FUNDAMENTALS *of* PLANT PHYSIOLOGY

FUNDAMENTALS *of* PLANT PHYSIOLOGY

SECOND EDITION

LINCOLN TAIZ

Professor Emeritus, University of California,
Santa Cruz, CA, USA

IAN MAX MØLLER

Professor Emeritus, Aarhus University, Denmark

ANGUS MURPHY

Professor, University of Maryland, College Park, MD, USA

WENDY A. PEER

Associate Professor, University of Maryland,
College Park, MD, USA

OXFORD
UNIVERSITY PRESS

OXFORD
UNIVERSITY PRESS

Oxford University Press is a department of the University of Oxford. It furthers the University's objective of excellence in research, scholarship, and education by publishing worldwide. Oxford is a registered trade mark of Oxford University Press in the UK and in certain other countries.

Published in the United States of America by Oxford University Press
198 Madison Avenue, New York, NY 10016, United States of America.

© 2025, 2018 by Oxford University Press

For titles covered by Section 112 of the US Higher Education Opportunity Act, please visit www.oup.com/us/he for the latest information about pricing and alternate formats.

All rights reserved. No part of this publication may be reproduced, stored in a retrieval system, or transmitted, in any form or by any means, without the prior permission in writing of Oxford University Press, or as expressly permitted by law, by license or under terms agreed with the appropriate reprographics rights organization. Inquiries concerning reproduction outside the scope of the above should be sent to the Rights Department, Oxford University Press, at the address above.

You must not circulate this work in any other form
and you must impose this same condition on any acquirer

Library of Congress Cataloging-in-Publication Data

Names: Taiz, Lincoln, author.
Title: Fundamentals of plant physiology / Lincoln Taiz, Professor Emeritus, University of California, Santa Cruz, USA, Ian Max Møller, Professor Emeritus, Aarhus University, Denmark, Angus Murphy, Professor, University of Maryland, USA, Wendy A. Peer, Associate Professor, University of Maryland, USA.
Description: Second edition. | New York : Oxford University Press, [2024] | Includes index.
Identifiers: LCCN 2024002173 | ISBN 9780197614167 (paperback) | ISBN 9780197614181 (epub)
Subjects: LCSH: Plant physiology—Textbooks.
Classification: LCC QK711.2 .T34 2024 | DDC 581.3—dc23/eng/20240214
LC record available at https://lccn.loc.gov/2024002173

Printed by Integrated Books International, United States of America

MIX
Paper
FSC FSC® C183721

About the Cover

The eleven plant images shown on the cover are (left to right, top row): Japanese cherry blossoms from a garden in Amsterdam, an agave leaf, and glassworts growing at the edge of the Mediterranean Sea; (second row) Japanese cheesewood growing in Sochi, Russia, and wheat growing in Japan; (third row) spadix with fruits of Western skunk cabbage, agave leaves, and sea kale (crambe); (bottom row) leaves and spathe of Western skunk cabbage, red samphire glassworts, and tropical coconut trees. Cherry blossoms exhibit the characteristic floral form found in members of the Rosaceae family cultivated for their fruits and flowers. Japanese cherry trees grown in public spaces are usually horticultural varieties that do not bear edible fruit. Agave grows in xeric environments and has succulent leaves and a shallow root system adapted to capturing and holding moisture. Cheesewood and other *Pittosporum* species are widely used as ornamental plants with growth habits that are highly responsive to light availability. Wheat and rice provide more food energy for the world population than any other crops. Sea kale is a halophytic member of the Brassicaceae family that grows in a variety of environments. Skunk cabbage is an unusual plant whose flower generates heat that helps it to emerge through the winter snow in early March; the heat also helps to volatilize the putrid smell that the flower generates to attract fly pollinators. Glassworts from the genus *Salicornia* are salt- and desiccation-tolerant species with succulent stems and small scalelike leaves that form a protective sheath around the stem. Coconut palm trees produce seeds that are highly adapted to multiple tropical environments. Coconut trees have many uses as food, drink, cosmetics, fuel, folk medicine, and building materials.

Brief Contents

Table of Contents

CHAPTER 4
Mineral Nutrition 85

CHAPTER 5
Assimilation of Inorganic Nutrients 117

Preface

It is with great pleasure that we present the Second Edition of *Fundamentals of Plant Physiology*.

Fundamentals was originally developed in response to requests from educators for a textbook that is both up-to-date and accessible. It was designed to be used as the primary textbook for undergraduate plant physiology and structure/function classes, where students may not yet have had extensive training in organic chemistry, genetics, plant anatomy, biochemistry, or molecular biology. *Fundamentals* presents plant physiology in the context of anatomy and growth in a manner that is particularly suited to programs in applied plant biology, such as horticulture and agronomy, or in ecology programs where training in physiology is a mandatory or optional component. Each chapter has been carefully tailored to achieve a discrete set of learning outcomes and overall competence to apply essential concepts of plant physiology to real-world problems.

Since the publication of the First Edition of *Fundamentals*, extraordinary advances in the plant sciences have been made. Driven by growing recognition of the existential threat of global climate change and increasing CO_2 levels, plant scientists, horticulturalists, and agronomists have achieved remarkable breakthroughs in plant stress physiology and in understanding of plant interactions with other organisms, how nutrient utilization is optimized, and how photosynthetic mechanisms respond to temperature and ozone. These advances have been combined with rapid molecular breeding and safer genome-editing techniques to produce crops that can continue to feed a growing global population in the face of farmland deterioration, decreased soil health, erratic climate changes, saltwater intrusion, rapid introduction of exotic pathogen and insect vectors, and devastating impacts of fire and flooding. Topics embedded throughout the Second Edition are strongly linked to these new realities.

The content of the Second Edition of *Fundamentals of Plant Physiology* is distilled from the Seventh Edition of the widely acclaimed upper-level *Plant Physiology and Development*, which was written and extensively reviewed by a consortium of leading plant scientists. The editors of the Second Edition of *Fundamentals* have transformed the most important elements of the cutting-edge material into a highly accessible textbook that is well-suited to students from a wide variety of backgrounds.

Another important pedagogical objective for the new Second Edition of *Fundamentals* has been to coordinate learning outcomes at different organizational levels and to better adapt them to student learning styles. To this end, section, chapter, and overall learning outcomes have been aligned with those communicated by instructors who teach plant physiology and plant structure/function courses. Self-study questions have also been greatly expanded and organized in a manner that allows them to be more in sync with examination and other evaluative processes. These questions were developed in consultation with instructors, teaching assistants, and students in actual plant physiology classes.

In terms of layout, the same single-column format with a running glossary in the margins that proved so successful in the First Edition has been continued in the Second Edition. While consistency of presentation enables instructors to retain existing course structure and design, many figures have been reconfigured or replaced to reduce complexity and to enhance understanding. Every effort has been made to ensure that, despite the updated content, courses taught using the Second Edition of *Fundamentals* can be completed in the same time frame as in the first edition. In short, our primary goal of *Fundamentals* has been to provide a concise, accessible, and focused treatment of the field of plant physiology, while maintaining the high standard of scientific accuracy and pedagogical richness for which *Plant Physiology and Development* is known.

It takes a village to produce a textbook, and *Fundamentals of Plant Physiology* is no exception. We thank all of the chapter authors of *Plant Physiology and Development*, Seventh Edition, and the staff at Oxford University Press who contributed in multiple ways to the publication of the Second Edition.

Lincoln Taiz   Angus Murphy

Ian Max Møller   Wendy A. Peer

August 2023

Editors

Ian Max Møller is Professor Emeritus of Molecular Biology and Genetics at Aarhus University, Denmark. He received his Ph.D. in Plant Biochemistry from Imperial College, London, UK, after which he worked at Lund University, Sweden, for 20 years before returning to Denmark. He has investigated plant mitochondria throughout his career, and his current interests include proteomics, oxidative stress, and posttranslational modification of proteins as a regulatory mechanism.

Angus Murphy is an Emeritus Professor of Plant Science and Landscape Architecture at the University of Maryland. He received his Ph.D. at the University of California, Santa Cruz and then moved to the faculty of Purdue University for 12 years before moving to the University of Maryland to serve as Chair of the Department of Plant Science and Landscape Architecture. His research explores the transport and metabolism of plant hormones and their role in development and environmental responses. Much of that research is focused on the functions of ATP binding cassette transporters and other proteins that are clustered in ordered membrane nanodomains.

Wendy A. Peer is an Associate Professor of Plant Biology and Chemical Ecology in the Department of Environmental Science and Technology at the University of Maryland. She received her Ph.D. at the University of California, Santa Cruz and then moved to the faculty of Purdue University before moving to the University of Maryland. Wendy Peer's research focuses on developmental, biotic, and abiotic interactions that affect seedling establishment and on identifying novel applications of microbial pairing with specific plant genotypes to enhance the resilience, quality, and nutrition of plant-based foods.

Lincoln Taiz is Professor Emeritus of Molecular, Cellular, and Developmental Biology at the University of California, Santa Cruz. He received his Ph.D. in Botany from the University of California, Berkeley. Dr. Taiz's primary research focus has been on the structure, function, and evolution of vacuolar H^+-ATPases. Other research interests include plant hormone transport and activity, cell wall mechanical properties, and heavy metal toxicity. Dr. Taiz is also the co-author with Lee Taiz of a popular science book on the discovery and denial of sex in plants, called *Flora Unveiled*.

Contributors

Andreas Blennow
Eduardo Blumwald
Frederica Brandizzi
Siobhan A. Braybrook
John J. Browse
John M. Christie
Asaph B. Cousins
José Feijó
Simon Gilroy
N. Michele Holbrook
Roger W. Innes
Alexander Jones
David M. Kramer
June M. Kwak
Pyung Ok Lim
Zhongchi Liu
Andreas Madlung
Ron Mittler
Sofía Otero
Yiping Qi
Simona Radutoiu
Allan G. Rasmusson
Nidhi Rawat
Sarah Robinson
Alexander Schulz
Claus Schwechheimer
Young Hun Song
Vijay Tiwari
Yi-Fang Tsay
Michael Udvardi
John M. Ward
Dolf Weijers
Christopher D. Whitewoods
Hye Ryun Wo

Digital Resources

to accompany *Fundamentals of Plant Physiology* 2e

Enhanced E-Book for the Student (ISBN 9780197614181)

Fundamentals of Plant Physiology is available as an e-book via RedShelf, VitalSource, and other leading higher education e-book vendors. The enhanced e-book provides student access to key resources:

- *Interactive Figurers:* Select text figures are presented in an interactive format that provides enhanced insight into key processes and cause-and-effect relationships.
- *Self-Study Questions:* Each section of the text provides self-study questions with which students can assess their mastery of content and concepts.
- *Flashcards:* This study aid assists students as they learn each chapter's essential vocabulary.

For the Instructor

(Available at oup.com/he/taiz-fundamentals2e)

Includes the following resources to aid instructors in developing their courses, delivering their lectures, and performing assessment:

- *Learning Objectives:* Outline the important take-aways of every major section.

  Learning objectives are placed immediately after section headings and are marked by a vertical brown bar. Learning outcome statements should be understood to begin with the unprinted phrase: "After reading this section, students should be able to:"
- *PowerPoint Resources:* A PowerPoint presentation for each chapter of the textbook includes all of the figures and tables from the chapter, with figure numbers and titles on each slide and complete captions in the Notes field.
- *Chapter Quizzes:* Brief quizzes assess student achievement of each chapter learning objective.
- *Interactive Figure Quizzes:* Brief quizzes to go with each interactive figure help confirm that students have acquired key concepts and content.

To learn more about any of these resources, or to get access, please contact your local OUP representative.

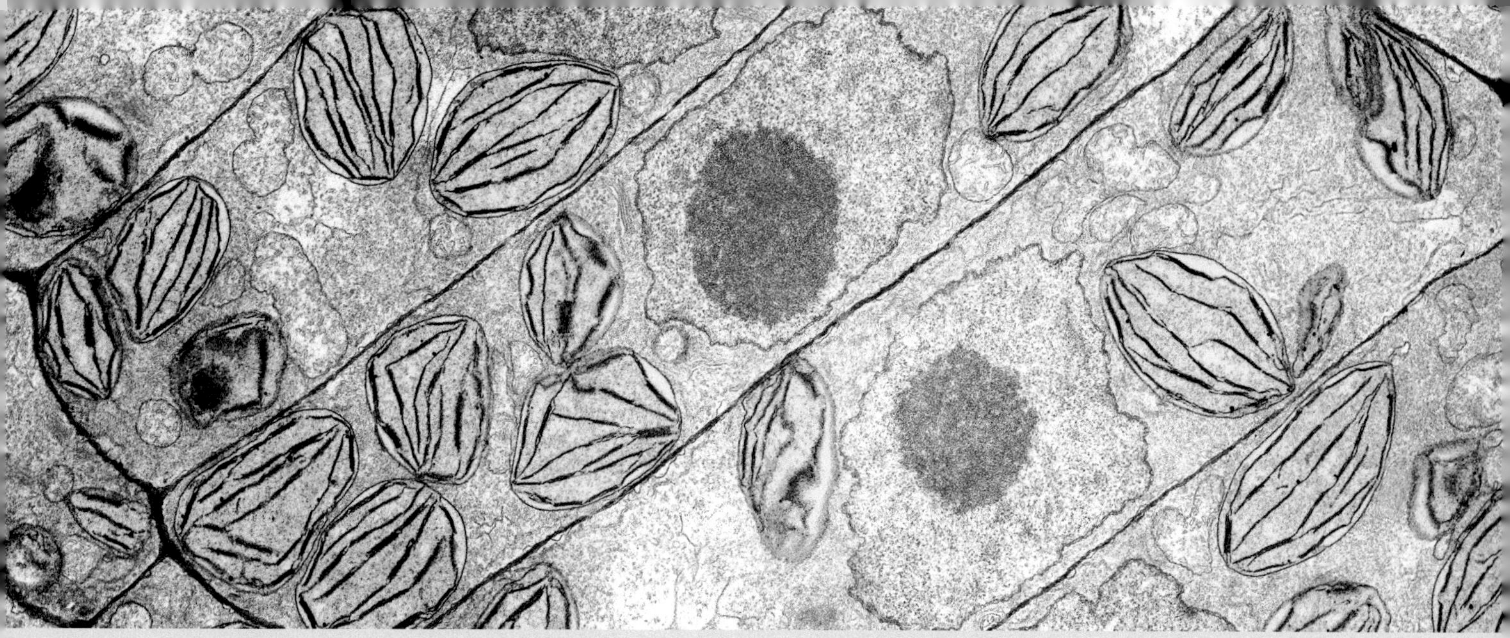

1 Plant and Cell Architecture

Plant physiology is the study of plant processes—how plants grow, develop, and function as they interact with their physical (abiotic) and living (biotic) environments. Although this book will emphasize the physiological, biochemical, and molecular functions of **plants**, it is important to recognize that, whether we are talking about gas exchange in the leaf, water conduction in the xylem, photosynthesis in the chloroplast, ion transport across membranes, signal transduction pathways involving light and hormones, or gene expression during development, all of these functions depend entirely on structures.

Plant functions occur across multiple scales. Tiny molecules recognize and bind each other to produce more elaborate structures that take on new functions. New leaves unfold as their cells multiply, diversify, and expand to capture light and exchange gases. Whole plants can function in communities as they compete to secure resources and reproduce, yet these landscape-scale functions continue to reflect structural interactions at the molecular and cellular scale.

The fundamental organizational unit of plants, and of all living organisms, is the cell. The term *cell* is derived from the Latin *cella*, meaning "storeroom" or "chamber." It was first used in biology in 1665 by the English scientist Robert Hooke to describe the individual units of the honeycomb-like structure he observed in cork under a compound microscope. The cork "cells" Hooke observed were actually the empty lumens of dead cells surrounded by cell walls, but the term is an apt one, because cells are the basic building blocks that define plant structure.

Moving outward from the cell, groups of specialized cells form specific tissues, and specific tissues arranged in particular patterns are the basis of three-dimensional organs. Just as plant anatomy, the study of the

plants All the families of plants inclusive of the non-seed, nonvascular plants.

macroscopic arrangements of cells and tissues within organs, received its initial impetus from improvements to the light microscope in the seventeenth century, so plant cell biology, the study of the interior of cells, was stimulated by the first application of the electron microscope to biological material in the mid-twentieth century. Subsequent improvements in microscopy and molecular biology have revealed astonishing variety and dynamics in the components that make up cells—the organelles, whose combined activities are required for the wide range of cellular and physiological functions that characterize biological organisms.

This chapter provides an overview of the basic anatomy and cell biology of plants, from the macroscopic structure of organs and tissues to the microscopic ultrastructure of cellular organelles. Subsequent chapters will treat these structures in greater detail from the perspective of their physiological and developmental functions at different stages of the plant life cycle.

1.1 Plant Life Processes: Unifying Principles

Describe unique features of plant life processes.

The spectacular diversity of plant size and form is familiar to everyone. Plants range in height from less than 1 cm to more than 100 m. Plant morphology, or form, is also surprisingly diverse. At first glance, the tiny plant duckweed (*Lemna*) seems to have little in common with a giant saguaro cactus or a redwood tree. No single plant shows the entire spectrum of adaptations to the range of environments that plants occupy on Earth, so plant physiologists often study **model organisms**, plants with short generation times and small **genomes** (the sum of their genetic information). In particular, the use of molecular tools to compare **phenotypes** with the **genotypes** of individual plants from the species *Arabidopsis thaliana* (Arabidopsis) has rapidly advanced the understanding of plant developmental programming and physiological responses. These models are useful because all plants, regardless of their specific adaptations, carry out fundamentally similar processes and are based on the same architectural plan.

We can summarize the major unifying principles of plants as follows:

- As Earth's primary producers, plants and green algae are the ultimate solar energy collectors. They harvest the energy of sunlight by converting light energy to chemical energy, which is stored in bonds formed when they synthesize carbohydrates from carbon dioxide and water.
- Other than certain reproductive cells, land plants do not move from place to place; they are sessile. As a substitute for motility, they have evolved the ability to grow toward essential resources, such as light, water, and mineral nutrients, throughout their life span.
- Plants are structurally reinforced to support their mass as they grow toward sunlight against the pull of gravity.
- Plants have mechanisms for moving water and minerals from the soil to the sites of photosynthesis and growth, as well as mechanisms for moving the products of photosynthesis to nonphotosynthetic organs and tissues.
- Plants lose water continuously by evaporation and have evolved mechanisms for avoiding desiccation.
- Plants develop from embryos that derive nutrients from the mother plant, and these additional food stores facilitate the production of large self-supporting structures on land.

model organism Organism that is particularly accessible and convenient for research and that provides information for hypothesis testing in other organisms.

genome Refers to all the genes in a haploid complement of eukaryotic chromosomes, in an organelle, a microbe, or the DNA or RNA content of a virus.

phenotype A set of observed or measured characteristics of an individual genotype found under a specific set of environmental conditions.

genotype The genetic constitution of an individual organism. Sometimes used to indicate a variant, cultivar, or ecotype carrying a unique set of genomic sequence variations.

Based on these principles, we can generally define terrestrial plants as sessile, multicellular organisms derived from embryos, adapted to land, and able to convert carbon dioxide into complex organic compounds through the process of photosynthesis. This broad definition includes a wide spectrum of organisms, from the hornworts to the **flowering plants**, as illustrated in the diagram, or cladogram, depicting evolutionary lineage as branches, or clades, on a tree (**Figure 1.1**). Plants share with (mostly aquatic) green algae the basal trait that is so important for photosynthesis in both clades: Their chloroplasts contain the pigments chlorophyll *a* and *b* as well as β-carotene. **Land plants**, or **embryophytes**, share the evolutionarily derived traits for surviving on land that are absent in the algae. Land plants include the **nonvascular plants**, or **bryophytes** (hornworts, mosses, and liverworts), and the **vascular plants**, or **tracheophytes**, which evolved from a common ancestor. The vascular plants consist of the **non-seed plants** (ferns and their relatives) and the **seed plants** (gymnosperms and angiosperms). Aquatic grasses are classified as land plants, as they are angiosperms that are evolutionarily adapted to periodic or continuous submersion in water.

flowering plant *See* angiosperm.

land plants (or **embryophytes)** All the families of plants inclusive of the non-seed, nonvascular plants.

nonvascular plants Plants that do not have vascular tissues, such as xylem and phloem.

vascular plant A plant that has xylem and phloem. Also called a tracheophyte.

bryophyte *See* nonvascular plants.

tracheophyte A plant that has xylem and phloem. Also called a vascular plant.

non-seed plants Plant families that do not produce a seed.

seed plants Plants in which the embryo is protected and nourished within a seed; gymnosperms and angiosperms.

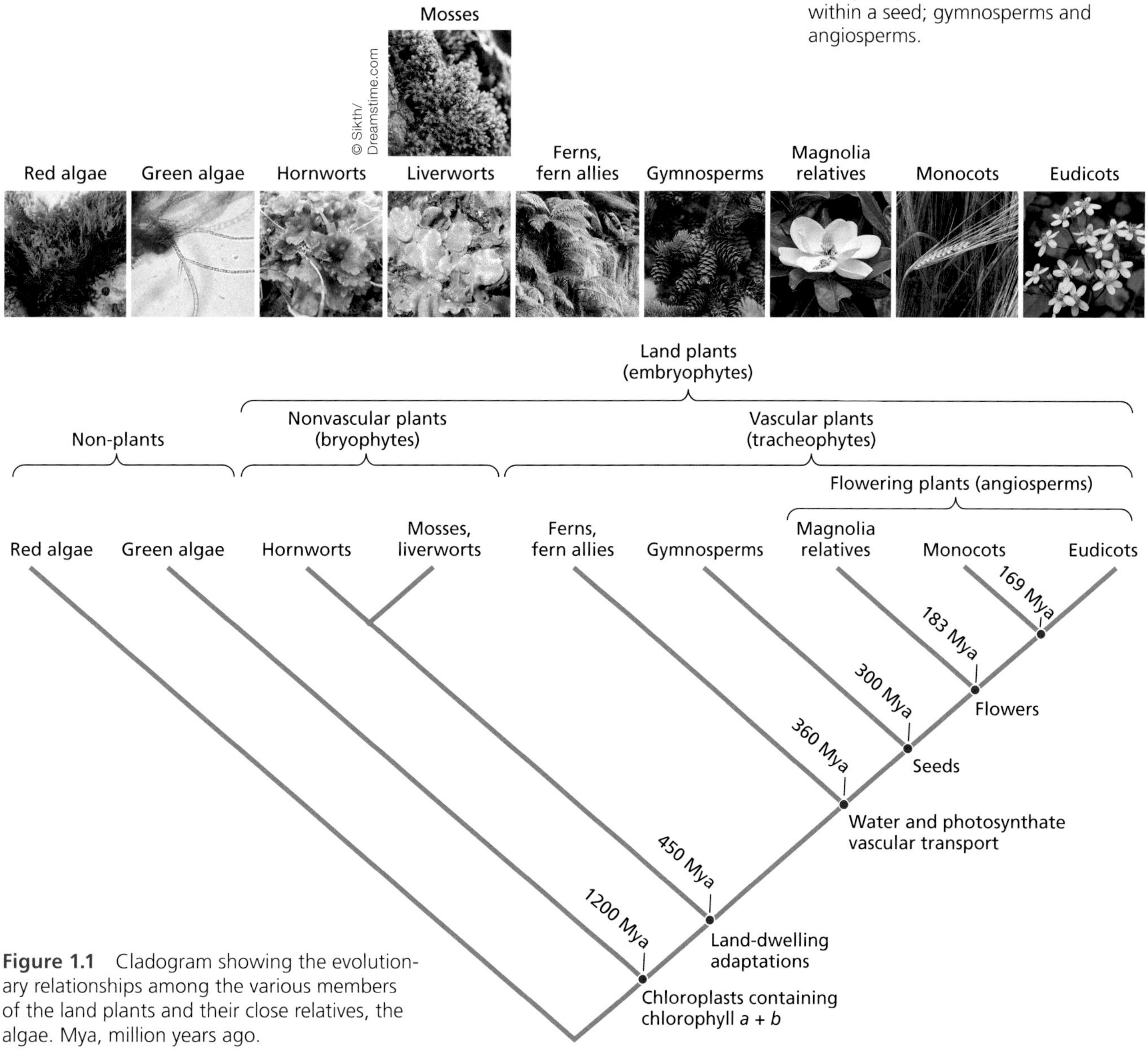

Figure 1.1 Cladogram showing the evolutionary relationships among the various members of the land plants and their close relatives, the algae. Mya, million years ago.

gymnosperms An early group of seed plants. Distinguished from angiosperms by having seeds borne unprotected (naked) in cones.

conifers Cone-bearing trees.

angiosperms The flowering plants. With their innovative reproductive organ, the flower, they are a more advanced type of seed plant and dominate the landscape. Distinguished from gymnosperms by the presence of a carpel that encloses the seeds.

monocot One of the two classes of flowering plants, characterized by a single seed leaf (cotyledon) in the embryo.

eudicot Abbreviated name for the eudicotyledons, one of two major classes of the angiosperms, and refers to the fact that plants in this class have two seed leaves (cotyledons).

alternation of generations The presence of two genetically distinct multicellular stages, one haploid and one diploid, in the plant life cycle. The haploid gametophyte generation begins with meiosis, while the diploid sporophyte generation begins with the fusion of sperm and egg.

diploid Cell with two copies of each chromosome and abbreviated as having 2*N* chromosomes.

haploid Cell with only one copy of each chromosome, abbreviated as 1*N*.

gamete A haploid (1*N*) reproductive cell.

meiosis The "reduction division" whereby two successive cell divisions produce four haploid (1*N*) cells from one diploid (2*N*) cell. In plants with alternation of generations, spores are produced by meiosis. In animals, which don't have alternation of generations, gametes are produced by meiosis.

spores Reproductive cells formed in plants by meiosis in the sporophyte generation. They give rise by mitotic divisions to the gametophyte generation.

sporophyte The diploid (2*N*) multicellular structure that produces haploid spores by meiosis. Contrast with *gametophyte*.

mitosis The ordered cellular process by which replicated chromosomes are distributed to daughter cells formed by cytokinesis.

gametophyte The haploid (1*N*) multicellular structure that produces haploid gametes by mitosis and differentiation. Contrast with sporophyte.

Because seed plants have many agricultural, industrial, timber, and medical uses, as well as an overwhelming dominance in terrestrial ecosystems, most research in plant biology has focused on the seed plants that have evolved in the last 300 million years (Figure 1.1). The **gymnosperms** (from the Greek for "naked seed") include the conifers, cycads, ginkgo, and gnetophytes (e.g., the medicinal plant *Ephedra*). About 800 species of gymnosperms are known. The **conifers** ("cone-bearers") are the largest group of gymnosperms and include such commercially important forest trees as pine, fir, spruce, and redwood. The **angiosperms** (from the Greek for "vessel seed") evolved about 183 million years ago and include three major groups: the **monocots**, **eudicots**, and so-called basal angiosperms (i.e., the Magnolia family and its relatives). The major anatomical innovation of the angiosperms is the flower; hence, they are referred to as *flowering plants*. Except for the northern boreal and temperate mountainous forests, angiosperms dominate the landscape. About 370,000 species are known, with an additional 17,000 undescribed species predicted by taxonomists using computer models.

Plant life cycles alternate between diploid and haploid generations

Plants, unlike animals, alternate between two distinct multicellular generations to complete their life cycle. This is called **alternation of generations**. One generation has **diploid** cells, cells with two copies of each chromosome and abbreviated as having 2N chromosomes, and the other generation has **haploid** cells, cells with only one copy of each chromosome, abbreviated as 1N. Each of these multicellular generations may be more or less physically and metabolically dependent on the other, depending on their evolutionary grouping.

When diploid (2*N*) animals, as represented by humans on the inner cycle in **Figure 1.2**, produce haploid **gametes**, egg (1*N*) and sperm (1*N*), they do so directly by the process of **meiosis**, cell division resulting in a reduction of the number of chromosomes from 2*N* to 1*N*. In contrast, the products of meiosis in diploid plants are **spores**, and diploid plant forms are therefore called **sporophytes**. Each spore is capable of undergoing **mitosis**, cell division that doesn't change the number of chromosomes in the daughter cells, to form a new haploid multicellular individual, the **gametophyte**, as shown by the outer cycles in Figure 1.2. The haploid gametophytes produce gametes, egg and sperm, by simple mitosis, whereas haploid gametes in animals are produced by meiosis. Once the haploid gametes fuse and **fertilization** takes place to create the 2*N* zygote, the life cycles of animals and plants are similar (Figure 1.2). The 2*N* zygote undergoes a series of mitotic divisions to produce the embryo, and eventually grows into the mature diploid adult.

Thus, all plant life cycles encompass two separate generations: the diploid, spore-producing sporophyte generation and the haploid, gamete-producing **gametophyte generation**. A line drawn between fertilization and meiosis divides these two separate stages of the generalized plant life cycle (Figure 1.2). Increasing the number of mitoses between fertilization and meiosis increases the size of the sporophyte generation and the number of spores that can be produced. Having more spores per fertilization event could compensate for low fertility when water becomes scarce on land. This could explain the marked tendency for the increase in size of the sporophyte generation, relative to the gametophyte generation, during the evolution of plants.

The sporophyte generation is dominant in the seed plants, the gymnosperms and angiosperms, and gives rise to different spores; the **megaspores** develop into the female gametophyte, and the **microspores** develop into the male gametophyte (Figure 1.2). The way the resulting male and female gametophytes are separated is quite diverse. In angiosperms, a single individual in a **monoecious** (from the

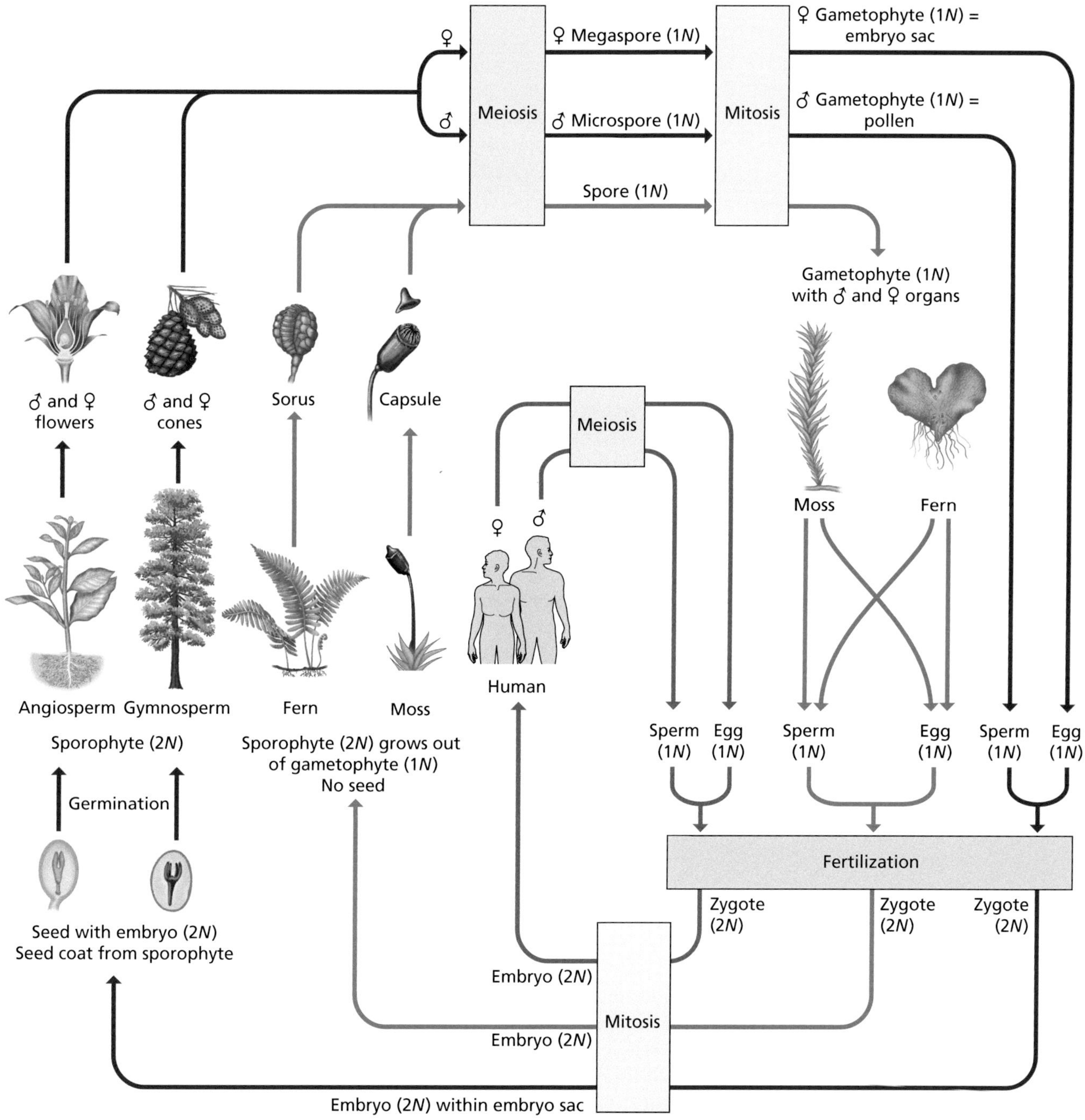

Figure 1.2 Diagram of the generalized life cycles of plants and animals. In contrast to animals, plants exhibit alternation of generations. Rather than producing gametes directly by meiosis as animals do, plants produce vegetative spores by meiosis. These 1*N* (haploid) spores divide to produce a second multicellular individual called the gametophyte. The gametophyte then produces gametes (sperm and egg) by mitosis. Following fertilization, the resulting 2*N* (diploid) zygote develops into the mature sporophyte generation, and the cycle begins again (discussed in Chapter 17).

Greek for "one house") species has flowers that produce both male and female gametophytes; both can occur in the single "perfect" flower as in tulips, or they can occur in separate male (staminate) and female (pistillate) flowers as in maize (corn; *Zea mays*). If male and female flowers occur on separate individuals, as in willow or poplar trees, then the species is **dioecious** (from the Greek for "two

fertilization The formation of a diploid (2*N*) zygote from the cellular and nuclear fusion of two haploid (1*N*) gametes, the egg and the sperm. In angiosperms, fertilization also involves fusion of a second sperm nucleus with the haploid nuclei (usually two) of the central cell to form the endosperm (usually triploid).

gametophyte generation The stage, or generation, in the life cycle of plants that produces gametes. It alternates with the sporophyte generation, in a process called alternation of generations.

houses”). In gymnosperms, ginkgos and cycads are dioecious, while conifers are monoecious. Conifers produce female cones, **megastrobili** (from the Greek for “large cones”; singular *megastrobilus*), usually higher up on the plant than the male cones, **microstrobili** (from the Greek for “small cones”; singular *microstrobilus*). Both megaspores and microspores produce gametophytes with only a few cells, compared with the sporophyte. The processes of gametophyte development are discussed in detail in Chapter 17.

1.2 Overview of Plant Structure

Describe the fundamental properties of plant structure.

Despite their apparent diversity, all seed plants have the same basic body plan (**Figure 1.3**). The vegetative body comprises three organs—the stem, the root, and the leaves—each with a different direction, or **polarity**, of growth. The **stem** grows upward and supports the aboveground part of the plant. The **root** anchors the plant and absorbs nutrients and water, and grows down below the ground. The **leaves**, whose primary function is photosynthesis, grow out laterally from the stem at the **nodes**. Variations in leaf arrangement (**phyllotaxis**) can give rise to many different forms of **shoots**, the term for the leaves and stem together. For example, leaf nodes can spiral around the stem, rotating by a fixed angle between each **internode** (the region between two nodes). Alternatively, leaves can arise oppositely or alternating on either side of the stem.

Organ position and shape are defined by directional patterns of growth that are determined by regulated cell division and expansion. The polarity of growth of the **primary plant axis** (the main stem and taproot) is vertical, whereas the typical leaf grows laterally at the margins to produce the flattened **leaf blade**. The **adaxial** (upper) surface of the leaf faces toward the stem, and the **abaxial** (lower) leaf surface faces away from the stem (Figure 1.3). The growth polarities of the organs are adapted to their functions: Leaves function in light absorption, stems elongate to lift the leaves toward sunlight, and roots elongate in search of water and nutrients from the soil. The cellular component that directly determines growth polarity in plants is the cell wall.

Plant cells are surrounded by rigid cell walls

The outer lipidic fluid boundary of the living cytoplasm of plant cells is the **plasma membrane** (also called plasmalemma) and is analogous to that found in animals, fungi, and bacteria. Within the plasma membrane, the highly crowded internal collection of hydrated ions, small and large molecules, organelles, and cytoskeletal components apart from the nucleus is called the **cytoplasm**. The liquid portion of the cytoplasm that can be physically separated from insoluble macromolecules, organelles, small compartments referred to as vesicles, and ribosomes is referred to as the **cytosol**. The internal compartment of the membrane-enclosed nucleus in eukaryotes is the nucleoplasm. The large internal compartment in most plant cells known as the **vacuole** is generally considered distinct from the cytoplasm as well, as it contains a less dense solution containing molecules and hydrolyzing enzymes. However, plant cells, unlike animal cells, are further enclosed by a rigid **cell wall** (**Figure 1.4**). Because of the absence of cell walls in animals, embryonic cells are able to migrate from one location to another; developing tissues and organs may thus contain cells that originated in different parts of the organism. In plants such cell migrations are prevented because each walled cell is cemented to its neighbors by a **middle lamella** composed mostly

megaspore The haploid (1*N*) spore that develops into the female gametophyte.

microspores The haploid (1*N*) cell that develops into the pollen tube or male gametophyte.

monoecious Refers to plants in which male and female flowers are found on the same individuals, such as cucumber (*Cucumis sativus*) and maize (corn; *Zea mays*). Contrast with *dioecious*.

dioecious Refers to plants in which male and female flowers are found on different individuals, such as spinach (*Spinacia*) and hemp (*Cannabis sativa*). Contrast with *monoecious*.

megastrobili The strobili or cones that contain the female gametophytic tissue.

microstrobili The strobili or cones that contain the male gametophytic tissue.

polarity (1) Property of some molecules, such as water, in which differences in the electronegativity of some atoms result in a partial negative charge at one end of the molecule and a partial positive charge at the other end. (2) Refers to the distinct ends and intermediate regions along an axis. Beginning with the single-celled zygote, the progressive development of distinctions along two axes: an apical–basal axis and a radial axis.

stem The typically aboveground primary axis of the shoot that bears leaves and buds. May also occur underground in the form of rhizomes, corms, and tubers.

root The organ, usually underground, that serves to anchor the plant in the soil, and to absorb water and mineral ions and conduct them to the shoot. In contrast to shoots, roots lack buds, leaves, or nodes.

leaves The main lateral appendages radiating out from stems and branches. Green leaves are usually the major photosynthetic organs of the plant.

node Position on the stem where leaves are attached.

shoots The organ, usually aboveground, that includes the stem, leaves, buds, and reproductive structures. Function in photosynthesis and reproduction.

phyllotaxy (phyllotaxis) The arrangement of leaves on the stem.

internode The portion of a stem between nodes.

primary plant axis The longitudinal axis of the plant defined by the positions of the shoot and root apical meristems.

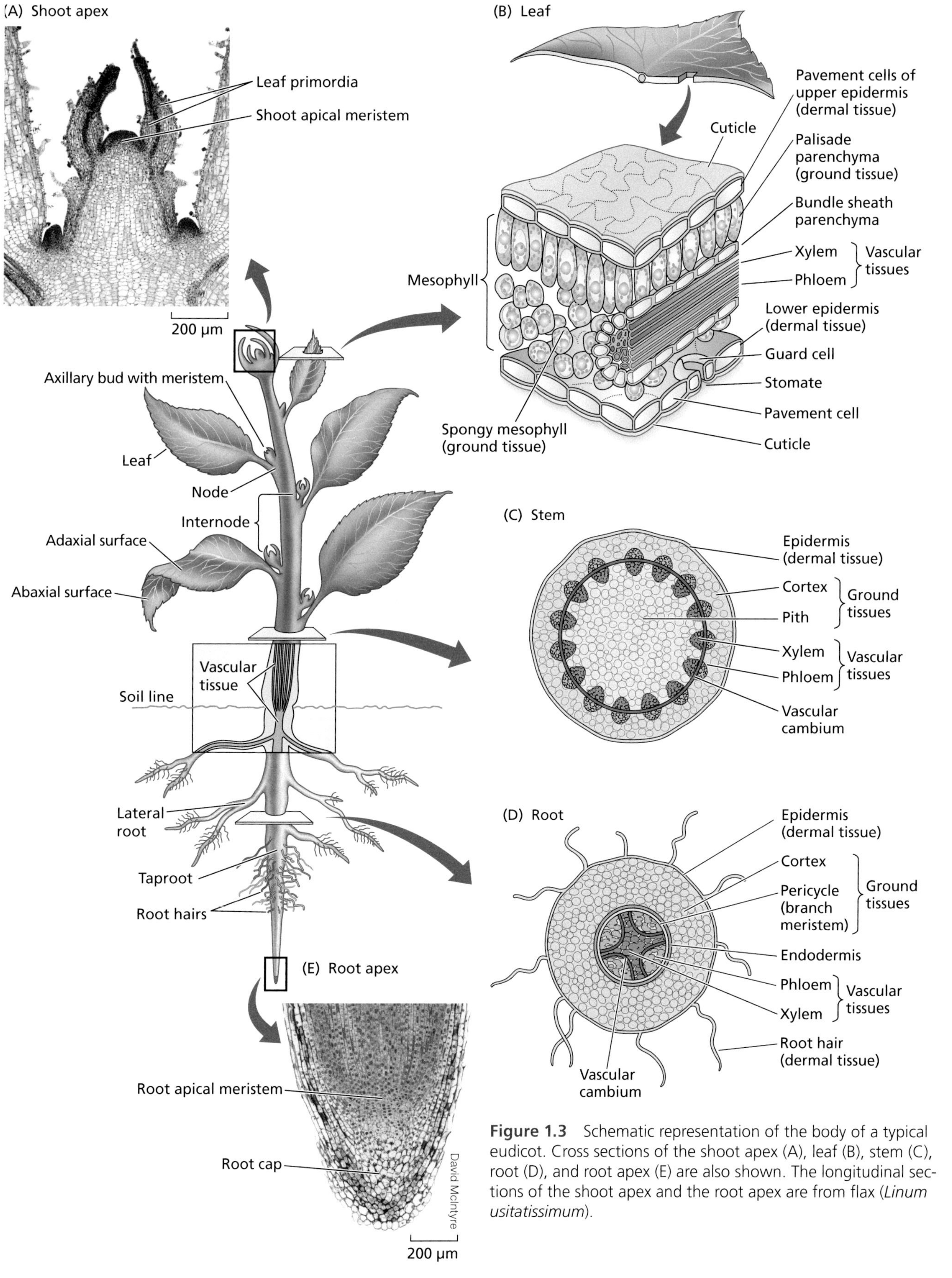

Figure 1.3 Schematic representation of the body of a typical eudicot. Cross sections of the shoot apex (A), leaf (B), stem (C), root (D), and root apex (E) are also shown. The longitudinal sections of the shoot apex and the root apex are from flax (*Linum usitatissimum*).

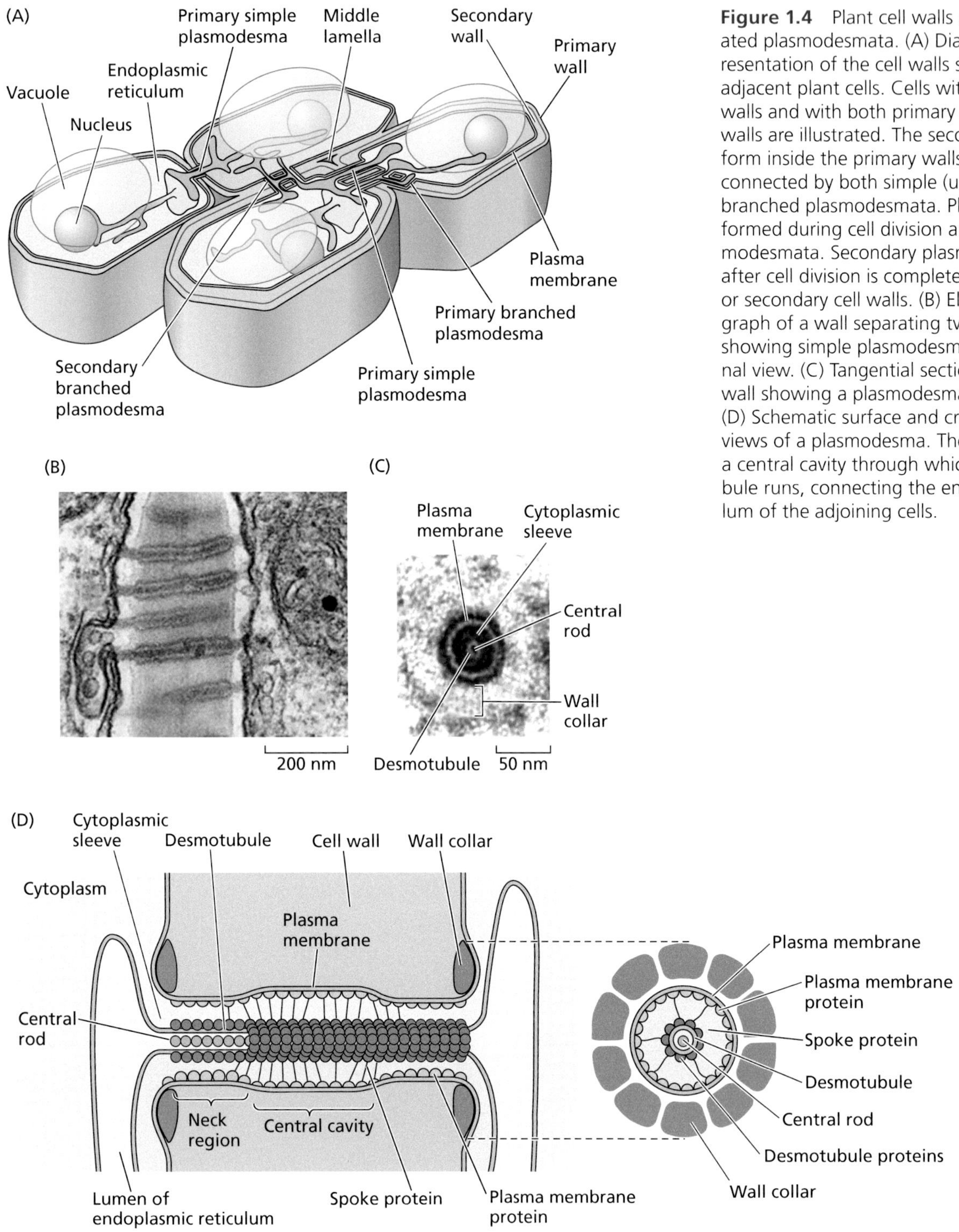

Figure 1.4 Plant cell walls and their associated plasmodesmata. (A) Diagrammatic representation of the cell walls surrounding four adjacent plant cells. Cells with only primary walls and with both primary and secondary walls are illustrated. The secondary walls form inside the primary walls. The cells are connected by both simple (unbranched) and branched plasmodesmata. Plasmodesmata formed during cell division are primary plasmodesmata. Secondary plasmodesmata form after cell division is completed, across primary or secondary cell walls. (B) Electron micrograph of a wall separating two adjacent cells, showing simple plasmodesmata in longitudinal view. (C) Tangential section through a cell wall showing a plasmodesma in cross section. (D) Schematic surface and cross-sectional views of a plasmodesma. The pore consists of a central cavity through which the desmotubule runs, connecting the endoplasmic reticulum of the adjoining cells.

leaf blade The broad, expanded area of the leaf; also called the lamina.

adaxial Referring to the upper surface of a leaf.

abaxial Referring to the lower surface of a leaf.

of **pectin**. As a consequence, plant development, unlike animal development, depends solely on patterns of cell division and cell enlargement.

Plant cells have two types of walls: primary and secondary (Figure 1.4A). **Primary cell walls** are typically thin (<1 μm) and are characteristic of young, growing cells. **Secondary cell walls** are thicker and stronger than primary walls and are deposited on the inner surface of the primary wall after most cell enlargement has ended. Secondary cell walls owe their strength and toughness to **lignin**, a brittle material. The evolution of lignified secondary cell walls provided plants with the structural reinforcement necessary to grow vertically above the soil and to colonize the land. Bryophytes lack lignified cell walls and are unable to grow more than a few centimeters above the ground.

Plasmodesmata allow the free movement of molecules between cells

The cytoplasm of neighboring cells is usually connected by **plasmodesmata** (singular *plasmodesma*), tubular structures 40 to 50 nm in diameter and formed by the connected membranes of adjacent cells (Figure 1.4). Plasmodesmata facilitate intercellular movement of proteins, nucleic acids, and other macromolecular signals that coordinate developmental processes among plant cells. Primary plasmodesmata are formed during cell division. Secondary plasmodesmata are formed after cells have divided. The core of each plasmodesma is the *desmotubule*, an endomembrane structure derived from both adjoining cells. Plant cells interconnected in this way form a continuous cytoplasmic space referred to as the **symplasm** (sometimes referred to as the **symplast**), and transport of small molecules through plasmodesmata is called **symplasmic transport** (discussed in Chapters 3 and 6). Transport through the permeable cell wall space outside the cells (the **apoplast**) is called **apoplastic transport**. Both forms of transport are important in the vascular system of plants (discussed in Chapter 6).

The symplasm can transport water, solutes, and macromolecules between cells without crossing the plasma membrane. However, there is a restriction on the size of molecules that can be transported via the symplasm; this restriction is called the size exclusion limit, and it varies with cell type, environment, and developmental stage. The transport can be followed by studying the movement of fluorescently labeled proteins or dyes between cells. The movement through plasmodesmata (Figure 1.4) is regulated, or gated, by deposition (closing) or degradation (opening) of callose within the plasmodesmatal opening.

New cells originate in dividing tissues called meristems

Plant growth is concentrated in localized regions of cell division called **meristems**. Most nuclear divisions (mitosis) and cell divisions (cytokinesis) occur in these meristematic regions. In a young plant, the most active meristems are the **apical meristems**, located at the tips of the stem and the root (Figure 1.3A and E). **Primary growth** is the phase of plant development that produces new organs for the basic plant form, the **primary plant body**. Primary growth results from the activity of apical meristems. Cell division in the meristem produces cuboidal cells about 10 μm on each edge. Division is followed by progressive cell expansion, and cells typically elongate to become much longer than they are wide (30–100 μm long, 10–25 μm wide—about half the width of a baby's fine hair and about 50 times the width of a typical bacterium). The increase in length produced by primary growth amplifies the plant's axial (top-to-bottom) polarity that is initially established in the embryo.

Meristematic tissue is also found along the length of the root and shoot. **Axillary buds** are meristems that develop at the node in the leaf axil—the region between the leaf and the shoot. Axillary buds become the apical meristems of branches. The branches of roots, the **lateral roots**, arise from meristematic cells in the **pericycle**, or root branch meristem (Figure 1.3D). This meristematic tissue then becomes the apical meristem of the lateral root.

Another set of meristematic cells, the **cambium**, gives rise to **secondary growth** and produces an increase in width or diameter of plants, having radial (inside-to-outside) polarity (**Figure 1.5**). The cambial layer that produces wood is called the **vascular cambium**. This meristem arises in the vascular system, between the xylem and the phloem of the primary plant body. The cells of the vascular cambium divide longitudinally to produce derivatives toward the inside

plasma membrane (plasmalemma) A bilayer of polar lipids (phospholipids or glycosylglycerides) and embedded proteins that together form a selectively permeable boundary around a cell.

cytoplasm The cellular matter enclosed by the plasma membrane exclusive of the nucleus.

cytosol The aqueous phase of the cytoplasm containing dissolved solutes but excluding supramolecular structures, such as ribosomes and components of the cytoskeleton.

vacuole A membrane-bound cellular organelle containing water and dissolved substances such as salts, sugars, enzymes, and amino acids. Vacuoles function in maintaining water balance and cellular turgor. Lytic vacuoles are characterized by a more acidic pH and can often occupy a majority of the space within a cell. Specialized protein storage vacuoles are found primarily in seeds and very young seedlings.

cell wall The rigid cell surface structure external to the plasma membrane that supports, binds, and protects the cell. Composed of cellulose and other polysaccharides and proteins.

middle lamella A thin layer of pectin-rich material at the junction where the primary walls of neighboring cells come into contact.

pectins A heterogeneous group of complex cell wall polysaccharides that form a gel in which the cellulose–hemicellulose network is embedded.

primary cell walls The thin (<1 μm), unspecialized cell walls that are characteristic of young, growing cells.

secondary cell wall Cell wall synthesized by nongrowing cells. Often multilayered and containing lignin, it differs in composition and structure from the primary wall.

lignin Highly branched phenolic polymer made up of phenylpropanoid alcohols that is deposited in secondary cell walls.

plasmodesmata (singular *plasmodesma*) Microscopic membrane-lined channel connecting adjacent cells through the cell wall and filled with cytoplasm and a central rod derived from the ER called the desmotubule.

symplasm The continuum of cellular cytoplasm connected by plasmodesmata.

symplast The continuous system of cell protoplasts interconnected by plasmodesmata.

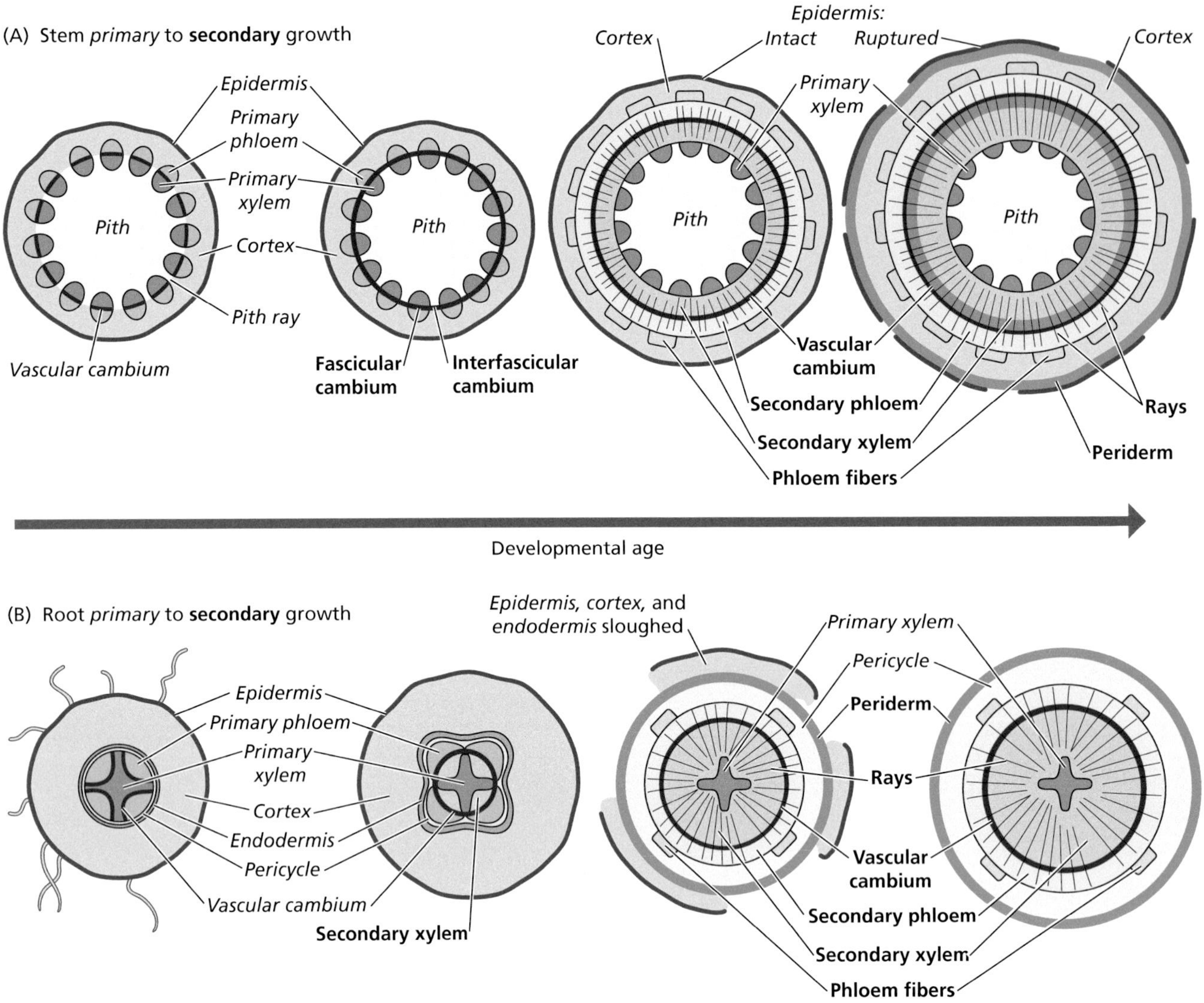

Figure 1.5 Secondary growth in stems and roots. Primary growth is labeled in italic text, secondary growth in bold. (A) Stem primary to secondary growth. The vascular cambium starts as separated growth regions in the vascular bundles, or fascia, of primary xylem and phloem. As the plant grows, the bundled, fascicular cambium becomes connected by interfascicular cambium between the bundles. Once the vascular cambium forms a continuous ring, it divides inward to generate secondary xylem and divides outward to generate the secondary phloem. Regions in the cortex develop into phloem fibers and the periderm, which contains the phellogen, or cork cambium, and the outer phelloderm. With growth, the epidermis ruptures and rays connect the inner and outer vasculature. (B) Root primary to secondary growth. The central vascular cylinder contains the primary phloem and primary xylem. As in the stem, the vascular cambium becomes connected and grows outward, generating secondary phloem and rays. As roots increase in girth, the pericycle generates the root periderm, while the outer epidermis, cortex, and endodermis are sloughed off. The pericycle produces the phloem fibers and rays as well as lateral roots. The vascular cambium produces secondary phloem and rings of secondary xylem.

symplasmic transport The intercellular transport of water and solutes through plasmodesmata.

apoplast The mostly continuous system of cell walls, intercellular air spaces, and xylem vessels in a plant.

or the outside of the stem or root. They also divide transversely to produce **rays** that transmit material radially outward. The inside derivatives differentiate into secondary xylem, which conducts water and nutrients from the soil upward to other plant organs. In temperate climates, summer wood is darker and denser than spring wood; alternating layers of summer and spring wood form **annual rings**. The vascular cambium derivatives displaced toward the outside of the secondary stem or root give rise to secondary phloem, which, like primary phloem, conducts the products of photosynthesis downward from the leaves to other organs of the plant. The associated **phloem fibers** add tensile strength to the stem, as do all fibers.

Finally, the **cork cambium**, or **phellogen**, is the cambial layer that produces the protective **periderm** (Figure 1.5) on the outside of woody plants. The cork cambium typically arises each year within the secondary phloem. The production of layers of water-resistant cork cells by the cork cambium isolates the outer primary tissues of the stem or root from their water supply, the xylem, causing them to shrivel and die. The **bark** of a woody plant is the collective term for several tissues—the secondary phloem, secondary phloem fibers, cortex (in stems), pericycle (in roots), and periderm—that can be peeled off as a unit at the soft layer of vascular cambium.

1.3 Plant Tissue Types

Distinguish the structural and functional differences of plant tissue types.

Three major tissue systems are present in all plant organs: dermal tissue, ground tissue, and vascular tissue (Figure 1.3B–D). **Dermal tissue** forms the outer protective layer of the plant and is called the **epidermis** in the primary plant body. **Ground tissue** fills out the three-dimensional bulk of the plant and includes the **pith** and **cortex** of primary stems and roots, and the **mesophyll** in leaves. **Vascular tissue** consists of two types of tissues: **xylem** and **phloem**, each of which consists of conducting cells, generalized parenchyma cells, and thick-walled fibers.

Dermal tissues cover the surfaces of plants

Three examples of leaf dermal cells are shown in **Figure 1.6**. Leaves have an upper and lower epidermis with different cell types in each (Figure 1.6A). The

apoplastic transport Movement of molecules through the cell wall continuum that is called the apoplast. Molecules may move through the linked cell walls of adjacent cells, and in that way move throughout the plant without crossing a plasma membrane.

meristems Localized regions of ongoing cell division that enable growth during postembryonic development.

apical meristems Localized regions made up of undifferentiated cells undergoing cell division without differentiation at the tips of shoots and roots.

primary growth The phase of plant development that gives rise to new organs and to the basic plant form.

primary plant body The part of the plant directly derived from the shoot and root apical meristems and primary meristems.

axillary buds Secondary meristems that are formed in the axils of leaves.

lateral roots Arise from the pericycle in mature regions of the root through establishment of secondary meristems that grow out through the cortex and epidermis, establishing a new growth axis.

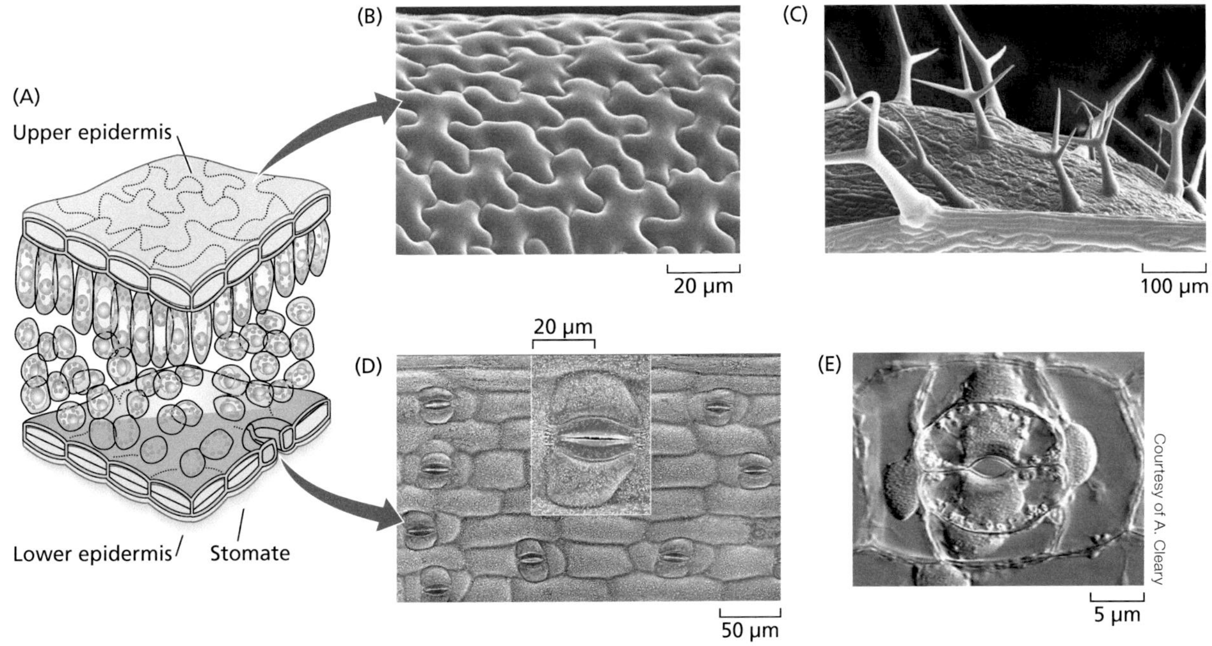

Figure 1.6 Dermal tissue of the leaf of a typical eudicot plant. (A) Generalized view of leaf structure. (B) Scanning electron micrograph of the epidermal cells of a *Galium aparine* leaf showing the puzzle-piece arrangement of pavement cells in the epidermis. (C) Scanning electron micrograph of an epidermis of a true leaf of Arabidopsis. The three-branched trichomes arise from a field of pavement cells and guard cell complexes. (D) Scanning electron micrograph with a slightly higher magnification inset of the stomatal complexes in a *Tradescantia* sepal. (E) Light micrograph of the stomatal complex of the *Tradescantia* sepal.

pericycle Meristematic cells forming the outermost layer of the vascular cylinder in the stem or root, interior to the endodermis.

cambium Layer of meristematic cells between the xylem and phloem that produces cells of these tissues and results in the lateral (secondary) growth of the stem or root.

secondary growth Growth, generally circumferential, that occurs after primary growth (stem and root elongation) is complete. It involves the vascular cambium (producing the secondary xylem and phloem) and the cork cambium (producing the periderm).

vascular cambium A lateral meristem consisting of fusiform and ray stem cells, giving rise to secondary xylem and phloem elements, as well as ray parenchyma.

rays Tissues of various height and width that radiate through the secondary xylem and phloem, and are formed from ray initials in the vascular cambium.

annual rings Alternating rings of spring and summer wood (formed by secondary growth of the xylem) seen in cross sections of stems of woody species.

phloem fiber Elongated, tapering sclerenchyma cell associated with the other cells in the phloem.

cork cambium A layer of lateral meristem that develops within mature cells of the cortex and the secondary phloem. Produces the secondary protective layer, the periderm. Also called phellogen.

phellogen *See* cork cambium.

periderm Tissue produced by the cork cambium that contributes to the outer bark of stems and roots during secondary growth of woody plants. Also replaces epidermis after wounding and in abscission layers after the shedding of plant parts.

bark Collective term for all the tissues outside the cambium of a woody stem or root, and composed of phloem, periderm, and dead tissue.

dermal tissue The tissue system that covers the outside of the plant body; the epidermis or periderm.

ground tissue The internal tissues of the plant, other than the vascular tissues.

epidermis The outermost layer of plant cells, typically one cell thick.

pith The ground tissue in the center of the stem or root.

cortex Ground tissue in the region of the primary stem or root located between the vascular tissue and the epidermis, mainly consisting of parenchyma.

epidermis includes the pavement cells, which are shaped like puzzle pieces in many flowering plants (Figure 1.6B). In Arabidopsis, pavement cells are the only cells in evidence on a seed leaf (cotyledon), but the epidermis of the true leaves differentiates into more cell types, becoming covered in three-pronged leaf hairs (trichomes) (Figure 1.6C). Many plants have relatively few chloroplasts in the epidermal tissue of true leaves, perhaps because chloroplast division is shut down. The exception to this is the guard cells (the amazing cells that form the "lips" of the mouths, or stomata, of the leaf; Figure 1.6D and E), which contain many chloroplasts. Once made, guard cells become cytoplasmically isolated from the rest of the leaf during the last cell division that forms them and have no plasmodesmata. The guard cell is a dynamic structure that responds rapidly to ambient light and water status (discussed in Chapters 3 and 6). In roots, root hairs differentiate from the epidermis.

Ground tissues form the bodies of plants

The leaf has ground tissue of two different types: the upper elongated *palisade mesophyll* and the irregularly shaped *spongy mesophyll* (**Figure 1.7**). The spongy mesophyll has large air spaces between the cells—they are not cemented all around their periphery with middle lamellae. This allows exchange of gases (carbon dioxide and oxygen) through the air spaces of the leaf during photosynthesis and respiration (discussed in Chapters 7 and 11). Both cell types have many chloroplasts (Figure 1.7), usually oriented at the cell periphery but able to move in response to light cues perceived by photoreceptors (discussed in Chapter 12).

Mesophyll cells of the leaf can differentiate into a variety of other cell types, and mesophyll is therefore considered a form of **parenchyma**, a ground tissue with thin primary walls (Figure 1.7). Parenchyma has the capacity to continue division and can differentiate into a variety of other ground tissues and vascular tissues after being produced by meristems. For example, parenchyma can differentiate into ground tissue that has thickened primary walls that can nevertheless continue to elongate (**Figure 1.8**). **Collenchyma** (e.g., as in the ribs of celery stalks) has very thick, layered walls (Figure 1.8A–D).

Parenchyma can also differentiate into the **sclerenchyma**, which has thick secondary walls (Figure 1.8 B and C). *Sclereids* come from parenchyma in leaves, fruit (e.g., pear), and flowers. Sclereid development in some tissues is dependent

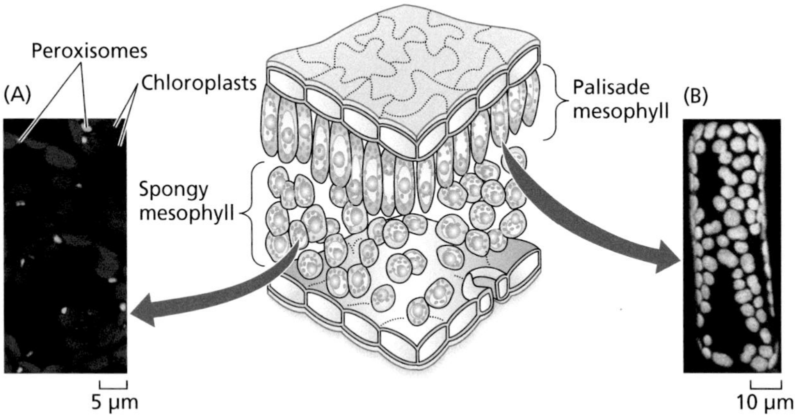

Figure 1.7 Spongy mesophyll and palisade cells in the leaf. (A) Fluorescence micrograph of spongy mesophyll cells showing peroxisomes (green) and chloroplasts (red). (B) Three-dimensional stereo view of chloroplast distribution in a palisade cell of a leaf.

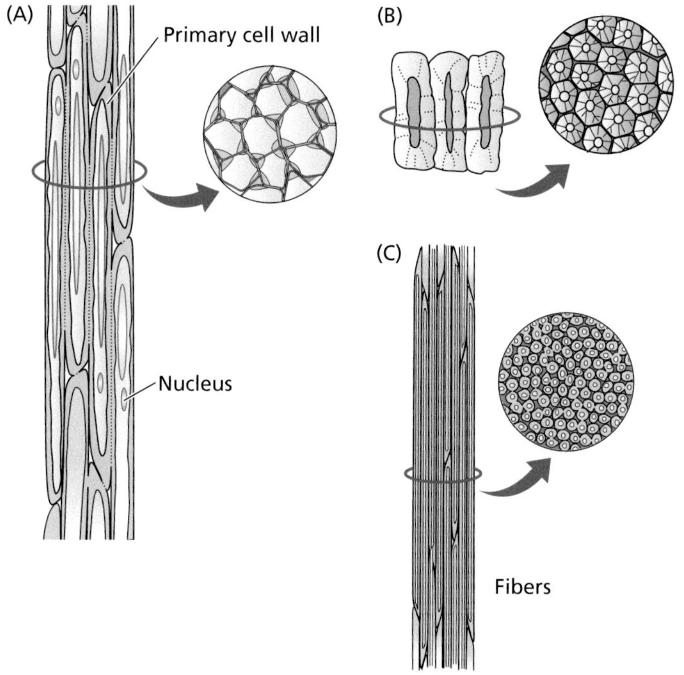

Figure 1.8 Ground tissue with thick primary walls. (A) Diagram of the collenchyma of celery in longitudinal view and cross section. (B) Diagram of sclereid cluster in longitudinal and cross-sectional views. (C) Diagram of fibers in longitudinal and cross-sectional views.

on exposure to environmental stress, such as wind and rain. **Fibers** develop from parenchyma and form elongated support structures with thickened secondary walls both in ground tissue (Figure 1.8C) and in vascular tissues (phloem fibers shown in Figure 1.5). They can elongate into the longest cells of higher plants; for example, individual fiber cells of the flax plants from the genus *Linum* can be 100 cm long! Flax fibers have very high tensile strength and are extensively used for textiles.

In the stem, the conductive tissues of the vascular cylinder are often filled with ground tissue including parenchymal cells that is referred to as *pith* (Figure 1.5A). In the root, the *cortex* is ground tissue found between the dermal and vascular tissues (Figure 1.5B). The boundary between the cortex and the vascular tissue is a specialized cell type called the **endodermis**, which has a lignin-impregnated **Casparian strip**. As we will describe in Chapter 3, the Casparian strip contributes to the separation of the cortical apoplast from the apoplast of the vascular tissue.

Vascular tissues form transport networks between different parts of the plant

Phloem cells conduct the products of photosynthesis from the leaves to roots, flowers, and fruits (discussed in Chapter 10), are living at maturity, and have nonlignified cell walls. They include **sieve cells** in gymnosperms, and **sieve tube elements** that stack to form sieve tubes in angiosperms (**Figure 1.9**). Within the sieve tubes, plasmodesmata can be seen in tangential sections of the forming sieve plate (Figure 1.9B). As with other plasmodesmatal arrays, callose is deposited at the sieve plate. The living sieve tube element is connected by plasmodesmata fields, or **sieve areas**, to adjacent cells, **companion cells** (in angiosperms) and **albuminous cells** (in gymnosperms). Also associated with phloem in some plants are fibers and storage parenchyma.

The xylem cells that conduct water and minerals from the root are nonliving at maturity and are known as **tracheary elements**. They consist of **tracheids** (found in all vascular plants) and shorter **vessel elements** (found mostly in

mesophyll Leaf tissue found between the upper and lower epidermal layers.

vascular tissues Plant tissues specialized for the transport of water (xylem) and photosynthetic products (phloem).

xylem The vascular tissue that transports water and ions from the root to the other parts of the plant.

phloem The tissue that transports the products of photosynthesis from mature leaves (or storage organs) to areas of growth and storage, including the roots.

parenchyma Metabolically active ground tissue consisting of thin-walled cells.

collenchyma A specialized parenchyma with irregularly thickened, pectin-rich, primary cell walls.

sclerenchyma Plant tissue composed of cells (sclereids and fibers), often dead at maturity, with thick, lignified secondary cell walls.

fiber An elongated, tapered sclerenchyma cell that provides support in vascular plants.

endodermis A specialized layer of cells surrounding the vascular tissue in roots and some stems.

Casparian strip A band in the cell walls of the endodermis that is impregnated with lignin. Prevents apoplastic movement of water and solutes into the stele.

sieve cells The relatively unspecialized sieve elements of gymnosperms.

sieve tube elements The highly differentiated sieve elements typical of the angiosperms.

sieve area A depression in the cell wall of a sieve tube element that contains a field of plasmodesmata.

companion cells In angiosperms, metabolically active cells that are connected to their sieve element by large, branched plasmodesmata and that take over many of the metabolic activities of the sieve element. In source leaves, they function in the transport of photosynthate into the sieve elements.

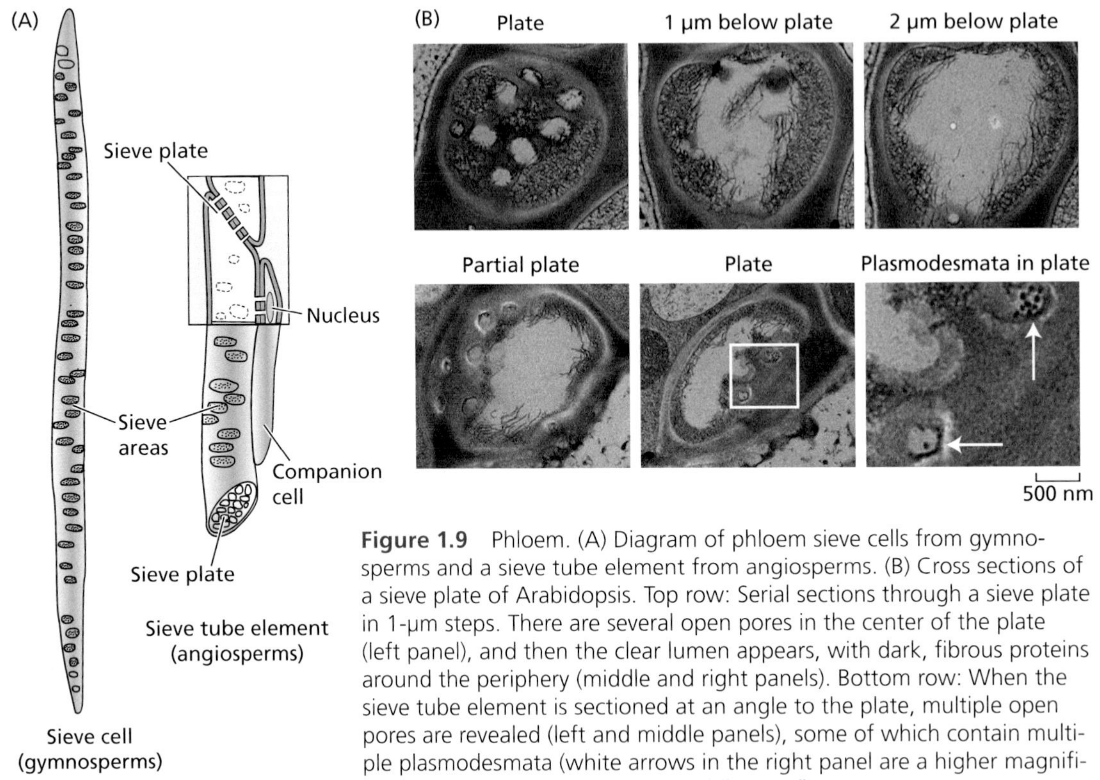

Figure 1.9 Phloem. (A) Diagram of phloem sieve cells from gymnosperms and a sieve tube element from angiosperms. (B) Cross sections of a sieve plate of Arabidopsis. Top row: Serial sections through a sieve plate in 1-µm steps. There are several open pores in the center of the plate (left panel), and then the clear lumen appears, with dark, fibrous proteins around the periphery (middle and right panels). Bottom row: When the sieve tube element is sectioned at an angle to the plate, multiple open pores are revealed (left and middle panels), some of which contain multiple plasmodesmata (white arrows in the right panel are a higher magnification of the boxed area in the middle panel).

angiosperms) (**Figure 1.10**). Vessel elements stack end to end to form wide (up to 0.7 mm) columns called vessels. Protoxylem cells with primary walls begin to differentiate into mature tracheary elements by laying down secondary cell walls with spiral bands of **cellulose** and strengthened with lignin. Once elongation has stopped, the top and bottom end walls develop large perforations. On the side walls, secondary walls continue to thicken, except in areas containing pits. The pits start as plasmodesmata fields and ultimately become channels through the walls shared between adjacent cells. The cells of the tracheids and vessels die, undergoing a process called programmed cell death (discussed in Chapter 15), leaving behind the hardened bundle of vessels made by the secondary walls and connected side to side by pits. The pits are important because flow through these narrow vessels depends on having a continuous liquid stream to conduct water and solutes from the roots to the shoots (discussed in Chapter 3). If an air bubble, or embolism, forms in a vessel element or tracheid, the stream can be diverted around the embolism through the pits into adjacent cells.

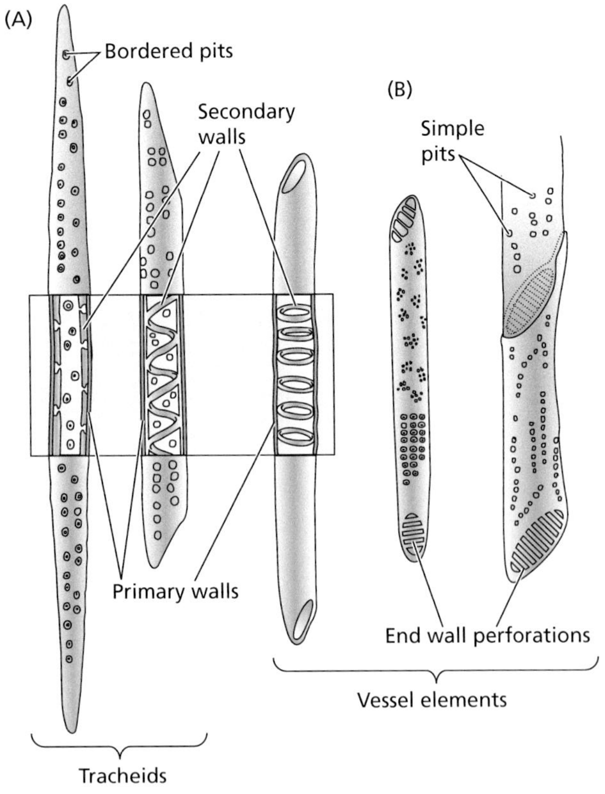

Figure 1.10 Xylem. (A) Diagram of two tracheids and a vessel element. The cutaway views (in the blue window) reveal thickenings of secondary wall in spiral (helical) and annular (ring-shaped) arrangements. (B) Diagram of two vessel elements, showing pits (bordered in tracheids, simple in vessel elements) and end wall perforations.

1.4 Plant Cell Compartments

Explain the biochemical structure and behavior of plant cell membranes and associated proteins.

All plant cells have the same basic organization: They contain cytoplasm, a nucleus, and other organelles, all enclosed by the plasma membrane and cell wall (**Figure 1.11**). All plant cells *begin* with a similar complement of organelles. Organelles fall into two main categories based on how they arise:

1. *The endomembrane system and peroxisomes*: the endoplasmic reticulum, nuclear envelope (which encapsulates the nucleus), Golgi apparatus, *trans* Golgi network, vacuole, and endosomes (smaller endomembrane compartments). Other organelles derived from the endomembrane system include oil bodies, peroxisomes, and specialized peroxisomes called glyoxysomes that function in lipid storage and carbon metabolism in seeds and leaves. With the exception of some peroxisomes,

albuminous cells Sieve element–associated cells in the phloem of gymnosperms. Although similar to companion cells in angiosperms, they have a different developmental origin. Also called *Strasburger cells*.

tracheary elements Water-transporting cells of the xylem.

tracheids Spindle-shaped, water-conducting cells with tapered ends and pitted walls without perforations found in the xylem of both angiosperms and gymnosperms.

vessel elements Nonliving water-conducting cells with perforated end walls found only in angiosperms and a small group of gymnosperms.

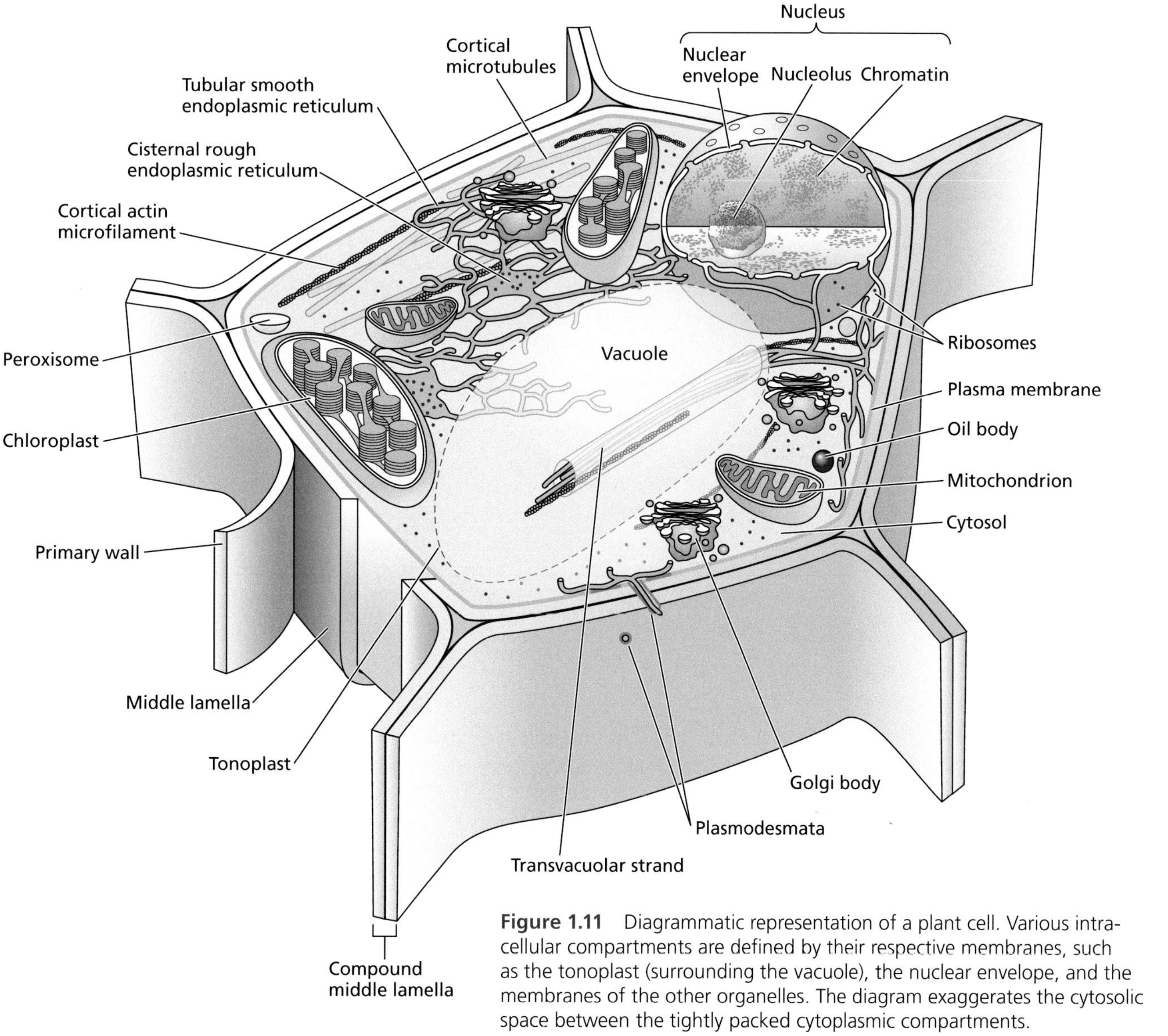

Figure 1.11 Diagrammatic representation of a plant cell. Various intracellular compartments are defined by their respective membranes, such as the tonoplast (surrounding the vacuole), the nuclear envelope, and the membranes of the other organelles. The diagram exaggerates the cytosolic space between the tightly packed cytoplasmic compartments.

cellulose A linear chain of (1,4)-linked β-D-glucose. The repeating unit is cellobiose.

fluid mosaic The common molecular lipid–protein structure for all biological membranes. A double layer (bilayer) of polar lipids (phospholipids or, in chloroplasts, glycosylglycerides) has a hydrophobic, fluid-like interior. When structural sterols and sphingolipids are present in the bilayer, the fluidity of the interior is decreased. Membrane proteins are embedded in the bilayer and may move laterally.

components of the endomembrane system are not formed by semiautonomous processes. The endomembrane system plays a central role in secretory processes, cell signaling, specialized metabolite and hormone production, membrane recycling, the cell cycle, and cell expansion.

2. *Semiautonomous (independently dividing) organelles of endosymbiotic origin*: plastids and mitochondria. These organelles function in energy metabolism and storage, and they synthesize a wide range of metabolites used in the biosynthesis of all cell components.

Because all of these cellular organelles are membranous compartments, we'll begin by describing membrane structure and function.

Biological membranes are lipid bilayers that contain proteins

All cells are enclosed in a membrane that serves as their outer boundary, separating the cytoplasm from the external environment. This plasma membrane allows the cell to take up and retain certain substances while excluding others. Various transport proteins embedded in the plasma membrane are responsible for this selective traffic of solutes—water-soluble ions and small, uncharged molecules—across the membrane. The accumulation or expulsion of ions or molecules in the cytoplasm through the action of transport proteins consumes metabolic energy. In plant cells, membranes enclose the genetic material, delimit the boundaries of other specialized internal organelles of the cell, and regulate the fluxes of ions and metabolites into and out of these compartments.

Biological membranes consist of a double layer (*bilayer*) of lipids in which proteins are embedded in what is referred to as a **fluid mosaic** (**Figure 1.12**). Membrane proteins may move laterally depending on the fluidity of the membrane that surrounds them as well as interactions between the membrane and associated proteins with the cytoskeleton and cell wall. Each layer of the bilayer is called a *leaflet*. In most membranes, proteins make up about half of the membrane's mass. However, the composition of the lipids and the properties of the proteins vary from membrane to membrane, conferring on each membrane its unique functional characteristics.

LIPIDS The most prominent membrane lipids found in plants are phospholipids, a class of lipids composed of glycerol covalently linked to two fatty acids and a phosphate group. Attached to the phosphate group in the phospholipid is a variable component, called the *head group*, such as choline, glycerol, or inositol (Figure 1.12B). The nonpolar hydrocarbon chains of the fatty acids form a hydrophobic region that excludes water. In contrast to the fatty acids, the head groups are highly polar; consequently, phospholipid molecules display both hydrophilic and hydrophobic properties (i.e., they are *amphipathic*). The fatty acid chains vary in length but usually consist of 16 to 24 carbons. If the carbons are linked by single bonds, the fatty acid chain is *saturated* (with hydrogen atoms), but if the chain includes one or more double bonds, it is *unsaturated*.

Double bonds in a fatty acid chain create a kink in the chain that prevents tight packing of the phospholipids in the bilayer (i.e., the bonds adopt a kinked *cis* configuration, as opposed to an unkinked *trans* configuration shown in Figure 1.12B). The kinks promote membrane fluidity and are critical for many membrane functions. Various phospholipids are distributed asymmetrically across the plasma membrane, giving the membrane sidedness; in terms of phospholipid composition, the outside leaflet of the plasma membrane that faces the outside of the cell is different from the inside leaflet that faces the

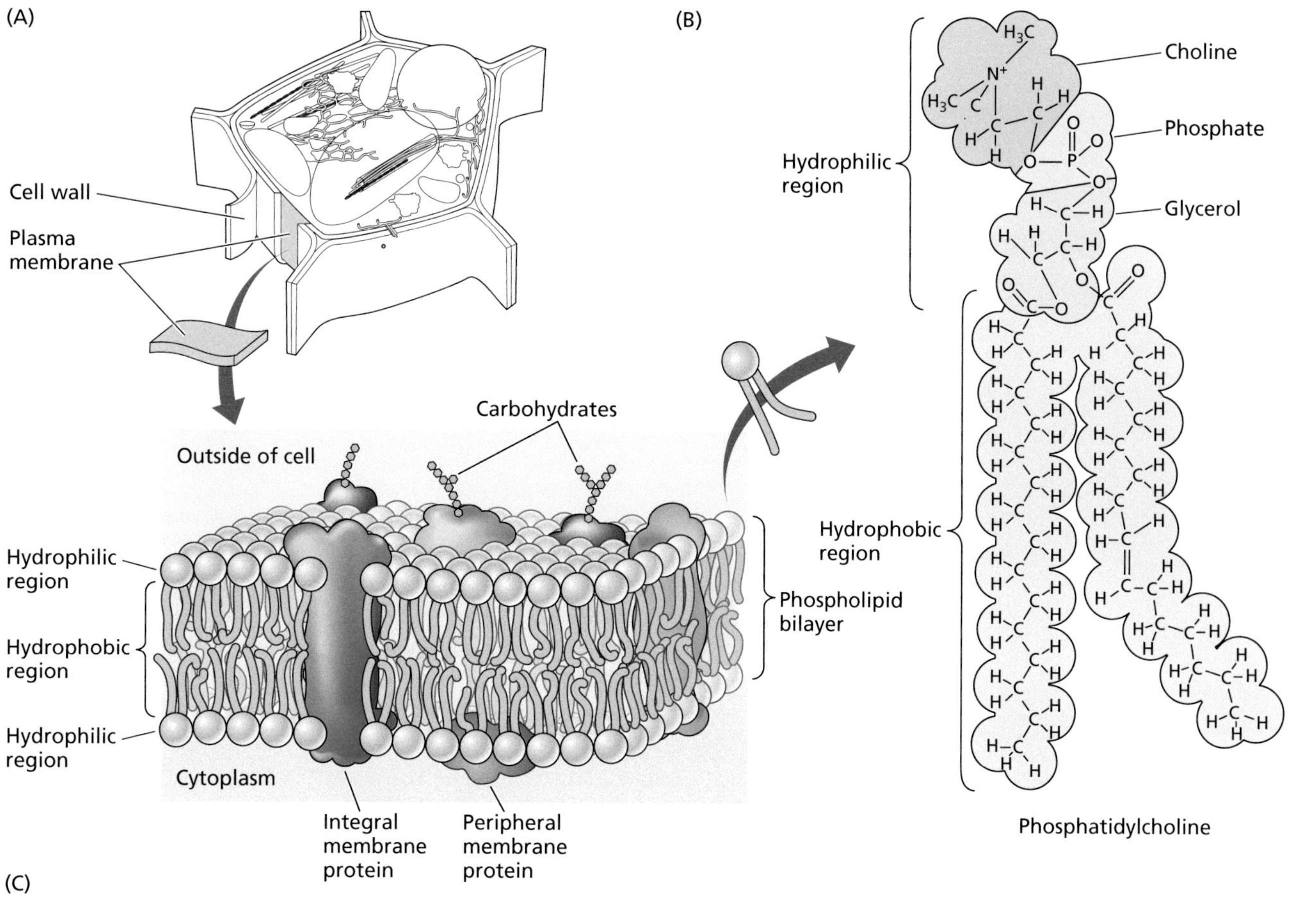

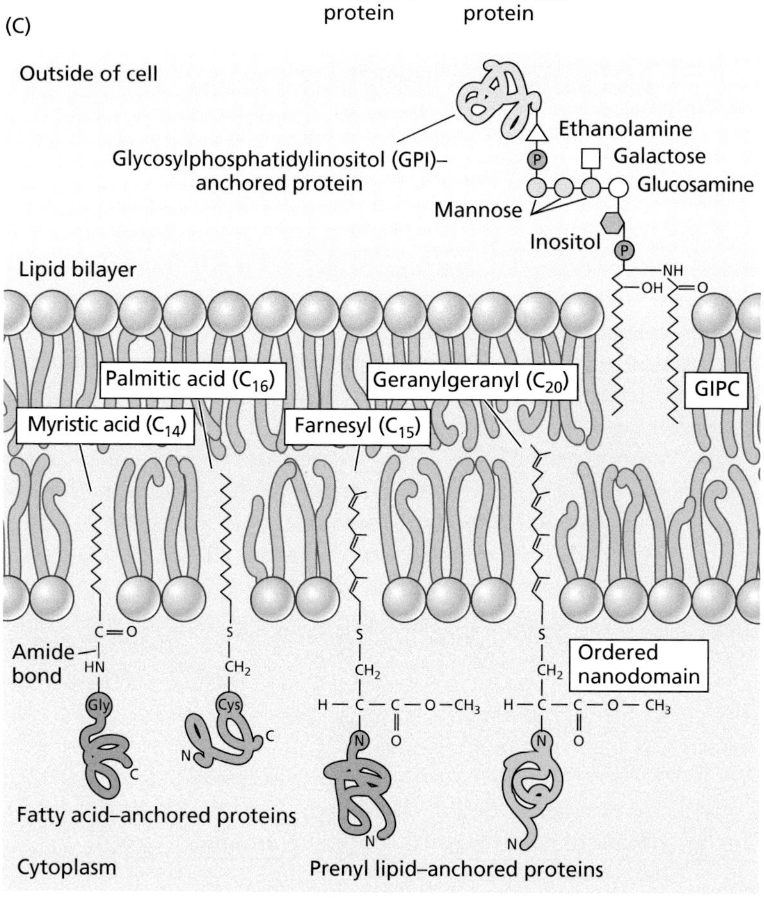

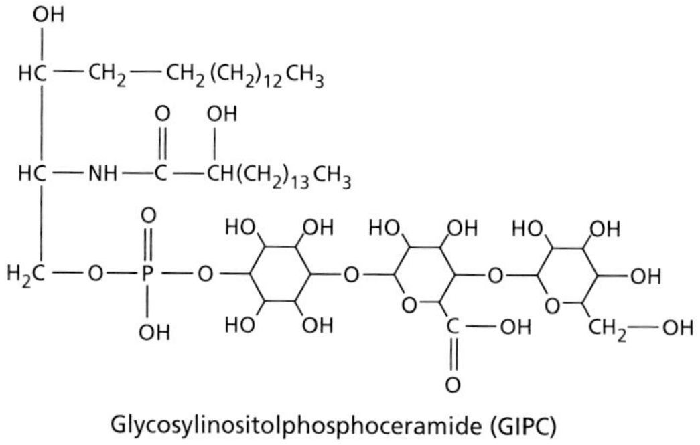

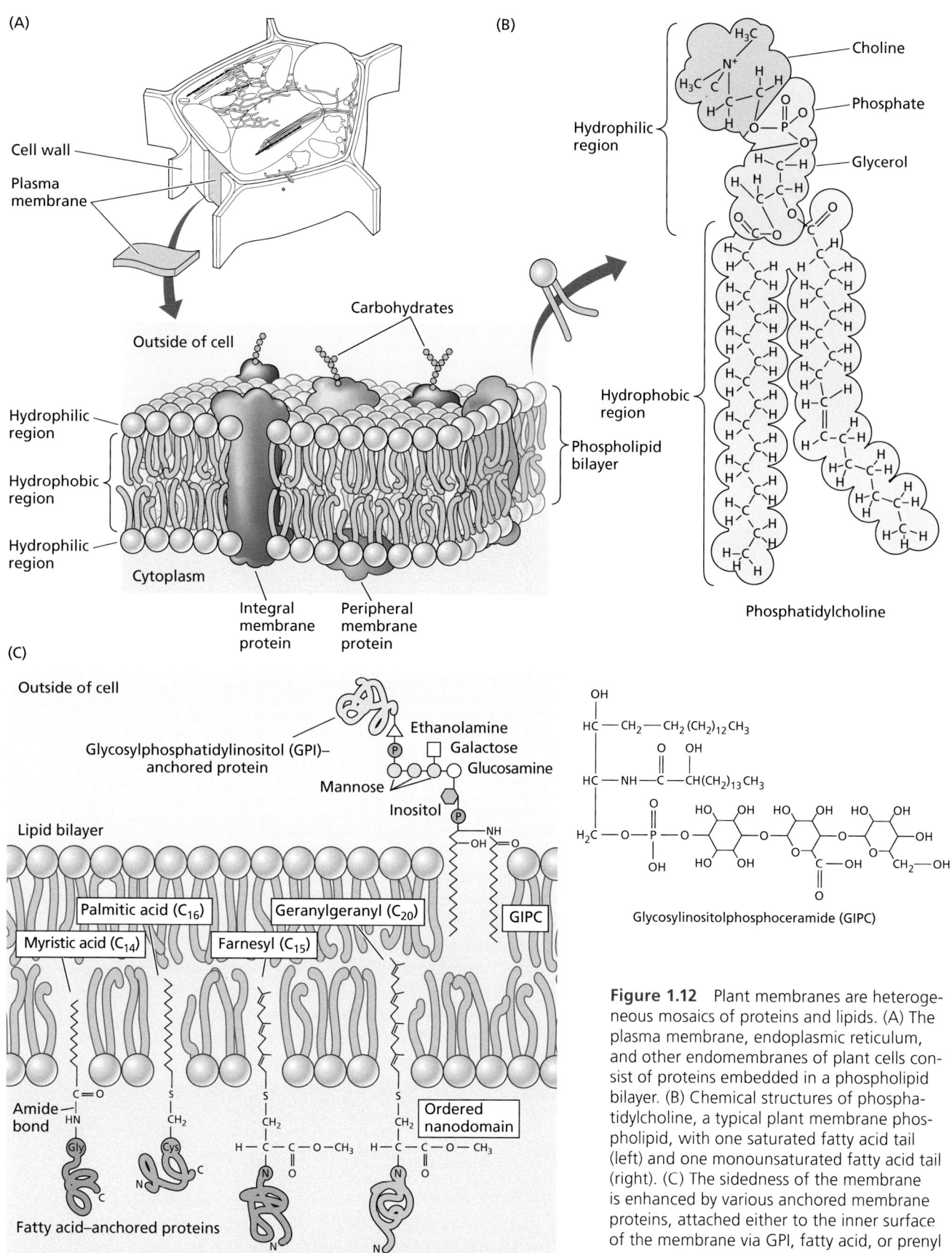

Figure 1.12 Plant membranes are heterogeneous mosaics of proteins and lipids. (A) The plasma membrane, endoplasmic reticulum, and other endomembranes of plant cells consist of proteins embedded in a phospholipid bilayer. (B) Chemical structures of phosphatidylcholine, a typical plant membrane phospholipid, with one saturated fatty acid tail (left) and one monounsaturated fatty acid tail (right). (C) The sidedness of the membrane is enhanced by various anchored membrane proteins, attached either to the inner surface of the membrane via GPI, fatty acid, or prenyl groups, or to the outer surface via glycosylinositolphosphorylceramide (GIPC).

cytoplasm. Turnover of membrane lipids is generally more rapid than that of membrane proteins.

In some membranes (particularly the plasma membrane), higher concentrations of sterols such as sitosterol and stigmasterol reduce the fluidity of membrane regions, resulting in the formation of relatively stable rigid lipid subdomains within the membrane. GIPC and glycosylceramide, known collectively as *sphingolipids*, are relatively hydrophobic lipids that limit lateral diffusion of proteins within the lipid phase, thereby further stabilizing lipid-ordered *nanodomains* that are important for clustering of signaling and transport protein complexes.

The membranes of plastids, the group of membrane-bound organelles to which chloroplasts belong, are unique in that their lipid component consists almost entirely of glycosylglycerides. The glycosyl polar head groups are derived from galactose, and galactolipids are derived from a common ancestor with cyanobacteria.

PROTEINS The proteins associated with the lipid bilayer are of two main types: integral and peripheral.

Integral membrane proteins are embedded in the lipid bilayer (Figure 1.12A). Most integral proteins span the entire width of the phospholipid bilayer, so one part of the protein interacts with the outside of the compartment, another part interacts with the hydrophobic core of the membrane, and a third part interacts with the interior (lumen) of the compartment. They may have one or many transmembrane domains.

Proteins that serve as membrane transporters (ion channels, carriers, and pumps; discussed in Chapter 6) are always integral membrane proteins, as are certain receptors that participate in signal transduction pathways (discussed in Chapter 12) and proteins in the endomembrane system that mediate membrane fusion. Some integral membrane proteins, such as those with extracellular arabinogalactan (carbohydrate) domains, interact directly with the cell wall.

Peripheral membrane proteins are attached to the membrane surface (Figure 1.12A) by noncovalent bonds and hydrophobic interactions, and can be dissociated from the membrane with high-salt solutions or chaotropic agents that break ionic and hydrogen bonds, respectively. Peripheral proteins serve a variety of functions in the cell. For example, some are involved in regulating intracellular membrane traffic or mediate interactions between the membranes and the major elements of the cytoskeleton.

Some peripheral membrane proteins are called **anchored proteins** as they are modified with hydrophobic prenyl, acyl, or myristol groups that enhance membrane interactions to varying extents. Such modifications are often reversible, allowing for dynamic interactions of the proteins with the membrane surface. Proteins can also be attached to the plasma membrane surface via a **glycosylphosphatidylinositol anchor** that is covalently attached to the protein in the endoplasmic reticulum (Figure 1.12C). Because this particular lipid anchor is asymmetrically distributed to the outer leaflets of the plasma membrane, it makes the two sides of the membrane even more distinct.

Specific lateral distribution of membrane components in the same plasma membrane is critical for development, growth, and nutrient transport. Polarized localization of proteins in the plasma membrane results in up to four distinct domains on each cell surface that are broadly designated as outer (peripheral), inner (central), apical, and basal. A combination of polar secretion, restricted lateral membrane diffusion, cytoskeleton corralling, and

integral membrane proteins Proteins that are embedded in the lipid bilayer of a membrane via at least one transmembrane domain.

peripheral membrane proteins Proteins that are bound to the membrane surface by noncovalent bonds, such as ionic bonds or hydrogen bonds.

anchored proteins Proteins that are bound to the membrane surface via lipid molecules, to which they are covalently attached.

glycosylphosphatidylinositol anchor In plants, a phosphoglyceride protein modification that anchors proteins to ordered plasma membrane domains.

endocytic recycling contributes to polarized distribution of plasma membrane proteins in these domains.

1.5 The Plant Cytoskeleton

Describe the key elements of nuclear architecture and explain their biological activity.

The cytoplasm is organized into a dynamic three-dimensional structure by a network of filamentous proteins called the **cytoskeleton**. This network provides the spatial organization for the organelles and serves as scaffolding for the movements of organelles and other cytoskeletal components. It also plays fundamental roles in endomembrane movement, mitosis, meiosis, cytokinesis, wall deposition, the maintenance of cell shape, and cell differentiation.

The plant cytoskeleton consists of microtubules and microfilaments

Two major types of cytoskeletal elements are observed in plant cells: microtubules and microfilaments. Each type is filamentous, having a fixed diameter and a variable length up to many micrometers.

Microtubules are polarized hollow cylinders with an outer diameter of 25 nm; they are composed of polymers of the protein **tubulin**. The tubulin monomer is a heterodimer composed of two similar polypeptide chains (α- and β-tubulin) (**Figure 1.13A**). A single microtubule consists of hundreds of thousands of tubulin monomers arranged in columns.

Microfilaments are solid, with a diameter of 7 nm. They are composed of the monomeric form of the protein **actin**, called globular actin, or G-actin. Activated by ATP, monomers of G-actin polymerize to form chains of actin subunits referred to as filamentous actin, or F-actin. A microfilament is helical, a shape resulting from the polarity of association of the G-actin monomers (**Figure 1.13B**).

Actin, tubulin, and their polymers are in constant flux in the living cell

In the cell, actin and tubulin subunits exist as pools of free proteins that are in dynamic equilibrium with their polymerized forms. Compared with animal cells, plants have larger pools of monomeric G-actin and strands made up of F-actin arrays. The cycle of polymerization–depolymerization is essential for cell life. Each of the monomers contains a bound nucleotide: ATP or ADP in the case of actin, GTP or GDP (guanosine tri- or diphosphate) in the case of tubulin. Both microtubules and microfilaments are polarized; that is, the two ends are different. The polarity is displayed by the different rates of growth of the two ends, with the plus end more active than the minus end.

In microfilaments, the polarity arises from the polarity of the actin monomer itself. ATP binds the nucleotide-binding cleft facing the fast-growing plus end. When ATP is hydrolyzed, then ADP is bound to the nucleotide-binding cleft and faces the slow-growing minus end. **Profilin** proteins assist in bringing monomers to the plus end of F-actin, where they are assembled by **formins**. The proteins **fimbrin** and **villin** function in bundling actin filaments into cables. The **actin-related protein 2/3 (Arp 2/3)** complex creates

cytoskeleton Composed of polarized microfilaments of actin or microtubules of tubulin, the cytoskeleton helps control the organization and polarity of organelles and cells during growth.

microtubule A component of the cell cytoskeleton and the mitotic spindle, and a player in the orientation of cellulose microfibrils in the cell wall. Made of tubulin.

tubulin A family of cytoskeletal GTP-binding proteins with three members, α-, β-, and γ-tubulin. α-tubulin forms heterodimers with β-tubulin, which polymerize to form microtubules.

microfilament A component of the cell cytoskeleton made of actin; it is involved in organelle motility within cells.

actin A major ATP-binding cytoskeletal protein. The monomeric, globular form of actin is called G-actin; the polymerized form in microfilaments is F-actin.

protofilament A column of polymerized tubulin monomers (α- and β-tubulin heterodimers) or a chain of polymerized actin subunits.

profilin Actin binding protein that keeps the depolymerized globular G-actin monomers charged with ATP, so that they can be rapidly re-integrated into F-actin. This protein also binds formins and thereby accelerates the formation of F-actin from formins.

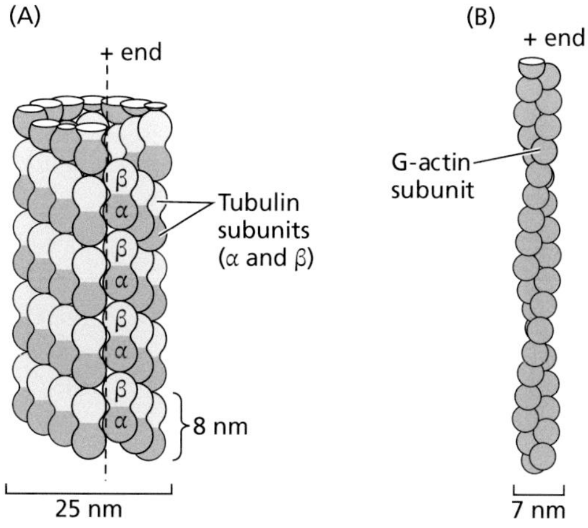

Figure 1.13 Cytoskeletal microtubules and microfilaments. (A) Drawing of a microtubule in longitudinal view. The organization of the α and β subunits is shown with the + end at the top. (B) Diagrammatic representation of a microfilament, showing an F-actin strand (**protofilament**) with a helical pitch based on the asymmetry of the monomers, the G-actin subunits. The + end of the microfilament is at the top.

formins Proteins that bind actin and actin–profilin complexes to initiate polymerization of the actin filament.

fimbrin An actin-binding protein that bundles F-actin filaments together into larger filamentous bundles.

villin An actin binding protein that bundles F-actin filaments.

actin-related protein 2/3 (Arp 2/3) Actin-related proteins 2 and 3, which bind to the side of a pre-existing actin filament and form a complex with actin to initiate growth of an actin filament branch.

treadmilling During interphase, a process by which microfilaments or microtubules in the cortical cytoplasm appear to migrate around the cell periphery due to addition of G-actin or tubulin heterodimers, respectively, to the plus end at the same rate as their removal from the minus end.

cytoplasmic streaming The coordinated movement of particles and organelles through the cytosol.

directed organelle movement Movement of an organelle in a particular direction, which can be driven by the interaction with molecular motors associated with the cytoskeleton.

pollen Small structures (microspores) produced by anthers of seed plants. Contain haploid male nuclei that will fertilize the egg in the ovule.

tip growth Localized growth at the tip of a plant cell, caused by localized secretion of new wall polymers. Occurs in pollen tubes, root hairs, some sclerenchyma fibers, and cotton fibers, as well as moss protonema and fungal hyphae. Contrast with *diffuse growth*.

branched actin filaments that are important for polarized growth of cells. Capping proteins can stabilize filament ends, and depolymerizing proteins enhance turnover. These events are highly dynamic, and are referred to as **treadmilling** when they occur simultaneously at the minus and plus ends of actin filaments.

In microtubules polarity arises from the tubular structure formed by α- and β-tubulin heterodimers (Figure 1.13A). The α-tubulin monomer exists only in the GTP form and is exposed on the minus end, and the β-tubulin can bind either GTP or GDP and is exposed on the plus end. Microfilaments and microtubules have half-lives, usually counted in minutes, determined by accessory proteins that regulate the dynamics of microfilaments and microtubules.

Cytoskeletal motor proteins mediate cytoplasmic streaming and directed organelle movement

Mitochondria, peroxisomes, and some endomembrane structures are extremely mobile in plant cells. These approximately 1-μm particles move at rates of about 1 to 10 μm s^{-1} in seed plants, depending on cell type and developmental stage. This movement is quite fast; it is equivalent to a 1-m object moving at 10 m s^{-1}, approximately the speed of the world's fastest human. Actin and its motor protein, myosin, act together to generate this movement using energy released by ATP hydrolysis and are therefore frequently referred to as the actomyosin cytoskeleton.

Cytoplasmic streaming refers to the coordinated flow of the cytoplasm with organelles inside the cell. The movement of individual organelles can be part of cytoplasmic streaming but is perhaps better called **directed organelle movement** because organelles can frequently move past each other in opposite directions on neighboring F-actin filaments. If directed organelle movement exerts sufficient viscous drag on the surrounding cytoplasm, then it will drive cytoplasmic streaming. Accordingly, there is almost no cytoplasmic streaming in small meristematic cells. As cells elongate and differentiate, rates of streaming increase. In the giant cells of the green algae *Chara* and *Nitella*, cytoplasmic streaming occurs in a helical path down one side of a cell and up the other side, at speeds of up to 75 μm s^{-1}. Similarly, in rapidly growing **pollen** tubes, the presence of a forward stream in the center and of return streams at the periphery creates fountain-like cytoplasmic streaming.

Molecular motors participate in all directed organelle movements. Plants have two types of motors, the myosins and the kinesins. The myosins are actin-binding proteins. The kinesins are associated with microtubules. When myosins and kinesins move along the cytoskeleton, they do so in a particular direction along the polar cytoskeletal polymers. Myosins generally move toward the plus end of the actin filaments, and a plant-specific isoform, myosin XI, plays an important role in **tip growth** of pollen tubes and root hairs. Kinesins can move along microtubules in both directions and can bind to chromatin or to other microtubules to help organize the spindle apparatus during mitosis.

All of the motors have separate head, neck, and tail domains (shown with myosin XI in **Figure 1.14**). In order for a motor to move an organelle, the motor dimerizes; two motor molecules interact and bind to the organelle at the cargo domain. The two heads of the dimer then alternately bind to the cytoskeleton and "walk" forward while the neck flexes as ATP is hydrolyzed (Figure 1.14C). In this way, the organelle (cargo) is moved along the cytoskeleton. Both polarized growth and cell expansion involve the cytoskeleton and cell wall. Cytoskeletal organization and modification of plant cell walls determine directional plant growth.

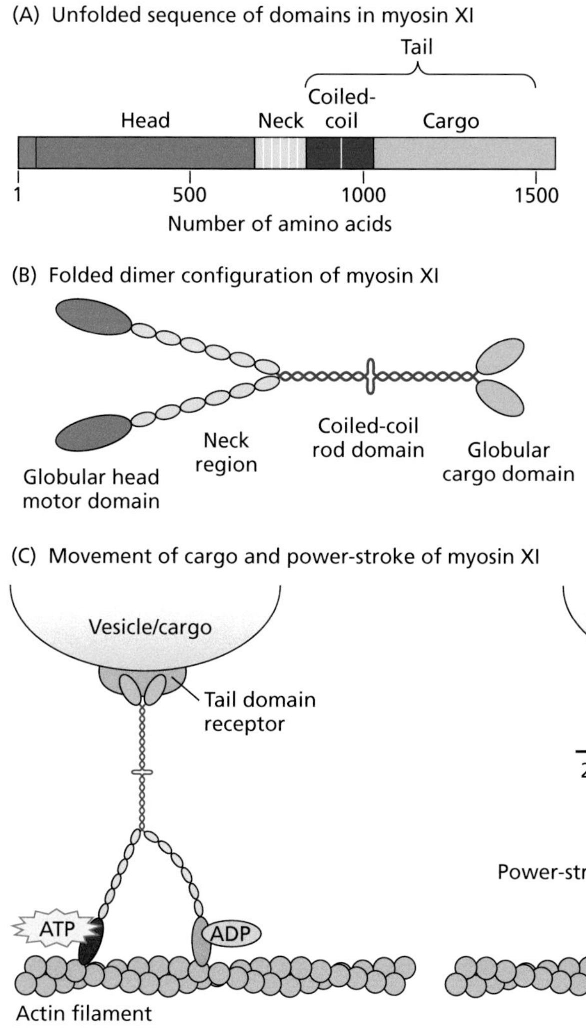

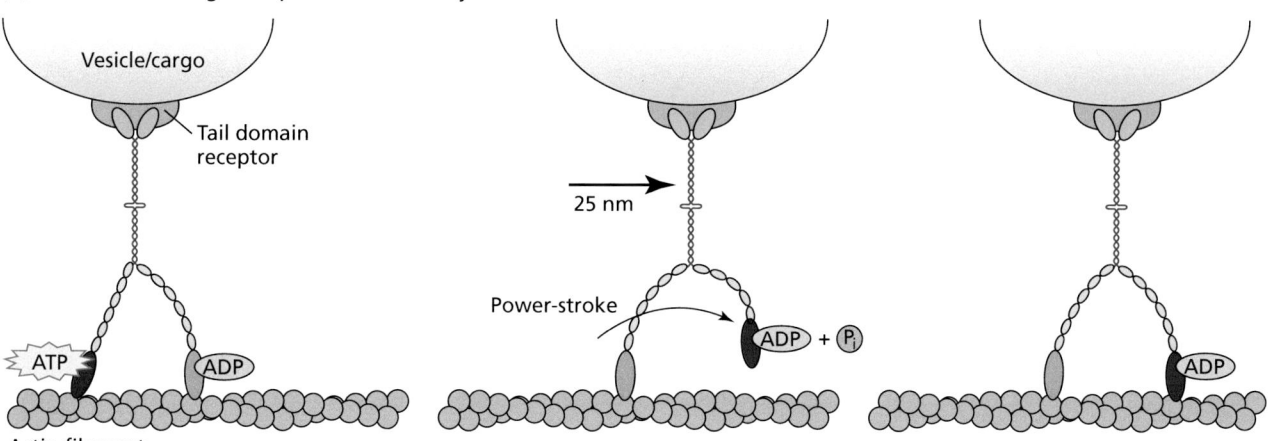

Figure 1.14 Myosin-driven movement of organelles. (A) Amino acid domains of a class XI myosin motor protein. (B) The globular head domain binds to the cytoskeleton reversibly, depending on the phosphorylation state of the nucleotide in the domain's ATPase active site. The neck region changes angle upon hydrolysis of ATP, flexing the head relative to the tail. The tail domain contains regions for dimerization, and the end of the globular tail domain binds to specific organelles or "cargo" and is called the *cargo domain*. (C) Movement and power-stroke of myosin XI. The two heads, shown in red and pink, have ATPase and motor activity that change the configuration of the neck region adjacent to the head so the motor "walks" along the actin filament during the power-stroke of the motor, when ATP is hydrolyzed to ADP and inorganic phosphate (Pi). The cargo moves about 25 nm with each step. What is now the trailing head domain in pink releases its ADP and binds ATP, allowing the process to repeat.

1.6 The Nucleus

Define the elements of the cytoskeleton and explain their biological activity.

The **nucleus** (plural *nuclei*) is the membrane-enclosed organelle that contains the genetic information primarily responsible for regulating the metabolism, growth, and differentiation of the cell. Collectively, these genes and their intervening sequences are referred to as the **nuclear genome**. The size of the nuclear genome in plants is highly variable, ranging from about 1.2×10^8 base pairs for the mustard relative *Arabidopsis thaliana* to 1×10^{11} base pairs for the lily *Fritillaria assyriaca*. For comparison, the human genome is about 3×10^9 base pairs. The remainder of the genetic information of the cell is contained in the two semiautonomous organelles—the plastid and the mitochondrion.

The nucleus consists of a complex matrix, the **nucleoplasm**, surrounded by a double membrane called the **nuclear envelope** (**Figure 1.15A**), which is a subdomain of the **endoplasmic reticulum** (ER). **Nuclear pores** form selective channels across both membranes, connecting the nucleoplasm with the cytoplasm (**Figure 1.15B**). There can be very few to many thousands of nuclear pores in an individual nuclear envelope.

nucleus (plural *nuclei*) The organelle that contains the genetic information primarily responsible for regulating cellular metabolism, growth, and differentiation.

nuclear genome The entire complement of DNA found in the nucleus.

nucleoplasm The soluble matrix of the nucleus in which the chromosomes and nucleolus are suspended.

nuclear envelope The double membrane surrounding the nucleus.

endoplasmic reticulum (ER) A continuous membrane system within the cytoplasm of eukaryotic cells that serves multiple functions, including the synthesis, modification, and intracellular transport of proteins.

nuclear pores Sites where the two membranes of the nuclear envelope join, forming a partial opening between the interior of nucleus and the cytosol.

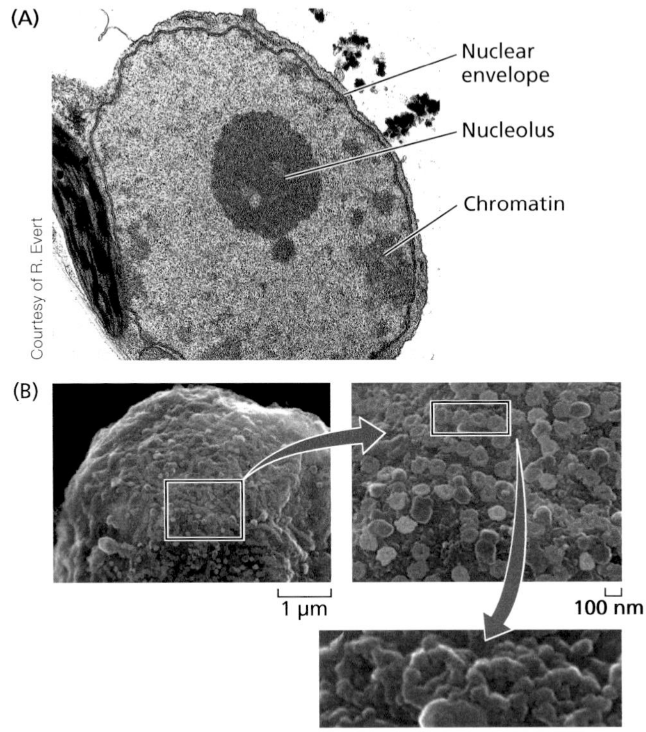

Figure 1.15 Pore complexes in the nuclear membrane. (A) Transmission electron micrograph of a plant nucleus, showing the nucleolus and the nuclear envelope. (B) Organization of the nuclear pore complexes (NPCs) on the nuclear surface in tobacco culture cells. The NPCs that touch each other are colored brown; the rest are colored blue. The first inset (top right) shows that most of the NPCs are closely associated, forming rows of 5 to 30 NPCs. The second inset (bottom right) shows the tight associations of the NPCs.

chromosome A threadlike or rod-shaped structure of DNA and protein found in the nucleus of most living cells, encoding genetic information in the form of genes.

chromatin The DNA–protein complex found in the interphase nucleus. Condensation of chromatin occurs during cell division to form the rod-shaped mitotic and meiotic chromosomes.

histones A family of proteins that interact with DNA and around which DNA is wound to form a nucleosome.

nucleosome A structure consisting of eight histone proteins around which DNA is coiled.

heterochromatin Chromatin that is densely packed and transcriptionally modified or suppressed. Compare with *euchromatin*.

euchromatin The dispersed, transcriptionally active form of chromatin. Compare with *heterochromatin*.

nucleolus (plural *nucleoli*) A densely granular region in the nucleus that is the site of ribosome synthesis.

ribosome The site of cellular protein synthesis; consists of RNA and protein.

nucleolar organizer region Site in the nucleolus where chromosomal regions that code for ribosomal RNA are clustered and are transcribed.

transcription The process by which the base sequence information in DNA is copied into an RNA molecule.

The nuclear "pore" is actually an elaborate structure composed of more than 100 different nucleoporin proteins arranged octagonally to form a 105-nm nuclear pore complex. The nucleoporins lining the 40-nm channel of the nuclear pore complex form a meshwork that acts as a supramolecular sieve. A specific amino acid sequence called the nuclear localization signal is required for a protein to gain entry into the nucleus when it exceeds the size of the nuclear pore opening.

The nucleus is the site of storage and replication of the **chromosomes**, composed of DNA and its associated proteins (**Figure 1.16**). Collectively, this DNA–protein complex is known as **chromatin**. The linear length of the entire DNA in any plant genome is about 2 m or 6 feet long. To solve the problem of packaging this chromosomal DNA within the nucleus, segments of the linear double helix of DNA are coiled twice around a solid cylinder of eight **histone** protein molecules, forming a **nucleosome**. Nucleosome assembly is also assisted by large protein complexes called condensins. Nucleosomes are arranged like beads on a string along the length of each chromosome. When the nucleus is not dividing, the chromosomes maintain their spatial independence; they do not "tangle," and instead remain quite discrete, giving rise to the possibility of separate regulation of each chromosome.

During mitosis, the chromatin condenses, first by coiling tightly into a 30-nm chromatin fiber, followed by further folding and packing (Figure 1.16). At interphase, two types of chromatin are distinguishable based on their degree of condensation: heterochromatin and euchromatin. **Heterochromatin** is a highly compact and transcriptionally inactive form of chromatin and accounts for about 10% of the DNA. Most of the heterochromatin is concentrated along the periphery of the nuclear membrane and is associated with regions of the chromosomes that contain few genes, such as telomeres and centromeres. The rest of the DNA consists of **euchromatin**, the dispersed, transcriptionally active form. Only about 10% of the euchromatin is transcriptionally active at any given time. The remainder is in an intermediate condensation state between transcriptionally active euchromatin and heterochromatin. During the cell cycle, chromatin undergoes dynamic structural changes. In addition to transient local changes that are required for transcription, heterochromatic regions can be converted to euchromatic regions, and vice versa.

Nuclei contain a densely granular region called the **nucleolus** (plural *nucleoli*) (Figure 1.15A), the site of **ribosome** synthesis. Typical cells have one nucleolus per nucleus; some cells have more. The nucleolus includes portions of one or more chromosomes where ribosomal RNA (rRNA) genes are clustered to form a structure called the **nucleolar organizer region**. Even though chromosomes remain largely separate within the nucleus, parts of several may come together to help form the nucleolus. The nucleolus assembles the ribosomal proteins (imported from the cytoplasm) and rRNA of the ribosome into a large and a small subunit, each exiting the nucleus separately through the nuclear pores. The two subunits unite in the cytoplasm to form a complete ribosome (**Figure 1.17A**, step 1). Assembled ribosomes are energy-consuming protein-synthesis machines that consume more than 50% of total cellular energy. Ribosomes synthesized inside the nucleus for cytoplasmic, "eukaryotic" protein synthesis, the 80S ribosomes (named for their centrifugal sedimentation coefficient), are larger than the ribosomes synthesized and functioning inside the endosymbiotic mitochondria and plastids for their separate program of "prokaryotic" protein synthesis, the 70S ribosomes.

Gene expression involves transcription, translation, and protein processing

The nucleus is the site of read-out, or **transcription**, of the cell's DNA (Figure 1.17A). Some of the DNA is transcribed by specific DNA-dependent RNA polymerases into messenger RNA (mRNA) that encodes proteins. Ribosomes in the cytoplasm then read mRNA exported from the nucleus in one direction, from the 5′ to the 3′ end (**Figure 1.17B**). Other regions of the DNA are transcribed into transfer RNA (tRNA) and rRNA to be used in **translation** (Figure 1.17A, steps 1–3). The RNA moves through nuclear pores to the cytoplasm, where "free" or cytoskeletally attached polyribosomes (groups of ribosomes translating a single RNA strand) translate the RNA into proteins destined for the cytoplasm and organelles that receive proteins independently of the endomembrane pathway. Stability or degradation of mRNAs is precisely regulated by many interacting proteins and is an important part of gene expression.

Most proteins that reside in or are trafficked through the endomembrane system enter the ER during the process of translation (co-translationally) on polyribosomes situated on the ER membrane. The mechanism of co-translational insertion of proteins into the ER is complex, involving the ribosomes, the mRNA that codes for the secretory protein, and a special protein-translocating pore, the **translocon**, in the ER membrane (Figure 1.17A, steps 4–7). Proteins that are produced by the endomembrane system are called secretory proteins and can either be secreted from the cell interior to the apoplast, retained in endomembrane organelles as either soluble or membrane-anchored proteins, or trafficked through the endomembrane system to the vacuole.

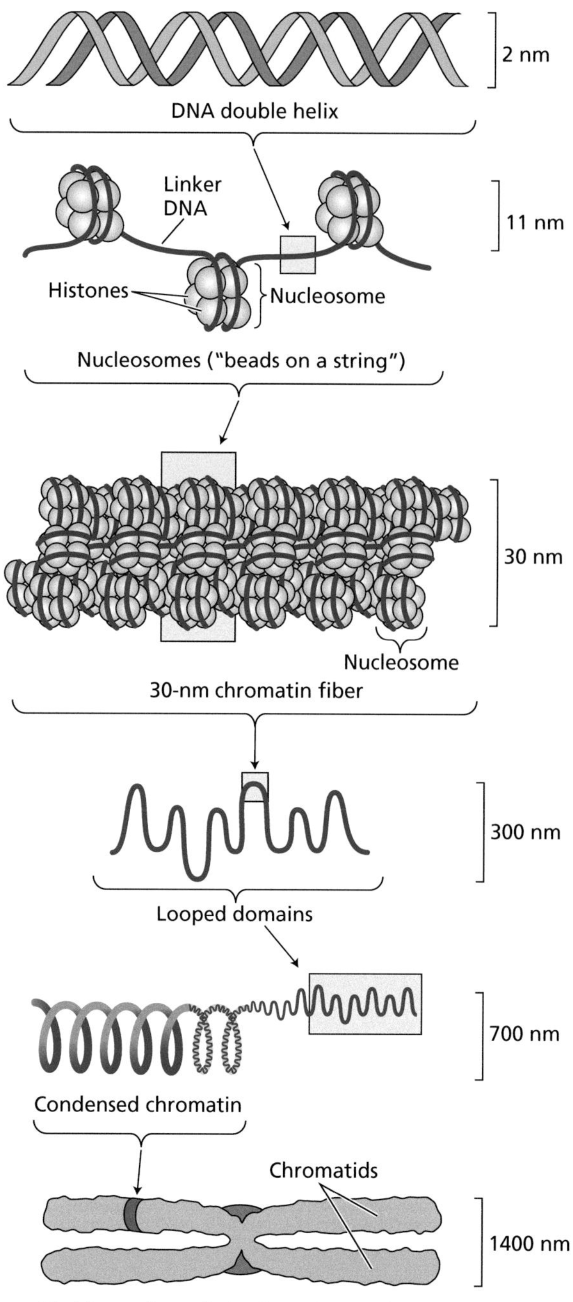

Figure 1.16 Packaging of DNA in a metaphase chromosome. The DNA is first aggregated into nucleosomes and then wound to form the 30-nm chromatin fibers. Further coiling leads to the condensed metaphase chromosome.

translation The process whereby a specific protein is synthesized on ribosomes according to the sequence information encoded by the mRNA.

translocons Pores in the rough endoplasmic reticulum that enable proteins synthesized on ribosomes to enter the ER lumen.

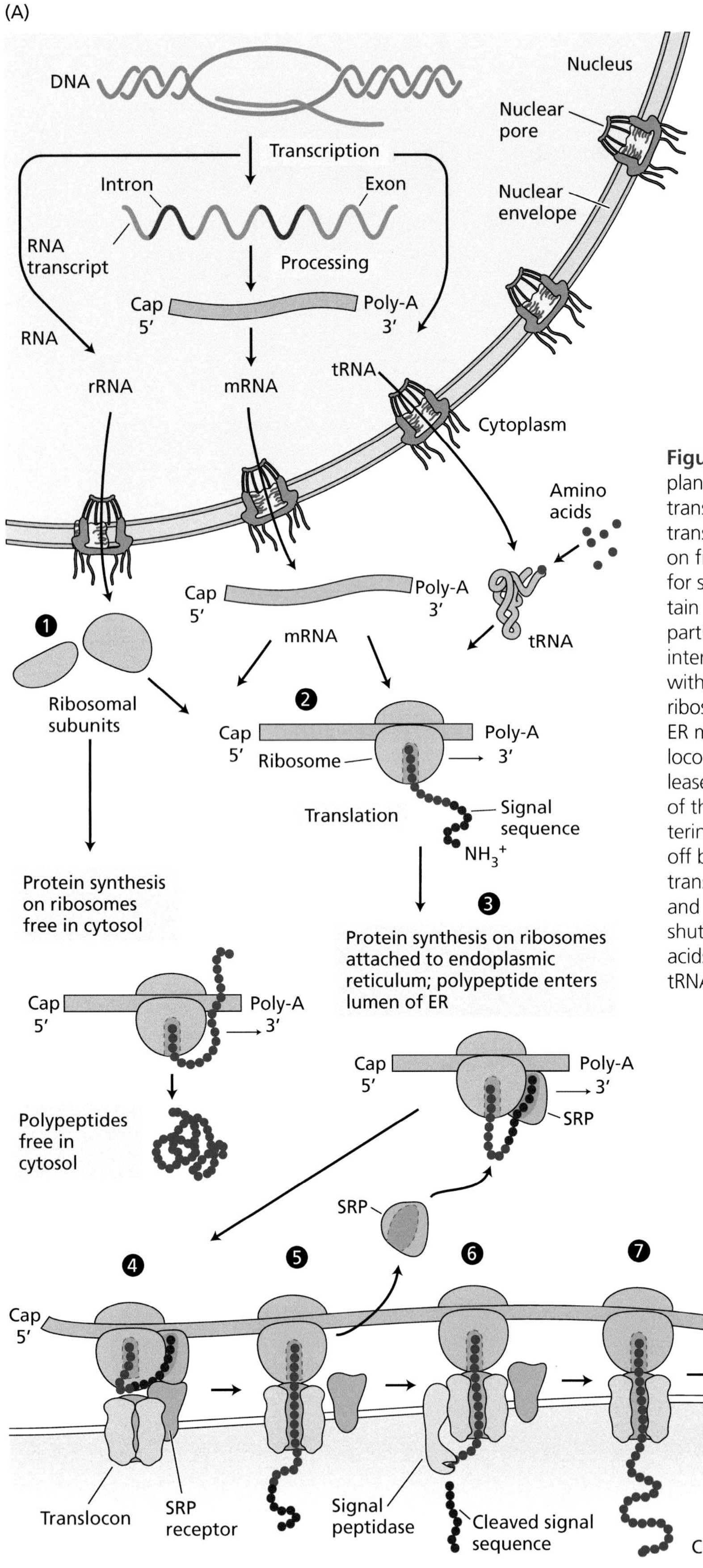

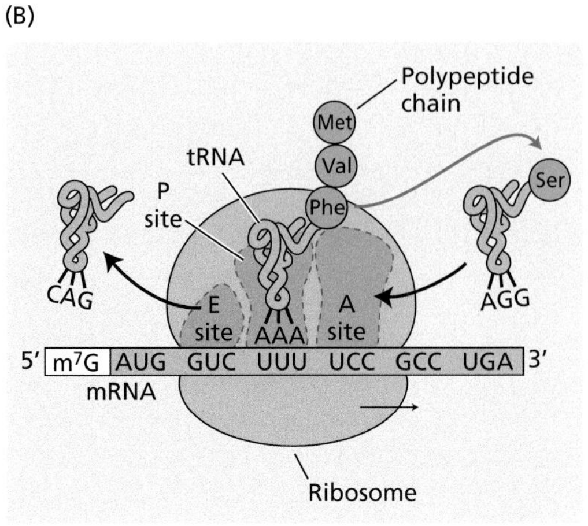

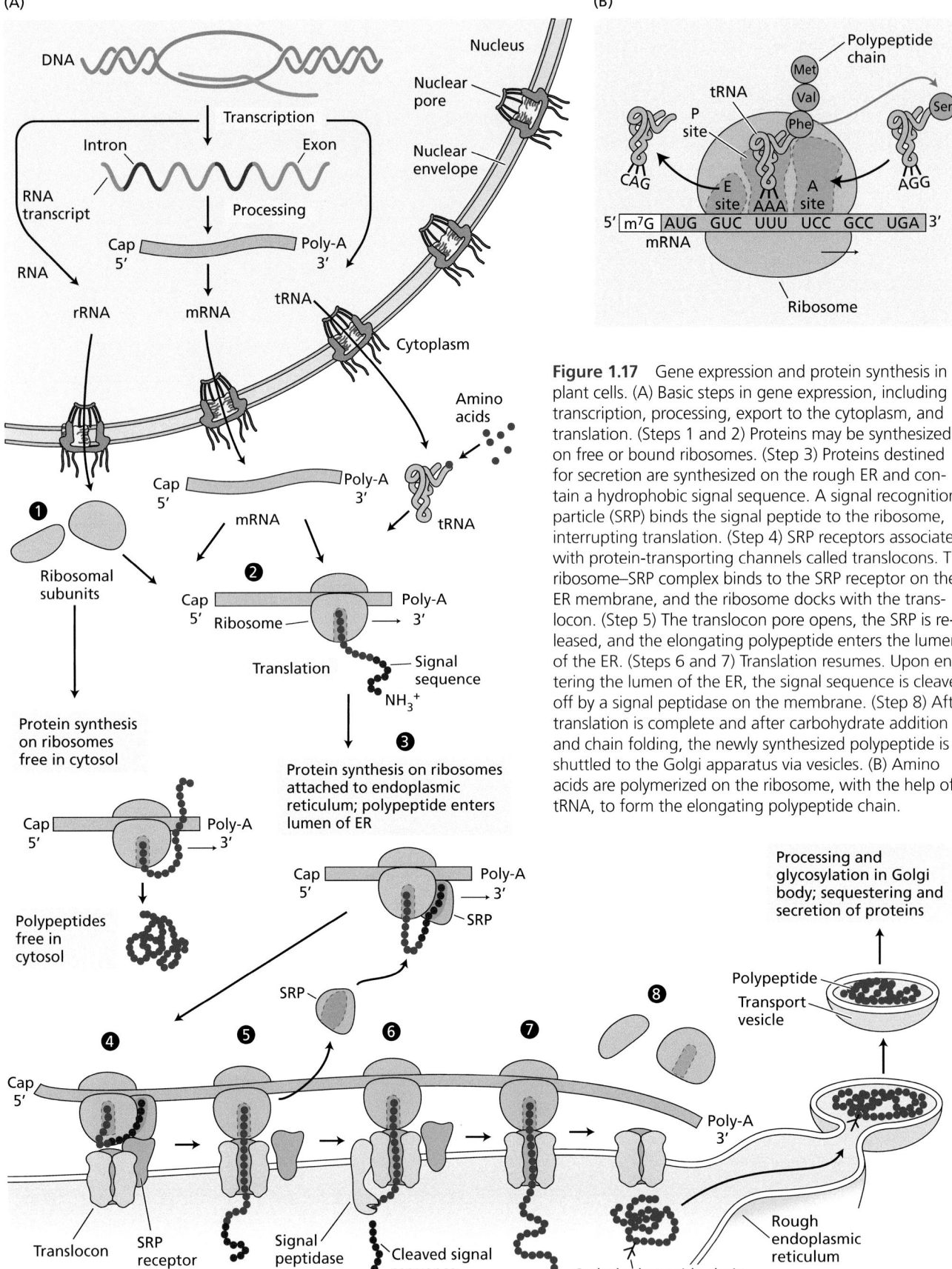

Figure 1.17 Gene expression and protein synthesis in plant cells. (A) Basic steps in gene expression, including transcription, processing, export to the cytoplasm, and translation. (Steps 1 and 2) Proteins may be synthesized on free or bound ribosomes. (Step 3) Proteins destined for secretion are synthesized on the rough ER and contain a hydrophobic signal sequence. A signal recognition particle (SRP) binds the signal peptide to the ribosome, interrupting translation. (Step 4) SRP receptors associate with protein-transporting channels called translocons. The ribosome–SRP complex binds to the SRP receptor on the ER membrane, and the ribosome docks with the translocon. (Step 5) The translocon pore opens, the SRP is released, and the elongating polypeptide enters the lumen of the ER. (Steps 6 and 7) Translation resumes. Upon entering the lumen of the ER, the signal sequence is cleaved off by a signal peptidase on the membrane. (Step 8) After translation is complete and after carbohydrate addition and chain folding, the newly synthesized polypeptide is shuttled to the Golgi apparatus via vesicles. (B) Amino acids are polymerized on the ribosome, with the help of tRNA, to form the elongating polypeptide chain.

Most secretory and integral membrane proteins have a **signal peptide**, a hydrophobic leader sequence of 18 to 30 amino acid residues at the amino-terminal end of the chain (Figure 1.17A). Early in translation, a **signal recognition particle** (**SRP**), made up of protein and RNA, binds both to this hydrophobic leader and to the ribosome, interrupting translation. The ER membrane contains **SRP receptors** that can associate with the translocons, through which the newly synthesized protein is threaded. During co-translational insertion into the ER, the mRNA–ribosome–SRP complex in the cytoplasm binds to the SRP receptor on the ER membrane surface, and the ribosome docks with the translocon. Docking opens the translocon pore, the SRP is released, translation resumes, and the elongating polypeptide enters the lumen of the ER. The signal peptide sequence is cleaved off by a signal peptidase associated with the ER membrane (Figure 1.17A). For integral membrane proteins, some parts of the polypeptide chain are translocated across the membrane while others are not. Completed integral membrane proteins are anchored to the membrane by one or more hydrophobic membrane-spanning domains. Transmembrane proteins that lack a signal peptide integrate into the membrane via internal hydrophobic and protein–protein interaction amino acid sequence motifs.

signal peptide A hydrophobic sequence of 18 to 30 amino acid residues at the amino-terminal end of a chain; it is found on all secretory proteins and most integral membrane proteins and permits their transit across the membrane of the rough ER.

The proteins synthesized on cytosolic ribosomes that are targeted to membrane organelles (plastids, mitochondria, or peroxisomes) after translation employ posttranslational insertion to cross the organelle membrane.

Posttranslational modification of proteins determines their location, activity, and longevity

Protein stability plays an important role in regulating a protein's longevity. Once synthesized, a protein has a finite life span in the cell, ranging from a few minutes to hours or days. Thus, steady-state levels of cellular proteins reflect an equilibrium between the synthesis and the degradation of those proteins, a balance referred to as **turnover**. In both plant and animal cells, there are two distinct pathways of protein turnover, one in specialized lytic vacuoles (called lysosomes in animal cells) and the other in the cytoplasm and nucleus.

Protein turnover in the cytoplasm and nucleus involves the ATP-dependent formation of a covalent bond between the protein that is to be degraded and a small, 76–amino acid polypeptide called **ubiquitin**. Addition of one or many ubiquitin molecules to a protein is called *ubiquitination*. Ubiquitination marks a protein for destruction by a large, ATP-dependent proteolytic complex called the **26S proteasome** that specifically recognizes such "ubiquitin-tagged" molecules (**Figure 1.18**). More than 90% of the short-lived proteins in eukaryotic cells are degraded via the ubiquitin pathway. There are a multitude of protein-specific ubiquitin ligases that control the turnover of specific target proteins to regulate

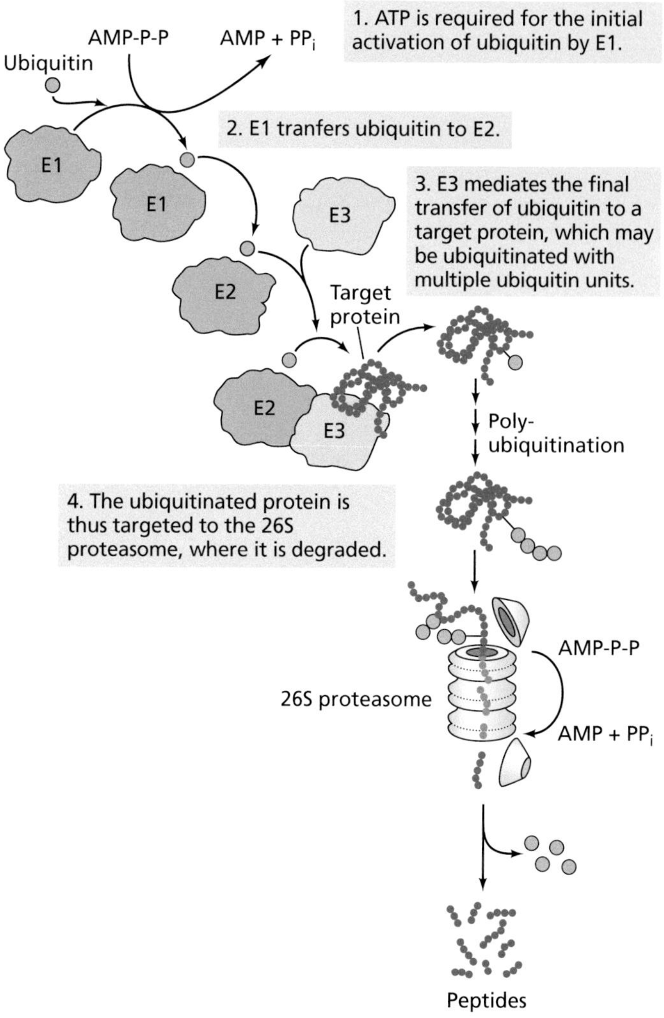
1. ATP is required for the initial activation of ubiquitin by E1.
Ubiquitin
AMP-P-P
AMP + PP$_i$
2. E1 tranfers ubiquitin to E2.
E1
E1
3. E3 mediates the final transfer of ubiquitin to a target protein, which may be ubiquitinated with multiple ubiquitin units.
E3
E2
Target protein
E2
E3
Poly-ubiquitination
4. The ubiquitinated protein is thus targeted to the 26S proteasome, where it is degraded.
26S proteasome
AMP-P-P
AMP + PP$_i$
Peptides

Figure 1.18 Generalized diagram of the cytoplasmic pathway of protein degradation. A similar pathway degrades nuclear proteins, membrane proteins, or proteins after delivery to the vacuole. Ubiquitination is initiated when the ubiquitin-activating enzyme (E1) catalyzes the ATP-dependent adenylation of the C terminus of ubiquitin. The adenylated ubiquitin is then transferred to a cysteine residue on a second enzyme, the ubiquitin-conjugating enzyme (E2). Proteins destined for degradation are bound by a third type of protein, a ubiquitin ligase (E3). The E2–ubiquitin complex then transfers its ubiquitin to a lysine residue of the protein bound to E3. This process can occur multiple times to form a polymer of ubiquitin. The ubiquitinated protein is then targeted to the proteasome for degradation.

development and environmental responses. Chapter 12 will discuss examples of this pathway in more detail in relation to plant hormone action.

1.7 The Endomembrane System

Describe the role of the endomembrane system components in plant cells.

The endomembrane system is a collection of related internal membranes that contributes to the synthesis of roughly one-third of all cellular proteins and allows the cell to communicate with the external environment via vesicular traffic at the plasma membrane. **Exocytosis** is the process that allows forward (anterograde) trafficking of biosynthetic cargo and membranes from the site of synthesis, the ER and the **Golgi apparatus**, to distal compartments of the endomembrane system, including the ***trans* Golgi network** (**TGN**), late endosomes, and ultimately the vacuole, plasma membrane, or cell wall. As exocytosis proceeds through successive endomembrane compartments, increasing luminal acidification contributes to its progression. **Endocytosis** is the opposite (retrograde) process of exocytosis. During endocytosis the plasma membrane forms invaginations that counterbalance the exocytic flow of membranes to the plasma membrane and allows the cell to incorporate internalized plasma membrane and extracellular material into the cell or recycle those components back to the plasma membrane. Under some conditions, an increase in the specialized form of endomembrane trafficking known as **autophagy** allows cells to consume their own organelles and other cytoplasmic components after transfer to the vacuole. A basal level of autophagy exists in cells to maintain homeostasis; however, in conditions of stress, autophagy is often increased.

The endoplasmic reticulum is a network of internal membranes

The **endoplasmic reticulum (ER)** is composed of an extensive network of tubules that is continuous with the nuclear envelope. The tubules form a network of flattened saccules called **cisternae** (singular *cisterna*) (**Figure 1.19**). The tubules spread throughout the cell, forming close associations with other organelles and the cytoskeleton, mainly actin. The ER network may therefore be a communication network among organelles within a cell, while also serving as a synthesis and delivery system for proteins and lipids. The ER is responsible for initiating the synthesis of secretory proteins; it also produces essential lipids for the cell and cellular organelles. Most of the ER lies just under the plasma membrane, residing in the outer layer of cytoplasm called the cell cortex. This portion of the ER is therefore called the **cortical ER**. The cortical ER is attached to the plasma membrane at membrane subdomains called ER–plasma membrane contact sites. ER contact sites are also formed with other organelles such as mitochondria, plastids, and peroxisomes. Because plant cells generally contain a large central vacuole, the cortical ER on one side of the cell connects with the cortical ER at the other side of the cell via the so-called transvacuolar strands of the ER. The ER is a highly dynamic organelle that remodels tubules into cisternae, and vice versa.

The region of the ER that has many membrane-bound ribosomes associated with co-translational insertion of membrane proteins is called the **rough ER** because the bound ribosomes described previously give the ER a rough appearance in electron micrographs (Figure 1.19A and B). Because the ER is the primary site of secretory protein synthesis, a sophisticated protein quality-control system is found inside the ER lumen to ensure the production of fully folded proteins. Once secretory proteins enter the ER, they undergo folding processes assisted by specialized resident ER proteins. High levels of calcium ions and oxidizing conditions contribute to folding and stabilization of the proteins via formation of cystine (cysteine-to-cysteine) linkages. Proteins that do not attain their properly

signal recognition particle (SRP) A ribonucleoprotein (protein–RNA complex) that recognizes and targets specific proteins to the endoplasmic reticulum in eukaryotes.

SRP receptor A receptor protein on the ER membrane that binds to the ribosome–SRP complex, permitting the ribosome to dock with the translocon pore through which the elongating polypeptide will enter the lumen of the ER.

turnover The balance between the rate of synthesis and the rate of degradation, usually applied to protein or RNA. An increase in turnover typically refers to an increase in degradation.

ubiquitin A small polypeptide that is covalently attached to proteins, and that serves as a recognition site for a large proteolytic complex, the proteasome.

26S proteasome A large proteolytic complex that degrades intracellular proteins marked for destruction by the attachment of one or more copies of the small protein, ubiquitin.

exocytosis The process by which vesicle contents are distributed to the plasma membrane or apoplast.

Golgi apparatus Endomembrane organelle that modifies and packages secreted proteins and some cell wall components. A site of key steps in protein glycosylation. Named for its discoverer, Camillo Golgi.

***trans* Golgi network (TGN)** in plants, a collective compartment of vesicles derived from the Golgi apparatus that are sorted to different subcellular destinations. The TGN also receives extracellular materials and recycled molecules from endocytic compartments.

endocytosis The formation of small vesicles from the plasma membrane, which detach and move into the cytosol, where they fuse with elements of the endomembrane system.

autophagy A catabolic mechanism that conveys cellular macromolecules and organelles via autophagosomes to lytic vacuoles where they are degraded and recycled.

endoplasmic reticulum (ER) A continuous membrane system within the cytoplasm of eukaryotic cells that serves multiple functions, including the synthesis, modification, and intracellular transport of proteins.

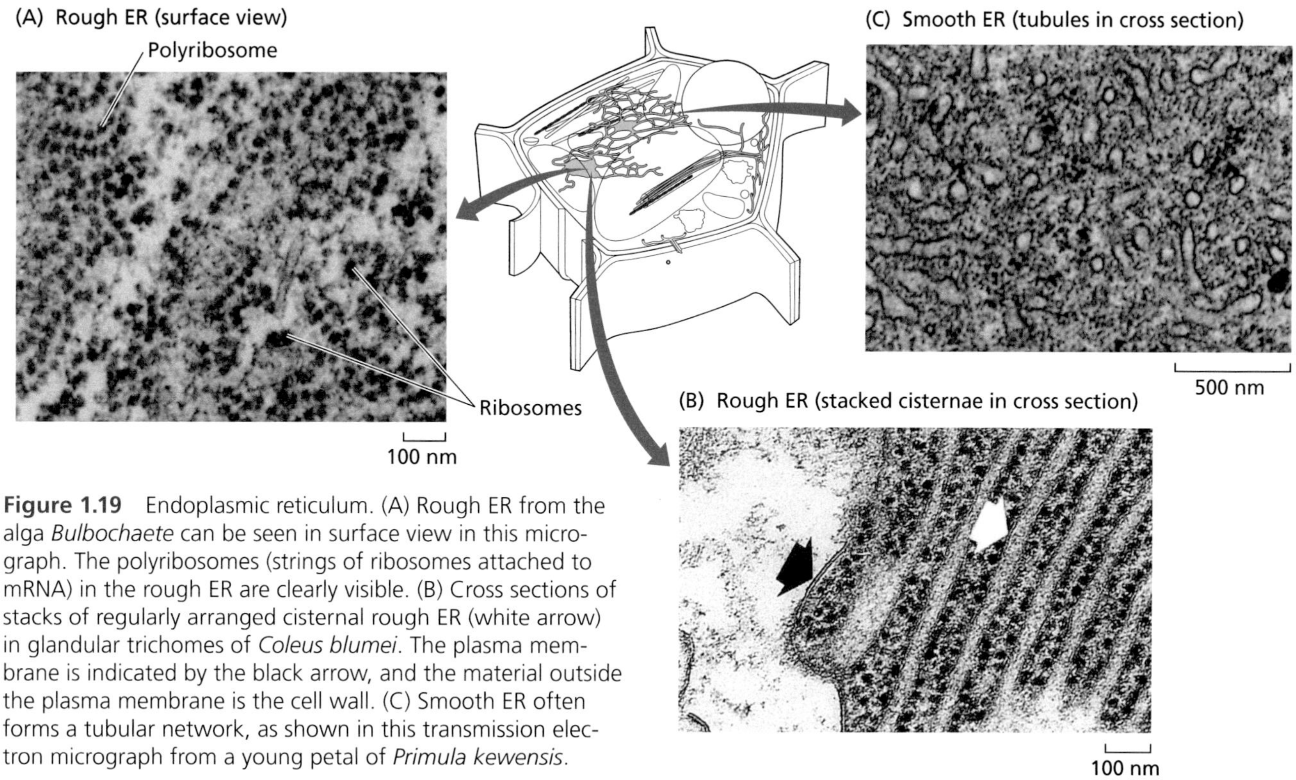

Figure 1.19 Endoplasmic reticulum. (A) Rough ER from the alga *Bulbochaete* can be seen in surface view in this micrograph. The polyribosomes (strings of ribosomes attached to mRNA) in the rough ER are clearly visible. (B) Cross sections of stacks of regularly arranged cisternal rough ER (white arrow) in glandular trichomes of *Coleus blumei*. The plasma membrane is indicated by the black arrow, and the material outside the plasma membrane is the cell wall. (C) Smooth ER often forms a tubular network, as shown in this transmission electron micrograph from a young petal of *Primula kewensis*.

folded conformation are retro-translocated via translocons situated in the ER membrane into the cytosol for degradation. Protein folding in the ER is monitored by conserved sensors that respond to the presence of hydrophobic domains of unfolded or misfolded proteins inside the ER. The activation of these sensors leads to an increased production of proteins that aid the folding and quality control of ER proteins. This process is collectively called the **unfolded protein response**.

The region of the ER without bound ribosomes is called **smooth ER** (Figure 1.19C) and is the major source of membrane phospholipids for the other compartments in the endomembrane system. Enzymes that initiate phospholipid and phospholipid head group synthesis on the cytosolic leaflet of the bilayer (i.e., the side of the membrane facing the cytosol) introduce an intrinsic lipid asymmetry in the endomembrane system and plasma membrane formed by exocytosis. Membrane asymmetry can, however, be adjusted by ATP-dependent enzymes called **flippases** that move phospholipids between membrane leaflets.

The ER is closely associated with the Golgi apparatus. Export of cargo proteins from the ER occurs primarily in membrane vesicles coated with coatomer type II (COPII) proteins. The COPII coat interacts with amino acid motifs in membrane-associated proteins at ER exit sites, which contact the acceptor *cis* cisternae of the Golgi apparatus. This anterograde flow of cargo is counterbalanced by retrograde traffic controlled by coatamer type I (COPI) protein-coated vesicles. This retrograde movement is important for retrieving escaping ER resident proteins and *cis* to *trans* cisternae–specific enzymes in the Golgi apparatus that participate in the continuous maturation and flow of the Golgi membranes.

Cell wall matrix polysaccharides, secretory proteins, and glycoproteins are processed in the Golgi apparatus

The Golgi apparatus is a polarized stack of cisternae, with thicker cisternae occurring on the *cis* side, or forming face. The Golgi accepts cargo transported in

cisternae (singular *cisterna*) A network of flattened saccules and tubules that compose the endoplasmic reticulum.

cortical ER The network of endoplasmic reticulum that lies just under the plasma membrane and is associated with the plasma membrane at specific contact points.

rough ER The endoplasmic reticulum to which ribosomes are attached. Rough ER synthesizes proteins that are transported by vesicles either to internal organelles or the plasma membrane.

unfolded protein response A cellular response that eliminates misfolded proteins in the lumen of the endoplasmic reticulum.

smooth ER The endoplasmic reticulum lacking attached ribosomes and usually consisting of tubules. Functions in lipid synthesis.

flippase Enzyme that "flips" newly synthesized phospholipids across the bilayer from the outer (cytoplasmic) face of the membrane to the inner leaflet, thereby ensuring symmetrical lipid composition of the membrane.

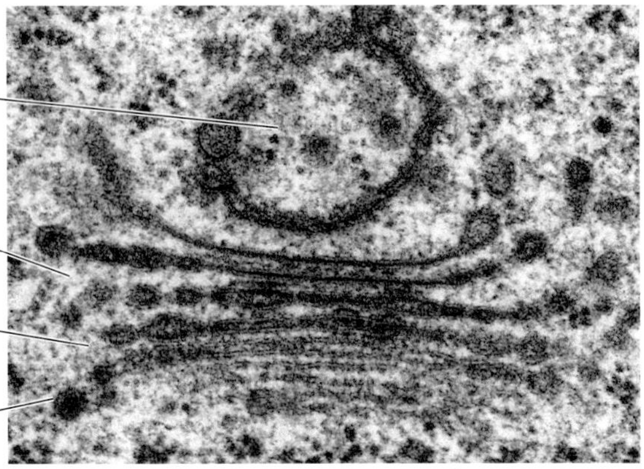

Figure 1.20 Electron micrograph of a Golgi apparatus in a tobacco (*Nicotiana tabacum*) root cap cell. The *cis*, medial, and *trans* cisternae are indicated. The *trans* Golgi network is associated with the *trans* cisternae and also functions as the early endosome in plant cells.

COPII carriers from the ER (**Figure 1.20**). The opposite, maturing face, or *trans* side, of the Golgi exhibits more flattened, thinner cisternae. Anterograde movement through the endomembrane system generally occurs via vesicles, but in the Golgi body it occurs by *cisternal maturation*, whereby *cis* cisternae formed from COPII vesicles mature stepwise into *trans* cisternae. The *trans* cisternae then mature into the *trans* Golgi network that then produces mostly uncoated secretory vesicles. The *trans* Golgi network also accepts vesicles and their cargo derived from endocytosis at the plasma membrane. In plants, the *trans* Golgi network functions in the sorting of membrane vesicles containing newly synthesized and recycled proteins and cell wall components known as **matrix polysaccharides** secreted to the plasma membrane. Specific membrane domains within the *trans* Golgi network are partially coated with **clathrin**, a protein that works in concert with adaptor proteins to select cargo and shape membranes into vesicles for transport into the vacuole. Proteins without a vacuolar localization signal are secreted to the cell wall/apoplast in what is often referred to as the *default* pathway.

Many of the proteins found in the lumen of the endomembrane system are **glycoproteins**—proteins with oligomeric sugar chains covalently attached—destined for secretion from the cell or delivery to the other endomembrane compartments. The enzymes that synthesize these sugar chains differ substantially from the enzymes that synthesize polysaccharides for incorporation into the cell wall. In the vast majority of cases, protein glycosylation begins in the ER with the attachment of a branched oligosaccharide chain made up of *N*-acetylglucosamine (GlcNAc), mannose (Man), and glucose (Glc) to one or more specific asparagine residues. These **N-linked glycoproteins** are trimmed in the ER and again during transition from the *cis* to the *trans* Golgi cisternae. Certain sugars, such as mannose, are removed from the oligosaccharide chains, and other sugars are added. In addition to these modifications, glycosylation of the –OH groups of hydroxyproline, serine, threonine, and tyrosine residues (**O-linked oligosaccharides**) occurs in the Golgi apparatus.

The plasma membrane has specialized regions involved in membrane recycling

Endocytotic membrane internalization at the plasma membrane occurs in small (100 nm) vesicles that are initially coated with clathrin, but clathrin-independent endocytosis has been documented in plant cells. Extracellular materials can be specifically recognized by receptors that are generally proteins spanning the plasma membrane. This process is called receptor-mediated endocytosis. Endocytosis proceeds through clathrin-mediated invagination of the plasma membrane progressing inward, followed by severing of the invagination neck. In plant cells, the difference in osmolarity (or more precisely water potential, discussed in Chapter 2) between the inside and the outside of the cell creates a large internal pressure (in the range of 0.2–1 MPa), called turgor pressure. Turgor pressure pushes the plasma membrane against the cell wall. Therefore, during endocytosis, plasma membrane invagination involves energy and must overcome opposing turgor pressure. After initial endocytosis, vesicles lose their clathrin coat to form endocytic vesicles that are then sorted through the *trans* Golgi network for recycling back to the plasma membrane or to **prevacuolar compartments** for trafficking to the vacuole. Prevacuolar compartment endosomes are also called

matrix polysaccharides Polysaccharides comprising the matrix of plant cell walls. In primary cell walls they consist of pectins, hemicelluloses, and proteins.

clathrin Proteins that have a unique *triskelion* structure that spontaneously assemble into 100-nm cages that coat vesicles associated with endocytosis at the plasma membrane and other cellular trafficking events.

glycoprotein Protein that has covalently attached sugar oligomers or polymers.

N-linked glycoprotein Glycan linked via a nitrogen atom to a protein. Formed by transfer of a 14-sugar glycan from the ER membrane–embedded dolichol diphosphate to the nascent polypeptide as it enters the lumen of the ER.

O-linked oligosaccharide Small polysaccharide that is covalently linked to the side chain hydroxyl of serine or threonine residues in a subgroup of plant glycoproteins. O-linked glycosylation occurs in the Golgi apparatus.

prevacuolar compartment A membrane compartment equivalent to the late endosome in animal cells where sorting occurs before cargo is delivered to the lytic vacuole. *See also* multivesicular body.

multivesicular bodies, as they generally contain smaller intraluminal vesicles.

Vacuoles have diverse functions in plant cells

The plant vacuole is a compartment enclosed by a membrane called the **tonoplast**. There are two primary types of vacuoles: lytic vacuoles found in vegetative cells and protein storage vacuoles found in developing seeds. The **vacuolar sap**, composed of water and solutes, is actively acidified (pH 5.5 and lower) by tonoplast-localized proton ATPases and pyrophosphatases in both lytic vacuoles and protein storage vacuoles during active growth. In general, vacuoles are small in young meristematic cells and increase in size during cell elongation and maturation, with the membrane being enlarged by fusion with membrane vesicles from the prevacuolar compartment. **Lytic vacuoles** (**Figure 1.21A**) can occupy 90 to 95% of plant cell volume, and function in calcium ion signaling, osmotic regulation, lytic digestion, xenobiotic disposal, and defense responses. Expansion of plant cells is driven primarily by increases in lytic vacuolar volume. Inorganic ions, sugars, organic acids, and pigments are some of the solutes that can accumulate in vacuoles in very high concentrations, or even as solid crystals, thanks to the presence of a variety of specific membrane transporters. Protein storage vacuoles (**Figure 1.21B**) are specialized compartments that accumulate large amounts of proteins and sugars in developing seeds.

Vacuoles can be inherited during cytokinesis of a mother cell but also are formed de novo in meristematic cells directly from the ER and from the fusion of endosomal compartments. Therefore, the tonoplast contains proteins and lipids that are originally synthesized in the ER.

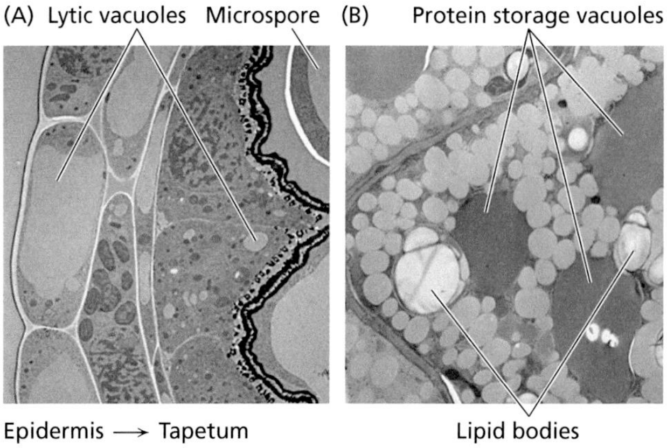

Figure 1.21 Transmission electron micrographs of the two primary types of vacuoles. (A) Cross section through a developing barley (*Hordeum vulgare* cv. Bowman) anther showing the four-layered wall (from outside to inside: epidermis, endothecium, middle layer, tapetum) that surrounds the developing microspores. The lytic vacuoles can be seen as electron-translucent spaces of various sizes and shapes in the different cell layers. The epidermis is characterized by a single prominent lytic vacuole that makes up most of the cell volume and restricts the cytoplasm to a thin layer pressed against the plasma membrane. By contrast, cells of the tapetum have a much denser cytoplasm and contain numerous small lytic vacuoles. (B) Section through castor bean (*Ricinus communis*) endosperm. The endosperm cytoplasm contains protein storage vacuoles of varying diameter that are characterized by an electron-opaque content. The small, globular, electron-translucent structures correspond to lipid bodies.

Oil bodies are lipid-storing organelles

Many plants synthesize and store large quantities of oil during seed development. These oils accumulate in organelles called **oil bodies** or **lipid bodies** (also known as **oleosomes**). Oil bodies are unique among the organelles in that they are surrounded by a phospholipid monolayer—derived from the ER. The phospholipids in the half–unit membrane are oriented with their polar head groups facing the aqueous phase of the cytosol and their hydrophobic fatty acid tails facing the lumen, where they interact with the stored **triacylglycerol** lipids. Oil bodies initially form as regions of differentiation within the ER membrane bilayer.

Peroxisomes play specialized metabolic roles in leaves and seeds

Peroxisomes are a class of organelles characterized by specialized metabolic functions and the ability to detoxify reactive oxygen species via the enzyme **catalase**. **Glyoxysomes** are peroxisomes specialized for the **β-oxidation** of fatty acids and the metabolism of glyoxylate, a two-carbon acid aldehyde, especially in germinating seeds. Glyoxysomes are associated with mitochondria and oil bodies (discussed in Chapter 11), while leaf peroxisomes function in photorespiration (discussed in Chapter 8). While peroxisomes can multiply by division much as plastids and mitochondria do, they are more often formed de novo directly from the ER.

multivesicular bodies Part of the prevacuolar sorting compartment that functions in the degradation of vacuoles and their membranes.

tonoplast The vacuolar membrane.

vacuolar sap The fluid contents of a vacuole, which may include water, inorganic ions, sugars, organic acids, and pigments.

lytic vacuoles Analogous to lysosomes in animal cells, plant lytic vacuoles contain hydrolytic enzymes that break down cellular macromolecules during senescence.

lipid body (also known as **oleosomes** or **oil bodies**) Organelle that accumulates and stores triacylglycerols. It is bounded by a single phospholipid leaflet ("half–unit membrane" or "phospholipid monolayer") derived from the endoplasmic reticulum.

triacylglycerols Three fatty acyl groups in ester linkage to the three hydroxyl groups of glycerol. Fats and oils.

peroxisome Organelle in which organic substrates are oxidized by O_2. These reactions generate H_2O_2 that is broken down to water by the peroxisomal enzyme catalase.

catalase An enzyme that breaks hydrogen peroxide down into water. When it is abundant in peroxisomes, it may form crystalline arrays.

glyoxysome An organelle found in the oil-rich storage tissues of seeds in which fatty acids are oxidized. A type of microbody.

β-oxidation Oxidation of fatty acids into fatty acyl-CoA, and the sequential breakdown of the fatty acids into acetyl-CoA units. NADH is also produced.

mitochondrion (plural *mitochondria*) The organelle that is the site for most reactions in the respiratory process in eukaryotes.

ATP synthase The multi-subunit protein complex that synthesizes ATP from ADP and phosphate (P_i). F_oF_1 and CF_0-CF_1 types are present in mitochondria and chloroplasts, respectively. Also called *ATPase* and *CF_0–CF_1 ATP synthase*.

cristae Folds in the inner mitochondrial membrane that project into the mitochondrial matrix.

matrix The aqueous, gel-like phase of a mitochondrion that occupies the internal space into which the cristae extend. Contains the DNA, ribosomes, and soluble enzymes required for the tricarboxylic acid cycle, oxidative phosphorylation, and other metabolic reactions.

tricarboxylic acid (TCA) cycle A cycle of reactions catalyzed by enzymes localized in the mitochondrial matrix that leads to the oxidation of pyruvate to CO_2. ATP and NADH are generated in the process. Also called the citric acid cycle.

plastids Cellular organelles found in eukaryotes, bounded by a double membrane, and sometimes containing extensive membrane systems. They perform many different functions: photosynthesis, starch storage, pigment storage, and energy transformations.

chloroplast The organelle that is the site of photosynthesis in photosynthetic eukaryotic organisms.

1.8 Independently Dividing Semiautonomous Organelles

Differentiate between the two main types of semiautonomous organelles in plants.

A typical plant cell has two primary types of semiautonomous organelles: mitochondria and plastids. Both types are separated from the cytoplasm by a double membrane (an outer and an inner membrane) and contain their own DNA and prokaryotic-type ribosomes as descendants of prokaryotic endosymbionts.

Mitochondria (singular *mitochondrion*) are the cellular sites of respiration, a process in which the energy released from sugar metabolism is used for the synthesis of ATP (adenosine triphosphate) from ADP (adenosine diphosphate) and inorganic phosphate (P_i).

Mitochondria are highly dynamic structures that can undergo both fission and fusion. Regardless of shape, all mitochondria have a smooth outer membrane and a highly convoluted inner membrane (**Figure 1.22**). The inner membrane contains an **ATP synthase** that uses a proton gradient to synthesize ATP for the cell. The proton gradient is generated through the cooperation of electron transporters called the electron transport chain, which is embedded in, and peripheral to, the inner membrane. The infoldings of the inner membrane are called **cristae** (singular *crista*). The compartment enclosed by the inner membrane, the mitochondrial **matrix**, contains the enzymes of the pathway of intermediary metabolism called the **tricarboxylic acid (TCA) cycle** (discussed in Chapter 11).

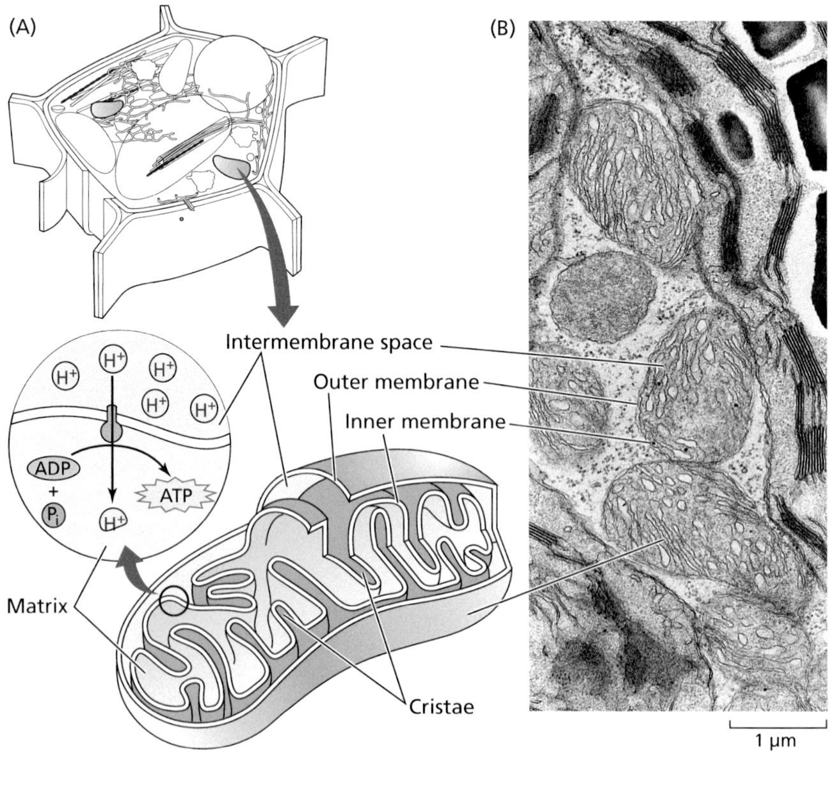

Figure 1.22 Mitochondria generate ATP via oxidative phosphorylation. (A) Diagrammatic representation of a mitochondrion, including the location of the ATP synthases involved in ATP synthesis on the inner membrane. (B) Electron micrograph of mitochondria from a leaf cell of Bermuda grass (*Cynodon dactylon*).

Plastids are another group of double-membrane–enclosed organelles derived from an ancestral endosymbiosis event with cyanobacteria. Plastids that function in photosynthesis are called **chloroplasts** (**Figure 1.23A**) (discussed in Chapter 7). In addition to their inner and outer envelope membranes (collectively referred to as the **chloroplast envelope**), chloroplasts possess a third system of internal membranes called **thylakoids**. A stack of thylakoids forms a **granum** (plural *grana*) (**Figure 1.23B**). Proteins and pigments (chlorophylls and carotenoids) that function in the photochemical events of photosynthesis are embedded in the thylakoid membrane. Adjacent grana are connected by unstacked membranes called **stroma lamellae** (singular *lamella*). The compartment surrounding the thylakoids, called the **stroma**, is analogous to the matrix of the mitochondrion and contains what may be Earth's most abundant protein, **ribulose 1,5-bisphosphate carboxylase/oxygenase**, called **Rubisco** for convenience. Rubisco is involved in capturing atmospheric carbon for conversion into organic acids during photosynthesis (discussed in Chapter 8). Although both the large and small subunits of Rubisco are encoded in the plastid genome in algae, in land plants the large subunit of Rubisco is encoded by the chloroplast genome while the small subunit is encoded by the nuclear genome.

chloroplast envelope The double-membrane system surrounding the chloroplast.

thylakoids The specialized, internal chlorophyll-containing membranes of the chloroplast where light absorption and the chemical reactions of photosynthesis take place.

granum (plural *grana*) In the chloroplast, a stack of thylakoids.

stroma lamellae Unstacked thylakoid membranes within the chloroplast.

stroma The fluid component surrounding the thylakoid membranes of a chloroplast.

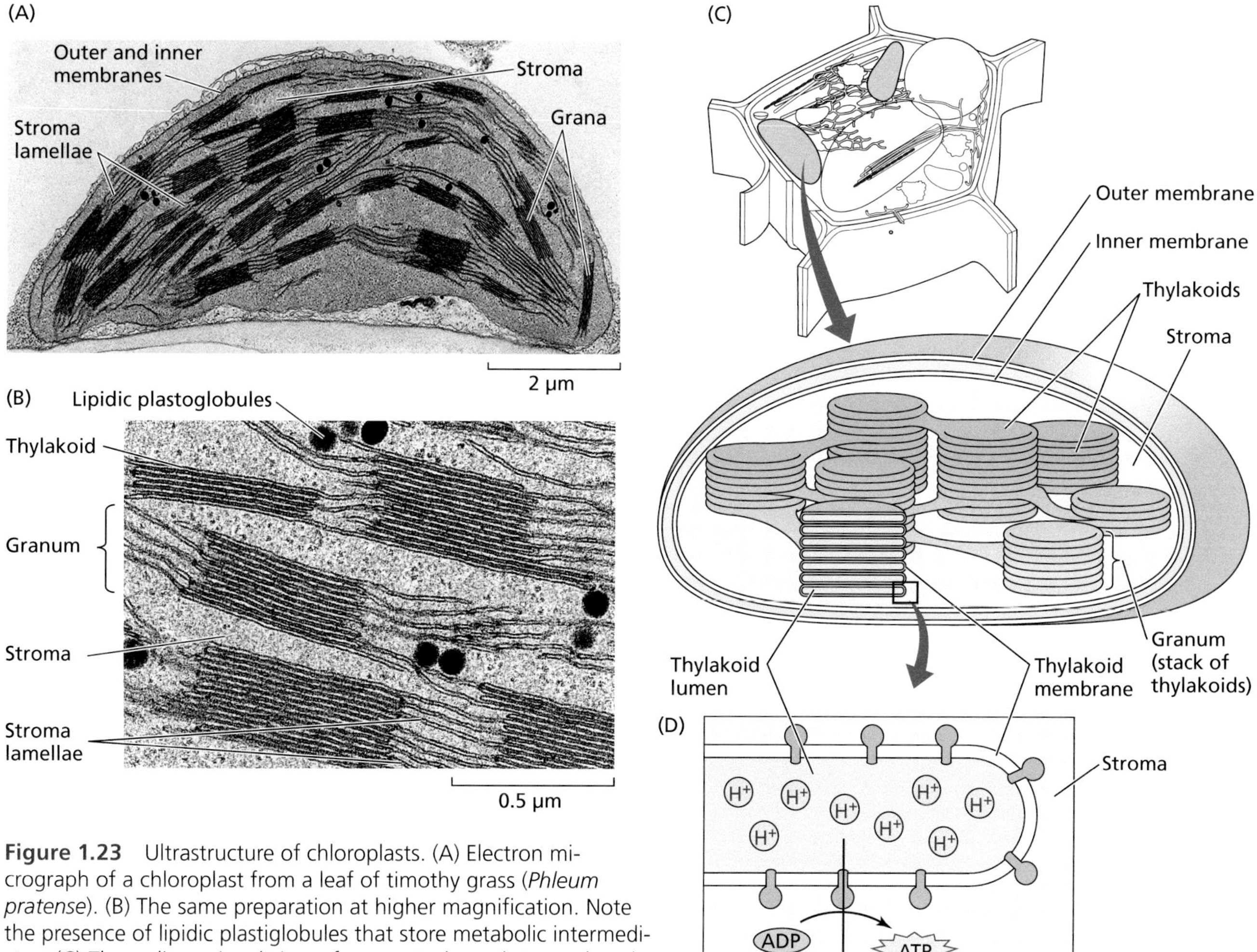

Figure 1.23 Ultrastructure of chloroplasts. (A) Electron micrograph of a chloroplast from a leaf of timothy grass (*Phleum pratense*). (B) The same preparation at higher magnification. Note the presence of lipidic plastiglobules that store metabolic intermediates. (C) Three-dimensional view of grana stacks and stroma lamellae, showing the complexity of the organization. (D) Diagrammatic representation of a thylakoid, showing the location of the ATP synthases in the thylakoid membrane.

Rubisco The acronym for the chloroplast enzyme *ribulose* 1,5-*bis*phosphate carboxylase/oxygenase. In a carboxylase reaction, Rubisco uses atmospheric CO_2 and ribulose 1,5-bisphosphate to form two molecules of 3-phosphoglycerate. It also functions as an oxygenase that adds O_2 to ribulose 1,5-bisphosphate to yield 3-phosphoglycerate and 2-phosphoglycolate. The competition between CO_2 and O_2 for ribulose 1, 5-bisphosphate limits net CO_2 fixation.

chromoplasts Plastids that contain high concentrations of carotenoid pigments, rather than chlorophyll.

leucoplasts Nonpigmented plastids, the most important of which is the amyloplast.

amyloplast A starch-storing plastid found abundantly in storage tissues of shoots and roots, and in seeds. Specialized amyloplasts in the root cap and shoot also serve as gravity sensors.

proplastid Type of immature, undeveloped plastid found in meristematic tissue.

etioplast Photosynthetically inactive form of chloroplast found in etiolated seedlings.

The various components of the photosynthetic apparatus are localized in different areas of the grana and the stroma lamellae. The ATP synthases of the chloroplast are located on the thylakoid membrane (**Figure 1.23C**). During photosynthesis, light-driven electron transfer reactions result in a proton gradient across the thylakoid membrane (**Figure 1.23D**) (see Chapter 7).

Plastids that contain high concentrations of carotenoid pigments, rather than chlorophyll, are called **chromoplasts**. Chromoplasts are responsible for the yellow, orange, and red colors of many fruits and flowers, as well as of autumn leaves (**Figure 1.24**). Nonpigmented plastids are called **leucoplasts**. Leucoplasts in specialized secretory tissues, such as the nectary, make monoterpenoids, volatile molecules (in essential oils) that often have a strong smell. The most important type of leucoplast is the **amyloplast**, a starch-storing plastid. Amyloplasts are abundant in the storage tissues of shoots and roots, and in seeds. Specialized amyloplasts in the root cap also serve as gravity sensors that enhance directional root growth downward into the soil (see Chapter 14).

Meristem cells contain **proplastids**, which have few internal membranes, no chlorophyll, and an incomplete complement of the enzymes necessary to carry out photosynthesis. While chloroplast development in gymnosperms occurs in both dark and light conditions, angiosperm chloroplast development from proplastids is triggered by light. Upon illumination, regulators of plastidial gene expression are imported into developing plastids to initiate expression of plastidial gene products that interact with nuclear-encoded functional proteins to initiate production of light-absorbing pigments, proliferation of internal membranes, formation of stroma lamellae and grana stacks, and assembly of light-harvesting complexes and other functional structures (Figure 1.23).

The maintenance of chloroplast structure depends on the presence of light; mature chloroplasts can revert to a simpler form termed **etioplasts** during extended periods of darkness. Likewise, under different environmental conditions or tissue differentiation, chloroplasts can be converted to chromoplasts, as in autumn leaves and ripening fruit, or to starch-containing amyloplasts in storage organs.

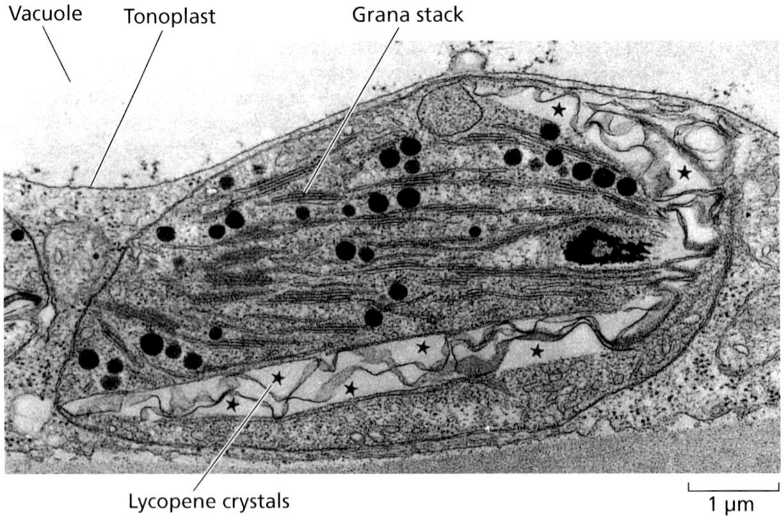

Figure 1.24 Electron micrograph of a chromoplast from tomato (*Solanum lycopersicum*) fruit at an early stage in the transition from chloroplast to chromoplast. Small grana stacks are still visible. Stars indicate crystals of lycopene, a type of carotenoid pigment, giving tomatoes their red color.

Plastidial and mitochondrial division are independent of nuclear division in land plants

Both mitochondria and plastids divide by the process of fission, consistent with their prokaryotic origins. Fission and organellar DNA replication are regulated independently of nuclear division. For example, the number of chloroplasts per cell volume depends on the cell's developmental history and its local environment. Accordingly, there are many more chloroplasts in the mesophyll cells in the interior of a leaf than in the epidermal cells forming the outer layer of the leaf.

Although the timing of the fission of chloroplasts and mitochondria is independent of the timing of cell division, these organelles require nuclear-encoded proteins to divide. In both bacteria and the semiautonomous organelles, fission is facilitated by proteins that form coordinated rings on the inner and outer membranes at the site of the future division plane. In plant cells, the genes that encode these proteins are located in the nucleus. Mitochondria and chloroplasts can also increase in size without dividing, in order to meet energy or photosynthetic demand. If, for example, the proteins involved in mitochondrial division are experimentally inactivated, the mitochondria become larger, allowing the cell to meet its energy needs. In plants with compromised chloroplast biogenesis and function, mitochondrial biogenesis and function are enhanced to compensate for energy deficiencies.

Both plastids and mitochondria can move around plant cells. In some plant cells the chloroplasts are anchored in the outer, cortical cytoplasm of the cell, but in others they are mobile, providing for optimal utilization of light for photosynthesis. Plastids and mitochondria are moved largely by myosins that motor along actin microfilaments.

1.9 Cell Cycle Regulation

Distinguish among the different molecular and cellular processes into their corresponding phase in the cell cycle.

The cell division cycle, or cell cycle, is the process by which cells reproduce themselves (**Figure 1.25**). The cell cycle consists of four phases: **gap 1 (G_1)**, **DNA synthesis phase (S, or S phase)**, **gap 2 (G_2)**, and **mitotic phase (M, or M phase)**. G_1 is the phase of variable duration when a newly formed daughter cell has not yet replicated its DNA. DNA is replicated during S phase. G_2 is the phase when a cell with replicated DNA has not yet proceeded to mitosis, which occurs during M phase. Collectively, G_1, S phase, and G_2 are referred to as **interphase**. M phase includes all the stages of mitosis, from metaphase to telophase. In vacuolated cells, the vacuole enlarges throughout interphase, and the plane of cell division bisects the vacuole during mitosis. Organelles distribute stochastically into future daughter cells.

The biochemical reactions governing the cell cycle are evolutionarily highly conserved in eukaryotes, and plants have retained the basic components of this mechanism. The key enzymes that control the transitions between the different phases of the cell cycle, and the entry of nondividing cells into the cell cycle, are the **cyclin-dependent kinases**, or **CDKs**. Protein kinases are enzymes that phosphorylate proteins using ATP. CDKs depend on regulatory subunits called **cyclins** for their activities. Specific cyclins increase in abundance at each cell cycle stage and are degraded before the transition to the next phase. Three cyclins have been shown to regulate the tobacco cell cycle, as shown in Figure 1.25:

1. Cyclin D, active late in G_1
2. Cyclin A, active in late S phase
3. Cyclin B, active just before M phase

gap 1 (G_1) The phase of the cell cycle preceding the synthesis of DNA.

DNA synthesis phase (S, or S phase) In the cell cycle, the stage during which DNA is replicated; it follows G_1 and precedes G_2.

gap 2 (G_2) The phase of the cell cycle following the synthesis of DNA.

mitotic phase (M, or M phase) Includes all the stages of mitosis, from metaphase to telophase.

interphase Collectively the G_1, S, and G_2 phases of the cell cycle.

cyclin-dependent kinases (CDKs) Protein kinases that regulate the transitions from G_1 to S, from S to G_2, and from G_2 to mitosis, during the cell cycle.

cyclins Regulatory proteins associated with cyclin-dependent kinases that play a crucial role in regulating the cell cycle.

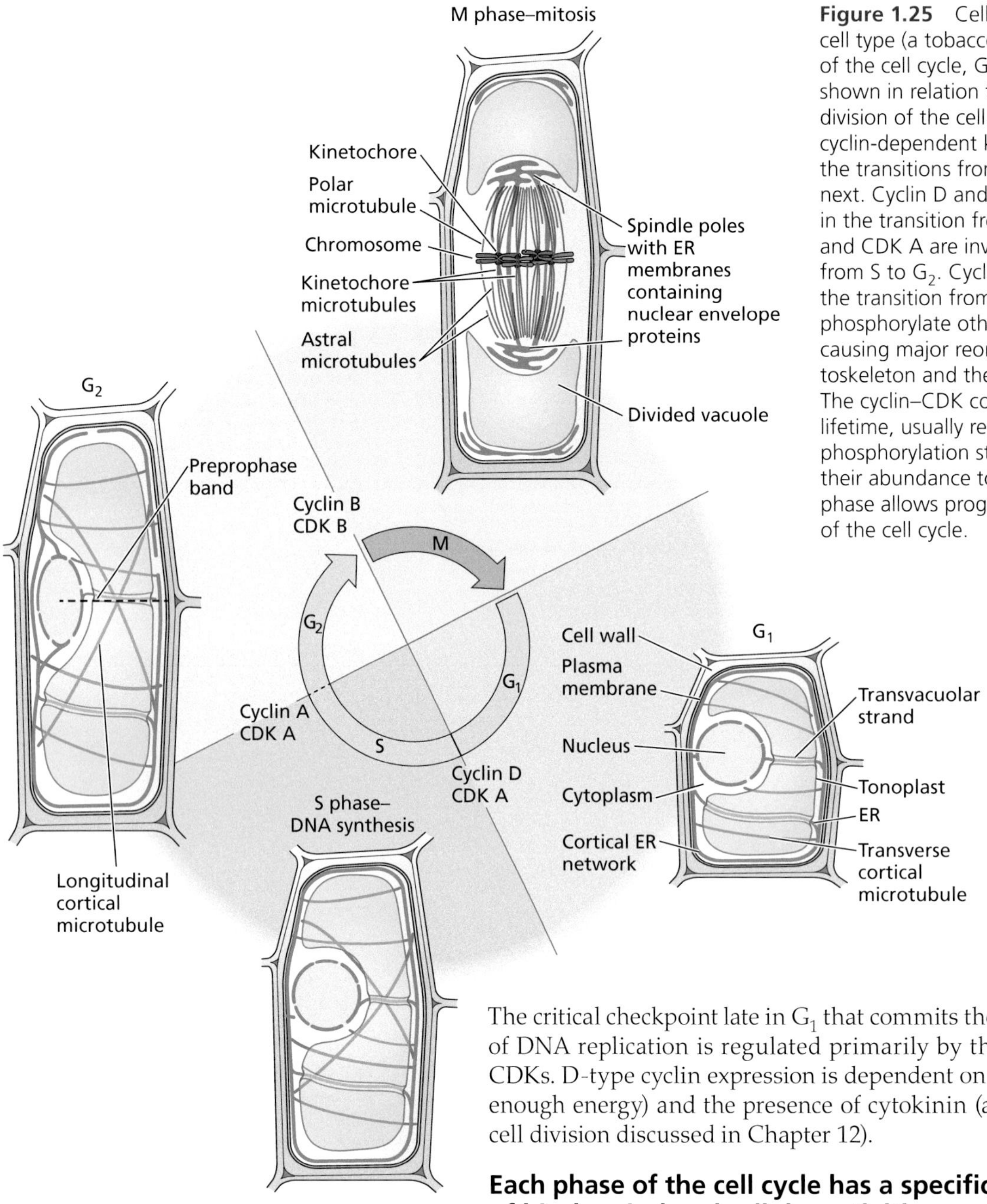

Figure 1.25 Cell cycle in a vacuolated cell type (a tobacco cell). The four phases of the cell cycle, G_1, S, G_2, and M, are shown in relation to elongation and division of the cell. Various cyclins and cyclin-dependent kinases (CDKs) regulate the transitions from one phase to the next. Cyclin D and CDK A are involved in the transition from G_1 to S. Cyclin A and CDK A are involved in the transition from S to G_2. Cyclin B and CDK B regulate the transition from G_2 to M. The kinases phosphorylate other proteins in the cell, causing major reorganization of the cytoskeleton and the membrane systems. The cyclin–CDK complexes have a finite lifetime, usually regulated by their own phosphorylation state; the decrease in their abundance toward the end of each phase allows progression to the next stage of the cell cycle.

The critical checkpoint late in G_1 that commits the cell to another round of DNA replication is regulated primarily by the D-type cyclins and CDKs. D-type cyclin expression is dependent on sugar availability (i.e., enough energy) and the presence of cytokinin (a hormone involved in cell division discussed in Chapter 12).

Each phase of the cell cycle has a specific set of biochemical and cellular activities

Nuclear DNA is prepared for replication in G_1 by the assembly of a prereplication complex at the origins of replication along the chromatin. DNA is replicated during S phase, and G_2 cells prepare for mitosis.

The whole architecture of the cell is altered as it enters mitosis. If the cell has a large central vacuole, this vacuole must first be divided in two by a coalescence of cytoplasmic transvacuolar strands that contain the nucleus; this becomes the region where nuclear division will occur. Golgi stacks and other organelles partition between the two halves of the cell, and the endomembrane system and cytoskeleton are extensively rearranged.

As a cell enters mitosis, the chromosomes change from their interphase state of organization within the nucleus and begin to condense to form the metaphase chromosomes (**Figure 1.26**). The metaphase chromosomes are held together by

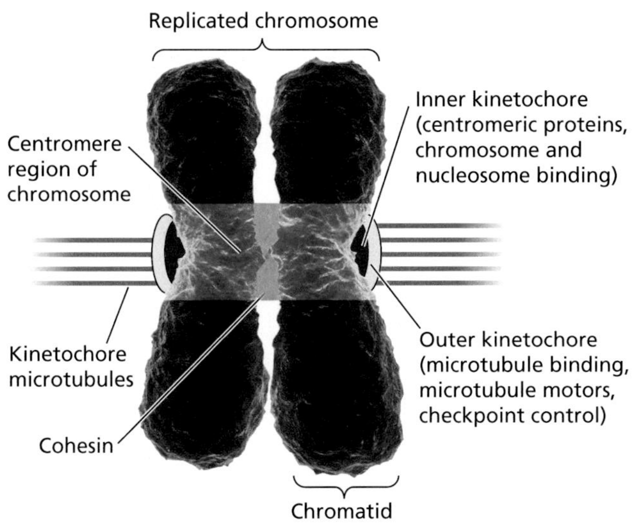

Figure 1.26 Structure of a metaphase chromosome. The centromeric DNA is highlighted, and the region where cohesion molecules bind the two chromosomes together is shown in orange. The kinetochore is a layered structure (the inner layer is shown in purple, the outer layer in yellow) that contains microtubule-associated proteins, including kinesins that help depolymerize the microtubules during shortening of kinetochore microtubules in anaphase.

special proteins called *cohesins*, which reside in the centromeric region of each chromosome pair. In order for the chromosomes to separate, these proteins must be cleaved by the enzyme *separase*. This occurs when the kinetochore attaches to spindle microtubules.

At a key regulatory point, or **checkpoint**, late in G_1 of the cell cycle, the cell becomes committed to the initiation of DNA synthesis. In mammalian cells, DNA replication and mitosis are linked—the cell division cycle, once begun, is not interrupted until all phases of mitosis have been completed. In contrast, plant cells can leave the cell division cycle either before or after replicating their DNA (i.e., during G_1 or G_2). Consequently, whereas most animal cells are diploid (having two sets of chromosomes), plant cells are frequently tetraploid (having four sets of chromosomes) or sometimes even polyploid (having many sets of chromosomes) after going through additional cycles of nuclear DNA replication without mitosis, a process called **endoreduplication**. Polyploidy, endoreduplication, and hybridization (genetic crosses between related species) has impacted the evolution of crop plants. Modern crop plants are largely derived from polyploidy and hybridization of wild relatives. The form and yields of modern crop plants reflect historical and artificial polyploidy and hybridization.

Mitosis and cytokinesis involve both microtubules and the endomembrane system

Mitosis is the process by which previously replicated chromosomes are aligned, separated, and distributed in an orderly fashion to daughter cells (**Figure 1.27**). Microtubule reorganizations are an integral part of mitosis and cytokinesis. The period immediately prior to prophase is called **preprophase**. During preprophase, microtubules are reorganized into a preprophase band of cortical microtubules (with a small number of microfilaments) around the nucleus at the site of the future cell plate—the precursor of the cross-wall. The position of the preprophase band, the underlying *cortical division site*, and the partition of cytoplasm that divides the central vacuoles determine the plane of cell division in plants, and thus play a crucial role in development.

At the start of **prophase**, the microtubules of the preprophase band start to depolymerize, and new microtubules polymerizing on the surface of the nuclear envelope begin to gather at two foci on opposite sides of the nucleus,

checkpoint One of several key regulatory points in the cell cycle. One checkpoint, late in G_1, determines if the cell is committed to the initiation of DNA synthesis.

endoreduplication Cycles of nuclear DNA replication without mitosis resulting in polyploidization.

preprophase In mitosis, the stage just before prophase during which the G_2 microtubules are completely reorganized into a preprophase band.

prophase The first stage of mitosis (and meiosis) prior to disassembly of the nuclear envelope, during which the chromatin condenses to form distinct chromosomes.

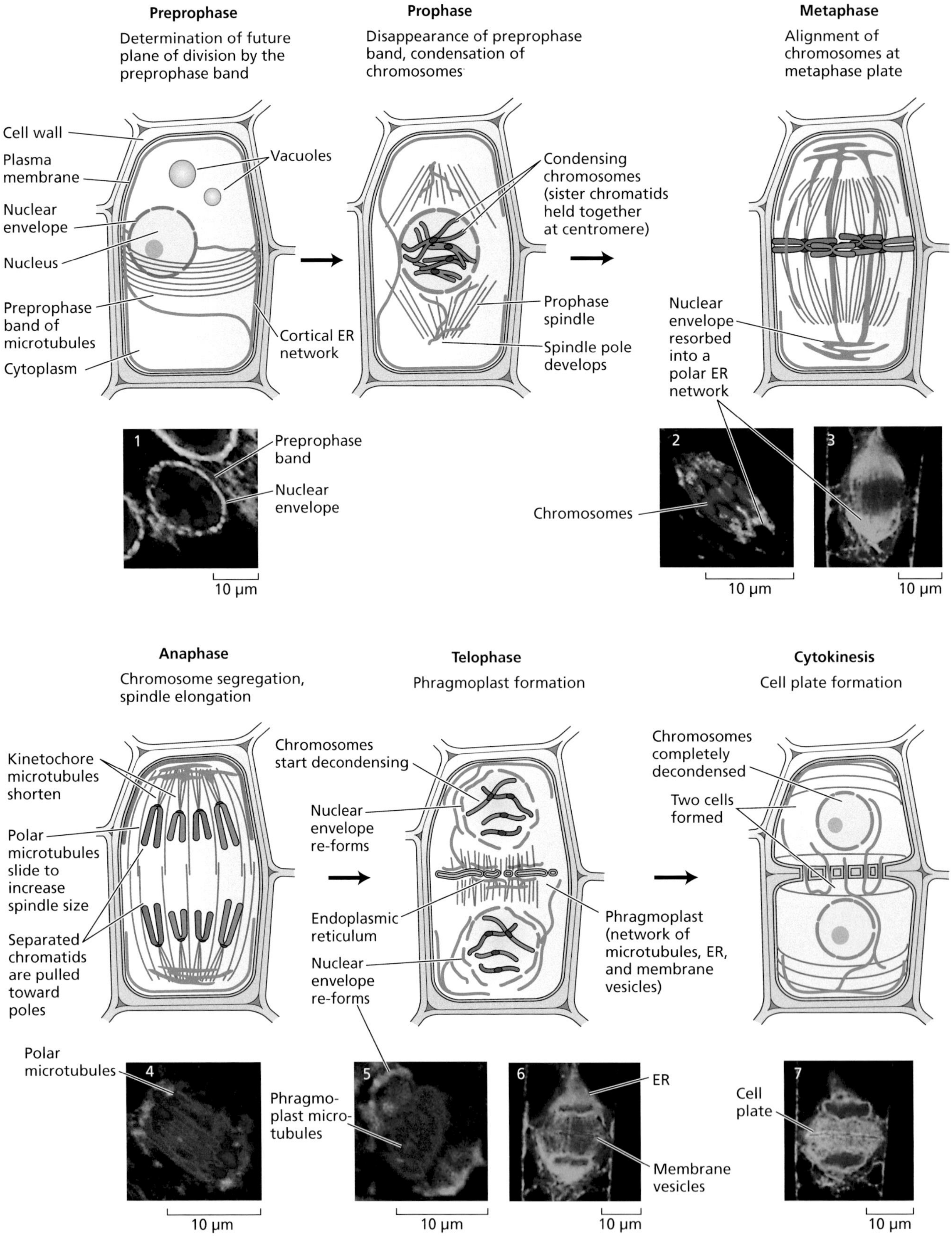
Preprophase
Determination of future plane of division by the preprophase band
Prophase
Disappearance of preprophase band, condensation of chromosomes
Metaphase
Alignment of chromosomes at metaphase plate
Cell wall
Plasma membrane
Nuclear envelope
Nucleus
Preprophase band of microtubules
Cytoplasm
Vacuoles
Cortical ER network
Condensing chromosomes (sister chromatids held together at centromere)
Prophase spindle
Spindle pole develops
Nuclear envelope resorbed into a polar ER network
1
Preprophase band
Nuclear envelope
10 μm
2
3
Chromosomes
10 μm
10 μm
Anaphase
Chromosome segregation, spindle elongation
Telophase
Phragmoplast formation
Cytokinesis
Cell plate formation
Kinetochore microtubules shorten
Polar microtubules slide to increase spindle size
Separated chromatids are pulled toward poles
Chromosomes start decondensing
Nuclear envelope re-forms
Endoplasmic reticulum
Nuclear envelope re-forms
Phragmoplast (network of microtubules, ER, and membrane vesicles)
Chromosomes completely decondensed
Two cells formed
Polar microtubules
4
10 μm
Phragmo-plast micro-tubules
5
10 μm
6
ER
Membrane vesicles
10 μm
Cell plate
7
10 μm

Figure 1.27 Changes in cellular organization that accompany mitosis in a meristematic (nonvacuolated) plant cell.

initiating spindle formation. Although not associated with the centrosomes (which plants, unlike animal cells, lack), these foci serve the same function in organizing microtubules. During prophase, the nuclear envelope remains intact, but it breaks down as the cell enters **metaphase** in a process that involves reorganization and reassimilation of the nuclear envelope into the ER. The nucleolus completely disappears during mitosis and gradually reassembles after mitosis, as the chromosomes decondense and reestablish their positions in the daughter nuclei.

In early metaphase, the preprophase band fully disappears and new microtubules polymerize, completing the **mitotic spindle**. The mitotic spindles of plant cells are more boxlike in shape than those in animal cells. The spindle microtubules in a plant cell arise from a diffuse zone consisting of multiple microtubule organizing centers at opposite ends of the cell and extend toward the middle in nearly parallel arrays.

The **centromere**, the region where the two chromatids are attached near the center of the chromosome, contains repetitive DNA, as does the **telomere**, which forms the chromosome ends that protect the chromosome from degradation. Some of the spindle microtubules bind to the chromosome in the special region of the centromere called the **kinetochore**, and the condensed chromosomes align at the metaphase plate.

The mechanism of chromosome separation during **anaphase** has two components. In early anaphase, the sister chromatids separate and begin to move toward their poles by regulated microtubule depolymerization. In late anaphase, the polar microtubules slide relative to each other and elongate, pushing the spindle poles farther apart. At the same time, the sister chromosomes are pushed to their respective poles.

At **telophase**, a new network of microtubules, F-actin, and ER called the **phragmoplast** appears (Figure 1.27). The phragmoplast organizes the region of the cytoplasm where cytokinesis takes place. The microtubules have now lost their spindle shape but retain polarity, with their minus ends still pointed toward the separated, now decondensing, chromosomes, where the nuclear envelope is in the process of re-forming. The plus ends of the microtubules point toward the midzone of the phragmoplast, where small vesicles produced by in the Golgi apparatus and *trans* Golgi network accumulate. Some content of these vesicles is also recycled from endocytosed parental cell plasma membranes.

Cytokinesis is the process that establishes the **cell plate**, a precursor of the cross-wall that will separate the two daughter cells. This cell plate, with its enclosing plasma membrane, is initiated by a large cluster of vesicles that fuse as an island in the center of the cell and then grows by additional vesicle fusion and expansion outward toward the parental wall. The forming cell plate fuses with the parental plasma membrane at a site determined by the location of the preprophase band (which disappeared earlier). Specific proteins recruited by preprophase band microtubules remain as landmarks within the cortical division zone after disassembly of the preprophase band. As the cell plate assembles, it traps ER tubules in plasma membrane–lined channels spanning the plate, thus connecting the two daughter cells (**Figure 1.28**). These cell plate–spanning ER tubules establish the sites for the primary plasmodesmata (Figure 1.4). After cytokinesis, microtubules re-form in the cell cortex. The new cortical microtubules have a transverse orientation relative to the future elongation axis of the cell, and this orientation determines the polarity of future cell extension.

metaphase A stage of mitosis during which the nuclear envelope breaks down and the condensed chromosomes align in the middle of the cell.

mitotic spindle The mitotic structure involved in chromosome movement. Polymerized from α- and β-tubulin monomers formed by the disassembly of the preprophase band in early metaphase.

centromere The constricted region on the mitotic chromosome where the kinetochore forms and to which spindle fibers attach.

telomeres Regions of repetitive DNA that form the chromosome ends and protect them from degradation.

kinetochore The site of spindle fiber attachment to the chromosome in anaphase.

anaphase The stage of mitosis during which the two chromatids of each replicated chromosome are separated and move toward opposite poles.

telophase Prior to cytokinesis, the final stage of mitosis (or meiosis) during which the chromatin decondenses, the nuclear envelope re-forms, and the cell plate extends.

phragmoplast An assembly of microtubules, membranes, and vesicles that forms during late anaphase or early telophase and precedes fusion of vesicles to form the cell plate.

cytokinesis In plant cells, following nuclear division, the separation of daughter nuclei by the formation of new cell wall.

cell plate A wall-like structure that separates newly divided cells. Formed by the phragmoplast and later becomes the cell wall.

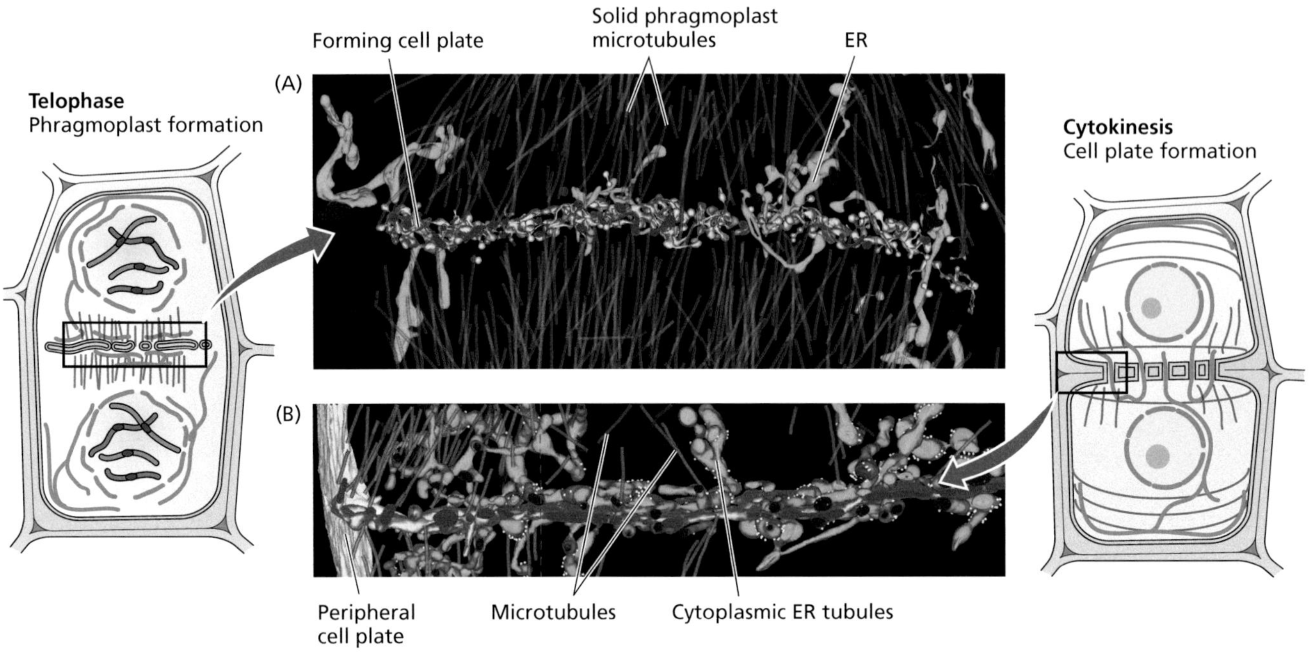

Figure 1.28 Changes in the organization of the phragmoplast and ER during cell plate formation. (A) The forming cell plate (yellow, seen from the side) in early telophase has only a few sites where it interacts with the ER (blue). The solid phragmoplast microtubules (purple) also have few ER cisternae among them. (B) Side view of the forming peripheral cell plate (yellow), showing that although many cytoplasmic ER tubules (blue) intermingle with microtubules (purple) in the peripheral growth zone, there is little direct contact between ER tubules and cell plate membranes. The small white dots are ER-bound ribosomes.

Summary

Despite their great diversity in form and size, all plants carry out similar physiological and biochemical processes. These processes depend entirely on structures that interact at every level of scale, from the molecular level to the anatomical level.

1.1 Plant Life Processes: Unifying Principles

- All land plants are sessile organisms and convert solar energy to chemical energy. They use growth instead of motility to obtain resources, and have vascular systems, rigid structures, and mechanisms to avoid desiccation on land. Plants develop from embryos sustained and protected by tissues from the mother plant.
- Plant classification is based on evolutionary relationships (**Figure 1.1**).
- Plant life cycles alternate between diploid and haploid generations (**Figure 1.2**).

1.2 Overview of Plant Structure

- All seed plants have the same basic vegetative body plan, consisting of stem, root, and leaves (**Figure 1.3**).
- New cells are produced in meristems that produce new organs such as roots, stems, leaves, and, after a transition from vegetative growth, flowers (**Figure 1.3**).
- Plant cells are surrounded by rigid, yet dynamic, cell walls that communicate with the cytoplasm (**Figure 1.4**).
- Primary cell walls are synthesized in actively growing cells, whereas secondary cell walls are deposited on the inner surface of the primary walls, after most cell expansion has ended (**Figure 1.4**).
- Because of rigid cell walls, plant development depends solely on patterns of cell division and directional cell enlargement.
- The cytoplasm of neighboring cells is connected by plasmodesmata, forming the symplasm, which allows not only water and small molecules but also proteins and RNAs to move between cells without crossing the plasma membrane (**Figure 1.4**).
- Secondary growth results in the increase in girth of roots and shoots through the action of the specialized meristems, the vascular and cork cambium (**Figure 1.5**).

(Continued)

Summary (*continued*)

1.3 Plant Tissue Types

- Dermal tissue, ground tissue, and vascular tissue are the three major tissue systems present in all plant organs (**Figure 1.3**).
- Dermal tissue includes the epidermis, which has multiple cell types, including pavement cells, stomatal guard cells, and trichomes (**Figure 1.6**).
- Parenchyma, collenchyma, and sclerenchyma are types of ground tissue. Parenchyma cells have the capacity to continue division and can differentiate into a variety of other ground tissues and vascular tissues (**Figures 1.7**, **1.8**).
- Vascular tissue has thickened secondary cell walls, perforated end walls, and pits that can develop into channels in the walls shared between adjacent cells (**Figures 1.9**, **1.10**).
- Mature xylem cells are optimized for water conductance and are dead at maturity (**Figure 1.10**).

1.4 Plant Cell Compartments

- All plant cells contain cytoplasm, a nucleus, and subcellular organelles, all enclosed by the plasma membrane and cell wall (**Figure 1.11**).
- All plants begin with a similar complement of organelles that arise either semiautonomously or from the endomembrane system.
- The endomembrane system plays a central role in secretory processes, cell signaling, specialized metabolite production, membrane recycling, the cell cycle, and cell expansion (**Figure 1.11**).
- Biological membranes are composed of lipids and both integral and peripheral membrane proteins. Peripheral membrane proteins with covalently linked membrane-embedded lipid moieties are called anchored proteins (**Figure 1.12**).
- The fluid-mosaic and ordered lipid domain structure of all plasma membranes permits regulation of transport into and out of the cell and between subcellular compartments (**Figure 1.12**).

1.5 The Plant Cytoskeleton

- A three-dimensional network of polymerizing and depolymerizing filamentous proteins—tubulin in microtubules, and actin in microfilaments—organizes the cytoplasm and provides a structural framework for interactions between endomembrane compartments (**Figure 1.13**).
- Molecular motors reversibly bind the cytoskeleton and direct organelle movement (**Figure 1.14**).
- During cytoplasmic streaming, bulk flow of the cytoplasm is driven by the viscous drag in the wake of myosin motor-driven organelles moving along F-actin.

1.6 The Nucleus

- The nucleus is the site of storage, replication, and transcription of the DNA in the chromatin, as well as the site for the synthesis of ribosomes and mRNA (**Figures 1.15–1.17**).
- The specialized membranes of the nuclear envelope are derived from and are continuous with the endoplasmic reticulum (ER), a component of the endomembrane system (**Figure 1.11**).
- Translation of mRNA into proteins occurs in the cytoplasm (**Figure 1.17**).
- Posttranslational regulation of gene expression involves mRNA and protein modification as well as turnover (**Figure 1.18**).

1.7 The Endomembrane System

- The ER is a network of membrane-bound tubules and cisternae that form a complex and dynamic structure that is to a large extent aligned with the F-actin cytoskeleton (**Figure 1.19**).
- The rough ER is involved in synthesis of proteins that enter the lumen of the ER; the smooth ER is the site of lipid biosynthesis (**Figure 1.19**).
- Secretion of proteins from cells begins at the rough ER (**Figure 1.19**).
- Glycoproteins destined for secretion are processed first in the ER and then in the Golgi apparatus (**Figure 1.20**).
- Vacuoles serve multiple functions in plant cells; in addition to being involved in the uptake of water and solutes, they play a role in cell expansion and in the storage of secondary metabolites that function in plant defense (**Figure 1.21**).
- Lipid bodies are derived from the ER and serve as a storage organelle for lipids known as triacylglycerols.
- Several types of plant peroxisomes are crucial organelles for detoxifying reactive oxygen species, degrading fatty acids, and other metabolic processes. Glyoxysomes participate in mobilizing lipids during seed germination.

1.8 Independently Dividing Semiautonomous Organelles

- Mitochondria and plastids each have an inner and an outer membrane (**Figures 1.22**, **1.23**).

(*Continued*)

Summary (*continued*)

- Chloroplasts have an internal thylakoid membrane system that contains photosynthetic proteins, chlorophylls and carotenoids, electron transport chains, and ATP synthases (**Figure 1.23**).
- Specialized plastids may contain high concentrations of pigments (chromoplasts) or starch (amyloplasts) (**Figure 1.24**).
- In plastids and mitochondria, fission and organellar DNA replication are regulated independently of nuclear division.

1.9 Cell Cycle Regulation

- The cell cycle, during which cells replicate their DNA and reproduce themselves, consists of four phases (**Figure 1.25**).
- Cyclins and cyclin-dependent kinases (CDKs) regulate the cell cycle, including the separation of paired metaphase chromosomes (**Figure 1.26**).
- Successful mitosis (**Figure 1.27**) and cytokinesis (**Figure 1.28**) require the coordinated participation of microtubules and the endomembrane system.

Suggested Reading

Burch-Smith, T. M., Stonebloom, S., Xu, M., and Zambryski, P. C. (2011) Plasmodesmata during development: Re-examination of the importance of primary, secondary, and branched plasmodesmata structure versus function. *Protoplasma* 248: 61–74.

Chapman, K. D., Dyer, J. M., and Mullen, R. T. (2012) Biogenesis and functions of lipid droplets in plants: Thematic Review Series: Lipid droplet synthesis and metabolism: from yeast to man. *J. Lipid Res.* 53: 215–226.

Gunning, B. E. S. (2009) *Plant Cell Biology on DVD*. Springer, New York, Heidelberg.

Hu, J., Baker, A., Bartel, B., Linka, N., Mullen, R. T., Reumann, S., and Zolman, B. K. (2012) Plant peroxisomes: Biogenesis and function. *Plant Cell* 24: 2279–2303.

Joppa, L. N., Roberts, D. L., and Pimm, S. L. (2011) How many species of flowering plants are there? *Proc. R. Soc. B* 278: 554–559.

Leroux, O. (2012) Collenchyma: A versatile mechanical tissue with dynamic cell walls. *Ann. Bot.* 110: 1083–1098.

McMichael, C. M., and Bednarek, S. Y. (2013) Cytoskeletal and membrane dynamics during higher plant cytokinesis. *New Phytol.* 197: 1039–1057.

Wasteneys, G. O., and Ambrose, J. C. (2009) Spatial organization of plant cortical microtubules: Close encounters of the 2D kind. *Trends Cell Biol.* 19: 62–71.

2 Water and Plant Cells

Water plays a crucial role in the life of all organisms, especially plants. Terrestrial plants have evolved mechanisms to prevent water loss. For example, leaf epidermal cells are protected by waxes to prevent water loss (discussed in Chapter 19). During photosynthesis, carbon dioxide from the atmosphere enters the leaf through the stomata (introduced in Chapter 1 and discussed in Chapters 3, 6, 8, and 9). At the same time water loss caused by transpiration during photosynthesis—and the threat of dehydration—occurs. To prevent leaf desiccation, water absorbed by the roots is transported through the plant body. Imbalances between the uptake and transport of water and water loss to the atmosphere result in water deficits that can impair many cellular processes. Thus, balancing the uptake, transport, and loss of water represents an important challenge for land plants.

A major difference between plant and animal cells is that plant cells have cell walls and animal cells do not (discussed in Chapter 1). Cell walls allow plant cells to build up large internal hydrostatic pressures, called **turgor pressure**. Well-watered plants have high turgor pressure and water-stressed plants have low turgor pressure and may appear wilted. In most cases turgor pressure can be restored by watering the plants. In this chapter we consider how water moves into and out of plant cells, emphasizing the molecular properties of water and the physical forces that influence water movement at the cellular level.

turgor pressure Force per unit of area in a liquid. In a plant cell, turgor pressure pushes the plasma membrane against the rigid cell wall and provides a force for cell expansion.

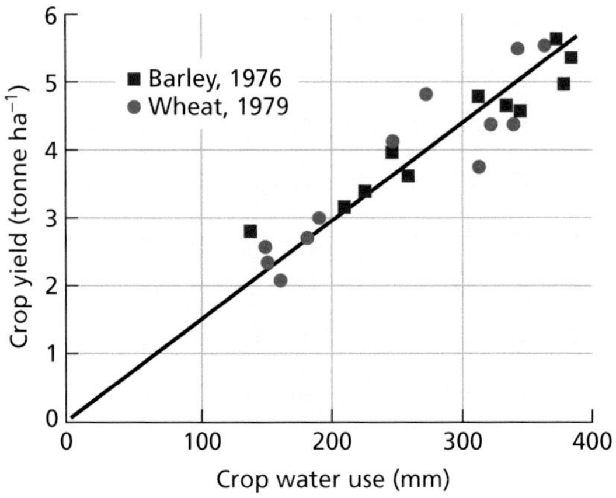

Figure 2.1 Grain yield as a function of water used is linear under a range of irrigation treatments for barley in 1976 and wheat in 1979 in southeastern England. Crop water use is the total amount of water used during the growth of the crop relative to the area over which the crop is grown (i.e., volume of water per area of land).

transpiration The evaporation of water from the surface of leaves and stems.

2.1 Water in Plant Life

Understand the role of water as a limiting resource for natural and agricultural ecosystems.

Of all the resources that plants need to grow and function, water is the most abundant and yet often the most limiting. Crop irrigation directly affects crop yield and reflects the fact that water is a key resource limiting agricultural productivity (**Figure 2.1**). Water availability also affects the productivity of natural ecosystems (**Figure 2.2**), leading to marked differences in vegetation type along precipitation gradients.

Water is frequently a limiting resource for plants, but much less so for animals, because plants use water in huge amounts. Most (~97%) of the water absorbed by a plant's roots is carried through the plant and evaporates from leaf surfaces. Such water loss is called **transpiration**. In contrast, only a small amount of the water absorbed by roots remains in the plant for growth (~2%) or is used in the biochemical reactions of photosynthesis and other metabolic processes (~1%).

Water loss to the atmosphere appears to be an inevitable consequence of carrying out photosynthesis on land. The uptake of CO_2 is coupled to the loss of water through a common pathway: As CO_2 diffuses into leaves, water vapor diffuses out. Because the driving gradient for water loss from leaves is much larger than that for CO_2 uptake, as many as 400 water molecules are lost for every CO_2 molecule gained. This unfavorable exchange has had a major influence on the evolution of plant form and function and explains why water plays such a key role in the physiology of plants.

We will begin our study of water by considering how its structure gives rise to some of its unique physical properties. We will then examine the physical basis for water movement, the concept of water potential, and the application of this concept to cell water relations.

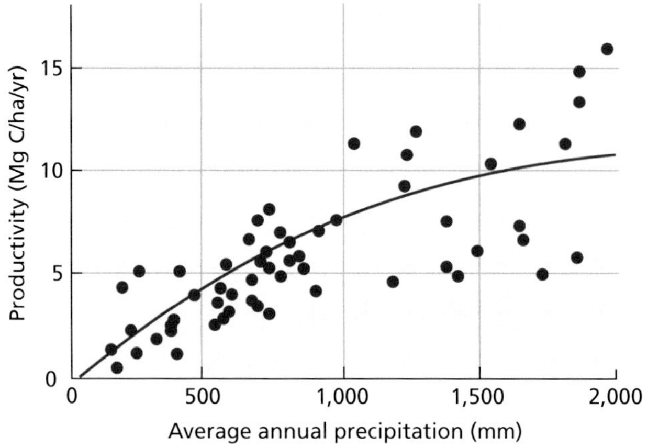

Figure 2.2 Productivity of various ecosystems worldwide as a function of annual precipitation is not linear. Above 2,000 to 3,000 mm yearly precipitation, productivity actually starts to decrease (not shown in graph). (Mg = 10^6 g)

2.2 The Structure and Properties of Water

Describe the physical and chemical properties that make water an excellent solvent.

The hydrogen bonding ability and polar structure of the water molecule give water special properties that enable it to act as a wide-ranging solvent and to be readily transported through the body of the plant. The formation of hydrogen bonds contributes to the high specific heat, surface tension, and tensile strength of water.

Water is a polar molecule that forms hydrogen bonds

The water molecule consists of an oxygen atom covalently bonded to two hydrogen atoms (**Figure 2.3A**). Because the oxygen atom is more **electronegative** than hydrogen, it tends to attract the electrons of the covalent bond. This attraction results in a partial negative charge at the oxygen end of the molecule and a partial positive charge at each hydrogen, making water a **polar** molecule. These partial charges are equal, so the water molecule carries no *net* charge.

Water molecules are tetrahedral in shape. Hydrogen atoms, each with a partial positive charge, are at two points of the tetrahedron. The other two points of the tetrahedron contain lone pairs of electrons, each with a partial negative charge (Figure 2.3A). Thus, each water molecule has two positive poles and two negative poles. These opposite partial charges create electrostatic attractions between water molecules, known as **hydrogen bonds** (**Figure 2.3B** and **C**).

Hydrogen bonds are formed only when highly electronegative atoms such as oxygen are covalently bonded to hydrogen. The small size of the hydrogen atom allows the partial positive charges to be more concentrated, and thus more effective in bonding electrostatically.

Hydrogen bonds are responsible for many of the unusual physical properties of water. Water can form up to four hydrogen bonds with adjacent water molecules, resulting in very strong intermolecular interactions. Hydrogen bonds can also

electronegative Having the capacity to attract electrons and thus producing a slightly negative electric charge.

polar A small difference in charge between two atoms in a molecule. An example is the water molecule where the oxygen atom has a partial negative charge relative to the two hydrogen atoms.

hydrogen bonds Weak chemical bonds formed between a hydrogen atom and an oxygen or nitrogen atom.

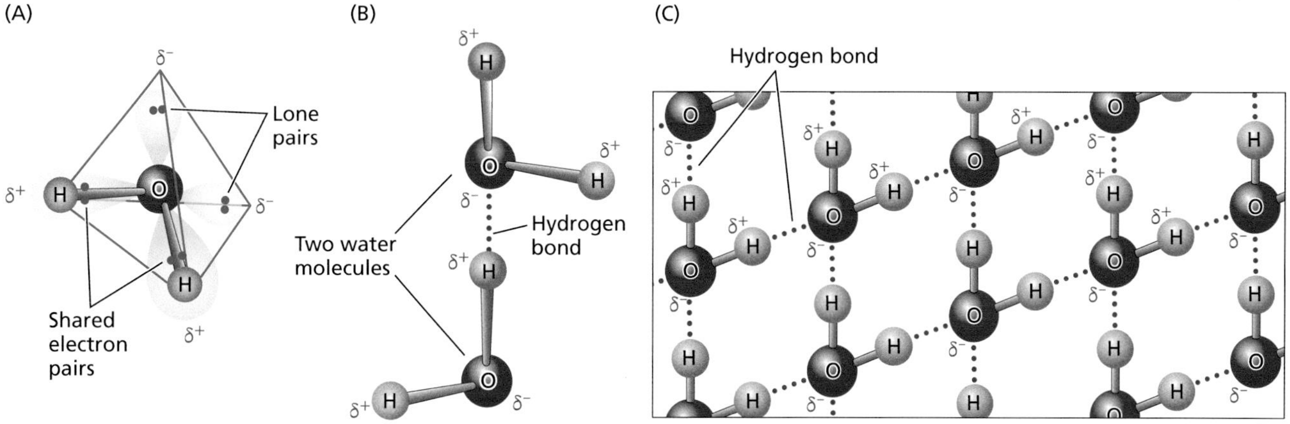

Figure 2.3 Structure of the water molecule. (A) Oxygen has six electrons in the outer shell; each hydrogen has one in the inner shell. The strong electronegativity of the oxygen atom (attracts electrons) means that the two electrons that form the covalent bond between oxygen and hydrogen are unequally shared, such that each hydrogen atom has a partial positive charge. Each of the two lone pairs of electrons of the oxygen atom produces a partial negative charge. (B) The opposite partial charges (δ^- and δ^+) on the water molecule lead to the formation of intermolecular hydrogen bonds with other water molecules. (C) Hydrogen bonds in water.

specific heat capacity Ratio of the heat capacity of a substance to the heat capacity of a reference substance, usually water. Heat capacity is the amount of heat needed to change the temperature of a unit mass by 1°C. The heat capacity of water is 1 calorie (4.184 Joule) per gram per degree Celsius.

latent heat of vaporization The energy needed to separate molecules from the liquid phase and move them into the gas phase at constant temperature.

surface tension A force exerted by water molecules at the air–water interface, resulting from the *cohesion* and *adhesion* properties of water molecules. This force minimizes the surface area of the air–water interface.

form between water and other molecules that contain electronegative atoms (O or N), especially when these are covalently bonded to H.

Water is an excellent solvent

Water dissolves greater amounts of a wider variety of substances than do other related solvents. Its versatility as a solvent is due in part to the small size of the water molecule. However, it is the hydrogen bonding ability of water and its polar structure that make it a particularly good solvent for ionic substances and for molecules such as sugars and proteins that contain polar –OH or $–NH_2$ groups.

Hydrogen bonding between water molecules and ions, and between water and polar solutes, lowers the free energy of the solute molecules and decreases electrostatic interactions, thereby increasing the solute solubility.

Water has distinctive thermal properties relative to its size

The extensive hydrogen bonding between water molecules results in water having both a high specific heat capacity and a high latent heat of vaporization.

Specific heat capacity is the heat energy required to raise the temperature of one unit of mass by one unit of temperature. Temperature is a measure of molecular kinetic energy (energy of motion). When the temperature of water is raised, the molecules vibrate faster and with greater amplitude. Hydrogen bonds act like rubber bands that absorb some of the energy from applied heat, leaving less energy available to increase motion. Thus, compared with other liquids, water requires a relatively large heat input to raise its temperature. This is important for plants because it helps buffer temperature fluctuations.

Latent heat of vaporization is the energy needed to separate molecules from the liquid phase and move them into the gas phase—a process that occurs during transpiration. The term *latent heat* refers to energy released or absorbed during a process at constant temperature, in this case a phase change. The latent heat of vaporization decreases as temperature increases, reaching a minimum at the boiling point (100°C). For water at 25°C, the heat of vaporization is 44 kJ mol^{-1}—the highest value known for any liquid. Most of this energy is used to break hydrogen bonds between water molecules.

As anyone who has exited a swimming pool can attest, evaporation cools the surface from which the liquid has evaporated. Only molecules with enough kinetic energy to overcome the intermolecular forces of the liquid phase are able to escape it and enter the gas phase. This reduces the average kinetic energy, and thus the temperature, of the liquid phase. Because water has such a high latent heat of vaporization, it moderates the temperature of transpiring leaves, which would otherwise increase due to the input of radiant energy from the sun.

Water has a high surface tension

Water molecules at an air–water interface form fewer hydrogen bonds than water molecules in the bulk liquid. Because hydrogen bonds help balance water's polar charge distribution, water molecules at the surface are in an energetically unfavorable state compared with those surrounded on all sides by other water molecules. As a result, the most stable (i.e., lowest-energy) configuration is one that minimizes the surface area of the air–water interface. To increase the area of an air–water interface, water molecules must move from the bulk liquid to the interface, which requires an input of energy to compensate for the reduction in the number of hydrogen bonds formed by molecules at the surface. The energy required to increase the surface area of a gas–liquid interface is known as **surface tension**.

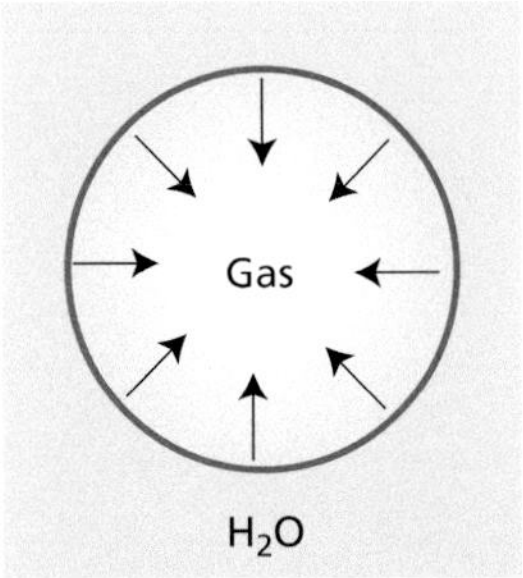

Surface tension of several liquids at 20°C (N/m)	
1% Gelatin	0.0083
Ethanol	0.0228
Phenol	0.0409
Water	0.0728

Figure 2.4 A gas bubble suspended in a liquid assumes a spherical shape such that its surface area is minimized. The magnitude of the pressure (force/area) exerted by the gas–liquid interface (arrows) is equal to $2T/r$, where T is the surface tension of the liquid (N/m) and r is the radius of the bubble (m). Water has an extremely high surface tension compared with other liquids at the same temperature.

Surface tension can be expressed in units of energy per area (J m^{-2}) but is generally expressed in the equivalent units of force per length (J m^{-2} = N m^{-1}). A joule (J) is the SI unit of energy, with units of force × distance (N m); a newton (N) is the SI unit of force, with units of mass × acceleration (kg m s^{-2}).

Water has a high surface tension, one of the highest of any liquid. Because hydrogen bonds are attractive forces that pull water molecules together, an air–water interface behaves like a stretched elastic membrane. Surface tension thus describes the tendency of a gas–liquid interface to contract. If the air–water interface is curved, surface tension produces a net force perpendicular to the surface that is directed inward (**Figure 2.4**). An important consequence of this is that an air bubble suspended in water will be squeezed by the surface tension of the water that surrounds it.

The extensive hydrogen bonding in water also gives rise to the property known as **cohesion**, the mutual attraction between molecules. A related property, called **adhesion**, is the attraction of water to a solid phase such as a cell wall or glass surface, which again is due primarily to the formation of hydrogen bonds. The degree to which water is attracted to the solid phase versus to itself can be quantified by measuring the **contact angle** (**Figure 2.5A**). Water tends to spread out (small contact angle) on a "wettable" substrate because the water can form hydrogen bonds with the solid. In contrast, water beads up on a nonwettable or hydrophobic substrate (large contact angle) because the water molecules have a more stable configuration when surrounded by other water molecules than when adjacent to the solid.

cohesion The mutual attraction between water molecules due to extensive hydrogen bonding.

adhesion The attraction of water to a solid phase such as a cell wall or glass surface, due primarily to the formation of hydrogen bonds.

contact angle A quantitative measure of the degree to which a water molecule is attracted to a solid phase versus to itself.

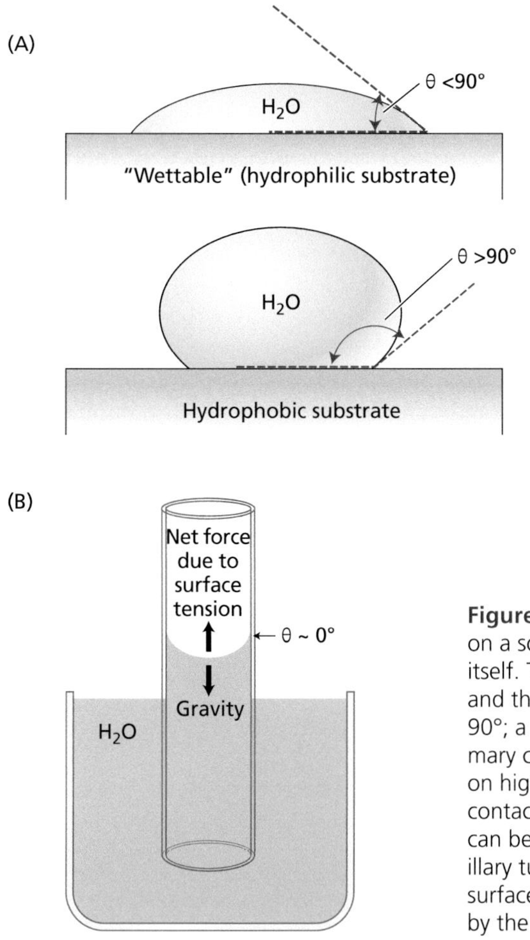

Figure 2.5 The interaction of water with surfaces. (A) The shape of a droplet placed on a solid surface reflects the relative attraction of the liquid to the solid versus to itself. The contact angle (θ) is the angle formed by the intersection of the liquid–vapor and the liquid–solid interface. "Wettable" surfaces have contact angles of less than 90°; a highly wettable (i.e., hydrophilic) surface (such as water on clean glass or primary cell walls) has a contact angle close to 0°. Water spreads out to form a thin film on highly wettable surfaces. In contrast, nonwettable (i.e., hydrophobic) surfaces have contact angles greater than 90°. Water "beads" up on such surfaces. (B) Capillarity can be observed when a liquid is supplied to the bottom of a vertically oriented capillary tube. If the walls are wettable (e.g., water on clean glass), the net force due to surface tension is upward. The water column rises until this upward force is balanced by the weight of the water column.

capillarity The movement of water for small distances up a glass capillary tube or within the cell wall, due to water's cohesion, adhesion, and surface tension.

tensile strength The ability to resist a pulling force. Water has a high tensile strength.

hydrostatic pressure Pressure generated by compression of water into a confined space. Measured in units called pascals (Pa) or, more conveniently, megapascals (MPa).

Cohesion, adhesion, and surface tension give rise to a phenomenon known as **capillarity** (**Figure 2.5B**). Consider a vertically oriented glass capillary tube with wettable walls (contact angle $<90°$). At equilibrium, the water level in the capillary tube will be higher than that of the water supply at its base. Water is drawn into the capillary tube because of (1) the attraction of water to the polar surface of the glass tube (adhesion) and (2) the surface tension of water. Together, adhesion and surface tension pull on the water molecules, causing them to move up the tube until this upward force is balanced by the weight of the water column. The narrower the tube, the higher the water level at equilibrium.

Water has a high tensile strength

Hydrogen bonds give water a high **tensile strength**, defined as the maximum force per unit area that a continuous column of water can withstand before breaking. We do not usually think of liquids as having tensile strength; however, such a property is evident in the rise of a water column in a capillary tube.

We can demonstrate the tensile strength of water by placing it in a clean glass syringe (**Figure 2.6**). When we *push* on the plunger, the water is compressed, and a positive **hydrostatic pressure** builds up. Pressure is measured in units called *pascals* (Pa) or, more conveniently, *megapascals* (MPa). One MPa equals approximately 9.9 atmospheres. Pressure is equivalent to a force per unit area ($1\ Pa = 1\ N\ m^{-2}$) and to an energy per unit volume ($1\ Pa = 1\ J\ m^{-3}$). **Table 2.1** compares units of pressure.

If instead of pushing on the plunger we *pull* on it, a tension, or *negative hydrostatic pressure*, develops as the water molecules resist being pulled apart. Negative pressures develop only when molecules are able to pull on one another. Hydrogen bonds between water molecules allow tensions to be transmitted through water, even though it is a liquid. In contrast, gases cannot develop negative pressures because the interactions between gas molecules are limited to elastic collisions.

How hard must we pull on the plunger before the water molecules are torn away from each other and the water column breaks? Careful studies have demonstrated that water can resist tensions greater than 20 MPa. The water column in a syringe, however, cannot sustain such large tensions because of the presence of microscopic gas bubbles in the bulk fluid or attached to the glass wall of the syringe. If the tension in the liquid, which exerts an outward-directed force on the gas–water interface, exceeds the inward-directed force resulting from surface tension, then the bubble will expand. This explains why gas bubbles interfere with the

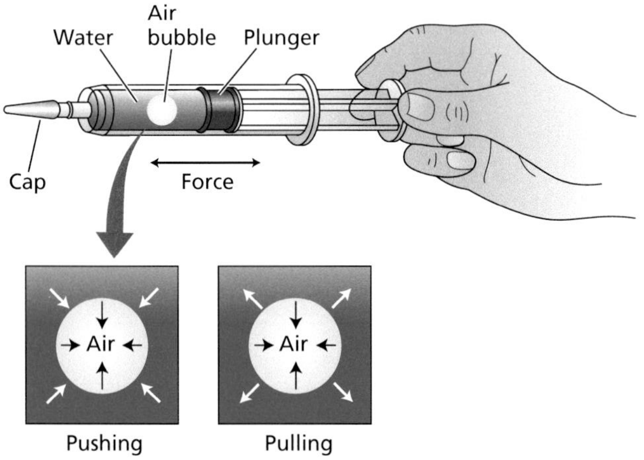

Figure 2.6 A sealed syringe can be used to create positive and negative pressures in fluids such as water. Pushing on the plunger causes the fluid to develop a positive, hydrostatic pressure (white arrows) that acts in the same direction as the inward force resulting from the surface tension of the gas–liquid interface (black arrows). Thus, a small air bubble trapped in the syringe will shrink as the pressure increases. Pulling on the plunger causes the fluid to develop a tension, or negative pressure. Air bubbles in the syringe will expand if the outward force on the bubble exerted by the fluid (white arrows) exceeds the inward force resulting from the surface tension of the gas–liquid interface (black arrows).

Table 2.1 Comparison of Units of Pressure

1 atmosphere	= 14.7 pounds per square inch
	= 760 mm Hg (at sea level, 45° latitude)
	= 1.013 bar
	= 0.1013 MPa
	= 1.013×10^5 Pa

A car tire is typically inflated to about 0.2 MPa.

The water pressure in home plumbing is typically 0.2–0.3 MPa.

The water pressure under 10 m (~30 feet) of water is 0.1 MPa.

ability of the water in the syringe to resist the pull exerted by the plunger. The expansion of gas bubbles due to tension in the surrounding liquid is known as **cavitation**. As we will see in Chapter 3, cavitation can have a devastating effect on water transport through the xylem.

cavitation The collapse of tension in a column of water resulting from the formation and expansion of tiny gas bubbles.

diffusion The movement of substances due to random thermal agitation from regions of high free energy to regions of low free energy (e.g., from high to low concentration).

osmosis The net movement of water across a selectively permeable membrane toward the region of more negative water potential, Ψ (lower concentration of water).

2.3 Diffusion and Osmosis

Explain the importance of diffusion and osmosis in transport over short (cellular) distances.

Cellular processes depend on the transport of molecules both to the cell and away from it. **Diffusion** is the net movement of substances from regions of higher to lower concentration. At the scale of a cell, diffusion is the dominant mode of transport. **Osmosis** is the net movement of a solvent across a selectively permeable barrier from a region of lower to higher solute concentration. Osmosis is a major factor governing the movement of water across selectively permeable cellular membranes.

Diffusion is the net movement of molecules by random thermal agitation

The molecules in a solution are not static; they are in continuous motion, colliding with one another and exchanging kinetic energy. The trajectory of any particular molecule after a collision is considered to be a random variable. Yet these random movements can result in the net movement of molecules.

Consider an imaginary plane dividing a solution into two equal volumes, A and B. As all the molecules undergo random motion, at each point in time there is some probability that any particular solute molecule will cross our imaginary plane. The number we expect to cross from A to B in any particular time will be proportional to the number at the beginning of the time point on side A, and the number crossing from B to A will be proportional to the number starting on side B.

If the initial concentration on side A is higher than that on side B, more solute molecules will be expected to cross from A to B than from B to A, and we will observe a net movement of solutes from A to B. Thus, diffusion results in the net movement of molecules from regions of high concentration to regions of low concentration, even though each molecule is moving in a random direction. The independent motion of each molecule explains why the system will evolve spontaneously—that is, without an input of energy—toward an equal number of A and B molecules on each side (**Figure 2.7**).

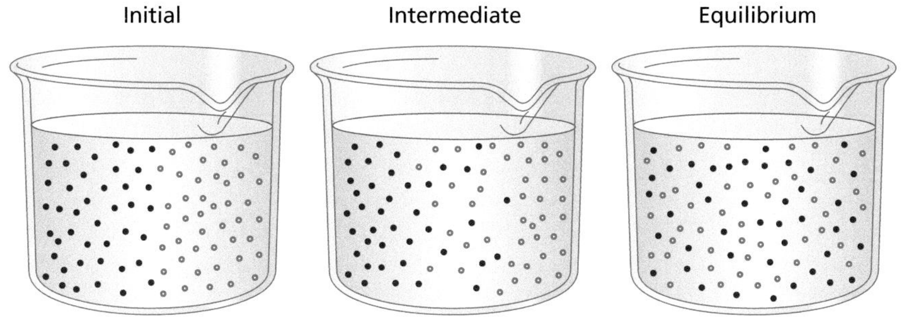

Figure 2.7 Thermal motion of molecules leads to diffusion—the gradual mixing of molecules and eventual dissipation of concentration differences. Initially, two materials containing different molecules are brought into contact. The materials may be gas or liquid. Diffusion is faster in gases and slower in liquids. With time, the mixing and randomization of the molecules diminish net movement. At equilibrium the two types of molecules are randomly (evenly) distributed.

flux density (J_s) The rate of transport of a substance s across a unit of area per unit of time. J_s may have units of moles per square meter per second (mol m^{-2} s^{-1}).

diffusion coefficient (D_s) The proportionality constant that measures how easily a specific substance s moves through a particular medium. The diffusion coefficient is a characteristic of the substance and depends on both the medium and the temperature.

selective permeability The membrane property that allows diffusion of some molecules across the membrane to a different extent than other molecules.

This tendency for a system to evolve toward an even distribution of molecules is synonymous with a decrease in free energy. Thus, diffusion represents the natural tendency of systems to move toward the lowest possible energy state.

In 1855, Adolf Fick showed that the rate of diffusion is directly proportional to the concentration gradient ($\Delta c_s/\Delta x$)—that is, to the difference in concentration of substance s (Δc_s) between two points separated by a very small distance Δx. In symbols, we write this relation as Fick's first law:

$$J_s = -D_s \frac{\Delta c_s}{\Delta x} \quad (2.1)$$

The rate of transport, expressed as the **flux density** (J_s), is the amount of substance s moving across a unit area per unit time (e.g., J_s may be expressed as moles per square meter per second [mol m^{-2} s^{-1}]). The **diffusion coefficient** (D_s, units of m^2 s^{-1}) is a proportionality constant that measures how easily substance s moves through a particular medium. The diffusion coefficient is a characteristic of the substance (larger molecules have smaller diffusion coefficients) and depends on both the medium (diffusion is typically 10,000 times faster in air than in a liquid, for example) and the temperature (substances diffuse faster at higher temperatures). The negative sign in the equation indicates that the flux moves down a concentration gradient.

Fick's first law says that a substance will diffuse faster when the concentration gradient becomes steeper (Δc_s is large) or when the diffusion coefficient is increased. Note that this equation accounts only for movement in response to a concentration gradient, and not for movement in response to other forces (e.g., pressure, electric fields).

Diffusion is most effective over short distances

Let's start a diffusion experiment with a mass of solute molecules initially concentrated around a position $x = 0$ (curve 1 in **Figure 2.8**). As the molecules move randomly, the average position of the solute molecules remains at $x = 0$ for all time points, but the distribution flattens out, as shown for two later time points (curves 2 and 3 in Figure 2.8). Comparing the distribution of the solutes at the three time points, we see that as the substance diffuses away from the starting point, the concentration gradient becomes less steep (Δc_s decreases). As a result, the rate at which solute molecules spread out from $x = 0$ is reduced.

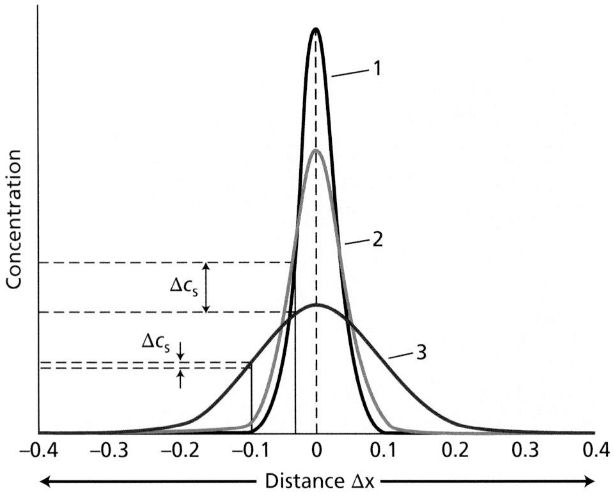

Figure 2.8 Graphical representation of the concentration gradient of a solute that is diffusing according to Fick's first law. The solute molecules were initially located in the plane indicated on the x-axis ("0"). Curve 1 shows the distribution of solute molecules shortly after placement at the plane of origin. Note how sharply the concentration drops off as the distance, x, from the origin increases. Curves 2 and 3 show the solute distribution at two later time points. The average distance of the diffusing molecules from the origin has increased, and the slope of the gradient has flattened out.

The average time required for a substance to diffuse (D) a given distance increases as the *square* of that distance (L), expressed as L^2/D_s. This scaling relationship reflects the "random walk" of diffusion—the fact that each molecule is as likely to move toward some point as away from it.

The diffusion coefficient for glucose in water is about 10^{-9} m^2 s^{-1}. Thus, the average time required for a glucose molecule to diffuse across a cell with a diameter of 50 μm is 2.5 s. However, the average time needed for the same glucose molecule to diffuse a distance of 1 m in water is approximately 32 years. These values show that diffusion in solutions can be effective within cellular dimensions but is far too slow to be effective over long distances.

Osmosis describes the net movement of water across a selectively permeable barrier

Membranes of plant cells are **selectively permeable**; that is, they allow water and other small, uncharged substances to move across the membrane but not larger solutes and charged substances. If the concentration of large or charged solutes is greater inside the cell than in the solution surrounding it, water will move spontaneously (i.e., without an input of energy) into the cell. The net movement of water across a selectively permeable barrier toward a region of higher solute concentration is referred to as *osmosis*.

Why does water move across a membrane into a region with higher solute concentration? As solutes "bounce" off the membrane, they hinder the movement of water molecules toward the membrane. This means that fewer water molecules pass through the membrane from the side with a higher solute concentration. On the side of the membrane without solutes, water molecules move randomly in all directions, including toward the membrane. As a result, the net movement of water across the membrane is toward the side with higher solute concentration.

Let's imagine what happens when we place a living cell in a beaker of pure water. The presence of a selectively permeable membrane means that water will move into the cell until one of two things happens: (1) The increase in cell volume causes the selectively permeable membrane to rupture, allowing solutes to diffuse freely, or (2) the increase in cell volume is mechanically constrained by the presence of a cell wall such that the driving force for water to enter the cell is balanced by a pressure exerted by the cell wall.

The first scenario describes what would happen to an animal cell, which lacks a cell wall. The second scenario is relevant to plant cells. The plant cell wall is very strong. The resistance of cells walls to deformation creates an inward force that raises the hydrostatic pressure within the cell.

2.4 Water Potential

Compute water potential with the formula $\Psi = \Psi_s + \Psi_p + \Psi_g$, and describe how each of the three main factors contributes to it.

All living things, including plants, require a continuous input of free energy to maintain and repair their highly organized structures, as well as to grow and reproduce. Processes such as biochemical reactions, solute accumulation, and long-distance transport of water and solutes are all driven by an input of free energy (the potential for performing work, including the work associated with moving water). In this section we examine how concentration, pressure, and gravity influence the free energy of water and thus affect the movement of water in plants.

The chemical potential of water represents the free-energy status of water

Chemical potential is a quantitative expression of the free energy associated with a substance. The unit of chemical potential is energy per mole of substance (J mol^{-1}). Note that chemical potential is a relative quantity: It represents the difference between the potential of a substance in a given state and the potential of the same substance in a standard state.

Historically, plant physiologists have most often used a related parameter called **water potential**, defined as the chemical potential of water divided by the partial molal volume of water (the volume of 1 mol of water): 18×10^{-6} m^3 mol^{-1}. Water potential is thus a measure of the free energy of water per unit volume (J m^{-3}). These units are equivalent to pressure units such as the pascal, which is the common measurement unit for water potential.

Water potential plays a central role in plant physiology because it describes the potential energy that can cause water to move from one part of the plant to another. Water flows spontaneously (i.e., without an input of energy) from

chemical potential The free energy associated with a substance that is available to perform work.

water potential (Ψ) A measure of the free energy associated with water per unit of volume (J m^{-3}). These units are equivalent to pressure units such as pascals. Ψ is a function of the solute potential, the pressure potential, and the gravitational potential: $\Psi = \Psi_s + \Psi_p + \Psi_g$. The term Ψ_g is often ignored because it is negligible for heights under 5 m.

regions of higher water potential to regions of lower water potential. Let's look more closely at the important concept of water potential.

Three major factors contribute to water potential

The major factors influencing the water potential in plants are *concentration, pressure,* and *gravity*. Water potential is symbolized by Ψ (the Greek letter psi), and the water potential of solutions may be dissected into individual components, usually written as the following sum:

$$\Psi = \Psi_s + \Psi_p + \Psi_g \quad (2.2)$$

The terms Ψ_s, Ψ_p, and Ψ_g denote the effects of solutes, pressure, and gravity, respectively, on the free energy of water. Water potentials are defined in relation to a reference, analogous to how the contour lines on a map specify the distance above sea level. The reference state most often used to define water potential is pure water at ambient temperature and standard atmospheric pressure. The reference height is generally set either at the base of the plant (for whole-plant studies) or at the level of the tissue under examination (for studies of water movement at the cellular level). Let's consider each of the terms on the right-hand side of Equation 2.2.

SOLUTES The term Ψ_s, called the **solute potential** or **osmotic potential**, represents the effect of dissolved solutes on water potential. Solutes reduce the free energy of water by diluting the water. This is primarily an entropy effect; that is, the mixing of solutes and water increases the disorder or entropy of the system and thereby lowers the free energy. This means that *the osmotic potential is independent of the specific nature of the solute*. For dilute solutions of nondissociating substances such as sucrose, the osmotic potential may be approximated by:

$$\Psi_s = -RTc_s \quad (2.3)$$

where R is the gas constant (8.32 J mol^{-1} K^{-1}), T is the absolute temperature (Kelvin, or K), and c_s is the solute concentration of the solution, expressed as **osmolarity** (moles of total dissolved solutes per volume of water [mol L^{-1}]). The minus sign indicates that dissolved solutes reduce the water potential of a solution relative to the reference state of pure water.

Equation 2.3 is valid for "ideal" solutions. Real solutions frequently deviate from the ideal, especially at high concentrations—for example, greater than 0.1 mol L^{-1}. Temperature also affects water potential; increasing the temperature will make the water potential more negative, and vice versa. We will describe water potential in terms of ideal solutions.

PRESSURE The term Ψ_p, called the **pressure potential**, represents the effect of hydrostatic pressure on the free energy of water. Positive pressures raise the water potential (indicating an increase in free energy); negative pressures reduce it. Both positive and negative pressures occur within plants. The positive hydrostatic pressure in cells is referred to as *turgor pressure*. Negative hydrostatic pressures, which frequently develop in xylem conduits, are referred to as **tension**. As we will see, tension is important in moving water long distances through the xylem of the plant.

Hydrostatic pressure is often measured as the deviation from atmospheric pressure. Remember that water in the reference state is at atmospheric pressure, where Ψ_p = 0 MPa for water in the standard state. Thus, the value of Ψ_p for pure water in an open beaker is 0 MPa, even though its absolute pressure is approximately 0.1 MPa (1 atmosphere).

GRAVITY Gravity causes water to move downward unless the force of gravity is opposed by an equal and opposite force. The **gravitational potential** (Ψ_g)

solute potential (Ψ_s) The effect of dissolved solutes on water potential. Also called osmotic potential (Ψ_s).

osmotic potential (Ψ_s) The effect of dissolved solutes on water potential. Also called solute potential.

osmolarity A unit of concentration expressed as moles of total dissolved solutes per liter of solution (mol L^{-1}). In biology, the solvent is usually water.

pressure potential (Ψ_p) The hydrostatic pressure of a solution in excess of ambient atmospheric pressure.

tension Negative hydrostatic pressure.

gravitational potential (Ψ_g) The part of the water potential caused by gravity. It is only of a significant size when considering water transport into trees and drainage in soils.

depends on the height (h) of the water above the reference-state water, the density of water (ρ_w), and the acceleration due to gravity (g). In symbols, we write the following:

$$\Psi_g = \rho_w g h \tag{2.4}$$

where $\rho_w g$ has a value of 0.01 MPa m^{-1}. Thus, raising water a distance of 10 m translates into a 0.1 MPa increase in water potential.

The gravitational potential (Ψ_g) is generally omitted in calculations of water transport at the cellular level, because differences in Ψ_g among neighboring cells are negligible compared with differences in the osmotic potential and the pressure potential. Thus, in these cases Equation 2.2 can be simplified as:

$$\Psi = \Psi_s + \Psi_p \tag{2.5}$$

Water potentials can be measured

Cell growth, photosynthesis, and crop productivity are all strongly influenced by water potential and its components. Plant scientists have thus expended considerable effort in devising accurate and reliable methods for evaluating the water status of plants.

The principal approaches for determining Ψ use either a psychrometer or a pressure chamber. Psychrometers take advantage of water's large latent heat of vaporization, which allows accurate measurements of (1) the vapor pressure of water in equilibrium with the sample or (2) the transfer of water vapor between the sample and a solution of known Ψ_s. The pressure chamber measures Y by applying external gas pressure to an excised leaf until water is forced out of the living cells.

In some cells, it is possible to measure Ψ_p directly by inserting a liquid-filled microcapillary that is connected to a pressure sensor into the cell. In other cases, Ψ_p is estimated as the difference between Ψ and Ψ_s. Solute concentrations (Ψ_s) can be determined using a variety of methods, including psychrometers and instruments that measure freezing point depression.

In discussions of water in dry soils and plant tissues with very low water contents, such as seeds, one often finds reference to the **matric potential**, Ψ_m. Under these conditions, water exists as a very thin layer, perhaps one or two molecules deep, bound to solid surfaces by electrostatic interactions. These interactions are not easily separated into their effects on Ψ_s and Ψ_p and are thus sometimes combined into a single term, Ψ_m.

2.5 Water Potential of Plant Cells

Describe the movement of water at the cellular level under different values of water potential.

Plant cells typically have water potentials of 0 MPa or less. A negative value indicates that the free energy of water in the cell is less than that of pure water at ambient temperature, atmospheric pressure, and equal height to the reference. As the water potential of the solution surrounding the cell changes, water will enter or leave the cell. Because water flows down its gradient in water potential (i.e., from higher to lower water potential), whether water enters or leaves the cell depends on the water potential of the cell and the solution surrounding it. In this section we illustrate the movement of water into and out of plant cells with some numeric examples.

Water enters the cell along a water potential gradient

First imagine an open beaker full of pure water at 20°C (**Figure 2.9A**). Because the water is open to the atmosphere, the pressure potential of the

matric potential (Ψ_m) The sum of osmotic potential (Ψ_s) + hydrostatic pressure (Ψ_p). Useful in situations (dry soils, seeds, and cell walls) where the separate measurement of Ψ_s and Ψ_p is difficult or impossible to obtain.

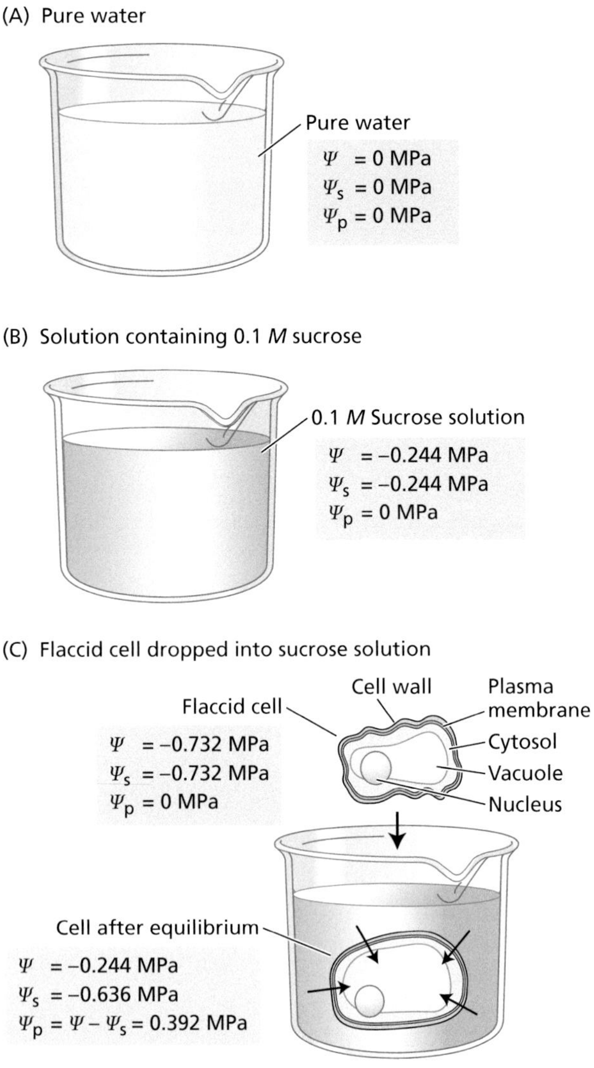

Figure 2.9 Water potential gradients can cause water to enter a cell. (A) Pure water. (B) A solution containing 0.1 *M* sucrose. (C) A flaccid cell (in air) with $\Psi_s = -0.732$ MPa (0.3 *M* dissolved solutes) is dropped into a 0.1 *M* sucrose solution. Because the starting water potential of the cell is less than the water potential of the solution, the cell takes up water. After equilibration, the water potential of the cell equals the water potential of the solution, and the result is a cell with a positive turgor pressure.

water is the same as atmospheric pressure ($\Psi_p = 0$ MPa). There are no solutes in the water, so $\Psi_s = 0$ MPa. Finally, because we will focus on transport processes occurring within the beaker, the reference height is equal to the level of the beaker, allowing us to ignore any contribution due to gravity (i.e., $\Psi_g = 0$ MPa). Therefore, the water potential is 0 MPa ($\Psi = \Psi_s + \Psi_p$).

Now imagine dissolving sucrose in the water to a concentration of 0.1 *M* (**Figure 2.9B**). This addition lowers the osmotic potential (Ψ_s) to −0.244 MPa and decreases the water potential (Ψ) to −0.244 MPa.

Next consider a plant cell that has been allowed to dry (i.e., a cell with no turgor pressure) until its cytoplasm is no longer pushing against the cell wall. We can describe a cell with zero turgor pressure as flaccid because its cell walls do not resist deformation. Let's also assume that the cell has a total internal solute concentration of 0.3 *M* (**Figure 2.9C**). This solute concentration gives an osmotic potential (Ψ_s) of −0.732 MPa. Because the cell is flaccid, the internal pressure is the same as atmospheric pressure, so the pressure potential (Ψ_p) is 0 MPa and the water potential of the cell is −0.732 MPa.

What happens if this cell is placed in the beaker containing 0.1 *M* sucrose (Figure 2.9C)? Because the water potential of the sucrose solution ($\Psi = -0.244$ MPa) is greater (less negative) than the water potential of the cell ($\Psi = -0.732$ MPa), water will move from the sucrose solution to the cell (from high to low water potential).

As water enters the cell, two things happen: The concentration of solutes decreases, and the cell wall is pushed outward. Both of these actions contribute to an increase in the cell's water potential. A lower solute concentration is associated with a lower (less negative) Ψ_s, while as the cell increases in volume, the cell wall extends a little but also resists deformation. This increases the pressure potential (Ψ_p) of the cell.

As the cell's water potential (Ψ) increases, the difference between the inside and outside water potentials ($\Delta\Psi$) is reduced. Eventually the cell Ψ reaches the same value as the Ψ of the sucrose solution. At this point, the cell is at equilibrium with the surrounding solution—$\Psi_{(cell)} = \Psi_{(solution)}$—and net water transport ceases ($\Delta\Psi = 0$ MPa).

Because the volume of the beaker is much larger than that of the cell, the tiny amount of water taken up by the cell does not significantly affect the solute concentration of the sucrose solution. Hence, Ψ_s, Ψ_p, and Ψ of the sucrose solution are not altered. Therefore, at equilibrium, $\Psi_{(cell)} = \Psi_{(solution)} = -0.244$ MPa.

Calculating cell Ψ_p and Ψ_s requires knowledge of the change in cell volume. In this example, let's assume that we know that the cell volume increased by 15%, such that the volume of the turgid cell is 1.15 times that of the flaccid cell. If we assume that the number of solutes in the cell remains constant as the cell hydrates, the final concentration of solutes will be diluted by 15%. The new Ψ_s can be calculated by dividing the initial Ψ_s by the relative increase in size of the hydrated cell: $\Psi_s = -0.732/1.15 = -0.636$ MPa. We can then calculate the pressure potential of the cell by rearranging Equation 2.5 as follows: $\Psi_p = \Psi - \Psi_s = (-0.244) - (-0.636) = 0.392$ MPa (Figure 2.9C).

Water can also leave the cell in response to a water potential gradient

If we now remove our plant cell from the 0.1 *M* sucrose solution (Figure 2.9C) and place it in a 0.3 *M* sucrose solution (**Figure 2.10A**), $\Psi_{(solution)}$ (–0.732 MPa) is more negative than $\Psi_{(cell)}$ (–0.244 MPa), and water will move from the turgid cell to the solution.

As water leaves the cell, the cell volume decreases. As the cell volume decreases, cell Ψ_p and Ψ decrease until $\Psi_{(cell)} = \Psi_{(solution)} = -0.732$ MPa. As before, we assume that the number of solutes in the cell remains constant as water flows from the cell. We know that Ψ_s in the flaccid cell was –0.732 MPa (Figure 2.9C) or precisely the same as the new $\Psi_{(solution)}$. This means that the cell volume will decrease back to that of the flaccid cell (from 115% to 100%), where $\Psi_s = -0.732$ MPa. This allows us to calculate that $\Psi_p = 0$ MPa using Equation 2.2.

In this example, the protoplast pulls away from the cell wall (i.e., the cell plasmolyzes), because sucrose molecules are able to diffuse through the water-filled pores of the cell walls. When this occurs, the difference in water potential between the cytoplasm and the solution is entirely across the plasma membrane, and thus the protoplast shrinks independently of the cell wall. Plant cells that are dehydrated osmotically with solutes that permeate the cell wall also undergo plasmolysis. In contrast, when a cell desiccates in air (e.g., as in the flaccid cell in Figure 2.9C), plasmolysis does not occur. Instead, the cell (cytoplasm + wall) shrinks as a unit, resulting in the cell wall being mechanically deformed as the cell loses volume (cytorrhysis).

If, instead of placing the turgid cell in the 0.3 *M* sucrose solution, we leave it in the 0.1 *M* solution and slowly squeeze it by pressing the cell between two plates (**Figure 2.10B**), we effectively raise the cell Ψ_p, consequently raising the cell Ψ and creating a $\Delta\Psi$ such that water now flows *out* of the cell. This is analogous

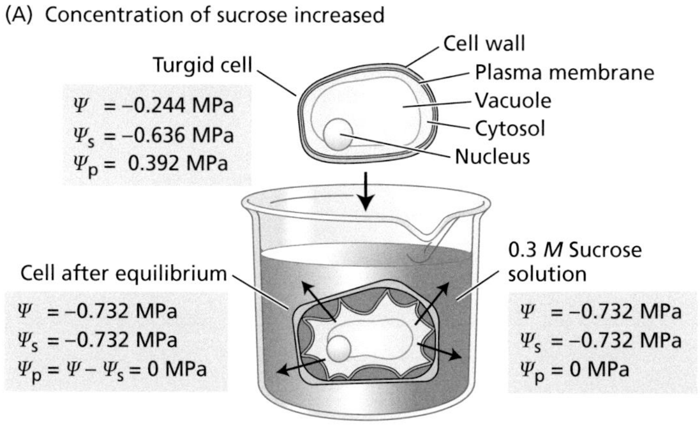

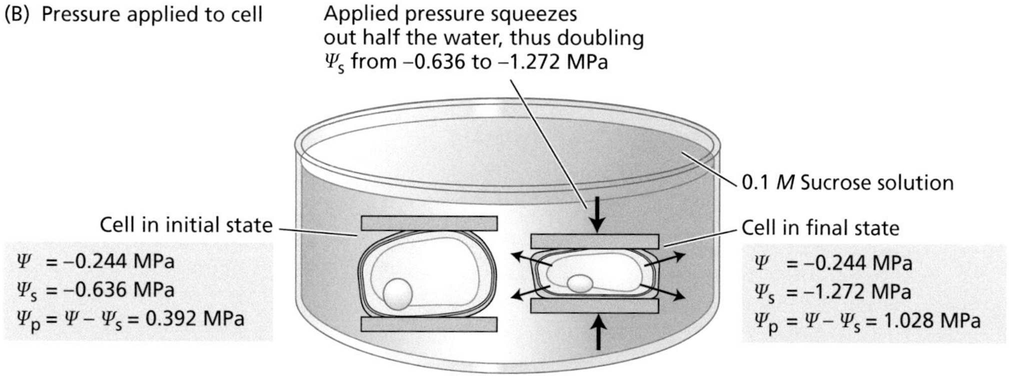

Figure 2.10 Water potential gradients can cause water to leave a cell. (A) If we move the turgid cell in Figure 2.9C to a solution with a higher sucrose concentration, the cell will lose water. The increased sucrose concentration lowers the solution water potential, draws water out of the cell, and thereby reduces the cell's turgor pressure. (B) Another way to make the cell lose water is to squeeze it slowly between two plates. In this case, half of the cell water is removed, so cell osmotic potential decreases (becomes more negative) by a factor of two.

to the industrial process of reverse osmosis in which externally applied pressure is used to separate water from dissolved solutes by forcing it across a semipermeable barrier. If we continue squeezing until half the cell's water is removed and then hold the cell in this condition, the cell will reach a new equilibrium. As in the previous example, at equilibrium $\Delta\Psi = 0$ MPa, the amount of water added to the external solution is so small that it can be ignored. The cell will thus return to the Ψ value that it had before the squeezing procedure. However, the components of the cell Ψ will be quite different.

Because half of the water was squeezed out of the cell while the solutes remained inside the cell (the plasma membrane is selectively permeable), the cell solution is concentrated twofold, and thus Ψ_s is lower (–0.636 MPa × 2 = –1.272 MPa). Knowing the final values for Ψ and Ψ_s, we can calculate the pressure potential, using Equation 2.5, as $\Psi_p = \Psi - \Psi_s$ = (–0.244 MPa) – (–1.272 MPa) = 1.028 MPa.

In our example we used an external force to change cell volume without a change in water potential. In nature, it is typically the water potential of the cell's environment that changes, and the cell gains or loses water until its Ψ matches that of its surroundings.

One point common to all these examples deserves emphasis: *Water flow across membranes is a passive process. That is, water moves in response to physical forces, toward regions of lower water potential or lower free energy.* There are no known metabolic "pumps" (e.g., reactions driven by ATP hydrolysis) that can be used to drive water across a semipermeable membrane against its free-energy gradient.

The only situation in which water can be said to move across a semipermeable membrane against its water potential gradient is when it is coupled to the movement of solutes. The transport of sugars, amino acids, or other small molecules by various membrane proteins can "drag" up to 260 water molecules across the membrane per molecule of solute transported.

Such transport of water can occur even when the movement is against the usual water potential gradient (i.e., toward a higher water potential), because the loss of free energy by the solute more than compensates for the gain of free energy by the water. The net change in free energy remains negative. The amount of water transported in this way is generally quite small compared with the passive movement of water down its water potential gradient.

Water potential and its components vary with growth conditions and location within the plant

In leaves of well-watered plants, Ψ ranges from –0.2 to about –1.0 MPa in herbaceous plants and to –2.5 MPa in trees and shrubs. Leaves of plants in arid climates can have much lower Ψ, down to below –10 MPa under the most extreme conditions.

Just as Ψ values depend on the growing conditions and the type of plant, so, too, the values of Ψ_s can vary considerably. In cells of well-watered garden plants (examples include lettuce, cucumber seedlings, and bean leaves), Ψ_s may be as high as –0.5 MPa (low cell solute concentration), although values of –0.8 to –1.2 MPa are more typical. In woody plants, Ψ_s tends to be lower (higher cell solute concentration), allowing the more negative midday Ψ typical of these plants to occur without a loss in turgor pressure.

Although Ψ_s *within* cells may be quite negative, the solution surrounding the cells—that is, in the cell walls and in the xylem, collectively referred to as the apoplast—is generally quite dilute. The Ψ_s of the apoplast is typically –0.1 to 0 MPa, although in certain tissues (e.g., developing fruits) and habitats (e.g., high-salinity environments) the concentration of solutes in the apoplast can be large.

Values for Ψ_p in cells of well-watered plants may range from 0.1 to as much as 3 MPa, depending on the value of Ψ_s inside the cell. A plant **wilts** when the

wilting Loss of rigidity, leading to a flaccid state, due to turgor pressure falling to zero.

turgor pressure inside the cells of its tissues falls toward zero. As more water is lost from the cell, the walls become mechanically deformed, and the cell may be damaged as a result.

2.6 Cell Wall and Membrane Properties

Explain the different physiological functions of cell wall and cell membrane with respect to water movement in plants.

Structural elements make important contributions to the water relations of plant cells. Cell wall elasticity defines the relation between turgor pressure and cell volume, while the permeability of the plasma membrane and the tonoplast (the vacuolar membrane) to water influences the rate at which cells exchange water with their surroundings. In this section we examine how wall and membrane properties influence the water status of plant cells.

Small changes in plant cell volume cause large changes in turgor pressure

Cell walls provide plant cells with a substantial degree of volume homeostasis relative to the large changes in water potential that they experience every day as a consequence of water losses caused by transpiration during photosynthesis. Because plant cells have fairly rigid walls, a change in cell Ψ is generally accompanied by a large change in Ψ_p, with relatively little change in cell (protoplast) volume, as long as Ψ_p is greater than 0.

This phenomenon is illustrated by the *pressure–volume curve* shown in **Figure 2.11**. As Ψ decreases from 0 to –1.2 MPa, the relative or percent water content is reduced by only slightly more than 5%. Most of the decrease in Ψ is due to a reduction in Ψ_p (by ~1.0 MPa); Ψ_s decreases by less than 0.2 MPa as a result of increased concentration of cell solutes.

Measurements of cell water potential and cell volume can be used to quantify how wall properties influence the water status of plant cells. Turgor pressure in most cells approaches zero as the cell volume decreases by 10 to 15%.

The volumetric elastic modulus, symbolized by ε (the Greek letter epsilon), can be determined by examining the relationship between Ψ_p and cell volume: ε is the change in Ψ_p for a given change in relative volume ($\varepsilon = \Delta\Psi_p/\Delta$[relative volume]). Cells with a large ε have stiff cell walls and thus experience larger changes in turgor pressure for the same change in cell volume than cells with a smaller ε and more elastic walls. The mechanical properties of cell walls vary among species and cell types, resulting in significant differences in the extent to which decreases in water potential affect cell volume.

A comparison of the cell water relations within stems of cacti illustrates the important role of cell wall properties. Cacti are stem succulent plants, typically found in arid regions. Their stems consist of an outer, photosynthetic layer that surrounds non-photosynthetic tissues that serve as a water storage reservoir (**Figure 2.12**). During drought, water is lost preferentially from these inner cells, despite the fact that the water potential of the two cell types remains in equilibrium (or very close to equilibrium). How does this happen?

Detailed studies of *Opuntia ficus-indica* demonstrate that the water storage cells are larger and have thinner walls than the photosynthetic cells, and are thus more flexible (have lower ε). For a given decrease in water potential, a water storage cell will lose a greater fraction of its water content than a photosynthetic cell. In addition, the solute concentration

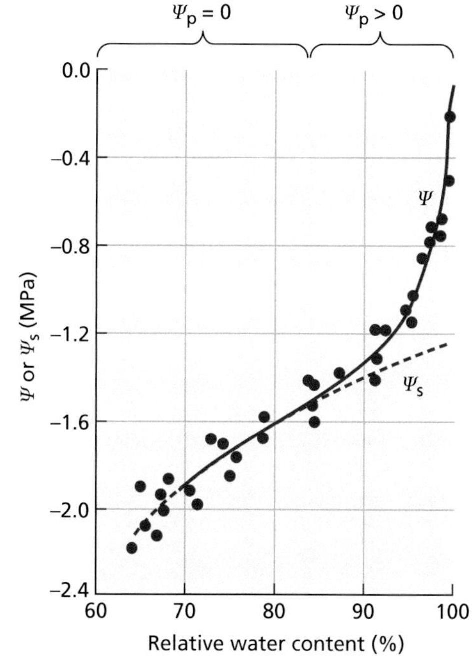

Figure 2.11 The relation among water potential (Ψ), solute potential (Ψ_s), and relative water content ($\Delta V/V$) in cotton (*Gossypium hirsutum*) leaves. Note that Ψ decreases steeply with the initial decrease in relative water content. In comparison, Ψ_s changes little. As cell volume decreases below 90% in this example, the situation reverses: Most of the change in water potential is due to a drop in cell Ψ_s, accompanied by relatively little change in turgor pressure.

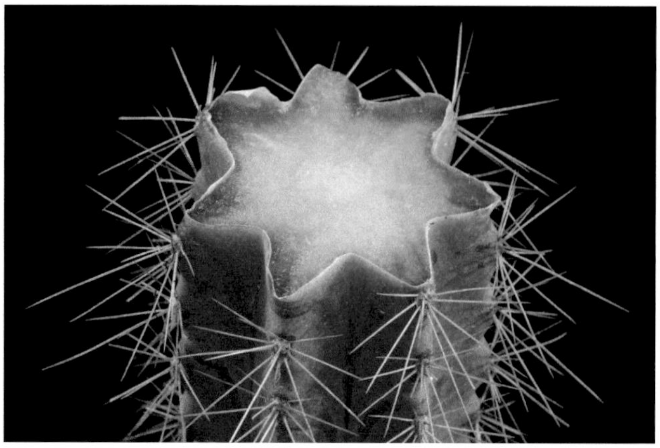

Figure 2.12 Cross section of a cactus stem, showing an outer, photosynthetic layer, and an inner, nonphotosynthetic tissue that functions in water storage. During drought, water is lost preferentially from nonphotosynthetic cells, so the water status of the photosynthetic tissue is maintained.

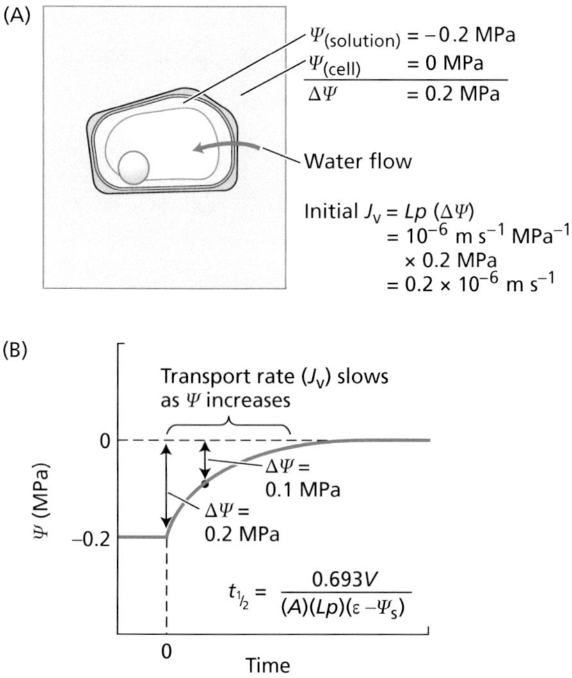

Figure 2.13 The rate of water transport into a cell depends on the magnitude of the water potential difference ($\Delta\Psi$) and the hydraulic conductivity of the plasma membranes (*Lp*). (A) In this example, the magnitude of the initial water potential difference is 0.2 MPa and *Lp* is 10^{-6} m s^{-1} MPa^{-1}. These values give an initial transport rate (J_V) of 0.2×10^{-6} m s^{-1}. (B) As water is taken up by the cell, the water potential difference decreases with time, leading to a slowing in the rate of water uptake. This effect follows an exponentially decaying time course with a half-time ($t_{1/2}$) that depends on the following cell parameters: volume (*V*), surface area (*A*), conductivity (*Lp*), volumetric elastic modulus (ε), and cell osmotic potential (Ψ_s).

hydraulic conductivity A measure of how readily water can move across a membrane; it is expressed in terms of volume of water per unit of area of membrane per unit of time per unit of driving force (i.e., m^3 m^{-2} s^{-1} MPa^{-1}).

of the water storage cells decreases during drought, in part due to the polymerization of soluble sugars into insoluble starch granules. A more typical plant response to drought is to accumulate solutes, in part to prevent water loss from cells (Chapter 19). However, in the case of cacti, the combination of more flexible cell walls and a decrease in solute concentration during drought allows water to be withdrawn preferentially from the water storage cells, thus helping maintain the hydration of the photosynthetic tissues.

The rate at which cells gain or lose water is influenced by plasma membrane hydraulic conductivity

So far, we have seen that water moves into and out of cells in response to a water potential gradient. The direction of flow is determined by the direction of the Ψ gradient, and the rate of water movement is proportional to the magnitude of the driving gradient. However, for a cell that experiences a change in the water potential of its surroundings (e.g., Figures 2.9 and 2.10), the movement of water across the plasma membrane will decrease with time as the internal and external water potentials converge (**Figure 2.13**). The rate approaches zero in an exponential manner. The time it takes for the rate to decline by half—its half-time, or $t_{1/2}$—is given by the following equation:

$$t_{1/2} = \left(\frac{0.693}{(A)(Lp)}\right)\left(\frac{V}{\varepsilon - \Psi_s}\right) \quad (2.6)$$

where *V* and *A* are, respectively, the volume and surface of the cell, and *Lp* is the **hydraulic conductivity** of the plasma membrane. Hydraulic conductivity describes how readily water can move across a membrane; it is expressed in terms of volume of water per unit area of membrane per unit time per unit driving force (i.e., m^3 m^{-2} s^{-1} MPa^{-1}).

A short half-time means a fast equilibration rate. Thus, cells with large surface-to-volume ratios, high membrane hydraulic conductivity, and stiff cell walls (large ε) will come rapidly into equilibrium with their surroundings. Cell half-times typically range from 1 to 10 s, although some are much shorter. Because of their short half-times, single cells reach water potential equilibrium with their surroundings in less than 1 min. For multicellular tissues, the half-times may be much longer.

Aquaporins facilitate the movement of water across membranes

For many years, plant physiologists were uncertain whether water movement into plant cells was limited to the diffusion of water molecules across the plasma membrane's lipid bilayer or if it also involved diffusion through protein-lined pores (**Figure 2.14**). Some studies suggested that diffusion directly across the lipid bilayer was not sufficient to account for observed rates of water movement across membranes, but the evidence in support of microscopic pores was not compelling.

This uncertainty was put to rest in 1991 with the discovery of **aquaporins**. Aquaporins are integral membrane proteins that form water-selective channels across the membrane. Because water diffuses much faster through such channels than through a lipid bilayer, aquaporins facilitate water movement into and out of plant cells.

Although aquaporins may alter the *rate* of water movement across the membrane, they do not change the *direction of transport* or the *driving force* for water movement. However, aquaporins can be reversibly "gated" (i.e., transferred between an open and a closed state) in response to physiological parameters such as intercellular pH and Ca^{2+} concentration. As a result, plants have the ability to regulate the permeability of their membranes to water (Chapters 3 and 6).

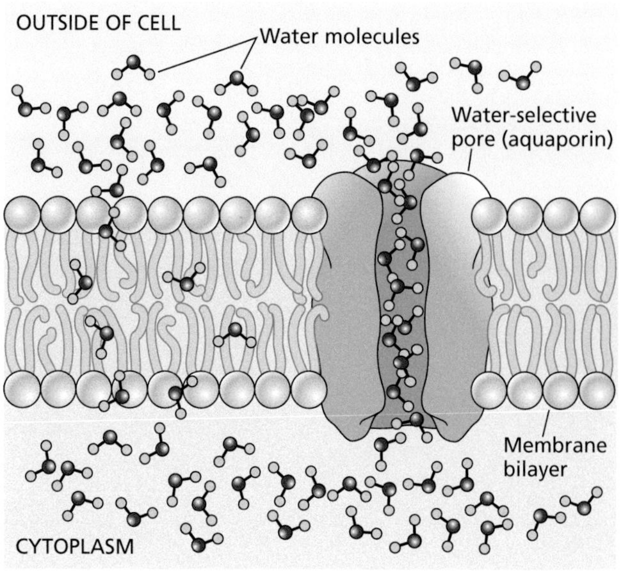

Figure 2.14 Water can cross biological membranes by diffusion. Individual water molecules can diffuse through the membrane bilayer, as shown on the left, and by linear diffusion of water molecules through water-selective pores formed by integral membrane proteins such as aquaporins.

2.7 Plant Water Status

Describe how water status affects many other functions in the plant.

The concept of water potential has two principal uses. First, water potential governs transport across plasma membranes, as we have described. Second, water potential is often used as a measure of the *water status* of a plant. In this section we discuss how the concept of water potential helps us evaluate the water status of a plant.

Physiological processes are affected by plant water status

Because of transpirational water loss to the atmosphere, plants are seldom fully hydrated. During periods of drought, they suffer from water deficits that lead to inhibition of plant growth and photosynthesis. **Figure 2.15** lists some of the physiological changes that occur as plants experience increasingly drier conditions.

The sensitivity of any particular physiological process to water deficits is, to a large extent, a reflection of that plant's strategy for dealing with the range of water availability that it experiences in its environment. The process that is most affected by water deficit is cell expansion (Figure 2.15). In many plants, reductions in water supply inhibit shoot growth and leaf expansion but *stimulate* root elongation. A relative increase in roots relative to leaves is an appropriate response to reductions in water availability, and thus the sensitivity of shoot growth to decreases in water availability can be seen as an adaptation to drought rather than a physiological constraint.

aquaporins Integral membrane proteins that form channels across a membrane, many of which are selective for water (hence the name). Such channels facilitate water movement across a membrane.

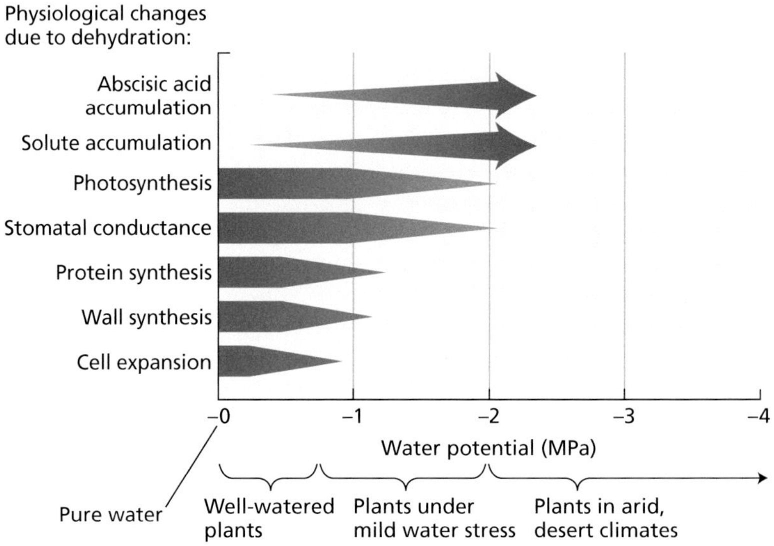

Figure 2.15 Sensitivity of various physiological processes to changes in water potential under various growing conditions. The thickness of the bars corresponds to the magnitude of the process. For example, cell expansion decreases as water potential falls (becomes more negative). Arrows indicate processes that are activated by water deficits. Abscisic acid is a hormone that induces stomatal closure during water stress (Chapters 3, 6, and 19).

However, what plants cannot do is to alter the availability of water in the soil. Figure 2.15 shows representative values for Ψ at various stages of water stress. Thus, drought does impose some absolute limitations on physiological processes, although the actual water potentials at which such limitations occur vary with species.

Solute accumulation helps cells maintain turgor and volume

The ability to maintain physiological activity as water becomes less available typically incurs some costs. The plant may spend energy to accumulate solutes to maintain turgor pressure, invest in the growth of nonphotosynthetic organs such as roots to increase water uptake capacity, or build xylem conduits capable of transporting water under large tensions. Thus, physiological responses to water availability reflect a trade-off between the benefits accrued by being able to carry out physiological processes (e.g., growth) over a wider range of environmental conditions and the carbon investment costs associated with such capability.

Plants that grow in saline environments, called **halophytes**, typically have very low values of Ψ_s. A low Ψ_s lowers cell Ψ enough to allow root cells to extract water from saline water without allowing excessive levels of salts to enter at the same time. Plants may also exhibit quite negative Ψ_s under drought conditions. Water stress typically leads to an accumulation of solutes in the cytoplasm and vacuole of plant cells, thus allowing the cells to maintain turgor pressure despite low water potentials.

Positive turgor pressure ($\Psi_p > 0$) is essential for many physiological processes. For example, the growth of plant cells requires turgor pressure to stretch the cell walls. The loss of turgor under water deficits can explain in part why cell growth is so sensitive to water stress, as well as why this sensitivity can be modified by varying the cell's osmotic potential. Turgor pressure also contributes to the mechanical rigidity of nonlignified cells and tissues, enables stomata to open (Chapter 3), and powers the movement of sugars through the phloem (Chapter 10).

However, the major benefit of maintaining positive turgor pressure may be in allowing the large fluctuations in water potential that occur each day to be accommodated primarily through changes in turgor pressure rather than cell volume. Large fluctuations in cell volume can impair metabolic processes through their impact on concentrations of key metabolites. The existence of mechanosensitive-activated signaling molecules in the plasma membrane (Chapter 12) suggests that plant cells may sense changes in their water status via changes in volume, rather than by responding directly to turgor pressure.

halophytes Plants that are native to saline soils and complete their life cycles in that environment. Contrast with *glycophytes*.

Summary

2.1 Water in Plant Life

- Water limits both agricultural and natural ecosystem productivity (**Figures 2.1**, **2.2**).
- About 97% of the water absorbed by roots is carried through the plant and is lost by transpiration from the leaf surfaces.
- The uptake of CO_2 is coupled to the loss of water through a common diffusional pathway.

2.2 The Structure and Properties of Water

- The polarity and tetrahedral shape of water molecules permit them to form hydrogen bonds that give water its unusual physical properties: It is an excellent solvent and has a high specific heat capacity, an unusually high latent heat of vaporization, and a high tensile strength (**Figures 2.3**, **2.6**).
- Cohesion, adhesion, and surface tension give rise to capillarity (**Figures 2.4**, **2.5**).

2.3 Diffusion and Osmosis

- The random thermal motion of molecules results in diffusion (**Figures 2.7**, **2.8**).
- Diffusion is important over short distances. The average time for a substance to diffuse a given distance increases as the square of that distance.
- Osmosis is the net movement of water across a selectively permeable barrier toward a region of higher solute concentration.

2.4 Water Potential

- Water's chemical potential is a measure of the free energy of water. It is quantified as the difference between the free energy of water in a given state and the free energy of water in a standard state.
- Concentration, pressure, and gravity contribute to water potential (Ψ) in plants.
- Ψ_s, the solute potential or osmotic potential, represents solutes' dilution of water and the reduction of the free energy of water.
- Ψ_p, the pressure potential, represents the effect of hydrostatic pressure on the free energy of water. Positive pressure (turgor pressure) raises the water potential; negative pressure (tension) reduces it.
- Ψ_g, the gravitational potential, represents the effect of gravity on the free energy of water. Gravity must be taken into account when considering water movement in trees but can be omitted when calculating cell water potential.

2.5 Water Potential of Plant Cells

- Plant cells typically have negative water potentials.
- Water enters or leaves a cell according to the water potential gradient.
- When a flaccid cell is placed in a solution that has a water potential greater (less negative) than the cell's water potential, water will move from the solution into the cell (from high to low water potential) (**Figure 2.9**).
- As water enters, the cell wall resists being stretched, increasing the turgor pressure (Ψ_p) of the cell.
- At equilibrium [$\Psi_{(cell)} = \Psi_{(solution)}$; $\Delta\Psi = 0$], net water movement ceases.
- When a turgid plant cell is placed in a sucrose solution that has a water potential more negative than the water potential of the cell, water will move from the turgid cell to the solution (**Figure 2.10**).
- If a cell is squeezed, its Ψ_p increases, resulting in a $\Delta\Psi$ such that water flows out of the cell (**Figure 2.10**).

2.6 Cell Wall and Membrane Properties

- Cell wall elasticity defines the relation between turgor pressure and cell volume, while the water permeability of the plasma membrane and tonoplast determines how fast cells exchange water with their surroundings.
- Because plant cells have fairly rigid walls, small changes in plant cell volume cause large changes in turgor pressure (**Figure 2.11**).
- For any non-zero initial $\Delta\Psi$, the net movement of water across the membrane will decrease with time as the internal and external water potentials converge (**Figure 2.13**).
- Aquaporins are water-selective membrane channels (**Figure 2.14**).

2.7 Plant Water Status

- During drought, photosynthesis and growth are inhibited, while concentrations of abscisic acid and solutes increase (**Figure 2.15**).
- During drought, plants must use energy to maintain turgor pressure by accumulating solutes, as well as to support root and vascular growth.
- Stretch-activated signaling molecules in the plasma membrane may permit plant cells to sense changes in their water status via changes in volume.

Suggested Reading

Bartlett, M. K., Scoffoni, C., and Sack, L. (2012) The determinants of leaf turgor loss point and prediction of drought tolerance of species and biomes: A global meta-analysis. *Ecol. Lett.* 15: 393–405.

Goldstein, G., Ortega, J. K. E., Nerd, A., and Nobel, P. S. (1991) Diel patterns of water potential components for the crassulacean acid metabolism plant *Opuntia ficus-indica* when well-watered or droughted. *Plant Physiol.* 95: 274–280.

Munns, R. (2002) Comparative physiology of salt and water stress. *Plant Cell Environ.* 25: 239–250.

Tardieu, F., Parent, B., Caldeira, C. F., and Welcker, C. (2014) Genetic and physiological controls of growth under water deficit. *Plant Physiol.* 164: 1628–1635.

Tyerman, S. D., McGaughey, S. A., Qiu, J., Yool, A. J., and Byrt, C. S. (2021) Adaptable and multifunctional ion-conducting aquaporins. *Annu. Rev. Plant Biol.* 72: 703–736. https://doi.org/10.1146/annurev-arplant-081720-013608.

Wheeler, T. D., and Stroock, A. D. (2008) The transpiration of water at negative pressures in a synthetic tree. *Nature* 455: 208–212.

3 Water Balance of Plants

Earth's atmosphere presents a formidable challenge to land plants. On the one hand, the atmosphere is the source of carbon dioxide (CO_2) needed for photosynthesis. On the other hand, the atmosphere is usually quite dry, leading to a net loss of water due to evaporation. Because plant surfaces do not allow the inward diffusion of CO_2 while preventing water loss, CO_2 uptake exposes plants to the risk of dehydration. This problem is compounded because the concentration gradient for CO_2 uptake is much smaller than the concentration gradient that drives water loss. To meet the contradictory demands of maximizing CO_2 uptake while limiting water loss, plants have evolved adaptations that control water loss from leaves, and that replace the water lost to the atmosphere with water drawn from the soil.

In this chapter we examine the mechanisms and driving forces of water transport within the plant and between the plant and its environment (**Figure 3.1**). We begin our examination of water transport by focusing on water in the soil. We then consider how water moves from the soil into the roots and from the roots up through specialized transport cells to the leaves, from which water is lost to the atmosphere. We end the chapter by considering the ways in which the leaf can control the loss of water, as well as the entry of CO_2, by regulating the opening and closing of stomata, the small openings through which CO_2 uptake and the majority of water loss occur.

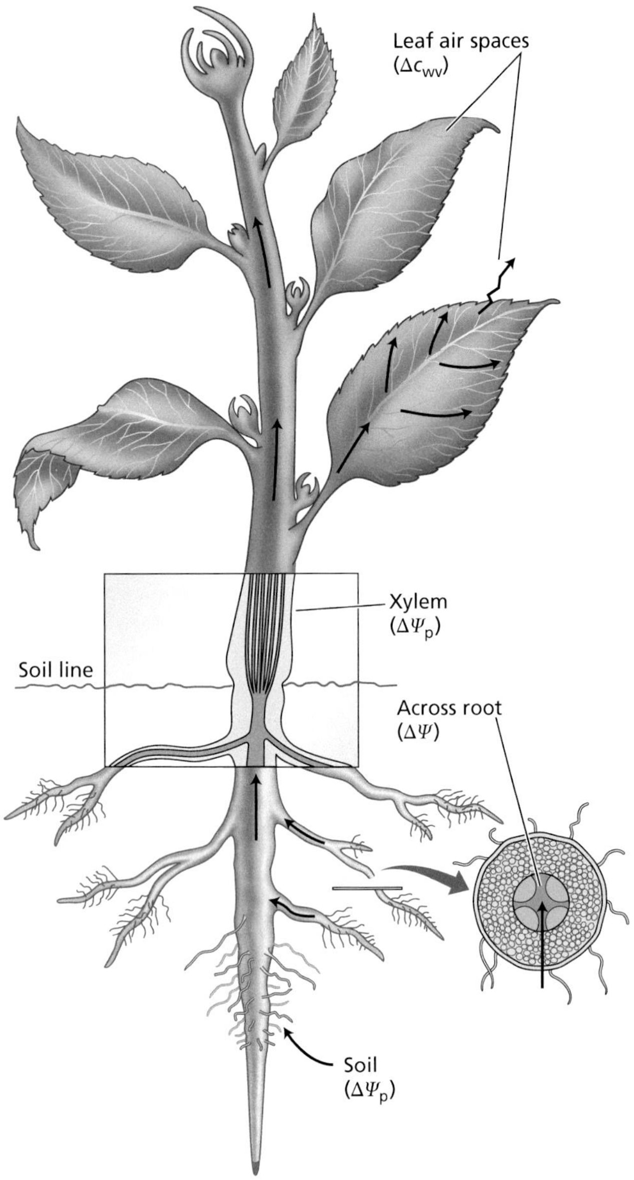

Figure 3.1 Main driving forces for water flow from the soil through the plant to the atmosphere. Differences in water vapor concentration (Δc_{wv}) between leaf and air are responsible for the diffusion of water vapor from the leaf to the air; differences in pressure potential ($\Delta\Psi_p$) drive the bulk flow of water through xylem conduits; and differences in water potential ($\Delta\Psi$) are responsible for the movement of water across the living cells in the root.

3.1 Water in the Soil

Explain the forces involved in water movement through the soil.

The water content and the rate of water movement in soils depend to a large extent on soil type and soil structure. At one extreme is sand, in which the soil particles may be 1 mm or more in diameter. Sandy soils have a relatively low surface area per gram of soil and have large spaces or channels between particles.

At the other extreme is clay, in which particles are smaller than 2 μm in diameter. Clay soils have much greater surface areas and smaller channels between particles. With the aid of organic substances such as humus (decomposing organic matter), clay particles may aggregate into "crumbs," allowing large channels to form that help improve soil aeration and infiltration of water.

When a soil is heavily watered by rain or by irrigation, the water percolates downward by gravity through the spaces between soil particles, partly displacing or trapping air in these channels. Because water is pulled into the spaces between soil particles by capillarity, the smaller channels are filled first. Depending on the amount of water available, water in the soil may exist as a film adhering to the surface of soil particles, it may fill the smaller but not the larger channels, or it may fill all of the spaces between particles.

In sandy soils, the spaces between particles are so large that water tends to drain from them and remain only on the particle surfaces and in the spaces where particles come into contact. In clay soils, the spaces between particles are so small that much water is retained against the force of gravity. A few days after a soaking rainfall, a clay soil might retain 40% water by volume. In contrast, sandy soils typically retain only about 15% water by volume after thorough wetting.

In this section we examine how the physical structure of the soil influences soil water potential, how water moves in the soil, and how roots absorb the water needed by the plant.

Soil water potential is affected by solutes, surface tension, and gravity

Like the water potential of plant cells, the water potential of soils may be dissected into three components: the osmotic potential, the pressure potential, and the gravitational potential. The **osmotic potential** (Ψ_s) (Chapter 2) of soil water is generally negligible, because except in saline soils, solute concentrations are low; a typical value might be −0.02 MPa. In soils that contain a substantial concentration of salts, however, Ψ_s can be quite negative, perhaps −0.2 MPa or lower.

The second component of soil water potential is the **pressure potential** (Ψ_p). For wet soils, Ψ_p is very close to zero. As soil dries out, Ψ_p decreases and can become quite negative. Where does the negative pressure potential in soil water come from, and why is Ψ_p related to the water content of the soil?

Water has a high surface tension that tends to minimize air–water interfaces independent of gravity (capillarity is discussed in Chapter 2). However, because

osmotic potential (Ψ_s) The effect of dissolved solutes on water potential. Also called solute potential.

pressure potential (Ψ_p) The hydrostatic pressure of a solution in excess of ambient atmospheric pressure.

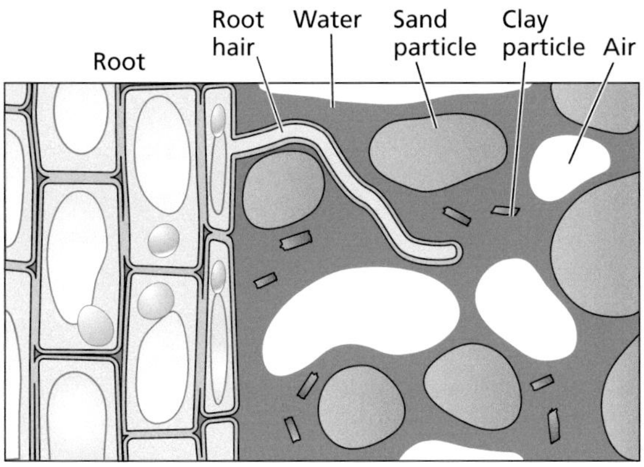

Figure 3.2 Root hairs make intimate contact with soil particles and greatly amplify the surface area used for water absorption by the plant. The soil is a mixture of particles (sand, clay, silt, and organic material), water, dissolved solutes, and air. When the soil is wet, the curvature of air–water interfaces is small. As water is absorbed by the plant, the soil solution recedes into smaller pockets, channels, and crevices between the soil particles. This recession causes the curvature of air–water interfaces in the soil to increase, resulting in a decrease in Ψ_p.

of adhesive forces, water also tends to cling to the surfaces of soil particles (**Figure 3.2**). As the water content of the soil decreases, the water recedes into the channels between soil particles, forming curved air–water interfaces. In Chapter 2 we learned that surface tension describes the tendency of a gas–liquid interface to contract. If the air–water interface is curved, surface tension produces a net force perpendicular to the surface that acts to reduce the area of the air–water interface. This force causes the Ψ_p of the water in the soil to decrease according to the Young–Laplace equation:

$$\Psi_p = \frac{-2T}{r} \quad (3.1)$$

where T is the surface tension of water (7.28×10^{-2} N m^{-1} at 20°C, equivalent to 7.28×10^{-8} MPa m), r is the radius of curvature (in meters) of the air–water interface, and the negative sign indicates that surface tension lowers the pressure in the water.

As soil dries out, water is lost from the largest spaces between soil particles first and subsequently from successively smaller spaces between and within soil particles. In this process, the value of Ψ_p in soil water can become quite negative because of the increasing curvature of air–water surfaces in pores of successively smaller diameter. For instance, a curvature of $r = 1$ μm (about the size of the largest clay particles) corresponds to a Ψ_p value of –0.15 MPa. The value of Ψ_p may easily reach –1 to –2 MPa as the air–water interface recedes into the smaller spaces between clay particles.

The third component of soil water potential is **gravitational potential** (Ψ_g). Gravity plays an important role in drainage. Water moves downwards because Ψ_g is proportional to elevation: higher at higher elevations, and vice versa. The gradient in Ψ_g is equal to 0.1 MPa over a distance of 10 m.

gravitational potential (Ψ_g) The part of the water potential caused by gravity. It is only of a significant size when considering water transport into trees and drainage in soils.

Water moves through the soil by bulk flow

Bulk or mass flow is the movement of molecules en masse, most often in response to a pressure or gravitational gradient. Common examples of bulk flow are water moving through a garden hose or down a river. Water moves through soils predominantly by bulk flow. Both gravity and pressure can drive bulk flow in soils. However, the movement of water toward roots is largely the result of gradients in Ψ_p ($\Delta\Psi_p$) (see Figure 3.1).

Water flows from regions of higher soil water content, where the water-filled spaces are larger and thus Ψ_p is less negative, to regions of lower soil water content, where the smaller size of the water-filled spaces is associated with more curved air–water interfaces and a more negative Ψ_p. Diffusion of water vapor from wetter to drier regions also accounts for some water movement, although its contribution is usually negligible except in very dry soils.

As plants absorb water from the soil, the soil is depleted of water near the surface of the roots. This depletion reduces Ψ_p near the root surface and establishes

soil hydraulic conductivity A measure of the ease with which water moves through a soil.

root hairs Microscopic extensions of root epidermal cells that greatly increase the surface area of the root for absorption.

a pressure gradient with respect to neighboring regions of soil that have higher Ψ_p values. Because the water-filled pore spaces in the soil are interconnected, water moves down the pressure gradient to the root surface by bulk flow through these channels.

The rate of water flow in soils depends on two factors: the size of the pressure gradient through the soil, and the hydraulic conductivity of the soil. **Soil hydraulic conductivity** is a measure of the ease with which water moves through the soil, and it varies with the type of soil and its water content. Sandy soils, which have large spaces between particles, have a large hydraulic conductivity when saturated, whereas clay soils, with only minute spaces between their particles, have an appreciably smaller saturated hydraulic conductivity.

As the water content of a soil decreases, the hydraulic conductivity falls dramatically. This decrease in soil hydraulic conductivity is due primarily to the replacement of water in the soil by air. When air moves into a soil channel previously filled with water, water movement through that channel is restricted to the periphery of the channel. As more of the soil spaces become filled with air, water flow is limited to fewer and narrower channels, and the hydraulic conductivity falls.

3.2 Water Absorption by Roots

Explain how water is absorbed by roots and how it reaches the xylem.

Contact between the surface of the root and the soil is essential for effective water absorption by the root. This contact provides the surface area needed for water uptake and is maximized by the growth of the root and of root hairs into the soil. **Root hairs** are filamentous outgrowths of root epidermal cells that greatly increase the surface area of the root, thus providing greater capacity for absorption of ions and water from the soil. When 4-month-old rye plants were examined, their root hairs were found to constitute about 60% of the surface area of the roots (Chapter 4).

Water enters the root most readily near the root tip. Mature regions of the root are less permeable to water because they have developed a modified epidermal layer that contains hydrophobic materials in its walls. Although it might at first seem counterintuitive that any portion of the root system should be impermeable to water, the older regions of the root must be sealed off if there is to be water uptake (and thus bulk flow of nutrients) from the regions of the root system that are actively exploring new areas in the soil (**Figure 3.3**).

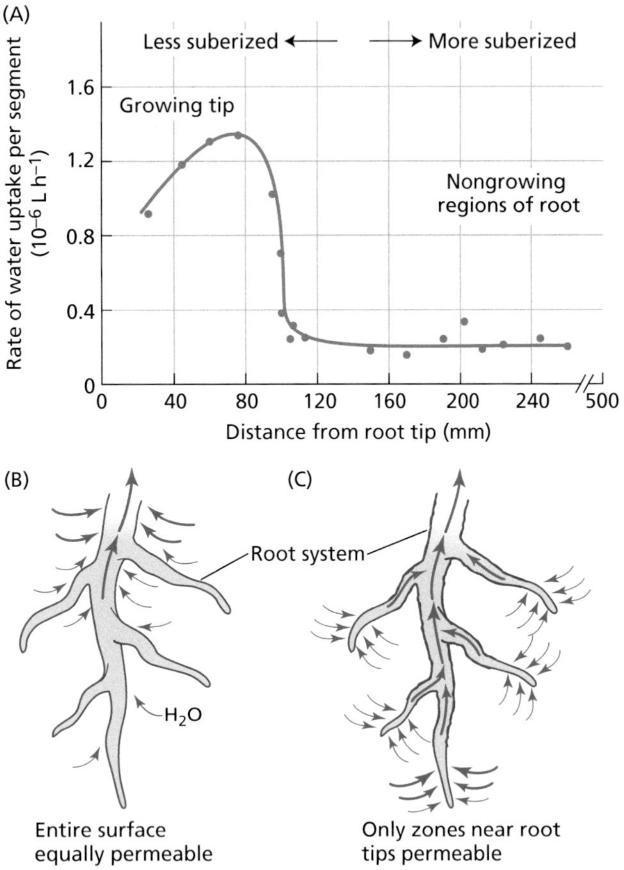

Figure 3.3 Water uptake by roots. (A) Rate of water uptake by short segments (3–5 mm) at various positions along an intact pumpkin (*Cucurbita pepo*) root. (B and C) Diagrams of water uptake in which the entire root surface is equally permeable (B) or is impermeable in older regions because of the deposition of suberin, a fatty acid-containing hydrophobic polymer, around the lignin that forms the Casparian strip. (C). When root surfaces are equally permeable, most of the water enters near the top of the root system, with more distal regions being hydraulically isolated as the suction in the xylem is relieved because of the inflow of water. Decreasing the permeability of older regions of the root allows xylem tensions to extend farther into the root system, allowing water uptake from distal regions of the root system.

The contact between the soil and the root surface is easily broken when the soil is disturbed. It is for this reason that newly transplanted seedlings and plants need to be protected from water loss for the first few days after transplantation. Thereafter, new root growth into the soil reestablishes soil–root contact, and the plant can better withstand water stress.

Let's consider how water moves within the root, and the factors that determine the rate of water uptake into the root.

Water moves in the root via the apoplast, symplasm, and transmembrane pathways

In the soil, water flows between soil particles. However, from the epidermis to the endodermis of the root, there are three pathways through which water can flow (**Figure 3.4**): the apoplast, the symplasm, and the transmembrane pathway.

1. The apoplast is the continuous system of cell walls, intercellular air spaces, and the lumens of nonliving cells (e.g., xylem conduits and fibers). In this pathway, water moves through cell walls and extracellular spaces without crossing any membranes as it travels across the root cortex.

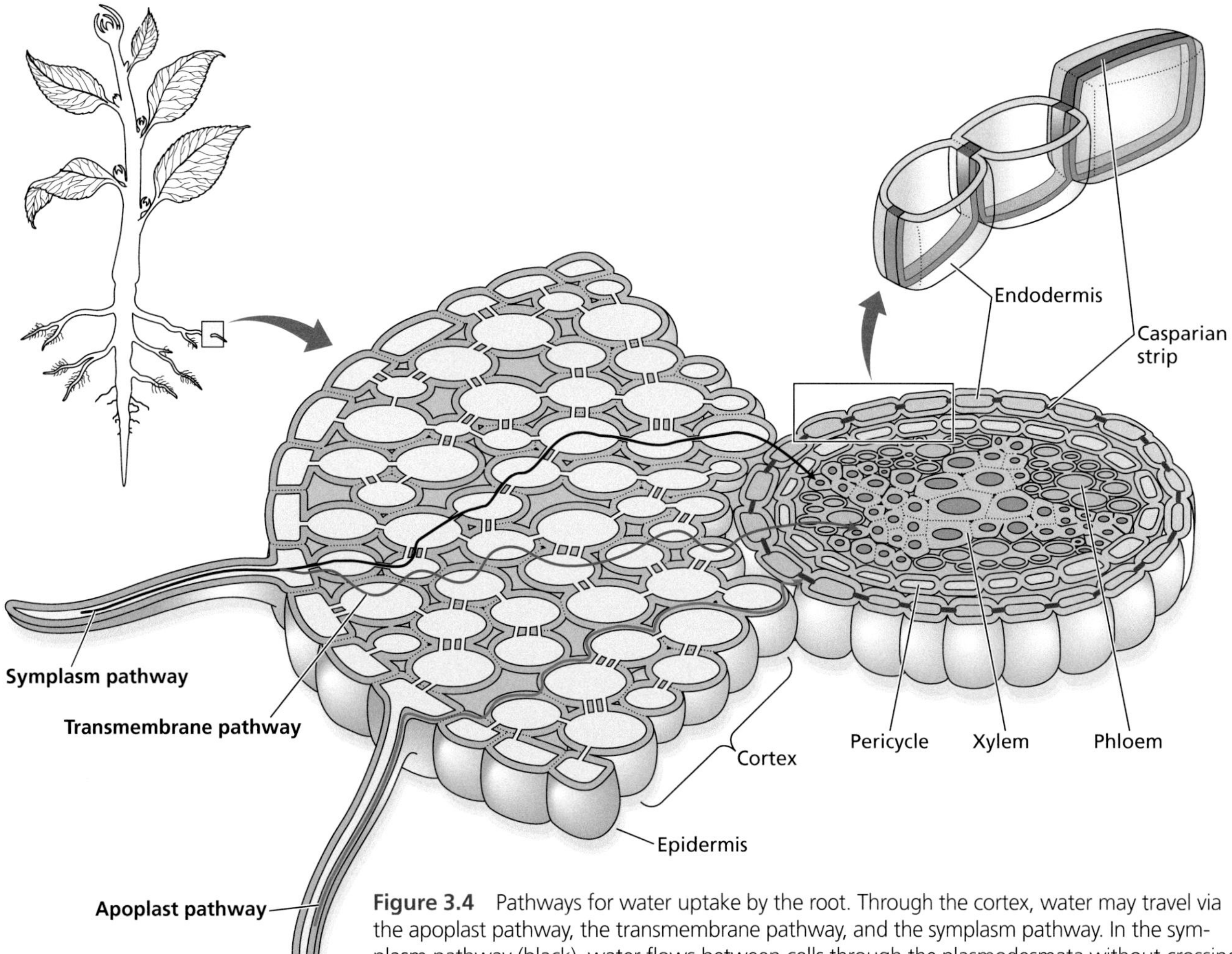

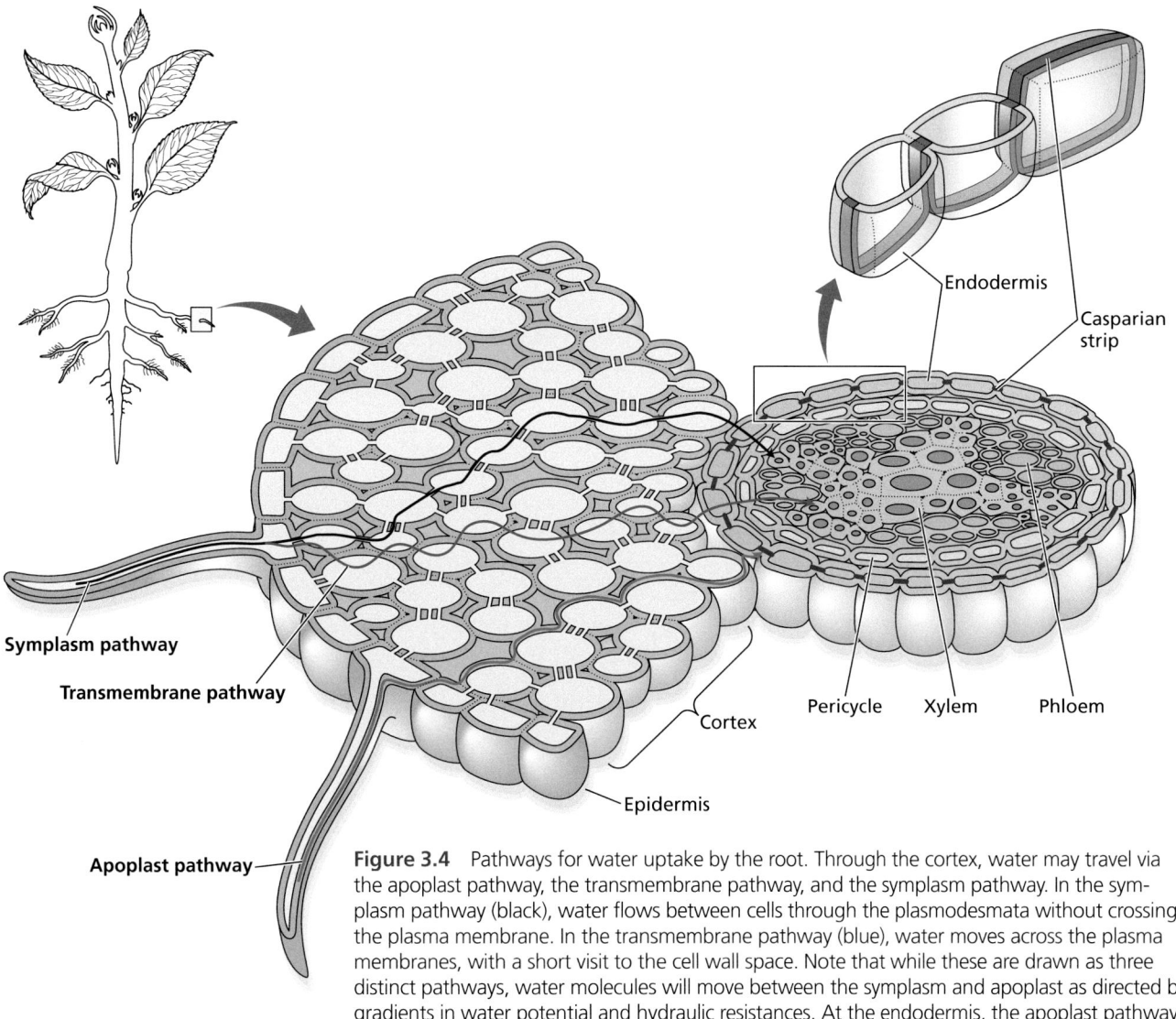

Figure 3.4 Pathways for water uptake by the root. Through the cortex, water may travel via the apoplast pathway, the transmembrane pathway, and the symplasm pathway. In the symplasm pathway (black), water flows between cells through the plasmodesmata without crossing the plasma membrane. In the transmembrane pathway (blue), water moves across the plasma membranes, with a short visit to the cell wall space. Note that while these are drawn as three distinct pathways, water molecules will move between the symplasm and apoplast as directed by gradients in water potential and hydraulic resistances. At the endodermis, the apoplast pathway is blocked by the Casparian strip. Some species also develop a Casparian strip in the outer cell layer of the root cortex, forming an exodermis in which flow through the apoplast pathway is blocked.

Casparian strip A band in the cell walls of the endodermis that is impregnated with lignin. Prevents apoplastic movement of water and solutes into the stele.

lignin Highly branched phenolic polymer made up of phenylpropanoid alcohols that is deposited in secondary cell walls.

root pressure A positive hydrostatic pressure in the xylem of roots that typically occurs at night in the absence of transpiration.

2. The symplasm consists of the entire network of cell cytoplasm interconnected by plasmodesmata. In this pathway, water travels across the root cortex via plasmodesmata (Chapter 1).
3. The transmembrane pathway is the route by which water enters a cell on one side, exits the cell on the other side, enters the next in the series, and so on. In this pathway, water crosses the plasma membrane of each cell in its path twice (once on entering and once on exiting). Transport across the tonoplast may also be involved.

While the relative importance of the apoplast, symplasm, and transmembrane pathways has not yet been fully established, it is important to remember that water moves not according to a single chosen path, but rather wherever the gradients and resistances direct it. A particular water molecule moving in the symplasm may cross the membrane and move in the apoplast for a moment, and then move across a membrane and back into the symplasm again.

At the endodermis, water movement through the apoplast pathway is obstructed by the Casparian strip (Figure 3.4). The **Casparian strip** is a band within the radial cell walls of the endodermis that is impregnated with **lignin** (Chapter 1). The Casparian strip breaks the continuity of the apoplast pathway, forcing water and solutes to pass through the plasma membrane in order to cross the endodermis. The plasma membrane can act selectively in terms of which solutes it allows to enter the cell and which it excludes. The Casparian strip allows the plant to control which solutes enter the xylem, something that would not be possible if water could move continuously through the apoplast from the soil to the xylem. The movement of water across plasma membranes explains why the driving gradient for water uptake by roots is the difference in soil and xylem water potential ($\Delta\Psi$) (Figure 3.1).

The requirement that water move symplasmically across the endodermis also helps explain why the permeability of roots to water depends strongly on the presence of aquaporins (Chapters 2 and 6). Down-regulating the expression of aquaporin genes markedly reduces the hydraulic conductivity of roots and can result in plants that wilt easily or that compensate by producing larger root systems.

Water uptake decreases when roots are subjected to low temperature or anaerobic conditions, or when treated with respiratory inhibitors. Decreased rates of respiration, in response to low temperature or anaerobic conditions such as occur in flooded soils, lead to a decrease in intracellular pH. In turn, the decrease in cytosolic pH alters the conductance of aquaporins in root cells, resulting in roots that are markedly less permeable to water. Thus, maintaining membrane permeability to water requires energy expenditure by root cells that is supplied by respiration (Chapter 11).

Solute accumulation in the xylem can generate "root pressure"

Plants sometimes exhibit a phenomenon referred to as **root pressure**. For example, if the stem of an actively growing plant is cut off just above the soil, the stump will often exude sap from the cut xylem for many hours. If a manometer is sealed over the stump, positive pressures as high as 0.2 MPa (and sometimes even higher) can be measured.

When transpiration is low or absent, positive hydrostatic pressure builds up in the xylem because roots continue to absorb ions from the soil and transport them into the xylem. The buildup of solutes in the xylem sap leads to a decrease in the xylem osmotic potential (Ψ_s) and thus to a decrease in the xylem water potential (Ψ). This lowering of the xylem Ψ provides a driving force for water absorption, which in turn leads to a positive hydrostatic pressure in the xylem. In effect, the multicellular root tissue behaves as an osmotic membrane does, building up a positive hydrostatic pressure in the xylem in response to the accumulation of solutes.

Root pressure is most likely to occur when soil water potentials are high and transpiration rates are low. As transpiration rates increase, water is transported through the plant and lost to the atmosphere so rapidly that a positive pressure resulting from ion uptake never develops in the xylem.

Plants that develop root pressure frequently produce liquid droplets on the edges of their leaves, a phenomenon known as **guttation** (**Figure 3.5**). Positive xylem pressure causes exudation of xylem sap through specialized pores called *hydathodes* that are associated with vein endings at the leaf margin. The "dew drops" that can be seen on the tips of grass leaves in the morning are actually guttation droplets exuded from hydathodes. Guttation is most noticeable when transpiration is suppressed and the relative humidity is high, such as at night. It is possible that root pressure reflects an unavoidable consequence of high rates of ion accumulation. However, the existence of positive pressures within the xylem at night can help dissolve gas bubbles, and thus play a role in reversing the deleterious effects of cavitation described in the next section.

Figure 3.5 Guttation in a leaf from lady's mantle (*Alchemilla vulgaris*). In the early morning, leaves secrete water droplets through the hydathodes, located at the margins of the leaves.

3.3 Water Transport through the Xylem

Describe the role of xylem in water transport and the forces involved in water movement through these cells.

In most plants, the xylem constitutes the longest part of the pathway of water transport. In a plant 1 m tall, more than 99.5% of the water transport pathway through the plant is within the xylem, and in tall trees the xylem represents an even higher percentage of the pathway. Compared with the movement of water through layers of living cells, the xylem is a pathway of low resistivity. In this section we examine how the structure of the xylem contributes to the movement of water from the roots to the leaves, and how negative pressures generated by transpiration pull water through the xylem.

The xylem consists of two types of transport cells

The conducting cells in the xylem have a specialized anatomy that enables them to transport large quantities of water with great efficiency. There are two main types of water-transporting cells in the xylem: tracheids and vessel elements (**Figure 3.6**). Vessel elements are found in angiosperms, a small group of gymnosperms called the Gnetales, and some ferns. Tracheids are present in both angiosperms and gymnosperms, as well as in ferns and other groups of vascular plants.

The maturation of both tracheids and vessel elements involves the production of secondary cell walls and the subsequent death of the cell—the loss of the cytoplasm and all of its contents. What remain are the thick, lignified cell walls, which form hollow tubes through which water can flow with relatively little resistance. However, because lignin is hydrophobic, it makes the secondary cell wall essentially impermeable to water. For this reason, tracheids and vessel elements develop specialized regions that allow water to flow from one water-transporting cell to another.

Tracheids are elongated, spindle-shaped cells (Figure 3.6A) that are arranged in overlapping vertical files (**Figure 3.7**). Water flows through the open tracheid lumen and between tracheids by means of the numerous **pits** in the tracheids' lateral walls (Figure 3.6B). Pits are microscopic regions where the lignified secondary wall is

guttation An exudation of liquid from the leaves due to root pressure.

tracheids Spindle-shaped, water-conducting cells with tapered ends and pitted walls without perforations found in the xylem of both angiosperms and gymnosperms.

pit A microscopic region where the secondary wall of a tracheary element is absent and the primary wall is thin and porous

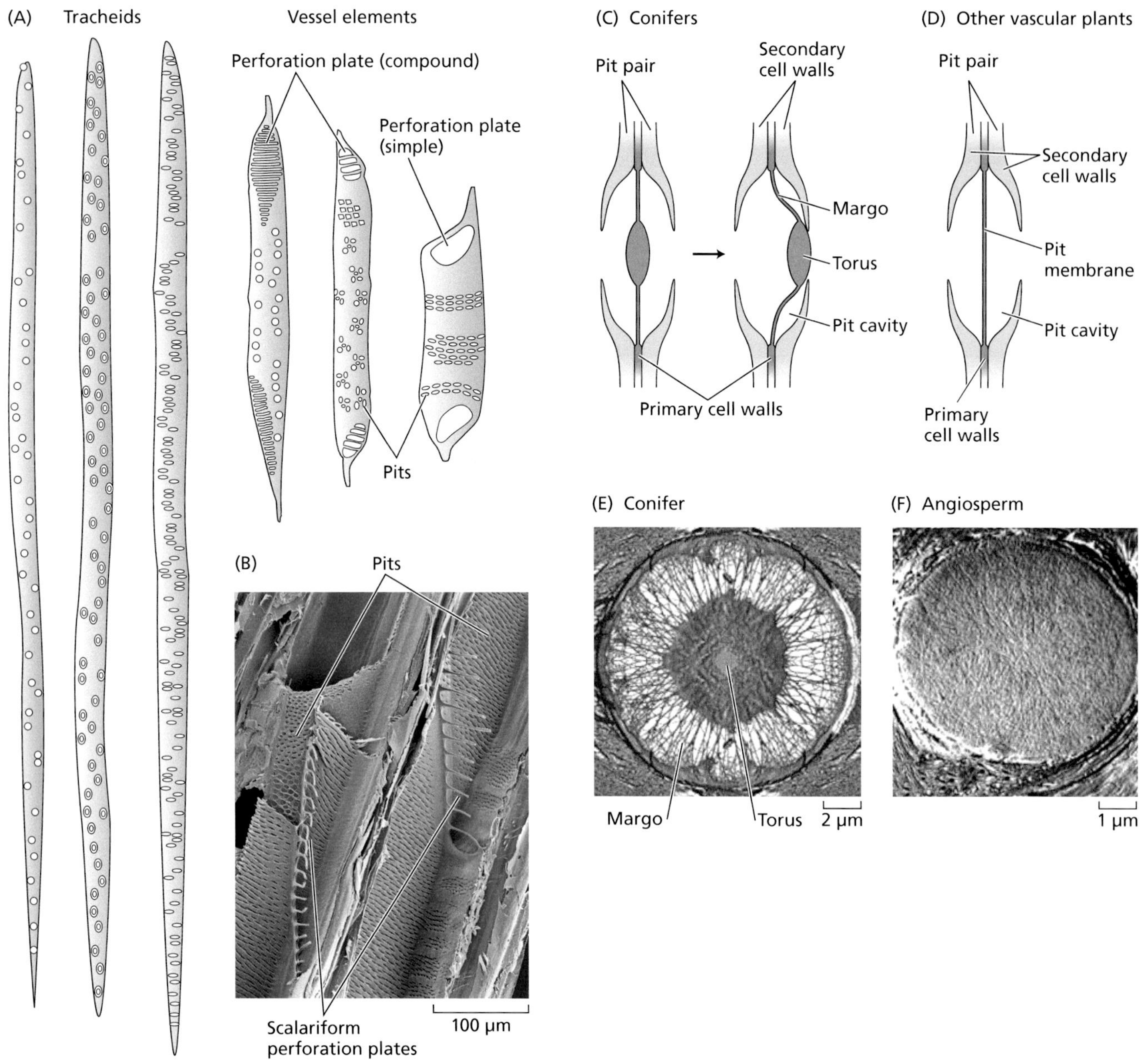

Figure 3.6 Xylem conduits and their interconnections. (A) Structural comparison of tracheids and vessel elements. Tracheids are elongated, hollow, dead cells with highly lignified walls. The walls contain numerous pits—regions where secondary wall is absent but primary wall remains. The shapes of pits and the patterns of wall pitting vary with species and organ type. Tracheids are present in all vascular plants. Vessels consist of a stack of two or more vessel elements. Like tracheids, vessel elements are dead cells and are connected to one another by perforation plates—regions of the wall where pores or holes have developed. Vessels are connected to other vessels and to tracheids through pits. Vessels are found in most angiosperms and are lacking in most gymnosperms. (B) Scanning electron micrograph showing two vessels (running diagonally from lower left to upper right). Pits are visible on the side walls, as are the scalariform (ladderlike) perforation plates between vessel elements. (C) Diagram of a coniferous bordered pit pair with the torus centered in the pit cavity (left) or lodged to one side of the cavity (right). When the pressure difference between two tracheids is small, the pit membrane lies close to the center of the bordered pit, allowing water to flow through the porous margo region of the pit membrane; when the pressure difference between two tracheids is large, such as when one has cavitated and the other remains filled with water under tension, the pit membrane is displaced such that the torus becomes lodged against the overarching walls, thereby preventing the gas-filled void (embolism) from propagating between tracheids. (D) In contrast, the pit membranes of angiosperms and other vascular plants are relatively homogeneous in their structure. These pit membranes have very small pores compared with those of conifers, which prevents the spread of embolism but also imparts a significant hydraulic resistance. (E and F) Scanning electron microscope images of pit membranes with the overarching secondary wall removed: the torus–margo membrane of a conifer tracheid (E); the homogenous pit membrane of an angiosperm vessel (F).

absent and only the primary wall is present (Figure 3.6C). Pits of one tracheid are located opposite pits of an adjoining tracheid, forming **pit pairs**. Pit pairs constitute a low-resistance path for water movement between tracheids. The water-permeable layer between pit pairs, consisting of two primary walls and a middle lamella, is called the **pit membrane**. In **bordered pits**, the secondary wall extends partway over the pit membrane, creating a pit cavity that is connected to the conduit lumen.

Vessel elements have end walls that are partially or completely open so they can be stacked end to end to form a multicellular conduit called a **vessel** (Figure 3.7). The end walls between adjacent vessel elements form a structure referred to as a **perforation plate**. Each perforation plate contains one or more openings through which water can flow. Simple perforation plates have only a single opening, and often the connection between adjacent vessel elements is completely open. Compound perforation plates have multiple openings, with the plate often consisting of ladderlike bars (scalariform perforation plates).

Vessel elements tend to be shorter and wider than tracheids, but the multicellular vessels formed from vessel elements can be much longer than tracheids. In some species they can be up to a meter or more in length. Like tracheids, vessel elements have pits on their lateral walls (Figure 3.6B) that allow water to flow from one conduit to another. The vessel elements found at the extreme ends of a vessel have tapered ends and are connected to neighboring vessels via pits.

It is important to emphasize that mature, water-transporting tracheids and vessels do not retain any cell membranes. Thus, the water-filled interiors of xylem conduits form part of the apoplast. Pit membranes, despite their name, do not function like cell membranes. Pit membranes are part of the cell wall, and because they are relatively porous they do not restrict the movement of solutes (i.e., they are not semipermeable and thus cannot cause water to flow by osmosis).

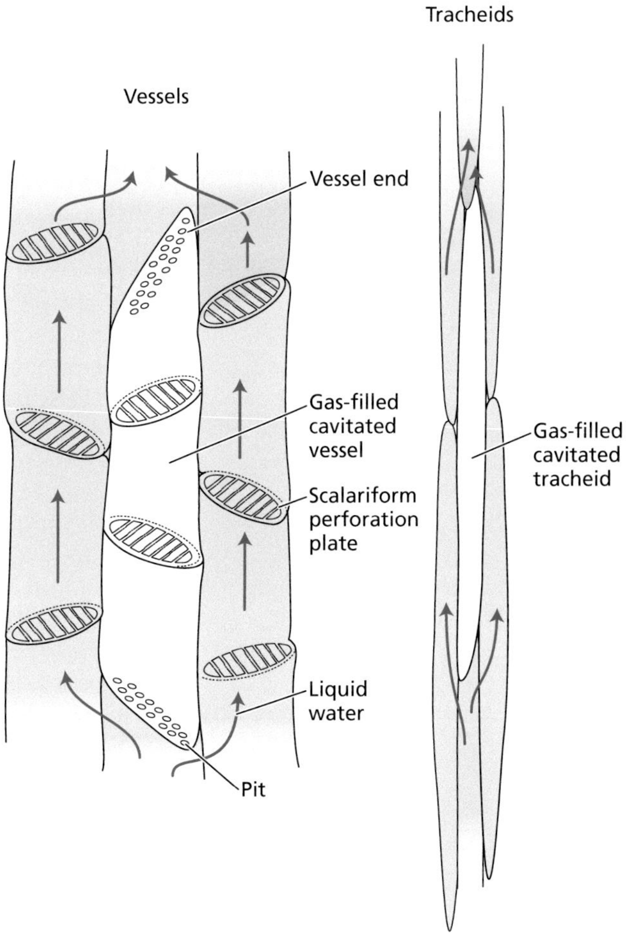

Figure 3.7 Vessels (left) and tracheids (right) form a series of parallel, interconnected pathways for water movement. Cavitation blocks water movement because of the formation of gas-filled (embolized) conduits. Because xylem conduits are interconnected through openings ("bordered pits") in their thick secondary walls, water can detour around the blocked vessel by moving through adjacent conduits. Pit membranes help prevent embolisms from spreading between xylem conduits. Thus, in the diagram on the right, the gas is contained within a single cavitated tracheid. In the diagram on the left, gas has filled the entire cavitated vessel, shown here as being made up of three vessel elements, each separated by scalariform perforation plates. In nature, vessels can be very long (up to several meters in length) and thus made up of many vessel elements.

Water moves through the xylem by pressure-driven bulk flow

Pressure-driven bulk flow of water is responsible for long-distance transport of water in the xylem (Figure 3.1). This pressure gradient must be large enough to overcome the gravitational gradient, which typically opposes the movement of water from the soil to the leaves, as well as the energy dissipated by viscosity. Viscosity describes a fluid's resistance to deformation; in the case of laminar flow through a tube, the relevant deformation arises because water is stationary at the walls of the tube and flows fastest at its center.

If we consider laminar flow through a tube, the rate of flow depends on the radius (r) of the tube, the viscosity (η) of the liquid, and the pressure gradient ($\Delta\Psi_p/\Delta x$) that drives the flow (Δx is a unit of distance along the flow path). Jean Léonard Marie Poiseuille (1797–1869) was a French physician and physiologist, and the relation just described is given by one form of Poiseuille's equation:

$$\text{Volume flow rate} = \left(\frac{\pi r^4}{8\eta}\right)\left(\frac{\Delta\Psi_p}{\Delta x}\right) \qquad (3.2)$$

pit pair Two pits occurring opposite one another in the walls of adjacent tracheids or vessel elements. Pit pairs constitute a low-resistance path for water movement between the conducting cells of the xylem.

pit membrane The porous layer in the xylem between pit pairs, consisting of two thinned primary walls and a middle lamella.

bordered pit A pit pair in which the pit chamber is over-arched by the cell wall, creating a larger pit chamber and smaller pit aperture.

vessel elements Nonliving water-conducting cells with perforated end walls found only in angiosperms and a small group of gymnosperms.

vessel A stack of two or more vessel elements in the xylem.

perforation plate The perforated end wall of a vessel element in the xylem.

torus A central thickening found in the pit membranes of tracheids in the xylem of most gymnosperms.

microfibril The major fibrillar component of the cell wall composed of layers of cellulose molecules packed tightly together by extensive hydrogen bonding.

margo A porous and relatively flexible region of the pit membranes in tracheids of conifer xylem, surrounding a central thickening, the torus.

expressed in cubic meters per second (m^3 s^{-1}). This equation tells us that pressure-driven bulk flow is extremely sensitive to the radius of the tube. If the radius is doubled, the volume flow rate increases by a factor of 16 (2^4). Vessel elements up to 500 μm in diameter, nearly an order of magnitude greater than that of the largest tracheids, occur in the stems of climbing species. These large-diameter vessels permit vines to transport large amounts of water despite the slenderness of their stems.

Equation 3.2 describes flow through a cylindrical tube and thus does not take into account that xylem conduits are of finite length, such that water must cross many pit membranes as it flows from the soil to the leaves. All else being equal, pit membranes should impede water flow through single-celled (and thus shorter) tracheids to a greater extent than through multicellular (and thus longer) vessels. How then is it possible that conifers are among the tallest trees in the world given that conifer xylem is formed entirely of tracheids?

The answer is that conifers (and the gymnosperm *Ginkgo biloba*) have pit membranes that are much more permeable to water than the pit membranes found in other plants. In conifers, the pit membrane has a central thickening, called a **torus** (plural *tori*), surrounded by a porous and relatively flexible region of **microfibrils** known as the **margo** (Figure 3.6C and E). The openings in the margo are large relative to the tiny pores in the pit membranes of other vascular plants. The strong dependence of pressure-driven flow on pore radius (Equation 3.2) explains why the hydraulic resistance to flow through a conifer pit is much less than in plants that lack torus–margo pit membranes.

The major role of pit membranes is to prevent the spread of gas bubbles, called *emboli*, within the xylem. Drought, which results in more negative water potentials, can cause xylem conduits to become air-filled, a potentially disastrous situation for a plant because embolized conduits can no longer transport water from the roots to the leaves. The torus–margo structure allows conifer pit membranes to function like valves: When they are centered in the pit cavity, the pits remain open; when they are lodged in the circular or oval wall thickenings bordering the pit, the pits are closed. Such lodging of the torus effectively prevents gas bubbles from spreading into neighboring tracheids (we will discuss this formation of bubbles, a process called cavitation, shortly). Plants that lack torus–margo pit membranes rely on the surface tension of water to prevent gas bubbles from spreading from one xylem conduit to another. For this reason, their pit membranes are constrained to have much smaller pores and thus a much larger hydraulic resistance.

Water movement through the xylem requires a smaller pressure gradient than movement through living cells

The xylem provides a low resistivity pathway for water movement. A comparison of the driving force required to move water through the xylem at a typical velocity with the driving force that would be needed to move water through a pathway made up of living cells at the same rate helps us appreciate xylem efficiency.

For the comparison, let's use 4 mm s^{-1} for the xylem transport velocity and 40 μm as the vessel radius. This is a high velocity for such a narrow vessel, so it will tend to exaggerate the pressure gradient required to support water flow in the xylem. Using a version of Poiseuille's equation (Equation 3.2), we can calculate the pressure gradient needed to move water at a velocity of 4 mm s^{-1} through an *ideal* tube with a uniform inner radius of 40 μm. The calculation gives a value of 0.02 MPa m^{-1}. Of course, *real* xylem conduits have irregular inner wall surfaces, and water flow through perforation plates and pits adds resistance to water transport. Such deviations from the ideal increase the viscous drag: Measurements show that the actual resistance is greater by approximately a factor of 2.

Let's now compare this value with the driving force that would be necessary to move water at the same velocity from cell to cell, crossing the plasma membrane each time. The estimated driving force needed to move water through a layer of cells at 4 mm s^{-1} is 2×10^8 MPa m^{-1}. This is ten orders of magnitude greater than the driving force needed to move water through our 40-μm-radius xylem vessel. The calculation clearly shows that water flow through the xylem is vastly more efficient than water flow across living cells. Nevertheless, the xylem can make a significant contribution to the total resistance to water flow through the plant.

What pressure difference is needed to lift water 100 m to a treetop?

With the previous example in mind, let's see what pressure gradient is needed to move water up to the top of a very tall tree. The tallest trees in the world are the coast redwoods (*Sequoia sempervirens*) of North America and the mountain ash (*Eucalyptus regnans*) of Australia. Individuals of both species can exceed 100 m.

If we think of the stem of a tree as a long pipe, we can estimate the pressure difference that is needed to overcome the viscous drag of moving water from the soil to the top of the tree by multiplying the pressure gradient needed to move the water by the height of the tree. The pressure gradients needed to move water through the xylem of very tall trees are on the order of 0.01 MPa m^{-1}, smaller than in our previous example. If we multiply this pressure gradient by the height of the tree (0.01 MPa $m^{-1} \times 100$ m), we find that the total pressure difference needed to overcome the resistance to water movement through the stem is equal to 1 MPa.

In addition to the hydraulic resistance due to viscosity, we must consider gravity. As described by Equation 2.4 ($\Psi_g = \rho_w gh$), for a height difference of 100 m, the difference in Ψ_g is approximately 1 MPa. That is, Ψ_g is 1 MPa higher at the top of the tree than at the ground level. Thus, the other components of water potential must be 1 MPa more negative at the top of the tree to counter the effects of gravity.

To allow transpiration to occur, the pressure gradient must be large enough to overcome the effects of gravity and the resistance to water movement through the xylem. Thus, we calculate that a pressure difference of roughly 2 MPa, from the base to the top branches, is needed to carry water up the tallest trees.

The cohesion–tension theory explains water transport in the xylem

In theory, the pressure gradient needed to move water through the xylem could result from the generation of positive pressures at the base of the plant. We mentioned previously that some roots can develop positive hydrostatic pressure in their xylem. However, root pressure is typically less than 0.1 MPa and disappears with transpiration or when soils are dry, so it is clearly inadequate to move water up a tall tree. Furthermore, because root pressure is generated by the accumulation of ions in the xylem, relying on this for transporting water would require a mechanism for dealing with these solutes once the water evaporates from the leaves.

Instead, transpiration causes the pressure in the xylem to decrease. It is the lowering of xylem pressures in leaves that creates the pressure gradient that causes water to flow upward from the roots. However, as our earlier calculations show, the $\Delta\Psi_p$ needed to move water through the xylem requires that pressures fall to values far below zero. Negative hydrostatic pressure means that the water molecules pull on one another (and the walls of the xylem conduit). Negative pressures in water are often referred to as tension.

In Chapter 2 we saw how the many hydrogen bonds that form between water molecules make water highly cohesive and give it a high tensile strength. It is these properties of water, coupled with the stiffness of the lignified conduit walls, that allow the xylem to develop and sustain large negative pressures. The idea that transpiration leads to a decrease in xylem pressure and that this tension pulls water through the xylem was first proposed toward the end of the nineteenth century. It is called the *cohesion–tension mechanism of sap ascent* because it requires the cohesive properties of water to sustain large tensions in the xylem water columns.

The xylem tensions needed to pull water from the soil develop in leaves because of transpiration. How does the loss of water vapor through open stomata result in the flow of water from the soil? When leaves open their stomata to obtain CO_2 for photosynthesis, water vapor diffuses out of the leaves. This causes water to evaporate from the surface of mesophyll cell walls inside the leaves. In turn, the loss of water from the cell walls causes the water potential in the walls to decrease as water is drawn into the interstices (spaces) of the cell wall where it forms curved air–water interfaces (**Figure 3.8**). Because water adheres to the cellulose microfibrils and other hydrophilic components of the cell wall, the curvature of these interfaces reduces Ψ_p, analogous to what happens

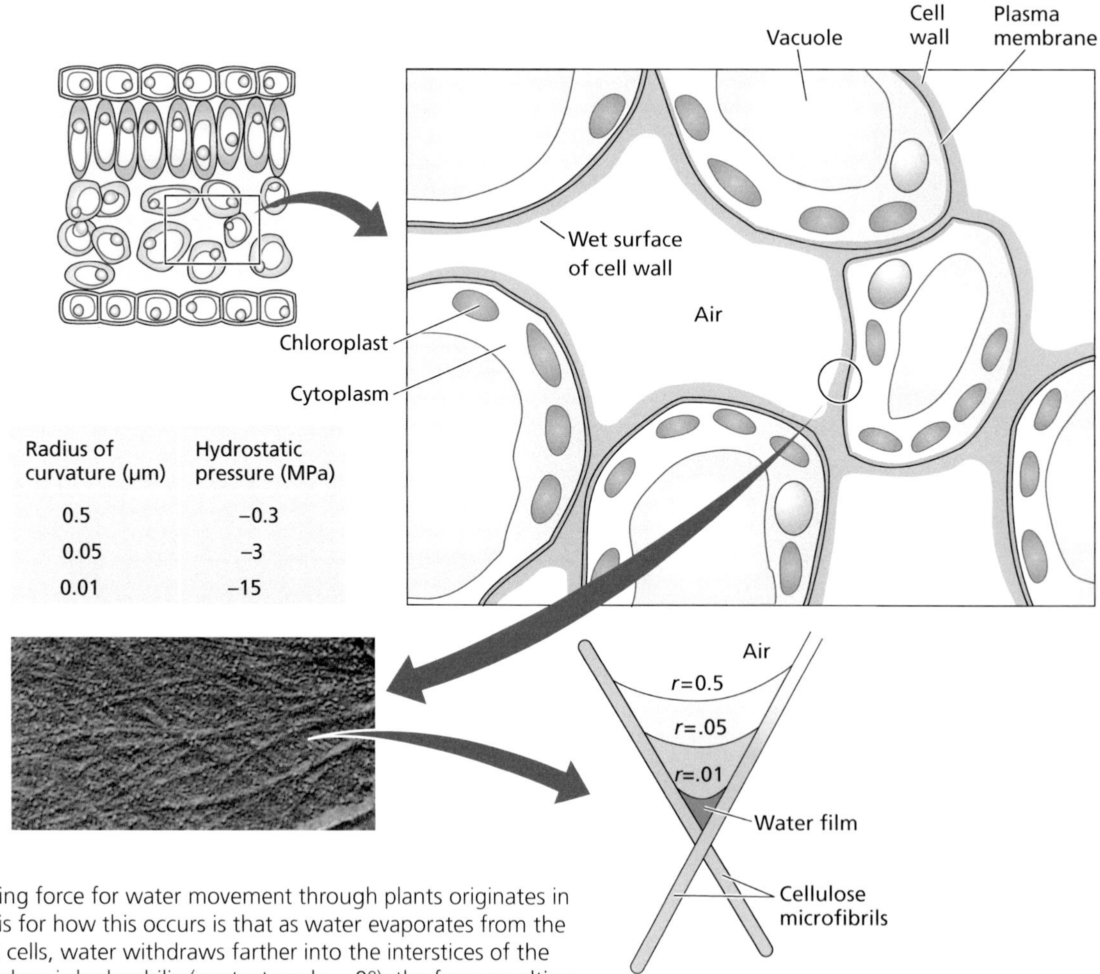

Figure 3.8 The driving force for water movement through plants originates in leaves. One hypothesis for how this occurs is that as water evaporates from the surfaces of mesophyll cells, water withdraws farther into the interstices of the cell wall. Because cellulose is hydrophilic (contact angle = 0°), the force resulting from surface tension causes a negative pressure in the liquid phase. As the radius of curvature of the air–water interfaces decreases, the hydrostatic pressure becomes more negative, as calculated from Equation 3.1.

in the soil. This creates a gradient in water potential that causes water to flow toward the sites of evaporation. Some of the water that flows toward the sites of evaporation comes from the protoplasts of adjacent cells. However, because leaves are connected to the soil via a low-resistance pathway—the xylem—most of what replaces water lost from the leaves due to transpiration comes from the soil. Water will flow from the soil when the water potential of the leaves is low enough to overcome the Ψ_p of the soil, as well as the resistance associated with moving water through the plant. Note that for water to be pulled from the soil requires that there be a continuous liquid-filled pathway extending from the sites of evaporation, down through the plant, and out into the soil.

The cohesion–tension mechanism explains how the substantial movement of water through plants can occur without the direct expenditure of metabolic energy: The energy that powers the movement of water through plants comes from the sun, which, by increasing the temperature of the leaf, drives the evaporation of water. However, water transport through the xylem is not "free." The plant must build xylem conduits capable of withstanding the large tensions needed to pull water from the soil. In addition, plants must accumulate enough solutes in their living cells that they are able to remain turgid even as water potentials decrease due to transpiration.

The cohesion–tension mechanism has been a controversial subject for more than a century and continues to generate lively debate. The main controversy surrounds the question of whether water columns in the xylem can sustain the large tensions (negative pressures) necessary to pull water up tall trees. Recently, water transport through a microfluidic device designed to function as a synthetic "tree" demonstrated the stable flow of liquid water at pressures lower (more negative) than –7.0 MPa.

Xylem transport of water in trees faces physical challenges

The large tensions that develop in the xylem of trees and other plants present significant physical challenges. First, the water under tension transmits an inward force to the walls of the xylem. If the cell walls were weak or pliant, they would collapse under this tension—analogous to what can happen to a drinking straw if you suck on it too hard. The secondary wall thickenings and lignification of tracheids and vessels are adaptations that offset this tendency to collapse. Plants that experience large xylem tensions tend to have dense wood, reflecting the mechanical stresses imposed on the wood by water under tension.

A second challenge is that water under such tensions is in a *physically metastable state*. Water is stable as a liquid when its hydrostatic pressure exceeds its saturated vapor pressure. When the hydrostatic pressure in liquid water becomes equal to its saturated vapor pressure, the water will undergo a phase change. We are all familiar with the idea of vaporizing water by increasing its temperature (raising its saturated vapor pressure). Less familiar, but still easily observed, is the fact that water can be made to boil at room temperature by placing it in a vacuum chamber (lowering the hydrostatic pressure of the liquid phase by reducing the pressure of the atmosphere).

Earlier we estimated that a pressure gradient of 2 MPa would be needed to supply water to leaves at the top of a 100-m-high tree. If we assume that the soil surrounding this tree is fully hydrated and lacks significant concentrations of solutes (i.e., $\Psi = 0$), the cohesion–tension theory predicts that the hydrostatic pressure of water in the xylem at the top of the tree will be –2 MPa. This value is substantially below the saturated vapor pressure (absolute pressure of ~0.002 MPa at 20°C). What maintains the water column in its liquid state?

Water in the xylem is in a metastable state because despite the existence of a thermodynamically lower energy state—the vapor phase—it remains a liquid. This situation occurs because (1) the cohesion and adhesion of water make the free-energy barrier for the liquid-to-vapor phase change very high, and (2) the structure of the xylem minimizes the presence of *nucleation sites*—sites that lower the energy barrier separating the liquid from the vapor phase.

The most important nucleation sites are gas bubbles because it takes less energy to expand the gas–liquid interface of an existing bubble than it does to form a bubble (gas void) de novo. Furthermore, the curvature of a bubble, and thus also the inward force resulting from surface tension, is a function of its size. At the same tension in the xylem, small bubbles remain stable or even contract, losing volume as they push gas back into solution, while large bubbles expand. If the tension in the xylem increases to the point where the inward force on a bubble resulting from surface tension is less than the outward force due to the negative pressure in the liquid phase, the bubble will expand. And once a bubble starts to expand, the force due to surface tension decreases, because the air–water interface of a larger bubble has less curvature. Thus, a bubble that exceeds the critical size for expansion will expand until it fills the entire conduit.

The absence of gas bubbles of sufficient size to destabilize the water column when under tension is partially because water must flow across the root endodermis to enter the xylem. The endodermis serves as a filter, preventing gas bubbles from entering the xylem. Pit membranes also function as filters as water flows from one xylem conduit to another. However, when pit membranes are exposed to air on one side—because of injury, leaf abscission, or the existence of a neighboring gas-filled conduit—pit membranes can serve as sites of entry for air. Air enters when the pressure difference across the pit membrane is sufficient either to allow air to penetrate the cellulose matrix of structurally homogeneous pit membranes (Figure 3.6D) or to dislodge the torus of a coniferous pit membrane (Figure 3.6C). This phenomenon is called *air seeding*.

A second mode by which bubbles can form in xylem conduits is freezing of the xylem tissues. Because water in the xylem contains dissolved gases and the solubility of gases in ice is very low, freezing of xylem conduits can lead to bubble formation.

In plant physiology, the phenomenon of bubble expansion is referred to as *cavitation*, and the resulting gas-filled void is referred to as an *embolism*. Its effect is similar to that of a vapor lock in the fuel line of an automobile or an embolism in a blood vessel. Cavitation breaks the continuity of the water column and prevents the transport of water under tension.

Vulnerability curves (**Figure 3.9**) provide a way of quantifying a species' susceptibility to cavitation and the impact of cavitation on flow through the xylem. A vulnerability curve plots the measured hydraulic conductivity (usually as a percent of maximum) of a branch, stem, or root segment versus the experimentally imposed level of xylem tension. Due to cavitation, xylem hydraulic conductivity decreases with increasing tensions until flow ceases entirely. However, the decrease in xylem hydraulic conductivity occurs at much lower tensions in species in moist habitats, such as cottonwood, than in species from more arid regions, such as sagebrush.

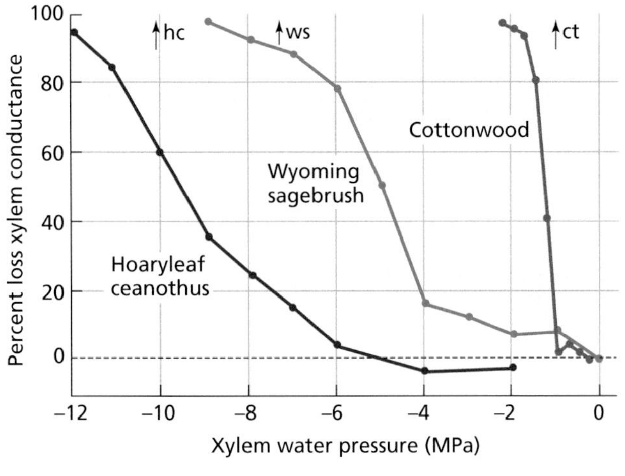

Figure 3.9 Xylem vulnerability curves represent the percentage loss of hydraulic conductance in stem xylem versus xylem water pressure, here in three species of contrasting drought tolerance. Data were obtained from excised branches subjected experimentally to increasing levels of xylem tension using a centrifugal force technique. Arrows on the upper axis indicate the minimum xylem pressure measured in the field for each species (hc: ceanothus, ws: sagebrush, ct: cottonwood).

Plants have several mechanisms to overcome losses of xylem conductivity caused by embolism

The impact of xylem cavitation on the plant can be minimized by several means. Pit membranes can prevent an expanding gas bubble from spreading to an adjacent conduit. In this way, pit membranes serve to contain embolism.

Most plants are able to tolerate small amounts of cavitation. Because the capillaries in the xylem are interconnected, water can detour around the embolized conduit by traveling through neighboring, water-filled conduits (Figure 3.7). Thus, the finite length of the tracheid and vessel conduits of the xylem, while resulting in an increased resistance to water flow, also provides a way to restrict the impact of cavitation. However, any reduction in xylem hydraulic capacity has the potential to affect productivity if it reduces the supply of water to the leaves. Soil drying results in increasingly negative xylem pressures and can lead to substantial amounts of embolism. Plants can mitigate the impact of soil drying by shedding leaves or reducing contact between roots and soil. During prolonged drought, however, xylem pressures will eventually reach levels that induce cavitation. Recent work shows that embolism is a major factor in drought-induced mortality of trees.

The major way that plants overcome embolism is by producing new xylem conduits or by growing new shoots. This highlights an important benefit of secondary growth (Chapter 15).

In some species, the development of root pressure allows embolized conduits to be refilled. Recall that root pressure results in the development of positive hydrostatic pressures in the xylem. These positive pressures in the liquid phase are transmitted to any gas bubbles in the xylem. In turn, the increase in pressure forces gas to dissolve and thus allows an embolized conduit to become filled with water.

3.4 Water Movement from the Leaf to the Atmosphere

Explain how water moves from the leaf to the atmosphere, and how the stomata regulate this process.

On its way from the leaf to the atmosphere, water is pulled from the xylem into the cell walls of the mesophyll, where it evaporates into the air spaces of the leaf (**Figure 3.10**). The water vapor then exits the leaf through the stomatal pore. The movement of liquid water through the living tissues of the leaf is controlled by gradients in water potential. However, transport in the vapor phase is by diffusion, so the final part of the transpiration stream is controlled by the *concentration gradient of water vapor.*

The waxy cuticle that covers the leaf surface is an effective barrier to water movement. It has been estimated that only about 5% of the water lost from leaves escapes through the cuticle. Almost all of the water lost from leaves is lost by diffusion of water vapor through the tiny stomatal pores. In most herbaceous species, stomata are present on both the adaxial and abaxial surfaces of the leaf, usually more abundant on the abaxial surface. In many tree species, stomata are located only on the abaxial surface of the leaf.

Leaves have a large hydraulic resistance

Although the distances that water must traverse within leaves are small relative to the entire soil-to-atmosphere pathway, the contribution of the leaf to the total hydraulic resistance is large. On average, leaves constitute 30% of the total liquid-phase resistance, and in some plants their contribution is much larger. This combination of short path length and large hydraulic resistance also occurs in roots, reflecting the fact that in both organs, water transport takes place across highly resistive living tissues as well as through the xylem.

Water enters leaves and is distributed across the leaf lamina in xylem conduits. Water must exit through the xylem walls and pass through multiple layers of living cells before it evaporates. Leaf hydraulic resistance thus reflects the number, distribution, and size of xylem conduits, as well as the hydraulic properties of leaf

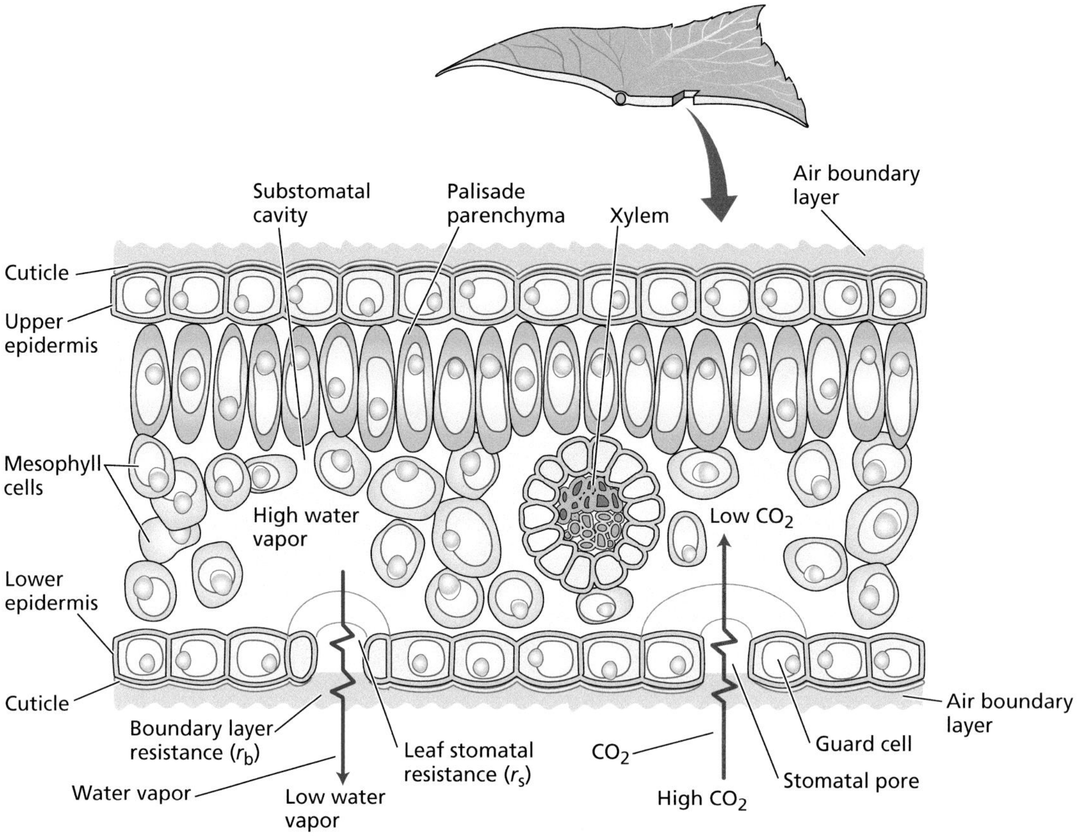

Figure 3.10 Water pathway through the leaf. Water is pulled from the xylem into the cell walls of the mesophyll, where it evaporates into the air spaces within the leaf. Water vapor then diffuses through the leaf air space, through the stomatal pore, and across the boundary layer of still air found next to the leaf surface. CO_2 diffuses into the leaf along its concentration gradient (low inside, higher outside).

mesophyll cells. The hydraulic resistance of leaves of diverse vein architectures varies as much as 40-fold. A large part of this variation appears to be due to the density of veins within the leaf and their distance from the evaporative leaf surface. Leaves with closely spaced veins tend to have lower hydraulic resistance and higher rates of photosynthesis, suggesting that the proximity of leaf veins to sites of evaporation exerts a significant impact on the rates of leaf gas exchange.

The hydraulic resistance of leaves varies in response to growth conditions and exposure to low leaf water potentials. For example, leaves of plants growing in shaded conditions exhibit greater resistance to water flow than do leaves of plants grown in higher light. Leaf hydraulic resistance also typically increases with leaf age. Over shorter time scales, decreases in leaf water potential lead to marked increases in leaf hydraulic resistance. The increase in leaf hydraulic resistance may result from decreases in the membrane permeability of mesophyll cells, cavitation of xylem conduits in leaf veins, or in some cases, the physical collapse of xylem conduits under tension.

difference in water vapor concentration (Δc_{wv}) The difference between the water vapor concentration of the air spaces inside the leaf and that of the air outside the leaf.

diffusional resistance (*r*) The restriction posed by the boundary layer and the stomata to the free diffusion of gases from and into the leaf.

The driving force for transpiration is the difference in water vapor concentration

Transpiration from the leaf depends on two major factors: (1) the **difference in water vapor concentration** between the leaf air spaces and the external bulk air (Δc_{wv}) and (2) the **diffusional resistance** (*r*) of this pathway. The difference in water vapor concentration is expressed as $c_{wv(leaf)} - c_{wv(air)}$. The water vapor concentration of air ($c_{wv[air]}$) can be readily measured, but that of the leaf ($c_{wv[leaf]}$) is more difficult to assess.

Whereas the volume of air space inside the leaf is small, the wet surface from which water evaporates is large. Air space volume is about 5% of the total leaf volume in pine needles, 10% in maize (corn; *Zea mays*) leaves, 30% in barley, and 40% in tobacco leaves. In contrast to the volume of the air space, the internal surface area from which water evaporates may be from 7 to 30 times the external leaf area. This high surface-to-volume ratio makes for rapid vapor equilibration inside the leaf. Thus, we can assume that the air space in the leaf is close to water potential equilibrium with the cell wall surfaces from which liquid water is evaporating.

Within the range of water potentials experienced by transpiring leaves (generally more positive than –2.0 MPa), the equilibrium water vapor concentration is within 2 percentage points of the saturation water vapor concentration. This allows one to estimate the water vapor concentration within a leaf from its temperature. Because the saturated water vapor content of air increases exponentially with temperature, leaf temperature has a marked impact on transpiration rates.

The concentration of water vapor, c_{wv}, changes at various points along the transpiration pathway. We see from **Table 3.1** that c_{wv} decreases at each step of the pathway from the cell wall surface to the bulk air outside the leaf. The important points to remember are that (1) the driving force for water loss from the leaf is the *absolute* concentration difference (difference in c_{wv}, in mol m^{-3}), and (2) this difference is markedly influenced by leaf temperature.

leaf stomatal resistance (r_s) The resistance to CO_2 diffusion imposed by the stomatal pores.

boundary layer resistance (r_b) The resistance to the diffusion of water vapor, CO_2, and heat due to the layer of unstirred air next to the leaf surface. A component of diffusional resistance.

Water loss is also affected by the pathway resistances

The second important factor governing water loss from the leaf is the diffusional resistance of the transpiration pathway, which consists of two varying components (Figure 3.10):

1. The resistance associated with diffusion through stomatal pores, the **leaf stomatal resistance** (r_s).
2. The resistance due to the layer of unstirred air next to the leaf surface through which water vapor must diffuse to reach the turbulent air of the atmosphere. This second resistance, r_b, is called the leaf **boundary layer resistance**. We will discuss this type of resistance before considering stomatal resistance.

Table 3.1 Representative Values for Relative Humidity, Absolute Water Vapor Concentration, and Water Potential for Four Points in the Pathway of Water Loss from a Leaf

		Water vapor	
Location	Relative humidity	Concentration (mol m^{-3}) (c_{wv})	Potential (MPa)[a]
Inner air spaces (25°C)	0.99	1.27	–1.38
Just inside stomatal pore (25°C)	0.97	1.21	–7.04
Just outside stomatal pore (25°C)	0.47	0.60	–103.7
Bulk air (20°C)	0.50	0.50	–93.6

Source: Adapted from P. S. Nobel. 1999. *Physiochemical and Environmental Plant Physiology*, 2nd ed. Academic Press, San Diego, CA.
Note: See Figure 3.10.
[a] Calculated using the equation $\Delta\Psi_p = (RT / \bar{V}) \ln (RH)$ with values for $RT / \bar{V}_w$ of 135 MPa at 20°C and 137.3 MPa at 25°C.

The thickness of the boundary layer is determined primarily by wind speed and leaf size. When the air surrounding the leaf is very still, the layer of unstirred air on the surface of the leaf may be so thick that it is the primary deterrent to water vapor loss from the leaf. Increases in stomatal apertures under such conditions have little effect on transpiration rate (**Figure 3.11**), although closing the stomata completely will still reduce transpiration.

When wind velocity is high, the moving air reduces the thickness of the boundary layer at the leaf surface, reducing the resistance of this layer. Under such conditions, stomatal resistance will largely control water loss from the leaf.

Various anatomical and morphological aspects of the leaf can influence the thickness of the boundary layer. Hairs on the surface of leaves can serve as microscopic windbreaks. Some plants have sunken stomata that provide a sheltered region outside the stomatal pore. The size and shape of leaves and their orientation relative to the wind direction also influence the way the wind sweeps across the leaf surface. Most of these factors, however, cannot be altered on an hour-to-hour or even a day-to-day basis. For short-term regulation of transpiration, control of stomatal apertures by the guard cells plays a crucial role in regulating leaf transpiration.

Some species are able to change the orientation of their leaves and thereby influence their transpiration and photosynthesis rates (Chapters 9 and 13). For example, when plants orient their leaves parallel to the sun's rays, leaf temperature is reduced and with it the driving force for transpiration, Δc_{wv}. Many grass leaves roll up as they experience water deficits, in this way reducing light interception. Even wilting can help ameliorate high transpiration rates by reducing the amount of radiation intercepted, resulting in lower leaf temperatures and a decrease in Δc_{wv}.

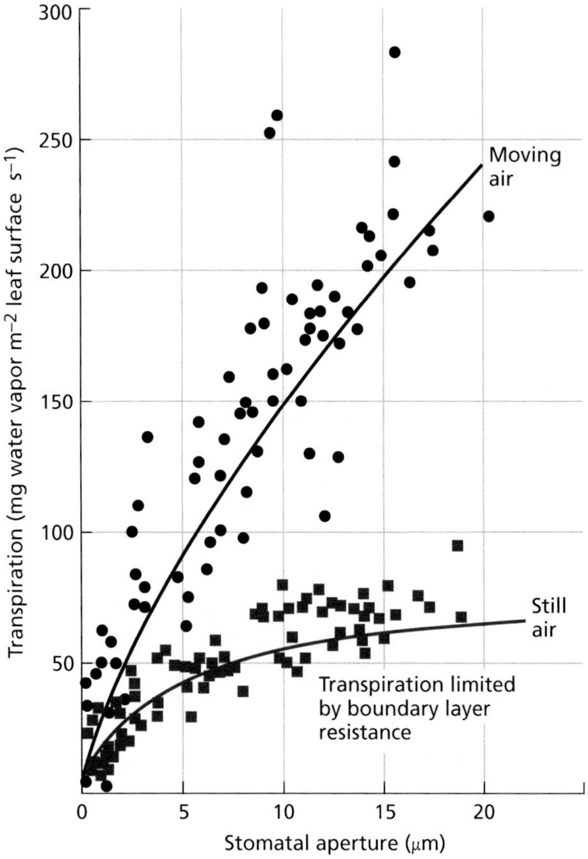

Figure 3.11 Dependence of transpiration on the stomatal aperture of zebra plant (*Zebrina pendula*) in still air and in moving air. The boundary layer is thicker and more rate-limiting in still air than in moving air. As a result, the stomatal aperture has less control over transpiration in still air.

Stomatal control couples leaf transpiration to leaf photosynthesis

Because the cuticle covering the leaf is nearly impermeable to water, most leaf transpiration results from the diffusion of water vapor through stomata (Figure 3.10). The microscopic stomatal pores provide a *variable-resistance pathway* for diffusional movement of gases across the epidermis and cuticle. Changes in stomatal resistance are important for regulating water loss by the plant and for controlling the rate of CO_2 uptake necessary for photosynthesis.

On a sunny morning when the supply of water is abundant and the solar radiation incident on the leaf favors high photosynthetic activity, the demand for CO_2 inside the leaf is large and the stomatal pores open wide, decreasing the stomatal resistance to CO_2 diffusion. Water loss by transpiration is substantial under these conditions, but since the water supply from the soil is plentiful, it is advantageous for the plant to accept an increased rate of water loss to obtain more products of photosynthesis, which are essential for growth and reproduction. By contrast, when soil water is less abundant, stomata open less or even remain closed despite abundant sunshine. At night, when there is no photosynthesis and thus no demand for CO_2 inside the leaf, stomata shut, preventing unnecessary loss of water.

The *temporal* regulation of stomatal apertures—open during the day, closed at night—occurs in the majority of plants. Only species that exhibit crassulacean acid metabolism (CAM) photosynthesis, which we will learn about in Chapter 8, exhibit an inverted pattern in which stomata open at night and close during the day. These plants conserve water by capturing and storing CO_2 at night when transpiration rates are low. They then draw on their stored CO_2 during the day, allowing them to photosynthesize despite closed stomata.

The cell walls of guard cells have specialized features

Plants regulate transpiration by opening and closing stomata. This biological control is exerted by a pair of specialized epidermal cells, the **guard cells**, which surround the stomatal pore (**Figures 3.12** and **3.13**).

Guard cells show considerable morphological diversity, but we can distinguish two main types. The most common type of guard cells have an elliptical contour (often called kidney-shaped) with the pore at their center (Figure 3.12A). In grasses (Figure 3.12B), guard cells have a characteristic dumbbell shape, with bulbous ends. The pore is a long slit located between the two "handles" of the dumbbells. These guard cells are always flanked by a pair of differentiated epidermal cells called **subsidiary cells**, which help the guard cells control the stomatal pore. Subsidiary cells are also found in some species with kidney-shaped stomata. The guard cells, subsidiary cells, and pore are collectively called the **stomatal complex**.

A distinctive feature of guard cells is the specialized structure of their walls. Portions of these walls are substantially thickened (Figure 3.13) and

guard cells A pair of specialized epidermal cells that surround the stomatal pore and regulate its opening and closing.

subsidiary cells Specialized epidermal cells that flank the guard cells and work with the guard cells in the control of stomatal apertures.

stomatal complex The guard cells, subsidiary cells, and stomatal pore, which together regulate leaf transpiration.

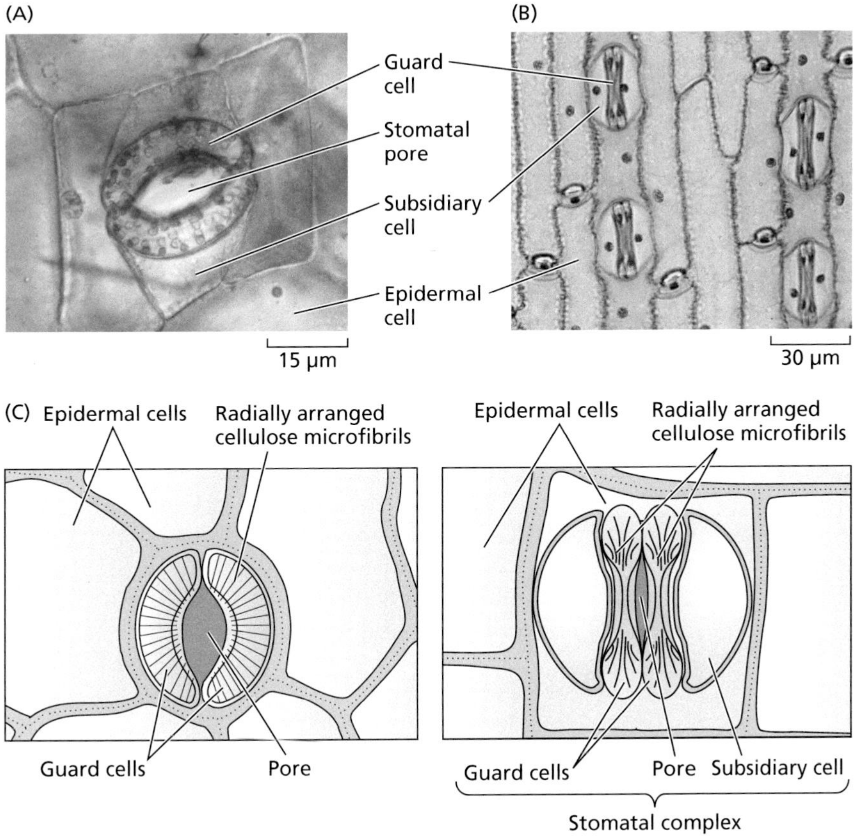

Figure 3.12 Stomata. (A) Most plants have kidney-shaped guard cells, as seen in this open stoma of *Tradescantia zebrina*. (B) Stomata of maize (*Zea mays*), showing the dumbbell-shaped guard cells typical of grasses. (C) Radial alignment of the cellulose microfibrils in guard cells and epidermal cells of a kidney-shaped stoma (left) and a grasslike stoma (right).

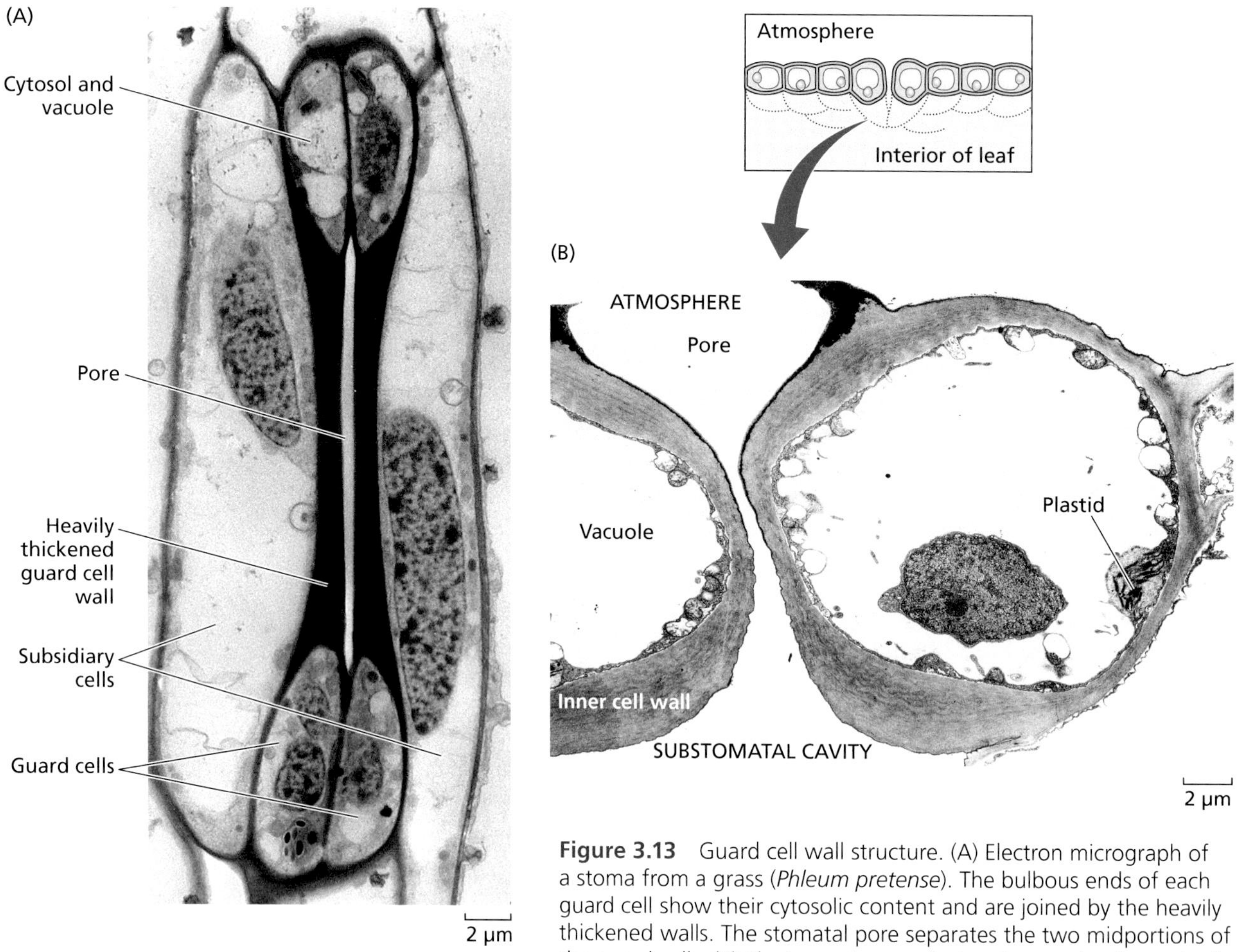

Figure 3.13 Guard cell wall structure. (A) Electron micrograph of a stoma from a grass (*Phleum pretense*). The bulbous ends of each guard cell show their cytosolic content and are joined by the heavily thickened walls. The stomatal pore separates the two midportions of the guard cells. (B) Electron micrograph showing a pair of guard cells from tobacco (*Nicotiana tabacum*). The section was made perpendicular to the main surface of the leaf. In both images, the stomatal pore is very small, indicating that the stomata are essentially closed. Note the uneven thickening pattern of the walls, which determines the asymmetric deformation of the guard cells when their volume increases during stomatal opening.

may be up to 5 μm across, in contrast to the 1 to 2 μm typical of epidermal cells. In kidney-shaped guard cells, a differential thickening pattern results in very thick inner and outer walls, a thin wall in contact with surrounding epidermal cells, and a somewhat thickened wall facing the pore. The portions of the outer wall in contact with the atmosphere often extend into well-developed ledges.

The alignment of **cellulose microfibrils**, which reinforce all plant cell walls and are an important determinant of cell shape (Chapter 1), plays an essential role in the opening and closing of the stomatal pore. In cylindrically shaped cells, cellulose microfibrils are oriented transversely to the long axis of the cell. As a result, the cell expands in the direction of its long axis, because the cellulose reinforcement offers the least resistance at right angles to its orientation.

In guard cells the microfibril organization is different. Kidney-shaped guard cells have cellulose microfibrils fanning out radially from the pore (Figure 3.12C). As a result, the wall facing the pore is much stronger than the

cellulose microfibril Thin, ribbon-like structure of indeterminate length and variable width composed of 1→4-linked β-D-glucan chains tightly packed in crystalline arrays alternating with less organized amorphous regions. Provides structural integrity to the cell walls of plants and determines the directionality of cell expansion.

wall in contact with epidermal cells. Thus, as a guard cell increases in volume, the side facing the epidermis expands more than the side facing the pore. This causes the guard cells to bow apart and the pore to open. In grasses, the dumbbell-shaped guard cells function like beams with inflatable ends. The orientation of the cellulose microfibrils is such that as the bulbous ends of the cells increase in volume, the beams move apart from each other and the slit between them widens (Figure 3.12C).

While the orientation of the cellulose microfibrils is a major determinant of guard cell expansion during stomatal opening, recent computer modeling studies and experiments with cell wall mutants have shown that the matrix polysaccharides (Chapter 1) play an important role as well. At the early stages of stomatal opening, the microfibril orientation dictates the pattern of opening. As the pressure in the guard cells increases, however, the cell wall pectins undergo strain-induced stiffening, and this property appears to be critical for proper opening of the stomata.

Changes in guard cell turgor pressure cause stomata to open and close

The early aspects of stomatal opening are ion uptake and other metabolic changes in the guard cells, which we will discuss in detail in Chapter 6. Here we will note the effect of decreases in osmotic potential (Ψ_s) resulting from ion uptake and from biosynthesis of organic molecules in the guard cells. The potassium ion concentration in guard cells increases severalfold when stomata open, from 100 m*M* in the closed state to 400 to 800 m*M* in the open state, depending on the species and the experimental conditions. In most species these large concentration changes in K^+ are electrically balanced by varying amounts of the anions Cl^- and malate^{2-}. However, in some species of the genus *Allium*, such as onion (*A. cepa*), K^+ is balanced solely by Cl^-.

Water relations in guard cells follow the same rules as in other cells. A decrease in Ψ_s causes the water potential to decrease, and water consequently moves into the guard cells. As water enters the cell, turgor pressure increases and the stomata open. Stomatal closure results from a decrease in guard cell turgor pressure as a result of solute efflux and polymerization and thus is essentially stomatal opening in reverse.

In grasses, stomatal opening involves the transfer of solutes out of subsidiary cells and into guard cells, resulting in an increase in the turgor pressure of guard cells and a decrease in the turgor pressure of subsidiary cells. As a result, the guard cells are able to move laterally into space previously occupied by subsidiary cells. The reciprocal exchange of solutes between guard cells and subsidiary cells is thought to play an important role in allowing grass stomata to open quickly and to achieve large apertures.

Internal and external signals regulate the osmotic balance of guard cells

Guard cells function as multisensory hydraulic valves. Environmental factors such as light quality and intensity, temperature, leaf water status, and intercellular CO_2 concentrations are sensed by guard cells, and these signals are integrated into well-defined stomatal responses.

Stomatal opening is triggered by both blue and red wavelengths of light. In guard cells, blue light is perceived by the phototropin photoreceptors (Chapter 13), which trigger responses that drive ion uptake (Chapter 6) and lower guard cell water potential. The result is water uptake and vacuolar expansion that produces rapid stomatal opening (Chapter 19 for details). Because the intensity of blue light needed to promote stomatal opening is extremely low, blue light is thought to be the principal signal for stomata to open at dawn. In most plants, higher intensities of red-enriched photosynthetically active radiation (PAR) (Chapter 7) induce stomatal opening during the day. The exception are CAM plants, which display the opposite response: Stomata open during the night. In both cases, the regulation of stomatal opening involves ion uptake. Opening responses also correlate with changes in sucrose signaling and are integrated with CO_2 sensing.

A potent signal for stomatal closure is the hormone abscisic acid, discussed in detail in Chapters 12 and 19. Abscisic acid produced by mesophyll cells in response to decreases in mesophyll cell turgor pressure is transported to guard cells, where it triggers the rapid release of solutes and thus a decrease in guard cell turgor pressure that causes stomata to close. Stomatal closure in response to water-deficit stress helps plants avoid developing xylem tensions associated with a high likelihood of cavitation.

The transpiration ratio measures the relationship between water loss and carbon gain

The effectiveness of plants in moderating water loss while allowing sufficient CO_2 uptake for photosynthesis can be assessed by a parameter called the **transpiration ratio**. This value is defined as the amount of water transpired by the plant divided by the amount of CO_2 assimilated by photosynthesis.

For plants in which the first stable product of carbon fixation is a three-carbon compound (C_3 plants; Chapter 8), as many as 400 molecules of water are lost for every molecule of CO_2 fixed by photosynthesis, giving a transpiration ratio of 400. (Sometimes the reciprocal of the transpiration ratio, called the *water-use efficiency*, is cited. Plants with a transpiration ratio of 400 have a water-use efficiency of 1/400, or 0.0025.)

The large ratio of H_2O efflux to CO_2 influx results from three factors:

1. The concentration gradient driving water loss is about 100 times larger than that driving the influx of CO_2. In large part, this difference is due to the low concentration of CO_2 in air (~0.042%) and the relatively high concentration of water vapor within the leaf.
2. CO_2 diffuses about 1.6 times more slowly through air than water does (the CO_2 molecule is larger than H_2O and has a smaller diffusion coefficient).
3. CO_2 must cross the plasma membrane, the cytoplasm, and the **chloroplast envelope** before it is assimilated in the chloroplast. These membranes add to the resistance of the CO_2 diffusion pathway.

Some plants use variations in the usual photosynthetic pathway for CO_2 fixation that substantially reduce their transpiration ratio. Plants in which a four-carbon compound is the first stable product of photosynthesis (C_4 plants; Chapter 8) generally transpire less water per molecule of CO_2 fixed than C_3 plants do; a typical transpiration ratio for C_4 plants is about 150. This is largely because C_4 photosynthesis results in a lower CO_2 concentration in the intercellular air space (Chapter 8), thus creating a larger driving force for the uptake of CO_2 and allowing these plants to operate with smaller stomatal apertures and thus lower transpiration rates.

Desert-adapted plants with CAM photosynthesis, in which CO_2 is initially fixed into four-carbon organic acids at night (Chapter 8), have even lower transpiration ratios; values of about 50 are not unusual. This is possible because their stomata have an inverted diurnal rhythm, opening at night and closing during the day. Transpiration is much lower at night, because the cool leaf temperature gives rise to only a very small Δc_{wv}.

transpiration ratio The ratio of water loss to photosynthetic carbon gain. Measures the effectiveness of plants in moderating water loss while allowing sufficient CO_2 uptake for photosynthesis.

chloroplast envelope The double-membrane system surrounding the chloroplast.

3.5 Overview: The Soil–Plant–Atmosphere Continuum

Differentiate the mechanisms of water transport from the soil through the plant to the atmosphere.

We have seen that movement of water from the soil through the plant to the atmosphere involves different mechanisms of transport:

- In the soil and the xylem, liquid water moves by bulk flow in response to a pressure gradient ($\Delta\Psi_p$).
- When liquid water is transported into and out of cells, the driving force is the water potential difference across the plasma membrane.
- In the vapor phase, water moves primarily by diffusion, at least until it reaches the outside air, where convection (a form of bulk flow) becomes dominant.

However, the key element in the transport of water from the soil to the leaves is the generation of negative pressures within the xylem due to the capillary forces within the cell walls of transpiring leaves. At the other end of the plant, soil water is also held by capillary forces. This results in a "tug-of-war" on a rope of water by capillary forces at both ends. As a leaf loses water due to transpiration powered by sunlight, water moves through the xylem driven by physical forces.

This simple mechanism makes for tremendous energetic efficiency, which is critical when as many as 400 molecules of water are being transported for every CO_2 molecule being taken up in exchange. Crucial elements that allow this transport system to function are a low-resistivity xylem flow path that is protected from cavitation, a high-surface-area root system for extracting water from the soil, and the ability to limit water loss from leaf surfaces when the soil is dry.

Summary

There is an inherent conflict between a plant's need for CO_2 uptake and its need to conserve water, resulting from water being lost through the same pores that let in CO_2. To manage this conflict, plants have evolved adaptations that control water loss from leaves, and that replace the water that is lost to the atmosphere with water from the soil.

3.1 Water in the Soil

- The water content and rate of movement in soils depend on soil type and structure, which influence the pressure gradient in the soil and its hydraulic conductivity.
- In soil, water may exist as a surface film on soil particles, or it may partially or completely fill the spaces between particles.
- Osmotic potential, pressure potential, and gravitational potential influence the movement of water from the soil through the plant to the atmosphere (**Figure 3.1**).
- The intimate contact between root hairs and soil particles greatly increases the surface area for water absorption (**Figure 3.2**).

3.2 Water Absorption by Roots

- Water uptake is mostly confined to regions near root tips (**Figure 3.3**).
- In the root, water may move via the apoplast, the symplasm, or the transmembrane pathway (**Figure 3.4**).
- Water movement through the apoplast is obstructed by the Casparian strip in the endodermis, which forces water to move symplasmically before it enters the xylem (**Figure 3.4**).
- When transpiration is low or absent, the continued transport of solutes into the xylem fluid leads to a decrease in Ψ_s and a decrease in Ψ, providing the force for water absorption and a positive Ψ_p, which yields a positive hydrostatic pressure in the xylem (**Figure 3.5**).

(Continued)

Summary *(continued)*

3.3 Water Transport through the Xylem

- Xylem conduits, which can be either single-celled tracheids or multicellular vessels, provide a low-resistance pathway for the transport of water (**Figure 3.6**).
- Elongated, spindle-shaped tracheids and stacked vessel elements have pits in lateral walls (**Figure 3.6**).
- Pressure-driven bulk flow moves water long distances through the xylem.
- The ascent of water through plants results from the decrease in water potential at the sites of evaporation within leaves (**Figure 3.8**).
- Cavitation breaks the continuity of the water column in the xylem and prevents the transport of water under tension (**Figures 3.7**, **3.9**).

3.4 Water Movement from the Leaf to the Atmosphere

- Water is pulled from the xylem into the cell walls of leaf mesophyll before evaporating into the leaf's air spaces (**Figure 3.10**).
- The hydraulic resistance of leaves is large and varies in response to growth conditions and exposure to low leaf water potentials.
- Transpiration depends on the difference in water vapor concentration between the leaf air spaces and the external air and on the diffusional resistance of this pathway, which consists of leaf stomatal resistance and boundary layer resistance (**Figure 3.11**, **Table 3.1**).
- Opening and closing of the stomatal pore are accomplished and controlled by guard cells (**Figures 3.12**, **3.13**).
- Stomatal movements are driven by changes in the solute concentration (Ψ_s) of guard cells. Blue and red wavelengths of light stimulate solute accumulation and thus stomatal opening, while the hormone abscisic acid triggers guard cells to decrease their solute concentration and thus to close.
- The effectiveness of plants in limiting water loss while allowing CO_2 uptake is given by the transpiration ratio.

3.5 Overview: The Soil–Plant–Atmosphere Continuum

- Physical forces, without the involvement of any metabolic pump, drive the movement of water from soil to plant to atmosphere, with the sun being the ultimate source for the energy.

Suggested Reading

Bramley, H., Turner, N. C., Turner, D. W., and Tyerman, S. D. (2009) Roles of morphology, anatomy and aquaporins in determining contrasting hydraulic behavior of roots. *Plant Physiol.* 150: 348–364.

Brodersen, C. R., Roddy, A. B., Wason, J. W., and McElrone, A. J. (2019) Functional status of xylem through time. *Annu. Rev. Plant Biol.* 70: 407–433. https://doi.org/10.1146/annurev-arplant-050718-100455.

Brodribb, T. J., McAdam, S. A. M., and Carins Murphy, M. R. (2017) Xylem and stomata, coordinated through space and time. *Plant Cell Environ.* 40: 872–880.

Buckley, T. N. (2019) How do stomata respond to water stress? *New Phytol.* 224: 21–33.

Cai, G., and Ahmed, M. A. (2022). The role of root hairs in water uptake: Recent advances and future perspectives. *J. Exp. Botany 73*(11): 3330–3338. https://doi.org/10.1093/jxb/erac114.

Calvo-Polanco, M., Ribeyre, Z., Dauzat, M., Reyt, G., Hidalgo-Shrestha, C., Diehl, P., Frenger, M., Simonneau, T., Muller, B., Salt, D. E., Franke, R. B., Maurel, C., and Boursiac, Y. (2021) Physiological roles of Casparian strips and suberin in the transport of water and solutes. *New Phytol. 232*(6): 2295–2307. https://doi.org/10.1111/nph.17765.

Choat, B., Brodribb, T. J., Brodersen, C. R., Duursma, R. A., López, R., and Medlyn, B. E. (2018) Triggers of tree mortality under drought. *Nature* 558: 531–539.

Hacke, U. G., Sperry, J. S., Pockman, W. T., Davis, S. D., and McCulloh, K. (2001) Trends in wood density and structure are linked to prevention of xylem implosion by negative pressure. *Oecologia* 126: 457–461.

Leakey, A. D. B., Ferguson, J. N, Pignon, C. P., Wu, A., Jin, Z., Hammer, G. L., and Lobell, D. B. (2019) Water use efficiency as a constraint and target for improving the resilience and productivity of C3 and C4 crops. *Annu. Rev. Plant Biol.* 70: 781–808. https://doi.org/10.1146/annurev-arplant-042817-040305.

Pittermann, J., Sperry, J. S., Hacke, U. G., Wheeler, J. K., and Sikkema, E. H. (2005) Torus-margo pits help conifers compete with angiosperms. *Science* 310: 1924.

Roelfsema, M. R. G., and Kollist, H. (2013) Tiny pores with a global impact. *New Phytol.* 197: 11–15.

Sack, L., and Scoffoni, C. (2013) Leaf venation: Structure, function, development, evolution, ecology and applications in past, present and future. *New Phytol.* 198: 983–1000.

Stroock, A. D., Pagay, V. V., Zwieniecki, M. A., and Holbrook, N. M. (2014). The physicochemical hydrodynamics of vascular plants. *Annu. Rev. Fluid Mech.* 46: 615–642.

4 Mineral Nutrition

Mineral nutrients are elements such as nitrogen, phosphorus, and potassium that plants acquire from the soil. Most mineral nutrients originate from weathered rocks and are released as the rock breaks down or is weathered. Although mineral nutrients continually cycle through all organisms, they enter the **biosphere** predominantly through the roots of plants; so, in a sense, plants act as "miners" of Earth's crust. The large surface area of roots and their ability to absorb nutrients, in elemental or compound form at low concentrations from the soil solution, increase the effectiveness of mineral acquisition by plants. After being absorbed by roots, the mineral elements or compounds that contain them are translocated to the different parts of the plant, where they serve in numerous biological functions. Other organisms, such as mycorrhizal fungi and nitrogen-fixing bacteria, often contribute to the plant's acquisition of mineral nutrients.

The study of how plants obtain and use mineral nutrients is called **mineral nutrition**. This area of research is central for improving modern agricultural practices and environmental protection, as well as for understanding plant ecological interactions in natural ecosystems. High agricultural yields often depend on fertilization with mineral nutrients. The application of a **limiting nutrient** will increase plant growth and yield unless another abiotic or biotic stress is also limiting. To meet the demand for food by the increasing world population, global fertilizer use has increased dramatically since 1940. Agricultural production is also increasing, especially grain and livestock production, in part because of fertilizer use but also because of improvements in crop genetics. Some of the environmental impacts of fertilizer use are discussed at the end of this chapter.

biosphere The parts of the surface and atmosphere of Earth that support life as well as the organisms living there.

mineral nutrition The study of how plants obtain and use mineral nutrients.

limiting nutrient A nutrient that is not sufficiently available to support plant growth. Its application will normally increase growth as long as other factors (abiotic or biotic) are not also limiting.

In this chapter we discuss the nutritional needs of plants, the symptoms of specific nutritional deficiencies, and the use of fertilizers to ensure proper plant nutrition. Then we examine how soil structure (the arrangement of solid, liquid, and gas phase components) and root morphology influence the transfer of nutrients from the environment into a plant. Finally, we introduce the topic of symbiotic mycorrhizal associations, which play key roles in nutrient acquisition in the majority of plants. Chapters 5 and 6 address additional aspects of nutrient assimilation and solute transport, respectively.

4.1 Essential Nutrients, Deficiencies, and Plant Disorders

Distinguish the roles of nutrients and how their deficiency impacts plant growth and development.

Only certain elements are essential for plants. An **essential element** is defined as one whose absence causes severe abnormalities in plant growth, development, or reproduction and may prevent a plant from completing its life cycle. Essential elements serve myriad functions in plants. They are components of cell walls, membranes, and other structural components, proteins that catalyze metabolism, nucleic acids that store and process genetic information, and molecular cascades involved in signal transduction, and are essential for osmotic homeostasis and other functions. If plants are given these essential elements, as well as water and energy from sunlight, they can synthesize all the compounds they need for normal growth. **Table 4.1** lists the elements that are considered to be essential for most, if not all, vascular plants. The first three elements—hydrogen, carbon, and oxygen—are not considered mineral nutrients because they are obtained primarily from water or atmospheric carbon dioxide.

Essential mineral elements are usually classified as *macronutrients* or *micronutrients* according to their relative concentrations in plants. In some cases, the differences in concentration between macronutrients and micronutrients are not as large as those indicated in Table 4.1. For example, some plant tissues, such as the leaf mesophyll, contain almost as much iron or manganese as sulfur or magnesium. Often elements are present in concentrations greater than the plant's minimum requirements, and the excess is stored.

While the classification of plant nutrients into macronutrients and micronutrients is a traditional approach, it is difficult to justify physiologically. Many scientists prefer to classify nutrients according to their biochemical role and physiological function. **Table 4.2** shows a classification in which plant nutrients have been divided into four basic groups:

1. Part of carbon compounds. Nitrogen, sulfur, and phosphorus form covalent bonds in carbon-containing compounds and macromolecules (e.g., nucleic acids, proteins, and lipids). The abundance of such crucial molecules explains why nitrogen, sulfur, and phosphorus are essential macronutrients. Plants assimilate nitrogen and sulfur via biochemical reactions involving oxidation and reduction reactions to form covalent bonds with carbon (Chapter 5).
2. Structural integrity. Silicon and boron are important for maintaining structural integrity. Silicon is a beneficial nutrient for most plants, but it is an essential nutrient only in Equisetaceae. Silicon is taken up as silicic acid and accumulates in cell walls as amorphous silica (SiO_2). Boron is an important structural component of pectin in cell walls (Chapter 1).

essential element A chemical element that is an intrinsic component of the structure or metabolism of a plant. When the element is in limited supply, a plant suffers abnormal growth, development, or reproduction.

Table 4.1 Tissue Concentrations of Essential Elements Required by Most Plants

Element	Chemical symbol	Concentration in dry matter (mg g^{-1} or µg g^{-1})[a]	Relative number of atoms with respect to molybdenum
Obtained from water or carbon dioxide			
Hydrogen	H	60	60,000,000
Carbon	C	450	40,000,000
Oxygen	O	450	30,000,000
Obtained from the soil			
Macronutrients			
Nitrogen	N	15	1,000,000
Potassium	K	10	250,000
Calcium	Ca	5	125,000
Magnesium	Mg	2	80,000
Phosphorus	P	2	60,000
Sulfur	S	1	30,000
Silicon	Si	1	30,000
Micronutrints			
Chlorine	Cl	100	3,000
Iron	Fe	100	2,000
Boron	B	20	2,000
Manganese	Mn	50	1,000
Sodium	Na	10	400
Zinc	Zn	20	300
Copper	Cu	6	100
Nickel	Ni	0.1	2
Molybdenum	Mo	0.1	1

Source: After E. Epstein. 1972. *Mineral Nutrition of Plants: Principles and Perspectives*. John Wiley and Sons, New York; E. Epstein. 1999. *Annu. Rev. Plant Physiol. Plant Mol. Biol.* 50: 641–664.

[a] The values for the nonmineral elements (H, C, O) and the macronutrients are in mg g^{-1}, while the values for micronutrients are expressed in μg g^{-1}.

3. Remain in ionic form. Both free ions dissolved in the plant water or ions electrostatically bound to substances such as the pectic acids present in the plant cell wall have important roles as enzyme cofactors, in regulating osmotic potentials, and in controlling membrane permeability.
4. Involved in oxidation–reduction (redox) reactions. Metals such as iron have important roles in reactions involving electron transfer and nitrogen fixation.

Please keep in mind that this classification is somewhat arbitrary because many elements serve several functional roles. For example, manganese is listed in group 4 as a metal involved in several key electron transfer reactions, yet it is a mineral element that remains in ionic form, which would place it in group 3.

Some naturally occurring elements, such as aluminum, selenium, and cobalt, are not essential elements yet can also accumulate in plants. Aluminum,

Table 4.2 Classification of Plant Mineral Nutrients According to Biochemical Function

Mineral nutrient	Functions
Group 1	**Nutrients that are part of carbon compounds**
N	Constituent of amino acids, amides, proteins, nucleic acids, nucleotides, coenzymes, hexosamines, etc.
S	Component of amino acids cysteine and methionine. Constituent of lipoic acid, coenzyme A, thiamine pyrophosphate, glutathione, biotin, 5′-adenylylsulfate, and 3′-phosphoadenosine.
P	Component of sugar phosphates, nucleic acids, nucleotides, coenzymes, phospholipids, phytic acid, etc. Has a key role in reactions that involve ATP.
Group 2	**Nutrients that are important for structural integrity**
Si	Deposited as amorphous silica in cell walls. Contributes to cell wall mechanical properties, including rigidity and elasticity.
B	Structural component of pectic polysaccharide RG II (rhamnogalacturonan II). Forms complexes with mannitol, mannan, polymannuronic acid, and other constituents of cell walls. Involved in cell elongation and nucleic acid metabolism.
Group 3	**Nutrients that remain in ionic form**
K	Required as a cofactor for more than 40 enzymes. Principal cation in establishing cell turgor and maintaining cell electroneutrality.
Ca	Constituent of the middle lamella of cell walls. Required as a cofactor by some enzymes involved in the hydrolysis of ATP and phospholipids. Acts as a second messenger in metabolic regulation.
Mg	Required by many enzymes involved in phosphate transfer. Constituent of the chlorophyll molecule.
Cl	Required for the photosynthetic reactions involved in O_2 evolution. Important anion for establishing cell turgor and maintaining cellular electroneutrality.
Zn	Constituent of alcohol dehydrogenase, glutamic dehydrogenase, carbonic anhydrase, etc.
Na	Involved in the regeneration of phosphoenolpyruvate in C_4 and CAM plants. Substitutes for potassium in some functions.
Group 4	**Nutrients that are involved in redox reactions**
Fe	Constituent of cytochromes and nonheme iron proteins involved in photosynthesis, N_2 fixation, and respiration.
Mn	Required for activity of some dehydrogenases, decarboxylases, kinases, oxidases, and peroxidases. Involved with other cation-activated enzymes and photosynthetic O_2 evolution.
Cu	Component of ascorbic acid oxidase, tyrosinase, monoamine oxidase, uricase, cytochrome oxidase, phenolase, laccase, and plastocyanin.
Ni	Constituent of urease. In N_2-fixing bacteria, constituent of hydrogenases.
Mo	Constituent of nitrogenase, nitrate reductase, and xanthine dehydrogenase.

Source: After H. J. Evans and G. J. Sorger. 1966. *Annu. Rev. Plant Physiol.* 17: 47–76; K. Mengel and E. A. Kirkby. 2001. *Principles of Plant Nutrition*, 5th ed. Kluwer Academic Publishers, Nordrecht, Netherlands.

for example, is not considered to be an essential element, although plants commonly contain from 0.1 to 500 μg aluminum per gram dry matter, and the addition of low amounts of aluminum to a nutrient solution may stimulate plant growth. Many species in the genera *Astragalus, Xylorhiza*, and *Stanleya* accumulate selenium (selenate is a sulfate analogue), although plants have not been shown to have a specific requirement for this element. Cobalt is part of cobalamin (vitamin B_{12} and its derivatives), a component of several enzymes in nitrogen-fixing microorganisms; thus, cobalt deficiency blocks the development and function of nitrogen-fixing nodules, but plants that are not fixing nitrogen do not require cobalt. Crop plants normally contain only relatively small amounts of such nonessential elements.

The following subsections describe the methods used to examine the roles of nutrient elements in plants.

Special techniques are used in nutritional studies

To demonstrate that an element is essential requires that plants be grown under experimental conditions in which only the element under investigation is absent. Such conditions are extremely difficult to achieve with plants grown in a complex medium such as soil. In the nineteenth century, scientists approached this problem by growing plants with their roots immersed in a **nutrient solution** containing only inorganic salts. Their demonstration that plants could grow normally with no soil or organic matter proved unequivocally that plants can fulfill all their needs from only mineral nutrient elements, water, air (CO_2), and sunlight.

The technique of growing plants with their roots immersed in a nutrient solution without soil is called **solution culture** or **hydroponics**. Successful hydroponic culture (**Figure 4.1A**) requires a large volume of nutrient solution or frequent adjustment of the nutrient solution to prevent nutrient uptake by roots from producing large changes in the mineral nutrient concentrations and pH of the solution. A sufficient supply of oxygen to the root system is also critical and may be achieved by vigorous bubbling of air through the solution. Hydroponics is used in the commercial production of many greenhouse and indoor crops, such as tomato (*Solanum lycopersicum*), cucumber (*Cucumis sativus*), and hemp (*Cannabis sativa*). In one form of commercial hydroponic culture, plants are grown in a supporting material such as sand, gravel, vermiculite, rockwool, polyurethane foams, or expanded clay. Nutrient solutions are then flushed through the supporting material, and old solutions are removed by leaching.

Another technique is to grow the plants in **aeroponics**. In this technique plants are grown with their roots suspended in air while being sprayed continuously with a nutrient solution (**Figure 4.1B**). This approach provides easy manipulation of the gaseous environment around the roots, but it requires higher concentrations of nutrients than hydroponic culture does to sustain rapid plant growth. Aeroponics is used in commercial food production, especially in indoor vertical agricultural facilities where temperature, humidity, light, and all nutrients are carefully controlled.

Nutrient solutions can sustain rapid plant growth

Many formulations have been used for nutrient solutions. Early formulations developed by Wilhelm Knop in Germany in the 1860s included only KNO_3, $Ca(NO_3)_2$, KH_2PO_4, $MgSO_4$, and an iron salt. At the time, this nutrient solution was believed to contain all the minerals required by plants; however, the chemicals were contaminated with other elements that are now known to be essential (such as boron and molybdenum). **Table 4.3** shows a more modern formulation for a nutrient solution called a modified **Hoagland solution**, named after Dennis R. Hoagland, a scientist who did pioneering work on hydroponics in the 1930s.

nutrient solution A solution containing only inorganic salts that supports the growth of plants in sunlight without soil or organic matter.

solution culture A technique for growing plants with their roots immersed in nutrient solution without soil. Also called hydroponics.

hydroponics A technique for growing plants with their roots immersed in nutrient solution without soil.

aeroponics The technique by which plants are grown with their roots suspended in air while being sprayed continuously with a nutrient solution.

Hoagland solution A type of nutrient solution for plant growth, originally formulated by Dennis R. Hoagland.

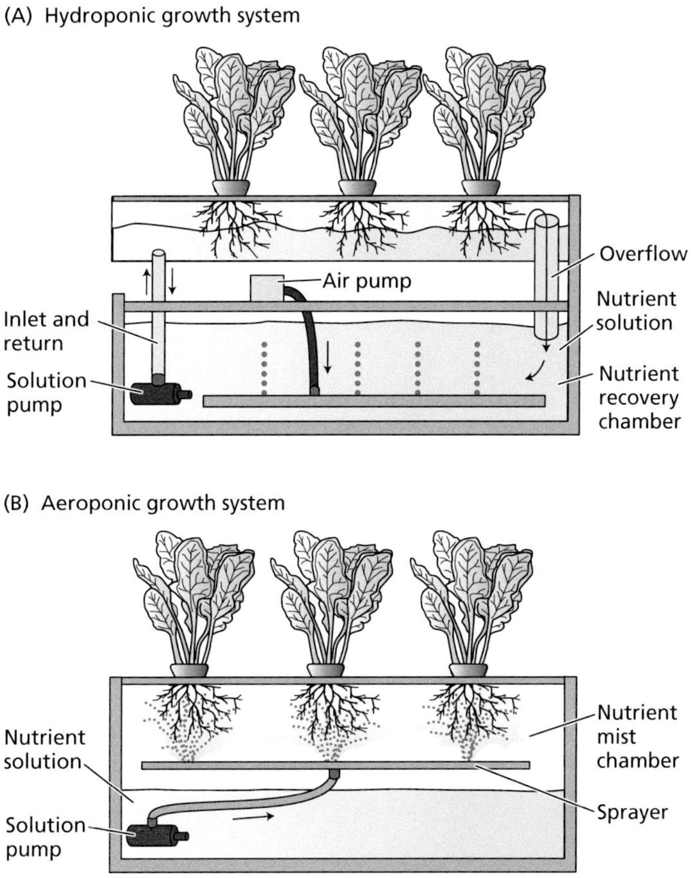

Figure 4.1 Solution culture systems. (A) In a hydroponic culture, plants are suspended by the base of the stem over a tank containing a nutrient solution. Air is pumped through a porous solid that generates a stream of small bubbles and keeps the solution fully saturated with oxygen. (B) In aeroponics, a high-pressure pump sprays nutrient solution on roots enclosed in a tank.

Table 4.3 Composition of a Modified Hoagland Nutrient Solution for Growing Plants

Compound	Molecular weight	Concentration of stock solution	Concentration of stock solution	Volume of stock solution per liter of final solution	Element	Final concentration of element	
	$g\ mol^{-1}$	mM	$g\ L^{-1}$	mL		μM	$\mu g\ g^{-1}$
Macronutrients							
KNO_3	101.10	1,000	101.10	6.0	N	16,000	224
$Ca(NO_3)_2{\cdot}4H_2O$	236.16	1,000	236.16	4.0	K	6,000	235
$NH_4H_2PO_4$	115.08	1,000	115.08	2.0	Ca	4,000	160
$MgSO_4{\cdot}7H_2O$	246.48	1,000	246.49	1.0	P	2,000	62
					S	1,000	32
					Mg	1,000	24
Micronutrients							
KCl	74.55	25	1.864	2.0 (combined, KCl through H_2MoO_4)	Cl	50	1.77
H_3BO_3	61.83	12.5	0.773		B	25	0.27
$MnSO_4{\cdot}H_2O$	169.01	1.0	0.169		Mn	2.0	0.11
$ZnSO_4{\cdot}7H_2O$	287.54	1.0	0.288		Zn	2.0	0.13
$CuSO_4{\cdot}5H_2O$	249.68	0.25	0.062		Cu	0.5	0.03
H_2MoO_4 (85% MoO_3)	161.97	0.25	0.040		Mo	0.5	0.05
NaFeDTPA	468.20	64	30.0	0.3–1.0	Fe	19.2–64.0	1.1–3.6
Optional[a]							
$NiSO_4{\cdot}6H_2O$	262.86	0.25	0.066	2.0	Ni	0.5	0.03
$Na_2SiO_3{\cdot}9H_2O$	284.20	1,000	284.20	1.0	Si	1,000	28

Source: After E. Epstein and A. J. Bloom. 2005. *Mineral Nutrition of Plants: Principles and Perspectives*, 2nd ed. Oxford University Press/Sinauer, Sunderland, MA.

Note: The macronutrients are added separately from stock solutions to prevent precipitation during preparation of the nutrient solution. A combined stock solution is made up containing all micronutrients except iron. Iron is added as sodium ferric diethylenetriaminepentaacetate; some plants, such as maize, require the higher concentration of iron shown in the table.

[a] Nickel is usually present as a contaminant of the other chemicals, so it may not need to be added explicitly. Silicon, if included, should be added first and the pH adjusted with HCl to prevent precipitation of the other nutrients.

A modified Hoagland solution contains all known mineral elements needed for rapid plant growth. The concentrations of these elements are set at the highest possible concentrations without producing toxicity symptoms or salinity stress (because nutrients are supplied as salts [e.g., sulfate, phosphate]), and thus may be several orders of magnitude higher than those found in the soil around plant roots. For example, whereas phosphorus is present in the soil solution at concentrations normally less than 0.06 $\mu g\ g^{-1}$ or 2 μ*M*, the modified Hoagland solution calls for 62 $\mu g\ g^{-1}$ or 2 m*M*. Such high initial concentrations permit plants to be grown in a medium for extended periods without replenishment of the nutrient, but they may injure young plants. Therefore, many researchers dilute their nutrient solutions severalfold and replenish them frequently to minimize fluctuations of nutrient concentration in the medium and in the plants.

Another important property of this modified Hoagland formulation is that nitrogen is supplied as both ammonium (NH_4^+) and nitrate (NO_3^-). Supplying nitrogen in a balanced mixture of cations (positively charged ions) and anions (negatively charged ions) tends to decrease the rapid rise in the pH of the medium that is commonly observed when the nitrogen is supplied solely as nitrate anion. Even when the pH of the medium is kept neutral, most plants grow better if they have access to both NH_4^+ and NO_3^-, because absorption and assimilation of the two nitrogen forms promote inorganic cation–anion balance in the plant.

A significant problem with nutrient solutions is maintaining the availability of iron. When supplied as an inorganic salt such as $FeSO_4$ or $Fe(NO_3)_2$, iron can precipitate out of solution as iron hydroxide, particularly under alkaline conditions. If phosphate salts are present, insoluble iron phosphate will also form. Precipitation of the iron out of solution makes it physically unavailable to plants, unless iron salts are added at frequent intervals. Earlier researchers solved this problem by adding iron together with citric acid or tartaric acid. Compounds such as these are called **chelators** because they form soluble complexes with cations such as ferric iron (Fe^{3+}) and Ca^{2+} in which the cation is held by ionic forces rather than by covalent bonds. **Figure 4.2** shows the structure of different species of Fe^{3+}-citrate complexes that are biologically relevant. In fact, plants produce and use organic chelators such as citrate to solubilize and transport iron into and around the organism. For example, iron is transported in the xylem (vascular tissue) as Fe^{3+}-citrate complexes.

More modern nutrient solutions use the chelators ethylenediaminetetraacetic acid (EDTA), diethylenetriaminepentaacetic acid (DTPA, or pentetic acid), or ethylenediamine-N,N′-bis(o-hydroxyphenylacetic) acid (o,o-EDDHA) to keep iron in solution where it is available to plants. In fact, plants use different mechanisms for iron uptake (Chapter 6). Eudicots use a reductase to reduce Fe^{3+} to ferrous iron (Fe^{2+}) at the root surface and then they take up Fe^{2+}. Monocots secrete chelators called phytosiderophores that bind Fe^{3+}, which is then transported into the plant complexed with the phytosiderophore. After uptake into the root, iron is kept soluble by chelation with organic compounds present in plant cells.

chelator A carbon compound (e.g., malic acid or citric acid) that can form a soluble noncovalent complex with certain cations, thereby facilitating their uptake.

Mineral deficiencies disrupt plant metabolism and function

An inadequate supply of an essential element results in a nutritional disorder manifested by characteristic deficiency symptoms. In hydroponic culture, withholding of an essential element can be readily correlated with a given set of symptoms. For example, a particular deficiency might elicit a specific pattern of leaf discoloration. Diagnosis of soil-grown plants can be more complex for the following reasons:

- Deficiencies of several elements may occur simultaneously.
- Deficiencies or excessive amounts of one element may induce deficiency or excessive accumulation of another element.
- Some virus-induced plant diseases may produce symptoms similar to those of nutrient deficiencies.

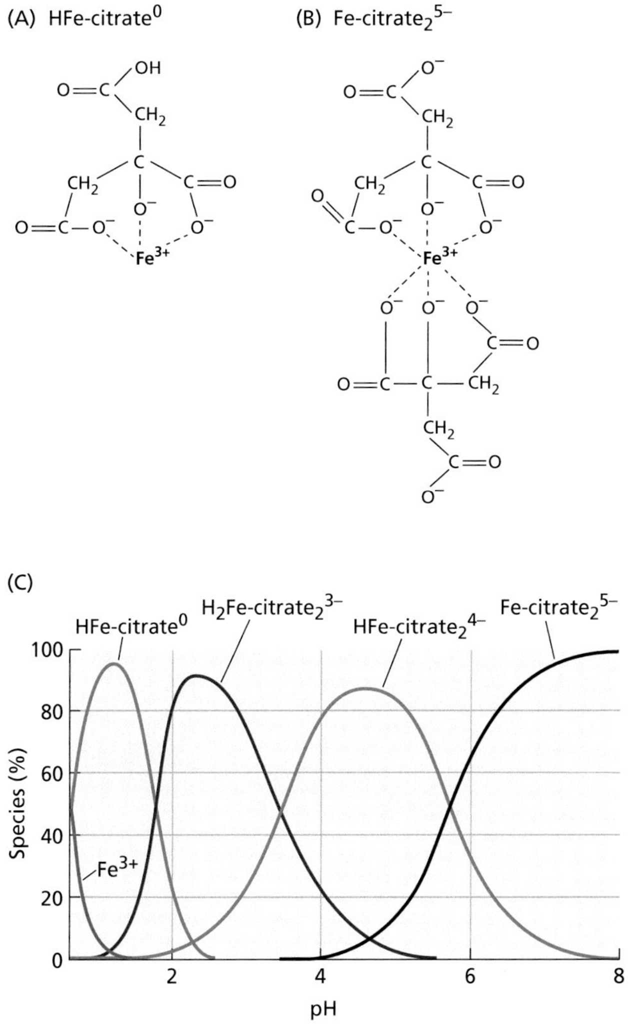

Figure 4.2 Examples of Fe^{3+}-citrate complexes in solution and the effect of pH. (A) HFe-citrate^0. (B) Fe-citrate$_2^{5-}$. (C) The predominance of different Fe^{3+}-citrate complexes in solution depends on pH. For example, Fe-citrate$_2^{5-}$, shown in (B), predominates at neutral pH.

Table 4.4 Mineral Elements Classified on the Basis of Their Mobility within a Plant and Their Tendency to Retranslocate during Deficiencies

Mobile	Intermediate	Immobile
Nitrogen	Sulfur	Calcium
Potassium	Iron	Boron
Magnesium	Zinc	
Phosphorus	Copper	
Chlorine		
Sodium		
Molybdenum		

Note: Within each group, elements are listed in the order of their abundance in the plant.

Nutrient deficiency symptoms in a plant are the expression of metabolic disorders resulting from the insufficient supply of an essential element. These disorders are related to the roles played by essential elements in normal plant metabolism and function (listed in Table 4.2).

Most essential elements have a variety of different physiological functions in plants, but some general statements are possible. The essential elements mainly function in plant structure, metabolism, signal transduction, and cellular osmoregulation. Research continues to reveal specific roles for these elements; for example, calcium ions function structurally in the cell wall and also regulate key enzymes, transporters, and ion channels in the cell.

An important clue in relating an acute deficiency symptom to a particular essential element is the extent to which an element can be recycled from older to younger leaves. Some elements, such as nitrogen, phosphorus, and potassium, can readily move from leaf to leaf, either because they are always primarily in soluble form (e.g., K^+) or because they are easily liberated from macromolecules such as proteins, nucleic acids, and lipids, via the action of proteases, nucleases, and lipases, respectively (e.g., NH_4^+ and HPO_4^{2-}) (Chapter 15). Other elements, such as boron and calcium, are relatively immobile in most plant species because they are associated with structures such as the cell wall that are more recalcitrant to decomposition (**Table 4.4**). If an essential element can be mobilized quickly, deficiency symptoms tend to appear first in older leaves. Conversely, deficiencies of immobile essential elements become evident first in younger leaves. Particular deficiency symptoms and functional roles of the essential elements (Table 4.2) are described below. Please keep in mind that many symptoms are highly dependent on plant species.

GROUP 1: DEFICIENCIES IN MINERAL NUTRIENTS THAT ARE PART OF CARBON COMPOUNDS This first group consists of nitrogen, phosphorus, and sulfur. Nitrogen and phosphorus availability in soils limits plant productivity in most natural and agricultural ecosystems. By contrast, soils generally contain sufficient sulfur due to atmospheric sulphate deposition primarily from coal-burning power plants. Coal use has been decreasing since 1980, and sulfur amendments to soil may become required. The oxidation–reduction states of nitrogen and sulfur range widely (Chapter 5). Plants take up highly oxidized inorganic forms of nitrate and sulfate and convert them into highly reduced forms found in organic compounds, such as amino acids. Phosphorus, by contrast, is not reduced after being taken up by plants as phosphate, and phosphate forms covalent bonds with carbon via oxygen in compounds such as ATP and other nucleotide phosphates, nucleic acids, phospholipids, phosphoproteins, and phosphorylated intermediates of metabolism.

NITROGEN Nitrogen is the mineral element that plants require in the greatest amounts (Table 4.1). It serves as a constituent of many plant cell components, including chlorophyll, proteins, and nucleic acids. Therefore, nitrogen deficiency rapidly limits plant growth. If such a deficiency persists, most species show leaf **chlorosis** (yellowing of the leaves), associated with deconstruction of the photosynthetic apparatus and of macromolecules more generally, especially in the older leaves near the base of the plant. Under severe nitrogen deficiency, these leaves become completely yellow (or tan) while mobilizing and exporting amino acids and nutrients to other organs, and then fall off the plant. Younger leaves may not show these symptoms initially because nitrogen can be mobilized from

chlorosis The yellowing of plant leaves such as occurs as a result of mineral deficiency. The leaves affected and the parts of the leaves that yellow can be diagnostic for the type of deficiency.

older leaves. Thus, a nitrogen-deficient plant may have light green upper leaves and yellow or tan lower leaves.

When nitrogen deficiency develops slowly, plants may have markedly slender and often woody stems. This woodiness may be due to a buildup of excess carbohydrates that cannot be used in the synthesis of amino acids or other nitrogen-containing compounds. Carbohydrates not used in nitrogen metabolism may also be used in anthocyanin synthesis, leading to accumulation of that pigment. This condition is revealed as a purple coloration in leaves, petioles, and stems of nitrogen-deficient plants of some species, such as tomato and certain varieties of maize (corn; *Zea mays*).

SULFUR Sulfur is found in certain amino acids (i.e., cysteine and methionine) and is a constituent of several coenzymes and vitamins, such as coenzyme A, *S*-adenosylmethionine, biotin, vitamin B_1, and pantothenic acid, which are essential for metabolism.

Many of the symptoms of sulfur deficiency are similar to those of nitrogen deficiency, including leaf chlorosis, stunting of growth, and anthocyanin accumulation. This similarity is not surprising, since sulfur and nitrogen are both constituents of proteins. The chlorosis caused by sulfur deficiency, however, generally arises initially in young and mature leaves, rather than in old leaves as in nitrogen deficiency, because sulfur, unlike nitrogen, is not as easily remobilized to the younger leaves in most species. Nonetheless, in some plant species sulfur chlorosis may occur simultaneously in all leaves, or even initially in older leaves.

PHOSPHORUS Phosphorus (as phosphate, PO_4^{3-}) is an integral component of important compounds in plant cells, including the sugar–phosphate intermediates of photosynthesis and respiration as well as the phospholipids that make up most plant membranes. The largest pool of organic phosphorus in cells is in nucleotides used in plant energy metabolism (such as ATP) and in DNA and RNA. Characteristic symptoms of phosphorus deficiency include stunted growth of the entire plant and a dark green coloration of the leaves, which may be malformed and contain small areas of dead tissue called **necrotic spots**.

As in nitrogen deficiency, some species may produce excess anthocyanins under phosphorus deficiency, giving the leaves a slight purple coloration. Unlike in nitrogen deficiency, the purple coloration of phosphorus deficiency is not associated with chlorosis. In fact, the leaves may be a dark greenish purple. Additional symptoms of phosphorus deficiency include the production of slender (but not woody) stems and the death of older leaves. Plant development, including flowering, may also be delayed.

GROUP 2: DEFICIENCIES IN MINERAL NUTRIENTS THAT ARE IMPORTANT FOR STRUCTURAL INTEGRITY This group consists of the macronutrient silicon and the micronutrient boron. These elements are usually present in plants as ester linkages between the element (X) and the carbon atom of an organic compound (i.e., R-C-OO-X).

SILICON Only members of the family Equisetaceae—called *scouring rushes* because at one time their ash, rich in gritty silica (SiO_2), was used to scour pots—require silicon to complete their life cycle. Nonetheless, many other species accumulate substantial amounts of silicon and show enhanced growth, fertility, and stress resistance when supplied with adequate amounts of silicon. For example, many grasses and rice (*Oryza* spp.) accumulate silicon in their leaves.

Plants deficient in silicon are more susceptible to lodging (the permanent displacement of crop stems from vertical alignment) and fungal infection. Silicon is deposited primarily in the endoplasmic reticulum, cell walls, and intercellular spaces as hydrated, amorphous silica ($SiO_2 \cdot nH_2O$). It also forms complexes with

necrotic spots Small spots of dead leaf tissue.

polyphenols and thus serves as a complement to lignin in the reinforcement of cell walls. In addition, silicon can lessen the toxicity of many metals, including aluminum and manganese.

BORON Boron serves as a structural component in the cell wall, where it cross-links RG II (rhamnogalacturonan II, a small pectic polysaccharide). It may also have additional physiological functions in plants; there is evidence for boron functioning in cell elongation, nucleic acid synthesis, hormone responses, membrane function, and cell cycle regulation. Boron-deficient plants may exhibit a wide variety of symptoms, depending on the species and the age of the plant.

A characteristic symptom of boron deficiency is black necrosis of young leaves and terminal buds. The necrosis of the young leaves occurs primarily at the base of the leaf blade. Stems may be unusually stiff and brittle. Apical dominance may also be lost, causing the plant to become highly branched; however, the terminal apices of the branches soon become necrotic because of inhibition of cell differentiation. Structures such as the fruits, fleshy roots, and tubers may exhibit necrosis or abnormalities related to the breakdown of internal tissues.

GROUP 3: DEFICIENCIES IN MINERAL NUTRIENTS THAT REMAIN IN IONIC FORM This group includes some of the most familiar mineral elements: the macronutrients potassium, calcium, and magnesium and the micronutrients chlorine, zinc, and sodium. These elements may be found as ions in solution in the cytosol or vacuoles, bound electrostatically or as ligands to larger, carbon-containing compounds.

POTASSIUM Potassium, present in plants as the cation K^+, plays an important role in regulation of the osmotic potential of plant cells. It also activates many enzymes involved in respiration and photosynthesis.

The first observable symptom of potassium deficiency is mottled or marginal chlorosis, which then develops into necrosis primarily at the leaf tips, at the margins, and between veins. In monocots, these necrotic lesions may eventually extend toward the leaf base. Because potassium can be mobilized to the younger leaves, these symptoms appear initially on the more mature leaves toward the base of the plant. The leaves may also curl and crinkle. The stems of potassium-deficient plants may be slender and weak, with abnormally short internodal regions. In potassium-deficient maize, the roots may have an increased susceptibility to root-rotting fungi present in the soil, and this susceptibility, together with effects on the stem, results in an increased tendency for the plant to lodge.

CALCIUM Calcium ions (Ca^{2+}) have two distinct roles in plants: (1) a structural/apoplastic role whereby Ca^{2+} binds to acidic groups of membrane lipids (phospho- and sulfolipids) and cross-links pectins, particularly in the middle lamellae that separate newly divided cells and (2) a signaling role whereby Ca^{2+} acts as a second messenger that initiates plant responses to environmental stimuli (Chapter 12). In its function as a second messenger, Ca^{2+} binds to proteins such as **calmodulin** (CaM), calmodulin-like proteins (CMLs), or calcineurin B–like proteins (CBLs) in the cytosol of plant cells. These Ca^{2+}-binding proteins regulate many other proteins such as kinases, phosphatases, and cytoskeletal proteins. Some targets of Ca^{2+} signaling bind Ca^{2+} directly, including ion channels, kinases, phospholipases, NADPH oxidases, and others. Ca^{2+} thereby regulates many cellular processes, including cell growth and differentiation and responses to hormones and environmental signals. While most essential nutrients are

calmodulin A Ca^{2+}-binding protein that regulates many cellular processes in a Ca^{2+}-dependent manner.

accumulated in plant cells, calcium ions are actively pumped out of the cytosol and into the cell wall space and vacuole, facilitating their role as a second messenger in signal transduction.

Characteristic symptoms of calcium deficiency include necrosis of young meristematic regions such as the tips of roots or young leaves, where cell division and cell wall formation are most rapid. Necrosis in slowly growing plants may be preceded by a general chlorosis and downward hooking of young leaves. Young leaves may also appear deformed. The root system of a calcium-deficient plant may appear brownish, short, and highly branched. Severe stunting may result if the meristematic regions of the plant die prematurely.

MAGNESIUM In plant cells, magnesium ions (Mg^{2+}) have a specific role in the activation of enzymes involved in respiration, photosynthesis, and the synthesis of DNA and RNA. Mg^{2+} is also part of the ring structure of the chlorophyll molecule. A characteristic symptom of magnesium deficiency is chlorosis between the leaf veins, occurring first in older leaves because of the high mobility of this cation. This pattern of chlorosis occurs because the chlorophyll in the vascular bundles remains unaffected longer than that in the cells between the bundles. If the deficiency is extensive, the leaves may become yellow or white. An additional symptom of magnesium deficiency may be senescence and premature leaf abscission.

CHLORINE The element chlorine is found in plants as the chloride ion (Cl^-). It is required for the water-splitting reaction of photosynthesis through which oxygen is produced. In addition, chlorine may be required for cell division in leaves and roots. Plants deficient in chlorine develop wilting of the leaf tips followed by general leaf chlorosis and necrosis. The leaves may also exhibit reduced growth. Eventually the leaves may take on a bronze-like color ("bronzing"). Roots of chlorine-deficient plants may appear stunted and thickened near the root tips.

Chloride ions are highly soluble and are generally available in soils because seawater is swept into the air by wind and delivered to soil when it rains. Therefore, chlorine deficiency is only rarely observed in plants grown in native or agricultural habitats. Most plants absorb chlorine at concentrations much higher than those required for normal growth and development.

ZINC Many enzymes require zinc ions (Zn^{2+}) for their activity, and zinc may be required for chlorophyll biosynthesis in some plants. Zinc deficiency is characterized by a reduction in internodal growth, and as a result plants display a rosette habit of growth in which the leaves form a circular cluster radiating at or close to the ground. The leaves may also be small and distorted, with leaf margins having a puckered appearance. These symptoms may result from loss of the ability to produce sufficient amounts of the auxin indole-3-acetic acid (IAA). Zinc deficiency results in a decrease in chlorophyll synthesis, and in some species (e.g., maize, sorghum, and beans) older leaves may show chlorosis between the leaf veins and then develop white necrotic spots.

SODIUM Species using the C_4 and crassulacean acid metabolism (CAM) pathways of carbon fixation (Chapter 8) may require sodium ions (Na^+). In these plants, Na^+ appears vital for regenerating phosphoenolpyruvate, the substrate for primary carboxylation in the C_4 and CAM pathways. Under sodium deficiency, the plants exhibit chlorosis and necrosis, or even fail to form flowers. Many C_3 species also benefit from exposure to low concentrations of Na^+. Sodium ions stimulate growth through enhanced cell expansion and can partly substitute for potassium ions as an osmotically active solute.

GROUP 4: DEFICIENCIES IN MINERAL NUTRIENTS THAT ARE INVOLVED IN REDOX REACTIONS This group of five micronutrients consists of the metals iron, manganese, copper, nickel, and molybdenum. Like the group 3 elements, these elements remain in ionic form. However, all of the group 4 elements can undergo reversible oxidation and reduction (e.g., $Fe^{2+} \leftrightarrow Fe^{3+}$), a property that is essential for their roles in electron transfer and energy transformation. They are usually found in association with larger molecules such as cytochromes, chlorophyll, and proteins (usually enzymes).

IRON Iron has an important role as a component of enzymes involved in the transfer of electrons (redox reactions), such as cytochromes (Chapters 7 and 11). In this role it is reversibly oxidized from Fe^{2+} to Fe^{3+} during electron transfer.

As in magnesium deficiency, a characteristic symptom of iron deficiency is interveinal chlorosis. This symptom, however, appears initially on younger leaves because iron, unlike magnesium, is not as readily mobilized from older leaves. Under conditions of extreme or prolonged deficiency, the veins may also become chlorotic, causing the whole leaf to turn white. The leaves become chlorotic because iron is required for the synthesis of some of the chlorophyll–protein complexes in the chloroplast. The low mobility of iron is probably due to its precipitation in the older leaves as insoluble oxides or phosphates. The precipitation of iron diminishes subsequent mobilization of the metal into the phloem for long-distance translocation.

MANGANESE Manganese ions (Mn^{2+}) activate several enzymes in plant cells. In particular, decarboxylases and dehydrogenases involved in the tricarboxylic acid cycle (Chapter 11) are specifically activated by manganese ions. The best-defined function of Mn^{2+} is in the photosynthetic reaction through which oxygen (O_2) is produced from water. The major symptom of manganese deficiency is interveinal chlorosis associated with the development of small necrotic spots. This chlorosis may occur on younger or older leaves, depending on plant species and growth rate.

COPPER Copper is involved in redox reactions in which it is reversibly oxidized from Cu^{+} to Cu^{2+}. An important copper-containing protein is plastocyanin, which functions as an electron carrier during the light reactions of photosynthesis (Chapter 7). The initial symptom of copper deficiency in many plant species is the production of dark green leaves, which may contain necrotic spots. The necrotic spots appear first at the tips of young leaves and then extend toward the leaf base along the margins. The leaves may also be twisted or malformed. Cereal plants exhibit white leaf chlorosis and necrosis with rolled tips. Under extreme copper deficiency, leaves may drop prematurely and flowers may be sterile.

NICKEL The only known requirement for nickel in vascular plants is for the nickel-containing (Ni^{2+}) enzyme urease. In addition, nitrogen-fixing microorganisms in legume root nodules require nickel (Ni^{+} through Ni^{4+}) for the enzyme that reprocesses some of the hydrogen gas generated during fixation (Chapter 5). Nickel-deficient plants accumulate urea in their leaves and consequently show leaf tip necrosis. Because plants require only minuscule amounts of nickel (Table 4.1), nickel deficiency in the field has been found in only one crop, pecan trees in the southeastern United States.

MOLYBDENUM Molybdenum ions (Mo^{4+} through Mo^{6+}) are components of several enzymes, including nitrate reductase, nitrogenase, xanthine dehydrogenase, aldehyde oxidase, and sulfite oxidase. Nitrate reductase catalyzes the reduction of nitrate to nitrite during its assimilation by the plant cell; nitrogenase converts nitrogen gas to ammonia in nitrogen-fixing microorganisms (Chapter 5). Because molybdenum is involved in both nitrate assimilation and nitrogen fixation, a molybdenum deficiency may bring about a nitrogen deficiency if the nitrogen source is primarily nitrate or if the plant depends on symbiotic nitrogen fixation. Consequently, the first indication of a molybdenum deficiency is general chlorosis between veins and necrosis of older leaves. In some plants, such as cauliflower and broccoli, the leaves may not become necrotic but instead may appear twisted and subsequently die (whiptail disease). Flowers may not develop, or the flowers may abscise prematurely.

Although plants require only tiny amounts of molybdenum (Table 4.1), some soils (e.g., acidic soils in Australia) supply inadequate concentrations. Small additions of molybdenum to such soils can greatly enhance crop or forage growth at negligible cost.

Plant tissue analysis reveals mineral deficiencies

Requirements for mineral elements change as a plant grows and develops. In crop plants, nutrient concentrations at certain stages of growth influence the yield of the economically important organs (tuber, grain, etc.). To optimize yields, farmers use analyses of nutrient concentrations in soil and in plants to determine fertilizer schedules.

Soil analysis is the chemical determination of the nutrient content in a soil sample from the root zone. The chemistry and the biology of soils are complex, and the results of soil analyses vary with sampling methods, storage conditions for the samples, and nutrient extraction techniques. Perhaps more important is that a particular soil analysis reflects the amount of nutrients *potentially* available to the plant roots from the soil, but soil analysis does not tell us how much of a particular mineral nutrient the plant actually needs or is able to absorb. This additional information is best determined by plant tissue analysis.

The traditional approach to **plant tissue analysis** is to determine the amount of a particular nutrient, usually in leaf samples, and determine if it is in a concentration range that is adequate for that species, the organ sampled, and the stage of growth. If the nutrient concentration is below the adequate range, it is in the **deficiency zone** and the application of that nutrient as a fertilizer or a soil amendment that would make the nutrient more available to plants, for example by changing soil pH, could improve growth and yield. Plant scientists recognize that this approach is overly simplified and that diagnosis of nutrient deficiency based on plant tissue analysis is complex. Knowing the concentration of some individual nutrients in the shoot is insufficient to identify a deficiency because plants tightly control the cellular concentration of some nutrients. The recent development of multi-element analysis (sometimes called ionomics) (**Box 4.1**) addresses some of this complexity by considering all nutrients in plant samples at the same time. This approach identifies multi-element patterns that can be used to diagnose plant nutrient deficiencies more accurately.

Because agricultural soils are often limited in the elements nitrogen, phosphorus, and potassium (NPK), many farmers routinely consider growth or yield responses for these elements. Fertilizers containing N, P, and K are often applied at planting or early in plant development before deficiency symptoms could occur. These applications are based on soil tests and the farmer's knowledge of the cropping history of the farm. During the growing season, if a nutrient deficiency is suspected, perhaps by agricultural drone monitoring, steps are taken to correct the deficiency before it reduces growth or yield.

soil analysis The chemical determination of the nutrient content in a soil sample, typically from the root zone.

plant tissue analysis In the context of mineral nutrition, the analysis of the concentrations of mineral nutrients in a plant sample.

deficiency zone In plant tissue, the range of concentrations of a mineral nutrient below the critical concentration, the highest concentration where reduced plant growth or yield is observed.

Box 4.1 Multi-Element Analysis: A Powerful Method to Study Mineral Nutrition

A plant's nutrient status, sometimes called the ionome or nutrient profile, can be described by its mineral nutrient and trace element composition. The elemental composition of an organism or genotype as it develops and responds to physiological stimuli can be determined by multi-element analysis (sometimes called ionomics). Interactions among nutrients have long been known to affect all aspects of plant nutrition, including nutrient availability in the soil, uptake by roots, and function in the plant. However, it wasn't until the early 2000s that patterns of elemental composition were treated as a phenotype. At that time, high-throughput multi-element analysis became possible with inductively coupled plasma mass spectrometry (ICP-MS), which can determine concentrations for 18 to 20 elements rapidly and simultaneously. New developments in plant genetics and genomics, as well as statistical methods to analyze large data sets, were also necessary to make this possible.

A particularly important finding has been that multiple element patterns provide a better identification of nutrient deficiency than does analysis of a single element. An important example of this involves iron deficiency. The concentration of iron in plants is very stable and provides almost no information to determine whether a plant is iron deficient. Multi-element analysis of iron-deficient Arabidopsis plants identified the pattern shown in **Figure A**. Iron-deficient plants show no change in iron content but have elevated manganese, cobalt, zinc, and cadmium and lower molybdenum, and this pattern is diagnostic for iron deficiency. Another example is phosphate deficiency. Low phosphorus concentration is diagnostic for phosphate deficiency. However, including concentrations of cobalt, zinc, boron, copper, and arsenic greatly improves the ability to diagnose phosphate deficiency.

Multi-element analysis has now been applied to many plant (and animal) species. One of the most important challenges is to identify genes that plants have used to adapt to different environmental conditions, including diverse natural soils. Alleles of such genes can be used in breeding projects to produce crop plants that are adapted to specific environmental conditions.

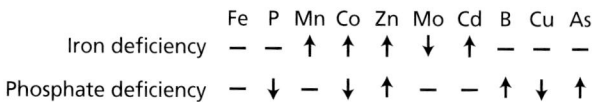

	Fe	P	Mn	Co	Zn	Mo	Cd	B	Cu	As
Iron deficiency	–	–	↑	↑	↑	↓	↑	–	–	–
Phosphate deficiency	–	↓	–	↓	↑	–	–	↑	↓	↑

Figure A Multi-element patterns can provide a more accurate diagnosis of nutrient deficiencies. Arabidopsis plants were grown in nutrient-sufficient or nutrient-deficient conditions. Arrows indicate elevated or decreased elemental concentrations that were predictive for nutrient deficiency. Dashes indicate elements that were not significantly different.

4.2 Treating Nutritional Deficiencies

Explain how nutritional deficiencies are ameliorated in agricultural systems.

Many traditional and subsistence farming practices promote the recycling of mineral elements. Crop plants absorb nutrients from the soil, humans and animals consume locally grown crops, and crop residues and manure from humans and animals return the nutrients to the soil. The main losses of nutrients from such agricultural systems are from leaching that carries dissolved ions, especially nitrate, away with drainage water. In acidic soils (pH < 7), leaching of nutrients other than nitrate may be decreased by the addition of lime—a mix of CaO, $CaCO_3$, and $Ca(OH)_2$—to make the soil more alkaline, because many mineral elements form less soluble compounds when the pH is higher than 7 (**Figure 4.3**). This decrease in leaching, however, may be gained at the expense of decreased availability of some nutrients, especially iron.

In the high-production agricultural systems of industrialized countries, a large proportion of crop biomass leaves the area of cultivation, and returning crop residues to the land where the crop was produced becomes difficult at best. Under most current agricultural practices, this unidirectional removal of nutrients from agricultural soils makes it necessary to restore the lost nutrients to these soils through the addition of fertilizers.

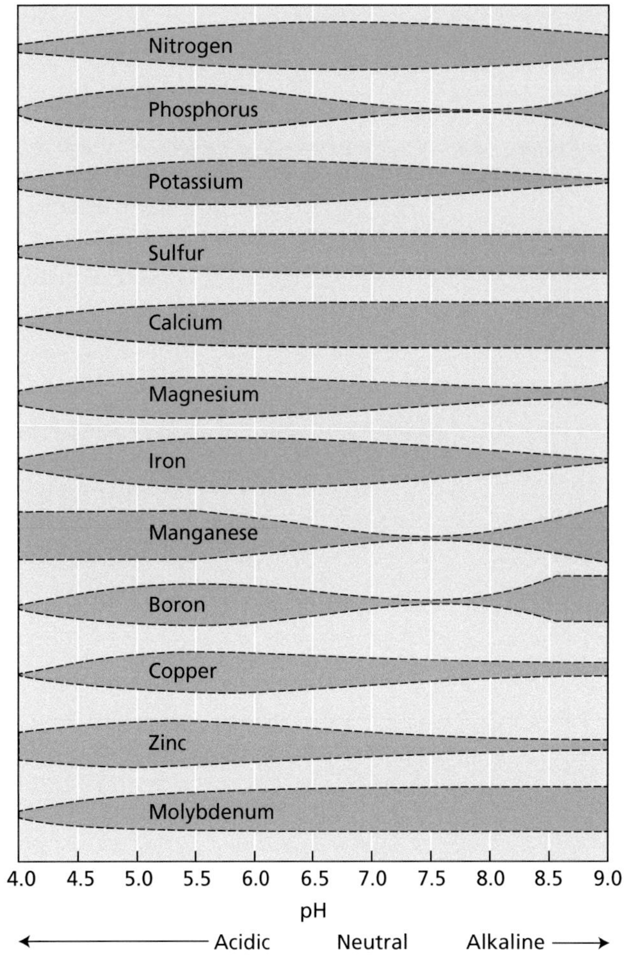

Figure 4.3 Influence of soil pH on the availability of nutrient elements in organic soils. The thickness of the horizontal bars indicates the degree of nutrient availability to plant roots. All of these nutrients are available in the pH range of 5.5 to 6.5.

Crop yields can be improved by the addition of fertilizers

Most inorganic fertilizers contain inorganic salts of the macronutrients nitrogen, phosphorus, and potassium. Fertilizers that contain only one of these three nutrients are termed *straight fertilizers*. Some examples of straight fertilizers are superphosphate, ammonium nitrate, and muriate of potash (potassium chloride). Fertilizers that contain two or more mineral nutrients are called *compound fertilizers* or *mixed fertilizers*, and the numbers on the package label, such as "10-14-10," refer to the weight percentages of N, P, and K, respectively, in the fertilizer.

With long-term agricultural production, consumption of micronutrients by crops can reach a point at which they, too, must be added to the soil as fertilizers. Adding micronutrients to the soil may also be necessary to correct a preexisting deficiency. For example, many acidic, sandy soils in humid regions are deficient in boron, copper, zinc, manganese, molybdenum, or iron and can benefit from nutrient supplementation.

Since soil pH affects the availability of all mineral nutrients (Figure 4.3), chemicals can be applied to the soil to modify soil pH. Addition of lime, as mentioned previously, can raise the pH of acidic soils (pH < 7); addition of elemental sulfur can lower the pH of alkaline soils (pH > 7). In the latter case, microorganisms absorb the sulfur and subsequently release sulfate and hydrogen ions that acidify the soil.

Fertilizers approved for organic agricultural practices originate from natural rock deposits such as sodium nitrate and rock phosphate (phosphorite) or residues from plants or animals. Natural rock deposits are chemically inorganic, but they are considered acceptable for use in organic agriculture. Plant and animal residues contain many nutrient elements in the form of organic compounds. Before crop plants can acquire the nutrient elements from these residues, complex organic compounds must be broken down to simple organic compounds such as amino acids or inorganic compounds such as ammonium and nitrate, usually by the action of soil microorganisms through a process called **mineralization**. Mineralization depends on many factors, including temperature, water and oxygen availability, pH, and the type and number of microorganisms present in the soil. As a consequence, rates of mineralization are highly variable, and nutrients from organic residues become available to plants over periods that range from days to months to years. Mineralization rates from organic residues are often low and limiting to plant growth. However, they may exceed crop demand at certain times—for example, before seedling emergence and early in the growing season as soils warm, increasing microbial activity. Asynchrony between mineral nutrient supply and plant demand is a major cause of nutrient loss to the surrounding environment in organic and traditional farming systems. Nonetheless, residues from **organic fertilizers** do improve the physical structure of most soils, enhancing water retention during drought and increasing drainage

mineralization The process of breaking down organic compounds, usually by soil microorganisms, and thereby releasing mineral nutrients in forms that can be assimilated by plants.

organic fertilizer A common usage describing fertilizer that contains nutrient elements derived from natural sources without any synthetic additions.

foliar application The application, and subsequent absorption, of mineral nutrients or other chemical compounds to leaves as sprays.

symbiosis The close association of two organisms in a relationship that may or may not be mutually beneficial. Often applied to beneficial (mutualistic) relationships.

in wet weather. In some developing countries, organic fertilizers are all that is available or affordable.

Some mineral nutrients can be absorbed by leaves

In addition to absorbing nutrients through their roots, most plants can absorb mineral nutrients applied to their leaves as a spray, a process known as **foliar application**. In some cases this method has agronomic advantages over the application of nutrients to the soil. Foliar application can reduce the lag time between application and uptake by the plant, which could be important during a phase of rapid growth. It can also circumvent the problem of restricted uptake of a nutrient from the soil. For example, foliar application of mineral nutrients such as iron, manganese, and copper may be more efficient than application through the soil, where these ions are bound to soil particles and hence are less available to the root system.

Nutrient uptake by leaves is most effective when the nutrient solution is applied to the leaf as a thin film. Production of a thin film often requires that the nutrient solutions be supplemented with surfactant chemicals, such as the detergent Tween 80® or organosilicon surfactants, that reduce surface tension. Nutrient movement into the plant seems to involve diffusion through the cuticle and uptake by leaf cells, although uptake through the stomatal pores may also occur.

For foliar nutrient application to be successful, damage to the leaves must be minimized. If foliar sprays are applied on a hot day, when evaporation is high, salts may accumulate on the leaf surface and cause burning or scorching. Spraying on cool days or in the evening helps alleviate this problem. Foliar application has proved economically successful mainly with tree crops and vines such as grapes, but it is also used with cereals. Nutrients applied to the leaves can save an orchard or vineyard when soil-applied nutrients would be too slow to correct a deficiency. In wheat (*Triticum aestivum*), nitrogen applied to the leaves during the later stages of growth enhances the protein content of the seeds.

4.3 Soil, Roots, and Microbes

Describe the interactions of the plant roots, soil, and soil microbes in plant nutrient acquisition.

Soil is complex physically, chemically, and biologically. It is a heterogeneous mixture of substances distributed in solid, liquid, and gas phases (Chapter 3). All of these phases interact with mineral elements. The inorganic particles of the solid phase provide a reservoir of potassium, phosphorus, calcium, magnesium, and iron. Also associated with this solid phase are organic compounds containing nitrogen, phosphorus, and sulfur, among other elements. The liquid phase of soil constitutes the soil solution that is held in pores between the soil particles. It contains dissolved mineral ions and organic compounds and serves as the medium for solute movement to the root surface. Gases such as oxygen, carbon dioxide, and nitrogen are dissolved in the soil solution, but roots exchange gases with soils predominantly through the air-filled pores between soil particles.

From a biological perspective, soil constitutes a diverse ecosystem in which plant roots and microorganisms interact. Many microorganisms play key roles in releasing (mineralizing) nutrients from organic sources, some of which are then made directly available to plants. Microbes also compete with the plants for mineral nutrients. Some specialized microorganisms, including mycorrhizal fungi and nitrogen-fixing bacteria, can form alliances with plants for their mutual benefit (**symbioses**, singular *symbiosis*). In this section we discuss the importance of soil properties, root structure, and symbiotic mycorrhizal relationships to plant mineral nutrition. Chapter 5 will address the symbiotic relationships of plants with nitrogen-fixing bacteria.

permanent charge Ionic charge of soil minerals caused by substitution of ions (for example, Al^{3+} for Si^{4+}) in the crystal structure during mineral formation.

Negatively charged soil particles affect the adsorption of mineral nutrients

Soils are composed of inorganic and organic particles, and both have predominantly negative charges on their surfaces. Many inorganic soil particles form crystal lattices, tetrahedral arrangements of the cationic forms of aluminum (Al^{3+}) and silicon (Si^{4+}) bound to oxygen atoms, thus forming aluminates and silicates. When cations of lesser charge (e.g., Mg^{2+}) replace Al^{3+} and Si^{4+} in the crystal lattice, these inorganic soil particles become negatively charged. This process is called isomorphous substitution. The negative charge of inorganic soil particles is called **permanent charge** and is largely pH independent.

Organic soil particles originate from dead plants, animals, and microorganisms that soil microorganisms have decomposed to various degrees. The negative surface charges of organic particles result from the dissociation of hydrogen ions from the carboxylic and hydroxyl groups present in this component of the soil. The negative charges on organic particles are predominately pH dependent, in contrast to inorganic soil particles.

Soils are categorized by particle size:

- Gravel consists of particles larger than 2 mm.
- Coarse sand consists of particles between 0.2 and 2 mm.
- Fine sand consists of particles between 0.02 and 0.2 mm.
- Silt consists of particles between 0.002 and 0.02 mm.
- Clay consists of particles smaller than 0.002 mm (2 μm).

The silicate-containing clay materials are further divided into three major groups—kaolinite, illite, and montmorillonite—based on differences in their structure and physical properties (**Table 4.5**). The kaolinite group is generally found in well-weathered soils; the montmorillonite and illite groups are found in less-weathered soils.

Mineral cations such as ammonium (NH_4^+) and potassium (K^+) are adsorbed to the negative surface charges of inorganic and organic soil particles or adsorbed within the lattices formed by soil particles. This cation adsorption is an important factor in soil fertility. Mineral cations adsorbed on the surface of soil particles are not readily leached when the soil is infiltrated by water and thus provide a nutrient reserve available to plant roots. Mineral nutrients adsorbed in this way can be replaced by other cations in a process known as

Table 4.5 Comparison of Properties of Three Major Types of Silicate Clays Found in the Soil

	Type of clay		
Property	Montmorillonite	Illite	Kaolinite
Size (μm)	0.01–1.0	0.1–2.0	0.1–5.0
Shape	Irregular flakes	Irregular flakes	Hexagonal crystals
Cohesion	High	Medium	Low
Water-swelling capacity	High	Medium	Low
Cation exchange capacity (milliequivalents 100 g^{-1})	80–100	15–40	3–15

Source: After N. C. Brady. 1974. *The Nature and Properties of Soils*, 8th ed. Macmillan, New York.

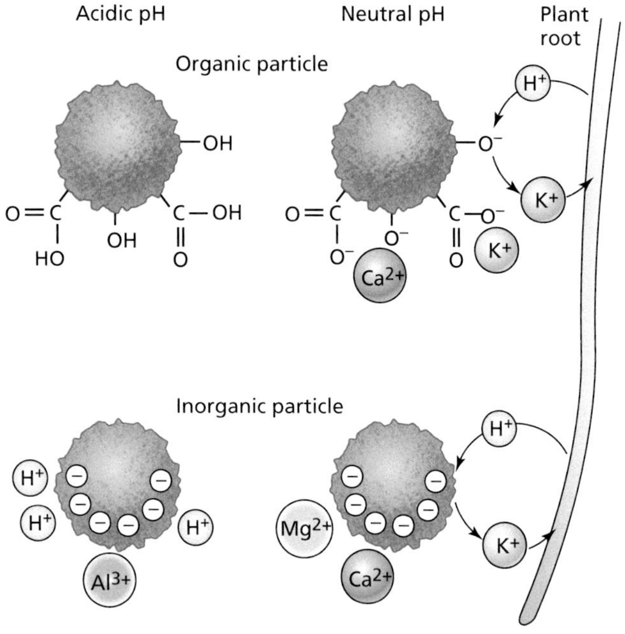

Figure 4.4 Cation exchange on the surface of soil particles. At acidic pH (left side), the cation exchange capacity (CEC) of organic soil particles is lower because organic acid sites (COOH, OH) are protonated. The CEC of inorganic particles is mainly pH independent and at lower pH tends to be occupied by H^+ and Al^{3+}. At neutral pH (right side), the CEC of organic particles is higher. Negatively charged sites on both organic and inorganic particles tend to be occupied by cationic nutrients such as Ca^{2+}, Mg^{2+}, and K^+. As roots grow, they release H^+ that can exchange on soil particles and release cationic nutrients for uptake by the root.

cation exchange (**Figure 4.4**). The degree to which a soil can adsorb and exchange ions is termed its *cation exchange capacity* (CEC) and is highly dependent on the soil type. A soil with a higher CEC generally has a larger reserve of mineral nutrients. CEC is due primarily to negative charges on clay and organic matter particles. Organic matter has a much higher CEC by weight compared with clay. Therefore, the CEC of a soil depends on its organic matter content. The CEC of inorganic particles, such as clay, is largely pH independent. However, the CEC of organic matter is highly dependent on soil pH: At low pH, organic acid groups are protonated and uncharged, whereas at higher pH they are deprotonated and carry a negative charge. As roots grow, they release protons into the soil, acidifying it. These protons can exchange cationic nutrients from soil particles, making them available for uptake.

Mineral anions such as nitrate (NO_3^-) and chloride (Cl^-) tend to be repelled by the negative charge on the surface of soil particles and remain dissolved in the soil solution. Thus, the anion exchange capacity of most agricultural soils is small compared with their CEC. Nitrate in particular remains mobile in the soil solution, where it is susceptible to leaching by water moving through the soil.

Phosphate solubility in soil is usually controlled by its association with Ca^{2+}, Fe^{3+}, and Al^{3+} to form insoluble inorganic compounds. As a result, phosphate is more soluble in soil solution at slightly acidic pH (Figure 4.3). However, the availability of phosphate is also controlled by its association with soil particles. Phosphate ions ($H_2PO_4^-$ and HPO_4^{2-}) bind to the edges of soil particles containing aluminum or iron because the positively charged iron and aluminum ions (Fe^{2+}, Fe^{3+}, and Al^{3+}) are associated with hydroxyl ion (OH^-) groups that become protonated and can bind phosphate. The lack of phosphate mobility or its low solubility can both limit its availability for uptake by plants. Formation of mycorrhizal symbioses helps overcome this lack of mobility. Additionally, the roots of some plants, such as those of lupine (*Lupinus albus*) and members of the Proteaceae (e.g., *Macadamia, Banskia, Protea, Hakea*), secrete large amounts of organic acids into the soil that release phosphate from iron, aluminum, and calcium phosphates.

Sulfate (SO_4^{2-}) in the presence of Ca^{2+} forms gypsum ($CaSO_4$). Gypsum is only slightly soluble, but it releases sufficient sulfate to support plant growth. Most nonacidic soils contain substantial amounts of Ca^{2+}; consequently, sulfate mobility in these soils is low, and sulfate is not highly susceptible to leaching.

Soil pH affects nutrient availability, soil microbes, and root growth

Hydrogen ion concentration (pH) is an important property of soils because it affects the growth of plant roots and soil microorganisms. Root growth is generally favored in slightly acidic soils, at pH values between 5.5 and 6.5. Fungi generally predominate in acidic soils (pH < 7); bacteria become more prevalent in alkaline soils (pH > 7). Soil pH determines the availability of soil nutrients (Figure 4.3). Acidity promotes the weathering of rocks that releases K^+, Mg^{2+}, Ca^{2+}, and Mn^{2+} and increases the solubility of carbonates, sulfates, and some phosphates. Increasing the solubility of nutrients enhances their availability to roots as concentrations increase in the soil solution.

cation exchange The replacement of mineral cations adsorbed to the surface of soil particles by other cations.

Major factors that lower the soil pH are the decomposition of organic matter, ammonium assimilation by plants and microbes, and the amount of rainfall. Carbon dioxide is produced as a result of the decomposition of organic matter and equilibrates with soil water in the following reaction:

$$CO_2 + H_2O \leftrightarrow H_2CO_3 \leftrightarrow H^+ + HCO_3^-$$

This releases hydrogen ions (H^+), lowering the pH of the soil. Microbial decomposition of organic matter also produces ammonia/ammonium (NH_3/NH_4^+) and hydrogen sulfide (H_2S) that can be oxidized in the soil to form the strong acids nitric acid (HNO_3) and sulfuric acid (H_2SO_4), respectively. As plant roots absorb ammonium ions from the soil and assimilate them into amino acids, the roots generate hydrogen ions that they secrete into the surrounding soil (Chapter 5). Hydrogen ions also displace K^+, Mg^{2+}, Ca^{2+}, and Mn^{2+} from the surfaces of soil particles. Leaching may then remove these ions from the upper soil layers, leaving a more acidic soil. By contrast, the weathering of rock in arid regions releases K^+, Mg^{2+}, Ca^{2+}, and Mn^{2+} into the soil, but because of the low rainfall, these ions do not leach from the upper soil layers, and the soil remains alkaline.

Excess mineral ions in the soil limit plant growth

When excess sodium salts are present in soil, the soil is said to be *saline*, and such soils may inhibit plant growth if the mineral ions reach concentrations that limit water availability or exceed the adequate levels for a particular nutrient. Sodium chloride and sodium sulfate are the most common salts in saline soils. Excess mineral ions in soils can be a major problem in arid and semiarid regions because rainfall is insufficient to leach them from the soil layers near the surface.

Irrigated agriculture fosters soil salinization if the amount of water applied is insufficient to leach the salt below the root zone. Irrigation water can contain 100 to 1,000 g of mineral ions per cubic meter. An average crop requires about 10,000 m^3 of water per hectare. Consequently, 1,000 to 10,000 kg of mineral ions per hectare may be added to the soil per crop, and over several growing seasons, high concentrations of mineral ions may accumulate in the soil.

In saline soils, plants encounter **salt stress**. Salt stress involves both water-deficit stress, as plant cells find it harder to extract water from the soil, and ionic stress, as ions (primarily Na^+ and Cl^-) accumulate to toxic levels within cells. Whereas many plants are affected adversely by the presence of relatively low concentrations of salt, other plants can survive (**salt-tolerant plants**) or even thrive (**halophytes**) at high salt concentrations. The mechanisms by which plants tolerate high salinity are complex (Chapter 19), involving expression of enzymes, biochemical synthesis of hydrophilic osmolytes that promote water uptake into cells, and membrane transport. Plants deal with excess mineral ions in different ways, either excluding them at the plasma membrane, taking them up and excreting them, or sequestering them. To prevent toxic buildup of mineral ions in the cytosol where crucial metabolism occurs, many plants sequester them in the vacuole. Efforts are underway to bestow salt tolerance on salt-sensitive crop species using both classic plant breeding and biotechnology.

Other mineral ions, including metal ions, can also accumulate to excessive amounts in soil and can cause severe toxicity in plants as well as humans. These metals include zinc, copper, cobalt, nickel, mercury, lead, cadmium, silver, and chromium.

Some plants develop extensive root systems

The ability of plants to obtain both water and mineral nutrients from the soil is related to their ability to develop an extensive root system and to various other traits such as flexible development in response to nutrient supply and demand, ability to secrete organic anions, and capacity to form mycorrhizal symbioses.

salt stress The adverse effects of excess minerals on plants.

salt-tolerant plants Plants that can survive or even thrive in high-salt soils. *See also* halophytes.

halophytes Plants that are native to saline soils and complete their life cycles in that environment. Contrast with *glycophytes*.

rhizosphere The immediate microenvironment surrounding the root.

primary root In monocots, a root generated directly by growth of the embryonic root or radicle.

nodal roots In monocots, adventitious roots that form after the emergence of primary roots.

taproot In eudicots, the main single root axis from which lateral roots develop.

In the late 1930s, H. J. Dittmer examined the root system of a single winter rye plant after 16 weeks of growth. He estimated that the plant had 13 million primary and lateral root axes, extending more than 500 km in length and providing 200 m^2 of surface area. This plant also had more than 10^{10} root hairs, providing another 300 m^2 of surface area. The total surface area of roots from a single rye plant equaled that of a professional basketball court. Other plant species may not develop such extensive root systems, which may limit their nutrient absorption capacity and increase their reliance on mycorrhizal symbioses.

In the desert, the roots of mesquite (genus *Prosopis*) may extend downward more than 50 m to reach groundwater. Annual crop plants have roots that usually grow between 0.1 and 2.0 m in depth and extend laterally to distances of 0.3 to 1.0 m. In orchards, the major root systems of trees planted 1 m apart reach a total length of 12 to 18 km per tree. The annual production of roots in natural ecosystems may easily surpass that of shoots, so in many respects the aboveground portions of a plant represent only "the tip of the iceberg." Nonetheless, making observations on root systems is difficult and usually requires special techniques.

Plant roots may grow continuously throughout the year if conditions are favorable. Their proliferation, however, depends on the availability of water and minerals in the immediate microenvironment surrounding the root, called the **rhizosphere**. If the rhizosphere is poor in nutrients or too dry, root growth is slow. As rhizosphere conditions improve, root growth increases. If fertilization and irrigation provide abundant nutrients and water, plants invest fewer resources into root growth and development and more into shoots, to maximize photosynthesis and growth, which decreases the root-to-shoot ratio. Under such luxurious conditions, which are rare in nature, a relatively small root system meets the nutrient needs of the whole plant. Fertilization and irrigation cause greater allocation of resources to the shoot and reproductive structures than to roots, and this shift in allocation pattern contributes to higher yield of aboveground parts.

Root systems differ in form but are based on common structures

The *form* of the root system differs greatly among plant species. In monocots, root development starts with the emergence of three to six **primary** (or *seminal*) **root** axes from the germinating seed. With further growth, the plant extends new adventitious roots, called **nodal roots** or *brace roots*. Over time, the primary and nodal root axes grow and branch extensively to form a complex *fibrous root system* (**Figure 4.5**). In fibrous root systems, all the roots generally have similar diameters (except where environmental conditions or pathogenic interactions modify the root structure), so it is impossible to distinguish a main root axis.

In contrast to monocots, eudicots develop root systems with a main single root axis, called a **taproot**, which may thicken as a result of secondary cambial activity (Chapter 15). From this main root axis, *lateral roots* develop to form an extensively branched root system. Nutrient availability also influences root development. Some plants, especially in the family Proteaceae, produce a proliferation of lateral roots termed *cluster roots* or *proteoid roots* in response to phosphate deficiency. An example of proteoid roots of *Hakea prostrata*

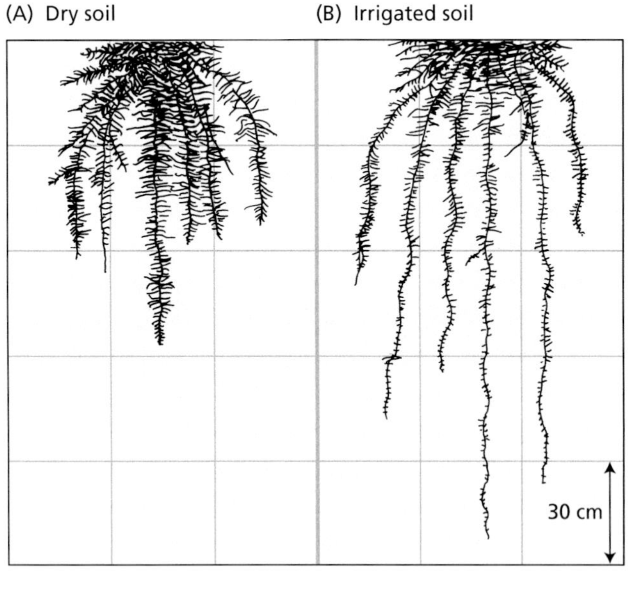

Figure 4.5 Fibrous root systems of wheat (a monocot). (A) The root system of a mature (3-month-old) wheat plant growing in dry soil. (B) The root system of a mature wheat plant growing in irrigated soil. The morphology of the root system is affected by the amount of water present in the soil. In a mature fibrous root system, the primary root axes are indistinguishable.

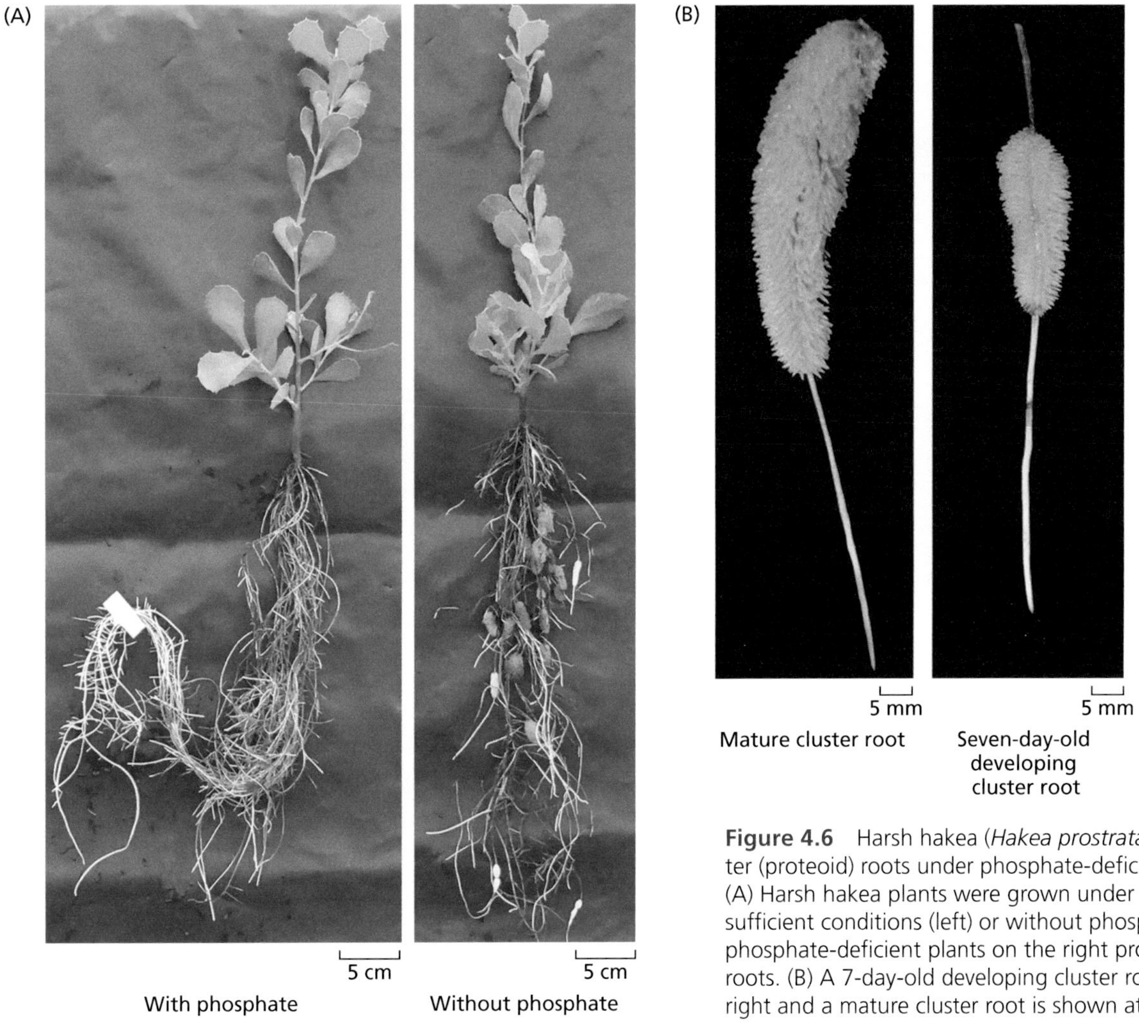

Figure 4.6 Harsh hakea (*Hakea prostrata*) produces cluster (proteoid) roots under phosphate-deficiency conditions. (A) Harsh hakea plants were grown under phosphate-sufficient conditions (left) or without phosphate (right). The phosphate-deficient plants on the right produced cluster roots. (B) A 7-day-old developing cluster root is shown at right and a mature cluster root is shown at left.

is shown in **Figure 4.6**. Proteoid roots secrete organic acids such as citrate or malate that lower the pH of the soil surrounding the root and solubilize calcium, aluminum, and iron phosphate, making phosphate available for uptake by the plant. The development of the root system in both monocots and eudicots depends on the activity of the root apical meristem and the production of lateral root meristems (Chapter 15). **Figure 4.7** is a generalized diagram of the apical region of a plant root and identifies three zones of activity: the meristematic, elongation, and maturation zones.

In the **meristematic zone** (Figure 4.7), cells divide both in the direction of the root base to form cells that will differentiate into the tissues of the functional root and in the direction of the root apex to form the **root cap**. The root cap protects the delicate meristematic cells as the root expands into the soil. It commonly secretes a gelatinous material containing polysaccharides and pectins called *mucigel*, which surrounds the root tip. The mucigel provides lubrication that eases the root's penetration of the soil, protects the root apex from desiccation, promotes the transfer of nutrients to the root, and affects interactions between the root and soil microorganisms. The root cap is central to the perception of gravity, the signal that directs the growth of roots downward. This process is termed the **gravitropic response** (Chapters 12 and 14).

Cell division in the root apex proper is relatively slow; this region is called the **quiescent center**. After a few generations of slow cell divisions, root cells

meristematic zone The region at the tip of the root containing the meristem that generates the body of the root. Located between the root cap and the elongation zone.

root cap Cells at the root apex that cover and protect the meristematic cells from mechanical injury as the root moves through the soil. Site for the perception of gravity and signaling for the gravitropic response in roots.

gravitropic response The growth initiated by the root cap's perception of gravity and the signal that directs the roots to grow downward.

quiescent center The central region of the root meristem where cells divide more slowly than surrounding cells, or do not divide at all. Serves as a reservoir of meristematic cells for tissue regeneration in case of wounding.

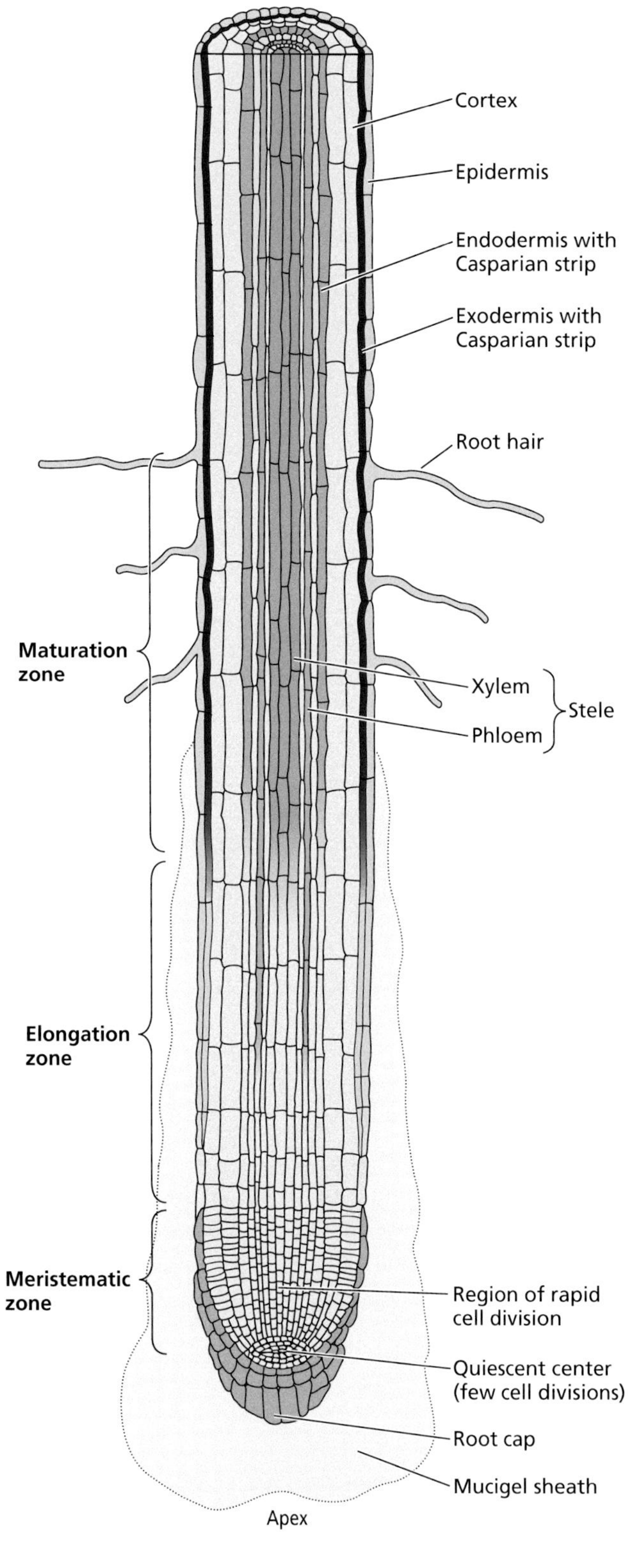

Figure 4.7 Diagrammatic longitudinal section of the apical region of the root. The meristematic cells are located near the tip of the root. These cells generate the root cap and the upper tissues of the root. In the elongation zone, cells differentiate to produce xylem, phloem, and cortex. Root hairs, formed in epidermal cells, first appear in the maturation zone.

displaced from the apex by about 0.1 mm begin to divide more rapidly. Cell division again tapers off at about 0.4 mm from the apex, and the cells expand equally in all directions.

The **elongation zone** begins approximately 0.7 to 1.5 mm from the apex (Figure 4.7). In this zone cells elongate rapidly. A final round of divisions of the innermost layer of cortex produces a central ring of cells called the **endodermis**. The radial walls of endodermal cells become thickened, and lignin is deposited there to form the Casparian strip. Formation of suberin lamellae along this strip results in a hydrophobic structure that prevents apoplastic movement of water or solutes across the root (Chapter 3). In addition to having an endodermis, 90% of angiosperm species also have an **exodermis**, the outermost layer of the cortex in roots. The exodermis is also suberized and is a barrier to apoplastic movement of water and solutes. The exodermis is absent in gymnosperms.

The area of the root interior to the endodermis is called the **stele** (Figure 4.7). The stele contains the vascular elements of the root: the **phloem**, which transports metabolites from the shoot to the root and to fruits and seeds, and the **xylem**, which transports water and solutes to the shoot (Chapters 1 and 3). Phloem develops more rapidly than xylem, attesting to the fact that phloem function is critical near the root apex. Large quantities of carbohydrates must flow through the phloem to the growing apical zones in order to support cell division and elongation. Carbohydrates provide rapidly growing cells with an energy source and with the carbon skeletons required to synthesize organic compounds. Six-carbon sugars (hexoses) also function as osmotically active solutes in the root. At the root apex, where the phloem is not yet developed, carbohydrate movement depends on symplasmic transport and is relatively slow.

Root hairs, with their large surface area for absorption of water and solutes and for anchoring the root to the soil, first appear in the **maturation zone** (Figure 4.7), and here the xylem develops the capacity to translocate substantial quantities of water and solutes to the shoot.

Different areas of the root absorb mineral ions differently

The precise point of entry of minerals into the root system has been a topic of considerable interest. Some researchers have claimed that nutrients are absorbed only at the apical regions of the root axes or branches; others claim that nutrients are absorbed over the entire root surface.

The Casparian strip of the endodermis and exodermis in angiosperms plays a major role in preventing nonselective entry of solutes into the xylem where they would be translocated to the rest of the plant. The mechanism of selective solute uptake that uses both transmembrane transporters and the Casparian strip is discussed in Chapters 3 and 6. But the Casparian strip does not extend to the root apex; the endodermis and exodermis mature, and each produces a Casparian strip in the late elongation zone or early maturation zone (Figure 4.7). This allows for a less-selective solute uptake mechanism at the root tip, and some solutes can be taken up predominantly at the root apex. For example, root absorption of calcium ions in barley (*Hordeum vulgare*) appears to be restricted to the apical region. The extent that particular nutrients (and other solutes) are taken up at the apex compared with the rest of the root depends on many factors, among them the nutrient's concentration in the soil solution, the demand for the nutrient by the plant, and the expression of uptake transporters in different parts of the root, especially in root hairs. Iron may be taken up either at the apical region, as in barley and other species, or over the entire root surface, as in maize. Potassium ions, nitrate, ammonium, and phosphate can be absorbed by all locations of the root surface, but in maize the elongation zone has the maximum rates of potassium ion accumulation and nitrate absorption. In maize and rice and in wetland species, the root apex absorbs ammonium more rapidly than the elongation zone does. Ammonium and nitrate uptake by conifer roots varies significantly across different regions of the root and may be influenced by rates of root growth and maturation. In several species, the root apex and root hairs are the most active in phosphate absorption. Hyphae of arbuscular mycorrhizal fungi also play a significant role in uptake of phosphate and other nutrients, and the development of this symbiosis can change the regions of the root involved in uptake.

The high rates of nutrient absorption in the apical root zones also result from the strong demand for nutrients in these tissues and the relatively high nutrient availability in the soil surrounding them. For example, cell elongation depends on the accumulation of solutes such as potassium ions, chloride, and nitrate to increase the osmotic pressure within the cell (Chapter 2). Ammonium is the preferred nitrogen source to support cell division in the meristem because meristematic tissues are often carbohydrate-limited and because assimilation of ammonium into organic nitrogen compounds consumes less energy than assimilation of nitrate (Chapter 5). The root apex and root hairs grow into fresh soil, where nutrients have not yet been depleted.

Within the soil, nutrients can move to the root surface both by mass flow and by diffusion. In mass flow, nutrients are carried by water moving through the soil toward the root. The amounts of nutrients provided to the root by mass flow depend on the nutrient concentrations in the soil solution and the rate of water flow through the soil toward the plant, which itself depends on transpiration rates. When both the rate of water flow and the concentrations of nutrients in the soil solution are high, mass flow can play an important role in nutrient supply. Hence, highly soluble nutrients such as nitrate are largely carried by mass flow, but this process is less important for nutrients with low solubility, such as phosphate and zinc ions.

In diffusion, mineral nutrients move from a region of higher concentration to a region of lower concentration. Nutrient uptake by roots lowers the concentrations of nutrients at the root surface, generating concentration gradients in the soil solution surrounding the root. Diffusion, or more precisely net flux, of nutrients down their concentration gradients, along with mass flow resulting from transpiration, can increase nutrient availability at the root surface.

When the rate of absorption of a nutrient by roots is high and the nutrient concentration in the soil solution is low, mass flow can supply only a small fraction of the total nutrient requirement. Under these conditions, plant nutrient absorption becomes independent of plant transpiration rates, and diffusion rates limit the movement of the nutrient to the root surface. When diffusion is too slow to maintain high nutrient concentrations near the root, a **nutrient depletion zone** forms adjacent to the root surface. This zone may extend from about 0.2 to 2.0 mm

elongation zone The region of rapid and extensive root cell elongation showing few, if any, cell divisions.

endodermis A specialized layer of cells surrounding the vascular tissue in roots and some stems.

Casparian strip A band in the cell walls of the endodermis that is impregnated with lignin. Prevents apoplastic movement of water and solutes into the stele.

exodermis A specialized layer of cells in most angiosperms, but not gymnosperms, containing a Casparian strip, in the outer layer of the cortex of the root.

stele In the root, the tissues located interior to the endodermis. The stele contains the vascular elements of the root: the phloem and the xylem.

phloem The tissue that transports the products of photosynthesis from mature leaves (or storage organs) to areas of growth and storage, including the roots.

xylem The vascular tissue that transports water and ions from the root to the other parts of the plant.

maturation zone The region of the root where differentiation occurs, including the production of root hairs and functional vascular tissue.

nutrient depletion zone The region surrounding the root surface showing diminished nutrient concentrations due to uptake into the roots and slow replacement by diffusion.

from the root surface, depending on the mobility of the nutrient in the soil. The nutrient depletion zone is particularly important for phosphate.

The formation of a depletion zone tells us something important about mineral nutrition. Because roots deplete the mineral supply in the rhizosphere, their effectiveness in mining minerals from the soil is determined not only by the rate at which they can remove nutrients from the soil solution but also by their continuous growth into undepleted soil. Without continuous growth, roots would rapidly deplete the soil adjacent to their surfaces. Optimal nutrient acquisition therefore depends both on the root system's capacity for nutrient uptake and on its ability to grow into fresh soil. The ability of the plant to form a mycorrhizal symbiosis is also critical in overcoming the effects of depletion zones, because the hyphae of the fungal symbionts grow beyond the depletion zone. These fungal structures take up nutrients far from the root (up to 25 cm in the case of arbuscular mycorrhizas) and translocate them rapidly to the roots, overcoming the slow diffusion in soil.

Nutrient availability influences root growth and development

Plants, which have limited mobility for most of their lives, must deal with changes in their local environment because they cannot move away from unfavorable conditions. Above the ground, light intensity, temperature, and humidity may fluctuate substantially during the day and across the canopy, but CO_2 and O_2 concentrations remain relatively uniform. In contrast, soil buffers the roots from temperature extremes, but the belowground concentrations of CO_2 and O_2, water, and nutrients are extremely heterogeneous, both spatially and temporally. For example, inorganic nitrogen concentrations in soil may range a thousandfold over a distance of centimeters or the course of hours. Given such heterogeneity, plants seek the most favorable conditions within their reach.

Roots sense the belowground environment and respond—through gravitropism, thigmotropism, chemotropism, and hydrotropism—by growing toward soil resources. Roots are able to sense nutrient availability and grow into areas of the soil with more favorable concentrations (**Figure 4.8**). Roots are also able to alter their metabolism and development to increase nutrient uptake under deficiency conditions.

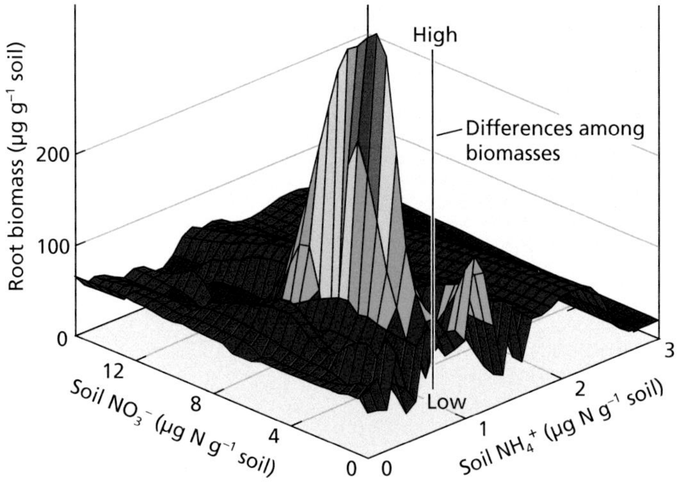

Figure 4.8 Root biomass as a function of extractable soil NH_4^+ and NO_3^-. The root biomass is shown (μg root dry weight g^{-1} soil) plotted against extractable soil NH_4^+ and NO_3^- (μg extractable N g^{-1} soil) for tomato (*Solanum lycopersicum* cv. T-5) growing in an irrigated field that had been fallow the previous 2 years. The colors emphasize the differences among biomasses, ranging from low (purple) to high (red).

Phosphate is generally trapped in surface soil horizons (layers), either tightly bound to iron and aluminum oxides or held as biological phosphorus in microorganisms. Phosphorus deficiency can trigger topsoil "foraging" or "mining" by plant roots. Some bean genotypes, for example, respond to phosphorus deficiency by producing more adventitious lateral roots, decreasing the growth angle of these roots (relative to the shoot) so they are more shallow, increasing the number of lateral roots emerging from the tap root, and increasing root hair density and length (**Figure 4.9**). These changes in root system architecture place more of the roots in the topsoil, where the majority of the phosphate resides. In some plants, these changes are manifest as cluster roots (Figure 4.6). Under continuous phosphorus deficiency, systemic signals, including miRNAs (Chapter 12) produced in the shoot, are transduced in the phloem to the roots, where they induce alterations in root architecture via auxin and other hormone signaling mechanisms (Chapter 15).

Nitrogen deficiency in the roots also results in both local and systemic responses. Under nitrogen

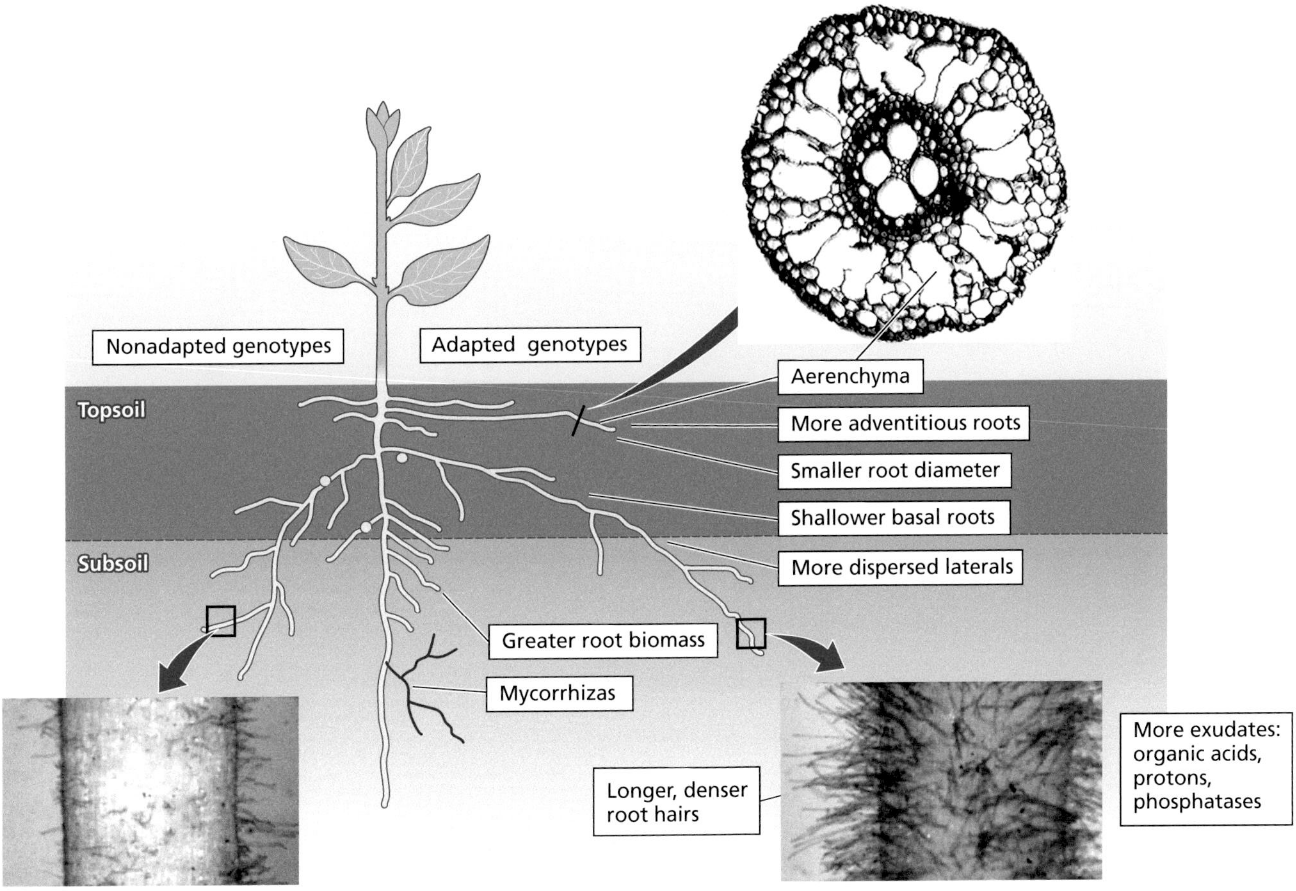

Figure 4.9 Topsoil foraging/mining of phosphorus by phosphorus-efficient bean genotypes. In response to low phosphate availability, some bean varieties, as shown on the right side, display modified root architecture with increased root density in the soil surface layer, increased root hair density, modified metabolism that results in secretion of protons and organic acids, and the ability to form mycorrhizal interactions. These adaptations all increase phosphate absorption by the root.

deficiency, the NRT1.1 nitrate transporter (Chapters 5 and 6) directly changes root auxin levels to reduce lateral root growth (Chapter 15). C-terminally Encoded Peptide 1 (CEP1) is also produced and further represses local lateral root growth. In split-root assays where part of the root system is exposed to nitrate while the rest experiences nitrate limitation, lateral root growth and nitrogen uptake are preferentially increased on the nitrate-sufficient side (**Figure 4.10A**), suggesting that systemic signaling is also involved. CEP is also xylem-mobile and functions together with cytokinins (Chapter 12) as a systemic nitrogen demand signal that travels to the shoot and is sensed in the phloem by CEP receptors 1 and 2. Activation of these receptors promotes expression of genes encoding two nonsecreted small proteins of the glutaredoxin family that travel in the phloem to the roots and promote nitrogen acquisition in areas of the root system with higher nitrate availability. **Figure 4.10B** shows a model for the regulation of nitrate acquisition via integrated local and systemic nitrogen-demand signaling. Local responses also modify root growth in response to high soil nutrient concentrations, as a few roots—as little as 4% of the root system in spring wheat and 12% in lettuce—are sufficient to supply all the nutrients required. Plants may thus diminish the allocation of their resources to roots while increasing allocations to the shoot and reproductive structures. This resource shifting is one mechanism through which fertilization stimulates crop yields.

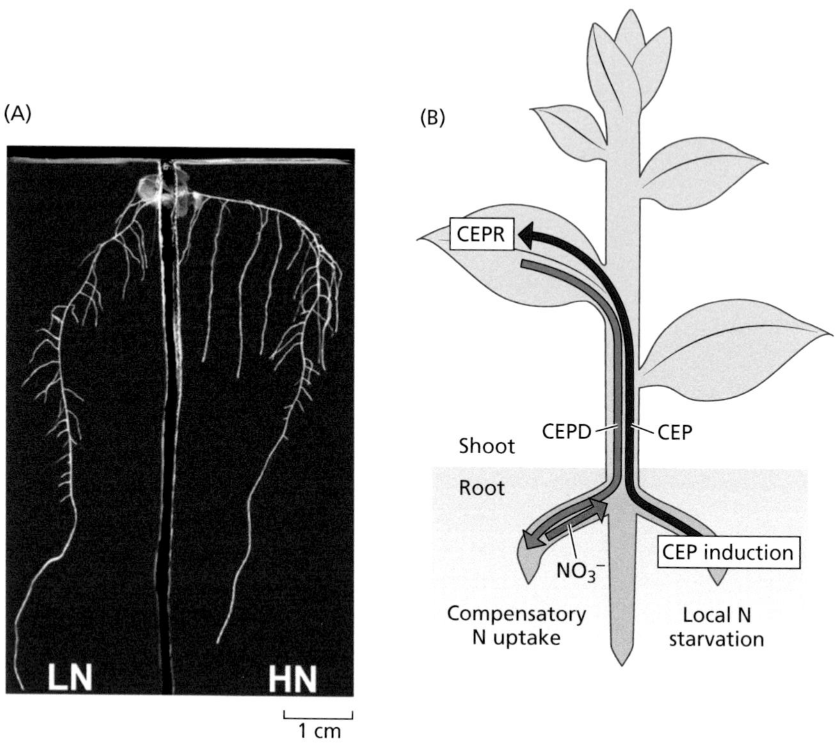

Figure 4.10 Root-to-shoot-to-root peptide signaling influences nitrate uptake in the context of uneven nitrogen supply to the root system. (A) Split-root experiment with half of an Arabidopsis root system growing on low nitrate (LN) and the other half growing on high nitrate (HN). After 5 days of growth in this setup, the high-nitrate side of the root system has longer lateral roots. (B) Model of how C-terminally Encoded Peptide 1 (CEP) signaling from low-nitrate roots stimulates nitrate uptake in high-nitrate roots via integration of nitrogen demand signals in the shoot. CEPD, CEP downstream; CEPR, CEP Receptor.

Mycorrhizal symbioses facilitate nutrient uptake by roots

Our discussion so far has centered on the direct acquisition of mineral elements and compounds by roots, but this process is usually modified by the association of **mycorrhizal fungi** with the root system to form a **mycorrhiza** (from the Greek words for "fungus" and "root"; plural *mycorrhizas* or *mycorrhizae*). The host plant supplies associated mycorrhizal fungi with carbohydrates and in return receives nutrients from the fungi. There are indications that drought (water-deficit stress) and disease tolerance may also be improved in the host plant.

Mycorrhizal symbioses of two main types—**arbuscular mycorrhizae** and **ectomycorrhizae**—are widespread in nature, occurring in about 90% of terrestrial plant species, including most major crops. The majority, perhaps 80%, are arbuscular mycorrhizas, which are symbioses between a phylum of fungi called the Glomeromycota and a broad range of angiosperms, gymnosperms, ferns, and liverworts. Their importance for herbaceous species and fruit trees makes arbuscular mycorrhizas vital to agricultural production, particularly in nutrient-poor soils. This is the most ancient type of mycorrhiza, occurring in fossils of the earliest land plants. This symbiosis preceded the evolution of true roots and therefore was probably crucial in facilitating plant establishment on land more than 450 million years ago.

By contrast, ectomycorrhizal symbioses evolved more recently. They are formed by far fewer plant species, notably trees in the families Pinaceae (pines, larches, Douglas fir), Fagaceae (beech, oak, chestnut), Salicaceae (poplar, aspen), Betulaceae (birch), and Myrtaceae (*Eucalyptus*). The fungal partners belong to either the Basidomycota or, less frequently, the Ascomycota. These symbioses play major roles in the nutrition of trees and therefore in the productivity of vast areas of boreal forest.

Some plant species, particularly those in the families Salicaceae (*Salix* [willow] and *Populus* [poplar, aspen]) and Myrtaceae (*Eucalyptus*), can form both arbuscular and ectomycorrhizal symbioses. Other plants are unable to form any kind of

mycorrhizal fungi Fungi that can form mycorrhizal symbioses with plants.

mycorrhiza (plural *mycorrhizae*) The symbiotic (mutualistic) association of certain fungi and plant roots. Facilitates the uptake of mineral nutrients by roots.

arbuscular mycorrhizae Symbioses between fungi in the phylum Glomeromycota and the roots of a broad range of angiosperms, gymnosperms, ferns, and liverworts. The hyphae of arbuscular mycorrhizae penetrate the cortical cells of the root.

ectomycorrhizae Symbioses in which the fungus typically forms a thick sheath, or mantle, around the roots. The root cells themselves are not penetrated by the fungal hyphae, and instead are surrounded by a network of hyphae called the Hartig net.

mycorrhiza. These include members of the families Brassicaceae, such as cabbage (*Brassica oleracea*) and the model plant *Arabidopsis thaliana*; Chenopodiaceae, such as spinach (*Spinacea oleracea*); and Proteaceae, such as macadamia nut (*Macadamia integrifolia*).

Certain agricultural practices may reduce or eliminate mycorrhiza formation in plants that normally form them. These practices include flooding (paddy rice does not form mycorrhizas, whereas upland rice does), extensive soil disturbance caused by plowing, application of high concentrations of fertilizer, and of course soil fumigation and application of some fungicides. Such practices may decrease yields in crops such as maize that are very dependent on mycorrhizae for nutrient uptake. Mycorrhizae (singular: mycorrhiza) also do not form in solution culture or in hydroponic cultivation. Nonetheless, for the majority of plants, mycorrhiza formation is the normal situation and the non-mycorrhizal state is essentially an artifact, brought about by particular agricultural practices.

Mycorrhizae modify the plant root system and influence plant mineral nutrient acquisition, but the way they do so varies between types. Arbuscular mycorrhizal fungi develop a highly branched system (mycelium) of hyphae (fine filamentous structures 2–10 µm in diameter) that extend into the soil beyond the root of their host (**Figure 4.11**). Different arbuscular mycorrhizal fungi vary considerably in their distance and intensity of soil exploration, but phosphate transfer from as far as 25 cm away from the root has been measured. The mycelium in soil also helps stabilize aggregates of soil particles, promoting good soil structure. The hyphae extend into soil well beyond the zone of depletion that develops around a root and thus can absorb an immobile nutrient such as phosphate from beyond the depletion zone. Hyphae also penetrate soil pores that are much narrower than those available to roots.

The root of the arbuscular mycorrhizal host plant looks almost the same as a non-mycorrhizal root, and the presence of the fungi can only be detected by staining and microscopy. Hyphae of arbuscular mycorrhizal fungi, growing from spores in the soil or roots of another plant, penetrate the root epidermis and colonize the root cortex. An exodermis is present only in angiosperms as the outermost cell layer of the cortex. Most angiosperms have an exodermis, and mycorrhizal fungi can gain access to the cortex by passing through the exodermis and invading the cortical cells to form either highly branched structures called **arbuscules** (Arum-type colonization; **Figure 4.12A**) or complex **hyphal coils** (Paris-type colonization; **Figure 4.12B**). The fungi are restricted to the cortex and never penetrate the endodermis or colonize the stele of the root. Arbuscules increase the area of contact between the symbionts and remain surrounded by a plant membrane that is involved in transferring nutrients from fungal to plant cells. The penetration process is genetically controlled by a pathway that nearly 400 million years later was partially coopted for the colonization of legume roots by nitrogen-fixing bacteria (Chapter 5).

Phosphate is delivered by the fungi directly to the root cortex. After export from the fungal arbuscules or coils, this phosphate is taken up by the plant cells. Some of the suite of plant **phosphate transporters** (Chapter 6) are specifically or preferentially expressed only in the plant membrane surrounding the arbuscules and coils in the root cortex and are not expressed in non-mycorrhizal roots. These transporters play a key role in the transfer of phosphate from fungus to plant. Plants regulate the level of fungal colonization in response to their phosphate status, indicating that plants

arbuscules Branched structures formed by mycorrhizal fungi inside cortical cells of the host plant root; the sites of nutrient transfer between the fungus and the plant.

hyphal coils Coiled structures formed by mycorrhizal fungi within the root cortical cells of its host plant; the sites of nutrient transfer between the fungus and the plant. Also called arbuscules.

phosphate transporter Protein in the plasma membrane specific for the uptake of phosphate by the cell.

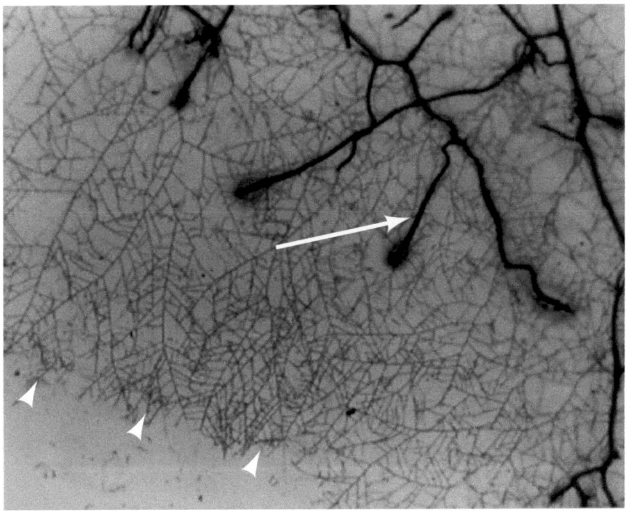

Figure 4.11 Visualization of the intact extraradical mycelium of *Glomus mosseae* spreading from colonized roots of cherry plum (*Prunus cerasifera*). The advancing front of the extraradical mycelium is indicated by arrowheads and the plant roots by an arrow. Note the differences in lengths and diameters of roots and hyphae.

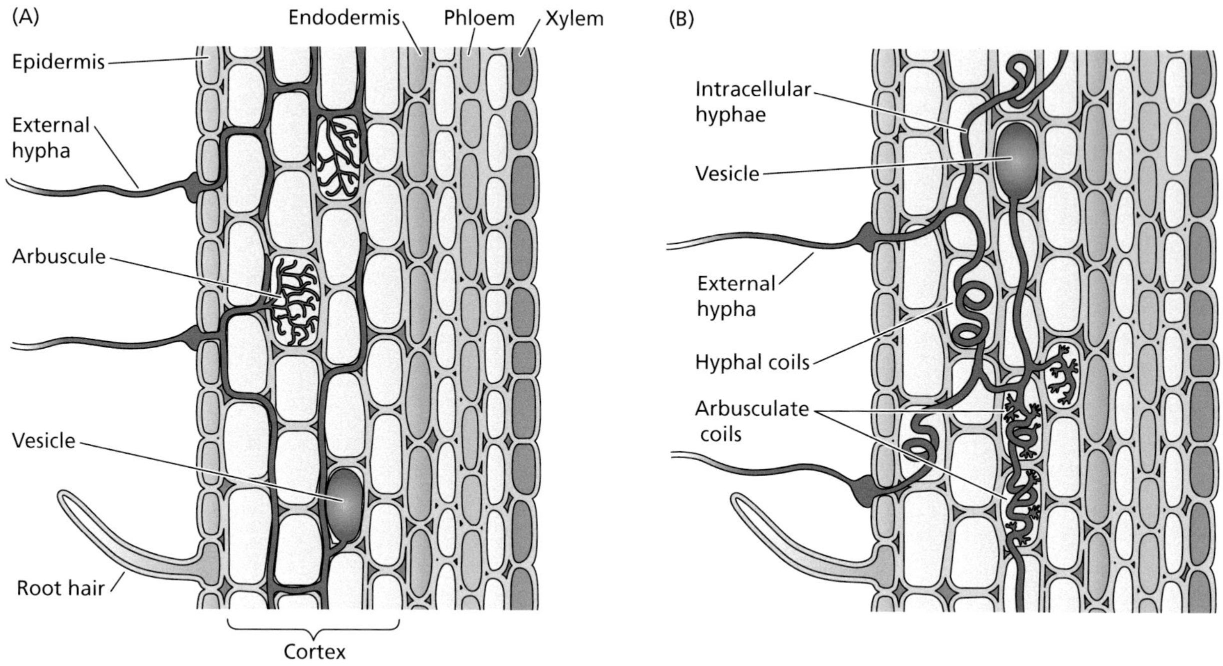

Figure 4.12 Diagrammatic representation of the two major forms of arbuscular mycorrhizal colonization of the root cortex. (A) Arum-type colonization is characterized by the formation of intracellular, highly branched arbuscules in root cortical cells. (B) Paris-type colonization is characterized by the formation of intracellular hyphal coils in root cortical cells, some of which (called arbusculate coils) bear small arbuscule-like branches.

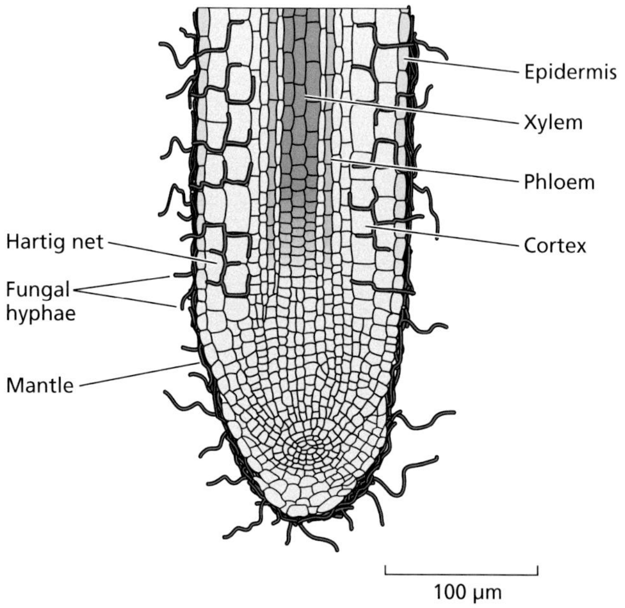

Figure 4.13 Diagrammatic representation of a longitudinal section of an ectomycorrhizal root. Fungal hyphae form a dense fungal mantle over the surface of the root and penetrate between epidermal cells, or epidermal and cortical cells, to form the Hartig net. Hyphae also grow extensively in soil, forming dense mycelium and/or mycelial strands.

control the amount of fixed carbon delivered to the fungus based on their phosphate requirement. We will need to know more about how the symbiotic partners interact to influence nutrient acquisition in order to harness arbuscular mycorrhizal symbiosis to optimize crop nutrition. In addition to their importance for phosphate uptake, arbuscular mycorrhizal fungi are known to be important in the uptake of nitrogen and zinc.

Roots colonized by ectomycorrhizal fungi can be clearly distinguished from non-mycorrhizal roots; they grow more slowly and often appear thicker and highly branched. The fungi typically form a thick sheath, or *mantle*, of mycelium around roots, and some of the hyphae penetrate between the epidermal and sometimes (in the case of conifers) the cortical cells (**Figure 4.13**). The root cells themselves are not penetrated by the fungal hyphae, but instead are surrounded by a network of hyphae called the **Hartig net**, which provides a large area of contact between the symbionts that is involved in nutrient transfers. The fungal mycelium also extends into the soil, away from the compact sheath, where it is present as individual hyphae, mycelial fans (**Figure 4.14**), or mycelial strands. The fans in particular play important roles in obtaining nutrients from the soil, especially soil organic matter.

Ectomycorrhizal fungi form interconnected networks among trees and produce many of the toadstools, puffballs, and truffles found in forests. Often the amount of fungal mycelium is so extensive that its total mass is much greater than that of the roots themselves. The arrangement and biochemical activities of the fungal structures in relation to the root tissues determine important aspects of nutrient acquisition by ectomycorrhizal roots and the form in which the nutrients pass from fungus to plant. In addition, all nutrients from the soil must pass through the fungal mantle covering the root epidermis before reaching the root cells themselves, giving the fungus a major role in uptake of all nutrients from the soil solution, including phosphate and inorganic forms of nitrogen (nitrate and ammonium). The extent to which fungi compete with the roots when nitrogen is in short supply is a matter of active research. The fungal mycelium that develops in soil proliferates extensively in soil organic matter patches (Figure 4.14). The hyphae have a marked ability to convert insoluble organic nitrogen and phosphorus to soluble forms and to pass these nutrients to the plants. In this way, ectomycorrhizal fungi enable their host plants to access organic sources of nutrients, avoid competition with free-living mineralizing organisms, and grow in highly organic forest soils that contain very low amounts of inorganic nutrients.

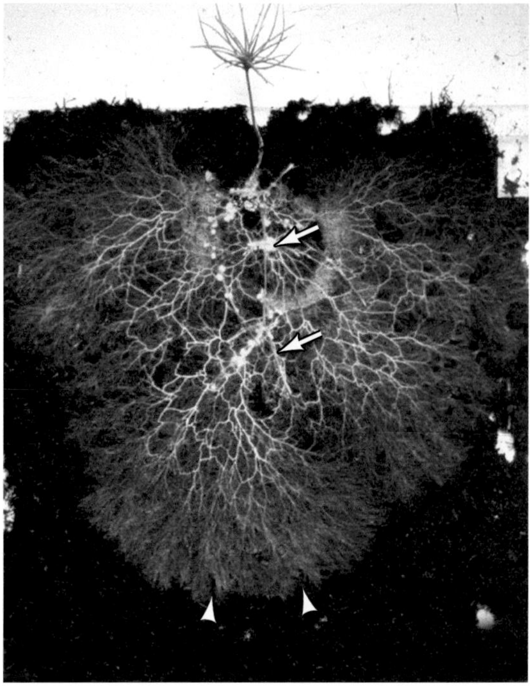

Figure 4.14 Pine seedling mycorrhizae. Seedling of pine (*Pinus*) showing mycorrhizal rootlets (upper arrow) colonized by an ectomycorrhizal fungus and grown in an observation chamber in forest soil. Note the differences between the mycelial front of dense hyphae advancing into the soil (arrowheads) and the aggregated mycelial strands (lower arrow).

Nutrients move between mycorrhizal fungi and root cells

Movement of nutrients from soil via a mycorrhizal fungus to root cells involves complex integration of structure and function in both fungus and plant symbionts. The interfaces where fungus and plant are juxtaposed are critical zones for transport and are composed of the plasma membranes of both organisms, plus variable amounts of cell wall material. Therefore, nutrient movements from fungus to plant are potentially under the control of these two membrane types and subject to regulatory transport processes as described in Chapter 6. Movement of nutrients from soil to the plant via a mycorrhizal fungus requires (at least) uptake of a nutrient from soil by the fungus, long-distance translocation of the nutrient through fungal hyphae (and mycelial strands when present), transmembrane efflux from the fungus to the apoplastic zone between the two membranes of the interface, and transport across the plant plasma membrane. Important issues to be resolved include the form of nutrient that is transferred and the mechanism and amounts of transfers. Mechanisms promoting efflux from the fungus to the interfacial apoplastic zone are poorly understood, while uptake into the plant has received more attention. In the case of phosphate, the plant uptake step is an active process requiring energy provided by a H^+-ATPase that generates a proton gradient (Chapter 6) and the presence of proton-coupled phosphate symporters in the plant membrane surrounding the intracellular fungal structures, which are specifically or preferentially expressed when the roots are mycorrhizal.

Transfer of nitrogen is more complex and more controversial. In ectomycorrhizae, for which a major role in plant nitrogen nutrition has long been accepted, organic nitrogen may move from fungus to plant, with the form (glutamine, glutamine and alanine, or glutamate) varying with fungal species and the selectivity of the efflux transporters in the fungus. Transfer of nitrogen as ammonium or ammonia also occurs.

Mycorrhizal networks can also transfer nutrients between plants. Stable isotope labeling and high-throughput DNA sequencing have been used to demonstrate distribution of nutrients based on plant source–sink relationships via ectomycorrhizal clusters associated with forest trees.

Hartig net A network of fungal hyphae that surround, but do not penetrate, the cortical cells of roots in ectomycorrhizal symbioses.

4.4 Fertilizer Production and Uses

Evaluate positive aspects of agricultural fertilizer production and use, and contrast positive factors with the negative impacts on environmental health and sustainability that increasingly motivate changes in modern agricultural practices.

More than half of the energy used in agriculture is expended on the production, distribution, and application of nitrogen fertilizers. Moreover, production of phosphorus fertilizers depends on nonrenewable resources that are likely to reach peak production during this century. Crop plants, however, typically use less than half of the fertilizer applied to the soils around them. The remaining minerals may leach into surface waters or groundwater, become associated with soil particles, or contribute to eutrophication (leading to algal blooms and fish kills), air pollution, or climate change after conversion to ammonia or nitrous oxide, for example.

As a consequence of fertilizer leaching, many water wells in the United States now exceed the federal standard for nitrate (NO_3^-) concentrations in drinking water of 10 mg/L. Infants have the greatest risk from nitrate poisoning (methemoglobinemia). In addition to nitrogen and phosphorus runoff from fertilizer application, nitrogen released to the environment by other human activities and deposited in the soil by rainwater, a process known as atmospheric nitrogen deposition, is altering ecosystems throughout the world. **Chemical fertilizers** are not the only contributors to environmental pollution. Extensive use of manures contributes to phosphorus and nitrogen runoff that stimulates eutrophic growth in large and small estuaries, with the result that waters become anoxic, toxic microorganisms proliferate, and marine life populations decline or disappear.

chemical fertilizers Fertilizers that provide nutrients in inorganic form.

Massive use of synthetic nitrogen (N) fertilizers beginning in the twentieth century, and now amounting to more than 120 teragrams (Tg; 1 Tg = 1 million metric tons) of N per year, underpinned the Green Revolution and ensures food security for much of humanity. However, the use of N fertilizers also has unfavorable consequences for the world in that reactive nitrogen species (RNS) adversely affect water and air quality and human and ecosystem health, taking us beyond a safe operating space for humanity. Human intervention in the global N cycle, primarily through the production and use of N fertilizers but also through the production of N oxides (NOx) from burning fossil fuels, has doubled the flux of N through the cycle, with additional impacts on Earth's climate. RNS compounds have many positive and negative effects on greenhouse gas (GHG) balance, although the net climate effect may be cooling, in part because of increased photosynthesis and CO_2 sequestration in the biosphere. Agriculture is the main source of ammonia emissions, which contribute the most to the cooling effect of RNS, mainly through aerosol formation and solar energy reflection and N deposition that stimulates carbon sequestration. Likewise, agriculture is the main source of the potent greenhouse gas N_2O, the primary RNS contributor to global warming. NOx have many interactions, leading to ozone (a GHG) and aerosol production, for example, which together lead to a small net cooling effect. Production and use of synthetic fertilizers have had other direct and indirect effects on climate, including the release of CO_2 from fossil fuels used to produce ammonia, which accounts for about 1% of global energy use. The doubling of the world population has led to a tripling of the consumption of fossil fuels and CO_2 emissions since the 1960s (**Figure 4.15**).

As a result, there is growing pressure to increase N-use efficiency (NUE) in agriculture through improved management practices that better match supply of fertilizer with crop demand for N, plant breeding to improve uptake and utilization of soil N, and the use of legumes that incorporate atmospheric di-nitrogen into

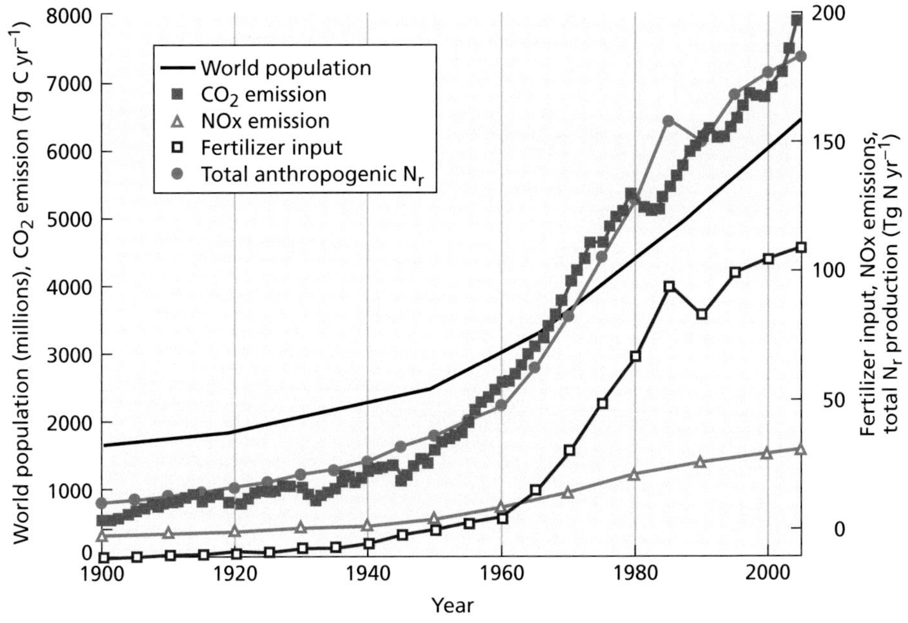

Figure 4.15 World trends in human population, N fertilizer consumption, N oxide (NOx) and CO_2 emissions, and human N fixation (total anthropogenic N_r). The rate of human N fixation is now about twice that of biological N fixation in terrestrial ecosystems. N_r, reactive nitrogen.

ammonia via symbiotic nitrogen fixation (Chapter 5). Globally, NUE, defined as nitrogen captured in harvested grain divided by total N inputs, stands at about 40% for the major cereals, wheat, maize, and rice, while it is 80% for the major legume, soybean. Thus, greater use of legumes for food production, or simply as a natural source of fixed N for agricultural systems when used as a cover crop in rotations or intercropped with a non-legume, could increase overall NUE of agriculture and reduce the need for synthetic N fertilizers. This would help ensure food security while reducing the negative environmental impacts of N fertilizers.

Summary

Plants are autotrophic organisms capable of using the energy from sunlight to synthesize all their components from carbon dioxide, water, and mineral elements. Although mineral nutrients continually cycle through all organisms, they enter the biosphere predominantly through the root systems of plants. After being absorbed by the roots, the mineral elements are translocated to the various parts of the plant, where they serve in numerous biological functions.

4.1 Essential Nutrients, Deficiencies, and Plant Disorders

- Studies of plant nutrition have shown that specific mineral elements are essential for plant life (**Tables 4.1, 4.2**).
- These elements are classified as macronutrients or micronutrients, depending on the relative amounts found in plants (**Table 4.1**).
- Nutritional disorders occur because nutrients have key roles within plants. They serve as components of organic compounds, in energy storage, in plant structures, as enzyme cofactors, and in electron transfer reactions.
- Mineral nutrition can be studied through solution media, which allows the characterization of specific nutrient requirements (**Figure 4.1**; **Table 4.3**).
- Certain visual symptoms are diagnostic for deficiencies in specific nutrients in vascular plants.
- Soil and plant tissue analysis can provide information on the nutritional status of the plant–soil system and can suggest corrective actions to avoid deficiencies or toxicities.

4.2 Treating Nutritional Deficiencies

- When crop plants are grown under modern high-production conditions, substantial amounts of nutrients are removed from the soil.
- To prevent the development of deficiencies, nutrients—particularly nitrogen, phosphorus, and potassium—can be added back to the soil in the form of fertilizers.

(Continued)

Summary (*continued*)

- Fertilizers that provide nutrients in inorganic forms are called inorganic fertilizers; those that derive from plant or animal residues or from natural rock deposits are considered organic fertilizers. In both cases, plants absorb the nutrients primarily as inorganic ions. Most fertilizers are applied to the soil, but some are sprayed on leaves.

4.3 Soil, Roots, and Microbes

- Soil is a complex substrate—physically, chemically, and biologically.
- Soil pH has a large influence on the availability of mineral elements to plants (**Figure 4.3**).
- The size of soil particles and the cation exchange capacity of the soil determine the extent to which a soil provides a reservoir for water and nutrients (**Table 4.5**; **Figure 4.4**).
- If mineral elements, especially sodium or metals, are present in excess in the soil, plant growth may be adversely affected. Certain plants are able to tolerate excess mineral elements, and a few species—for example, halophytes in the case of sodium—may thrive under these extreme conditions.
- To obtain nutrients from the soil, plants develop extensive root systems (**Figures 4.5**, **4.6**), form symbioses with mycorrhizal fungi, and produce and secrete protons or organic anions into the soil.
- Roots continually deplete the nutrients from the immediate soil around them.
- Nutrient availability influences root growth (**Figures 4.8–4.10**).
- The majority of plants have the ability to form symbioses with mycorrhizal fungi.
- The fine hyphae of mycorrhizal fungi extend the reach of roots into the surrounding soil and facilitate the acquisition of nutrients (**Figures 4.11–4.14**). Arbuscular mycorrhizas increase uptake of mineral nutrients, particularly phosphorus, whereas ectomycorrhizas play a significant role in obtaining nitrogen from organic sources.
- In return, plants provide carbohydrates to the mycorrhizal fungi.

4.4 Fertilizer Production and Use

- The production and use of fertilizers, in particular nitrogen fertilizers, has increased many-fold during the twentieth century and has helped ensure food security for most of humankind.
- However, the use of N fertilizers also has unfavorable consequences for the world in that reactive nitrogen species (RNS) adversely affect water and air quality and human and ecosystem health.
- There is therefore a need to increase N-use efficiency (NUE) in agriculture through improved management practices, through plant breeding to improve uptake and utilization of soil N, and through the increased use of legumes with symbiotic nitrogen fixation.

Suggested Reading

Baxter, I. R., Vitek, O., Lahner, B., Muthukumar, B., Borghi, M., Morrissey, J., Guerinot, M. L., and Salt, D. E. (2008) The leaf ionome as a multivariable system to detect a plant's physiological status. *Proc. Natl. Acad. Sci. USA* 105: 12081–12086.

Bennett, A. E., and Groten, K. (2022) The costs and benefits of plant–arbuscular mycorrhizal fungal interactions. *Annu. Rev. Plant Biol.* 73(1): 649–672.

Erisman, J. W., Galloway, J., Seitzinger, S., Bleeker, A., and Butterbach-Bahl, K. (2011) Reactive nitrogen in the environment and its effect on climate change. *Curr. Opin. Environ. Sustain.* 3: 281–290.

Godfray, H. C., Beddington, J. R., Crute, I. R., Haddad, L., Lawrence, D., Muir, J. F., Pretty, J., Robinson, S., Thomas, S. M., and Toulmin, C. (2010) Food security: The challenge of feeding 9 billion people. *Science* 327: 812–818.

Hose, E., Clarkson, D. T., Steudle, E., Schreiber, L., and Hartung, W. (2001) The exodermis: A variable apoplastic barrier. *J. Exp. Botany* 52: 2245–2264.

Huang, X. Y., and Salt, D. E. (2016) Plant ionomics: From elemental profiling to environmental adaptation. *Mol. Plant* 9: 787–797.

Jackson, R. B., Canadell, J., Ehleringer, J. R., Mooney, H. A., Sala, O. A., and Schulze, E.-D. (1996) A global analysis of root distributions for terrestrial biomes. *Oecologia* 108: 389–411.

Lambers, H. (2022) Phosphorus acquisition and utilization in plants. *Annu. Rev. Plant Biol.* 73(1): 17–42.

Liu, Q., Wu, K., Song, W., Zhong, N., Wu, Y., and Fu, X. (2022) Improving crop nitrogen use efficiency toward sustainable Green Revolution. *Annu. Rev. Plant Biol.* 73(1): 523–551.

Vukosav, P., Mlakar, M., and Tomišić, V. (2012) Revision of iron (III)-citrate speciation in aqueous solution: Voltammetric and spectrophotometric studies. *Anal. Chim. Acta* 745: 85–91.

5 Assimilation of Inorganic Nutrients

Higher plants are autotrophic organisms that can synthesize all of their organic molecular components out of inorganic nutrients obtained from their surroundings. For many inorganic nutrients, this process involves absorption from the soil by the roots and incorporation into the organic compounds that are essential for growth and development. This incorporation of inorganic nutrients into organic substances such as pigments, enzyme cofactors, lipids, nucleic acids, and amino acids is termed **nutrient assimilation**.

Assimilation of some nutrients—particularly nitrogen and sulfur—involves a complex series of biochemical reactions that are among the most energy-consuming reactions in living organisms:

- In nitrate (NO_3^-) assimilation, the nitrogen in NO_3^- is converted to a higher-energy (more reduced) form in nitrite (NO_2^-), then to a yet-higher-energy (even more reduced) form in ammonium (NH_4^+), and finally into the amide nitrogen of the amino acid glutamine. This process consumes the equivalent of 12 ATPs per amide nitrogen.
- Plants such as legumes form symbiotic relationships with nitrogen-fixing bacteria to convert molecular nitrogen (N_2) into ammonia (NH_3). Ammonia is the first stable product of natural fixation; at physiological pH, however, ammonia is protonated to form the ammonium ion (NH_4^+). The process of biological nitrogen fixation, together with the subsequent assimilation of NH_3 into an amino acid, consumes the equivalent of about 16 ATPs per amide nitrogen.
- Sulfate (SO_4^{2-}), the most oxidized form of sulfur, is reduced to sulfite (SO_3^{2-}) and then sulfide (S^{2-}) in a series of steps before it is assimilated into the amino acid cysteine, and the process consumes about

nutrient assimilation The incorporation of mineral nutrients into carbon compounds such as pigments, enzyme cofactors, lipids, nucleic acids, or amino acids.

14 ATPs. The sulfate intermediate 5′-adenylylsulfate is incorporated into glucosinolates and other sulfur-containing compounds.

- For some perspective on the enormous energies involved, consider that if these reactions run rapidly in reverse—say, from NH_4NO_3 (ammonium nitrate) to N_2—they become explosive, liberating vast amounts of energy as motion, heat, and light. Nearly all explosives, including nitroglycerin, TNT (trinitrotoluene), and gunpowder, are based on the rapid oxidation of nitrogen or sulfur compounds.

Assimilation of other nutrients, especially the macronutrient and micronutrient cations (Chapter 4), involves the formation of complexes with organic compounds. For example, Mg^{2+} associates with chlorophyll pigments, Ca^{2+} associates with pectates within the cell wall, and Mo^{6+} associates with enzymes such as nitrate reductase and nitrogenase. These complexes are highly stable, and removal of the nutrient from the complex may result in total loss of function.

This chapter outlines the primary reactions through which the major nutrients (nitrogen, sulfur, phosphate, and oxygen) are assimilated and discusses the organic products of these reactions. We emphasize the physiological implications of the required energy expenditures and introduce the topic of symbiotic nitrogen fixation. Plants serve as the major conduit through which nutrients pass from slower geophysical domains into faster biological ones; this chapter thus highlights the vital role of plant nutrient assimilation in the human diet.

5.1 Nitrogen in the Environment

Summarize the biogeochemical nitrogen cycle and effects of different nitrogen forms in plant tissues.

Many prominent biochemical compounds in plant cells contain nitrogen. For example, nitrogen is found in the nucleotides and amino acids that form the building blocks of nucleic acids and proteins, respectively. Only the elements oxygen, carbon, and hydrogen are more abundant in plants than nitrogen. Most natural and agricultural ecosystems show dramatic gains in productivity after fertilization with inorganic nitrogen, attesting to the importance of this element and to the fact that it is present in suboptimal amounts.

In this section we discuss the biogeochemical cycle of nitrogen, the crucial role of nitrogen fixation in the conversion of molecular nitrogen into ammonium and nitrate, and the fate of ammonium and nitrate in plant tissues.

Nitrogen passes through several forms in a biogeochemical cycle

Nitrogen is present in many forms in the biosphere. The atmosphere contains vast quantities (about 78% by volume) of molecular nitrogen (N_2). For the most part, this large reservoir of nitrogen is not directly available to living organisms. Acquisition of nitrogen from the atmosphere requires the breaking of an exceptionally stable triple covalent bond between two nitrogen atoms ($N\equiv N$) to produce ammonia (NH_3) or nitrate (NO_3^-). These reactions, known as **nitrogen fixation**, occur through both industrial and natural processes.

Industrial processes use elevated temperature (about 200°C), high pressure (about 200 atmospheres, or 20 MPa), and a metal catalyst (usually iron) to combine N_2 with hydrogen and form ammonia. These extreme conditions are required to overcome the high activation energy of the reaction. This nitrogen fixation reaction, called the *Haber–Bosch process*, is a starting point for the manufacture

nitrogen fixation The natural or industrial processes by which atmospheric nitrogen N_2 is converted to ammonia (NH_3) or nitrate (NO_3^-).

of nitrogen fertilizers, which globally exceeds 110 million metric tons per year (110 Mt y^{-1}, or 110×10^{12} g y^{-1}). It has been estimated that half of the nitrogen in the human body comes from this process.

The following natural processes fix about 260 Mt y^{-1} (± 100 Mt y^{-1}) of nitrogen (**Table 5.1**):

- *Lightning.* Lightning is responsible for about 2% of the nitrogen fixed by natural processes. Lightning converts water vapor and oxygen into highly reactive hydroxyl free radicals, free hydrogen atoms, and free oxygen atoms that attack molecular nitrogen (N_2) to form nitric acid (HNO_3). This nitric acid subsequently falls to Earth with rain.
- *Biological nitrogen fixation.* The remaining 98% results from biological nitrogen fixation, in which bacteria or cyanobacteria (blue-green algae) fix N_2 into ammonia (NH_3). This ammonia dissolves in water to form ammonium (NH_4^+):

$$NH_3 + H_2O \rightarrow NH_4^+ + OH^- \tag{5.1}$$

The high cost of fertilizers, especially for poor farmers, combined with negative impact of fertilizers on the environment, has resulted in intense efforts to better understand and implement biological nitrogen fixation, as well as to improve the nitrogen use efficiency of crops.

Once fixed into ammonia or nitrate, nitrogen enters a biogeochemical cycle and passes through several organic or inorganic forms before it eventually returns to molecular nitrogen (**Figure 5.1** and Table 5.1). The ammonium (NH_4^+) and nitrate (NO_3^-) ions in the soil solution that are generated through fixation or released through decomposition of soil organic matter become the object of intense competition among plants and microorganisms. To be competitive, plants have evolved mechanisms for scavenging these ions rapidly from the soil

Table 5.1 The Major Processes of the Biogeochemical Nitrogen Cycle

Process	Definition	Rate (Mt y^{-1}, 10^{12}g y^{-1})[a]
Industrial fixation	Industrial conversion of molecular nitrogen to ammonia	110–120
Atmospheric fixation	Lightning and photochemical conversion of molecular nitrogen to nitrate	2–10
Biological fixation	Prokaryotic conversion of molecular nitrogen to ammonia	150–380
Plant acquisition	Plant absorption and assimilation of ammonium or nitrate	1,000–1,200
Immobilization	Microbial absorption and assimilation of ammonium or nitrate	N/C
Ammonification	Bacterial and fungal catabolism of soil organic matter to ammonium	N/C
Anammox	Anaerobic ammonium oxidation: bacterial conversion of ammonium and nitrite to molecular nitrogen	N/C
Nitrification	Bacterial (*Nitrosomonas* spp.) oxidation of ammonium to nitrite and subsequent bacterial (*Nitrobacter* spp.) oxidation of nitrite to nitrate	N/C
Mineralization	Bacterial and fungal catabolism of soil organic matter to mineral nitrogen through ammonification or nitrification	N/C
Volatilization	Physical loss of gaseous ammonia to the atmosphere	60–80
Ammonium fixation	Physical embedding of ammonium into soil particles	10
Denitrification	Bacterial conversion of nitrate to nitrous oxide and molecular nitrogen	100–280
Nitrate leaching	Physical flow of nitrate dissolved in groundwater out of the topsoil and eventually into the oceans	40–70

Source: W. H. Schlesinger. 1997. *Biogeochemistry: An Analysis of Global Change*, 2nd ed., Academic Press, San Diego, CA.

[a] N/C, not calculated.

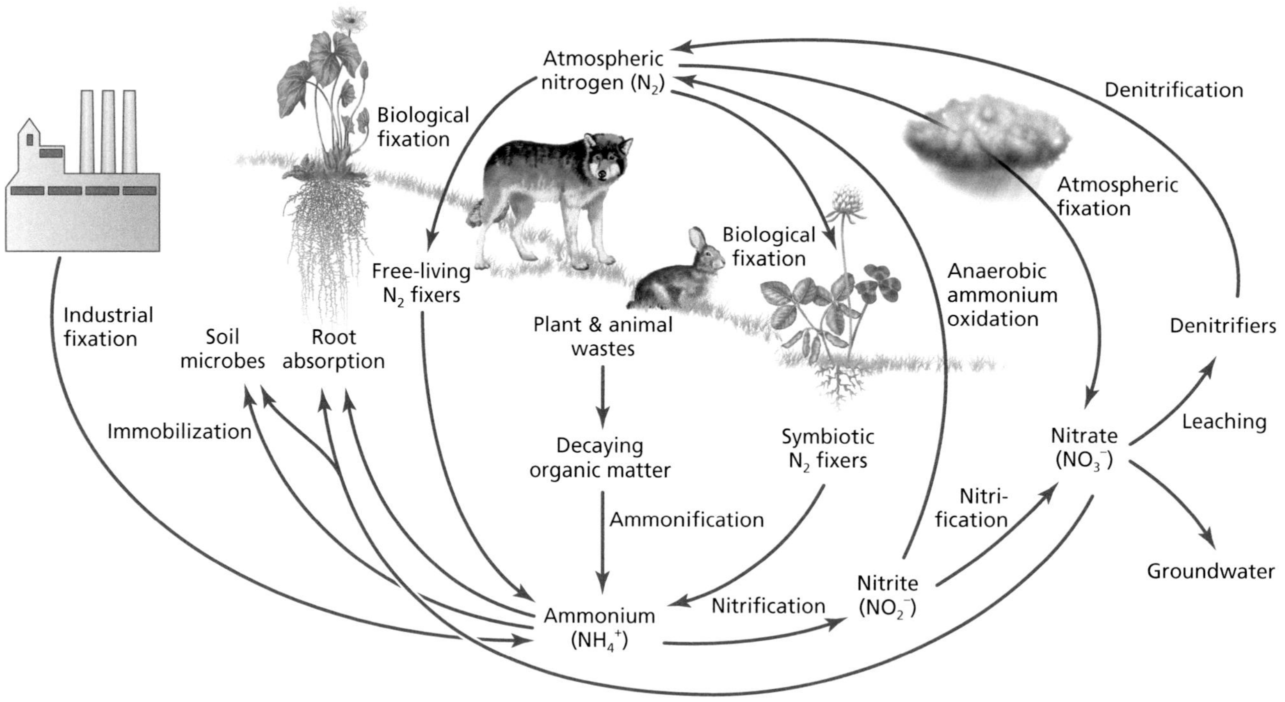

Figure 5.1 The nitrogen cycle. Nitrogen cycles through the atmosphere, changing from a gas to soluble ions before it is incorporated into organic compounds in living organisms. Some of the steps involved in the nitrogen cycle are shown.

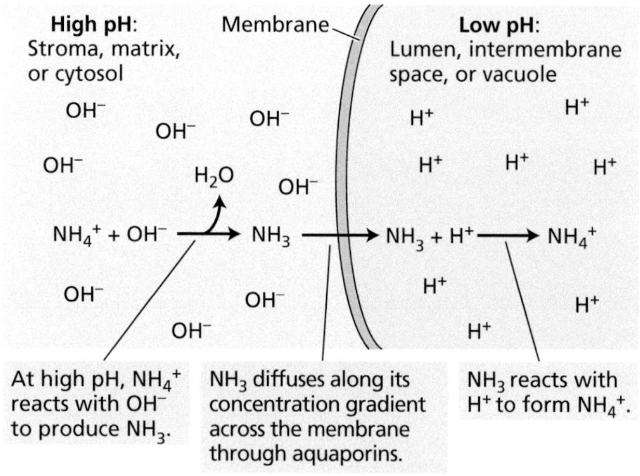

Figure 5.2 NH_4^+ toxicity derives from dissipation of pH gradients. The left side represents the stroma, matrix, or cytosol, where the pH is high; the right side represents the lumen, intermembrane space, or vacuole, where the pH is lower; and the membrane represents the thylakoid, inner mitochondrial, or tonoplast membrane for a chloroplast, mitochondrion, or root cell, respectively. The net result of the reaction shown is that both the OH^- concentration on the left side and the H^+ concentration on the right side have been diminished; that is, the pH gradient has been dissipated. The movement of ammonia across membranes can occur through aquaporins (Chapter 6).

solution. Under the elevated soil concentrations that occur after fertilization, the absorption of ammonium and nitrate by the roots may exceed the capacity of a plant to assimilate these ions, leading to their accumulation within the plant's tissues.

Unassimilated ammonium or nitrate may be dangerous

If it accumulates to high concentrations in living tissues, ammonium is toxic to both plants and animals. Ammonium dissipates the transmembrane proton gradients (**Figure 5.2**) that are required for photosynthetic and respiratory electron transport (Chapters 7 and 11), for sequestering metabolites in the vacuole, and for transporting nutrients across biological membranes (Chapter 6). Since high concentrations of ammonium are dangerous, animals have developed a strong aversion to its smell. For example, the active ingredient in smelling salts, a medicinal vapor released under the nose to revive a person who has fainted, is ammonium carbonate. To avoid the toxic effects of ammonium, plants suppress ammonium uptake by a phosphorylation-dependent allosteric feedback mechanism involving the ammonium transporter AMT1;1, they assimilate ammonium near the site of absorption or generation, and they rapidly store any excess in their vacuoles.

Unlike the case with ammonium, plants can store high concentrations of nitrate, and they can translocate it from tissue to tissue without deleterious effect. Yet if livestock or humans consume plant material (or other foods like cured meat) that is high in nitrate, they may suffer methemoglobinemia, a disease in which the liver reduces nitrate to nitrite, which combines with hemoglobin and renders hemoglobin unable to bind oxygen. Humans and other animals may also convert nitrate into nitrosamines, which are potent carcinogens, or into nitric oxide, a potent signaling molecule involved in many physiological processes in humans, such as widening of blood vessels, and for signaling in plants (Chapters 12 and 14).

nitrate reductase An enzyme located in the cytosol that reduces nitrate (NO_3^-) to nitrite (NO_2^-). It catalyzes the first step by which nitrate absorbed by roots is assimilated into organic form.

5.2 Nitrate Assimilation

Describe the mechanisms that plants use to assimilate nitrate and transport it from roots and shoots.

Plant roots actively absorb nitrate from the soil solution via several low- and high-affinity nitrate–proton cotransporters in the NRT1 and NRT2 family (Chapter 6). Plants eventually assimilate most of this nitrate into organic nitrogen compounds. The first step in this process is the conversion of nitrate to nitrite in the cytosol, a reduction reaction (for redox properties, see Chapters 7 and 11) that involves the transfer of two electrons. The enzyme **nitrate reductase** (**Figures 5.3** and **5.4**) catalyzes this reaction:

$$NO_3^- + NAD(P)H + H^+ \rightarrow NO_2^- + NAD(P)^+ + H_2O \qquad (5.2)$$

where NAD(P)H indicates either NADH or NADPH. The most common form of nitrate reductase uses only NADH as an electron donor; another form of the enzyme that is found predominantly in nongreen tissues such as roots can use either NADH or NADPH.

The nitrate reductases of higher plants are composed of two identical subunits, each containing three prosthetic groups: flavin adenine dinucleotide (FAD),

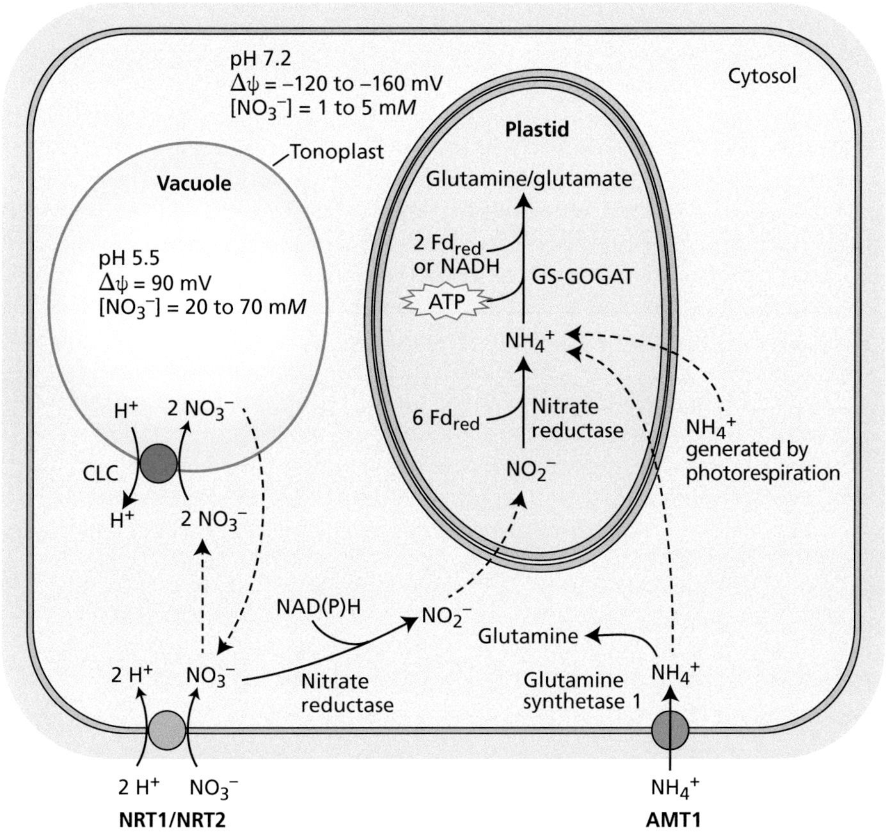

Figure 5.3 Overview of nitrogen assimilation in plant cells. Nitrate is taken up across the plasma membrane by proton-coupled symporters, nitrate transporters NRT1 or NRT2, using the proton gradient as the driving force. Nitrate is then converted to ammonium in a two-step process by nitrate reductase in the cytosol and nitrite reductase in the plastids. Nitrate can be transported across the tonoplast by the proton-nitrate antiporter CLC and stored in the vacuole up to a concentration of 70 m*M*. Ammonium, taken up by AMT1 transporters or produced intracellularly by nitrate assimilation or photorespiration, can be assimilated into glutamine and glutamate by glutamine synthetase (GS) and glutamate synthase, which has the systematic name glutamine: 2-oxoglutarate aminotransferase (GOGAT). $\Delta\psi$, membrane potential; Fd, ferredoxin.

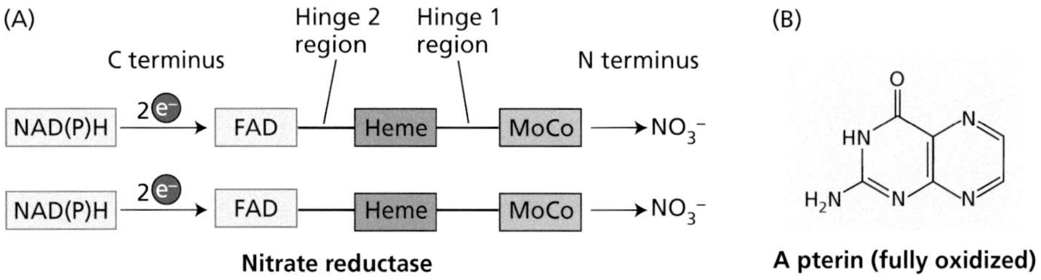

Figure 5.4 Model of the nitrate reductase dimer. Model of the nitrate reductase dimer, illustrating the three binding domains whose polypeptide sequences are similar in eukaryotes: FAD, heme, and molybdenum complex (MoCo). (A) The NAD(P)H binds at the FAD-binding region of each subunit and initiates a two-electron transfer from the carboxyl (C) terminus, through each of the electron transfer components, to the amino (N) terminus. Nitrate is reduced at the MoCo near the amino terminus. The polypeptide sequences of the hinge regions are highly variable among species. (B) The pterin of the MoCo.

heme, and a molybdenum ion complexed to an organic molecule called a *pterin*. Nitrate reductase is the main molybdenum-containing protein in vegetative tissues; one symptom of molybdenum deficiency is the accumulation of nitrate that results from diminished nitrate reductase activity.

X-ray crystallography and comparison of the amino acid sequences for nitrate reductase from several species with the sequences of other well-characterized proteins that bind FAD, heme, or molybdenum ions have led to a multiple-domain model for nitrate reductase; a simplified three-domain model is shown in Figure 5.4. The FAD-binding domain accepts two electrons from NADH or NADPH. The electrons then pass through the heme domain to the molybdenum complex, where they are transferred to nitrate.

Many factors regulate nitrate reductase

Nitrate, light, and carbohydrates influence the abundance and activity of nitrate reductase at the transcriptional, translational, and posttranslational levels. Nitrate reductase mRNA in roots is induced very quickly after addition of nitrate, and maximum levels can be reached within 30 min. In contrast to the rapid accumulation of mRNA, there is a gradual linear increase in nitrate reductase activity upon addition of nitrate, reflecting that the synthesis of the nitrate reductase protein requires the presence of the nitrate reductase mRNA.

The posttranslational modification of the nitrate reductase protein involves a reversible phosphorylation (Chapter 12) and is analogous to the regulation of pyruvate dehydrogenase (Chapter 11). Carbohydrate concentrations, light, and other environmental factors stimulate a protein phosphatase that dephosphorylates a key serine residue in the hinge 1 region of nitrate reductase (between the molybdenum complex and heme-binding domains; Figure 5.4) and thereby activates the enzyme. Operating in the reverse direction, darkness and Mg^{2+} stimulate a protein kinase that phosphorylates the same serine residues, which then interact with a 14-3-3 inhibitor protein and thereby inactivate nitrate reductase. *Regulation of nitrate reductase activity through phosphorylation and dephosphorylation provides more rapid control than can be achieved through synthesis or degradation of the enzyme (minutes vs. hours).*

Nitrite reductase converts nitrite to ammonium

Nitrite (NO_2^-) is a highly reactive, and potentially toxic, ion. Plant cells immediately transport the nitrite generated by nitrate reduction (Reaction 5.2) from the cytosol into the plastid, where nitrite is rapidly converted into ammonium

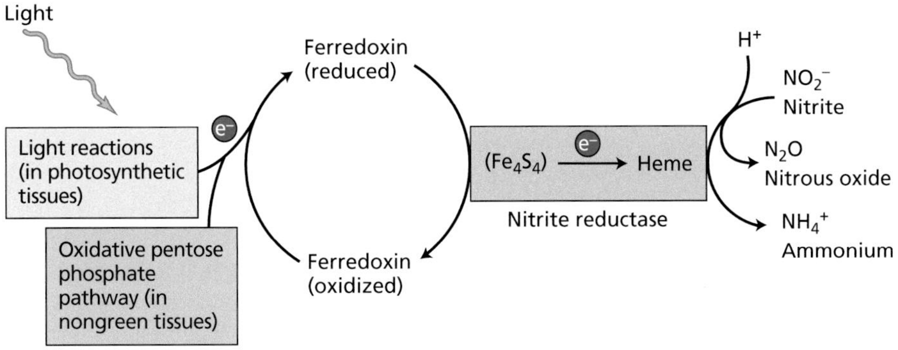

Figure 5.5 Model for coupling of photosynthetic electron flow to nitrate reduction. Model for coupling of photosynthetic electron flow, via ferredoxin, to the reduction of nitrite by nitrite reductase. The enzyme contains two prosthetic groups, an iron–sulfur cluster (Fe_4S_4) and heme, which participate in the reduction of nitrite to ammonium.

by the enzyme nitrite reductase. Nitrite reduction involves the transfer of six electrons, according to the following overall reaction:

$$NO_2^- + 6\,Fd_{red} + 8\,H^+ \rightarrow NH_4^+ + 6\,Fd_{ox} + 2\,H_2O \qquad (5.3)$$

where Fd is ferredoxin and the subscripts *red* and *ox* stand for *reduced* and *oxidized*, respectively. Reduced ferredoxin is derived from photosynthetic electron transport in the chloroplasts (Chapter 7) and from NADPH generated by the oxidative pentose phosphate pathway in the plastids of nongreen tissues (Chapter 11).

In some species, chloroplasts and root plastids contain different forms of nitrite reductase, but both forms consist of a single polypeptide containing two prosthetic groups: an iron–sulfur cluster (Fe_4S_4) and a specialized heme. These groups act together to bind nitrite and reduce it to ammonium. Although no nitrogen compounds of intermediate redox states accumulate, a small percentage (0.02–0.2%) of the nitrite reduced is released as nitrous oxide (N_2O), a greenhouse gas. The electron flow through ferredoxin, Fe_4S_4, and heme can be represented as in **Figure 5.5**.

Nitrite reductase is encoded in the nucleus and synthesized in the cytoplasm with an N-terminal transit peptide that targets it to the plastids. Elevated concentrations of NO_3^- or exposure to light induce the transcription of nitrite reductase mRNA. Accumulation of the end products in the process—asparagine and glutamine—represses this induction.

Both roots and shoots assimilate nitrate

In many plants, nitrate is reduced primarily in the roots when soil nitrate is low. As the supply of nitrate increases, a greater proportion of the absorbed nitrate is translocated via the xylem to the shoot and assimilated there. Even under similar conditions of nitrate supply, the balance between root and shoot nitrate metabolism—as indicated by the proportion of nitrate reductase activity in each of the two organs or by the relative concentrations of nitrate and reduced nitrogen in the xylem sap—varies among species.

In plants such as cocklebur (*Xanthium strumarium*), nitrate metabolism is restricted to the shoot, whereas in other plants, such as white lupine (*Lupinus albus*), most nitrate is metabolized in the roots (**Figure 5.6**). Generally, species

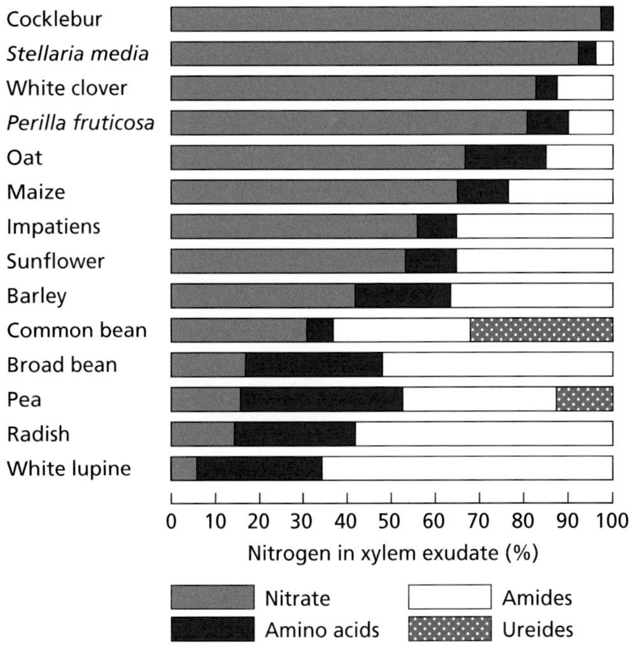

Figure 5.6 Relative amounts of nitrate and other nitrogen compounds in the xylem sap of various plant species. The plants were grown with their roots exposed to nitrate solutions, and xylem sap was collected by severing the stem. Note the presence of ureides in common bean and pea; only legumes of tropical origin export nitrogen in such compounds.

native to temperate regions rely more heavily on nitrate assimilation by the roots than do species in tropical or subtropical regions. Apart from species-based differences, partitioning of nitrate assimilation between the root and the shoot is also affected by environmental conditions such as salt stress and elevated CO_2, among others.

Nitrate can be transported in both xylem and phloem

Proper partitioning of nitrate assimilation among various tissues is important to maximize nitrogen use efficiency and to optimize plant growth. Root-to-shoot nitrate transport, a key factor in determining the partitioning of nitrate assimilation between root and shoot, occurs via the xylem, using transpiration as the driving force.

Root-to-shoot nitrate transport is controlled mainly at the step of xylem loading. In Arabidopsis, nitrate transporter 1.5 (NRT1.5), expressed in the xylem pole pericycle, is responsible for loading nitrate into the xylem. In addition to xylem transport, phloem nitrate transport, driven by osmotic gradients and from source to sink tissues, is responsible for redistributing nitrate to nitrogen-demanding tissues. Nitrate can be stored in large quantities in the vacuole, up to a concentration of 70 m*M* (Figure 5.3), from which it can be retrieved when needed. Nitrate storage capacity and retrieval efficiency are well correlated with yield in grain crops. When the external nitrate supply is low, nitrate stored in old leaves is retrieved and loaded into phloem by the nitrate transporter NRT1.7 to feed growing tissues such as young leaves and seeds (**Figure 5.7**). This phloem-mediated remobilization can help plants sustain growth under nitrogen deficiency. In contrast, when soil nitrate supply is high, nitrate from the roots is directed via the xylem stream to mature leaves where transpiration is greatest.

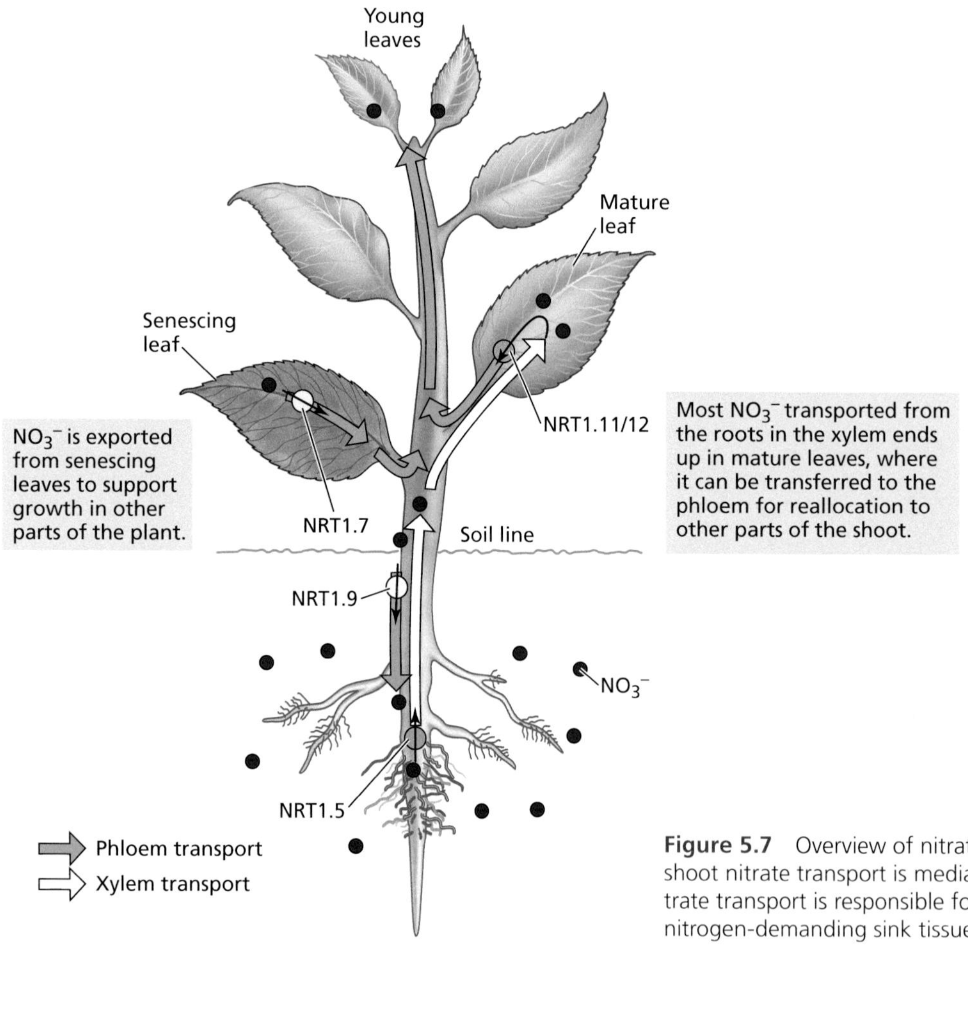

Figure 5.7 Overview of nitrate transport in plants. Root-to-shoot nitrate transport is mediated by xylem, while phloem nitrate transport is responsible for distributing nitrate into nitrogen-demanding sink tissues. NRT, nitrate transporter.

In this case, xylem-borne nitrate can be transferred into the phloem by the nitrate transporters NRT1.11 and NRT1.12 in the major vein of mature leaves to satisfy the high nitrogen demand of young leaves. Plants rely on phloem nitrate transport to reallocate xylem-borne and stored nitrate under nitrogen-sufficient and nitrogen-deficient conditions, respectively, to feed growing tissues. Similar to NRT1.11 and NRT1.12, root-expressed NRT1.9 also mediates xylem-to-phloem transfer of nitrate, serving as a negative regulatory module for root-to-shoot nitrate transport. Depending on external and internal nitrogen status, xylem and phloem nitrate transport work together to ensure that nitrate is properly distributed and assimilated in various tissues.

Transceptor contributes to nitrate signaling

NRT1.1 is a dual-affinity nitrate transporter involved in nitrate uptake (Chapter 6), and it also functions as a nitrate transceptor (transporter with receptor function). Nitrate rapidly induces the expression of approximately 500 nitrate-related genes—including those encoding nitrate reductase, nitrite reductase, and nitrate transporters—in a coordinated manner. This rapid transcriptional response is termed the primary nitrate response, and it prepares plants to acquire and assimilate nitrate when external nitrate is available. Expression levels of nitrate-related genes are stringently regulated according to the external nitrate concentration and display a biphasic saturation pattern (**Figure 5.8A**). As part of the primary nitrate response, plants use dual-affinity binding and altered phosphorylation of

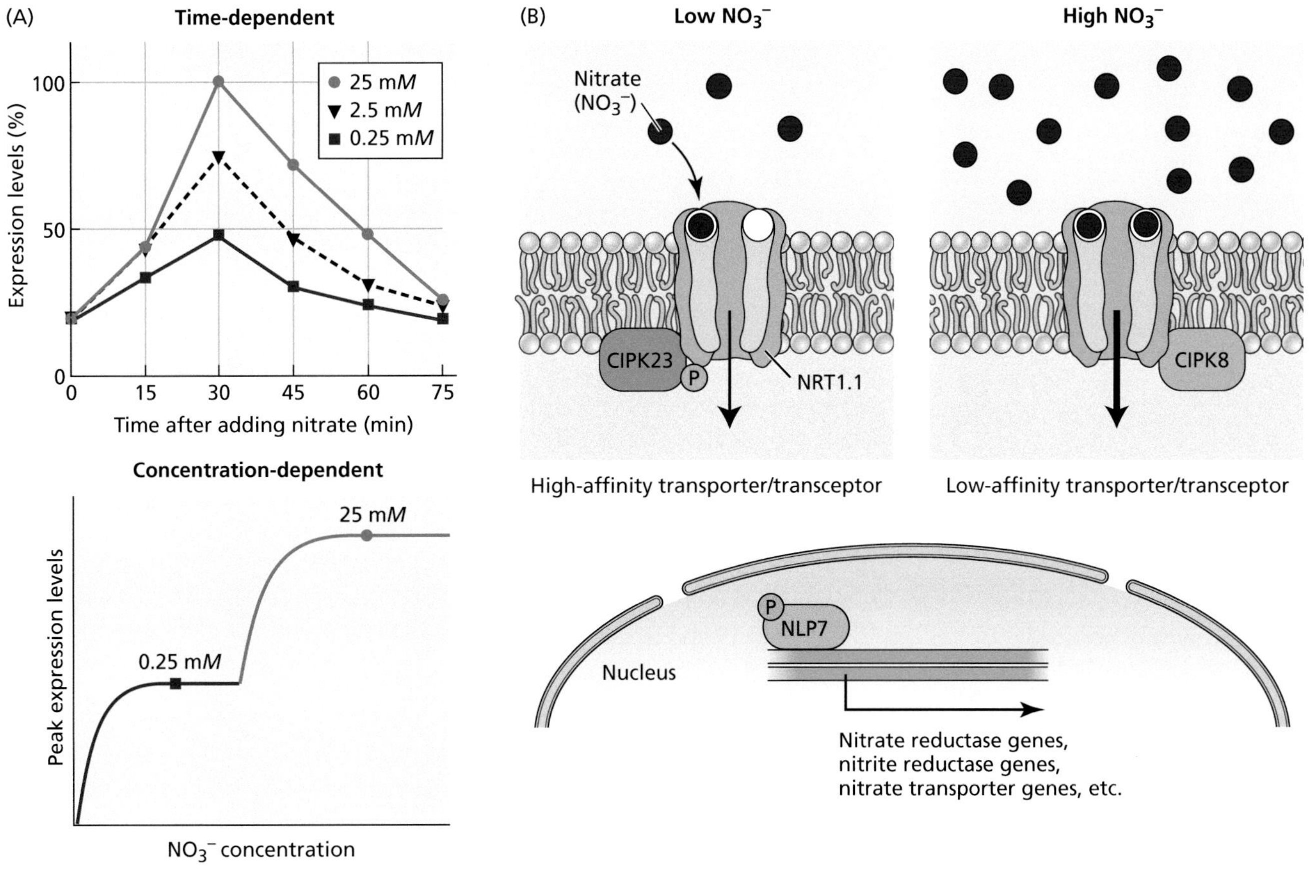

Figure 5.8 The primary nitrate response. (A) Time-dependent (top) and concentration-dependent (bottom) expression patterns of the nitrate-related genes including nitrate reductase. (B) The transceptor NRT1.1 uses dual-affinity binding to sense a wide range of changes in soil nitrate concentration with the help of two protein kinases, CIPK23 and CIPK8 (Chapter 12), which operate in the high- and low-affinity phases, respectively. Transcription factor NLP7 can then bind the promoters of nitrate-related genes to elicit their expression.

the transceptor NRT1.1 to sense wide-ranging variations of the external nitrate levels, and then use the transcription factor NLP7 to bind to the promoters of the nitrate reductase and nitrite reductase genes (**Figure 5.8B**). To ensure that nitrate acquisition is well coordinated with C assimilation and internal N status, nitrate transporter gene NRT2.1 is also fine-tuned by CEP1-mediated systemic nitrogen-demand signaling and a light-responsive transcription factor.

5.3 Ammonium Assimilation

> Describe the different processes used by plants to assimilate ammonium into amino acids.

Plant cells avoid ammonium toxicity by rapidly converting the ammonium generated from nitrate assimilation or photorespiration (Chapter 8) into amino acids. The primary pathway for this conversion involves the sequential actions of glutamine synthetase (GS) and glutamate synthase, which has the systematic name glutamine:2-oxoglutarate aminotransferase (GOGAT). In this section we discuss the enzymatic processes that mediate the assimilation of ammonium into essential amino acids, and the role of amides in the regulation of nitrogen and carbon metabolism.

Converting ammonium to amino acids requires two enzymes

Glutamine synthetase (**GS**) combines ammonium with glutamate to form glutamine (**Figure 5.9A**):

$$\text{Glutamate} + NH_4^+ + \text{ATP} \rightarrow \text{glutamine} + \text{ADP} + P_i \quad (5.4)$$

This reaction requires the hydrolysis of one ATP and involves a divalent cation such as Mg^{2+}, Mn^{2+}, or Co^{2+} as cofactor. Plants have two classes of GS, GS1 in the cytosol and GS2 in chloroplasts/plastids.

Cytosolic GS1 isozymes, expressed specifically in phloem and predominantly in the vascular bundles of roots, function in primary ammonia assimilation and produce glutamine for long-distance nitrogen transport. The GS2 in root plastids generates amide nitrogen for local consumption, whereas the GS2 in the chloroplasts of mesophyll cells is involved in both primary ammonium assimilation and the reassimilation of photorespiratory NH_4^+ (Chapter 8). Loss-of-function *gs2* mutants die under photorespiratory conditions because of the accumulation of toxic concentrations of NH_4^+, but *gs1* mutants do not display this phenotype. Light and carbohydrate concentrations alter the expression of the plastid forms of the enzyme GS2, but they have little effect on the cytosolic forms GS1.

Elevated plastid levels of glutamine stimulate the activity of **glutamate synthase** (also known as *glutamine:2-oxoglutarate aminotransferase*, or **GOGAT**). This enzyme transfers the amide group of glutamine to 2-oxoglutarate, yielding two molecules of glutamate (Figure 5.9A). Plants contain two types of GOGAT; one accepts electrons from NADH, and the other accepts electrons from ferredoxin (Fd):

$$\text{Glutamine} + \text{2-oxoglutarate} + \text{NADH} + H^+ \rightarrow 2\ \text{glutamate} + \text{NAD}^+ \quad (5.5)$$

$$\text{Glutamine} + \text{2-oxoglutarate} + 2\ \text{Fd}_{red} \rightarrow 2\ \text{glutamate} + 2\ \text{Fd}_{ox} \quad (5.6)$$

In plastids of nonphotosynthetic tissues such as roots or the vascular bundles of developing leaves, glutamate synthase (NADH-GOGAT) uses NADH as coenzyme (Reaction 5.5). In roots, NADH-GOGAT functions in the assimilation of NH_4^+ absorbed from the rhizosphere (the soil near the surface of the roots); in vascular bundles of developing leaves, NADH-GOGAT assimilates glutamine translocated from roots or senescing leaves.

In chloroplasts, Fd-GOGAT (Reaction 5.6) functions in both primary ammonium assimilation and photorespiratory nitrogen metabolism. Both the

glutamine synthetase (GS) An enzyme that catalyzes the condensation of ammonium and glutamate to form glutamine. The reaction is critical for the assimilation of ammonium into essential amino acids. Two forms of GS exist—one in the cytosol and one in chloroplasts.

glutamate synthase (GOGAT) An enzyme that transfers the amide group of glutamine to 2-oxoglutarate, yielding two molecules of glutamate. Also known as glutamine:2-oxoglutarate aminotransferase.

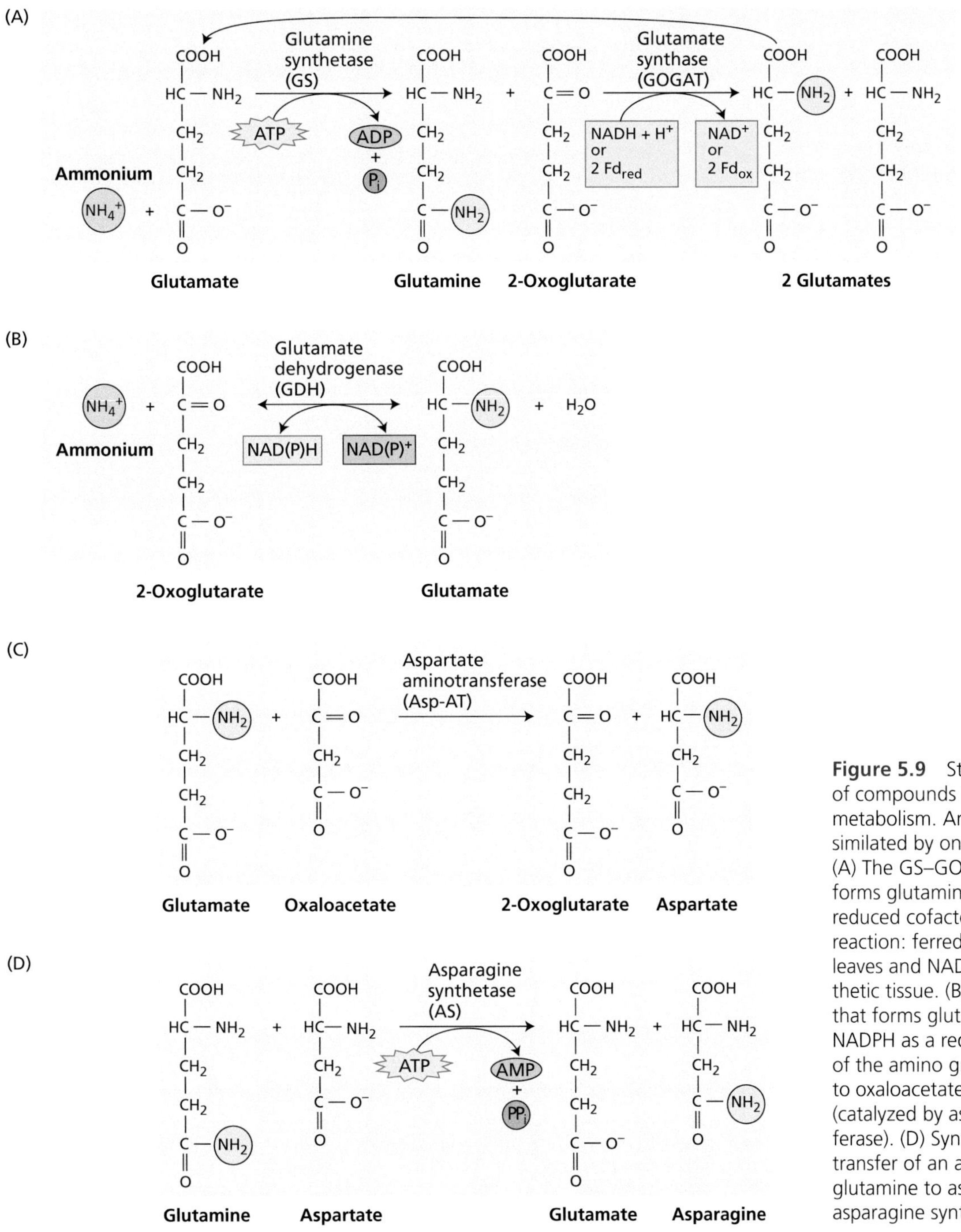

Figure 5.9 Structure and pathways of compounds involved in ammonium metabolism. Ammonium can be assimilated by one of several processes. (A) The GS–GOGAT pathway that forms glutamine and glutamate. A reduced cofactor is required for the reaction: ferredoxin (Fd) in green leaves and NADH in nonphotosynthetic tissue. (B) The GDH pathway that forms glutamate using NADH or NADPH as a reductant. (C) Transfer of the amino group from glutamate to oxaloacetate to form aspartate (catalyzed by aspartate aminotransferase). (D) Synthesis of asparagine by transfer of an amino acid group from glutamine to aspartate (catalyzed by asparagine synthetase).

amount of protein and its activity increase with light levels. Roots, particularly those under nitrate nutrition, have Fd-GOGAT in their plastids. Fd-GOGAT in the roots presumably functions to incorporate the glutamine generated during nitrate assimilation. Electrons to reduce Fd in roots are generated by the oxidative pentose phosphate pathway (Chapter 11).

Ammonium can be assimilated via an alternative pathway

Glutamate dehydrogenase (**GDH**) catalyzes a reversible reaction that synthesizes or deaminates glutamate (**Figure 5.9B**):

$$\text{2-Oxoglutarate} + NH_4^+ + NAD(P)H \rightarrow \text{glutamate} + H_2O + NAD(P)^+ \qquad (5.7)$$

glutamate dehydrogenase (GDH) An enzyme that catalyzes a reversible reaction that synthesizes or deaminates glutamate as part of the nitrogen assimilation process.

aspartate aminotransferase An aminotransferase that transfers the amino group from glutamate to the carboxyl atom of oxaloacetate to form aspartate.

asparagine synthetase (AS) An enzyme that transfers nitrogen as an amino group from glutamine to aspartate, forming asparagine.

An NADH-dependent form of GDH is found in mitochondria, and an NADPH-dependent form is localized in the chloroplasts of photosynthetic organs. Although both forms are relatively abundant, they cannot substitute for the GS–GOGAT pathway to assimilate ammonium. GDH has a very low affinity for ammonium (K_m of 10–80 mM), which is much higher than the ammonium concentration typically found in plant tissue (0.2–1.0 mM). Therefore, the primary function of GDH is to deaminate glutamate and thereby provide five-carbon skeletons (2-oxoglutarate) to plants in darkness or to germinating seeds (see Figure 5.9B).

Transamination reactions transfer nitrogen

Once assimilated into glutamine and glutamate, nitrogen can be further incorporated into other amino acids via transamination reactions. The enzymes that catalyze these reactions are called aminotransferases. An example is **aspartate aminotransferase**, which catalyze the following reaction (**Figure 5.9C**):

$$\text{Glutamate} + \text{oxaloacetate} \rightarrow \text{2-oxoglutarate} + \text{aspartate} \quad (5.8)$$

in which the amino group of glutamate is transferred to the carboxyl group of oxaloacetate to produce aspartate. Aspartate is a metabolically reactive amino acid that serves as a nitrogen donor in numerous aminotransferase reactions to produce lysine, threonine, and methionine. Aspartate also participates in the malate–aspartate shuttle to transfer reducing equivalents from the mitochondrion and chloroplast into the cytosol in C_3 plants and in the transport of carbon from the mesophyll to the bundle sheath for C_4 carbon fixation in some C_4 plants (Chapter 8). All transamination reactions require pyridoxal phosphate (vitamin B_6) as a cofactor.

Aminotransferases are found in the cytosol, chloroplasts, mitochondria, and peroxisomes. The aminotransferases in chloroplasts may have a significant role in amino acid biosynthesis, because plant leaves or isolated chloroplasts exposed to radioactively labeled carbon dioxide rapidly incorporate the label into glutamate, aspartate, alanine, serine, and glycine.

Asparagine and glutamine link carbon and nitrogen metabolism

Asparagine, isolated from asparagus as early as 1806, was the first amide to be identified. It serves not only as a component of proteins but also as a key compound for nitrogen transport and storage because of its stability and high nitrogen-to-carbon ratio (2 N to 4 C for asparagine compared with 2 N to 5 C for glutamine and 1 N to 5 C for glutamate).

The major pathway for asparagine synthesis involves the transfer of the amide nitrogen from glutamine to aspartate (**Figure 5.9D**):

$$\text{Glutamine} + \text{aspartate} + \text{ATP} \rightarrow \text{glutamate} + \text{asparagine} + \text{AMP} + \text{PP}_i \quad (5.9)$$

Asparagine synthetase (**AS**), the enzyme that catalyzes this reaction, is found in the cytosol of leaves and roots and in nitrogen-fixing nodules (Section 5.5). In maize (corn; *Zea mays*) roots, particularly those under potentially toxic levels concentrations of ammonia, ammonium may replace glutamine as the source of the amide group.

High levels of light and carbohydrate, conditions that stimulate plastid GS2 (Reaction 5.4) and Fd-GOGAT (Reactions 5.5 and 5.6), inhibit the expression of genes coding for AS and the activity of the enzyme. Thus, they favor nitrogen assimilation into glutamine and glutamate, compounds that are rich in carbon and participate in the synthesis of new plant materials.

In contrast, energy-limited conditions inhibit GS2 and GOGAT, stimulate AS, and thus favor nitrogen assimilation into asparagine, a compound that is rich in nitrogen and sufficiently stable for long-distance transport or long-term storage.

5.4 Amino Acid Biosynthesis

Describe the pathways by which amino acids are synthesized.

Humans and most animals are heterotrophic for nitrogen, relying on organic nitrogen in the foods they eat. Furthermore, they cannot synthesize certain amino acids—histidine, isoleucine, leucine, lysine, methionine, phenylalanine, threonine, tryptophan, valine, and, in the case of young humans, arginine (adult humans can synthesize arginine)—and thus must obtain these so-called essential amino acids from their diet. In contrast, plants acquire inorganic nitrogen from the soil and can synthesize all of the 20 amino acids that are common in proteins. The nitrogen-containing amino group, as discussed in Section 5.3, is derived from transamination reactions with glutamine or glutamate. The carbon skeletons for amino acids derive from 3-phosphoglycerate, phosphoenolpyruvate, or pyruvate generated during glycolysis, or from 2-oxoglutarate or oxaloacetate generated in the tricarboxylic acid cycle (**Figure 5.10**). Parts of

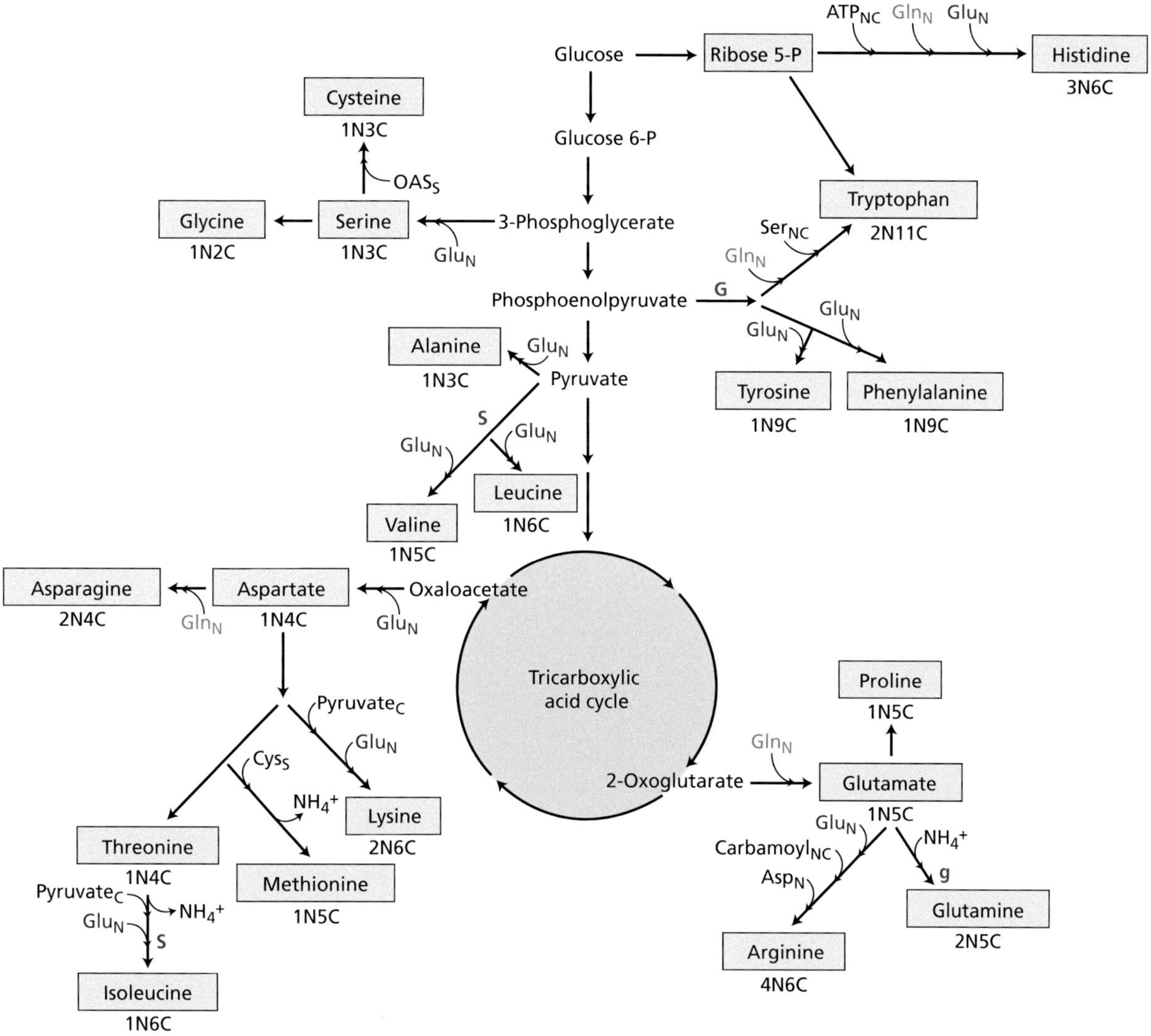

Figure 5.10 Biosynthetic pathways for the carbon skeletons of the 20 standard amino acids. N sources, S sources, and additional C sources in the biosynthetic pathways are also indicated. Asp_N, aspartate (N source); ATP_{NC}, ATP (N and C source); $carbamoyl_{NC}$, carbamoyl (N and C source); Cys_S, cysteine (S-source); Gln_N (blue), glutamine (N source); Glu_N (red), glutamate (N source); OAS_S, O-acetyl-serine (S-source); Pyr_C, pyruvate (C source); Ser_{NC}, serine (N and C source). Herbicides can block steps in amino acid synthesis: G (bold red), glyphosate; g (bold red), glufosinate, S (bold red), sulfonylureas.

these pathways required for synthesis of the essential amino acids are missing in animals and they are therefore appropriate targets for herbicides. Thus, substances that block these pathways, such as glyphosate (Roundup®), glufosinate, and sulfonylureas, are lethal to plants, but at low concentrations do not appear to injure animals.

5.5 Biological Nitrogen Fixation

Explain biological nitrogen fixation and the importance of the symbiotic relationship between plants and nitrogen-fixing bacteria.

Biological nitrogen fixation accounts for most of the conversion of atmospheric N_2 into ammonium, and thus serves as the key entry point of molecular nitrogen into the biogeochemical cycle of nitrogen (Figure 5.1). Bacterial nitrogen fixation involves the symbiotic and associative relationships between nitrogen-fixing organisms and higher plants; nodules, the specialized plant-produced structures infected by nitrogen-fixing bacteria; the genetic and signaling interactions that regulate nitrogen fixation by symbiotic prokaryotes and their hosts; and the properties of the nitrogenase enzymes that fix nitrogen.

Free-living and symbiotic bacteria fix nitrogen

Some bacteria can convert atmospheric nitrogen into ammonia (**Table 5.2**). Most of these nitrogen-fixing prokaryotes, also called diazotrophs, live in the soil, generally independent of other organisms. Several form symbiotic associations with higher plants in which the prokaryote directly provides the host plant with fixed nitrogen in exchange for other nutrients and carbohydrates. Such symbioses occur in nodules that form on the roots or sometimes stems of the plant and contain the nitrogen-fixing bacteria.

The most common type of symbiosis occurs between members of the plant family Fabaceae (Leguminosae) and soil bacteria of the genera *Azorhizobium*,

Table 5.2 Examples of Organisms That Can Carry Out Nitrogen Fixation

Symbiotic nitrogen fixation	
Host plant	**N-fixing symbionts**
Leguminous: legumes, *Parasponia* (nonlegume)	*Azorhizobium*, *Bradyrhizobium*, *Mesorhizobium*, *Rhizobium*, *Sinorhizobium*
Actinorhizal: alder (tree), *Ceanothus* (shrub), *Casuarina* (tree), *Datisca* (shrub)	*Frankia*
Gunnera	*Nostoc*
Azolla (water fern)	*Anabaena*
Sugarcane	*Acetobacter*
Miscanthus	*Azospirillum*
Free-living nitrogen fixation	
Type	**N-fixing genera**
Cyanobacteria (blue-green algae)	*Anabaena*, *Calothrix*, *Nostoc*
Other bacteria	
Aerobic	*Azospirillum*, *Azotobacter*, *Beijerinckia*, *Derxia*, *Gloeothece*
Facultative	*Bacillus*, *Klebsiella*
Anaerobic	
Nonphotosynthetic	*Clostridium*, *Methanococcus* (archaebacterium)
Photosynthetic	*Chromatium*, *Rhodospirillum*

Bradyrhizobium, Mesorhizobium, Rhizobium, and *Sinorhizobium* (collectively called **rhizobia**; **Figure 5.11** and **Table 5.3**). Another common type of symbiosis occurs between several woody plant species, such as alder trees, and soil bacteria of the genus *Frankia*; these plants are known as **actinorhizal** plants. Still other types of nitrogen-fixing symbioses involve the South American herb *Gunnera* and the tiny water fern *Azolla,* which form associations with the cyanobacteria *Nostoc* and *Anabaena,* respectively (**Figure 5.12**). Finally, several types of nitrogen-fixing bacteria are associated with grasses or cereals.

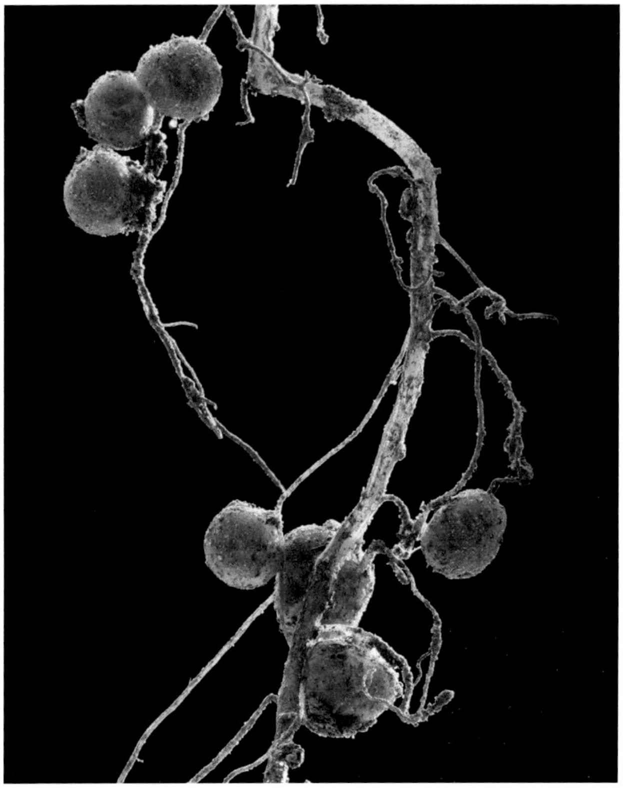

Figure 5.11 Root nodules on a common bean (*Phaseolus vulgaris*). The nodules, the spherical structures, are a result of infection by *Rhizobium* sp.

Nitrogen fixation requires microanaerobic or anaerobic conditions

Because nitrogen fixation involves the expenditure of large amounts of energy, the nitrogenase enzymes that catalyze these reactions have sites that facilitate the high-energy exchange of electrons. Oxygen, being a strong electron acceptor, can damage these sites and irreversibly inactivate nitrogenase, so nitrogen must be fixed under anaerobic conditions. Each of the nitrogen-fixing organisms listed in Table 5.2 either functions under natural anaerobic conditions or creates an internal, local anaerobic (microanaerobic) environment separated from the oxygen in the atmosphere or the solution that surrounds it.

In filamentous cyanobacteria, nitrogen fixation occurs in specialized cells called *heterocysts* (Figure 5.12). The low oxygen concentration in heterocysts is achieved by the elimination of oxygen-producing photosystem II activity (Chapter 7), an increased respiration rate, and synthesis of a heterocyst-specific peptidoglycan layer that thickens the cell wall and minimizes oxygen diffusion. Heterocysts differentiate from vegetative cells when cyanobacteria are deprived of NH_4^+. As a result of nitrogen starvation, 2-oxoglutarate accumulates in vegetative cells, inducing the synthesis of the transcription factors nitrogen control gene (NtcA) and heterocyst differentiation master regulator (HetR), which act as a global nitrogen regulator and a specific master regulator, respectively, and leading to heterocyst differentiation. HetR interacts with heterocyst commitment protein (HetP), a protein containing no domain of known function, to commit the cell to irreversibly differentiate into a heterocyst. The number of heterocysts in filaments depends on the environmental conditions. Usually one heterocyst cell is present

rhizobia A collective term for the genera of soil bacteria that form symbiotic (mutualistic) relationships with members of the plant family Fabaceae (Leguminosae).

actinorhizal Pertaining to several woody plant species, such as alder trees, in which symbiosis occurs with soil bacteria of the nitrogen-fixing genus *Frankia*.

Table 5.3 Associations Between Host Plants and Rhizobia

Plant host	Rhizobial symbiont
Parasponia (a nonlegume)	Bradyrhizobium spp.
Soybean (*Glycine max*)	*Bradyrhizobium japonicum* (slow-growing type); *Sinorhizobium fredii* (fast-growing type)
Alfalfa (*Medicago sativa*)	*Sinorhizobium meliloti*
Sesbania (aquatic)	*Azorhizobium* (forms both root and stem nodules; the stems have adventitious roots)
Bean (*Phaseolus*)	*Rhizobium leguminosarum* bv. *phaseoli*; *R. tropicii*; *R. etli*
Clover (*Trifolium*)	*Rhizobium leguminosarum* bv. *trifolii*
Pea (*Pisum sativum*)	*Rhizobium leguminosarum* bv. *viciae*
Aeschynomene (aquatic)	Photosynthetic *Bradyrhizobium* clade (photosynthetically active rhizobia that form stem nodules, probably associated with adventitious roots)

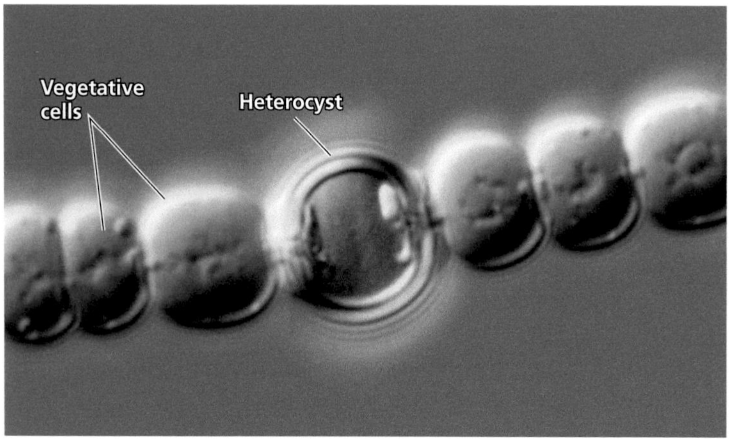

Figure 5.12 A heterocyst of a cyanobacterium. A heterocyst in a filament of the nitrogen-fixing cyanobacterium *Anabaena*, which forms associations with the water fern *Azolla*. The thick-walled heterocysts, interspersed among vegetative cells, have an anaerobic inner environment that allows cyanobacteria to fix nitrogen in aerobic conditions.

for every ten vegetative cells, enabling an efficient use of energy resources for carbon and nitrogen metabolism in the two types of cells.

Cyanobacteria can fix nitrogen under anaerobic conditions such as those that occur in flooded fields. In Asian countries, nitrogen-fixing cyanobacteria of both the heterocyst and nonheterocyst types are a major means for maintaining an adequate nitrogen supply in the soil of rice fields. Cyanobacteria fix nitrogen when the fields are flooded and die as the fields dry, releasing the fixed nitrogen to the soil. Another important source of available nitrogen in flooded rice fields is the *Azolla–Anabaena* association, which can fix as much as 0.5 kg of atmospheric nitrogen per hectare per day, a rate of fertilization that is sufficient to attain moderate rice yields. **Box 5.1** discusses how plants can obtain fixed nitrogen in future agriculture.

Free-living bacteria that are capable of fixing nitrogen are aerobic, facultative, or anaerobic (Table 5.2):

- *Aerobic* nitrogen-fixing bacteria such as *Azotobacter* are thought to maintain a low oxygen concentration (microaerobic conditions) through their high levels of respiration. Others, such as *Gloeothece*, evolve O_2 photosynthetically during the day and fix nitrogen during the night when respiration lowers oxygen concentrations.
- *Facultative* organisms are able to grow under both aerobic and anaerobic conditions and generally fix nitrogen only under anaerobic conditions.
- Obligate *anaerobic* nitrogen-fixing bacteria that grow in environments devoid of oxygen can be either photosynthetic (e.g., *Rhodospirillum*) or nonphotosynthetic (e.g., *Clostridium*).

Box 5.1 Challenges and Solutions for Solving Nitrogen Deficiency in Future Agriculture

Nutrient acquisition, especially of phosphorus and nitrogen, is crucial for balancing plant growth and reproduction. New agricultural technologies and practices are required for supporting an increasing human population. In the coming decades the challenge will be to meet future food demands while reducing the environmental impact of intensive agriculture, currently using high inputs of industrial fertilizers. Acquisition of nutrients through beneficial plant–microbe associations is recognized as a sustainable solution to this challenge. Identifying microbes that can efficiently provide these nutrients to plant hosts in return for carbon resources, without inducing a pathogenic effect, is now a prime target for research.

In the legume–*Rhizobium* symbiosis, the nitrogen-fixing bacteria have the capacity to provide most of the nitrogen required by the plant host. Consequently, the areas cultivated with legumes as food and cover or forage crops are increasing to meet societal demands for protein production in a sustainable manner. Elite nitrogen-fixing rhizobial strains are used as inoculum, but to take full advantage of their symbiotic capacities we need to understand how to increase bacterial persistence across diverse soils, and to ensure their effective selection by the legume hosts.

Cereals lack this endosymbiotic capacity and are therefore dependent on nitrogen fertilizers. Identification of root-associative diazotrophs and knowledge of how to manipulate and adapt diazotrophs to agricultural settings can reduce dependence on nitrogen fertilizers. Plant genotype, the isolate, and the environment all have a major effect on the establishment of root–diazotroph associations, as well as on the amount of fixed nitrogen received by the host. Biotechnological approaches target genetic engineering of cereals by expression of the nitrogenase complex in the chloroplasts and mitochondria of cereal cells, or by transferring the ability to recognize and accommodate the nitrogen-fixing bacteria through symbiotic association. Once successful, these biotechnological approaches will have the potential to create a new agricultural revolution.

Cereals can develop associative relationships with nitrogen-fixing bacteria, but in these interactions root nodules are not produced. Instead, the nitrogen-fixing bacteria attach to the root surfaces, mainly around the elongation zone and the root hairs, or live as endophytes, colonizing the apoplastic space of plant tissues without causing disease. For example, *Acetobacter diazotrophicus* and *Herbaspirillum* species live in the apoplast of stem tissues in sugarcane and may provide their host with about 30% of its nitrogen. Recent studies of indigenous maize native cultivars grown in nitrogen-poor, high-humidity environments of Mexico's highlands showed that their aerial roots secrete carbohydrate-rich mucigel and associate with nitrogen-fixing bacteria. Under these conditions, the fixation of atmospheric nitrogen contributed 30 to 80% of the nitrogen nutrition of these maize genotypes. The potential for associative and endophytic nitrogen-fixing bacteria to supplement the nitrogen nutrition of cereals has been explored, but the large diversity of bacterial species, and the variety of plant responses to these bacteria, have impeded progress.

Symbiotic nitrogen fixation occurs in specialized structures

Some symbiotic nitrogen-fixing prokaryotes dwell within **nodules**, the special organs of the plant host that enclose the nitrogen-fixing bacteria (Figure 5.11). In the case of *Gunnera*, these organs are preexisting stem glands that develop independently of the symbiont. In the case of legumes and actinorhizal plants, bacteria induce the plant to form the nodules.

Legumes and actinorhizal plants regulate gas permeability in their nodules, maintaining oxygen concentrations of 20 to 40 nanomolar (n*M*) within the nodule (about 10,000 times lower than equilibrium concentrations in water). These concentrations can support respiration but are sufficiently low to avoid inactivation of the nitrogenase. Gas permeability increases in the light and decreases under drought (water-deficit stress) or upon exposure to nitrate. The mechanism for regulating gas permeability is not yet known, but it may involve potassium ion fluxes into and out of infected cells.

Nodules contain oxygen-binding heme proteins called **leghemoglobins** that have a high affinity for oxygen (a K_m of about 10 n*M*), about ten times higher than that of the β chain of human hemoglobin. Leghemoglobins are the most abundant proteins in nodules, giving them a heme-pink color, and are crucial for symbiotic nitrogen fixation.

Leghemoglobins increase the rate of oxygen transport to the respiring symbiotic bacteria, decreasing substantially the steady-state concentration of oxygen in infected cells. To continue aerobic respiration under such conditions, the bacteroid uses a specialized electron transport chain (Chapter 11) in which the terminal oxidase has an affinity for oxygen even higher than that of leghemoglobins, a K_m of about 7 n*M*.

Establishing symbiosis requires an exchange of signals

The symbiosis between legumes and rhizobia is not obligatory. Legume seedlings germinate without any association with rhizobia, and they may remain unassociated throughout their life cycle. Rhizobia also occur as free-living organisms in the soil. Under nitrogen-limited conditions, however, an elaborate exchange of signals takes place: The plant secretes specific signals that are recognized by potential endosymbionts, which in turn secrete signals that the plant recognizes. This signaling, the subsequent infection process, and the development of nitrogen-fixing nodules involve specific genes in both the host and the symbionts.

Plant genes up-regulated during root nodule symbiosis have been generically called **nodulin genes**. A set of rhizobial genes called **nodulation (*nod*) genes** is responsible for the biosynthesis and secretion of specific signaling molecules, the Nod factors. The *nod* genes are classified as common *nod* genes or host-specific

nodules Specialized organs of a plant host that contain symbiotic nitrogen-fixing bacteria.

leghemoglobin An oxygen-binding heme protein produced by legumes during nitrogen-fixing symbioses with rhizobia. Found in the cytoplasm of infected nodule cells, it facilitates the diffusion of oxygen to the respiring symbiotic bacteria.

nodulin genes Plant genes specific to nodule formation.

nodulation (*nod*) genes Rhizobial genes, the products of which participate in nodule formation.

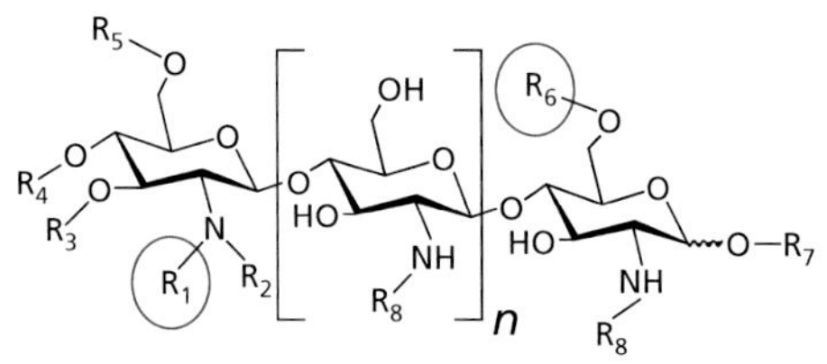

R1: fatty acid: C18:0; C18:1, C16:0; C16:1
R2: methyl, hydrogen
R3: hydrogen, carbamoyl
R4: hydrogen, carbamoyl
R5: hydrogen, carbamoyl, acetyl
R6: hydrogen, fucose, sulfate, acetyl, 4-*O*-acetyl fucose, 2-*O*-methyl fucose
R7: hydrogen, glycerol, mannosyl
R8: acetyl
n=1, 2, 3

Figure 5.13 Nod factors are lipochitin oligosaccharides. The fatty acid chain typically has 16 to 18 carbons. The number of repeated middle sections (*n*) is usually two or three. The number before the colon gives the total number of carbons in the fatty acyl chain, and the number after the colon gives the number of double bonds.

nod genes. The common *nod* genes—*nodA*, *nodB*, and *nodC*—are found in all rhizobial strains; the host-specific *nod* genes—such as *nodP*, *nodQ*, and *nodH*; or *nodF*, *nodE*, and *nodL*—may be present or not in different rhizobial species and determine the host range (the plants that can be infected). Only one of the *nod* genes, the regulatory *nodD*, is constitutively expressed, and as we will explain in detail, its protein product (NodD) regulates the transcription of the other *nod* genes.

The first stage in the formation of the symbiotic relationship between the nitrogen-fixing bacteria and their host is migration of the bacteria toward the roots of the host plant. This migration is a chemotactic response mediated by chemical attractants, especially (iso)flavonoids and betaines, secreted by the roots. These attractants activate the rhizobial NodD protein, which then induces transcription of the other *nod* genes. The promoter region of all *nod* operons, except that of *nodD*, contains a highly conserved sequence called the *nod* box. Binding of the activated NodD to the *nod* box induces transcription of the other *nod* genes.

Nod factors produced by bacteria act as signals for symbiosis

Nod factors are lipochitin oligosaccharide signal molecules, all of which have a chitin β-1,4-linked *N*-acetyl-D-glucosamine backbone (varying in length from three to six sugar units) and a fatty acid chain on the C-2 position of the nonreducing sugar (**Figure 5.13**).

Three of the *nod* genes (*nodA*, *nodB*, and *nodC*) encode enzymes (NodA, NodB, and NodC, respectively) that are required for synthesizing the basic lipochitinous structure:

1. NodA is an *N*-acyltransferase that catalyzes the addition of a fatty acyl chain.
2. NodB is a chitin-oligosaccharide deacetylase that removes the acetyl group from the terminal nonreducing sugar.
3. NodC is a chitin-oligosaccharide synthase that links *N*-acetyl-D-glucos amine monomers.

The NOD proteins, whose presence varies among rhizobial species, are primarily involved in the modification of the length and saturation of the fatty acyl chain (NodE and NodF) or the addition of reducing or nonreducing sugar groups (NodS, NodH, and NodZ) important in determining host specificity.

Legume hosts use Nod factor receptors (NFRs) to recognize and respond to specific Nod factors (**Figure 5.14**). These receptors have extracellular regions with three sugar-binding LysM domains (for lysin motif, a widespread protein module originally identified in enzymes that degrade bacterial cell walls but also present in many other proteins), a single-pass transmembrane domain, and an intracellular protein kinase domain. Binding of Nod factors to the NFR extracellular domains activates the intracellular protein kinase domain, which then initiates a signaling cascade. Some components of this signaling cascade, such as symbiosis receptor-like kinase (SYMRK), nucleoporins, cation channels, and cyclic nucleotide–gated channels, are also crucial for plant symbiosis with arbuscular mycorrhizal fungi (Chapters 4 and 18) and are required for activation of oscillations in the Ca^{2+} concentration in the nuclei of root epidermal cells (Figure 5.14; Chapter 12). Nuclear Ca^{2+} oscillations are interpreted by a Ca^{2+}/calmodulin-dependent protein kinase (CCaMK) that associates with a transcriptional activator that, together with specific transcription factors, activates the expression

Nod factors Lipochitin oligosaccharide signal molecules active in regulating gene expression during nitrogen-fixing nodule formation. All Nod factors have a chitin β-1,4-linked *N*-acetyl-D-glucosamine backbone (varying in length from three to six sugar units) and a fatty acid chain on the C-2 position of the nonreducing sugar.

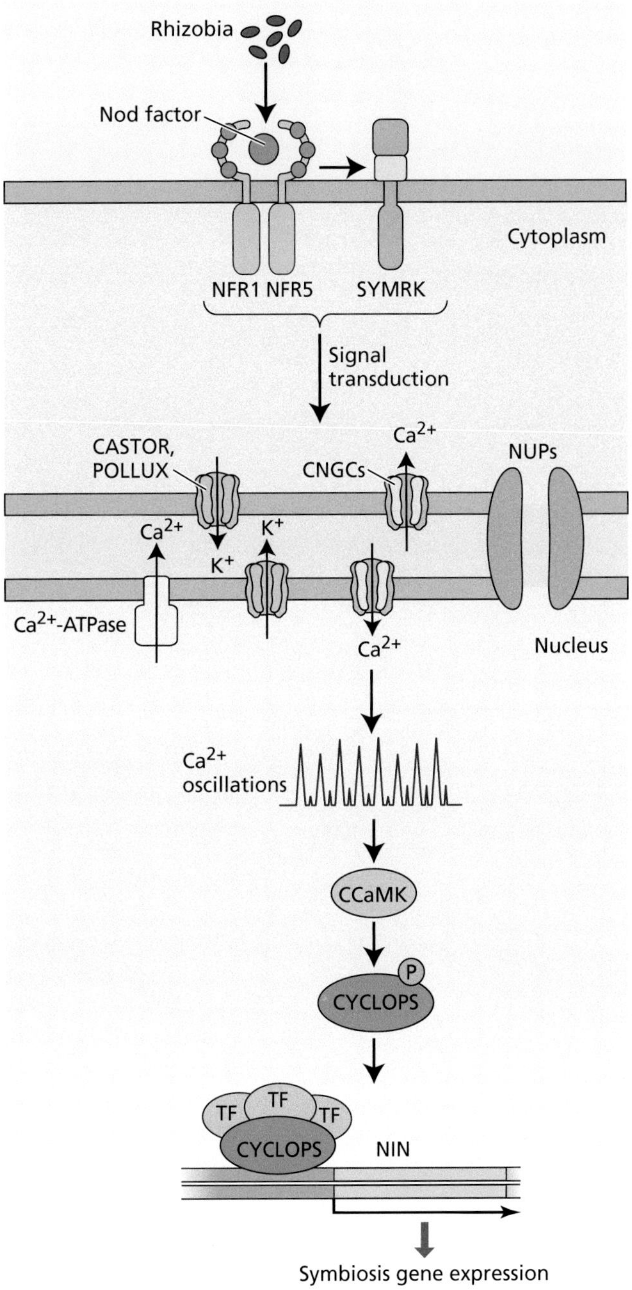

Figure 5.14 Simplified intracellular nodulation signaling cascade. Nodulation processes begin with binding of compatible Nod factors to Nod factor receptors (NFR1 and NFR5). CASTOR/POLLUX, Ca^{2+}/K^+ cation channels; CCaMK, Ca^{2+}/calmodulin-dependent protein kinase; CNGCs, cyclic nucleotide–gated channels; CYCLOPS, a transcriptional activator; NIN, nodule inception protein, controlling both nodulation and infection; NUPs, nucleoporins; SYMRK, symbiotic receptor kinase; TF, transcription factor.

of early nodulin genes, leading to root nodule formation and bacterial infection.

Nodule formation involves phytohormones

Infection and nodule organogenesis occur simultaneously during root nodule symbiosis. Nod factor–producing rhizobia attach to developing root hairs, which then reorient their tip growth and curl around the bacteria (**Figure 5.15A** and **Figure 5.15B**). The rhizobia enclosed in the curl multiply and form a microcolony. The cell wall of the curling root hair undergoes major remodeling in response to Nod factors, and an apoplastic infection chamber is formed. The next step is formation of the **infection thread** (**Figure 5.15C**), an internal tubular extension of the plasma membrane that is produced by the fusion of Golgi-derived membrane vesicles at the site of infection. The thread grows at its tip by the fusion of secretory vesicles to the end of the tube. In parallel with infection thread formation in the root hairs, cells deeper within the root cortex dedifferentiate and start dividing, forming a distinct area called a *nodule primordium*, from which the nodule will develop. The nodule primordia form opposite the protoxylem poles of the root vascular bundle.

Different signaling compounds, acting either positively or negatively, control the development of nodule primordia. Nod factors activate localized cytokinin signaling in the root cortex and pericycle, leading to the localized suppression of polar auxin transport, inducing localized cell division. Ethylene is synthesized in the region of the pericycle, diffuses into the cortex, and blocks cell division opposite the phloem poles of the root (Chapter 15).

The infection thread filled with proliferating rhizobia elongates through the root hair and cortical cell layers, in the direction of the nodule primordium, where it branches, leading to increased access to multiple plant cells. When the infection thread reaches the cells of the nodule primordium, its tip fuses with the plasma membrane of a host cell (**Figure 5.15D**) and bacterial cells surrounded by host-derived membrane are released into the cytoplasm, forming the *symbiosomes* (**Figure 5.15E** and **Figure 5.15F**).

At first the bacteria within symbiosomes continue to divide, and the surrounding symbiosome membrane (also called the *peribacteroid membrane*) increases in surface area to accommodate this growth by fusing with smaller vesicles. Soon thereafter, upon an undetermined signal from the plant, the bacteria stop dividing and begin to differentiate into nitrogen-fixing **bacteroids**.

The mature nodule develops a vascular system (which facilitates the exchange of fixed nitrogen produced by the bacteroids for nutrients contributed

infection thread An internal tubular extension of the plasma membrane of root hairs through which rhizobia enter root cortical cells.

bacteroids Endosymbiotic bacteria that have differentiated into a nondividing nitrogen-fixing state.

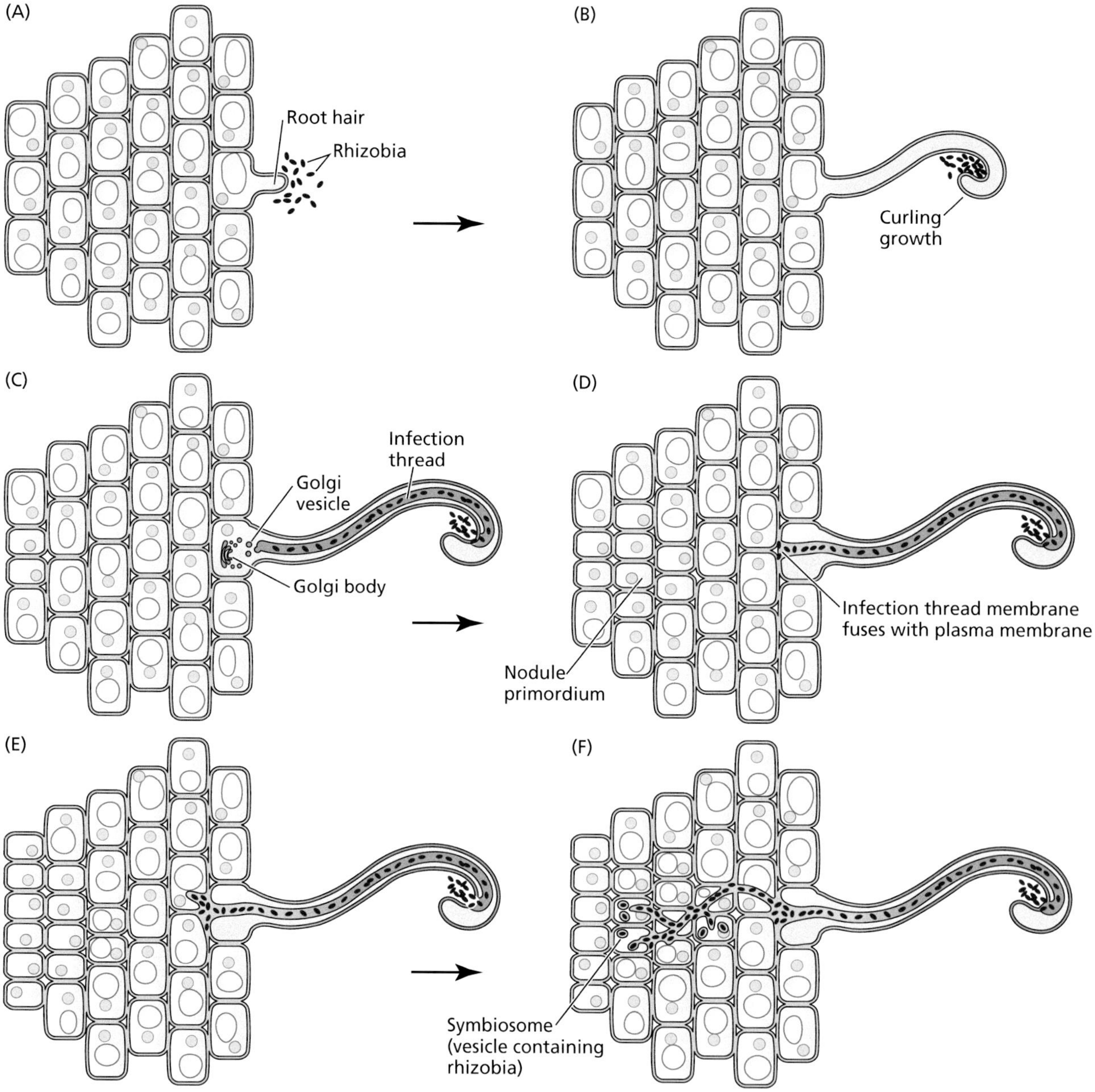

Figure 5.15 The infection process during nodule organogenesis. (A) Rhizobia bind to an emerging root hair in response to chemical attractants sent by the plant. (B) In response to factors produced by the bacteria, the root hair exhibits abnormal curling growth, and rhizobia cells proliferate within the coils. (C) Localized degradation of the root hair wall leads to infection and formation of the infection thread from Golgi secretory vesicles of root cells. (D) The infection thread reaches the end of the cell, and its membrane fuses with the plasma membrane of the root hair cell. (E) Rhizobia are released into the apoplast and penetrate the compound middle lamella to the subepidermal cell plasma membrane, leading to the initiation of a new infection thread, which forms an open channel with the first. (F) The infection thread extends and branches until it reaches target cells, where vesicles (symbiosomes) composed of plant membrane that enclose bacterial cells are released into the cytosol.

by the plant) and a layer of cells to limit diffusion of O_2 into the root nodule interior. In some temperate legumes (e.g., pea), the nodules are elongated and cylindrical because of the presence of a *nodule meristem* at the tip. The nodules of tropical legumes, such as soybean and peanut, lack a persistent meristem and are spherical.

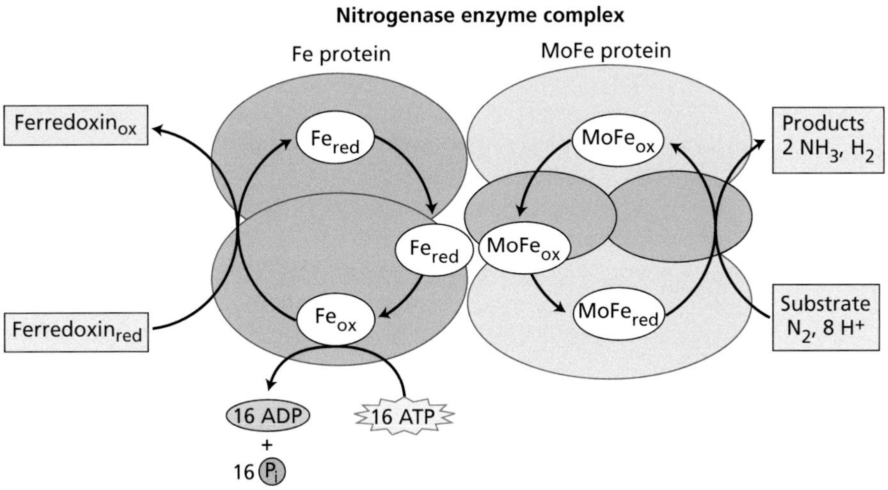

Figure 5.16 The reaction catalyzed by nitrogenase. Ferredoxin reduces the Fe protein. Binding and hydrolysis of ATP to the Fe protein are thought to cause a conformational change of the Fe protein that facilitates the redox reactions. The Fe protein reduces the MoFe protein, and the MoFe protein reduces the N_2.

The nitrogenase enzyme complex fixes N_2

Biological nitrogen fixation, like industrial nitrogen fixation, produces ammonia from molecular nitrogen. The overall reaction is

$$N_2 + 8e^- + 8H^+ + 16\,ATP \rightarrow 2NH_3 + H_2 + 16\,ADP + 16\,P_i \qquad (5.10)$$

nitrogenase enzyme complex The two-component protein complex that catalyzes the biological nitrogen fixation reaction in which ammonia is produced from molecular nitrogen.

Note that the reduction of N_2 to two NH_3, a six-electron transfer, is coupled to the reduction of two protons to evolve H_2. The **nitrogenase enzyme complex** catalyzes this reaction.

The nitrogenase enzyme complex has two components (the Fe protein and the MoFe protein), neither of which has catalytic activity by itself (**Figure 5.16**). For one dinitrogen molecule to be reduced to two ammonia molecules, the Fe protein and the MoFe protein have to interact eight times:

- The Fe protein, known as the reductase, participates in the redox reactions that convert N_2 to NH_3. It is a homodimeric protein with two iron–sulfur clusters (four Fe^{2+}/Fe^{3+} and four S^{2-}) and two ATP-binding sites. The Fe protein is irreversibly inactivated by O_2 with a typical half-decay time of 30 to 45 s.
- The MoFe protein has four subunits, each with two Mo–Fe–S clusters. The MoFe protein is also inactivated by O_2, with a half-decay time in air of 10 min.

In the overall nitrogen reduction reaction, ferredoxin serves as an electron donor to the Fe protein, which in turn hydrolyzes ATP and reduces the MoFe protein. Although the MoFe protein can reduce numerous substrates (**Table 5.4**), under natural conditions it reacts only with N_2 and H^+. Researchers use the reduction of acetylene to ethylene to estimate nitrogenase activity.

The energetics of nitrogen fixation are complex. The triple bond of N_2 is very stable and requires a large investment of energy to break it to produce ammonia in both industrial and biological NH_3 production. The industrial Haber-Bosch process occurs under high temperature

Table 5.4 Reactions Catalyzed by Nitrogenase

$N_2 \rightarrow NH_3$	Molecular nitrogen fixation
$2H^+ \rightarrow H_2$	H_2 production
$N_2O \rightarrow N_2 + H_2O$	Nitrous oxide reduction
$N_3^- \rightarrow N_2 + NH_3$	Azide reduction
$C_2H_2 \rightarrow C_2H_4$	Acetylene reduction
$ATP \rightarrow ADP + P_i$	ATP hydrolytic activity

Note: The reactions in the first two rows occur under natural conditions.

Source: R. H. Burris. 1976. Nitrogen fixation. In *Plant Biochemistry*, 3rd ed., J. Bonner and J. Varner, eds., Academic Press, New York, pp. 887–908.

and pressure with a catalyst to produce ammonia. Calculations based on the carbohydrate metabolism of legumes show that a plant respires 7 to 12 moles of CO_2 per mole of N_2 fixed. The enzymatic reaction is limited by the slow operation (number of N_2 molecules reduced per unit time is about 5 s^{-1}) of the nitrogenase complex. The bacteroid synthesizes large amounts of nitrogenase (up to 20% of the total protein in the cell) to optimize the rate of nitrogen fixation (Reaction 5.10).

Under natural conditions, substantial amounts of H^+ are reduced to H_2 gas, and this process can compete with N_2 reduction for electrons from nitrogenase. In rhizobia, 30 to 60% of the energy supplied to nitrogenase may be lost as H_2, diminishing the efficiency of nitrogen fixation. Some rhizobia, however, contain hydrogenase, an enzyme that can split the H_2 formed and generate electrons for N_2 reduction, thus improving the efficiency of nitrogen fixation.

Amides and ureides are the transported forms of nitrogen

The symbiotic nitrogen-fixing prokaryotes release ammonia that, to avoid toxicity, must be rapidly converted into organic forms in the root nodules before being transported to the shoot via the xylem. Nitrogen-fixing legumes can be classified as amide exporters or ureide exporters, depending on the composition of the xylem sap. Amides (principally the amino acids asparagine or glutamine) are exported by temperate-region legumes, such as pea (*Pisum*), clover (*Trifolium*), broad bean (*Vicia*), and lentil (*Lens*) (Figure 5.6).

Ureides are exported by legumes of tropical origin, such as soybean (*Glycine*), common bean (*Phaseolus*), peanut (*Arachis*), and southern pea (*Vigna*) (Figure 5.6). The three major ureides are allantoin, allantoic acid, and citrulline (**Figure 5.17**). Allantoin is synthesized in peroxisomes from uric acid, and allantoic acid is synthesized from allantoin in the endoplasmic reticulum. The site of citrulline synthesis from the amino acid ornithine has not yet been determined. All three compounds are ultimately released into the xylem and transported to the shoot, where they are rapidly catabolized to ammonium. This ammonium enters the assimilation pathway described earlier.

5.6 Sulfur Assimilation

Describe the mechanisms in which plants can assimilate sulfur and synthesize sulfur-containing amino acids.

Sulfur is among the most versatile elements in living organisms. Disulfide bridges in proteins play structural and regulatory roles (Chapter 8). Sulfur participates in electron transport through iron–sulfur clusters (Chapters 7 and 11). The catalytic sites for several enzymes and coenzymes, such as urease and coenzyme A, contain sulfur. Specialized metabolites (compounds that are not involved in primary pathways of growth and development) that contain sulfur range from the rhizobial Nod factors to glucosinolate plant defense compounds (Chapter 18) that are also anticarcinogenic to the antiseptic alliin in garlic.

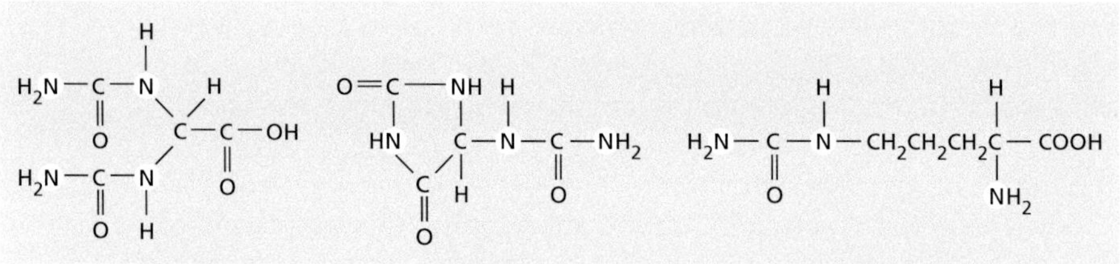

Allantoic acid   Allantoin   Citrulline

Figure 5.17 The major ureide compounds in nitrogen transport. The major ureide compounds used to transport nitrogen from sites of fixation to sites where their deamination will provide nitrogen for amino acid and nucleoside synthesis.

The versatility of sulfur derives in part from the property that it shares with nitrogen: *multiple stable oxidation states*. In this section we discuss sulfur assimilation into the two sulfur-containing amino acids, cysteine and methionine.

Sulfate is the form of sulfur transported into plants

Most of the sulfur in higher-plant cells derives from sulfate (SO_4^{2-}) transported via an H^+–SO_4^{2-} symporter (Chapter 6) from the soil solution. Sulfate in the soil comes predominantly from the weathering of parent rock material. Industrialization, however, adds an additional source of sulfate: atmospheric pollution. The burning of fossil fuels releases several gaseous forms of sulfur, including sulfur dioxide (SO_2) and hydrogen sulfide (H_2S), which find their way to the soil in rain. The practice of burning hard coal for fuel has been decreasing since the 1980s, decreasing the deposition of atmospheric sulfur in the soil.

In the gas phase, sulfur dioxide reacts with a hydroxyl radical and oxygen to form sulfur trioxide (SO_3). SO_3 dissolves in water to become sulfuric acid (H_2SO_4), a strong acid, which is the major cause of acid rain. Plants can metabolize sulfur dioxide taken up in the gaseous form through their stomata. Nonetheless, prolonged exposure (>8 h) to high atmospheric concentrations (>0.3 ppm) of SO_2 causes extensive tissue damage because of the formation of sulfuric acid inside the plant leaf.

Sulfate assimilation occurs mostly in leaves

Sulfur assimilation is similar to nitrogen assimilation in many ways, as the sulfate is reduced to sulfite and then reduced to sulfide for synthesis of the sulfur-containing amino acid cysteine (Figure 5.10). The main difference from N assimilation is that sulfate is very stable and thus needs to be activated before any subsequent reactions may proceed. Activation begins with a reaction between sulfate and ATP to form adenosine-5′-phosphosulfate (APS) and pyrophosphate (PP_i), after which it can be reduced or phosphylated. The reduction of sulfate to cysteine changes the oxidation number of sulfur from +6 to −2, thus involving the transfer of eight electrons. Glutathione, ferredoxin, NAD(P)H, or *O*-acetylserine may serve as electron donors at various steps of the pathway. Leaves are generally much more active than roots in sulfur assimilation, presumably because photosynthesis provides reduced ferredoxin (Chapter 7), and photorespiration generates serine (Chapter 8), which may stimulate the production of *O*-acetylserine that is part of the cysteine synthesis pathway. Sulfur assimilated in leaves is exported via the phloem to sites of protein synthesis (shoot and root apices, and fruits) mainly as glutathione:

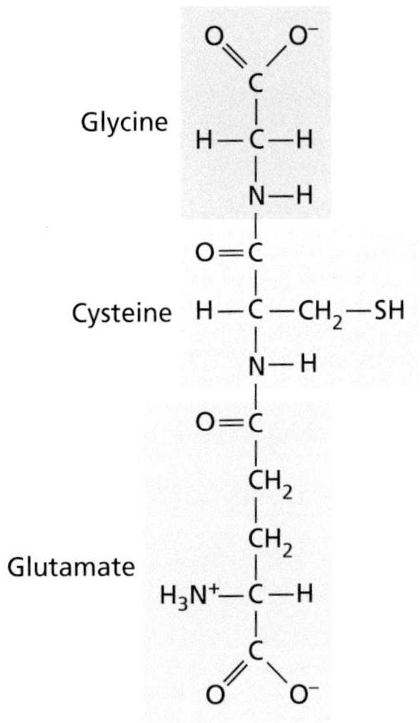

Reduced glutathione

Methionine is synthesized from cysteine

Methionine, the other sulfur-containing amino acid found in proteins, is synthesized in plastids from cysteine. After cysteine and methionine are synthesized, sulfur can be incorporated into proteins and a variety of other compounds, such as acetyl-CoA and *S*-adenosylmethionine. The latter compound is important in the synthesis of ethylene (Chapters 12, 17, and 19).

5.7 Phosphate Assimilation

Describe the role of ATP during phosphate assimilation and the relationship between phosphate and sulfate signaling.

Phosphate (HPO_4^{2-}) in the soil solution is readily transported into plant roots via an H^+–HPO_4^{2-} symporter (Chapter 6) and incorporated into a variety of organic compounds, including sugar phosphates, phospholipids, and nucleotides. The main entry point of phosphate into assimilatory pathways occurs during the formation of ATP, the energy "currency" of the cell. In the overall reaction for this process, inorganic phosphate is added to the second phosphate group in adenosine diphosphate to form a phosphate ester bond.

In mitochondria, the energy for ATP synthesis derives from the oxidation of NADH or succinate by oxidative phosphorylation (Chapter 11). ATP synthesis is also driven by light-dependent photophosphorylation in the chloroplasts (Chapter 7). In addition, glycolysis incorporates inorganic phosphate into 1,3-bisphosphoglyceric acid, forming a high-energy acyl phosphate group. This phosphate can be donated to ADP to form ATP in a substrate-level phosphorylation reaction (Chapter 11). Once incorporated into ATP, the phosphate group may be transferred via many different reactions to form the various phosphorylated compounds found in plant cells.

miRNAs contribute to phosphate and sulfate signaling

Mediated by the transcription factor PHR1, phosphate deficiency quickly induces the expression of miR399 in shoots, and later in roots. Shoot-derived miR399 is transported via phloem transport into roots (Chapter 10), where it facilitates cleavage of *PHOSPHATE 2* (*PHO2*) mRNA that encodes a ubiquitin-conjugating E2 enzyme (Chapter 12). This action relieves the degradation of PHO1 and PHOSPHATE TRANSPORTER 1 (PHT1) to enhance root-to-shoot phosphate translocation and phosphate uptake, respectively, under conditions of phosphate deficiency.

In a similar process, mediated by the transcription factor SULFUR LIMITATION 1 (SLIM1), sulfate deficiency induces an accumulation of miR395 in phloem, thereby limiting the expression of its target *SULFATE TRANSPORTER 2;1* (*SULTR2;1*) in xylem parenchyma to enhance root-to-shoot sulfate transport. Additional targets of miR395 include three ATP sulfurylase isoforms (*ATPS1*, *ATPS3*, and *ATPS4*) that modulate not only sulfate assimilation flux but also sulfate translocation.

5.8 Oxygen Assimilation

Assess the importance of oxygen assimilation in plant metabolism.

Respiration accounts for the bulk (~90%) of the O_2 assimilated by plant cells (Chapter 11). Another major pathway for the assimilation of O_2 into organic compounds involves the incorporation of O_2 from water (Chapter 8). A small proportion of oxygen can be directly assimilated into organic compounds in the process of *oxygen fixation* via enzymes known as *oxygenases*. The most prominent oxygenase in plants is ribulose 1,5-bisphosphate carboxylase/oxygenase (Rubisco), which during photorespiration incorporates oxygen into an organic compound and releases energy (Chapter 8).

5.9 The Energetics of Nutrient Assimilation

Explain how plants couple nutrient assimilation to photosynthetic electron transport to produce energy.

Nutrient assimilation—particularly of nitrogen and sulfur—requires large amounts of energy to convert stable, low-energy, highly oxidized inorganic compounds into high-energy, highly reduced organic compounds. The nitrate (NO_3^-) assimilation process consumes the equivalent of 17 ATP per amide nitrogen. This means that about 25% of the total energy expenditures in both roots and shoots is used to assimilate nitrogen, a constituent that accounts for less than 2% of the total dry weight of the plant. For comparison, the assimilation of sulfate (SO_4^{2-}) into the amino acid cysteine consumes about 14 ATP per S atom assimilated.

Nitrogen and carbon assimilation are closely linked. Many of the assimilatory reactions for nitrogen occur in the stroma of the chloroplast where they have ready access to powerful reducing agents, such as NADPH, thioredoxin, and ferredoxin, generated during photosynthetic electron transport. The process of coupling nutrient assimilation to photosynthetic electron transport is called **photoassimilation**.

photoassimilation The coupling of nutrient assimilation to photosynthetic electron transport.

Photoassimilation and the C_3 carbon fixation cycle occur in the same compartment. However, only when photosynthetic electron transport generates reductant in excess of the needs of the C_3 carbon fixation cycle—for example, under conditions of high light and low CO_2—does photoassimilation proceed. High concentrations of CO_2 inhibit nitrate assimilation in the shoots of C_3 plants but not in C_4 plants, for complex reasons (Chapter 8 describes C_3 and C_4 photosynthesis). If, as expected, atmospheric levels of CO_2 increase to about 700 ppm during this century (Chapter 9), CO_2 inhibition of shoot nitrate assimilation will increasingly affect plant–nutrient relations. Food quality of C_3 crops such as wheat has already suffered losses and will decline even more during the coming decades. Breeding crops for enhanced root nitrate and ammonium assimilation has the potential to mitigate such losses in food quality, but this approach is yet untapped.

Summary

Nutrient assimilation is the often energy-requiring process by which plants incorporate inorganic nutrients into the carbon constituents necessary for growth and development.

5.1 Nitrogen in the Environment

- When nitrogen is fixed into ammonia (NH_3) or nitrate (NO_3^-), it passes through several organic or inorganic forms before it eventually returns to molecular nitrogen (N_2) (**Table 5.1**; **Figure 5.1**).
- At high concentrations, ammonium (NH_4^+) is toxic to living tissues, but nitrate can be safely stored and translocated in plant tissues (**Figures 5.2**, **5.3**).

5.2 Nitrate Assimilation

- Plant roots actively absorb nitrate, then reduce it to nitrite (NO_2^-) in the cytosol by nitrate reductase, and further reduce it to ammonium in the plastids by nitrite reductase (**Figure 5.3**).
- Nitrate reductase contain three domains that bind FAD, heme, and molybdenum ions, respectively (**Figure 5.4**).
- Nitrate, light, and carbohydrate levels affect the transcription and translation of nitrate reductase.
- Darkness and Mg^{2+} can inactivate nitrate reductase. Such inactivation is faster than regulation by decreased synthesis or degradation of the enzyme.
- In chloroplasts and root plastids, the enzyme nitrite reductase reduces nitrite to ammonium (**Figures 5.2**, **5.5**).
- Both roots and shoots assimilate nitrate (**Figures 5.6**, **5.7**).
- Nitrate can be transported in both the xylem and phloem (**Figure 5.7**).
- Nitrate transceptor (transporter with receptor function) NRT1.1 not only helps the plant acquire nitrate from soil but also senses nitrate concentration differences in the soil and elicits different levels of gene expression (**Figure 5.8**).

5.3 Ammonium Assimilation

- Plant cells avoid ammonium toxicity by rapidly converting ammonium into amino acids (**Figure 5.9**).
- Nitrogen is incorporated into other amino acids via transamination reactions involving glutamine and glutamate.
- The amino acid asparagine is a key compound for nitrogen transport and storage.

5.4 Amino Acid Biosynthesis

- The carbon skeletons for amino acids derive from intermediates of glycolysis and the tricarboxylic acid cycle (**Figure 5.10**).

(Continued)

Summary *(continued)*

5.5 Biological Nitrogen Fixation

- Biological nitrogen fixation accounts for most of the ammonia formed from atmospheric N_2 (**Figure 5.1**; **Table 5.1**).
- Several types of nitrogen-fixing bacteria form symbiotic associations with higher plants (**Figures 5.11**, **5.12**; **Tables 5.2**, **5.3**).
- Nitrogen fixation requires anaerobic or microanaerobic conditions.
- Symbiotic nitrogen-fixing prokaryotes function within nodules, specialized structures formed by the plant host (**Figure 5.11**).
- The symbiotic relationship is initiated by the migration of the nitrogen-fixing bacteria toward the roots of the host plant, which is mediated by chemical attractants secreted by the roots.
- Attractants activate the rhizobial NodD protein, which then induces the biosynthesis of Nod factors that act as signals for symbiosis (**Figures 5.13**, **5.14**).
- Nod factors induce root hair curling, sequestration of rhizobia, cell wall degradation, and bacterial access to the root hair plasma membrane, from which an infection thread forms (**Figure 5.15**).
- Filled with proliferating rhizobia, the infection thread elongates through root tissue in the direction of the developing nodule, which arises from cortical cells (**Figure 5.15**).
- In response to a signal from the plant, the bacteria in the nodule stop dividing and differentiate into nitrogen-fixing bacteroids.
- The reduction of N_2 to NH_3 is catalyzed by the nitrogenase enzyme complex (**Figure 5.16**).
- Fixed nitrogen is transported as amides or ureides (**Figures 5.6**, **5.17**).

5.6 Sulfur Assimilation

- Most assimilated sulfur derives from sulfate (SO_4^{2-}) absorbed from the soil solution, but plants can also metabolize gaseous sulfur dioxide (SO_2) entering via stomata.
- Synthesis of sulfur-containing organic compounds begins with the reduction of sulfate to the amino acid cysteine.
- Sulfate is mainly assimilated in leaves, involving three subcellular compartments of cytosol, chloroplast, and mitochondria.

5.7 Phosphate Assimilation

- Roots absorb phosphate (HPO_4^{2-}) from the soil solution, and its assimilation occurs with the formation of ATP.
- From ATP, the phosphate group can be transferred to many different carbon compounds in plant cells.

5.8 Oxygen Assimilation

- Respiration and the oxygenase activity of Rubisco account for most of the O_2 assimilated by plant cells, but direct oxygen fixation is also catalyzed by other oxygenases.

5.9 The Energetics of Nutrient Assimilation

- Energy-requiring nutrient assimilation is coupled to photosynthetic electron transport, which generates powerful reducing agents.
- Photoassimilation operates only when photosynthetic electron transport generates reductant in excess of the needs of the C_3 carbon fixation cycle.
- Rising levels of atmospheric CO_2 inhibit nitrate assimilation in the shoots of C_3 plants.

Suggested Reading

Chiou, T.-J., and Lin, S.-I. (2011) Signaling network in sensing phosphate availability in plants. *Annu. Rev. Plant Physiol.* 62: 185–206.

Liu, Q., Wu, K., Song, W., Zhong, N., Wu, Y., and Fu, X. (2022) Improving crop nitrogen use efficiency toward sustainable Green Revolution. *Annu. Rev. Plant Biol.* 73(1): 523–551.

Rudgers, J. A., Afkhami, M. E., Bell-Dereske, L., Chung, Y. A., Crawford, K. M., Kivlin, S. N., Mann, M. A., and Nuñez, M. A. (2020) Climate disruption of plant–microbe interactions. *Annu. Rev. Ecol. Evol. System.* 51(1): 561–586.

Santi, C., Bogusz, D., and Franche, C. (2013) Biological nitrogen fixation in non-legume plants. *Ann. Bot.* 111: 743–767.

Schroeder, J. I., Delhaize, E., Frommer, W. B., Guerinot, M. L., Harrison, M. J., Herrera-Estrella, L., Horie, T., Kochian, L. V., Munns, R., Nishizawa, N. K., Tsay, Y. F., and Sanders, D. (2013) Using membrane transporters to improve crops for sustainable food production. *Nature* 497: 60–66.

Takahashi, H., Kopriva, S., Giordano, M., Saito, K., and Hell, R. (2011) Sulfur assimilation in photosynthetic organisms: Molecular functions and regulations of transporters and assimilatory enzymes. *Annu. Rev. Plant Biol.* 62: 157–184.

Wang, Y.-Y., Cheng, Y.-H., Chen, K.-E., and Tsay, Y. F. (2018) Nitrate transport, signaling, and use efficiency. *Annu. Rev. Plant Biol.* 69: 85–122.

Zhang, Y., Gallant, É., Park, J. D., and Seyedsayamdost, M. R. (2022) The small-molecule language of dynamic microbial interactions. *Annu. Rev. Microbiol.* 76(1): 641–660.

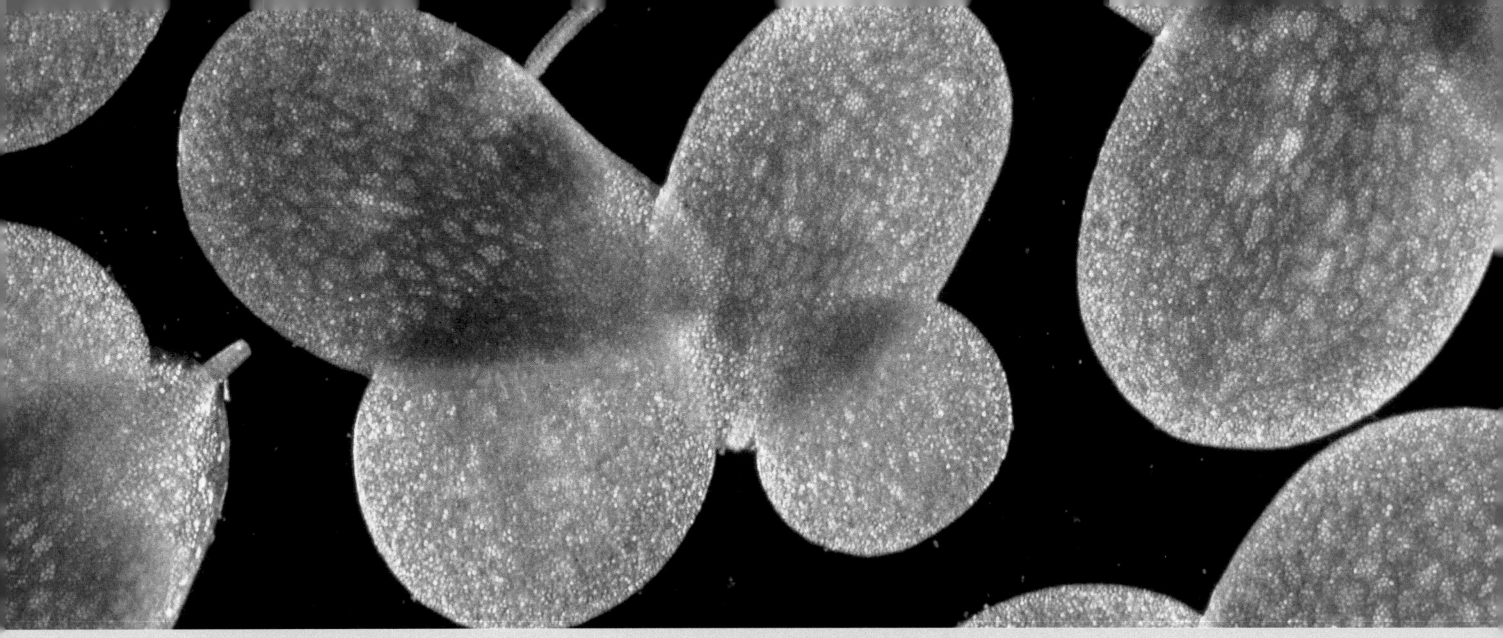

6 Solute Transport

The interior of a plant cell is separated from the external environment by a plasma membrane that is only two lipid molecules thick. This thin layer separates a relatively constant internal environment from variable external surroundings. In addition to forming a hydrophobic barrier to diffusion, the membrane must facilitate and continuously regulate the inward and outward traffic of selected molecules and ions as the cell takes up nutrients, exports solutes, and regulates its turgor pressure. Similar functions are performed by the internal membranes that separate the various compartments within each cell. The plasma membrane also detects information about the environment, about molecular signals from other cells, and about the presence of invading pathogens. Often these signals are relayed by changes in ion fluxes across the membrane.

The movement of ions and other solutes from one location to another within an organism is known as **transport**. Transport of solutes into or within cells is regulated mainly by membrane proteins. Transport between plant organs, or between plant and environment, is also controlled by membrane transport proteins at the cellular level. For example, the transport of sucrose from leaf to root via the phloem, referred to as **translocation**, is driven and regulated by membrane transport proteins into the phloem cells of the leaf and from the phloem into the storage cells of the root (Chapter 10).

This chapter focuses on the physical and chemical principles that govern the movement of molecules in solution and how these principles apply to membranes and biological systems. We discuss the molecular mechanisms of transport in living cells and the great variety of membrane transport proteins that are responsible for the particular transport properties of plant

transport Molecular or ionic movement from one location to another; may involve crossing a diffusion barrier such as one or more membranes.

translocation The movement of photosynthate from sources to sinks in the phloem.

passive transport Diffusion across a membrane. The spontaneous movement of a solute across a membrane in the direction of a gradient of (electro)chemical potential (from higher to lower potential). Downhill transport.

active transport The use of energy to move a solute across a membrane against a concentration gradient, a potential gradient, or both (electrochemical potential gradient). Uphill transport.

chemical potential The free energy associated with a substance that is available to perform work.

cells. Because transported substances, including carbohydrates, amino acids, and metals such as iron and zinc, are vital for human nutrition, understanding and manipulating solute transport in plants can contribute solutions to sustainable food production.

6.1 Passive and Active Transport

Differentiate the physical bases of passive and active transport.

According to Fick's first law (Equation 2.1), the movement of molecules by diffusion always proceeds spontaneously, down a gradient of free energy or chemical potential, until equilibrium is reached. The spontaneous "downhill" movement of molecules across a biological membrane is called **passive transport**. At equilibrium, no further net movement of solutes can occur without the application of a driving force.

The movement of substances across a biological membrane against a gradient of chemical potential, or "uphill," is termed **active transport**. It is not spontaneous, and it requires that work be done on the system by the application of cellular energy. One common way of accomplishing this task is to couple transport to the hydrolysis of ATP.

In Chapter 2 we calculated the driving force for diffusion or, conversely, the energy input necessary to move substances against a gradient by measuring the potential-energy gradient. For uncharged solutes this gradient is often a simple function of the difference in concentration. Biological transport can be driven by four major forces: concentration, hydrostatic pressure, gravity, and electrical fields. (However, recall from Chapter 2 that in small-scale biological systems, gravity seldom contributes substantially to the force that drives transport.)

The **chemical potential** for any solute is defined as the sum of the concentration, electrical, and hydrostatic potentials (and the associated chemical potential under standard conditions). *The importance of the concept of chemical potential is that it sums all the forces that may act on a molecule to drive net transport.*

$$\tilde{\mu}_j = \mu_j^* + RT\ln C_j + z_jFE + \bar{V}_jP \qquad (6.1)$$

$\tilde{\mu}_j$	μ_j^*	$RT\ln C_j$	z_jFE	$\bar{V}_jP$
Chemical potential for a given solute, j	Chemical potential of j under standard conditions	Concentration (activity) component	Electrical-potential component	Hydrostatic-pressure component

Here $\tilde{\mu}_j$ is the chemical potential of the solute species j in joules per mole (J mol^{-1}), μ_j^* is its chemical potential under standard conditions (a correction factor that will cancel out in future equations and so can be ignored), R is the universal gas constant, T is the absolute temperature, and C_j is the concentration of j (more accurately the activity, but they are nearly the same for relatively dilute solutions).

The electrical term, z_jFE, applies only to ions; z is the electrostatic charge of the ion (+1 for monovalent cations, –1 for monovalent anions, +2 for divalent cations, and so on), F is Faraday's constant (96,500 Coulombs, equivalent to the electrical charge on 1 mol of H^+), and E is the overall electrical potential of the solution (with respect to ground). The final term, $\bar{V}_jP$, expresses the contribution of the partial molal volume of $j(\bar{V}_j)$ and pressure (P) to the chemical potential of j. (The partial molal volume of j is the change in volume per mole of substance j added to the system, for an infinitesimal addition.)

This final term, $\bar{V}_jP$, makes a much smaller contribution to $\tilde{\mu}_j$ than the concentration and electrical terms, except in the very important case of osmotic water movements. As discussed in Chapter 2, when we consider water movement at the cellular scale, the chemical potential of water (i.e., the water potential) depends on the concentration of dissolved solutes and the hydrostatic pressure on the system.

In general, molecules always move energetically downhill from areas of higher chemical potential to areas of lower chemical potential by diffusion (passive transport). Movement of molecules against a chemical-potential gradient is indicative of active transport (**Figure 6.1**).

We can approximate the chemical potential of a neutral solute diffusing in any compartment by the concentration term alone (unless a solution is concentrated, causing hydrostatic pressure to build up within the plant cell). From Equation 6.1, the chemical potential of sucrose, which is uncharged, inside a cell can be described as follows (in the next four equations, the subscript *s* stands for sucrose and the superscripts *i* and *o* stand for inside and outside, respectively):

$$\underset{\substack{\text{Chemical potential}\\\text{of sucrose solution}\\\text{inside the cell}}}{\tilde{\mu}_s^i} = \underset{\substack{\text{Chemical potential}\\\text{of sucrose solution}\\\text{under standard}\\\text{conditions}}}{\mu_s^*} + \underset{\substack{\text{Concentration}\\\text{components}}}{RT \ln C_s^i} \tag{6.2}$$

The chemical potential of sucrose outside the cell is calculated in the same way:

$$\tilde{\mu}_s^o = \mu_s^* + RT \ln C_s^o \tag{6.3}$$

We can calculate the difference in the chemical potential of sucrose between the solutions inside and outside the cell, $\Delta \tilde{\mu}_s$, regardless of the mechanism of transport. To get the signs right, remember that for inward transport, sucrose is being removed (–) from outside the cell and added (+) to the inside, so the change in free energy in joules per mole of sucrose transported will be as follows:

$$\Delta \tilde{\mu}_s = \mu_s^i - \tilde{\mu}_s^o \tag{6.4}$$

Substituting the terms from Equations 6.2 and 6.3 into Equation 6.4, we get the following:

$$\Delta \tilde{\mu}_s = (\mu_s^* + RT \ln C_s^i) - (\mu_s^* + RT \ln C_s^o) = RT(\ln C_s^i - \ln C_s^o) = RT \ln \frac{C_s^i}{C_s^o} \tag{6.5}$$

If this difference in chemical potential is negative, sucrose can diffuse inward spontaneously (provided the membrane is permeable to sucrose). In other words,

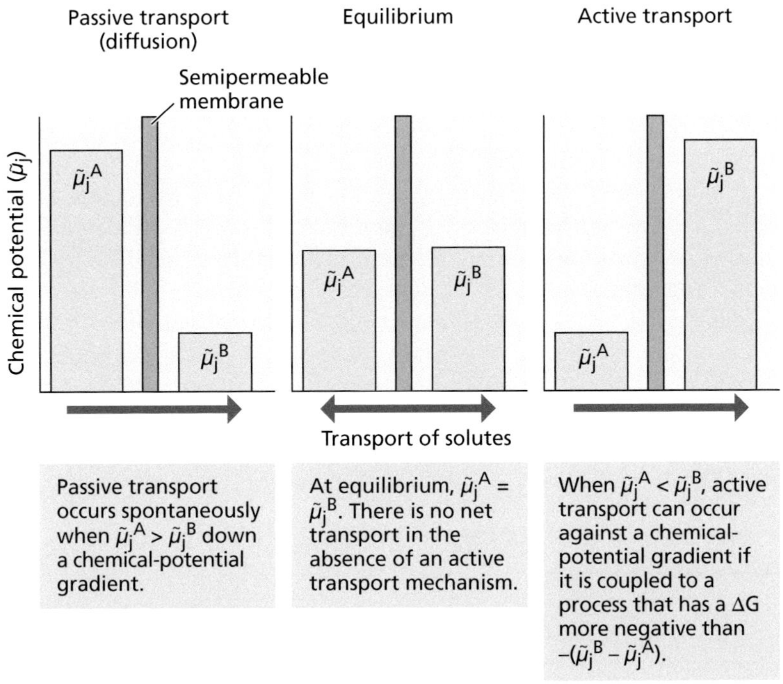

Figure 6.1 Relationship between chemical potential, $\tilde{\mu}$, and the transport of molecules across a semipermeable membrane. The net movement of molecular species *j* between compartments A and B depends on the relative magnitude of the chemical potential of *j* in each compartment, represented here by the size of the boxes. Movement down a chemical gradient occurs spontaneously and is called passive transport; movement against, or up, a gradient requires energy and is called active transport.

the driving force ($\Delta \tilde{\mu}_s$) for solute diffusion is related to the magnitude of the concentration gradient (C_s^i / C_s^o).

If the solute carries an electrical charge (for example, the potassium ion), the electrical component of the chemical potential must also be considered. Suppose the membrane is permeable to K^+ and Cl^- rather than to sucrose. Because the ionic species (K^+ and Cl^-) diffuse independently, each has its own chemical potential. Thus, for inward K^+ diffusion,

$$\Delta \tilde{\mu}_K = \tilde{\mu}_K^i - \tilde{\mu}_K^o \tag{6.6}$$

Substituting the appropriate terms from Equation 6.1 into Equation 6.6, we get

$$\Delta \tilde{\mu}_s = (RT \ln [K^+]^i + zFE^i) - (RT \ln [K^+]^o + zFE^o) \tag{6.7}$$

and because the electrostatic charge of K^+ is +1, z = +1, and

$$\Delta \tilde{\mu}_K = RT \ln \frac{[K^+]^i}{[K^+]^o} + F(E^i - E^o) \tag{6.8}$$

The magnitude and sign of this expression will indicate the driving force and direction for K^+ diffusion across the membrane. A similar expression can be written for Cl^- (but remember that for Cl^-, $z = -1$).

Equation 6.8 shows that ions, such as K^+, diffuse in response to both their concentration gradients ($[K^+]^i/[K^+]^o$) and any electrical-potential difference between the two compartments ($E^i - E^o$). One important implication of this equation is that ions can be driven passively against their concentration gradients if an appropriate voltage (electrical field) is applied between the two compartments. Because of the importance of electrical fields in the biological transport of any charged molecule, $\tilde{\mu}$ is often called the **electrochemical potential**, and $\Delta \tilde{\mu}$ is the difference in electrochemical potential between two compartments.

6.2 Transport of Ions across Membrane Barriers

Define the factors that influence the distribution of ions across membranes.

If two ionic solutions are separated by a biological membrane, diffusion is complicated by the fact that the ions must move through the membrane as well as across the open solutions. The extent to which a membrane permits the movement of a substance is called **membrane permeability**. Permeability depends on the composition of the membrane as well as on the chemical nature of the solute. In a loose sense, permeability can be expressed in terms of a diffusion coefficient for the solute through the membrane. However, permeability is influenced by several additional factors, such as the ability of a substance to enter the membrane, that are difficult to measure.

Despite its theoretical complexity, we can readily measure permeability by determining the rate at which a solute passes through a membrane under a specific set of conditions. Generally, the membrane will hinder diffusion and thus reduce the speed with which equilibrium is reached. For any particular solute, however, the permeability or resistance of the membrane itself cannot alter the final equilibrium conditions. Equilibrium occurs when $\Delta \tilde{\mu}_j = 0$.

electrochemical potential The chemical potential of an electrically charged solute.

membrane permeability The extent to which a membrane permits or restricts the movement of a substance.

Different diffusion rates for cations and anions produce diffusion potentials

When salts diffuse across a membrane, an electrical membrane potential (voltage) can develop. Consider the two KCl solutions separated by a

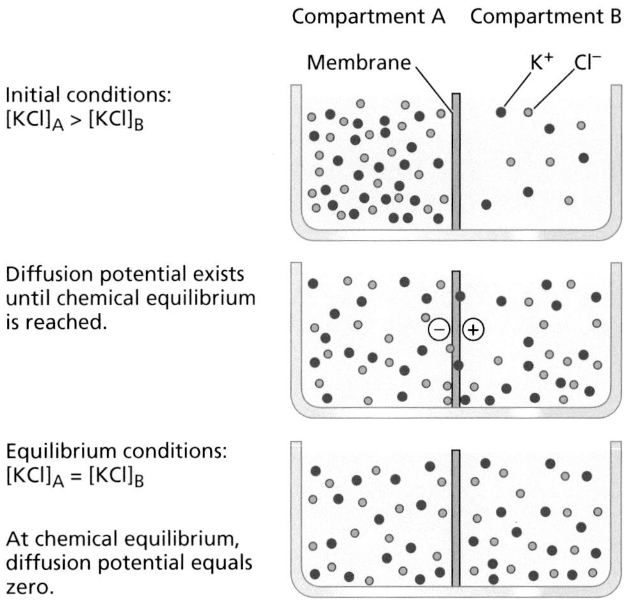

Figure 6.2 Development of a diffusion potential and a charge separation between two compartments separated by a membrane that is preferentially permeable to potassium ions. If the concentration of potassium chloride is higher in compartment A ($[KCl]_A > [KCl]_B$), potassium and chloride ions will diffuse into compartment B. If the membrane is more permeable to potassium ions than to chloride, potassium ions will diffuse faster than chloride ions, and a charge separation (+ and –) will develop, resulting in establishment of a diffusion potential.

membrane in **Figure 6.2**. The K^+ and Cl^- ions will permeate the membrane independently as they diffuse down their respective gradients of electrochemical potential. And unless the membrane is very porous, its permeability to the two ions will differ.

As a consequence of these different permeabilities, K^+ and Cl^- will initially diffuse across the membrane at different rates. The result is a slight separation of charge, which instantly creates an electrical potential across the membrane. In biological systems, membranes are usually more permeable to K^+ than to Cl^-. Therefore, K^+ will diffuse out of the cell (compartment A in Figure 6.2) faster than Cl^-, causing the cell to develop a negative electrical charge with respect to the extracellular medium. A potential that develops as a result of diffusion is called a **diffusion potential**.

The principle of electrical neutrality must always be kept in mind when the movement of ions across membranes is considered: Bulk solutions always contain equal numbers of anions and cations. The existence of a membrane potential implies that the distribution of charges across the membrane is uneven; however, the actual number of unbalanced ions is negligible in chemical terms. For example, the generation of a membrane potential of –100 millivolts (mV), like that found across the plasma membranes of many plant cells, is due to ions moving across the membrane and results in the presence of only 1 extra anion out of every 100,000 within the cell—a concentration difference of only 0.001%! As Figure 6.2 shows, these extra anions are found immediately adjacent to the surface of the membrane; there is no charge imbalance throughout the bulk of the cell.

In our example of KCl diffusion across a membrane, electrical neutrality is preserved, because as K^+ moves ahead of Cl^- in the membrane, the resulting diffusion potential retards the movement of K^+ and speeds that of Cl^-. Ultimately, both ions diffuse at the same rate, but the diffusion potential persists and can be measured. As the system moves toward equilibrium and the concentration gradient collapses, the diffusion potential also collapses.

How does membrane potential relate to ion distribution?

Because the membrane in the preceding example is permeable to both K^+ and Cl^- ions, equilibrium will not be reached for either ion until the concentration gradients decrease to zero. However, if the membrane were permeable only to K^+, diffusion of K^+ would carry charges across the membrane until the membrane potential balanced the concentration gradient. Because a change in potential requires very few ions, this balance would be reached very quickly. Potassium ions would then be at equilibrium, even though the change in the concentration gradient for K^+ would be negligible.

When the distribution of any solute across a membrane reaches equilibrium, the passive flux, J (i.e., the amount of solute crossing a unit area of membrane

diffusion potential The potential (voltage) difference that develops across a semipermeable membrane as a result of the differential permeability of solutes with opposite charges (for example, K^+ and Cl^-).

per unit time), is the same in the two directions—outside to inside and inside to outside:

$$J_{o\to i} = J_{i\to o}$$

Fluxes are related to $\Delta\tilde{\mu}$; thus, at equilibrium the electrochemical potentials will be the same even if the concentrations are quite different:

$$\tilde{\mu}_j^{o} = \mu_j^{i}$$

and for any given ion (the ion is symbolized here by the subscript *j*),

$$\mu_j^* + RT\ln C_j^o + z_jFE^o = \mu_j^* + RT\ln C_j^i + z_jFE^i \quad (6.9)$$

By rearranging Equation 6.9, we obtain the difference in electrical potential between the two compartments at equilibrium ($E^i - E^o$):

$$E^i - E^o = \frac{RT}{z_jF}\left(\ln\frac{C_j^o}{C_j^i}\right)$$

This electrical-potential difference is known as the **Nernst potential** (ΔE_j) for that ion,

$$\Delta E_j = E^i - E^o$$

and

$$\Delta E_j = \frac{RT}{z_jF}\left(\ln\frac{C_j^o}{C_j^i}\right) \quad (6.10)$$

or

$$\Delta E_j = \frac{2.3RT}{z_jF}\left(\log\frac{C_j^o}{C_j^i}\right)$$

This relationship, known as the *Nernst equation*, states that at equilibrium, the difference in concentration of an ion between two compartments is balanced by the voltage difference between the compartments. The Nernst equation can be further simplified for a univalent cation at 25°C:

$$\Delta E_j = 59\text{mV}\log\frac{C_j^o}{C_j^i} \quad (6.11)$$

Note that a tenfold difference in concentration corresponds to a Nernst potential of 59 mV ($C_o/C_i = 10/1$; log 10 = 1). That is, a membrane potential of 59 mV would maintain a tenfold concentration gradient of an ion whose movement across the membrane is driven by passive diffusion. Similarly, if a tenfold concentration gradient of an ion existed across the membrane, passive diffusion of that ion down its concentration gradient (if it were allowed to come to equilibrium) would result in a difference of 59 mV across the membrane.

All living cells exhibit a membrane potential that is due to selective ion movement across the plasma membrane. We can determine these membrane potentials by inserting a microelectrode into the cell and measuring the voltage difference between the inside of the cell and the extracellular medium (**Figure 6.3**).

The Nernst equation can be used at any time to determine whether a given ion is at equilibrium across a membrane. However, a distinction must be made between equilibrium and steady state. *Steady state* is the condition in which influx and efflux of a given solute are equal, and therefore the ion concentrations are constant over time. Steady state is not necessarily the same as equilibrium (Figure 6.1); in steady state, the existence of active transport across the membrane prevents many diffusive fluxes from ever reaching equilibrium.

Nernst potential The electrical potential described by the Nernst equation.

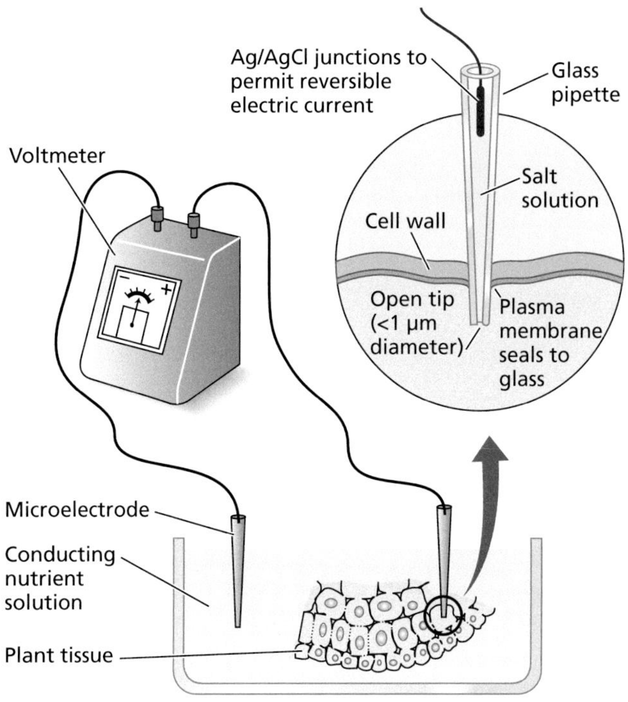

Figure 6.3 Diagram of a pair of microelectrodes used to measure membrane potentials across a plasma membrane. One of the glass micropipette electrodes is inserted into the cell compartment under study (usually the vacuole or the cytoplasm), while the other is kept in an electrically conducting solution that serves as a reference. The microelectrodes are connected to a voltmeter, which records the electrical-potential difference between the cell compartment and the solution. Typical membrane potentials across plant plasma membranes range from −100 to −200 mV. The magnification (circle and dark arrow) shows how electrical contact with the interior of the cell is made through the open tip of the glass micropipette, which contains an electrically conducting salt solution.

The Nernst equation distinguishes between active and passive transport

Table 6.1 shows how experimental measurements of ion concentrations at steady state in pea root cells compare with predicted values calculated from the Nernst equation. In this example, the concentration of each ion in the external solution bathing the tissue and the measured membrane potential were substituted into the Nernst equation, and the internal concentration of each ion was predicted.

The Nernst equation calculations assume passive ionic distribution, but notice that, of all the ions shown in Table 6.1, only K^+ is at or near equilibrium. The anions NO_3^-, Cl^-, $H_2PO_4^-$, and SO_4^{2-} all have higher internal concentrations than predicted, indicating that their uptake is active. The cations Na^+, Mg^{2+}, and Ca^{2+} have lower internal concentrations than predicted; therefore, these ions are actively exported.

The example shown in Table 6.1 is an oversimplification because plant cells have several internal compartments, each of which can differ in its ionic composition from the others. The cytosol and the vacuole are the most important intracellular compartments in determining the ionic relations of plant cells. In most mature plant cells, the central vacuole occupies 90% or more of the cell volume, and the cytosol is restricted to a thin layer around the periphery of the cell.

Because of its small volume, the cytosol of most angiosperm cells is difficult to assay chemically. In brief:

- Potassium ions accumulate passively in the cytosol. When extracellular K^+ concentrations are very low, K^+ may be taken up actively.
- Anions are actively taken up into the cytosol.
- Sodium ions are actively pumped out of the cytosol into the extracellular space and vacuole.
- Protons are actively extruded from the cytosol. This process helps maintain the cytosolic pH near neutrality, while the vacuole

Table 6.1 Comparison of Observed and Calculated Ion Concentrations in Pea Root Tissue

Ion	Concentration in external medium (mmol L^{-1})	Internal concentration[a] (mmol L^{-1})	
		Calculated	Observed
K^+	1	74	75
Na^+	1	74	8
Mg^{2+}	0.25	1,340	3
Ca^{2+}	1	5,360	2
NO_3^-	2	0.0272	28
Cl^-	1	0.0136	7
$H_2PO_4^-$	1	0.0136	21
SO_4^{2-}	0.25	0.00005	19

Source: Data from N. Higinbotham et al. 1967. *Plant Physiol.* 42: 37–46.

Note: The membrane potential was measured as −110 mV.

[a] Internal concentration values were derived from ion content of hot water extracts of 1- to 2-cm intact root segments.

Goldman equation An equation that predicts the diffusion potential across a membrane, as a function of the concentrations and permeabilities of all ions (e.g., K^+, Na^+, and Cl^-) that permeate the membrane.

transport proteins Transmembrane proteins that are involved in the movement of molecules or ions from one side of a membrane to the other side.

channels Transmembrane proteins that function as selective pores for passive transport of ions or water across the membrane.

carriers Membrane transport proteins that bind to a solute, undergo conformational change, and release the solute on the other side of the membrane.

pumps Membrane proteins that carry out primary active transport across a biological membrane. Most pumps transport ions, such as H^+ or Ca^{2+}.

and the extracellular medium are generally more acidic by one or two pH units.

- Calcium ions are actively transported out of the cytosol at both the plasma membrane and the vacuolar membrane, which is called the tonoplast.

Many different ions cross the membranes of living cells simultaneously, but K^+ has the highest concentrations in plant cells, and it exhibits high permeabilities. A modified version of the Nernst equation, the **Goldman equation**, includes all permeant ions (all ions for which mechanisms of transmembrane movement are known) and therefore gives a more accurate value for the diffusion potential. The diffusion potential calculated from the Goldman equation is termed the *Goldman diffusion potential.*

Proton transport is a major determinant of the membrane potential

In most eukaryotic cells, K^+ has both the greatest internal concentration and the highest membrane permeability, so the diffusion potential may approach E_K, the Nernst potential for K^+. In some cells of some organisms—particularly in some mammalian cells such as neurons—the normal resting potential of the cell may also be close to E_K. This is not the case with plants and fungi, however, which often show experimentally measured membrane potentials that are much more negative (often –200 to –100 mV) than those calculated from the Goldman equation, which are usually only –80 to –50 mV. Thus, the membrane potential must have a second component. The additional voltage is provided by the plasma membrane H^+-ATPase, an *electrogenic pump* that transports H^+ out of plant cells unaccompanied by a balancing charge.

Energy for active transport by the plasma membrane H^+-ATPase is provided by the hydrolysis of ATP. We can study the dependence of the plasma membrane potential on ATP by observing the effect of cyanide (CN^-) on the membrane potential (**Figure 6.4**). Cyanide rapidly poisons the mitochondria, and ATP within cells becomes depleted. As ATP synthesis is inhibited, the membrane potential falls to the level of the Goldman diffusion potential.

Thus, the membrane potentials of plant cells have two components: a diffusion potential and a component resulting from active, electrogenic ion transport—the net transport of a charge across the membrane. When cyanide inhibits electrogenic ion transport, the pH of the external medium increases while the cytosol becomes acidic, consistent with the active transport of protons out of the cell that is electrogenic.

A change in membrane potential caused by an electrogenic pump will change the driving forces for diffusion of all ions that cross the membrane. For example, the outward transport of H^+ can create an electrical driving force for the passive diffusion of K^+ into the cell. Protons are transported electrogenically across the plasma membrane not only in plants but also in bacteria, algae, fungi, and some animal cells, such as those of the stomach epithelia.

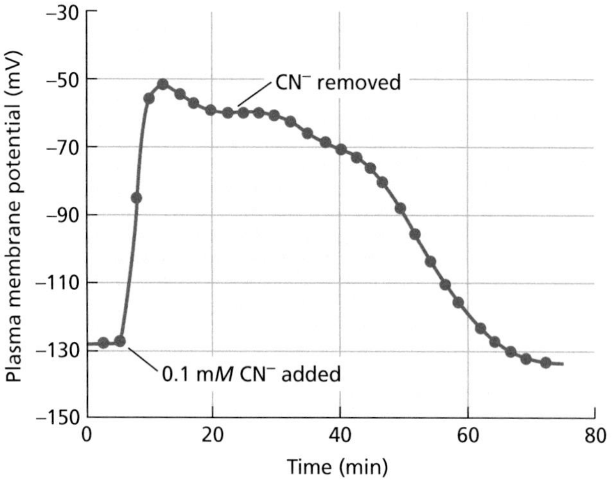

Figure 6.4 The plasma membrane potential of a pea cell collapses when cyanide (CN^-) is added to the bathing solution. Cyanide blocks ATP production in the cell by poisoning the mitochondria. The collapse of the membrane potential upon addition of cyanide indicates that an ATP supply is necessary for maintenance of the potential. Washing the cyanide out of the tissue results in a slow recovery of ATP production and restoration of the membrane potential.

ATP synthesis in mitochondria and chloroplasts also depends on a H^+-ATPase. In these organelles, this transport protein is usually called an *ATP synthase* because it forms ATP rather than hydrolyzing it.

6.3 Membrane Transport Processes

Explain different types of transport proteins and their role in membrane functions.

Artificial membranes made of pure phospholipids have been used extensively to study membrane permeability. When the permeability of artificial phospholipid bilayers to ions and molecules is compared with that of biological membranes, important similarities and differences become evident (**Figure 6.5**).

Biological and artificial membranes have similar permeabilities to nonpolar molecules and many small polar molecules. However, biological membranes are much more permeable to ions, to some large polar molecules, such as sugars, and to water than artificial bilayers. This is because biological membranes contain **transport proteins** that facilitate the passage of selected ions and other molecules. The general term *transport proteins* encompasses three main categories of proteins: **channels**, **carriers**, and **pumps** (**Figure 6.6**), each of which we describe in more detail.

Transport proteins exhibit specificity for the solutes they transport, so cells require a great diversity of transport proteins. The simple prokaryote *Haemophilus influenzae*, the first organism for which the complete genome was sequenced, has only 1,743 genes, yet more than 200 of those genes (>10% of the genome) encode various proteins involved in membrane transport. In Arabidopsis, out of a predicted 33,602 protein-coding genes, as many as 1,800 may encode transport proteins.

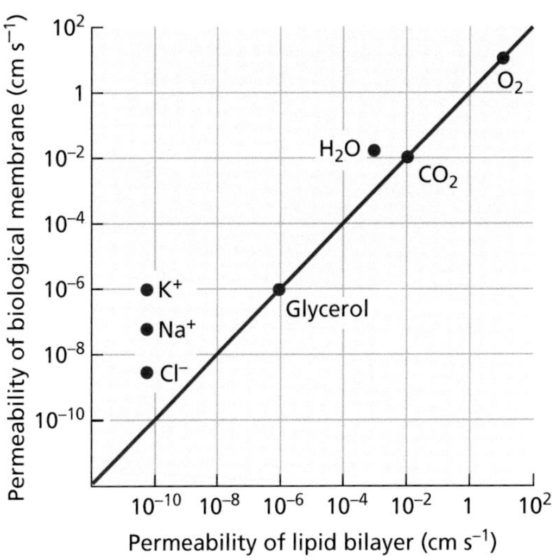

Figure 6.5 Typical values for the permeability of a biological membrane to various substances compared with those for an artificial phospholipid bilayer. For nonpolar molecules such as O_2 and CO_2, and for some small uncharged molecules such as glycerol, permeability values are similar in both systems. For ions and selected polar molecules, including water, the permeability of biological membranes is increased by one or more orders of magnitude because of the presence of transport proteins. Note the logarithmic scale.

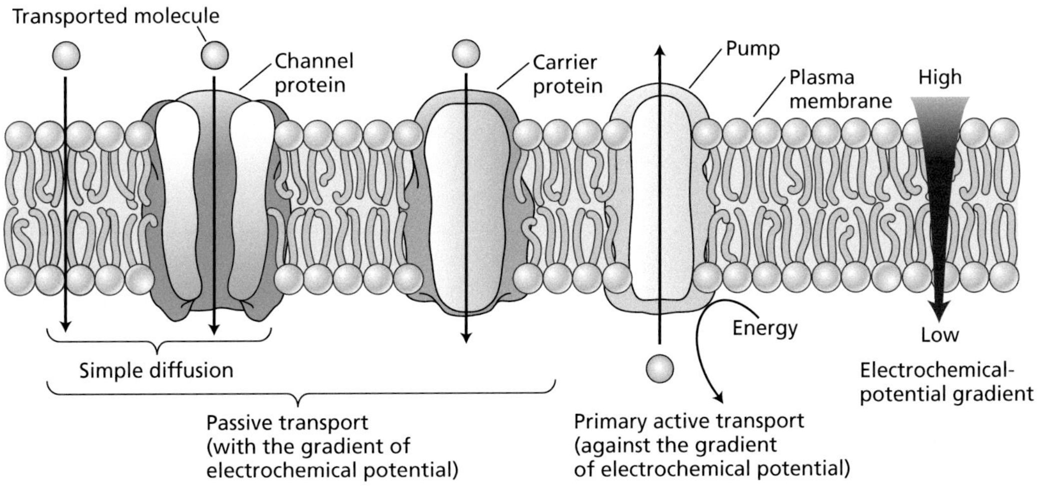

Figure 6.6 Three classes of membrane transport proteins: channels, carriers, and pumps. Channels and carriers can mediate the passive transport of a solute across a membrane down the electrochemical potential gradient of the solute. Channel proteins act as membrane pores, and their specificity is determined primarily by the biophysical properties of the channel. Carrier proteins bind the transported molecule on one side of the membrane and release it on the other side. Primary active transport is carried out by pumps that use energy directly, usually from ATP hydrolysis, to transport solutes against their electrochemical potential gradient.

gate A structural domain of the channel protein that opens or closes the channel in response to external signals such as voltage changes, hormone binding, or light.

Transport proteins are usually specific for the substances they transport, but they may transport other solutes with a similar size and charge. For example, a K^+ transporter in the plasma membrane may transport K^+, Rb^+, and Na^+ with different preferences. However, most K^+ transporters are completely ineffective in transporting anions such as Cl^- or uncharged solutes such as sucrose. Similarly, most sucrose transporters transport a range of uncharged glycosides, but no other substances.

Channels enhance diffusion across membranes

Channels are transmembrane proteins that function as selective pores through which ions, and in some cases neutral molecules, can diffuse across the membrane. The size of a pore and the density and nature of the surface charges on its interior lining determine the transport specificity of a channel. Transport through channels is always passive, and the specificity of transport depends on the pore size and electrical charge more than on selective binding (**Figure 6.7**).

As long as the channel pore is open, substances that can enter the pore will diffuse through it extremely rapidly: about 10^8 ions per second through an ion channel. Channel pores are not open all the time, however. Channel proteins contain regions called **gates** that open and close the pore in response to signals. Signals that can regulate channel activity include membrane potential changes, ligands, hormones, light, and posttranslational protein modifications such as phosphorylation (Chapter 12). For example, voltage-gated channels open or close in response to changes in the membrane potential (Figure 6.7B). Another intriguing regulatory signal is mechanical force, which changes the conformation and thereby controls the gating of mechanosensitive ion channels in plants and other organisms.

Individual ion channels can be studied in detail by an electrophysiological technique called patch clamping that detects the electric current carried by ions diffusing through a single open channel or a collection of channels. Patch clamp studies reveal that for a given ion, such as K^+, a membrane has a variety of different channels. These channels may open over different voltage ranges, or in response to different signals, which may include K^+ or Ca^{2+} concentrations, pH, reactive oxygen species, and so on. The ion permeability of a membrane depends on which channels are open and on the substrate specificity of those channels. This allows ion transport to be fine-tuned to the prevailing conditions.

As we saw in the experiment presented in Table 6.1, the distribution of most ions is not close to equilibrium across the membrane. Therefore, we know that channels for most ions are usually closed. Plant cells generally accumulate more anions than could occur via a strictly passive mechanism. Therefore, when anion channels open, anions flow out of

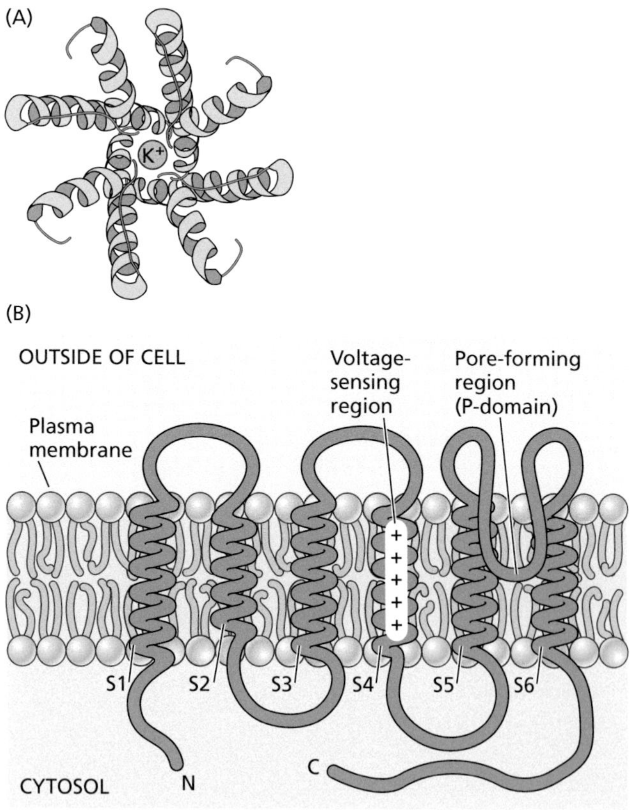

Figure 6.7 Models of K^+ channels in plants. (A) Top view of a channel, looking through the pore of the protein. Membrane-spanning helices of four subunits come together in an inverted teepee with the pore at the center. The pore-forming regions of the four subunits dip into the membrane, forming a K^+ selectivity finger region at the outer part of the pore. (B) Side view of an inwardly rectifying K^+ channel, showing a polypeptide chain of one subunit, with six membrane-spanning helices (S1–S6). The fourth helix contains positively charged amino acids and acts as a voltage sensor. The pore-forming region (P-domain) is a loop between helices 5 and 6.

the cell, and then active mechanisms are required for anion uptake. Calcium ion channels are tightly regulated and essentially open only during signal transduction (Chapter 12). Calcium ion channels function only to allow Ca^{2+} flux into the cytosol, and Ca^{2+} must be expelled from the cytosol by active transport. In contrast, K^+ can diffuse either inward or outward through channels, depending on whether the membrane potential is more negative or more positive than E_K, the potassium ion equilibrium potential.

K^+ channels that open only at potentials more negative than the Nernst potential for K^+ are specialized for inward diffusion of K^+ and are **inwardly rectifying**, or simply *inward* K^+ channels. Conversely, K^+ channels that open only at potentials more positive than the Nernst potential for K^+ are **outwardly rectifying**, or *outward* K^+ channels (**Figure 6.8**). Guard cells have both types of K^+ channels to control stomatal opening and closing.

Carriers bind and transport specific substances

Unlike channels, carrier proteins do not have pores that extend completely across the membrane. In transport mediated by a carrier, the substance being transported is initially bound to a specific site on the carrier protein. This allows the carrier to be highly selective for a particular substrate to be transported. Binding of the specific substrate causes a conformational change in the protein, which exposes the substrate to the solution on the other side of the membrane. Transport is complete when the substrate dissociates from the transport protein.

Because a conformational change in the protein is required to transport an individual molecule or ion, the rate of transport by a carrier is many orders of magnitude slower than that through a channel. Typically, carriers may transport 100 to 1,000 ions or molecules per second, while millions of ions can pass through an open ion channel per second. The binding and release of molecules at a specific site on a carrier protein are similar to the binding and release of molecules by an enzyme in an enzyme-catalyzed reaction. While channels function only in passive transport, carriers can function in either passive transport or secondary active transport. Passive transport via a carrier is sometimes called **facilitated diffusion**, although it resembles diffusion only in that it transports substances down their gradient of electrochemical potential, without an additional input of energy. (The term "facilitated diffusion" might seem more appropriately applied to transport through channels, but historically it has not been used in that way.)

Primary active transport requires energy

To carry out active transport, a carrier must couple the energetically uphill transport of a solute with another, energy-releasing event so that the overall free-energy change is negative. **Primary active transport** is coupled directly to a source of energy other than $\Delta \tilde{\mu}_j$, such as ATP hydrolysis, an oxidation–reduction reaction, or the absorption of light by the carrier protein (such as bacteriorhodopsin in halobacteria).

Membrane proteins that carry out primary active transport are called **pumps** (Figure 6.6). Most pumps transport inorganic ions, such as H^+ or Ca^{2+}. However, pumps belonging to the ATP-binding cassette (ABC) family of transporters can carry large organic molecules.

Ion pumps can be further characterized as either electrogenic or electroneutral. In general, **electrogenic transport** refers to ion transport involving the net movement of charge across the membrane. An example of electrogenic transport is the auxin/proton symporter AUX1 that moves one negative ion (auxin, indole-3-acetic acid, IAA^-) and two H^+ into the cell. In contrast, **electroneutral transport**, as the name implies, involves no net movement of charge. The cation/proton antiporter CHX exports one cation (e.g., Na^+) and imports one proton, resulting in electroneutral transport.

inwardly rectifying Refers to ion channels that open only at potentials more negative than the prevailing Nernst potential for a cation, or more positive than the prevailing Nernst potential for an anion, and thus mediate inward current.

outwardly rectifying Refers to ion channels that open only at potentials more positive than the prevailing Nernst potential for a cation, or more negative than the prevailing Nernst potential for an anion, and thus mediate outward current.

facilitated diffusion Passive transport across a membrane using a carrier.

primary active transport The direct coupling of a metabolic energy source such as ATP hydrolysis, oxidation–reduction reaction, or light absorption to active transport by a carrier protein.

pumps Membrane proteins that carry out primary active transport across a biological membrane. Most pumps transport ions, such as H^+ or Ca^{2+}.

electrogenic transport Active ion transport involving the net movement of charge across a membrane.

electroneutral transport Active ion transport that involves no net movement of charge across a membrane.

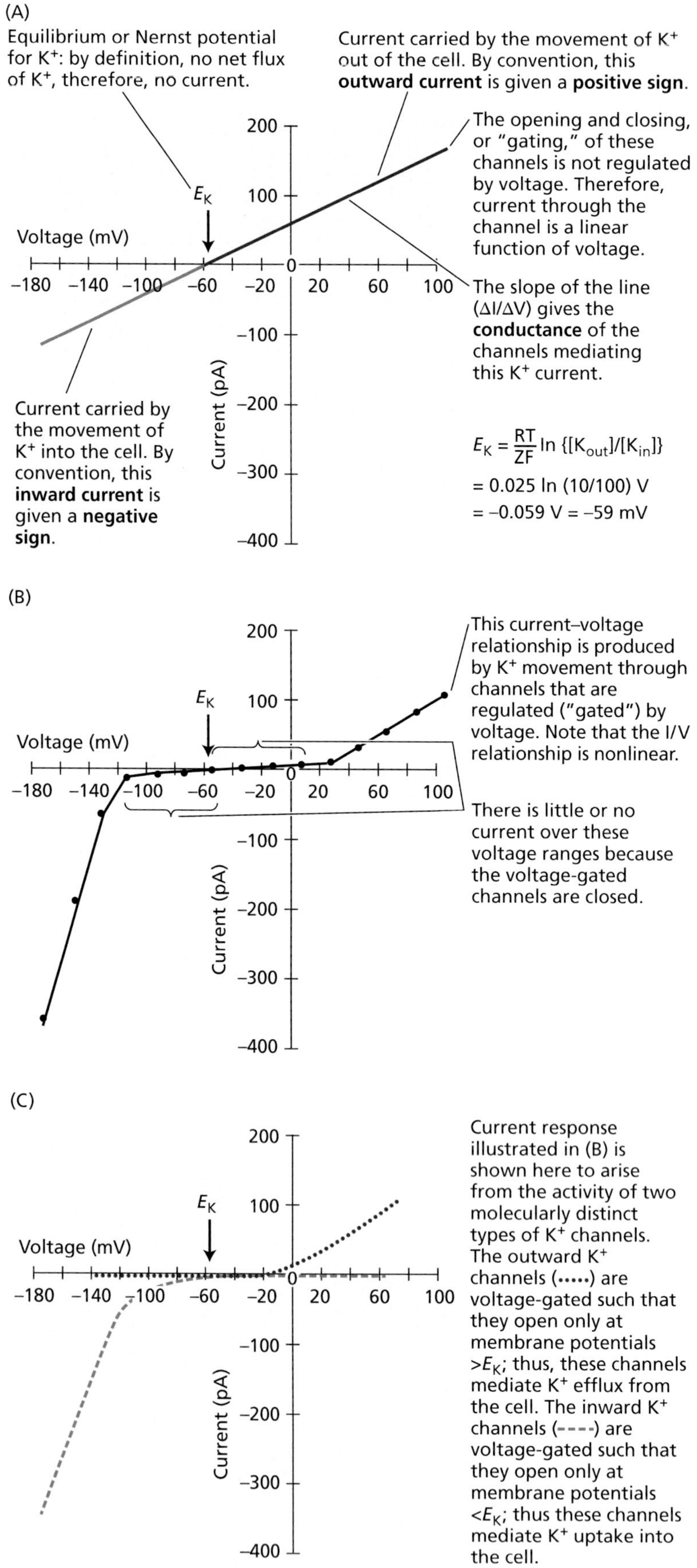
(A)
Equilibrium or Nernst potential for K+: by definition, no net flux of K+, therefore, no current.
Current carried by the movement of K+ out of the cell. By convention, this **outward current** is given a **positive sign**.
The opening and closing, or "gating," of these channels is not regulated by voltage. Therefore, current through the channel is a linear function of voltage.
The slope of the line (ΔI/ΔV) gives the **conductance** of the channels mediating this K+ current.
Current carried by the movement of K+ into the cell. By convention, this **inward current** is given a **negative sign**.
E_K
Voltage (mV)
Current (pA)
200
100
0
−100
−200
−300
−400
−180 −140 −100 −60 −20 20 60 100
$E_K = \frac{RT}{ZF} \ln \{[K_{out}]/[K_{in}]\}$
= 0.025 ln (10/100) V
= −0.059 V = −59 mV
(B)
This current–voltage relationship is produced by K+ movement through channels that are regulated ("gated") by voltage. Note that the I/V relationship is nonlinear.
There is little or no current over these voltage ranges because the voltage-gated channels are closed.
E_K
Voltage (mV)
Current (pA)
(C)
Current response illustrated in (B) is shown here to arise from the activity of two molecularly distinct types of K+ channels. The outward K+ channels (•••••) are voltage-gated such that they open only at membrane potentials >E_K; thus, these channels mediate K+ efflux from the cell. The inward K+ channels (----) are voltage-gated such that they open only at membrane potentials <E_K; thus these channels mediate K+ uptake into the cell.
E_K
Voltage (mV)
Current (pA)

Figure 6.8 Current–voltage relationships. (A) Diagram showing the current that would result from K^+ flux through a set of hypothetical plasma membrane K^+ channels that were not voltage-regulated, given a K^+ concentration in the cytosol of 100 m*M* and an extracellular K^+ concentration of 10 m*M*. Note that the current would be linear, and that there would be zero current at the equilibrium (Nernst) potential for K^+ (E_K). (B) Actual K^+ current data from an Arabidopsis guard cell protoplast, with the same intracellular and extracellular K^+ concentrations as in (A). These currents result from the activities of voltage-regulated K^+ channels. Note that, again, there is zero net current at the equilibrium potential for K^+. However, there is also zero net current over a broader voltage range because the channels are closed over this voltage range in these conditions. When the channels are closed, no K^+ can flow through them, hence zero current is observed over this voltage range. (C) The current–voltage relationship in (B) results from the activities of two sets of channels (the inwardly rectifying K^+ channels and the outwardly rectifying K^+ channels), which together produce the current–voltage relationship.

For the plasma membrane of plants, fungi, and bacteria, as well as for plant tonoplast and other plant and animal endomembranes, H^+ is the principal ion that is electrogenically pumped across the membrane. The **plasma membrane H^+-ATPase** generates the gradient of electrochemical potential of H^+ across the plasma membrane, while the **vacuolar H^+-ATPase** (usually called the **V-ATPase**) and the **H^+-pyrophosphatase** (**H^+-PPase**) electrogenically pump protons into the lumen of the vacuole and the Golgi cisternae.

Secondary active transport is driven by ion gradients

Solutes are also actively transported across a membrane against their gradient of electrochemical potential by coupling the uphill transport of one solute to the downhill transport of another. This type of carrier-mediated cotransport is termed **secondary active transport**. In plants and fungi, secondary active transporters mainly use the downhill transport of H^+ to drive the uptake of nutrients such as NO_3^-, SO_4^{2-}, and $H_2PO_4^-$; the uptake of many metabolites such as amino acids, peptides, and sucrose; and the export of Na^+, which at high concentrations is toxic to plant cells. **Figure 6.9** shows how secondary active transport may involve the binding of a substrate (S) and an ion (usually H^+) to a carrier protein and a conformational change in that protein.

Secondary active transport is driven indirectly by pumps. In plant cells, protons are extruded from the cytosol by electrogenic H^+-ATPases operating in the plasma membrane and at the vacuolar membrane. Consequently, a membrane potential and a pH gradient are created at the expense of ATP hydrolysis.

plasma membrane H^+-ATPase An ATPase that pumps H^+ across the plasma membrane energized by ATP hydrolysis.

vacuolar H^+-ATPase (V-ATPase) A large, multi-subunit enzyme complex, related to F_oF_1-ATPases, present in endomembranes (tonoplast, Golgi). Acidifies the vacuole and provides the proton motive force for the secondary transport of a variety of solutes into the lumen. V-ATPases also function in the regulation of intracellular protein trafficking.

H^+-pyrophosphatase (H^+-PPase) An electrogenic pump that moves protons into the vacuole, energized by the hydrolysis of pyrophosphate. H^+-PPases are also thought to function in reverse to regenerate pyrophosphate when cytosolic levels are depleted.

secondary active transport Active transport that uses energy stored in the proton motive force or other ion gradient and operates by symport or antiport.

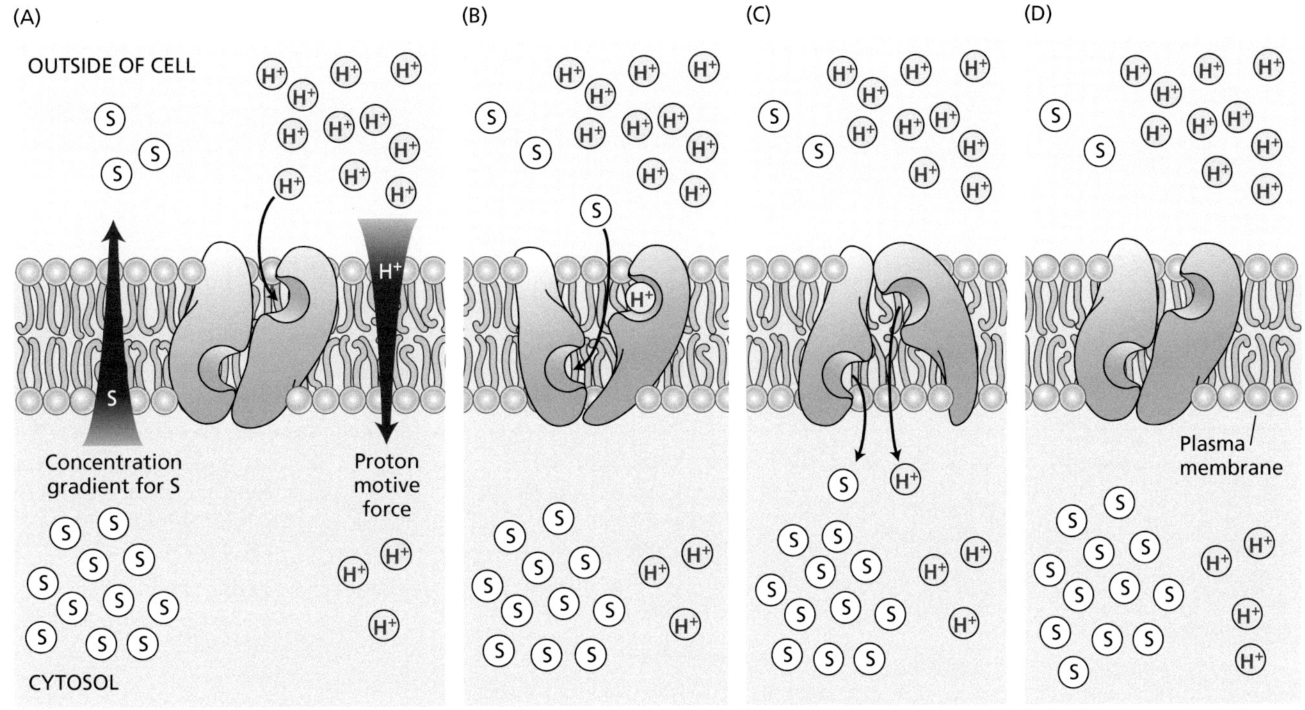

Figure 6.9 Hypothetical model of secondary active transport. In secondary active transport, the energetically uphill transport of one solute is driven by the energetically downhill transport of another solute. In the illustrated example, energy that was stored as proton motive force ($\Delta \tilde{\mu}_{H^+}$, symbolized by the red arrow on the right in [A]) is being used to take up a substrate (S) against its concentration gradient (red arrow on the left). (A) In the initial conformation, the binding sites on the protein are exposed to the outside environment and can bind a proton. (B) This binding results in a conformational change that permits a molecule of S to be bound. (C) The binding of S causes another conformational change that exposes the binding sites and their substrates to the inside of the cell. (D) Release of a proton and a molecule of S to the interior of the cell restores the original conformation of the carrier and allows a new pumping cycle to begin.

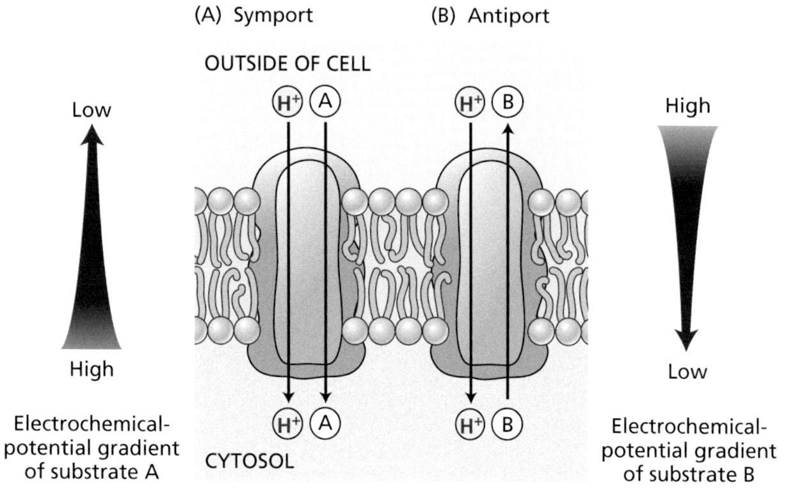

Figure 6.10 Two examples of secondary active transport coupled to a primary proton gradient. (A) In symport, the energy dissipated by a proton moving back into the cell is coupled to the uptake of one molecule of a solute (e.g., a sugar) into the cell. (B) In antiport, the energy dissipated by a proton moving back into the cell is coupled to the active transport of a substrate (e.g., a sodium ion) out of the cell. In both cases, the substrate under consideration is moving against its electrochemical potential gradient. Both neutral and charged substrates can be transported by such secondary active transport processes.

proton motive force (PMF) The energetic effect of the electrochemical H^+ gradient across a membrane, expressed in units of electrical potential.

symport A type of secondary active transport in which two substances are moved in the same direction across a membrane.

antiport A type of secondary active transport in which the passive (downhill) movement of protons or other ions drives the active (uphill) transport of a solute in the opposite direction.

This gradient of electrochemical potential for H^+, referred to as $\Delta \tilde{\mu}_{H^+}$ or (when expressed in other units) the **proton motive force (PMF)**, represents stored free energy in the form of the H^+ gradient.

There are two types of secondary active transport: symport and antiport. In **symport**, facilitated by *symporter* proteins, two substances move in the same direction through the membrane (Figures 6.9 and **Figure 6.10A**). In **antiport**, facilitated by *antiporter* proteins, two substances move in opposite directions (**Figure 6.10B**). Considering the direction of the H^+ gradient, proton-coupled symporters generally function in solute uptake into the cytosol while proton-coupled antiporters function to export solutes out of the cytosol.

In both types of secondary transport, the ion or solute being transported simultaneously with the protons is moving against its gradient of electrochemical potential, so its transport is active. However, the energy driving this transport is provided by the proton motive force rather than directly by ATP hydrolysis.

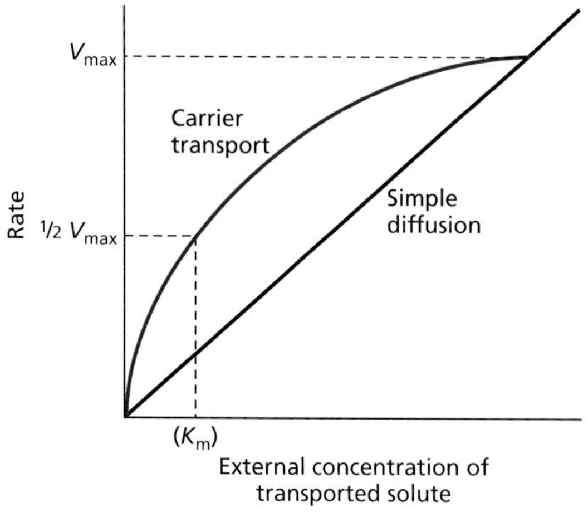

Figure 6.11 Carrier transport often shows enzyme kinetics, including saturation (V_{max}). In contrast, simple diffusion through open channels is ideally directly proportional to the concentration of the transported solute or, for an ion, to the difference in electrochemical potential across the membrane.

Kinetic analyses can elucidate transport mechanisms

Thus far we have described the energetics or thermodynamics of transport, which determine the direction of transport. It is equally important to understand the kinetics, or rate, of transport. Enzyme kinetics are often used to study transport because both involve the binding and dissociation of molecules at binding sites on proteins. Kinetic experiments provide new insights into the regulation of transport.

In kinetic experiments, the effects of solute concentrations on transport rates are measured, and the transport rates can then be used to distinguish between different transporters. The maximum rate (V_{max}) of carrier- or channel-mediated transport cannot be exceeded, regardless of the concentration of substrate (**Figure 6.11**). V_{max} is approached when the substrate-binding site on the carrier is always occupied or when flux through the channel is maximal. The concentration of transporter, not the concentration of solute, becomes rate-limiting. Thus, V_{max} is an indicator of the number of molecules of the specific transport protein that are functioning in the membrane.

The constant K_m (which is numerically equal to the solute concentration that yields half the maximum rate of transport) tends to reflect the properties of the binding site. Low K_m values indicate high binding affinity for the transported substance. Higher values of K_m indicate a lower affinity for the solute. The affinity can be so low that in practice V_{max} is never reached.

Cells or tissues often display complex kinetics for the transport of a solute. Complex kinetics usually indicate the presence of more than one type of transport mechanism—for example, both high- and low-affinity transporters. **Figure 6.12** shows the rate of sucrose uptake by soybean cotyledon protoplasts as a function of the external sucrose concentration. Uptake increases sharply with concentration and begins to saturate at about 10 m*M*, indicating that sucrose uptake at low concentrations is an energy-dependent, carrier-mediated process (H^+–sucrose symport). At concentrations above 10 m*M*, uptake becomes linear and nonsaturable within the concentration range tested and sucrose enters the cells by transport down its concentration gradient and is therefore insensitive to metabolic poisons. Inhibition of ATP synthesis with metabolic poisons blocks the saturable component, but not the linear one (not shown). Consistent with these data, both H^+–sucrose symporters and sucrose facilitators (i.e., transport proteins that mediate transmembrane sucrose flux down its free-energy gradient) have been identified at the molecular level.

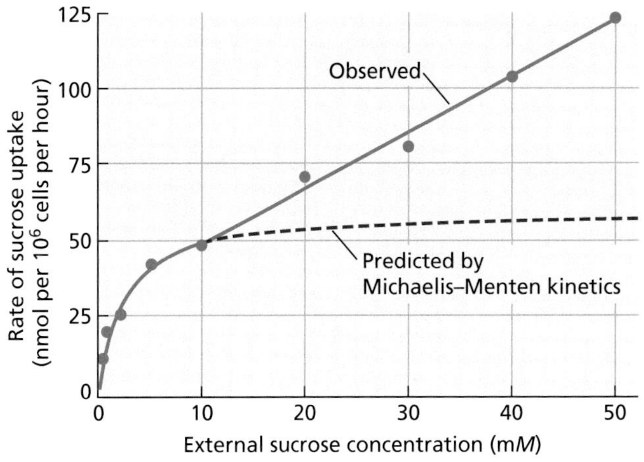

Figure 6.12 The transport properties of a solute can change with solute concentrations. For example, at low concentrations (1–10 m*M*), the rate of uptake of sucrose by soybean cells shows saturation kinetics typical of carriers. A curve fitted to these data is predicted to approach a maximum rate (V_{max}) of 57 nmol per 10^6 cells per hour. Instead, at higher sucrose concentrations, the uptake rate continues to increase linearly over a broad range of concentrations, consistent with the existence of additional facilitated transport mechanisms for sucrose uptake.

6.4 Membrane Transport Proteins

Describe the transporter families important for the movement of different groups of molecules.

Numerous representative transport proteins located in the plasma membrane and the tonoplast are illustrated in **Figure 6.13**. Typically, transport across a biological membrane is energized by one primary active transport system coupled to ATP hydrolysis. In plants and fungi, the H^+-ATPase is the primary pump that generates the ion gradient and an electrochemical potential. Many other ions or neutral organic solutes can then be transported by a variety of secondary active transport proteins, which energize the transport of their substrates by simultaneously carrying one or two H^+ down their energy gradient. Thus, protons circulate across the membrane, outward through the primary active transport proteins, and back into the cell through the secondary active transport proteins. Most of the ion gradients across membranes of higher plants are generated and maintained by electrochemical-potential gradients of H^+, which are generated by electrogenic proton pumps.

In plants, Na^+ is transported out of cells by a Na^+–H^+ antiporter while Cl^-, NO_3^-, $H_2PO_4^-$, sucrose, amino acids, and other substances enter the cell via specific H^+ symporters. What about potassium ions? Potassium ions are mainly taken up from the apoplast through inwardly rectifying K^+ channels (Figure 6.8). K^+ uptake through channels is driven by the negative membrane potential. When passive K^+ uptake is not energetically favorable, for example if extracellular K^+ is too low, K^+ can be taken up by symport with H^+. K^+ uptake via a symport mechanism can accumulate more K^+ in cells relative to the extracellular K^+ concentration because transport is driven by both the negative membrane potential and the transmembrane pH gradient, both of which are generated by H^+-ATPases in the plasma membrane. K^+ efflux from cells occurs through outwardly rectifying K^+ channels and is driven by membrane potentials more positive than E_K (Figure 6.8), which can be achieved by efflux of Cl^- or other anions through anion channels.

Transporter gene identification, genome sequencing, and transport assays using heterologous expression systems have revolutionized the study of transport proteins. As the number of sequenced genomes has increased, it has become common to identify putative transporter genes by phylogenetic analysis, in which

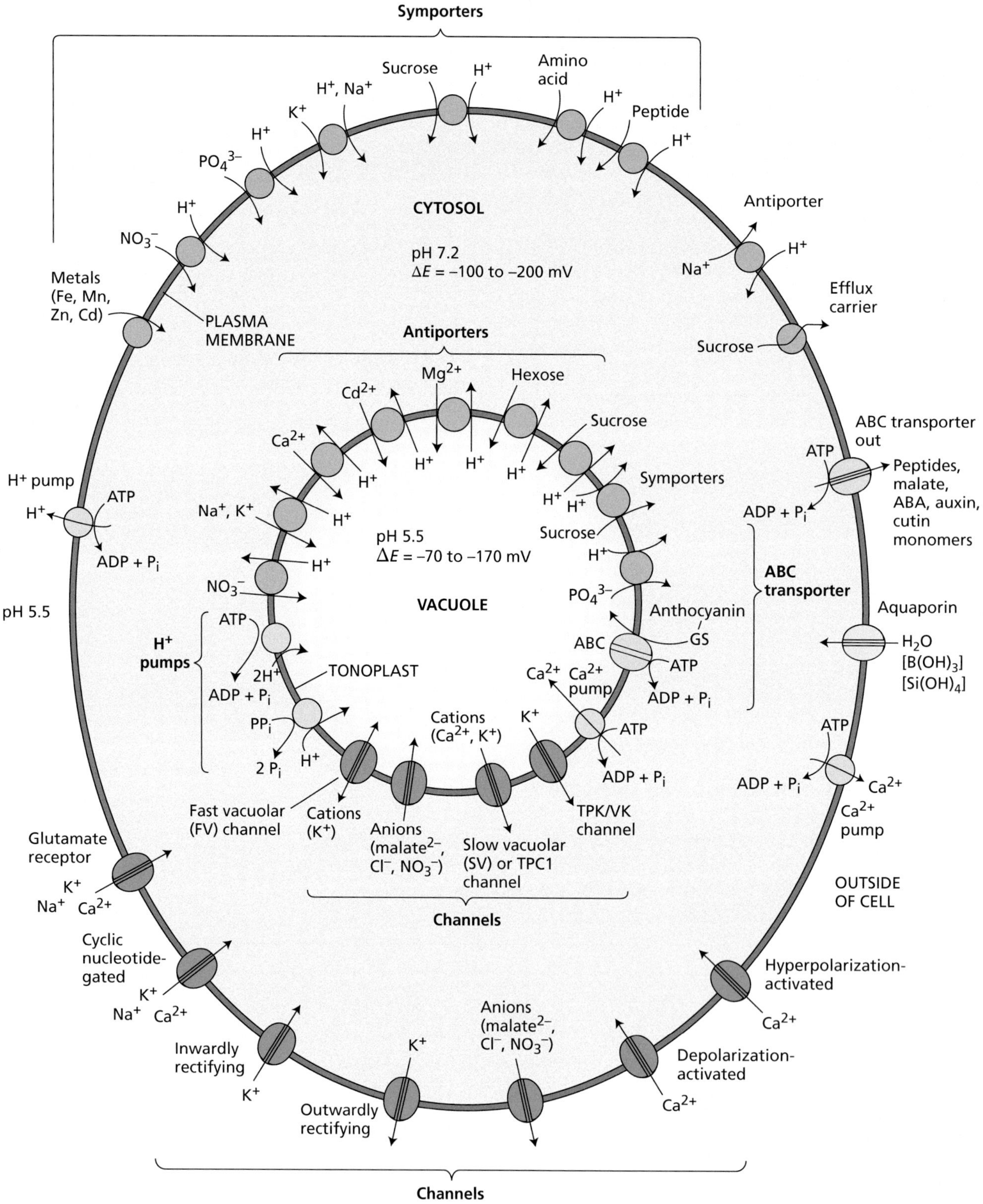

Figure 6.13 Overview of the various transport proteins in the plasma membrane and tonoplast of plant cells. The electrical potentials of the cytosol and vacuole are given using the extracellular space as a reference. Thus, the vacuole is 30 mV less negative than the cytosol.

sequence comparison with genes encoding transporters of known function in other organisms allows one to predict function in the organism of interest. It has become evident that gene families, rather than individual genes, exist in plant genomes for most transport functions. Within a gene family, variations in transport kinetics, in modes of regulation, in subcellular localization, and in differential tissue expression give plants a remarkable plasticity to respond to a broad range of environmental conditions.

Transporters exist for diverse nitrogen-containing compounds

Nitrogen, one of the macronutrients, can be present in the soil solution as nitrate (NO_3^-), ammonia (NH_3), or ammonium (NH_4^+), and in some soils in organic form such as amino acids. Plants have both electrogenic and electroneutral ammonium transporters, and both types bind NH_4^+, deprotonate it, and transport NH_3. In addition, electrogenic ammonium transporters also transport H^+, and therefore are NH_3–H^+ symporters. The cellular uptake of anions such as NO_3^- requires energy, and NO_3^- is taken up by symport with H^+. Plant NO_3^- transporters are of particular interest because of their complexity. For example, the nitrate transporter NRT1.1 encodes an inducible NO_3^-–H^+ symporter that functions as a dual-affinity carrier, with its mode of action (high affinity or low affinity) being switched by its phosphorylation status. Remarkably, this transporter also functions as a NO_3^- sensor that regulates NO_3^--induced gene expression (Chapter 5).

Once nitrogen has been incorporated into organic molecules (Chapter 5), there are a variety of mechanisms that distribute it throughout the plant. Peptide transporters provide one such mechanism to mobilize nitrogen reserves during seed germination and senescence. In the carnivorous pitcher plant *Nepenthes alata*, high levels of expression of a peptide transporter are found in the pitcher, where the transporter presumably mediates uptake of peptides from digested insects into the internal tissues.

Some peptide transporters operate by coupling with the H^+ electrochemical gradient. Other peptide transporters are members of the ABC family of proteins, which directly use energy from ATP hydrolysis for transport; thus, this transport does not depend on a primary electrochemical gradient. The ABC family is an extremely large protein family, and its members transport diverse substrates, ranging from small inorganic ions to macromolecules. For example, large metabolites such as flavonoids, anthocyanins, and metabolic products are sequestered in the vacuole via the action of specific ABC transporters, while other ABC transporters mediate the transmembrane transport of the hormones abscisic acid and auxin.

The plasma membrane amino acid transporters of eukaryotes are large and the majority rely on the proton gradient for coupled amino acid uptake and are present in plants. In general, amino acid transporters can provide high- or low-affinity transport, and they have overlapping substrate specificities. Many amino acid transporters show distinct tissue-specific expression patterns, suggesting specialized functions in different cell types, including vascular tissues for long-distance transport of nitrogen in plants.

Amino acid and peptide transporters have important roles in addition to their function as distributors of nitrogen resources. Because plant hormones are frequently found conjugated with amino acids and peptides, transporters for those molecules are involved in the distribution of hormone conjugates within the cell and throughout the plant. The hormone auxin is derived from tryptophan, and the genes encoding some auxin transporters are related to amino acid transporters (Chapters 12 and 14). In another example, proline accumulates under salt stress and lowers the water potential of the cell, thereby promoting cellular water retention under stress conditions (Chapter 19).

Cation transporters are diverse

Cations are transported by pumps, carriers, and channels. The relative contributions of each type of transport mechanism differ depending on the membrane, cell type, and prevailing conditions.

CATION CHANNELS Approximately 50 genes in the Arabidopsis genome encode channels mediating cation uptake across the plasma membrane or intracellular membranes such as the tonoplast. Some of these channels are highly selective for specific ionic species, such as potassium ions. Others allow passage of a variety of cations, sometimes including Na^+, even though this ion is toxic when over-accumulated. Cation channels are categorized into five types based on their deduced structures and cation selectivity (**Figure 6.14**).

Of the five types of plant cation channels, the voltage-gated K^+ channels have been the most thoroughly characterized. Plant voltage-gated channels are highly K^+-selective and can be either inwardly or outwardly rectifying or weakly rectifying (Figure 6.8). Some members of this family can:

- Mediate K^+ uptake or efflux across the guard cell plasma membrane.
- Provide a major conduit for K^+ uptake from the soil.
- Participate in K^+ release to xylem vessels from the living stelar cells.
- Play a role in K^+ uptake in pollen, a process that promotes water influx and pollen tube elongation.

Some voltage-gated channels, such as those in roots, can mediate high-affinity K^+ uptake, allowing passive K^+ uptake at micromolar external K^+ concentrations as long as the membrane potential is sufficiently hyperpolarized to drive this uptake.

However, not all ion channels are as strongly regulated by membrane potential. For example, the TPK channel is not voltage-regulated (Figure 6.14A).

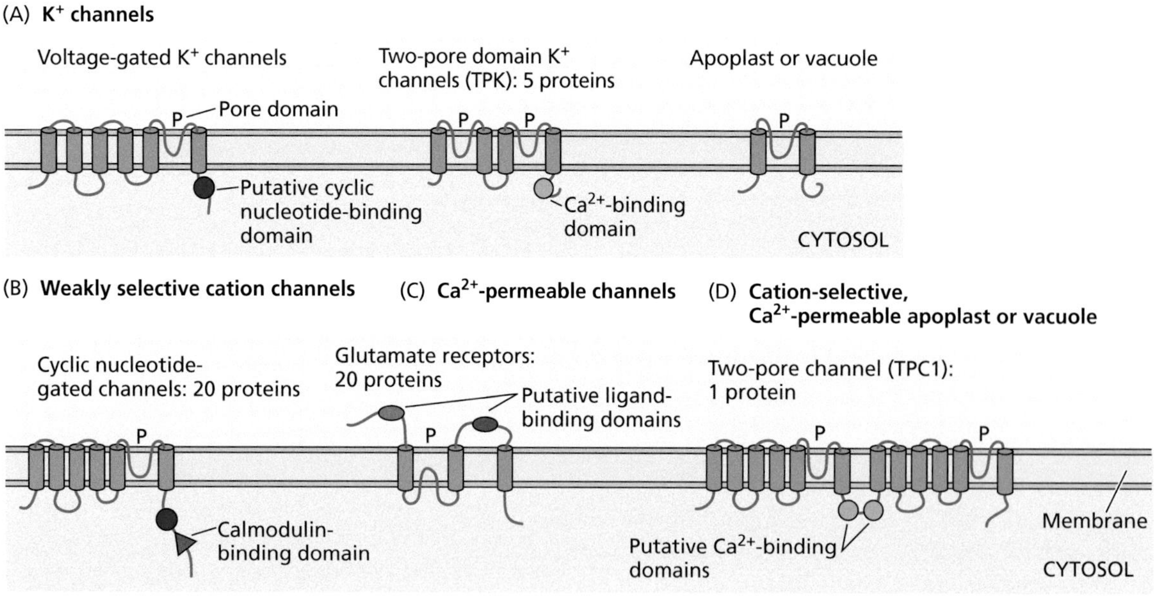

Figure 6.14 Five families of Arabidopsis cation channels. Some channels have been identified from sequence homology with channels of animals, while others have been experimentally verified. (A) Two families of K^+-selective channels: voltage-gated K^+ channels and two-pore domain K^+ channels (TPK) that are not voltage-gated. (B) Weakly selective cation channels with activity regulated by binding of cyclic nucleotides. (C) Glutamate receptor Ca^{2+} channels; these channels are activated by extracellular glutamate and are Ca^{2+}-permeable. (D) Two-pore channel: TPC1 is the sole two-pore channel of this type encoded in the Arabidopsis genome. TPC1/SV is permeable to mono- and divalent cations, including Ca^{2+}.

Ligand-gated channels are activated when they bind a ligand. For example, cyclic nucleotide-gated cation channels are activated by the binding of the Ca^{2+}-binding protein calmodulin and cyclic nucleotide ligands, such as cGMP. These channels exhibit weak selectivity with permeability to K^+, Na^+, and Ca^{2+}. Cyclic nucleotide-gated cation channels are involved in diverse physiological processes, including disease resistance, senescence, temperature sensing, and pollen tube growth and viability. Another interesting set of ligand-gated channels are the glutamate receptor channels (Figure 6.14C). Plant glutamate receptor channels are permeable to Ca^{2+}, K^+, and Na^+ to varying extents, but they have been particularly implicated in Ca^{2+} uptake and signaling in nutrient acquisition in roots and in guard cell and pollen tube physiology.

Ion fluxes must also occur in and out of the vacuole, and both cation- and anion-permeable channels have been characterized in the vacuolar membrane. Plant vacuolar cation channels include the Ca^{2+}-activated TPC1 cation channel (Figures 6.13 and 6.14D) and most TPK channels (Figure 6.13), which are highly selective K^+ channels that are activated by Ca^{2+}. In addition, Ca^{2+} efflux from internal storage sites such as the vacuole plays an important signaling role. For a more detailed description of these signal transduction pathways, see Chapter 12.

CATION CARRIERS A variety of carriers also move cations into plant cells (Figure 6.13). One family of transporters that specializes in K^+ transport across plant membranes is the HAK/KT/KUP family of high-affinity K^+ transporters in plants (HAK for short). They contains both high-affinity and low-affinity transporters, some of which also mediate Na^+ influx at high external Na^+ concentrations. High-affinity HAK transporters are thought to take up K^+ via H^+–K^+ symport, and these transporters are particularly important for K^+ uptake from the soil when soil K^+ concentrations are low. A second family, the cation–proton antiporters (CPAs), mediates electroneutral exchange of H^+ and other cations, including K^+. A third family consists of the Trk/Ktr/HKT, another high-affinity K^+ transporter family (HKT for short). They can operate as K^+–H^+ or K^+–Na^+ symporters, or as Na^+ channels under high external Na^+ concentrations. HKT transporters are central elements in plant tolerance of saline conditions.

Irrigation increases soil salinity, and salinization of croplands is an increasing problem worldwide. Although halophytic plants, such as those found in salt marshes, are adapted to a high-salt environment, such environments are deleterious to other, glycophytic, plant species, including the majority of crop species. Plants have evolved mechanisms to extrude Na^+ across the plasma membrane, to sequester salt in the vacuole, and to redistribute Na^+ within the plant body.

At the plasma membrane, a Na^+–H^+ antiporter was uncovered in a screen to identify Arabidopsis mutants that showed enhanced sensitivity to salt; hence, this antiporter was named Salt Overly Sensitive, or SOS1. SOS1-type antiporters in the root extrude Na^+ from the plant, thereby lowering internal concentrations of this toxic ion (Chapter 19).

In the tonoplast, electroneutral H^+-coupled antiporters for Na^+ or K^+ in the Na^+/H^+ Antiporter (NHX) family function to accumulate Na^+ or K^+ in the vacuole. NHX transporters couple the energetically downhill movement of H^+ into the cytosol with Na^+ or K^+ uptake into the vacuole. When the Arabidopsis *AtNHX1* antiporter gene is overexpressed, it greatly increases salt tolerance in both Arabidopsis and crop species such as maize (corn; *Zea mays*), wheat, and tomato.

Whereas SOS1 and NHX antiporters reduce cytosolic Na^+ concentrations, HKT1 transporters transport Na^+ from the apoplast into the cytosol. However, Na^+ uptake by HKT1 transporters in the plasma membrane of root xylem parenchyma cells retrieves Na^+ from the transpiration stream, thereby reducing Na^+ concentrations and attendant toxicity in photosynthetic tissues. Transgenic

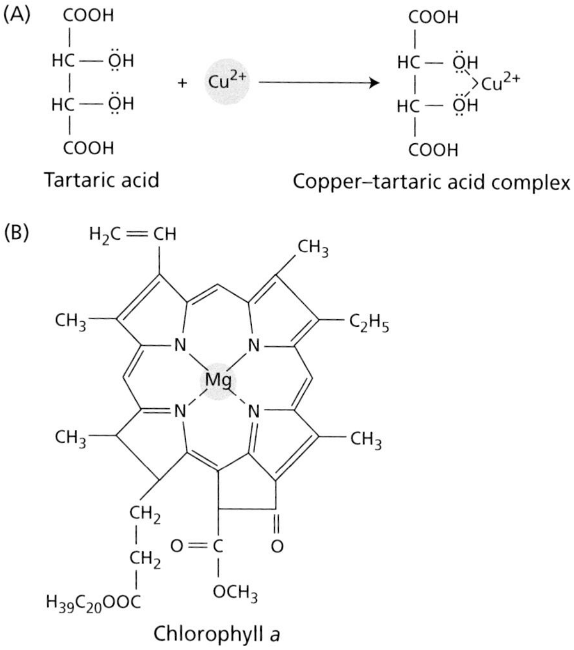

Figure 6.15 Examples of coordination complexes. Coordination complexes form when oxygen or nitrogen atoms of a carbon compound donate shared electron pairs (represented by dots) to form a bond with a cation. (A) Copper ions share electrons with the hydroxyl oxygens of tartaric acid. (B) Magnesium ions share electrons with nitrogen atoms in chlorophyll *a*. Dashed lines represent a coordination bond between electrons from the nitrogen atoms and the magnesium cation.

expression of a HKT1 transporter in a durum (pasta) variety of wheat greatly increases grain yield in wheat grown on saline soils.

Calcium ions are actively pumped out of the cytoplasm by Ca^{2+}-ATPases in plasma membrane and some endomembranes such as the tonoplast and endoplasmic reticulum (Figure 6.13). Much of the Ca^{2+} inside the cell is stored in the central vacuole, where it is sequestered via Ca^{2+}-ATPases and Ca^{2+}–H^+ antiporters. The large free-energy gradient for Ca^{2+} favors its entry into the cytosol from both the apoplast and intracellular stores via Ca^{2+}-permeable channels. Calcium ion concentrations in the apoplast are usually in the millimolar range; in contrast, free cytosolic Ca^{2+} concentrations are kept in the hundreds of nanomolar (10^{-9} *M*) to 1 micromolar (10^{-6} *M*) range.

CATION FUNCTION IN CELLS The noncovalent bonds formed between cations and carbon compounds are either coordination bonds or electrostatic bonds. In the formation of a coordination complex, several oxygen or nitrogen atoms of a carbon compound donate unshared electrons to form a bond with the cation mineral nutrient. As a result, the positive charge on the cation is neutralized or distributed.

Coordination bonds typically form between polyvalent cations and carbon compounds—for example, complexes between copper ions and a chelator such as tartaric acid (**Figure 6.15A**) or between magnesium ions and chlorophyll *a* (**Figure 6.15B**). The nutrients that are assimilated as coordination complexes include copper, zinc, iron, and magnesium ions. Electrostatic bonds form because of the attraction of a positively charged cation for a negatively charged group such as a carboxylate ($-COO^-$) on a carbon compound. In contrast, the cation in an electrostatic bond retains its positive charge. Monovalent cations such as potassium ions can form electrostatic bonds with many organic acids (**Figure 6.16A**). Nonetheless, many of the potassium ions accumulated by plant cells and function in osmotic regulation and enzyme activation remain in the cytosol and vacuole as free ions. Divalent ions such as the calcium ion form electrostatic bonds with the carboxylic groups of polygalacturonic acid of pectins in the cell wall (**Figure 6.16B**) (Chapter 1).

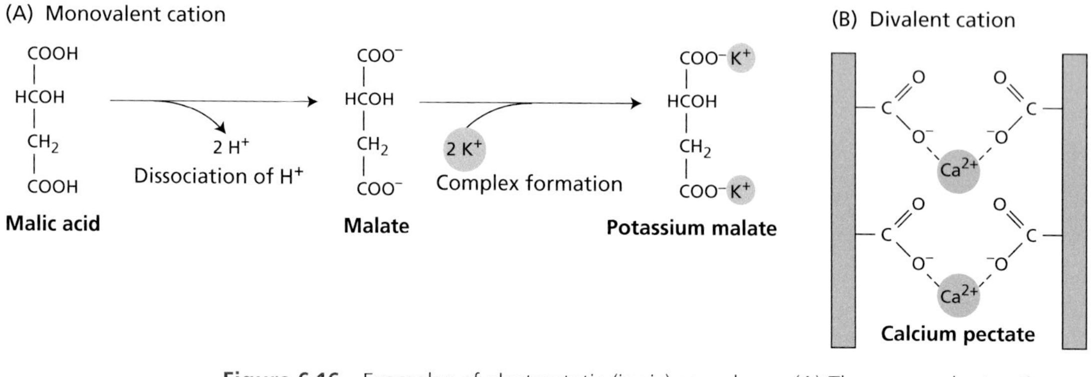

Figure 6.16 Examples of electrostatic (ionic) complexes. (A) The monovalent cation K^+ and malate form the complex potassium malate. (B) The divalent cation Ca^{2+} and carboxylic acid groups from pectate form the complex calcium pectate to cross-links between parallel pectin strands (brown). Calcium ion cross-links play a structural role in the cell walls.

Anion transporters have been identified

Nitrate (NO_3^-), chloride (Cl^-), sulfate (SO_4^{2-}), and phosphate ($H_2PO_4^-$) are the main inorganic ions in plant cells, and malate^{2-} is a major organic anion. The free-energy gradient for all of these anions is in the direction of passive efflux, which means that anion efflux occurs if an anion channel opens in the plasma membrane and that energy is required for anion accumulation in cells. Several types of plant anion channels have been characterized by electrophysiological techniques, and most anion channels appear to be permeable to a variety of anions. In particular, several anion channels, each with differential voltage dependence and anion permeabilities, are important for anion efflux from guard cells.

Anion–H^+ symporters mediate the energetically uphill transport of anions into plant cells and are specific for particular anions such as nitrate, chloride, phosphate, sulfate, and organic anions such as malate and citrate. In Arabidopsis, a family of about nine high-affinity and low-affinity plasma membrane phosphate transporters are proton symporters. These transporters are expressed primarily in the root epidermis and root hairs, and their expression is induced upon phosphate starvation. Other phosphate–H^+ symporters are localized to membranes of intracellular organelles such as the tonoplast, mitochondria, and plastids, including the inner plastid envelope, where they mediate exchange of inorganic phosphate with phosphorylated carbon compounds.

Transporters for metal and metalloid ions transport essential micronutrients

Several metals are essential nutrients for plants, although they are required in only trace amounts (Chapter 4). More than 25 zinc- and iron-regulated (ZIP) transporters mediate the uptake of iron, manganese, and zinc ions into plants, and other transporter families mediate the uptake of copper and molybdenum ions. Metal ions are usually present at low concentrations in the soil solution, so these transporters are typically high-affinity transporters. To absorb sufficient iron from the soil solution, roots have evolved several mechanisms that increase the solubility of iron and therefore its availability for plants (**Figure 6.17**). These mechanisms include:

- Soil acidification, which increases the solubility of ferric iron, followed by the reduction of ferric iron to the more soluble ferrous form (Fe^{2+}). This mechanism is used by eudicots and non-graminaceous monocots.
- Release of compounds that form stable, soluble complexes with iron, the so-called siderophores. This mechanism is used by graminaceous monocots (grasses).

In the soil, iron is present primarily as ferric iron (Fe^{3+}), in oxides such as $Fe(OH)^{2+}$, $Fe(OH)_3$, and $Fe(OH)_4^-$. At neutral pH, ferric iron is highly insoluble. Roots generally acidify the soil around them: They pump out protons, respiration in the roots releases CO_2 that forms carbonic acid in the soil, the uptake and assimilation of ammonium cause the release of excess protons, and the roots secrete organic acids that enhance iron and phosphate availability. Iron deficiency also stimulates proton export in roots. Plasma membranes in roots contain an enzyme called *iron-chelate reductase*, which reduces ferric iron to ferrous iron, with cytosolic NADH or NADPH serving as the electron donor (Figure 6.17A). The activity of this enzyme increases under iron deficiency.

Several compounds secreted by roots form stable chelates with iron. Examples include malic acid, citric acid, phenolics, and piscidic acid. Grasses produce a special class of iron chelators called *siderophores*. Siderophores are composed of amino acids not found in proteins, such as mugineic acid,

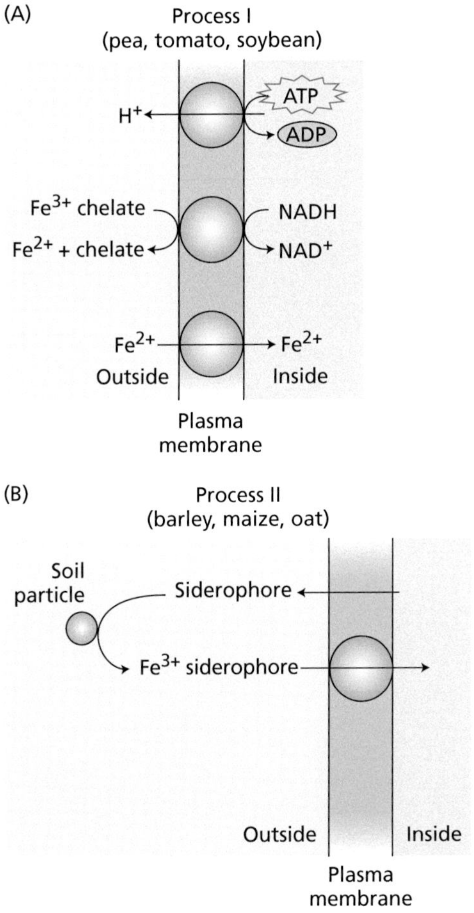

Figure 6.17 Two mechanisms for iron uptake by plants. (A) A process used by eudicots such as pea, tomato, and soybean. The chelators include organic compounds such as malic acid, citric acid, phenolics, and piscidic acid. (B) A process used by grasses such as barley, maize, and oat. After the grass secretes a siderophore that removes iron from soil particles, the whole iron–chelate complex is transported into root cells.

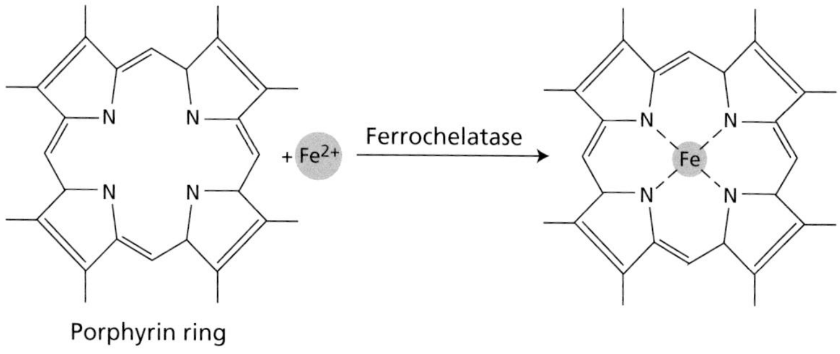

Figure 6.18 The ferrochelatase reaction. The enzyme ferrochelatase catalyzes the insertion of iron into the porphyrin ring to form a coordination complex for cytochromes in plants and hemoglobin in N-fixing bacteria.

and form highly stable complexes with Fe^{3+}. Root cell plasma membranes in grasses have Fe^{3+}–siderophore transport systems to take up the chelated iron. Under iron deficiency, grass roots release more siderophore into the soil and increase their capacity for Fe^{3+}–siderophore transport (Figure 6.17B). Iron deficiency is the most common human nutritional disorder worldwide, so an increased understanding of how plants accumulate iron aid efforts to improve the nutritional value of crops.

Once in the plant, metal ions, usually as chelates, must be transported into the xylem for distribution throughout the plant body via the transpiration stream, and metals must also be sent to their appropriate subcellular destinations. Ferrous iron absorbed by the roots is oxidized to ferric iron in the cytoplasm. Much of the iron taken up in the roots is transported to leaves in a complex with citrate or nicotianamine. Once in the leaves, the enzyme ferrochelatase inserts an iron cation into the porphyrin precursor of heme groups found in the cytochromes located in the chloroplasts and mitochondria (**Figure 6.18**) (Chapters 7 and 11). While most of the iron in plants is found in heme groups, iron–sulfur proteins of the electron transport chain also in chloroplasts and mitochondria contain nonheme iron covalently bound to the sulfur atoms of cysteine residues. Iron is also found in Fe_2S_2 centers, which contain two irons (each complexed with the sulfur atoms of cysteine residues) and two inorganic sulfides.

Free iron (iron not complexed with carbon compounds) may interact with oxygen to form highly damaging hydroxyl radicals, $HO^{\bullet}$. Plant cells may limit such damage by storing iron in an iron–protein complex called **ferritin**. Ferritins are essential for protection against oxidative damage but not for either seedling development or proper functioning of the photosynthetic apparatus. Ferritin consists of a protein shell with 24 identical subunits forming a hollow sphere that has a molecular mass of about 480 kDa. Within this sphere is a core of 5,400 to 6,200 iron atoms present as a ferric oxide–phosphate complex. The mechanism of iron release from ferritin is unknown but involves the breakdown of the protein shell. The concentration of free iron in plant cells regulates the de novo biosynthesis of ferritin.

Metalloids are elements that have properties of both metals and nonmetals. Boron and silicon are two metalloids that play important roles in cell wall structure—boron by cross-linking cell wall polysaccharides, and silicon by increasing structural rigidity. Boron (as boric acid [$B(OH)_3$; also written H_3BO_3]) and silicon (as silicic acid [$Si(OH)_4$; also written H_4SiO_4]) both enter cells via aquaporin-type channels and are exported via efflux transporters, probably by secondary active transport. Because of similarities in chemical structure, arsenite (an arsenic oxide) can also enter plant roots via the silicon channel and be exported to the transpiration stream via the silicon exporter. Rice is particularly efficient at taking up arsenite, and as a result, arsenic poisoning from human consumption of rice is a significant problem in regions of southeast Asia. The metalloid selenium can enter the roots as selenate through the sulfate transporter. It can be incorporated into compounds in place of sulfur, thereby disrupting protein function. In some plants, such as *Astragulus bisculatus* (locoweed), selenium is incorporated into methylselenocysteine and moves via the transpiration stream to stomata where it is volatilized. Plants that accumulate selenium can poison livestock.

ferritin A protein that functions in cellular iron storage in multiple compartments, including plastids and mitochondria.

aquaporins Integral membrane proteins that form channels across a membrane, many of which are selective for water (hence the name). Such channels facilitate water movement across a membrane.

Aquaporins have diverse functions

Aquaporins are an abundant class of membrane transporters (Chapters 2 and 3) with approximately 35 members in Arabidopsis. As the name implies, many aquaporin proteins function as water channels, while some aquaporin proteins mediate the influx of mineral nutrients (e.g., boric acid and silicic acid). Aquaporins can also act as conduits for carbon dioxide, ammonia (NH_3), and hydrogen peroxide (H_2O_2) movement across the plasma membrane and other plant membranes.

Aquaporin activity is regulated by phosphorylation as well as by pH, Ca^{2+} concentration, heteromerization, and reactive oxygen species (Chapter 12). Such regulation may account for the ability of plant cells to quickly alter their permeability to water in response to circadian rhythm and to stresses such as salt, chilling, drought (water-deficit stress), and flooding (anoxia). Regulation also occurs at the level of gene expression. Aquaporin genes are highly expressed in epidermal and endodermal cells and in the xylem parenchyma, which may be critical points for control of water movement.

Plasma membrane H^+-ATPases are highly regulated P-type ATPases

The active transport of protons across the plasma membrane to the apoplast creates pH and electrical potential gradients that drive the transport of many ions and uncharged solutes through the various secondary active transporters and channels. H^+-ATPase activity is also important for the regulation of cytosolic pH and for the control of cell turgor, which drives organ (leaf and flower) movement, stomatal opening, and cell growth. **Figure 6.19** illustrates how a membrane H^+-ATPase might work.

Plant and fungal plasma membrane H^+-ATPases and Ca^{2+}-ATPases are members of a class known as P-type ATPases, which are phosphorylated as part of the catalytic cycle that hydrolyzes ATP. Plant plasma membrane H^+-ATPases are encoded by a family of about a dozen genes. Some H^+-ATPases exhibit cell-specific patterns of expression. For example, several H^+-ATPases are expressed in guard cells, where they energize the plasma membrane to drive solute uptake during stomatal opening.

In general, H^+-ATPase expression is high in cells with key functions in nutrient movement, including root endodermal cells and cells involved in nutrient uptake

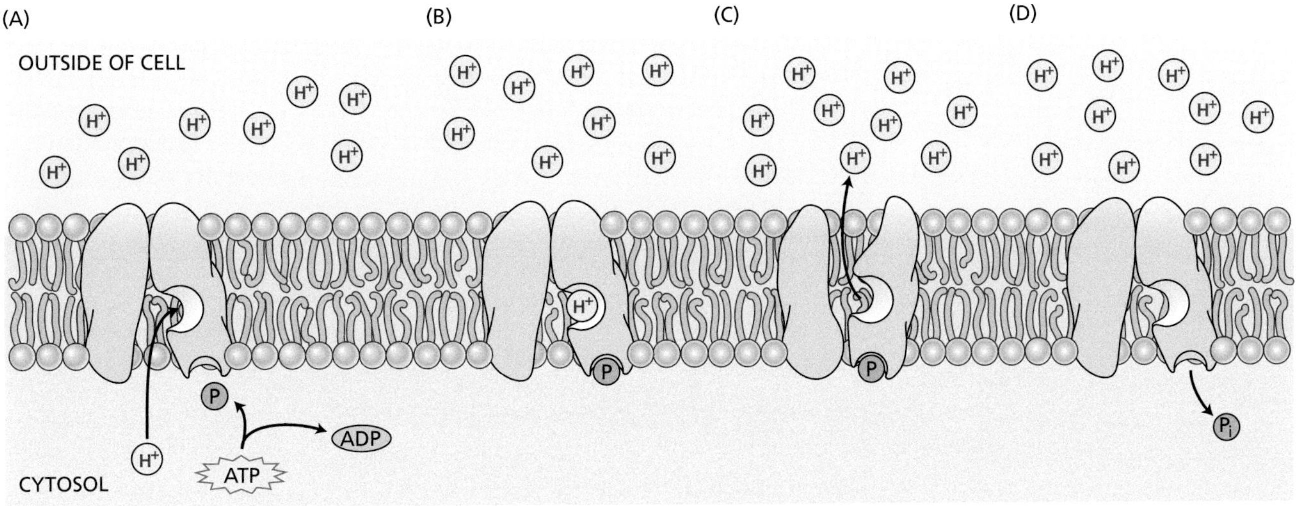

Figure 6.19 Hypothetical steps in the transport of a proton against its chemical gradient by H^+-ATPase. (A) The pump, embedded in the membrane, binds the proton on the inside of the cell and (B) is phosphorylated by ATP. (C) This phosphorylation leads to a conformational change that exposes the proton to the outside of the cell and makes it possible for the proton to diffuse away. (D) Release of the phosphate ion (P_i) from the pump into the cytosol restores the initial configuration of the H^+-ATPase and allows a new pumping cycle to begin.

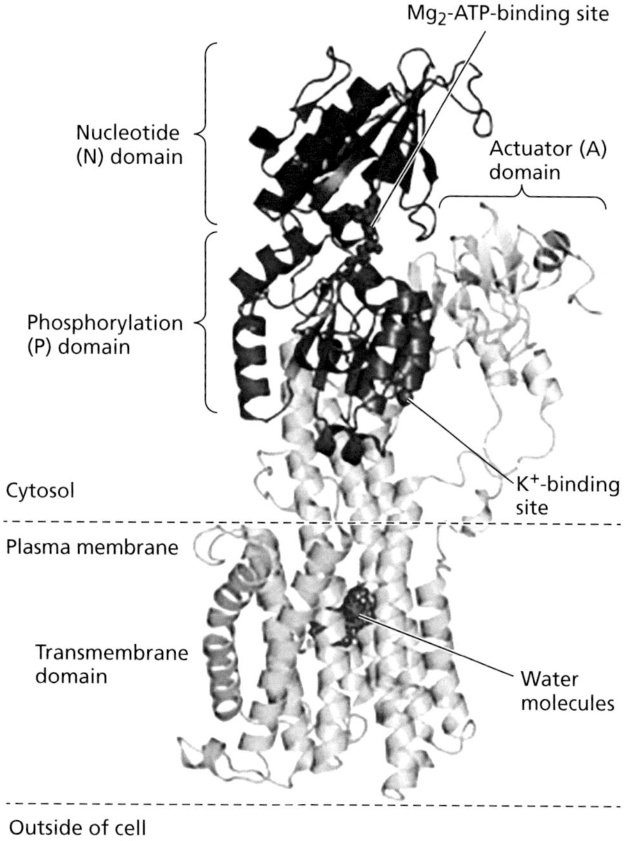

Figure 6.20 Three-dimensional representation of the plasma membrane H^+-ATPase AtAHA2 from *Arabidopsis thaliana*. The nucleotide (N) binding domain, phosphorylation (P) domain, and actuator (A) domain all protrude into the cytosol, while the transmembrane domain contains ten transmembrane helices that anchor the enzyme in the membrane and contain the transmembrane pathway for the translocation of protons. The solvent cavity within the transmembrane domain contains eight to ten water molecules. Mg_2-ATP- and K^+-binding sites are indicated. Posttranslational modifications lead to displacement of the C-terminal autoinhibitory domain (not shown here), resulting in H^+-ATPase activation.

from the apoplast that surrounds the developing seed. In cells in which multiple H^+-ATPases are co-expressed, they may be differentially regulated or may function redundantly, perhaps providing a "fail-safe" mechanism to this all-important transport function.

The plasma membrane H^+-ATPase AtAHA2 from *Arabidopsis thaliana* has ten membrane-spanning helices that loop back and forth across the membrane (**Figure 6.20**). Some of the membrane-spanning domains make up the pathway through which protons are pumped. The nucleotide (N) domain binds Mg_2-ATP, and the phosphorylation (P) domain contains the aspartic acid residue that becomes phosphorylated during the catalytic cycle.

Like other enzymes, the plasma membrane H^+-ATPase is regulated by the concentration of substrate (ATP), pH, temperature, and other factors. In addition, H^+-ATPase molecules can be reversibly activated or deactivated by specific signals, such as light, hormones, or pathogen attack. This type of regulation is mediated by a specialized autoinhibitory domain at the C-terminal end of the polypeptide chain, which acts to regulate the activity of the H^+-ATPase. If the autoinhibitory domain is removed by a protease, the enzyme becomes irreversibly activated.

The autoinhibitory effect of the C-terminal domain is also regulated by protein kinases and phosphatases that add phosphate groups to or remove them from serine or threonine residues on this domain (Chapter 12). Phosphorylation promotes binding of 14-3-3 proteins, which are soluble proteins that participate in many signal transduction pathways and function via protein–protein interaction; 14-3-3 proteins bind to the phosphorylated region of the H^+-ATPase and displace the autoinhibitory domain, leading to H^+-ATPase activation. The fungal toxin **fusicoccin**, which is a strong activator of the H^+-ATPase, activates this pump by increasing 14-3-3 binding affinity even in the absence of phosphorylation (Chapter 18). The effect of fusicoccin on the guard cell H^+-ATPases is so strong that it can lead to irreversible stomatal opening, wilting, and even plant death.

The tonoplast H^+-ATPase drives solute accumulation in vacuoles

Plant cells increase their size primarily by taking up water into a large central vacuole. Therefore, the water potential of the vacuole must be kept sufficiently low for water to enter from the cytosol. The tonoplast regulates the traffic of ions and metabolites between the cytosol and the vacuole, just as the plasma membrane regulates their uptake into the cell. The tonoplast has a diversity of anion and cation channels including the vacuolar H^+-ATPase (V-ATPase), which transports protons into the vacuole (Figure 6.13).

fusicoccin A fungal toxin that induces acidification of plant cell walls by activating H^+-ATPases in the plasma membrane. Fusicoccin stimulates rapid acid growth in stem and coleoptile sections. It also stimulates stomatal opening by stimulating proton pumping at the guard cell plasma membrane.

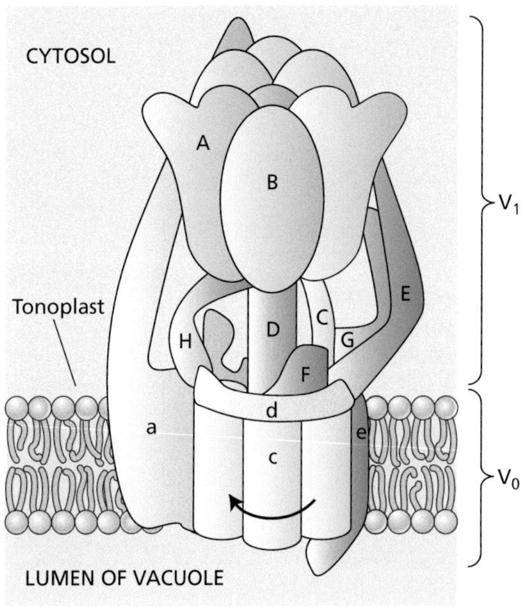

Figure 6.21 Model of the V-ATPase rotary motor. Many polypeptide subunits come together to make up this complex enzyme. The V_1 catalytic complex, which is easily dissociated from the membrane, contains the nucleotide-binding and catalytic sites required for ATP hydrolysis. Components of V_1 are designated by uppercase letters. The integral membrane complex mediating H^+ transport is designated V_0, and its subunits are designated by lowercase letters. It is proposed that ATPase reactions catalyzed by each of the A subunits, acting in sequence, drive the rotation of the shaft (D) and the six c subunits. The rotation of the c subunits relative to the a subunit is thought to drive the transport of H^+ across the membrane.

Vacuolar ATPases belong to a general class of ATPases that are present on the endomembrane systems of all eukaryotes (**Figure 6.21**). The vacuolar H^+-ATPase differs both structurally and functionally from the plasma membrane H^+-ATPase. The V-ATPase is more closely related to the F-ATPases of mitochondria and chloroplasts, and the V-ATPase, unlike the plasma membrane ATPases discussed earlier, does not form a phosphorylated intermediate during ATP hydrolysis. Because of their similarities to F-ATPases, V-ATPases are assumed to operate like tiny rotary motors.

Vacuolar ATPases are electrogenic proton pumps that generate a proton motive force across the tonoplast. Electrogenic proton pumping accounts for the fact that the tonoplast has a membrane potential of –20 to –30 mV (cytosol relative to the vacuole lumen). Electrophysiologists describe membrane potentials relative to the cytoplasm, so the vacuolar lumen is considered equivalent to the extracellular space. Anions such as Cl^- and malate^{2-} are transported from the cytosol into the vacuole through channels in the tonoplast. Relatively few protons need to be pumped across the membrane to generate a membrane potential. The activity of tonoplast anion channels neutralizes the membrane potential and allows the vacuolar H^+-ATPase to generate a large concentration gradient of protons (pH gradient) across the tonoplast resulting in the acidic vacuolar sap (pH 5.5) in contrast to the cytosol (pH 7.0–7.5). Whereas the electrical component of the proton motive force drives the uptake of anions into the vacuole, the electrochemical-potential gradient for H^+ ($\tilde{\mu}_{H^+}$) is harnessed to drive the uptake of cations and sugars into the vacuole via secondary transport (antiporter) systems (Figure 6.13).

H^+-pyrophosphatases and P-type H^+-ATPases also pump protons at the tonoplast

Another type of proton pump, a H^+-pyrophosphatase (H^+-PPase), works in parallel with the vacuolar ATPase to create a proton gradient across the tonoplast (Figure 6.13). This enzyme consists of a single polypeptide that harnesses energy from the hydrolysis of inorganic pyrophosphate (PP_i) to drive H^+ transport.

The free energy released by PP_i hydrolysis is less than that from ATP hydrolysis. However, the H^+-PPase transports only one H^+ ion per PP_i molecule hydrolyzed, whereas the V-ATPase appears to transport two H^+ ions per ATP hydrolyzed. Thus, the energy available per H^+ transported appears to be approximately the same, and the two enzymes seem to be able to generate comparable proton gradients. Interestingly, the plant H^+-PPase is not found in animals or in yeast, although similar enzymes are present in some bacteria and protists.

Both the V-ATPase and the H^+-PPase are found in other compartments of the endo-membrane system where the V-type ATPases also regulate vesicle trafficking and secretion. Crop plants overexpressing H^+-PPase are resistant to salt and drought stress and have increased biomass. These phenotypes may have several causes, including increased H^+ transport into vacuoles, localization

water potential (Ψ) A measure of the free energy associated with water per unit of volume (J m^{-3}). These units are equivalent to pressure units such as pascals. Ψ is a function of the solute potential, the pressure potential, and the gravitational potential: $\Psi = \Psi_s + \Psi_p + \Psi_g$. The term Ψ_g is often ignored because it is negligible for heights under 5 m.

phototropins Blue-light photoreceptors that primarily regulate phototropism, chloroplast movements, and stomatal opening. Phototropins are autophosphorylating protein kinases that are stimulated by blue light interactions with a flavin co-factor.

of H^+-PPase in membranes other than the tonoplast, decreased cytosolic pyrophosphate concentration, and up-regulation of plasma membrane H^+-ATPases.

Although the pH of most plant vacuoles is mildly acidic (~5.5), the pH of the vacuoles of some species is much lower—a phenomenon termed *hyperacidification*. Vacuolar hyperacidification is responsible for the sour taste of certain fruits (lemons) and vegetables (rhubarb) and flower color in petunia. Hyperacidification of the vacuoles of petunia petals and citrus fruit has been shown to require the activity of P-type H^+-ATPases in the tonoplast. Since P-type H^+-ATPases transport one H^+ per ATP hydrolyzed, they are capable of generating a larger pH gradient than either V-type ATPases or H^+-PPases.

6.5 Transport in Stomatal Guard Cells

Describe the biological and biophysical principles of solute transport that mediate stomatal opening and closing.

Stomatal guard cells (**Figure 6.22**), found in the epidermis of leaves and other green tissues, control the aperture of stomatal openings that allow for gas exchange between a plant and the atmosphere—primarily water vapor efflux into the atmosphere and CO_2 uptake into leaves (Chapters 3 and 9). Guard cells are an important model for studying transport processes for several reasons: Guard cells reversibly change their shape using an osmotic mechanism, they respond to many environmental and internal stimuli, and they are easily accessible and visible in the epidermis. The osmotic mechanism that guard cells use to change shape, and thereby regulate stomatal aperture, depends on pumps, transporters, and channels in the plasma membrane and vacuolar membrane of guard cells.

Blue light induces stomatal opening

Opening of the stomatal pore requires guard cells surrounding the pore to increase in volume, which they do by taking in water from the extracellular space. Water movement across membranes is driven by a difference in **water potential** (Chapter 2), and guard cells generate a more negative water potential by accumulating K^+ salts. An important physiological stimulus that triggers stomatal opening is blue light. As shown in **Figure 6.23A**, blue light is detected by **phototropins**, plasma membrane protein kinases that use a flavin cofactor to absorb blue light (Chapter 13). Phototropin activation initiates a signaling pathway that leads to phosphorylation and activation of plasma membrane H^+-ATPases. Phototropin does not phosphorylate H^+-ATPases directly—that action is accomplished by another protein kinase that has not been identified yet. Activation of plasma membrane H^+-ATPases causes hyperpolarization of the membrane potential, a key regulator of transport activities in the cell. Hyperpolarization activates inwardly rectifying K^+ channels, allowing K^+ uptake into the guard cells. Anion uptake is accomplished by H^+-coupled anion transporters that are stimulated

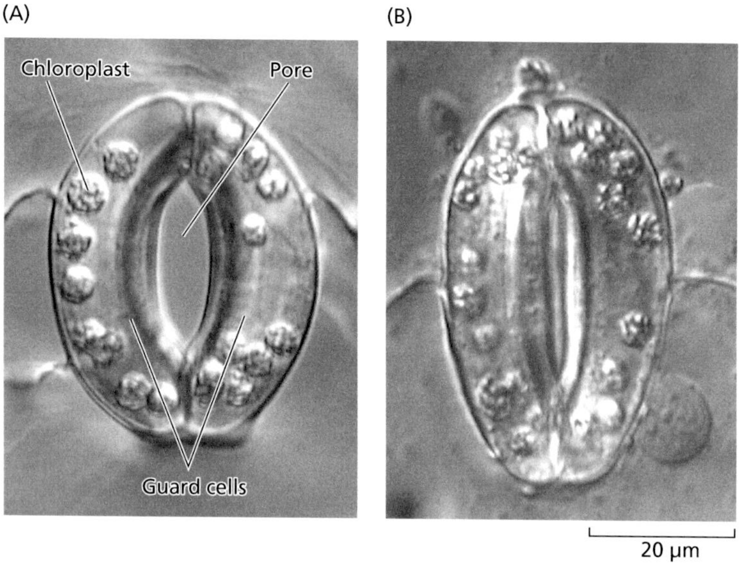

Figure 6.22 Micrograph of a light-stimulated stomatal opening in detached epidermis of *Vicia faba*. An open, light-treated stoma is shown in (A), and a closed, dark-treated stoma is shown in (B). Stomatal aperture is quantified by measuring the width of the stomatal pore. Courtesy of E. Raveh

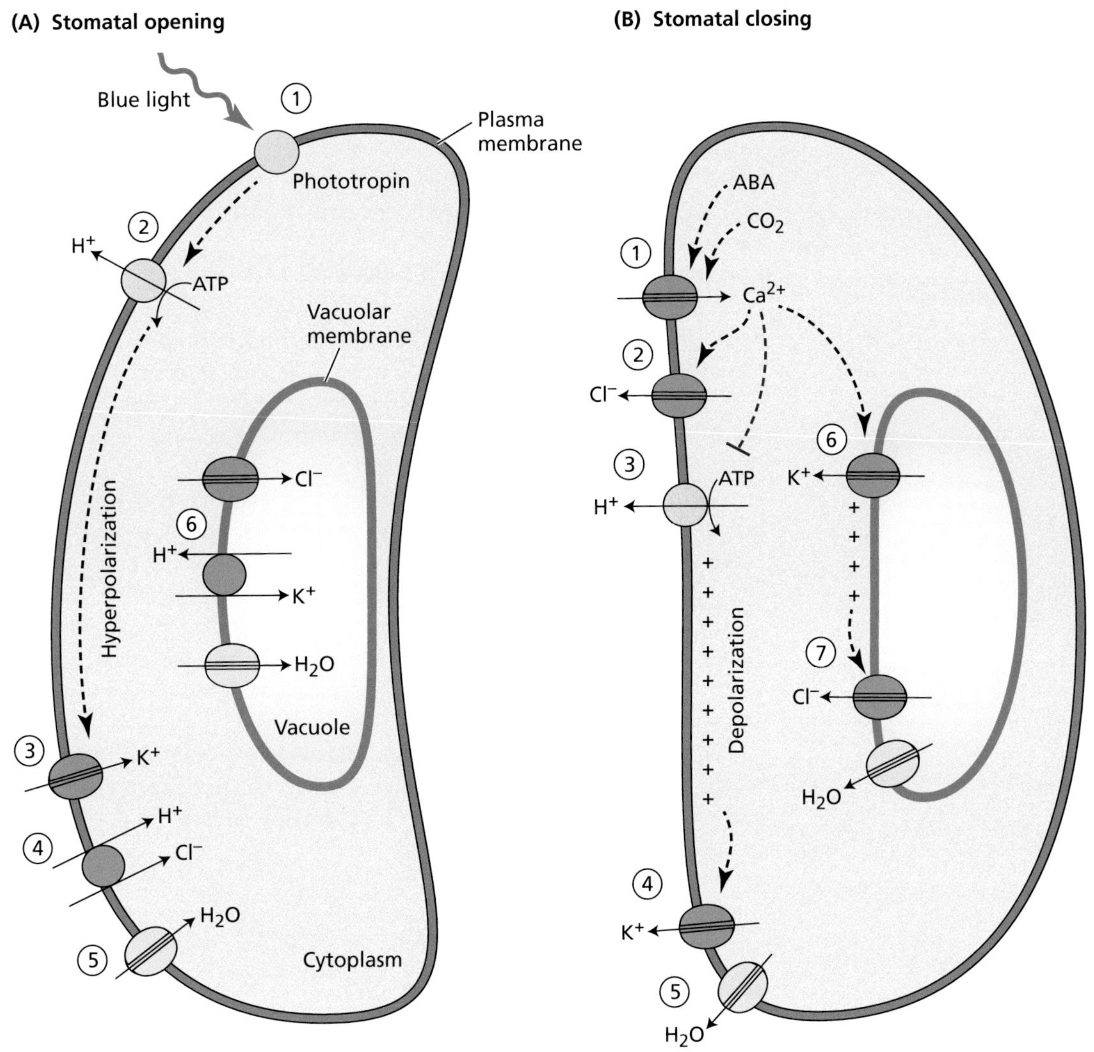

Figure 6.23 Transport activities in guard cells during stomatal opening and closing. (A) Stomatal opening. ① Phototropins, plasma membrane blue-light receptors, receive a blue light signal. ② Signaling results in activation of plasma membrane H^+-ATPases, which hyperpolarizes the membrane potential and increases the transmembrane pH gradient. ③ Inwardly rectifying K^+ channels are activated by hyperpolarization, and ④ anions such as Cl^- are accumulated via H^+-coupled transporters. ⑤ The resulting accumulation of K^+ salts drives water uptake via aquaporins. ⑥ The vacuole also accumulates K^+ salts, but this mechanism involves K^+–H^+ antiporters and anion channels. (B) Stomatal closing. ① ABA or elevated CO_2 leads to the activation of Ca^{2+} channels and elevated cytoplasmic Ca^{2+}, which is a central regulator for stomatal closing. ② Elevated Ca^{2+} results in activation of anion efflux channels and ③ inhibition of plasma membrane H^+-ATPases. ④ The resulting plasma membrane depolarization activates outwardly rectifying K^+ channels, allowing K^+ efflux. ⑤ The efflux of K^+ salts drives water efflux from the cells via aquaporins. ⑥ Elevated cytoplasmic Ca^{2+} activates TPK K^+ channels in the tonoplast, allowing K^+ efflux from the vacuole. ⑦ The resulting tonoplast depolarization drives anion efflux from the vacuole via anion channels. Dashed lines indicate stimulatory or inhibitory signal transduction steps.

by the increased transmembrane pH gradient caused by the activation of plasma membrane H^+-ATPases. The more negative **osmotic potential** due to uptake of K^+ salts drives water uptake into the guard cells via aquaporins, thus increasing guard cell volume and turgor and causing a change in the shape of the cells, which opens the stomatal pore (Chapter 3).

osmotic potential (Ψ_s) The effect of dissolved solutes on water potential. Also called solute potential.

The volume of the guard cell vacuole also increases during stomatal opening. Proton pumping from the vacuolar H^+-PPase or V-ATPase does not appear to be required. Anion uptake through anion channels and K^+ uptake via K^+–H^+

antiporters (Figure 6.23A) increases K^+ salt concentration in the vacuole and drives water uptake into the vacuole via aquaporins.

Abscisic acid and high CO_2 induce stomatal closing

The mechanism of stomatal closing is the opposite of stomatal opening in that it is driven by depolarization at the plasma membrane, efflux of K^+ salts from guard cells, loss of water via aquaporins, a decrease in guard cell volume and turgor, and a change in cell shape that results in closing of the stomatal pore. The hormone abscisic acid (ABA) (Chapter 12) and elevated intracellular CO_2 both induce stomatal closing. Elevated cytoplasmic Ca^{2+} in guard cells is an important signaling intermediate in stomatal closing (**Figure 6.23B**). ABA receptor activation triggers the activation of Ca^{2+} channels in the plasma membrane, allowing Ca^{2+} influx. The elevated cytosolic Ca^{2+} in turn leads to activation of plasma membrane anion channels and inhibition of plasma membrane H^+-ATPases, both of which lead to membrane depolarization.

Membrane depolarization, due to anion efflux through plasma membrane anion channels, activates outwardly rectifying K^+ channels in the plasma membrane and drives K^+ efflux. This results in the net efflux of K^+ salts from guard cells, making the osmotic potential of the cells less negative and driving water efflux via aquaporins. The cell volume and turgor decrease, and the shape of the guard cells changes to close the stomatal pore. Note that in both stomatal opening and closing, K^+ movement across the plasma membrane is driven by changes in membrane potential.

The volume of the guard cell vacuole also decreases during stomatal closure. It is expected that K^+ efflux from vacuoles, as well as of anions such as Cl^-, NO_3^-, and malate, is important, but the mechanism is not fully understood. Elevated cytosolic Ca^{2+} activates highly selective TPK channels that may provide a pathway for K^+ efflux from the vacuole. The resulting vacuolar membrane depolarization may drive anion efflux from the vacuole via anion channels (Figure 6.23B).

6.6 Ion Transport in Roots

Define the cell types that function in xylem loading and the pathways by which it occurs.

Mineral nutrients absorbed by the root are carried to the shoot by the transpiration stream moving through the xylem (Chapter 3). Both the initial uptake of nutrients and water and the subsequent movement of these substances from the root surface across the cortex and into the xylem are highly specific, well-regulated processes.

Ion transport across the root obeys the same biophysical laws that govern cellular transport. However, as we have seen in the case of water movement (Chapter 3), the anatomy of roots imposes special constraints on the pathway of ion movement. In this section we discuss the pathways and mechanisms involved in the radial movement of ions from the root surface to the tracheary elements of the xylem.

Solutes move through both apoplast and symplasm

Thus far, our discussion of cellular ion transport has not included the cell wall. In terms of the transport of small molecules, the cell wall is a fluid-filled lattice of polysaccharides through which mineral nutrients diffuse readily. Because all plant cells are separated by cell walls, ions can diffuse across a tissue (or be carried passively by water flow) entirely through the cell wall space without ever entering a living cell. This continuum of cell walls is called the **extracellular space**,

extracellular space In plants, the space continuum outside the plasma membrane made up of interconnecting cell walls through which water and mineral nutrients readily diffuse.

or **apoplast**. Typically, 5 to 20% of the plant tissue volume is occupied by cell walls.

Just as the cell walls form a continuous phase, so do the cytoplasms of neighboring cells, collectively referred to as the **symplasm**. Plant cells are interconnected by cytoplasmic bridges called plasmodesmata (singular *plasmodesma*) (Chapter 1), cylindrical pores 20 to 60 nm in diameter (**Figure 6.24**). Each plasmodesma is lined with plasma membrane and contains a narrow tubule, the *desmotubule*, which is a continuation of the endoplasmic reticulum.

In tissues where significant amounts of intercellular transport occur, neighboring cells contain large numbers of plasmodesmata, up to 15 per square micrometer of cell surface. Specialized secretory cells, such as floral nectaries and leaf salt glands, have high densities of plasmodesmata.

Inorganic ions, water, and small organic molecules can move from cell to cell through these pores. Because each plasmodesma is partly occluded by the desmotubule and its associated proteins (Chapter 1), the movement of large molecules such as proteins through plasmodesmata requires special mechanisms. Ions, by contrast, appear to move symplasmically through the plant by simple diffusion through plasmodesmata (Chapter 3).

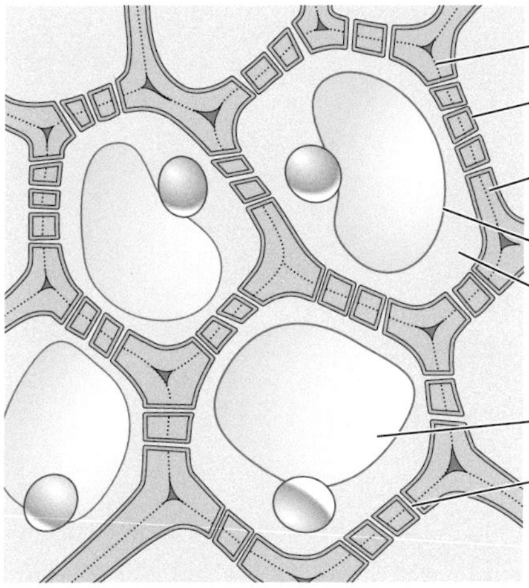

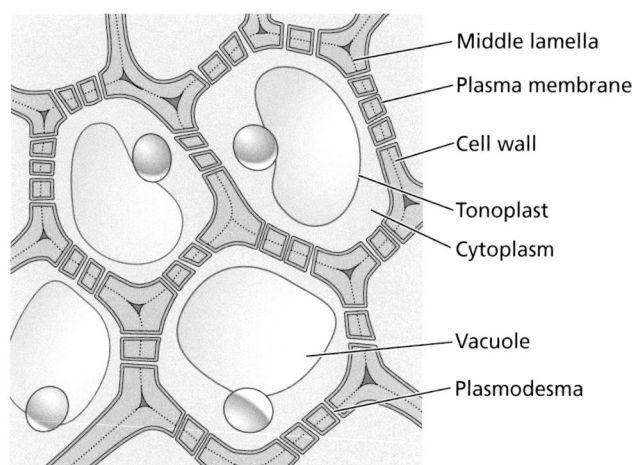

Figure 6.24 Plasmodesmata connect the cytoplasm of neighboring cells, thereby facilitating cell-to-cell communication.

Ions cross both symplasm and apoplast

Absorption of most ions by the root (Chapter 6) is more pronounced in the root hair zone than in the meristematic and elongation zones. Cells in the root hair zone have completed their elongation but have not yet begun secondary growth (Chapter 15). The root hairs are modifications of specific epidermal cells that greatly increase the surface area available for ion absorption (Chapter 14).

An ion that enters a root may immediately enter the symplasm by crossing the plasma membrane of an epidermal cell, or it may enter the apoplast and diffuse between the epidermal cells through the cell walls. From the apoplast of the cortex, an ion (or other solute) may either be transported across the plasma membrane of a cortical cell, thus entering the symplasm, or diffuse radially all the way to the endodermis via the apoplast. The apoplast forms a continuous phase from the root surface through the cortex. However, in all cases, ions must enter the symplasm before they can enter the stele, because of the presence of the Casparian strip. The Casparian strip is a lignified or suberized layer that forms as rings around the specialized cells of the endodermis (**Figure 6.25**; Chapters 1, 2, and 4) and effectively blocks the entry of water and solutes into the stele via the apoplast.

The stele consists of dead tracheary elements surrounded by living pericycle and xylem parenchyma cells. Once an ion has entered the stele through the symplasmic connections across the endodermis, it continues to diffuse through the living cells. Finally, the ion is exported into the apoplast and diffuses into the conducting cells of the xylem—because these cells are dead, their interiors are continuous with the apoplast. The Casparian strip selectively allows symplasmic ions and solutes to enter the xylem and be transported to the shoot. The Casparian strip also prevents ions from diffusing back out of the root through the apoplast. Thus, the presence of the Casparian strip allows the plant to maintain a higher ion concentration in the xylem than exists in the soil water surrounding the roots. The roots of most eudicots have an additional cell layer with a Casparian strip called the exodermis. When present, the exodermis is the outermost cell layer of the cortex and serves a function similar to that of the endodermis.

apoplast The mostly continuous system of cell walls, intercellular air spaces, and xylem vessels in a plant.

symplasm The continuum of cellular cytoplasm connected by plasmodesmata.

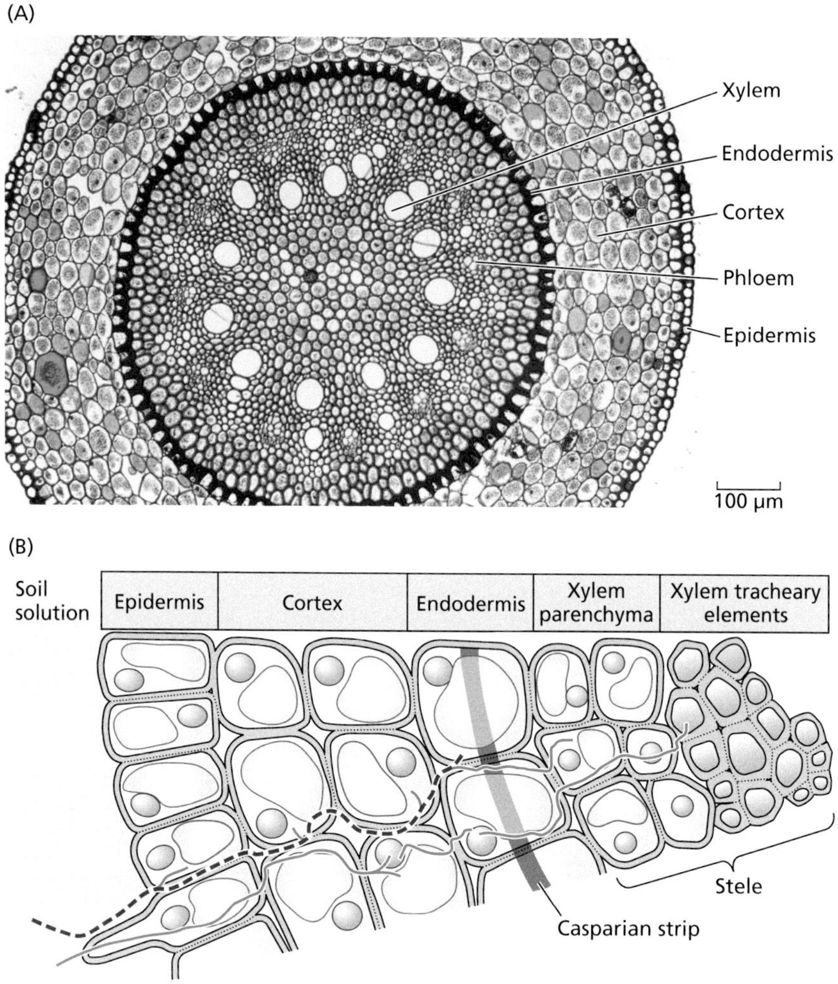

Figure 6.25 Tissue organization in roots. (A) Cross section through a root of carrion flower (genus *Smilax*), a monocot, showing the epidermis, cortex parenchyma, endodermis, xylem, and phloem. (B) Schematic diagram of a root cross section, illustrating the cell layers through which solutes pass from the soil solution to the xylem tracheary elements. The apoplastic pathway for solutes (dashed red line) stops at the Casparian strip; solutes in the apoplast can enter the symplasm (blue lines). The symplasmic pathway for solutes crosses the Casparian strip into the stele (blue lines).

Xylem parenchyma cells participate in xylem loading

The process whereby ions exit the symplasm of a xylem parenchyma cell and enter the conducting cells of the xylem for translocation to the shoot is called **xylem loading**. Xylem loading is a highly regulated process. Xylem parenchyma cells, like other living plant cells, maintain plasma membrane H^+-ATPase activity and a negative membrane potential. Transporters that specifically function in the unloading of solutes to the tracheary elements have been identified by electrophysiological and genetic approaches. The plasma membranes of xylem parenchyma cells contain proton pumps, aquaporins, and a variety of ion channels and carriers specialized for influx or efflux.

xylem loading The process whereby ions exit the symplast and enter the conducting cells of the xylem.

The stelar outwardly rectifying K^+ channel (SKOR) is expressed in cells of the pericycle and xylem parenchyma, where it functions as an efflux channel, transporting K^+ from the living cells into the tracheary elements. In mutant Arabidopsis plants lacking the SKOR channel protein, or in plants in which

SKOR has been pharmacologically inactivated, K^+ transport from the root to the shoot is severely reduced, confirming the function of this channel protein.

Several types of anion-selective channels have also been identified that participate in unloading of Cl^- and NO_3^- from the xylem parenchyma. Drought, abscisic acid (ABA) treatment, or elevation of cytosolic Ca^{2+} concentrations (which often occurs as a response to ABA) all reduce the activity of SKOR and anion channels of the root xylem parenchyma, a response that could help maintain cellular hydration in the root under desiccating conditions.

Other, less-selective, ion channels found in the plasma membrane of xylem parenchyma cells are permeable to K^+, Na^+, and anions. Other transport proteins mediate loading of boron (as boric acid [$B(OH)_3$] or borate [$B(OH)_4^-$]), Mg^{2+}, and $H_2PO_4^-$. Thus, the flux of ions from the xylem parenchyma cells into the xylem tracheary elements is under tight metabolic control through the regulation of plasma membrane H^+-ATPases, ion efflux channels, and carriers.

Summary

The movement of ions and other solutes from one location to another within an organism is known as transport. Transport processes are predominantly mediated by membrane proteins. Plants exchange solutes with their local environment, among their tissues and organs, and within their cells. Transport proteins control cellular transport as well as long-distance transport processes such as metabolite transport in the phloem. Ion transport in plants is vital to their mineral nutrition and stress tolerance, and modulation of plant transport components and properties has potential to improve the nutritive value, stress tolerance, and yield of crops.

6.1 Passive and Active Transport

- Concentration gradients and electrical-potential gradients, the main forces that drive transport across biological membranes, are integrated by a term called the electrochemical potential (**Equation 6.8**).
- Movement of solutes across membranes down their free-energy gradient is facilitated by passive transport mechanisms, whereas movement of solutes against their free-energy gradient is known as active transport and requires energy input (**Figure 6.1**).

6.2 Transport of Ions across Membrane Barriers

- The extent to which a membrane permits the movement of a substance is a property known as membrane permeability (**Figure 6.5**).
- Permeability depends on the membrane proteins that facilitate the transport of specific substances, on the lipid composition of the membrane, and on the chemical properties of the solutes.
- The Nernst equation describes the distribution of a single permeant ionic species across a membrane at equilibrium (**Equation 6.10**).
- Transport of H^+ across the plant plasma membrane by H^+-ATPases is a major determinant of the membrane potential (**Figure 6.20**).

6.3 Membrane Transport Processes

- Biological membranes contain specialized proteins—channels, carriers, and pumps—that facilitate solute transport (**Figure 6.6**).
- The net result of membrane transport processes is that most ions are maintained in disequilibrium with their surroundings (**Table 6.1**).
- Channels are regulated protein pores that, when open, greatly enhance fluxes of ions and, in some cases, neutral molecules across membranes (**Figures 6.6, 6.7**).
- Organisms have a great diversity of ion channel types. Depending on the channel type, channels can be nonselective or highly selective for just one ionic species. Channels can be regulated by many parameters, including voltage, intracellular signaling molecules, ligands, hormones, and light (**Figures 6.8, 6.13, 6.14**).
- Carriers bind specific substances and transport them at a rate several orders of magnitude lower than that of channels (**Figures 6.6, 6.11**).
- Pumps require energy for transport. Active transport of H^+ and Ca^{2+} across plant plasma membranes is mediated by pumps (**Figures 6.13, 6.19–6.21**).
- Secondary active transporters in plants harness energy from energetically downhill movement of protons to mediate energetically uphill transport of another solute (**Figure 6.9**).

(Continued)

Summary (*continued*)

- In symport, both transported solutes move in the same direction across the membrane, whereas in antiport, the two solutes move in opposite directions (**Figure 6.10**).

6.4 Membrane Transport Proteins

- Many channels, carriers, and pumps of the plant plasma membrane and tonoplast have been identified at the molecular level (**Figure 6.13**) and characterized using electrophysiological (**Figures 6.3**, **6.8**) and biochemical techniques.
- Transporters exist for diverse nitrogenous compounds, including ammonium, nitrate, amino acids, and peptides.
- Plants have a great variety of cation channels that can be classified according to their ionic selectivity and regulatory mechanisms (**Figure 6.14**).
- Several different classes of cation carriers mediate K^+ uptake into the cytosol (**Figure 6.13**).
- Na^+–H^+ and K^+–H^+ antiporters in the tonoplast accumulate Na^+ and K^+ in the vacuole (**Figure 6.13**).
- Ca^{2+} is an important second messenger in signal transduction cascades, and its cytosolic concentration is tightly regulated. Ca^{2+} enters the cytosol passively, via Ca^{2+}-permeable channels, and is actively removed from the cytosol by Ca^+ pumps and Ca^{2+}–H^+ antiporters (**Figure 6.13**).
- Selective carriers that mediate NO_3^-, Cl^-, SO_4^{2-}, and $H_2PO_4^-$ uptake into the cytosol and anion channels that nonselectively mediate anion efflux from the cytosol regulate cellular concentrations of these macronutrients (**Figure 6.13**).
- Both essential and toxic metal ions are transported by high-affinity ZIP transport proteins.
- Aquaporins act as water channels, mediate the influx of mineral nutrients (e.g., boron and silicic acid), and act as conduits for the movement of CO_2, NH_3, and H_2O_2 across the plasma membrane and other plant membranes. Regulation of aquaporin opening allows plant cells to rapidly change their permeability to water in response to environmental stimuli.
- Plasma membrane H^+-ATPases are encoded by a multigene family, and their activity is reversibly controlled by an autoinhibitory domain (**Figure 6.20**).
- Like the plasma membrane, the tonoplast also contains both cation and anion channels, as well as a diversity of other transporters (**Figure 6.13**).
- Three types of proton pumps are found in the vacuolar membrane. Vacuolar H^+-ATPases (V-ATPases), H^+-pyrophosphatases (H^+-PPases), and in some cells P-type H^+-ATPases regulate the proton motive force across the tonoplast, which in turn drives the movement of other solutes across this membrane via antiport and symport mechanisms (**Figures 6.13**, **6.21**).

6.5 Transport in Stomatal Guard Cells

- Stomatal opening involves activation of P-type H^+-ATPases in the guard cell plasma membrane. The resulting membrane hyperpolarization drives the uptake of K^+ salts, a more negative osmotic potential, cellular uptake of water, and stomatal opening (**Figures 6.22**, **6.23**).
- Stomatal closing is driven by anion channel activation that depolarizes the guard cell plasma membrane. K^+ efflux follows via outward K^+ channels. The osmotic potential becomes less negative and water efflux results in a decrease in guard cell volume and stomatal closure (**Figures 6.22**, **6.23**).

6.6 Ion Transport in Roots

- Solutes such as mineral nutrients move between cells either through the extracellular space (the apoplast) or from cytoplasm to cytoplasm (via the symplasm). The cytoplasm of neighboring cells is connected by plasmodesmata, which facilitate symplasmic transport (**Figure 6.24**).
- When a solute enters the root, it may be taken up into the cytosol of an epidermal cell, or it may diffuse through the apoplast into the root cortex and then enter the symplasm through a cortical or endodermal cell.
- The presence of the Casparian strip prevents apoplastic diffusion of solutes into the stele. Solutes enter the stele via diffusion from endodermal cells to pericycle and xylem parenchyma cells.
- During xylem loading, solutes are released from xylem parenchyma cells to the conducting cells of the xylem, and then move to the shoot in the transpiration stream (**Figure 6.25**).

Suggested Reading

Barbier-Brygoo, H., Vinauger, M., Colcombet, J., Ephritikhine, G., Frachisse, J., and Maurel, C. (2000) Anion channels in higher plants: Functional characterization, molecular structure and physiological role. *Biochim. Biophys. Acta* 1465: 199–218.

Geldner, N. (2013) The endodermis. *Annu. Rev. Plant. Biol.* 64: 531–558.

Jammes, F., Hu, H. C., Villiers, F., Bouten, R., and Kwak, J. M. (2011) Calcium-permeable channels in plant cells. *FEBS J.* 278: 4262–4276.

Lawson, T., and Matthews, J. (2020) Guard cell metabolism and stomatal function. *Annu. Rev. Plant. Biol.* 71(1): 273–302.

Palmgren, M. G., and Nissen, P. (2011) P-type ATPases. *Annu. Rev. Biophys.* 40: 243–266.

Pantoja, O. (2021) Recent advances in the physiology of ion channels in plants. *Annu. Rev. Plant. Biol.* 72(1): 463–495.

Wang, Y., and Wu, W.-H. (2013) Potassium transport and signaling in higher plants. *Annu. Rev. Plant Biol.* 64: 451–476.

Yamaguchi, T., Hamamoto, S., and Uozumi, N. (2013) Sodium transport system in plant cells. *Front. Plant Sci.* 4: 410.

7 Photosynthesis: The Light Reactions

Life on Earth ultimately depends on energy derived from the sun. Photosynthesis is the biological process that harvests this energy. A large fraction of the planet's energy resources results from photosynthetic activity in either recent or ancient times (fossil fuels).

The term *photosynthesis* means "synthesis using light." Oxygenic photosynthetic organisms (like plants that produce O_2 as a byproduct) use solar energy to synthesize complex carbon compounds. More specifically, light energy drives the synthesis of carbohydrates and the generation of oxygen from carbon dioxide and water. Energy stored in these carbohydrate molecules can be used later to power cellular processes in the plant and can serve as the energy source for all forms of life.

This chapter focuses the role of light in photosynthesis, the structure of the photosynthetic apparatus, and the processes that begin with the excitation of chlorophyll by light and culminate in the synthesis of ATP and NADPH.

7.1 Photosynthesis in Green Plants

Explain the general principles of photosynthesis.

The most active photosynthetic tissue in vascular plants is the mesophyll of leaves. Mesophyll cells have many chloroplasts, which contain the specialized light-absorbing green pigments, the **chlorophylls**. In photosynthesis, the plant uses solar energy to oxidize water, thereby releasing oxygen, and to reduce carbon dioxide, thereby forming large carbon compounds, primarily sugars. The complex series of reactions that culminate in the reduction of CO_2 include the **thylakoid reactions**, also called the **light reactions**, and the carbon fixation reactions.

chlorophylls A group of light-absorbing green pigments active in photosynthesis.

thylakoid reactions (or **light reactions)** The chemical reactions of photosynthesis that occur in the specialized internal membranes of the chloroplast (called thylakoids). Include photosynthetic electron transport and ATP synthesis.

The light-dependent reactions (also called the "light reactions") of photosynthesis take place in the specialized internal membranes of the chloroplast called thylakoids (Chapter 1). The end products of the light-dependent reactions are the high-energy compounds ATP and NADPH that are used for the synthesis of sugars in the **carbon fixation reactions** or just **carbon reactions**. These synthetic processes take place in the *stroma* of the chloroplast, the aqueous region that surrounds the thylakoids. The light-dependent (light harvesting) reactions are the subject of this chapter; the carbon fixation reactions will be discussed in Chapter 8.

In the chloroplast, light energy is converted into chemical energy by two different functional units called *photosystems*. The absorbed light energy is used to power the transfer of electrons through a series of compounds that act as electron donors and electron acceptors. The majority of electrons are extracted from H_2O, which is oxidized to O_2, and ultimately reduce $NADP^+$ to NADPH. Light energy is also used to generate a proton motive force (Chapter 6) across the thylakoid membrane; this proton motive force is used to synthesize ATP.

carbon fixation reactions (or **carbon reactions)** The synthetic reactions occurring in the stroma of the chloroplast that use the high-energy compounds ATP and NADPH for the incorporation of CO_2 into carbon compounds.

wavelength (λ) A unit of measurement for characterizing light energy. The distance between successive wave crests. In the visible spectrum, it corresponds to a color.

frequency (ν) A unit of measurement that characterizes waves, in particular light energy. The number of wave crests that pass an observer in a given time.

photon A discrete physical unit of radiant energy.

quantum (plural *quanta*) A discrete packet of energy contained in a photon.

7.2 General Concepts

Describe the physical and chemical aspects of photosynthesis.

The essential concepts for understanding photosynthesis include the nature of light, the properties of pigments, and the various roles of pigments.

Light consists of photons with characteristic energies

Light has properties of both particles and waves. A light wave (**Figure 7.1**) is characterized by a **wavelength**, denoted by the Greek letter lambda (λ), which is the distance between successive wave crests. The **frequency**, represented by the Greek letter nu (ν), is the number of wave crests that pass an observer in a given time. A simple equation relates the wavelength, the frequency, and the speed of any wave:

$$c = \lambda \nu \qquad (7.1)$$

where c is the speed of the wave—in the present case, the speed of light (3.0×10^8 m s^{-1}). The light wave is a transverse (side-to-side) electromagnetic wave, in which both electric and magnetic fields oscillate perpendicularly to the direction of propagation of the wave and at 90° with respect to each other.

A light particle, known as a **photon**, contains an amount of energy that is not continuous but rather is delivered at discrete levels, called **quanta**. The energy (E) of a photon depends on the frequency of the light according to a relation known as Planck's law:

$$E = h \nu \qquad (7.2)$$

where h is Planck's constant (6.626×10^{-34} J s). How photons interact with matter depends strongly on their frequency or energy, and only a narrow range of frequencies in the visible spectrum are useful for photosynthesis (**Figure 7.2**). Lower-frequency photons, with energies below that of red light, and higher-frequency photons, with energies above that of violet light, cannot drive photosynthesis.

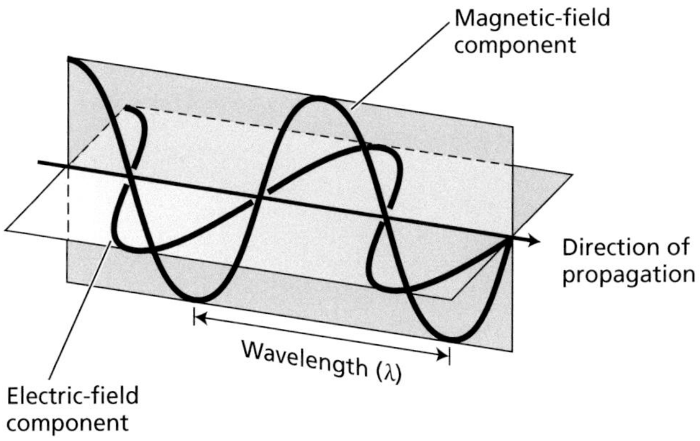

Figure 7.1 Wave properties of light. Light is a transverse electromagnetic wave, consisting of oscillating electric and magnetic fields that are perpendicular to each other and to the direction of propagation of the light. Light moves at a speed of 3.0×10^8 m s^{-1}. The wavelength (λ) is the distance between successive crests of the wave.

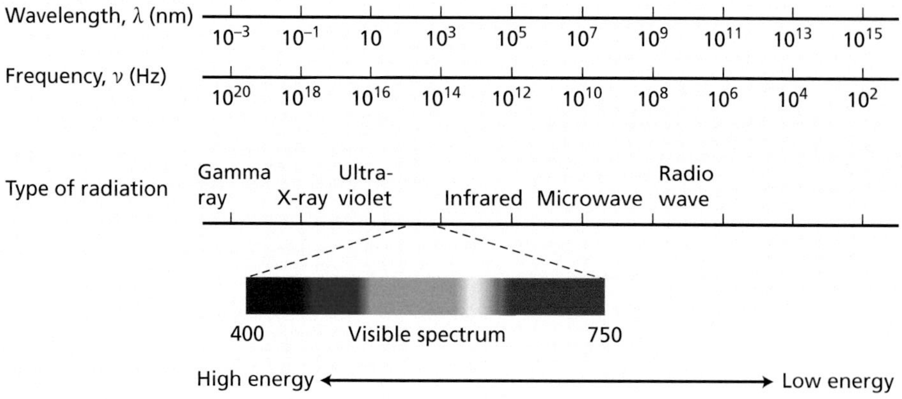

Figure 7.2 Electromagnetic spectrum. Wavelength (λ) and frequency (ν) are inversely related. Our eyes are sensitive to only a narrow range of wavelengths of radiation, the visible region, which extends from about 400 nm (violet) to about 700 nm (red). Short-wavelength (high-frequency) light has a high energy content; long-wavelength (low-frequency) light has a low energy content.

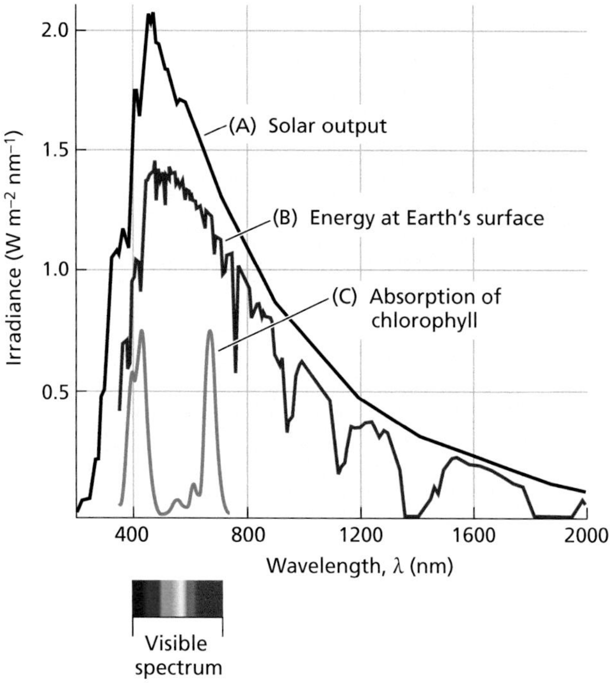

Figure 7.3 The solar spectrum and its relation to the absorption spectrum of chlorophyll. Curve A is the energy output of the sun as a function of wavelength. Curve B is the energy that strikes Earth's surface. The sharp valleys in the infrared region beyond 700 nm represent the absorption of solar energy by molecules in the atmosphere, chiefly water vapor. Curve C is the absorption spectrum of chlorophyll, which absorbs strongly in the blue (~430 nm) and red (~660 nm) portions of the spectrum. Because the green light in the middle of the visible spectrum is not efficiently absorbed, some of it is reflected into our eyes and gives plants their characteristic green color.

Absorption of photosynthetically active light changes the electronic states of chlorophylls

For light energy to power photosynthesis, it must be captured and converted into electrical or chemical forms. The first step in this process is the absorption of light by pigments associated with the photosynthetic apparatus. Different energies (or frequencies) of light interact with molecules in different ways, inducing different types of transitions. For light to be absorbed, the absorbing pigment must have an energy transition that matches the photon's energy. If the photon's energy does not precisely match such a transition (i.e., it is too low or too high), the light will not be absorbed. Thus, although sunlight that strikes Earth's surface contains photons with a wide range of different energies, only a small fraction of this light can be used for photosynthesis (**Figure 7.3**). Plant photosynthesis can use light energy between about 400 and 700 nm ranging from violet to red and far-red wavelengths, nearly matching the visible spectrum (Figure 7.2). Plant biologists refer to this part of the spectrum as photosynthetically active radiation (PAR).

This range of photon energies can excite electrons in plant pigments, mostly chlorophylls, from low-energy molecular orbitals to higher-energy "excited states" (**Figure 7.4**). Such excited states in chlorophyll molecules are the essential starting points for the light reactions. By contrast, lower-energy infrared radiation tends to excite vibrational motions or rotations of chemical bonds, the energy of which is rapidly dissipated as heat and thus is not useful for driving photosynthesis. Photons with higher energy, including deeper UV light, tend to directly ionize molecules, essentially knocking electrons off the molecules, thereby generating radicals that can damage biological materials.

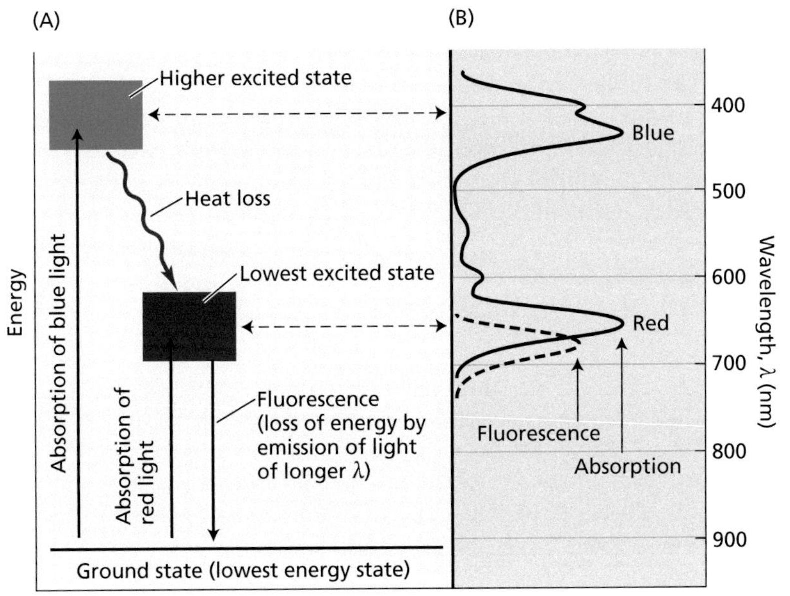

Figure 7.4 Light absorption and emission by chlorophyll. (A) Energy level diagram. Absorption or emission of light is indicated by vertical arrows that connect the ground state with excited electron states. The blue and red absorption bands of chlorophyll (which absorb blue and red photons, respectively) correspond to the upward vertical arrows, signifying that energy absorbed from light causes the molecule to change from the ground state to an excited state. The downward-pointing arrow indicates fluorescence, in which the molecule goes from the lowest excited state to the ground state while re-emitting energy as a photon. (B) Spectra of absorption and fluorescence. The long-wavelength (red) absorption band of chlorophyll corresponds to light that has the energy required to cause the transition from the ground state to the first excited state. The short-wavelength (blue) absorption band corresponds to a transition to a higher excited state.

The **absorption spectrum** (plural *spectra*) of a typical chlorophyll found in chloroplasts is shown by the blue curve in Figure 7.3. An absorption spectrum provides a measure of the amount of **light energy** taken up or absorbed by a molecule or substance as a function of the wavelength of the light. The absorption spectrum for a particular substance can be determined using a spectrophotometer, as illustrated in **Figure 7.5**. In many cases, absorption spectra are plotted as the absorption of light (on the y-axis) as a function of the wavelength of light (on the x-axis). However, it is also informative to plot absorption as a function of frequency, which, because of the relationship described by Equation 7.2, gives us the dependence of absorption on the energy of the photons.

absorption spectrum A graphic representation of the amount of light energy absorbed by a substance plotted against the wavelength of the light.

light energy The energy associated with photons.

Chlorophyll appears green to our eyes because it absorbs light mainly in the red and blue parts of the spectrum, so the light that is not absorbed is enriched in green wavelengths (~550 nm) (Figures 7.3 and 7.4). The absorption of light is represented by Reaction 7.3, in which chlorophyll (Chl) in its lowest energy, or ground, state absorbs a photon (represented by $h\nu$) and makes a transition to a higher energy, or excited, state (Chl*):

$$Chl + h\nu \rightarrow Chl^* \qquad (7.3)$$

The distribution of electrons in the excited molecule is different from the distribution in the ground-state molecule. Absorption of red light excites the chlorophyll in the ground state to a higher-energy "excited state" (the "Lowest

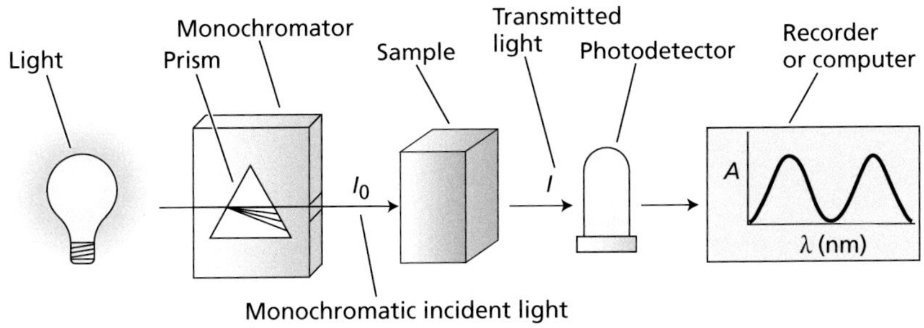

Figure 7.5 Schematic diagram of a spectrophotometer. The instrument consists of a light source, a monochromator that contains a wavelength selection device such as a prism, a sample holder, a photodetector, and a recorder or computer. The output wavelength of the monochromator can be changed by rotation of the prism; the graph of absorbance (*A*) versus wavelength (λ) is called a spectrum.

lowest excited state The excited state with the lowest energy attained when a chlorophyll molecule in a higher energy state gives up some of its energy to the surroundings as heat.

fluorescence Following light absorption, the emission of light at a slightly longer wavelength (lower energy) than the wavelength of the absorbed light.

energy transfer In the light reactions of photosynthesis, the direct transfer of energy from an excited molecule, such as β-carotene, to another molecule, such as chlorophyll. Energy transfer can also take place between chemically identical molecules, as in chlorophyll-to-chlorophyll transfer.

photochemistry The very rapid chemical reactions in which light energy absorbed by a molecule causes a chemical reaction to occur.

bacteriochlorophylls Light-absorbing pigments active in photosynthesis in anoxygenic photosynthetic organisms.

carotenoids Linear polyenes arranged as a planar zigzag chain, with conjugated double bonds. These orange pigments serve both as antenna pigments and photoprotective agents.

excited state" in Figure 7.4) by promoting an electron from a low-energy orbital to a higher-energy orbital. Blue light can also be absorbed by chlorophyll, initially producing an excited state of even higher energy (the "Higher excited state" in Figure 7.4) than that produced by red light. However, this higher-energy excited state is extremely unstable and rapidly decays to the **lowest excited state**, which is the starting point for reactions that drive photosynthesis. The initial photosynthetic reactions must be extremely rapid to outcompete the decay of the lowest excited state; if the energy in the excited state is not used by these reactions, it will last only several nanoseconds (1 nanosecond = 10^{-9} s) at most.

The lowest excited state of chlorophyll can decay by several alternative pathways:

1. Re-emit a photon and thereby return to its ground state—a process known as **fluorescence**. When it does so, the wavelength of fluorescence is slightly longer (and of lower energy) than the wavelength of absorption, because a portion of the excitation energy is converted into heat before the fluorescent photon is emitted. Chlorophylls fluoresce in the red region of the spectrum (see Figure 7.4).
2. Return to its ground state by directly converting its excitation energy into heat, with no emission of a photon.
3. Participate in **energy transfer**, during which the chlorophyll transfers its energy to another molecule, including other chlorophylls. In this way, hundreds of chlorophyll molecules can transfer energy over long distances of 50 to 100 nm.
4. Form a highly reactive state called triplet chlorophyll, which can react with oxygen to form toxic byproducts.
5. Initiate **photochemistry**, in which the energy of the excited state causes chemical reactions to occur, ultimately storing energy in photosynthetic products. The photochemical reactions of photosynthesis are among the fastest known chemical reactions. This speed is necessary for photochemistry to outcompete the non–energy-storing decay pathways.

Photosynthetic pigments absorb the light that powers photosynthesis

The energy of sunlight is first absorbed by the pigments of the plant. All pigments active in photosynthesis are found in the chloroplast. Structures and absorption spectra of several photosynthetic pigments are shown in **Figure 7.6** and **Figure 7.7**, respectively. The chlorophylls and **bacteriochlorophylls** (pigments found in certain bacteria) are the common pigments of photosynthetic organisms. Chlorophylls *a* and *b* are abundant in green plants, and chlorophylls *c*, *d*, and *f* are found in some protists and cyanobacteria. Several different types of bacteriochlorophylls have been found; type *a* is the most widely distributed.

All chlorophylls have a complex ring structure that is chemically related to the heme groups found in hemoglobin and cytochromes (Figure 7.6A). A long hydrocarbon tail is almost always attached to the ring structure. The tail anchors the chlorophyll in the thylakoid membrane. The ring structure contains some loosely bound electrons and is the part of the molecule involved in electronic transitions and redox (reduction–oxidation) reactions.

The different types of **carotenoids** found in photosynthetic organisms are molecules with multiple conjugated double bonds (see Figure 7.6B). Absorption bands in the 400 to 500 nm region give carotenoids their characteristic orange color. The color of carrots, for example, is due to the carotenoid β-carotene, whose structure and absorption spectrum are shown in Figures 7.6 and 7.7, respectively.

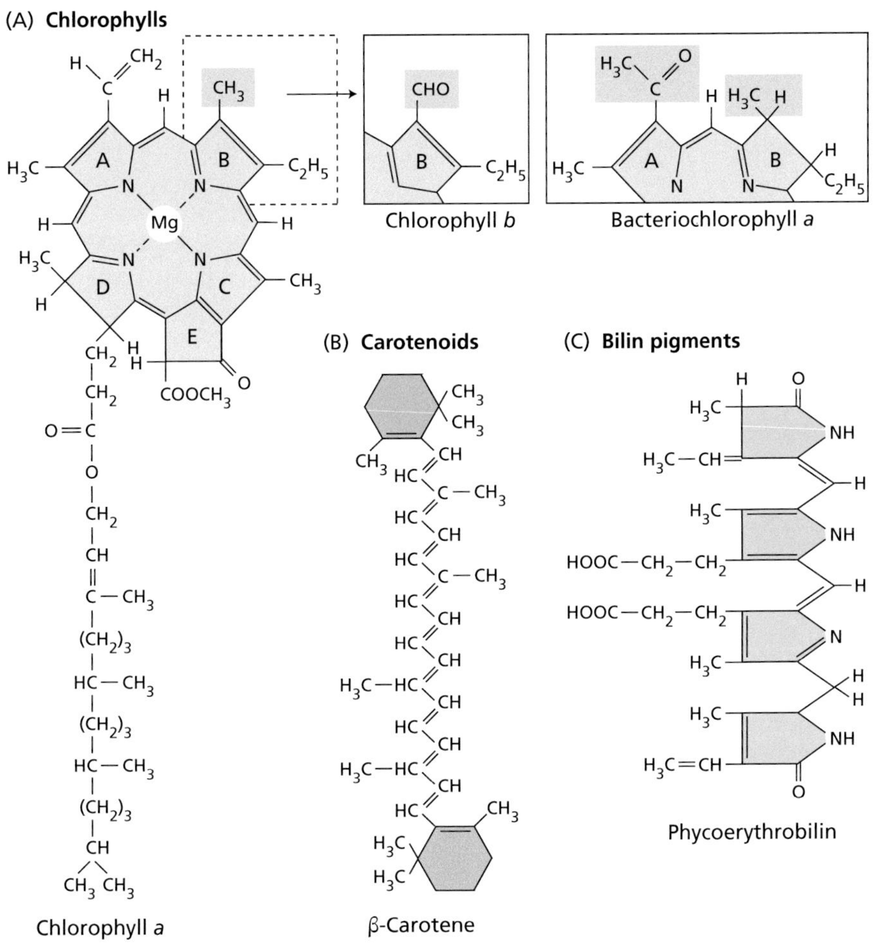

Figure 7.6 Molecular structure of some photosynthetic pigments. (A) The chlorophylls have a porphyrin-like ring structure (green) with a magnesium ion (Mg) coordinated in the center and a long hydrophobic hydrocarbon tail that anchors them in the photosynthetic membrane. The porphyrin-like ring is the site of the electron rearrangements that occur when the chlorophyll is excited, and of the unpaired electrons when it is either oxidized or reduced. Various chlorophylls differ chiefly in the substituents around the rings and the pattern of double bonds. (B) Carotenoids are linear polyenes that serve as both antenna pigments and photoprotective agents. (C) Bilin pigments are open-chain tetrapyrroles found in antenna structures known as phycobilisomes that occur in cyanobacteria and red algae.

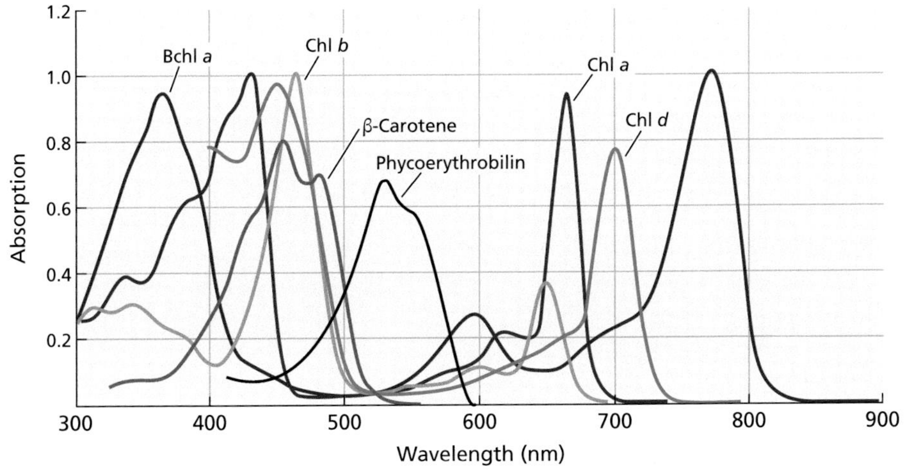

Figure 7.7 Absorption spectra of some photosynthetic pigments. These include β-carotene, chlorophyll *a* (Chl *a*), chlorophyll *b* (Chl *b*), bacteriochlorophyll *a* (Bchl *a*), chlorophyll *d* (Chl *d*), and phycoerythrobilin. The absorption spectra shown are for pure pigments dissolved in nonpolar solvents, except phycoerythrin, a protein from cyanobacteria that contains a phycoerythrobilin chromophore covalently attached to the peptide chain. In many cases the spectra of photosynthetic pigments in vivo are substantially affected by the environment of the pigments in the photosynthetic membrane.

Carotenoids are found in all known natural photosynthetic organisms. Carotenoids are integral constituents of the thylakoid membrane and are usually associated intimately with many of the proteins that make up the photosynthetic apparatus. The light energy absorbed by the carotenoids can be transferred to chlorophyll for photosynthesis; because of this role they are called **accessory pigments**. Carotenoids also help protect the organism from damage caused by light, by "quenching" reactive intermediates such as triplet chlorophylls (Chapter 9).

accessory pigments Light-absorbing molecules in photosynthetic organisms that work with chlorophyll *a* in the absorption of light used for photosynthesis. They include carotenoids, other chlorophylls, and phycobiliproteins.

action spectrum A graphic representation of the magnitude of a biological response to light as a function of wavelength.

quantum yield (φ) The ratio of the yield of a particular product of a photochemical process to the total number of quanta absorbed.

antenna complex A group of pigment molecules that cooperate to absorb light energy and transfer it to a reaction center complex.

reaction center complex A group of electron transfer proteins that receive energy from the antenna complex and convert it into chemical energy using oxidation–reduction reactions.

7.3 Key Experiments in Understanding Photosynthesis

Explain the principles of photosynthesis using the early experimental approaches.

Establishing the overall chemical equation of photosynthesis required several hundred years and contributions by many scientists. In 1771, Joseph Priestley observed that a sprig of mint growing in air in which a candle had burned out regenerated something in the air so that another candle could burn. He had discovered oxygen evolution by plants. A Dutch biologist, Jan Ingenhousz, documented the essential role of light in photosynthesis in 1779.

Other scientists established the roles of CO_2 and H_2O and showed that organic matter, specifically carbohydrate, is a product of photosynthesis along with oxygen. By the end of the nineteenth century, the balanced overall chemical reaction for photosynthesis could be written as follows:

$$6\,CO_2 + 6\,H_2O \xrightarrow{\text{Light, plant}} C_6H_{12}O_6 + 6\,O_2 \qquad (7.4)$$

where $C_6H_{12}O_6$ represents a simple sugar such as glucose. As we will discuss in Chapter 8, glucose is not the actual product of the carbon fixation reactions, so this part of the reaction should not be taken literally. However, the energetics for the actual reaction are approximately the same as represented here.

The chemical reactions of photosynthesis are complex. At least 50 intermediate reaction steps have now been identified, and additional steps will undoubtedly be discovered. An early clue to the chemical nature of the essential chemical process of photosynthesis came in the 1920s from investigations of photosynthetic bacteria that did not produce oxygen as an end product. From his studies on these bacteria, C. B. van Niel concluded that photosynthesis is a redox process. This conclusion has served as a fundamental concept on which all subsequent research on photosynthesis has been based.

Action spectra relate light absorption to photosynthetic activity

The use of action spectra has been central to the development of our current understanding of photosynthesis. An **action spectrum** describes the effectiveness of different wavelengths of light in promoting a biological response. For example, an action spectrum for photosynthesis can be constructed by measuring the rate of oxygen evolution as a function of increasing light intensities (the number of photons per unit of area per second) at different wavelengths (**Figure 7.8**). The efficiency with which a particular wavelength intensity of light induces photosynthesis will be higher when it is more effectively absorbed by photosynthetic

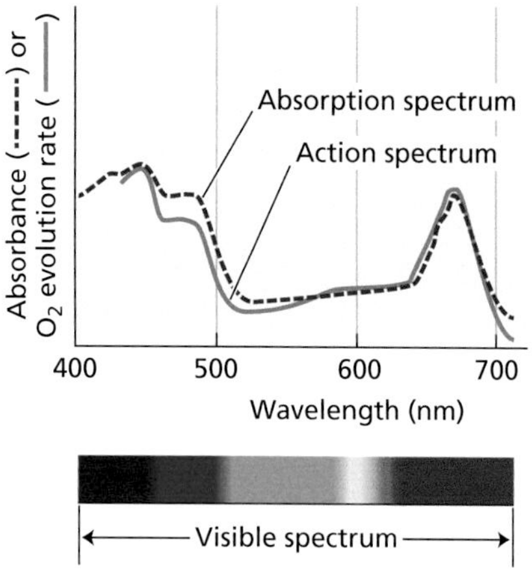

Figure 7.8 Action spectrum compared with an absorption spectrum. The absorption spectrum is measured as shown in Figure 7.5. Because photosynthesis is saturated at high light, the rate of oxygen evolution should be measured at very low light or estimated by the half-saturation point. An action spectrum is measured by plotting a response to light, such as oxygen evolution, as a function of wavelength. If the pigment used to obtain the absorption spectrum is the same as that which causes the response, the absorption and action spectra will match. In the example shown here, the action spectrum for oxygen evolution matches the absorption spectrum of intact chloroplasts quite well, indicating that light absorption by the chlorophylls mediates oxygen evolution. Discrepancies are found in the region of carotenoid absorption, from 450 to 550 nm, indicating that energy transfer from carotenoids to chlorophylls is not as effective as energy transfer between chlorophylls.

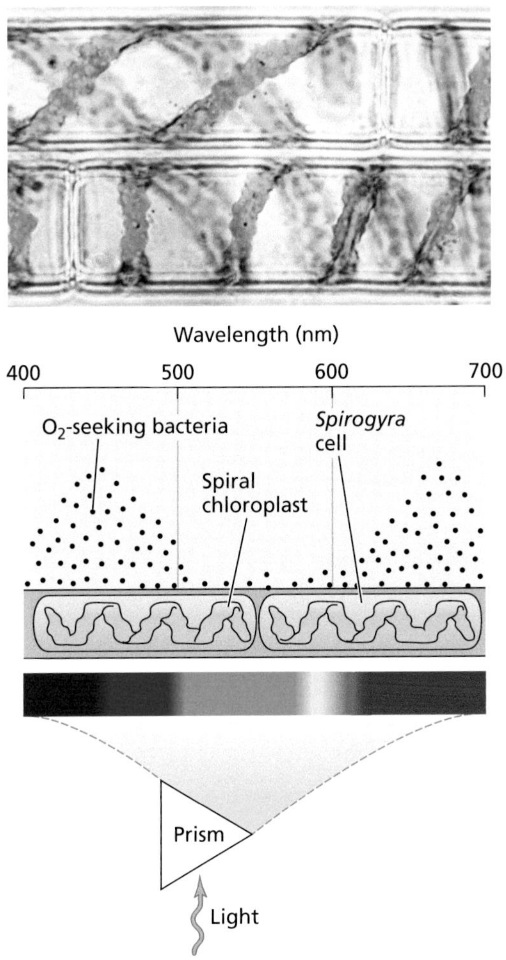

Figure 7.9 Schematic diagram of the action spectrum measurements by T. W. Engelmann. Engelmann projected a spectrum of light onto the spiral chloroplast of the filamentous green alga *Spirogyra* and observed that O_2-seeking bacteria introduced into the system collected in the region of the spectrum where chlorophyll pigments absorb. This action spectrum gave the first indication of the effectiveness of light absorbed by pigments in driving photosynthesis. © Biophoto Associates/Science Source

pigments. Thus, an action spectrum can identify the *chromophores* (pigments) responsible for a particular light-induced phenomenon.

Some of the first action spectra were measured by T. W. Engelmann in the late 1800s (**Figure 7.9**). Engelmann used a prism to disperse sunlight into a rainbow that was allowed to fall on an aquatic algal filament. A population of O_2-seeking bacteria was introduced into the system. The bacteria congregated in the regions of the filaments that evolved the most O_2. These were the regions illuminated by blue light and red light, which are strongly absorbed by chlorophyll. Today, action spectra can be measured in room-sized spectrographs in which a huge monochromator bathes the experimental samples in monochromatic light. The technology is more sophisticated, but the principle is the same as that of Engelmann's experiments.

A special version of the action spectrum measures the effectiveness of the light that is actually absorbed by the plant, rather than the total incident light. In this case, the **quantum yield** of photosynthesis—the fraction of absorbed light that is actually used to drive productive photosynthesis—can be calculated. The quantum yield of photosynthesis under low light can be close to 1.0, which means nearly every photon that is absorbed is used to drive photosynthesis.

Photosynthesis takes place in complexes containing light-harvesting antennas and photochemical reaction centers

A portion of the light energy absorbed by chlorophylls and carotenoids is eventually stored as chemical energy via the formation of chemical bonds. This conversion of energy from one form to another is a complex process that depends on cooperation between many pigment molecules and a group of electron transfer proteins.

The majority of the pigments serve as an **antenna complex**, collecting light and transferring the energy to the **reaction center complex**, where the chemical oxidation and reduction reactions leading to long-term energy storage take place (**Figure 7.10**).

How does the plant benefit from this division of labor between antenna and reaction center pigments? Even in bright sunlight, a single chlorophyll molecule absorbs only a few photons each second. If there were a reaction center associated with each chlorophyll molecule, the reaction center enzymes would be idle most of the time, only occasionally being activated by photon absorption. However, if a reaction center receives energy from many pigments at once, the system is kept active a large fraction of the time.

In 1932, Robert Emerson and William Arnold performed a key experiment that provided the first evidence for the cooperation of many chlorophyll molecules in energy conversion

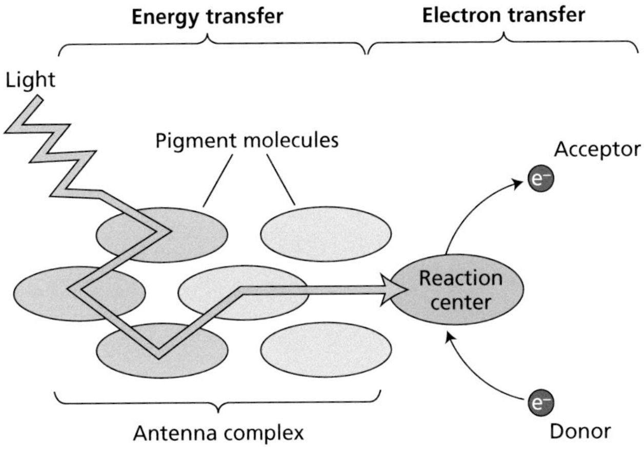

Figure 7.10 Basic concept of energy transfer during photosynthesis. Many pigments together serve as an antenna, collecting light and transferring its energy to the reaction center, where chemical reactions store some of the energy by transferring electrons from a chlorophyll pigment to an electron acceptor molecule. An electron donor then reduces the chlorophyll again. The transfer of energy in the antenna is a purely physical phenomenon and involves no chemical changes.

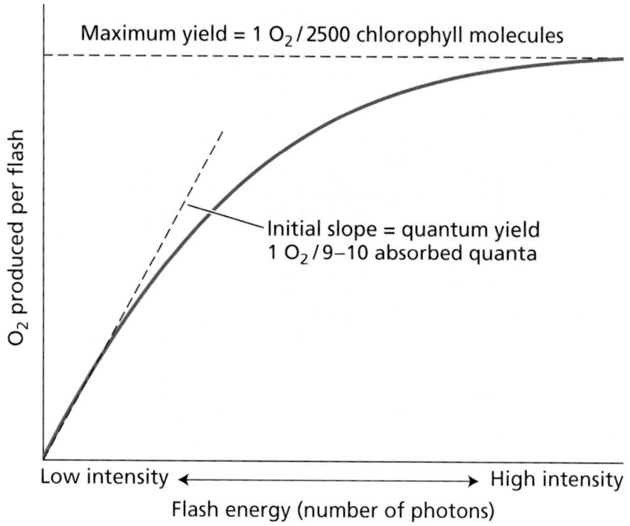

Figure 7.11 Relationship of oxygen production to flash energy. Emerson and Arnold showed the first evidence for the interaction between the antenna pigments and the reaction center. They found that, at saturating energies, the maximum amount of O_2 produced is 1 molecule per 2,500 chlorophyll molecules. In their measurement of the relationship of oxygen production to flash energy, Emerson and Arnold were surprised to find that under saturating conditions, only 1 molecule of oxygen was produced for each 2,500 chlorophyll molecules in the sample. We know now that several hundred pigments are associated with each reaction center and that each reaction center must operate four times to produce 1 molecule of oxygen—hence the value of 2,500 chlorophylls per O_2.

during photosynthesis. They delivered very brief (10^{-5} s) flashes of light to a suspension of the green alga *Chlorella pyrenoidosa* and measured the amount of oxygen produced. The flashes were spaced about 0.1 s apart, a time that Emerson and Arnold had determined in earlier work was long enough for the enzymatic steps of the process to be completed before the arrival of the next flash. The investigators varied the energy of the flashes and found that at high energies the oxygen production did not increase when a more intense flash was given: The photosynthetic system was saturated with light (**Figure 7.11**).

The reaction centers and most of the antenna complexes are integral components of the thylakoid membrane. In eukaryotic photosynthetic organisms, these membranes are found within the chloroplast; in photosynthetic prokaryotes, the site of photosynthesis is the plasma membrane or membranes derived from it.

The chemical reaction of photosynthesis is driven by light

It is important to realize that the chemical reaction shown in Equation 7.4 is energetically uphill, meaning that it cannot proceed without a large input of energy. The equilibrium constant for Equation 7.4, calculated from tabulated free energies of formation for each of the compounds involved, is about 10^{-500}. This number is so close to zero that one can be quite confident that in the entire history of the universe no molecule of glucose has formed spontaneously from H_2O and CO_2 without external energy being provided. The energy needed to drive the photosynthetic reaction comes from light. Here's a simpler form of Equation 7.4:

$$CO_2 + H_2O \xrightarrow{\text{Light, plant}} (CH_2O) + O_2 \qquad (7.5)$$

where (CH_2O) is one-sixth of a glucose molecule. About ten photons of light are required to drive the reaction of Equation 7.5.

Light drives the reduction of NADP+ and the formation of ATP

The overall process of photosynthesis is a redox chemical reaction, in which electrons are removed from one chemical species, thereby oxidizing it, and added to another species, thereby reducing it. In 1937, Robert Hill found that in the light, isolated chloroplast thylakoids reduce a variety of compounds, such as iron salts. These compounds serve as oxidants in place of CO_2, as the following equation shows:

$$4\,Fe^{3+} + 2\,H_2O \rightarrow 4\,Fe^{2+} + O_2 + 4\,H^+ \qquad (7.6)$$

Many compounds have since been shown to act as artificial electron acceptors in what has come to be known as the Hill reaction. Artificial electron acceptors have been invaluable in elucidating the reactions that precede carbon reduction and have provided the first evidence that oxygen evolution could occur in the absence of carbon dioxide. This led to the now accepted and proven idea that the oxygen in photosynthesis originates from water, not from carbon dioxide.

We now know that during the normal functioning of the photosynthetic system, light reduces $NADP^+$, which in turn serves as the reducing agent for carbon fixation in the Calvin–Benson cycle (Chapter 8). ATP is also formed during the electron flow from water to $NADP^+$, and it too is used in carbon reduction.

The chemical reactions in which water is oxidized to oxygen, $NADP^+$ is reduced to NADPH, and ATP is formed all take place in the thylakoid membranes. Carbon fixation and reduction reactions primarily occur in the stroma, the aqueous region of the chloroplast.

photosystem I (PSI) A system of photoreactions that absorbs maximally far-red light (700 nm), oxidizes plastocyanin, and reduces ferredoxin.

photosystem II (PSII) A system of photoreactions that absorbs maximally red light (680 nm), oxidizes water, and reduces plastoquinone. Operates very poorly under far-red light.

Oxygen-evolving organisms have two photosystems that operate in series

The light reactions of photosynthesis involve two types of photochemical reaction centers, contained within **photosystems I** and **II** (**PSI** and **PSII**), which operate in series to carry out the early energy storage reactions of photosynthesis. PSI preferentially absorbs far-red light of wavelengths greater than 680 nm; PSII preferentially absorbs red light of 680 nm and is driven very poorly by far-red light. Another difference between the photosystems is that:

- PSI produces a strong reductant, capable of reducing $NADP^+$, and a weak oxidant.
- PSII produces a very strong oxidant, capable of oxidizing water, and a weaker reductant than the one produced by PSI.

The reductant produced by PSII re-reduces the oxidant produced by PSI. These properties of the two photosystems are shown schematically in **Figure 7.12**. The scheme of photosynthesis, called the *Z scheme* (because it looks like the letter Z), has become the basis for understanding O_2-evolving (oxygenic) photosynthetic organisms. It accounts for the operation of two physically and chemically distinct photosystems (I and II), each with its own pool of antenna pigments and photochemical reaction center. The two photosystems are linked by an electron transport chain.

Almost all the energy for life on our planet comes from photosynthesis, and thus its efficiency controls the maximum overall productivity of our ecosystems. Two distinct parameters determine the efficiency of photosynthesis: the quantum yield and the energy conversion efficiency.

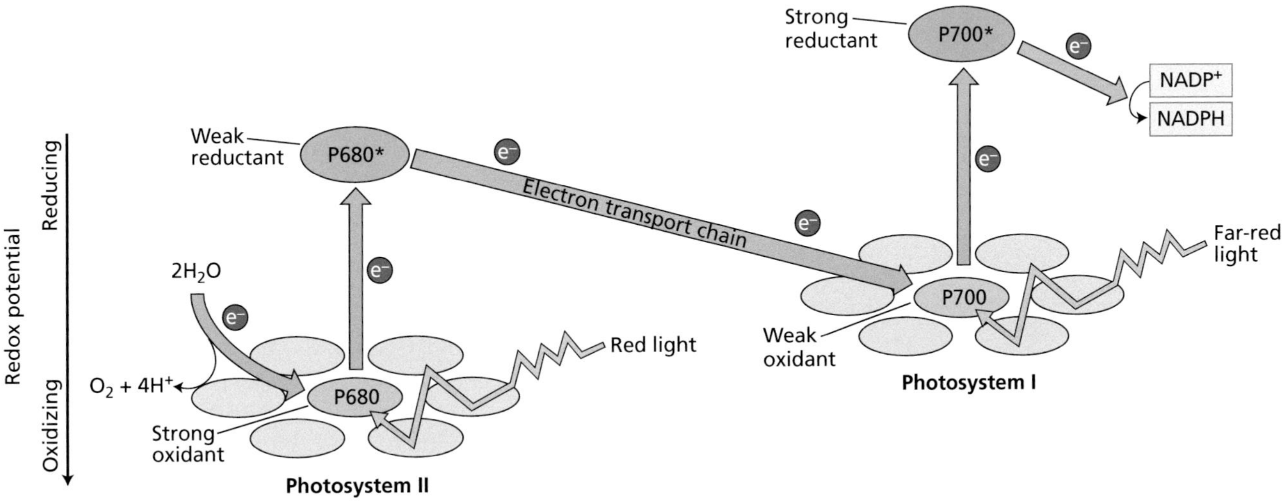

Figure 7.12 Z scheme of photosynthesis. Red light absorbed by photosystem II (PSII) produces a strong oxidant and a weak reductant. Far-red light absorbed by photosystem I (PSI) produces a weak oxidant and a strong reductant. The strong oxidant generated by PSII oxidizes water, while the strong reductant produced by PSI reduces $NADP^+$. This scheme is basic to an understanding of photosynthetic electron transport. P680 and P700 refer to the wavelengths of maximum absorption of the reaction center chlorophylls in PSII and PSI, respectively. Where asterisks are added, as in P680* and P700*, the photosystem is in the excited state.

energy conversion efficiency The fraction of energy from photons that is stored in photochemical reactions. Energy is lost in making reactions go forward, which prevents back reactions. This is distinct from the quantum yield, which describes the fraction of light that induces stable photochemistry.

thylakoids The specialized, internal chlorophyll-containing membranes of the chloroplast where light absorption and the chemical reactions of photosynthesis take place.

stroma The fluid component surrounding the thylakoid membranes of a chloroplast.

grana lamellae Stacked thylakoid membranes within the chloroplast. Each stack is called a granum.

The graph shown in Figure 7.11 permits us to calculate the quantum yield of photochemistry (Φ), defined as follows:

$$\Phi = \frac{\text{Number of photochemical products}}{\text{Total number of quanta absorbed}} \tag{7.7}$$

At low light intensities, the curve shows its highest, and nearly linear, increase in oxygen evolution with increasing light intensity. Over this range, the quantum yield of photochemistry can be as high as 0.95, meaning that 95% of the photons absorbed by chlorophylls are used in photochemistry. Because the formation of the stable photosynthetic products O_2 and fixed CO_2 (sugars) requires multiple photochemical events, it has a lower quantum yield of formation than the photochemical quantum yield. It takes about ten photons to produce one molecule of O_2, so the quantum yield of O_2 production is about 0.1, even though the quantum yield for each photochemical step in the process is nearly 1.0.

Although the photochemical quantum yield under optimal conditions is nearly 1.0, the fraction of energy that is stored by photosynthesis—that is, the **energy conversion efficiency**—is much less. One major reason for this loss of energy is that a photon's energy is absorbed by generating a series of intermediates in the light reactions, ultimately forming O_2, NADPH, and ATP. The equilibrium constant for each step has a large decrease in free energy, which ensures that the forward reactions are much faster than the reverse reactions. In this way, the energy from the photon is "trapped," preventing it from being lost by reverse reactions, thereby increasing the quantum yield of light capture, but at the cost of a loss in overall energy storage.

When 680 nm red light is absorbed, the total energy input (Equation 7.2) is 1,760 kJ per mole of oxygen formed. This amount of energy is more than enough to drive Reaction 7.5, which has a standard-state free-energy change of +467 kJ mol^{-1}. The maximum efficiency of conversion of light energy from 680 nm red photons (the optimal wavelength) into chemical energy is therefore about 27%. Blue light is also strongly absorbed by the photosystems, with similar quantum efficiency. Blue photons have about 50% higher energy content than red photons, but once absorbed this extra energy is lost extremely rapidly, resulting in a lower energy conversion efficiency for the formation of stable products. White light from the sun is composed of photons that span the range of photosynthetically active radiation, from about 400 to 700 nm, and the average energy conversion efficiency is intermediate between that of red and blue photons.

At higher light intensities, photosynthesis becomes saturated and is limited by downstream processes, such as the fixation of CO_2, leading to lower quantum efficiencies (Figure 7.11). While Figure 7.11 shows the light saturation curve for a short flash of light, the general shape of the curve under constant light is similar. The higher the light intensity, the smaller the slope of the response of photosynthesis to further increases in light intensity. This flattening of the curve means that the quantum efficiency of the light reactions decreases at high light intensities—that is, a higher fraction of the energy from photons is lost as heat. The system is "light-limited" when the photons arrive more slowly than the maximum rates at which the system can use them. But when light intensity increases, photons arrive more rapidly than can be used. Too much light can lead to buildup of reactive intermediates that can lead to photodamage. To prevent these situations, the capture of light is down-regulated by the plant, dissipating more energy as heat.

Most of the energy produced by the light reactions is stored in the form of fixed carbon. A large fraction of this fixed carbon is subsequently used for cellular maintenance processes, with a smaller fraction being used to generate new biomass (Chapter 9). Thus, the overall energy conversion efficiency for making new plant matter is only a few percent, far below the theoretical maximum, limiting the energy available (and the rate of CO_2 uptake) in our environment. Decreasing these large energy losses is a target of efforts to improve crop productivity, though it is not clear to what extent this can be engineered while still maintaining the robustness of plants in their environments.

7.4 Organization of the Photosynthetic Apparatus

> Describe how the photosynthetic apparatus is organized physically and how that regulates photosynthesis.

The photosynthetic apparatus and the structure of its components have a specific architecture that allows them to efficiently harness light energy into energy stored in chemical bonds. The molecular structure of the system leads to its functional characteristics.

The chloroplast is the site of photosynthesis

In photosynthetic eukaryotes, photosynthesis takes place in the subcellular organelle known as the chloroplast (Chapter 1). **Figure 7.13** shows a transmission electron micrograph of a thin section from a pea chloroplast. The most striking aspect of the structure of the chloroplast is the extensive system of internal membranes known as **thylakoids**. All the chlorophyll is contained within this membrane system, which is the site of the light reactions of photosynthesis.

The carbon reduction reactions, which are catalyzed by water-soluble enzymes, take place in the **stroma**, the region of the chloroplast outside the thylakoids. Most of the thylakoids appear to be very closely associated with each other. These stacked membranes are known as **grana lamellae** (singular *lamella*; each stack is called a *granum*), and the exposed membranes in which stacking is absent are known as **stroma lamellae**.

Two separate membranes, each composed of a lipid bilayer and together known as the **envelope**, surround most types of chloroplasts (**Figure 7.14**). This double-membrane system contains a variety of metabolite transport systems.

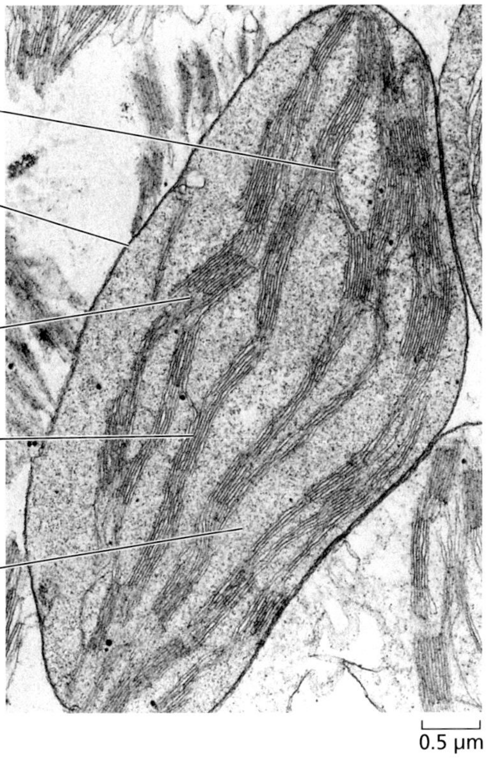

Figure 7.13 Transmission electron micrograph of a chloroplast from pea (*Pisum sativum*) fixed in glutaraldehyde and OsO_4, embedded in plastic resin, and thin-sectioned with an ultramicrotome. Courtesy of J. Swafford

stroma lamellae Unstacked thylakoid membranes within the chloroplast.

envelope The double-membrane system surrounding the chloroplast or the nucleus. The outer membrane of the nuclear envelope is continuous with the endoplasmic reticulum.

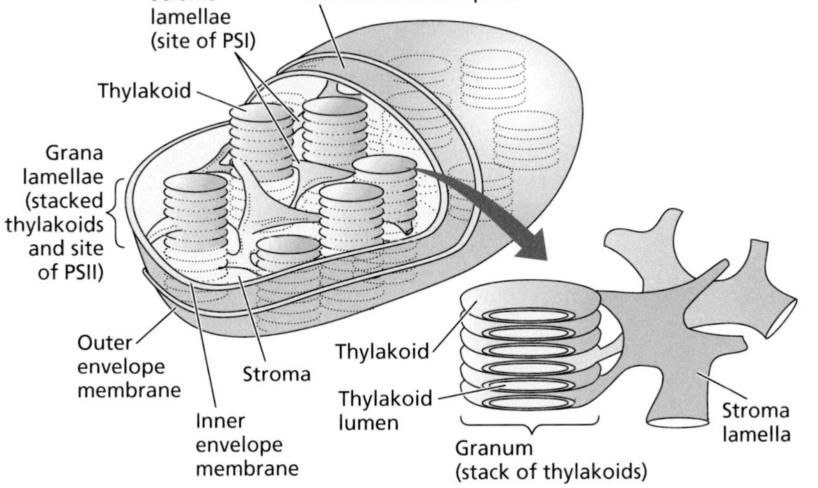

Figure 7.14 Schematic picture of the overall organization of the membranes in the chloroplast. The chloroplast of vascular plants is surrounded by the inner and outer membranes (the envelope). The region of the chloroplast that is inside the inner membrane and surrounds the thylakoid membranes is known as the stroma. It contains the enzymes that catalyze carbon fixation and other biosynthetic pathways. The thylakoid membranes are highly folded and appear in many pictures to be stacked like coins (the granum), although in reality they form one or a few large interconnected membrane systems, with a well-defined interior and exterior with respect to the stroma.

integral membrane proteins Proteins that are embedded in the lipid bilayer of a membrane via at least one transmembrane domain.

The chloroplast also contains its own DNA, RNA, and ribosomes. Some of the chloroplast proteins are products of transcription and translation within the chloroplast itself, whereas most of the others are encoded by nuclear DNA, synthesized on cytoplasmic ribosomes, and then imported into the chloroplast.

Thylakoids contain integral membrane proteins

A wide variety of proteins essential to photosynthesis are embedded in the thylakoid membranes. In many cases, portions of these proteins extend into the aqueous regions on both sides of the thylakoids. These **integral membrane proteins** contain a large proportion of hydrophobic amino acids and are therefore much more stable in a non-aqueous medium such as the hydrocarbon portion of the membrane.

The reaction centers, the antenna pigment–protein complexes, and most of the electron carrier proteins are all integral membrane proteins. In all known cases, integral membrane proteins have a unique orientation within the membrane. Thylakoid integral membrane proteins have one region pointing toward the stromal side of the membrane and the other oriented toward the interior space of the thylakoid, known as the *lumen* (Figure 7.14).

The chlorophylls and carotenoids in the thylakoid membrane are associated in a noncovalent, but highly specific, way with proteins, thereby forming pigment–protein complexes that are structurally organized to optimize energy transfer to, and subsequent electron transfer by, the reaction centers.

Photosystems I and II are spatially separated in the thylakoid membrane

The PSII reaction center, along with its antenna chlorophylls and associated electron transport proteins, is located predominantly in the grana lamellae (**Figure 7.15**). The PSI reaction center and its associated antenna pigments and electron transfer proteins, as well as the ATP synthase enzyme that catalyzes the formation of ATP, are found almost exclusively in the stroma lamellae and at the edges of the grana lamellae. The cytochrome b_6f complex of the electron transport chain that connects the two photosystems is distributed between stroma and grana lamellae. (The organization can also be seen in Figure 7.18.)

The two photochemical events that take place in O_2-evolving photosynthesis are spatially separated. This separation requires that mobile carriers move electrons between the two photosystems, from the grana region to the stroma region of the thylakoids. These diffusible carriers are the redox cofactor plastoquinone (PQ) and the blue-colored copper protein plastocyanin (PC), which carry electrons from PSII to the cytochrome b_6f complex, and from the cytochrome b_6f complex to PSI, respectively. In addition, water oxidation by PSII releases protons into the lumen of the grana, which must diffuse to the stroma lamellae to reach the ATP synthase, where they are used to drive the synthesis of ATP. The functional significance of this large separation (many tens of nanometers) between PSI and PSII is not entirely clear, but it is thought to allow the two photosystems to self-organize and thus improve the efficiency of energy distribution between the two photosystems. Spatial separation also allows for uncoupling of PSI from PSII under high light condition to prevent photooxidation of PSI.

There are more PQ and PC molecules than photo-centers in the thylakoids, and these "pools" of electron carriers can interact with multiple PSI or PSII complexes. This allows the photosystems to act cooperatively without requiring a strict one-to-one stoichiometry between the two photosystems or their excitation by light. Instead, PSII reaction centers feed reducing equivalents into a common intermediate pool of lipid-soluble electron carriers (plastoquinone). The PSI reaction centers remove the reducing equivalents from the common pool, rather than from any specific PSII reaction center complex. In some species, including

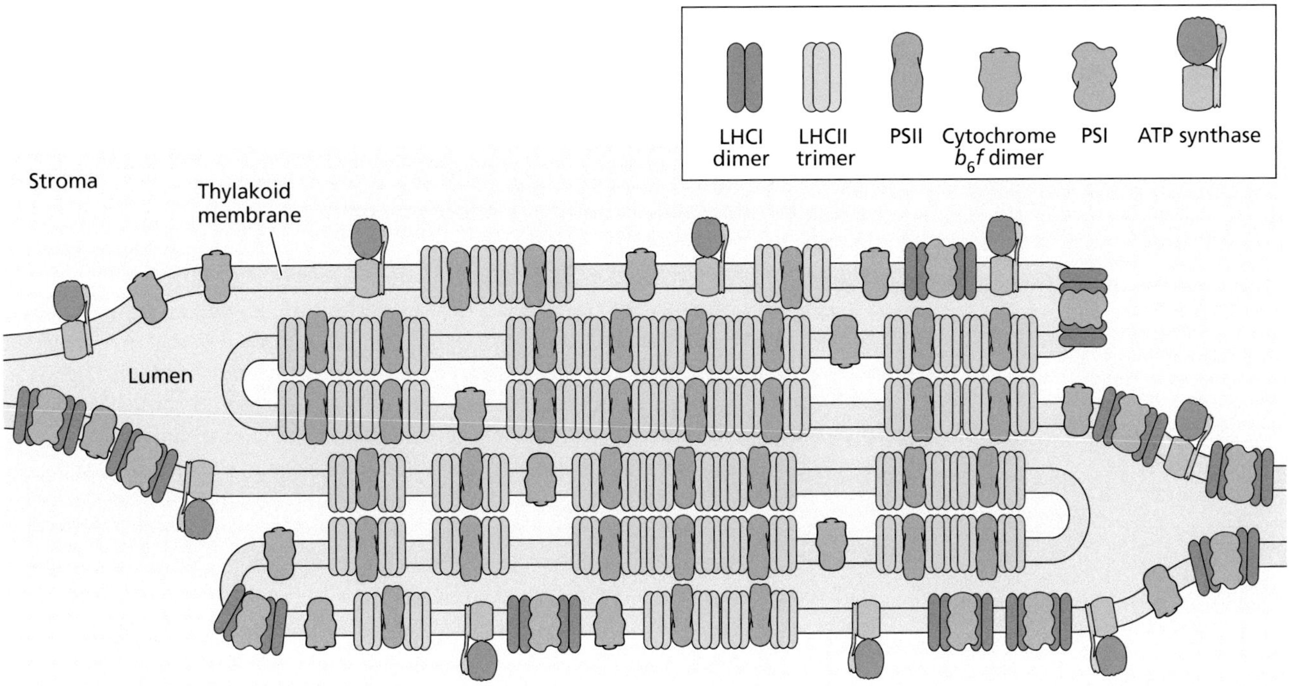

Figure 7.15 Organization of the four major protein complexes of the thylakoid membrane. PSII is located predominantly in the stacked regions of the thylakoid membrane; PSI and ATP synthase are found in the unstacked regions protruding into the stroma. Cytochrome b_6f complexes are evenly distributed. This lateral separation of the two photosystems requires that electrons and protons produced by PSII be transported a considerable distance before they can be acted on by PSI and ATP synthase, respectively. The light-harvesting antenna complexes, LHCI and LCHII, associate with PSI and PSII, respectively.

many vascular plants, there is relatively more PSII than PSI in the thylakoids, but in cyanobacteria there is often a higher proportion of PSI. The reasons for these different stoichiometries are not entirely understood, but they are thought to help prevent the buildup of reactive intermediates and balance the relative needs for ATP and NADPH by controlling the relative rates of linear and cyclic electron flow.

7.5 Organization of Light-Absorbing Antenna Systems

Describe the structure and function of the light-absorbing antenna systems.

The antenna systems of different classes of photosynthetic organisms are remarkably varied, in contrast to the reaction centers, which appear to be similar even in distantly related organisms. The variety of antenna complexes reflects evolutionary adaptation to the diverse environments in which different organisms live, as well as the need in some organisms to balance energy input to the two photosystems.

Antenna systems contain chlorophyll and are membrane-associated

Antenna systems function to deliver energy efficiently to the reaction centers with which they are associated. The size of the antenna system varies considerably in different organisms, ranging from a low of 20 to 30 bacteriochlorophylls per reaction center in some photosynthetic bacteria, to generally 200 to 300 chlorophylls per reaction center in vascular plants, to a few thousand pigments per

fluorescence resonance energy transfer (FRET) The physical mechanism by which excitation energy is conveyed from a molecule that absorbs light to an adjacent molecule.

reaction center in some types of algae and bacteria. The molecular structures of antenna pigments are also quite diverse, although all of them are associated in some way with the photosynthetic membrane. In almost all cases, the antenna pigments are associated with proteins to form pigment–protein complexes.

The physical mechanism by which excitation energy is conveyed from the chlorophyll that absorbs the light to the reaction center occurs predominantly through **fluorescence resonance energy transfer**, often abbreviated as **FRET**. By this mechanism the excitation energy is transferred from one molecule to another by a nonradiative process.

A useful analogy for resonance transfer is the transfer of energy between a tuning fork and a piano soundboard. After the tuning fork is struck and properly placed on the soundboard, it begins to vibrate. The efficiency of energy transfer between the tuning fork and the soundboard depends on their distance from each other and their relative orientation, as well as on their vibrational frequencies, or pitches. Similar parameters affect the efficiency of energy transfer in antenna complexes, with energy substituted for pitch.

Energy transfer in antenna complexes is usually very efficient: Approximately 95 to 99% of the photons absorbed by the antenna pigments have their energy transferred to the reaction center, where it can be used for photochemistry. There is an important difference between energy transfer among pigments in the antenna and the electron transfer that occurs in the reaction center: Whereas energy transfer is a purely physical phenomenon, electron transfer involves chemical (redox) reactions.

The antenna funnels energy to the reaction center

Pigments within the antenna that funnel absorbed energy toward the reaction center have absorption maxima that progressively shift toward longer red wavelengths (**Figure 7.16**). This red shift in absorption maximum means that the

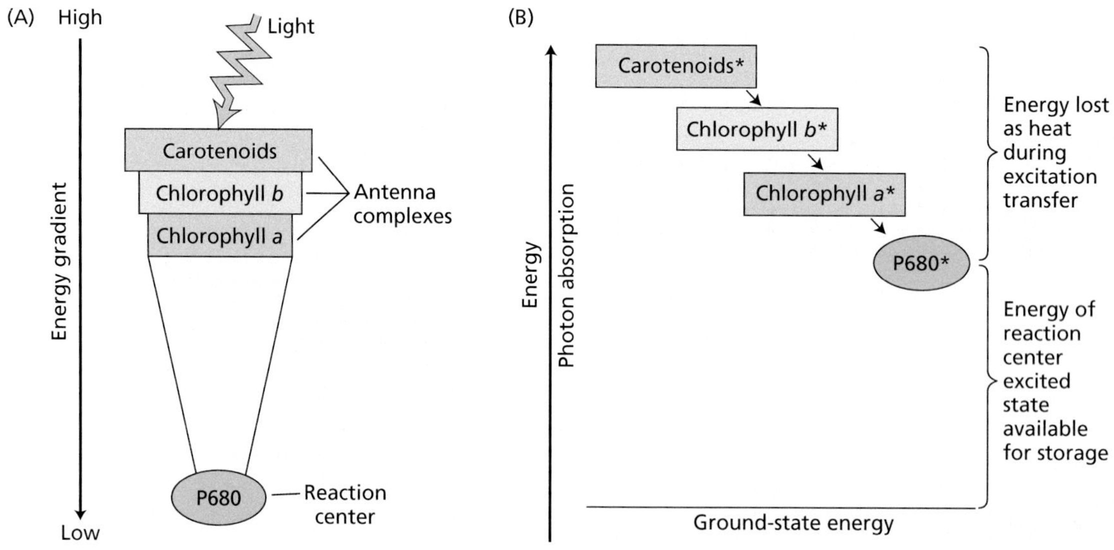

Figure 7.16 Funneling of excitation from the antenna system toward the reaction center. (A) The excited-state energy of pigments increases with distance from the reaction center; that is, pigments closer to the reaction center are lower in energy than those farther from the reaction center. This energy gradient ensures that excitation transfer toward the reaction center is energetically favorable and that excitation transfer back out to the peripheral portions of the antenna is energetically unfavorable. (B) Some energy is lost as heat to the environment by this process, but under optimal conditions almost all the excitation energy absorbed in the antenna complexes can be delivered to the reaction center. The asterisks denote excited states.

energy of the excited state is somewhat lower nearer the reaction center than in the more peripheral portions of the antenna system. This loss of energy helps drive flow of the remaining energy toward the reaction centers, where it can initiate the photochemical light reactions.

As a result, when excitation is transferred, for example, from a chlorophyll *b* molecule absorbing maximally at 650 nm to a chlorophyll *a* molecule absorbing maximally at 670 nm, the difference in energy between these two excited chlorophylls is lost to the environment as heat. If the excitation energy were to be transferred back to the chlorophyll *b*, then the energy lost as heat would have to be resupplied. The probability of reverse transfer is therefore small because thermal energy is not sufficient to make up the deficit between the lower-energy and higher-energy pigments. This effect gives the energy-trapping process a degree of directionality or irreversibility and makes the delivery of excitation to the reaction center more efficient. In essence, the system sacrifices some energy from each quantum so that nearly all of the quanta can be trapped by the reaction center.

light-harvesting complex I (LHCI) One of a large family of the most abundant antenna proteins, associated primarily with photosystem II.

light-harvesting complex II (LHCII) One of a large family of the most abundant antenna proteins, associated primarily with photosystem II.

chlorophyll *a/b* antenna proteins Chlorophyll-containing proteins associated with one or the other of the two photosystems in eukaryotic organisms. Also known as light-harvesting complex proteins (LHC proteins).

Many antenna pigment–protein complexes have a common structural motif

In all eukaryotic photosynthetic organisms that contain both chlorophyll *a* and chlorophyll *b*, the most abundant antenna proteins are members of a large family of structurally related proteins. Some of these proteins are associated primarily with PSII and are called **light-harvesting complex II (LHCII)** proteins; others are associated with PSI and are called **light-harvesting complex I (LHCI)** proteins. These antenna complexes are also known as **chlorophyll *a/b* antenna proteins**.

The LHC proteins are almost entirely embedded in the thylakoid membrane. LHCII contains three α-helical regions and binds 14 chlorophyll *a* and *b* molecules, as well as four carotenoids. The structure of the LHCI proteins is similar to that of the LHCII proteins. All of these proteins have significant sequence similarity and are almost certainly descendants of a common ancestral protein.

Light absorbed by carotenoids or chlorophyll *b* in the LHC proteins is rapidly transferred to chlorophyll *a* and then to other chlorophyll-containing antenna pigments that are intimately associated with the respective reaction center. The LHCII complex is also involved in regulatory processes to optimize photosynthesis and prevent damage to the photosystems.

7.6 Mechanisms of Electron Transport

Describe the chemical reactions and the functions of the proteins in electron transfer during photosynthesis.

The chemical reactions involved in electron transfer during photosynthesis include excitation of chlorophyll by light and the reduction of the first electron acceptor, the flow of electrons through photosystems II and I, the oxidation of water as the primary source of electrons, and the reduction of the final electron acceptor ($NADP^+$), as well as the chemiosmotic mechanism that mediates ATP synthesis.

Electrons from chlorophyll travel through the carriers organized in the Z scheme

Figure 7.17 shows a simplified version of the Z scheme, in which all the electron carriers known to function in electron flow from H_2O to $NADP^+$ are arranged vertically at their midpoint redox potentials, which can be used to estimate the energy stored in those states compared with the unexcited (or ground) state. Components known to react with each other are connected by arrows, so the Z scheme is really a synthesis of both kinetic and thermodynamic information. The

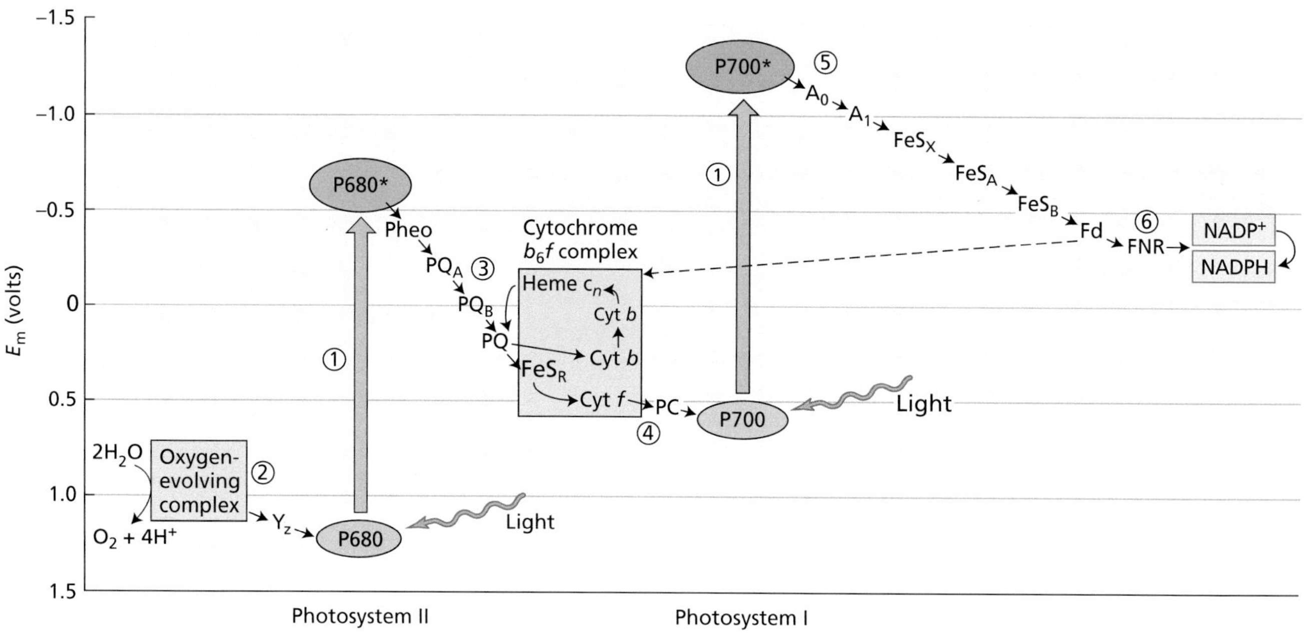

Figure 7.17 Detailed Z scheme for O_2-evolving photosynthetic organisms. The redox carriers are placed at their midpoint redox potentials (at pH 7). ① The vertical arrows represent photon absorption by the reaction center chlorophylls: P680 for photosystem II (PSII) and P700 for photosystem I (PSI). The excited PSII reaction center chlorophyll, P680*, transfers an electron to pheophytin (Pheo). ② On the oxidizing side of PSII (to the left of the vertical arrow joining P680 with P680*), P680 oxidized by light is re-reduced by Y_z, which has received electrons from oxidation of water. ③ On the reducing side of PSII (to the right of the arrow joining P680 with P680*), pheophytin transfers electrons to the acceptors PQ_A and PQ_B, which are plastoquinones. ④ The cytochrome b_6f complex transfers electrons to plastocyanin (PC), a soluble protein, which in turn reduces $P700^+$ (oxidized P700). ⑤ The acceptor of electrons from P700* (A_0) is thought to be a chlorophyll, and the next acceptor (A_1) is a quinone. A series of membrane-bound iron–sulfur proteins (FeS_X, FeS_A, and FeS_B) transfers electrons to soluble ferredoxin (Fd). ⑥ The soluble flavoprotein ferredoxin–$NADP^+$ reductase (FNR) reduces $NADP^+$ to NADPH, which is used in the Calvin–Benson cycle to reduce CO_2 (Chapter 8). The dashed line indicates cyclic electron flow around PSI.

large vertical arrows represent the input of light energy into the system. Note that this "energy diagram" view of the Z scheme does not represent changes in the positions of the various states or imply that they physically move in these directions. A spatial representation of these reactions is shown in **Figure 7.18**.

When photons excite the specialized chlorophyll of the reaction centers (P680 for PSII; P700 for PSI), an electron is ejected. The electron then passes through a series of electron carriers and eventually reduces P700 (for electrons from PSII) or $NADP^+$ (for electrons from PSI). Much of the following discussion describes the movements of these electrons, and how these result in the storage of energy in the final products.

Almost all the chemical processes that make up the light reactions of photosynthesis are carried out by four major protein complexes: PSII, the cytochrome b_6f complex, PSI, and ATP synthase. These four integral membrane complexes are vectorially oriented in the thylakoid membrane to function as follows (Figures 7.18):

- PSII oxidizes water to O_2 in the thylakoid lumen and in the process releases protons into the lumen. The reduced product of photosystem II is plastohydroquinone (PQH_2). The overall reaction results in transfer of electrons from the lumen to the stromal side of the thylakoid membrane, as well as the release of protons into the stroma (from water oxidation) and uptake of protons from the stroma (by PQ reduction to PQH_2).
- Cytochrome b_6f oxidizes PQH_2 molecules that were reduced by PSII and delivers electrons to PSI via the soluble copper protein plastocyanin. The

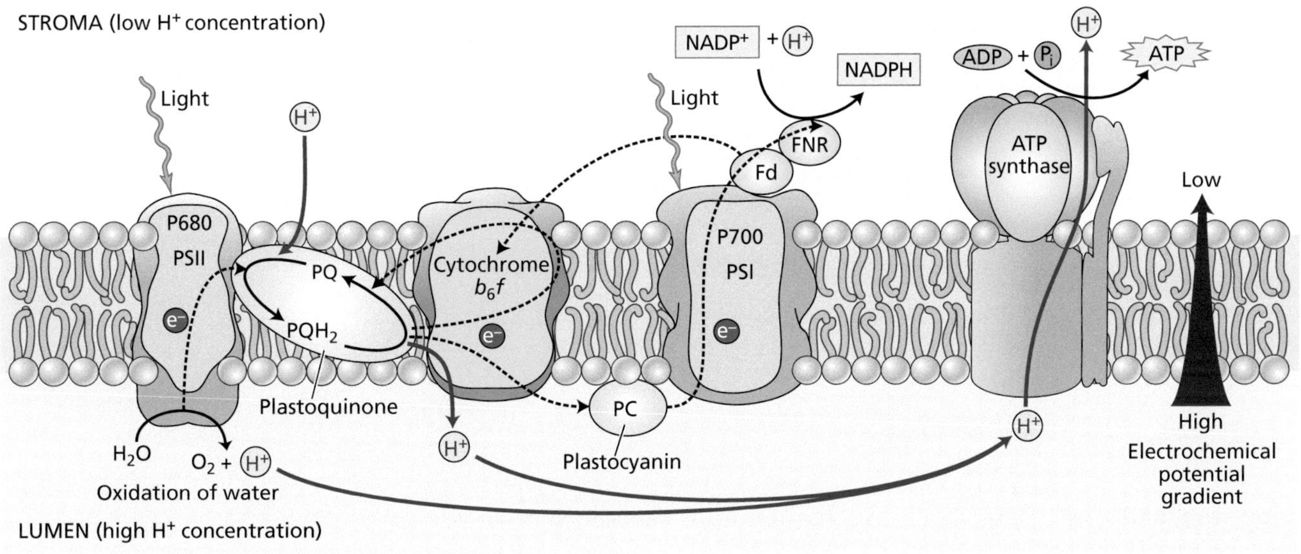

Figure 7.18 Transfer of electrons and protons in the thylakoid membrane is carried out vectorially by four protein complexes. Water is oxidized and protons are released in the lumen by PSII. PSI reduces $NADP^+$ to NADPH in the stroma, via the action of ferredoxin (Fd) and the flavoprotein ferredoxin–$NADP^+$ reductase (FNR). Protons are also transported into the lumen by the action of the cytochrome b_6f complex and contribute to the electrochemical proton gradient. These protons must then diffuse to the ATP synthase enzyme, where their diffusion down the electrochemical potential gradient is used to synthesize ATP in the stroma. Reduced plastoquinone (PQH_2) and plastocyanin transfer electrons to cytochrome b_6f and to PSI, respectively. The dashed lines represent electron transfer; solid blue lines represent proton movement.

oxidation is a complex process, called the *Q cycle*, that involves a series of electron and proton transfer reactions that not only provide electrons to PSI but also contribute to the thylakoid proton motive force.

- PSI reduces $NADP^+$ to NADPH in the stroma by the action of ferredoxin (Fd) and the flavoprotein ferredoxin–$NADP^+$ reductase (FNR). The overall reaction results in the transfer of electrons from the lumen to the stromal side of the thylakoid membrane.
- ATP synthase synthesizes ATP from ADP and inorganic phosphate (P_i) as protons diffuse back through it from the lumen into the stroma. The overall reaction results in the transfer of protons from the lumen to the stromal side of the thylakoid membrane.

Energy is captured when an excited chlorophyll reduces an electron acceptor molecule

The function of light is to excite a specialized chlorophyll in the reaction center, either by direct absorption or, more frequently, via energy transfer from an antenna pigment. This excitation process can be envisioned as the promotion of an electron from the highest-energy filled orbital of the chlorophyll to the lowest-energy unfilled orbital (**Figure 7.19**). The higher orbital has higher free energy, with the electron only loosely bound to the chlorophyll and easily lost if a nearby molecule can accept the electron.

This process can occur because of the excited-state chlorophyll that allows it to act as a strong electron donor (reductant) and a strong electron acceptor (oxidant). As illustrated in Figure 7.19, the ground-state chlorophyll has a pair of electrons in one of its orbitals (designated by the up and down arrows). Since the orbital is occupied by two electrons of different spins, this state is stable and does not easily lose or gain electrons.

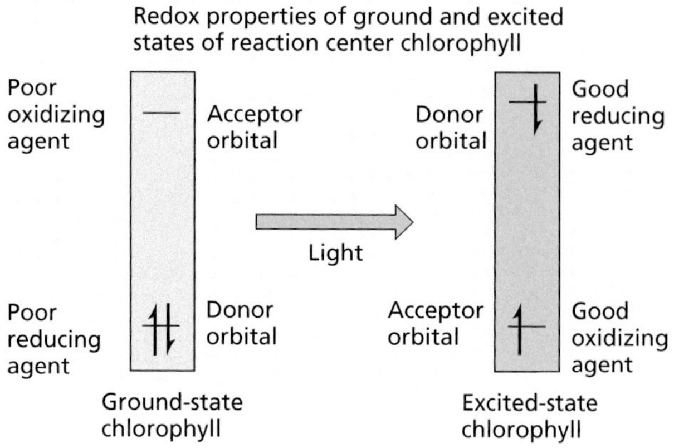

Figure 7.19 Orbital occupation diagram for the ground and excited states of reaction center chlorophyll. In the ground state the molecule is a poor reducing agent (loses electrons from a low-energy orbital) and a poor oxidizing agent (accepts electrons only into a high-energy orbital). In the excited state the situation is markedly different, and an electron can be lost from the high-energy orbital, making the molecule an extremely powerful reducing agent. This is the reason for the extremely negative excited-state redox potential shown by P680* and P700* in Figure 7.17. The excited state can also act as a strong oxidant by accepting an electron into the lower-energy orbital, although this pathway is not significant in reaction centers.

charge-separated state State produced in photosynthetic reaction centers after excitation by light in which an electron is moved from the lumen side to the stromal side. This movement of electrons forms an electric field across the thylakoid.

Exciting this chlorophyll promotes one of the two electrons to a higher orbital; in this state, the two orbitals are only partially filled and are therefore unstable. If this excited chlorophyll cannot react with other molecules, it will re-form the ground state and lose the extra energy as either fluorescence (light) or heat. However, if the excited chlorophyll can interact with nearby molecules, it can perform secondary photochemistry because the two partially filled orbitals are reactive. One of these orbitals can readily give up an electron, making it a strong reductant, while the other can readily accept an electron, making it a strong oxidant (Figure 7.19).

In the reaction centers, the first photochemical reaction involves the transfer of an excited electron on the chlorophyll to an acceptor molecule, forming a **charge-separated state** in which the reduced electron acceptor has a negative charge and the chlorophyll has a positive charge, but the overall charge of the reaction center remains the same because the electrons are simply rearranged, not lost or gained.

The excited state of chlorophyll decays in a few nanoseconds, losing its energy as heat or fluorescence. Therefore, the initial steps of photosynthesis must be extremely rapid: They occur within a few picoseconds (1 picosecond = 10^{-12} s). Such extremely rapid electron transfer reactions require that the donor and acceptor molecules be packed together very closely and have energy levels that are optimized for rapid electron transfer.

Going from the excited chlorophyll to the charge-separated state results in some loss of energy, but the first charge-separated state is still highly unstable. The acceptor transfers its extra electron to a secondary acceptor and so on down a chain of electron carriers. We refer to this chain as the acceptor-side electron carriers because they accept electrons from the excited reaction center chlorophyll. In parallel, the oxidized chlorophyll can extract an electron from nearby electron donors. The donor-side reactions produce the fully reduced chlorophyll, thus preventing the return of electrons from the acceptor-side carriers.

Each of the electron transfer steps (on both the acceptor and donor sides) progressively stabilizes the charge-separated states of the reaction centers. However, because energy is conserved, this stabilization results in a loss of energy, so that the more stable the state, the less energy it contains. This trade-off partly explains the differences in quantum yield and energy conversion. Because the subsequent charge-separated states are progressively more stable, the following forward reactions can also be slower. This is important because the biochemical reactions that are ultimately powered by the light reactions are much slower than the initial photochemistry, on a scale of milliseconds to seconds.

The reaction center chlorophylls of the two photosystems absorb at different wavelengths

PSI and PSII have distinct absorption characteristics (Figures 7.12 and 7.17). For example, the reduced and oxidized states of the reaction center chlorophylls have distinct absorption spectra that can be probed using a spectrophotometer that measures the amount of light of different wavelengths absorbed by a sample (Figure 7.5). In the oxidized state, chlorophylls lose their characteristic strong light absorbance in the red region of the spectrum; they become **bleached**. It is therefore possible to monitor the redox state of these chlorophylls by time-resolved optical absorbance measurements in which this bleaching is monitored directly.

Using such techniques, it was found that the reaction center chlorophyll of PSI absorbs maximally at 700 nm in its reduced (ground) state. Accordingly, this chlorophyll is named **P700** (the P stands for *pigment*). The analogous optical transient of PSII is at 680 nm, so its reaction center chlorophyll is known as **P680**. The primary donors of PSI (P700) and PSII (P680) are composed of two chlorophyll *a* molecules, though they may not act as functional dimers. In the oxidized state, reaction center chlorophylls contain an unpaired electron.

bleached The loss of chlorophyll's characteristic absorbance due to its conversion into another structural state, often by oxidation.

P700 The chlorophyll of the photosystem I reaction center that absorbs maximally at 700 nm in its neutral state. The P stands for pigment.

P680 The chlorophyll of the photosystem II reaction center that absorbs maximally at 680 nm in its neutral state. The P stands for pigment.

oxygen-evolving complex (OEC) The complex associated with photosystem II that oxidizes water and produces molecular oxygen (O_2).

The PSII reaction center is a multi-subunit pigment–protein complex

PSII is a multi-subunit protein supercomplex. In vascular plants, the multi-subunit protein supercomplex has two complete reaction centers and some antenna complexes. The core of the reaction center consists of two membrane proteins known as D1 and D2, as well as other proteins (**Figure 7.20**).

The primary donor chlorophyll, additional chlorophylls, carotenoids, pheophytins, and plastoquinones (two electron acceptors described below) are bound to the membrane proteins D1 and D2. Other proteins serve as antenna complexes or are involved in oxygen evolution. Some, such as cytochrome b_{559}, have no known function but may be involved in a protective cycle around PSII.

Water is oxidized to oxygen by PSII

Water is oxidized according to the following chemical reaction:

$$2\,H_2O \rightarrow O_2 + 4\,H^+ + 4\,e^- \qquad (7.8)$$

This reaction indicates that four electrons are removed from two water molecules, generating an oxygen molecule and four hydrogen ions (protons).

Water is a very stable molecule. Oxidation of water to form molecular oxygen requires the formation of an extremely strong oxidant. The photosynthetic **oxygen-evolving complex** (**OEC**) is the only known biochemical system that carries out this reaction, and it is the source of almost all the oxygen in Earth's atmosphere.

The OEC contains a catalytic cluster with four manganese (Mn) ions as well as Cl^- and Ca^{2+} ions. The D1 protein of PSII contains a special tyrosine residue, called Y_z, that acts as the primary electron donor to the excited chlorophyll P680 (Figure 7.17), forming a free radical oxidized tyrosine residue that, in turn, extracts electrons from the OEC. Each successive excitation of PSII results in an additional oxidation of the OEC, generating a progressive series of oxidation states—known as the *S states* and labeled S_0, S_1, S_2, S_3, and S_4.

When four oxidizing equivalents have accumulated in the OEC, it extracts four electrons from two H_2O molecules to generate O_2. The oxidation of water also releases four protons

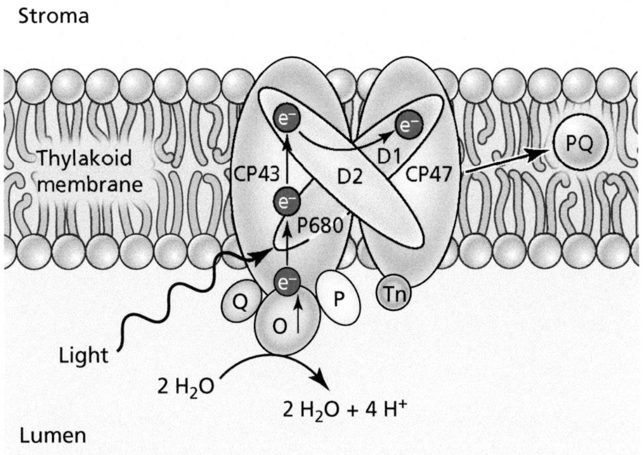

Figure 7.20 Simplified structure of one complete reaction center of the dimeric multi-subunit PSII supercomplex from spinach as determined by electron microscopy. The simplified core subunits of PSII are the reaction center (D1, D2), inner light-harvesting antennas (CP43, CP47), and the oxygen-evolving complex (PsbO, PsbP, PsbQ, PsbTn). LHCII are not shown. The arrows show the path of the electrons; PSII passes electrons to plastoquinone (PQ).

pheophytin A chlorophyll in which the central magnesium atom has been replaced by two hydrogen atoms.

plastohydroquinone (PQH_2) The fully reduced form of plastoquinone.

cytochrome b_6f complex A large multi-subunit protein complex containing two *b*-type hemes, one *c*-type heme (cytochrome *f*), and a Rieske iron–sulfur protein. A relatively immobile complex distributed equally between the grana and the stroma regions of the thylakoid membranes.

cytochrome *f* A subunit in the cytochrome b_6f complex that plays a role in electron transport between photosystems I and II.

Rieske iron–sulfur protein A protein subunit in the cytochrome b_6f complex, in which two iron atoms are bridged by two sulfur atoms, with two histidine and two cysteine ligands.

into the lumen of the thylakoid, and these protons are ultimately transferred from the lumen to the stroma by translocation through the ATP synthase (Figure 7.18).

Pheophytin and two quinones accept electrons from PSII

Pheophytin, a chlorophyll in which the central magnesium ion has been replaced by two hydrogen ions, acts as an early acceptor in PSII. The structural change gives pheophytin chemical and spectral properties that are slightly different from those of Mg-containing chlorophylls. Pheophytin passes electrons to a complex of two plastoquinones in close proximity to an iron ion.

The two plastoquinones, PQ_A and PQ_B, are bound to the reaction center and receive electrons from pheophytin in a sequential fashion. PQ_A can accept only one electron at a time, so it acts as a relay, transferring electrons from pheophytin to PQ_B. The function of PQ_B is more complex. After one excitation of PSII, PQ_B can accept one electron, forming a stable, tightly bound free radical intermediate (plastosemiquinone) (**Figure 7.21**). A second PSII excitation results in the uptake of two protons from the stromal side of the thylakoid and the formation of the fully reduced, protonated **plastohydroquinone (PQH_2)**. The PQH_2 then dissociates from the reaction center complex and enters the hydrophobic portion of the membrane, where it transfers its electrons to the cytochrome b_6f complex. The empty Q_B site on PSII can be repopulated with an oxidized PQ, re-forming PQ_B. Unlike the large protein complexes of the thylakoid membrane, PQH_2 is a small, nonpolar molecule that diffuses readily in the nonpolar core of the membrane bilayer in both its fully oxidized (PQ) and fully reduced (PQH_2) forms. This allows the PQ/PQH_2 system to act as a transmembrane shuttle for electrons and protons.

Electron flow through the cytochrome b_6f complex also transports protons

The **cytochrome b_6f complex** is a large multi-subunit protein with several prosthetic groups (**Figure 7.22**). The complex is a functional dimer and contains two *b*-type hemes and one *c*-type heme (**cytochrome *f***). In *c*-type cytochromes the heme is covalently attached to the protein; in *b*-type cytochromes the chemically similar protoheme group is not covalently attached. The complex contains a **Rieske iron–sulfur protein** (named for the scientist who discovered it), in which two iron ions are bridged by two sulfide ions. The cytochrome b_6f complex also contains additional cofactors, including an additional heme group (called heme c_n), a chlorophyll, and a carotenoid, whose functions are yet to be fully resolved.

The cytochrome b_6f complex and the related cytochrome bc_1 complex in the mitochondrial electron transport chain operate through a mechanism known as

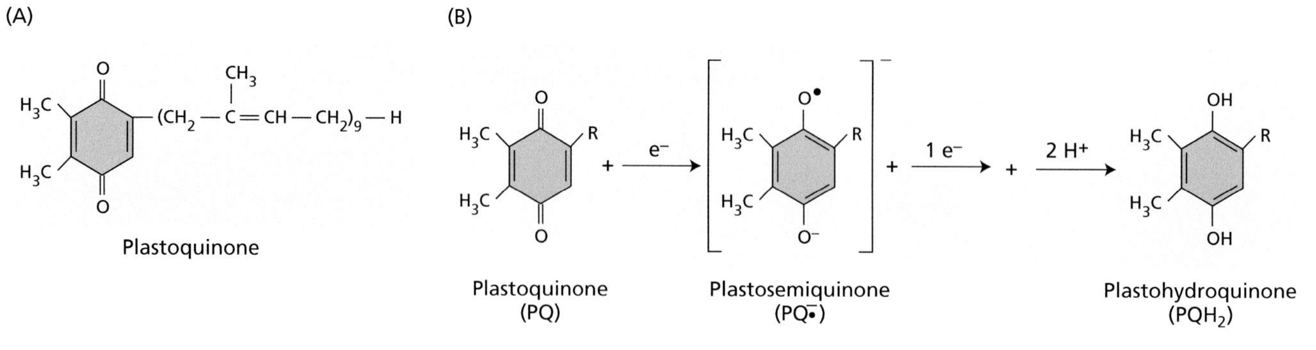

Figure 7.21 Structure and reactions of plastoquinones that operate in PSII. (A) The plastoquinone consists of a quinoid head and a long nonpolar tail that anchors it in the membrane. (B) Redox reactions of plastoquinone. The fully oxidized plastoquinone (PQ), anionic plastosemiquinone ($PQ^{\bullet-}$), and reduced plastohydroquinone (PQH_2) forms are shown; R represents the side chain.

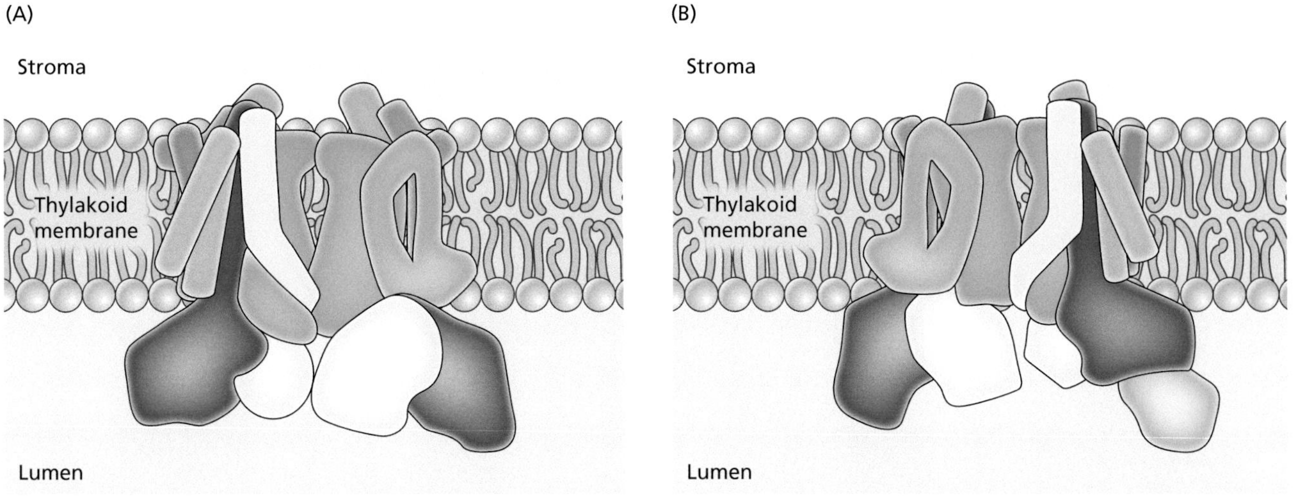

Figure 7.22 Structure of the cytochrome b_6f dimer complex from spinach. (A) Arrangement of the proteins in the complex. The major subunits: cytochrome b_6 protein (green), cytochrome *f* protein (purple), Rieske iron–sulfur protein (yellow), and other, smaller subunits in lavender, pink, orange, and teal. (B) Shows the complex rotated 180 degrees from A in the plane of the thylakoid membrane. Cytochrome b_6f dimer (from A) and plastocyanin (PC; blue) are shown in relation to each other.

the **Q cycle**, which was first proposed by Peter Mitchell in 1975 and later modified by many other researchers (**Figure 7.23**). In this mechanism, PQH_2, formed by light excitation of PSII or other processes, binds to a site called Q_o that faces the lumenal side of the thylakoid. One of the electrons on PQH_2 is transferred to the Rieske iron–sulfur center, then to cytochrome *f*, PC, and finally P700 of PSI. The extraction of one electron from the PQH_2 bound at Q_o results in the formation of plastosemiquinone, which is highly reactive and delivers an electron to the nearby *b*-type heme. This electron is then transferred across the thylakoid, via the *b* and c_n hemes, to a second binding site called Q_i. Note that the reduction of the *b*-type hemes appears to be uphill in terms of free energy, but is driven forward by subsequent reactions. The overall reaction has a negative free energy.

This process results in the transfer of one electron across the thylakoid membrane and the release of two protons into the lumen. A second turnover of the Q_o site introduces a second electron into the cytochrome *b* chain, allowing a PQ bound to the Q_i site to be reduced to PQH_2, with the uptake of protons from the stromal side of the thylakoid membrane. The resulting PQH_2 can then diffuse to Q_o, where it will be oxidized, delivering the protons to the lumen. Overall, the Q cycle results in two protons being deposited into the lumen for each electron that is transferred from PQH_2 to P700, increasing the number of protons available for ATP synthesis.

Q cycle A mechanism for oxidation of plastohydroquinone (reduced plastoquinone, also called plastoquinol) in chloroplasts and of ubihydroquinone (reduced ubiquinone, also called ubiquinol) in mitochondria.

Plastocyanin carries electrons between the cytochrome b_6f complex and photosystem I

The location of the two photosystems at different sites on the thylakoid membranes (Figure 7.15) requires that at least one component is capable of moving along or within the membrane in order to deliver electrons produced by PSII to PSI. The cytochrome b_6f complex is distributed equally between the grana and the stroma regions of the membranes, but its large size makes it unlikely that it is a mobile carrier of electrons between the photosystems. Plastoquinone acts as the mobile carrier of electrons from PSII to the cytochrome b_6f complex. Electrons are then transferred between the cytochrome b_6f complex and P700 by **plastocyanin (PC)**, a small (10.5 kDa), water-soluble, copper-containing protein. This protein is found in the lumenal space (Figure 7.23).

plastocyanin (PC) A small (10.5 kDa), water-soluble, copper-containing protein that transfers electrons between the cytochrome b_6f complex and P700. This protein is found in the lumenal space.

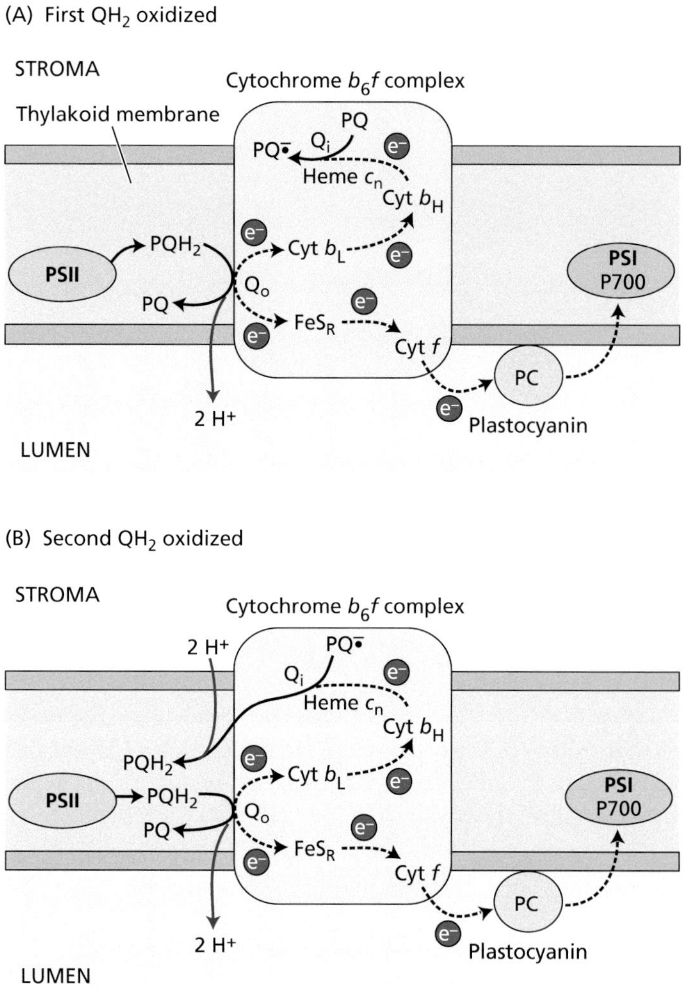

Figure 7.23 Mechanism of electron and proton transfer in the cytochrome b_6f complex. This complex contains two *b*-type cytochromes (Cyt *b*), a *c*-type cytochrome (Cyt *c*, historically called cytochrome *f*), a Rieske Fe–S protein (**FeS_R**), and two quinone oxidation–reduction sites. (A) Oxidation of the first PQH_2: A plastohydroquinone (PQH_2) molecule produced by the action of PSII (Figure 7.21) binds to a site called Q_o near the lumenal side of the complex and is oxidized in a special process in which one of its two electrons is transferred to the FeS_R and the other to heme b_L (a low-potential or strongly reducing *b* heme). Two protons from PQH_2 are released into the lumen during this process. The electron transferred to FeS_R is passed to cytochrome *f* (Cyt *f*) and then to plastocyanin (PC), which reduces oxidized P700 of PSI. The reduced b_L heme transfers an electron to the higher-potential b_H heme. The oxidized PQH_2, called plastoquinone (PQ), is released from the Q_o site into the thylakoid membrane. (B) Oxidation of the second PQH_2 results in a cyclic process: A second PQH_2 is oxidized at the Q_o site, with one electron going from FeS_R to PC and finally to P700. The second electron goes through the two *b*-type hemes and cyt c_n heme. The two electrons accumulated on the hemes cooperate to reduce PQ to PQH_2, at the Q_i site, which is located near the stromal side of the membrane, taking up two protons from the stroma. The PQH_2 released from Q_i into the thylakoid membrane can be oxidized at the Q_o site. Overall, for each electron that is passed from PQH_2 at the Q_o site to PC, two protons are released into the lumen.

FeS_R An iron- and sulfur-containing subunit of the cytochrome b_6f complex, involved in electron and proton transfer. See also Rieske iron–sulfur protein.

Fe–S centers Prosthetic groups consisting of inorganic iron and sulfur that are abundant in proteins in respiratory and photosynthetic electron transport.

FeS_X, FeS_A, FeS_B Membrane-bound iron–sulfur proteins that transfer electrons between photosystem I and ferredoxin.

ferredoxin (Fd) A small, water-soluble iron–sulfur protein involved in electron transport in photosystem I.

ferredoxin–NADP⁺ reductase (FNR) A membrane-associated flavoprotein that receives electrons from photosystem I and reduces $NADP^+$ to NADPH.

photophosphorylation The formation of ATP from ADP and inorganic phosphate (P_i), catalyzed by the CF_0F_1-ATP synthase and using light energy stored in the proton gradient across the thylakoid membrane.

The PSI reaction center oxidizes PC and reduces ferredoxin, which transfers electrons to $NADP^+$

The PSI reaction center complex is a large multi-subunit complex (**Figure 7.24**). In contrast to PSII, in which the antenna chlorophylls are associated with the reaction center but present on separate pigment–proteins, a core antenna consisting of about 100 chlorophylls is an integral part of the PSI reaction center. The core antenna and P700 are bound to two proteins, PsaA and PsaB, with molecular masses in the range of 66 to 70 kDa. The PSI reaction center complex from pea contains four LHCI complexes in addition to the core structure. The total number of chlorophyll molecules in this complex is nearly 200. The vast majority of these chlorophylls act as part of an antenna, funneling light energy into the core of the reaction center, where photochemistry takes place.

An early electron acceptor is a chlorophyll molecule, and another is a quinone species, phylloquinone, also known as vitamin K_1. Additional electron acceptors include a series of three membrane-associated iron–sulfur proteins, also known as **Fe–S centers: FeS_X, FeS_A, and FeS_B** (Figures 7.17 and 7.24). FeS_X is part of the P700-binding protein; FeS_A and FeS_B reside on PsaC, which is part of the PSI reaction center complex. Electrons are transferred through FeS_A and FeS_B to **ferredoxin (Fd)**, a small, water-soluble iron–sulfur protein. The membrane-associated flavoprotein **ferredoxin–NADP⁺ reductase (FNR)** oxidizes the reduced ferredoxin and reduces $NADP^+$ to NADPH, thus completing the sequence of electron flow that begins with the oxidation of water.

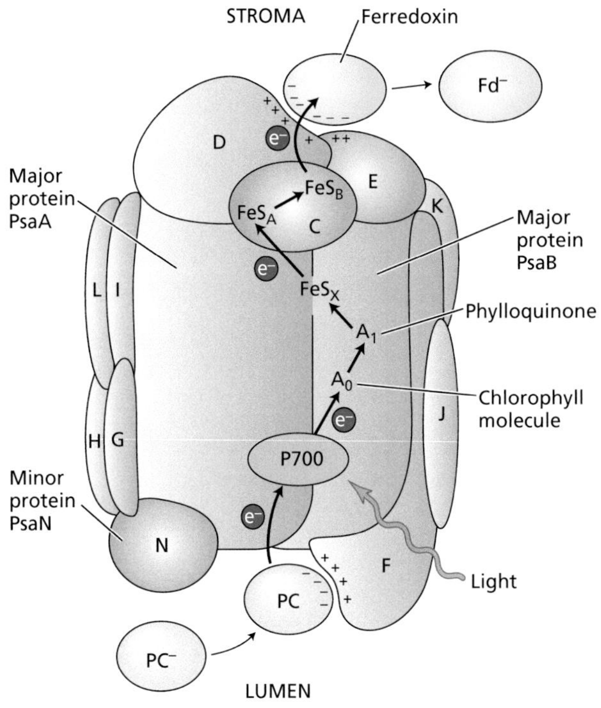

Figure 7.24 Structure of PSI. Structural model of the PSI reaction center from vascular plants. Components of the PSI reaction center are organized around two major core proteins, PsaA and PsaB. The minor proteins PsaC to PsaN are labeled C to N. Electrons are transferred from plastocyanin (PC) to P700 (Figures 7.17 and 7.18) and then to a chlorophyll molecule (A_0), to phylloquinone (A_1), to the Fe–S centers FeS_X, FeS_A, and FeS_B, and finally to the soluble iron–sulfur protein ferredoxin (Fd).

In addition to the reduction of $NADP^+$, reduced ferredoxin produced by PSI has several other functions in the chloroplast, such as the supply of reductants to reduce nitrite (Chapter 5) and the regulation of some of the carbon-fixation enzymes (Chapter 8).

Some herbicides block photosynthetic electron flow

The use of herbicides to kill unwanted plants is widespread in modern agriculture. Many different classes of herbicides have been developed. Some act by blocking amino acid, carotenoid, or lipid biosynthesis or by disrupting cell division. Other herbicides, such as dichlorophenyldimethylurea (DCMU, also known as diuron) and paraquat, block photosynthetic electron flow (**Figure 7.25**).

DCMU blocks electron flow at the quinone acceptors of PSII, by competing for the binding site of plastoquinone that is normally occupied by PQ_B. Paraquat accepts electrons from the early acceptors of PSI and then reacts with oxygen to form superoxide, $O_2^{\bullet -}$, a species that is very damaging to chloroplast components. In this way paraquat siphons electrons off the electron transport and prevents the formation of NADPH.

7.7 Proton Transport and ATP Synthesis in the Chloroplast

Explain how ATP is synthesized in the chloroplast.

Light energy is used to reduce $NADP^+$ to NADPH. Several steps in this sequence of electron transfer reactions also store energy in the form of an electrochemical gradient of protons that in turn drives **photophosphorylation**, the light-dependent synthesis of ATP. This process was discovered by Daniel Arnon and his coworkers in the 1950s. Under normal cellular conditions, photophosphorylation requires electron flow. Electron flow without accompanying phosphorylation is said to be **uncoupled**.

It is widely accepted that photophosphorylation works via the chemiosmotic mechanism, proposed in the 1960s by Peter Mitchell. The same general mechanism drives phosphorylation during aerobic respiration in bacteria and mitochondria, as well as the transfer of many ions and metabolites across membranes. Chemiosmosis appears to be a unifying aspect of membrane processes in all forms of life.

In Chapter 6 we discussed the role of ATPases in chemiosmosis and ion transport at the cell's plasma membrane. The ATP

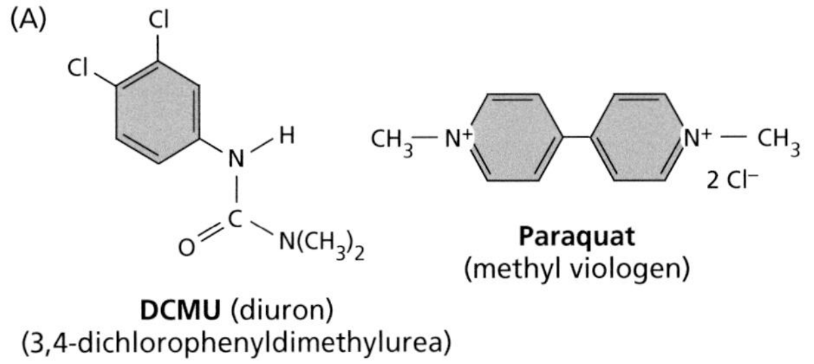

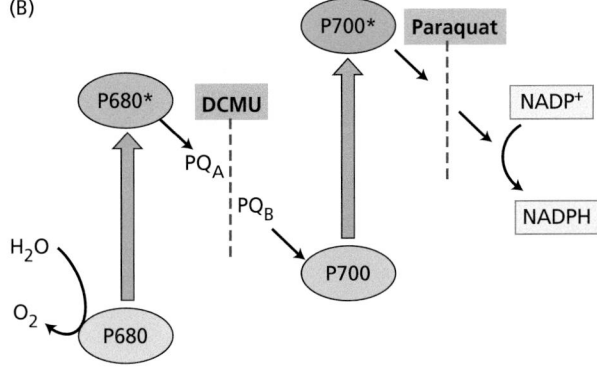

Figure 7.25 Chemical structure and mechanism of action of two important herbicides. (A) Chemical structure of 3,4-dichlorophenyldimethylurea (DCMU, also known as diuron) and methyl viologen (paraquat), two herbicides that block photosynthetic electron flow. (B) Sites of action of the two herbicides. DCMU blocks electron flow at the plastoquinone acceptors of PSII by competing for the binding site of plastoquinone. Paraquat acts by accepting electrons from the early acceptors of PSI. Interference by DCMU and paraquat is indicated by dashed lines.

uncoupled A process in which protons translocated into the lumen as part of the light reactions can escape to the stroma and bypass the ATP synthase so that ATP is not synthesized. Similar uncoupled reactions occur in the mitochondria when protons that are translocated out of the matrix as a result of the electron transport activity of complexes I, III, and IV return to the matrix and bypass the F_oF_1-ATP synthase.

chemical potential The free energy associated with a substance that is available to perform work.

proton motive force (PMF) The energetic effect of the electrochemical H^+ gradient across a membrane, expressed in units of electrical potential.

used by the plasma membrane ATPase is synthesized by photophosphorylation in the chloroplast and oxidative phosphorylation in the mitochondrion. Here we are concerned with chemiosmosis and transmembrane proton concentration differences used to make ATP in the chloroplast.

The basic principle of chemiosmosis is that ion concentration differences and electrical potential differences across membranes are sources of free energy that can be used by the cell. Differences in **chemical potential** of any molecular species whose concentrations are not the same on opposite sides of a membrane provide such a source of energy (Chapter 6).

The asymmetric nature of the photosynthetic membrane and the electron transfer systems leads to the storage of energy in two forms. First, electrons flow through the photosystems across the thylakoid membrane from the lumen (where water is split) to the stroma (where NADPH is produced). Second, protons flow from one side of the membrane to the other during electron flow, as discussed earlier. The direction of proton translocation is such that the stroma becomes more alkaline (fewer H^+ ions) and the lumen becomes more acidic (more H^+ ions) as a result of electron transport (Figures 7.18 and 7.23).

Some of the early evidence supporting a chemiosmotic mechanism of photosynthetic ATP formation was provided by an elegant experiment using isolated thylakoid membranes carried out by André Jagendorf and coworkers (**Figure 7.26**). Using pH changes and dark conditions, they found that ATP was formed from ADP and P_i by this process, with no light input or electron transport. This result supports the predictions of the chemiosmotic mechanism.

Mitchell proposed that the total energy available for ATP synthesis, which he called the **proton motive force** (Δp), is the sum of a proton chemical potential and a transmembrane electrical potential. These two components of the proton motive force from the outside of the membrane to the inside are given by the following equation:

$$\Delta p = \Delta E - 59\ mV \times (\text{pH}_\text{i} - \text{pH}_\text{o}) \tag{7.9}$$

where ΔE is the transmembrane electrical potential, and $\text{pH}_\text{i} - \text{pH}_\text{o}$ (or ΔpH) is the pH difference across the membrane. The constant of proportionality (at 25°C) is 59 mV per pH unit, so a transmembrane pH difference of one pH unit

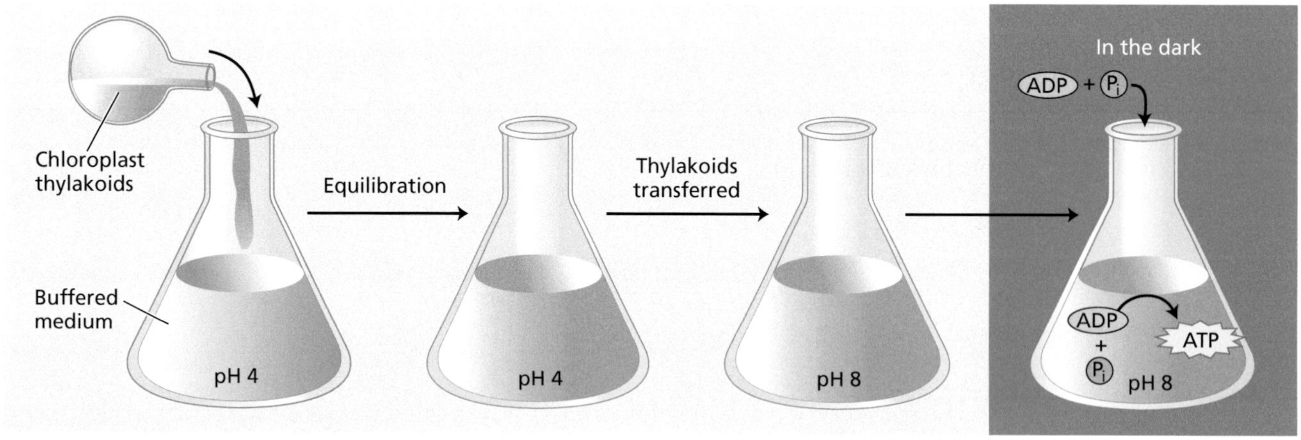

Figure 7.26 Summary of the experiment carried out by Jagendorf and coworkers. Isolated chloroplast thylakoids kept previously at pH 8 were equilibrated in an acid medium at pH 4. The thylakoids were then transferred to a buffer at pH 8 that contained ADP and P_i. The proton gradient generated by this manipulation provided a driving force for ATP synthesis in the absence of light. This experiment verified a prediction of the chemiosmotic model stating that a chemical potential across a membrane can provide energy for ATP synthesis.

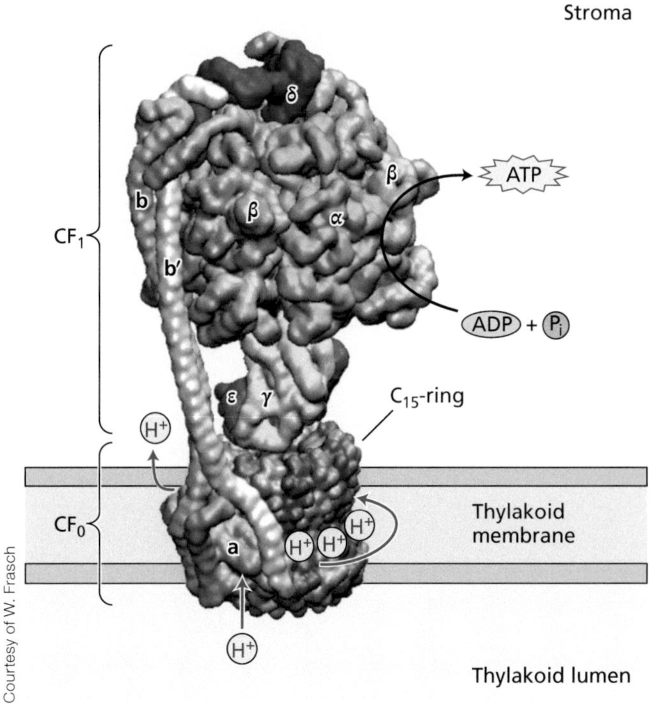

Figure 7.27 Structure of chloroplast F_0F_1-ATP synthase. This enzyme consists of a large multi-subunit complex, CF_1, attached on the stromal side of the membrane to an integral membrane portion, known as CF_0. CF_1 consists of five different polypeptides, with stoichiometry of $\alpha_3\ \beta_3\ \gamma\ \delta\ \varepsilon$. CF_0 contains probably four different polypeptides, with a stoichiometry of a b b′ c_{14}. Protons from the lumen are transported by the rotating *c* polypeptide and are ejected on the stromal side. The structure closely resembles that of the mitochondrial F_0F_1-ATP synthase (Chapters 6 and 11) and the vacuolar V-type ATPase (Chapter 6). Courtesy of W. Frasch

is equivalent to a membrane potential of 59 mV. While it is thought that mitochondria store Δp almost exclusively as electrical potential, the chloroplast also stores some of this energy as a pH gradient, which causes the lumen of the thylakoid to become more acidic with respect to the stroma, which in turn plays key roles in regulating light capture and electron transfer.

In the thylakoids, ATP is synthesized by an enzyme complex (mass ~400 kDa) known by several names: **ATP synthase**, **ATPase** (after the reverse reaction of ATP hydrolysis), and **CF_0-CF_1 ATP synthase**. This enzyme consists of two parts: a hydrophobic membrane-bound portion called CF_0 and a portion that sticks out into the stroma called CF_1 (**Figure 7.27**). CF_0 appears to form a channel across the membrane through which protons can pass. CF_1 is made up of several peptides, including three copies of each of the α and β peptides arranged alternately much like the sections of an orange. Whereas the catalytic sites are located largely on the β polypeptide, many of the other peptides are thought to have primarily regulatory functions. CF_1 is the portion of the complex that synthesizes ATP.

The molecular structure of the mitochondrial ATP synthase has been determined by X-ray crystallography and cryo-electron microscopy methods. Although there are significant differences between the chloroplast and mitochondrial enzymes, they have the same overall architecture and probably nearly identical catalytic sites. In fact, there are remarkable similarities in the way electron flow is coupled to proton translocation in chloroplasts and mitochondria (**Figure 7.28**). Another remarkable aspect of the mechanism of the ATP synthase is that the internal stalk and probably much of the CF_0 portion of the enzyme rotate during catalysis. The enzyme is actually a tiny molecular motor. Three molecules of ATP are synthesized for each rotation of the enzyme.

Cyclic electron flow augments the output of ATP to balance the chloroplast energy budget

The linear electron flow pathway produces ATP and NADPH, but at a fixed ratio that is too low to support CO_2 fixation and other processes that require ATP. Therefore, under certain conditions, the light reactions provide additional sources of ATP. One major source is a process called **cyclic electron flow**, in which electrons flow from the reducing side of PSI back to P700 via plastohydroquinone and the cytochrome b_6f complex. This cyclic electron flow is coupled to proton pumping from the stroma to the lumen and can be used for ATP synthesis without water oxidation or $NADP^+$ reduction (Figure 7.18). Cyclic electron flow is especially important as an ATP source in the **bundle sheath** chloroplasts of some plants that carry out C_4 carbon fixation (Chapter 8).

ATP synthase The multi-subunit protein complex that synthesizes ATP from ADP and phosphate (P_i). F_0F_1 and CF_0-CF_1 types are present in mitochondria and chloroplasts, respectively. Also called *ATPase* and *CF_0–CF_1 ATP synthase*.

ATPase The multi-subunit protein complex that synthesizes ATP from ADP and phosphate (P_i). F_0F_1 and CF_0-CF_1 types are present in mitochondria and chloroplasts, respectively. Also called *ATP synthase* and *CF_0-CF_1 ATP synthase*.

CF_0CF_1-ATP synthase The multi-subunit protein complex that synthesizes ATP from ADP and phosphate (P_i). F_0F_1 and CF_0-CF_1 types are present in mitochondria and chloroplasts, respectively. Also called ATP synthase and ATPase.

cyclic electron flow In photosystem I, the flow of electrons from the electron acceptors through the cytochrome b_6f complex and back to P700, coupled to proton pumping into the lumen. This electron flow energizes ATP synthesis but does not oxidize water or reduce $NADP^+$.

bundle sheath One or more layers of closely packed cells surrounding the small veins of leaves and the primary vascular bundles of stems.

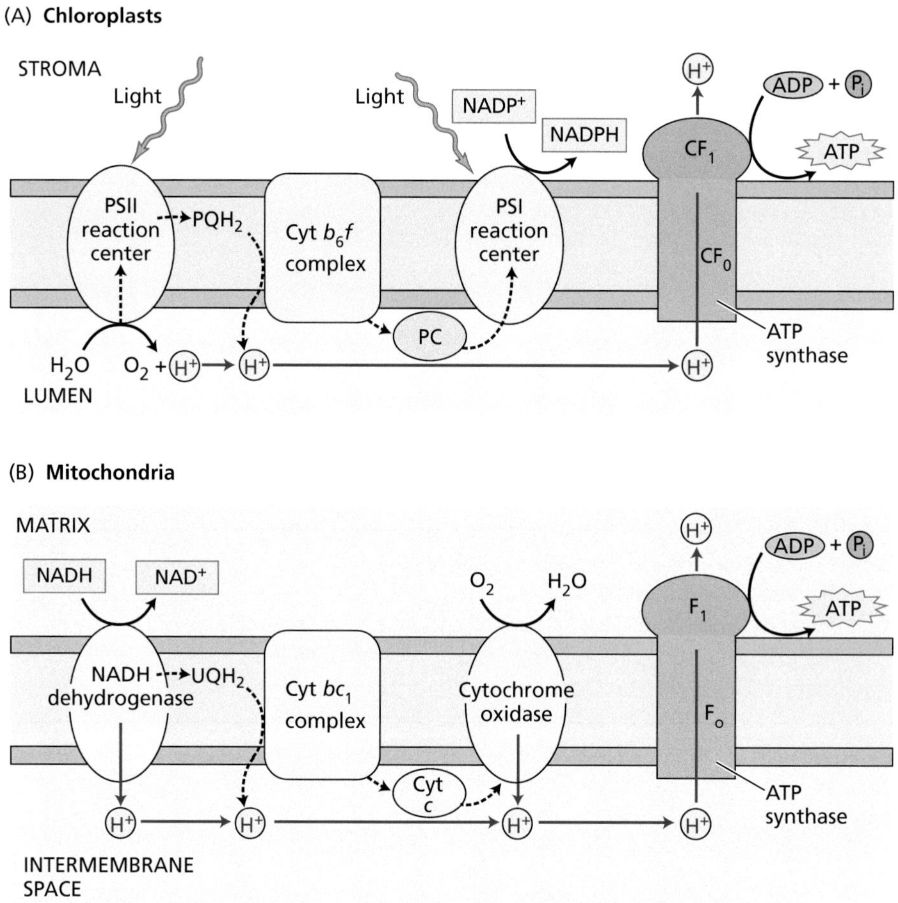

Figure 7.28 Similarities of photosynthetic and respiratory electron flow in chloroplasts and mitochondria. Electron flow is coupled to proton translocation, creating a transmembrane proton motive force (Δp). The energy in the proton motive force is then used by ATP synthase to synthesize ATP. (A) Chloroplasts carry out noncyclic electron flow, oxidizing water and reducing $NADP^+$. Protons are produced by the oxidation of water and by the oxidation of PQH_2 by the cytochrome b_6f complex. (B) Mitochondria oxidize NADH to NAD^+ and reduce oxygen to water. Protons are pumped by the enzyme NADH dehydrogenase, the cytochrome bc_1 complex, and cytochrome oxidase. The ATP synthases in the three systems are very similar in structure. UQH_2, ubiquinol.

Summary

Photosynthesis in plants captures light energy for the synthesis of carbohydrates and the generation of oxygen from carbon dioxide and water. Energy stored in carbohydrates is used to power cellular processes in the plant and can serve as the energy resource for all forms of life.

7.1 Photosynthesis in Green Plants

- Within chloroplasts, chlorophylls absorb light energy for the oxidation of water, releasing oxygen and generating NADPH and ATP (thylakoid reactions or light reactions).
- NADPH and ATP are used to reduce carbon dioxide to form sugars (carbon fixation reactions, discussed in **Chapter 8**).

7.2 General Concepts

- Light behaves as both a particle and a wave, delivering energy as photons, some of which are absorbed and used by plants (**Figures 7.1–7.3**).
- Light-energized chlorophyll may fluoresce or transfer energy to another molecule to drive chemical reactions (**Figures 7.4**, **7.10**).

(Continued)

Summary (*continued*)

- All photosynthetic organisms contain a mixture of pigments with distinct structures and light-absorbing properties (**Figures 7.6**, **7.7**).

7.3 Key Experiments in Understanding Photosynthesis

- An action spectrum for photosynthesis shows algal oxygen evolution at certain wavelengths (**Figures 7.8**, **7.9**).
- Antenna pigment–protein complexes collect light energy and transfer it to the reaction center complexes (**Figures 7.10**, **7.16**).
- Light drives the reduction of $NADP^+$ and the formation of ATP. Oxygen-evolving organisms have two photosystems (PSI and PSII) that operate in series (**Figure 7.12**).

7.4 Organization of the Photosynthetic Apparatus

- Within the chloroplast, thylakoid membranes contain the reaction centers, the light-harvesting antenna complexes, and most of the electron carrier proteins. PSI and PSII are spatially separated in thylakoids (**Figures 7.13–7.15**).

7.5 Organization of Light-Absorbing Antenna Systems

- The antenna system funnels energy to the reaction center (**Figure 7.16**).
- Light-harvesting proteins of both photosystems are structurally similar.

7.6 Mechanisms of Electron Transport

- The Z scheme illustrates the flow of electrons from H_2O to $NADP^+$ through carriers in PSII and PSI (**Figures 7.12**, **7.17**).
- Three large protein complexes transfer electrons: PSII, the cytochrome b_6f complex, and PSI (**Figure 7.18**).
- PSI reaction center chlorophyll absorbs maximally at 700 nm; PSII reaction center chlorophyll absorbs maximally at 680 nm.
- The PSII reaction center is a multi-subunit protein–pigment complex (**Figure 7.21**).
- Manganese ions are required to oxidize water.
- Two hydrophobic plastoquinones accept electrons from PSII (**Figures 7.18**, **7.20**).
- Protons are transported into the thylakoid lumen when electrons pass through the cytochrome b_6f complex (**Figures 7.18**, **7.22**).
- Plastoquinone and plastocyanin carry electrons between PSII and PSI (**Figures 7.18**, **7.23**).
- $NADP^+$ is reduced by the PSI reaction center, using Fe–S centers and ferredoxin as electron carriers (**Figure 7.24**).
- Herbicides may block photosynthetic electron flow (**Figure 7.25**).

7.7 Proton Transport and ATP Synthesis in the Chloroplast

- In vitro transfer of pH 4–equilibrated chloroplast thylakoids to a pH 8 buffer resulted in the formation of ATP from ADP and P_i, supporting the chemiosmotic mechanism (**Figure 7.26**).
- Protons move down an electrochemical gradient (proton motive force), passing through an ATP synthase and forming ATP (**Figure 7.27**).
- During catalysis, the CF_0 portion of the ATP synthase rotates like a miniature motor.
- Proton translocation in chloroplasts and mitochondria shows significant similarities (**Figure 7.28**).
- Cyclic electron flow generates ATP, but no NADPH, by proton pumping (**Figure 7.15**).

Suggested Reading

Armbruster, U., Correa Galvis, V., Kunz, H.-H., and Strand, D. D. (2017) The regulation of the chloroplast proton motive force plays a key role for photosynthesis in fluctuating light. *Curr. Opin. Plant Biol.* 37: 56–62.

Croce, R., and van Amerongen, H. (2020) Light harvesting in oxygenic photosynthesis: Structural biology meets spectroscopy. *Science* 369: eaay2058.

Davis, G. A., Rutherford, A. W., and Kramer, D. M. (2017) Hacking the thylakoid proton motive force for improved photosynthesis: Modulating ion flux rates that control proton motive force partitioning into $\Delta\psi$ and ΔpH. *Philos Trans. R. Soc. Lond. B. Biol. Sci.* 372: 20160381.

Hohmann-Marriott, M. F., and Blankenship, R. E. (2011) Evolution of photosynthesis. *Annu. Rev. Plant Biol.* 62: 515–548.

Kaiser, E., Morales, A., and Harbinson, J. (2018) Fluctuating light takes crop photosynthesis on a rollercoaster ride. *Plant Physiol.* 176: 977–989.

Kanazawa, A., Neofotis, P., Davis, G. A., Fisher, N., and Kramer, D. M. (2020) Diversity in photoprotection and energy balancing in terrestrial and aquatic phototrophs. In A. W. D. Larkum, A. R. Grossman, and J. A. Raven, eds., *Photosynthesis in Algae: Biochemical and Physiological Mechanisms*. Springer International Publishing, Cham, Switzerland, pp. 299–327.

Kramer, D. M., Nitschke, W., and Cooley, J. W. (2009) The cytochrome *bc1* and related *bc* complexes: The Rieske/cytochrome *b* complex as the functional core of a central electron/proton transfer complex. In C. N. Hunter, F. Daldal, M. C. Thurnauer, and J. T. Beatty, eds., *The Purple Phototrophic Bacteria*. Springer Netherlands, Dordrecht, pp. 451–473.

Kromdijk, J., Głowacka, K., Leonelli, L., Gabilly, S. T., Iwai, M., Niyogi, K. K., and Long, S. P. (2016) Improving photosynthesis and crop productivity by accelerating recovery from photoprotection. *Science* 354: 857–861.

Lubitz, W., Chrysina, M., and Cox, N. (2019) Water oxidation in photosystem II. *Photosynth. Res.* 142: 105–125.

8 Photosynthesis: The Carbon Reactions

The initial source of energy for nearly all life on our planet comes from solar radiant energy that strikes Earth's surface. Aquatic and terrestrial photosynthetic organisms capture approximately 3×10^{21} Joules per year of sunlight energy and use it to fix approximately 2×10^{11} tonnes of carbon per year. In Chapter 7, we learned how photosynthetic organisms harvest light energy and convert it into chemical energy. How do plants capture carbon dioxide (CO_2) and convert it into sugars?

More than 1 billion years ago, heterotrophic cells acquired the ability to convert sunlight into chemical energy through primary endosymbiosis of ancient photosynthetic cyanobacteria, which gave rise to plastids. Phylogenic analyses of plastids, cyanobacteria, and eukaryotes have determined that the Archaeplastidae, the autotrophic eukaryotic descendants of this ancient monophyletic endosymbiotic event, comprises two major lineages: Chloroplastidae (Viridiplantae: green algae, land plants) and Rhodophyceae (red algae) (Figure 1.1). The integration of cyanobacterial genes into the host's genome dramatically reduced the size of the plastid genome (plastome). This transfer of genes required the development of complex transport mechanisms across the outer and inner membrane of the plastid to target nucleus-encoded proteins to the endosymbiont, and plastid-encoded proteins to the host, as well as the exchange of numerous compounds between the plastid and the rest of the cell. The ancestral endosymbiont conveyed the ability not only to carry out photosynthesis but also to synthesize novel compounds, such as *starch*, and is now essential for nitrogen and sulfur assimilation, amino acid and fatty acid synthesis, and many other specialized functions.

In Chapter 7 we learned about the light-dependent reactions of photosynthesis: The photochemical oxidation of water to molecular oxygen

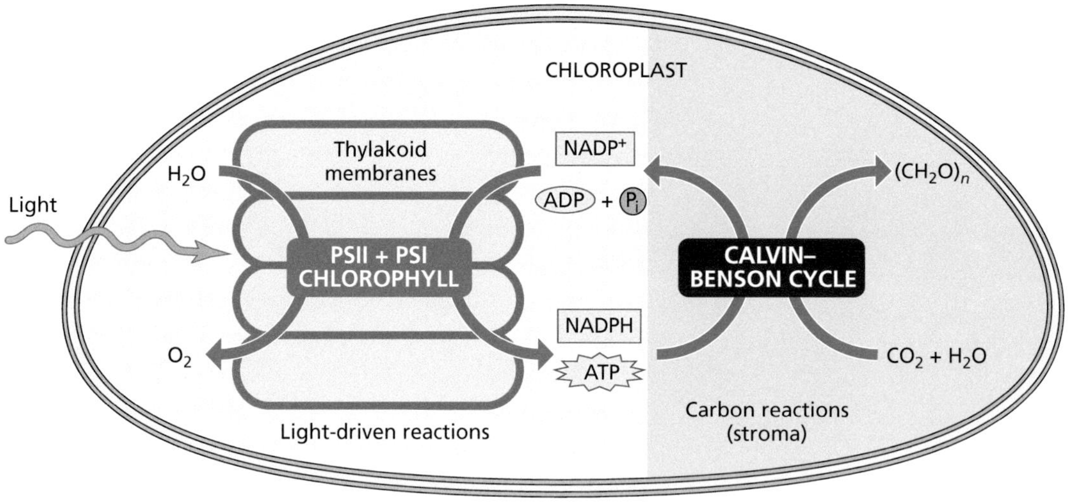

Figure 8.1 Light and carbon reactions of photosynthesis in chloroplasts of land plants. In thylakoid membranes, the excitation of chlorophyll in the photosynthetic electron transport system (photosystem II [PSII] + photosystem I [PSI]) by light elicits the formation of ATP and NADPH (Chapter 7). In the stroma, both ATP and NADPH are consumed by the Calvin–Benson cycle in a series of enzyme-driven reactions that reduce atmospheric CO_2 to carbohydrates (triose phosphates).

generates ATP, reduced ferredoxin, and NADPH. These products are used to drive the enzyme-catalyzed reduction of atmospheric CO_2 to triose phosphates, ultimately leading to the synthesis of carbohydrates and other cell components (**Figure 8.1**). The reduction of CO_2 within the stroma of chloroplasts is commonly referred to as the *carbon reactions of photosynthesis* because not only do products of photochemistry (e.g., ATP and NADPH) provide the energy for these reactions, but light is required to activate many of the enzymes required for carbon assimilation.

We begin this chapter by describing the metabolic cycle that incorporates atmospheric CO_2 into organic compounds: the **Calvin–Benson cycle**. Next, we consider how the unavoidable reaction of ribulose 1,5-bisphosphate carboxylase/oxygenase (Rubisco) with molecular oxygen decreases the efficiency of photosynthetic CO_2 assimilation, and how photorespiration releases CO_2 while recycling the byproducts of this oxygenation reaction that would otherwise be unusable by the cell. We also examine two CO_2-concentrating mechanisms that land plants have evolved for minimizing the oxygenation of Rubisco and the loss of energy and CO_2 during photorespiration: C_4 photosynthesis and crassulacean acid metabolism (CAM). Finally, we discuss the two major products of the photosynthetic CO_2 fixation: **starch**, the reserve polysaccharide that accumulates transiently in chloroplasts; and **sucrose**, the disaccharide that is generated outside the chloroplast and is exported from leaves (*sources*) to developing tissues and storage organs (*sinks*) of the plant.

Calvin–Benson cycle The biochemical pathway for the reduction of CO_2 to carbohydrate. The cycle involves three phases: the carboxylation of ribulose 1,5-bisphosphate with atmospheric CO_2, catalyzed by Rubisco; the reduction of the formed 3-phosphoglycerate to triose phosphates; and the regeneration of ribulose 1,5-bisphosphate through the concerted action of ten enzymatic reactions.

starch A polyglucan consisting of long chains of 1,4-linked glucose molecules and branch points where 1,6-linkages are used. Starch is the carbohydrate storage form in most plants.

sucrose A disaccharide consisting of a glucose and a fructose molecule linked via an ether bond between C-1 on the glucosyl subunit and C-2 on the fructosyl unit. Sucrose is the carbohydrate transport form (e.g., in the phloem between source and sink).

8.1 The Calvin–Benson Cycle

Describe the three phases of Calvin–Benson cycle and the light-dependent mechanisms that influence the cycle.

Fixation of atmospheric CO_2 into organic compounds ultimately provides the substrates and energy needed for cellular metabolism, and thus life on earth.

These endergonic (energy-requiring) transformations are driven by the energy from the photochemical reactions (Chapter 7). The predominant pathway of autotrophic CO_2 fixation, the Calvin–Benson cycle, is found in many prokaryotes and in all photosynthetic eukaryotes, from algae to angiosperms. This pathway decreases the oxidation state of carbon, and the Calvin–Benson cycle is also aptly named the *reductive pentose phosphate cycle* and the *photosynthetic carbon reduction cycle*. How is CO_2 fixed via Rubisco and assimilated by the Calvin–Benson cycle using ATP and NADPH generated by the light reactions (Figure 8.1), and how is the cycle is regulated?

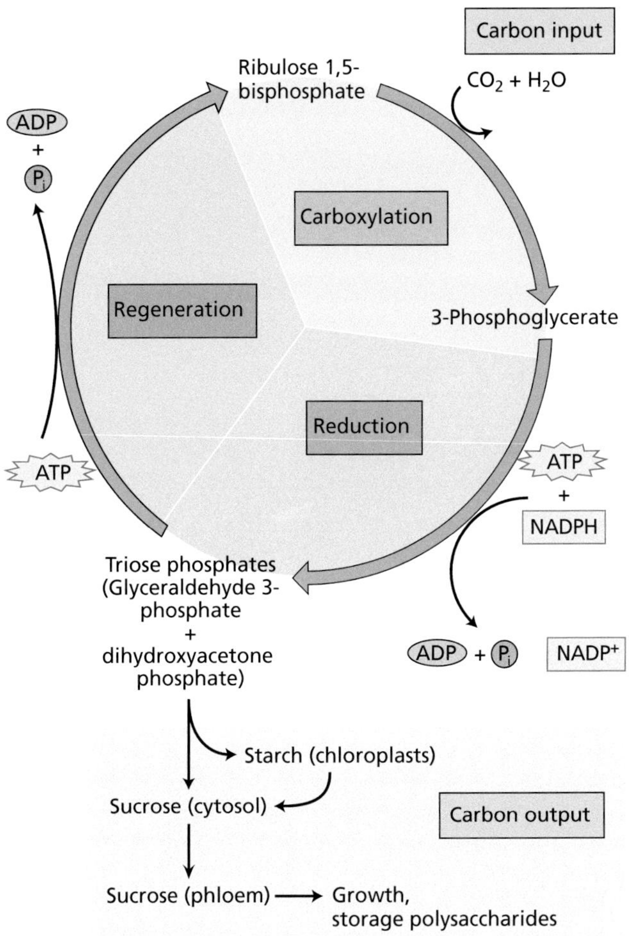

Figure 8.2 The Calvin–Benson cycle proceeds in three phases: (1) *Carboxylation*, which covalently links atmospheric carbon (CO_2) to a carbon skeleton; (2) *reduction*, which forms a carbohydrate (triose phosphate) at the expense of photochemically generated ATP and reducing equivalents in the form of NADPH; and (3) *regeneration*, which restores the CO_2 acceptor ribulose 1,5-bisphosphate. At steady state, the input of CO_2 equals the output of triose phosphates. The latter either serve as precursors of starch biosynthesis in the chloroplast or are exported to the cytosol for sucrose biosynthesis and other metabolic reactions. Sucrose is loaded into the phloem sap and used for growth or polysaccharide biosynthesis in other parts of the plant.

The Calvin–Benson cycle has three phases: carboxylation, reduction, and regeneration

In the 1950s, a series of ingenious experiments carried out by M. Calvin, A. Benson, J. A. Bassham, and their colleagues provided convincing evidence for the Calvin–Benson cycle. The Calvin–Benson cycle can be separated into three phases that are highly coordinated in the chloroplast (**Figures 8.2** and **8.3**):

1. *Carboxylation* of the CO_2 acceptor molecule. Rubisco catalyzes the reaction of CO_2 and water with a five-carbon acceptor molecule (ribulose 1,5-bisphosphate) to generate two molecules of a three-carbon Calvin–Benson cycle intermediate (3-phosphoglycerate).
2. *Reduction* of 3-phosphoglycerate. Two successive enzymatic reactions phosphorylate and reduce 3-phosphoglycerate to the triose phosphate glyceraldehyde 3-phosphate using photochemically generated ATP and NADPH.
3. *Regeneration* of the CO_2 acceptor ribulose 1,5-bisphosphate. The cycle is completed by regeneration of ribulose 1,5-bisphosphate through a series of ten enzyme-catalyzed reactions, consuming additional ATP.

The majority of the triose phosphate generated in the Calvin–Benson cycle is used to regenerate ribulose 1,5-bisphosphate. However, the input of atmospheric CO_2 generates additional triose phosphates that can be used by several other metabolic pathways, including for the generation of high-energy storage compounds such as starch in the chloroplast, or for export to the cytosol for the formation of sucrose or for respiration in cytosol and mitochondria (Chapter 11).

In summary, triose phosphates are formed in the carboxylation and reduction phases of the Calvin–Benson cycle using energy (ATP) and reducing equivalents (NADPH) generated by the illuminated photosystems of chloroplast thylakoid membranes:

$$3\ CO_2 + 3 \text{ ribulose 1,5-bisphosphate} + 3\ H_2O + 6\ \text{NADPH} + 6\ H^+ + 6\ \text{ATP}$$
$$\downarrow$$
$$6 \text{ Triose phosphates} + 6\ \text{NADP}^+ + 6\ \text{ADP} + 6\ P_i$$

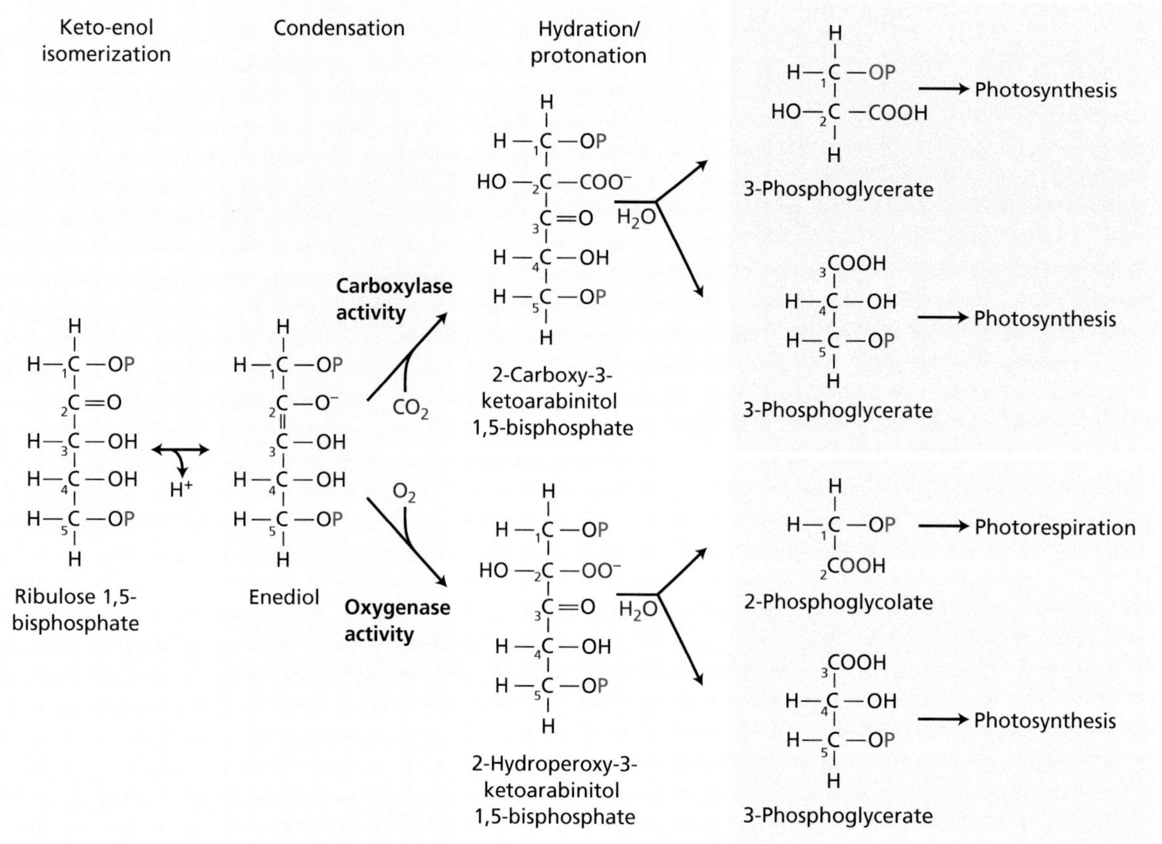

Figure 8.3 Carboxylation and oxygenation of ribulose 1,5-bisphosphate catalyzed by Rubisco. The binding of ribulose 1,5-bisphosphate to Rubisco facilitates the formation of an enzyme-bound enediol intermediate that can be attacked by CO_2 or O_2 at carbon 2. With CO_2, the product is a six-carbon intermediate (2-carboxy-3-ketoarabinitol 1,5-bisphosphate); with O_2, the product is a five-carbon reactive intermediate (2-hydroperoxy-3-ketoarabinitol 1,5-bisphosphate). The hydration of these intermediates at carbon 3 triggers the cleavage of the carbon–carbon bond between carbons 2 and 3, yielding two molecules of 3-phosphoglycerate (carboxylase activity) or one molecule of 2-phosphoglycolate and one molecule of 3-phosphoglyerate (oxygenase activity).

From these six triose phosphates, five are used in the regeneration phase that restores the CO_2 acceptor (ribulose 1,5-bisphosphate) for the continuous functioning of the Calvin–Benson cycle:

$$\text{5 Triose phosphates} + 3\ \text{ATP} + 2\ H_2O$$
$$\downarrow$$
$$3\ \text{ribulose 1,5-bisphosphate} + 3\ \text{ADP} + 2\ P_i$$

The sixth triose phosphate represents the net synthesis of an organic compound from CO_2 that is used as a building block for stored carbon or for other metabolic processes. Hence, the fixation of three CO_2 molecules generates one new triose phosphate (glyceraldehyde 3-phosphate) using 6 NADPH and 9 ATP:

$$3\ CO_2 + 5\ H_2O + 6\ \text{NADPH} + 9\ \text{ATP}$$
$$\downarrow$$
$$\text{Glyceraldehyde 3} - \text{phosphate} + 6\ \text{NADP}^+ + 9\ \text{ADP} + 8\ P_i$$

Stated another way, the Calvin–Benson cycle uses two molecules of NADPH and three molecules of ATP to assimilate each molecule of CO_2.

An induction period precedes the steady state of photosynthetic CO_2 assimilation

In the dark, the enzymes of the Calvin–Benson cycle and most of the triose phosphates are needed to restore adequate concentrations of metabolic intermediates for the light-dependent reactions when leaves receive light again. The rate of CO_2 fixation increases with time in the first few minutes after the onset of illumination—a time lag called the **induction period**. The acceleration of the photosynthetic rate is due to both the activation of enzymes by light (discussed later) and an increase in the concentration of intermediates of the Calvin–Benson cycle. In short, the six triose phosphates formed in the carboxylation and reduction phases of the Calvin–Benson cycle during the induction period are used mainly to build up the intermediates needed to regenerate the CO_2 acceptor ribulose 1,5-bisphosphate.

When photosynthesis reaches a steady state, five of the six triose phosphates formed continue to contribute to the regeneration of the CO_2 acceptor ribulose 1,5-bisphosphate, while the sixth triose phosphate can be used in the chloroplast for starch formation, in the cytosol for sucrose synthesis, and for other metabolic processes, such as respiration.

induction period The period of time (time lag) between the perception of a signal and the activation of the response. In the Calvin–Benson cycle, the time elapsed between the onset of illumination and the full activation of the cycle.

Many mechanisms regulate the Calvin–Benson cycle

The efficient use of energy in the Calvin–Benson cycle requires the existence of specific regulatory mechanisms that ensure not only that all intermediates in the cycle are present at adequate concentrations in the light but also that the cycle is turned off in the dark. To produce the necessary metabolites in response to environmental stimuli, chloroplasts achieve the appropriate rates of biochemical transformations through the modification of enzyme levels (μmoles of enzyme/chloroplast) and catalytic activities (moles of substrate converted/[minute × mole of enzyme]).

Gene expression and protein biosynthesis determine the concentrations of enzymes in cell compartments. The amounts of enzymes present in the chloroplast stroma are regulated by the coordinated expression of nuclear and plastid genomes. This coordination is controlled in part by pentatricopeptide repeat (PPR) proteins, which are nucleus-encoded chloroplast RNA-binding proteins that help regulate expression of plastid genes. Nucleus-encoded enzymes are translated on 80S ribosomes in the cytosol and subsequently transported into the plastid. Alternatively, plastid-encoded proteins are translated in the stroma on prokaryotic-like 70S ribosomes.

Light modulates the expression of stromal enzymes encoded by the nuclear genome via specific photoreceptors (e.g., phytochrome and blue-light receptors; Chapters 12 and 13). However, nuclear gene expression needs to be synchronized with the expression of other components of the photosynthetic apparatus in the organelle. Most of the regulatory signaling between nucleus and plastids is anterograde—that is, the products of nuclear genes control the transcription and translation of plastid genes. Such is the case, for example, in the assembly of stromal Rubisco from eight nucleus-encoded small subunits (S) and eight plastid-encoded large subunits (L). However, in some cases (e.g., the synthesis of proteins associated with chlorophyll), regulation can be retrograde—that is, the regulatory signal flows from the plastid to the nucleus.

In contrast to slow changes in catalytic rates caused by variations in the concentration of enzymes, posttranslational modifications rapidly change the specific activity of chloroplast enzymes (moles of substrate converted/[minute × mole of enzyme]). Two general mechanisms accomplish the light-mediated modification of the kinetic properties of stromal enzymes:

1. Change in covalent bonds that result in a chemically modified enzyme, such as the carbamylation of amino groups [Enz–NH_2 + CO_2 ↔ Enz–NH–CO_2^- + H^+], the reduction of disulfide bonds [Enz–$(S)_2$ + Prot–$(SH)_2$ ↔ Enz–$(SH)_2$ + Prot–$(S)_2$], and (de)phosphorylation by kinases and phosphatases
2. Modification of noncovalent interactions caused by changes in (i) ionic composition of the cellular milieu (e.g., pH, Mg^{2+}), (ii) binding of enzyme effectors, (iii) close association with regulatory proteins in supramolecular complexes, or (iv) interaction with thylakoid membranes

In our further discussion of regulation in this section, we examine light-dependent mechanisms that regulate the specific activity of five pivotal enzymes within minutes of the dark-to-light transition: Rubisco, fructose 1,6-bisphosphatase, sedoheptulose 1,7-bisphosphatase, phosphoribulokinase, and NADP–glyceraldehyde-3-phosphate dehydrogenase.

Rubisco activase regulates the catalytic activity of Rubisco

Because Rubisco has such a central role in carbon fixation, its activity is very tightly regulated. Rubisco that has been inactivated (e.g., at night) is reactivated when conditions become favorable for photosynthesis. First, the enzyme Rubisco activase removes bound sugar phosphates from Rubisco, and then Rubisco is activated by binding of a CO_2 molecule (activator CO_2). After this activation, the enzyme can start its catalytic cycle, whereby CO_2 reacts with ribulose 1,5-bisphosphate. The activity of Rubisco activase is, in turn, regulated by the ferredoxin–thioredoxin system.

Light regulates the Calvin–Benson cycle via the ferredoxin–thioredoxin system

Light regulates the catalytic activity of four enzymes of the Calvin–Benson cycle directly via the **ferredoxin–thioredoxin system**. This mechanism uses ferredoxin reduced by the photosynthetic electron transport chain together with two chloroplast proteins (ferredoxin–thioredoxin reductase and thioredoxin) to regulate fructose 1,6-bisphosphatase, sedoheptulose 1,7-bisphosphatase, phosphoribulokinase, and NADP–glyceraldehyde-3-phosphate dehydrogenase (**Figure 8.4**).

Light transfers electrons from water to ferredoxin via the photosynthetic electron transport system (Chapter 7). Reduced ferredoxin converts the disulfide bond of the regulatory protein thioredoxin (—S—S—) to the reduced state (—SH HS—) with the iron–sulfur enzyme ferredoxin–thioredoxin reductase. Subsequently, reduced thioredoxin cleaves a specific disulfide bridge (oxidized cysteines) of the target enzyme, forming free (reduced) cysteines. The cleavage of the target enzyme disulfide bonds causes a conformational change that increases catalytic activity (Figure 8.4). Deactivation of thioredoxin-activated enzymes takes place when darkness relieves the "electron pressure" from photosynthetic electron transport, but the precise details of the deactivation process are unknown.

Proteomic studies have shown that the ferredoxin–thioredoxin system regulates enzymes functional in numerous chloroplast processes other than carbon fixation. Additionally, thioredoxin protects proteins against damage caused by reactive oxygen species, such as hydrogen peroxide (H_2O_2), the superoxide anion ($O_2^{\bullet -}$), and the hydroxyl radical ($^{\bullet}OH$).

ferredoxin–thioredoxin system Three chloroplast proteins (ferredoxin, ferredoxin–thioredoxin reductase, thioredoxin). The three proteins use reducing power from the photosynthetic electron transport chain to reduce protein disulfide bonds in a cascade of thiol–disulfide exchanges. As a result, light controls the activity of several enzymes of the Calvin–Benson cycle.

Light-dependent ion movements modulate enzymes of the Calvin–Benson cycle

Upon illumination, the electron transport chain leads to the flow of protons from the stroma into the thylakoid lumen and generates the proton motive force, which leads to the release of Mg^{2+} from the thylakoid lumen to the stroma. These

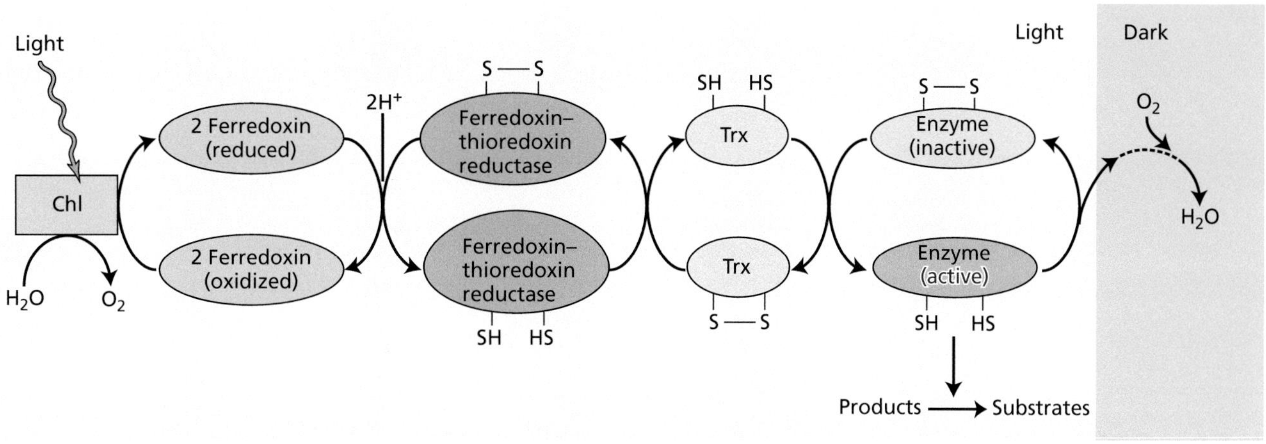

Figure 8.4 Ferredoxin–thioredoxin system. The ferredoxin–thioredoxin system links the redox status of the thylakoid membranes to the activity of enzymes in the chloroplast stroma. The activation of enzymes of the Calvin–Benson cycle starts in the light with the reduction of ferredoxin by the photosynthetic electron transport chain (Chl) (Chapter 7). Reduced ferredoxin, together with two protons, is used to reduce a catalytically active disulfide bond (—S—S—) of the iron–sulfur enzyme ferredoxin–thioredoxin reductase, which in turn reduces the unique disulfide bond (—S—S—) of the regulatory protein thioredoxin (Trx). The reduced form of thioredoxin (—SH HS—) then reduces regulatory disulfide bonds of the target enzyme, triggering its conversion to the catalytically active state that catalyzes the transformation of substrates into products. Darkness halts the electron flow from ferredoxin to the enzyme, and thioredoxin becomes oxidized through an unknown mechanism involving molecular oxygen. Next, the unique disulfide bond (—S—S—) of thioredoxin brings the reduced form (—SH HS—) of the enzyme back to the oxidized form (—S—S—) with the concurrent loss of the catalytic capacity. In contrast to thioredoxin-activated enzymes, a chloroplast enzyme of the oxidative pentose phosphate cycle, glucose 6-phosphate dehydrogenase, does not operate in the light but is functional in the dark because thioredoxin reduces the disulfide critical for the activity of the enzyme.

light-actuated ion fluxes decrease the stromal concentration of protons (the pH increases from 7 to 8) and increase Mg^{2+} from 2 to 5 m*M*. The light-mediated increase of pH and the concentration of Mg^{2+} activate enzymes of the Calvin–Benson cycle that require Mg^{2+} for catalysis and are more active at pH 8 than at pH 7: Rubisco, fructose 1,6-bisphosphatase, sedoheptulose 1,7-bisphosphatase, and phosphoribulokinase. Modifications of the ionic composition within the chloroplast stroma are rapidly reversed upon darkening.

8.2 The Oxygenation Reaction of Rubisco and Photorespiration

Define the process of photorespiration and distinguish among the cellular compartments where it occurs.

Rubisco catalyzes both the carboxylation and the oxygenation of ribulose 1,5-bisphosphate (Figure 8.3). Carboxylation yields two molecules of 3-phosphoglycerate, and oxygenation produces one molecule each of 3-phosphoglycerate and 2-phosphoglycolate. 2-Phosphoglycolate is an inhibitor of two chloroplast enzymes (triose phosphate isomerase and phosphofructokinase) and cannot be directly processed by the Calvin–Benson cycle. To avoid both the drain of carbon out of the Calvin–Benson cycle and enzyme inhibition, 2-phosphoglycolate is metabolized through the oxidative photosynthetic carbon cycle. This network of coordinated enzymatic reactions, also known as **photorespiration**, occurs in chloroplasts, leaf peroxisomes, and mitochondria (**Figure 8.5**).

Recent studies have shown that photorespiration is an essential component of photosynthesis that not only salvages part of the assimilated carbon but also links to other important pathways (e.g., amino acid biosynthesis, nitrogen assimilation, 1-carbon metabolism) of contemporary land plants.

photorespiration Uptake of atmospheric O_2 with a concomitant release of CO_2 by illuminated leaves. Molecular oxygen serves as substrate for Rubisco, producing 2-phosphoglycolate that enters the photorespiratory cycle (the oxidative photosynthetic carbon cycle). The activity of the cycle recovers some of the carbon found in 2-phosphoglycolate, but some is lost to the atmosphere.

(A)

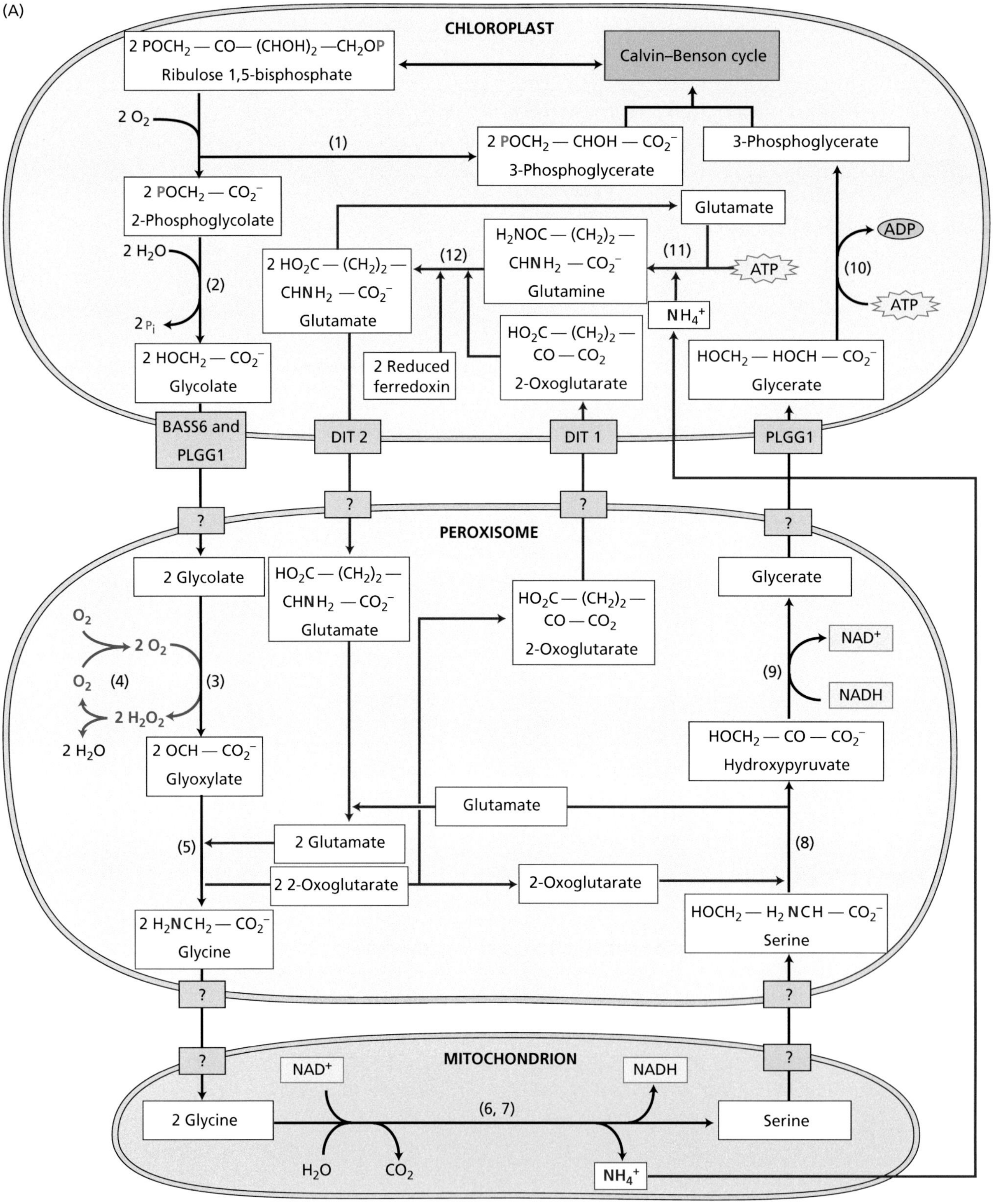
CHLOROPLAST
2 POCH2 — CO— (CHOH)2 —CH2OP
Ribulose 1,5-bisphosphate
Calvin–Benson cycle
2 O2
(1)
2 POCH2 — CHOH — CO2−
3-Phosphoglycerate
3-Phosphoglycerate
2 POCH2 — CO2−
2-Phosphoglycolate
Glutamate
2 H2O
(2)
2 Pi
H2NOC — (CH2)2 —
CHNH2 — CO2−
Glutamine
(12)
(11)
ATP
ADP
(10)
ATP
2 HO2C — (CH2)2 —
CHNH2 — CO2−
Glutamate
NH4+
HO2C — (CH2)2 —
CO — CO2
2-Oxoglutarate
2 HOCH2 — CO2−
Glycolate
2 Reduced
ferredoxin
HOCH2 — HOCH — CO2−
Glycerate
BASS6 and
PLGG1
DIT 2
DIT 1
PLGG1
?
PEROXISOME
2 Glycolate
HO2C — (CH2)2 —
CHNH2 — CO2−
Glutamate
HO2C — (CH2)2 —
CO — CO2
2-Oxoglutarate
Glycerate
O2
2 O2
O2
(4)
(3)
2 H2O2
2 H2O
NAD+
(9)
NADH
2 OCH — CO2−
Glyoxylate
HOCH2 — CO — CO2−
Hydroxypyruvate
Glutamate
(5)
2 Glutamate
(8)
2 2-Oxoglutarate
2-Oxoglutarate
2 H2NCH2 — CO2−
Glycine
HOCH2 — H2NCH — CO2−
Serine
MITOCHONDRION
NAD+
NADH
(6, 7)
2 Glycine
Serine
H2O
CO2
NH4+

(B)

Reactions of photorespiration

Reaction number (see Part A) and enzyme name	**Reaction**[a]
1. Rubisco	2 Ribulose 1,5-bisphosphate + 2 O_2 → 2 2-phosphoglycolate + 2 3-phosphoglycerate
2. Phosphoglycolate phosphatase	2 2-Phosphoglycolate + 2 H_2O → 2 glycolate + 2 P_i
3. Glycolate oxidase	2 Glycolate + 2 O_2 → 2 glyoxylate + 2 H_2O_2
4. Catalase	2 H_2O_2 → 2 H_2O + O_2
5. Glutamate:glyoxylate aminotransferase	2 Glyoxylate + 2 glutamate → 2 glycine + 2 2-oxoglutarate
6. Glycine decarboxylase complex (GDC)	Glycine + NAD^+ + [GDC] → CO_2 + NH_4^+ + NADH + [GDC-THF-CH_2]
7. Serine hydroxymethyl transferase	[GDC-THF-CH_2] + glycine + H_2O → serine + [GDC]
8. Serine:2-oxoglutarate aminotransferase	Serine + 2-oxoglutarate → hydroxypyruvate + glutamate
9. Hydroxypyruvate reductase	Hydroxypyruvate + NADH + H^+ → glycerate + NAD^+
10. Glycerate kinase	Glycerate + ATP → 3-phosphoglycerate + ADP
11. Glutamine synthetase	Glutamate + NH_4^+ + ATP → glutamine + ADP + Pi
12. Ferredoxin-dependent glutamate synthase (GOGAT)	2-Oxoglutarate + glutamine + 2 Fd_{red} + 2 H^+ → 2 glutamate + 2 Fd_{oxid}

[a]Locations: Chloroplasts; peroxisomes; mitochondria. Fd: ferredoxin.

(C) Net reaction of photorespiration

2 Ribulose 1,5-bisphosphate + 3 O_2 + H_2O + glutamate
↓ **(reactions 1 to 9)**
Glycerate + 2 3-phosphoglycerate + NH_4^+ + CO_2 + 2 P_i + 2-oxoglutarate

Two reactions in the chloroplasts restore the molecule of glutamate:

2-Oxoglutarate + NH_4^+ + [(2 Fd_{red} + 2 H^+), ATP]
↓ **(reactions 11 and 12)**
Glutamate + H_2O + [(2 Fd_{oxid}), ADP + P_i]

and the molecule of 3-phosphoglycerate:

Glycerate + ATP
↓ **(reaction 10)**
3-Phosphoglycerate + ADP

Hence, the consumption of three molecules of atmospheric oxygen in the C_2 oxidative photosynthetic carbon cycle (two in the oxygenase activity of Rubisco and one in peroxisomal oxidations) elicits

- the release of one molecule of CO_2 and
- the consumption of two molecules of ATP and two molecules of reducing equivalents (2 Fd_{red} + 2 H^+)

for

- incorporating a three-carbon skeleton back into the Calvin–Benson cycle, and
- restoring glutamate from NH_4^+ and 2-oxoglutarate.

Figure 8.5 Operation of photorespiration. (A) Enzymatic reactions are distributed in three organelles: chloroplasts, peroxisomes, and mitochondria. In *chloroplasts*, the oxygenase activity of Rubisco reacting with two molecules of oxygen yields two molecules of 2-phosphoglycolate, which are converted into two molecules of glycolate. The glycolates are transported from the chloroplast and imported into the peroxisome where they are oxidized by O_2 to glyoxylate. The glyoxylate is converted into glycine, which moves from peroxisomes and into the mitochondria. In *mitochondria*, two molecules of glycine (four carbons) yield a molecule of serine (three carbons) with the concurrent release of CO_2 (one carbon) and NH_4^+. The amino acid serine is then transported back to the peroxisome and transformed to glycerate (three carbons), which return to chloroplasts in a process that recovers part of the carbon (three carbons) and all the nitrogen released by the GDC cycle. For details, see the text. The violet boxes at the membranes are transporters. BASS6, the bile acid sodium symporter; DIT 1, 2-oxoglutarate/malate translocator; DIT 2, glutamate/malate translocator; PLGG1, the plastidial glycolate/glycerate transporter; both in the inner envelope membrane of the chloroplast. (B) A description of the numbered reactions of photorespiration shown in (A). (C) Net reaction of the photorespiratory cycle.

The oxygenation of ribulose 1,5-bisphosphate sets in motion photorespiration

In evolutionary terms, Rubisco appears to have evolved from an ancient enolase in the archaeal methionine-salvage pathway. Billions of years ago, oxygenation of ribulose 1,5-bisphosphate was insignificant in non-oxygenic prokaryotes because of the lack of O_2 and high levels of CO_2 in the ancient atmosphere. The high concentrations of O_2 and low levels of CO_2 in the extant atmosphere boost the oxygenase activity of Rubisco, making inevitable the formation of the toxic 2-phosphoglycolate. All Rubiscos catalyze the incorporation of O_2 into ribulose 1,5-bisphosphate. Even homologs from anaerobic autotrophic bacteria exhibit the oxygenase activity, demonstrating that the oxygenase reaction is intrinsically linked to the active site of Rubisco and not an adaptive response to the appearance of O_2 in the biosphere.

Photorespiration is shown in Figure 8.5. The oxygenation of the 2,3-enediol isomer of ribulose 1,5-bisphosphate with one molecule of O_2 yields an unstable intermediate that rapidly splits into one molecule each of 3-phosphoglycerate and 2-phosphoglycolate (reaction 1). In chloroplasts of land plants, 2-phosphoglycolate phosphatase catalyzes the rapid hydrolysis of 2-phosphoglycolate to glycolate (reaction 2). The subsequent transformations of glycolate take place in the peroxisomes and mitochondria. Glycolate leaves the chloroplasts through specific transporters in the envelope inner membrane and diffuses to the peroxisomes (Figure 8.5A). In peroxisomes, glycolate oxidase catalyzes the oxidation of glycolate by O_2, producing H_2O_2 and glyoxylate (reaction 3). The peroxisomal catalase breaks down the H_2O_2, releasing O_2 and H_2O (reaction 4). The glutamate:glyoxylate aminotransferase catalyzes the transamination of glyoxylate with glutamate, yielding the amino acid glycine (reaction 5).

Glycine exits the peroxisomes and enters the mitochondria, where a multienzyme complex of glycine decarboxylase (GDC) and serine hydroxymethyltransferase (SHMT) catalyzes the conversion of two molecules of glycine and one molecule of NAD^+ into one molecule each of serine, NADH, NH_4^+, and CO_2 (reactions 6 and 7). This sequence of reactions is initiated by GDC, using one molecule of NAD^+ for the oxidative decarboxylation of one molecule of glycine, yielding one molecule each of NADH, NH_4^+, and CO_2 and the activated one-carbon unit methylene tetrahydrofolate (THF) bound to GDC (GDC-THF-CH_2):

$$\text{Glycine} + \text{NAD}^+ + \text{GDC-THF} \rightarrow \text{NADH} + \text{NH}_4^+ + \text{CO}_2 + \text{GDC-THF-CH}_2$$

Next, SHMT catalyzes the addition of the methylene unit to a second molecule of glycine, forming serine and regenerating GDC-THF:

$$\text{Glycine} + \text{GDC-THF-CH}_2 \rightarrow \text{Serine} + \text{GDC-THF}$$

The oxidation of carbon atoms (two molecules of glycine [oxidation states C1: +3; C2: –1] → serine [oxidation states C1: +3; C2: 0; C3: –1] and CO_2 [oxidation state C: +4]) drives the reduction of oxidized pyridine nucleotide:

$$\text{NAD}^+ + \text{H}^+ + 2\,\text{e}^- \rightarrow \text{NADH}$$

The products of the GDC and SHMT reactions are metabolized at different locations in leaf cells. NADH is oxidized to NAD^+ in the mitochondria. NH_4^+ and CO_2 diffuse out of the mitochondria and can be refixed in the chloroplasts, where they are assimilated to form glutamate and 3-phosphoglycerate, respectively.

The newly formed serine diffuses from the mitochondria back to the peroxisomes for donation of its amino group to 2-oxoglutarate via transamination catalyzed by serine:2-oxoglutarate aminotransferase, forming glutamate and hydroxypyruvate (reaction 8). Next, an NADH-dependent reductase catalyzes the transformation of hydroxypyruvate into glycerate (reaction 9). Finally, glycerate reenters the chloroplast, where it is phosphorylated by glycerate kinase using ATP to yield 3-phosphoglycerate and ADP (reaction 10). Hence, the formation of 2-phosphoglycolate (via Rubisco) and the phosphorylation of glycerate (via glycerate kinase) metabolically link the Calvin–Benson cycle to photorespiration.

The NH_4^+ released in the oxidation of glycine diffuses from the matrix of the mitochondria to the chloroplasts (Figure 8.5A). In the chloroplast stroma, glutamine synthetase catalyzes the ATP-dependent incorporation of NH_4^+ into glutamate, yielding glutamine, ADP, and inorganic phosphate (reaction 11). Subsequently, glutamine and 2-oxoglutarate are substrates of ferredoxin-dependent glutamate synthase (GOGAT) for the production of two molecules of glutamate (reaction 12). The reassimilation of NH_4^+ released from the photorespiratory cycle restores glutamate for the action of the peroxisomal glutamate:glyoxylate aminotransferase in the conversion of glyoxylate to glycine (reaction 5).

Carbon, nitrogen, and oxygen atoms circulate through photorespiration (**Figure 8.6**).

- In the *carbon* cycle, the chloroplasts transfer two molecules of glycolate (four carbon atoms) to peroxisomes and recover one molecule of glycerate (three carbon atoms). The mitochondria release one molecule of CO_2 (one carbon atom).
- In the *nitrogen* cycle, the chloroplasts transfer one molecule of glutamate (one nitrogen atom) and recover one molecule of NH_4^+ (one nitrogen atom).

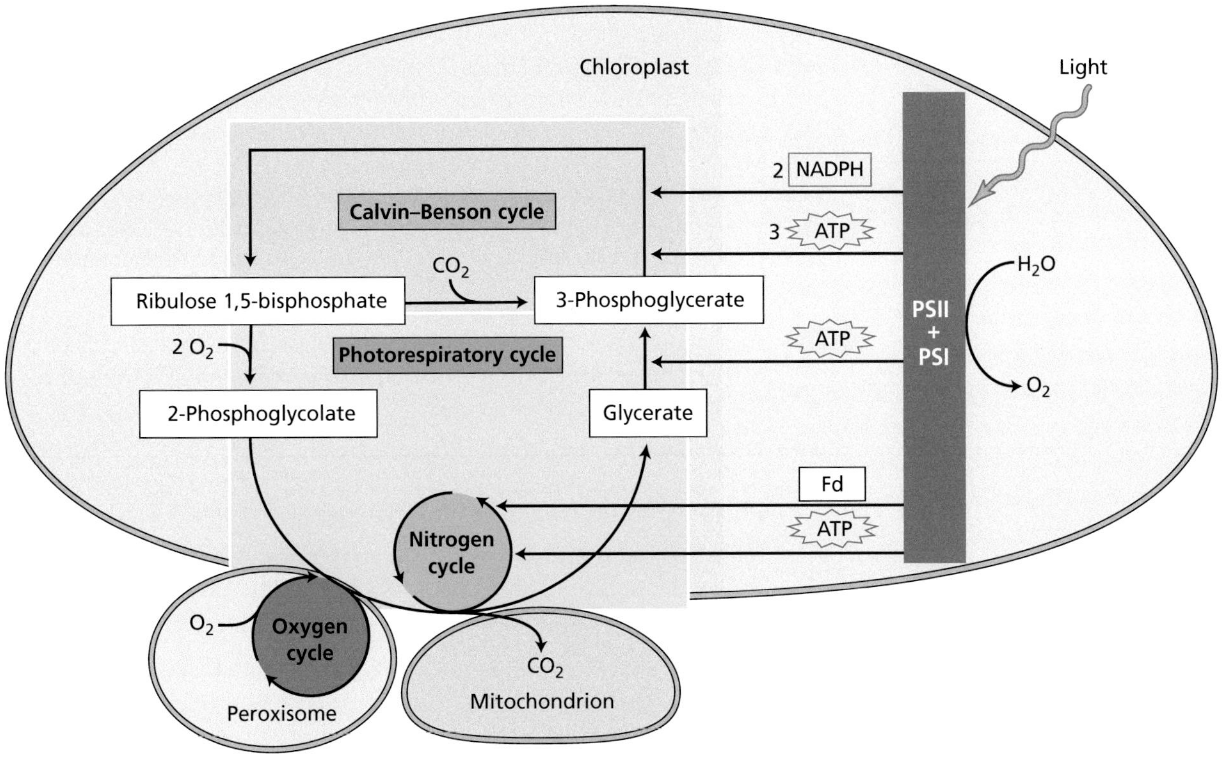

Figure 8.6 Dependence of photorespiration on chloroplast metabolism. The supply of ATP and reducing equivalents from light reactions in thylakoid membranes is needed for the functioning of photorespiration in three compartments: chloroplasts, peroxisomes, and mitochondria. The *carbon cycle* uses (1) NADPH and ATP to maintain the adequate level of ribulose 1,5-bisphosphate in the Calvin–Benson cycle and (2) ATP to convert glycerate to 3-phosphoglycerate in photorespiration. The *nitrogen cycle* employs ATP and reducing equivalents to restore glutamate from NH_4^+ and 2-oxoglutarate coming from the photorespiratory cycle. In the peroxisome, the *oxygen cycle* contributes to the removal of H_2O_2 formed in the oxidation of glycolate by O_2.

- In the *oxygen* cycle, Rubisco and glycolate oxidase catalyze the incorporation of two molecules of O_2 each (eight oxygen atoms) when two molecules of ribulose 1,5-bisphosphate enter photorespiration (Figure 8.5, reactions 1 and 3). However, catalase releases one molecule of O_2 from two molecules of H_2O_2 (two oxygen atoms) (Figure 8.5, reaction 4). Hence, three molecules of O_2 (six oxygen atoms) are reduced in the photorespiratory cycle.
- In vivo, three aspects regulate the distribution of metabolites between the Calvin–Benson cycle and photorespiration: (i) the kinetic properties of Rubisco (inherent to the plant), (ii) the availability of atmospheric CO_2 and O_2, and (iii) the temperature.

The specificity factor ($S_{c/o}$) estimates the preference of Rubisco for CO_2 versus O_2:

$$S_{c/o} = [V_C/K_C]/[V_O/K_O] \tag{8.1}$$

where V_C and V_O are the maximum velocities for carboxylation and oxygenation, respectively, and K_C and K_O are the Michaelis–Menten constants for CO_2 and O_2, respectively. $S_{c/o}$ determines the ratio of the velocity of carboxylation (v_C) to that of oxygenation (v_O) at given concentrations of CO_2 and O_2:

$$S_{c/o} = v_C / v_O \times [(O_2)/(CO_2)] \tag{8.2}$$

When the concentration of CO_2 around the active site is equal to that of O_2 [$(O_2)/(CO_2) = 1$], $S_{c/o}$ estimates the relative capacity of Rubisco for carboxylation and oxygenation (v_C/v_O). $S_{c/o}$ at a given temperature is generally constant but tends to decrease as temperature increases. Additionally, $S_{c/o}$ for Rubisco from different organisms varies: The $S_{c/o}$ of Rubisco from cyanobacteria (~40) is lower than that from C_3 plants (~82–90) and from C_4 species (~70–82) (C_3 and C_4 photosynthesis are discussed below). There is generally a trade-off in the affinity of Rubisco for CO_2 and the catalytic speed of the carboxylation reaction such that enzymes that evolved under high-CO_2 conditions tend to have a low $S_{c/o}$ and those that evolved under low-CO_2 availability tend to have a high $S_{c/o}$.

The ambient temperature exerts an important influence over $S_{c/o}$ and the concentrations of O_2/CO_2 around the active site of Rubisco. Warmer environments have the effect of:

- Increasing the oxygenase activity of Rubisco more than the carboxylase activity, suggesting that a greater increase of K_C for CO_2 than of K_O for O_2 diminishes $S_{c/o}$.
- Lowering the solubility of CO_2 to a greater extent than O_2. The increase of $[O_2]/[CO_2]$ decreases the v_C/v_O ratio; that is, the oxygenase activity of Rubisco prevails over the carboxylase activity.
- Reducing the stomatal aperture to conserve water. The stomatal closure lowers the uptake of atmospheric CO_2, thereby decreasing CO_2 at the active site of Rubisco (Chapter 3).

Overall, warmer environments significantly limit the efficiency of photosynthetic carbon assimilation because the progressive increase in temperature tilts the balance away from carboxylation (photosynthesis) and toward oxygenation (photorespiration) (Chapter 9).

Photorespiration is linked to the photosynthetic electron transport system

Photosynthetic carbon metabolism in intact leaves reflects the competition for Rubisco between two mutually opposing cycles, the Calvin–Benson cycle and photorespiration. These cycles are interlocked with the photosynthetic electron transport system for the supply of ATP and reducing equivalents (reduced ferredoxin and NADPH) (Figure 8.6). For salvaging two molecules of 2-phosphoglycolate by conversion to one molecule of 3-phosphoglycerate, photophosphorylation provides one molecule of ATP necessary for the transformation of glycerate to 3-phosphoglycerate (Figure 8.5, reaction 10), while the consumption of NADH by hydroxypyruvate reductase (Figure 8.5, reaction 9) is counterbalanced by its production by glycine decarboxylase (Figure 8.5, reaction 6).

In photorespiration, nitrogen:

- *Enters* into the peroxisome through the transamination step catalyzed by the glutamate:glyoxylate aminotransferase (two nitrogen atoms) (Figure 8.5, reaction 5), and
- *Leaves* (1) from the mitochondria as NH_4^+ (one nitrogen atom), in the reaction catalyzed by the glycine decarboxylase–serine hydroxymethyltransferase complex (Figure 8.5, reactions 6 and 7) (2) from the peroxisomes in the transamination step catalyzed by serine:2-oxoglutarate aminotransferase (one nitrogen atom) (Figure 8.5, reaction 8).

The photosynthetic electron transport system supplies one molecule of ATP and two molecules of reduced ferredoxin needed for salvaging one molecule of NH_4^+ through its incorporation into glutamate via glutamine synthetase (Figure 8.5, reaction 11) and ferredoxin-dependent glutamate synthase (GOGAT) (Figure 8.5, reaction 12).

In summary, each oxygenation reaction produces one molecule each of 3-phosphoglycerate and 2-phosphoglycolate such that the net reaction of two oxygenation reactions is:

$$\text{2 Ribulose 1,5-bisphosphate} + 3\,O_2 + H_2O + \text{ATP} + [\text{2 ferredoxin}_{red} + 2\,H^+ + \text{ATP}]$$
$$\downarrow$$
$$\text{3 3-Phosphoglycerate} + CO_2 + 2\,P_i + \text{ADP} + [\text{2 ferredoxin}_{oxid} + \text{ADP} + P_i]$$

Because of the additional provision of ATP and reducing power for the operation of the photorespiratory cycle, the quantum requirement for CO_2 fixation under photorespiratory conditions (high $[O_2]$ and low $[CO_2]$) is higher than under nonphotorespiratory conditions (low $[O_2]$ and high $[CO_2]$). **Box 8.1** discusses genetically engineered photorespiratory bypasses in land plants. The quantum requirement for CO_2 fixation is further discussed in Chapter 9.

Box 8.1 Production of Biomass May Be Enhanced by Engineering Photorespiration

Solutions to the current food and energy shortages depend on the degree to which land plants can be altered for higher CO_2 assimilation. When O_2 outcompetes CO_2, the oxygenase activity of Rubisco lowers the amount of carbon that enters the Calvin–Benson cycle. Therefore, to understand how to engineer leaf cells for improving the photosynthetic efficiency, scientists are tackling various aspects of the oxygenation reaction of Rubisco and photorespiration, from modifying the active site of Rubisco to genetically engineering bypasses of the traditional photorespiratory pathways. The engineering of catalytically more efficient Rubiscos in plants has been challenging, but future progress is likely with the rapid development of new genetic tools and a better understanding of Rubisco kinetic constraints. An attractive alternative possibility is to incorporate more efficient mechanisms for retrieving the carbon atoms of the 2-phosphoglycolate that is generated by the reaction of Rubisco with O_2 and the wasteful release of ammonia by the GDC in the mitochondria.

There have been several engineering approaches to lower the flux of photorespiratory metabolites through peroxisomes and mitochondria, releasing photorespired CO_2 in the chloroplast where it can be directly refixed. One of the first approaches was to introduce a bacterial (*E. coli*) glycolate catabolic pathway into chloroplasts of a land plant (Arabidopsis). Chloroplasts of these transgenic plants have an entirely functional photorespiratory cycle formed from the bacterial enzymes glycolate dehydrogenase, tartronic semialdehyde synthase, and tartronic semialdehyde reductase. The engineered plants grow faster, have increased biomass, and contain higher levels of soluble sugars under short-day growth conditions.

(*Continued*)

Box 8.1 *(continued)*

A second approach used the overexpression of three enzymes in the chloroplast stroma of Arabidopsis—glycolate oxidase, catalase, and malate synthase—to release CO_2 from glycolate. First, the oxidation of glycolate by the novel chloroplast glycolate oxidase yields glyoxylate and H_2O_2, and catalase catalyzes the subsequent decomposition of H_2O_2 [2 glycolate + 2 O_2 → 2 glyoxylate + 2 H_2O_2; 2 H_2O_2 → 2 H_2O + O_2]. Next, the successive action of two enzymes converts two molecules of glyoxylate (two carbon atoms) into pyruvate (three carbon atoms) and CO_2 (one carbon atom):

- Malate synthase catalyzes the condensation of glyoxylate with acetyl-CoA [CoA-S~CO-CH_3], yielding malate [glyoxylate + CoA-S~CO-CH_3 → malate + CoA-SH].
- The chloroplastic NADP–malic enzyme catalyzes the decarboxylation of malate to pyruvate with the concurrent formation of NADPH [malate + $NADP^+$ → pyruvate + CO_2 + NADPH + H^+].

Finally, chloroplastic pyruvate dehydrogenase catalyzes the conversion of pyruvate into acetyl-CoA, yielding NADH and another molecule of CO_2 [pyruvate + CoA-SH + NAD^+ → CoA-S~CO-CH_3 + CO_2 + NADH + H^+]. As a result of this alternative cycle, one molecule of glycolate (two carbon atoms) is converted to two molecules of CO_2 (two carbon atoms). The oxidation of carbon atoms yields reducing power in the form of NADPH and NADH. These Arabidopsis plants also had increased biomass production compared with control plants under short-day growth conditions.

These novel pathways depart from plant photorespiration in sidestepping the mitochondrial and peroxisomal reactions. Therefore, the shift of glycolate from plant photorespiration to the engineered pathways releases CO_2 in the immediate vicinity of Rubisco, allowing rapid refixation of the CO_2, and at the same time avoids the use of energy (ATP and reductant) required to recover NH_4^+.

Recently, similar photorespiratory bypasses have been combined with repression of the chloroplast glycolate exporter (PLGG1) (Figure 8.5), which leads to further yield increases in field-grown tobacco. There are still questions regarding the mechanisms that confer the yield benefits of these photorespiratory bypasses, but these initial studies are encouraging and there are continued efforts in this area of research (see https://ripe.illinois.edu).

Enzymes of plant photorespiration derive from different ancestors

The complete genomes of different organisms have demonstrated that all photorespiratory enzymes are present in plants and in red and green algae. Moreover, these phylogenetic studies suggest that the distribution of enzymes in plants correlates with the origin of compartments involved in photorespiration. Chloroplast enzymes (e.g., glycerate kinase) evolved from a cyanobacterial endosymbiont, while mitochondrial enzymes (e.g., glycine decarboxylase) have a proteobacterial ancestor.

Photorespiration interacts with many metabolic pathways

Early research suggested that photorespiration served to salvage carbon diverted by the oxygenase activity of Rubisco and to protect plants from stressful conditions such as high light, drought (water deficit stress), and salt stress. The negative impact of lacking photorespiration on photosynthetic CO_2 assimilation was first demonstrated by plant mutants that do not survive in air (21% O_2; 0.04% CO_2) but resume their normal growth in high-CO_2 environments (2% CO_2). This *photorespiratory phenotype* served to identify previously unknown enzymes and reactions of photorespiration. For example, Arabidopsis mutants lacking glycerate kinase accumulate glycerate and are incapable of growing in normal air but are viable in atmospheres with elevated levels of CO_2.

Studies have also revealed a tight connection between photorespiration and other pathways of plant metabolism. For example, key steps of photorespiration interact with:

- *Nitrogen metabolism at multiple levels*: Photorespiration reassimilates NH_4^+ formed in the mitochondria, uses glutamate in peroxisomal transaminations, and produces amino acids (serine, glycine) for other metabolic pathways.
- *Cell redox homeostasis*: H_2O_2 formed by the peroxisomal glycolate oxidase regulates the redox state of leaves. The formation of H_2O_2 induces

suicide programs in catalase-deficient barley plants that exhibit the photorespiratory phenotype. Although H_2O_2 damages key cellular molecules such as DNA and lipids, the current view recognizes this reactive oxygen species as a signaling molecule linked to hormone and stress responses.

- *C_1 metabolism*: 5,10-methylene tetrahydrofolate is the cofactor required by glycine decarboxylase–serine hydroxymethyltransferase in the conversion of glycine to serine in the mitochondria. Reactions mediated by folates transfer one-carbon units in the synthesis of precursors for proteins, nucleic acids, lignin, and alkaloids.
- *Expression of transcription factors*: More than 200 transcription factors are differentially expressed when plants are transferred from atmospheres with high levels of CO_2 to normal air. Photorespiratory conditions (high [O_2], low [CO_2]) enhance the expression of genes encoding the components of cyclic electron flow pathways, consistent with the additional demand for energy by the photorespiratory pathway. Photorespiratory conditions decrease the expression of genes encoding proteins involved in synthesis of starch and sucrose and in the metabolism of nitrogen and sulfur.

8.3 Inorganic Carbon-Concentrating Mechanisms

Define the carbon-concentrating mechanisms that have evolved in land plants as an adaptation to lower atmospheric CO_2 concentrations.

Except for some photosynthetic bacteria, all other photo-autotrophic organisms in the biosphere use the Calvin–Benson cycle to assimilate atmospheric CO_2. The pronounced reduction in atmospheric CO_2 and rise in O_2 levels that commenced about 350 million years ago triggered a series of adaptations to handle an environment that promoted photorespiration in photosynthetic organisms. These adaptations include various strategies for active uptake of CO_2 and HCO_3^- from the surrounding environment and accumulation of inorganic carbon near Rubisco. The immediate consequence of higher levels of CO_2 around Rubisco is a decrease in the oxygenation reaction. Plasma membrane CO_2 and HCO_3^- pumps have been studied extensively in prokaryotic cyanobacteria, eukaryotic algae, and aquatic plants but have not been identified in terrestrial plants.

In land plants, the diffusion of CO_2 from the atmosphere to the chloroplast plays a crucial role in net photosynthesis. During C_3 photosynthesis, once CO_2 is within the leaf it has to cross four barriers (the cell wall, plasma membrane, cytosol, and chloroplast envelope) before reaching Rubisco; C_3 plants only fix CO_2 into three-carbon acids (3-phosphogylycerate). Recent evidence has revealed that pore-forming membrane proteins (aquaporins) and leaf anatomy probably both influence the diffusion of CO_2 within the leaf to the initial site of carboxylation. However, the combined effects of relatively low atmospheric CO_2 availability (e.g., low stomatal conductance because of drought) and resistance to CO_2 movement within the leaf can limit rates of C_3 photosynthesis and decrease leaf-level water use efficiency (Chapter 9).

Land plants evolved two carbon-concentrating mechanisms for increasing the concentration of CO_2 at the site of Rubisco carboxylation: C_4 photosynthetic carbon fixation (C_4) and crassulacean acid metabolism (CAM). The uptake of atmospheric CO_2 by these carbon-concentrating mechanisms precedes CO_2 assimilation through the Calvin–Benson cycle.

C_4 photosynthesis Photosynthetic carbon metabolism in which the initial fixation of CO_2 is catalyzed by phospho-*enol*pyruvate carboxylase (not by Rubisco as in C_3 photosynthesis), producing a four-carbon compound (oxaloacetate). The fixed carbon is subsequently released and refixed by the Calvin–Benson cycle.

8.4 Inorganic Carbon-Concentrating Mechanisms: C_4 Photosynthetic Carbon Fixation

Describe the C_4 photosynthetic carbon cycle and the associated Kranz anatomy.

C_4 photosynthesis has evolved as one of the major carbon-concentrating mechanisms used by terrestrial plants to compensate for limitations associated with low availability of atmospheric CO_2. This CO_2-concentrating mechanism evolved in more than 60 plant lineages, and some of the most productive crops on the planet (e.g., maize [corn; *Zea mays*], sugarcane, sorghum) use this mechanism to enhance the carboxylation capacity of Rubisco. In this section we examine:

- The biochemical and anatomical attributes of C_4 photosynthesis that minimize the oxygenation reaction of Rubisco oxygenation and the rate of photorespiration
- The requirements of the spatial separation of the C_4 biochemistry and variation in decarboxylation of inorganic carbon
- The light-mediated regulation of enzyme activities
- The importance of C_4 photosynthesis for sustaining plant growth in many tropical areas

Malate and aspartate are the primary carboxylation products of the C_4 cycle

In the late 1950s, H. P. Kortschack and Y. Karpilov independently observed that ^{14}C label appeared initially in the four-carbon acids malate and aspartate when $^{14}CO_2$ was provided to leaves of sugarcane and maize in the light. This finding was unexpected because a three-carbon acid, 3-phosphoglycerate, is the first labeled product in the Calvin–Benson cycle. In subsequent experiments, M. D. Hatch and C. R. Slack used "pulse-chase" labeling with $^{14}CO_2$ followed by unlabeled CO_2 to demonstrate the transfer of carbon from four-carbon acids to the Calvin–Benson cycle. They found that (1) malate and aspartate are the first stable intermediates of photosynthesis and (2) the carbon 4 of these four-carbon acids subsequently becomes carbon 1 of 3-phosphoglycerate. These transformations take place in two morphologically distinct cell types—the mesophyll and bundle sheath cells—that are separated by their respective walls and membranes ("Diffusion barrier" in **Figure 8.7**). This pathway is named the *C_4 photosynthetic carbon cycle* (also known as the Hatch–Slack cycle or the C_4 cycle).

Figure 8.7 The C_4 photosynthetic carbon cycle involves five successive stages in two distinct cell types. (A) In the mesophyll cells, the enzyme phosphoenolpyruvate carboxylase (PEPCase) catalyzes the reaction of HCO_3^-, provided by the uptake of atmospheric CO_2 (reaction 1) with phosphoenolpyruvate, a three-carbon compound (reaction 2). The reaction product, oxaloacetate, a four-carbon compound, is further transformed to malate by the action of NADP–malate dehydrogenase (reaction 3). The four-carbon acid moves to the bundle sheath cell, close to vascular connections. The decarboxylating enzyme (here, NADP–malic enzyme in the chloroplast; reaction 5a) releases the CO_2 from the four-carbon acid, yielding a three-carbon acid (e.g., pyruvate). The released CO_2 in the bundle sheath chloroplast builds up a large excess of CO_2 over O_2 around Rubisco, thereby facilitating the assimilation of CO_2 through the Calvin–Benson cycle. The residual three-carbon acid (pyruvate) flows back to the mesophyll cell. To complete the C_4 cycle, the enzyme pyruvate–phosphate dikinase (reaction 8) catalyzes the regeneration of phosphoenolpyruvate, the acceptor of the HCO_3^-, for another turn of the cycle. The consumption of two molecules of ATP per molecule of fixed CO_2 (reactions 8–10) drives the C_4 cycle in the direction of the arrows, thus pumping CO_2 from the atmosphere to the Calvin–Benson cycle. The assimilated carbon leaves the chloroplast and, after transformation to sucrose in the cytosol, enters the phloem for translocation to other parts of the plant. (B) The reactions of the C4 photosynthetic carbon cycle, some of which are shown in (A). ▶

(A)

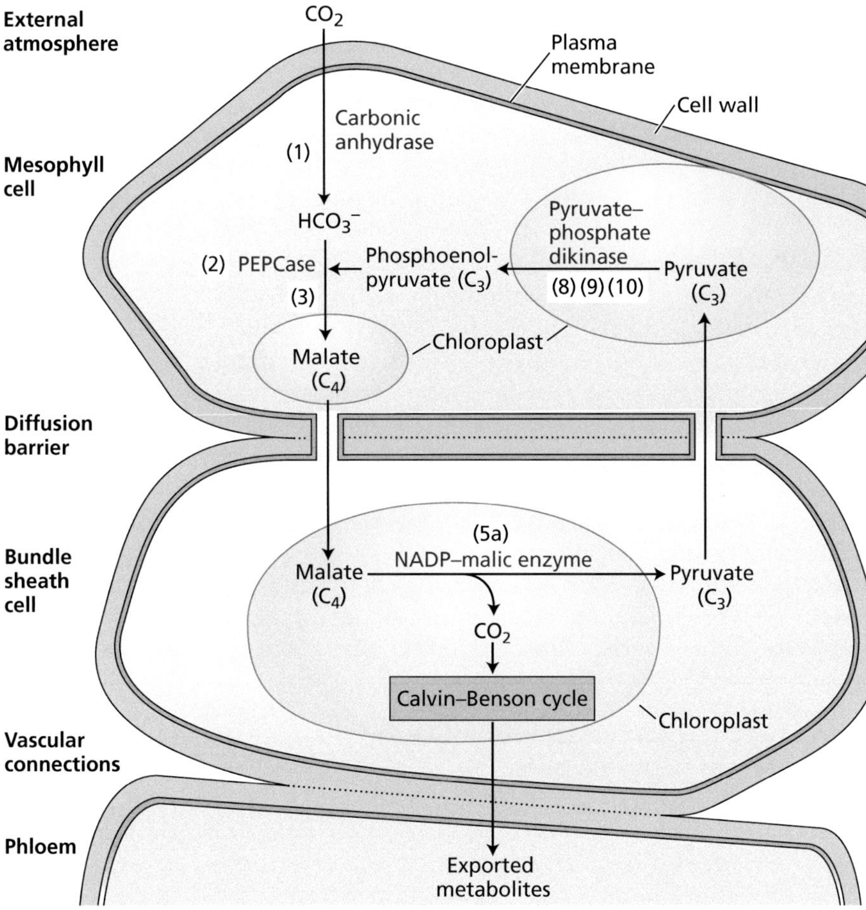

(B)

Reactions of C_4 photosynthesis

Reaction numbers (see Part A) and enzyme name	Reaction
1. Carbonic anhydrase (cytosol)	$CO_2 + H_2O \rightarrow H_2CO_3 \rightarrow H^+ + HCO_3^-$
2. PEPCase (cytosol)	Phosphoenolpyruvate + HCO_3^- → oxaloacetate + P_i
3. NADP–malate dehydrogenase (chloroplast)	Oxaloacetate + NADPH + H^+ → malate + $NADP^+$
4. Aspartate aminotransferase (chloroplast/mitochondrion)	Oxaloacetate + glutamate → aspartate + 2-oxoglutarate
Decarboxylating enzymes	
5a. NADP–malic enzyme (chloroplast)	Malate + $NADP^+$ → pyruvate + CO_2 + NADPH + H^+
5b. NAD–malic enzyme (mitochondrion)	Malate + NAD^+ → pyruvate + CO_2 + NADH + H^+
6. Phosphoenolpyruvate carboxykinase (cytosol)	Oxaloacetate + ATP → phosphoenolpyruvate + CO_2 + ADP
7. Alanine aminotransferase (cytosol)	Pyruvate + glutamate → alanine + 2-oxoglutarate
8. Pyruvate–phosphate dikinase (chloroplast)	Pyruvate + P_i + ATP → phosphoenolpyruvate + AMP + PP_i
9. Adenylate kinase (chloroplast)	AMP + ATP → 2 ADP
10. Pyrophosphatase (chloroplast)	$PP_i + H_2O \rightarrow 2\ P_i$

Note: P_i and PP_i stand for inorganic phosphate and pyrophosphate, respectively.

Kranz anatomy (German *kranz*, "wreath") The wreathlike arrangement of mesophyll cells around a layer of bundle sheath cells. The two concentric layers of photosynthetic tissue surround the vascular bundle. This anatomical feature is typical of leaves of many C_4 plants.

During C_4 photosynthesis, the enzyme phosphoenolpyruvate carboxylase (PEPCase), rather than Rubisco, catalyzes the initial carboxylation in mesophyll cytosol close to the external atmosphere (Figure 8.7, reaction 1). The carboxylation by PEPCase uses bicarbonate (HCO_3^-) instead of CO_2, and unlike with Rubisco, this reaction is not inhibited by O_2. The four-carbon acids formed in mesophyll cells flow across the diffusion barrier via the plasmodesmata to the bundle sheath cells, where they are decarboxylated, releasing CO_2 that is refixed by Rubisco via the Calvin–Benson cycle. Although all C_4 plants share primary carboxylation via PEPCase, the other enzymes used to release CO_2 around Rubisco vary among different C_4 biochemical subtypes.

Since the seminal studies of the 1950s and the 1960s, the C_4 cycle has been associated with a particular leaf structure, called **Kranz anatomy** (*Kranz*, German for "wreath"). Typical Kranz anatomy exhibits an inner ring of bundle sheath cells around vascular tissues and an outer layer of mesophyll cells (Figure 8.7A and B). This particular leaf anatomy generates a diffusion barrier that (1) separates the uptake of atmospheric carbon by PEPCase in mesophyll cells from CO_2 assimilation by Rubisco in bundle sheath cells and (2) limits the leakage of CO_2 from bundle sheath to mesophyll cells. However, there are now clear examples of single-cell C_4 photosynthesis in several green algae, diatoms, and aquatic and land plants (Figure 8.7C). In summary, a diffusion gradient is needed to concentrate CO_2 around Rubisco, but this barrier can be *between* or *within* cells as long as the resistance is sufficient to minimize energy-wasteful overcycling and leakage of CO_2 from the compartmentalized area.

Kranz-type C_4 plants assimilate CO_2 by the concerted action of two different types of cells

The key features of the C_4 cycle were initially described in leaves of plants such as maize whose vascular tissues are surrounded by two distinctive photosynthetic cell types. In this anatomical context, the assimilation of CO_2 from the external atmosphere in the bundle sheath cells proceeds through five successive stages described in reactions in Figure 8.7:

(a) Hydration of CO_2 by carbonic anhydrase (reaction 1), to generate HCO_3^- that is used to carboxylate phosphoenolpyruvate by PEPCase in the mesophyll cell cytosol (reaction 2). The reaction product, oxaloacetate, is subsequently reduced to malate by NADP–malate dehydrogenase in the mesophyll chloroplasts (reaction 3) or converted to aspartate by transamination with glutamate in the cytosol, depending on the C_4 subtype (reaction 4).

(b) Diffusion of the four-carbon acids (malate or aspartate) through the plasmodesmata into the bundle sheath cells surrounding the vascular bundles.

(c) Decarboxylation of the four-carbon acids and generation of CO_2, which is then reduced to triose phosphate via the Calvin–Benson cycle. Prior to this reaction, an aspartate aminotransferase catalyzes the conversion of aspartate back to oxaloacetate in some C_4 plants (reaction 4). Different types of C_4 plants make use of different decarboxylases to release CO_2 for the effective suppression of the oxygenase reaction of Rubisco (reactions 5a, 5b, and 6).

(d) Diffusion back to the mesophyll cells of the three-carbon backbone (pyruvate or alanine) formed by the decarboxylation step.

(e) Regeneration of phosphoenolpyruvate, the HCO_3^- acceptor, which consumes ATP and inorganic phosphate and releases AMP and pyrophosphate. Two molecules of ATP are consumed in the conversion of pyruvate to phosphoenolpyruvate: one in the reaction catalyzed by pyruvate–phosphate dikinase (reaction 8) and another in the formation of AMP from ADP by adenylate kinase (reaction 9). When alanine is the three-carbon acid returned from the bundle sheath cells, the formation of pyruvate by alanine aminotransferase (reaction 7) precedes phosphorylation by pyruvate–phosphate dikinase.

The compartmentation of enzymes ensures that inorganic carbon from the surrounding atmosphere can be taken up initially by mesophyll cells, fixed subsequently by the Calvin–Benson cycle of bundle sheath cells, and finally exported to the phloem (Figure 8.7 and Chapter 10).

The C_4 subtypes use different mechanisms to decarboxylate four-carbon acids transported to bundle sheath cells

The C_4 biochemical subtypes employ different mechanisms to decarboxylate the four-carbon acids in the bundle sheath cells, and shuttle different three-carbon compounds from bundle sheath cells back to the mesophyll cells (Figure 8.7B). Additionally, the four-carbon acid transported to the bundle sheath cells can be malate (C_3 plants) or aspartate (C_4 plants), produced in the chloroplasts or cytosol of mesophyll cells, respectively.

In the NADP–malic enzyme (NADP–ME) type of C_4 photosynthesis, malate enters the chloroplast of bundle sheath cells, where it is decarboxylated by NADP–ME (Figure 8.7, reaction 5a).

In the NAD–malic enzyme (NAD–ME) and PEP carboxykinase (PEPCK) types of C_4 photosynthesis, cytosolic aspartate aminotransferase of the bundle sheath cells catalyzes the conversion of aspartate back to oxaloacetate [aspartate + pyruvate → oxaloacetate + alanine]. In NAD–ME type C_4 plants, the oxaloacetate is reduced to malate in the mitochondria of bundle sheath cells via NAD–malate dehydrogenase and subsequently decarboxylated by NAD–ME (Figure 8.7, reaction 5b). In PEPCK type C_4 plants, the majority of the oxaloacetate is decarboxylated to phosphoenolpyruvate in the cytosol of bundle sheath cells (Figure 8.7, reaction 6), but these plants can also use NAD–ME to decarboxylate malate provided by the mesophyll cells. The released CO_2 diffuses from the mitochondria or cytosol to the chloroplasts of bundle sheath cells for fixation by Rubisco and reduction by the Calvin–Benson cycle.

The CO_2 released by the three decarboxylation reactions increases the concentration of CO_2 around the active site of Rubisco in the chloroplasts of bundle sheath cells, thereby minimizing the inhibition by O_2. Pyruvate (from NADP–ME type C_4 photosynthesis) and alanine (from NAD–ME and PEPCK types) are transported from bundle sheath cells to the mesophyll cells for the regeneration of phosphoenolpyruvate. Although traditionally C_4 plants were classified as one subtype or another, data suggest that in some C_4 plants there is flexibility and overlap between these pathways.

Bundle sheath cells and mesophyll cells exhibit anatomical and biochemical differences

Originally described for tropical grasses and *Atriplex*, the C_4 cycle is now known to occur in at least 62 independent lineages of angiosperms distributed across 21 different families. C_4 plants evolved from C_3 ancestors around 30 million years ago in response to multiple environmental stimuli such as atmospheric changes (decline of CO_2 concentration, increase of O_2 concentration), modification of global weather, periods of drought, and intense solar radiation. The transition from C_3 to C_4 plants requires the coordinated modification of genes that affect leaf anatomy, cell ultrastructure, metabolite transport, and regulation of metabolic enzymes. The analyses of C_3 and C_4 genomes and transcriptomes indicate that convergent evolution underlies the multiple origins of C_4 plants.

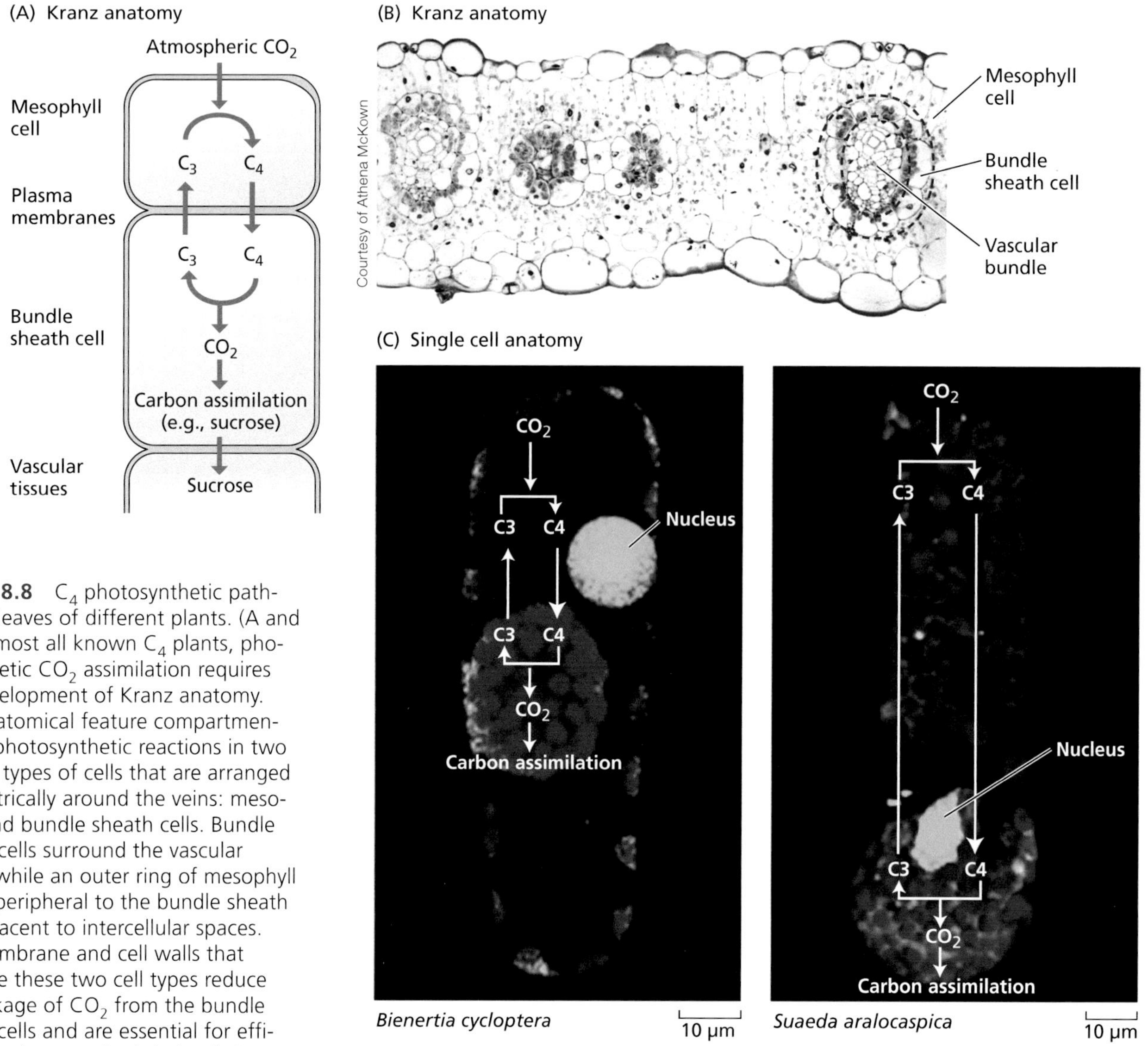

Figure 8.8 C_4 photosynthetic pathway in leaves of different plants. (A and B) In almost all known C_4 plants, photosynthetic CO_2 assimilation requires the development of Kranz anatomy. This anatomical feature compartmentalizes photosynthetic reactions in two distinct types of cells that are arranged concentrically around the veins: mesophyll and bundle sheath cells. Bundle sheath cells surround the vascular tissue, while an outer ring of mesophyll cells is peripheral to the bundle sheath and adjacent to intercellular spaces. The membrane and cell walls that separate these two cell types reduce the leakage of CO_2 from the bundle sheath cells and are essential for efficient C_4 photosynthesis in land plants. (B) Kranz anatomy. Light micrograph of transverse section of leaf blade of *Flaveria australasica* (NAD–malic enzyme type C_4 photosynthesis). (C) Single-cell C_4 photosynthesis occurs in organisms that contain the equivalent of the C_4 compartmentation in a single cell. Studies on the key photosynthetic enzymes of these plants indicate two dimorphic chloroplasts located in different cytoplasmic locations functioning analogously to mesophyll and bundle sheath cells in Kranz-type C_4 plants. Here, diagrams of the C_4 cycle are superimposed on fluorescent staining and immunofluorescence of *Bienertia cycloptera* (left) and *Suaeda aralocaspica* (right).

The distinctive Kranz anatomy increases the concentration of CO_2 in bundle sheath cells to almost tenfold higher than that in the external atmosphere (**Figure 8.8A,B**). The efficient accumulation of CO_2 in the vicinity of Rubisco reduces the rate of photorespiration to 2 to 3% of photosynthesis. Mesophyll and bundle sheath cells show large biochemical differences. PEPCase and Rubisco are located in mesophyll and bundle sheath cells, respectively, while the decarboxylases are found in different intracellular compartments of bundle sheath cells: NADP–ME in chloroplasts, NAD–ME in mitochondria, and PEPCK in the cytosol. In addition, mesophyll cells contain randomly arranged chloroplasts with stacked thylakoid membranes, while in some C_4 subtypes (primarily NADP–ME) the chloroplasts in bundle sheath cells are concentrically arranged and exhibit unstacked thylakoids. These chloroplast characteristics correlate with the energy requirements of the type of C_4 photosynthesis. For example, C_4 species of the NADP–ME subtype, in which malate is shuttled from mesophyll chloroplasts to bundle sheath cells, exhibit functional photosystems II and I in mesophyll chloroplasts, whereas bundle sheath chloroplasts are deficient in or have low levels of photosystem II and the associated oxygen-evolving complex (i.e., O_2 production from photosystem II is generally low in the bundle sheath cells of NADP–ME plants). In NAD–ME and PEPCK subtypes, by contrast, the bundle sheath chloroplasts and the mesophyll chloroplasts have similar levels of photosystem II.

The C_4 cycle also concentrates CO_2 in single cells

The finding of C_4 photosynthesis in terrestrial plants devoid of Kranz anatomy disclosed a much greater diversity in modes of C_4 carbon fixation than had previously been thought to exist. Four plants in the Amaranthaceae that grow in the Middle East and parts of Asia, *Suaeda aralocaspica* and three *Bienertia* species, perform complete C_4 photosynthesis within single chlorenchyma cells (**Figure 8.8C**). The external region, proximal to the external atmosphere, carries out the initial carboxylation and regeneration of phosphoenolpyruvate, whereas the internal region functions in the decarboxylation of four-carbon acids and the refixation of the liberated CO_2 via Rubisco. The chlorenchyma cells of these species have dimorphic chloroplasts with different subsets of enzymes, similar to the two types of chloroplasts found in mesophyll versus bundle sheath cells of Kranz-type C_4 plants. In the single-cell C_4 plants, the large vacuole separating the two types of chloroplasts provides the diffusion resistance that allows the CO_2 concentration to build up around Rubisco.

Aquatic photosynthetic organisms, which account for approximately half of the global primary productivity, also use a single-cell CO_2-concentrating mechanism to suppress the Rubisco oxygenation reaction.

Light regulates the activity of key C_4 enzymes

In addition to supplying ATP and NADPH for the operation of the C_4 cycle, light is essential for the regulation of several participating enzymes. Variations in photon flux density elicit changes in the activities of NADP–malate dehydrogenase, PEPCase, and pyruvate–phosphate dikinase by two different mechanisms: thiol–disulfide exchange [Enz–(Cys-S)$_2$ $\leftrightarrow$ Enz–(Cys-SH)$_2$] and phosphorylation–dephosphorylation of specific amino acid residues (e.g., serine, Enz–Ser-OH $\leftrightarrow$ Enz–Ser-OP).

NADP–malate dehydrogenase is regulated via the ferredoxin–thioredoxin system as in C_3 plants (Figure 8.4). The enzyme is reduced (activated) by thioredoxin when leaves are illuminated, and it is oxidized (deactivated) in the dark. The diurnal phosphorylation of PEPCase by a specific kinase, named PEPCase kinase, is thought to increase the uptake of ambient CO_2, and the nocturnal dephosphorylation by protein phosphatase 2A brings PEPCase back to low activity. A highly unusual enzyme regulates the dark–light activity of pyruvate–phosphate dikinase (PPDK). PPDK is modified by a bifunctional threonine kinase–phosphatase that catalyzes both ADP-dependent phosphorylation and P_i-dependent dephosphorylation of PPDK. Darkness promotes the phosphorylation of PPDK by the regulatory kinase–phosphatase [$(PPDK)_{active}$ + ADP $\rightarrow$ $(PPDK\text{-}P)_{inactive}$ + AMP], causing the loss of enzyme activity. The phosphorolytic cleavage of the phosphoryl group in the light by the same enzyme restores the catalytic capacity of PPDK [$(PPDK\text{-}P)_{inactive} + P_i \rightarrow (PPDK)_{active} + PP_i$].

Photosynthetic assimilation of CO_2 in C_4 plants requires more transport processes than in C_3 plants

The chloroplasts export part of the fixed carbon to the cytosol during active photosynthesis while importing the phosphate released from biosynthetic processes to replenish ATP and other phosphorylated metabolites in the stroma. In C_3 plants, the major factors that modulate the partitioning of assimilated carbon between the chloroplast and cytosol are the relative concentrations of triose phosphates and inorganic phosphate. Triose phosphate isomerases rapidly interconvert dihydroxyacetone phosphate and glyceraldehyde 3-phosphate in the plastid and cytosol. The triose phosphate translocator—a protein complex in the inner membrane of the chloroplast envelope—exchanges chloroplast triose phosphates for cytosol phosphate. Thus, C_3 plants require one transport process across the chloroplast envelope to export triose phosphates (three molecules of CO_2 assimilated) from the chloroplasts to the cytosol.

In C_4 plants, the distribution of photosynthetic CO_2 assimilation over two different cells entails a massive flux of metabolites between mesophyll cells and bundle sheath cells. Moreover, three different pathways accomplish the assimilation of inorganic carbon in C_4 photosynthesis. In this context, different metabolites flow from the cytosol of leaf cells to chloroplasts, mitochondria, and conducting tissues. Therefore, the composition and the function of translocators in organelles and plasma membrane of C_4 plants depend on the pathway used for CO_2 assimilation. For example, mesophyll cells of NADP–ME type C_4 photosynthesis use four transport steps across the chloroplast envelope to fix one molecule of atmospheric CO_2: (1) import of cytosolic pyruvate (Na^+-dependent pyruvate transporter); (2) export of stromal phosphoenolpyruvate (phosphoenolpyruvate phosphate translocator); (3) import of cytosolic oxaloacetate (dicarboxylate transporter); and (4) export of stromal malate (dicarboxylate transporter).

In hot, dry climates, the C_4 cycle reduces photorespiration

Elevated temperatures limit the rate of photosynthetic CO_2 assimilation in C_3 plants by decreasing the solubility of CO_2 and the ratio of the carboxylation to oxygenation reactions of Rubisco. Because of the decrease in the photosynthetic activity of Rubisco, the energy demands associated with photorespiration increase in warmer areas of the world. In C_4 plants, two features help overcome the deleterious effects of high temperature:

- First, atmospheric CO_2 enters the cytoplasm of mesophyll cells where carbonic anhydrase rapidly and reversibly converts CO_2 into bicarbonate [$CO_2 + H_2O \rightarrow HCO_3^- + H^+$] ($K_{eq} = 1.7 \times 10^{-4}$). The equilibrium between CO_2 and HCO_3^- is determined by the pH of the cytoplasm, but carbonic anhydrase speeds up the time to reach equilibration. This reaction by carbonic anhydrase is essential to maintain the high rates of C_4 photosynthesis, particularly in warm climates that tend to decrease the levels of CO_2 within the leaf. PEPCase also has a high affinity for HCO_3^- and thus high activity even when HCO_3^- is low. This enables C_4 plants to reduce their stomatal aperture at high temperatures and thereby conserve water while fixing CO_2 at rates equal to or greater than those of C_3 plants.
- Second, the high concentration of CO_2 in bundle sheath chloroplasts minimizes the oxygenation reaction of Rubisco and the operation of photorespiration.

The response of net CO_2 assimilation to temperature in part influences the distribution of C_3 and C_4 species on Earth. The optimal photosynthetic efficiency of C_3 species generally occurs at temperatures lower than for C_4 species: approximately 20 to 25°C and 25 to 35°C, respectively. By enabling more efficient assimilation of CO_2 at higher temperatures, C_4 species become more abundant in the tropics and subtropics and less abundant at latitudes farther from the equator. Although C_4 photosynthesis is commonly dominant in warm environments, a group of perennial grasses (*Miscanthus*, *Spartina*) are chilling-tolerant C_4 species that thrive in areas where the weather is moderately cold (Chapter 9).

8.5 Inorganic Carbon-Concentrating Mechanisms: Crassulacean Acid Metabolism (CAM)

Compare and contrast CAM photosynthesis and C_4 photosynthesis.

Another mechanism for concentrating CO_2 around Rubisco is present in approximately 350 genera and 35 terrestrial plant families that inhabit arid environments with seasonal water availability, including commercially important plants such as pineapple (*Ananas comosus*), agave (*Agave* spp.), cacti (Cactaceae), and orchids (Orchidaceae). This photosynthetic carbon fixation is named **crassulacean acid metabolism (CAM)**, to recognize its initial observation in *Kalanchoe pinnata*, a succulent member of the Crassulaceae. Like the C_4 mechanism, CAM appears to have originated during the last 35 million years in habitats where reductions in atmospheric CO_2 coupled with reduced stomatal conductance due to water limitations provided the selective pressure. The leaves of CAM plants have traits, such as thick cuticles and stomata with small apertures, that minimize water loss but also minimize the access to atmospheric CO_2. The carbon-concentrating mechanism in CAM plants is temporal spacing between night and day reactions (**Figure 8.9**), which differs from that of C_4 plants, in which the carbon-concentrating mechanism is physical separation between the mesophyll and bundle sheath cells.

In all CAM plants, cytosolic PEPCase fixes HCO_3^- from atmospheric and respiratory CO_2 into oxaloacetate using phosphoenolpyruvate formed via the glycolytic breakdown of stored carbohydrates (Figure 8.9 and Figure 8.7, reaction 2). A cytosolic NADP–malate dehydrogenase converts the oxaloacetate to malate (Figure 8.7, reaction 3), which is stored in the vacuoles for the remainder of the night. During the day, the stored malate exits the vacuole for decarboxylation by cytosolic NADP–ME, mitochondrial NAD–ME, or cytosolic PEPCK, the same enzymes used in the three C_4 biochemical subtypes (Figure 8.7B). The released CO_2 is made available to chloroplasts for fixation via Rubisco, while the three-carbon acid byproduct is converted to triose phosphates and subsequently to starch or sucrose via gluconeogenesis (Figure 8.9).

Changes in the rate of carbon uptake and in enzyme regulation throughout the day create a 24-h CAM cycle. Four distinct phases define the temporal control of C_4 and C_3 carboxylations within the same cellular environment: phase I (night), phase II (early morning), phase III (daytime), and phase IV (late afternoon). During the nocturnal phase I, when stomata are open, CO_2 is captured in the form of HCO_3^- by PEPCase and stored as malate in the vacuole. During diurnal phase III, when stomata are closed and the Calvin–Benson cycle is activated by light, the stored malate is decarboxylated and provides high concentrations of CO_2 around the active site of Rubisco, thereby minimizing photorespiration. The transient phases II and IV shift the metabolism in preparation for phases III and I, respectively. In phase II, Rubisco activity increases, while it decreases in phase IV. In contrast, the activity of PEPCase increases in phase IV and declines in phase II. The contribution of each phase to the overall carbon balance varies considerably among different CAM plants and is sensitive to environmental conditions. Constitutive CAM plants use the nocturnal uptake of CO_2 at all times. Facultative CAM plants resort to the CAM pathway only when induced by water deficit stress or salt stress; otherwise, they operate as C_3 plants.

Whether the triose phosphates produced by the Calvin–Benson cycle are stored as starch in chloroplasts or used for the synthesis of sucrose depends on the plant species. However, these carbohydrates ultimately ensure not only plant growth but also the supply of substrates for the next nocturnal carboxylation

crassulacean acid metabolism (CAM) A biochemical process for concentrating CO_2 at the carboxylation site of Rubisco. Found in the family Crassulaceae (*Crassula*, *Kalanchoe*, *Sedum*) and numerous other families of angiosperms. In CAM, CO_2 uptake and initial fixation take place at night, and decarboxylation and reduction of the internally released CO_2 occur during the day.

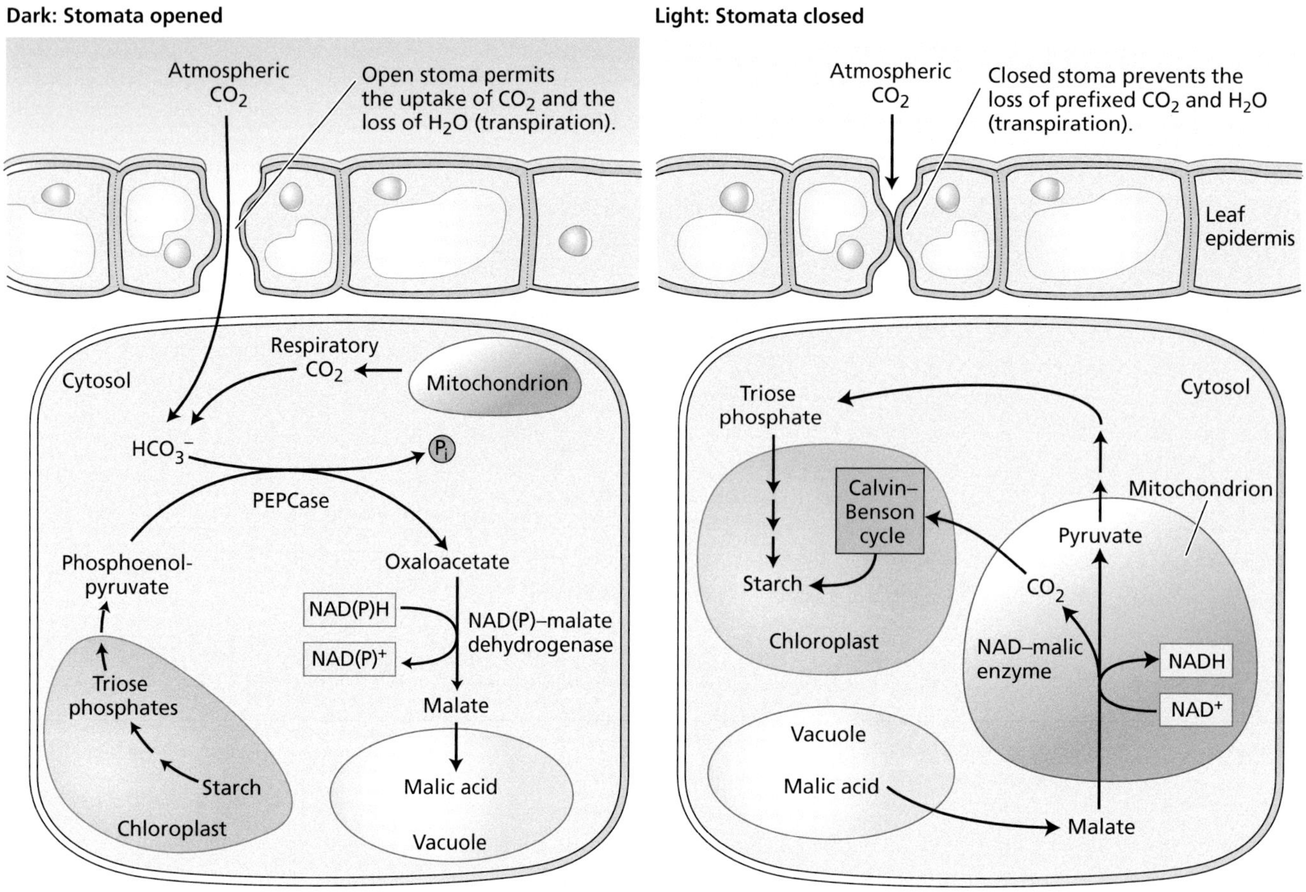

Figure 8.9 Crassulacean acid metabolism (CAM) in *Kalanchoe*. In CAM metabolism, CO_2 uptake is separated temporally from refixation via the Calvin–Benson cycle. The uptake of atmospheric CO_2 takes place at night when stomata are open. At this stage, gaseous CO_2 in the cytosol, coming from the external atmosphere and mitochondrial respiration, increases the levels of HCO_3^- ($CO_2 + H_2O \leftrightarrow HCO_3^- + H^+$). Then cytosolic PEPCase catalyzes a reaction between HCO_3^- and phosphoenolpyruvate provided by the nocturnal breakdown of chloroplast starch. The four-carbon acid formed, oxaloacetate, is reduced to malate, which in turn diffuses into the acidic vacuole and becomes protonated to malic acid. During the day, the malic acid stored in the vacuole flows back to the cytosol and dissociates back to malate. Mitochondrial NAD–malic enzyme decarboxylates the malate, releasing CO_2 that is refixed into carbon skeletons by the Calvin–Benson cycle. In essence, the diurnal accumulation of starch in the chloroplast constitutes the net gain of the nocturnal uptake of inorganic carbon. The adaptive advantage of the stomatal closure during the day is that it prevents not only water loss by transpiration but also the diffusion of internal CO_2 to the external atmosphere. Figure 8.7B describes the reactions. Note that different CAM species can have different contributions of NAD–ME, NADP–ME, and PEPCK as the decarboxylating enzyme.

phase. To sum up, the temporal separation of nocturnal initial carboxylation from diurnal decarboxylation increases the concentration of CO_2 near Rubisco and reduces photorespiration, thereby increasing the efficiency of photosynthesis.

Different mechanisms regulate C_4 PEPCase and CAM PEPCase

Comparative analysis of photosynthetic PEPCases provides a remarkable example of the adaptation of enzyme regulation to particular metabolisms. Phosphorylation of plant PEPCases by PEPCase kinase converts the inactive nonphosphorylated form into the active phosphorylated counterpart:

$$\text{PEPCase}_{\text{inactive}} + \text{ATP} \rightarrow \text{PEPCase} - \text{P}_{\text{active}} + \text{ADP}$$

Dephosphorylation of PEPCase by protein phosphatase 2A brings the enzyme back to the inactive form. PEPCase in C_4 plants functions during the day and has reduced activity at night, while PEPCase in CAM plants operates at night and has reduced activity in the daytime. The contrasting responses of photosynthetic PEPCases to light are conferred by regulatory elements that control the synthesis and degradation of PEPCase kinases. The synthesis of PEPCase kinase is mediated by light-sensing mechanisms in C_4 leaves and by endogenous circadian rhythms in CAM leaves.

CAM is a versatile mechanism sensitive to environmental stimuli

The high-water-use efficiency in CAM plants probably accounts for their extensive diversification and speciation in water-limited environments. CAM plants that grow in deserts, such as cacti, open their stomata during the cool nights and close them during the hot, dry days. The potential advantage of terrestrial CAM plants in arid environments is well illustrated by the unintentional introduction of the American prickly pear (*Opuntia stricta*) into the Australian ecosystem. From a few plants in 1840, the population of *O. stricta* expanded to occupy 25 million ha in less than a century.

Closing the stomata during the day minimizes the loss of water in CAM plants, but because H_2O and CO_2 share the same diffusion pathway, CO_2 must then be taken up by the open stomata at night (Figure 8.9). The availability of light mobilizes the reserves of vacuolar malic acid for the action of specific decarboxylating enzymes—NAD–ME, NADP–ME, and PEPCK—and the assimilation of the resulting CO_2 via the Calvin–Benson cycle. CO_2 released by decarboxylation does not escape from the leaf because stomata are closed during the day. As a consequence, the internally generated CO_2 is fixed by Rubisco and converted to carbohydrates by the Calvin–Benson cycle. Thus, stomatal closure during the day not only helps conserve water but also assists in the buildup of the elevated internal concentration of CO_2 that enhances the photosynthetic carboxylation of ribulose 1,5-bisphosphate.

Genotypic attributes and environmental factors modulate the extent to which the biochemical and physiological capacity of CAM plants is expressed. Although many species in the family Crassulaceae (e.g., *Kalanchoe*) are obligate CAM plants that exhibit circadian rhythmicity, others (e.g., *Clusia*) show C_3 photosynthesis and CAM simultaneously in distinct leaves. The proportion of CO_2 taken up by PEPCase at night or by Rubisco during the day (net CO_2 assimilation) is adjusted by (1) stomatal behavior, (2) fluctuations in organic acid and storage carbohydrate accumulation, (3) the activities of primary (PEPCase) and secondary (Rubisco) carboxylating enzymes, (4) the activities of decarboxylating enzymes, and (5) the rates of synthesis and breakdown of three-carbon skeletons.

Many CAM representatives are able to adjust their pattern of CO_2 uptake in response to longer-term variations of environmental conditions. The ice plant (*Mesembryanthemum crystallinum* L.), agave, and *Clusia* are among the plants that use CAM when water is scarce but undergo a gradual transition to C_3 when water is sufficient. Other environmental conditions, such as salinity, temperature, and light, also contribute to the extent of CAM induction in these species. This form of regulation requires the expression of numerous CAM genes in response to stress signals.

The water-conserving closure of stomata in arid lands may not be the unique basis of CAM evolution, because paradoxically, CAM species are also found among aquatic plants. Perhaps this mechanism also enhances the acquisition of inorganic carbon (as HCO_3^-) in aquatic habitats, where the high resistance to gas diffusion in liquid restricts the availability of CO_2.

8.6 Accumulation and Partitioning of Photosynthates—Starch and Sucrose

Differentiate sucrose and starch mobilization and storage during the day and night.

Metabolites accumulated in the light—photosynthates—become the ultimate source of energy for plant growth, maintenance, and development. The photosynthetic assimilation of CO_2 by most leaves yields sucrose in the cytosol and starch in the chloroplasts. During the day, sucrose flows from the leaf cytosol to heterotrophic sink tissues, while starch accumulates as dense, insoluble granules in chloroplasts (**Figure 8.10**). The onset of darkness not only stops the assimilation of CO_2 but also initiates the degradation of chloroplast starch. The content of starch in the chloroplasts falls through the night because breakdown

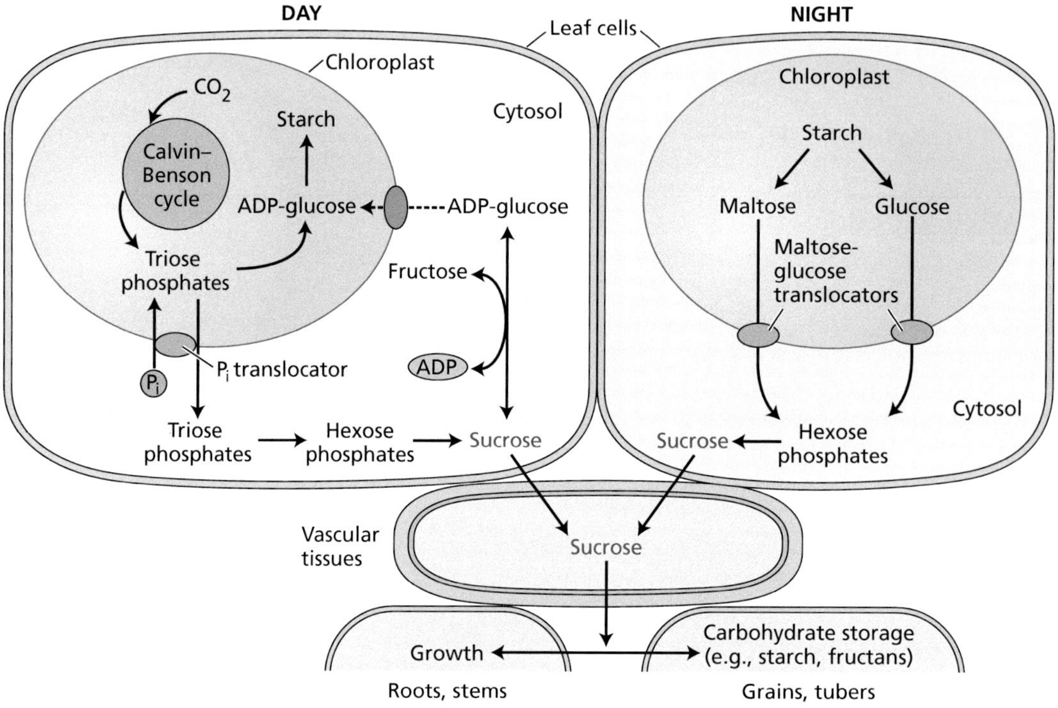

Figure 8.10 Carbon mobilization in land plants. During the day, carbon assimilated photosynthetically is either used for the formation of starch in the chloroplast or is exported to the cytosol for the synthesis of sucrose. External and internal stimuli control the partitioning between starch and sucrose. Triose phosphates from the Calvin–Benson cycle may be used for (1) the synthesis of chloroplast ADP-glucose (the glucosyl donor for starch synthesis) or (2) translocation to the cytosol for the synthesis of sucrose. At night, the cleavage of the glycosidic linkages of starch releases both maltose and glucose, which flow across the chloroplast envelope to supplement the hexose phosphate pool and contribute to sucrose synthesis. Transport across the chloroplast envelope, carried out by translocators for phosphate and maltose and glucose, conveys information between the two compartments. As a consequence of the diurnal synthesis and nocturnal breakdown, the levels of chloroplast starch are maximal during the day and minimal at night. This transitory starch serves as the nocturnal energy reserve that provides an adequate supply of carbohydrates to land plants, and also as a diurnal overflow that accepts the carbon excess when photosynthetic CO_2 assimilation proceeds faster than the synthesis of sucrose. Daily, sucrose links the assimilation of inorganic carbon (CO_2) in leaves to the use of organic carbon for growth and storage in nonphotosynthetic parts of the plant.

products flow to the cytosol to sustain the export of sucrose to other organs. The large fluctuation of stromal starch in the light versus the dark is why the polysaccharide stored in chloroplasts is called *transitory starch*. Transitory starch functions as (1) an overflow mechanism that stores photosynthate when the synthesis and transport of sucrose are limited during the day and (2) an energy reserve to provide an adequate supply of carbohydrate at night when sugars are not formed by photosynthesis. Plants vary widely in the extent to which they accumulate starch and sucrose in leaves (Figure 8.10). In some species (e.g., soybean, sugar beet, Arabidopsis), the ratio of starch to sucrose in the leaf is almost constant throughout the day. In others (e.g., spinach, French beans), starch accumulates when the amount of sucrose exceeds the storage capacity of the leaf or the demand of sink tissues.

The carbon metabolism of leaves also responds to the requirements of sink tissues for energy and growth. Regulatory mechanisms ensure that physiological processes in the chloroplast are synchronized not only with the cytoplasm of the leaf cell but also with other parts of the plant during the day–night cycle. An abundance of sugars in leaves promotes plant growth and carbohydrate storage in reserve organs, while low levels of sugars in sink tissues stimulate the rate of photosynthesis. Sucrose transport links the availability of carbohydrates in source leaves to the use of energy and the formation of reserve polysaccharides in sink tissues (Chapter 10).

Summary

Sunlight ultimately provides energy for the assimilation of inorganic carbon into organic material (autotrophy). The Calvin–Benson cycle is the predominant pathway for this conversion in many prokaryotes and in all plants.

8.1 The Calvin–Benson Cycle

- NADPH and ATP generated by light energy in chloroplast thylakoids drive the endergonic fixation of atmospheric CO_2 through the Calvin–Benson cycle in the chloroplast stroma (**Figure 8.1**).
- The Calvin–Benson cycle has three phases: (1) carboxylation of ribulose 1,5-bisphosphate with CO_2 catalyzed by Rubisco, yielding 3-phosphoglycerate; (2) reduction of the 3-phosphoglycerate to triose phosphates using ATP and NADPH; and (3) regeneration of the CO_2 acceptor molecule ribulose 1,5-bisphosphate (**Figure 8.2**).
- CO_2 and O_2 compete in the carboxylation and oxygenation reactions catalyzed by Rubisco (**Figure 8.3**).
- Rubisco activase controls the activity of Rubisco, wherein CO_2 functions as both activator and substrate.
- Light regulates the activity of Rubisco activase and four enzymes of the Calvin–Benson cycle via the ferredoxin–thioredoxin system and changes in Mg^{2+} concentration and pH (**Figure 8.4**).

8.2 The Oxygenation Reaction of Rubisco and Photorespiration

- Photorespiration minimizes the loss of fixed CO_2 by the oxygenase activity of Rubisco (**Figure 8.5**).
- Chloroplasts, peroxisomes, and mitochondria participate in the movement of carbon, nitrogen, and oxygen atoms through photorespiration (**Figures 8.5**, **8.6**).
- Kinetic properties of Rubisco, temperature, and the availability of atmospheric CO_2 and O_2 control the balance between the Calvin–Benson cycle and photorespiration.

8.3 Inorganic Carbon-Concentrating Mechanisms

- Terrestrial plants have two carbon-concentrating mechanisms that precede CO_2 assimilation through the Calvin–Benson cycle: C_4 photosynthetic carbon fixation (C_4) and crassulacean acid metabolism (CAM).

8.4 Inorganic Carbon-Concentrating Mechanisms: C_4 Photosynthetic Carbon Fixation

- The C_4 photosynthetic carbon cycle fixes atmospheric CO_2 via PEPCase into carbon skeletons in one compartment. The four-carbon acid products flow to another

(Continued)

Summary *(continued)*

compartment where CO_2 is released and refixed via Rubisco (**Figure 8.7**).

- The C_4 cycle may be driven by diffusion gradients between mesophyll and bundle sheath cells (Kranz anatomy) as well as by gradients within a single cell (**Figure 8.8**).
- Light regulates the activity of key C_4 cycle enzymes: NADP–malate dehydrogenase, PEPCase, and pyruvate–phosphate dikinase.
- The CO_2-concentrating mechanism in C_4 plants reduces the oxygenation reaction of Rubisco and water loss in hot, dry climates.

8.5 Inorganic Carbon-Concentrating Mechanisms: Crassulacean Acid Metabolism (CAM)

- CAM photosynthesis captures atmospheric CO_2 at night and scavenges respiratory CO_2 behind closed stomata during the day (**Figure 8.9**).
- CAM is generally associated with anatomical features that minimize water loss and is typically found in plants growing in arid environments.
- Genetics and environmental factors determine CAM expression.

8.6 Accumulation and Partitioning of Photosynthates—Starch and Sucrose

- In most leaves, sucrose in the cytosol and starch in chloroplasts are the end products of photosynthetic CO_2 assimilation (**Figure 8.10**).
- During the day, sucrose flows from the leaf cytosol to sink tissues, while starch accumulates as granules in chloroplasts. At night, the starch content of chloroplasts falls to provide carbon skeletons for sucrose synthesis in the cytosol to nourish heterotrophic tissues.

Suggested Reading

Balsera, M., Uberegui, E., Schürmann, P., and Buchanan, B. B. (2014) Evolutionary development of redox regulation in chloroplasts. *Antioxid. Redox Signal.* 21: 1327–1355.

Barett, J., Girr, P., and Mackinder, L. C. M. (2021) Pyrenoids: CO_2 fixing phase separated liquid organelles. *Biochim. Biophys. Acta.* 1868: 118949.

Bräutigam, A., Schlüter, U., Eisenhut, M., and Gowik., U. (2017) On the evolution of CAM photosynthesis. *Plant Physiol.* 174: 473–477.

Edwards, E. J. (2014) The inevitability of C_4 photosynthesis. *eLife* 3: e03702.

Ermakova, M., Danila, F. R., Furbank, R. T., and von Caemmerer, S. (2020) On the road to C_4 rice: Advances and perspectives. *Plant J.* 101: 940–950.

Flamholz, A. I., Prywes, N., Moran, U., Davidi, D., Bar-On, Y. M., Oltroffe, L. M., Alves, R., Savage, D., and Milo, R. (2019) Revisiting trade-offs between Rubisco kinetic parameters *Biochemistry* 58: 3365–3376.

Lawson, T., and Matthews, J. (2020) Guard cell metabolism and stomatal function. *Annu. Rev. Plant Biol.* 71: 273–302.

Michelet, L., Zaffagnini, M., Morisse, S., Sparla, F., Pérez-Pérez, M. E., Francia, F., Danon, A., Marchand, C. H., Fermani, S., Trost, P., et al. (2013) Redox regulation of the Calvin-Benson cycle: Something old, something new. *Front. Plant Sci.* 4: 1–21.

Mueller-Cajar, O. (2017) The diverse AAA+ machines that repair inhibited Rubisco active sites. *Front. Mol. Biol.* 4: 1–31.

Sage, R. F., Christin, P. A., and Edwards, E. J. (2011) The C_4 plant lineages of planet Earth. *J. Exp. Bot.* 62: 3155–3169.

Schlüter, U., and Weber, A. P. M. (2020) Regulation and evolution of C_4 photosynthesis. *Annu. Rev. Plant Biol.* 71: 183–215.

Weber, A. P. M., and Bar-Even, A. (2019) Update: Improving the efficiency of photosynthetic carbon reactions. *Plant Physiol.* 179: 803–812.

9 Photosynthesis: Physiological and Ecological Considerations

The conversion of solar energy to the chemical energy of organic compounds is a complex process that includes electron transport and photosynthetic carbon metabolism (Chapters 7 and 8). This chapter addresses some of the photosynthetic responses of the intact leaf to its environment. Additional photosynthetic responses to different types of stress will be covered in Chapter 19. When discussing photosynthesis in this chapter, we are referring to the rate of net photosynthesis, the difference between photosynthetic carbon assimilation and the loss of CO_2 via photorespiration and mitochondrial respiration in the light.

The impact of the environment on photosynthesis is of broad interest, especially to physiologists, ecologists, evolutionary biologists, climate change scientists, and agronomists. From a physiological standpoint, we wish to understand the direct responses of photosynthesis to environmental factors such as light, atmospheric CO_2 concentrations, and temperature, as well as the indirect responses (mediated through the effects of stomatal control) to environmental factors such as humidity and soil moisture. The dependence of photosynthetic processes on environmental conditions is also important to agronomists because plant productivity, and hence crop yield, depends strongly on prevailing photosynthetic rates in a dynamic environment. To the ecologist, photosynthetic variation across plant lineages in different environments is of great interest in terms of adaptation and evolution.

In studying the environmental dependence of photosynthesis, a central question arises: How many environmental factors can limit photosynthesis at one time? The British plant physiologist F. F. Blackman hypothesized in 1905 that, under any particular set of conditions, the rate of photosynthesis is limited by the slowest step in the process, the so-called *limiting factor*. The implication of this hypothesis is that at any given time, photosynthesis can be limited either by light or by CO_2 concentration, for instance, but not by both factors. This hypothesis has had a marked influence on the approach used by plant physiologists to study photosynthesis—that is, varying one factor and keeping all other environmental conditions constant. Instrumental to this analysis was the development of a biochemical model of C_3 photosynthesis published in 1980 by Graham Farquhar, Susanne von Caemmerer, and Joe Berry. This model has been expanded over the years to describe the three major metabolic processes that have been identified as important for photosynthetic performance:

1. Rubisco capacity
2. Regeneration of ribulose bisphosphate (RuBP)
3. Metabolism of the triose phosphates

Graham Farquhar and Tom Sharkey pointed out that we can think of the controls on the overall net photosynthetic rates of leaves in economic terms, considering "supply" and "demand" functions for carbon dioxide. The major metabolic processes just referred to take place in the palisade cells and spongy mesophyll of the leaf (**Figure 9.1**). These biochemical activities describe the "demand" for CO_2 by photosynthetic metabolism in the cells. However, the rate of CO_2 "supply" to these cells is largely determined by diffusion limitations resulting from boundary layer conductance, stomatal regulation, and subsequent resistance in the mesophyll (also Chapter 3). The coordinated actions of "demand" by photosynthetic cells and "supply" of atmospheric CO_2 affect leaf photosynthetic rates of net CO_2 uptake.

In this chapter we focus on how naturally occurring variation in light and temperature influences photosynthesis in leaves and how leaves in turn adjust or acclimate to such variation. We also explore how atmospheric CO_2 and temperature influence photosynthesis, an especially important consideration in a world where CO_2 concentrations and temperature are rapidly increasing as humans continue to burn fossil fuels for energy production.

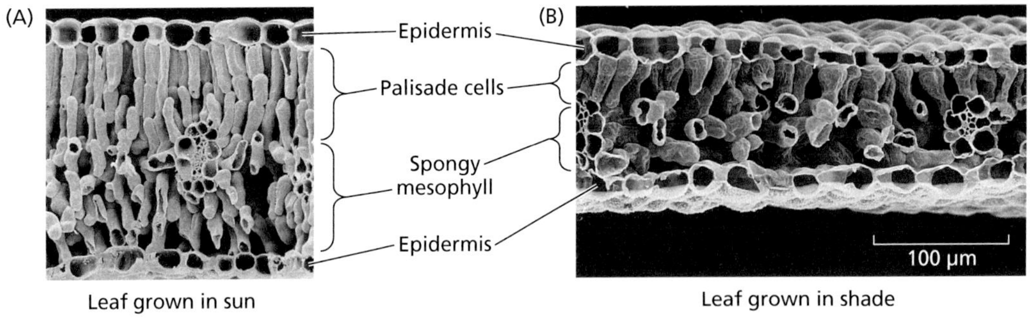

Figure 9.1 Scanning electron micrographs of the leaf anatomy of a legume (*Thermopsis montana*) grown in different light environments. Note that the sun leaf (A) is much thicker than the shade leaf (B), and that the palisade (column-like) cells are much longer in the leaves grown in sunlight. Layers of spongy mesophyll cells can be seen below the palisade cells.

9.1 Photosynthesis Is Influenced by Leaf Properties

Explain the relationship between leaf characteristics and light and the subsequent effect on photosynthesis.

Scaling up from the chloroplast (the focus of Chapters 7 and 8) to the leaf adds new levels of complexity to photosynthesis. At the same time, the structural and functional properties of the leaf add additional levels of regulation to CO_2 and H_2O exchange between leaves and the atmosphere.

We will start by examining the capture of light, and how leaf anatomy and leaf orientation maximize light absorption for photosynthesis. Then we will describe how leaves acclimate to their light environment. We will see that the photosynthetic response of leaves grown under different light conditions also reflects the ability of a plant to grow under different light environments. However, there are limits to the extent to which photosynthesis in a species can acclimate to very different light environments. For example, in some situations photosynthesis is limited by an inadequate supply of light. In other situations, absorption of too much light would cause severe problems if special mechanisms did not protect the photosynthetic system from excessive light. While plants have multiple levels of control over photosynthesis that allow them to grow successfully in constantly changing environments, there are ultimately limits to what is possible.

Consider the many ways in which leaves are exposed to different spectra (qualities) and quantities of light. Plants grown outdoors are exposed to sunlight, and the spectrum of that light will depend on whether it is measured in full sunlight or under the shade of a canopy. Plants grown indoors may receive incandescent, fluorescent, or LED lighting, each of which differs in intensity and spectral composition from sunlight. To account for these differences in spectral quality and quantity, we need to define how we measure and express the light availability for photosynthesis.

The light reaching the plant is a flux, and that flux can be measured in either energy or photon flux units. **Irradiance** is the amount of energy that falls on a flat surface of known area per unit time, expressed in watts per square meter ($W\ m^{-2}$). Recall that time (seconds) is contained within the term watt: $1\ W = 1$ joule (J) s^{-1}. Quantum flux, or photon flux density (PFD), is the number of incident **quanta** (singular *quantum*) striking the leaf, expressed in moles per square meter per second ($mol\ m^{-2}\ s^{-1}$), where *moles* refers to the number of photons (1 mol of light $= 6.02 \times 10^{23}$ photons, Avogadro's number). Quanta and energy units for sunlight can be interconverted relatively easily, provided that the wavelength of the light, λ, is known. The energy of a photon is related to its wavelength as follows:

$$E = \frac{hc}{\lambda} \tag{9.1}$$

where c is the speed of light ($3 \times 10^8\ m\ s^{-1}$), h is Planck's constant (6.63×10^{-34} J s), and λ is the wavelength of light, usually expressed in nanometers ($1\ nm = 10^{-9}\ m$). From Equation 9.1 it can be calculated that a photon at 400 nm has twice the energy of a photon at 800 nm.

When considering photosynthesis and light, it is appropriate to express light as photosynthetic photon flux density (PPFD)—the flux of light (usually expressed as micromoles per square meter per second [$\mu mol\ m^{-2}\ s^{-1}$]) within the photosynthetically active range (400–700 nm). How much light is there on a sunny day? Under direct sunlight on a clear day, PPFD is about 2,000 $\mu mol\ m^{-2}\ s^{-1}$ at the top of a dense forest canopy, but it may be only 10 $\mu mol\ m^{-2}\ s^{-1}$ at the bottom of the canopy because of light absorption by the leaves overhead.

Leaf anatomy and canopy structure optimize light absorption

On average, about 340 W ($J\ s^{-1}$) of radiant energy from the sun reach each square meter of Earth's surface. When this sunlight strikes the vegetation, only 5%

irradiance The amount of energy that falls on a flat surface of known area per unit of time. Expressed as watts per square meter ($W\ m^{-2}$). Time (seconds) is contained within the term watt: 1 W = 1 joule (J) s^{-1}, or as micromoles of quanta per square meter per second ($\mu mol\ m^{-2}\ s^{-1}$), also referred to as fluence rate.

quantum (plural *quanta*) A discrete packet of energy contained in a photon.

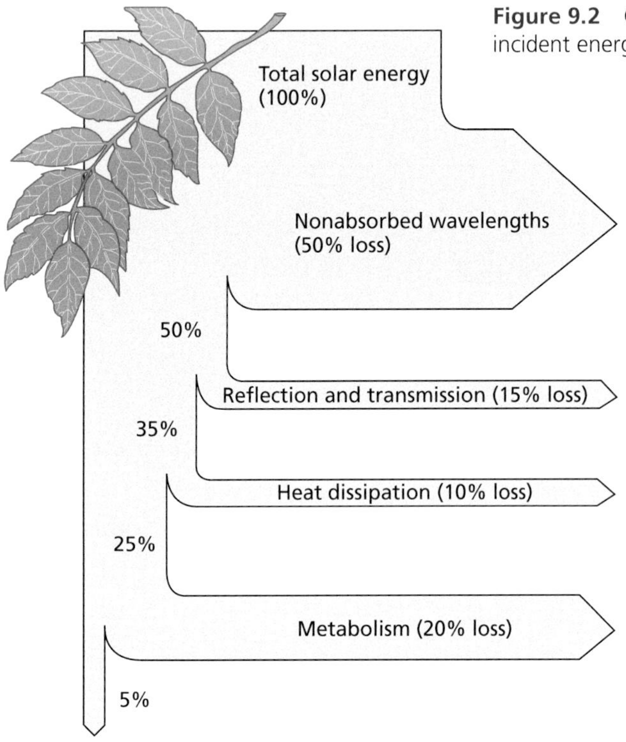

Figure 9.2 Conversion of solar energy into carbohydrates by a leaf. Of the total incident energy, only 5% is converted into carbohydrates.

of the energy is ultimately converted into carbohydrates by photosynthesis (**Figure 9.2**). The reason this percentage is so low is that a major percentage of the light is either too short or too long a wavelength to be absorbed by the photosynthetic pigments (**Figure 9.3**). Furthermore, of the photosynthetically active radiation (PAR) (400–700 nm) that is incident on a leaf, a small percentage is transmitted through the leaf and some is also reflected from its surface. Because chlorophyll absorbs strongly in the blue and red regions of the spectrum (Figure 7.3), green wavelengths are primarily transmitted and reflected (Figure 9.3)—hence the green color of vegetation. Lastly, a percentage of the photosynthetically active radiation that is initially absorbed by the leaf is consumed by metabolism and a smaller amount is lost as heat (Chapter 7).

The anatomy of the leaf is highly specialized for light absorption. The outermost cell layer, the epidermis, is typically transparent to visible light, and the individual cells are often convex. Convex epidermal cells can act as lenses and focus light so that the amount of light reaching some of the chloroplasts can be greater than the amount of incident light. Epidermal focusing is common among herbaceous plants and is especially prominent among tropical plants that grow in the forest understory, where the incident light levels are low.

Below the epidermis, the top layers of photosynthetic cells in eudicots are called **palisade cells**; they are shaped like pillars that stand in parallel columns one to three layers deep (Figure 9.1). Some leaves have several layers of columnar palisade cells, and we may wonder how efficient it is for a plant to invest energy in developing multiple cell layers when the high chlorophyll content of the first layer would appear to allow little transmission of incident light to the leaf interior. In fact, more light than might be expected penetrates the first layer of palisade cells because of the *sieve effect* and *light channeling*.

The **sieve effect** occurs because chlorophyll is not uniformly distributed throughout cells but instead is confined to the chloroplasts. This packaging of chlorophyll results in shading between the chlorophyll molecules and creates gaps between the chloroplasts where light is not absorbed—hence the reference to a sieve. Because of the sieve effect, the total absorption of light by a given amount of chlorophyll in chloroplasts of a palisade cell is less than the light that would be absorbed by the same amount of chlorophyll were it uniformly distributed in solution.

Light channeling occurs when some of the incident light is propagated through the central vacuoles of the palisade cells and through the air spaces between the cells, an arrangement that results in the transmission of light into the leaf interior. In the interior, below the palisade layers, is the **spongy mesophyll**,

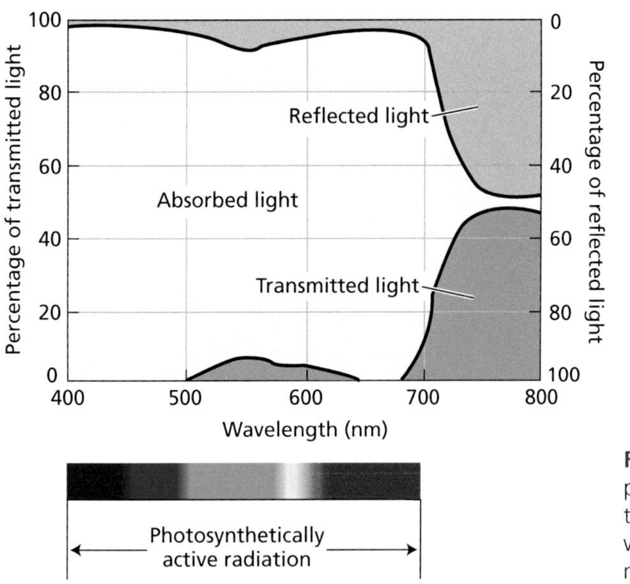

Figure 9.3 Optical properties of a bean leaf. Shown here are the percentages of light absorbed, reflected, and transmitted, as a function of wavelength. The transmitted and reflected green light in the wave band at 500 to 600 nm gives leaves their green color. Note that most of the light above 700 nm is not absorbed by the leaf.

where the cells are very irregular in shape and are surrounded by large air spaces (Figure 9.1). The large air spaces generate many interfaces between air and water that reflect and refract the light, thereby randomizing its direction of travel. This phenomenon is called **interface light scattering**.

Light scattering is especially important in leaves because the multiple refractions between cell–air interfaces greatly increase the length of the path over which photons travel, thereby increasing the probability of absorption. In fact, photon path lengths within leaves are commonly four times longer than the thickness of the leaf. Thus, the palisade cell properties that allow light to pass through, and the spongy mesophyll cell properties that are conducive to light scattering, result in more uniform light absorption throughout the leaf.

In some environments, such as deserts, there is so much light that it is potentially harmful to the photosynthetic machinery of leaves. In these environments, leaves often have special anatomical features, such as hairs, salt glands, and epicuticular wax, that increase the reflection of light from the leaf surface, thereby reducing light absorption. Such adaptations can decrease light absorption by as much as 60%, thereby reducing overheating and other problems associated with the absorption of too much solar energy.

At the whole-plant level, leaves at the top of a canopy absorb most of the sunlight, and reduce the amount of radiation that reaches leaves lower down in the canopy. Leaves that are shaded by other leaves experience lower light levels and different light quality than the leaves above them and have much lower photosynthetic rates. However, like the layers of an individual leaf, the structure of most plants, and especially of trees, represents an outstanding adaptation for light interception. The elaborate branching structure of trees vastly increases the interception of sunlight. In addition, leaves at different levels of the canopy have varied morphology and physiology that help improve light capture throughout the canopy. The result is that very little PAR penetrates all the way to the bottom of the forest canopy; almost all of the PAR is absorbed by leaves before reaching the forest floor (**Figure 9.4**).

The deep shade of a forest floor thus makes for a challenging growth environment for plants. However, in many shady habitats **sunflecks** are a common

palisade cells Below the leaf upper epidermis, the top one to three layers of pillar-shaped photosynthetic cells.

sieve effect The penetration of photosynthetically active light through several layers of cells due to the gaps between chloroplasts permitting the passage of light.

light channeling In photosynthetic cells, the propagation of some of the incident light through the central vacuole of the palisade cells and through the air spaces between the cells.

spongy mesophyll Mesophyll cells of very irregular shape located below the palisade cells and surrounded by large air spaces. They function in photosynthesis and gas exchange.

interface light scattering The randomization of the direction of photon movement within plant tissues due to the reflecting and refracting of light from the many air–water interfaces. Greatly increases the probability of photon absorption within a leaf.

sunflecks Patches of sunlight that pass through openings in a forest canopy to the forest floor. A major source of incident radiation for plants growing under the forest canopy.

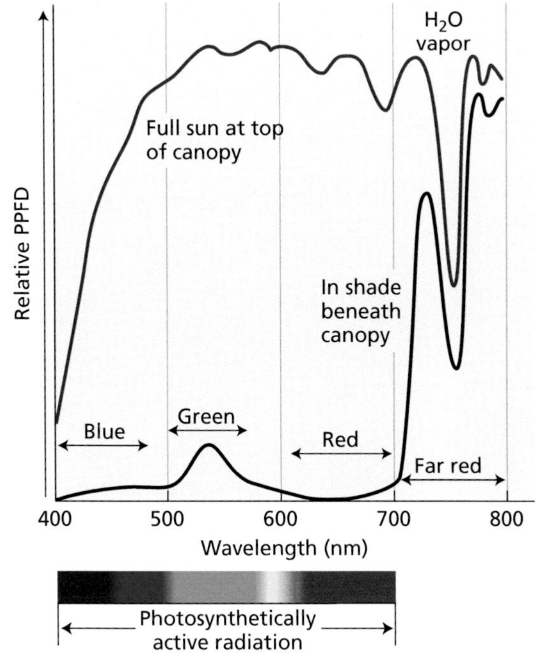

Figure 9.4 Relative spectral distributions of sunlight at the top of a canopy and in the shade under the canopy. Most photosynthetically active radiation is absorbed by leaves in the canopy.

solar tracking The movement of leaf blades throughout the day so that the planar surface of the blade remains perpendicular to the sun's rays.

environmental feature that brings high light levels deep into the canopy. These are patches of sunlight that pass through small gaps in the leaf canopy; as the sun moves, the sunflecks move across the normally shaded leaves. In spite of the short, ephemeral nature of sunflecks, the photons in them can comprise nearly 50% of the total light energy available during the day. In a dense forest, sunflecks can change the sunlight impinging on a shade leaf by more than tenfold within seconds. This critical energy is available for only a few seconds to minutes, and in a very high dose. Many deep-shade species that experience sunflecks have physiological mechanisms for taking advantage of this burst of light when it occurs. Sunflecks also play an important role in the carbon metabolism of densely planted crops, where the lower leaves are shaded by leaves higher up on the plant.

Leaf angle and leaf movement can control light absorption

The angle of the leaf relative to the sun determines the amount of sunlight incident on it. Incoming sunlight can strike a flat leaf surface at a variety of angles depending on the time of day and the orientation of the leaf. Maximum incident radiation occurs when sunlight strikes a leaf perpendicular to its surface. When the rays of light deviate from perpendicular, however, the incident sunlight on a leaf is proportional to the angle at which the rays hit the surface.

Under natural conditions, leaves exposed to full sunlight at the top of the canopy tend to have steep leaf angles, so less than the maximum amount of sunlight is incident on the leaf blade; this allows more sunlight to penetrate into the canopy. For this reason, it is common to see the angle of leaves within a canopy decrease (become more horizontal) with increasing depth in the canopy.

Some leaves maximize light absorption by **solar tracking**; that is, they continuously adjust the orientation of their blades (laminae) such that they remain perpendicular to the sun's rays (**Figure 9.5**). Many species, including alfalfa, cotton, soybean, bean, and lupine, have leaves capable of solar tracking.

Solar-tracking leaves present a nearly vertical position at sunrise, facing the eastern horizon. Individual leaf blades then begin to track the rising sun, following its movement across the sky with an accuracy of $\pm 15°$ until sunset, when the blades are again nearly vertical, facing west. During the night the leaves take a horizontal position and reorient just before dawn so that they face the eastern horizon, ready for another sunrise. Leaves track the sun only on clear days, and they stop moving when a cloud obscures the sun. In the case of intermittent cloud cover, some leaves can reorient as rapidly as 90° per hour and thus can catch up with the new solar position when the sun emerges from behind a cloud.

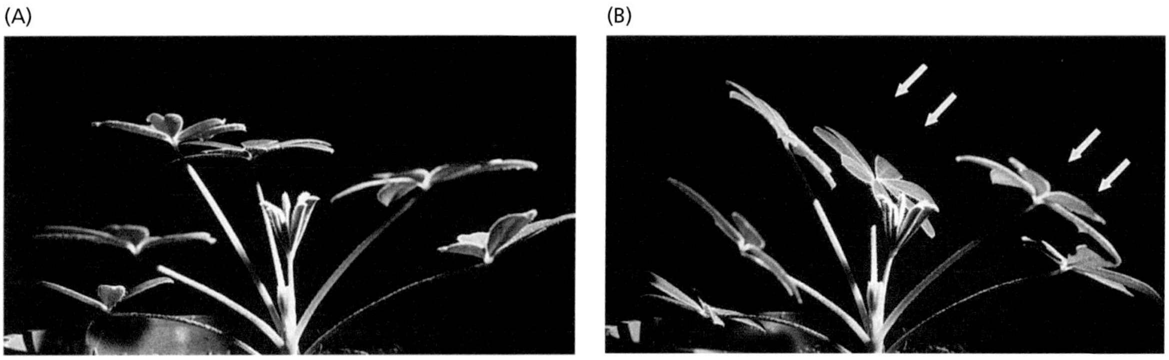

Figure 9.5 Leaf movement in a sun-tracking plant. (A) Initial leaf orientation in the lupine *Lupinus succulentus*, with no direct sunlight. (B) Leaf orientation 4 h after exposure to oblique light. Arrows indicate the direction of the light beam. Movement is generated by asymmetric swelling of a pulvinus, found at the junction between the lamina and the petiole. In natural conditions, the leaves track the sun's trajectory in the sky.

Solar tracking is a blue-light response (Chapter 13), and the sensing of blue light in solar-tracking leaves occurs in specialized regions of the leaf or stem. In species of *Lavatera* (Malvaceae), the photosensitive region is located in or near the major leaf veins, but in many species, notably legumes, leaf orientation is controlled by a specialized organ called the **pulvinus** (plural *pulvini*), found at the junction between the blade and the petiole. In lupines (*Lupinus*, Fabaceae), for example, leaves consist of five or more leaflets, and the photosensitive region is in a pulvinus located at the basal part of each leaflet lamina (Figure 9.5). The pulvinus contains motor cells that change their osmotic potential and generate mechanical forces that determine laminar orientation. In other plants, leaf orientation is controlled by small mechanical changes along the length of the petiole and by movements of the younger parts of the stem.

Heliotropism is another term used to describe leaf orientation by solar tracking. Leaves that maximize light interception by solar tracking are referred to as *diaheliotropic*. Some solar-tracking plants can also move their leaves so that they *avoid* full exposure to sunlight, thus minimizing heating and water loss. These sun-avoiding leaves are called *paraheliotropic*. Some plant species, such as soybean, have leaves that can display diaheliotropic movements when they are well watered and paraheliotropic movements when they experience water stress.

pulvinus (plural *pulvini*) A turgor-driven organ found at the junction between the blade and the petiole of the leaf that provides the mechanical force for leaf movements.

heliotropism The movements of leaves toward or away from the sun.

acclimation The increase in plant stress tolerance due to exposure to prior stress. May involve changes in gene expression. Contrast with *adaptation*.

plasticity The ability to adjust morphologically, physiologically, and biochemically in response to changes in the environment.

Leaves acclimate to sun and shade environments

Acclimation is a developmental process in which leaves express a set of biochemical and morphological adjustments that are suited to the particular environment in which the leaves are exposed. Acclimation can occur in both mature leaves and newly developing leaves. **Plasticity** is the term we use to define how much adjustment can take place. Many plants have developmental plasticity to respond to a range of light regimes, growing as sun plants in sunny areas and as shade plants in shady habitats. The ability to acclimate is important, given that shady habitats can receive less than 20% of the PAR available in an exposed habitat, and deep-shade habitats receive less than 1% of the PAR incident at the top of the canopy.

In some plant species, individual leaves that develop under very sunny or deep shady environments are often unable to persist when transferred to the other type of habitat. In such cases, the mature leaf will abscise (senesce and fall off) and a new leaf will develop that is better suited for the new environment. You may notice this if you take a plant that developed indoors and transfer it outdoors; after some time, if it is the right type of plant, a new set of leaves will develop that are better suited to high sunlight. However, some plant species are not able to acclimate when transferred from a sunny to a shady environment, or vice versa. The lack of acclimation indicates that these plants have lower plasticity and are specialized for either a sunny or a shady environment. When plants adapted to deep-shade conditions are transferred into full sunlight, the leaves experience chronic photoinhibition and leaf bleaching, and they eventually die. We will discuss photoinhibition later in this chapter.

Sun and shade leaves have contrasting biochemical and morphological characteristics:

- Shade leaves increase light capture by having more total chlorophyll per reaction center, a higher ratio of chlorophyll *b* to chlorophyll *a*, and usually thinner laminae than sun leaves.
- Sun leaves have less total chlorophyll per reaction center, increase CO_2 assimilation by having more Rubisco, and can dissipate excess light energy by having a large pool of xanthophyll-cycle components. Morphologically they have thicker leaves and a thicker palisade layer than shade leaves (Figure 9.1).

light compensation point The amount of light reaching a photosynthesizing leaf at which photosynthetic CO_2 uptake exactly balances respiratory CO_2 release.

These morphological and biochemical modifications are associated with specific acclimation responses to the *amount* of sunlight in a plant's habitat, but light *quality* can also influence such responses. For example, far-red light, which is absorbed primarily by photosystem I (PSI), is proportionally more abundant in shady habitats than in sunny ones (Chapter 15). To better balance the flow of energy through PSII and PSI, the adaptive response of some shade plants is to produce a higher ratio of PSII to PSI reaction centers, compared with that found in sun plants. Other shade plants, rather than altering the ratio of PSII to PSI reaction centers, add more antenna chlorophyll to PSII to increase absorption by this photosystem. These changes appear to enhance light absorption and optimize energy transfer in shady environments.

9.2 Effects of Light on Photosynthesis in the Intact Leaf

Describe the relationship between light and photosynthetic rate and the different mechanisms by which plants respond to light changes.

Light is a critical resource that often limits plant growth, but at times leaves can be exposed to too much rather than too little light. Photosynthetic responses to light can help explain contrasting physiological properties between sun and shade plants, and between C_3 and C_4 species (Chapter 8). Responses to excess light help protect the photosystems from damage.

Photosynthetic light-response curves reveal differences in leaf properties

Measuring net CO_2 fixation in intact leaves across varying PPFD levels generates light-response curves (**Figure 9.6**). In near darkness there is little photosynthetic carbon assimilation, but because mitochondrial respiration continues, CO_2 is given off by the leaf (Chapter 11). The uptake of CO_2 is negative in this part of the light-response curve. At higher PPFD levels, photosynthetic CO_2 assimilation eventually reaches a point where CO_2 uptake exactly balances CO_2 evolution, the **light compensation point**. The PPFD at which different leaves reach the light compensation point can vary among species and developmental conditions. One of the more interesting differences is found between plants that normally grow in full sunlight and those that grow in the shade (**Figure 9.7**). Light compensation points of sun plants range from 10 to 20 μmol m^{-2} s^{-1} PPFD, whereas corresponding values for shade plants are 1 to 5 μmol m^{-2} s^{-1} PPFD.

Why are light compensation points lower for shade plants? For the most part, this is because respiration rates in shade plants are very low; therefore, only low rates of photosynthesis

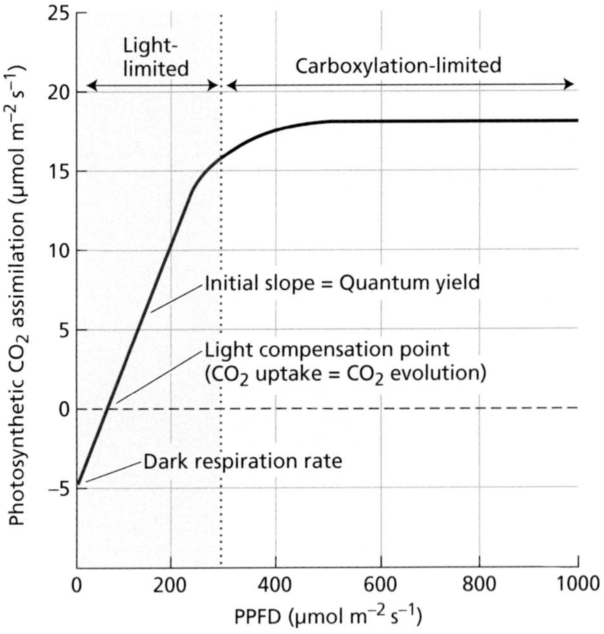

Figure 9.6 Response of photosynthesis to light in a C_3 plant. In darkness, respiration causes a net efflux of CO_2 from the plant. The light compensation point is reached when photosynthetic CO_2 assimilation equals the amount of CO_2 evolved by respiration. Increasing light above the light compensation point proportionally increases photosynthesis, indicating that photosynthesis is limited by the rate of electron transport, which in turn is limited by the amount of available light. This portion of the curve is referred to as light-limited. Further increases in photosynthesis are eventually limited by the carboxylation capacity of Rubisco or the metabolism of triose phosphates. This part of the curve is referred to as carboxylation-limited.

Figure 9.7 Light-response curves of photosynthetic carbon fixation in sun and shade plants. Triangle orache (*Atriplex triangularis*) is a sun plant, and wild ginger (*Asarum caudatum*) is a shade plant. Typically, shade plants have a low light compensation point and have lower maximum photosynthetic rates than sun plants. The dashed red line has been extrapolated from the measured part of the curve.

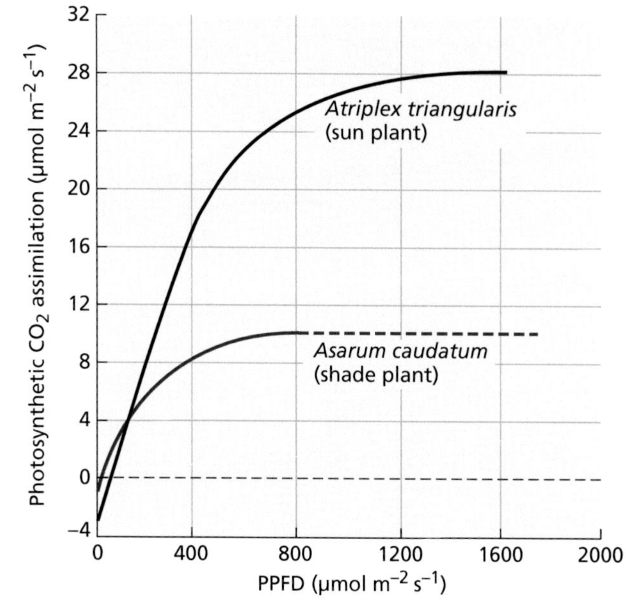

are necessary to bring the net rates of CO_2 exchange to zero. Low respiratory rates are a consequence of slow growth rates of shade plants and allow them to survive in light-limited environments through their ability to achieve positive CO_2 uptake rates at lower PPFD values compared with sun plants.

A linear relationship between PPFD and photosynthetic rate persists at light levels above the light compensation point (Figure 9.6). Throughout this linear portion of the light-response curve, photosynthesis is light-limited; more light stimulates proportionally more photosynthesis. When corrected for light absorption, the slope of this linear portion of the curve provides the **maximum quantum yield** of photosynthesis for the leaf. Leaves of sun and shade plants show very similar quantum yields despite their different growth habitats. This is because the basic biophysical and biochemical processes that determine quantum yield are the same for these two types of plants. But quantum yield can vary among plants with different photosynthetic pathways (e.g., between C_3 and C_4 plants).

Quantum yield is the ratio of a given light-dependent product to the number of absorbed photons (see Equation 7.7). Photosynthetic quantum yield can be expressed on either a CO_2 or an O_2 basis, and as explained in Chapter 7, the quantum yield of photochemistry is about 0.95. However, the maximum photosynthetic quantum yield of an integrated process such as photosynthesis is lower than the theoretical yield when measured in chloroplasts (organelles) or whole leaves. Based on the biochemistry discussed in Chapter 8, we expect the theoretical maximum quantum yield for photosynthesis to be 0.125 for C_3 plants (one CO_2 molecule fixed per eight photons absorbed). But under today's atmospheric conditions (417 ppm CO_2, 21% O_2), the quantum yields for CO_2 of C_3 and C_4 leaves vary between 0.04 and 0.07 mole of CO_2 per mole of photons.

In C_3 plants the reduction from the theoretical maximum is caused primarily by energy loss through photorespiration. In C_4 plants the reduction is caused by the additional energy requirements of the CO_2-concentrating mechanism and the potential cost of refixing CO_2 that has diffused out from within the bundle sheath cells. If C_3 leaves are exposed to low O_2 concentrations, photorespiration is minimized, and the maximum quantum yield increases to about 0.09 mole of CO_2 per mole of photons. In contrast, if C_4 leaves are exposed to low O_2 concentrations, the quantum yields for CO_2 fixation remain constant at about 0.06 to 0.07 mole of CO_2 per mole of photons. This is because the carbon-concentrating mechanism in C_4 photosynthesis eliminates nearly all the oxygenation reaction of Rubisco and consequently the CO_2 evolved via photorespiration.

At higher PPFD along the light-response curve, the photosynthetic response to light starts to level off (Figures 9.6 and 9.7) and eventually approaches *saturation*. Beyond the light saturation point, net photosynthesis no longer increases, indicating that factors other than incident light, such as electron transport rate, Rubisco activity, or the metabolism of triose phosphates, limit photosynthesis. Light saturation levels for shade plants are substantially lower than those for

maximum quantum yield The ratio between photosynthetic product and the number of photons absorbed by a photosynthetic tissue. In a graphic plot of photon flux and photosynthetic rate, the maximum quantum yield is given by the slope of the linear portion of the curve.

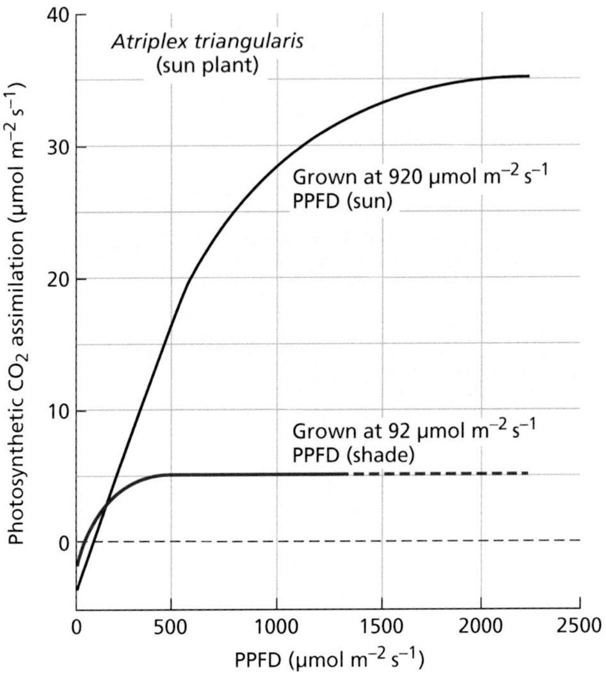

Figure 9.8 Light-response curve of photosynthesis of a sun plant grown under sun versus shade conditions. The upper curve represents an *A. triangularis* leaf grown at a PPFD ten times higher than that of the lower curve. In the plant grown at the lower light levels, photosynthesis saturates at a substantially lower PPFD, indicating that the photosynthetic properties of a leaf depend on its growing conditions. The dashed red line has been extrapolated from the measured part of the curve.

sun plants (Figure 9.7). This is also true for leaves of the same plant when grown in sun versus shade (**Figure 9.8**). These levels usually reflect the maximum PPFD to which a leaf was exposed during growth.

The light-response curve of most leaves saturates between 500 and 1,000 μmol m^{-2} s^{-1}, well below full sunlight (which is ~2,000 μmol m^{-2} s^{-1}). An exception to this is well-fertilized crop leaves, which often saturate above 1,000 μmol m^{-2} s^{-1}. Although individual leaves are rarely able to use full sunlight, whole plants usually consist of many leaves that shade each other. Thus, at any given time of the day only a small proportion of the leaves are exposed to full sun, especially in plants with dense canopies. The rest of the leaves receive subsaturating photon fluxes that come from sunflecks that pass through gaps in the leaf canopy, diffuse light, and light transmitted through other leaves.

Because the photosynthetic response of the intact plant is the sum of the photosynthetic activity of all the leaves, only rarely is photosynthesis light-saturated at the level of the whole plant (**Figure 9.9**). It is for this reason that crop productivity is usually related to the total amount of light received during the growing season, rather than to single-leaf photosynthetic capacity. Given enough water and nutrients, the more light a crop receives, the higher the biomass produced.

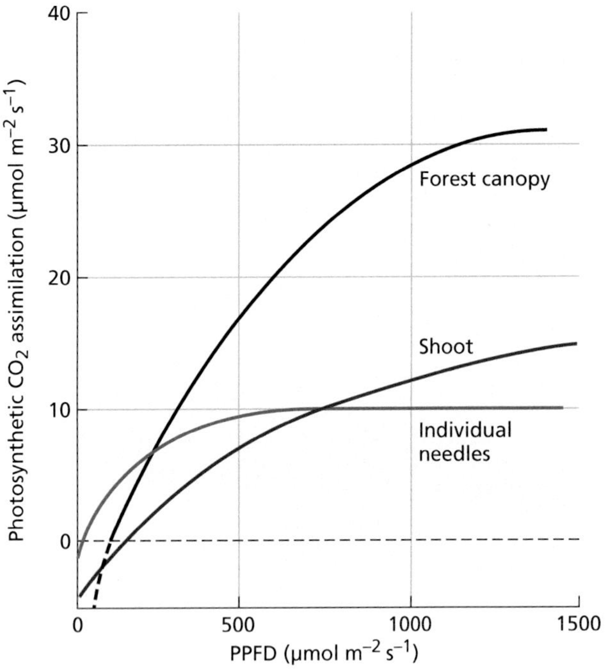

Figure 9.9 Changes in photosynthesis (expressed on a per-square-meter basis) in individual needles, a complex shoot, and a forest canopy of Sitka spruce (*Picea sitchensis*) as a function of PPFD. Complex shoots consist of groupings of needles that often shade each other, similar to the situation in a canopy where branches often shade other branches. As a result of shading, much higher PPFD levels are needed to saturate photosynthesis. The dashed portion of the forest canopy trace has been extrapolated from the measured part of the curve.

Leaves must dissipate excess light energy as heat

When exposed to excess light, leaves must dissipate the surplus absorbed light energy to prevent damage to the photosynthetic apparatus (**Figure 9.10**). There are several routes for energy dissipation that involve *nonphotochemical quenching*, the quenching of chlorophyll fluorescence by mechanisms other than photochemistry. The most important example involves the transfer of absorbed light energy away from electron transport toward heat production through the xanthophyll cycle.

THE XANTHOPHYLL CYCLE The xanthophyll cycle, which comprises the three carotenoids violaxanthin, antheraxanthin,

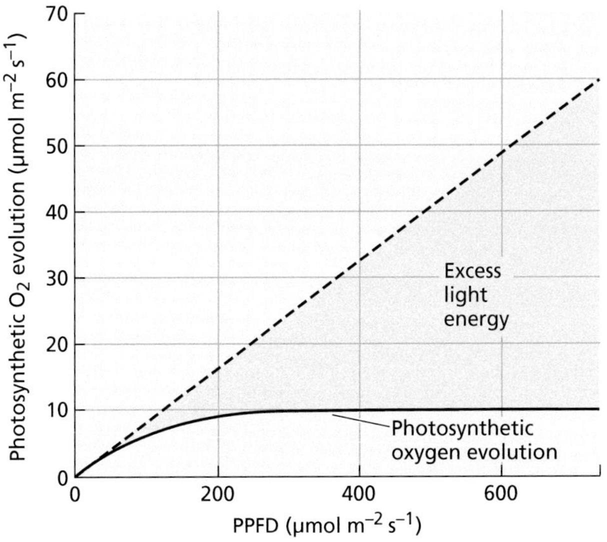

Figure 9.10 Excess light energy in relation to a light-response curve of photosynthetic oxygen evolution in a shade leaf. The dashed line shows theoretical oxygen evolution in the absence of any rate limitation to photosynthesis. At PPFD levels up to 150 μmol m^{-2} s^{-1}, a shade plant is able to use the absorbed light. Above 150 μmol m^{-2} s^{-1}, however, photosynthesis saturates, and an increasingly larger amount of the absorbed light energy must be dissipated. At higher PPFD levels there is a large difference between the fraction of light used by photosynthesis versus that which must be dissipated (excess light energy). The differences are much greater in a shade plant than in a sun plant.

and zeaxanthin, provides a mechanism to dissipate excess light energy in the leaf (**Figure 9.11**). Under high light, violaxanthin is reduced to antheraxanthin and then to zeaxanthin in the thylakoid lumen (the reverse reaction occurs in the stroma), and the reaction forms water at each step. Both of the aromatic rings of violaxanthin have a bound oxygen atom. In antheraxanthin only one of the two rings has a bound oxygen, and in zeaxanthin neither does. Zeaxanthin is the most effective of the three xanthophylls in heat dissipation, and antheraxanthin is only half as effective. Whereas the antheraxanthin content remains relatively constant throughout the day, the zeaxanthin content increases at high PPFD and decreases at low PPFD.

In leaves growing under full sunlight, zeaxanthin and antheraxanthin can make up 40% of the total xanthophyll-cycle pool at maximum PPFD levels attained at midday (**Figure 9.12**). In these conditions a substantial amount of excess light energy absorbed by the thylakoid membranes can be dissipated as heat, thus preventing damage to the photosynthetic machinery of the chloroplast. Leaves that grow in full sunlight contain a substantially larger xanthophyll pool than do shade leaves, so they can dissipate higher amounts of excess light energy.

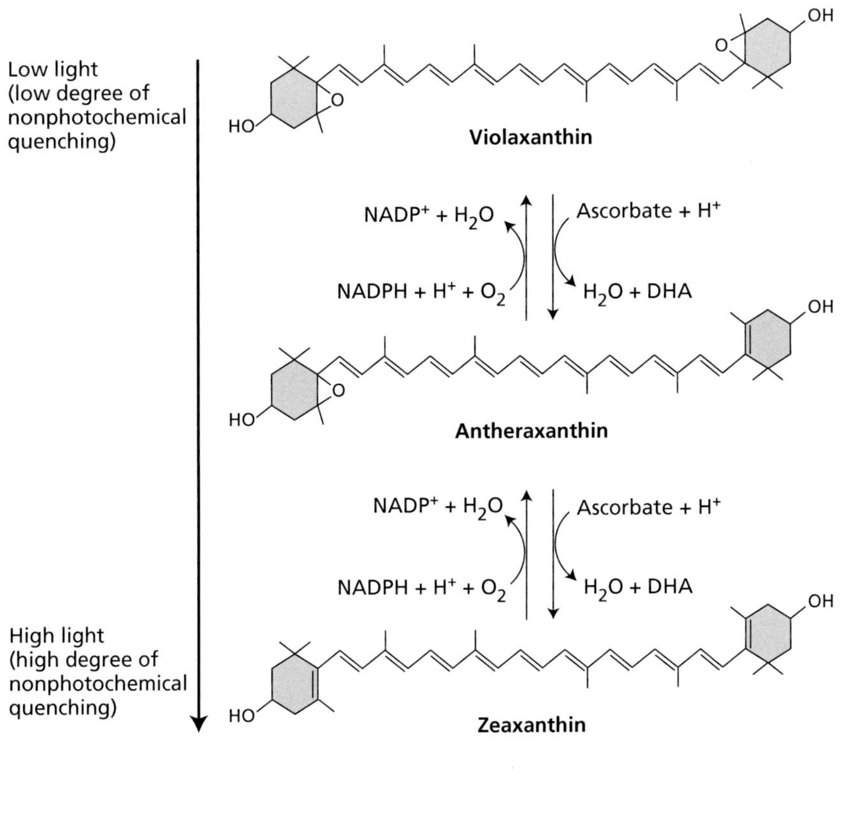

Figure 9.11 Chemical structure of violaxanthin, antheraxanthin, and zeaxanthin. The highly quenched state of PSII is associated with zeaxanthin, the unquenched state with violaxanthin. Enzymes interconvert these two carotenoids, with antheraxanthin as the intermediate, in response to changing conditions, especially changes in light intensity. Zeaxanthin formation uses ascorbate as a cofactor, and violaxanthin formation requires NADPH. DHA, dehydroascorbate.

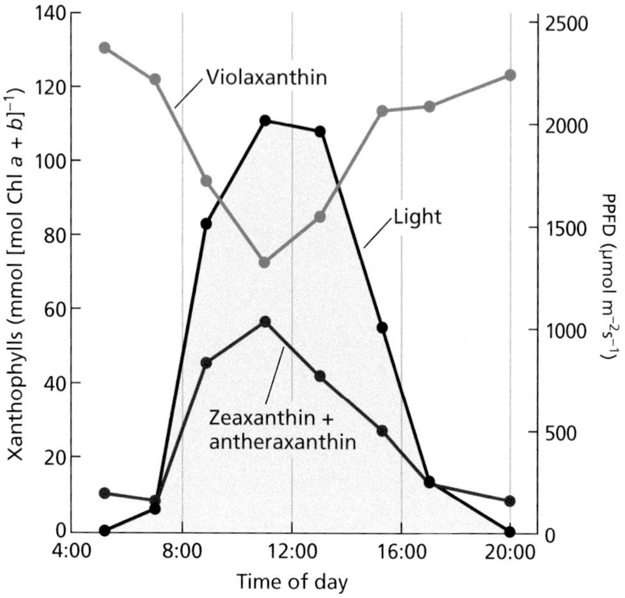

Figure 9.12 Diurnal changes in xanthophyll content as a function of PPFD in sunflower (*Helianthus annuus*). As the amount of light incident to a leaf increases, a greater proportion of violaxanthin is converted to antheraxanthin and zeaxanthin, thereby dissipating excess excitation energy and protecting the photosynthetic apparatus.

Nevertheless, the xanthophyll cycle also operates in plants that grow in the low light of the forest understory, where they are occasionally exposed to sunflecks. Exposure to just one sunfleck results in the conversion of much of the violaxanthin in the leaf to zeaxanthin.

The xanthophyll cycle is also important in species that remain green during winter, when photosynthetic rates are low yet light absorption remains high. Unlike in the diurnal cycling of the xanthophyll pool observed in the summer, zeaxanthin levels remain high all day during the winter. This mechanism maximizes dissipation of light energy, thereby protecting the leaves against photooxidation when winter cold prevents carbon assimilation.

CHLOROPLAST MOVEMENTS An alternative means of reducing the absorption of excess light energy is to move the chloroplasts within the cell so that they are no longer exposed to high light. Chloroplast movement is widespread among algae, mosses, and leaves of higher plants. If chloroplast orientation and location are controlled, leaves can regulate how much incident light is absorbed. Under low light, chloroplasts gather at the cell surfaces parallel to the plane of the leaf so that they are aligned perpendicularly to the incident light—a position that maximizes absorption of light (**Figure 9.13A** and **Figure 9.13B**; Chapter 13).

Under high light, the chloroplasts move to the cell surfaces that are parallel to the incident light, thus minimizing absorption of light (**Figure 9.13C**). Such chloroplast rearrangement can decrease the amount of light absorbed by the leaf

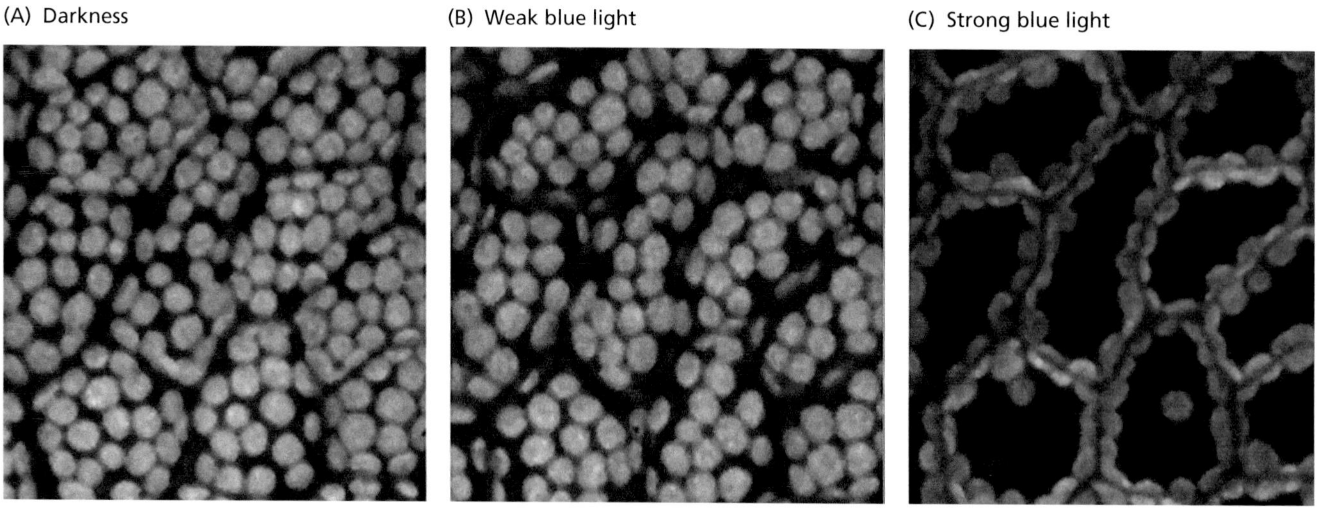

Figure 9.13 Chloroplast distribution in photosynthesizing cells of the duckweed *Lemna*. These surface views show the same cells under three conditions: (A) darkness, (B) weak blue light, and (C) strong blue light. In (A) and (B), chloroplasts are positioned near the upper surface of the cells, where they can absorb maximum amounts of light. When the cells are irradiated with strong blue light (C), the chloroplasts move to the side walls, where they shade each other, thus minimizing the absorption of excess light.

by about 15%. Chloroplast movement in leaves is a typical blue-light response (Chapter 13). Blue light also controls chloroplast orientation in many of the lower plants, but in some algae, chloroplast movement is controlled by phytochrome. In leaves, chloroplasts move along actin microfilaments in the cytoplasm, and calcium ions regulate their movement.

LEAF MOVEMENTS Plants have also evolved responses that reduce the excess radiation load on whole leaves during high sunlight periods, especially when transpiration and its cooling effects are reduced because of water stress. These responses often involve changes in the leaf orientation relative to the incoming sunlight. For example, heliotropic leaves of both alfalfa and lupine track the sun, but at the same time can reduce incident light levels by folding leaflets together so that the leaf blade become nearly parallel to the sun's rays (paraheliotropic), as discussed earlier. Another common response is mild wilting around solar noon, as seen in many sunflowers, whereby a leaf droops to a vertical orientation, again effectively reducing the incident heat load and reducing transpiration and incident light levels. Many grasses are able to effectively "twist" through loss of turgor in bulliform cells, resulting in reduced incident PPFD.

Absorption of too much light can lead to photoinhibition

When leaves are exposed to more light than they can use (Figure 9.10), the reaction center of PSII is inactivated and often damaged in a phenomenon called **photoinhibition**. The characteristics of photoinhibition in the intact leaf depend on the amount of light to which the plant is exposed. The two types of photoinhibition are dynamic photoinhibition and chronic photoinhibition.

Under moderate excess light, **dynamic photoinhibition** is observed. Quantum yield decreases, but the maximum photosynthetic rate remains unchanged. Dynamic photoinhibition is caused by the diversion of absorbed light energy toward photoprotective heat dissipation—hence the decrease in quantum yield. This decrease is often temporary, and quantum yield can return to its initial higher value when PPFD decreases below saturation levels. **Figure 9.14** shows how photons from sunlight are allocated to photosynthetic reactions versus being thermally dissipated as excess energy over the course of a day under favorable and stressed environmental conditions.

photoinhibition The inhibition of photosynthesis by excess light.

dynamic photoinhibition Photoinhibition of photosynthesis in which quantum efficiency decreases but the maximum photosynthetic rate remains unchanged. Occurs under moderate, not high, excess light.

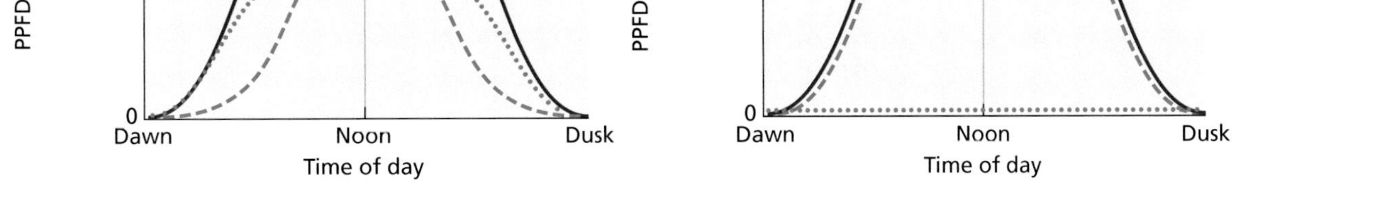

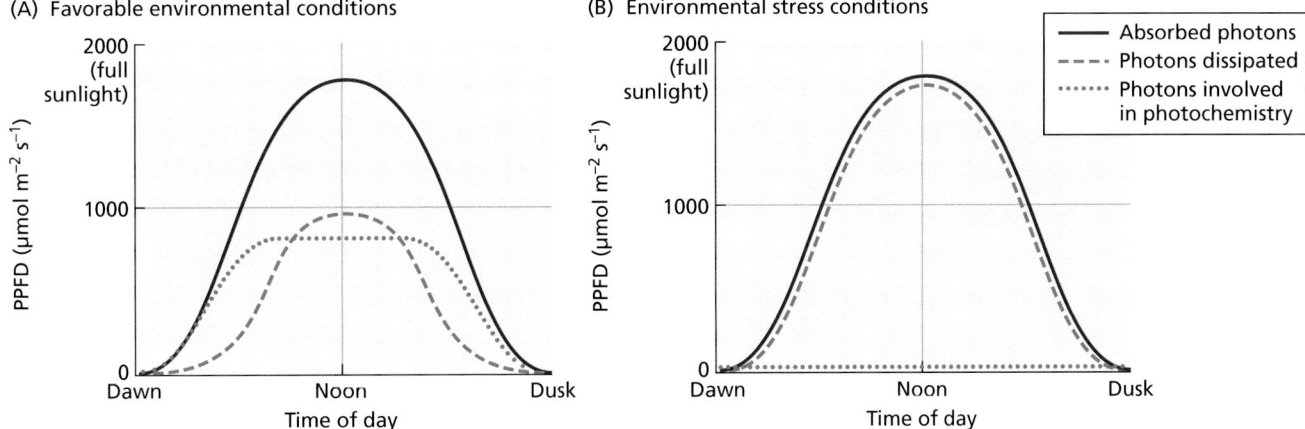

Figure 9.14 Changes over the course of a day in the allocation of photons absorbed by sunlight. Shown here are contrasts in how the photons striking a leaf are either involved in photochemistry or thermally dissipated as excess energy under favorable conditions (A) and stress conditions (B).

chronic photoinhibition Photoinhibition of photosynthentic activity in which both quantum efficiency and the maximum rate of photosynthesis are decreased. Occurs under high levels of excess light.

Chronic photoinhibition results from exposure to high levels of excess light that damage the photosynthetic system and decrease both instantaneous quantum yield and maximum photosynthetic rate. This would happen if the stress condition in Figure 9.14B persisted because photoprotection was not possible. In contrast to the effects of dynamic photoinhibition, the effects of chronic photoinhibition are relatively long-lasting, persisting for weeks or months.

Early researchers of photoinhibition interpreted decreases in quantum yield as damage to the photosynthetic apparatus. It is now recognized that short-term decreases in quantum yield reflect protective mechanisms, whereas chronic photoinhibition represents actual damage to the chloroplast resulting from excess light and a failure of the protective mechanisms.

How significant is photoinhibition in nature? Dynamic photoinhibition appears to occur daily, when leaves are exposed to maximum amounts of light and there is a corresponding reduction in carbon fixation. Photoinhibition is more pronounced at low temperatures and becomes chronic under more extreme climatic conditions.

9.3 Effects of Temperature on Photosynthesis in the Intact Leaf

Explain how temperature and CO_2 affect photosynthesis.

Photosynthesis (CO_2 uptake) and transpiration (H_2O loss) share a common pathway. That is, CO_2 diffuses into the leaf, and H_2O diffuses out, through the stomatal opening regulated by the guard cells (Chapter 3). While these are independent processes, vast quantities of water are lost during photosynthetic periods, with the molar ratio of H_2O loss to CO_2 uptake often exceeding 250. The high water-loss rate also removes heat from leaves through evaporative cooling, keeping leaves relatively cool even under full sunlight conditions. Thus, in addition to allowing CO_2 uptake for photosynthesis, transpiration helps prevent leaves from overheating (e.g., high temperature stress), but this water loss comes at a cost, especially in arid and semiarid ecosystems.

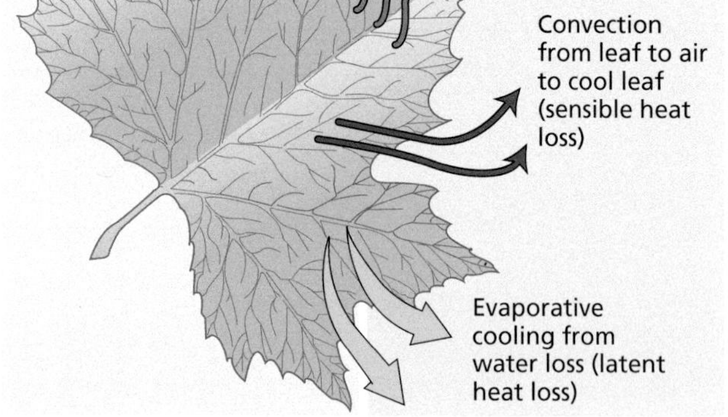

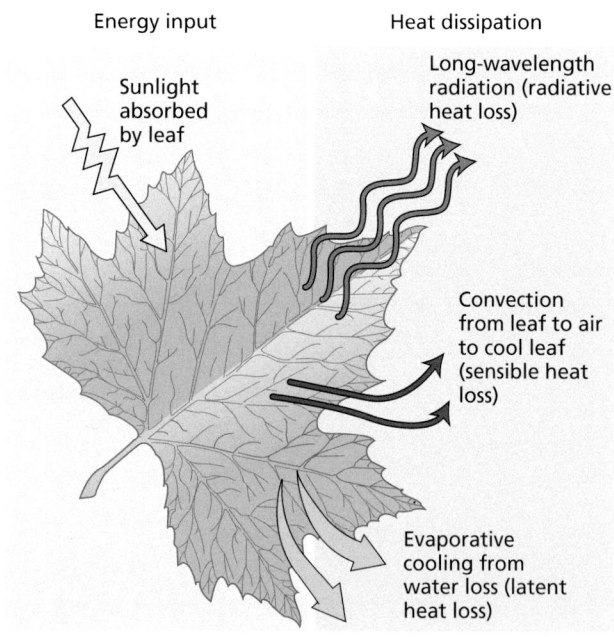

Figure 9.15 Absorption and dissipation of energy from sunlight by the leaf. The imposed heat load must be dissipated in order to avoid damage to the leaf. The heat load is dissipated by emission of long-wavelength radiation, by sensible heat loss to the air surrounding the leaf, and by the evaporative cooling caused by transpiration.

Leaves must dissipate vast quantities of heat

The heat load on a leaf exposed to full sunlight is very high. In fact, under normal sunny conditions with moderate air temperatures, a leaf would warm up to a dangerously high temperature if all incident solar energy were absorbed and none of the heat was dissipated. However, this does not occur because leaves absorb only about 50% of the total solar energy (300–3,000 nm), with most of the absorption occurring in the visible portion of the spectrum (Figures 9.2 and 9.3). This amount of energy absorption is still large, and a leaf can typically dissipate this heat load through three processes (**Figure 9.15**):

1. Radiative heat loss: All objects emit long-wave radiation (at about 10,000 nm) in proportion to their temperature (in absolute temperature, K) to the fourth power. However, the maximum emitted wavelength is inversely proportional to the leaf temperature, and leaf temperatures are low enough that the wavelengths emitted are not visible to the human eye.

2. Sensible heat loss: If the temperature of the leaf is higher than that of the air circulating around the leaf, the heat is convected (transferred) away from the leaf to the air. The size and shape of a leaf influence the amount of sensible heat loss, with smaller leaves being more efficient in heat loss than larger leaves.
3. Latent heat loss: Because the evaporation of water requires energy, when water evaporates from a leaf (transpiration), it removes large amounts of heat from the leaf and thus cools it. The human body is cooled by the same principle, commonly known as perspiration.

Bowen ratio The ratio of sensible heat loss to evaporative heat loss, the two most important processes in the regulation of leaf temperature.

Sensible heat loss and evaporative heat loss are the most important processes in the regulation of leaf temperature, and the ratio of the two fluxes is called the **Bowen ratio**:

$$\text{Bowen ratio} = \frac{\text{Sensible heat loss}}{\text{Evaporative heat loss}} \tag{9.2}$$

In well-watered crops, transpiration (Chapter 3), and hence water evaporation from the leaf, is high, so the Bowen ratio is low. Conversely, when evaporative cooling is limited, the Bowen ratio is high. For example, in a water-stressed crop, partial stomatal closure reduces evaporative cooling, and the Bowen ratio is increased. The amount of evaporative heat loss (and thus the Bowen ratio) is influenced by the degree to which stomata remain open.

Plants with very high Bowen ratios conserve water, but consequently may also experience high leaf temperatures. However, the temperature difference between the leaf and the air does increase the amount of sensible heat loss. Reduced growth is usually correlated with high Bowen ratios, because a high Bowen ratio is indicative of at least partial stomatal closure and reduced CO_2 availability for photosynthesis.

There is an optimal temperature for photosynthesis

Maintaining favorable leaf temperatures is crucial to plant growth because maximum photosynthesis occurs within a relatively narrow temperature range. The peak photosynthetic rate across a range of temperatures is the *photosynthetic thermal optimum*. When a plant is either above or below its optimal temperature, photosynthetic rates decrease. The photosynthetic thermal optimum reflects biochemical, genetic (adaptation), and environmental (acclimation) components.

Species adapted to different thermal regimes usually have an optimal temperature range for photosynthesis that reflects the temperatures of the environment in which they evolved. For example, there is an especially clear contrast between the C_3 plant *Atriplex glabriuscula*, which commonly grows in cool coastal environments, and the C_4 plant *Tidestromia oblongifolia*, from a hot desert environment (**Figure 9.16**). The ability to acclimate or biochemically adjust to temperature can also be found within species. When plants of the same species are grown at different temperatures and then tested for their photosynthetic response, they generally show photosynthetic thermal optima that correlate with the temperature at which they were grown. That is, plants of the same species grown at low temperatures have higher photosynthetic rates at low temperatures, whereas those same plants grown at high temperatures have higher photosynthetic rates at high

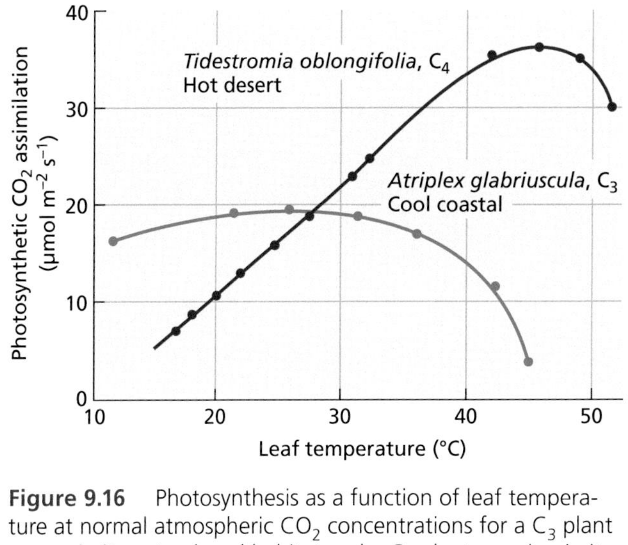

Figure 9.16 Photosynthesis as a function of leaf temperature at normal atmospheric CO_2 concentrations for a C_3 plant grown in its natural cool habitat and a C_4 plant growing in its natural hot habitat.

temperatures. Plants with a high thermal plasticity are capable of growing over a wide range of temperatures.

Changes in photosynthetic rates in response to temperature play an important role in plant adaptations to different environments and contribute to plants being productive even in some of the most extreme thermal habitats. In the lower temperature range, plants growing in alpine areas of Colorado and arctic regions in Alaska are capable of net CO_2 uptake at temperatures close to 0°C. At the other extreme, plants living in Death Valley, California, one of the hottest places on Earth, can achieve positive photosynthetic rates at temperatures approaching 50°C (Figure 9.16).

Photosynthesis is sensitive to both high and low temperatures

When photosynthetic rates are plotted as a function of temperature, the temperature-response curve has an asymmetric bell-type shape (Figure 9.16). In spite of some differences in shape, the temperature-response curve of photosynthesis among and within species has many common features. The ascending portion of the curve represents a temperature-dependent stimulation of enzymatic activities; the flat top is the temperature range that is optimal for photosynthesis; and the descending portion of the curve is associated with temperature-sensitive deleterious effects, some of which are reversible while others are not.

What factors are associated with the decline in photosynthesis above the photosynthetic temperature optimum? Temperature affects all biochemical reactions of photosynthesis as well as membrane integrity in chloroplasts, so it is not surprising that the responses to temperature are complex. Cellular respiration rates also increase as a function of temperature, but they are not the primary reason for the sharp decrease in net photosynthesis at high temperatures. A major impact of high temperature is on membrane-bound electron transport processes, which can become uncoupled or unstable at high temperatures. This cuts off the supply of reducing power and ATP needed to fuel net photosynthesis and leads to a sharp overall decrease in photosynthesis.

Under ambient CO_2 concentrations and with favorable light and soil moisture conditions, the photosynthetic thermal optimum is often limited by the activity of Rubisco. In leaves of C_3 plants, the response to increasing temperature reflects conflicting processes: an increase in carboxylation rate and a decrease in the affinity of Rubisco for CO_2 with a corresponding increase in the oxygenation rate and photorespiration (Chapter 8). There is also evidence that Rubisco activity decreases because of negative heat effects on Rubisco activase at higher (>35°C) temperatures. The reduction in the affinity for CO_2 and the increase in photorespiration attenuate the potential temperature response of photosynthesis under ambient CO_2 concentrations. By contrast, in plants with C_4 photosynthesis, the CO_2 surrounding Rubisco is saturating, or nearly so, and the negative effect of high temperature on Rubisco's affinity for CO_2 is not realized. This is one reason that leaves of C_4 plants tend to have a higher photosynthetic temperature optimum than do leaves of C_3 plants (Figure 9.16).

At low temperatures, the oxygenation rate catalyzed by Rubisco is decreased, but C_3 photosynthesis can become limited by factors such as phosphate availability in the chloroplast. When triose phosphates are exported from the chloroplast to the cytosol, an equimolar amount of inorganic phosphate is taken up via translocators in the chloroplast membrane. If the rate of triose phosphate used in the cytosol decreases, phosphate uptake into the chloroplast is inhibited and photosynthesis becomes phosphate-limited. Starch synthesis and sucrose synthesis decrease rapidly with decreasing temperature as the demand by sink tissues diminishes, reducing the consumption of triose phosphates and causing the phosphate limitation observed at low temperatures. C_4 plants are typically warm-adapted species, and their photosynthetic rates decrease sharply at low

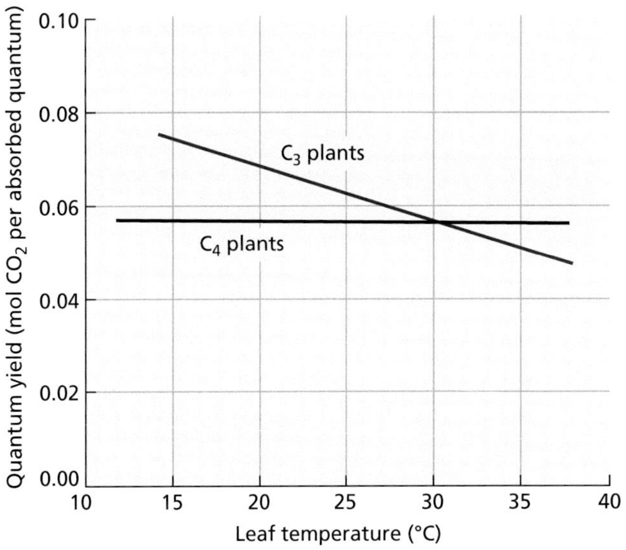

Figure 9.17 Quantum yield of photosynthetic carbon fixation in C_3 and C_4 plants as a function of leaf temperature. Photorespiration increases with temperature in C_3 plants, and the energy cost of net CO_2 fixation increases accordingly. This higher energy cost is reflected in lower quantum yields at higher temperatures. In contrast, photorespiration is very low in C_4 plants and the quantum yield does not show a temperature dependence. Note that at lower temperatures the quantum yield of C_3 plants is higher than that of C_4 plants, indicating that C_3 photosynthesis is more efficient at lower temperatures.

temperatures. However, some C_4 species, such as *Miscanthus*, appear to be less cold-sensitive than others, such as the warm-adapted maize (corn; *Zea mays*).

Photosynthetic efficiency is temperature-sensitive

Photorespiration (Chapter 8) and the quantum yield (light-use efficiency) differ between C_3 and C_4 photosynthesis, with changes particularly noticeable as temperatures vary. **Figure 9.17** illustrates quantum yield for photosynthesis as a function of leaf temperature in C_3 and C_4 plants in today's atmosphere of 417 ppm CO_2. In the C_4 plants, the quantum yield remains constant with temperature, reflecting low rates of photorespiration. In the C_3 plants, the quantum yield decreases with temperature, reflecting a stimulation of photorespiration by temperature and an ensuing higher energy cost for net CO_2 fixation.

The reduced quantum yield due to increased photorespiration leads to expected differences in the photosynthetic capacities of C_3 and C_4 plants in habitats with different temperatures. **Figure 9.18** shows the predicted relative rates of primary productivity of C_3 and C_4 grasses along a latitudinal transect in the Great Plains of North America from southern Texas in the United States to Manitoba in Canada. The decline in C_4 relative to C_3 productivity moving northward closely parallels the shift in abundance of plants with these pathways in the Great Plains: C_4 species are more common below 40°N, and C_3 species dominate above 45°N.

9.4 Effects of Carbon Dioxide on Photosynthesis in the Intact Leaf

Describe the relationship between different photosynthetic lifestyles and the atmospheric CO_2 concentration.

Atmospheric CO_2 diffuses into leaves—first through stomata, then through the intercellular air spaces, and ultimately into cells and chloroplasts. In the presence of adequate amounts of light, higher CO_2 concentrations typically support higher photosynthetic rates. The reverse is also true: Low CO_2 concentrations can limit the amount of photosynthesis, particularly in C_3 plants.

In this section we discuss the concentration of atmospheric CO_2 in recent history, and its availability for carbon-fixing processes. Then we consider the limitations that CO_2 places on photosynthesis and the impact of the CO_2-concentrating mechanisms of C_4 and CAM plants.

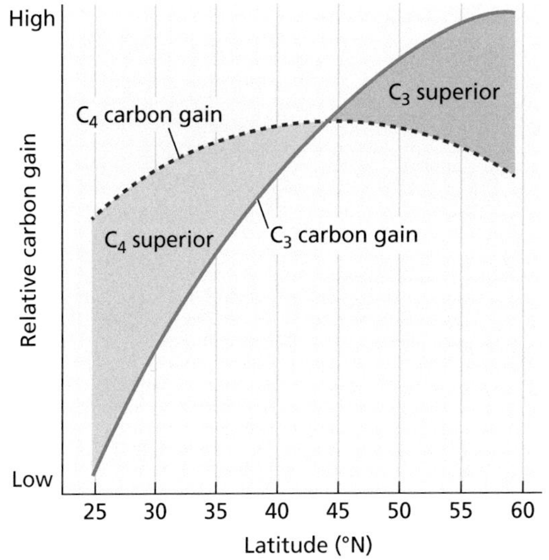

Figure 9.18 Relative rates of photosynthetic carbon gain predicted for identical C_3 and C_4 grass canopies as a function of latitude across the Great Plains of North America.

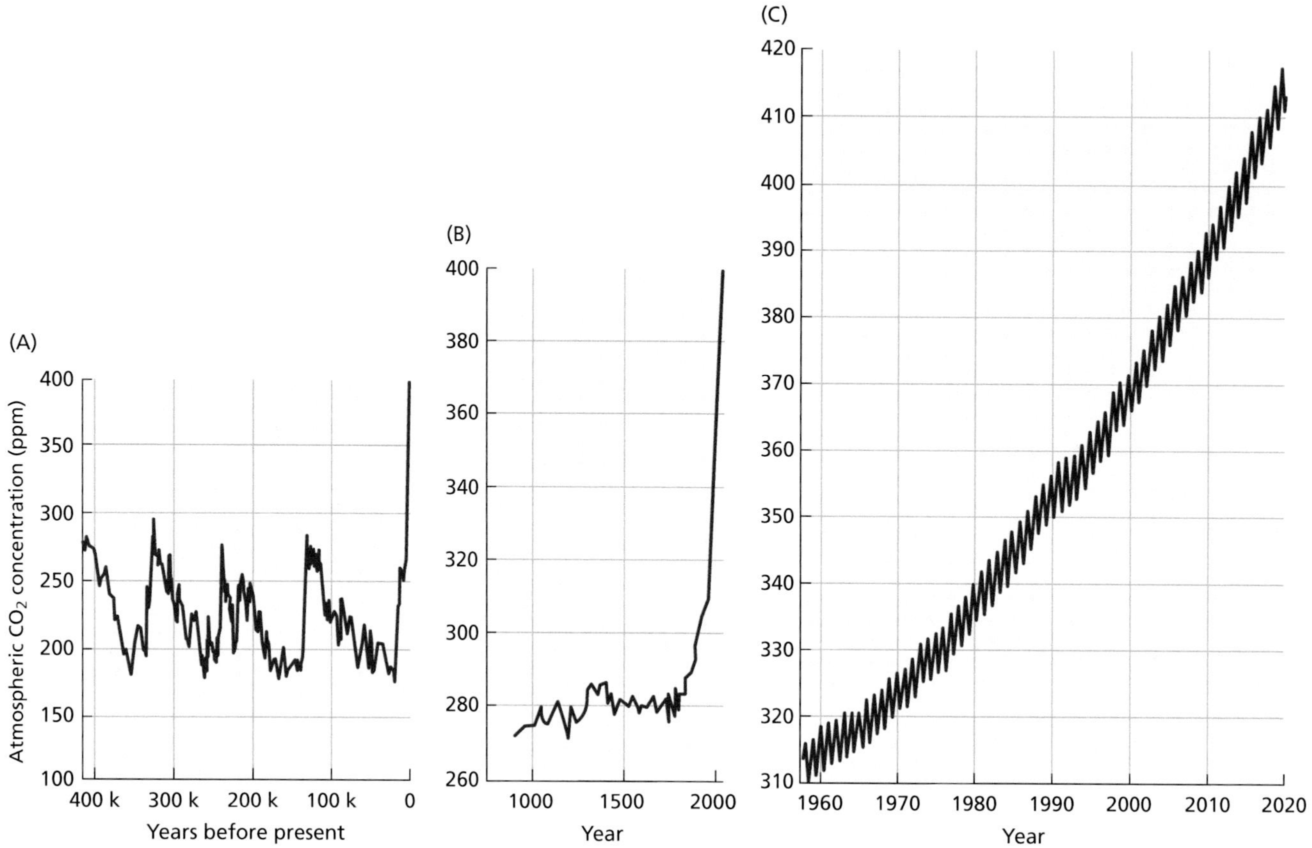

Figure 9.19 Concentration of atmospheric CO_2 from 420,000 years ago to the present. (A) Past atmospheric CO_2 concentrations, determined from bubbles trapped in glacial ice in Antarctica, were much lower than current levels. (B) In the last 1,000 years, the rise in atmospheric CO_2 concentration coincides with the Industrial Revolution and the increased burning of fossil fuels. (C) Current atmospheric concentrations of CO_2 measured at Mauna Loa, Hawaii, continue to rise. The wavy nature of the trace is caused by change in atmospheric CO_2 concentrations associated with seasonal changes in relative balance between photosynthesis and respiration rates as well as seasonal forest fires. Each year the highest CO_2 concentration is observed in May, just before the Northern Hemisphere growing season, and the lowest concentration is observed in October.

Atmospheric CO_2 concentration keeps rising

Carbon dioxide presently accounts for about 0.0417%, or 417 ppm, of atmospheric air. The partial pressure of ambient CO_2 (c_a) varies with atmospheric pressure and is approximately 40 pascals (Pa) at sea level. Water vapor usually accounts for up to 2% of the atmosphere and O_2 for about 21%. The largest constituent in the atmosphere is diatomic nitrogen, at about 78%.

Today the atmospheric concentration of CO_2 is almost twice the concentration that prevailed over the last 400,000 years, as measured from air bubbles trapped in glacial ice in Antarctica (**Figure 9.19A** and **Figure 9.19B**), and it is higher than any experienced on Earth in the last 2 million years. Most extant plant taxa are therefore thought to have evolved in a low-CO_2 world (~180–280 ppm CO_2). Only when one looks back more than 35 million years does one find much higher CO_2 concentrations (>1,000 ppm). Thus, the geologic trend over these many millions of years was one of decreasing atmospheric CO_2 concentrations.

Currently, the CO_2 concentration of the atmosphere is increasing by about 1 to 3 ppm each year, primarily because of the burning of fossil fuels (e.g., coal, oil, and natural gas) and deforestation (**Figure 9.19C**). Since 1958, when C. David

Keeling began systematic measurements of CO_2 in the clean air at Mauna Loa, Hawaii, atmospheric CO_2 concentrations have increased by more than 30%. By year 2100 the atmospheric CO_2 concentration could reach 600 to 750 ppm unless fossil fuel emissions and deforestation are diminished.

boundary layer resistance (r_b) The resistance to the diffusion of water vapor, CO_2, and heat due to the layer of unstirred air next to the leaf surface. A component of diffusional resistance.

stomatal resistance A measurement of the limitation to the free diffusion of gases from and into the leaf posed by the stomatal pores. The inverse of stomatal conductance.

CO_2 diffusion to the chloroplast is essential to photosynthesis

For photosynthesis to occur in C_3 plants, CO_2 must diffuse from the atmosphere into the leaf and to the carboxylation site of Rubisco. The diffusion rate depends on the CO_2 concentration gradient in the leaf (Chapters 3 and 8) and on resistances along the diffusion pathway, which comprises both gas and liquid routes, or phases. The cuticle that covers the leaf is nearly impermeable to CO_2, so the main port of entry of CO_2 into the leaf is the stomatal pore. (The same path is traveled in the reverse direction by H_2O.) CO_2 diffuses through the pore into the substomatal cavity and into the intercellular air spaces between the mesophyll cells. This portion of the diffusion path of CO_2 is in the gas phase. The remainder of the diffusion path to the chloroplast is in the liquid phase, which begins at the water layer that wets the walls of the mesophyll cells and continues through the plasma membrane, the cytosol, and the chloroplast.

The sharing of the stomatal entry pathway by CO_2 and H_2O presents the plant with a functional dilemma. In air of high relative humidity, the diffusion gradient that drives water loss is about 50 times larger than the gradient that drives CO_2 uptake. In drier air, this difference can be much larger. Therefore, a decrease in stomatal resistance through the opening of stomata facilitates higher CO_2 uptake but is unavoidably accompanied by substantial water loss. Not surprisingly, many adaptive features help counteract this water loss in plants in arid and semiarid regions of the world.

Each portion of the CO_2 diffusion pathway imposes a resistance to CO_2 diffusion, so the supply of CO_2 for photosynthesis meets a series of different points of resistance. The gas phase of CO_2 diffusion into the leaf can be divided into three components (the boundary layer, the stomata, and the intercellular spaces of the leaf), each of which can impose a significant resistance to CO_2 diffusion (**Figure 9.20**). An evaluation of the magnitude of each point of resistance is helpful for understanding CO_2 limitations to photosynthesis.

The boundary layer consists of relatively unstirred air near the leaf surface, and its resistance to diffusion is called the **boundary layer resistance**. The boundary layer resistance affects all diffusive processes, including water and CO_2 diffusion as well as sensible heat loss, discussed earlier. The boundary layer resistance decreases with smaller leaf size and greater wind speed. Smaller leaves thus have a lower resistance to CO_2 and water diffusion and tend to have higher sensible heat loss. Leaves of desert plants are usually small, facilitating sensible heat loss. In contrast, large leaves are often found in the humid tropics, especially in the shade. These leaves have large boundary layer resistances, but they can dissipate the radiation heat load by evaporative cooling made possible by the abundant water supply in these habitats.

After diffusing through the boundary layer, CO_2 enters the leaf through the stomatal pores, which imposes the next type of resistance in the diffusion pathway, **stomatal resistance**.

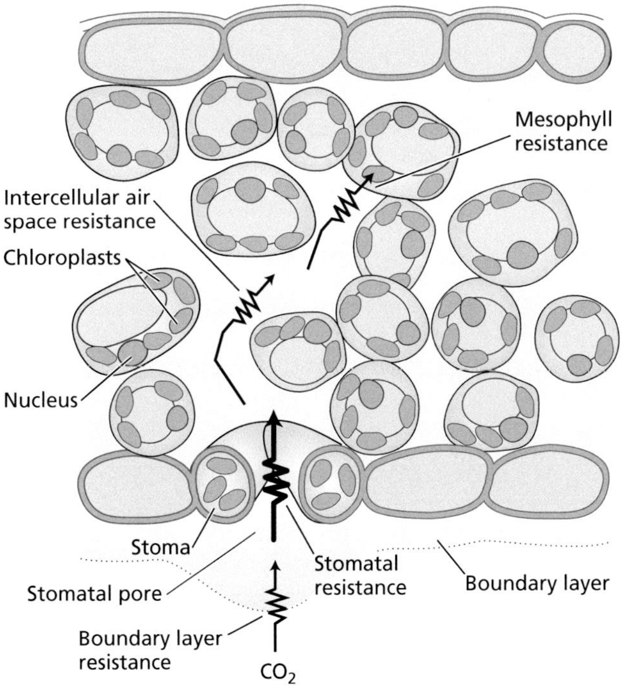

Figure 9.20 Points of resistance to the diffusion and fixation of CO_2 from outside the leaf to the chloroplasts. The stomatal aperture is the major point of resistance to CO_2 diffusion into the leaf. In some cases, the nuclei and chloroplasts are depicted to be in front of the vacuole, not inside it (Figure 9.13).

intercellular air space resistance The resistance or hindrance that slows down the diffusion of CO_2 inside a leaf, from the substomatal cavity to the walls of the mesophyll cells.

mesophyll resistance The resistance to CO_2 diffusion imposed by the liquid phase inside leaves. The liquid phase includes diffusion from the intercellular leaf spaces to the carboxylation sites in the chloroplast.

CO_2 compensation point The CO_2 concentration at which the rate of respiration balances the photosynthetic rate.

Under most conditions in nature, in which the air around a leaf is seldom completely still, the boundary layer resistance is much smaller than the stomatal resistance, and the main limitation to CO_2 diffusion into the leaf is imposed by the stomatal resistance.

There are two additional resistances within the leaf. The first is resistance to CO_2 diffusion in the air spaces that separate the substomatal cavity from the walls of the mesophyll cells. This is called the **intercellular air space resistance**. The second is the **mesophyll resistance**, which is resistance to CO_2 diffusion in the liquid phase. In C_3 plants, localization of chloroplasts near the cell periphery minimizes the distance that CO_2 must diffuse through liquid to reach carboxylation sites within the chloroplast. The mesophyll resistance to CO_2 diffusion can have a similar limitation to photosynthesis as stomatal resistance. Because the stomatal guard cells can impose a variable and potentially large resistance to CO_2 influx and water loss in the diffusion pathway, regulating the stomatal aperture provides the plant with an effective way to control gas exchange between the leaf and the atmosphere. Alternatively, decreasing mesophyll resistance may increase CO_2 availability for photosynthesis without negatively affecting leaf water loss because mesophyll resistance is not thought to influence the movement of water out of the leaf. Traits such as mesophyll cell wall thickness and the mesophyll cell surface area exposed to the intercellular air space can both influence mesophyll resistance. These traits have potentially important implications for breeding increased photosynthetic water-use efficiency in agronomically important plants, particularly when soil water availability is low.

CO_2 supply imposes limitations on photosynthesis

For C_3 plants growing with adequate light, water, and nutrients, CO_2 enrichment above natural atmospheric concentrations results in increased photosynthesis and enhanced productivity. Expressing photosynthetic rate as a function of the partial pressure of CO_2 in the intercellular air space (c_i) within the leaf makes it possible to evaluate limitations to photosynthesis imposed by CO_2 supply. At low c_i concentrations, photosynthesis is strongly limited by the low CO_2. In the absence of photosynthesis, leaves release CO_2 because of mitochondrial respiration (Chapter 11).

Increasing c_i to the concentration at which photosynthesis and respiration balance each other defines the **CO_2 compensation point**. This is the point at which net assimilation of CO_2 by the leaf is zero (**Figure 9.21**). This concept is analogous to that of the light compensation point discussed earlier (Figure 9.6). The CO_2 compensation point reflects the balance between photosynthesis uptake of CO_2 and the CO_2 released by photorespiration and respiration under constant PPFD, whereas the light compensation point reflects that balance as a function of PPFD under constant CO_2 concentration.

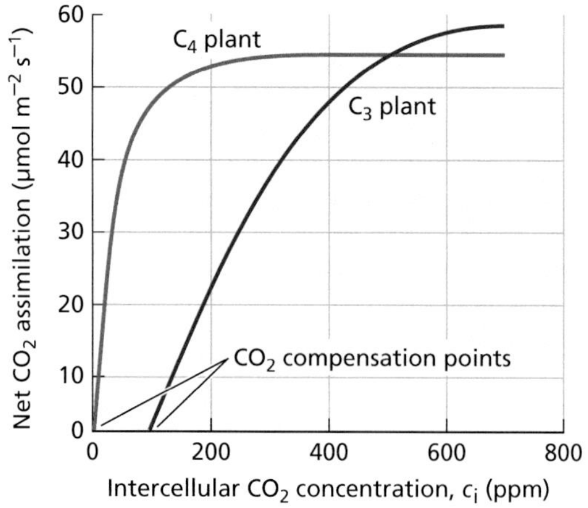

Figure 9.21 Changes in photosynthesis as a function of intercellular CO_2 concentrations in Arizona honeysweet (*Tidestromia oblongifolia*), a C_4 plant, and creosote bush (*Larrea tridentata*), a C_3 plant. Photosynthetic rate is plotted against calculated intercellular CO_2 concentration inside the leaf. The intercellular CO_2 concentration at which net CO_2 assimilation is zero defines the CO_2 compensation point.

C_3 VERSUS C_4 PLANTS In C_3 plants, increasing c_i above the compensation point increases photosynthesis over a wide concentration range (Figure 9.21). At low to intermediate CO_2 concentrations, photosynthesis is limited by the carboxylation capacity of Rubisco. At higher c_i concentrations, photosynthesis begins to saturate as the net photosynthetic rate becomes limited by another factor. At these higher c_i concentrations, net photosynthesis becomes limited by the capacity of the light reactions to generate sufficient NADPH and ATP to regenerate the acceptor molecule ribulose 1,5-bisphosphate. Most leaves appear to regulate their c_i values by controlling stomatal opening, so that c_i remains at an intermediate subambient concentration between the limits imposed by carboxylation capacity and the capacity to regenerate ribulose 1,5-bisphosphate. In this way, both the light and dark reactions of photosynthesis are co-limiting. A plot of net CO_2 assimilation as a function of c_i tells us how photosynthesis is regulated by CO_2, independent of the functioning of stomata (Figure 9.21).

Comparing such a plot for C_3 and C_4 plants reveals interesting differences between the two pathways of carbon metabolism (Chapter 8):

- In C_4 plants, photosynthetic rates saturate at CO_2 concentrations in the intercellular air space (c_i) of about 100 to 200 ppm, reflecting the effective CO_2-concentrating mechanisms operating in these plants.
- In C_3 plants, increasing c_i concentrations continue to stimulate photosynthesis over a much broader CO_2 range than for C_4 plants.
- In C_4 plants, the CO_2 compensation point is nearly zero, reflecting their very low levels of photorespiration.
- In C_3 plants, the CO_2 compensation point is about 50 to 100 ppm at 25°C, reflecting CO_2 produced by photorespiration.

These responses reveal that C_3 plants will be more likely than C_4 plants to benefit from today's ongoing increases in atmospheric CO_2 concentrations (Figure 9.21). Because photosynthesis in C_4 plants is saturated at low CO_2 concentrations, C_4 plants do not benefit as much from increases in atmospheric CO_2 concentrations. However, the response of photosynthesis in both C_3 and C_4 plants to changes in CO_2 concentrations is strongly influenced by leaf temperature and soil water availability, which are also expected to change as atmospheric CO_2 concentrations rise.

From an evolutionary perspective, the ancestral photosynthetic pathway is C_3 photosynthesis, and C_4 photosynthesis evolved from it. During earlier geologic time periods (>35 million years ago) when atmospheric CO_2 concentrations were much higher than they are today, CO_2 diffusion through stomata into leaves would have resulted in higher c_i values and therefore high photosynthetic rates and reduced photorespiration in C_3 plants. However, the lower CO_2 concentrations in the atmosphere that have occurred more recently would have led to reduced photosynthetic rates and higher photorespiration. The evolution of C_4 photosynthesis is one biochemical adaptation to a CO_2-limited atmosphere. Our current understanding is that the earliest C_4 plants evolved recently in geologic terms, approximately 25 to 30 million years ago, when the atmospheric CO_2 concentration was relatively low.

C_4 photosynthesis first became a prominent component of terrestrial ecosystems in the warmest growing regions of Earth when global CO_2 concentrations decreased below a critical threshold (**Figure 9.22**), and stomatal conductance was low due to soil aridity and low relative humidity in the atmosphere. The negative impacts of high photorespiration and CO_2 limitation on C_3 photosynthesis would have been greatest under these warm to hot growing conditions, drought, and low atmospheric CO_2. These conditions were likely key factors favoring the evolution and ultimately geographic expansion of C_4 plants.

Because of the CO_2-concentrating mechanisms in C_4 plants, the CO_2 concentration at the site of Rubisco carboxylation within C_4 chloroplasts is typically close to saturating for Rubisco carboxylation activity. As a

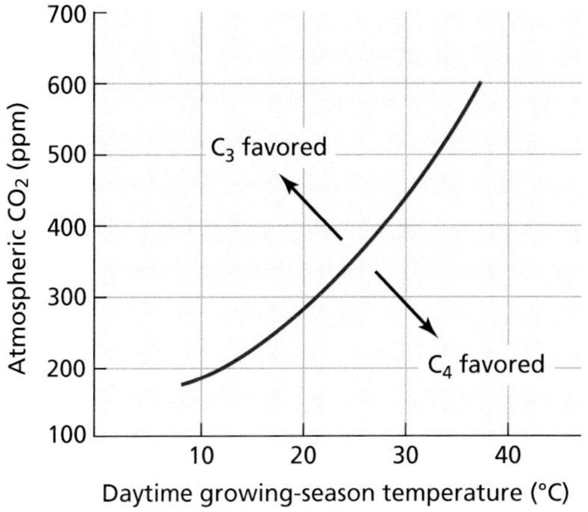

Figure 9.22 Combination of global atmospheric CO_2 concentrations and daytime growing-season temperatures that are predicted to favor C_3 versus C_4 grasses. At any point in time, Earth is at a single atmospheric CO_2 concentration, resulting in the expectation that C_4 plants would be most common in habitats with the warmest growing seasons.

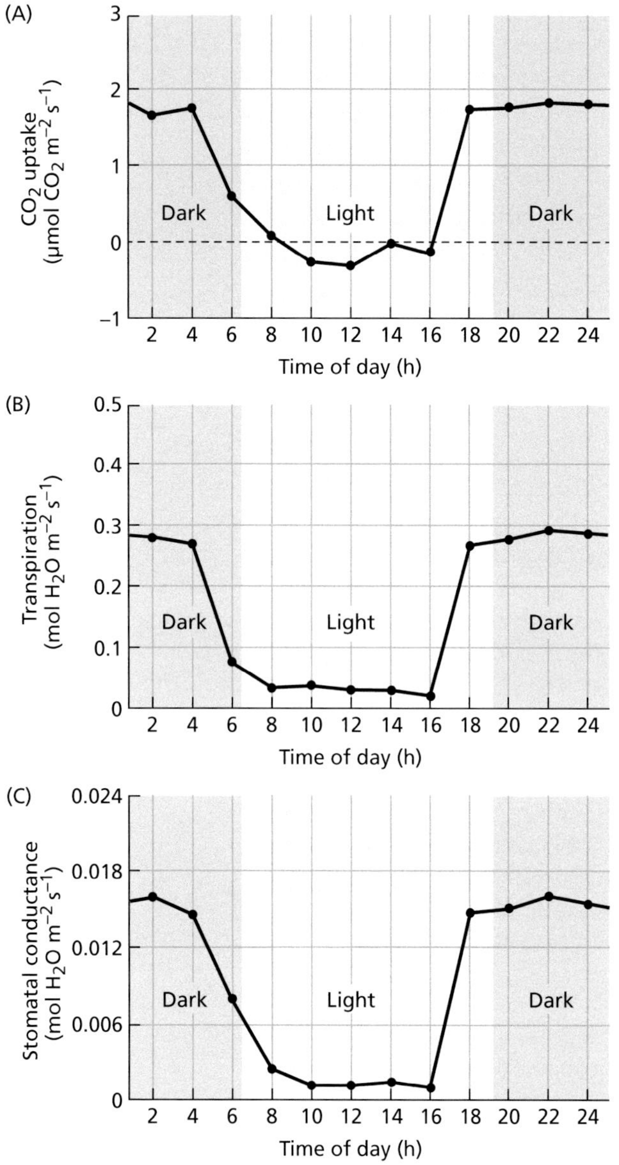

Figure 9.23 Photosynthetic net CO_2 uptake, H_2O transpiration, and stomatal conductance of a CAM plant, the orchid *Doritaenopsis*, during a 24-h period. Whole plants were kept in a gas-exchange chamber in the laboratory. The dark period is indicated by dark areas. Three parameters were measured over the study period: (A) photosynthetic rate, (B) water loss, and (C) stomatal conductance. In contrast to plants with C_3 or C_4 metabolism, CAM plants open their stomata and fix CO_2 at night.

result, plants with C_4 metabolism require less Rubisco than C_3 plants to achieve a given rate of photosynthesis, and thus use less nitrogen for growth. In addition, the CO_2-concentrating mechanism allows the leaf to maintain high photosynthetic rates at lower c_i values. This permits stomata to remain relatively closed, resulting in less water loss for a given rate of photosynthesis. Thus, the CO_2-concentrating mechanism helps C_4 plants use water and nitrogen more efficiently than C_3 plants. However, the additional energy cost required by the CO_2-concentrating mechanism (Chapter 8) reduces the light-use efficiency of C_4 photosynthesis. This is probably one reason that most shade-adapted plants in temperate regions are not C_4 plants and that C_4 plants are typically found in open habitats with high light.

CAM PLANTS Plants with crassulacean acid metabolism (CAM), including many cacti, orchids, bromeliads, and other succulents, have stomatal activity patterns that contrast with those found in C_3 and C_4 plants. CAM plants open their stomata at night and close them during the day, exactly the opposite of the pattern observed in leaves of C_3 and C_4 plants (**Figure 9.23**). At night, atmospheric CO_2 diffuses into CAM plants where it is combined with phosphoenolpyruvate and fixed into oxaloacetate, which is reduced to malate (Chapter 8). Then during the day CO_2 is released from malate and taken up by Rubisco behind closed stomata. Because stomata are open primarily at night, when lower temperatures and higher humidity reduce transpiration demand, the ratio of water loss to CO_2 uptake is much lower in CAM plants than it is in either C_3 or C_4 plants.

The main photosynthetic constraint on CAM metabolism is that the capacity to store malic acid is limited, and this limitation restricts the total amount of CO_2 uptake. However, the daily cycle of CAM photosynthesis can be very flexible. Some CAM plants are able to enhance total photosynthesis during wet conditions by directly fixing atmospheric CO_2 via the Calvin–Benson cycle at the beginning and end of the day, when temperature gradients are less extreme. Other plants may use CAM as a survival mechanism during severe water limitations. For example, cladodes (flattened stems) of cacti can survive after detachment from the plant for several months without water. Their stomata are closed all the time, and the CO_2 released by respiration is refixed into malate. This process, which has been called *CAM idling*, also allows the intact plant to survive for prolonged drought periods while losing remarkably little water.

How will photosynthesis and respiration change in the future under elevated CO_2 conditions?

The consequences of increasing global atmospheric CO_2 are under intense scrutiny by scientists and government agencies, particularly because of the evidence that the greenhouse effect is altering the world's climate. The **greenhouse effect** refers to the warming of Earth's climate that is caused by the trapping of long-wavelength radiation by the atmosphere.

A greenhouse roof transmits visible light, which is absorbed by plants and other surfaces inside the greenhouse. Some of the absorbed light energy is converted to heat, and some of it is reemitted as long-wavelength radiation. Because glass poorly transmits long-wavelength radiation, this radiation cannot leave the greenhouse through the glass roof, and the greenhouse heats up. Certain gases in the atmosphere, particularly CO_2 and methane, play a role similar to that of the glass roof in a greenhouse. The increased CO_2 concentrations, and elevated temperatures associated with the greenhouse effect, affect photosynthesis and plant growth in a number of ways. At today's atmospheric CO_2 concentrations, photosynthesis in C_3 plants is CO_2-limited. However, other factors, such as temperature, water availability, and nutrient cycling are expected to have a greater impact on photosynthesis as atmospheric CO_2 concentrations continue to rise. At extremes, all of these factors will have negative impacts, but moderate increases in leaf temperatures and decreased stomatal conductance would be expected to marginally increase photosynthetic rates.

greenhouse effect The warming of Earth's climate, caused by the trapping of long-wavelength radiation by CO_2 and other gases in the atmosphere. The term is derived from the heating of a greenhouse that results from the penetration of long-wavelength radiation through the glass roof, the conversion of the long-wave radiation to heat, and the blocking of the heat escape by the glass roof.

A central question in plant physiology today is: How will photosynthesis and respiration differ by the year 2100 when global CO_2 concentrations have reached 600 ppm, 700 ppm, or even higher? This question is particularly relevant as humans continue to add CO_2 derived from fossil fuel combustion to Earth's atmosphere. Under well-watered and highly fertilized laboratory conditions, most C_3 plants grow about 30% faster when the CO_2 concentration reaches 600 to 750 ppm than they do today; above that atmospheric CO_2 concentration, the growth rate becomes more limited by the nutrients available to the plant. To study this question in the field, scientists need to be able to create realistic simulations of future environments. A promising approach to the study of plant physiology and ecology in environments with elevated CO_2 concentrations has been the use of *Free Air CO_2 Enrichment* (FACE) experiments. This is an ongoing long-term experiment spanning different biomes and continents.

For FACE experiments, entire fields of plants or natural ecosystems are encircled by emitters that add CO_2 to the air to create the high-CO_2 environment we might expect to have 20 to 30 years from now. **Figure 9.24** shows FACE experiments in a deciduous forest and a crop canopy.

FACE experiments have provided key new insights into how plants and ecosystems will respond to the CO_2 concentrations expected in the future. One key observation is that plants with the C_3 photosynthetic pathway are much more responsive than C_4 plants under well-watered conditions, with net photosynthetic rates increasing 20% or more in C_3 plants and little to not at all in C_4 plants. Photosynthesis increases in C_3 plants because c_i concentrations increase (Figure 9.21). At the same time, there is a down-regulation of photosynthetic capacity manifested in reduced activity of the enzymes associated with the enzymatic reactions of photosynthesis. The FACE experiments have also demonstrated that under field conditions, which may have many additional limitations to plant growth, the stimulation of crop yield is generally lower than would be predicted from experiments conducted in greenhouse or growth chambers, which often minimize the influence of these other limiting factors.

Elevated CO_2 concentrations will affect many plant processes. For instance, leaves tend to keep their stomata more closed under elevated CO_2 concentrations. As a direct consequence of reduced transpiration, leaf temperatures are higher (Figure 9.24C), which may feedback on both photosynthesis, as previously discussed, and mitochondrial respiration. From FACE studies, it has become increasingly clear that acclimation occurs under higher CO_2 concentrations in

(A)

Figure 9.24 Free Air CO_2 Enrichment (FACE) experiments are used to study how plants and ecosystems will respond to future CO_2 concentrations. Shown here are FACE experiments in (A) a deciduous forest and (B) a crop canopy. (C) Infrared image of a crop canopy from (B). Under elevated CO_2 concentrations, leaf stomata are more closed, resulting in higher leaf temperatures.

(B)

(C)

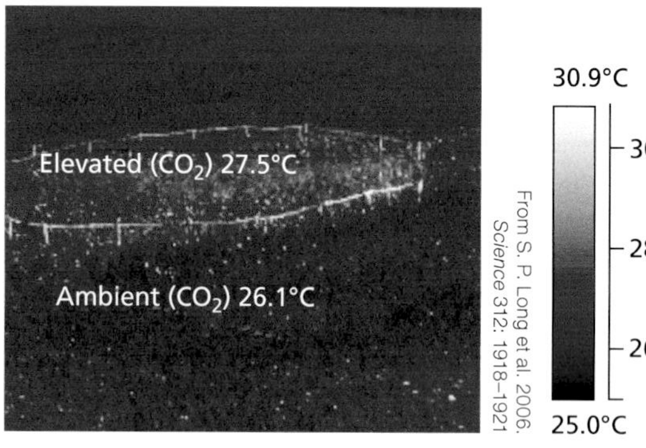

which photosynthesis and respiration rates are different than they would be under today's atmospheric conditions, but not as high as would have been predicted without the down-regulation of acclimation response.

While CO_2 is indeed important for photosynthesis and respiration, other factors are important for growth under elevated CO_2. For example, a common FACE observation is that plant growth under elevated CO_2 concentrations quickly becomes constrained by nutrient availability (remember Blackman's rule of limiting factors). A second observation is that the presence of pollutant trace gases, such as ozone, can reduce the net photosynthetic response below the maximum values predicted from initial FACE and greenhouse studies of nearly four decades ago.

As a result of increased atmospheric CO_2 concentrations, warmer and drier conditions are also predicted to occur in the near future, which we are already starting to see. The interaction of CO_2, temperature, and water availability has been demonstrated to have important implications for predicting current and future plant productivity, in both managed and natural ecosystems. Of particular importance is that the stimulation of photosynthesis and yield by increases in atmospheric CO_2 concentrations is often partially reduced by increases in temperature and restricted stomatal conductance under drought conditions. Important progress is being made by studying how growth of irrigated and fertilized crops compares with that of plants in natural ecosystems in a world with elevated CO_2. Understanding these responses is crucial as society looks for increased agricultural outputs to support rising human populations and to provide raw materials for biofuels, while minimizing the need for resource inputs.

Summary

In considering photosynthetic performance, both the limiting factor hypothesis and an "economic perspective" emphasizing CO_2 "supply" and "demand" have guided research.

9.1 Photosynthesis Is Influenced by Leaf Properties

- Leaf anatomy is highly specialized for light absorption (**Figure 9.1**).
- About 5% of the solar energy reaching Earth is converted into carbohydrates by photosynthesis. Much absorbed light is lost by reflection and transmission, metabolism, and heat (**Figures 9.2**, **9.3**).
- In dense forests, almost all photosynthetically active radiation is absorbed by leaves (**Figure 9.4**).
- Leaves of some plants maximize light absorption by solar tracking (**Figure 9.5**).
- Some plants respond to a range of light regimes. However, sun and shade leaves have contrasting morphological and biochemical characteristics.
- To increase light absorption, some shade plants produce a higher ratio of PSII to PSI reaction centers, while others add antenna chlorophyll to PSII.

9.2 Effects of Light on Photosynthesis in the Intact Leaf

- Light-response curves show the PPFD where photosynthesis is limited by light or by carboxylation capacity. The slope of the linear portion of the light-response curve measures the maximum quantum yield (**Figure 9.6**).
- Light compensation points for shade plants are lower than for sun plants because respiration rates in shade plants are low (**Figures 9.7**, **9.8**).
- Beyond the saturation point, factors other than incident light, such as electron transport, Rubisco activity, or triose phosphate metabolism, limit photosynthesis. Rarely is an entire plant light-saturated (**Figure 9.9**).
- The xanthophyll cycle dissipates excess absorbed light energy to avoid damaging the photosynthetic apparatus (**Figures 9.10**, **9.11**, **9.12**). Chloroplast movements also limit excess light absorption (**Figure 9.13**).
- Dynamic photoinhibition temporarily diverts excess light absorption to heat but maintains maximum photosynthetic rate (**Figure 9.14**). Chronic photoinhibition is irreversible.

9.3 Effects of Temperature on Photosynthesis in the Intact Leaf

- Leaf absorption of light energy generates a heat load that must be dissipated (**Figure 9.15**).
- Plants are remarkably plastic in their adaptations to temperature. Optimal photosynthetic temperatures have strong biochemical, genetic (adaptation), and environmental (acclimation) components.
- Temperature-response curves identify (a) a temperature range where enzymatic events are stimulated, (b) a range for optimal photosynthesis, and (c) a range where deleterious events occur (**Figure 9.16**).
- Below 30°C, the quantum yield of C_3 plants is generally higher than that of C_4 plants; above 30°C, the situation is reversed (**Figure 9.17**). Because of photorespiration, the quantum yield is strongly dependent on temperature in C_3 plants but is nearly independent of temperature in C_4 plants.
- Reduced quantum yield due to photorespiration leads to differences in the photosynthetic capacities of C_3 and C_4 plants and results in a shift of species dominance across different latitudes (**Figure 9.18**).

9.4 Effects of Carbon Dioxide on Photosynthesis in the Intact Leaf

- Atmospheric CO_2 concentrations have been increasing since the Industrial Revolution because of human use of fossil fuels and deforestation (**Figure 9.19**).
- Concentration gradients drive the diffusion of CO_2 from the atmosphere to the carboxylation site in the leaf, using both gas and liquid routes. There are multiple resistances along the CO_2 diffusion pathway, but under most conditions stomatal resistance has the greatest effect on CO_2 diffusion into a leaf (**Figure 9.20**).

(Continued)

Summary (*continued*)

- Enrichment of CO_2 above natural atmospheric concentrations results in increased photosynthesis and productivity (**Figure 9.21**), but this can be attenuated by increased leaf temperature and reduced stomatal conductance.
- C_4 photosynthesis may have become prominent due to high rates of photorespiration in Earth's warmest and driest regions when global atmospheric CO_2 concentrations fell below a threshold value (**Figure 9.22**).
- Opening at night and closing during the day, the stomatal activity of CAM plants contrasts with that found in C_3 and C_4 plants (**Figure 9.23**).
- Free Air CO_2 Enrichment (FACE) experiments (**Figure 9.24**) show that C_3 plants are more responsive to elevated CO_2 than are C_4 plants; however, the response of C_3 plants under FACE conditions is often less than that predicted from experiments in greenhouse and growth chambers, as other factors can limit plant growth under field conditions.

Suggested Reading

Ainsworth, E.A., and Long, S. P. (2020) 30 years of free-air carbon dioxide enrichment (FACE): What have we learned about future crop productivity and its potential for adaptation? *Glob. Change Biol.* 27: 27–49. https://doi.org/10.1111/gcb.15375.

Bassi, R., and Dall'Osto, L. (2021) Dissipation of light energy absorbed in excess: The molecular mechanisms. *Annu. Rev. Plant Biol.* 72: 47–76.

Blankenship, R. E. (2014) *Molecular Mechanisms of Photosynthesis*, 2nd ed. Blackwell Science Ltd, Oxford.

Demmig-Adams, B., and Adams, W. W. (2006) Photoprotection in an ecological context: The remarkable complexity of thermal energy dissipation. *New Phytol.* 172: 11–21.

Dodd, A. N., Borland A. M., Haslam R. P., et al. (2002) Crassulacean acid metabolism: Plastic, fantastic. *J. Exp. Bot.* 53: 569–580.

Evans, J. R. (2020) Mesophyll conductance: Walls, membranes and spatial complexity. *New Phytol.* 229: 1864–1876.

Long, S. P., Ainsworth, E. A., Leakey, A. D., et al. (2006) Food for thought: Lower-than-

Murchie, E. H., and Niyogi, K. K. (2011) Manipulation of photoprotection to improve plant photosynthesis. *Plant Physiol.* 155: 86–92.

Osborne, C. P., and Sack, L. (2012) Evolution of C_4 plants: A new hypothesis for an interaction of CO_2 and water relations mediated by plant hydraulics. *Philos. Trans. R. Soc. Lond. B. Biol. Sci.* 367: 583–600.

Terashima, I., Hanba, Y. T., Tholen, D., and Niinemets, Ü. (2011) Leaf functional anatomy in relation to photosynthesis. *Plant Physiol.* 155: 108–116.

Zhu, X. G., Long, S. P., and Ort, D. R. (2010) Improving photosynthetic efficiency for greater yield. *Annu. Rev. Plant Biol.* 61: 235–261.

10 Translocation in the Phloem

Survival on land poses some serious challenges to terrestrial plants. Foremost among these challenges is the need to acquire and retain water. In response to such environmental pressures, plants evolved roots that anchor the plant and absorb water and nutrients. Plants also need to photosynthesize, and plants evolved leaves and mechanisms to reduce water loss from leaves during gas exchanges associated with photosynthesis. As plants increased in size, the roots and leaves became increasingly separated from each other in space. Thus, systems evolved for long-distance transport that allowed the shoot and the root to efficiently exchange products of absorption and assimilation.

You will recall from Chapters 3 and 6 that the xylem is the tissue that transports water and minerals from the root system to the aerial portions of the plant. The **phloem** is the tissue that transports (*translocates*) the products of photosynthesis—particularly sugars—from mature leaves to areas of growth and storage, including the roots.

Along with the sugars, signals (regulatory molecules), water, and various compounds are redistributed throughout the plant body via the phloem. All of these molecules appear to move with the transported sugars. The compounds to be redistributed, some of which initially arrive in the mature leaves via the xylem, can be either transferred out of the leaves without modification or metabolized before redistribution. The fluid that flows through the phloem—the water plus all its solutes—is called *phloem sap*. (*Sap* is a general term used to refer to the fluid contents of plant cells.)

In this chapter we first examine the patterns and pathway of translocation and focus on the loading of sugars in mature leaves. We then follow the sugar movement in the phloem from source to sink and discuss the

phloem The tissue that transports the products of photosynthesis from mature leaves (or storage organs) to areas of growth and storage, including the roots.

source Any organ that is capable of exporting photosynthetic products in excess of its own needs, such as a mature leaf or a storage organ. Contrast with *sink*.

sink Any organ that imports photosynthate, including nonphotosynthetic organs and organs that do not produce enough photosynthetic products to support their own growth or storage needs, such as roots, tubers, developing fruits, and immature leaves. Contrast with *source*.

collection phloem Sieve elements of sources.

transport phloem Sieve elements of the pathway connecting source and sinks such as the phloem of midribs and stem internodes.

release phloem Sieve elements of sinks where sugars and other photosynthetic products are unloaded into sink tissues.

photosynthate A carbon-containing product of photosynthesis.

mechanism of phloem transport, including the rate of movement, magnitude of pressure gradients between sources and sinks, materials translocated in the phloem, and its wound response. The destination of the sugar molecules' journey through the plant is their unloading in sink organs and subsequent consumption and storage. Finally, we explore the role of the phloem in coordinating plant growth by the allocation and partitioning of photosynthetic products and by offering a rapid pathway for signaling.

Throughout, we will primarily describe translocation in the phloem of angiosperms, because most of the research has been conducted on that group of plants. The translocation mechanism in gymnosperms differs in some respects from angiosperms but is less well understood.

10.1 Patterns of Translocation: Source to Sink

Describe the general components and dynamics of source-to-sink translocation.

The two long-distance transport pathways—the phloem and the xylem—extend throughout the plant body. Unlike the xylem, the phloem does not translocate materials exclusively in either an upward or a downward direction. Rather, sap is translocated from areas of supply, called **sources**, to areas of metabolism or storage, called **sinks** (**Figure 10.1**). The terms **collection phloem**, **transport phloem**, and **release phloem** are often used to characterize the prevalent phloem functions in sources, connecting pathways, and sinks, respectively.

Sources include exporting organs, typically mature leaves that produce photosynthate in excess of their own needs. The term **photosynthate** refers to products of photosynthesis (Chapter 8). Another type of source is a storage organ during the exporting phase of its development. For example, the storage root of the biennial wild beet (*Beta maritima*) is a sink during the growing season of the first year, when it accumulates sugars received from the source leaves. During the second growing season, the same root becomes a source; the sugars are remobilized and used to produce a new shoot, which ultimately becomes reproductive.

Sinks include all nonphotosynthetic organs of the plant and organs that do not produce enough photosynthetic products to support their own growth or

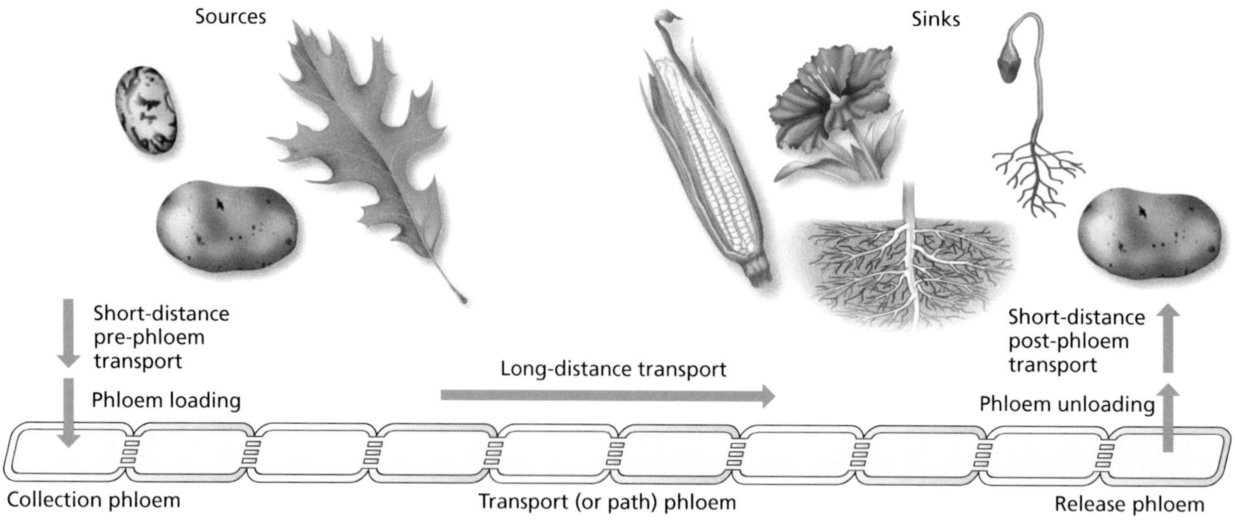

Figure 10.1 Phloem translocates materials from sources to sinks.

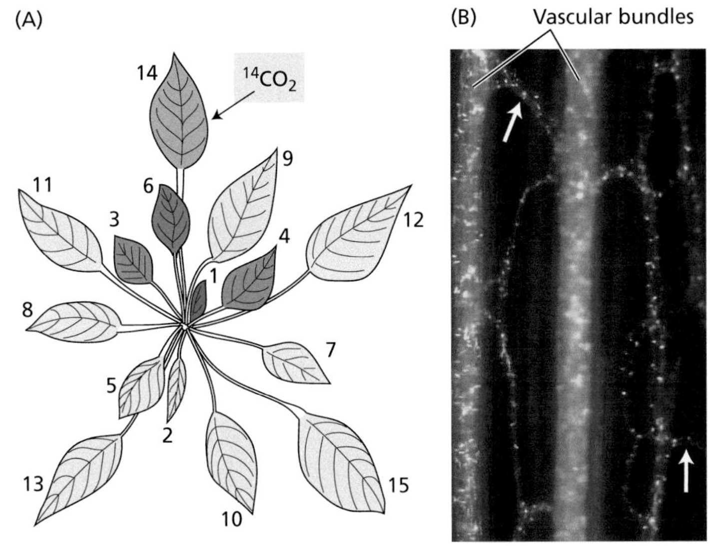

Figure 10.2 Source-to-sink patterns of phloem translocation. (A) Distribution of radioactivity from a single labeled source leaf in an intact plant. The distribution of radioactivity in leaves of a sugar beet plant (*Beta vulgaris*) was determined 1 week after $^{14}CO_2$ was supplied for 4 h to a single source leaf (leaf 14, arrow). The degree of radioactive labeling is indicated by the intensity of shading of the leaves. Leaves are numbered according to their age; the youngest, newly emerged leaf is designated 1. The ^{14}C label was translocated mainly to the sink leaves directly above the source leaf (that is, sink leaves with the most direct vascular connections to the source; for example, leaves 1 and 6 are sink leaves directly above source leaf 14). (B) Longitudinal view of the typical three-dimensional structure of the phloem in a thick section (from an internode of dahlia [*Dahlia pinnata*]), shown here after clearing, staining with aniline blue (which binds to callose and fluoresces yellow), and viewing under an epifluorescence microscope. The sieve plates of the phloem sieve elements (Figure 10.7) are seen as numerous small dots because of their callose content. Two large longitudinal vascular bundles are prominent. This staining reveals the delicate sieve tubes forming the phloem network; two phloem anastomoses (vascular interconnections) are marked by arrows.

storage needs. Roots, tubers, developing fruits, and immature leaves, which must import carbohydrate for normal development, are all examples of sink tissues. Labeling studies support the source-to-sink pattern of translocation in the phloem (**Figure 10.2A**).

Although the overall pattern of transport in the phloem can be stated simply as source-to-sink movement, the specific pathways involved are often more complex, depending on proximity, development, vascular interconnections (**Figure 10.2B**), and modification of translocation pathways. Not all sources supply all sinks on a plant; rather, certain sources preferentially supply specific sinks.

10.2 Pathways of Translocation

Explain the structures and mechanisms by which translocation occurs.

The phloem is generally found on the outer side of both primary and secondary vascular tissues (**Figure 10.3** and **Figure 10.4**). In plants with secondary growth, the phloem constitutes the inner bark. Although phloem is commonly located external to the xylem, it is *also* found on the inner side in many eudicot families. In these families, the phloem in the two positions is called external and internal phloem, respectively.

The small veins of leaves, the primary vascular bundles of stems, and the root stele are each separated from the ground parenchyma by a layer of compactly

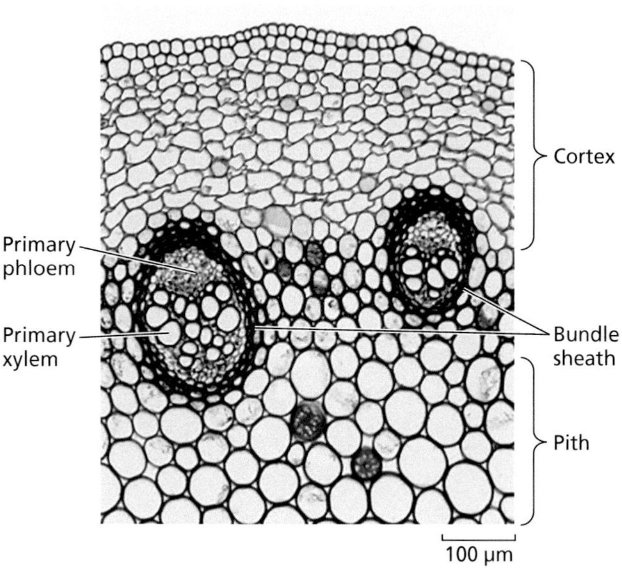

Figure 10.3 Transverse section of a vascular bundle of buttercup (*Ranunculus repens*). The primary phloem is toward the outside of the stem. Both the primary phloem and the primary xylem are surrounded by a bundle sheath of thick-walled sclerenchyma cells, which isolate the vascular tissue from the ground tissue. Fibers and xylem vessels are stained red.

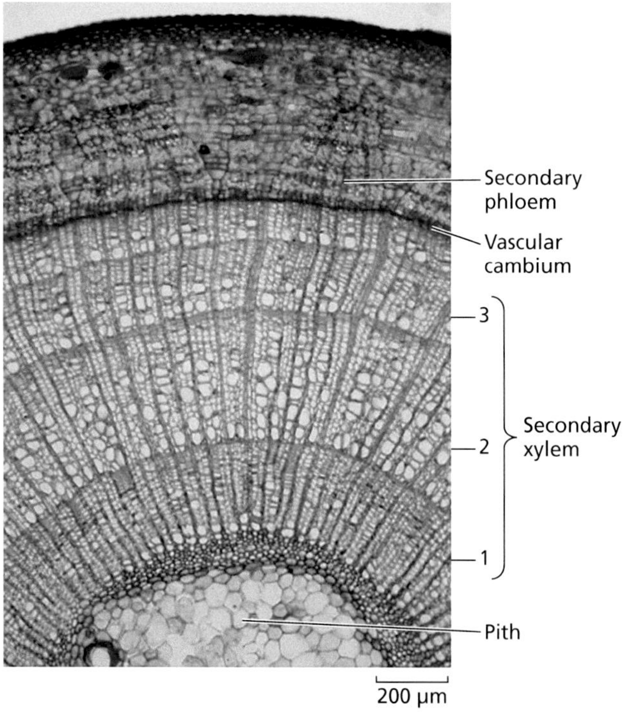

Figure 10.4 Transverse section of a 4-year-old stem of a basswood (*Tilia*) tree. The numbers 1, 2, and 3 indicate growth rings in the secondary xylem. The old (outer) secondary phloem is pushed outward by expansion of the xylem. Only the most recent (innermost) layer of secondary phloem is functional.

arranged cells, called the bundle sheath, starch sheath, and endodermis, respectively. The bundle sheath cells involved in C_4 metabolism were discussed in Chapter 8. In the vascular tissue of leaves, the bundle sheath surrounds the small veins all the way to their ends, isolating the veins from the intercellular air spaces of the leaf (Figure 10.3).

We begin our discussion of translocation pathways with the experimental evidence demonstrating that *sieve elements* are the conducting cells in the phloem. Then we examine their structure and physiology.

Sugar is translocated in phloem sieve elements

Early experiments on phloem transport date back to the nineteenth century, indicating the importance of long-distance transport in plants. These classic experiments demonstrated that removal of a ring of bark around the trunk of a tree, which removes the secondary phloem (Figure 10.4), effectively stops sugar transport from the leaves to the roots without altering water transport through the xylem. When radioactively labeled compounds became available, $^{14}CO_2$ was used to show that sugars made in the photosynthetic process are translocated through the phloem sieve elements.

Mature sieve elements are living cells specialized for translocation

The cells of the phloem that conduct sugars and other organic materials throughout the plant are called **sieve elements**. *Sieve element* is a comprehensive term that includes both the **sieve tube elements** typical of the angiosperms and the *sieve cells* of gymnosperms. In addition to sieve elements, the phloem tissue contains companion cells and parenchyma cells, which store and release food molecules. In some cases the phloem tissue also includes fibers and sclereids (for protection and strengthening of the tissue) and laticifers (latex-containing cells). However, only the sieve elements are directly involved in translocation.

Mature sieve elements are unique among living plant cells (**Figure 10.5** and **Figure 10.6**). They lack many structures normally found in living cells, even in the undifferentiated cells from which they are formed. Although sieve elements retain their plasma membrane, they lose their nuclei and vacuoles during differentiation. Microtubules, Golgi bodies, and ribosomes are also generally absent from the mature cells. Organelles that are retained include somewhat modified mitochondria, plastids, and smooth endoplasmic reticulum. The walls are nonlignified, though they are secondarily thickened in some cases.

Sieve elements of most angiosperms are rich in specific structural proteins called P-proteins (Figure 10.5B and 10.6C). P-protein occurs in several different forms (tubular, fibrillar, granular, and crystalline), depending on the species and maturity of the cell, and is involved in sealing off damaged sieve tubes.

In conclusion, the cellular structure of sieve elements is different from that of tracheary elements of the xylem,

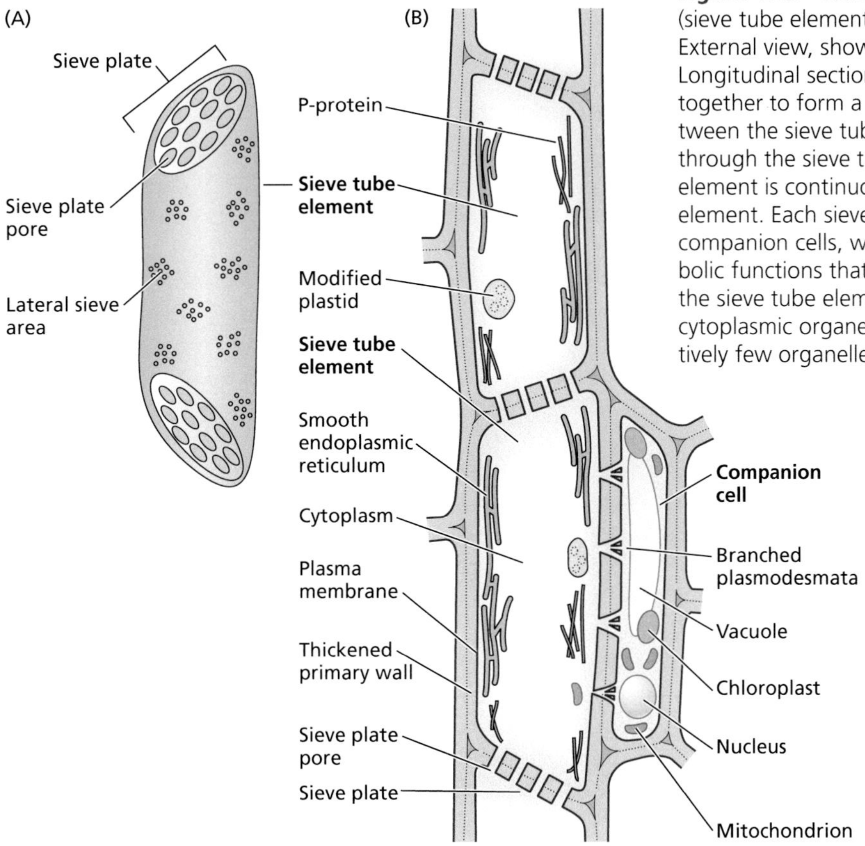

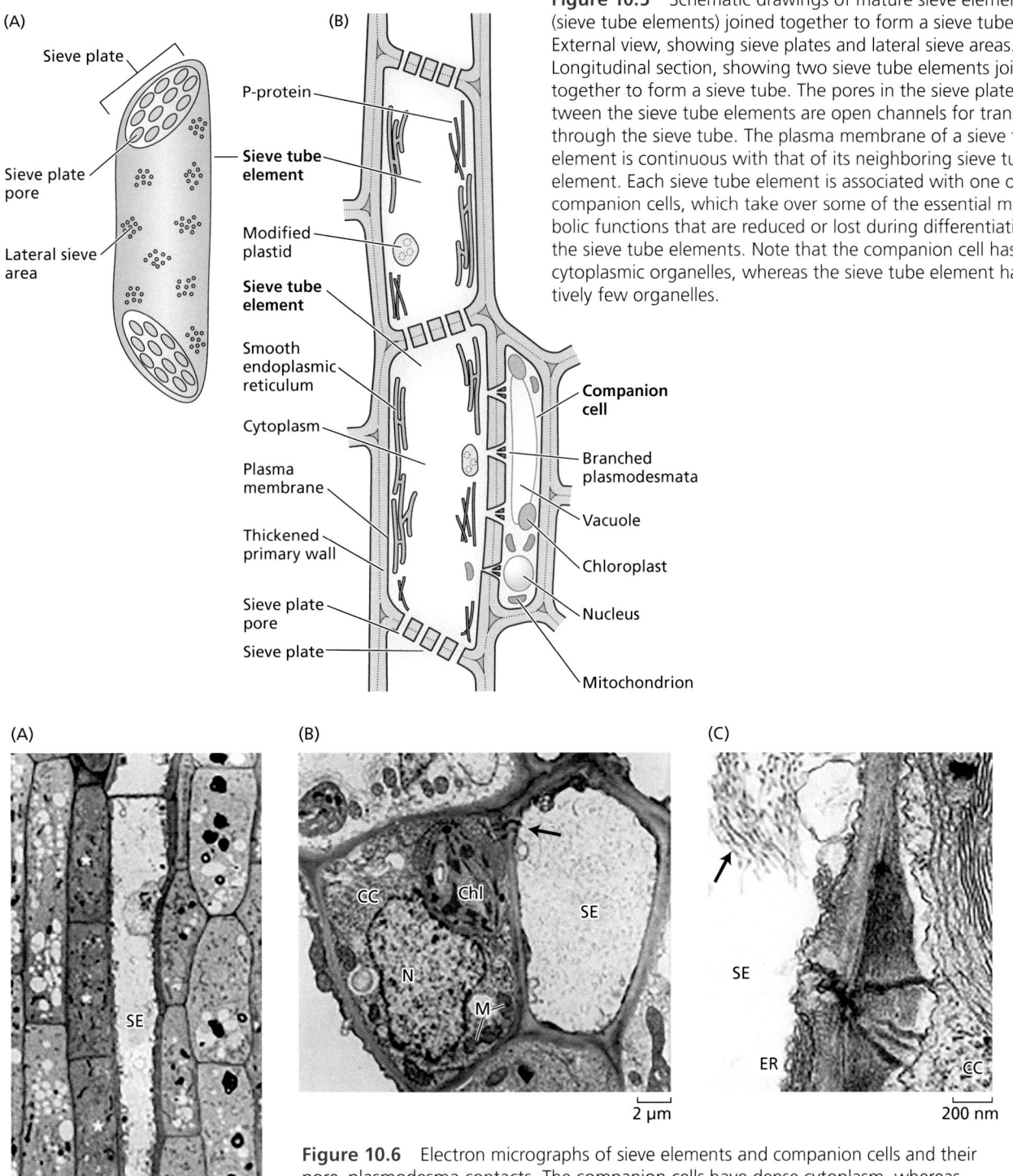

Figure 10.5 Schematic drawings of mature sieve elements (sieve tube elements) joined together to form a sieve tube. (A) External view, showing sieve plates and lateral sieve areas. (B) Longitudinal section, showing two sieve tube elements joined together to form a sieve tube. The pores in the sieve plates between the sieve tube elements are open channels for transport through the sieve tube. The plasma membrane of a sieve tube element is continuous with that of its neighboring sieve tube element. Each sieve tube element is associated with one or more companion cells, which take over some of the essential metabolic functions that are reduced or lost during differentiation of the sieve tube elements. Note that the companion cell has many cytoplasmic organelles, whereas the sieve tube element has relatively few organelles.

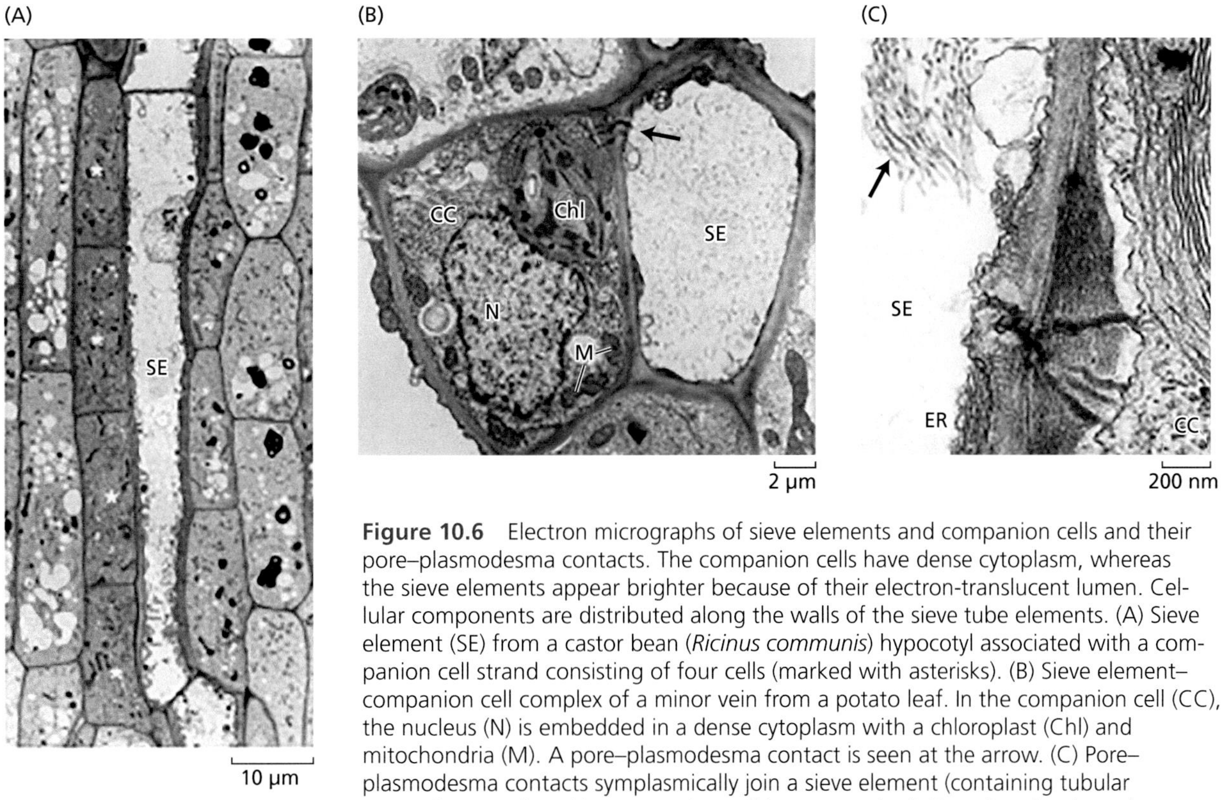

Figure 10.6 Electron micrographs of sieve elements and companion cells and their pore–plasmodesma contacts. The companion cells have dense cytoplasm, whereas the sieve elements appear brighter because of their electron-translucent lumen. Cellular components are distributed along the walls of the sieve tube elements. (A) Sieve element (SE) from a castor bean (*Ricinus communis*) hypocotyl associated with a companion cell strand consisting of four cells (marked with asterisks). (B) Sieve element–companion cell complex of a minor vein from a potato leaf. In the companion cell (CC), the nucleus (N) is embedded in a dense cytoplasm with a chloroplast (Chl) and mitochondria (M). A pore–plasmodesma contact is seen at the arrow. (C) Pore–plasmodesma contacts symplasmically join a sieve element (containing tubular P-protein, arrow) and its companion cell in a potato leaf. Note that the sieve element plasma membrane is lined with smooth ER cisternae (ER).

sieve tube elements The highly differentiated sieve elements typical of the angiosperms.

sieve elements Cells of the phloem that conduct sugars and other organic materials throughout the plant. Refers to both sieve tube elements (angiosperms) and sieve cells (gymnosperms).

sieve area A depression in the cell wall of a sieve tube element that contains a field of plasmodesmata.

sieve plates Sieve areas found in angiosperm sieve tube elements; they have larger pores (sieve plate pores) than other sieve areas and are generally found in end walls of sieve tube elements.

sieve tube Tube formed by the joining together of individual sieve tube elements at their end walls.

which lack a plasma membrane, have lignified secondary walls, and are dead at maturity. As we will see, persistence of the sieve element plasma membrane is critical to the mechanism of translocation in the phloem.

Large pores in cell walls are the prominent feature of sieve elements

Sieve elements (sieve cells and sieve tube elements) have characteristic **sieve areas** in their cell walls, where pores interconnect the conducting cells (**Figure 10.7**). The sieve area pores range in diameter from less than 1 μm to around 10 μm and develop from plasmodesmata. Unlike sieve areas of gymnosperms, the sieve areas of angiosperms can differentiate into **sieve plates** (Figure 10.7 and **Table 10.1**). Sieve plates have larger pores than the other sieve areas in the cell and are generally found on the end walls of sieve tube elements, where the individual cells are joined together to form a longitudinal series called a **sieve tube** (Figure 10.5).

The distribution of sieve tube contents in functional sieve plate is a critical question when considering the mechanism of phloem transport. Early micrographs showed blocked or occluded pores, thought to be artifacts due to damage as the tissues were prepared for observation. Less invasively prepared phloem depicts the sieve plate pores as open channels that appear to allow unfettered transport (Figure 10.7).

In contrast to pores in sieve tube elements of angiosperms, sieve areas in gymnosperms do not appear to be open channels. All of the sieve areas in gymnosperms such as conifers are structurally similar, though they can be more numerous on the overlapping end walls of sieve cells. The pores of gymnosperm sieve areas meet in large median cavities in the middle of the wall. Smooth endoplasmic reticulum (SER) covers the sieve areas and is continuous through the sieve pores and median cavity. Such obstructed pores would result in a high

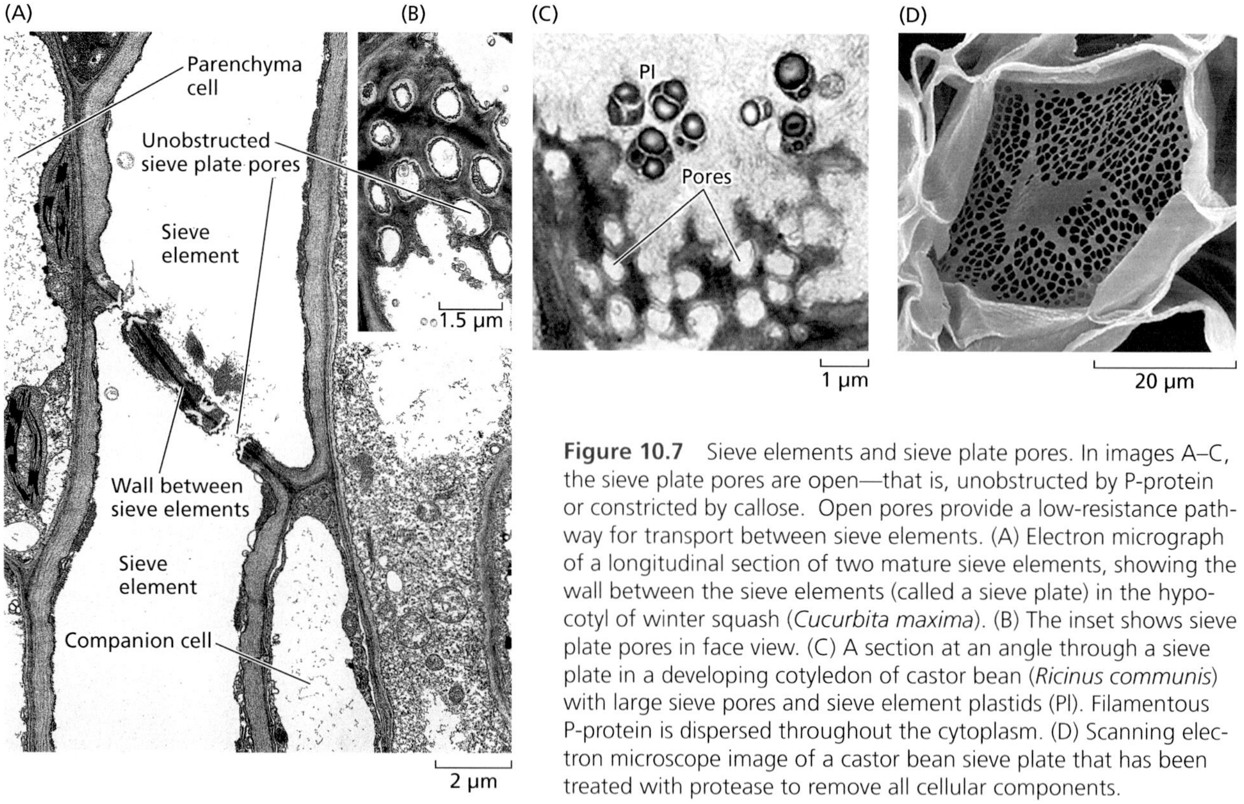

Figure 10.7 Sieve elements and sieve plate pores. In images A–C, the sieve plate pores are open—that is, unobstructed by P-protein or constricted by callose. Open pores provide a low-resistance pathway for transport between sieve elements. (A) Electron micrograph of a longitudinal section of two mature sieve elements, showing the wall between the sieve elements (called a sieve plate) in the hypocotyl of winter squash (*Cucurbita maxima*). (B) The inset shows sieve plate pores in face view. (C) A section at an angle through a sieve plate in a developing cotyledon of castor bean (*Ricinus communis*) with large sieve pores and sieve element plastids (Pl). Filamentous P-protein is dispersed throughout the cytoplasm. (D) Scanning electron microscope image of a castor bean sieve plate that has been treated with protease to remove all cellular components.

Table 10.1 Characteristics of Sieve Elements in Seed Plants

Sieve tube elements found in angiosperms
1. Some sieve areas are differentiated into sieve plates; individual sieve tube elements are joined together into a sieve tube.
2. Sieve plate pores are open channels.
3. P-protein is present in all eudicots and many monocots.
4. Companion cells are sources of ATP and perhaps other compounds. In some species they serve as transfer cells or intermediary cells
Sieve tube elements found in gymnosperms
1. There are no sieve plates; all sieve areas are similar.
2. Pores in sieve areas appear blocked with membranes.
3. There is no P-protein.
4. Strasburger cells have similar functions as companion cells.

resistance to mass flow between sieve cells. Observation of living material with confocal laser scanning microscopy confirms that the observed distribution of SER is not an artifact of fixation.

Table 10.1 lists characteristics of sieve tube elements and sieve cells.

Companion cells aid the highly specialized sieve elements

Each sieve tube element is usually associated with one or more **companion cells** (Figures 10.5 and 10.6). As a rule, the division of a single mother cell forms the sieve tube element and the companion cell in metaphloem and secondary phloem. Numerous specialized plasmodesmatal connections penetrate the walls between sieve tube elements and their companion cells; the plasmodesmata are complex and branched on the companion cell side, while forming a sieve pore on the sieve element side (Figure 10.6B and C). The presence of these **pore–plasmodesma contacts** suggests a close functional relationship between a sieve element and its companion cell, an association that has been demonstrated by the rapid exchange of fluorescent dyes, and even proteins, between the two cells.

Companion cells differ from sieve elements in having dense, ribosome-rich cytoplasm and many mitochondria. They have taken over some of the critical metabolic functions of sieve elements, such as protein and lipid synthesis, that are reduced or lost during sieve element differentiation. In addition to nursing sieve elements, companion cells are involved in coordinating essential plant functions, such as flowering, tuber development, and emergence of lateral roots.

Companion cells play a role in phloem loading, where they are involved in the last step of the transport of photosynthetic products from mature leaves to the sieve elements in the minor (small) veins of the leaf. The evolution of different phloem-loading strategies in seed plants is reflected by the existence of at least three different types of companion cells in the minor veins of mature, exporting leaves: **"ordinary" companion cells** that are indistinguishable from those in transport phloem, transfer cells, and intermediary cells.

Gymnosperm sieve cells are also tightly coupled with pore–plasmodesma contacts to specialized cells with dense cytoplasm and abundant mitochondria. In contrast to companion cells, these *Strasburger cells* are not ontogenetically related (not sister cells) to sieve elements. Like companion cells, they are involved in phloem loading.

companion cells In angiosperms, metabolically active cells that are connected to their sieve element by large, branched plasmodesmata and that take over many of the metabolic activities of the sieve element. In source leaves, they function in the transport of photosynthate into the sieve elements.

pore–plasmodesma contacts Specific symplasmic contacts between a sieve element and its companion cell, consisting of sieve pores and branched plasmodesmata on the sieve element and companion cell side, respectively.

ordinary companion cell A type of companion cell with a variable number of plasmodesmata connecting it to neighboring cells other than its associated sieve element.

pre-phloem transport The movement of photosynthates from to mesophyll to bundle sheath and phloem parenchyma, prior to their uptake by sieve element–companion cell complexes.

phloem loading The movement of photosynthetic products into the sieve elements of mature leaves. *See also* phloem unloading.

export The movement of photosynthate in sieve elements away from the source tissue.

long-distance transport Translocation through the phloem to the sink.

10.3 Phloem Loading

Describe the three mechanisms by which phloem is loaded.

Photosynthates travel from their origin in mesophyll cells to the minor veins, where they are loaded into the sieve elements. Phloem loading of sugar molecules—often sucrose—provides the driving force for long-distance transport.

Several steps are involved in the transport of photosynthate from the mesophyll chloroplasts to the sieve elements of mature leaves:

1. Triose phosphate formed by photosynthesis during the day is transported from the chloroplast to the cytosol, where it is converted to sucrose. During the night, carbon from stored starch exits the chloroplast primarily in the form of maltose and is converted to sucrose.
2. Sucrose moves from producing cells in the mesophyll to cells in the vicinity of the sieve elements in the smallest veins of the leaf (**Figure 10.8**). This **pre-phloem transport** usually covers a distance of only a few cell diameters.
3. In the process called **phloem loading**, sugars are transported into the sieve elements and companion cells. Note that sieve elements and companion cells are often considered a functional unit, called the *sieve element–companion cell complex*. Once inside the sieve elements, sucrose and other solutes are translocated away from the source, a process known as **export**. Translocation through the vascular system to a sink is referred to as **long-distance transport**.

Phloem loading can occur via the apoplast or symplasm

Photosynthates in source leaves must move from the photosynthesizing cells in the mesophyll to the sieve elements. Studies using fluorescent tracers of the molecular size of sucrose suggest that the initial pre-phloem pathway is symplasmic and follows the sugar concentration gradient. However, some sugars might be transported to the apoplast prior to phloem loading (**Figure 10.9A**), or they might move entirely through the symplasm to the sieve elements via the plasmodesmata (**Figure 10.9B**). (Chapter 1 describes the symplasm and apoplast.) One of the two routes, apoplastic or symplasmic, is predominant in a given plant species; many species, however, show evidence of being able to use more than one loading mechanism. Three distinct mechanisms for phloem loading are now recognized: apoplastic loading, symplasmic loading with oligomer trapping, and passive symplasmic loading. For simplicity's sake, we will initially consider the routes separately, then return to the subject of loading diversity.

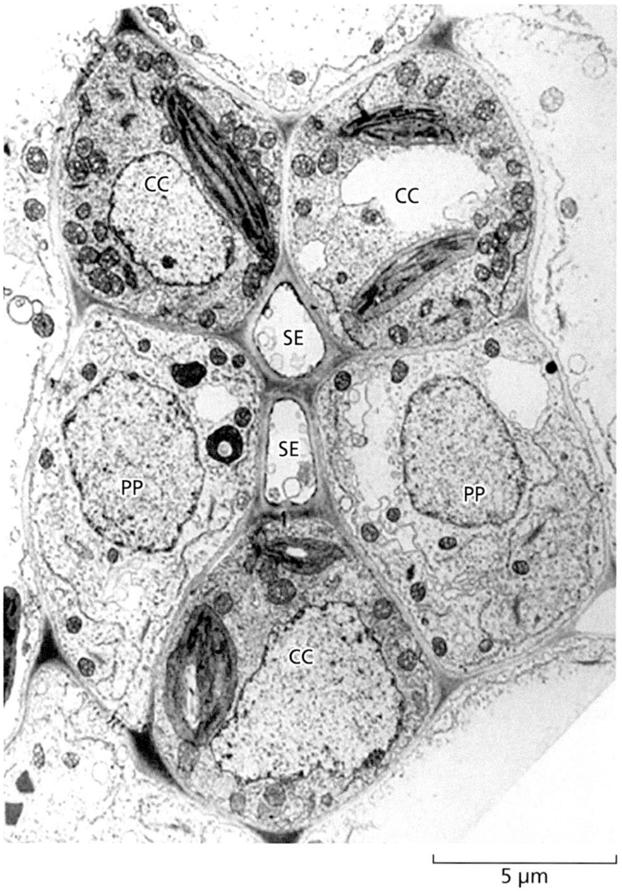

Figure 10.8 Electron micrograph showing the relationships among the phloem cell types of the smallest vein in a source leaf of potato. The image shows two sieve elements (SE), three large companion cells (CC), and two phloem parenchyma cells (PP), all surrounded by the bundle sheath. Photosynthates from the mesophyll are loaded from the bundle sheath or the phloem parenchyma into the sieve element–companion cell complexes.

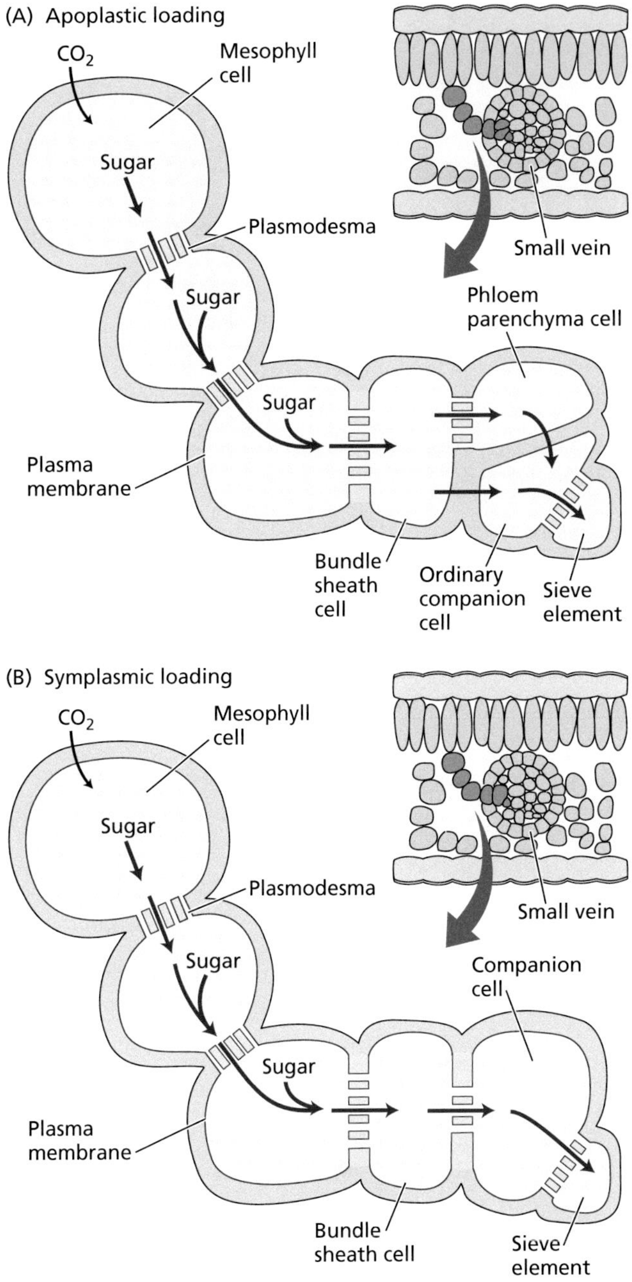

Figure 10.9 Schematic diagram of pathways of phloem loading in source leaves. (A) In the partly apoplastic pathway, sugars initially move through the symplasm but enter the apoplast before loading into the companion cells and sieve elements. Sugars loaded into the companion cells are thought to move through plasmodesmata into the sieve elements. (B) In the entirely symplasmic pathway, sugars move from one cell to another through plasmodesmata, all the way from the mesophyll to the sieve elements. Note that symplasmically loading plants typically do not contain phloem parenchyma cells.

We will discuss apoplastic loading first and then introduce the two types of symplasmic loading (oligomer trapping and passive), following the order in which their importance was recognized.

Apoplastic loading is characteristic of many herbaceous species

In the case of apoplastic loading, the sugars enter the apoplast surrounding the sieve element–companion cell complex. The sucrose transporters that mediate the efflux of sucrose, most likely from the phloem parenchyma bundle sheath cells to the apoplast near the sieve element–companion cell complexes, have been identified—first in Arabidopsis and rice—as a subfamily of the SWEET transporters. Sugars are then actively taken up from the apoplast into the sieve element–companion cell complex by a selective transporter located in the plasma membranes of these cells.

The apoplastic phloem-loading model leads to three basic predictions:

1. Transported sugars should be found in the apoplast.
2. In experiments in which sugars are supplied to the apoplast, the exogenously supplied sugars should accumulate in sieve elements and companion cells.
3. Inhibition of sugar efflux from the phloem parenchyma or of uptake from the apoplast should result in inhibition of export from the leaf.

Many studies devoted to testing these predictions have provided solid evidence for apoplastic loading in several species.

Sucrose loading in the apoplastic pathway requires metabolic energy

In many of the species initially studied, sugars become more concentrated in the sieve elements and companion cells than in the mesophyll. This difference in solute concentration can be demonstrated through measurement of the osmotic potential (Ψ_s) of the various cell types in the leaf (Chapters 2 and 3).

In sugar beet, the osmotic potential of the mesophyll is approximately –1.3 MPa, and the osmotic potential of the sieve elements and companion cells is about –3.0 MPa. Most of this difference in osmotic potential is thought to result from accumulated sugar, specifically sucrose, which is the major transport sugar in this species. Experimental studies have also demonstrated that both externally supplied sucrose and sucrose derived from photosynthate accumulate in the phloem of the minor veins of sugar beet source leaves (**Figure 10.10**).

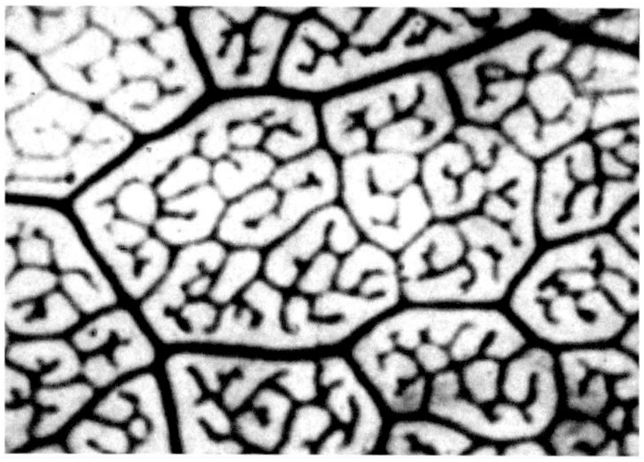

Figure 10.10 Autoradiograph shows that labeled sugar moves against its concentration gradient from the apoplast into sieve elements and companion cells of a sugar beet source leaf. A solution of ^{14}C-labeled sucrose was applied for 30 min to the adaxial surface of a sugar beet leaf that had previously been kept in darkness for 3 h. The leaf cuticle was removed to allow penetration of the labeled sucrose solution to the interior of the leaf (the apoplast). The sieve elements and companion cells of the small veins in the source leaf contain high concentrations of labeled sugar, shown by the black accumulations, indicating that sucrose is actively transported against its concentration gradient (from the apoplast into the leaf veins).

transfer cell A type of companion cell similar to an ordinary companion cell, but with fingerlike projections of the cell wall that greatly increase the surface area of the plasma membrane and increase the capacity for solute transport across the membrane from the apoplast.

The fact that sucrose is at a higher concentration in the sieve element–companion cell complex than in surrounding cells indicates that sucrose is actively transported against its chemical-potential gradient. The dependence of sucrose accumulation on active transport is supported by the fact that treating source tissue with respiratory inhibitors both decreases the ATP concentration and inhibits loading of exogenous sugar.

Plants that load sugars apoplastically into the phloem may also load amino acids and sugar alcohols (sorbitol and mannitol) actively. In contrast, other metabolites, such as organic acids and hormones, may enter sieve elements passively.

Phloem loading in the apoplastic pathway involves a sucrose–H$^+$ symporter

A sucrose–H$^+$ symporter appears to mediate the transport of sucrose from the apoplast into the sieve element–companion cell complex. Chapter 6 describes that symport is a secondary transport process that uses the energy generated by the proton pump, in this case a companion cell–specific isoform. The energy dissipated by protons moving back into the cell is coupled to the uptake of a substrate, in this case sucrose (**Figure 10.11**). While various sugars are present in the apoplast, the sucrose–H$^+$ symporter selects the one sugar species that is transported long-distance in the phloem, in this case sucrose.

Many sucrose–H$^+$ symporters have been cloned and localized in the phloem. SUT1 and SUC2 and their homologs appear to be the major sucrose transporters in phloem loading. They are located in the plasma membrane of either companion cells or sieve elements.

Transfer cells are companion cells that are specialized for membrane transport

Some apoplastically loading plants, such as broad bean (*Vicia faba*), have transfer cells in minor veins. Transfer-type companion cells have fingerlike wall ingrowths, particularly on the cell walls that face away from source sieve elements (**Figure 10.12A**). These wall ingrowths greatly increase the surface area of the plasma membrane, thus increasing the potential for solute transfer across the membrane. Relatively few plasmodesmata connect this type of companion cell to any of the surrounding cells except its own sieve element. Xylem parenchyma cells can also be modified as **transfer cells**, serving to retrieve and reroute solutes moving in the xylem, which is also part of the apoplast. Transfer cells occur most frequently at nodes in transport phloem, as well as in source phloem and in post–sieve element unloading pathways.

A paucity or virtual absence of plasmodesmata between the sieve element–companion cell complex and neighboring cells (bundle sheath or phloem parenchyma) seems to be correlated with apoplastic loading. The function of the remaining plasmodesmata is not known. The fact that

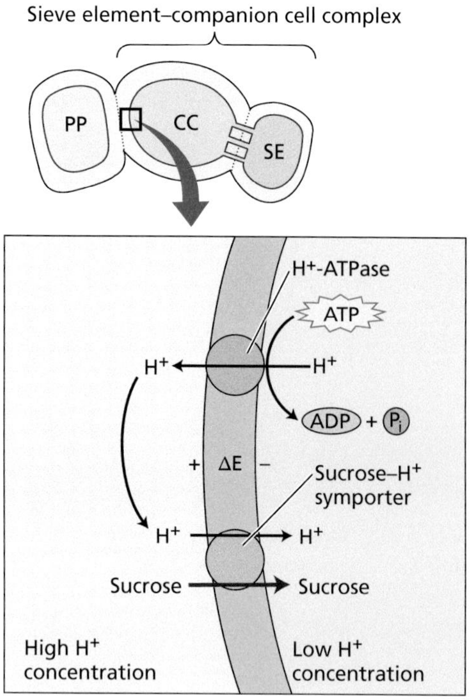

Figure 10.11 ATP-dependent sucrose transport in apoplastic sieve element loading. In the cotransport model of sucrose loading into the symplasm of the sieve element–companion cell complex, the plasma membrane ATPase pumps protons out of the cell into the apoplast, establishing a higher proton concentration in the apoplast and a membrane potential of approximately −120 mV (ΔE). The energy in the electrochemical proton gradient is then used to drive the transport of sucrose into the symplasm of the sieve element–companion cell complex through a sucrose–H$^+$ symporter. CC, companion cell; PP, phloem parenchyma cell; SE, sieve element.

they are present indicates that they must have a function, and an important one, because the cost of having them is high: They are the avenues by which viruses spread systemically in a plant.

intermediary cell A type of companion cell with numerous plasmodesmatal connections to surrounding cells, particularly to the bundle sheath cells.

Phloem loading is symplasmic in some species

Apoplastic phloem loading is prevalent in species that transport only sucrose and have few plasmodesmata leading into the minor vein phloem. However, many other species have numerous plasmodesmata at the interface between the sieve element–companion cell complex and the surrounding cells (**Figure 10.12B**). The operation of a symplasmic pathway requiring the presence of open plasmodesmata is implied in these species, since phloem isolation seems to be essential for apoplastic loading.

The oligomer-trapping model explains symplasmic loading in plants with intermediary-type companion cells

The sugars tri- and tetrasaccharides raffinose and stachyose (Figure 10.15 shows structures) move through the symplasm of some species, and, in addition to the disaccharide sucrose, are transported in the phloem. These species have intermediary cells in the minor veins and abundant plasmodesmata leading into the minor veins. Some examples of such species are common coleus (*Coleus blumei*), pumpkin and squash (*Cucurbita pepo*), and melon (*Cucumis melo*). **Intermediary cells** are companion cells that have numerous plasmodesmata connecting them to bundle sheath cells (Figure 10.12B). Although the presence of many plasmodesmatal connections to surrounding cells is the most characteristic feature of intermediary cells, they are also distinctive in having numerous small vacuoles, as well as poorly developed thylakoids and a lack of starch grains in the chloroplasts.

Two major questions emerge concerning symplasmic loading:

1. How are raffinose and stachyose selected for transport in the phloem? In apoplastic phloem loading, the involvement of symporters provides a clear mechanism for selectivity because symporters are specific for certain sugar molecules. Symplasmic loading, by contrast, depends on the diffusion of sugars from the mesophyll to the sieve elements via the plasmodesmata. How can diffusion through plasmodesmata during symplasmic loading be selective for certain sugars?
2. Data from several species showing symplasmic loading indicate that sieve elements and companion cells have a higher osmolyte content (more negative

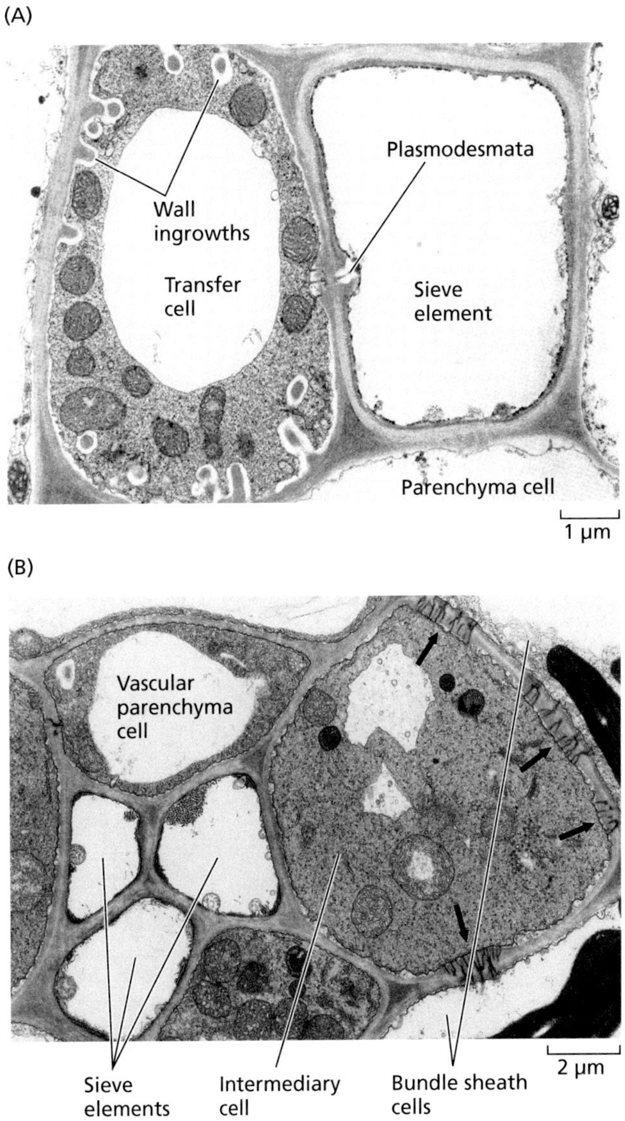

Figure 10.12 Electron micrographs of specialized companion cells in minor veins of mature leaves. (A) A sieve element adjacent to a transfer-type companion cell with numerous wall ingrowths in pea (*Pisum sativum*). (B) A typical intermediary-type companion cell in the minor vein phloem of heartleaf mask flower (*Alonsoa warscewiczii*) with numerous fields of plasmodesmata (arrows) connecting it to neighboring bundle sheath cells. These plasmodesmata are branched on both sides, but the branches are longer and narrower on the intermediary cell side.

oligomer-trapping model A model that explains the active accumulation of tri-, tetra-, and pentasaccharides in sieve element–companion cell complexes of symplasmically loading species.

osmotic potential) than the mesophyll. How can diffusion-dependent symplasmic loading account for the observed selectivity for transported molecules and the accumulation of sugars against a concentration gradient?

The **oligomer-trapping model** (**Figure 10.13**) addresses these questions. The model states that the sucrose synthesized in the mesophyll diffuses from the bundle sheath cells into the intermediary cells through the abundant plasmodesmata that connect the two cell types. In the intermediary cells, raffinose and stachyose (oligomers made of three and four hexose sugars, respectively) are synthesized from the transported sucrose and from galactinol (a metabolite of galactose). Because of their relatively large size, raffinose and stachyose cannot diffuse back into the bundle sheath cells, but they can move on into the sieve element. Sugar concentrations in the sieve elements of these plants can reach concentrations equivalent to those in plants that load apoplastically. Sucrose can continue to diffuse into the intermediary cells, because its synthesis in the mesophyll and its use in the intermediary cells maintain the concentration gradient (**Figure 10.14**). Like apoplastic loading, oligomer trapping also requires metabolic energy in the companion cells—not for membrane transport, but for the synthesis of tri- and tetrasaccharides.

The oligomer-trapping model makes three predictions:

1. Sucrose should be more concentrated in the mesophyll than in the intermediary cells.
2. The enzymes for raffinose and stachyose synthesis should be preferentially located in the intermediary cells.
3. The plasmodesmata linking the bundle sheath cells and the intermediary cells should exclude molecules larger than sucrose.

Several studies verify the predictions and support the oligomer-trapping model of symplasmic loading in some species. However, recent modeling results suggest

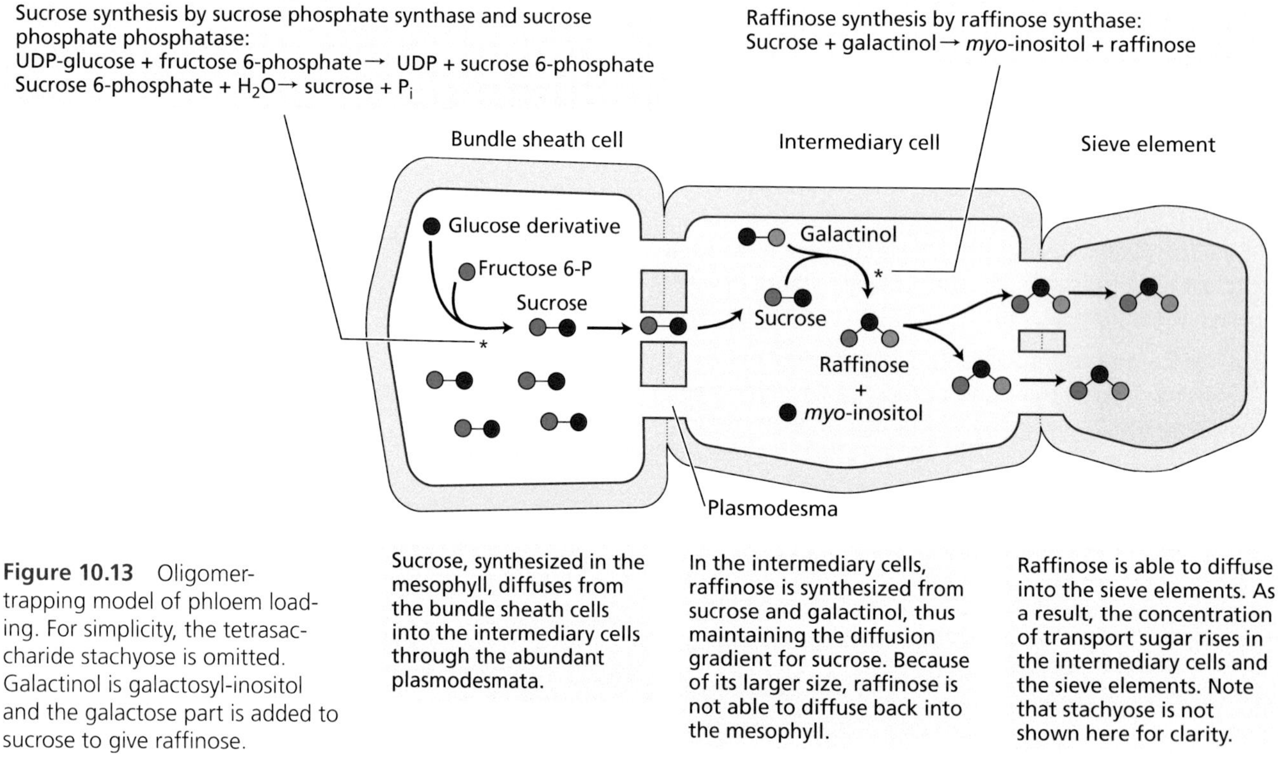
Sucrose synthesis by sucrose phosphate synthase and sucrose phosphate phosphatase:
UDP-glucose + fructose 6-phosphate → UDP + sucrose 6-phosphate
Sucrose 6-phosphate + H_2O → sucrose + P_i
Raffinose synthesis by raffinose synthase:
Sucrose + galactinol → *myo*-inositol + raffinose
Bundle sheath cell
Intermediary cell
Sieve element
Glucose derivative
Fructose 6-P
Sucrose
Galactinol
Sucrose
Raffinose
+
myo-inositol
Plasmodesma
Sucrose, synthesized in the mesophyll, diffuses from the bundle sheath cells into the intermediary cells through the abundant plasmodesmata.
In the intermediary cells, raffinose is synthesized from sucrose and galactinol, thus maintaining the diffusion gradient for sucrose. Because of its larger size, raffinose is not able to diffuse back into the mesophyll.
Raffinose is able to diffuse into the sieve elements. As a result, the concentration of transport sugar rises in the intermediary cells and the sieve elements. Note that stachyose is not shown here for clarity.

Figure 10.13 Oligomer-trapping model of phloem loading. For simplicity, the tetrasaccharide stachyose is omitted. Galactinol is galactosyl-inositol and the galactose part is added to sucrose to give raffinose.

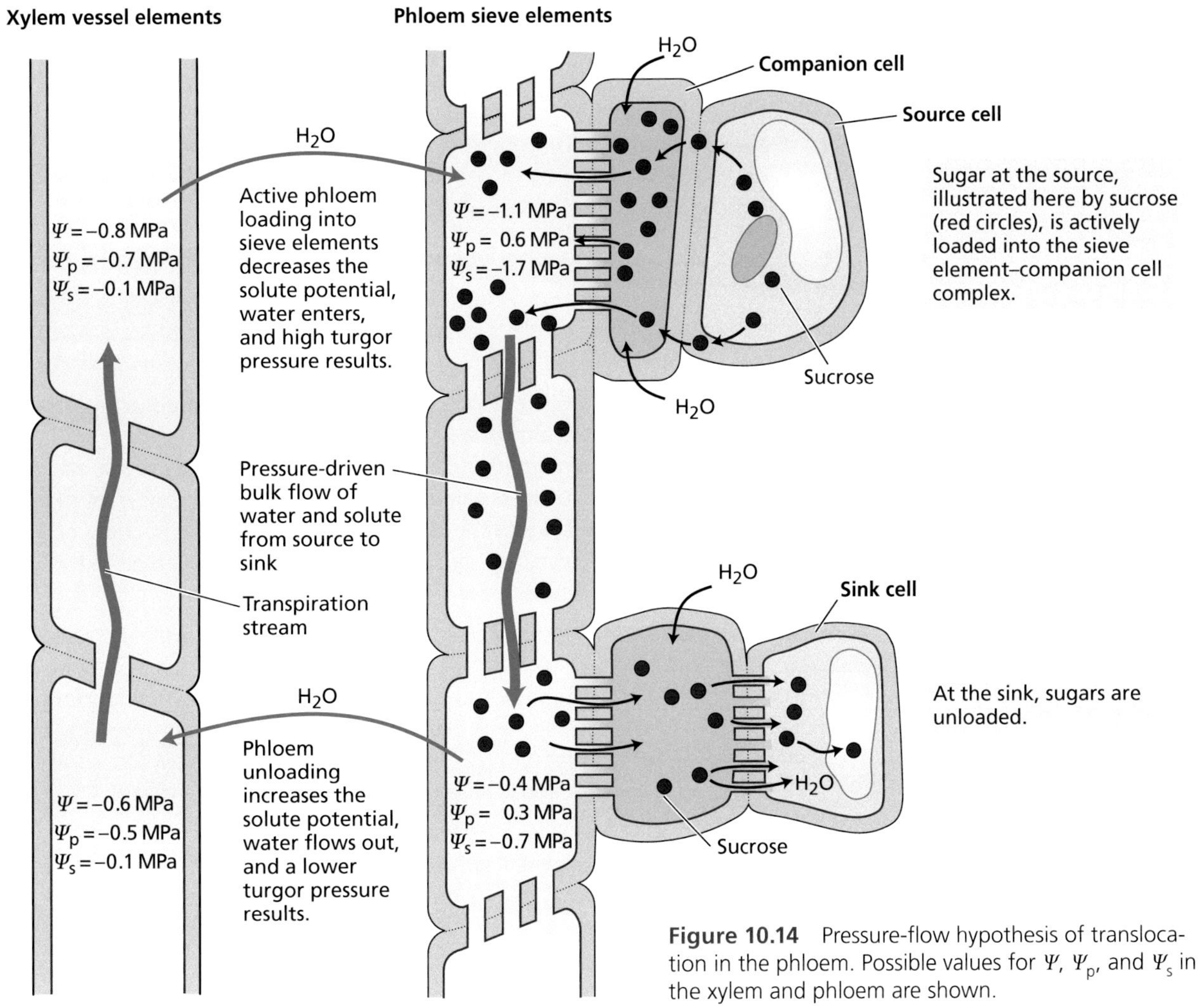
Xylem vessel elements
Phloem sieve elements
H_2O
Companion cell
Source cell
Sugar at the source, illustrated here by sucrose (red circles), is actively loaded into the sieve element–companion cell complex.
H_2O
Active phloem loading into sieve elements decreases the solute potential, water enters, and high turgor pressure results.
$\Psi = -0.8$ MPa
$\Psi_p = -0.7$ MPa
$\Psi_s = -0.1$ MPa
$\Psi = -1.1$ MPa
$\Psi_p = 0.6$ MPa
$\Psi_s = -1.7$ MPa
Sucrose
H_2O
Pressure-driven bulk flow of water and solute from source to sink
Transpiration stream
H_2O
Sink cell
At the sink, sugars are unloaded.
H_2O
Phloem unloading increases the solute potential, water flows out, and a lower turgor pressure results.
$\Psi = -0.6$ MPa
$\Psi_p = -0.5$ MPa
$\Psi_s = -0.1$ MPa
$\Psi = -0.4$ MPa
$\Psi_p = 0.3$ MPa
$\Psi_s = -0.7$ MPa
H_2O
Sucrose

Figure 10.14 Pressure-flow hypothesis of translocation in the phloem. Possible values for Ψ, Ψ_p, and Ψ_s in the xylem and phloem are shown.

that the diffusion of oligosaccharides, such as raffinose and stachyose, back into the mesophyll is prevented not only by the physical size of plasmodesmata but also by the onset of opposing mass flow from the bundle sheath.

Phloem loading is passive in several tree species

Passive symplasmic phloem loading is widespread among tree species. While the data supporting this mechanism are relatively recent, passive symplasmic loading was actually part of Münch's original conception of pressure flow.

It has become apparent that several tree species possess abundant plasmodesmata between the sieve element–companion cell complex and surrounding cells, but do not have intermediary-type companion cells and do not transport raffinose and stachyose. Willow (*Salix babylonica*) and apple (*Malus domestica*) trees are among the species that fall into this category. These plants have no concentrating step in the pathway from the mesophyll into the sieve element–companion cell complex. Since a concentration gradient from the mesophyll into the phloem drives diffusion along the pre-phloem pathway, the absolute concentrations of sugars in the source leaves of these species must be high to maintain the required high turgor pressures in the sieve elements. Although there is wide variation and considerable overlap among groups of plants with different loading mechanisms, source leaf sugar concentrations are generally higher in the tree species that load passively. The feasibility of this mode of phloem loading was recently demonstrated.

Gymnosperms seem to load photosynthates passively and have a high frequency of functional plasmodesmata in each interface between mesophyll and sieve elements. Gymnosperm needles have xerophytic adaptations that involve more cells in the pre-phloem pathway. However, water relations in this complex pathway indicate that mass flow starts inside the bundle sheath and drives photosynthates via several parenchyma and Strasburger cells into the sieve elements.

The type of phloem loading is correlated with several significant characteristics

As already discussed, the operation of apoplastic and symplasmic phloem-loading pathways is correlated with several defining characteristics, listed in **Table 10.2**.

- Species that have *apoplastic phloem loading* as their dominant loading strategy translocate sucrose almost exclusively and have either intermediary-type companion cells or transfer cells in the minor veins. These species usually possess few connections between the sieve element–companion cell complex and the surrounding cells. Active carriers in the sieve element–companion cell complex concentrate sucrose in the phloem cells and generate the driving force for long-distance transport.
- Species that use *symplasmic phloem loading with oligomer trapping* translocate oligosaccharides such as raffinose in addition to sucrose. They have intermediary-type companion cells in the minor veins, with abundant connections between the sieve element–companion cell complex and the surrounding cells. Oligomer trapping probably concentrates transport sugars in the phloem cells and generates the driving force for long-distance transport.
- Species that have *passive symplasmic phloem loading* translocate sucrose and sugar alcohols and have ordinary companion cells in the minor

Table 10.2 Patterns in Apoplastic and Symplasmic Loading

Feature	Apoplastic loading	Symplasmic oligomer trapping	Passive symplasmic loading
Transport sugar	Sucrose	Raffinose and stachyose in addition to sucrose	Sucrose and sugar alcohols
Characteristic companion cells	Ordinary companion cells or transfer cells	Intermediary cells	Ordinary companion cells
Number and conductivity of plasmodesmata connecting the sieve-element–companion cell complex to surrounding cells	Low	High	High
Dependence on active carriers in sieve element–companion cell complex	Transporter driven	Independent of transporters	Independent of transporters
Overall concentration of transport sugars in source leaves	Low	Low	High
Cell type in which driving force for long-distance transport is generated	Sieve element–companion cell complex	Intermediary cells	Mesophyll
Growth habit	Mainly herbaceous	Herbs and woody species	Mainly trees

Sources: Y. V. Gamalei. 1985. *Fiziologiya Rasenii* (Moscow) 32: 886–875; A. J. E. van Bel et al. 1992. *Acta Bot. Neerl.* 41: 121–141; E. A. Rennie and R. Turgeon. 2009. *Proc. Natl. Acad. Sci. USA* 106: 14163–14167.

Note: Plants using all three mechanisms of phloem loading may also transport sugar alcohols. In addition, some species may load both apoplastically and symplasmically, since different types of companion cells can be found within the veins of a single species.

veins. These species also possess abundant connections all the way from mesophyll to the sieve element–companion cell complex. Species with passive symplasmic loading are characterized by high overall sugar concentration in the source leaves, which maintains a concentration gradient between the mesophyll and the sieve element–companion cell complex. The high sugar concentrations in the pre-phloem pathway might already—before the sieve-element–companion cell complex—generate the driving force for long-distance transport. Many of the species with passive symplasmic loading are trees.

Our discussion thus far has considered apoplastic loading, symplasmic loading with oligomer trapping, and passive loading separately. However, increasing evidence shows that many, if not all, plants are capable of more than one loading mechanism. For example, both structural and physiological data indicate that some oligomer-trapping plants are also capable of apoplastic loading.

Plasmodesmatal frequencies suggest that the passive loading strategy is ancestral in the angiosperms, while apoplastic loading and oligomer trapping evolved later. However, the ability to load by multiple mechanisms may have been present even in the earliest angiosperms. Multiple loading mechanisms may allow plants to adapt quickly to abiotic stresses, such as low temperatures. Certainly, the evolution of different loading types and the environmental pressures related to their evolution will continue to be important research areas in the future, as loading pathways are clarified in more species.

10.4 Long-Distance Transport: A Pressure-Driven Mechanism

Explain the mechanism underlying the main theory of long-distance transport.

Once the main transport sugar is loaded into the minor vein sieve elements, it follows the continuous low-resistance pathway of the transport phloem to the organs consuming it (sink) (Figure 10.1). The transport phloem can span long distances: In giant redwoods (*Sequoiadendrum giganteum*) it reaches more than 100 m from the needles at the top of the tree to its finest roots. This long-distance pathway is continuous and comprises first the primary phloem of leaves, petioles, and primary stems; then the secondary phloem in the inner bark of stem and root; and finally, the primary phloem of the fine root system. Independent of whether they originated in procambium or cambium, all sieve elements have the same ultrastructure, though cell width and sieve pore size are generally larger in the secondary phloem.

An important, but often neglected, aspect of the transport phloem is the unloading and reloading of photosynthates along the pathway. Retrieval of photosynthates might just be to compensate for the leakage of photosynthates caused by the steep sugar gradient between sieve elements and the apoplast. In trees, however, unloading and reloading regularly take place between the transport phloem and the surrounding tissue, such as for storage of proteins and starch in the ray parenchyma. Remobilization of these stored compounds supplies the axial buds (sinks) with nutrients needed for new growth during spring. Companion cells in the transport phloem are instrumental in these processes.

Mass transfer is much faster than diffusion

The rate of movement of materials in the sieve elements can be expressed in two ways: as **velocity**, the linear distance traveled per unit time, or as **mass transfer rate**, the quantity of material passing through a given cross section of phloem or sieve elements per unit time. Mass transfer rates based on the cross-sectional

velocity The linear distanced traveled per unit time.

mass transfer rate The quantity of material passing through a given cross section of phloem or sieve elements per unit of time.

area of the sieve elements are preferred because the sieve elements are the conducting cells of the phloem. Values for mass transfer rate range from 1 to 15 g h^{-1} cm^{-2} of sieve elements (in SI units, 2.8–41.7 μg s^{-1} mm^{-2}).

Velocities and mass transfer rates can both be measured with radioactive tracers. In the simplest type of experiment for measuring velocity, ^{11}C- or ^{14}C-labeled CO_2 is applied for a brief period of time to a source leaf (pulse labeling), and the arrival of label at a sink tissue or at a particular point along the pathway is monitored with an appropriate detector.

In general, velocities measured by a variety of techniques average about 1.0 m h^{-1} (0.28 mm s^{-1}) and range from 0.3 to 1.5 m h^{-1} (in SI units, 0.08–0.42 mm s^{-1}). Transport values exceed the rate of diffusion by many orders of magnitude. Any proposed mechanism of phloem translocation must account for these high velocities.

The pressure-flow model is a passive mechanism for phloem transport

The most widely accepted mechanism of phloem translocation in angiosperms is the pressure-flow model. The pressure-flow model explains phloem translocation as a flow of solution (mass flow or bulk flow) driven by an osmotically generated pressure gradient between source and sink. The pressure-flow model predicts mass flow, and data both support and challenge the model.

The pressure is osmotically generated

Diffusion is far too slow to account for the velocities of solute movement observed in the phloem. Translocation velocities average 1 m h^{-1}; the rate of diffusion would be 1 m per 62 years for sucrose at 25°C! (Chapter 2 discusses diffusion velocities and the distances over which diffusion is an effective transport mechanism.)

The **pressure-flow model**, first proposed by Ernst Münch in 1930, states that a flow of solution in the sieve elements is driven by an osmotically generated *pressure gradient* between source and sink. Phloem loading at the source and phloem unloading at the sink establish the pressure gradient.

As we discussed earlier, three different mechanisms are recognized to generate high concentrations of sugars in the sieve elements of the source: active membrane transport, oligomer trapping, and passive symplasmic transport. Let's use the water potential to examine sucrose transport: $\Psi = \Psi_s + \Psi_p$; that is, $\Psi_p = \Psi - \Psi_s$ (Equation 2.5). In source tissues, an accumulation of sugars in the sieve elements generates a low (negative) solute potential (Ψ_s) and causes a steep drop in the water potential (Ψ). In response to the water-potential gradient, water enters the sieve elements and causes the turgor pressure (Ψ_p) to increase. Water enters the sieve elements through water channels, called **aquaporins**.

At the receiving end of the translocation pathway, **phloem unloading** leads to a lower sugar concentration in the sieve elements, generating a higher (less negative) solute potential in the sieve elements of sink tissues. As the water potential of the phloem rises above that of the xylem, water tends to leave the phloem in response to the water-potential gradient, causing a decrease in turgor pressure in the sieve elements of the sink. Figure 10.14 illustrates the *pressure-flow hypothesis*; the figure specifically shows the case of apoplastic phloem loading.

The phloem sap moves by mass flow down the pressure gradient. Mass flow means that solutes move at the same rate as the water molecules, and this is possible where no membranes are crossed during intercellular transport. Since this is the case between sieve tube elements, mass flow can occur from a source organ with a lower water potential to a sink organ with a higher water potential, or vice versa, depending on the identities of the source and sink organs. In fact, Figure 10.14 illustrates an example in which the flow is against the water-potential gradient. Such water movement does not violate the laws of thermodynamics,

pressure-flow model A widely accepted model of phloem translocation in angiosperms. It states that transport in the sieve elements is driven by a pressure gradient between source and sink. The pressure gradient is osmotically generated and results from the loading at the source and unloading at the sink.

aquaporins Integral membrane proteins that form channels across a membrane, many of which are selective for water (hence the name). Such channels facilitate water movement across a membrane.

phloem unloading The movement of photosynthates from the sieve elements to neighboring cells that store or metabolize them or pass them on to other sink cells via short-distance transport. *See also* phloem loading.

because mass flow is driven by a pressure gradient, as opposed to osmosis, which is driven by a water-potential gradient.

According to the pressure-flow hypothesis, movement in the translocation pathway is driven by transport of solutes and water *into* source sieve elements and *out of* sink sieve elements. The passive, pressure-driven, long-distance translocation in the sieve tubes ultimately depends on the mechanisms involved in phloem loading and unloading. These mechanisms are responsible for setting up the pressure gradient.

Some predictions of pressure flow have been confirmed, while others require further experimentation

Some important predictions emerge from the model of phloem translocation we have just described:

- No true bidirectional transport (i.e., simultaneous transport in both directions) in a single sieve tube can occur. A mass flow of solution precludes such bidirectional movement because a solution can flow in only one direction in a pipe at any one time. Furthermore, water and solutes must move at the same velocity in a flowing solution.
- No great expenditures of energy are required to drive translocation in the tissues along the path. Therefore, treatments that restrict the supply of ATP in the path, such as low temperature, anoxia, and metabolic inhibitors, should not stop translocation. However, the sieve tube lumen and the sieve plate pores must be largely unobstructed. If P-protein or other materials block the pores, the resistance to flow of the sieve element sap might be too great.
- The pressure-flow hypothesis predicts the presence of a positive pressure gradient, with turgor pressure higher in sieve elements of sources than in those of sinks.
- The pressure difference must be large enough to overcome the resistance of the pathway and to maintain flow at the observed velocities. Thus, pressure gradients should be larger in long transport pathways, for example in trees, than in short transport pathways, as in herbaceous plants.

There are plenty of observations consistent with the first two predictions. Here we focus on the last two predictions.

Functional sieve plate pores appear to be open channels

Ultrastructural studies of sieve elements are challenging because of the high internal pressure in these cells. When the phloem is excised or killed slowly with chemical fixatives, the turgor pressure in the sieve elements is released. The contents of the cells surge toward the point of pressure release and, in the case of sieve tube elements, accumulate on the sieve plates. This accumulation is probably the reason that many earlier electron micrographs show sieve plates that are obstructed, particularly with P-protein or organelles.

Newer, rapid freezing and fixation techniques provide reliable pictures of undisturbed sieve elements and are confirmed by live microscopy. When young Arabidopsis plants are rapidly frozen by plunging them into slush nitrogen, then freeze-substituted and fixed, sieve plate pores are often unobstructed in the tissue (Figure 10.7A). The sieve plate pores of living, translocating sieve elements of many eudicots have also been observed to be mostly open. Live cell microscopy, as well as careful preparation for electron microscopy, has observed the sieve element organelles (mitochondria, plastids, and ER) typically at the sieve element periphery. They seem to be anchored to each other or to the sieve element plasma membrane by minute protein "clamps." The open pores and anchored organelles to prevent their surging toward sieve plates seen in many species (Figures 10.6 and 10.7) are consistent with mass flow.

What about the distribution of P-protein in the sieve tube lumen? Electron micrographs of sieve tube members prepared by rapid freezing and fixation have often shown P-protein along the periphery of the sieve elements or as a fine meshwork in the lumen of the cell. Furthermore, the sieve plate pores often contain P-protein in similar positions, lining the pore or in a loose network (Figures 10.6C and 10.7C).

Are the pressure gradients in the sieve elements sufficient to drive phloem transport in trees?

Mass flow or bulk flow is the combined movement of all the molecules in a solution, driven by a pressure gradient. What are the pressure values in sieve elements, and how can they be determined? Does a pressure gradient exist between sources and sinks, and if so, is the gradient modest or substantial? Do large plants, such as trees, have proportionally higher pressures in the phloem than small herbaceous species?

Turgor pressure in sieve elements can either be calculated from the water potential and solute potential ($\Psi_p = \Psi - \Psi_s$) or measured directly. A well-established technique uses phloem-sucking aphids. Aphids are small insects that feed by inserting their mouthparts, consisting of four tubular stylets, into a single sieve element of a leaf or stem. Sap can be collected from aphid stylets cut from the body of the insect, usually with a laser, after the aphid has been anesthetized with CO_2. After stylectomy, micromanometers or pressure transducers are sealed over exuding aphid stylets. The data obtained are accurate because aphids pierce only a single sieve element, and the plasma membrane apparently seals well around the aphid stylet. Pressures measured using the aphid stylet technique range from approximately 0.7 to 1.5 MPa in both herbaceous plants and small trees. These values fit well with direct measurements of the turgor pressures in functional sieve tubes of morning glory (*Ipomoea nil*); in a 7-m stem of this vine, turgor pressure averaged 1.1 MPa.

This study of *Ipomoea* assessed every factor that is necessary to test the feasibility of pressure flow through passive microtubes (= sieve tubes) as described by the Poiseuille equation (Chapter 3):

$$Velocity = \frac{k \cdot \Delta\Psi p}{\eta \cdot \Delta x} \tag{10.1}$$

where k is the conductivity of the tube (m^2), $\Delta\Psi_p$ is the pressure differential (Pa), η is the sap viscosity (Pa s), and Δx is the length of the tube (m).

The specific conductivity of sieve tubes was derived from the geometry of sieve elements and dimensions of sieve pores along the transport phloem. The study concluded that a pressure of 0.21 MPa is sufficient to drive transport through 1 m of *Ipomoea* primary phloem. This study confirmed earlier studies that derived turgor pressures from phloem sap concentrations and calculated that the observed pressure gradients are sufficient to drive mass flow. In trees, however, systematic studies of turgor gradients in sieve tubes are lacking. The data are critical to any evaluation of the pressure-flow hypothesis.

One observation is certain: Turgor pressures in trees are not proportionally higher than those in herbaceous plants. One study compared calculated turgor pressures (the technique often used in trees) and pressures measured using aphid stylets in small willow saplings. The two techniques yielded comparable values, averaging 0.6 MPa for the calculated pressures and 0.8 MPa for the measured pressures. Calculated pressures were as high as 2.0 MPa in large white ash trees. These values are not substantially different from those measured in herbaceous plants, as previously noted. (Herbaceous plants and trees do often differ in their phloem-loading strategies in a way that is consistent with the relatively low pressures in trees.)

Although the relatively low turgor pressures measured in the phloem of trees are inconsistent with one of the predictions of the pressure-flow theory, the theory is still the best available to describe the experimental observations.

10.5 Materials Translocated in the Phloem

Distinguish between the molecules that move in the phloem sap and those that seal off damaged phloem to limit sap loss.

Water is the most abundant substance in the phloem. Dissolved in the water are the translocated solutes, including carbohydrates, amino acids, hormones, some inorganic ions, proteins and RNAs, and some specialized metabolites involved in defense and protection. Carbohydrates are the most significant and concentrated solutes in phloem sap (**Table 10.3**), with sucrose being the sugar most commonly transported in sieve elements. There is always some sucrose in sieve element sap, and it can reach concentrations of 0.3 to 0.9 *M*. Sugars, potassium ions, and amino acids and their amides are the principal molecules contributing to the osmotic potential of the phloem (**Figure 10.15**).

Table 10.3 The Composition of Phloem Sap, Collected as an Exudate from Cuts in the Phloem

	mg mL^{-1} [m*M*]	m*M*		
Component	*Ricinus* (Hall & Baker, 1972)	*Ricinus* (Peuke, 2010)	*Lupinus* low-saline soil (Pate, 1989)	*Lupinus* high-saline soil (Pate, 1989)
Sugars	80.0–106.0 [234–309]	433	652	600
Amino acids	5.2 [28]	67.5	41	110
Organic acids	2.0–3.2	8.02[a]	60	56
Protein	1.45–2.20	–	–	–
K^+	2.3–4.4 [59–112]	67.1	66.9	52.6
Na^+	–	6.96	8.1	92.6
Cl^-	0.355–0.675 [10–19]	12	7.9	68
SO_4^{2-}	–	1.29	4.3	5.5
PO_4^{3-}	0.350–0.550 [3.7–5.8]	6.56	10	12.6
NO_3^-	–	0.59	–	–
Mg^{2+}	0.109–0.122 [4.5–5.0]	3.71	3.4	2.7
Ca^{2+}	–	1.21	1.5	0.91

Sources: S. M. Hall and D. A. Baker. 1972. *Planta* 106: 131–140; J. S. Pate. 1989. In *Transport of Photoassimilates*, D. A. Baker and J. A. Milburn (eds.). Longman Scientific & Technical, Harlow, UK. pp. 138–166; A. D. Peuke. 2010. *J. Exp. Bot.* 61: 635–655

[a] Malate only.

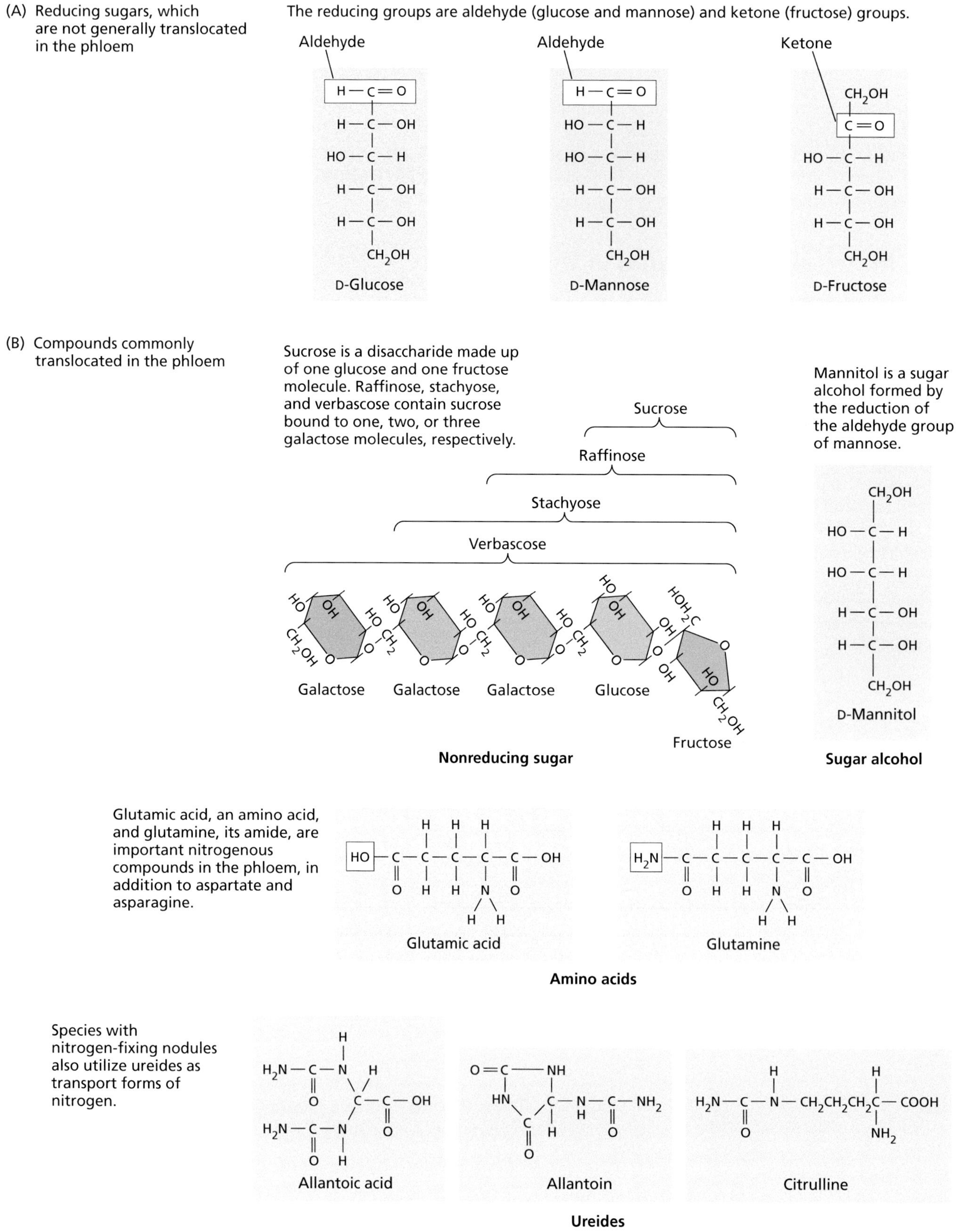

(A) Reducing sugars, which are not generally translocated in the phloem
The reducing groups are aldehyde (glucose and mannose) and ketone (fructose) groups.
Aldehyde
Aldehyde
Ketone
D-Glucose
D-Mannose
D-Fructose
(B) Compounds commonly translocated in the phloem
Sucrose is a disaccharide made up of one glucose and one fructose molecule. Raffinose, stachyose, and verbascose contain sucrose bound to one, two, or three galactose molecules, respectively.
Sucrose
Raffinose
Stachyose
Verbascose
Galactose
Galactose
Galactose
Glucose
Fructose
Nonreducing sugar
Mannitol is a sugar alcohol formed by the reduction of the aldehyde group of mannose.
D-Mannitol
Sugar alcohol
Glutamic acid, an amino acid, and glutamine, its amide, are important nitrogenous compounds in the phloem, in addition to aspartate and asparagine.
Glutamic acid
Glutamine
Amino acids
Species with nitrogen-fixing nodules also utilize ureides as transport forms of nitrogen.
Allantoic acid
Allantoin
Citrulline
Ureides

Figure 10.15 Structures of (A) compounds not normally translocated in the phloem and (B) compounds commonly translocated in the phloem.

Complete identification of solutes that are mobile in the phloem and that have a significant function has been difficult; no one method of sampling phloem sap is free of artifacts or provides a complete picture of mobile solutes. The collection of phloem sap is experimentally challenging because of the high turgor pressure in the sieve elements and the wound reactions that plug sieve plates. Because of processes that plug sieve plate pores, only a few species exude phloem sap from wounds that sever sieve elements.

A preferable approach for collecting exuded sap is to use an aphid stylet as a "natural syringe," as we described. The high turgor pressure in the sieve element forces the cell contents through the stylet to the cut end, where they can be collected. However, quantities of collected sap are small, and the method is technically difficult. Nonetheless, this method is thought to yield relatively pure sap from the sieve elements and companion cells and to provide a fairly accurate picture of the composition of phloem sap.

Sugars are translocated in a nonreducing form

Results from many analyses of collected sap indicate that the translocated carbohydrates are nonreducing sugars, like sucrose. Reducing sugars, such as the hexoses glucose and fructose, contain an exposed aldehyde or ketone group, so they are more reactive than nonreducing sugars (Figure 10.15A). In a nonreducing sugar, such as sucrose, the ketone or aldehyde group is reduced to an alcohol or combined with a similar group on another sugar. Translocated sugar alcohols include mannitol and sorbitol (Figure 10.15B).

Most researchers believe that the nonreducing sugars are the major compounds translocated in the phloem because they are less reactive than their reducing counterparts. Reducing sugars, such as hexoses, are quite reactive and may be as much of a problem as reactive oxygen and nitrogen species. Animals can transport glucose in their blood because it is present in fairly low concentrations, but the very high concentrations of sugars in the phloem make it necessary to transport nonreducing sugars.

Other small organic and inorganic solutes are translocated in the phloem

Nitrogen is found in the phloem mainly in amino acids—especially glutamate and aspartate—and their respective amides, glutamine and asparagine. Reported concentrations of amino acids and organic acids vary widely, even for the same species, but they are usually low compared with those of carbohydrates (Table 10.3). Like sugars, the phloem-mobile amino acids are actively loaded into the phloem, as confirmed by the localization of amino acid transporters in the plasma membrane of companion cells.

Some inorganic ions are also translocated in the phloem, including K^+, NO_3^-, Mg^{2+}, PO_4^{3-}, and Cl^-. By contrast, Ca^{2+} and SO_4^{2-} are relatively immobile in the phloem. Potassium ions appear to serve as a mobile energy source in the phloem by assisting the proton pump in sucrose uptake and retrieval. Small cytosolic solutes in companion cells might easily move into the neighboring sieve element, which is well connected by pore–plasmodesma contacts (Figure 10.6B and C). From here, they are swept along with the phloem sap. It is not settled, however, whether the pore–plasmodesma contacts have a mechanism to retain valuable solutes in the companion cells. Size does not seem to be critical; passage of up to 70 kDa macromolecules was reported using GFP-tagged proteins as tracers. By contrast, the passage of ions and small organic molecules is challenging to test. Simple occurrence in the phloem is not sufficient evidence since collection of phloem sap might lead to displacement of the solute in question. In what may have been a response to sampling, nucleotide phosphates have been found in the phloem sap. It is difficult to imagine that companion cells lose these energy-rich compounds constitutively.

Hormones do not seem to be retained in the companion cell. Almost all the endogenous plant hormones, including auxin, gibberellins, cytokinins, and abscisic acid, have been found in sieve elements. The long-distance transport of hormones, especially auxin, is thought to occur at least partly in the sieve elements. Hormones might indeed enter the sieve tubes via the pore–plasmodesma contacts, as has been shown for auxin.

Phloem-mobile macromolecules often originate in companion cells

Macromolecules found in the phloem sap include structural P-proteins (involved in the sealing of wounded sieve elements in cucurbit species), as well as several water-soluble proteins and RNAs. The function of many of the proteins commonly found in phloem sap is related to stress and defense reactions. The possible roles of RNAs and proteins as signal molecules will be discussed later.

Damaged sieve elements are sealed off

As detailed earlier in this section, sieve element sap is rich in sugars and other organic molecules. These molecules represent an energy investment for the plant, and their loss must be prevented when sieve elements are damaged. Short-term sealing mechanisms involve sap proteins (P-proteins), while the principal long-term mechanism for preventing sap loss entails closing sieve plate pores with callose, a glucose polymer.

When a sieve tube is cut or punctured, the release of pressure causes the contents of the sieve elements to surge toward the cut end, from which the plant could lose much sugar-rich phloem sap if there were no sealing mechanism. When surging occurs, however, P-protein is trapped on the sieve plate pores, helping seal the sieve element and prevent further loss of sap (**Figure 10.16A**). Direct support for the sealing function of P-protein has been found in both tobacco and Arabidopsis, in which mutants lacking P-protein lose significantly more transport sugar by sap exudation after wounding than do wild-type plants. No visible phenotypic differences were observed between the mutant and wild-type plants.

A longer-term solution to sieve tube damage is the production of **callose** in the sieve pores (**Figure 10.16B**). Callose, a β-1,3-**glucan**, is synthesized by an enzyme in the plasma membrane (callose synthase) and is deposited between the plasma membrane and the cell wall. Callose is synthesized in functioning sieve elements in response to damage and other stresses, such as mechanical stimulation and high temperatures, or in preparation for normal developmental events, such as dormancy. The deposition of **wound callose** in the sieve pores efficiently seals off damaged sieve elements from surrounding intact tissue, with complete occlusion occurring about 10 min after injury (**Figure 10.16C**). In all cases, as sieve elements recover from damage or break dormancy, the callose disappears from the sieve pores; its dissolution is mediated by a callose-hydrolyzing enzyme (Figure 10.16C). While Arabidopsis and tobacco mutants lacking P-protein show no visible phenotypic changes, Arabidopsis mutants lacking a specific callose synthase show lack of wound-induced callose at sieve plates and disturbance of sieve pore development.

callose A β-1,3-glucan synthesized in the plasma membrane and deposited between the plasma membrane and the cell wall. Synthesized by sieve elements in response to damage or stress, or as part of a normal developmental process.

glucan A polysaccharide made from glucose units.

wound callose Callose deposited in the sieve pores of damaged sieve elements that seals them off from surrounding intact tissue. As sieve elements recover, the callose disappears from pores.

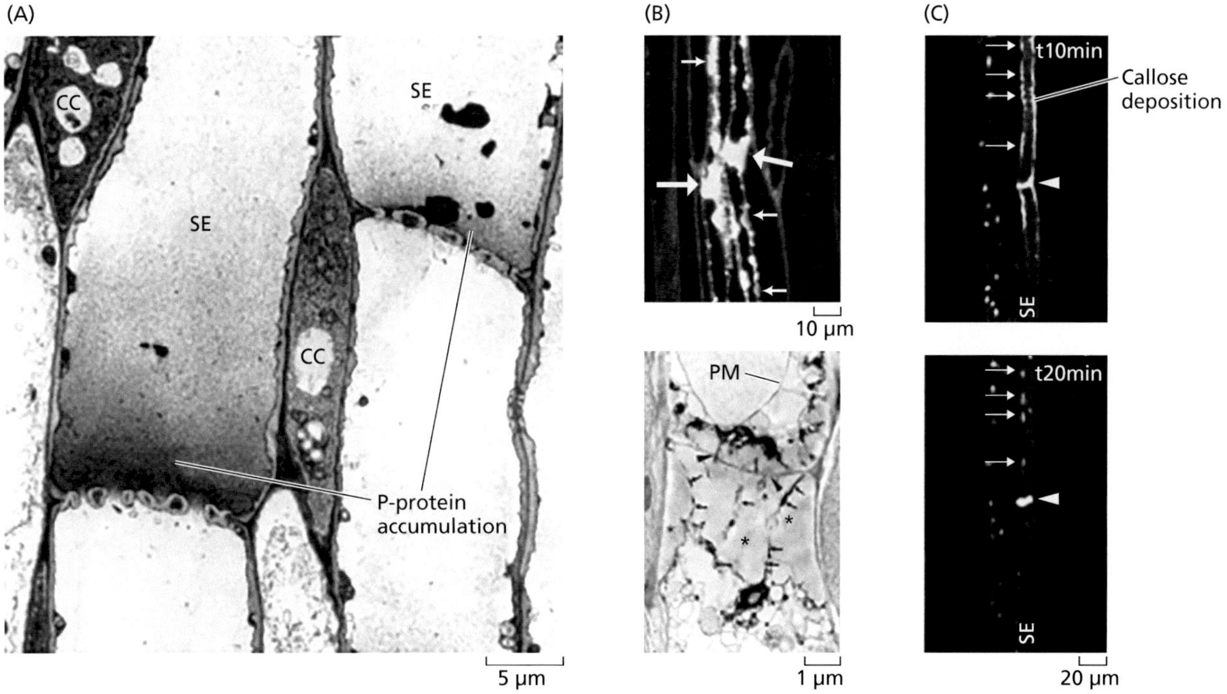

Figure 10.16 Sealing of sieve elements by P-protein and callose. (A) P-protein filaments surge toward a phloem cut (~0.1 mm below this image) and plug the sieve plates of squash (*Cucurbita pepo*). (B) Callose covers sieve plates and lateral walls of pea (*Pisum satium*) root phloem 48 h after severing the stele, as seen by bright fluorescence (thick and thin arrows in top image, respectively) and transmission electron microscopy (bottom image). Cytoplasmic compounds (arrows) that initially plug the sieve pores (arrowheads) are subsequently enclosed by callose (asterisks in bottom image). (C) Callose deposition (yellow) in leaf phloem is induced by heating the leaf tip 30 mm away at sieve plates (arrowhead) and at lateral sieve pores (arrows). The callose is degraded over time. Top image, 10 min after heating. Bottom image, 20 min after heating. CC, companion cell; PM, plasma membrane; SE, sieve element.

10.6 Phloem Unloading and Sink-to-Source Transition

Describe the mechanisms of phloem unloading and how organs transition from sink to source.

Now we have reached the destination of our journey from source to sink. The driving force for phloem translocation is the loading of sugars in source leaves, leading to osmotic pressure in source phloem. The transport phloem forms a continuous pathway to the organs consuming or storing the photosynthates. The plant ensures integrity of this pathway by wound responses that plug the sieve plate pores and seal them with callose. At sink sites, sugars are released from the phloem, thus reducing its osmotic pressure. Without sugar release, transport would stop since the pressure differential between source and sink would equilibrate (Figure 10.14). Thus, the rate of phloem transport depends not only on the rate of loading in source leaves but also on the rate of sugar consumption and removal in the sink organs. Inhibitor studies have shown that import into sink tissue is energy-dependent. Accordingly, phloem unloading is coupled to **sink activity**.

Let's take a closer look at **import** into sinks such as developing roots, tubers, and reproductive structures. In many ways the events in sink tissues are simply

sink activity The rate of uptake of photosynthate per unit of weight of sink tissue.

import The movement of photosynthate in sieve elements into sink organs.

the reverse of the events in sources. The following steps are involved in the import of sugars into sink cells:

1. *Phloem unloading.* This is the process by which imported sugars leave the sieve elements of sink tissues.
2. *Post-phloem transport.* After unloading, the sugars are transported to cells in the sink by means of a short-distance transport pathway.
3. *Storage and metabolism.* In the final step, sugars are stored or metabolized in sink cells.

How does all of this work together? Are phloem unloading and short-distance transport symplasmic or apoplastic? Is sucrose hydrolyzed during the process? Do phloem unloading and subsequent steps require energy? Finally, we examine the process by which a young, importing leaf becomes an exporting source leaf.

Phloem unloading and short-distance transport can occur via symplasmic or apoplastic pathways

In sink organs, sugars move from the sieve elements to the cells that store or metabolize them. Sinks vary from growing vegetative organs (root tips and young leaves) to storage tissues (roots and stems) to organs of reproduction and dispersal (fruits and seeds). Because sinks vary so greatly in structure and function, there is no single mechanism of phloem unloading and short-distance transport. Differences in import pathways due to differences in sink types are emphasized in this section; however, the pathways often switch during sink development as well.

As in sources, the sugars may move entirely through the symplasm via the plasmodesmata in sinks, or they may enter the apoplast at some point. **Figure 10.17** diagrams the several possible pathways in sinks. Both unloading and the short-distance pathway appear to be completely symplasmic in meristematic and elongating regions of primary root tips and in some young eudicot leaves, such as those of sugar beet and tobacco (Figure 10.17A). Loading and unloading strategies seem to be independent; plant species that load sugars apoplastically release them at root tips symplasmically.

Symplasmic unloading supplies growing vegetative sinks

Meristematic regions of the root and shoot and elongating regions of plant organs are heterotrophic—that is, they depend on phloem import, as they are not able to meet their energy needs through photosynthesis. This is also the case for young leaves when photosynthesis is not yet meeting their energy needs.

In symplasmic unloading, no membranes are crossed during sugar uptake into the sink cells, and transport is passive: Transport sugars move from a high concentration in the sieve elements to a low concentration in the sink cells via specific, funnel-shaped plasmodesmata, and the process involves both diffusion and bulk flow. Phloem-mobile macromolecules appear to be sorted out at the next post-phloem interface, the pericycle–endodermis interface, through which small solutes and proteins up to 27 kDa, but no larger molecules, pass readily.

While symplasmic unloading is passive, it depends on energy consumption in the sink. Transport sugars are used as substrates for respiration and are metabolized into storage polymers and into compounds needed for growth. Metabolic energy is required in these sink organs mainly for respiration and for biosynthetic reactions. Sucrose metabolism thus results in a low sucrose concentration in the sink cells, extending the source-to-sink sugar gradient in the phloem via the post-phloem pathway right into the sink cells.

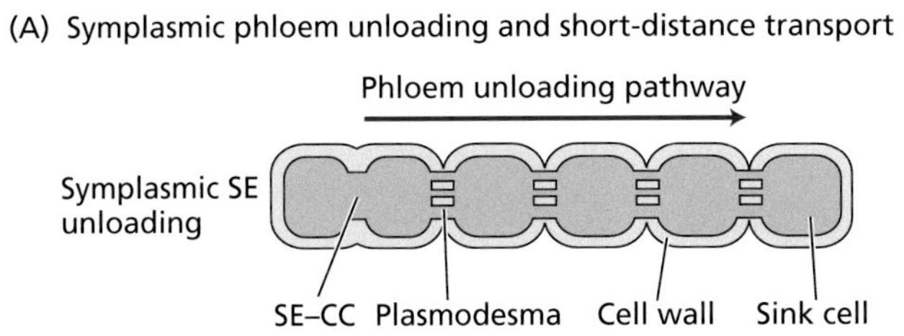
(A) Symplasmic phloem unloading and short-distance transport
Phloem unloading pathway
Symplasmic SE unloading
SE–CC
Plasmodesma
Cell wall
Sink cell

Figure 10.17 Pathways for phloem unloading and short-distance transport. The sieve element–companion cell complex (SE–CC) is considered a single functional unit. The presence of plasmodesmata is assumed to provide functional symplasmic continuity. An absence of plasmodesmata between cells indicates an apoplastic transport step. (A) Symplasmic phloem unloading and short-distance transport. All steps are symplasmic. (B) Apoplastic phloem unloading and short-distance transport.

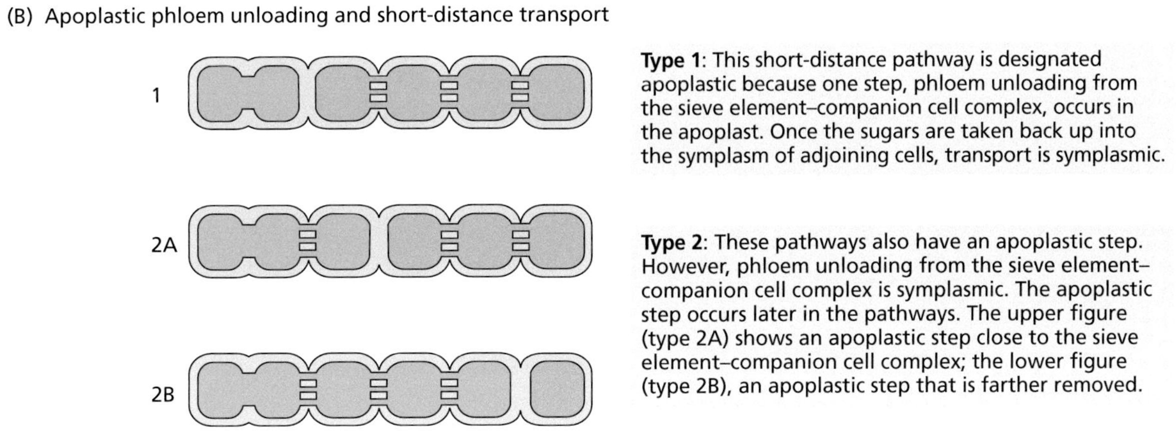
(B) Apoplastic phloem unloading and short-distance transport
1
Type 1: This short-distance pathway is designated apoplastic because one step, phloem unloading from the sieve element–companion cell complex, occurs in the apoplast. Once the sugars are taken back up into the symplasm of adjoining cells, transport is symplasmic.
2A
Type 2: These pathways also have an apoplastic step. However, phloem unloading from the sieve element–companion cell complex is symplasmic. The apoplastic step occurs later in the pathways. The upper figure (type 2A) shows an apoplastic step close to the sieve element–companion cell complex; the lower figure (type 2B), an apoplastic step that is farther removed.
2B

Import into seeds, fruits, and storage organs often involves an apoplastic step

While symplasmic unloading predominates in most sink tissues, part of the short-distance pathway is apoplastic in some sink organs at some stages of development—for example, in fruits, seeds, and other storage organs that accumulate high concentrations of sugars. Sugars exit the sieve elements (phloem unloading) via a symplasmic pathway and are transferred from the symplasm to the apoplast at a point distal from the sieve element–companion cell complex (type 2 in Figure 10.17B).

The pathway can switch between symplasmic and apoplastic in these sinks, with an apoplastic step being required when sink sugar concentrations are high. The apoplastic step can be located at the site of unloading itself (type 1 in Figure 10.17B) or can be farther away from the sieve elements (type 2). The type 2 arrangement, typical of developing seeds, appears to be the most common in apoplastic pathways. Early in ovule development, the phloem end at the chalaza (Chapter 17) is symplasmically isolated and unloading is apoplastic, but unloading switches to a symplasmic pathway at the time of fertilization (**Figure 10.18**). In developing seeds, an apoplastic step occurs next to the embryo sink because there are no symplasmic connections between the maternal tissues and the embryo. This apoplastic step permits membrane control over the substances that enter the embryo, because two membranes must be crossed in the process.

When an apoplastic step occurs in the import pathway, the transport sugar can be partly metabolized in the apoplast, or it can cross the apoplast unchanged. For example, sucrose can be hydrolyzed into glucose and fructose in the apoplast by invertase, a sucrose-splitting enzyme, and glucose and/or fructose would then enter the sink cells. Sucrose splitting doubles the osmolarity (more negative Ψs) and drags water out of the sucrose-releasing cell. Thus, sucrose-cleaving enzymes play a role in the control of phloem transport by sink tissues.

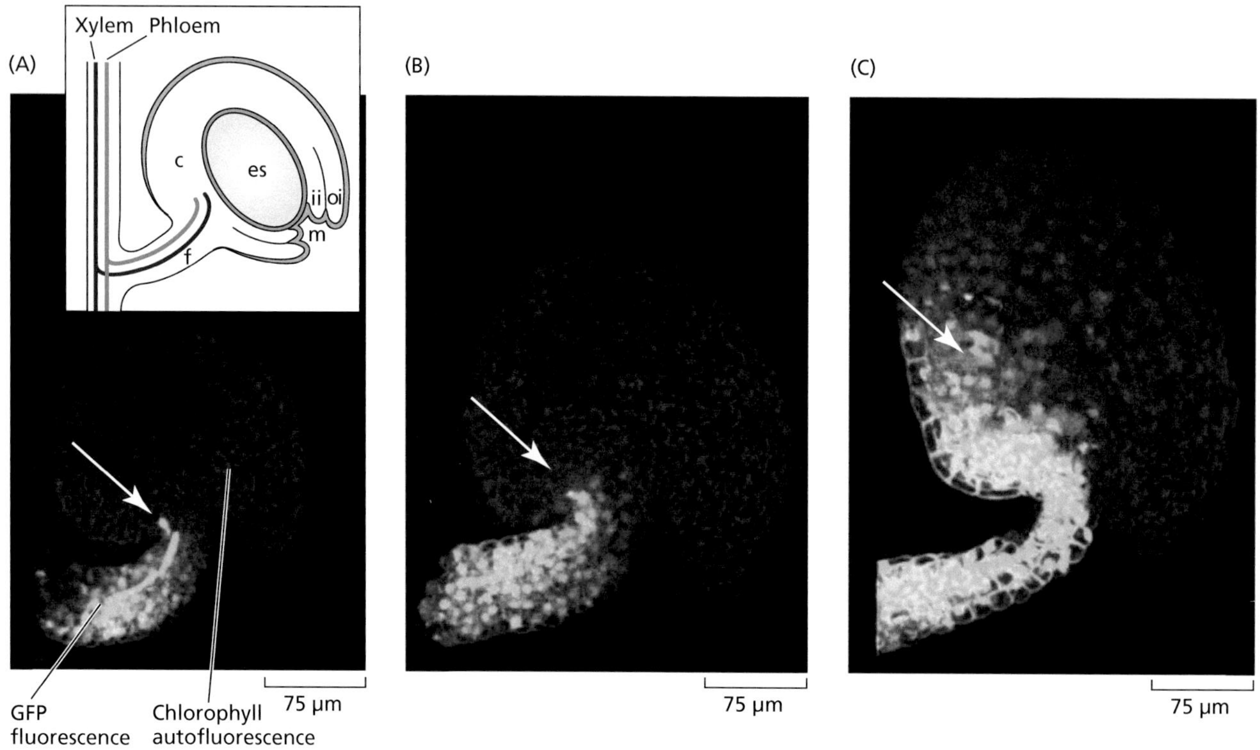

Figure 10.18 The unloading pathway can switch from an apoplastic to a symplasmic pathway. GFP marks the switch from apoplastic to symplasmic unloading into developing and fertilized ovules. GFP fluorescence is shown in green, chlorophyll autofluorescence in red. (A) Developing ovule of a closed flower. The GFP (arrow) is confined to the phloem because there are no functional plasmodesmata connecting the ends of the sieve tubes to surrounding cells (unloading is apoplastic). The inset shows a drawing of an ovule at this stage. Red line = xylem, green line = phloem. c, chalaza; es, embryo sac; f, funiculus; ii, inner integument; m, micropyle; oi, outer integument. (B) Mature ovule of a flower that is about to open. GFP is still not being unloaded into the integuments (arrow). (C) A fertilized ovule. The arrow marks GFP that has moved into the integuments via the symplasm.

Apoplastic import is active and requires metabolic energy

In apoplastic import, sugars must cross at least two membranes: the plasma membrane of the cell that is releasing the sugar and the plasma membrane of the sink cell. Transport across membranes may be energy-dependent. While some evidence indicates that both efflux and uptake of sucrose can be active transport, the transporters have yet to be completely characterized.

Some of the same sucrose transporters described earlier for sucrose loading could also be involved in sucrose unloading; the direction of transport would depend on the sucrose gradient, the pH gradient, and the membrane potential. Furthermore, symporters important in phloem loading have been found in some sink tissues—for example, SUT1 in potato tubers. The symporter may function in sucrose retrieval from the apoplast, in import into sink cells, or in both. Monosaccharide transporters must be involved in uptake into sink cells when sucrose is hydrolyzed in the apoplast by cell wall invertases.

The transition of a leaf from sink to source is gradual

Leaves of eudicots such as tomato and bean begin their development as sink organs. A transition from sink to source status occurs later in development, when the leaf is approximately 25% expanded, and is usually complete when the leaf is 40 to 50% expanded. Export from the leaf begins at the tip or apex of the blade and progresses toward the base until the whole leaf becomes a sugar exporter.

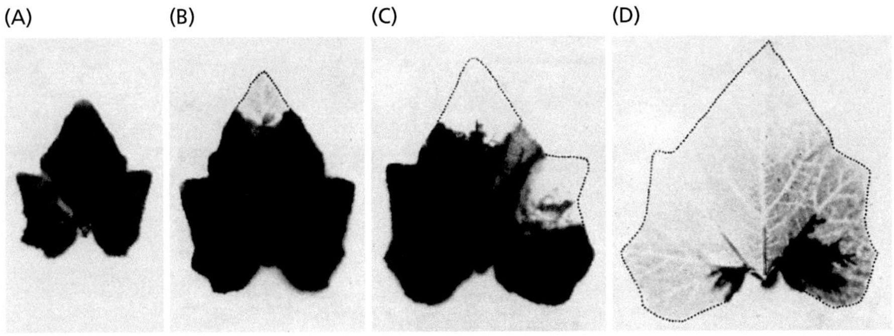

Figure 10.19 Autoradiographs of a leaf of summer squash (*Cucurbita pepo*), showing the transition of the leaf from sink to source status. In each case, the leaf imported ^{14}C from the source leaf on the plant for 2 h. Label is visible as black accumulations. (A) The entire leaf is a sink, importing sugar from the source leaf. (B–D) The base is still a sink. As the tip of the leaf loses the ability to unload and stops importing sugar (as shown by the loss of black accumulations), it gains the ability to load and to export sugar.

During the transition period, the tip exports sugar, while the base still imports it from the other source leaves (**Figure 10.19**).

The maturation of leaves is accompanied by functional and anatomical changes, resulting in a reversal of transport direction from importing to exporting. In general, the cessation of import and the initiation of export are independent events. In albino leaves of tobacco, which have no chlorophyll and therefore are incapable of photosynthesis, import stops at the same developmental stage as in green leaves, even though export is not possible. Therefore, a developmental switch besides the initiation of export must occur in developing leaves of tobacco that causes them to cease importing sugars.

Sugars are unloaded and loaded almost entirely via different veins in tobacco (**Figure 10.20**), contributing to the conclusion that import cessation and export initiation are two separate events. The minor veins that are eventually responsible for most of the loading in tobacco and other *Nicotiana* species do not mature until about the time import ceases, as was shown with uptake of radioactively labeled sucrose.

The direction of phloem translocation was also studied by following the expression pattern of the SUC2 sucrose–H^+ symporter (**Figure 10.21**). The ability to accumulate exogenous sucrose in the sieve element–companion cell complex

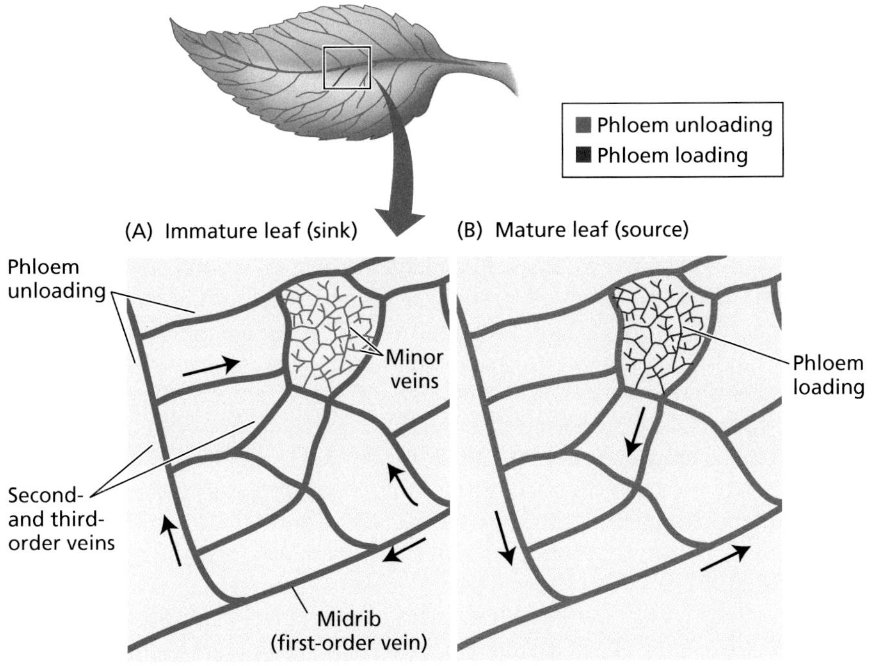

Figure 10.20 Division of labor in the veins of a tobacco leaf. (A) When the leaf is immature and still in its sink phase, photosynthate is imported from mature leaves and distributed (arrows) throughout the blade (lamina) via the larger, major veins (thicker lines). The major veins are labeled, with the midrib being the first-order vein. The imported photosynthate unloads from the same major veins into the mesophyll. The smallest, minor veins are found within the areas enclosed by the third-order veins. The minor veins do not function in import and unloading because they are immature. Figure is drawn to scale from an autoradiograph. (B) In a mature source leaf, import has ceased and export has begun. Photosynthate loads into the minor veins, while the larger veins serve only in export (arrows); they can no longer unload. Figure is not drawn to scale or in correct proportions, since the blade (lamina) grows considerably as the leaf matures.

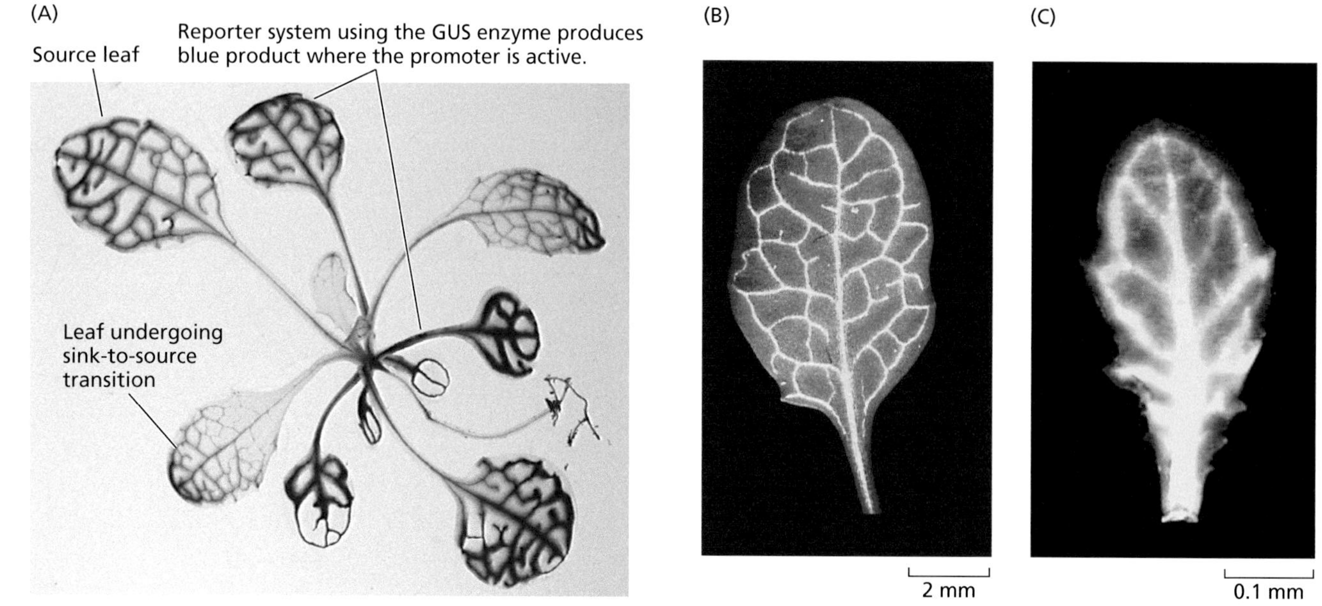

Figure 10.21 Export from source tissue depends on the placement and activity of active sucrose transporters. (A) Arabidopsis rosette transformed with a construct consisting of a reporter gene under control of the *AtSUC2* promoter. SUC2, a sucrose–H^+ symporter, is one of the major sucrose transporters functioning in phloem loading. The reporter system used (GUS) forms a visible product (blue) where the promoter is active. Staining is visible only in the vascular tissue of source leaves and in the tips of leaves undergoing the sink-to-source transition. (B,C) GFP fluorescence (yellow here because of a background red chlorophyll fluorescence) in source and sink leaves from transgenic Arabidopsis plants expressing GFP under control of the *SUC2* promoter indicates that GFP moves through plasmodesmata from companion cells into sieve elements of source leaves and from sieve elements into the surrounding mesophyll of sink leaves. GFP is localized by its fluorescence after excitation with blue light. (B) GFP is synthesized in companion cells and moves into the sieve elements of the source, as indicated by the bright fluorescence in the veins. (C) Free GFP is imported into sink leaves and moves into the surrounding mesophyll. Because GFP has moved into surrounding tissues, the veins are no longer distinctly delineated, and GFP fluorescence is more diffuse. Note that the scales in (B) and (C) are different; even though the source leaf in (B) appears to be the same size as the sink leaf in (C), the source leaf is actually much larger (note the different scale bars).

is acquired as the leaves undergo the sink-to-source transition, suggesting that the symporter required for loading has become functional. In developing Arabidopsis leaves, expression of the symporter begins in the tip and proceeds to the base during a sink-to-source transition (Figure 10.21A). This is the same pattern (basipetal) that is seen in the development of export capacity.

Live microscopy was used to visualize the veins responsible for loading in source leaves, and for unloading in sink leaves. Plants were transformed with the gene for green fluorescent protein (GFP) from jellyfish, under control of the *SUC2* promoter from Arabidopsis. The SUC2 sucrose–H^+ symporter is synthesized within the companion cells, so proteins expressed under the control of its promoter, including GFP, are also synthesized in the companion cells. GFP moves through plasmodesmata from companion cells into sieve elements of source leaves (Figure 10.21B) and migrates within the phloem to sink tissues. In young leaves, GFP is imported through the large veins, from which it spreads out into the mesophyll (Figure 10.21C).

The change that stops import during sink-to-source transition must involve blockage of unloading from the large veins at some point in the development of mature leaves. Factors that could account for the cessation of unloading include plasmodesmatal closure and a decrease in plasmodesmatal frequency. Recently it was shown that cessation of phloem unloading from large veins is initiated by activation of the TOR (TARGET OF RAPAMYCIN) regulatory complex. If this metabolic complex is inhibited, GFP spreading into the mesophyll continues beyond the sink-to-source transition.

Export of sugars begins when apoplastic loading is activated and has accumulated sufficient photosynthate in the sieve elements to drive translocation out of the leaf. The following events precede photosynthate export:

- The leaf is synthesizing photosynthate in sufficient quantity that some is available for export. The sucrose-synthesizing genes are being expressed.
- Minor veins responsible for loading have matured. A regulatory element (enhancer) has been identified in the DNA of Arabidopsis that acts as part of a cascade of events leading to minor vein maturation. The enhancer can activate a reporter gene fused to a companion cell–specific promoter and does so in the same tip-to-base pattern as in the sink-to-source transition.
- The sucrose–H^+ symporter is expressed and in place in the plasma membrane of the sieve element–companion cell complex. Binding sites for transcription factors have been identified in the *SUC2* promoter that mediate this source-specific and companion cell–specific gene expression.

allocation The regulated diversion of photosynthate into storage, utilization, and/or transport.

partitioning The differential distribution of photosynthate to multiple sinks within the plant.

10.7 Photosynthate Distribution: Allocation and Partitioning

Describe allocation and partitioning of photosynthate and how they are regulated.

The photosynthetic rate determines the total amount of fixed carbon available to the leaf, and the amount of fixed carbon available for translocation depends on subsequent metabolic events. The regulation of the distribution of fixed carbon into various metabolic pathways is called **allocation** (**Figure 10.22**).

The vascular bundles in a plant form a system of "pipes" that can direct the flow of photosynthates to various sinks: young leaves, stems, roots, fruits, or seeds. The vascular system is highly interconnected, forming an open network that allows source leaves to communicate with multiple sinks. The differential distribution of photosynthates within the plant is called **partitioning** (Figure 10.22).

What determines the volume of flow to any given sink? How do sinks compete? How do sources and sink communicate with each other?

Allocation includes storage, utilization, and transport

The carbon fixed in a source cell can be used for storage, metabolism, and transport:

- *Synthesis of storage compounds.* Starch is synthesized and stored in chloroplasts and, in most species, is the primary storage form that is mobilized for translocation during the night. Plants that store carbon primarily as starch are called *starch storers.*
- *Metabolic utilization.* Fixed carbon can be used within various compartments of the photosynthesizing cell to meet the energy needs of the cell or to provide carbon skeletons for the synthesis of other compounds required by the cell.
- *Synthesis of transport compounds.* Fixed carbon can be incorporated into transport sugars for export to various sink tissues. A portion of the transport sugar can also be stored temporarily in the vacuole.

Figure 10.22 Allocation and partitioning determine the distribution of photosynthate within a plant.

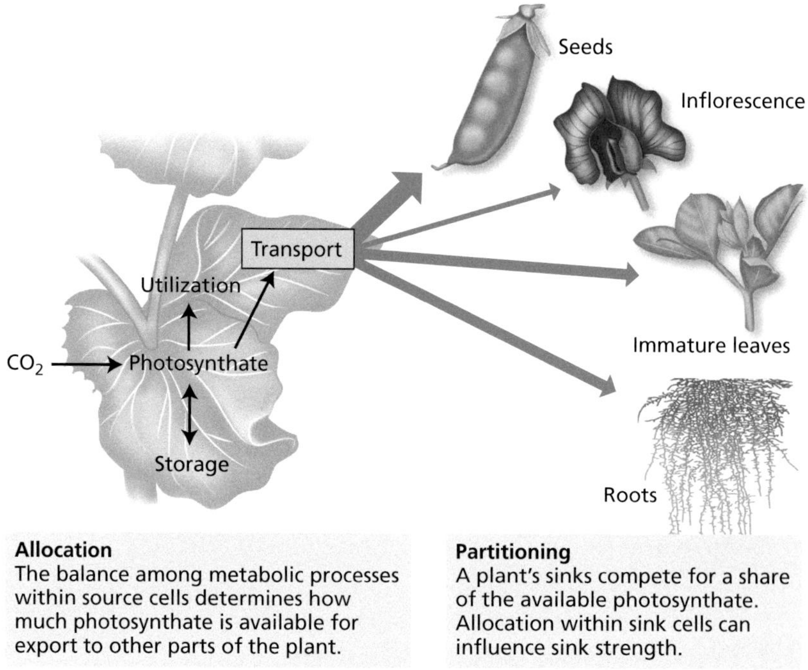
Seeds
Inflorescence
Transport
Utilization
CO_2
Photosynthate
Storage
Immature leaves
Roots
Allocation
The balance among metabolic processes within source cells determines how much photosynthate is available for export to other parts of the plant.
Partitioning
A plant's sinks compete for a share of the available photosynthate. Allocation within sink cells can influence sink strength.

Allocation is also a key process in sink tissues. Once the transport sugars have been unloaded and enter the sink cells, they can remain as such or can be transformed into various other compounds. In storage sinks, fixed carbon can be accumulated as sucrose or hexose in vacuoles or as starch in amyloplasts. In growing sinks, sugars can be used for respiration and for the synthesis of other molecules required for growth.

Source leaves regulate allocation

Increases in the rate of photosynthesis in a source leaf generally result in an increase in the rate of translocation from the source. Control points for the allocation of photosynthate include the distribution of triose phosphates to the following processes:

- Regeneration of intermediates in the C_3 photosynthetic carbon reduction cycle (the Calvin–Benson cycle)
- Starch synthesis
- Sucrose synthesis, as well as distribution of sucrose between transport and temporary storage pools

Various enzymes operate in the pathways that process the photosynthate. During the day, the rate of starch synthesis in the chloroplast must be coordinated with sucrose synthesis in the cytosol. Triose phosphates produced in the chloroplast by the Calvin–Benson cycle can be used for either starch or sucrose synthesis or in respiration. Sucrose synthesis in the cytosol diverts triose phosphate away from starch synthesis and storage. The control of these steps is complex and may be different in plants with different loading strategies.

However, there is a limit to the amount of carbon that normally can be diverted from starch synthesis in species that store carbon primarily as starch. Studies of allocation between starch and sucrose under different conditions suggest that a fairly steady rate of translocation throughout the 24-h period is a priority for most plants.

Various sinks partition transport sugars

Sinks compete for the photosynthate exported by the sources. Competition determines the partitioning of transport sugars among the various sink tissues of the plant, at least in the short term. The allocation of sugar within a sink (storage or metabolism) affects its ability to compete for available sugars. In this way, the processes of partitioning and allocation interact.

Of course, events in sources and sinks must be synchronized. Partitioning determines the patterns of growth, and growth must be balanced between shoot growth (photosynthetic productivity) and root growth (water and mineral uptake) in such a way that the plant can respond to the challenges of a variable environment. The goal is *not* a constant root-to-shoot ratio but one that secures a supply of carbon and mineral nutrients appropriate to the needs of the plant. So an additional level of control lies in the interaction between areas of supply and demand.

sink strength The ability of a sink organ to mobilize assimilates toward itself. Depends on two factors: sink size and sink activity.

Sink tissues compete for available translocated photosynthate

Translocation to sink tissues depends on the position of the sink in relation to the source and on the vascular connections between source and sink. Another factor determining the pattern of transport is competition between sinks—for example, between terminal sinks (e.g., seeds) or between terminal sinks and axial sinks (e.g., axial parenchyma in sugarcane or in trees for seasonal growth) along the transport pathway. For example, young leaves might compete with roots for photosynthates in the translocation stream. Competition has been shown by numerous experiments in which removal of a sink tissue from a plant generally results in increased translocation to alternative, and hence competing, sinks. Conversely, increased sink size, for example increased fruit load, decreases translocation to other sinks, especially the roots. Pruning of grape or fruit trees uses gardeners' traditional knowledge of these processes.

In the reverse type of experiment, the source supply can be altered while the sink tissues are left intact. When the supply of photosynthates from sources to competing sinks is suddenly and drastically reduced by shading of all the source leaves but one, the sink tissues become dependent on a single source. In sugar beet and bean plants, the rates of photosynthesis and export from the single remaining source leaf usually do not change over the short term (approximately 8 h). However, the roots receive less sugar from the single source, while the young leaves receive relatively more. Presumably, the young leaves can deplete the sugar content of the sieve elements more readily and thus increase the pressure gradient and the rate of translocation toward themselves.

Treatments such as making the sink water potential more negative increase the pressure gradient and enhance transport to the sink. Treatment of the root tips of pea (*Pisum sativum*) seedlings with mannitol solutions increased the import of sucrose over the short term by more than 300%, possibly because of a turgor decrease in the sink cells. Longer-term experiments show the same trend. Moderate water stress induced with polyethylene glycol treatment of the roots increased the proportion of photosynthates transported to the roots of apple plants over a period of 15 days, but decreased the proportion transported to the shoot apex. This contrasts with shading treatments just discussed, in which source limitation diverts more sugar to the young leaves.

Sink strength depends on sink size and activity

The ability of a sink to mobilize photosynthate toward itself is often described as **sink strength**. Sink strength depends on two factors—sink size and sink activity—as follows:

$$\text{Sink strength} = \text{sink size} \times \text{sink activity}$$

sink size The total weight of the sink.

sink activity The rate of uptake of photosynthate per unit of weight of sink tissue.

Sink size is the total biomass of the sink tissue, and **sink activity** is the rate of uptake of photosynthates per unit biomass of sink tissue. Altering either the size or the activity of the sink results in changes in translocation patterns. For example, the ability of a pea pod to import carbon depends on the dry weight of that pod as a proportion of the total number of pods.

Changes in sink activity depend on a balance between processes such as the unloading from the sieve elements, metabolism in the cell wall, uptake from the apoplast, and photosynthate utilization for either growth or storage.

Experimental treatments to manipulate sink strength, such as cooling the sink tissue, inhibit all activities that require metabolic energy and typically result in a decrease in the velocity of transport toward the sink. More specific are experiments that take advantage of our ability to over- or underexpress enzymes such as those involved in sucrose metabolism. The two major enzymes that split sucrose are invertase and sucrose synthase, both of which can catalyze the first step in sucrose utilization. Their activity increases phloem unloading and thereby sink strength.

The source adjusts over the long term to changes in the source-to-sink ratio

If all but one of the source leaves of a soybean plant are shaded for an extended period (e.g., 8 days), many changes occur in the single remaining source leaf. These changes include a decrease in starch concentration and increases in photosynthetic rate, Rubisco activity, sucrose concentration, transport from the source, and orthophosphate concentration. Thus, the observed short-term changes in the distribution of photosynthate among different sinks are followed, over a longer term, by adjustments in the source leaf's metabolism.

Photosynthetic rate (the net amount of carbon fixed per unit leaf area per unit time) often increases over several days when sink demand increases, and it decreases when sink demand decreases. An accumulation of photosynthate (sucrose or hexoses) in the source leaf can account for the linkage between sink demand and photosynthetic rate in starch-storing plants.

Sugars act as signaling molecules that regulate many metabolic and developmental processes in plants (Chapter 15). In general, carbohydrate depletion enhances the expression of genes for photosynthesis, reserve mobilization, and export processes, while abundant carbon resources favor genes for storage and utilization.

Sucrose or hexoses that would accumulate in leaves as a result of decreased sink demand are well known to repress photosynthetic genes. Interestingly, the genes for invertase and sucrose synthase, both of which can catalyze the first step in sucrose utilization, and genes for sucrose–H^+ symporters, which play a key role in apoplastic loading, are also among those regulated by carbohydrate supply.

10.8 Transport of Signaling Molecules

Describe the various signals transmitted by the phloem and their roles in physiology and development.

Besides its major function in the long-distance transport of photosynthate, the phloem is also a conduit for the transport of signaling molecules from one part of the organism to another. Such long-distance signals coordinate the activities of sources and sinks and regulate plant growth and development, for example flowering (Chapter 16). As indicated earlier, the signals between sources and sinks might be physical or chemical. Physical signals such as turgor change are transmitted rapidly via the interconnecting system of sieve elements. Sieve tubes can propagate electrical signals such as those generated from herbivory (Chapter

18). Depolarization of the negative membrane potential of sieve elements can trigger callose deposition and thus plugging of sieve pores. Molecules traditionally considered to be chemical signals, such as proteins and plant hormones, are found in the phloem sap, as are messenger RNAs and noncoding small RNAs. Other mobile mRNAs and proteins travel cell to cell via the plasmodesmata. The translocated carbohydrates themselves may also act as signals for shoot branching (Chapter 15).

Turgor pressure and chemical signals coordinate source and sink activities

Turgor pressure may play a role in coordinating the activities of sources and sinks. Increase of phloem unloading due to rapid sugar utilization in the sink leads to reduction of turgor pressures in the release phloem to be transmitted to the sources. If sieve element turgor were involved in controlling loading, this signal from the sinks would increase loading. The opposite response would be seen when unloading was slow in the sinks. Loading of sugars from storage along the axial pathway—such as from ray parenchyma in trees—also responds to changes in solute demand. Some data suggest that cell turgor can modify the activity of the proton-pumping ATPase at the plasma membrane and thereby alter sucrose loading rates.

Shoots produce growth regulators such as auxin that can be rapidly transported to the roots via the phloem in addition to the xylem parenchyma (Chapter 14). Roots produce cytokinins, which move to the shoots through the xylem. Gibberellins (GA) and abscisic acid (ABA) are also transported throughout the plant in the vascular system. Plant hormones play a role in regulating source–sink relationships. They affect photosynthate partitioning in part by controlling sink growth, leaf senescence, and other developmental processes. Plant defense responses against herbivores and pathogens can also change allocation and partitioning of photosynthates, with plant defense hormones such as jasmonates mediating the responses.

Hormones might regulate apoplastic loading and unloading by influencing the number of active transporters in plasma membranes. For example, transport of auxin and sucrose in the phloem increases sucrose transporter expression in rose petals to stimulate sucrose unloading, thereby inhibiting petal abscission (Chapter 15). Together, sucrose and ABA signaling enhance starch biosynthesis in maize endosperm (*Zea mays*, corn) (Chapter 14). Other potential sites of hormone regulation of unloading include tonoplast transporters, enzymes for metabolism of incoming sucrose, wall extensibility, and plasmodesmatal permeability in the case of symplasmic unloading, making the target tissue accessible for mobile RNAs and proteins.

As indicated earlier, carbohydrate levels can influence the expression of genes encoding photosynthesis components, as well as genes involved in sucrose hydrolysis. Many genes have been shown to be responsive to sugar depletion and abundance. Thus, not only is sucrose transported in the phloem, but sucrose or its metabolites can also act as signals that modify the activities of sources and sinks. For example, sucrose–H^+ symporter mRNA declines in sugar beet source leaves fed exogenous sucrose through the xylem. The decline in symporter mRNA is accompanied by a loss of symporter activity in plasma membrane vesicles isolated from the leaves. A working model includes the following steps:

1. Decreased sink demand leads to high sucrose levels in the vascular tissue.
2. High sucrose levels lead to down-regulation of the symporter gene in the source.
3. Decreased loading results in increased sucrose concentrations in the source.

This model is supported by the observation that the down-regulation of the sucrose loading by antisense inhibition of the sucrose–H^+ symporter SUT1 leads to increased sucrose concentrations and starch accumulation, and a lowered photosynthetic rate in source leaves, as compared with the wild type.

In summary, sugars and other metabolites have been shown to interact with hormonal signals to control and integrate many plant processes. Gene expression in some source–sink systems responds to both sugar and hormonal signals.

Mobile RNAs function as signal molecules to regulate growth and development

Endogenous RNA molecules and proteins have been found in phloem sap, and at least some of these can function as signal molecules or generate phloem-mobile signals. The endogenous RNA molecules travel as complexes with specific proteins (ribonucleo-proteins [RNPs]) in the phloem sap. Analyses of phloem exudates and of graft transmission suggest that all classes of RNA molecules—tRNA, rRNA, mRNA, short interfering RNA (siRNA), and microRNA (miRNA)—move with the phloem sap to distant tissues. It appears that most of them are produced in companion cells. After being unloaded in sink tissues, they are able to modify the function of specific cells. For example, it was shown by grafting that silencing spreads in the phloem systemically to young leaves, flowers, and primary roots and follows the same symplasmic unloading pathway as photosynthates do. miRNAs are involved in coordinating growth and development such as nutrient starvation response or potato tuberization. rRNA and tRNA fragments in the phloem sap are not part of a functional protein biosynthesis machinery in sieve elements, as previously assumed, but are able to inhibit translation.

Mobile proteins also function as signal molecules to regulate growth and development

A classic example of phloem signaling is the FLOWERING LOCUS T (FT) protein, which is a significant component of the floral stimulus that moves from source leaf to apex and that induces flowering at the apex in response to inductive conditions (Chapter 16). The FT protein has been shown to move from companion cells of source leaves, where it is expressed, into the sieve elements, assisted by an interacting protein that is necessary for passage of the pore–plasmodesma contacts. Long-distance transport of FT protein into the apical tissues is facilitated by yet another protein and takes some 8 h, as has been demonstrated using heat shock–induced FT expression. Unloading from the phloem and transport to the shoot apical meristem, where FT promotes flowering, takes another 4 h. FT mutant analysis indicates that both processes, uploading to and unloading from the phloem, are actively regulated.

During vegetative growth, phloem-mobile transcription factors and peptide hormones were recently shown to integrate plant growth and fine-tune nitrogen acquisition. A light-responsive transcription factor of 18 kDa moves from shoot to root via the phloem to coordinate shoot and root growth with carbon assimilation and nitrogen acquisition. Both nitrogen acquisition and nitrate transport are fine-tuned by three complementary, phloem-mobile peptide hormones, which act downstream of leaf-localized receptor kinases that respond to small xylem-mobile peptides moving from root to shoot.

Plasmodesmata function in phloem signaling

Plasmodesmata are involved in nearly every aspect of phloem translocation, from loading to long-distance transport (pores in sieve areas and sieve plates are modified plasmodesmata) to allocation and partitioning. What role might plasmodesmata play in macromolecular signaling in the phloem?

The mechanism of plasmodesmatal transport (called trafficking) can be either passive (nontargeted) or selective and regulated. When a molecule moves passively, its size must be smaller than the **size exclusion limit** of the plasmodesmata. Phloem plasmodesmata typically have a wider size exclusion limit than those in nonvascular tissues. As indicated earlier, GFP (27 kDa) moves passively through plasmodesmata in the phloem. When expressed in the companion cells, it passes the pore–plasmodesma contacts and is swept along with the phloem sap (Figure 10.21C). By contrast, FT protein (20 kDa) does not enter the phloem stream in the absence of the FT interacting protein. Generally, when a molecule moves in a selective fashion, it must possess a trafficking signal or interact in some other way with the plasmodesmata. The transport of some developmental transcription factors and of viral **movement proteins** appears to occur by means of a selective mechanism. Viral movement proteins interact directly with plasmodesmata to allow the passage of viral nucleic acids between cells; viruses also move in the phloem. Once at the plasmodesmata, movement proteins act to increase the size exclusion limit of the plasmodesmata to allow the viral genome to move between cells. Endogenous proteins are thought to carry out similar functions for endogenous macromolecules such as FT protein and some P-proteins. Interaction with components at or within the plasmodesmata, such as chaperones, is also required.

It is fitting to end this chapter with research topics that will continue to engage plant physiologists in the future: regulation of growth and development via the transport of endogenous RNA and protein signals, the nature of the proteins that facilitate the transport of signals through plasmodesmata, and the possibility of targeting signals to specific sinks in contrast to mass flow. Many other potential areas of inquiry have been indicated as well, such as the mechanism of phloem transport in gymnosperms, the nature and role of proteins in the lumen of the sieve elements, and the magnitude of pressure gradients in the sieve elements, especially in trees. As always in science, an answer to one question generates more questions!

From a more applied angle there are many possibilities for applying scientific knowledge about phloem translocation to the improvement of crop yield and resistance to abiotic and biotic stress factors (**Box 10.1**).

size exclusion limit (SEL) The restriction on the size of molecules that can be transported via the symplast. It is imposed by the width of the cytoplasmic sleeve that surrounds the desmotubule in the center of the plasmodesma.

movement proteins Nonstructural proteins encoded by a virus's genome that facilitate movement of that virus through the symplast.

Box 10.1 Relevance of Phloem Translocation and Signaling for Climate Change and Biotechnology

Phloem translocation and phloem signaling rapidly respond to abiotic and biotic changes of the environment. These include changes in CO_2 concentration, temperature, and light-induced ROS production, as well as attacks by caterpillars, fungi, and bacteria. The signaling network linking translocation in the phloem with the activity in neighboring tissues is thus of considerable basic agricultural importance. A thorough understanding of the mechanisms regulating photosynthate translocation and phloem signaling should provide the basis for technology aimed at enhancing crop productivity by increasing the accumulation of photosynthate in edible sink tissues, such as cereal grains.

Flower induction is an example of the complexity of phloem signaling. In cells abutting the conducting phloem of mature leaves, the specific flower-inducing protein FLOWERING LOCUS T (FT) is expressed once the day length is appropriate. FT transmission to the apical meristem is regulated by several subsequent checkpoints: It can enter the phloem stream only in the presence of an interacting protein, it can travel only as rapidly as the phloem sap, and its phloem exit has to be accepted by cell contacts in the unloading pathway of the shoot apical meristem. All of these checkpoints are interesting targets for biotechnological and breeding approaches to control flowering.

Also, the sugar levels in source and sink organs are involved in coordinating whole-plant physiology: Sucrose or

(Continued)

Box 10.1 *(continued)*

hexoses that would accumulate because of decreased sink demand are well known to repress photosynthetic genes. Among the genes regulated by the carbohydrate supply are those for invertase and sucrose synthase, both of which can catalyze the first step in sucrose utilization, and genes for sucrose–H^+ symporters, which play a key role in apoplastic loading. Regulation of photosynthesis by sink demand suggests that sustained increases in photosynthesis in response to elevated CO_2 in the atmosphere may depend on increasing sink strength. Research should aim at increasing the strength of existing sinks or developing new sinks. These are interesting targets for biotechnological and breeding approaches to control fruit set and seed filling.

Attaining higher yields of crop plants is one goal of research on photosynthate allocation and partitioning. Whereas grains and fruits are examples of edible yields, total yield includes inedible portions of the shoot. The harvest index, the ratio of economical yield (edible grain) to total aboveground biomass, has increased over the years largely because of the efforts of plant breeders. One goal of modern plant physiology is to further increase yield based on a fundamental understanding of metabolism, development, and, in the present context, partitioning.

However, allocation and partitioning in the whole plant must be coordinated such that increased transport to edible tissues does not occur at the expense of other essential processes and structures. Crop yield may also be improved if photosynthates that are normally "lost" by the plant are retained. For example, losses due to nonessential respiration or exudation from roots could be reduced. In the latter case, care must be taken not to disrupt essential processes outside the plant, such as growth of beneficial microbial species in the vicinity of the root that obtain nutrients from the root exudate.

Developmental regulators show feedback mechanisms between the activity of meristems and the arrival of hormones and carbohydrates through the phloem. At the same time, the conducting phloem elements keep pace with organ growth. If we want to improve crop plants in order to cope with increasing threat levels, such as by drought, flooding, heat, and pests, and to raise productivity, we need to understand and manipulate this regulatory interplay.

Summary

Phloem translocation moves the products of photosynthesis from mature leaves to areas of growth and storage. It also transmits chemical signals and redistributes ions and other substances throughout the plant body.

10.1 Patterns of Translocation: Source to Sink

- Phloem does not translocate materials exclusively in either an upward or a downward direction. Sap is translocated from sources to sinks, and the pathways involved are often complex (**Figures 10.1**, **10.2**).

10.2 Pathways of Translocation

- Sieve elements of the phloem conduct sugars and other organic materials throughout the plant (**Figures 10.3–10.5**).
- During development, sieve elements retain their plasma membrane but lose many organelles, retaining only modified mitochondria, plastids, and smooth endoplasmic reticulum (**Figures 10.5**, **10.6**).
- Sieve elements are interconnected through pores in their cell walls (**Figure 10.7**).
- Companion cells aid transport of photosynthetic products to the sieve elements. They also supply proteins to the sieve elements (**Figures 10.5**, **10.6**).
- In gymnosperms, smooth ER covers the sieve areas and is continuous through the sieve pores and median cavity (**Table 10.1**).

10.3 Phloem Loading

- The export of sugars from sources involves allocation of photosynthate to transport sugars, pre-phloem transport, and phloem loading in minor veins (**Figure 10.8**).
- Phloem loading can occur by way of the symplasm or apoplast (**Figure 10.9**).
- Sucrose is actively transported into the sieve element–companion cell complex in the apoplastic pathway (**Figures 10.10**, **10.11**).
- The oligomer-trapping model holds that oligomers are synthesized from sucrose in the intermediary cells; the larger oligosaccharides can diffuse only into the sieve elements (**Figures 10.12B**, **10.13**).
- Apoplastic and symplasmic phloem-loading pathways have defining characteristics (**Table 10.2**).

(Continued)

Summary (*continued*)

10.4 Long-Distance Transport: A Pressure-Driven Mechanism

- The transport phloem forms a passive, low-resistance pathway but is actively involved in unloading and reloading of sugars to axial sinks.
- Transport velocities in the phloem are high and exceed the rate of diffusion over long distances by many orders of magnitude.
- The pressure-flow model explains phloem translocation as a bulk flow of solution driven by an osmotically generated pressure gradient between source and sink (**Figure 10.14**).
- Phloem loading at the source and phloem unloading at the sink establish the pressure gradient for passive, long-distance bulk flow.
- Pressure gradients in the phloem sieve elements may be modest; pressures in herbaceous plants and trees appear to be similar.
- The pressure-flow model in its present form does not fully explain phloem transport in trees.

10.5 Materials Translocated in the Phloem

- The composition of sap has been determined; nonreducing sugars are the main transported molecules (**Table 10.3**, **Figure 10.15**).
- Sap includes proteins, many of which may have functions related to stress and defense reactions.
- P-proteins and callose seal off damaged phloem to limit loss of sap (**Figure 10.16**).

10.6 Phloem Unloading and Sink-to-Source Transition

- The import of sugars into sink cells involves phloem unloading, short-distance transport, and storage or metabolism.
- Phloem unloading and short-distance transport may operate by symplasmic or apoplastic pathways in different sinks (**Figure 10.17**).
- Transport into sink tissues is energy-dependent.
- Import cessation and export initiation are separate events, and there is a gradual transition from sink to source (**Figures 10.18–10.20**).
- The transition from sink to source requires several changes, including the expression and localization of the sucrose–H^+ symporter (**Figure 10.21**).

10.7 Photosynthate Distribution: Allocation and Partitioning

- Allocation in source leaves includes synthesis of storage compounds, metabolic utilization, and synthesis of transport compounds (**Figure 10.22**).
- The regulation of allocation must thus control the distribution of fixed carbon to the Calvin–Benson cycle, starch synthesis, sucrose synthesis, and respiration.
- A variety of chemical and physical signals are involved in partitioning resources among the various sinks.
- In competing for photosynthate, sink strength depends on sink size and sink activity.
- In response to altered conditions, short-term changes alter the distribution of photosynthate among different sinks, while long-term changes take place in source metabolism and alter the amount of photosynthate available for transport.

10.8 Transport of Signaling Molecules

- Turgor pressure, cytokinins, gibberellins, and abscisic acid have signaling roles in coordinating source and sink activities.
- Some hormone peptides, transcription factors, and other proteins can move from shoot to root companion cells in the phloem to coordinate carbon and nitrate acquisition with growth.
- Proteins and RNAs transported in the phloem can alter cellular functions.
- Interaction between mobile molecules and pore–plasmodesma contacts may control what passes through plasmodesmata.

Suggested Reading

Braun, D. M. (2022) Phloem loading and unloading of sucrose: What a long, strange trip from source to sink. *Annu. Rev. Plant Biol.* 73(1): 553–584.

Comtet, J., Turgeon, R., and Stroock, A. (2017) Phloem loading through plasmodesmata: A biophysical analysis. *Plant Physiol.* 175: 904–915.

Jensen, K. H., Berg-Sørensen, K., Bruus, H., Holbrook, N. M., Liesche, J., Schulz, A., Zwieniecki, M. A., and Bohr, T. (2016) Sap flow and sugar transport in plants. *Rev. Mod. Phys.* 88: 1–63.

Kehr, J., and Kragler, F. (2018) Long-distance RNA movement. *New Phytol.* 218: 29–40.

Knoblauch, M., Knoblauch, J., Mullendore, D. L., Savage, J. A., Babst, B. A., Beecher, S. D., Dodgen, A. C., Jensen, K. H., and Holbrook, N. M. (2016) Testing the Munch hypothesis of long-distance phloem transport in plants. *eLife* 5: e15341.

Liesche, J., and Schulz, A. (2018) Phloem transport in gymnosperms: A question of pressure and resistance. *Curr. Opin. Plant Biol.* 43: 36–42.

Otero, S., and Helariutta, Y. (2017) Companion cells: A diamond in the rough. *J. Exp. Bot.* 68: 71–78.

Patrick, J. W. (2013) Does Don Fisher's high-pressure manifold model account for phloem transport and resource partitioning? *Front. Plant Sci.* 4: 184.

Ross-Elliott, T. J., Jensen, K. H., Haaning, K. S., Wagner, B. M., Knoblauch, J., Howell, A. H., Mullendore, D. L., Monteith, A. G., Paultre, D., Yan, D. W., et al. (2017) Phloem unloading in Arabidopsis roots is convective and regulated by the phloem pole pericycle. *eLife* 6: e15341.

Schulz, A. (2015) Diffusion or bulk flow: How plasmodesmata facilitate pre-phloem transport of assimilates. *J. Plant Res.* 128: 49–61.

11 Respiration and Lipid Metabolism

Photosynthesis provides the organic building blocks that plants (and nearly all other organisms) depend on. Respiration, with its associated carbon metabolism, releases the energy stored in carbon compounds for cellular use. At the same time, it generates many carbon precursors for biosynthesis.

We begin this chapter by reviewing respiration in its metabolic context, emphasizing connections between the pathways involved and special features that are peculiar to plants. We also relate this to our understanding of the biochemistry and molecular biology of plant mitochondria and respiratory fluxes in intact plant tissues. Then we describe the pathways of lipid synthesis that lead to the accumulation of fats and oils, which plants use both for energy and carbon storage, and as a principal component of cellular membranes. Finally, we discuss the catabolic pathways involved in the breakdown of lipids and the conversion of their degradation products into sugars that occurs during the germination of fat-storing seeds.

11.1 Overview of Plant Respiration

Describe the processes involved in cellular respiration and the redox reactions that oxidize the carbon product of photosynthesis.

Aerobic (oxygen-requiring) respiration is common to most eukaryotic organisms, and the respiratory process in plants is similar to that found in other aerobic eukaryotes. However, some specific aspects of plant respiration distinguish it, especially from its animal counterpart. **Aerobic respiration** is a biological process by which reduced organic compounds are oxidized in a controlled manner. During respiration, energy is released and transiently carried by a compound, **adenosine triphosphate (ATP)**, which is used by the cell for maintenance and development.

aerobic respiration The complete oxidation of carbon compounds to CO_2 and H_2O, using oxygen as the final electron acceptor. Energy is released and conserved as ATP.

adenosine triphosphate (ATP) The major carrier of chemical energy in the cell, which by hydrolysis is converted to adenosine diphosphate (ADP) or adenosine monophosphate (AMP) with release of energy.

Respiration also plays a crucial role in generating organic acids, or their anions, that are used for biosynthesis, nitrogen assimilation, and secretion. The anabolic role of respiration is particularly important in actively growing tissues.

Glucose and fatty acids are usually cited as the substrates for respiration. In plant cells, however, the substrates (reduced carbon) are mainly derived from sucrose, other sugars, organic acids, triose phosphates from photosynthesis, and to some extent the products from lipid and protein degradation (**Figure 11.1**).

From a chemical standpoint, plant respiration can be expressed as the oxidation of the 12-carbon molecule sucrose and the reduction of 12 molecules of O_2:

$$C_{12}H_{22}O_{11} + 13H_2O \rightarrow 12CO_2 + 48H^+ + 48e^- \quad (11.1)$$

$$12O_2 + 48H^+ + 48e^- \rightarrow 24H_2O \quad (11.2)$$

giving the following net reaction:

$$C_{12}H_{22}O_{11} + 12O_2 \rightarrow 12CO_2 + 11H_2O \quad (11.3)$$

This reaction is the reversal of the photosynthetic process; it represents a coupled redox reaction in which sucrose is completely oxidized to CO_2 while oxygen serves as the ultimate electron acceptor and is reduced to water in the process. The controlled release of this free energy, along with its coupling to the synthesis of ATP, is the primary, although by no means only, role of respiratory metabolism.

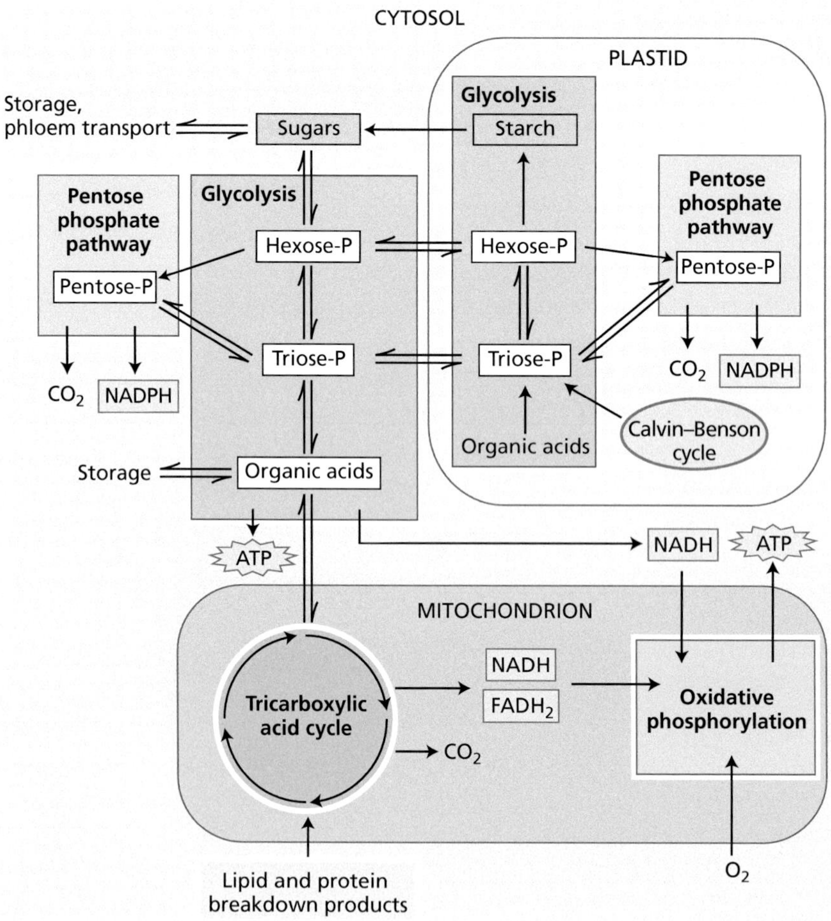

Figure 11.1 Overview of respiration. Substrates for respiration are generated by other cellular processes and enter the respiratory pathways. Glycolysis and the oxidative pentose phosphate pathways in the cytosol and plastids convert sugars into organic acids such as pyruvate, via hexose phosphates and triose phosphates, generating NADH or NADPH, and ATP. The organic acids are oxidized in the mitochondrial tricarboxylic acid (TCA) cycle, and the NADH and $FADH_2$ produced provide the energy for ATP synthesis by the electron transport chain and ATP synthase in oxidative phosphorylation. In gluconeogenesis, carbon from lipid breakdown in the glyoxysomes is metabolized in the TCA cycle, and then used to synthesize sugars in the cytosol by reverse glycolysis.

To extract energy, the cell oxidizes sucrose in a series of step-by-step reactions, where the steps with the largest energy release are coupled to energy conservation. The reactions can be grouped into four major processes: glycolysis, the oxidative pentose phosphate pathway, the tricarboxylic acid (TCA) cycle, and oxidative phosphorylation. These pathways do not function in isolation, but rather exchange metabolites at several levels. The substrates of respiration enter the respiratory process at different points in the pathways, as summarized in Figure 11.1:

- **Glycolysis** involves a series of reactions catalyzed by enzymes located in the cytosol and the plastids. A sugar—for example, sucrose—is partly oxidized via six-carbon sugar phosphates (hexose phosphates) and three-carbon sugar phosphates (triose phosphates) to produce an organic acid—mainly pyruvate. The process yields a small amount of energy as ATP and reducing power in the form of a reduced nicotinamide nucleotide, NADH.
- In the **oxidative pentose phosphate pathway**, also located in both the cytosol and the plastids, the six-carbon glucose 6-phosphate is initially oxidized to the five-carbon ribulose 5-phosphate. Carbon is lost as CO_2, and reducing power is conserved in the form of another reduced nicotinamide nucleotide, NADPH. In subsequent near-equilibrium reactions of the pentose phosphate pathway, ribulose 5-phosphate is converted into sugar phosphates containing three to seven carbon atoms. These intermediates can be used in biosynthetic pathways or re-enter glycolysis.
- The combined reactions of pyruvate dehydrogenase and the **tricarboxylic acid (TCA) cycle** completely oxidize pyruvate to 3 CO_2, via stepwise oxidations of organic acids in the innermost compartment of the mitochondrion—the matrix. This process mobilizes the major amount of reducing power (16 NADH + 4 $FADH_2$ per sucrose) and a small amount of energy (ATP) from the breakdown of sucrose.
- In **oxidative phosphorylation**, electrons are transferred along an electron transport chain consisting of a series of protein complexes embedded in the inner of the two mitochondrial membranes. This system transfers electrons from NADH (and related species)—produced by glycolysis, the oxidative pentose phosphate pathway, and the TCA cycle—to oxygen. This electron transfer releases a large amount of free energy, much of which is conserved through the synthesis of ATP from ADP and P_i (inorganic phosphate), catalyzed by the enzyme ATP synthase. Collectively, the redox reactions of the electron transport chain and the synthesis of ATP are called oxidative phosphorylation.

Nicotinamide adenine dinucleotide (NAD^+/NADH) is a soluble organic cofactor (coenzyme) associated with many enzymes that catalyze cellular redox reactions. NAD^+ is the oxidized form that undergoes a reversible two-electron reduction to yield NADH (**Figure 11.2**). The standard reduction potential for the NAD^+/NADH redox couple is about –320 mV. This tells us that NADH is a relatively strong reductant (i.e., electron donor), which can conserve the free energy carried by the electrons released during the stepwise oxidations of glycolysis and the TCA cycle. A related soluble compound, nicotinamide adenine dinucleotide phosphate ($NADP^+$/NADPH), has a similar function in photosynthesis (Chapters 7 and 8) and the oxidative pentose phosphate pathway, and also takes part in mitochondrial metabolism. Similarly, the predominantly tightly bound redox couples FAD/$FADH_2$ and FMN/$FMNH_2$ (Figure 11.2) play important roles in respiration and photosynthesis.

The oxidation of NADH by oxygen via the electron transport chain releases free energy (220 kJ mol^{-1}) that drives the synthesis of approximately 60 ATP

glycolysis A series of reactions in which a sugar is oxidized to produce two molecules of pyruvate. A small amount of ATP and NADH is produced.

oxidative pentose phosphate pathway A cytosolic and plastidic pathway that oxidizes glucose and produces NADPH and several sugar phosphates.

tricarboxylic acid (TCA) cycle A cycle of reactions catalyzed by enzymes localized in the mitochondrial matrix that leads to the oxidation of pyruvate to CO_2. ATP and NADH are generated in the process. Also called the citric acid cycle.

oxidative phosphorylation The transfer of electrons to oxygen in the mitochondrial electron transport chain that is coupled to ATP synthesis from ADP and phosphate by ATP synthase.

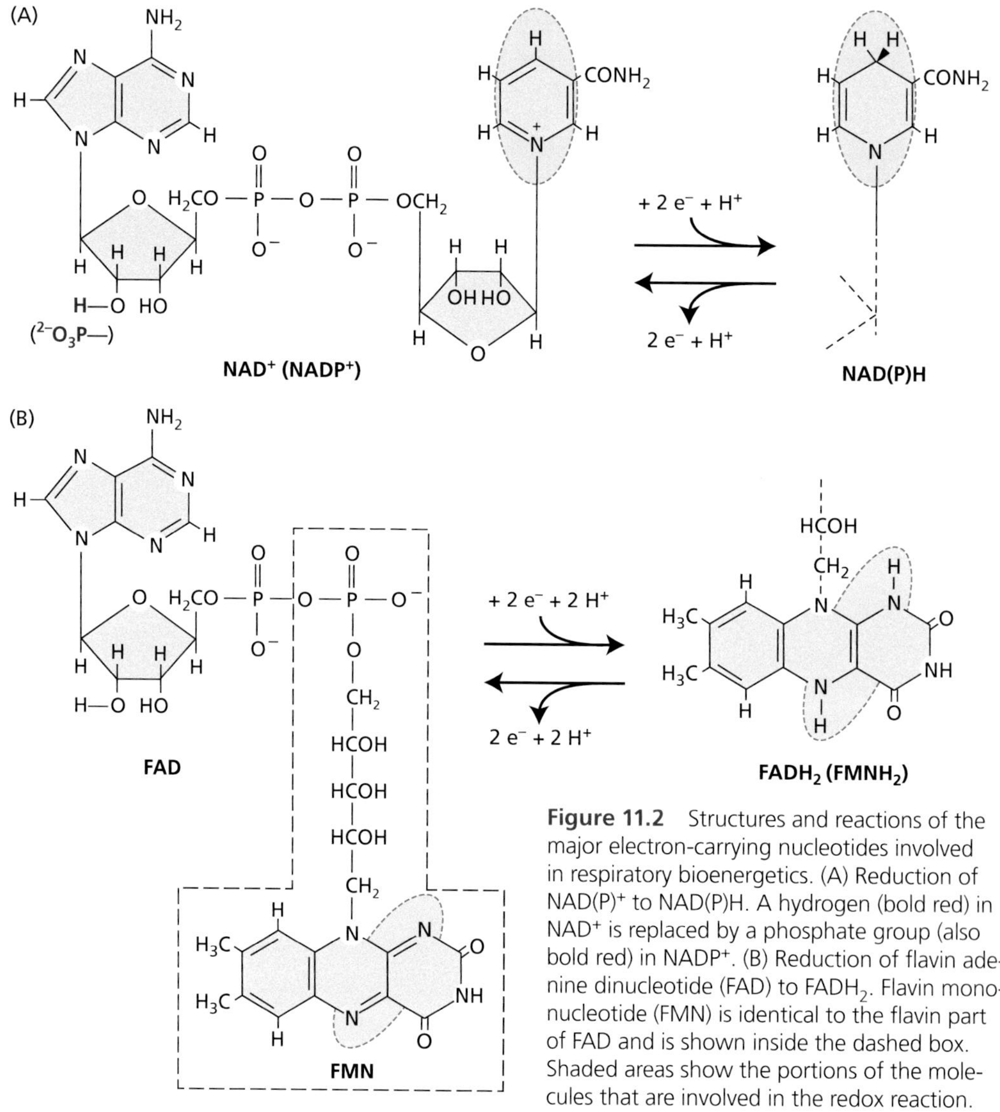

Figure 11.2 Structures and reactions of the major electron-carrying nucleotides involved in respiratory bioenergetics. (A) Reduction of $NAD(P)^+$ to NAD(P)H. A hydrogen (bold red) in NAD^+ is replaced by a phosphate group (also bold red) in $NADP^+$. (B) Reduction of flavin adenine dinucleotide (FAD) to $FADH_2$. Flavin mononucleotide (FMN) is identical to the flavin part of FAD and is shown inside the dashed box. Shaded areas show the portions of the molecules that are involved in the redox reaction.

per sucrose. We can now formulate a more complete picture of respiration as related to its role in cellular energy metabolism by coupling the following two reactions:

$$C_{12}H_{22}O_{11} + 12\,O_2 \rightarrow 12\,CO_2 + 11\,H_2O \tag{11.3}$$

$$60\,ADP + 60\,P_i \rightarrow 60\,ATP + 60\,H_2O \tag{11.4}$$

Recall that not all the carbon that enters the respiratory pathway is fully oxidized to CO_2. Many respiratory carbon intermediates are the starting points for pathways that synthesize amino acids, nucleotides, lipids, and numerous other compounds.

11.2 Glycolysis

Describe glycolysis, gluconeogenesis, the functions of PEP carboxylase and pyruvate kinase, and the fate of pyruvate in subsequent oxidation and fermentation pathways.

In the early steps of glycolysis (from the Greek words *glykos*, "sugar," and *lysis*, "splitting"), carbohydrates are converted into hexose phosphates, each of which is then split into two triose phosphates. In a subsequent energy-conserving phase, each triose phosphate is oxidized and rearranged to yield one molecule of pyruvate, an organic acid. Besides preparing the substrate for oxidation in the TCA cycle, glycolysis yields a small amount of chemical energy in the form of ATP and NADH.

When molecular oxygen is unavailable—for example, in plant roots in flooded soils—glycolysis can be the main source of energy for cells. For this to work, the *fermentative pathways*, which are carried out in the cytosol, must reduce pyruvate to recycle the NADH produced by glycolysis. In this section we describe the basic glycolytic and fermentative pathways, emphasizing features that are specific to plant cells.

Glycolysis metabolizes carbohydrates from several sources

Glycolysis occurs in the cytosol of all organisms (prokaryotes and eukaryotes). The principal reactions associated with the classic glycolytic pathway in plants are almost identical to those in animal cells. However, plant glycolysis has unique regulatory features, alternative enzymatic routes for several cytosolic steps, and a parallel glycolytic pathway in plastids (**Figure 11.3**).

In animals, the substrate of glycolysis is glucose, and the end product is pyruvate. Because sucrose is the major translocated sugar in most plants and is therefore the form of carbon that most nonphotosynthetic tissues import, sucrose (not glucose) can be argued to be the true sugar substrate for plant glycolysis. The end products of plant glycolysis are pyruvate and another organic acid anion, malate.

In the early steps of glycolysis, sucrose is split into its two 6-carbon sugars (hexoses), glucose and fructose, which can readily enter the glycolytic pathway.

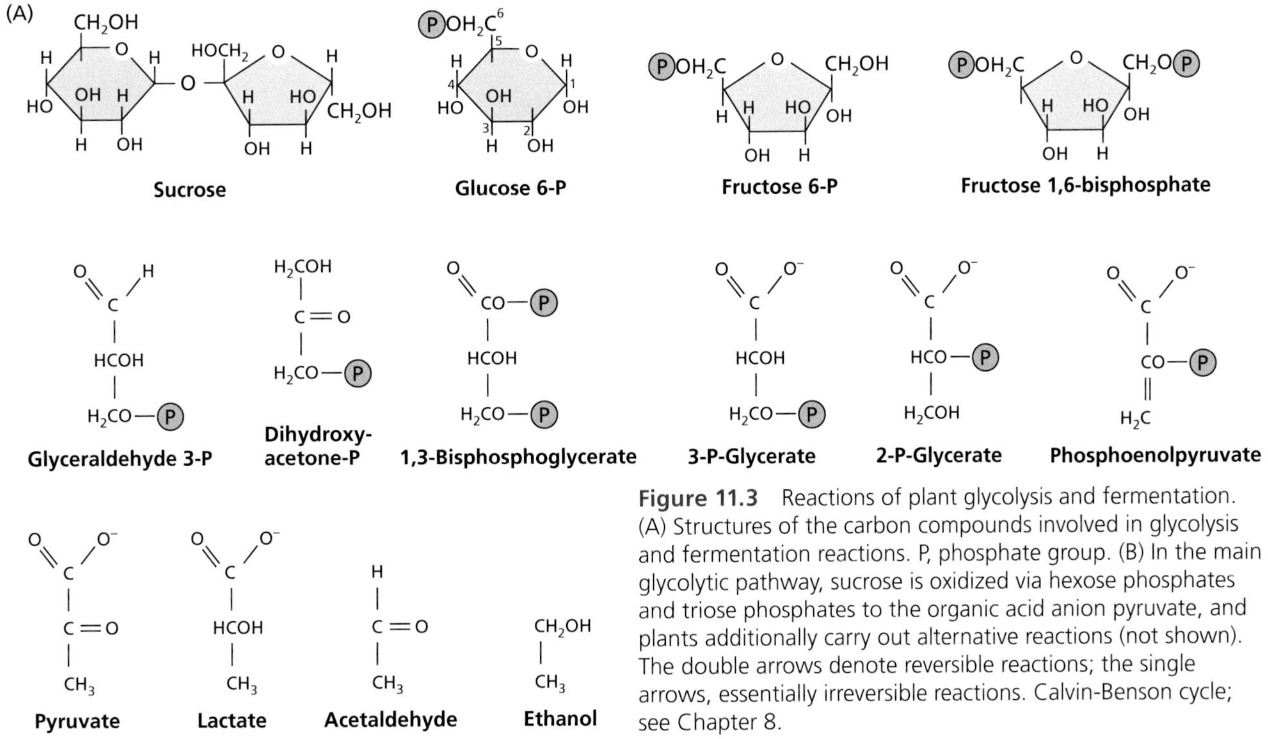

Figure 11.3 Reactions of plant glycolysis and fermentation. (A) Structures of the carbon compounds involved in glycolysis and fermentation reactions. P, phosphate group. (B) In the main glycolytic pathway, sucrose is oxidized via hexose phosphates and triose phosphates to the organic acid anion pyruvate, and plants additionally carry out alternative reactions (not shown). The double arrows denote reversible reactions; the single arrows, essentially irreversible reactions. Calvin-Benson cycle; see Chapter 8.

(B)

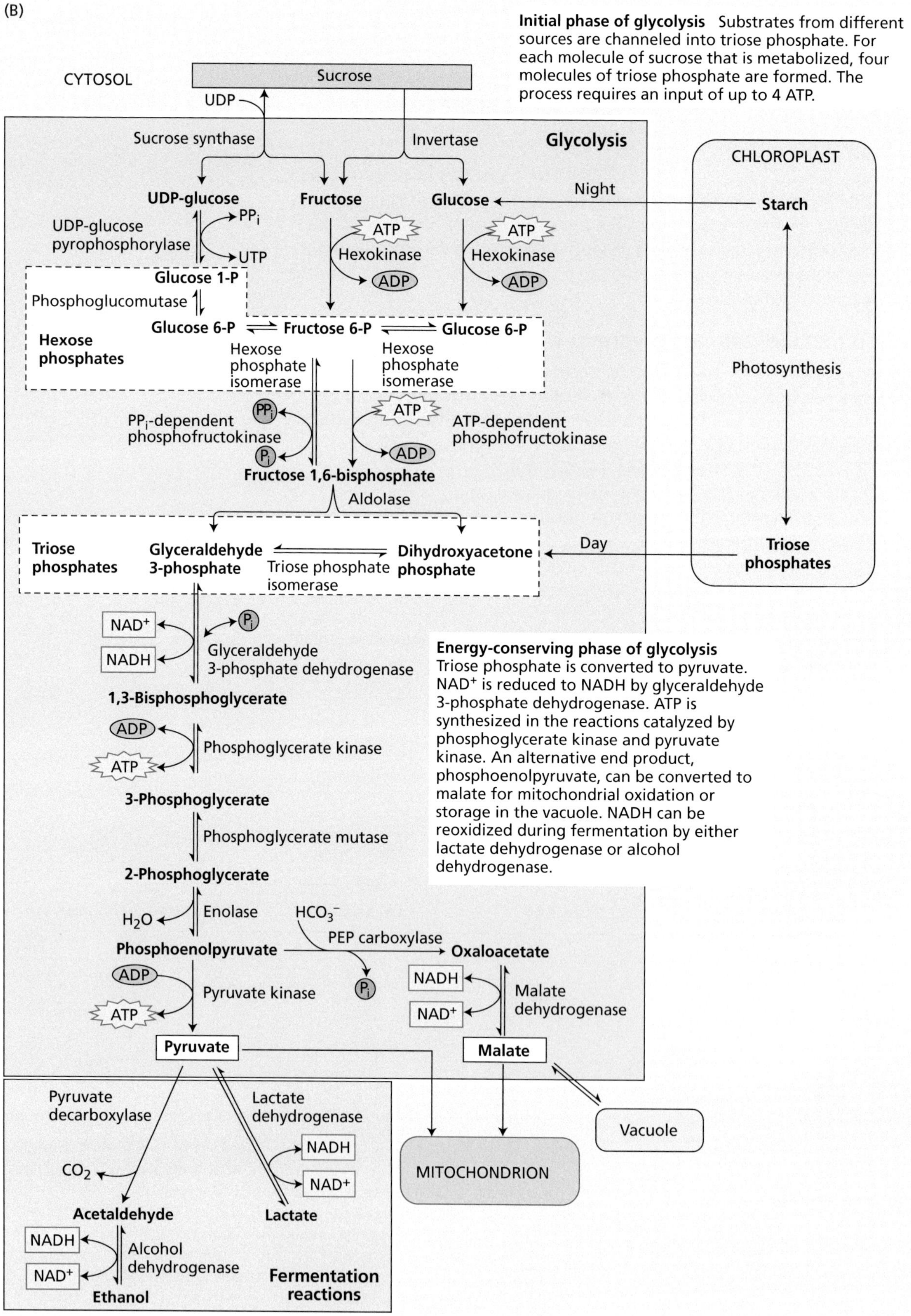
Initial phase of glycolysis Substrates from different sources are channeled into triose phosphate. For each molecule of sucrose that is metabolized, four molecules of triose phosphate are formed. The process requires an input of up to 4 ATP.
CYTOSOL
Sucrose
UDP
Sucrose synthase
Invertase
Glycolysis
CHLOROPLAST
UDP-glucose
Fructose
Glucose
Night
Starch
UDP-glucose pyrophosphorylase
PP_i
UTP
ATP
Hexokinase
ADP
ATP
Hexokinase
ADP
Glucose 1-P
Phosphoglucomutase
Glucose 6-P
Fructose 6-P
Glucose 6-P
Hexose phosphates
Hexose phosphate isomerase
Hexose phosphate isomerase
Photosynthesis
PP_i-dependent phosphofructokinase
PP_i
P_i
ATP
ADP
ATP-dependent phosphofructokinase
Fructose 1,6-bisphosphate
Aldolase
Triose phosphates
Glyceraldehyde 3-phosphate
Triose phosphate isomerase
Dihydroxyacetone phosphate
Day
Triose phosphates
NAD^+
NADH
P_i
Glyceraldehyde 3-phosphate dehydrogenase
1,3-Bisphosphoglycerate
ADP
ATP
Phosphoglycerate kinase
3-Phosphoglycerate
Phosphoglycerate mutase
2-Phosphoglycerate
Energy-conserving phase of glycolysis
Triose phosphate is converted to pyruvate. NAD^+ is reduced to NADH by glyceraldehyde 3-phosphate dehydrogenase. ATP is synthesized in the reactions catalyzed by phosphoglycerate kinase and pyruvate kinase. An alternative end product, phosphoenolpyruvate, can be converted to malate for mitochondrial oxidation or storage in the vacuole. NADH can be reoxidized during fermentation by either lactate dehydrogenase or alcohol dehydrogenase.
H_2O
Enolase
HCO_3^-
PEP carboxylase
Phosphoenolpyruvate
Oxaloacetate
ADP
ATP
Pyruvate kinase
P_i
NADH
NAD^+
Malate dehydrogenase
Pyruvate
Malate
Pyruvate decarboxylase
Lactate dehydrogenase
NADH
NAD^+
CO_2
MITOCHONDRION
Vacuole
Acetaldehyde
Lactate
NADH
NAD^+
Alcohol dehydrogenase
Ethanol
Fermentation reactions

Two pathways for the splitting of sucrose are known in plants, both of which take part in the use of sucrose from phloem unloading (Chapter 10): the invertase pathway and the sucrose synthase pathway.

Invertases in the cell wall, vacuole, and cytosol hydrolyze sucrose into glucose and fructose. The hexoses are then phosphorylated in the cytosol by a hexokinase that uses ATP to form **hexose phosphates**. Alternatively, *sucrose synthase* combines sucrose with UDP to produce fructose and UDP-glucose in the cytosol. UDP-glucose pyrophosphorylase can then convert UDP-glucose and pyrophosphate (PP_i) into UTP and glucose 6-phosphate (Figure 11.3). While the sucrose synthase reaction is close to equilibrium, the invertase reaction is essentially irreversible, driving the flux in the forward direction.

Through studies of transgenic plants lacking specific invertases or sucrose synthase, each enzyme has been found to be essential for specific life processes, but differences are observed among plant tissues and species. For example, sucrose synthase and cell wall invertase are especially involved in fruit development in several crop species (Chapter 17), whereas the cytosolic invertase is important for growth, cell wall integrity, and respiration. Both sucrose synthase and invertases can degrade sucrose for glycolysis, and if one of the enzymes is absent, the other enzyme(s) can still maintain respiration. The existence of several enzymes or pathways that serve a similar function and can replace each other is called **metabolic redundancy**. In plastids, there is in many cases a parallel glycolytic pathway that produces metabolites, ATP, and reducing power for plastidial reactions that synthesize, for example, fatty acids, tetrapyrroles, and aromatic amino acids. Starch is synthesized and catabolized only in plastids, and carbon obtained from starch degradation (at night) enters the glycolytic pathway in the cytosol primarily as glucose (Chapter 8). In the light, photosynthetic products can enter the glycolytic pathway directly as triose phosphate. In overview, glycolysis works like a funnel, with an initial phase collecting carbon from different carbohydrate sources, depending on the physiological situation.

In the initial phase of glycolysis, each hexose unit is phosphorylated twice and then split, producing two molecules of **triose phosphate**. This series of reactions consumes zero to four molecules of ATP per sucrose unit, depending on whether the sucrose is split by sucrose synthase or invertase and whether the ATP- or PPi-dependent phosphofructokinases are active. The initial phase also includes several essentially irreversible reactions of the glycolytic pathway, which are catalyzed by hexokinase and phosphofructokinase (Figure 11.3). As we will see later in this section, the phosphofructokinase reaction is one of the control points of glycolysis in both plants and animals.

hexose phosphates Six-carbon sugars with phosphate groups attached.

metabolic redundancy A common feature of plant metabolism in which different pathways serve a similar function. They can therefore replace each other without apparent loss in function.

triose phosphate A three-carbon sugar phosphate.

The energy-conserving phase of glycolysis produces pyruvate, ATP, and NADH

The reactions discussed thus far convert carbon from the various substrate pools to triose phosphates, *glyceraldehyde 3-phosphate*, and its isomer dihydroxyacetone phosphate, which are efficiently interconverted by the triose phosphate isomerase. From this step onward, the glycolytic pathway can extract usable energy in the energy-conserving phase.

Glyceraldehyde 3-phosphate dehydrogenase catalyzes the oxidation of the aldehyde to a carboxylic acid, reducing NAD^+ to NADH. This reaction releases sufficient free energy to allow the phosphorylation (using inorganic phosphate) of glyceraldehyde 3-phosphate to produce 1,3-bisphosphoglycerate. The phosphorylated carboxylate on carbon 1 of 1,3-bisphosphoglycerate is a strong donor of P_i groups (Figure 11.3A).

Phosphoglycerate kinase catalyzes the next step where the phosphate on carbon 1 is transferred to a molecule of ADP, yielding ATP and 3-phosphoglycerate. For

substrate-level phosphorylation A process that involves the direct transfer of a phosphate group from a substrate molecule to ADP to form ATP.

PEP carboxylase A cytosolic enzyme that forms oxaloacetate by the carboxylation of phospho*enol*pyruvate.

gluconeogenesis The synthesis of carbohydrates through the reversal of glycolysis.

each sucrose entering glycolysis, four ATPs are generated by this reaction—one for each molecule of 1,3-bisphosphoglycerate.

This type of ATP synthesis, traditionally referred to as **substrate-level phosphorylation**, involves the direct transfer of a phosphate group from a substrate molecule to ADP to form ATP. ATP synthesis by substrate-level phosphorylation is mechanistically distinct from ATP synthesis by the oxidative phosphorylation in mitochondria or by photophosphorylation in chloroplasts.

In the subsequent two reactions, the phosphate on 3-phosphoglycerate is transferred to carbon 2, and then a molecule of water is removed, yielding the compound *phosphoenolpyruvate* (*PEP*). The phosphate group on PEP makes it an extremely strong substrate for ATP formation. *Pyruvate kinase* catalyzes a second substrate-level phosphorylation by converting ADP and PEP into ATP and pyruvate. This final step, which is the third essentially irreversible step in glycolysis, yields four additional molecules of ATP for each sucrose molecule that enters the pathway.

At the end of the glycolytic process, plants have alternative pathways for metabolizing PEP. In one pathway PEP is carboxylated by the ubiquitous cytosolic enzyme **PEP carboxylase** to form the organic acid anion oxaloacetate. (The same reaction is responsible for the initial CO_2 fixation in CAM plants; Chapter 8.) The oxaloacetate is then reduced to malate by the action of *malate dehydrogenase*, which uses NADH as a source of electrons (Figure 11.3). The resulting malate can be exported to the vacuole for storage or to the mitochondrion for degradation in the TCA cycle. Thus, the action of pyruvate kinase and PEP carboxylase can produce pyruvate or malate for mitochondrial respiration, although pyruvate dominates in most tissues.

Plants have alternative glycolytic reactions

In glycolysis, sugars are degraded to the organic acid pyruvate, and the glycolytic reaction catalyzed by *ATP-dependent phosphofructokinase* is essentially irreversible (Figure 11.3). However, many organisms can also operate a similar pathway in the opposite direction. This process, to synthesize sugars from organic acids, is known as **gluconeogenesis**.

Gluconeogenesis is particularly important in plants (such as the castor oil plant *Ricinus communis* and sunflower) that store carbon in the form of oils (triacylglycerols) in the seeds. Plants cannot transport lipids, so when such a seed germinates, the oil is converted by gluconeogenesis into sucrose, which is transported to the growing cells in the seedling to fuel their energy requirements. In the initial phase of glycolysis, gluconeogenesis overlaps with the pathway for synthesis of sucrose from photosynthetic triose phosphate, which is typical of leaf cells.

The enzyme *fructose 1,6-bisphosphatase* converts fructose 1,6-bisphosphate irreversibly into fructose 6-phosphate and P_i during gluconeogenesis. Therefore, ATP-dependent phosphofructokinase and fructose 1,6-bisphosphate phosphatase represent major control points of carbon flux through the glycolytic and gluconeogenic pathways of both plants and animals as well as in photosynthetic sucrose synthesis in plants.

In plants, the interconversion of fructose 6-phosphate and fructose 1,6-bisphosphate is made more complex by the presence of an additional (cytosolic) enzyme, *PP_i-dependent phosphofructokinase* (pyrophosphate:fructose 6-phosphate 1-phosphotransferase), which catalyzes the following reversible reaction (Figure 11.3):

$$\text{Fructose 6-P} + PP_i \rightleftarrows \text{fructose 1,6-bisphosphate} + P_i \qquad (11.5)$$

where -P represents bound phosphate. PP_i-dependent phosphofructokinase is found in the cytosol of most plant tissues at activities that can be higher than those of ATP-dependent phosphofructokinase. The reaction catalyzed by PP_i-dependent phosphofructokinase is readily reversible, but it is unlikely to operate in sucrose synthesis. Suppression of PP_i-dependent phosphofructokinase in transgenic plants has shown that it contributes to glycolytic flux but that it is not essential for plant survival. The three enzymes that interconvert fructose 6-phosphate and fructose 1,6-bisphosphate are all regulated to match the plant demands for both respiration and synthesis of sucrose and polysaccharides. As a consequence, the operation of the glycolytic pathway in plants has several unique characteristics.

fermentation The metabolism of pyruvate in the absence of oxygen, leading to the oxidation of the NADH generated in glycolysis to NAD^+. Allows glycolytic ATP production to function in the absence of oxygen.

Fermentation regenerates the NAD^+ needed for glycolytic ATP production in the absence of oxygen

Oxidative phosphorylation does not function in the absence of oxygen. Glycolysis then cannot continue because the cell's limited supply of NAD^+ becomes tied up in the reduced state (NADH), and the glyceraldehyde 3-phosphate dehydrogenase comes to a halt. To overcome this limitation, plants and other organisms can further metabolize pyruvate by carrying out one or more forms of **fermentation** (Figure 11.3).

Alcoholic fermentation (widely known from brewer's yeast) is common in plants. Pyruvate decarboxylase oxidizes pyruvate to CO_2 and acetaldehyde, and alcohol dehydrogenase reduces the latter to ethanol by oxidizing NADH. In lactic acid fermentation (common in mammalian muscle but also in plants), the enzyme lactate dehydrogenase uses NADH to reduce pyruvate to lactate, thus regenerating NAD^+.

Plant tissues may be subjected to low (hypoxic) or zero (anoxic) concentrations of ambient oxygen. Compact and highly active organs such as tubers, seeds, and apical meristems can have highly hypoxic interiors, but the best-studied example involves waterlogged soils in which the diffusion of oxygen is reduced and roots become hypoxic, forcing the tissues to carry out fermentative metabolism. In maize (corn; *Zea mays*), the initial metabolic response to low oxygen concentrations is lactic acid fermentation, but the subsequent response is alcoholic fermentation. Ethanol is thought to be a less toxic end product of fermentation because it can diffuse out of the cell, whereas lactate accumulation promotes deleterious acidification of the cytosol.

It is important to consider the efficiency of fermentation. *Efficiency* is defined here as the energy conserved as ATP relative to the energy potentially available in a molecule of sucrose. Normal glycolysis leads to a net synthesis of four ATP molecules for each sucrose molecule converted into pyruvate. The efficiency of fermentation is only about 4% because most of the energy available in sucrose remains in the ethanol or lactate. In contrast, the pyruvate produced by glycolysis during aerobic respiration is completely oxidized into CO_2 by mitochondria, resulting in a much more efficient use of the free energy available in sucrose.

Changes in the glycolytic pathway under oxygen deficiency can increase the ATP yield of fermentation. An example of this is when sucrose is degraded via sucrose synthase instead of invertase, avoiding ATP consumption by the hexokinase, or by engaging PP_i-dependent phosphofructokinase. Such modifications emphasize the importance of energetic efficiency for plant survival in the absence of oxygen.

Because of the low energy recovery of fermentation, an increased rate of glycolytic flux is needed to sustain the ATP production necessary for cell survival. The increase in glycolytic rate in the absence of O_2 is called the *Pasteur effect* after the French microbiologist Louis Pasteur, who first noted it in yeast. Glycolysis is up-regulated by changes in metabolite levels and by the induced synthesis of the glycolytic and fermentation enzymes in response to lower O_2 and energy status.

ribulose 5-phosphate In the pentose phosphate pathway, the initial five-carbon product of the oxidation of glucose 6-phosphate; in subsequent reactions, it is converted into sugars containing three to seven carbon atoms.

11.3 The Oxidative Pentose Phosphate Pathway

Summarize the steps of the oxidative pentose phosphate pathway and how it is regulated.

In plants, oxidation of sugars can occur via the glycolytic pathway in the cytosol (Figure 11.3) or the oxidative pentose phosphate pathway. The oxidative pentose phosphate pathway reactions are carried out by soluble enzymes present in both the cytosol and plastids (Figure 11.1 and **Figure 11.4**). Under most conditions, the pathway in plastids predominates over that in the cytosol.

The first two reactions convert the six-carbon molecule glucose 6-phosphate into the five-carbon unit **ribulose 5-phosphate**, with loss of a CO_2 molecule and generation of two molecules of NADPH (not NADH). The remaining reactions convert ribulose 5-phosphate into the glycolytic intermediates glyceraldehyde 3-phosphate and fructose 6-phosphate. These products can be further metabolized by glycolysis to yield pyruvate. Alternatively, glucose 6-phosphate can be regenerated from glyceraldehyde 3-phosphate and fructose 6-phosphate by glycolytic enzymes as in the reaction:

$$\begin{aligned} &6\,\text{Glucose 6-P} + 12\,\text{NADP}^+ + 7\,\text{H}_2\text{O} \rightarrow \\ &\quad 5\,\text{glucose 6-P} + 6\,\text{CO}_2 + \text{P}_\text{i} + 12\,\text{NADPH} + 12\,\text{H}^+ \end{aligned} \tag{11.6}$$

The net result is the complete oxidation of one glucose 6-phosphate molecule to CO_2 with the concomitant synthesis of 12 NADPH molecules, while five glucose 6-phosphate molecules are regenerated.

Studies of the release of CO_2 from isotopically labeled glucose indicate that the pentose phosphate pathway accounts for 10 to 25% of the glucose breakdown, with the rest occurring mainly via glycolysis. However, the contribution of the pentose phosphate pathway changes during development and with changes in growth conditions as the plant's requirements for specific products vary.

The oxidative pentose phosphate pathway produces NADPH and biosynthetic intermediates

The oxidative pentose phosphate pathway plays several roles in plant metabolism:

- *NADPH supply in the cytosol.* NADPH drives reductive steps associated with biosynthetic and stress defense reactions and is a substrate for reactions that remove reactive oxygen species (ROS). The pentose phosphate pathway may therefore also contribute to cellular energy metabolism; that is, electrons from NADPH may end up reducing O_2 and generating ATP through oxidative phosphorylation.
- *NADPH supply in plastids.* In nongreen plastids, such as amyloplasts in the root, and chloroplasts functioning in the dark, the pentose phosphate pathway is the major supplier of NADPH. The NADPH is used for biosynthetic reactions such as lipid synthesis and nitrogen assimilation. The formation of NADPH by glucose 6-phosphate oxidation in amyloplasts may also signal sugar status to the thioredoxin system for control of starch synthesis.
- *Supply of substrates for biosynthetic processes.* Another intermediate in the pentose phosphate pathway, the four-carbon erythrose 4-phosphate, combines with PEP in the initial reaction of the shikimic acid pathway that produces plant phenolic compounds, including the aromatic amino acids and the precursors of lignin and flavonoids. This role of the pentose phosphate pathway is supported by the observation that its enzymes are induced by stress conditions such as wounding, under which aromatic compounds are needed for reinforcing and protecting the tissue.

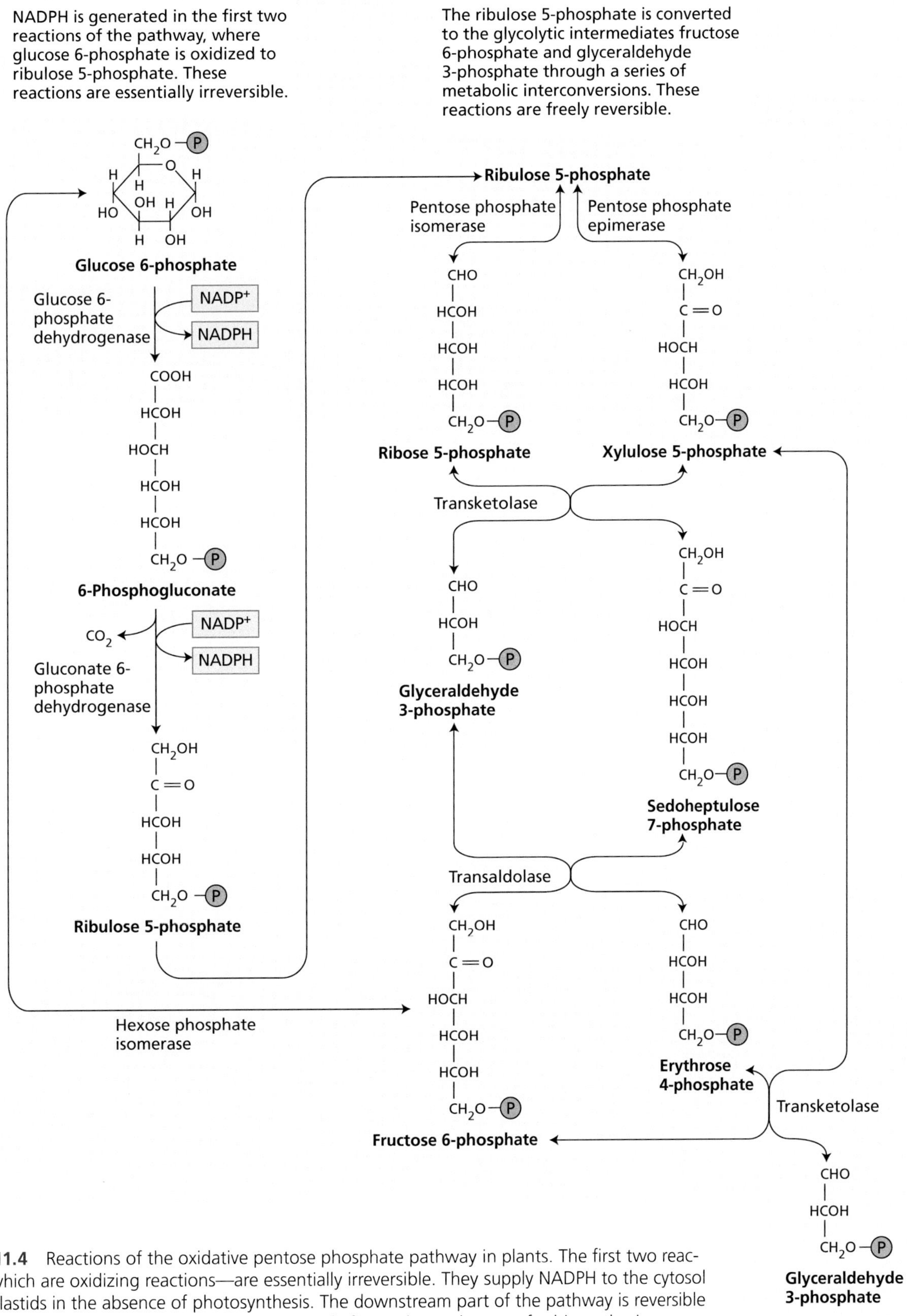

NADPH is generated in the first two reactions of the pathway, where glucose 6-phosphate is oxidized to ribulose 5-phosphate. These reactions are essentially irreversible.
The ribulose 5-phosphate is converted to the glycolytic intermediates fructose 6-phosphate and glyceraldehyde 3-phosphate through a series of metabolic interconversions. These reactions are freely reversible.
Glucose 6-phosphate
Glucose 6-phosphate dehydrogenase
NADP+
NADPH
6-Phosphogluconate
CO2
Gluconate 6-phosphate dehydrogenase
Ribulose 5-phosphate
Hexose phosphate isomerase
Ribulose 5-phosphate
Pentose phosphate isomerase
Pentose phosphate epimerase
Ribose 5-phosphate
Xylulose 5-phosphate
Transketolase
Glyceraldehyde 3-phosphate
Sedoheptulose 7-phosphate
Transaldolase
Erythrose 4-phosphate
Transketolase
Fructose 6-phosphate
Glyceraldehyde 3-phosphate

Figure 11.4 Reactions of the oxidative pentose phosphate pathway in plants. The first two reactions—which are oxidizing reactions—are essentially irreversible. They supply NADPH to the cytosol and to plastids in the absence of photosynthesis. The downstream part of the pathway is reversible (as denoted by bidirectional arrows), so it can supply five-carbon substrates for biosynthesis even when the oxidizing reactions are inhibited—for example, in chloroplasts in the light.

glucose 6-phosphate dehydrogenase A cytosolic and plastidic enzyme that catalyzes the initial reaction of the oxidative pentose phosphate pathway.

mitochondrion (plural *mitochondria*) The organelle that is the site for most reactions in the respiratory process in eukaryotes.

outer mitochondrial membrane The outer of the two mitochondrial membranes, which appears to be freely permeable to all small molecules.

inner mitochondrial membrane The inner of the two mitochondrial membranes, containing the electron transport chain, the F_oF_1-ATP synthase, and numerous transporters.

The oxidative pentose phosphate pathway is controlled by cellular redox status

Each enzymatic step in the oxidative pentose phosphate pathway is catalyzed by a group of isoenzymes that vary in their abundance and regulatory properties among plant cell types. The initial reaction of the pathway, catalyzed by **glucose 6-phosphate dehydrogenase**, is inhibited by a high ratio of NADPH to $NADP^+$.

In the light, little operation of the oxidative pentose phosphate pathway occurs in chloroplasts. Plastidic glucose 6-phosphate dehydrogenase is inhibited by a reductive inactivation involving the *ferredoxin–thioredoxin system* (Chapter 8) and by the NADPH:$NADP^+$ ratio. Moreover, the end products of the pathway, fructose 6-phosphate and glyceraldehyde 3-phosphate, are being synthesized by the Calvin–Benson cycle. Thus, mass action will drive the nonoxidative reactions of the pathway in the reverse direction. In this way, synthesis of erythrose 4-phosphate can be maintained in the light. In nongreen plastids, the glucose 6-phosphate dehydrogenase is less sensitive to inactivation by reduced thioredoxin and NADPH and can therefore reduce $NADP^+$ to maintain a relatively high reduction level of plastid components in the absence of photosynthesis.

11.4 The Tricarboxylic Acid Cycle

Summarize the steps of the tricarboxylic acid cycle, how it is regulated, and the role of malate in this process.

During the nineteenth century, biologists discovered that in the absence of air, glucose-fed yeast cells produce ethanol or lactic acid, whereas in the presence of air, cells consume O_2 and produce CO_2 and H_2O. In 1937 the German-born British biochemist Hans A. Krebs reported the discovery of the citric acid cycle—most commonly known as the *tricarboxylic acid (TCA) cycle*. The elucidation of the TCA cycle explained how pyruvate is broken down into CO_2 and H_2O, and also highlighted the key concept of cycles in metabolic pathways. For his discoveries, Hans Krebs was awarded the Nobel Prize in physiology or medicine in 1953.

Mitochondria are semiautonomous organelles

The breakdown of sucrose into pyruvate releases less than 25% of the total energy in sucrose; the remaining energy is stored in the four molecules of pyruvate. The next two stages of respiration (the TCA cycle and oxidative phosphorylation) take place within an organelle enclosed by a double membrane, the **mitochondrion** (plural *mitochondria*).

Plant mitochondria are usually spherical or rodlike and range from 0.5 to 1.0 µm in diameter and up to 3 µm in length (**Figure 11.5A, B**), but they can also be branched or reticulated. The number and sizes of mitochondria in a cell can vary dynamically as a result of mitochondrial fission, fusion, and degradation (Figure 11.5C) while keeping up with cell division. Like chloroplasts, they are semiautonomous organelles, because they contain ribosomes, RNA, and DNA, which encode a limited number of mitochondrial proteins. Plant mitochondria are thus able to carry out protein synthesis and transmit their genetic information. Metabolically active cells usually contain more mitochondria than less active cells, reflecting the mitochondrial role in energy metabolism. Tapetum cells in the anthers (Chapter 1 and 16), for example, show a marked increase in the number of mitochondria during pollen development.

The ultrastructural features of plant mitochondria are similar to those of mitochondria in other eukaryotes (Figure 11.5A and B). Plant mitochondria have two membranes: A smooth **outer mitochondrial membrane** completely surrounds a highly invaginated **inner mitochondrial membrane**. The invaginations of the

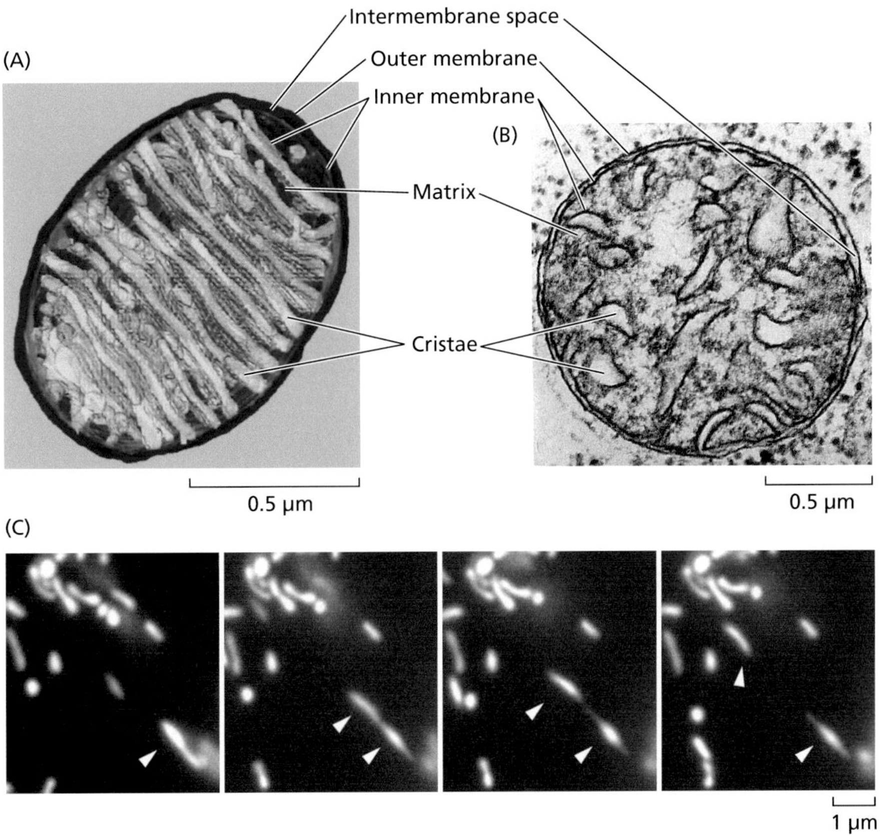

Figure 11.5 Structure of mitochondria from animals and plants. (A) Three-dimensional tomography picture of a chicken brain mitochondrion, showing the invaginations of the inner membrane, called cristae, as well as the locations of the matrix and intermembrane space. (B) Electron micrograph of a mitochondrion in a mesophyll cell of broad bean (*Vicia faba*). (C) Time-lapse pictures showing a dividing mitochondrion in an Arabidopsis epidermal cell (arrowheads). All the visible organelles are mitochondria labeled with green fluorescent protein. The pictures shown were taken 2 s apart.

inner membrane are known as **cristae** (singular *crista*). As a consequence of its greatly enlarged surface area, the inner membrane can contain 50% of the total mitochondrial protein. The region between the two mitochondrial membranes is known as the **intermembrane space**. The compartment enclosed by the inner membrane is referred to as the mitochondrial **matrix**. It has a very high content of macromolecules, approximately 50% by weight. Because there is little water in the matrix, mobility is restricted, and it is likely that matrix proteins are organized into multienzyme complexes (so-called metabolons) to facilitate substrate channeling.

Intact mitochondria are osmotically active; that is, they take up water and swell when placed in a hypoosmotic medium. Ions and polar molecules are generally unable to diffuse freely through the inner membrane, which functions as the osmotic barrier. The outer membrane is permeable to solutes that have a molecular mass of less than approximately 6,000 Da—that is, most cellular metabolites and ions, but not proteins. The lipid fraction of both membranes is primarily made up of phospholipids, 80% of which are either phosphatidylcholine or phosphatidylethanolamine. About 15% of the inner mitochondrial membrane lipids is diphosphatidylglycerol (also called cardiolipin), which occurs only in that membrane.

cristae Folds in the inner mitochondrial membrane that project into the mitochondrial matrix.

intermembrane space The fluid-filled space between the two mitochondrial membranes or between the two chloroplast envelope membranes.

matrix The aqueous, gel-like phase of a mitochondrion that occupies the internal space into which the cristae extend. Contains the DNA, ribosomes, and soluble enzymes required for the tricarboxylic acid cycle, oxidative phosphorylation, and other metabolic reactions.

Pyruvate enters the mitochondrion and is oxidized via the TCA cycle

The TCA cycle got its name because of the importance of the tricarboxylic acids citric acid (citrate) and isocitric acid (isocitrate) as intermediates (**Figure 11.6**). This cycle constitutes the second stage in respiration and takes place in the mitochondrial matrix. Its operation requires that the pyruvate generated in

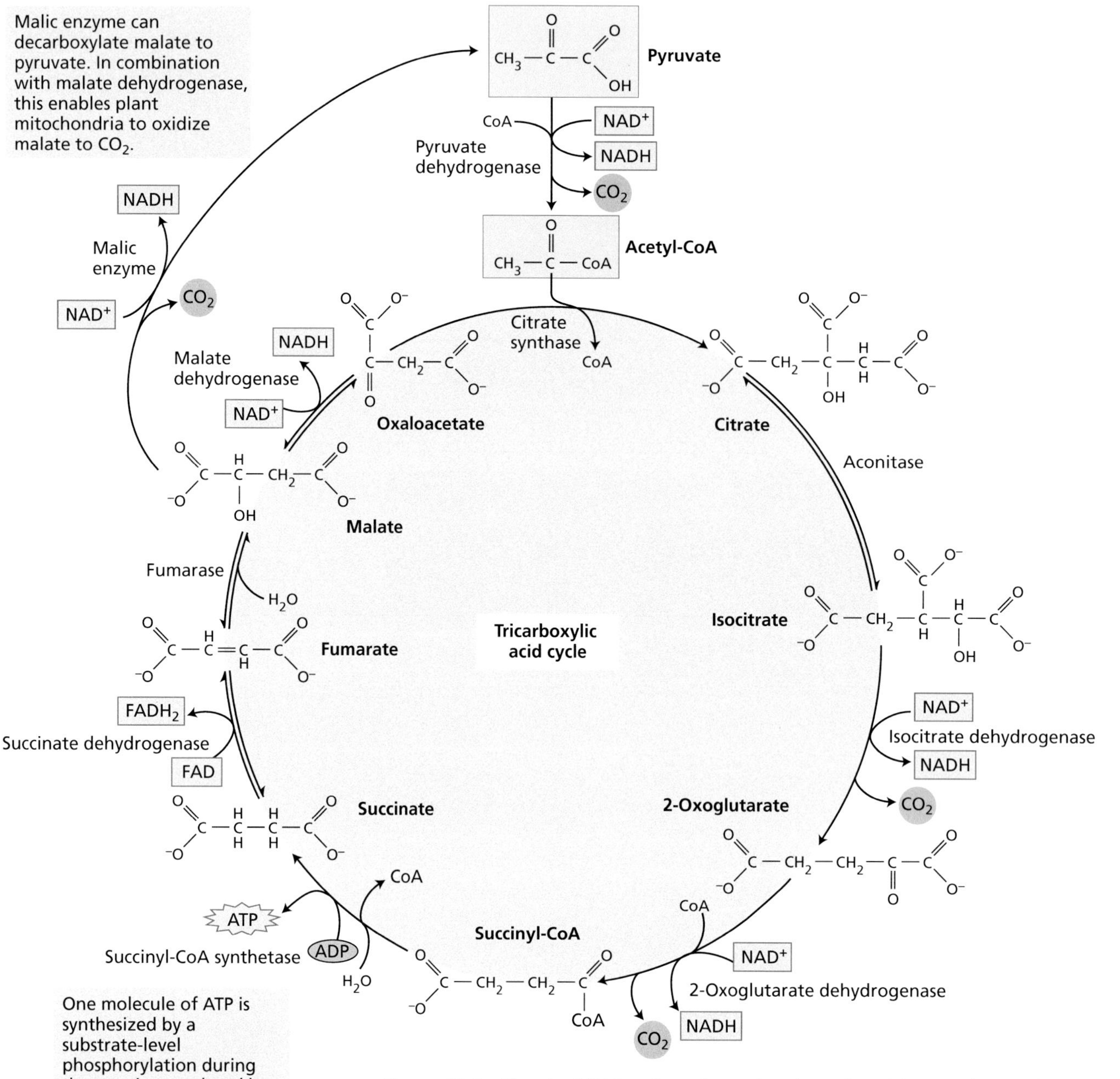

Figure 11.6 The plant TCA cycle and associated reactions. The reactions and enzymes of the TCA cycle are displayed, along with the associated reactions of pyruvate dehydrogenase and malic enzyme. Pyruvate is completely oxidized to three molecules of CO_2, and malic enzyme enables plant mitochondria to completely oxidize malate. The electrons released during these oxidations are used to reduce four molecules of NAD^+ to NADH and one molecule of FAD to $FADH_2$. Malic enzyme is not found in mammalian mitochondria.

the cytosol during glycolysis be transported through the inner mitochondrial membrane barrier via a specific transport protein.

Once inside the mitochondrial matrix, pyruvate is decarboxylated in an oxidation reaction catalyzed by **pyruvate dehydrogenase**, a large complex consisting of several enzymes. The products are NADH, CO_2, and acetyl-CoA, in which the acetyl group derived from pyruvate is linked by a thioester bond to a soluble cofactor, coenzyme A (CoA) (Figure 11.6).

pyruvate dehydrogenase An enzyme in the mitochondrial matrix that decarboxylates pyruvate, producing NADH (from NAD^+), CO_2, and acetic acid in the form of acetyl-CoA (acetic acid bound to coenzyme A).

In the next reaction, citrate synthase, formally the first enzyme in the TCA cycle, combines the acetyl group of acetyl-CoA with a four-carbon dicarboxylic acid (*oxaloacetate*) to give a six-carbon tricarboxylic acid (citrate). Citrate is then isomerized to isocitrate by the enzyme aconitase.

The following two reactions are successive oxidative decarboxylations, each of which produces one NADH and releases one molecule of CO_2, yielding a four-carbon product bound to CoA, succinyl-CoA. At this point, all three carbons that entered the mitochondrion as pyruvate have been oxidized into CO_2.

In the remainder of the TCA cycle, succinyl-CoA is oxidized to oxaloacetate, allowing the continued operation of the cycle. Initially the large amount of free energy available in the thioester bond of succinyl-CoA is conserved through the synthesis of ATP from ADP and P_i via a substrate-level phosphorylation catalyzed by *succinyl-CoA synthetase*. The resulting succinate is oxidized to fumarate by *succinate dehydrogenase*, which is the only membrane-associated enzyme of the TCA cycle and also part of the electron transport chain.

The electrons and protons removed from succinate end up not on NAD^+, but on another cofactor involved in redox reactions: **flavin adenine dinucleotide** (**FAD**). FAD is covalently bound to the active site of succinate dehydrogenase and undergoes a reversible two-electron reduction to produce $FADH_2$ (Figure 11.2B).

In the final two reactions of the TCA cycle, fumarate is hydrated to produce malate, which is subsequently oxidized by *malate dehydrogenase* to regenerate oxaloacetate and produce another molecule of NADH. The oxaloacetate produced is now able to react with another acetyl-CoA and continue the cycling.

In total, the stepwise oxidation of one molecule of pyruvate (3C) in the mitochondrion gives rise to three molecules of CO_2, and much of the free energy released during these oxidations is conserved in the form of four NADH and one $FADH_2$. In addition, one molecule of ATP is produced by a substrate-level phosphorylation.

flavin adenine dinucleotide (FAD) A riboflavin-containing cofactor that undergoes a reversible two-electron reduction to produce $FADH_2$.

malic enzyme An enzyme that catalyzes the oxidation of malate to pyruvate, permitting plant mitochondria to oxidize malate or citrate to CO_2 without involving pyruvate generated by glycolysis.

The TCA cycle of plants has unique features

The TCA cycle reactions outlined in Figure 11.6 are not all identical to those carried out by animal mitochondria. For example, the step catalyzed by succinyl-CoA synthetase produces GTP or ATP in animals, but exclusively ATP in plants. These nucleotides are energetically equivalent.

A special feature of the plant TCA cycle is the presence of large amounts of **malic enzyme** in the mitochondrial matrix of plants. This enzyme catalyzes the oxidative decarboxylation of malate:

$$\text{Malate} + \text{NAD}^+ \rightarrow \text{pyruvate} + \text{CO}_2 + \text{NADH} \quad (11.7)$$

The activity of malic enzyme enables plant mitochondria to operate alternative pathways for the metabolism of PEP derived from glycolysis. As already described, malate can be synthesized from PEP in the cytosol via the enzymes PEP carboxylase and malate dehydrogenase (Figure 11.3). For degradation, malate is transported into the mitochondrial matrix, where malic enzyme can oxidize it to pyruvate. This reaction makes possible the oxidation of TCA cycle intermediates such as malate (**Figure 11.7A**) or citrate (**Figure 11.7B**) to CO_2. Many plant tissues, not only those that carry out crassulacean acid metabolism (Chapter 8), store significant amounts of malate or other organic acids in their vacuoles. Degradation of malate via mitochondrial malic enzyme is important for regulating the levels of organic acids in cells—for example, during fruit ripening.

Instead of being degraded, the malate produced via PEP carboxylase can replenish TCA cycle intermediates used in biosynthesis. For example, export of 2-oxoglutarate for nitrogen assimilation in the chloroplast leads to a shortage of oxaloacetate for the citrate synthase reaction. This malate can be replaced through

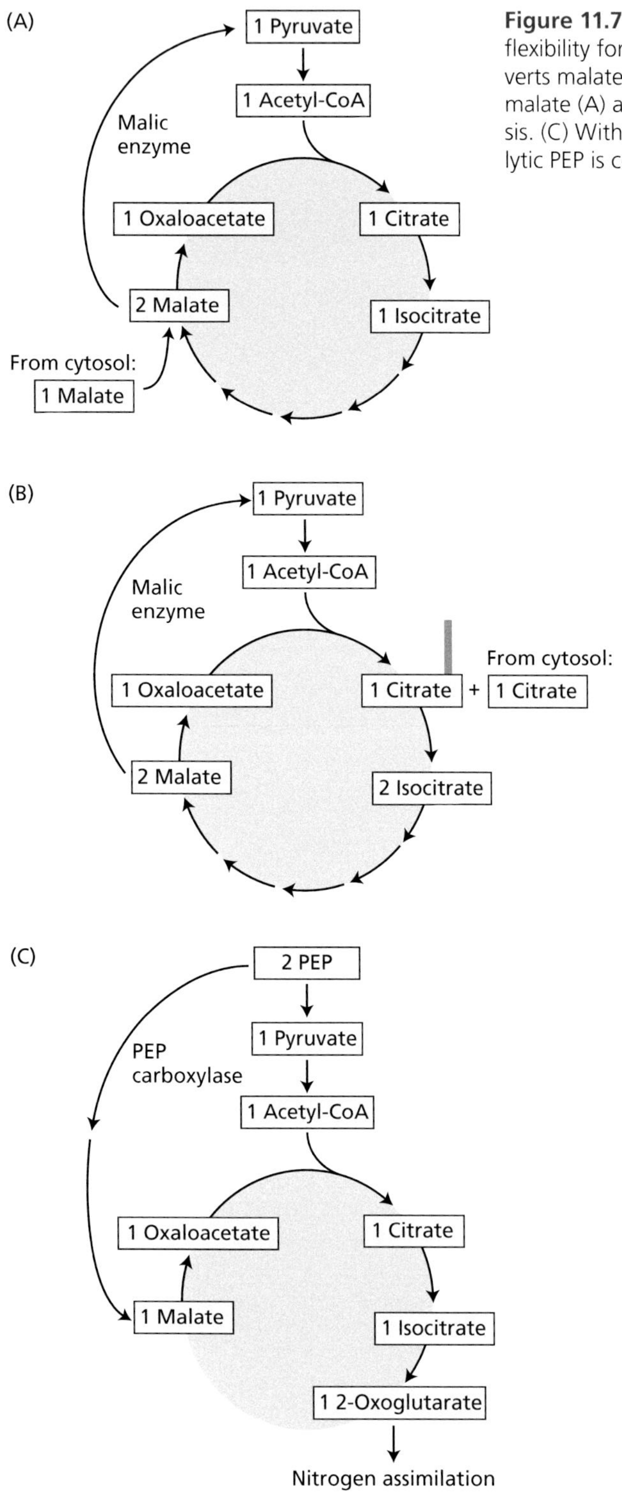

Figure 11.7 Malic enzyme and PEP carboxylase provide plants with metabolic flexibility for the metabolism of PEP and pyruvate. (A and B) Malic enzyme converts malate into pyruvate and thus allows plant mitochondria to oxidize both malate (A) and citrate (B) to CO_2 without involving pyruvate delivered by glycolysis. (C) With the added action of PEP carboxylase to the standard pathway, glycolytic PEP is converted into 2-oxoglutarate, which is used for nitrogen assimilation.

the PEP carboxylase pathway (**Figure 11.7C**). Reactions that replenish intermediates in a metabolic cycle are called *anaplerotic*.

Gamma-aminobutyric acid (GABA) is an amino acid that accumulates under several stress conditions in plants and that may have a role as a signal. GABA is synthesized from 2-oxoglutarate and degraded into succinate by the so-called **GABA shunt**, which bypasses TCA cycle enzymes that may be inhibited under plant stress.

11.5 Oxidative Phosphorylation

Describe oxidative phosphorylation with its different molecular control points and mechanisms for producing ATP.

ATP is the energy transmitter used by cells to drive life processes, so chemical energy conserved during the TCA cycle in the form of NADH and $FADH_2$ must be converted into ATP to perform useful work in the cell. This O_2-dependent process, called oxidative phosphorylation, occurs in the inner mitochondrial membrane.

Oxidative phosphorylation is fundamentally similar in all aerobic cells, yet the mitochondrial electron transport chain of vertebrate and arthropod animals lacks non-energy-conserving NAD(P)H dehydrogenases and alternative oxidase that are present in most eukaryotes, and that are especially rich in variants in plants.

The electron transport chain catalyzes a flow of electrons from NADH to O_2

For each molecule of sucrose oxidized through glycolysis and the TCA cycle, 4 molecules of NADH are generated in the cytosol, and 16 molecules of NADH plus 4 molecules of $FADH_2$ (associated with succinate dehydrogenase) are generated in the mitochondrial matrix. These reduced compounds must be reoxidized, or the entire respiratory process will rapidly come to a halt.

The electron transport chain catalyzes a transfer of two electrons from NADH (or $FADH_2$) to oxygen, the final electron acceptor of the respiratory process. For the oxidation of NADH, the reaction can be written as:

$$\text{NADH} + \text{H}^+ + \tfrac{1}{2}\text{O}_2 \rightarrow \text{NAD}^+ + \text{H}_2\text{O} \qquad (11.8)$$

From the reduction potentials for the NADH–NAD^+ pair (–320 mV) and the H_2O–½ O_2 pair (+810 mV), it can be calculated that the standard free energy released during this overall reaction ($-nF\Delta E^{0'}$) is about 220 kJ per mole of NADH. Because the succinate–fumarate reduction potential is higher (+30 mV), only 152 kJ per mole of succinate is released. The role of the electron transport chain

GABA shunt A pathway supplementing the tricarboxylic acid cycle with the ability to form and degrade GABA.

is to bring about the oxidation of NADH (and $FADH_2$) and, in the process, use some of the free energy released to generate an electrochemical proton gradient, $\Delta \tilde{\mu}_{H^+}$, across the inner mitochondrial membrane.

The electron transport chain of plants contains the same set of electron carriers found in the mitochondria of other organisms (**Figure 11.8**). The individual electron transport proteins are organized into four transmembrane multiprotein complexes (identified by roman numerals I through IV), all of which are localized in the inner mitochondrial membrane. Three of these complexes (I, III, and IV) are engaged in proton pumping.

COMPLEX I (NADH DEHYDROGENASE) Electrons from NADH generated by the TCA cycle in the mitochondrial matrix are oxidized by complex I (an NADH dehydrogenase). The electron carriers in complex I include a tightly bound cofactor (**flavin mononucleotide** [**FMN**], which is chemically similar to FAD; Figure 11.2B) and several iron–sulfur centers. Complex I then transfers these electrons to ubiquinone. Four protons are pumped from the matrix into the intermembrane space for every electron pair passing through the complex.

NADH dehydrogenase (complex 1) A multi-subunit protein complex in the mitochondrial electron transport chain that catalyzes oxidation of NADH and reduction of ubiquinone linked to the pumping of protons from the matrix to the intermembrane space.

flavin mononucleotide (FMN) A riboflavin-containing cofactor that undergoes a reversible one- or two-electron reduction to produce FMNH or $FMNH_2$. FAD and FMN occur as cofactors in flavoproteins.

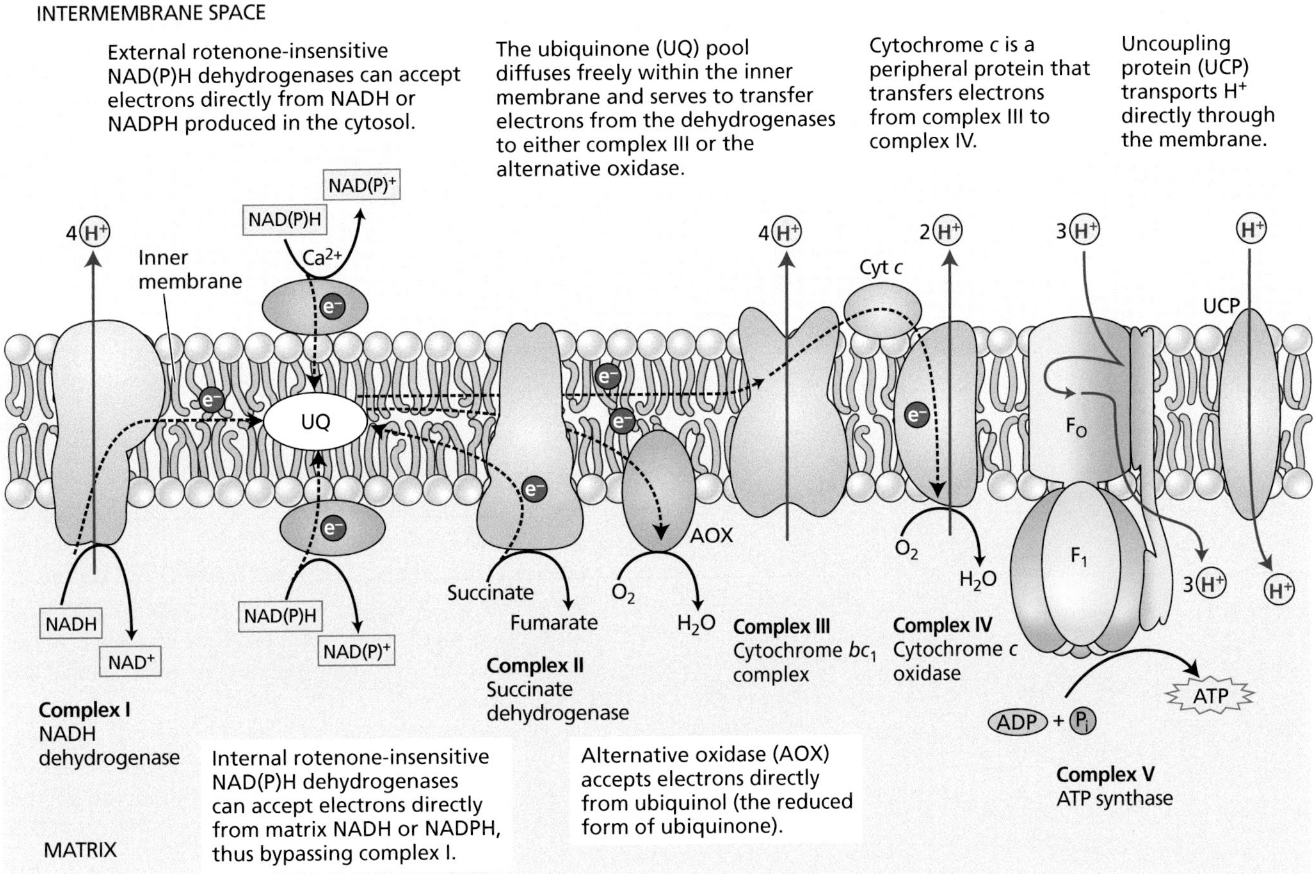
INTERMEMBRANE SPACE
External rotenone-insensitive NAD(P)H dehydrogenases can accept electrons directly from NADH or NADPH produced in the cytosol.
The ubiquinone (UQ) pool diffuses freely within the inner membrane and serves to transfer electrons from the dehydrogenases to either complex III or the alternative oxidase.
Cytochrome c is a peripheral protein that transfers electrons from complex III to complex IV.
Uncoupling protein (UCP) transports H+ directly through the membrane.
NAD(P)+
NAD(P)H
4 H+
Inner membrane
Ca2+
e−
UQ
4 H+
2 H+
3 H+
H+
Cyt c
UCP
Fo
F1
AOX
O2
H2O
NADH
NAD+
NAD(P)H
NAD(P)+
Succinate
Fumarate
O2
H2O
Complex III
Cytochrome bc1 complex
Complex IV
Cytochrome c oxidase
3 H+
H+
ATP
ADP + Pi
Complex I
NADH dehydrogenase
Complex II
Succinate dehydrogenase
Internal rotenone-insensitive NAD(P)H dehydrogenases can accept electrons directly from matrix NADH or NADPH, thus bypassing complex I.
Alternative oxidase (AOX) accepts electrons directly from ubiquinol (the reduced form of ubiquinone).
Complex V
ATP synthase
MATRIX

Figure 11.8 Organization of the electron transport chain and ATP synthesis in the inner membrane of the plant mitochondrion. Mitochondria from nearly all eukaryotes contain the four standard protein complexes: I, II, III, and IV. The structures of all complexes have been determined, but they are shown here as simplified shapes. The electron transport chain of the plant mitochondrion contains additional enzymes that do not pump protons, external and internal rotenone-insensitive NAD(P)H dehydrogenases and the alternative oxidase (in green). Additionally, uncoupling proteins directly bypass the ATP synthase by allowing passive proton influx. This multiplicity of bypasses in plants, where vertebrate animals have only the uncoupling protein, gives greater flexibility to plant energy coupling.

ubiquinone A mobile electron carrier of the mitochondrial electron transport chain. Chemically and functionally similar to plastoquinone in the photosynthetic electron transport chain.

Q cycle A mechanism for oxidation of plastohydroquinone (reduced plastoquinone, also called plastoquinol) in chloroplasts and of ubihydroquinone (reduced ubiquinone, also called ubiquinol) in mitochondria.

cytochrome *c* A peripheral, mobile component of the mitochondrial electron transport chain that oxidizes complex III and reduces complex IV.

NAD(P)H dehydrogenases A collective term for membrane-bound enzymes that oxidize NADH or NADPH, or both, and reduce quinone. Several are present in the electron transport chain of mitochondria; for example, the proton-pumping complex I but also simpler non–proton-pumping enzymes.

alternative oxidase An enzyme in the mitochondrial electron transport chain that reduces oxygen to water and oxidizes ubiquinol (ubihydroquinone).

Ubiquinone, a small lipid-soluble electron and proton carrier, is localized within the inner membrane. It is not tightly associated with any protein, and it can diffuse within the hydrophobic core of the membrane bilayer.

COMPLEX II (SUCCINATE DEHYDROGENASE) Oxidation of succinate in the TCA cycle is catalyzed by this complex, and the reducing equivalents are transferred via the cofactors, $FADH_2$ and several iron–sulfur centers, to ubiquinone. Complex II does not pump protons.

COMPLEX III (CYTOCHROME *BC*$_1$ COMPLEX) Complex III oxidizes reduced ubiquinone (ubiquinol) and transfers the electrons via an iron–sulfur center, two *b*-type cytochromes (b_{565} and b_{560}), and a membrane-bound cytochrome c_1 to cytochrome *c*. Four protons per electron pair are pumped out of the matrix by complex III using a mechanism called the **Q cycle**. Both structurally and functionally, ubiquinone and the cytochrome bc_1 complex are very similar to plastoquinone and the cytochrome b_6f complex, respectively, in the photosynthetic electron transport chain (Chapter 7).

Cytochrome *c* is a small protein loosely attached to the outer surface of the inner membrane and serves as a mobile carrier to transfer electrons between complexes III and IV, similar to the role of plastocyanin in photosynthetic electron transport (see Figure 7.28).

COMPLEX IV (CYTOCHROME *C* OXIDASE) Complex IV contains two copper centers (Cu_A and Cu_B) and cytochromes *a* and a_3. This complex is the terminal oxidase and brings about the four-electron reduction of O_2 to two molecules of H_2O. Two protons are pumped out of the matrix per electron pair (Figure 11.8).

The electron transport chain may be more complex than the description here implies. Plant respiratory complexes contain several plant-specific subunits whose functions are still unknown. Several of the complexes contain subunits that participate in functions other than electron transport, such as protein import. Finally, the proton-pumping complexes have been observed to form a single supercomplex (called a respirasome), in contrast to being individually mobile in the membrane. However, the functional significance of supercomplexes is still unclear.

The electron transport chain has supplementary branches

In addition to the set of protein complexes just described, the plant electron transport chain contains components not found in mammalian mitochondria (Figure 11.8). Especially, additional non-energy-conserving **NAD(P)H dehydrogenases** and a so-called **alternative oxidase** are bound to the inner membrane. In contrast to the proton-pumping complexes I, III, and IV, these enzymes do not pump protons, so the energy released from oxidation of NADH is not conserved as ATP, but instead is turned into heat.

Plant mitochondria have two pathways for oxidizing matrix NADH. Electron flow through complex I, described in the previous subsection, is sensitive to inhibition by several compounds, including rotenone and piericidin. Plant mitochondria also have a rotenone-insensitive dehydrogenase, ND_{in}(NADH), on the matrix surface of the inner mitochondrial membrane. This enzyme oxidizes NADH derived from the TCA cycle and may also be a bypass engaged when complex I is overloaded, as we will see later in this section. An NADPH dehydrogenase, ND_{in}(NADPH), is also present on the matrix surface, but very little is known about this enzyme.

Rotenone-insensitive NAD(P)H dehydrogenases, mostly Ca^{2+}-dependent, are also attached to the outer surface of the inner membrane facing the intermembrane space. They oxidize either NADH or NADPH from the cytosol.

Electrons from these external NAD(P)H dehydrogenases—ND_{ex}(NADH) and ND_{ex}(NADPH)—enter the main electron transport chain at the level of the ubiquinone pool.

Plants have an additional respiratory pathway for the oxidation of ubiquinol and reduction of oxygen. This pathway involves the alternative oxidase, which, unlike cytochrome *c* oxidase, is insensitive to inhibition by cyanide, carbon monoxide, and the signal molecule nitric oxide.

Some additional electron transport chain dehydrogenases present in plant mitochondria directly perform important carbon conversions. A *proline dehydrogenase* oxidizes the amino acid proline. Proline accumulates during drought or salt stress (Chapter 19), and it is degraded by this mitochondrial pathway when water status returns to normal. An electron transfer flavoprotein:quinone oxidoreductase mediates the degradation of several amino acids that are used by plants as a reserve under carbon starvation conditions induced by light deprivation. A third example is a galactono-gamma-lactone dehydrogenase, specific to plants, that performs the last step in the major pathway for synthesis of the antioxidant *ascorbic acid* (also known as vitamin C). The enzyme uses cytochrome *c* as its electron acceptor, in competition with normal respiration.

chemiosmotic mechanism The mechanism whereby the electrochemical gradient of protons established across a membrane by an electron transport process is used to either drive energy-requiring ATP synthesis (mitochondria and chloroplasts) or carrier-mediated anion efflux at the plasma membrane (as in polar auxin transport).

electrochemical proton gradient The sum of the electrical charge gradient and the pH gradient across the membrane, resulting from a concentration gradient of protons.

ATP synthesis in the mitochondrion is coupled to electron transport

In oxidative phosphorylation, the transfer of electrons to oxygen via complexes I, III, and IV is coupled to the synthesis of ATP from ADP and P_i via the F_oF_1-ATP synthase (complex V). The number of ATPs synthesized depends on the electron donor.

In experiments conducted on isolated mitochondria, electrons donated to complex I (e.g., generated by malate oxidation) give ADP:O ratios (the number of ATPs synthesized per two electrons transferred to oxygen) of 2.4 to 2.7 (**Table 11.1**). Electrons donated to complex II (from succinate) and to the external NADH dehydrogenase give values in the range of 1.6 to 1.8, while electrons donated directly to cytochrome *c* oxidase (complex IV) via artificial electron carriers give values of 0.8 to 0.9. Results such as these have led to the general concept that there are three sites of energy conservation along the electron transport chain, at complexes I, III, and IV.

The experimental ADP:O ratios agree quite well with the values calculated on the basis of the number of H^+ pumped by complexes I, III, and IV and the cost of 4 H^+ for producing 1 ATP (Table 11.1). For instance, electrons from external NADH pass only complexes III and IV, so a total of 6 H^+ are pumped, giving 1.5 ATP (when the alternative oxidase pathway is not used).

The mechanism of mitochondrial ATP synthesis is based on the **chemiosmotic mechanism** described in Chapter 7, which was first proposed in 1961 by Nobel laureate Peter Mitchell as a general mechanism of energy conservation across biological membranes. According to the chemiosmotic mechanism, the orientation of electron carriers within the inner mitochondrial membrane allows for the transfer of protons across the inner membrane during electron flow (Figure 11.8).

Because the inner mitochondrial membrane is highly impermeable to protons, an **electrochemical proton gradient** can build up. The free-energy difference associated with the electrochemical

Table 11.1 Theoretical and Experimental ADP:O Ratios in Isolated Plant Mitochondria

	ADP:O ratio	
Electrons feeding into	Theoretical[a]	Experimental
Complex I	2.5	2.4–2.7
Complex II	1.5	1.6–1.8
External NADH dehydrogenase	1.5	1.6–1.8
Complex IV	1.0[b]	0.8–0.9

[a] It is assumed that complexes I, III, and IV pump 4, 4, and 2 H^+ per 2 electrons, respectively; that the cost of synthesizing one ATP and exporting it to the cytosol is 4 H^+; and that the non-energy-conserving pathways are not active.

[b] Cytochrome *c* oxidase (complex IV) pumps only 2 protons. However, 2 electrons move from the outer surface of the inner membrane (where the electrons are donated) across the inner membrane to the inner, matrix side. As a result, 2 H^+ are consumed on the matrix side. This means that the net movement of H^+ and charges is equivalent to the movement of a total of 4 H^+, giving an ADP:O ratio of 1.0.

F_oF_1-ATP synthase A multi-subunit protein complex associated with the inner mitochondrial membrane that couples the passage of protons across the membrane to the synthesis of ATP from ADP and phosphate. The subscript "o" in F_o refers to the binding of the inhibitor oligomycin. Similar to CF_0CF_1-ATP synthase in photophosphorylation, to which oligomycin does not bind and inhibit (hence the subscript is "0").

F_o The integral membrane part of the F_oF_1-ATP synthase.

F_1 The ATP-binding, matrix-facing part of the F_oF_1-ATP synthase.

proton gradient, $\Delta\tilde{\mu}_{H^+}$ (expressed in kJ mol^{-1}), is also referred to as the *proton motive force*, Δp, when expressed in units of volts. The transfer of an H^+ from the intermembrane space to the matrix will release energy due to the Δp, which is the sum of an electrical transmembrane potential component (ΔE) and a chemical-potential component (ΔpH), according to the approximate equation:

$$\Delta p = \Delta E - 59 \text{ mV} \times \Delta \text{pH (at 25°C)} \tag{11.9}$$

where

$$\Delta E = E_{\text{inside}} - E_{\text{outside}} \tag{11.10}$$

and

$$\Delta \text{pH} = \text{pH}_{\text{inside}} - \text{pH}_{\text{outside}} \tag{11.11}$$

ΔE results from the asymmetric distribution of charged species (H^+ and other ions) across the membrane, and ΔpH is due to the H^+ concentration difference across the membrane. Because protons are translocated from the mitochondrial matrix to the intermembrane space, the resulting ΔE across the inner mitochondrial membrane has a negative value. Under normal conditions, the ΔpH is approximately 0.5 and the ΔE approximately –200 mV. Because the membrane is only 7 to 8 nm thick, this ΔE corresponds to an electric field of at least 25 million V/m (or ten times the field generating a lightning flash in a thunderstorm), emphasizing the enormous forces involved in electron transport.

As Equation 11.9 shows, both ΔE and ΔpH contribute to the proton motive force in plant mitochondria, although ΔpH constitutes the smaller part, probably because of the large buffering capacity of both cytosol and matrix, which prevents large pH changes. This situation contrasts with that in the chloroplast, where almost all of the proton motive force across the thylakoid membrane is due to ΔpH (Chapter 7).The free-energy input required to generate $\Delta\tilde{\mu}_{H^+}$ comes from the free energy released during electron transport. This is done by asymmetric uptake and release of H^+ from reactions or via redox-gated H^+ channels. Because of the low permeability (conductance) of the inner membrane to protons, their electrochemical gradient can be consumed to carry out chemical work (ATP synthesis). The $\Delta\tilde{\mu}_{H^+}$ is coupled to the synthesis of ATP by an additional protein complex associated with the inner membrane, the F_oF_1-ATP synthase.

The **F_oF_1-ATP synthase** (also called *complex V*) consists of two major components, F_o and F_1 (Figure 11.8). **F_o** (subscript "o" for oligomycin-sensitive) is an integral membrane protein complex containing at least three different polypeptides. They form the channel through which protons cross the inner membrane. The other component, **F_1**, is a peripheral membrane protein complex that is composed of at least five different subunits and contains catalytic sites for converting ADP and P_i into ATP. This complex is attached to the matrix side of F_o.

The passage of protons through the channel is coupled to the catalytic cycle of the F_1 component of the ATP synthase, allowing the ongoing synthesis of ATP and the simultaneous use of the $\Delta\tilde{\mu}_{H^+}$. One ATP is synthesized for each 3 H^+ passing through the F_o component from the intermembrane space to the matrix, down the electrochemical proton gradient.

A high-resolution structure for the F_1 component of the mammalian ATP synthase provided evidence for a mechanism in which a part of F_o rotates relative to F_1 to couple H^+ transport to ATP synthesis. The structure and function of the CF_0CF_1-ATP synthase in chloroplasts are similar to those of the mitochondrial ATP synthase, except that the chloroplast complex does not bind oligomycin, and is therefore designated with the subscript zero instead of 'o' (Chapter 7).

The operation of a chemiosmotic mechanism of ATP synthesis has several implications. First, the true site of ATP formation on the inner mitochondrial membrane is the ATP synthase, not complex I, III, or IV. These complexes serve as sites of energy conservation whereby electron transport is coupled to the generation of a $\Delta\tilde{\mu}_{H^+}$. The synthesis of ATP decreases the $\Delta\tilde{\mu}_{H^+}$ and, as a consequence, its restriction on the electron transport complexes. Electron transport is therefore stimulated by a large supply of ADP.

The chemiosmotic mechanism also explains the action mechanism of **uncouplers**. These are a wide range of chemically unrelated, artificial compounds (including 2,4-dinitrophenol and *p*-trifluoromethoxycarbonylcyanide phenylhydrazone [FCCP]) that decrease mitochondrial ATP synthesis but stimulate the rate of electron transport. All uncouplers make the inner membrane leaky to protons and prevent the buildup of a sufficiently large $\Delta\tilde{\mu}_{H^+}$ to drive ATP synthesis or restrict electron transport. The inner membrane of plant mitochondria, like the mammalian one, has a natural uncoupling protein that can directly lower the $\Delta\tilde{\mu}_{H^+}$ in response to regulatory cues (Figure 11.8).

uncoupler A chemical compound that increases the proton permeability of membranes and thus uncouples the formation of the proton gradient from ATP synthesis.

Transporters exchange substrates and products

The electrochemical proton gradient also plays a role in the movement of the organic acid anions of the TCA cycle, and of the substrates and products of ATP synthesis, into and out of mitochondria (**Figure 11.9**). Although ATP is synthesized in the mitochondrial matrix, most of it is used outside the mitochondrion, so an efficient mechanism is needed for moving ADP (and P_i) into and ATP out of the organelle.

The ADP/ATP (adenine nucleotide) transporter performs the active exchange of ADP and ATP across the inner membrane. The movement of the more negatively charged ATP^{4-} out of the mitochondrion in exchange for ADP^{3-}—that is, one net negative charge out—is driven by the electrical-potential gradient (ΔE, positive outside) generated by proton pumping.

The uptake of inorganic phosphate (P_i) involves an active phosphate transporter protein that uses the chemical-potential component (ΔpH) of the proton motive force to drive the electroneutral exchange of P_i^- (in) for OH^- (out). As long as a ΔpH is maintained across the inner membrane, the P_i content within the matrix remains high. Similar reasoning applies to the uptake of pyruvate, which is driven by the electroneutral exchange of pyruvate for OH^-, leading to efficient uptake of pyruvate from the cytosol (Figure 11.9).

The total energetic cost of taking up one P_i and one ADP into the matrix and exporting one ATP is equivalent to the movement of one H^+ from the intermembrane space into the matrix:

- Moving one OH^- out in exchange for P_i^- is equivalent to one H^+ in, so this electroneutral exchange consumes the ΔpH, but not the ΔE.
- Moving one negative charge out (ADP^{3-} entering the matrix in exchange for ATP^{4-} leaving) is the same as moving one positive charge in, so this transport lowers only the ΔE.

This proton, which drives the exchange of ATP for ADP and P_i, should also be included in our calculation of the cost of synthesizing one ATP. Thus, the total cost is 3 H^+ used by the ATP synthase plus 1 H^+ for the exchanges across the membrane, or a total of 4 H^+.

The inner membrane also contains transporters for dicarboxylic acids (malate or succinate) exchanged for P_i^{2-} and for the tricarboxylic acids (citrate, aconitate, or isocitrate) exchanged for dicarboxylic acids (Figure 11.9).

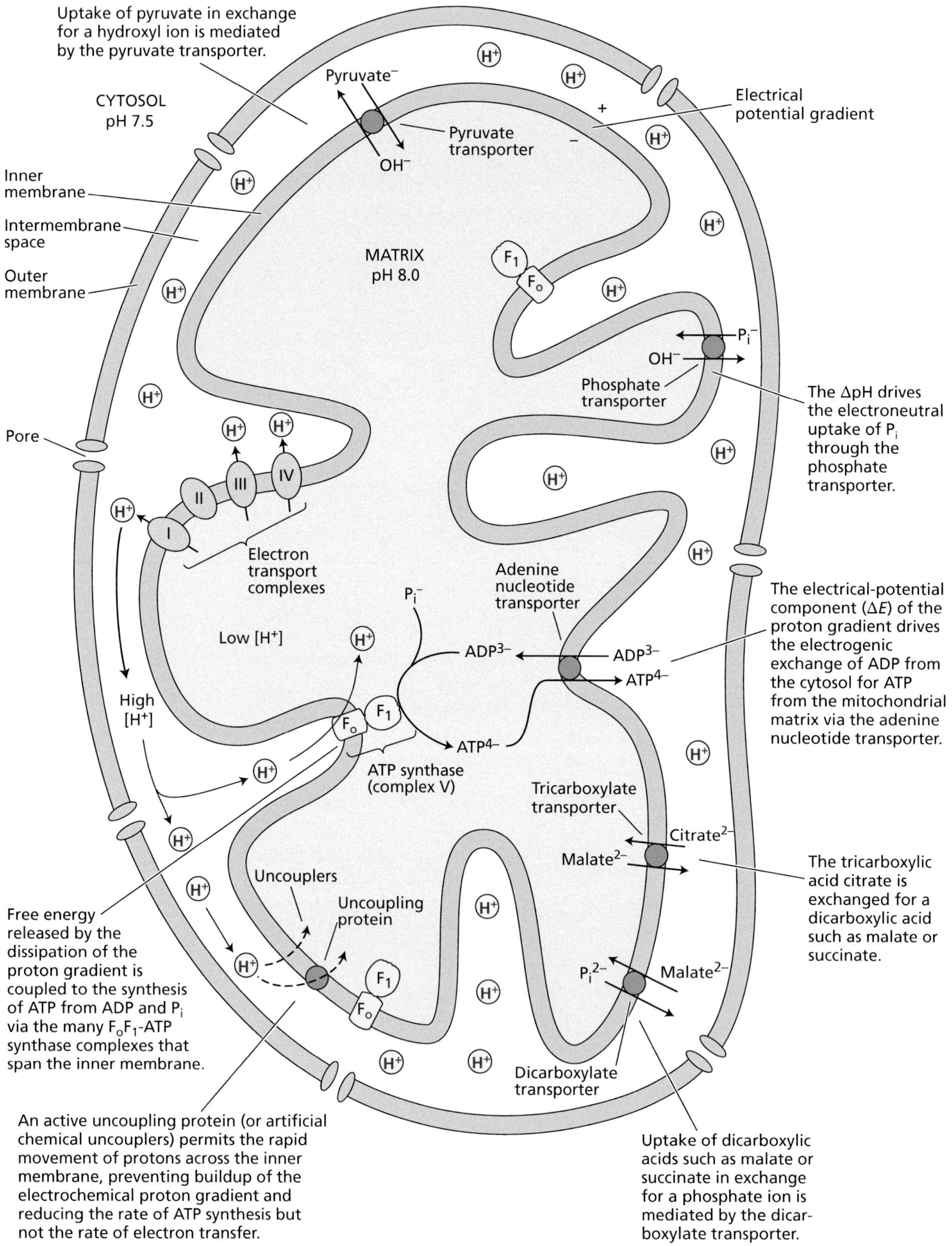

Figure 11.9 Transmembrane transport in plant mitochondria. An electrochemical proton gradient, $\Delta\tilde{\mu}_{H^+}$, consisting of an electrical-potential component (ΔE, −200 mV, negative inside) and a chemical-potential component (ΔpH, alkaline inside), is established across the inner mitochondrial membrane during electron transport. The $\Delta\tilde{\mu}_{H^+}$ is used by specific transporters that move metabolites across the inner membrane.

Aerobic respiration yields about 60 molecules of ATP per molecule of sucrose

The complete oxidation of a sucrose molecule leads to the net formation of:

- Eight molecules of ATP by substrate-level phosphorylation (four from glycolysis and four from the TCA cycle)
- Four molecules of NADH in the cytosol
- Sixteen molecules of NADH plus four molecules of $FADH_2$ (via succinate dehydrogenase) in the mitochondrial matrix

Based on theoretical ADP:O values (Table 11.1), we can estimate that 52 molecules of ATP will be generated per molecule of sucrose by oxidative phosphorylation. The complete aerobic oxidation of sucrose (including substrate-level phosphorylation) results in a total of about 60 ATP synthesized per sucrose molecule (**Table 11.2**).

Using 50 kJ mol^{-1} as the actual free energy of formation of ATP in vivo, we find that about 3,010 kJ mol^{-1} of free energy is conserved in the form of ATP per mole of sucrose oxidized during aerobic respiration. This amount represents about 52% of the standard free energy available from the complete oxidation of sucrose; the rest is lost as heat. It also represents a vast improvement over fermentative metabolism, in which only 4% of the energy available in sucrose is converted into ATP.

Table 11.2 The Maximum Yield of Cytosolic ATP from the Complete Oxidation of Sucrose to CO_2 via Aerobic Glycolysis and the TCA Cycle

Part reaction	ATP per sucrose[a]
Glycolysis	
4 substrate-level phosphorylations	4
4 NADH	4 × 1.5 = 6
TCA cycle	
4 substrate-level phosphorylations	4
4 $FADH_2$	4 × 1.5 = 6
16 NADH	16 × 2.5 = 40
Total	60

Source: Adapted from M. D. Brand. 1994. *Biochem.* (*Lond.*) 16: 20–24.

Note: Glycolysis is here assumed to use invertase and ATP-phosphofructokinases, but the ATP yield can be higher if other enzymes are active. Cytosolic NADH is assumed to be oxidized by the external NADH dehydrogenase. The other non-energy-conserving pathways (e.g., the alternative oxidase) are assumed not to be engaged.

[a] Calculated using the theoretical ADP:O values from Table 11.1.

Plants have several mechanisms that lower the ATP yield

As we have seen, a complex machinery is required for conserving energy in oxidative phosphorylation. So it is perhaps surprising that plant mitochondria have several functional proteins that reduce this efficiency. Plants are probably less limited by energy supply (sunlight) than by other factors in the environment (e.g., access to water and nutrients). As a consequence, metabolic flexibility may be more important to them than energetic efficiency.

THE ALTERNATIVE OXIDASE Most plants display a capacity for *cyanide-resistant respiration* that is comparable to the capacity of the cyanide-sensitive cytochrome *c* oxidase pathway. The cyanide-resistant oxygen uptake is catalyzed by the alternative oxidase. Electrons feed off the main electron transport chain into this alternative pathway at the level of the ubiquinone pool (Figure 11.8). The alternative oxidase, the only component of the alternative pathway, catalyzes a four-electron reduction of oxygen to water. When electrons pass to the alternative pathway from the ubiquinone pool, two sites of proton pumping (at complexes III and IV) are bypassed. Because there is no energy-conservation site in the alternative pathway between ubiquinone and oxygen, the free energy that would normally be conserved as ATP is lost as heat when electrons are shunted through this pathway.

How can a process as seemingly energetically wasteful as the alternative pathway contribute to plant metabolism? One example of the functional usefulness of the alternative oxidase is in so-called thermogenic flowers of several plant families—for example, the voodoo lily (*Sauromatum guttatum*). In such flowers, a combination of very high respiratory rate and very high alternative oxidase activity leads to massive heat production that is important for pollination. Thermogenesis is also important for reproduction in eastern skunk cabbage

uncoupling protein A protein that increases the proton permeability of the inner mitochondrial membrane and thereby decreases energy conservation.

(*Symplocarpus foetidus*), the sacred lotus (*Nelumbo nucifera*), and Macrozamia cycads. However, in most plants the respiratory rates are too low to generate sufficient heat to raise the temperature significantly, so it must have other roles.

The transcription of the alternative oxidase is often induced by reactive oxygen species (ROS) generated from biotic and abiotic stresses that can inhibit mitochondrial respiration. These include phosphate deficiency, chilling, drought, and osmotic stress (Chapter 19). The up-regulation of alternative oxidase expression is an example of *retrograde regulation,* in which nuclear gene expression responds to changes in organellar status by the action of several stress-associated transcription factors.

The activity of the alternative oxidase, which functions as a dimer, is regulated by reversible oxidation–reduction of an intermolecular sulfhydryl bridge, by the reduction level of the ubiquinone pool, and by pyruvate. The first two factors ensure that the enzyme is most active under reducing conditions, while the last factor ensures that the enzyme has high activity when there is plenty of substrate for the TCA cycle.

If the cell consumes relatively little ATP, ADP levels will be low and $\Delta\tilde{\mu}_{H^+}$ will be high. This will restrict the respiration rate, and the reduction level in the mitochondrial electron transport chain will rise and activate the alternative oxidase present. The increased reduction level will also lead to increased formation of ROS, which can cause damage and act as a signal for the activation of alternative oxidase expression (Figure 11.10). The increased amount of alternative oxidase can prevent overreduction by draining off electrons from the ubiquinone pool (Figure 11.8), thus limiting the production of ROS and acting as damage control. In this way, the alternative oxidase allows the mitochondrion to adjust the relative rates of ATP production and synthesis of carbon skeletons for use in biosynthetic reactions.

THE UNCOUPLING PROTEIN The inner membrane of mitochondria has an **uncoupling protein** that increases the proton permeability of the membrane and thus acts as an uncoupler. This protein is induced by stress and stimulated by ROS. In knockout mutants, photosynthetic carbon assimilation and growth were decreased consistent with the interpretation that the uncoupling protein, like the alternative oxidase, functions to prevent overreduction of the electron transport chain and formation of ROS.

ROTENONE-INSENSITIVE NAD(P)H DEHYDROGENASES Multiple plant mitochondrial NADH or NADPH dehydrogenases are insensitive to the inhibitor rotenone (Figure 11.8). The internal, rotenone-insensitive NADH dehydrogenase [ND_{in}(NADH)] may work as a non-proton-pumping bypass when complex I is overloaded. These NADH dehydrogenases and the NADPH dehydrogenases are likely to make plant respiration more flexible and allow control of specific redox homeostasis of NADH and NADPH in mitochondria and cytosol (**Figure 11.10**).

Respiration is an integral part of a redox and biosynthesis network

In addition to the 2-oxoglutarate provision for nitrogen assimilation (Figure 11.7C), glycolysis, the oxidative pentose phosphate pathway, and the TCA cycle produce the building blocks for the synthesis of many central plant metabolites, including amino acids, lipids, nucleotides, and hormones (**Figure 11.11**). These are in turn linked to pathways for synthesis of specialized metabolites (Chapter 19). Indeed, much of the reduced carbon that is metabolized by glycolysis and the TCA cycle is diverted to biosynthetic purposes and not oxidized to CO_2.

Mitochondria are also a part of the cellular redox network. Variations in consumption or production of redox and energy-carrying compounds such as NAD(P)H and organic acids affect metabolic pathways in the cytosol and in plastids. Of special importance is the synthesis of ascorbic acid, a central redox

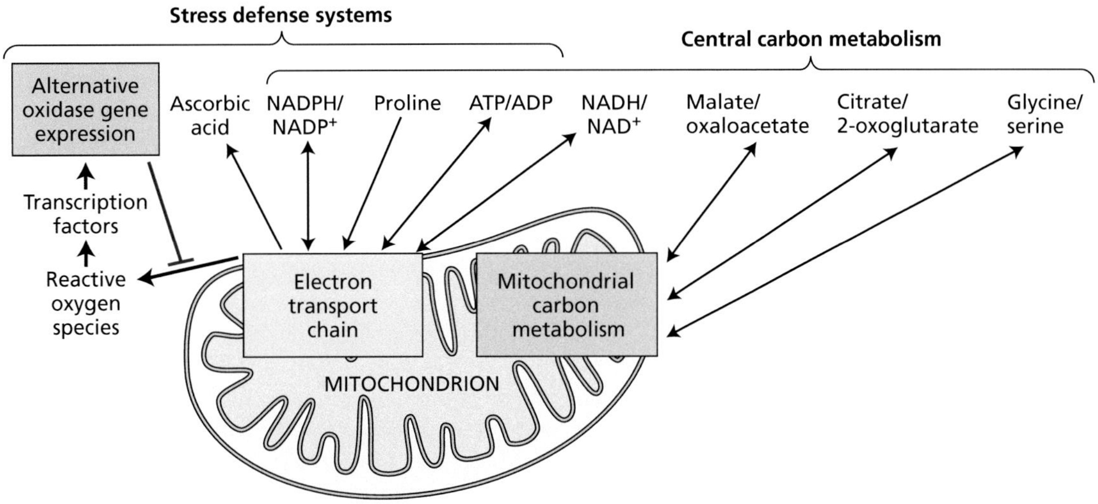

Figure 11.10 Metabolic interactions between mitochondria and cytosol. Mitochondrial activities can influence the cytosolic levels of redox and energy molecules involved in stress defense and in central carbon metabolism (such as growth processes and photosynthesis). An exact distinction between stress defense and carbon metabolism cannot be made because they have components in common. Arrows denote effects caused by changes in mitochondrial synthesis [e.g., reactive oxygen species (ROS), ATP, or ascorbic acid] or degradation [e.g., NAD(P)H, proline, or glycine]. The ROS-mediated activation of expression of nuclear genes for the alternative oxidase is an example of retrograde regulation.

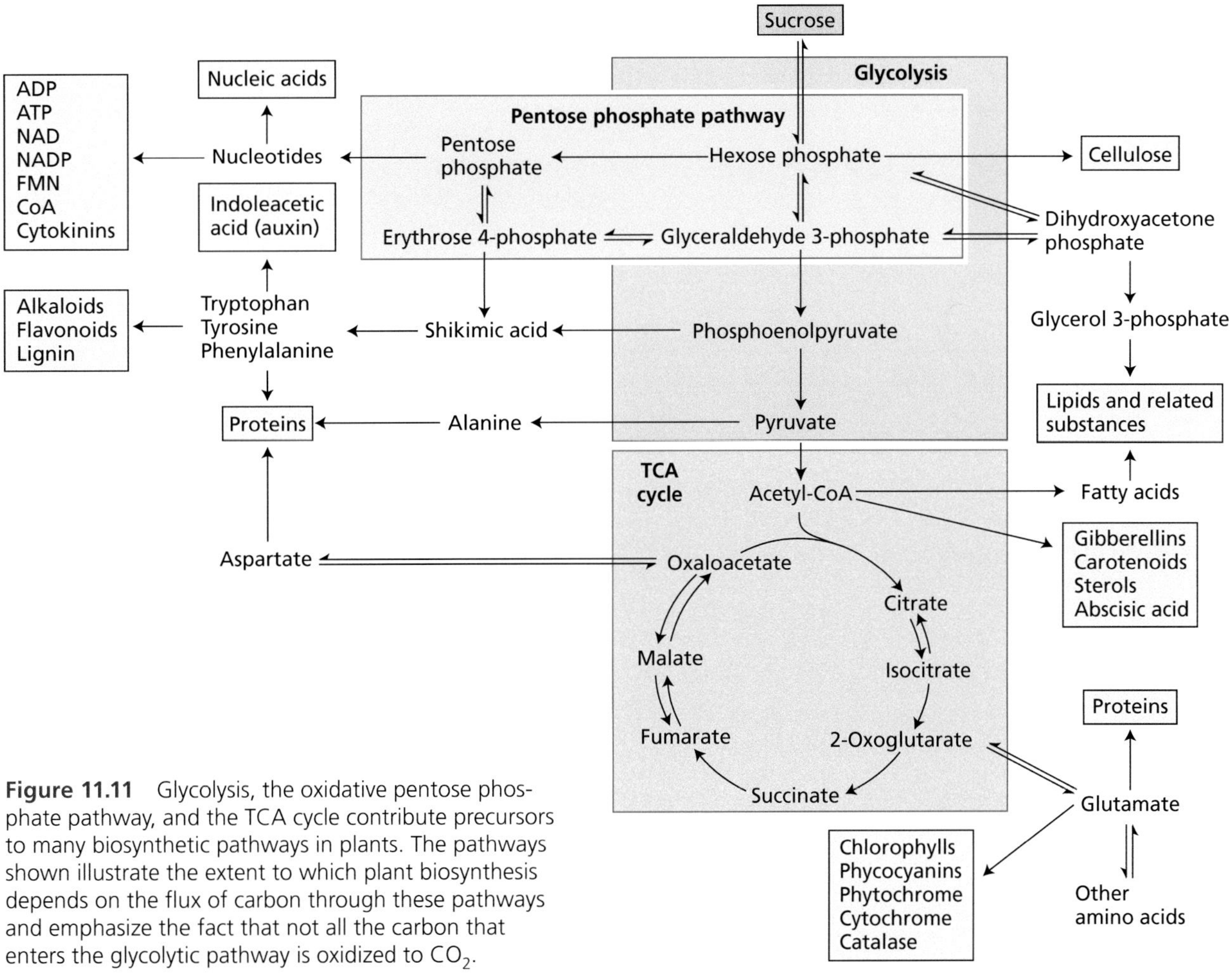

Figure 11.11 Glycolysis, the oxidative pentose phosphate pathway, and the TCA cycle contribute precursors to many biosynthetic pathways in plants. The pathways shown illustrate the extent to which plant biosynthesis depends on the flux of carbon through these pathways and emphasize the fact that not all the carbon that enters the glycolytic pathway is oxidized to CO_2.

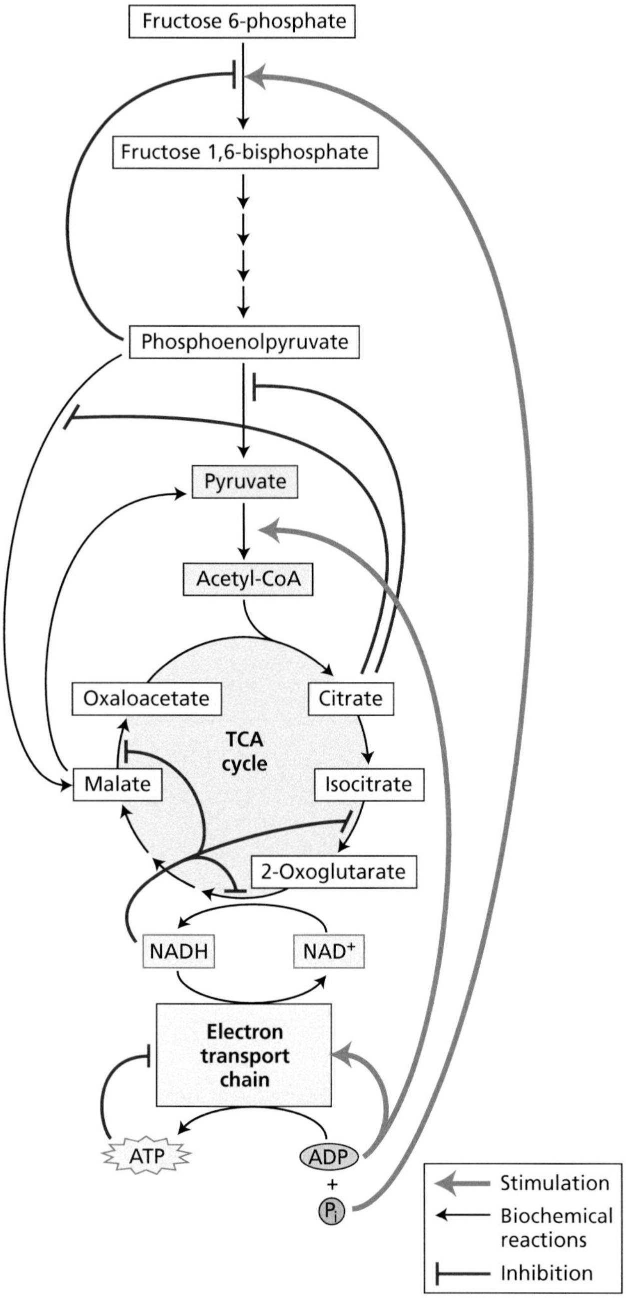

Figure 11.12 Model of bottom-up regulation of plant respiration. Directly or indirectly, several substrates for respiration (e.g., ADP) stimulate while products of respiration (e.g., ATP) inhibit upstream reactions in a stepwise fashion. The image shows a selected set of effects that illustrates the stepwise bottom-up principle. For instance, ATP inhibits the electron transport chain, leading to an accumulation of NADH. NADH inhibits TCA cycle enzymes such as isocitrate dehydrogenase and 2-oxoglutarate dehydrogenase. TCA cycle intermediates such as citrate then inhibit the PEP-metabolizing enzymes in the cytosol. Finally, PEP inhibits the conversion of fructose 6-phosphate into fructose 1,6-bisphosphate and restricts carbon flow into glycolysis. In this way, respiration can be up- or down-regulated in response to changing demands for any of its products: ATP, organic acids, and glycolytic intermediates.

and stress defense molecule in plants, by the electron transport chain (Figure 11.10). Mitochondria also carry out steps in the biosynthesis of coenzymes necessary for many metabolic enzymes throughout the cell.

Respiration is controlled at multiple levels

Gene expression establishes the protein levels for the respiratory enzymes. The levels change during development and, for some enzymes, such as alternative oxidase and NAD(P)H dehydrogenases, also in response to external factors such as stress, light, and nitrogen supply. However, most respiratory proteins are present in higher abundance than the basic respiration rate requires. Therefore, their activities can be either up- or down-regulated in response to metabolite or redox changes by protein modification or dynamic binding of molecules, and respiration can be rapidly controlled by the demands for its products.

Energy-limited shade plants, for example, have been shown to minimize their respiratory rate by minimizing ATP consumption. To achieve this, the plant respiratory rate is under "bottom-up" control (**Figure 11.12**). The substrates of ATP synthesis—ADP and P_i—are key short-term regulators of the rates of oxidative phosphorylation and therefore, via a stepwise series of effects, also of the TCA cycle and glycolysis. Multiple control points exist in all three processes; here we give a brief overview of some major features of the dynamic control of respiration. In contrast to plant cells, animal cells regulate respiration from the "top down," with a primary activation occurring at the phosphofructokinase and secondary activation at the pyruvate kinase.

If a plant cell's demand for ATP in the cytosol decreases relative to the supply, less ADP becomes available for ATP synthesis in the mitochondria, and the electron transport chain will operate at a reduced rate (Figure 11.12). This slowdown leads to an increase in matrix NADH, which inhibits the activity of pyruvate dehydrogenase and several TCA cycle dehydrogenases, and in matrix NADPH, which is required for thioredoxin function. As in photosynthetic carbon fixation (Chapter 8), redox levels control TCA cycle enzymes via thioredoxin-mediated removal of cysteine residue dimers. Thus, TCA cycle inhibition results in a buildup of TCA cycle intermediates and their derivatives that directly or indirectly down-regulate the action of upstream steps such as cytosolic pyruvate kinase and PEP carboxylase. This increases the cytosolic PEP concentration, which in turn reduces the rate of conversion of fructose 6-phosphate into fructose 1,6-bisphosphate, thus inhibiting glycolysis (Figure 11.12). This bottom-up inhibitory effect of PEP on ATP-dependent phosphofructokinase activity is strongly decreased by the activator P_i, making the cytosolic ratio of PEP to P_i a critical factor in the control of plant glycolytic activity.

One benefit of the bottom-up control is that it permits plants to regulate the net glycolytic flux to pyruvate in the cytosol independently of processes such as the photosynthetic sucrose–triose phosphate–starch interconversion. Another benefit is that glycolysis can adjust to the demands for both ATP and biosynthetic precursors from the TCA cycle, such as 2-oxoglutarate needed for nitrogen assimilation (Chapter 5). The two glycolytic enzymes metabolizing PEP in plant cells—pyruvate kinase and PEP carboxylase—are regulated by similar metabolites, which means that their joint action required for amino acid biosynthesis can be coordinately controlled.

A consequence of bottom-up control of glycolysis is that its rate can influence cellular concentrations of sugars, in combination with sugar-supplying processes such as phloem transport. Glucose and sucrose are potent signaling molecules that help the plant adjust its growth and development to its carbohydrate status. For example, hexokinase functions not only as a glycolytic enzyme but also as a glucose receptor in the nucleus. This induces sugar-dependent gene expression and works to adjust plant growth and metabolism to the carbohydrate status.

Pyruvate dehydrogenase is an important control point, especially for reducing TCA cycle activity during photosynthesis. This enzyme is posttranslationally phosphorylated by a *regulatory protein kinase* and dephosphorylated by a *protein phosphatase*. Pyruvate dehydrogenase is inactive in the phosphorylated state, yet the regulatory protein kinase is inhibited by pyruvate, stimulating pyruvate dehydrogenase activity when its substrate is available. In combination with multiple other effectors (**Figure 11.13**), this regulation adjusts the input of substrate to the TCA cycle to the cellular demand.

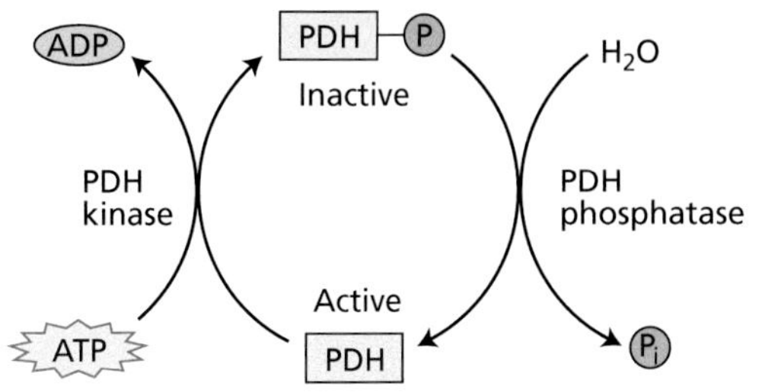

Pyruvate + CoA + NAD^+ ⟶ Acetyl-CoA + CO_2 + NADH + H^+

Effect on PDH activity	Mechanism
Activating	
Pyruvate	Inhibits kinase
ADP	Inhibits kinase
Mg^{2+} (or Mn^{2+})	Stimulates phosphatase
Inactivating	
NADH	Inhibits PDH Stimulates kinase
Acetyl-CoA	Inhibits PDH Stimulates kinase
NH_4^+	Inhibits PDH Stimulates kinase

Figure 11.13 Metabolic regulation of pyruvate dehydrogenase (PDH) activity, directly and by reversible phosphorylation. Upstream and downstream metabolites control PDH activity by direct actions on the enzyme itself or by regulating its protein kinase or protein phosphatase.

11.6 Respiration in Intact Plants and Tissues

Describe the relationship between photosynthesis and respiration and the factors that affect the respiratory rates of different organs.

Many rewarding studies of plant respiration and its regulation have been carried out on organelles and enzymes isolated from plant tissues. But how does this knowledge relate to the function of the whole plant in a natural or agricultural setting? What happens when green organs are exposed to light, and respiration and photosynthesis operate simultaneously? How are they functionally integrated in the cell?

Plants respire roughly half of the daily photosynthetic yield

Many factors can affect the respiration rate of an intact plant or of its individual organs. Relevant factors include the species and growth habit of the plant, the type and age of the specific organ, and environmental variables such as light, external O_2 and CO_2 concentrations, temperature, and nutrient and water supply (Chapter 19). By measuring different oxygen isotopes, it is possible to measure the in vivo activities of the alternative oxidase and cytochrome *c* oxidase simultaneously. Therefore, we know that a significant part of respiration in most tissues takes place via the "energy-wasting" alternative pathways.

Whole-plant respiration rates, particularly when considered on a fresh-weight basis, are generally lower than respiration rates reported for animal tissues. The reason is simply that plant tissues generally contain fewer mitochondria per gram fresh weight due to the presence of a large vacuole and a cell wall, neither of which contains mitochondria. Nonetheless, respiration rates in some plant tissues are as high as those observed in actively respiring animal tissues, so the respiratory process in plants is not inherently slower than in animals. In fact, isolated plant mitochondria respire as fast as or faster than mammalian mitochondria.

The effect of respiration on the overall carbon economy of the plant can be substantial. Whereas only green tissues photosynthesize, all tissues respire, and they do so 24 h a day. Even in photosynthetically active tissues, respiration, if integrated over the entire day, uses a substantial fraction of the gross photosynthesis. A survey of several herbaceous species indicated that 30 to 60% of the daily gain in photosynthetic carbon is lost to respiration, although these values tend to decrease in older plants. Trees respire a similar fraction of their photosynthetic production, but their respiratory loss increases with age as the ratio of photosynthetic to nonphotosynthetic tissue decreases. In general, unfavorable growth conditions will increase respiration relative to photosynthesis, and thus lower the overall carbon yield available for growth.

Respiratory processes operate during photosynthesis

Mitochondria are involved in the metabolism of photosynthesizing leaves in several ways. Mitochondria are major suppliers of ATP to the cytosol in the light (e.g., for driving biosynthetic pathways), especially under conditions where photosynthetically produced ATP is used in the chloroplasts for carbon fixation. Furthermore, the glycine generated by photorespiration is oxidized to serine in the mitochondrion in a reaction involving mitochondrial oxygen consumption (Chapter 8). Given that rates of photorespiration often reach 20 to 40% of the gross photosynthetic rate, photorespiration is the largest daytime provider of NADH for the mitochondrial oxidative phosphorylation. By comparison, mitochondria in photosynthesizing tissue do operate some respiration via the TCA cycle, yet far slower than the gross photosynthetic rate, generally by a factor of 6- to 20-fold. The activity of pyruvate dehydrogenase, one of the ports of entry into the TCA cycle, decreases in the light to 25% of its activity in darkness. A basal flux through the respiratory pathways is, however, needed during photosynthesis to supply precursors for biosynthetic reactions, such as the 2-oxoglutarate needed for nitrogen assimilation (Figures 11.7C and 11.11) or as a cofactor for 2-oxoglutarate-dioxygenase enzymes that activate or inactive gibberellins or jasmonates, for example (Chapter 12).

Additional evidence for the involvement of mitochondrial respiration in photosynthesis has been obtained in studies with mitochondrial mutants defective in respiratory complexes. Compared with the wild type, these plants have disturbed leaf development and photosynthesis because changes in levels of redox-active metabolites are communicated between mitochondria and chloroplasts, negatively affecting photosynthetic function.

Different tissues and organs respire at different rates

Respiration is often considered to have two components of comparable magnitude. **Maintenance respiration** is needed to support the function and turnover of the tissues already present. **Growth respiration** provides the energy needed for converting sugars into the building blocks that make up new tissues. A useful rule of thumb is that the greater the overall metabolic activity of a given tissue, the higher its respiration rate. Developing buds usually show very high rates of respiration, and respiration rates of vegetative organs usually decrease from the point of growth (e.g., the leaf tip in eudicots and the leaf base in monocots) to more differentiated regions. A well-studied example is the growing barley leaf.

In mature vegetative organs, stems generally have the lowest respiration rates, whereas leaf and root respiration varies with the plant species and the

maintenance respiration The respiration needed to support the function and turnover of existing tissue. Contrast with *growth respiration*.

growth respiration The respiration that provides the energy needed for converting sugars into the building blocks that make up new tissue. Contrast with *maintenance respiration*.

environmental conditions. Low availability of soil nutrients, for example, increases the demand for respiratory ATP production in the root. This increase reflects increased energy costs for active ion uptake and root growth in search of nutrients. **Box 11.1** discusses how crop yield is affected by changes in respiration rates.

When a plant organ reaches maturity, its respiration rate either remains roughly constant or decreases slowly as the tissue ages and ultimately senesces. Exceptions to this pattern include the marked rise in respiration, known as the *climacteric,* that accompanies the onset of ripening in many fruits (e.g., avocado, apple, and banana), and senescence in detached leaves and flowers. During fruit ripening, massive conversion of, for example, starch (banana) or organic acids (tomato and apple) into sugars occurs, accompanied by a rise in the concentration of the hormone ethylene (Chapter 17) and in the activity of the alternative oxidase pathway.

Box 11.1 Modifying Respiration for Future Needs

To keep up with the growing human population, we need to increase agricultural food production by 70% by the year 2060. This must take place without an increase in the cultivated area, with limited water resources in many places, and preferably using fewer agrochemicals. Therefore, we need to develop plants that give a greater net carbon yield by growing more efficiently under future climate conditions.

In plants, a large part (30–60%) of the photosynthetic carbon gain is lost through respiratory processes, and several investigations have shown that respiration can limit the growth rate of plants. Plant respiration is therefore a relevant target for modification to increase plant productivity.

A simplistic approach would be to remove genes for the non-energy-conserving enzymes NAD(P)H dehydrogenases or alternative oxidase in the mitochondrial electron transport chain. Electrons passing through those enzymes yield 30 to 100% less ATP, so efficiency could potentially be gained by preventing their activity. Unfortunately, suppression of dehydrogenases oxidizing internal NADH or external NADPH in Arabidopsis and suppression of the alternative oxidase in soybean have all given reduced growth under standard greenhouse conditions. This shows that these enzymes have important and nonredundant functions in normal plant growth.

J. S. Amthor and colleagues identified several strategies for specifically decreasing respiratory carbon loss by decreasing the cellular demand for ATP. Here are some examples:

- In plant cells, several ATP-consuming metabolic cycles, so-called futile cycles, take place, such as the simultaneous synthesis and degradation of sucrose and fructose 1,6-bisphosphate in glycolysis. In both cases there is a synthesis enzyme using ATP and another enzyme that performs the reverse reaction, but without ATP recovery (Figure 11.3). These reactions take place simultaneously, simply consuming ATP without apparent function. If the enzymes could be modified, without creating side effects, to prevent the cycling, ATP could be saved and less respiration would be needed.
- Reschedule the expression of the alternative oxidase so that it is active only during the day, where its activity is required to balance the dissipation of excess reducing power generated by photosynthesis.
- Relocate nitrate assimilation mechanisms (Chapter 5) from root to shoot so that the nitrate assimilation in the cytosol and chloroplasts can use the excess reducing power generated by photosynthesis in a more productive way instead of "burning it off" through the alternative oxidase. Modifications of nitrogen signaling mechanisms (Chapter 12) are another approach to improve utilization in shoot tissues.

To estimate the improvement in net carbon yield that can be achieved, Amthor and colleagues calculated that if all futile cycles could be eliminated (which is quite unrealistic), this could give a reduction in maintenance respiration of 15%, which would give a biomass gain of about 5%. This may not seem like much, but the effects of the various strategies should be additive.

An important issue is to breed plants that are adapted for optimal function at future higher atmospheric CO_2 concentrations and associated higher temperatures and drought. CO_2 release by plant respiration is approximately six times that of anthropogenic release. However, it has been difficult to predict the impact of atmospheric changes on respiration because it strongly interacts with photosynthesis, photorespiration, and nitrogen assimilation. Therefore, respiratory components of prediction models for future plant growth and the plants' impact on the future atmosphere are less certain.

Another respiratory factor important for global warming is the aerenchyma in flooded plants (Chapter 19). This structure forms the dominant pathway for emission of methane, the second most important greenhouse gas, from flooded soils, especially from rice paddies.

respiratory quotient (RQ) The ratio of CO_2 evolution to O_2 consumption.

temperature coefficient (Q_{10}) The increase in the rate of a process (e.g., respiration) for every 10°C increase in temperature.

Different tissues can use different substrates for respiration. Sugars dominate overall, but organic acids contribute to maturation in apples and lemons, amino acids in leaves during light deprivation, and reductant (NADH) generated from photorespiration in illuminated leaves and lipid degradation in germinating lipid-containing seeds (discussed later in this chapter). These compounds have different ratios of carbon to oxygen atoms, so the ratio of CO_2 release to O_2 consumption, called the **respiratory quotient (RQ)**, varies with the substrate oxidized. Since RQ can be determined in the field, it is an important parameter in analyses of carbon metabolism on a larger scale. Alcoholic fermentation releases CO_2 without consuming O_2, so a high RQ is also a marker for fermentation.

Environmental factors alter respiration rates

Many environmental factors can alter the operation of metabolic pathways and change respiratory rates. Disadvantageous growth conditions will generally increase respiration to supply energy for protective reactions. However, environmental O_2, temperature, and CO_2 can either stimulate or suppress respiration.

OXYGEN Plant respiration can be limited by the supply of oxygen, the terminal substrate in the overall respiratory process. Respiration rates typically decrease if the atmospheric O_2 concentration is below 5% for whole organs or below 2 to 3% for tissue slices. Many plant species, such as tomato and the dogwood tree *Cornus florida*, can survive only in well-drained, aerated soils, because of the need for a constant supply of oxygen to their roots from the soil. Oxygen diffuses slowly in aqueous solutions, both inside and outside the cells. Compact organs such as seeds and potato tubers have a significant O_2 concentration gradient, decreasing from the surface to the center, which can restrict respiration and therefore also the ATP:ADP ratio. The diffusion limitation is most significant in seeds with a thick seed coat and in plant organs submerged in water. The problem of oxygen supply is particularly important in plants growing in very wet or recurrently flooded soils, where limitation in oxygen supply to the root can be severe (Chapter 19). Such roots must survive on anaerobic (fermentative) metabolism (Figure 11.3) or develop structures that can supply oxygen to the roots.

TEMPERATURE Respiration operates over a wide temperature range. Over short time, it typically increases with temperatures between 0 and 30°C and reaches a plateau at 40 to 50°C. At higher temperatures, it again decreases because of inactivation of the respiratory machinery. The increase in respiration rate for every 10°C increase in temperature is called the **temperature coefficient (Q_{10})**. This ratio varies with plant development and external factors. On a longer timescale, plants acclimate to low temperatures by increasing their respiratory capacity so that ATP production can be maintained.

Low temperatures are used to retard postharvest respiration during the storage of fruits and vegetables, but those temperatures must be adjusted with care. For instance, when potato tubers are stored at temperatures above 10°C, respiration and ancillary metabolic activities are sufficient to allow sprouting. However, below 5°C, stored starch is converted into sucrose (which protects against frost damage), imparting an unwanted sweetness to the tubers. Therefore, potatoes are best stored at 7 to 9°C, which prevents the breakdown of starch while minimizing respiration and germination.

CARBON DIOXIDE It is common in commercial practice to store fruits at low temperatures under 2 to 3% O_2. The reduced temperature lowers the respiration rate, as does the reduced O_2 level. Low levels of oxygen, instead of anoxic conditions, are used in order to avoid fermentative metabolism. Additionally, an artificially elevated CO_2 concentration (3–5%) is used because it has a direct inhibitory effect on respiration.

The atmospheric CO_2 concentration is currently (2023) about 417 ppm, but because of human activities it is projected to reach about 700 ppm before year 2100 (Chapter 9). Plant CO_2 uptake by photosynthesis and release by plant respiration are both much larger processes than the release of CO_2 caused by burning of fossil fuels. Therefore, small effects of elevated CO_2 concentrations on plant respiration will affect future global atmospheric changes.

Within the range relevant for atmospheric changes, there is no direct effect of CO_2 concentration on plant respiration. Over longer time, changes in carbon substrate availability, allocation, and partitioning (Chapter 10), and in gene expression will affect plant respiration, yet this is highly variable among plant species, making global predictions difficult. The CO_2-induced increase in temperature (global warming) is expected to generate an increase in respiration that is, however, smaller than the increase in photosynthesis. There is no present consensus that elevations in atmospheric CO_2 would be attenuated by a feedback inhibition of respiration.

11.7 Lipid Metabolism

Describe the different types of lipids and where and how they are synthesized and metabolized.

Lipids are essential for life because of their roles in membrane structure and function as well as for energy and carbon storage. All plant cell membranes have distinct lipid compositions, such as the inner membrane of mitochondria in which cardiolipin contributes to cristae architecture and interacts with electron transfer chain components. Distinct lipids that include sterols, sphingolipids, glycolipids, and phospholipids are also required for the structure and function of the endomembranes and the plasma membrane (Chapter 1), membrane anchoring, modification, and function of some membrane proteins (Chapter 1), signaling and signal transduction (Chapter 12), membrane transporter activity, and membrane fluidity at different temperatures (Chapters 1 and 19, later this chapter). Other lipids important for plant structure are cutin and waxes, which make up the protective cuticle that reduces water loss from exposed plant tissues, and terpenoids (also known as isoprenoids), which include carotenoids involved in photosynthesis and sterols present in many plant membranes, as well as defense compounds (Chapter 18). Fats and oils are important storage forms of reduced carbon in many seeds, including those of agriculturally important species such as soybean, sunflower, canola, peanut, and cotton. Some fruits, such as olives and avocados, also store fats and oils.

Here, we focus on two types of glycerolipids: the *triacylglycerols*, which make up the fats and oils stored in seeds, and the *polar glycerolipids*, which form the lipid bilayers of cellular membranes (**Figures 11.14** and **11.15**). We also

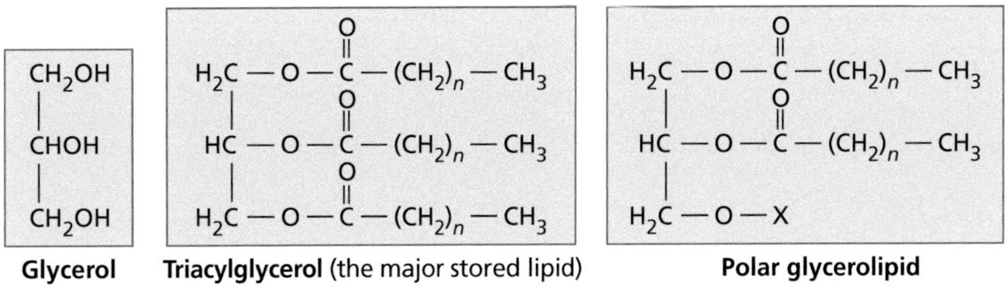

X = H	Diacylglycerol (DAG)
X = HPO_3^-	Phosphatidic acid
X = $PO_3^- — CH_2 — CH_2 — \overset{+}{N}(CH_3)_3$	Phosphatidylcholine
X = $PO_3^- — CH_2 — CH_2 — NH_2$	Phosphatidylethanolamine
X = galactose	Galactolipids

Figure 11.14 Structural features of triacylglycerols and polar glycerolipids in higher plants. The carbon chain lengths of the fatty acids, which always have an even number of carbons, range from 8 to 30 but are typically 16 or 18. Thus, the value of *n* is usually 14 or 16.

triacylglycerols Three fatty acyl groups in ester linkage to the three hydroxyl groups of glycerol. Fats and oils.

Figure 11.15 Major polar glycerolipid classes found in plant membranes: (A) Glyceroglycolipids and a sphingolipid and (B) glycerophospholipids. Two of at least six different fatty acids may be attached to the glycerol backbone. One of the more common molecular species is shown for each glycerolipid class. The numbers given below each name refer to the number of carbons (number before the colon) and the number of double bonds (number after the colon). ▶

examine the complex process by which germinating seeds obtain carbon skeletons and metabolic energy from the oxidation of fats and oils.

Fats and oils store large amounts of energy

Fats and oils belong to the general class termed *lipids*, a structurally diverse group of hydrophobic compounds that are soluble in organic solvents and highly insoluble in water. Lipids represent a more reduced form of carbon than carbohydrates, so the complete oxidation of 1 g of fat or oil (which contains ~40 kJ of energy) can produce considerably more ATP than the oxidation of 1 g of starch (~15.9 kJ). Conversely, the biosynthesis of lipids requires a correspondingly large investment of metabolic energy.

Triacylglycerols are stored in oil bodies

Fats and oils exist mainly in the form of **triacylglycerols** (*acyl* refers to the fatty acid portion), in which fatty acid molecules are linked by ester bonds to the three hydroxyl groups of glycerol (Figure 11.14).

The fatty acids in plants are usually straight-chain carboxylic acids having an even number of carbon atoms. The carbon chains can be as short as 8 units and as long as 30 or more, but most commonly are 16 or 18 carbons long. Fatty acids are often indicated by a numeric abbreviation that contains their number of carbons and number of double bonds, separated by a colon (e.g., 16:0, 18:1). *Oils* are liquid at room temperature, primarily because of the presence of carbon–carbon double bonds (unsaturation) in their component fatty acids; *fats*, which have a higher proportion of saturated fatty acids, are solid at room temperature. The major fatty acids in plant lipids are shown in **Table 11.3**. However, several plant species synthesize unusual, industrially useful fatty acids in their seed oils. **Box 11.2** discusses why plant lipids are a focus of biotechnology.

Table 11.3 Common Fatty Acids in Higher Plant Tissues

Name[a]	Structure
Saturated fatty acids	
Lauric acid (12:0)	$CH_3(CH_2)_{10}CO_2H$
Myristic acid (14:0)	$CH_3(CH_2)_{12}CO_2H$
Palmitic acid (16:0)	$CH_3(CH_2)_{14}CO_2H$
Stearic acid (18:0)	$CH_3(CH_2)_{16}CO_2H$
Unsaturated fatty acids	
Oleic acid (18:1)	$CH_3(CH_2)_7CH{=}CH(CH_2)_7CO_2H$
Linoleic acid (18:2)	$CH_3(CH_2)_4CH{=}CH{-}CH_2{-}CH{=}CH(CH_2)_7CO_2H$
Linolenic acid (18:3)	$CH_3CH_2CH{=}CH{-}CH_2{-}CH{=}CH{-}CH_2{-}CH{=}CH{-}(CH_2)_7CO_2H$

[a] Each fatty acid has a numeric abbreviation. The number before the colon represents the total number of carbons; the number after the colon is the number of double bonds.

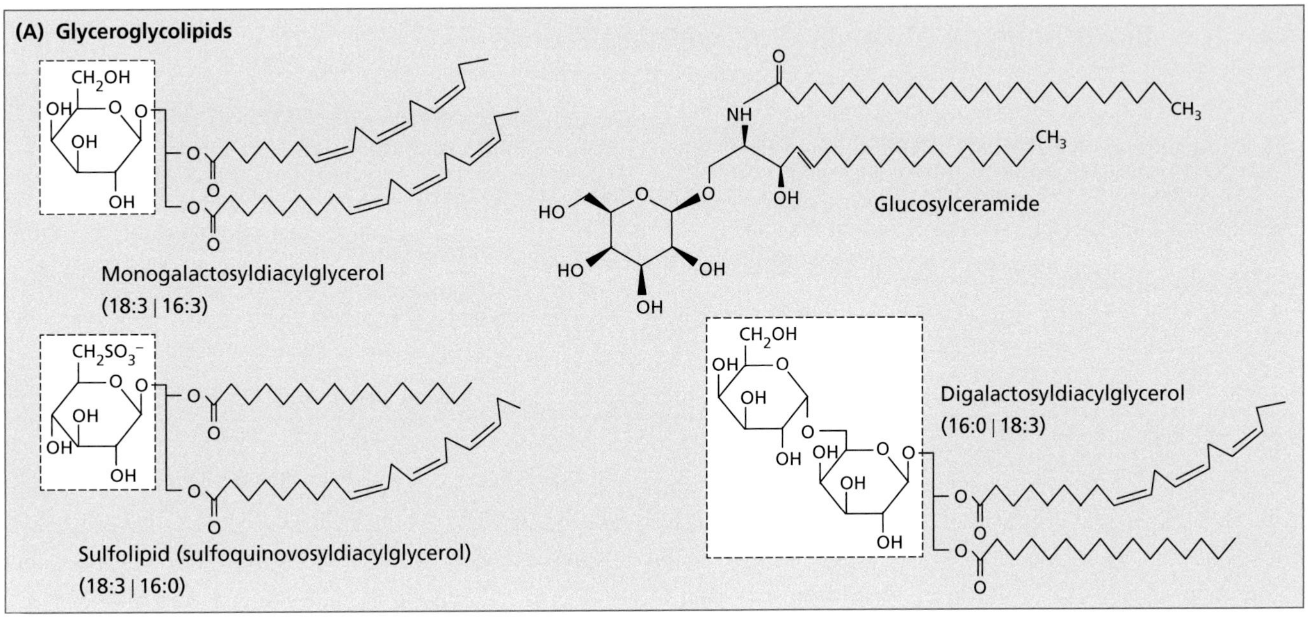
(A) Glyceroglycolipids
Monogalactosyldiacylglycerol
(18:3 | 16:3)
Glucosylceramide
Sulfolipid (sulfoquinovosyldiacylglycerol)
(18:3 | 16:0)
Digalactosyldiacylglycerol
(16:0 | 18:3)

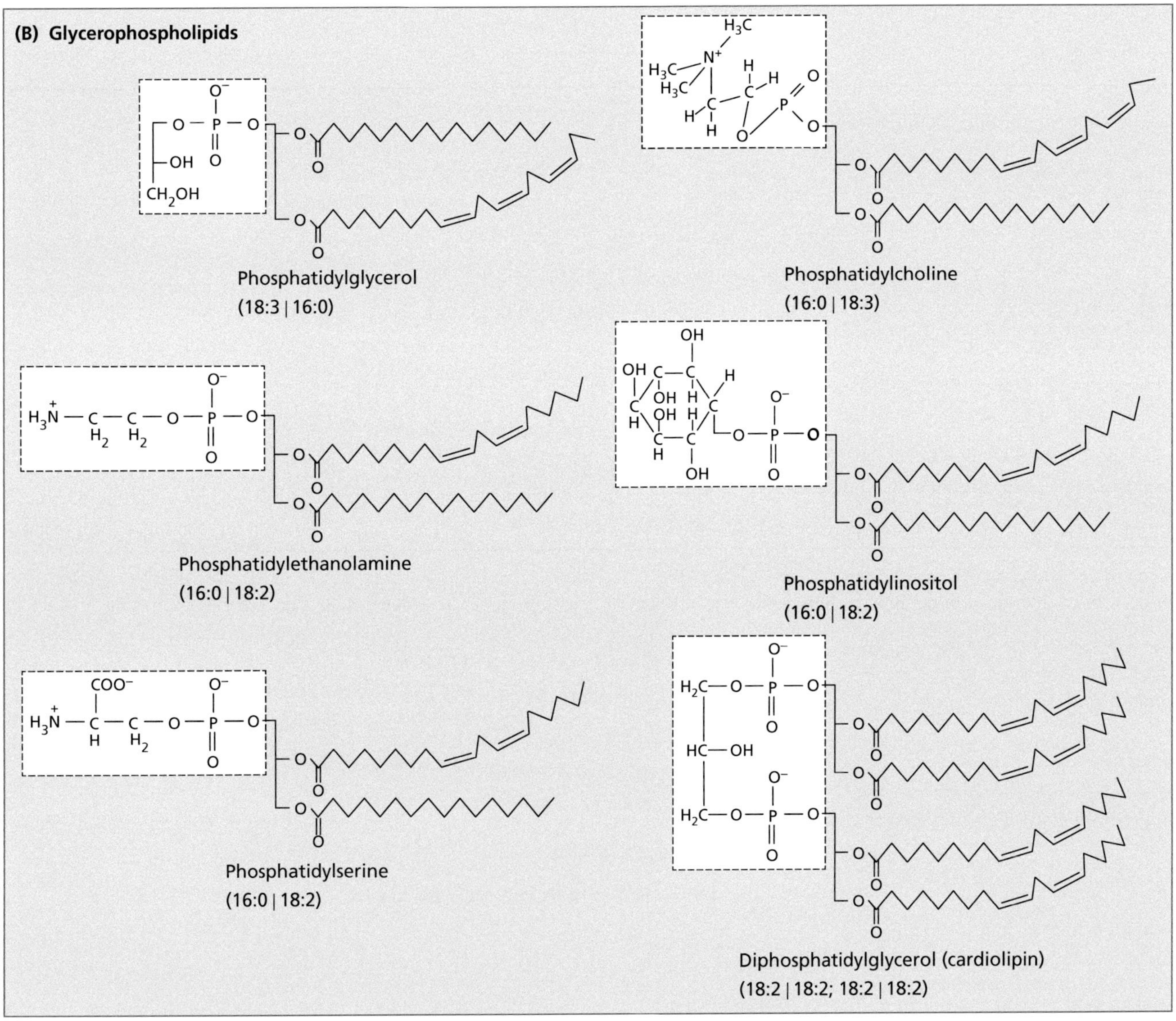
(B) Glycerophospholipids
Phosphatidylglycerol
(18:3 | 16:0)
Phosphatidylcholine
(16:0 | 18:3)
Phosphatidylethanolamine
(16:0 | 18:2)
Phosphatidylinositol
(16:0 | 18:2)
Phosphatidylserine
(16:0 | 18:2)
Diphosphatidylglycerol (cardiolipin)
(18:2 | 18:2; 18:2 | 18:2)

Box 11.2 Biotechnology of Lipids in a Changing World

The glycerolipids in cellular membranes are integral to temperature responses of plants. As the climate transition takes hold, our knowledge of how the biochemistry and membrane physiology of lipids affect the temperature responses of photosynthesis, respiration, and other processes will be used to strengthen the adaptability of crops and other plants to the altered environments in which they will be grown.

Plant lipids, especially vegetable oils produced by soybean, canola, oil palm, and other crops, have been a focus of biotechnology for several decades. The most successful use of biotechnology has been the production of high-oleic soybean (Qualisoy) and canola varieties that have allowed for the elimination of unhealthy trans fats from food. Researchers are also exploring possibilities for producing oil in leaves of crop plants to provide for energy-dense biofuels. Finally, transferring the pathways for synthesis of unusual fatty acids found in some wild plant species to crop plants will provide chemical feedstocks for the synthesis of plastics, resins, and other products for society.

The proportions of fatty acids in plant oils vary with the plant species. For example, peanut oil is about 9% palmitic acid, 59% oleic acid, 21% linoleic acid, and 11% minor components while soybean oil is 13% palmitic acid, 7% oleic acid, 51% linoleic acid, 23% linolenic acid, and 6% minor components. In most seeds, triacylglycerols are stored in the cytoplasm of either cotyledon or endosperm cells in organelles known as **oil bodies** (Chapter 1). The oil body membrane is a single layer of phospholipids (i.e., a half-bilayer) with the hydrophilic ends of the phospholipids exposed to the cytosol and the hydrophobic acyl hydrocarbon chains facing the triacylglycerol interior. The oil body is stabilized by the presence of specific proteins, called oleosins, that coat its outer surface and prevent the phospholipids of adjacent oil bodies from coming in contact and fusing.

The unique membrane structure of oil bodies results from the pattern of triacylglycerol biosynthesis. Triacylglycerol synthesis is completed by enzymes located in the membranes of the endoplasmic reticulum (ER), and the resulting fats accumulate between the two monolayers of the ER membrane bilayer. The bilayer swells apart as more fats are added to the growing structure, while the oleosins are synthesized by ribosomes attached to the ER, and ultimately a mature oil body coated with oleosins buds off from the ER.

oil body Organelle that accumulates and stores triacylglycerols. It is bounded by a single phospholipid leaflet ("half-unit membrane" or "phospholipid monolayer") derived from the endoplasmic reticulum.

polar glycerolipids The main structural lipids in membranes, in which the hydrophobic portion consists of two 16-carbon or 18-carbon fatty acid chains esterified to positions 1 and 2 of a glycerol.

glyceroglycolipids Glycerolipids in which sugars form the polar head group. Glyceroglycolipids are the most abundant glycerolipids in chloroplast membranes.

glycerophospholipids Polar glycerolipids in which the hydrophobic portion consists of two 16-carbon or 18-carbon fatty acid chains esterified to positions 1 and 2 of a glycerol backbone. The phosphate-containing polar head group is attached to position 3 of the glycerol.

Polar glycerolipids are the main structural lipids in membranes

As described in Chapter 1, each membrane in the cell is a bilayer of *amphipathic* (i.e., having both hydrophilic and hydrophobic regions) lipid molecules in which a polar head group interacts with the aqueous environment while hydrophobic fatty acid chains form the core of the membrane. This hydrophobic core prevents unregulated diffusion of solutes between cell compartments and thereby allows the metabolism of the cell to be organized.

The main structural lipids in membranes are the **polar glycerolipids** (Figures 11.14 and 11.15), in which the hydrophobic portion consists of two 16-carbon or 18-carbon fatty acid chains esterified to positions 1 and 2 of a glycerol backbone. The polar head group is attached to position 3 of the glycerol. There are two categories of polar glycerolipids:

1. **Glyceroglycolipids**, in which sugars form the head group (Figure 11.15A)
2. **Glycerophospholipids**, in which the head group contains phosphate (Figure 11.15B)

Plant membranes have additional structural lipids, including sphingolipids and sterols, which are structural and functional components of the plasma membrane (Chapter 1). Other lipids perform specific roles in photosynthesis and

Table 11.4 Glycerolipid Components of Cellular Membranes

Lipid	Lipid composition (mole percentages of total)		
	Chloroplast	Endoplasmic reticulum	Mitochondrion
Phosphatidylcholine	4	47	43
Phosphatidylethanolamine	—	34	35
Phosphatidylinositol	1	17	6
Phosphatidylglycerol	7	2	3
Diphosphatidylglycerol (cardiolipin)	—	—	13
Monogalactosyldiacylglycerol	55	—	—
Digalactosyldiacylglycerol	24	—	—
Sulfolipid	8	—	—

other processes. Included among these lipids are chlorophylls, plastoquinone, ubiquinone, carotenoids, and tocopherols, which together account for about one-third of the lipids in plant leaves.

Figure 11.15 shows the nine major polar glycerolipid classes in plants, each of which can be associated with many different fatty acid combinations. The structures shown illustrate some of the more common molecular species.

Chloroplast membranes, which account for 70% of the membrane lipids in photosynthetic tissues, are dominated by glyceroglycolipids; other membranes of the cell contain glycerophospholipids (**Table 11.4**). In nonphotosynthetic tissues, glycerophospholipids are the major membrane glycerolipids. Interestingly, under phosphate starvation, many plant species partially replace glycerophospholipids with glyceroglycolipids, to allow some recovery of phosphate for the synthesis of RNA and other phosphorus-rich compounds.

Glycerolipids are synthesized in the plastids and the ER

In plants, fatty acids (16:0 and 18:1) are synthesized primarily in the plastids, while in animals they are synthesized primarily in the cytosol. In the plastids, the fatty acids can be incorporated into phosphatidic acid, a phosphoglycerol molecule with two acyl chains, which is a precursor of glyceroglycolipids. They can also be exported to the ER, where they are linked with glycerol to form triacylglycerol or glycerophospholipids. Fatty acids may undergo further modification after they are linked with glycerol. Additional double bonds are placed in the 16:0 and 18:1 fatty acids by a series of desaturase isozymes. *Desaturase isozymes* are integral membrane proteins found in the chloroplast and the ER. Each desaturase inserts a double bond at a specific position in the fatty acid chain, and the enzymes act sequentially to produce the final 18:3 and 16:3 products.

The first steps of glycerolipid synthesis attach two fatty acid molecules to glycerol-3-phosphate to form **phosphatidic acid**. The action of a specific phosphatase produces **diacylglycerol** (**DAG**) from phosphatidic acid. Phosphatidic acid can also be converted directly into phosphatidylinositol or phosphatidylglycerol; DAG can give rise to phosphatidylethanolamine or phosphatidylcholine (Chapter 12).

Lipid composition influences membrane function

A central question in membrane biology is the functional reason behind lipid diversity. Each membrane system of the cell has a characteristic and distinct complement of lipid types, and within a single membrane, each class of lipids has a distinct fatty acid composition (Table 11.4). Why is lipid diversity needed?

phosphatidic acid (PA) A diacylglycerol that has a phosphate on the third carbon of the glycerol backbone.

diacylglycerol (DAG) A molecule consisting of the three-carbon glycerol molecule to which two fatty acids are covalently attached by ester linkages.

One aspect of membrane biology that might offer answers to this central question is the relationship between lipid composition and the ability of organisms to adjust to temperature changes. For example, chill-sensitive plants experience sharp reductions in growth rate and development at temperatures between 0 and 12°C (Chapter 19). Many economically important crops, such as cotton, soybean, maize, rice, and many tropical and subtropical fruits, are classified as chill-sensitive. In contrast, most plants that originate from temperate regions can grow and develop at chilling temperatures and are classified as chill-resistant plants.

It has been suggested that, because of the decrease in lipid fluidity at lower temperatures, the primary event of chilling injury is *a transition from a liquid-crystalline phase to a gel phase* in cellular membranes. According to this hypothesis, such a transition results in alterations in the metabolism of chilled cells and would lead to injury and death of the chill-sensitive plants. Several lines of evidence indicate that high levels of saturated fatty acids in phosphatidylglycerol are specifically associated with chilling sensitivity.

Other research, however, suggests that the relationship between membrane unsaturation and plant responses to temperature is more subtle and complex. The responses of Arabidopsis mutants with increased saturation of fatty acids to low temperatures are not what is predicted by the chill-sensitivity hypothesis, suggesting that normal chilling injury may not be strictly related to the level of unsaturation of membrane lipids.

Membrane lipids are precursors of important signaling compounds

Plants, animals, and microbes all use membrane lipids as precursors for compounds that are used for intracellular or long-range signaling. For example, the hormone jasmonate—derived from linolenic acid (18:3)—activates plant defenses against insects and many fungal pathogens (Chapter 18). In addition, jasmonates regulate other aspects of plant growth, including flower and seed development (Chapters 12, 16, and 17).

Phosphatidylinositol 4,5-bisphosphate (**PIP_2**) is the most important of several phosphorylated derivatives of phosphatidylinositol known as *phosphoinositides*. Receptor-mediated activation of phospholipase C leads to the hydrolysis of PIP_2 into inositol trisphosphate ($InsP_3$) and diacylglycerol, both of which act as intracellular secondary messengers (Chapter 12).

The action of $InsP_3$ in releasing Ca^{2+} into the cytoplasm (through Ca^{2+}-sensitive channels in the tonoplast and other membranes), and thereby regulating cellular processes, has been demonstrated in several plant systems, including the stomatal guard cells. Information about other types of lipid signaling in plants is becoming available through biochemical and molecular genetic studies of phospholipases and other enzymes involved in the generation of these signals.

Storage lipids are converted into carbohydrates in germinating seeds

During germination (Chapter 14), embryos of oil-containing seeds metabolize stored triacylglycerols by converting them into sucrose. Plants are not able to transport fats from the cotyledons to other tissues of the germinating seedling, so they must convert stored lipids into a more mobile form of carbon, generally sucrose. This process involves several steps that are located in different cellular compartments: oil bodies, glyoxysomes, mitochondria, and the cytosol.

phosphatidylinositol 4,5-bisphosphate (PIP_2) A group of phosphorylated derivatives of phosphatidylinositol.

OVERVIEW: LIPIDS TO SUCROSE In oilseeds, the conversion of lipids into sucrose is triggered by germination. It begins with the hydrolysis of triacylglycerols

(A)

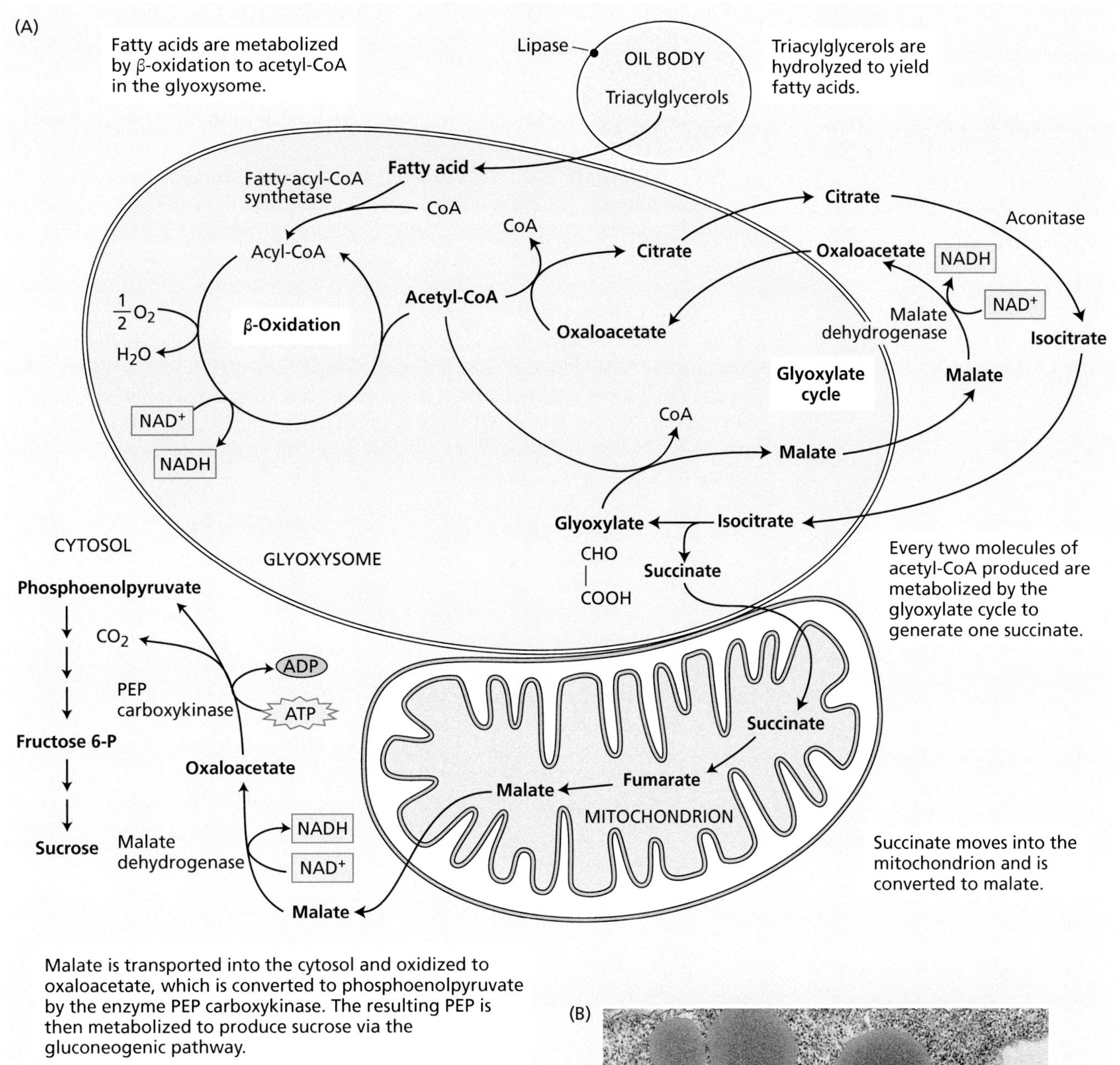

(B)

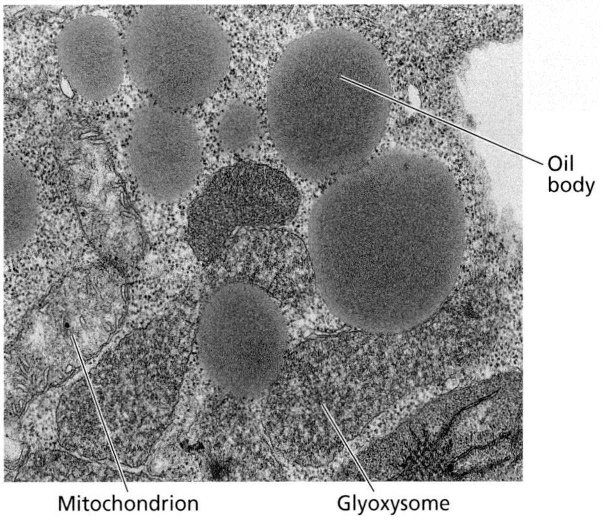

Figure 11.16 Conversion of fats into sugars during germination in oil-storing seeds. (A) Carbon flow during fatty acid breakdown and gluconeogenesis (Figures 11.3 and 11.6 show chemical structures). (B) Electron micrograph of a cell from the oil-storing cotyledon of a cucumber seedling, showing glyoxysomes, mitochondria, and oil bodies.

stored in oil bodies into free fatty acids, followed by oxidation of those fatty acids in glyoxysomes to produce acetyl-CoA (**Figure 11.16**). In most oilseeds, approximately 30% of the acetyl-CoA is used for energy production via respiration, and the rest is converted into sucrose through the glyoxylate cycle and gluconeogenesis.

β-oxidation Oxidation of fatty acids into fatty acyl-CoA, and the sequential breakdown of the fatty acids into acetyl-CoA units. NADH is also produced.

glyoxylate cycle The sequence of reactions that convert two molecules of acetyl-CoA to succinate in the glyoxysome.

LIPASE-MEDIATED HYDROLYSIS The initial step in the conversion of lipids into carbohydrates is the breakdown of triacylglycerols stored in oil bodies by a lipase, which hydrolyzes triacylglycerols into three fatty acid molecules and one molecule of glycerol. During the breakdown of lipids, oil bodies and glyoxysomes are generally in close physical association (Figure 11.16B).

Β–OXIDATION OF FATTY ACIDS The fatty acid molecules enter the glyoxysome, where they are activated by conversion into fatty-acyl-CoA by *fatty-acyl-CoA synthetase.* Fatty-acyl-CoA is the initial substrate for the **β-oxidation** series of reactions, in which C_n fatty acids (fatty acids composed of *n* carbons) are sequentially broken down into *n*/2 molecules of acetyl-CoA (Figure 11.16A). This reaction sequence involves the reduction of ½ O_2 to H_2O and the formation of one NADH for each β-oxidation cycle. In plant seed storage tissues, the four enzymes associated with β-oxidation are located exclusively in the glyoxysome or the equivalent organelle in vegetative tissues, the peroxisome (Chapter 1). β-oxidation is also important for the conversion of indole-3-butric acid, a storage form of auxin used in induce rooting in cuttings, into active indole-3-acetic acid (auxin) (Chapter 12).

THE GLYOXYLATE CYCLE The function of the **glyoxylate cycle** is to convert two molecules of acetyl-CoA into succinate. The acetyl-CoA produced by β-oxidation is metabolized in the glyoxysome through a series of reactions that make up the glyoxylate cycle (Figure 11.16A). Initially, the acetyl-CoA reacts with oxaloacetate to give citrate, which is then transferred to the cytoplasm for isomerization to isocitrate by aconitase. Isocitrate is reimported into the glyoxysome and converted into malate by two reactions that are unique to the glyoxylate cycle:

1. First, isocitrate (C_6) is cleaved by the enzyme isocitrate lyase to give succinate (C_4) and glyoxylate (C_2). The succinate is exported to the mitochondria.
2. Next, malate synthase combines a second molecule of acetyl-CoA with glyoxylate to produce malate (C_4).

Malate is then transferred to the cytoplasm and converted into oxaloacetate by the cytoplasmic isozyme of malate dehydrogenase. Oxaloacetate is reimported into the glyoxysome and combines with another acetyl-CoA to continue the cycle (Figure 11.16A). The glyoxylate produced keeps the cycle operating, but the succinate is exported to the mitochondria for further processing.

THE MITOCHONDRIAL ROLE Moving from the glyoxysomes to the mitochondria, the succinate is converted into malate by the two corresponding TCA cycle reactions. The resulting malate can be exported from the mitochondria in exchange for succinate via the dicarboxylate transporter located in the inner mitochondrial membrane (Figure 11.9). Malate is then oxidized to oxaloacetate by malate dehydrogenase in the cytosol, and the resulting oxaloacetate is converted into carbohydrates by the reversal of glycolysis (gluconeogenesis). This conversion requires circumventing the irreversibility of the PEP carboxylase reaction (Figure 11.3) and is facilitated by PEP carboxykinase, which uses ATP as a phosphate group donor to convert oxaloacetate into PEP and CO_2 (Figure 11.16A).

From PEP, gluconeogenesis can proceed to the production of glucose. Sucrose is the final product of this process and is the primary form of reduced carbon translocated from the endosperm or cotyledons to the growing seedling tissues. Not all seeds quantitatively convert fat into sugar.

Summary

Using the building blocks provided by photosynthesis, respiration releases the energy stored in carbon compounds in a controlled manner for cellular use. At the same time, it generates many carbon precursors for biosynthesis.

11.1 Overview of Plant Respiration

- In plant respiration, reduced cellular carbon generated by photosynthesis is oxidized to CO_2 and water, and this oxidation is coupled to the synthesis of ATP.
- Respiration takes place via four main processes: glycolysis, the oxidative pentose phosphate pathway, the TCA cycle, and oxidative phosphorylation (the electron transport chain and ATP synthesis) (**Figure 11.1**).

11.2 Glycolysis

- In glycolysis, carbohydrates are converted into pyruvate in the cytosol, and a small amount of ATP is synthesized via substrate-level phosphorylation. NADH is also produced (**Figure 11.3**).
- Plant glycolysis has alternative enzymes for several steps. These allow differences in substrates used, products made, and the direction of the pathway.
- When insufficient O_2 is available, fermentation regenerates NAD^+ for glycolysis. Only a minor fraction of the energy available in sugars is conserved by fermentation (**Figure 11.3**).
- Plant glycolysis is regulated from the bottom up by its products (**Figure 11.12**).

11.3 The Oxidative Pentose Phosphate Pathway

- Carbohydrates can be oxidized via the oxidative pentose phosphate pathway, which provides building blocks for biosynthesis and reducing power in the form of NADPH (**Figure 11.4**).

11.4 The Tricarboxylic Acid Cycle

- In the mitochondrial matrix, pyruvate is oxidized into CO_2 via the TCA cycle, generating a large number of reducing equivalents in the form of NADH and $FADH_2$ (**Figures 11.5**, **11.6**).
- In plants, the TCA cycle is involved in alternative pathways that allow oxidation of malate or citrate and export of intermediates for biosynthesis (**Figures 11.6**, **11.7**).

11.5 Oxidative Phosphorylation

- Electron transport from NADH and $FADH_2$ to oxygen is coupled by enzyme complexes to proton transport across the inner mitochondrial membrane. This generates an electrochemical proton gradient used for powering synthesis and export of ATP (**Figures 11.8–11.10**).
- During aerobic respiration, up to 60 molecules of ATP are produced per molecule of sucrose (**Tables 11.1** and **11.2**).
- Typical for plant respiration is the presence of several proteins (alternative oxidase, NAD[P]H dehydrogenases, and uncoupling protein) that lower the energy recovery but also give increased metabolic flexibility (**Figure 11.8**).
- The main products of the respiratory process are ATP and metabolic intermediates used in biosynthesis. The cellular demand for these compounds regulates respiration via control points in the electron transport chain, the TCA cycle, and glycolysis (**Figures 11.10–11.13**).

11.6 Respiration in Intact Plants and Tissues

- More than 50% of the daily photosynthetically fixed carbon may be respired by a plant.
- Many factors can affect the respiration rate observed at the whole-plant level. These factors include the nature and age of the plant tissue and environmental factors such as light, temperature, nutrient and water supply, and O_2 and CO_2 concentrations.

11.7 Lipid Metabolism

- Fatty acids are synthesized in plastids. Fatty acids from the plastids can be transported to the ER, where they are further modified.
- Triacylglycerols (fats and oils) are an efficient form for storage of reduced carbon, particularly in seeds. Triacylglycerols are synthesized in the ER and accumulate within the phospholipid bilayer, forming oil bodies.
- Polar glycerolipids, a primary structural components of membranes (**Figures 11.14**, **11.15**; **Tables 11.3**, **11.4**), are synthesized in the ER and in the plastids.
- The function of a membrane may be influenced by its lipid composition. The degree of unsaturation of the fatty acids influences the sensitivity of plants to chilling temperatures, but does not seem to be involved in typical chilling injury.
- Some lipid derivatives, such as jasmonates, are important plant hormones.
- During germination in oil-storing seeds, the stored lipids are metabolized to carbohydrates in a series of reactions that include the glyoxylate cycle. The glyoxylate cycle takes place in glyoxysomes and the cytosol, and subsequent steps occur in the mitochondria (**Figure 11.16**).
- Acetyl-CoA generated during lipid breakdown in the glyoxysomes is used for respiration or converted into carbohydrates in the cytosol by gluconeogenesis (**Figure 11.16**).

Suggested Reading

Amthor, J. S., Bar-Even, A., Hanson, A. D., Millar, A. H., Stitt, M., Sweetlove, L. J., and Tyerman, S. D. (2019) Engineering strategies to boost crop productivity by cutting respiratory carbon loss. *Plant Cell* 31: 297–314.

Atkin, O. K., Meir, P., and Turnbull, M. H. (2014) Improving representation of leaf respiration in large-scale predictive climate-vegetation models. *New Phytol.* 202: 743–748.

Dusenge, M. E., Duarte, A. G., and Way, D. A. (2019) Plant carbon metabolism and climate change: Elevated CO_2 and temperature impacts on photosynthesis, photorespiration and respiration. *New Phytol.* 221: 32–49.

Holzl, G., and Doermann, P. (2019) Chloroplast lipids and their biosynthesis. *Annu. Rev. Plant Biol.* 70: 51–81.

Møller, I. M., Igamberdiev, A. U., Bykova, N. V., Finkemeier, I., Rasmusson, A. G., and Schwarzländer, M. (2020) Matrix redox physiology governs the regulation of plant mitochondrial metabolism through post-translational protein modifications. *Plant Cell* 32: 573–594.

Møller, I. M., Rasmusson, A. G., and Van Aken, O. (2021) Plant mitochondria: Past, present and future. *Plant J.* 108: 912–959.

Nakamura, Y. (2017) Plant phospholipid diversity: Emerging functions in metabolism and protein–lipid interactions. *Trends Plant Sci.* 22: 1027–1040.

Rasmusson, A. G., Geisler, D. A., and Møller, I. M. (2008) The multiplicity of dehydrogenases in the electron transport chain of plant mitochondria. *Mitochondrion* 8: 47–60.

Sweetlove, L. J., Beard, K. F. M., Nunes-Nesi, A., Fernie, A. R., and Ratcliffe, R. G. (2010) Not just a circle: Flux modes in the plant TCA cycle. *Trends Plant Sci.* 15: 462–470.

Vanlerberghe, G. C. (2013) Alternative oxidase: A mitochondrial respiratory pathway to maintain metabolic and signaling homeostasis during abiotic and biotic stress in plants. *Int. J. Mol. Sci.* 14: 6805–6847.

12 Signals and Signal Transduction

As sessile organisms, plants constantly adjust their responses to the surrounding environment to optimize growth and ensure survival. To facilitate such adjustments, plants have evolved sophisticated sensory systems to optimize water and nutrient usage; to monitor light quantity, quality, and directionality; and to defend themselves from biotic and abiotic threats. Some of the earliest documented experiments analyzing the responses of seedlings to light and gravity were conducted by Charles and Francis Darwin. They observed that a unidirectional light source was perceived at the coleoptile tip of grasses, yet the bending response took place farther back along the shoot tissue. This led the Darwins to conclude that there must be a mobile signal that transferred information from the site of perception to the region of bending. The mobile signal was later identified as the growth hormone auxin, indole-3-acetic acid.

In general, an environmental input that initiates one or more plant responses is referred to as a **signal**, and the physical component that biochemically responds to that signal is designated a receptor. Receptors are either proteins or, in the case of light receptors, proteins associated with light-absorbing pigments. Once receptors sense their specific signal, they must *transduce* the signal (i.e., convert it from one form to another) in order to amplify the signal and trigger the cellular response. Receptors often do this by modifying the activity of other proteins or by employing intracellular signaling molecules called **second messengers**; these molecules then alter cellular processes such as gene transcription. Hence, all signal transduction pathways typically involve the following chain of events:

$$\text{Signal} \rightarrow \text{receptor} \rightarrow \text{signal transduction} \rightarrow \text{response}$$

signal An environmental input that initiates one or more plant responses.

second messenger A small intracellular molecule (e.g., cyclic AMP, cyclic GMP, calcium, IP_3, or diacylglycerol) whose concentration increases or decreases in response to the activation of a receptor by an external signal, such as hormones or light. It then diffuses intracellularly to the target enzymes or intracellular receptor to produce and amplify the physiological response.

secondary signal A signaling component that is produced or released by activation of a primary signaling pathway. Often a hormone that acts on sites distal to the site of activation.

signal transduction pathway A sequence of biochemical processes by which an extracellular signal (typically light or a hormone) interacts with a receptor, causing a change in the level of a second messenger and ultimately a change in cell function.

In many cases the initial response produces or mobilizes **secondary signals** that are often small molecules such as Ca^{2+} or hormones that then activate "downstream" physiological responses. Most of the specific events and intermediate steps involved in plant signal transduction have been characterized and described collectively as **signal transduction pathways**.

We begin this chapter by providing a brief overview of the types of external cues that direct plant growth. Next we discuss how plants employ signal transduction pathways to regulate gene expression and posttranslational responses. Signals are often amplified via second messenger or protein modification cascades and may be mobilized to coordinate responses throughout the plant. Finally, we examine how individual stimulus-response cascades are often integrated with other signaling pathways, termed *cross-regulation*, to shape plant responses to their environment in time and space.

12.1 Temporal and Spatial Aspects of Signaling

Differentiate spatial and temporal responses in terms of fast and slow signaling mechanisms.

Plant signal transduction mechanisms may be relatively rapid or extremely slow (**Figure 12.1**). When some carnivorous plants, most notably Venus flytrap (*Dionaea muscipula*), catch insects, they use modified leaf traps that close within milliseconds after touch stimulation. Similarly, the sensitive plant (*Mimosa pudica*) folds its leaflets rapidly upon being touched. Young seedlings reorient themselves with respect to gravity minutes after being placed horizontally. Such rapid response mechanisms generally involve initial electrochemical responses to transduce signals. In contrast, plants attacked by insect herbivores may emit volatiles to attract insect predators within a few hours. Processes occurring on this timescale often involve new gene transcription and translation activity and often involve transduction of signals from one organ to another.

Longer-term environmental responses modify developmental programs to shape plant architecture over the entire life of the plant. Examples of long-term responses include modulation of root branching in response to nutrient availability, growth of sun or shade leaves to adjust to light conditions, and activation of lateral bud outgrowth when the shoot apex is damaged by grazing animals. Long-term plant responses can operate over timescales of months or years and often involve signal transduction at the organismal level.

12.2 Signal Perception and Amplification

Describe how different signals are perceived and amplified, and how signaling mechanisms are conserved across kingdoms.

Although highly varied in nature and makeup, all signal transduction pathways share common features: An initial stimulus is perceived by a receptor and transmitted via intermediate processes to sites where physiological responses are initiated (**Figure 12.2**). The stimulus may derive from developmental programming or from the external environment. When the response mechanism reaches an optimal point, feedback mechanisms attenuate the processes and reset the sensor mechanism. Receptors can be located at the plasma membrane, cytosol, endomembrane system, or nucleus, as exemplified by hormone and touch receptors (**Figure 12.3**).

Figure 12.1 Timing of plant responses to the environment ranges from very rapid to extremely slow. (A) Insect movements on modified leaves of a Venus flytrap (*Dionaea muscipula*) activate trigger hairs, inducing rapid closure of the leaf lobes. (B) The leaves of sundew plants (*Drosera anglica*) capture insects in a sticky fluid produced by stalked glands called tentacles, and then roll up to secure the prey and begin digestion. (C) A hawthorn tree (*Crataegus* spp.) subjected to prevailing onshore winds responds slowly by growing away from the wind. (D) Tree trunks and branches can respond slowly to mechanical stress by producing reaction wood. In this case the tree is an angiosperm, which produces *tension wood* on the adaxial surface. Gymnosperms produce *compression wood* on the abaxial surface. (E) Cross section through a gymnosperm tree branch with compression wood, creating an asymmetrical ring structure.

Protein kinases are major signal transduction components

One of the most conserved mechanisms in all organisms is modification of proteins, lipids, or nucleic acids by **kinase** enzymes that add a phosphate group from ATP to modify the properties of those molecules. Protein kinases transduce and amplify signals by phosphorylating serine, threonine, tyrosine, or histidine residues of target proteins to alter their biological activity or association with other proteins and lipids (**Figure 12.4A**). When a protein functions as a receptor and also transduces that signal by phosphorylating another molecule, it is called a **receptor kinase**.

Receptor serine/threonine kinases (RKs; historically known as "receptor-like kinases") are a prominent family of kinases that phosphorylate serine or, less commonly, threonine protein residues. The name "receptor-like" was given to these proteins before their function as receptors was clearly established, and the name has persisted. One large class of RKs comprises plasma membrane–spanning

kinases Enzymes that have the capacity to transfer phosphate groups from ATP to other molecules.

receptor kinase A protein in a signaling pathway that detects the presence of a ligand, such as a hormone, by phosphorylating itself or another protein.

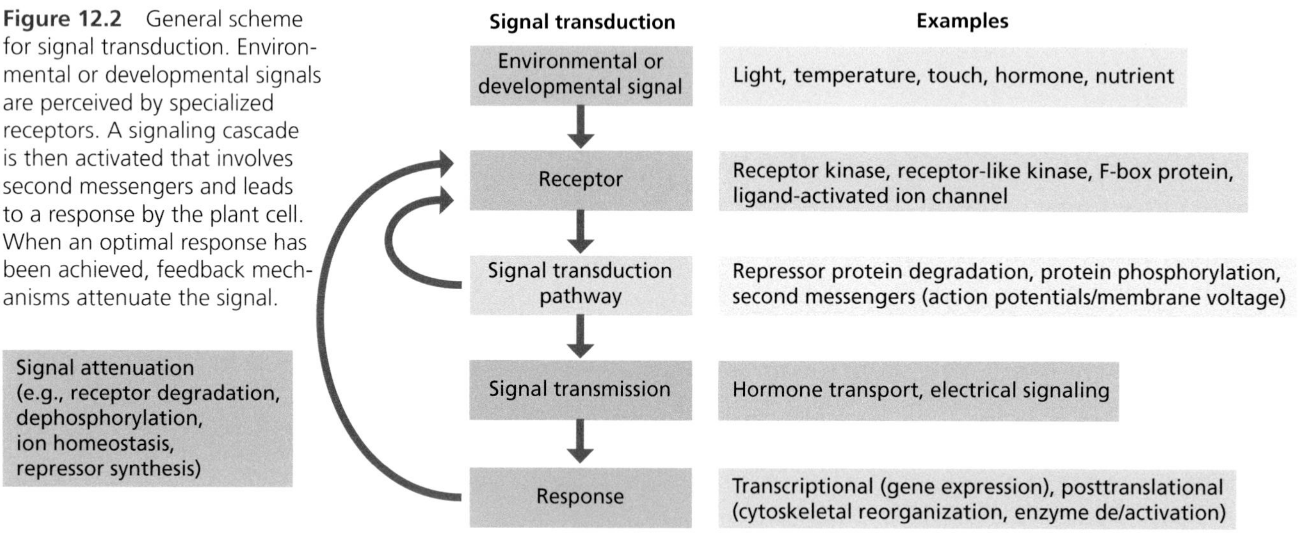

Figure 12.2 General scheme for signal transduction. Environmental or developmental signals are perceived by specialized receptors. A signaling cascade is then activated that involves second messengers and leads to a response by the plant cell. When an optimal response has been achieved, feedback mechanisms attenuate the signal.

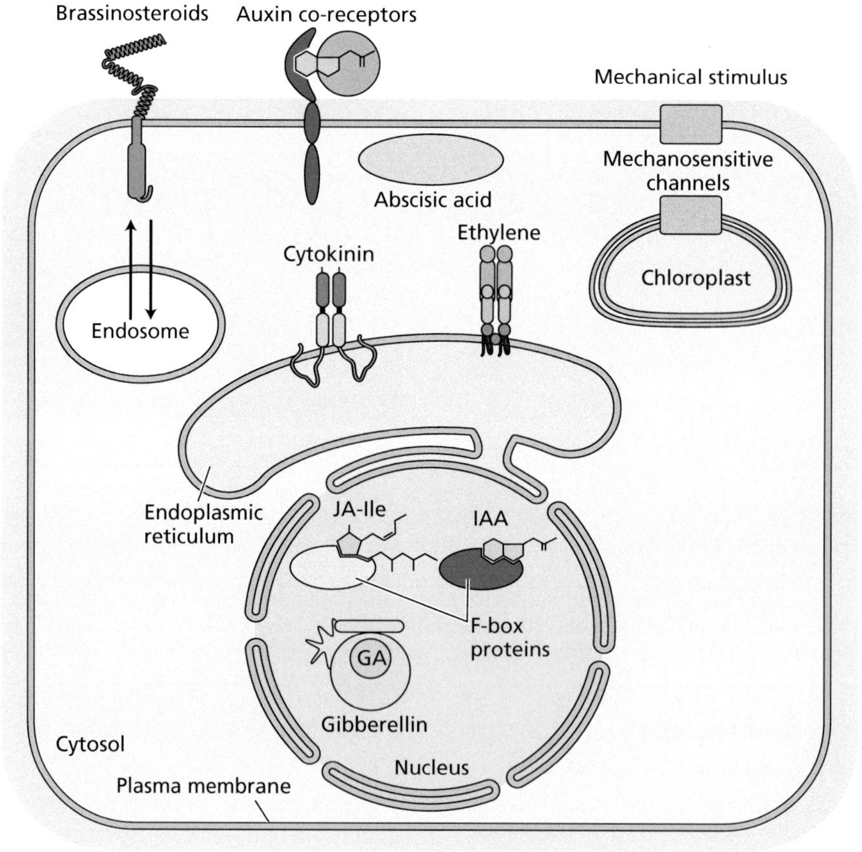

Figure 12.3 Primary locations of plant receptors and mechanosensitive receptors in the cell. The individual receptors are discussed individually below. GA, gibberellin; IAA, indole-3-acetic acid; JA-Ile, jasmonic acid-isoleucine.

receptors with extracellular leucine rich repeat (LRR) ligand-binding domains and an intracellular serine/threonine kinase domain for signal transmission (**Figure 12.4B**). Ligand binding to an RK receptor results in dimerization with co-receptors to initiate phosphorylation events in the intracellular kinase domains of both RKs. These, in turn, initiate multiple phosphorylation-dependent signaling processes. An example is the FLAGELLIN SENSITIVE 2 receptor that binds to the flg22 peptide of pathogenic bacteria and interacts with BRI1 ASSOCIATED KINASE (BAK1)/SOMATIC EMBRYOGENESIS RECEPTOR KINASE (SERK)

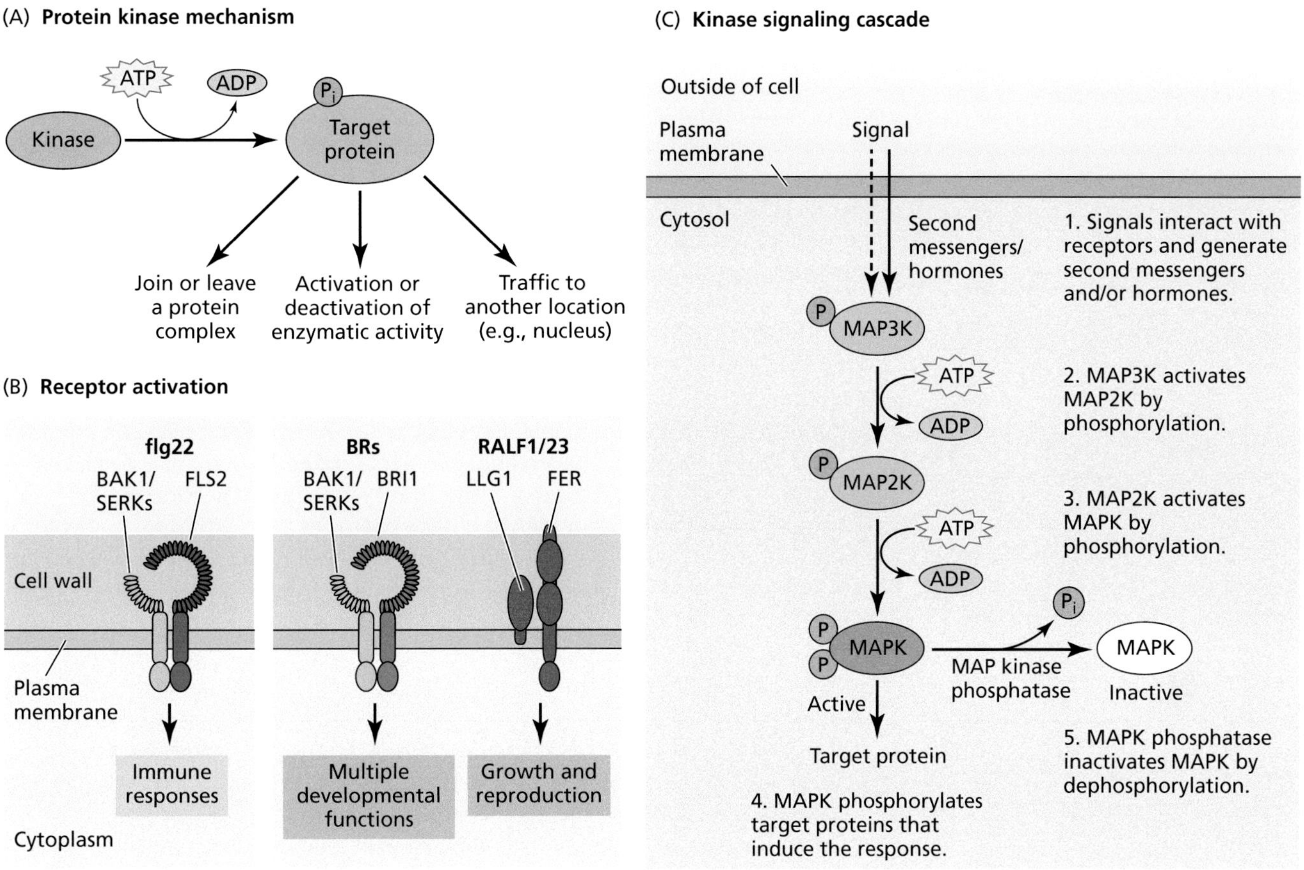

Figure 12.4 Protein kinases phosphorylate amino acid residues to modify protein activity, location, and/or complex formation. (A) Kinases activate transfer of a phosphate group from ATP to a serine, threonine, tyrosine, or less commonly, histidine residue. (B) Receptor-like serine/threonine kinases (RKs) bind to extracellular ligands and heteromerize with co-receptors to initiate cellular signal transduction. Ligand binding to a receptor results in dimerization and protein phosphorylation to initiate defense, hormone signaling, or other developmental responses. Some receptor components have multiple names when they function in more than one response depending on their co-receptor protein. Ligands can be derived from pathogens such as the flg22 elicitor peptide from flagellin, plant signaling peptides (e.g., RAPID ALKALINIZATION FACTORS [RALFs]), or hormones (brassinosteroids [BRs]). BAK1, BRASSINOSTEROID ACTIVATED KINASE 1; BRI1, BRASSINOSTEROID INSENSITIVE 1; SERK, SOMATIC EMBRYOGENESIS RECEPTOR KINASE; FER, FERONIA; FLS2, Flagellin Sensitive 2; LLG1, LORELEI 1. (C) Mitogen-activated protein kinase (MAPK) pathways amplify signals to achieve a rapid and massive response to an environmental or developmental stimulus. The solid and dotted arrows leading from the signal indicate direct and indirect activation, respectively.

co-receptors to initiate immune responses (Figure 12.4B). These responses are discussed in Chapter 19.

BAK1 co-receptors also function in intercellular communication that regulates plant growth and development. They heteromerize with the BRASSINOSTEROID INSENSITIVE 1 (BRI1) receptor when it is bound to brassinosteroid hormones to regulate multiple gene transcription events (Figure 12.4B). In many of the LRR RK-dependent signaling events, related cytoplasmic kinases play a role in downstream signaling.

Another group of RKs exemplified by the FERONIA/LORELEI co-receptor group functions in monitoring of cell wall integrity via a Glc2-N-glycan–binding malectin domain that is associated with the cell wall–facing outer leaflet of the plasma membrane. FERONIA binds secreted RAPID ALKALINIZATION FACTOR (RALF) peptides and stimulates an increase in apoplastic pH via transient increases in cytosolic Ca^{2+} and differential phosphorylation of multiple proteins during pollination and other rapid growth events.

Mitogen Activated Protein (MAP) kinase A large family of eukaryotic serine/threonine protein kinases that function in diverse environmental and developmental processes. MAP kinases generally initiate protein kinase cascades that rapidly amplify signals to activate cellular responses.

AGC kinase A member of a large subfamily of protein kinases that are activated by environmental stimuli of second messengers. AGC kinases share a conserved catalytic domain and a regulatory C-terminal domain.

phototropism The alteration of plant growth patterns in response to the direction of incident radiation, especially blue light.

phosphatases Enzymes that remove a phosphate group from a protein.

Kinases also function in signal amplification

If a receptor is considered the gateway through which a signal enters a signaling network, the receptor location to some extent prescribes the length of the subsequent signaling pathway; such pathways can consist of either a few signaling steps or an elaborate cascade of signaling events. The perception of signals at the plasma membrane often activates transduction pathways with many intermediates. In the case of signaling pathways that must eventually reach the nucleus to regulate gene expression, signal strength along the pathway will dissipate unless it is reinforced by signal amplification events. In the absence of amplification, any activated signaling intermediate that must traverse the cytosol to translocate to the nucleus will become diluted because of its diffusion and deactivation mechanisms (for example, by dephosphorylation, degradation, or sequestration). Furthermore, many chemical signals are present at very low concentrations, and receptors can similarly occur at very low density, such that the initial signal may be quite weak. Signal amplification cascades serve to adjust, maintain, or even enhance signal strength over larger distances.

Protein kinase cascades are an important signal amplification mechanism. **Mitogen Activated Protein (MAP) kinase cascades** play an important role in multiple plant responses. As shown in **Figure 12.4C**, signals perceived by receptors can produce a series of phosphorylation events that rapidly amplify the signal. Within minutes of signal perception, the MAP kinase cascade can trigger a suite of downstream responses.

Other kinases control the timing of signaling events by linkage to the cycle of cellular reproduction. The cyclin-dependent kinases function only when associated with cyclin regulators of the cell cycle. Association of different cyclins with a particular cyclin-dependent kinase regulates the phosphorylation of discrete regulators at each step of the cell cycle (discussed in Chapter 1).

Plants also have a large group of **AGC kinases** that are evolutionarily related to animal cyclic adenosine monophosphate (A-class), cyclic guanosine monophosphate (G-class), and Ca^{2+}-dependent (C-class) protein. AGC kinases similar to the phototropin blue-light receptors that function in **phototropism**, chloroplast movement, and stomatal responses to blue light (Chapter 13) are present in algae and vascular plants. In higher plants, members of the AGC family also regulate polar auxin transport streams and phosphorylate lipids to create signaling intermediates that function in multiple cellular responses.

Ca^{2+}-binding protein kinases represent another important group of protein kinase families in plants. These kinases function as a component of Ca^{2+} signaling and are discussed in the context of Ca^{2+} second messengers later in this section.

Phosphatases are the "off switch" of protein phosphorylation

Protein phosphorylation can have a multitude of effects on protein function, localization, and abundance. The effects of phosphorylation can be reversed by **phosphatases** that remove the phosphate group from substrate proteins. Phosphatases can be grouped by structural similarity, by the specific amino acid residues that they dephosphorylate, or by molecular interactions that stimulate their dephosphorylation activity. The cytosolic serine/threonine-specific protein phosphatases of the 2A (PP2A) and 2C (PP2C) subfamilies are the primary protein phosphatases that function in plants. The dephosphorylation of phosphorylated proteins by phosphatases is often part of a negative feedback regulatory circuit, and many of the protein kinase–dependent phosphorylation events mentioned earlier are antagonized by phosphatases.

Other protein modifications can reconfigure cellular processes

Interactions and cellular distribution of proteins can be regulated by modifications other than phosphorylation. As discussed in Chapter 1, *glycosylation* of proteins in the Golgi apparatus and the endoplasmic reticulum (ER) can change their activity and destination within, or even outside, the cell. Other protein modifications modify interactions with other proteins, association with membranes, or even activity of the protein itself. The most common modifications are those that target cytosolic proteins to membrane complexes and surfaces. These include (1) *N-myristoylation*, whereby a 14-carbon myristoyl group is added to the N-terminal glycine residue of a protein; (2) *prenylation*, whereby a farnesyl phosphate or geranylgeranyl molecule derived from terpenoid biosynthesis is attached to cysteine residues at a signature C-terminal cysteine-containing motif; (3) *S-acylation*, a reversible process that takes place at membrane surfaces whereby a transferase enzyme attaches a C-16 or C-18 fatty acid to a thiol residue to stabilize membrane interactions; and (4) *nitrosylation* of cysteine residues in proteins under conditions where nitric oxide added to cysteine residues prevents formation of cysteine linkages or inhibits enzymatic activity. Protein nitrosylation functions in both programmed development and stress responses.

Small-molecule second messengers rapidly amplify signals

Second messengers represent another strategy to enhance or propagate signals. These small molecules and ions are rapidly produced or mobilized at relatively high levels after signal perception, and can modify the activity of target signaling proteins. Probably the most ubiquitous second messenger in all eukaryotes is the divalent calcium cation Ca^{2+}, which in plants is involved in a vast number of signaling pathways. Cytosolic free Ca^{2+} levels increase rapidly when Ca^{2+}-permeable ion channels open to allow passive Ca^{2+} influx from Ca^{2+} stores into the cytosol (**Figure 12.5**). Channel activity must be tightly regulated to maintain precise control over the timing and duration of cytosolic Ca^{2+} elevation. Generally, ion channels are gated, meaning the channel pores are opened or closed by changes in transmembrane electrical potential, membrane tension, posttranslational modification, or binding of a ligand (Chapter 6). Several families of Ca^{2+}-permeable channels have been identified in plants; these include plasma membrane–localized glutamate-like receptors (GLRs) and cyclic nucleotide-gated channels (CNGCs). Electrophysiological and other evidence supports the presence of Ca^{2+}-permeable channels at the tonoplast and the ER.

Once receptor-mediated signaling activates Ca^{2+}-permeable channels, Ca^{2+} sensor proteins play a pivotal role as signaling intermediaries, linking Ca^{2+} signals to changes in cellular activities. Most plant genomes contain four major multigene families of Ca^{2+} sensors: the calmodulin (CaM) and calmodulin-like proteins, the Ca^{2+}-dependent protein kinases (CDPKs), the Ca^{2+}/calmodulin-dependent protein kinases (CCaMKs), and the calcineurin-B-like proteins (CBLs), which function in concert with CBL-interacting protein kinases (CIPKs). Members of these sensor families modulate the activity of target proteins either by binding to CaM or by phosphorylating the target protein in a Ca^{2+}-dependent manner. Target proteins include transcription factors, various protein kinases, Ca^{2+}-ATPases, enzymes producing reactive oxygen species (ROS), and ion channels. Finally, Ca^{2+} pumps and Ca^{2+} exchangers in organelle and plasma membranes actively remove Ca^{2+} from the cytosol to terminate Ca^{2+} signaling (Figure 12.5).

Ca^{2+} can also modulate extra- and intracellular pH changes generated by plasma membrane H^+ pumps. As we will discuss later, the pH of the cell wall regulates the protonation status of weakly acidic small molecules such as plant hormones and consequently affects their ability to enter cells by diffusion. Cell wall pH regulation may thus represent a mechanism to fine-tune hormone uptake and signaling. Evidence also exists for pH-dependent gating of potassium ion channels and aquaporins.

Signaling molecules derived from cellular lipids (diacylglycerol, inositol 1,4,5-trisphosphate, and phosphatidic acid; **Figure 12.6**) are enzymatically generated from membrane lipids and are thought to also regulate Ca^{2+} fluxes

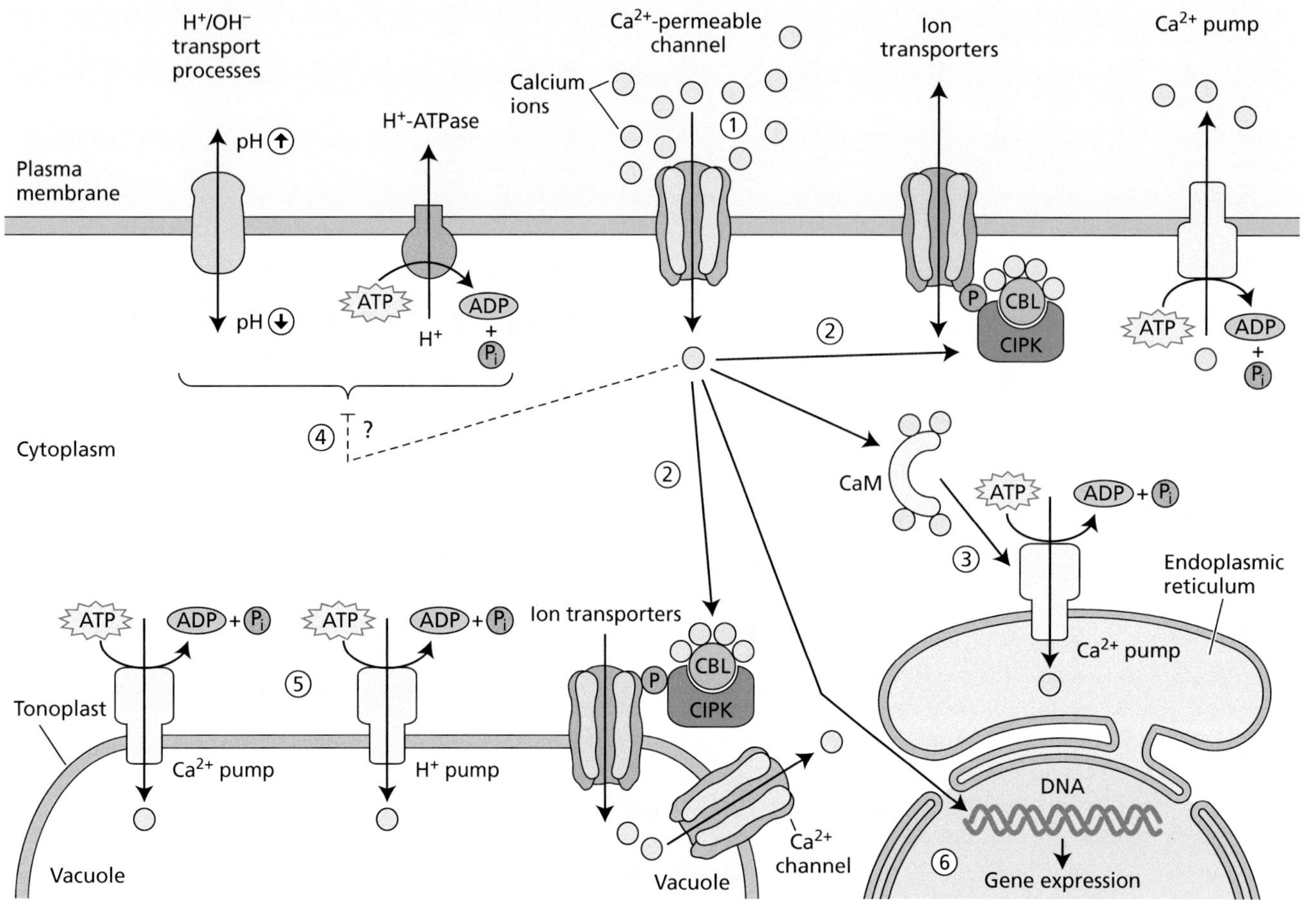

1. Signal-induced activation of Ca^{2+}-permeable ion channels leads to an increase in the concentration of free cytosolic Ca^{2+}.
2. Activated CBL/CIPK Ca^{2+} sensor proteins interact with ion transporters on the plasma membrane and the vacuolar membrane.
3. Activated calmodulin (CaM) stimulates Ca^{2+} pumps on the endoplasmic reticulum membrane.
4. Ca^{2+} may affect membrane transporters, triggering changes in the intracellular and extracellular pH.
5. Pumps on vacuolar membrane create H^+ and Ca^{2+} gradients between cytoplasm and vacuole.
6. Increased Ca^{2+} in nucleus activates Ca^{2+}-dependent transcription.

Figure 12.5 Calcium ions, pH, and reactive oxygen species (ROS) function as second messengers that amplify signals and regulate the activity of target signaling proteins to trigger physiological responses. An increase in the $[Ca^{2+}]_{cyt}$ activates Ca^{2+} sensor proteins (calmodulins [CaMs], Ca^{2+}-dependent protein kinases [CDPKs], and calcineurin-B-like proteins/CBL-interacting protein kinases [CBL/CIPKs]), which are located at different subcellular sites. Six types of activation are shown in the figure.

phosphatidic acid (PA) A diacylglycerol that has a phosphate on the third carbon of the glycerol backbone.

reactive oxygen species (ROS) These include the superoxide anion ($O_2^{\bullet-}$), hydrogen peroxide (H_2O_2), the hydroxyl radical (HO•), and singlet oxygen. They are generated in several cell compartments and can act as signals or cause damage to cellular components.

in response to environmental cues. Diacylglycerol is also rapidly converted by lipid kinases to **phosphatidic acid** (**PA**). PA can also be formed by enzymatic processing of phosphatidyl choline. PA is a cone-shaped lipid that promotes vesicle budding, docking, and fusion by increasing local negative membrane curvature and enhancing some interactions of membrane-associated proteins.

Reactive oxygen species (**ROS**) can be both cytotoxic byproducts of metabolic processes such as respiration and photosynthesis and signaling molecules that regulate plant responses to environmental and endogenous signals. ROS are formed in mitochondria, plastids, and peroxisomes, but plasma membrane–localized NADPH oxidases make up the best-understood family of ROS-producing enzymes involved in signal transduction. NADPH oxidases (or respiratory burst oxidase homologs [RBOHs]) transfer electrons from the cytosolic electron donor NADPH across the membrane to reduce extracellular molecular oxygen. The resulting ROS, superoxide, can dismutate to hydrogen peroxide, a more membrane-permeant ROS that also enters cells through specific aquaporin channels. NADPH oxidase is regulated by phosphorylation of its N-terminal amino acids and by Ca^{2+} binding.

Targets of ROS signaling include the thiol side chain of cysteine amino acid residues, which can be oxidatively modified to form intramolecular (within the polypeptide/protein) or intermolecular (oxidative cross-linking of different [poly]peptides/proteins) disulfide bonds. Such disulfide bonds usually change the conformation of the affected proteins and thereby their activities. Oxidation of lipids by ROS also functions in stress responses (discussed in Chapter 19), and redox regulation has been shown to alter the DNA-binding activity or cellular localization of several transcription factors and transcriptional activators.

12.3 Hormones and Plant Development

Explain the roles of hormones in plant development and environmental responses.

The form and function of multicellular organisms would not be possible without efficient communication among cells, tissues, and organs. In higher plants, regulation and coordination of metabolism, growth, and morphogenesis often depend on chemical signals from one part of the plant to another. Hormones are chemical messengers that are produced in one cell and modulate cellular processes in another cell by interacting with specific proteins that function as receptors linked to cellular signal transduction pathways. Most plant hormones are capable of activating responses in target cells at very low concentrations.

Although the details of hormonal control of development are quite diverse, hormonal pathways share common features (**Figure 12.7**). Environmental cues and/or developmental programming often result in increases or decreases in hormone biosynthesis. The hormone is then transported to a site of action. Perception of the hormone by a receptor results in transcriptional or posttranscriptional (e.g., phosphorylation, protein turnover, ion extrusion) events that ultimately induce a physiological or developmental response. The response can then be attenuated by negative feedback repression of hormone synthesis and by other processes that restore basal active hormone levels. In this way the plant reacquires the ability to respond to the next signal input.

Plant development is regulated by ten major hormones: **auxins**, **gibberellins (GAs)**, **cytokinins**, **ethylene**, **abscisic acid (ABA)**, **brassinosteroids (BRs)**, **jasmonates**

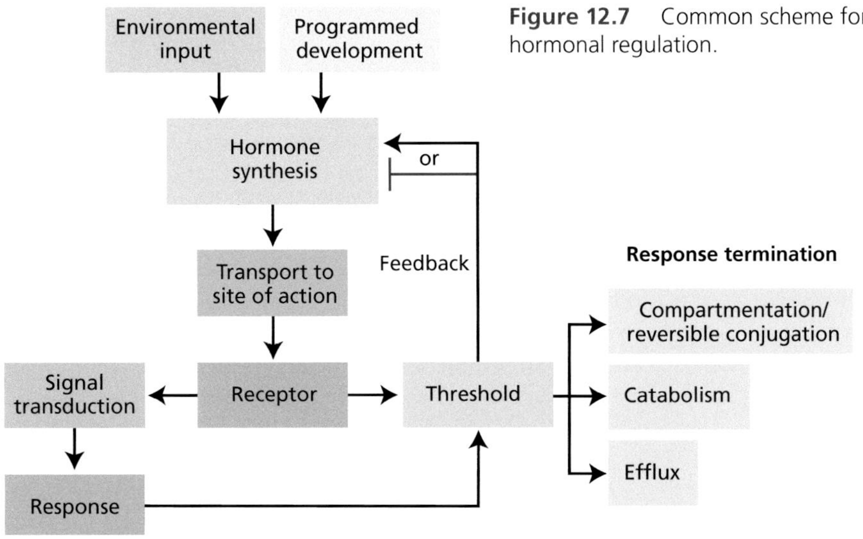

Figure 12.7 Common scheme for hormonal regulation.

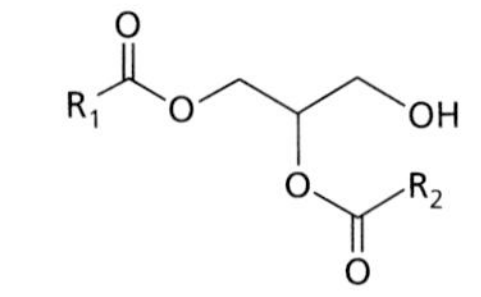

1,2-diacylglycerol

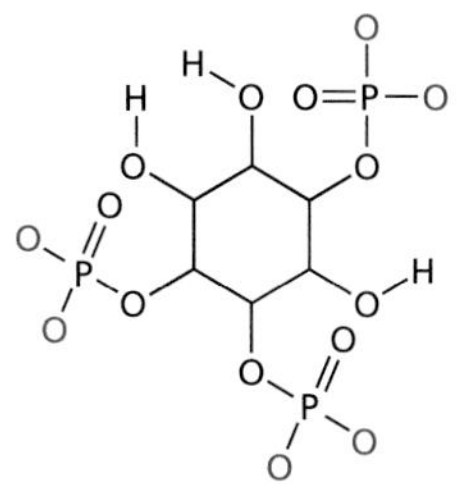

Inositol 1,4,5-triphosphate

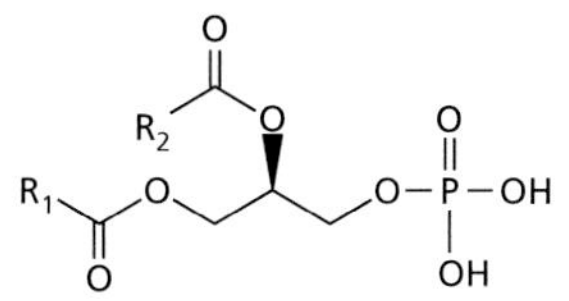

Phosphatidic acid

Figure 12.6 Signaling molecules derived from membrane phospholipids. Lipid-modifying enzymes remodel cellular membranes and produce lipid signaling molecules. R_1 and R_2 represent acyl side chains.

auxins Major plant hormones involved in numerous developmental processes, including cell elongation, organogenesis, apical–basal polarity, differentiation, apical dominance, and tropic responses. The most abundant chemical form is indole-3-acetic acid.

gibberellins (GAs) A large group of chemically related plant hormones synthesized by a branch of the terpenoid pathway and associated with the promotion of stem growth (especially in dwarf and rosette plants), seed germination, and many other functions.

cytokinins A class of plant hormones that function in cell division and differentiation, as well as axillary bud growth, and leaf senescence. Cytokinins are adenine derivatives, the most common form of which is zeatin.

ethylene A gaseous plant hormone involved in fruit ripening, abscission, and the growth of etiolated seedlings. The chemical formula of ethylene is C_2H_4.

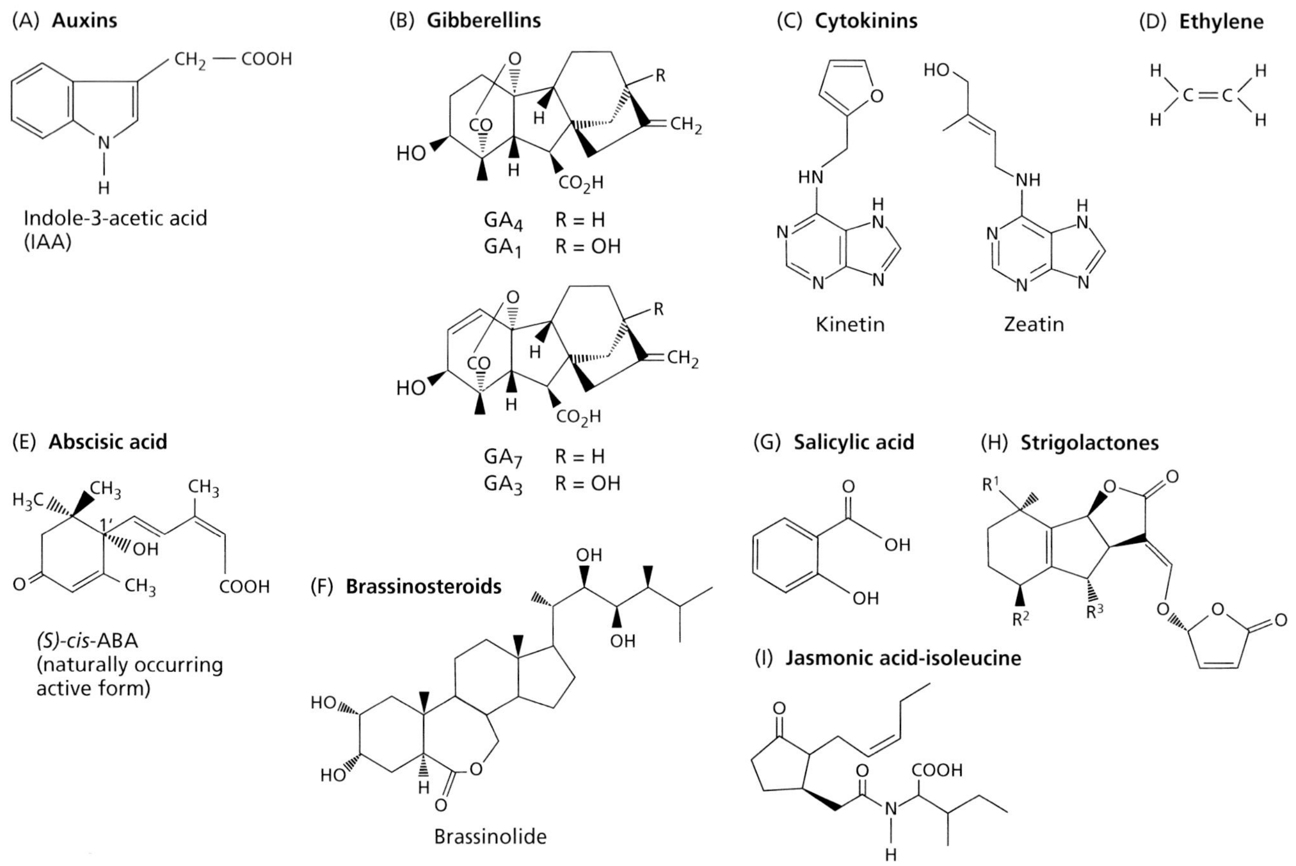

Figure 12.8 Chemical structures of phytohormones.

(primarily jasmonic acid-isoleucine, JA-Ile), **salicylic acid (SA)**, **pipecolates** (not shown in Figure 12.8), and **strigolactones** (**Figure 12.8**). In addition, several peptides trigger responses in adjacent or distal cells to regulate growth, nutrient assimilation, or biotic interactions. Here we briefly introduce auxins, gibberellins, cytokinins, ethylene, abscisic acid, brassinosteroids, and strigolactones. The roles of jasmonates, salicylic acid, and pipecolates in biotic interactions will be discussed in Chapter 18.

abscisic acid (ABA) A plant hormone that functions in regulation of seed dormancy as well as in stress responses such as stomatal closure during water deficit and responses to cold and heat. ABA is derived from carotenoid precursors.

brassinosteroids (BRs) A group of plant steroid hormones that play important roles in many developmental processes, including cell division and cell elongation in stems and roots, photomorphogenesis, reproductive development, leaf senescence, and stress responses.

jasmonates Plant hormones that function in plant defenses to biotic and abiotic stresses as well as some other aspects of development. Jasmonic acid is derived from the octadecanoid pathway. Can function as a volatile signal when methylated (methyl jasmonate) and is primarily active when conjugated to the amino acid isoleucine (JA-Ile).

Auxin was discovered in early studies of phototropism

Auxin is essential to plant growth and regulates virtually every aspect of plant development. As described at the beginning of this chapter, Charles and Francis Darwin suggested the presence of a transmissible signal such as auxin after observing the bending of seedling hypocotyls and leaf sheaths (coleoptiles) in response to light. Subsequently, it was shown that the signal was a chemical that could diffuse through gelatin blocks (**Figure 12.9**). Plant physiologists named the chemical signal auxin from the Greek *auxein*, meaning "to increase" or "to grow," and identified indole-3-acetic acid (IAA; Figure 12.8A) as the primary plant auxin. In some species, 4-chloro-IAA, phenylacetic acid, and indole-3-butyric acid (IBA) function as natural auxins, but IAA is the most abundant and physiologically important form. Auxins are also produced by a variety of microbes that live in, on, or near plants (the phytobiome) and are deployed to promote plant growth that benefits the microbes. Because the structure of IAA is relatively simple, researchers were quickly able to synthesize a wide array of molecules with auxin activity. Some of these compounds, such as 1-naphthalene acetic acid (NAA),

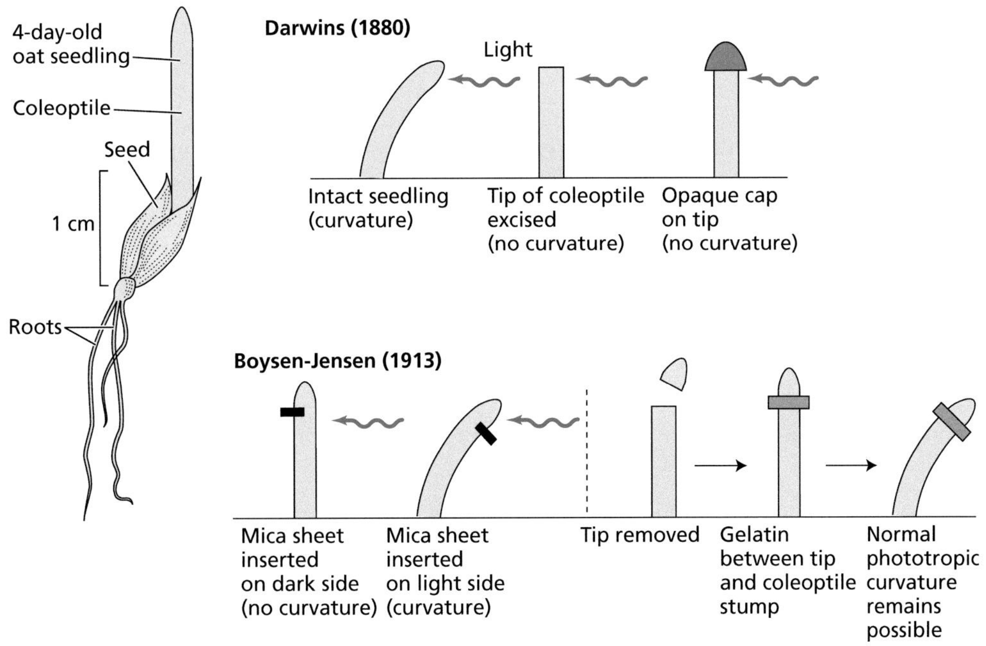

From experiments on coleoptile phototropism, the Darwins concluded in 1880 that a growth stimulus is produced in the coleoptile tip and is transmitted to the growth zone.

In 1913, P. Boysen-Jensen discovered that the growth stimulus passes through gelatin but not through water-impermeable barriers such as mica.

Figure 12.9 Early experiments on the chemical nature of auxin.

2,4-dichlorophenoxyacetic acid (2,4-D), and 2-methoxy-3,6-dichlorobenzoic acid (dicamba), are now used widely as growth regulators and herbicides in horticulture and agriculture.

Gibberellins promote stem growth and were discovered in studies of the "foolish seedling disease" of rice

A second group of plant hormones are the gibberellic acid compounds known collectively as gibberellins (GAs), which are numbered in the chronological sequence of their discovery. This group comprises a large number of compounds, all of which are tetracyclic (four-ringed) diterpenoid acids, but only a few of which have intrinsic biological activity (Figure 12.8B). Gibberellins function in the induction of internode elongation in dwarf plants, promotion of seed germination (Chapter 14), the transition to flowering and pollen development (Chapter 16), and fruit development (Chapter 17).

Gibberellins were first recognized in the early twentieth century as natural products in the fungus *Gibberella/Fusarium fujikuroi*, from which the hormones derive their unusual name. Rice plants infected with this fungus grow abnormally tall, which leads to lodging (falling over) and reduced yield. Such overgrowth can be reproduced by applying gibberellins to uninfected rice seedlings. *Fusarium fujikuroi* produces several different gibberellins, the most abundant of which is GA_3, which is used as a growth regulator in horticultural production. GA_3 sprayed on grape vines is routinely used to produce the large, seedless grapes we purchase in the grocery store (**Figure 12.10A**). GA_3 can also be sprayed to induce stem elongation of dwarf maize (*Zea mays*) (**Figure 12.10B**) and many rosette plants (**Figure 12.10C**). GAs were subsequently discovered in a majority of plants studied to date, with GA_1 and/or GA_4 having the primary "hormonal" function.

Cytokinins were discovered as cell division–promoting factors in tissue-culture experiments

Cytokinins were discovered in a search for factors that stimulated plant cells to divide (i.e., undergo cytokinesis). A small molecule was identified that could, in the presence of auxin, stimulate tobacco pith parenchyma tissue to proliferate in

salicylic acid (SA) A benzoic acid derivative that functions as a local endogenous signal in coordination with systemically distributed pipecolates for systemic acquired resistance.

pipecolates hormones derived from pipecolic acid that function as long distance signals in systemic acquired resistance responses

strigolactones Carotenoid-derived plant hormones that inhibit shoot branching. They also play roles in the soil by stimulating the growth of arbuscular mycorrhizae and the germination of parasitic plant seeds, such as those of *Striga* (the source of their name).

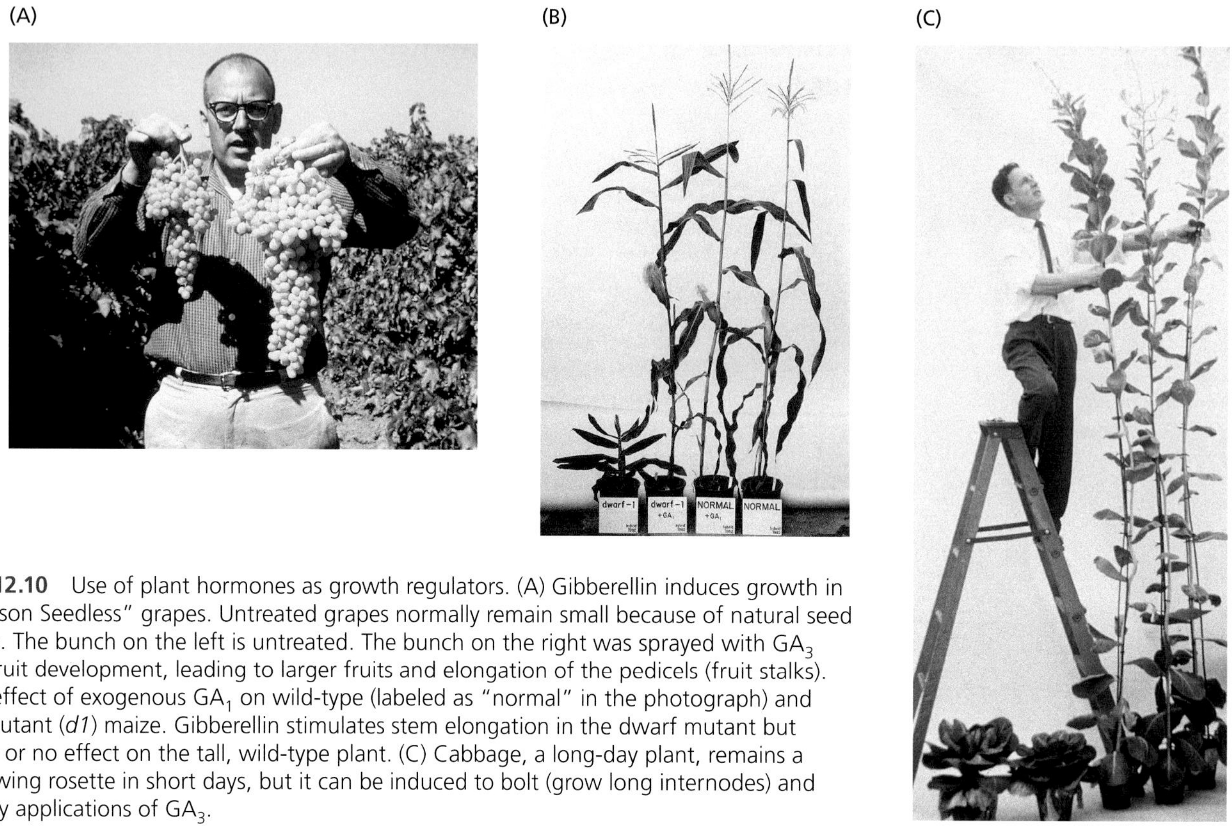

Figure 12.10 Use of plant hormones as growth regulators. (A) Gibberellin induces growth in "Thompson Seedless" grapes. Untreated grapes normally remain small because of natural seed abortion. The bunch on the left is untreated. The bunch on the right was sprayed with GA_3 during fruit development, leading to larger fruits and elongation of the pedicels (fruit stalks). (B) The effect of exogenous GA_1 on wild-type (labeled as "normal" in the photograph) and dwarf mutant (*d1*) maize. Gibberellin stimulates stem elongation in the dwarf mutant but has little or no effect on the tall, wild-type plant. (C) Cabbage, a long-day plant, remains a low-growing rosette in short days, but it can be induced to bolt (grow long internodes) and flower by applications of GA_3.

© Sylvan Wittwer/Visuals Unlimited

culture (**Figure 12.11A**). The cytokinesis-inducing molecule was named *kinetin*. While kinetin is a synthetic cytokinin, its structure is similar to that of naturally occurring cytokinins such as zeatin (Figure 12.8C).

Cytokinins regulate many physiological and developmental processes, including leaf senescence, axillary branch outgrowth, vascular development, breaking of bud dormancy, and formation of apical meristems (Chapter 15). In addition, cytokinins play important roles in nutrient and abiotic stress responses. Cytokinins also function in interactions with symbiotic nitrogen-fixing bacteria and arbuscular mycorrhizal fungi, as well as pathogenic bacteria, fungi, nematodes, and viruses (**Figure 12.11B**).

Ethylene is a gaseous hormone that promotes fruit ripening and other developmental processes

Ethylene is a gas with a simple chemical structure (Figure 12.8D) and was first identified as a plant growth regulator in 1901. Subsequently, ethylene was identified as a natural product synthesized by plant tissues. Ethylene regulates a wide range of processes in plants, including seedling growth, cell expansion and differentiation (Chapter 14), leaf and flower senescence and abscission (Chapter 15), and stress responses (Chapters 18 and 19). Phytobiome bacteria have been shown to influence plant growth and stress responses by manipulating levels of ethylene and its metabolic precursors.

Abscisic acid regulates seed maturation and stomatal closure in response to water stress

Abscisic acid (ABA) is a ubiquitous hormone in vascular plants and has also been found in mosses, some phytopathogenic fungi, and a wide range of metazoans. ABA is a terpenoid (Figure 12.8E) that was identified in the 1960s as a

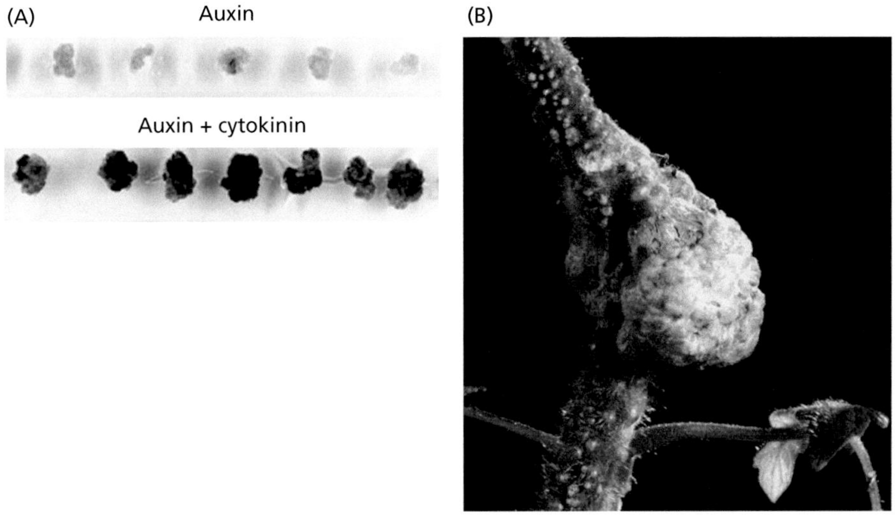

Figure 12.11 Cytokinin enhances cell division and greening. (A) Wild-type Arabidopsis leaf explants were induced to form callus (undifferentiated cells) by culturing in the presence of auxin alone (top) or auxin plus cytokinin (bottom). Cytokinin was required for callus growth and greening in the presence of light. (B) Tumor that formed on a tomato stem expressing increased cytokinin levels after being infected with the crown gall bacterium, *Agrobacterium tumefaciens*. Two months before this photo was taken, the stem was wounded and inoculated with a virulent strain of the crown gall bacterium.

growth-inhibiting compound associated with the onset of bud dormancy and promotion of cotton fruit abscission. However, later work showed that ABA promotes senescence, the process preceding abscission, rather than abscission itself. Since then, ABA has also been shown to regulate salinity, dehydration, and temperature stress responses, including stomatal closure (**Figure 12.12**). ABA also promotes seed maturation and dormancy (Chapter 14) and regulates the growth of roots and shoots, heterophylly (the production of different leaf types on an individual plant) (Chapter 15), and flowering (Chapter 16), as well as some responses to pathogens (Chapter 18).

Brassinosteroids regulate photomorphogenesis, floral sex identity, and other developmental processes

Brassinosteroids are ubiquitous polyhydroxylated steroids that were first discovered as growth-promoting substances present in the pollen of *Brassica napus* (rape plant). The most bioactive brassinosteroid in eudicots is **brassinolide** (Figure 12.8F). Its immediate precursor, castasterone, is also an active hormone in some monocot species. In the angiosperms, brassinosteroids are found at low

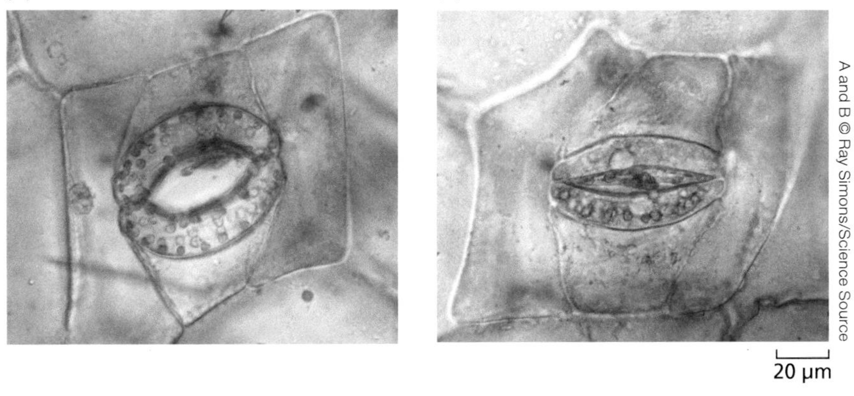

Figure 12.12 Stomatal closure in response to ABA. (A) Stomata are open in the light for gas exchange with the environment. (B) ABA treatment closes stomata in the light. This reduces water loss during the day under drought stress conditions (Chapter 19). Regulation of stomatal opening is also discussed in Chapters 3, 6, and 13.

brassinolide A plant steroid hormone with growth-promoting activity, first isolated from *Brassica napus* pollen. One of a group of plant hormones with similar structures and activities, called brassinosteroids.

Figure 12.13 Loss of the active brassinosteroid in the *nana1* mutant of maize results in dwarfism and altered floral sex determination. The staminate wild-type tassel is shown on the left, and the feminized *nana1* tassel is shown on the right.

levels in various organs (e.g., flowers, leaves, roots) and at higher relative levels in pollen, immature seeds, and fruits.

Brassinosteroids play pivotal roles in multiple aspects of plant development, including germination, elongation growth, photomorphogenesis (Chapter 14), reproductive development (Chapter 17), leaf senescence (Chapter 16), and stress responses (Chapter 18). Mutants deficient in brassinosteroid synthesis exhibit dwarfism, reduced apical dominance, de-etiolated growth in the dark, and feminized male flowers and tassels in some species (**Figure 12.13**).

Strigolactones suppress branching and promote rhizosphere interactions

Strigolactones are a group of terpenoid lactones (Figure 12.8H) that were originally discovered as host-derived germination stimulants for root parasitic plants, such as the witchweeds (*Striga* spp.) and broomrapes (*Orobanche* and *Phelipanche* spp.) as well as attractants for arbuscular mycorrhizal fungi (Chapter 4). Strigolactones suppress shoot branching and stimulate cambial activity and secondary growth (Chapter 15). Strigolactones and other species-specific apocarotenoids have analogous functions in roots, where they reduce adventitious and lateral root formation and promote root hair growth. Karrikins, found in smoke from forest and chaparral fires, are similar to strigolactones and stimulate seed germination (Chapter 14). Rhizosphere microbial communities also produce similar carotenoids to modify root growth.

12.4 Phytohormone Metabolism and Homeostasis

Understand hormone metabolism and homeostasis and how it regulates developmental and environmental responses.

To be effective signals, the concentrations of plant hormones must be tightly regulated. In the simplest terms, a hormone's concentration in any given tissue or cell is determined by the balance between the rate of increase in its concentration (e.g., by local synthesis or by import from elsewhere in the plant) and the rate of decrease in its concentration (e.g., by inactivation, degradation, sequestration, or efflux) (**Figure 12.14**). In this section we discuss mechanisms for modulating hormone concentrations locally (within a cell or tissue). We cover hormone transport between different parts of a plant in the following section.

Indole-3-pyruvate is the primary intermediate in auxin biosynthesis

IAA is synthesized in a two-step process using the amino acid tryptophan as a precursor. Tryptophan is converted by a tryptophan aminotransferase (TAA) to

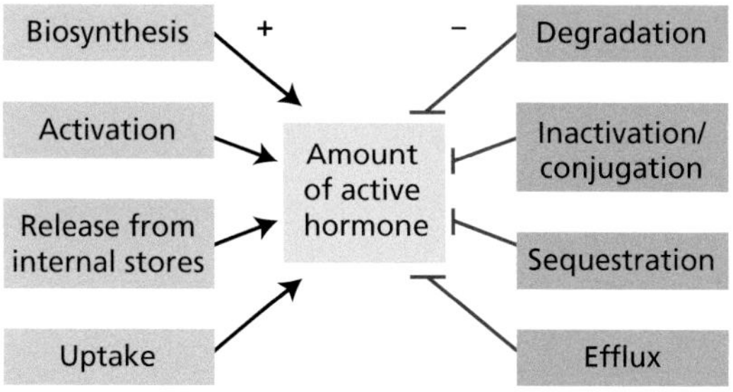

Figure 12.14 Homeostatic regulatory mechanisms that influence hormone concentration. Both positive and negative factors work in concert to maintain hormone homeostasis.

indole-3-pyruvate (IPyA), which is then converted to IAA by members of the YUCCA flavin monooxygenase family (**Figure 12.15A**). The YUCCA enzymes were named for the yucca-like tall inflorescences and narrow, epinastic leaves observed in Arabidopsis when the enzyme is overproduced.

IAA biosynthesis is associated with rapidly dividing and growing tissues. Although virtually all plant tissues appear to be capable of producing low levels of IAA, shoot and root apical meristems, young leaves, and young fruits are the primary sites of auxin synthesis. In plants that produce indole glucosinolate defense compounds, IAA can also be synthesized from tryptophan via a pathway with indole acetonitrile as an intermediate. In maize kernels, IAA also appears to be synthesized by a tryptophan-independent pathway.

Auxin is toxic at high cellular concentrations, and auxin is reversibly conjugated to amino acids for storage. When IAA is conjugated to Asp or Glu, the IAA conjugate is oxidized by a dioxygenase, processed by amidohydrolases, and then further catabolized for complete inactivation. In some species, IAA is also reversibly converted to methyl-IAA via an indole-3-acetate O-methyltransferase. Indole-3-butyric acid (IBA) is an endogenous auxin in some plants that is also routinely used in horticulture to promote rooting of cuttings. IBA is rapidly converted to IAA in the peroxisome by β-oxidation. Auxin and auxin conjugates can also be sequestered in the endoplasmic reticulum by homeostatic processes.

The well-documented toxicity of exogenously applied auxin, especially on eudicot species, forms the basis for a family of synthetic auxins, such as 2,4-dichlorophenoxyacetic acid (2,4-D) and dicamba, that have long been used as herbicides for eudicot weeds. Synthetic auxins are effective herbicides as they are generally much less subject to degradation, conjugation, transport, and sequestration than natural auxins. This class of herbicides is widely used in turfgrass weed control and row crop agriculture to kill germinating dicot weeds, as they are more readily inactivated by monocot crops compared to eudicot weeds. The artificial auxin 1-naphthylene acetic acid is less toxic to plants and is used as a growth regulator to thin fruit in apple orchards to ensure that the size of the remaining fruit will yield a high market value.

Gibberellins are synthesized by oxidation of the diterpene *ent*-kaurene

Gibberellins are synthesized in several parts of a plant, including developing and germinating seeds, young leaves, and elongating internodes. The biosynthetic pathway (Figure 12.15B), which starts in plastids, leads to the production of geranylgeranyl diphosphate (GGPP), which gets converted into *ent*-kaurene. This compound is oxidized sequentially by enzymes associated with the ER, leading to GA_{12}. Dioxygenase enzymes in the cytosol oxidize GA_{12} to active gibberellins. GA is catabolized by oxidation (via GA 2-oxidases), methylation, and glycosylation. Genetic modulation of these pathways plays an important role in plant development.

Gibberellin antagonists are used as plant growth regulators. Uniconazole and paclobutrazol are used to inhibit GA biosynthesis and, thus, decrease the stature of the plants to reduce lodging or reduce internode and petiole elongation in horticultural and ornamental crops.

Cytokinins are adenine derivatives with isoprene side chains

Cytokinins are derived from adenine and have isoprenoid side chains. Zeatin, dihydrozeatin (DHZ), and isopentenyl adenine (iP) are the most abundant cytokinins in higher plants. A simplified schematic of the cytokinin biosynthetic pathway is shown in Figure 12.8C.

Cytokinin levels are regulated at the level of synthesis but are also present in the plant as inactive ribosides, ribotides, and **glycosides**. Active cytokinins are also irreversibly cleaved by cytokinin oxidases.

Cytokinins are widely used as growth regulators in tissue culture and vegetative plant propagation.

glycosides Compounds containing an attached sugar or sugars.

ACC synthase The enzyme that catalyzes the synthesis of 1-aminocyclopropane-1-carboxylic acid (ACC) from *S*-adenosylmethionine.

ACC oxidase The enzyme that catalyzes the conversion of 1-aminocyclopropane-1-carboxylic acid (ACC) to ethylene, the last step in ethylene biosynthesis.

Ethylene is synthesized from methionine via ACC

Ethylene can be produced in almost all plant tissues depending on the stage of development and environmental conditions. For example, mature fruits of certain species undergo a respiratory burst in response to ethylene, and ethylene levels increase in these fruits at the time of ripening (Chapter 17). Ethylene is derived from the amino acid methionine and the intermediate *S*-adenosylmethionine, which is generated in the Yang cycle (Figure 12.8D). The first committed step in ethylene biosynthesis is the conversion of *S*-adenosylmethionine to 1-aminocyclopropane-1-carboxylic acid (ACC) by the enzyme **ACC synthase**. ACC is then converted to ethylene by **ACC oxidases**. ACC functions directly as a signaling compound in mosses, and there is some evidence that ACC interacts with glutamate-like receptors in angiosperms.

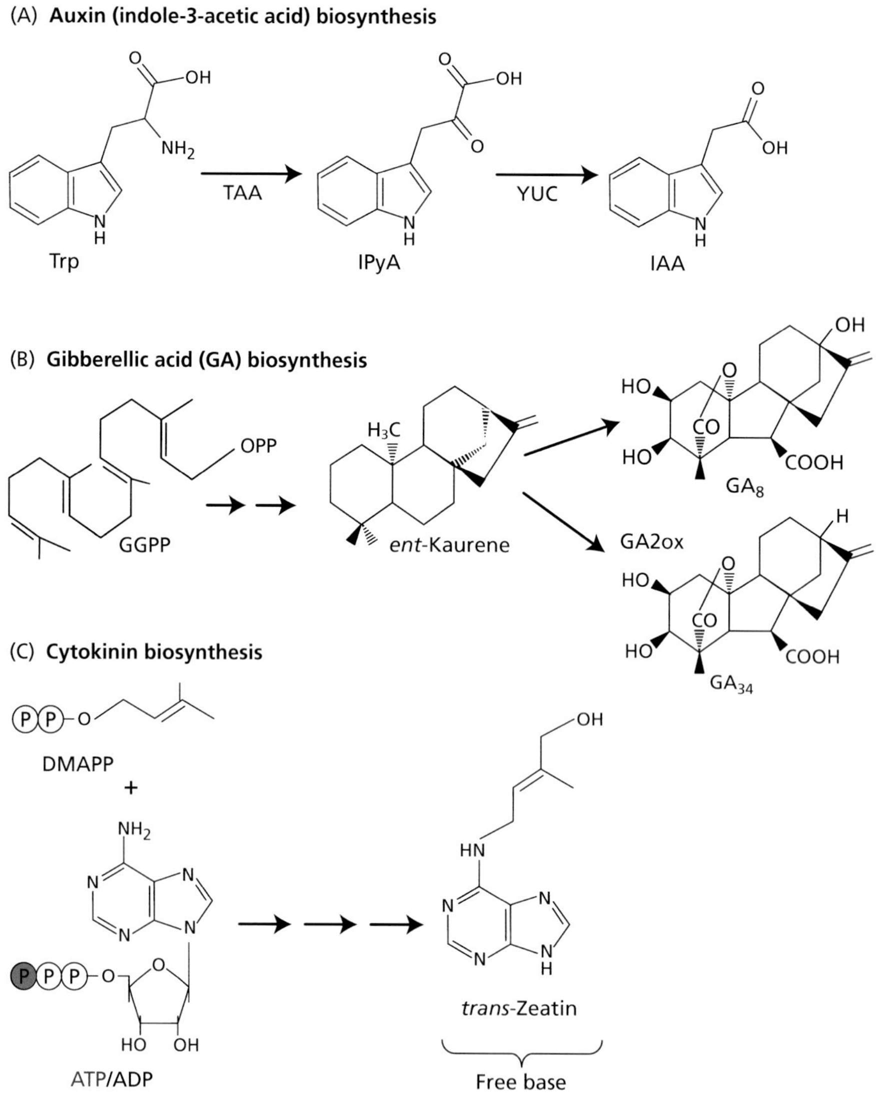

Figure 12.15 Continued

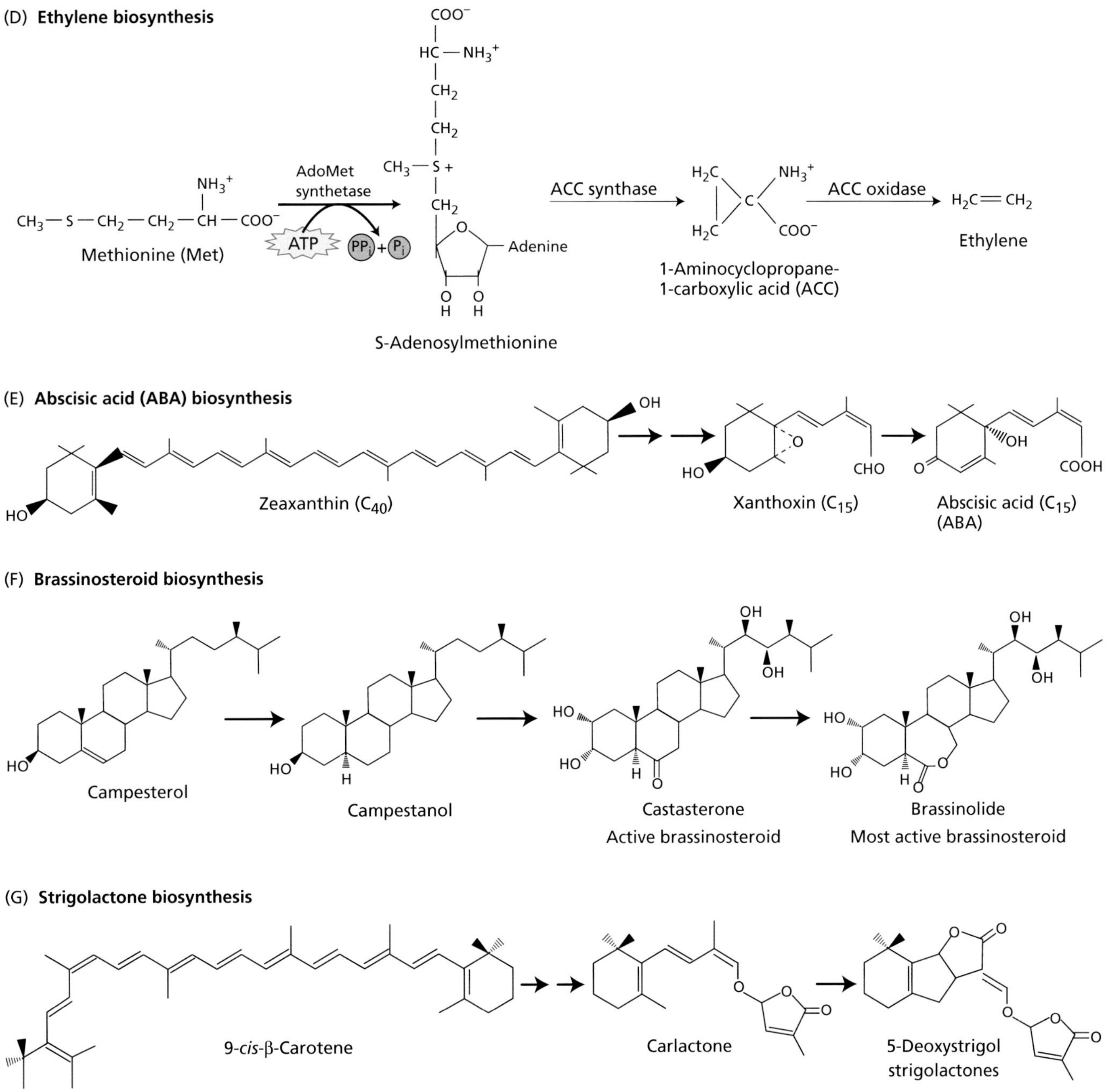

Figure 12.15 Abbreviated biosynthetic pathways of phytohormones. (A) Auxin biosynthesis from tryptophan (Trp). In the first step, Trp is converted to indole-3-pyruvate (IPyA) by the TAA family of tryptophan amino transferases. Subsequently, IAA is produced from IPyA by the YUCCA (YUC) family of flavin monooxygenases. (B) Gibberellin (GA) biosynthesis. In the plastid, geranylgeranyl diphosphate (GGPP) is converted to *ent*-kaurene. After this committed step, *ent*-kaurene is converted in multiple steps to active forms of GA. (C) Cytokinin biosynthesis. An isopentenyl side chain from dimethylallyl diphosphate (DMAPP) is transferred by isopentenyl transferase to an adenosine moiety (ATP or ADP) and eventually is converted to zeatin. (D) Ethylene biosynthesis. The amino acid methionine is converted to *S*-adenosylmethionine, which is converted to 1-aminocyclopropane-1-carboxylic acid (ACC) by the enzyme ACC synthase. Oxidation of ACC via ACC oxidase produces ethylene. (E) ABA biosynthesis via the terpenoid pathway. The initial stages occur in plastids, where isopentenyl diphosphate (IPP) is converted to zeaxanthin, which is modified and cleaved to form xanthoxin. Xanthoxin then is converted to ABA in the cytosol. (F) Brassinosteroid synthesis. The primary precursor for brassinosteroid biosynthesis is campesterol. In different branches of the pathway, cholesterol and sitosterol can also serve as precursors. Active brassinosteroids, castasterone and brassinolide, are derived from the immediate precursor campestanol after multiple C-6 oxidation steps. (G) Strigolactone biosynthesis. Cleavage of 9-*cis*-ß-carotene produces the intermediate carlactone. The final conversion to strigolactones occurs in the cytosol.

Because ethylene is a gas, it diffuses out of plant tissues, and therefore the level decreases rapidly when biosynthesis is interrupted. Ethylene is used to promote fruit ripening, particularly when used with ethylene-deficient cultivars.

Abscisic acid is synthesized from a carotenoid intermediate

ABA is synthesized in almost all cells that contain plastids and has been detected in every major organ and tissue. ABA is a 14-carbon terpenoid, or sesquiterpenoid, which is synthesized in plants via carotenoid intermediates (Figure 12.8E). Enzymatic cleavage of the carotenoid is a rate-limiting step in ABA synthesis that produces the 14-carbon precursor molecule xanthoxin, which subsequently moves into the cytosol, where a series of oxidative reactions converts it to ABA. Further oxidation by ABA-8′-hydroxylases leads to ABA inactivation. ABA can also be inactivated by reversible conjugation.

ABA concentrations can fluctuate dramatically in specific tissues during development or in response to changing environmental conditions. In developing seeds, for example, ABA levels can increase 100-fold within a few days, reaching average concentrations in the micromolar range, and then decline to very low levels as maturation proceeds (Chapter 14). Under conditions of water stress (i.e., dehydration stress), ABA in the leaves can increase 50-fold within 4 to 8 h (Chapter 19).

Brassinosteroids are derived from the sterol campesterol

Brassinosteroids are synthesized from the sterol campesterol, which is similar in structure to cholesterol. Members of the **cytochrome P450 monooxygenase** (**CYP**) enzyme family that are associated with the ER catalyze most of the reactions in the brassinosteroid biosynthetic pathway (Figure 12.15F). Bioactive brassinosteroid levels are also modulated by regulation of gene expression, epimerization, oxidation, hydroxylation, sulfonation, and conjugation to glucose or lipids.

Propiconazole is a commonly used fungicide, because it inhibits the activity of cytochrome P450 enzymes that synthesize the structural sterol ergosterol in fungal pathogens. When used at higher concentrations, this fungicide is used as a growth regulator to reduce the height of turf grasses and maize. These observations led to the discovery that the maize *nana1* dwarf and floral sex mutant (Figure 12.13) is deficient in brassinosteroid biosynthesis.

Strigolactones are synthesized from β-carotene

Strigolactones and other apocarotenoid signaling molecules are derived from carotenoid precursors in plastids in a pathway that is conserved up to synthesis of the intermediate carlactone. Synthesis of strigolactones begins with cleavage of the carotenoid precursors (Figure 12.15G). Diverse cytochrome P450 isoforms produce strigolactones and similar signaling molecules that vary among species. Carotenoid cleavage can also occur via nonenzymatic oxidative processes. The strigolactone signaling pathway will be discussed in Chapter 18.

12.5 Movement of Hormones within the Plant

Describe how hormones are transported throughout the plant.

So far we have looked at the mechanisms that control the synthesis, turnover, and intracellular distribution of plant hormones. In this section we focus on the movement of hormones between the cells, tissues, and organs of the plant. Lipophilic hormones such as ABA and strigolactones can diffuse across membranes, but in some tissues they are also actively transported by ATP-binding cassette (ABC) transporters. ABA and cytokinins can move over long distances in the xylem transpiration stream (Chapter 3) and have been shown to be actively transported into the vascular system in the root. Gibberellins that are synthesized

cytochrome P450 monooxygenase (CYP) A generic term for a large number of related, but distinct, mixed-function oxidative enzymes localized on the endoplasmic reticulum. CYPs participate in a variety of oxidative processes, including steps in the biosynthesis of gibberellins and brassinosteroids.

in the root epidermis can also be taken up in those cells via a nitrate/peptide transporter. An active transport mechanism controls gibberellin levels in root vascular tissues, resulting in accumulation of this growth hormone in expanding endodermal cells that control root elongation. As a gaseous compound, ethylene is more soluble in lipid bilayers than in the aqueous phase and can freely pass through the plasma membrane. In contrast, its precursor, ACC, is water-soluble and is thought to be transported via the xylem to shoot tissues. It is currently unknown whether brassinosteroids are active locally in tissues where they synthesized. Brassinosteroids do not seem to undergo root-to-shoot and shoot-to-root translocation, as experiments in pea and tomato indicate that reciprocal stock/scion grafting of wild-type to brassinosteroid-deficient mutants does not rescue the phenotype of the latter. However, brassinosteroids are thought to be moved locally from intracellular sites of synthesis to cell surface receptors via plasma membrane ABCB transporters. Auxins and cytokinins can also move with source–sink fluxes in the phloem (Chapter 10).

Plant polarity is maintained by polar auxin streams

Despite the existence of diverse hormone transport mechanisms in plants, auxin is the only plant hormone that has been clearly shown to function via highly regulated vectorial streams. These **polar auxin transport** streams are found in almost all plants, including bryophytes. Long-distance polar auxin transport through the vascular parenchyma from sites of synthesis in apical tissues and young leaves to the root tip regulates stem elongation, apical dominance, and lateral branching (Chapter 15). Auxin flows redirected at the root apex into the root epidermis are necessary for root gravitropic responses (Chapter 14). Localized polar auxin streams are essential components of organogenesis (Chapters 15–17).

By convention, auxin transport from the shoot and root apices to the root–shoot transition zone is referred to as a **basipetal** flow, whereas downward auxin flow in the root is referred to as **acropetal** transport. As this terminology can be confusing, a newer terminology assigns the term *rootward* transport to all auxin flows toward the root apex and the term *shootward* transport to any directional flow away from the root apex. Both shootward and rootward polar auxin transport are primary mechanisms for effecting programmed and plastic directional growth.

Polar transport proceeds in a cell-to-cell fashion, rather than via the symplasm; that is, auxin exits a cell through the plasma membrane, diffuses across the cell wall, and enters the next cell through its plasma membrane (**Figure 12.16**). The overall process requires metabolic energy, as evidenced by the sensitivity of polar transport to O_2 deprivation, sucrose depletion, and metabolic inhibitors. The velocity of polar auxin transport is faster than diffusion but much slower than phloem translocation rates (Chapter 10). Polar transport is observed with all natural and some synthetic auxins; other weak organic acids, inactive auxin analogs, and IAA conjugates are poorly transported. Although polar auxin concentration gradients in the embryo appear to be initially established by localized auxin synthesis, they are amplified and extended by specific transporter proteins in the plasma membrane. The cellular processes that use the proton gradients generated by plasma membrane ATPases to generate polar auxin streams are described by a **chemiosmotic mechanism** of auxin transport.

AUXIN UPTAKE IAA is a weak acid (pK_a 4.75). In the apoplast, where plasma membrane H^+-ATPases normally maintain a cell wall solution of pH 5 to 5.5, 15 to 25% of the auxin is present in a lipophilic, undissociated form (IAAH) that diffuses passively across the plasma membrane down a concentration gradient. Therefore, auxin may enter the cell from any side. Auxin uptake is accelerated by secondary active transport (Chapter 6) of the amphipathic, anionic IAA^- present in the apoplast via AUXIN1/LIKE AUXIN1 (AUX1/LAX) symporters

polar auxin transport Directional auxin movement that functions in programmed development and plastic growth responses. Long-distance polar auxin transport maintains the overall polarity of the plant apical–basal axis and supplies auxin for direction into localized streams.

basipetal From the growing tip of a shoot or root toward the base (junction of the root and shoot).

acropetal From the base to the tip of an organ, such as a stem, root, or leaf.

chemiosmotic mechanism The mechanism whereby the electrochemical gradient of protons established across a membrane by an electron transport process is used to either drive energy-requiring ATP synthesis (mitochondria and chloroplasts) or carrier-mediated anion efflux at the plasma membrane (as in polar auxin transport).

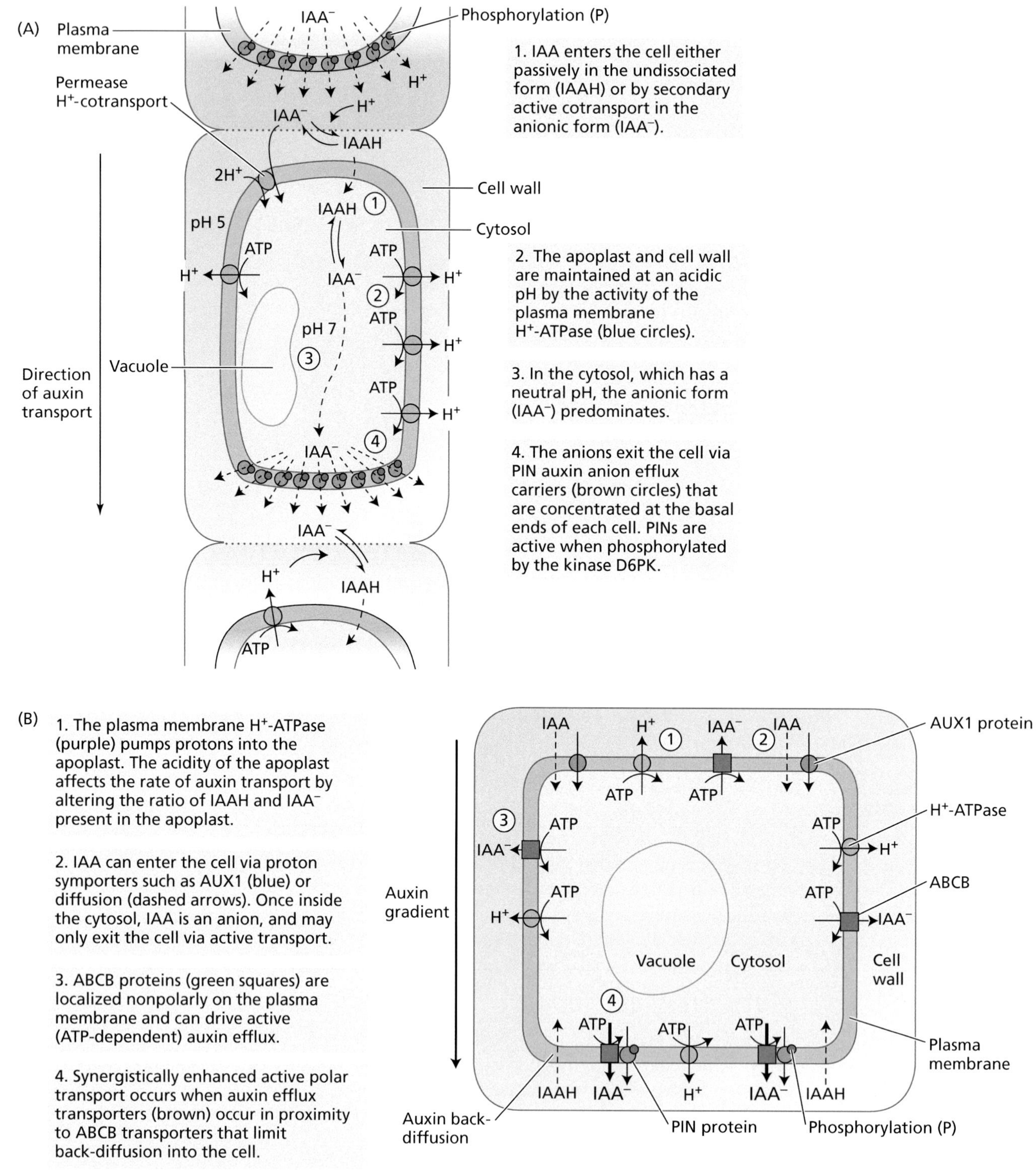
(A)
Plasma membrane
Permease H+-cotransport
IAA−
Phosphorylation (P)
H+
IAA−
H+
IAAH
2H+
Cell wall
IAAH ①
pH 5
Cytosol
ATP
ATP
H+
IAA−
H+
②
ATP
pH 7
H+
③
Vacuole
Direction of auxin transport
ATP
H+
④
IAA−
IAA−
H+
IAAH
ATP
1. IAA enters the cell either passively in the undissociated form (IAAH) or by secondary active cotransport in the anionic form (IAA−).
2. The apoplast and cell wall are maintained at an acidic pH by the activity of the plasma membrane H+-ATPase (blue circles).
3. In the cytosol, which has a neutral pH, the anionic form (IAA−) predominates.
4. The anions exit the cell via PIN auxin anion efflux carriers (brown circles) that are concentrated at the basal ends of each cell. PINs are active when phosphorylated by the kinase D6PK.
(B)
1. The plasma membrane H+-ATPase (purple) pumps protons into the apoplast. The acidity of the apoplast affects the rate of auxin transport by altering the ratio of IAAH and IAA− present in the apoplast.
2. IAA can enter the cell via proton symporters such as AUX1 (blue) or diffusion (dashed arrows). Once inside the cytosol, IAA is an anion, and may only exit the cell via active transport.
3. ABCB proteins (green squares) are localized nonpolarly on the plasma membrane and can drive active (ATP-dependent) auxin efflux.
4. Synergistically enhanced active polar transport occurs when auxin efflux transporters (brown) occur in proximity to ABCB transporters that limit back-diffusion into the cell.
IAA
H+ ①
IAA− ②
IAA
AUX1 protein
ATP
ATP
③
ATP
ATP
H+-ATPase
IAA−
H+
ABCB
Auxin gradient
ATP
ATP
H+
IAA−
Vacuole
Cytosol
Cell wall
④
ATP
ATP
ATP
Plasma membrane
IAAH
IAA−
H+
IAA−
IAAH
Auxin back-diffusion
PIN protein
Phosphorylation (P)

Figure 12.16 The cellular basis of polar auxin transport is described by a well-tested chemiosmotic mechanism. (A) Simplified chemiosmotic mechanism of polar auxin transport. Shown here is one elongated cell in a column of auxin-transporting cells. Additional export mechanisms contribute to transport by preventing reuptake of IAA at sites of export and in adjoining cell files. PIN anion efflux carriers are active when phosphorylated (indicated by red dot) by the AGC kinase D6 PROTEIN KINASE (D6PK). (B) Model for polar auxin transport in small cells with significant back-diffusion of auxin due to a high surface-to-volume ratio. ABCB proteins are thought to maintain polar streams by preventing reuptake of auxin exported at carrier sites. In larger cells, ABCB transporters appear to prevent movement of auxin out of polar streams into adjoining cell files.

that cotransport two protons along with the auxin anion. This secondary active transport of auxin allows for greater auxin accumulation than does simple diffusion because anionic auxin is driven across the membrane by the proton motive force (i.e., the high H^+ concentration in the apoplastic solution). The most important contribution of AUX1 is its role in creating cellular sinks that drive polar auxin transport streams. Shootward auxin flows in the *aux1* mutant of Arabidopsis are completely disrupted, resulting in agravitropic root growth; expression of *AUX1* under the control of a promoter associated with the lateral root cap completely restores gravitropic growth in this mutant. The compound 1-naphthoxyacetic acid is often used as an inhibitor of the auxin uptake activity of AUX1/LAX proteins. The NRT1.1 and 1.5 nitrate transporters, which are up-regulated under nitrogen-deficient conditions, also transport auxin, adding to the auxin uptake capacity of root cells.

AUXIN EFFLUX At the neutral pH of the cytosol, the anionic form of auxin, IAA^-, predominates. Transport of IAA^- out of the cell is driven by the negative membrane potential inside the cell. However, because the lipid bilayer of the membrane is impermeable to the anion, auxin export out of the cell must occur via transport proteins in the plasma membrane. Where **PIN auxin efflux carrier proteins** are polarly localized—that is, present on the plasma membrane at only one end of a cell—auxin uptake into the cell and subsequent efflux via PIN give rise to a net polar transport (Figure 12.16A). The PIN family of proteins is named after the pin-shaped inflorescences formed by the *pin1* mutant of Arabidopsis (**Figure 12.17A**). Different PIN family members mediate auxin efflux in each tissue, and *pin* mutants exhibit phenotypes consistent with this function in these tissues. Of the PIN proteins, PIN1 is the most studied, as it is essential to virtually every aspect of polar development and organogenesis in plant shoots.

A subset of ATP-dependent transporters from the large superfamily of ATP-binding cassette (ABC) integral membrane transporters amplifies efflux and prevents reuptake of exported auxin, especially in small cells where auxin concentrations are high. Defective *ABCB* (ABC"B" class) genes in Arabidopsis, maize, and sorghum result in dwarf phenotypes of varying severity and in altered gravitropism and reduced auxin efflux (**Figure 12.17B**). In general, ABCBs are uniformly, rather than polarly, distributed in the plasma membranes of cells in shoot and root apices. However, when specific ABCB and PIN proteins co-occur in the same location in the cell, the directionality of auxin transport is enhanced; PINs function synergistically with ABCBs to stimulate polar auxin transport (Figure 12.16B). The compound *N*-1-naphthylphthalamic acid (NPA) binds to ABCB and PIN auxin transport proteins and is used as an inhibitor of both ABCB and PIN auxin efflux activity.

Auxin transport is regulated by multiple mechanisms

As would be expected for such an important function, auxin transport is regulated by both transcriptional and posttranscriptional mechanisms. Auxin itself regulates expression of the genes encoding auxin transporters to increase or decrease their abundance and thus to regulate auxin levels.

Auxin transporters are also regulated by AGC kinases that regulate the polar localization and activity of PIN proteins and the efflux activity of ABCB19 in phototropic responses (Chapter 14). Membrane composition and cell wall structure also regulate transporter activity, as both PIN and ABCB plasma membrane localization is dependent on structural sterols or sphingolipids, and PIN1 polar localization is abolished in protoplasts and in **cellulose synthase**–deficient mutants of Arabidopsis. Furthermore, some natural compounds, primarily flavonoids, function as auxin efflux inhibitors.

Regulation of the cellular trafficking of auxin transport proteins to and from the plasma membrane plays a particularly important role in plant development.

PIN auxin efflux carrier proteins Plasma membrane transport proteins that amplify localized, directional auxin streams associated with embryonic development, organogenesis, and tropic growth.

cellulose synthase Enzyme that catalyzes the synthesis of individual (1,4)-linked β-D-glucans that make up cellulose microfibrils.

Figure 12.17 Phenotypes associated with the deletion of auxin efflux proteins. (A) PIN1 in Arabidopsis. (Left) Localization of the PIN1 protein at the basal ends of conducting cells (yellow) in Arabidopsis inflorescences as seen by immunofluorescence microscopy. (Middle) The *pin1* mutant of Arabidopsis. A normal wild-type Arabidopsis plant is on the right. (B) The *BR2* (*Brachytic 2*) gene encodes an ABCB required for normal auxin transport in maize, and *br2* mutants have short internodes. The *br2* mutants have compact lower stalks (middle and right) but normal tassels and ear (left and middle).

Specific chaperone proteins are required for successful direction of AUX1, PIN, and ABCB auxin transporters to the plasma membrane. However, the polarized trafficking of a subset of PIN proteins is the most notable.

Polar localization of the PIN auxin efflux proteins is thought to involve three processes:

- Initial isotropic (nondirectional) trafficking to the plasma membrane
- **Transcytosis** and concentration in polarized plasma membrane domains. Polar localization and activity of PINs is regulated by the activity of the D6PK and PINOID AGC kinases. The name PINOID derives from the phenotypic similarity of the *pinoid* and *pin1* mutants of Arabidopsis.
- Stabilization via interactions with the cell wall

transcytosis Redirection of a secreted protein from one membrane domain within a cell to another, polarized domain.

12.6 Hormonal Signaling Pathways

Describe the types of hormone receptors and their signaling mechanisms and electrical signaling.

Hormones act on cells that possess specific receptors that bind the hormone and initiate signal transduction events. Plants employ large numbers of receptor kinases and signal transduction kinases to bring about the physiological responses in hormone target cells. In this section we examine the types of receptors and signal transduction pathways that are associated with the primary plant hormones.

The cytokinin and ethylene signal transduction pathways derive from bacterial two-component regulatory systems

In bacteria, **two-component regulatory systems** mediate a wide range of responses to environmental stimuli. The two components of this signaling system consist of a membrane-bound histidine kinase **sensor protein** and a soluble **response regulator** protein. Sensor proteins receive the input signal, undergo autophosphorylation on a histidine residue, and transfer the phosphoryl group to a conserved aspartate residue on the response regulator. The activated response regulators primarily function as transcription factors that activate cellular responses. Modifications of this two-component system function in plant cytokinin and ethylene signal transduction.

Cytokinin signaling is mediated by a phospho-relay system consisting of a transmembrane cytokinin receptor, a phosphotransfer protein, and a nuclear response regulator (**Figure 12.18A**). The cytokinin receptors are hybrid sensor histidine kinases, as they contain the sensor input and histidine kinase (transmitter) domains but also have a receiver domain similar to the bacterial response regulator protein. The majority of the cytokinin receptors are in the ER membrane, with the hormone binding domain in the ER lumen and the transmitter and receiver domains in the cytosol (Figure 12.18A). Cytokinin binding triggers autophosphorylation of a histidine residue on the transmitter domain, followed by transfer of this same phosphate to the aspartate residue on the receiver domain. The phosphate is then transferred to a histidine phosphotransfer (HPT) protein that then transfers the phosphate to nuclear transcriptional regulators to regulate their activity.

Ethylene receptors are also similar to bacterial two-component histidine kinases, but do not exhibit phospho-relay activity. Only two of the ethylene receptor protein family in Arabidopsis function as histidine kinases and this activity is not essential to their function. The ethylene receptor ETR1 is also located in the ER membrane and interacts with the soluble serine/threonine kinase CONSTITUTIVE TRIPLE RESPONSE 1 (CTR1) and the transmembrane protein ETHYLENE-INSENSITIVE 2 (EIN2) (**Figure 12.18B**). When ethylene is not present, the cytosolic C-terminal domain of EIN2 is phosphorylated by CTR1 and remains in the cytosol. When ethylene binds to the ETR1 receptor dimer, CTR1 is "switched off," and the dephosphorylated C-terminus of EIN2 is cleaved by proteases and migrates into the nucleus. In the nucleus, the C-terminal fragment of EIN2 associates with a complex made up of Ethylene Insensitive 3 (EIN3) and an Ethylene Insensitive Like (EIL) transcription factor to activate histone acetyl transferase (HAT) activity. HAT activation relaxes chromatin-histone interactions of ethylene-responsive genes to enhance gene transcription. Genes that are repressed by ethylene signaling have a transcription factor called EIN2-NUCLEAR ASSOCIATED PROTEIN1(ENAP1) associated with their promoter. Interaction of an EIN3/EIL complex with ENAP1 recruits a SRT1/2 histone deactylase (HDAC) complex to enhance chromatin-histone interactions and, thus, repress gene expression. In this way, ethylene can activate one set of genes and inactivate another set.

two-component regulatory systems Signaling pathways common in prokaryotes. They typically involve a membrane-bound histidine kinase sensor protein that senses environmental signals and a response regulator protein that mediates the response. Although rare in eukaryotes, systems resembling bacterial two-component systems are involved in both ethylene and cytokinin signaling.

sensor proteins Bacterial receptor proteins that perceive external or internal signals as part of two-component regulatory systems. They consist of two domains, an *input domain*, which receives the environmental signal, and a *transmitter domain*, which transmits the signal to the response regulator.

response regulator One component of the two-component regulatory systems that are composed of a histidine kinase sensor protein and a response regulator protein. Response regulators have a *receiver domain*, which is phosphorylated by the sensor protein, and an *output domain*, which carries out the response.

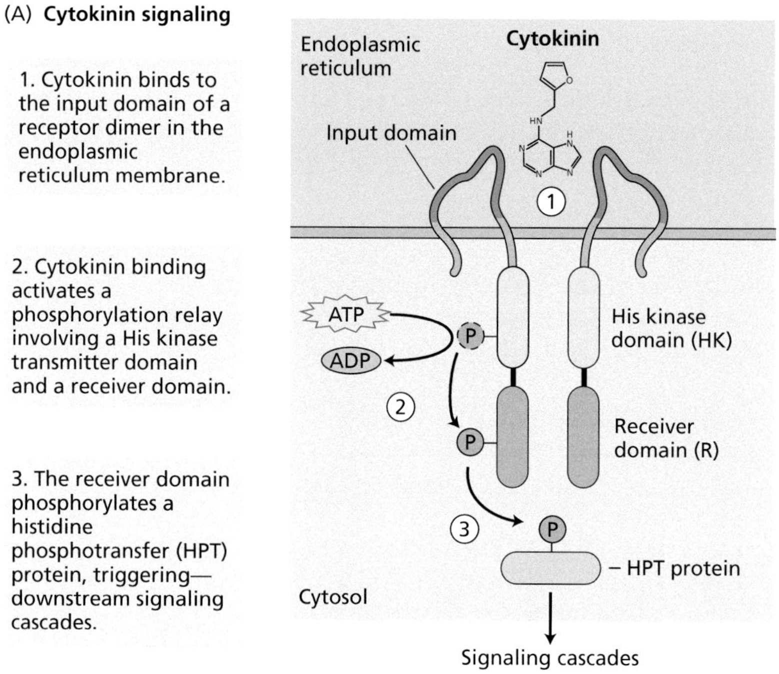
(A) Cytokinin signaling
1. Cytokinin binds to the input domain of a receptor dimer in the endoplasmic reticulum membrane.
2. Cytokinin binding activates a phosphorylation relay involving a His kinase transmitter domain and a receiver domain.
3. The receiver domain phosphorylates a histidine phosphotransfer (HPT) protein, triggering— downstream signaling cascades.
Endoplasmic reticulum
Cytokinin
Input domain
1
ATP
ADP
P
2
P
3
P
His kinase domain (HK)
Receiver domain (R)
– HPT protein
Cytosol
Signaling cascades

Figure 12.18 Continued

Phosphorylation also regulates brassinosteroid, ABA, and auxin perception

Binding of active brassinosteroids to a plasma membrane–localized receptor initiates signal transduction pathways that amplify the signal and inactivate transcriptional repressors in the nucleus. As discussed earlier in this chapter, brassinosteroid binding to the BRASSINOSTEROID-INSENSITIVE 1 (BRI1) receptor-like kinase dimer results in recruitment and phosphorylation of the BRI1-ASSOCIATED RECEPTOR KINASE1 (BAK1). Phosphorylation of BAK1 results in activation of specific transcription factors to enhance expression of brassinosteroid-responsive genes (**Figure 12.19A**).

ABA signaling involves the activity of both kinases and phosphatases. ABA binding to its receptor regulates a subset of phosphatases to regulate the protein kinase activity of a group of **SNF1-Related Kinase2** (SnRK2) proteins. In the absence of ABA, these phosphatases bind to SnRK2s to prevent their kinase activity (**Figure 12.19B**). ABA binding changes the conformation of the receptor protein to permit or enhance interaction with the phosphatase and release the SnRK2 kinases from inhibition. SnRK2 proteins are then free to phosphorylate their many target proteins, including ion channels regulating stomatal aperture and the transcription factors that bind ABA response elements in gene promoters to activate ABA-responsive gene expression. The phosphatases also interact with other proteins involved in cellular ABA responses such as stomatal closing. ABA signal transduction is therefore based on reversing the balance between phosphatase and kinase activities.

A similar phosphorylation mechanism mediates a subset of very fast auxin responses at the cell surface (**Figure 12.19C**). In Arabidopsis, auxin binding to Zn^{2+} binding proteins from the AUXIN Binding Protein 1 (ABP1) / ABP1-like (ABL) family activates their co-receptors called TRANSMEMBRANE KINASES. Activated TMKs initiate rapid phosphorylation of a large number of cytosolic and plasma membrane proteins. Most prominently, TMKs then activate the plasma

SNF1-Related Kinase2 (SnRK2) A family of kinases that includes ABA-activated protein kinases or stress-activated protein kinases.

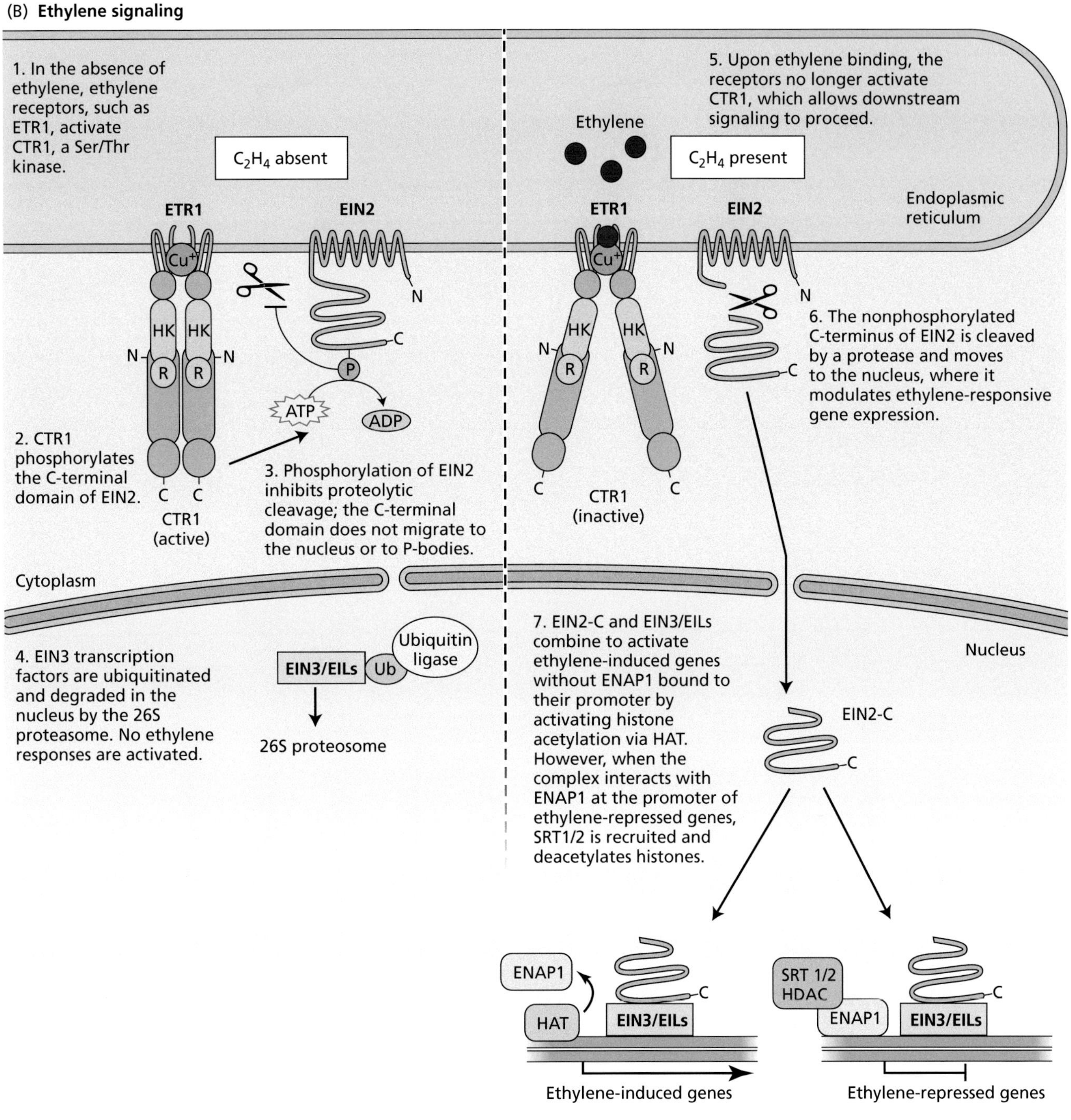
(B) Ethylene signaling
1. In the absence of ethylene, ethylene receptors, such as ETR1, activate CTR1, a Ser/Thr kinase.
C2H4 absent
ETR1
EIN2
Cu+
HK HK
N R R N
C
N
P
ATP
ADP
2. CTR1 phosphorylates the C-terminal domain of EIN2.
C C
CTR1 (active)
3. Phosphorylation of EIN2 inhibits proteolytic cleavage; the C-terminal domain does not migrate to the nucleus or to P-bodies.
Cytoplasm
4. EIN3 transcription factors are ubiquitinated and degraded in the nucleus by the 26S proteasome. No ethylene responses are activated.
EIN3/EILs
Ub
Ubiquitin ligase
26S proteosome
Ethylene
5. Upon ethylene binding, the receptors no longer activate CTR1, which allows downstream signaling to proceed.
C2H4 present
ETR1
EIN2
Endoplasmic reticulum
CTR1 (inactive)
6. The nonphosphorylated C-terminus of EIN2 is cleaved by a protease and moves to the nucleus, where it modulates ethylene-responsive gene expression.
7. EIN2-C and EIN3/EILs combine to activate ethylene-induced genes without ENAP1 bound to their promoter by activating histone acetylation via HAT. However, when the complex interacts with ENAP1 at the promoter of ethylene-repressed genes, SRT1/2 is recruited and deacetylates histones.
Nucleus
EIN2-C
ENAP1
HAT
EIN3/EILs
SRT 1/2 HDAC
ENAP1
EIN3/EILs
Ethylene-induced genes
Ethylene-repressed genes

Figure 12.18 Cytokinin and ethylene signaling in Arabidopsis. (A) Cytokinin binding to the input domain of the receptor dimer inside the ER to activate cytosolic signaling cascades via phosphorylation of the HPT protein. (B) Ethylene binds to ETHYLENE RECEPTOR 1 (ETR1) dimer inside the ER via a copper ion cofactor, which is assembled into the ethylene receptors. In the absence of ethylene, the kinase CONSTITUTIVE TRIPLE RESPONSE 1 (CTR1) phosphorylates ETHYLENE-INSENSITIVE 2 (EIN2) to stabilize it and prevent cleavage of the EIN2 C terminus (EIN2-C). EIN2-C then migrates to the nucleus where it forms a complex with ETHYLENE INSENSITIVE 3 and one of a group of ETHYLENE INSENSITIVE LIKE transcription factors. The complex enhances ethylene-responsive gene expression by increasing Histone Acetyl Transferase activity. The complex reduces expression of ethylene-repressed genes via interaction with EIN2 NUCLEAR ASSOCIATED PROTEIN 1 (ENAP1) and a sirtuin (SRT)1/2 histone deacetylase (HDAC) that increases chromatin-histone interactions.

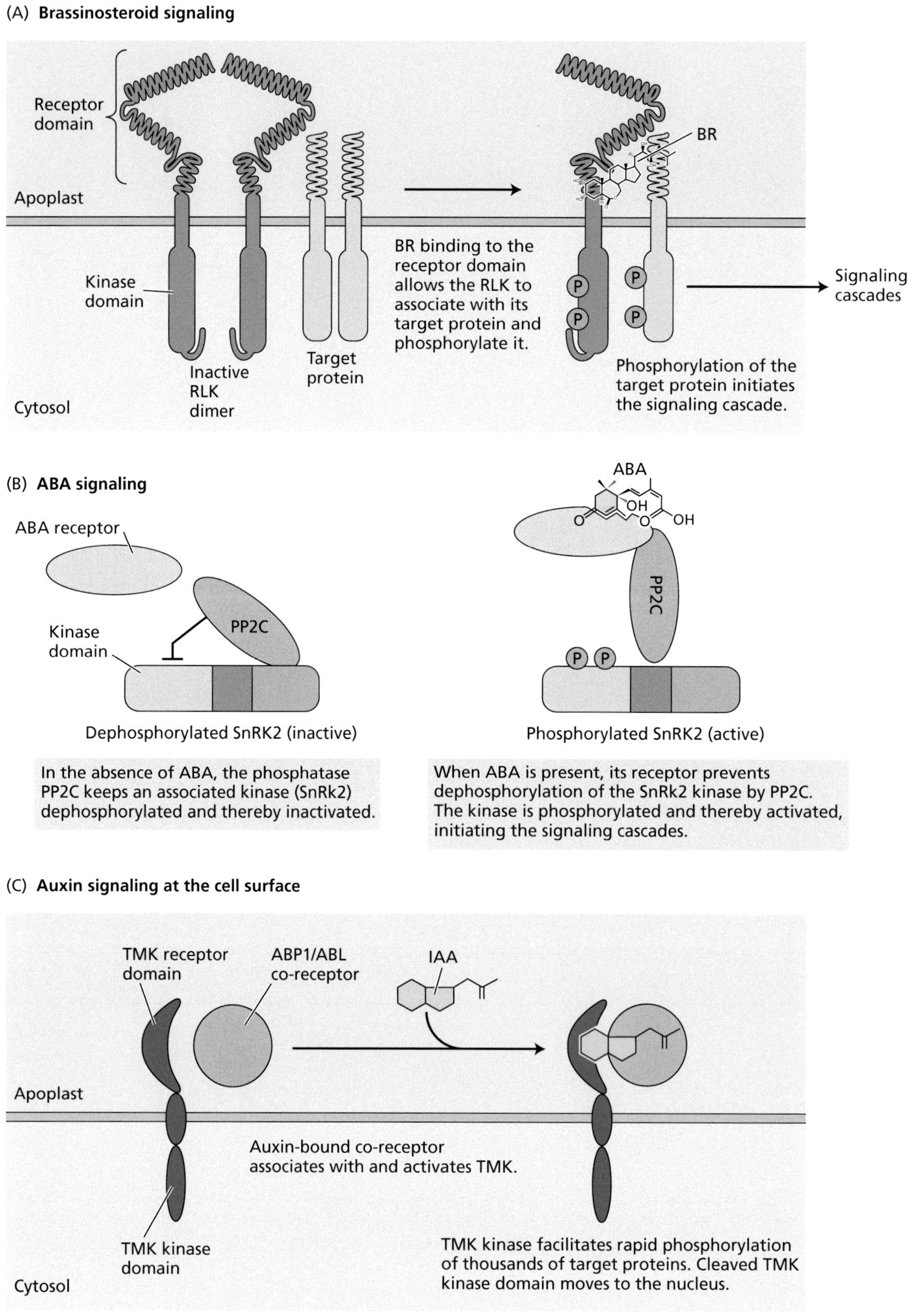
(A) Brassinosteroid signaling
Receptor domain
Apoplast
Kinase domain
Inactive RLK dimer
Target protein
Cytosol
BR
BR binding to the receptor domain allows the RLK to associate with its target protein and phosphorylate it.
P
Signaling cascades
Phosphorylation of the target protein initiates the signaling cascade.
(B) ABA signaling
ABA receptor
Kinase domain
PP2C
Dephosphorylated SnRK2 (inactive)
In the absence of ABA, the phosphatase PP2C keeps an associated kinase (SnRk2) dephosphorylated and thereby inactivated.
ABA
OH
O
Phosphorylated SnRK2 (active)
When ABA is present, its receptor prevents dephosphorylation of the SnRk2 kinase by PP2C. The kinase is phosphorylated and thereby activated, initiating the signaling cascades.
(C) Auxin signaling at the cell surface
TMK receptor domain
ABP1/ABL co-receptor
IAA
Apoplast
Auxin-bound co-receptor associates with and activates TMK.
TMK kinase domain
Cytosol
TMK kinase facilitates rapid phosphorylation of thousands of target proteins. Cleaved TMK kinase domain moves to the nucleus.

Figure 12.19 Brassinosteroids (A), ABA (B), and auxin (C) also trigger kinase/phosphatase signaling cascades.

membrane H^+-ATPase to increase apoplastic acidification and cellular elongation and to regulate the developmental responses of etiolated (dark-grown) seedlings. TMKs also directly activate nuclear repressors of auxin signaling.

Multiple plant hormone receptors use ubiquitination to remove repressor proteins

Protein degradation of transcriptional repressor proteins was first described as part of the auxin signaling pathway. Since then, the **ubiquitin–proteasome pathway** has also been shown to be central to the jasmonate and gibberellin signaling pathways. In brief, a small protein called ubiquitin is first activated by an *E1 ubiquitin-activating* enzyme and then attached to a target protein via an *E2 ubiquitin-conjugating* enzyme and a *ubiquitin E3 ligase* complex. The ubiquitin E3 ligase complexes are highly conserved among eukaryotes and are called **S-phase kinase associated protein 1** (**Skp1**)**/Cullin/F-box** (**SCF**) **complexes**, based on the names of the components of the complex that was first described in studies of animal signal transduction. A superscript term, usually derived from a mutant screen, is applied to an E3 ligase name (e.g., SCF^{TIR1}) to indicate which F-box protein the complex contains. F-box proteins typically recruit target proteins to the SCF complex so that they can be tagged with multiple copies of ubiquitin by the E3 ligase to target the protein for degradation by the 26S proteasome (Chapter 1).

Several of these F-box proteins are components of hormone receptor complexes, and the proteins targeted for degradation are transcriptional repressors (**Figure 12.20**). This occurs in the signaling pathways of auxin, jasmonates, and gibberellins (all small organic acids), demonstrating the adaption of a common mechanism to specific signals.

In the auxin nuclear (occurring in the nucleus) signaling pathway, IAA acts as a "molecular glue" to bring together TIR1/Auxin F-box proteins with **AUXIN/INDOLE-3-ACETIC ACID** (**AUX/IAA**) repressors of **auxin response factor** (**ARF**) transcription factors. ARFs are generally transcriptional activators of genes with **auxin response element** (**AuxRE**) in their promoter regions. When auxin concentrations are low, AUX/IAA repressor proteins form heterodimers with the ARFs that repress transcriptional activation. The AUX/IAA/ARF complexes also recruit chromatin remodeling factors via an interaction with the TOPLESS (TPL) protein, which recruits histone deacetylases (HDACs) to maintain the chromatin in a transcriptionally inactive state. Ubiquitination and degradation of AUX/IAAs in the presence of auxin allow the ARFs to dimerize or even oligomerize recruit histone acetyltransferases (HATs) that loosen chromatin to activate gene expression (Figure 12.20A). Among the many target genes of this pathway are genes encoding auxin-metabolizing enzymes and AUX/IAA repressors, which eventually feedback on the system to reduce active auxin levels and terminate ARF-dependent signaling. Synthetic AuxRE promoters fused to reporter proteins are used to visualize cellular auxin levels. Disappearance of fluorescent proteins fused to AUX/IAA "degron" domains is used to negatively report rapid changes in intracellular auxin levels.

A small subset of ARF transcription factors contain unique protein domains that directly bind auxin. When activated by auxin binding, these ARFs function in auxin-dependent leaf and flower development. One unique auxin F-box protein functions in canonical nuclear auxin receptor complexes but also functions as a cytosolic auxin receptor. This specialized AFB converts cytosolic ATP to the second messenger cyclic adenosine monophosphate (cAMP), which, in turn, activates rapid Ca^{2+} release in Arabidopsis roots that is required for gravitropic bending (Chapter 14).

Jasmonates and gibberellins also promote the interaction between an F-box protein of an SCF complex and target proteins (Figure 12.20B, Figure 12.20C). The F-box protein CORONATINE-INSENSITIVE1 (COI1) functions as a JA-Ile receptor (Figure 12.20B). Like auxin, JA-Ile promotes the interaction between COI1 and

ubiquitin–proteasome pathway A mechanism for the specific degradation of cellular proteins involving two discrete steps: the polyubiquitination of proteins via the E3 ubiquitin ligase and the degradation of the tagged protein by the 26S proteasome.

S-phase kinase associated protein 1 (Skp1)/Cullin/F-box (SCF) complexes Large protein complexes that function as E3 ubiquitin ligases in the signaling pathways of several plant hormones.

AUXIN/INDOLE-3-ACETIC ACID (AUX/IAA) A family of short-lived small proteins that combine with the TIR1/AFB proteins to form the primary auxin receptor. This family of short-lived small proteins in Arabidopsis regulates auxin-induced gene expression by binding to ARF protein that is bound to DNA. If the specific ARF is a transcriptional activator, the Aux/IAA binding represses transcription.

auxin response factors (ARFs) A family of proteins that regulate the transcription of specific genes involved in auxin responses; they are inhibited by association with specific Aux/IAA repressor proteins, which are degraded in the presence of auxin.

auxin response element (AuxRE) A DNA promoter sequence that modulates gene expression when bound by auxin-responsive transcription factors.

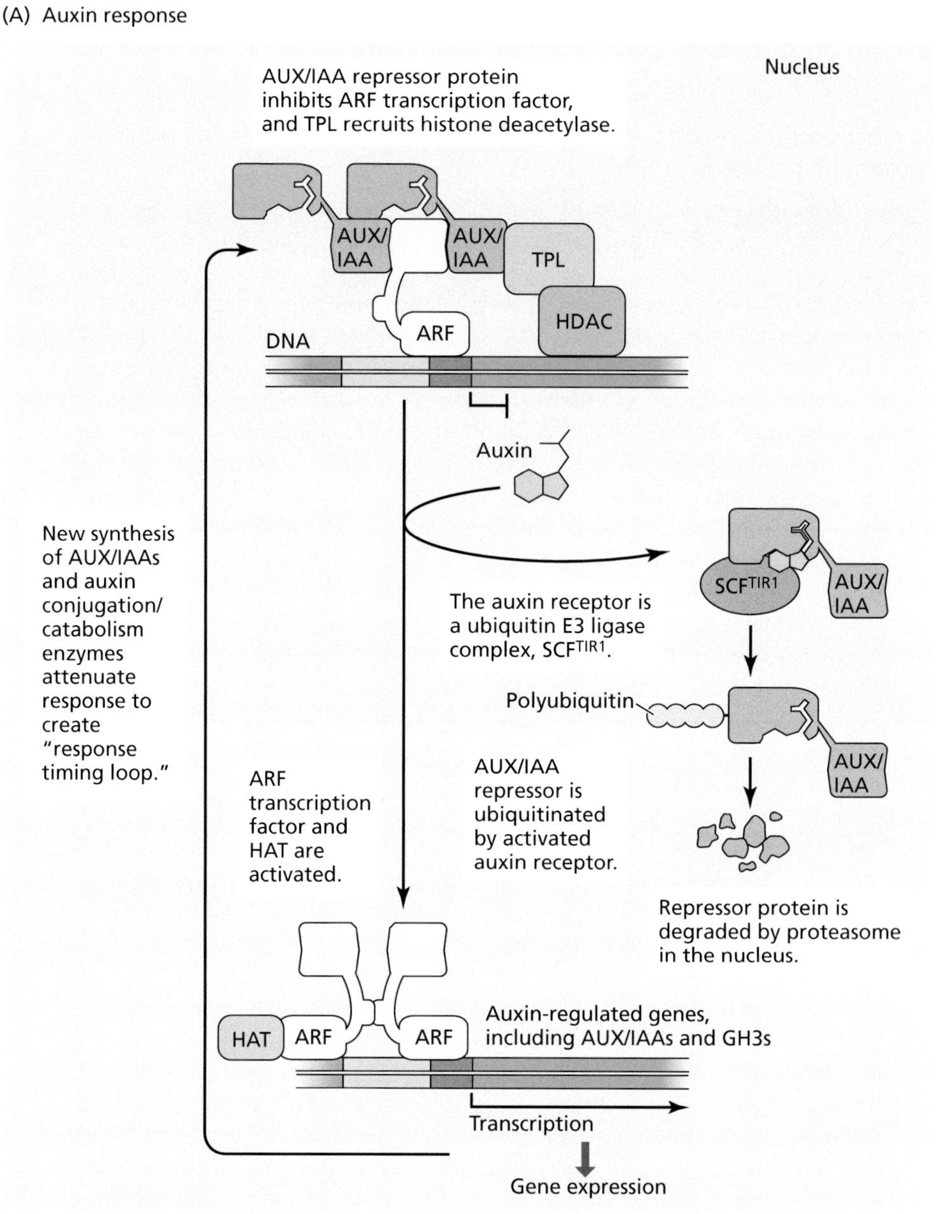

Figure 12.20 Signal transduction pathways in plants often function by inactivating repressor proteins. Auxin, JA-Ile (JA), and gibberellin (GA) signal by promoting interaction between components of the SCF ubiquitination machinery and repressor proteins operating in each hormone's signal transduction pathway. Auxin (A) and the jasmonate JA-Ile (B) directly promote interaction between the SCF^{TIR1} and SCF^{COI1}complexes and the AUX/IAA and JAZ repressors, respectively. The TOPLESS (TPL) protein functions as a co-repressor with AUX/IAAs and JAZ. In JAZ signaling, TPL functions as a co-repressor after co-phytochrome-interacting recruitment with NINJA. TPL recruits histone deacetylase (HDAC) to the complex to keep the associated chromatin in a repressed state. Release of AUX/IAA and JAZ repression after hormone binding releases TPL and allows activation of the chromatin via histone acetyltransferase (HAT). The structural characteristics of the ARF and AUX/IAA proteins that function in auxin signaling have been determined by X-ray crystallography and are reflected in the figure. The structural characteristics of the JAZ repressor protein are only partially known. (C) In contrast, gibberellin additionally requires a separate receptor protein, GID1, to form the complex between SCF^{SLY1} and DELLA repressor proteins. In all three cases, addition of multiple ubiquitins (polyubiquitin) marks the repressor proteins for degradation. This triggers the activation of ARF, MYC2, and PIF3/4 transcription factors, resulting in auxin-, JA-Ile-, and gibberellin-induced changes in gene expression.

repressors of JA-Ile-induced gene expression called **JA-ILE ZIM-DOMAIN (JAZ)** proteins, thereby targeting JAZ proteins for degradation. Analogous to AUX/IAA proteins, JAZ repressor proteins suppress transcription of jasmonate-responsive genes, but in this case by binding to and repressing basic helix-loop-helix (bHLH)

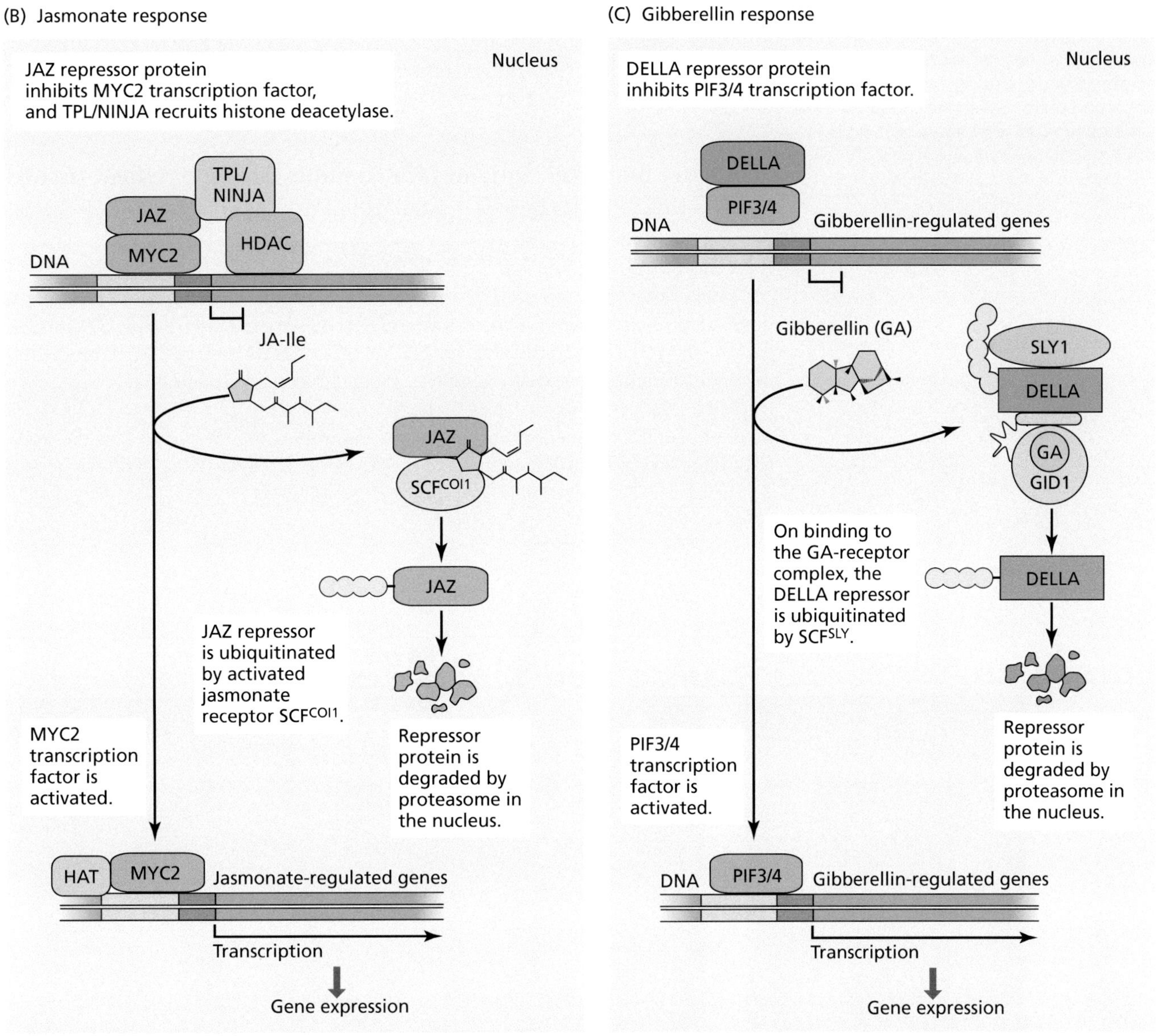

MYC transcription factors via interactions with TPL and a novel interacting protein, NINJA. JA-Ile-induced ubiquitin-dependent degradation of JAZ repressor proteins results in the release and activation of these transcription factors, triggering the induction of jasmonate-responsive gene expression.

Gibberellin signaling involves components of the SCF complex (Figure 12.20C). However, the gibberellin receptor **GIBBERELLIN INSENSITIVE DWARF 1** (**GID1**) does not itself function as an F-box protein. Instead, when GID1 binds gibberellin, the receptor undergoes a conformational change that promotes the binding of repressor proteins known as DELLAs (for a signature Asp-Glu-Leu-Leu-Ala amino acid sequence). This in turn induces a conformational change in the DELLA protein and facilitates the interaction of GID1-bound DELLA to an SCF^{SLY1} complex to trigger degradation of the DELLA proteins by the 26S proteasome. DELLA degradation results in the release and activation of **phytochrome interacting factor** (**PIF**) and other transcription factors to effect changes in gene expression.

JA-ILE ZIM-DOMAIN (JAZ) A transcriptional repressor that serves as a switch for jasmonate signaling. In the presence of JA, JAZ is degraded, allowing positive transcriptional regulators to activate JA-induced genes.

GIBBERELLIN INSENSITIVE DWARF 1 (GID1) Gibberellin receptor protein in rice.

phytochrome interacting factors (PIFs) A family of phytochrome-interacting proteins that may activate and repress gene transcription; some PIFs are targets for phytochrome-mediated degradation.

action potential A transient event in which the membrane potential difference rapidly decreases (depolarizes) and abruptly increases (hyperpolarizes). Action potentials, which are triggered by the opening of ion channels, can be self-propagating along linear files of cells, especially in the vascular systems of plants.

In all of the signaling mechanisms described in this section, there are multiple isoform combinations present in each cell type and at each phase of plant development. This allows for a diversity of responses appropriate to that cell or tissue type. These responses will be discussed in detail in subsequent chapters.

Plants use electrical signaling for communication between tissues

Plants employ long-distance electrical signaling to communicate between distant parts of the plant body. The most common type of electrical signaling in plants is the **action potential**, the transient depolarization of the plasma membrane of a cell generated by voltage-gated ion channels (Chapter 6). Action potentials have been shown to mediate touch-induced leaflet closure in sensitive plant (*Mimosa pudica*), as well as the rapid closure (~0.1 s) of Venus flytrap, which occurs when an insect touches the touch-sensitive trigger hairs on the adaxial sides of the trap-like leaf lobes (**Figure 12.21**). Each touch of a trigger hair is detected by a mechanosensitive receptor that activates an initial Ca^{2+} influx and an action potential (electrical signal) characterized by K^+ efflux. To avoid accidental triggering

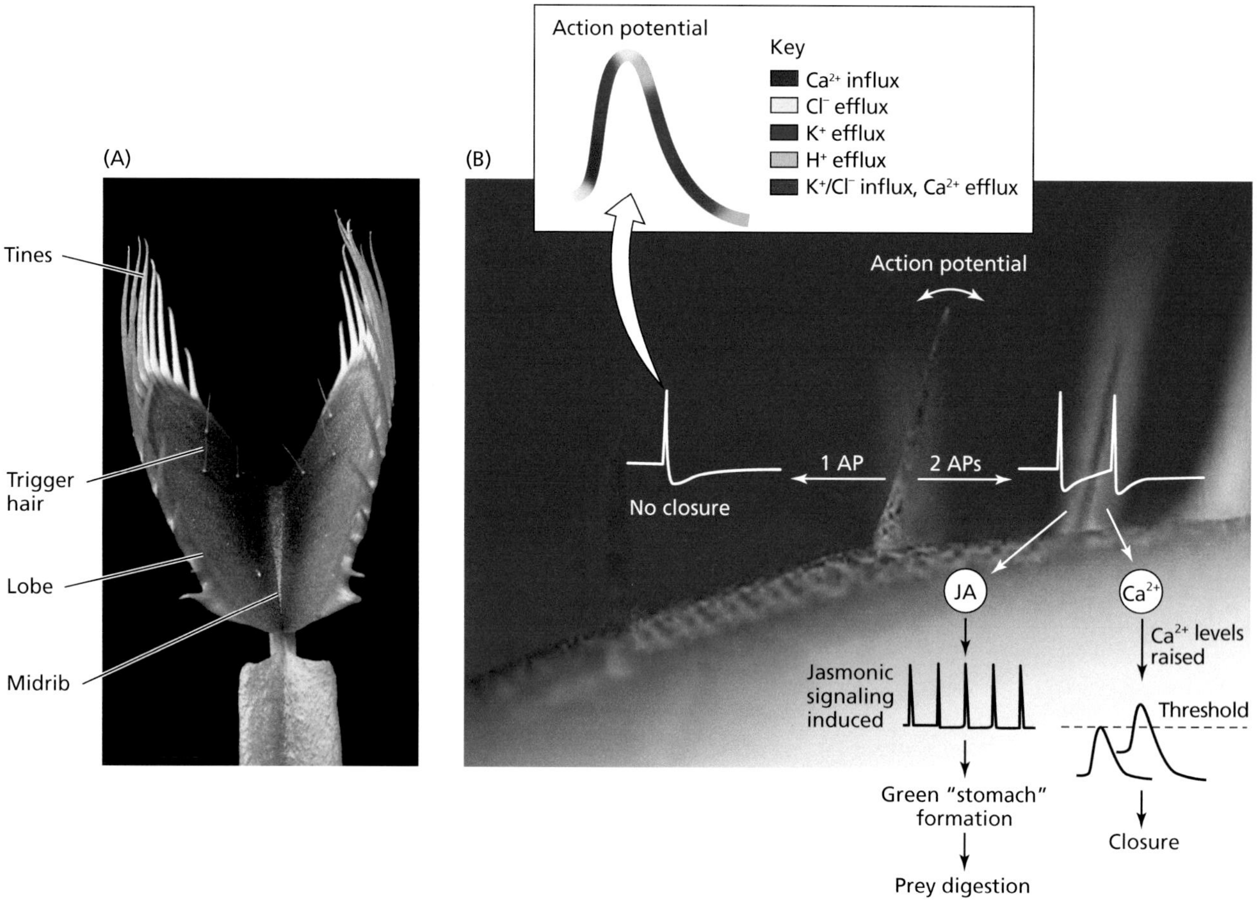

Figure 12.21 Electrical signaling in Venus flytrap (*Dionaea muscipula*). (A) Photograph of the snap-trap leaves with needlelike tines and touch-sensitive trigger hairs. (B) Action potentials (APs) associated with Ca^{2+} influx and K^+ efflux are triggered by movement of mechanosensitive channels in trigger hairs. Mechanical stimulation of the trigger hairs twice by a prey raises Ca^{2+} levels to a threshold that results in flytrap closure. Five triggering events are sufficient to induce jasmonate signaling that stimulates production of digestive enzymes and compounds that create the "green stomach" that digests the prey.

of flytrap closure, either two hairs must be touched within 20 s of each other, or one hair must be touched twice in rapid succession to reach a threshold Ca^{2+} level. A total of five action potentials are required to stimulate jasmonate signaling and the resultant production of digestive enzymes and other components of the "green stomach" that digests the prey.

Electrical signaling has also been shown to facilitate rapid communication between remote parts of plants in response to various types of stresses. As we will discuss in Chapter 18, electrical signals can be propagated throughout the plant via the vasculature in response to damage caused by chewing insects. Despite some similarities, electrical signaling in plants lacks structures similar to animal synapses that secrete neurotransmitters.

Cross-regulation allows signal transduction pathways to be integrated

Within plant cells, signal transduction pathways never function in isolation, but operate as part of a complex web of signaling interactions. These interactions account for the fact that plant hormones often exhibit *agonistic* (additive or positive) or *antagonistic* (inhibitory or negative) interactions with other signals. Classic examples include the antagonistic interaction between gibberellin and ABA in the control of seed germination (Chapter 14). When signaling pathways intersect, the interaction is termed **cross-regulation**, and three categories have been proposed (**Figure 12.22**):

1. **Primary cross-regulation** involves distinct signaling pathways regulating a shared transduction component in a positive or a negative manner.
2. **Secondary cross-regulation** involves the output of one signaling pathway regulating the abundance or perception of a second signal.
3. **Tertiary cross-regulation** involves the outputs of two distinct pathways influencing each other.

cross-regulation The interaction of two or more signaling pathways.

primary cross-regulation Involves distinct signaling pathways regulating a shared transduction component in a positive or a negative manner.

secondary cross-regulation Regulation by the output of one signal pathway of the abundance or perception of a second signal.

tertiary cross-regulation Regulation that involves the outputs of two distinct signaling pathways exerting influences on one another.

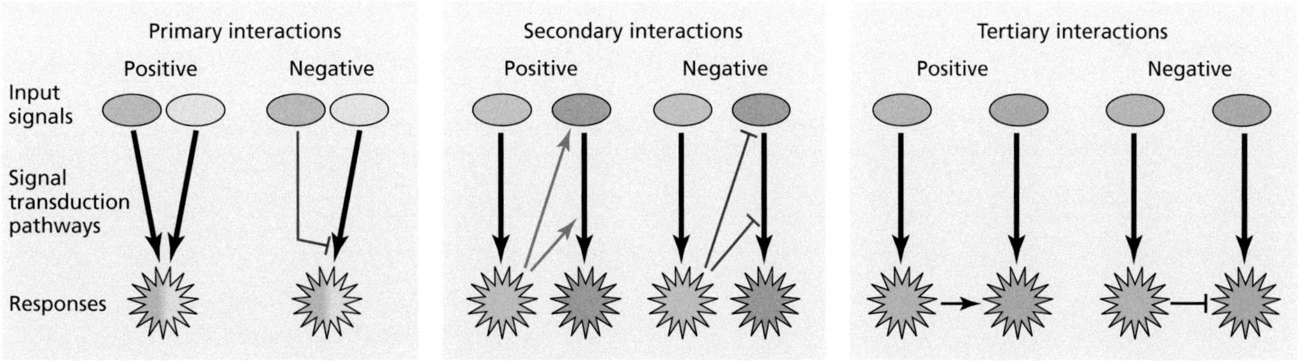

Two input pathways regulate a single shared protein or multiple shared proteins controlling a response. Both pathways have the same effect on the response.

Two input pathways converge on shared protein(s), but one of the pathways inhibits the effect of the other.

Two input pathways regulate separate responses. In addition, one pathway enhances the input levels or perception of the other pathway.

As in the positive interaction, except that one pathway represses the input levels or perception of the other pathway.

The response of one of the signaling pathways promotes the response of the other pathway.

The response of one of the signaling pathways inhibits the response of the other pathway.

Figure 12.22 Signal transduction pathways operate as part of a complex web of signaling interactions. Three types of cross-regulation have been proposed: primary, secondary, and tertiary. Input signals are shown here as ovals, signal transduction pathways are indicated by heavy arrows, and responses (pathway outputs) are shown as stars. Green (positive) or red (negative) colored lines indicate where one pathway influences the other pathway. The three types of cross-regulation can be either positive or negative.

Summary

Both short- and long-term physiological responses to external and internal signals arise from the transformation (transduction) of signals into mechanistic pathways. In order to activate areas that may be distal from the initial signaling location, signaling intermediates are amplified before dissemination (transmission). Once in play, signaling pathways often overlap into complex signaling networks, a phenomenon termed cross-regulation, to coordinate integrated physiological responses.

12.1 Temporal and Spatial Aspects of Signaling

- Plants use signal transduction to coordinate both rapid and slow responses to stimuli (**Figures 12.1**, **12.2**).

12.2 Signal Perception and Amplification

- Receptors are present throughout cells and are conserved among bacteria, plant, animal, and fungal kingdoms (**Figure 12.3**).
- Protein modification by phosphorylation is used to transduce and amplify signals (**Figure 12.4**).
- Signals can also be amplified by second messengers such as Ca^{2+}, H^+, reactive oxygen species (ROS), and modified lipids (lipid signaling molecules), although it can be challenging to parse their signaling targets (**Figure 12.5**).
- Modified lipids regulate growth and stress responses (**Figure 12.6**).

12.3 Hormones and Plant Development

- Hormones are conserved chemical messengers that can, at very low concentrations, transmit signals between cells and initiate physiological responses (**Figure 12.7**).
- Except for ethylene, phytohormones contain ring structures (**Figure 12.8**).
- The first growth hormone to be identified was auxin, during studies of coleoptile bending due to phototropism (**Figure 12.9**).
- The gibberellin group of growth hormones can be used as growth regulators (**Figure 12.10**).
- Tissue-culture experiments revealed the role of cytokinins as cell division–promoting factors (**Figure 12.11**).
- Ethylene is a gaseous hormone that promotes fruit ripening and other developmental processes.
- Abscisic acid regulates seed maturation and stomatal closure in response to water stress (**Figure 12.12**).
- Brassinosteroids are lipid-soluble hormones that regulate many processes, including photomorphogenesis, germination, and floral sex identity (**Figure 12.13**).
- Strigolactones are derived from enzymatic cleavage of carotenoids and regulate branching from the primary root-shoot axis.

12.4 Phytohormone Metabolism and Homeostasis

- The concentration of hormones is tightly regulated so that signals produce timely responses without compromising sensitivity to the same signal in the future (**Figure 12.14**).
- Plant hormones are synthesized from amino acid, nucleoside, terpene, and sterol precursors (**Figure 12.15**).

12.5 Movement of Hormones within the Plant

- Auxin gradients and polar transport are central components of programmed and plastic development. Polar transport of auxin is driven by chemiosmotic gradients and is primarily regulated by polarized efflux carriers (**Figure 12.16**).
- Polar auxin transport regulates organogenesis and maintains the polarity of plant form (**Figure 12.17**).

12.6 Hormonal Signaling Pathways

- Cytokinin and ethylene pathways use derived two-component regulatory systems, which involve membrane-bound sensor proteins and soluble response regulator proteins (**Figures 12.18**).
- Brassinosteroids and abscisic acid pathways utilize phosphorylation to transduce signals (**Figure 12.19**).
- Many plant hormone pathways employ negative regulators (inactivating repressors), allowing faster activation of downstream response genes (**Figures 12.20**).
- Plants can also employ fast-acting, long-distance electrical signaling using action potentials (**Figure 12.21**).
- Integration of signal transduction pathways is accomplished through cross-regulation (**Figure 12.22**).

Suggested Reading

Binder, B. (2020) Ethylene signaling in plants. *J. Biol. Chem.* 295(22): 7710–7725.

Blasquez, M. A., Nelson, D. C., and Weijer, D. (2020) Evolution of plant hormone response pathways. *Annu. Rev. Plant Biol.* 71: 327–353. doi:10.1146/annurev-arplant-050718-100309.

Hwang, I., Sheen, J., and Müller, B. (2012) Cytokinin signaling networks. *Annu. Rev. Plant Biol.* 63: 353–380.

Kim, E.-J., and Russanova, E. (2020) Brassinosteroid signalling. *Current Biol.* 30(7): R294–R298. doi:10.1016/j.cub.2020.02.011.

Suarez-Rodriguez, M. C., Petersen, M., and Mundy, J. (2010) Mitogen-activated protein kinase signaling in plants. *Annu. Rev. Plant Biol.* 61: 621–649.

Wang, X., Devaiah, S. P., Zhang, W., and Welti, R. (2006) Signaling functions of phosphatidic acid. *Prog. Lipid Res.* 45: 250–278.

13 Signals from Sunlight

Sunlight serves not only as an energy source for photosynthesis, but also as a signal that regulates various developmental processes, from seed germination to fruit development and senescence (**Figure 13.1**). Sunlight also provides directional cues for plant growth as well as nondirectional cues for plant movements. We have already touched on several light-sensing mechanisms in the preceding chapters. In Chapter 9 we saw that chloroplasts move within leaf palisade cells in response to blue light to achieve optimal light absorption. We also learned that the leaves of many species are able to bend toward the sun during its progress across the sky, a phenomenon known as **solar tracking**. As discussed in Chapters 3 and 6, stomata use blue light as a signal for opening, a sensory response that enables CO_2 to enter the leaf and initiate transpiration.

In later chapters we will encounter examples of light-regulated plant development. For example, many seeds require light to germinate. Sunlight inhibits stem growth and stimulates leaf expansion in growing seedlings, two of several light-induced phenotypic changes collectively referred to as **photomorphogenesis** (**Figure 13.2**). Most of us are familiar with the observation that the branches of houseplants placed near a window grow toward the incoming light. This phenomenon, called **phototropism**, is an example of how plants alter their growth patterns in response to the direction of incident radiation. This response is stimulated primarily by blue light (**Figure 13.3**) and is particularly important for seedling establishment (Chapter 14). In some species the leaves fold up at night (**nyctinasty**) and open at dawn (**photonasty**). Photonastic movements are plant movements in response to nondirectional light. As we will discuss in Chapter 16, many plants flower at specific times of the year in response to changing day length, a phenomenon called **photoperiodism**.

solar tracking The movement of leaf blades throughout the day so that the planar surface of the blade remains perpendicular to the sun's rays.

photomorphogenesis The influence and specific roles of light on plant development. In the seedling, light-induced changes in gene expression that support aboveground growth in the light rather than belowground growth in the dark.

phototropism The alteration of plant growth patterns in response to the direction of incident radiation, especially blue light.

nyctinasty Sleep movements of leaves. Leaves extend horizontally to face the light during the day and fold together vertically at night.

photonasty Plant movements in response to nondirectional light.

photoperiodism A biological response to the length and timing of day and night, making it possible for an event to occur at a particular time of year.

Figure 13.1 Sunlight exerts multiple influences on plants. Plants expose their leaves to sunlight to transform solar energy into chemical energy, and they also use sunlight for a wide range of developmental signals that optimize photosynthesis and detect seasonal changes.

In addition to visible light, sunlight also contains infrared and potentially damaging ultraviolet (UV) radiation (**Figure 13.4**). Many plants can sense the presence of UV radiation and protect themselves against cellular damage by synthesizing simple phenolics, flavonoids, and carotenoid compounds that act as sunscreens and remove damaging oxidants and free radicals generated by high-energy UV photons.

All of the photoresponses to visible light noted above, as well as the responses to UV radiation, involve receptors that detect specific wavelengths of light and induce developmental or physiological changes. As we saw in Chapter 12, signal transduction involves a chain of reactions beginning with a receptor and ending with a physiological response. The receptor molecules that plants use

Figure 13.2 Comparison of seedlings grown in the light versus the dark. (Left) Cress seedlings grown in the light. (Right) Cress seedlings grown in the dark. The light-grown seedlings exhibit photomorphogenesis. The dark-grown seedlings undergo etiolation, characterized by elongated hypocotyls and a lack of chlorophyll.

photoreceptors Proteins that sense the presence of light and initiate a response via a signaling pathway.

Figure 13.3 Time-lapse photograph of a maize (corn; *Zea mays*) coleoptile growing toward unilateral blue light given from the right. In the first image on the left, the coleoptile is about 3 cm long. The consecutive exposures were made 30 min apart. Note the increasing angle of curvature as the coleoptile bends. (Courtesy of M. A. Quiñones.)

to detect sunlight are termed **photoreceptors**. Photoreceptors undergo a conformational change when irradiated by a particular wavelength of light (perceived by the eye as color) to initiate signaling reactions that typically involve second messengers and phosphorylation cascades collectively referred to as photoresponses.

In this chapter we discuss the signaling mechanisms involved in light-regulated growth and development, focusing primarily on the receptors for red light (620–700 nm), far-red light (710–850 nm), blue light (350–500 nm), and UV-B radiation (290–320 nm).

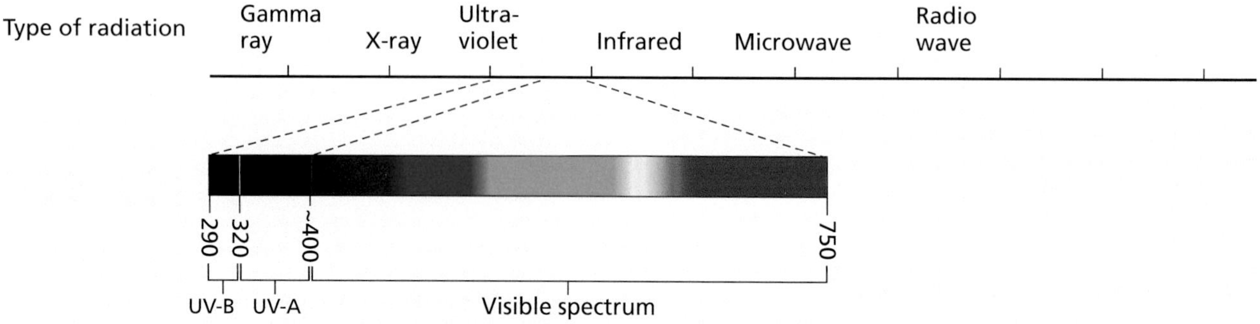

Figure 13.4 Plants can use visible light and UV-A and UV-B radiation as developmental signals (all wavelengths in nm).

13.1 Plant Photoreceptors

Distinguish among the types of photoreceptors and how they respond to wavelength, fluence, and irradiance.

Pigments, such as chlorophyll and the accessory pigments of photosynthesis (Chapter 7), are molecules that absorb visible light at specific wavelengths, and reflect or transmit the nonabsorbed wavelengths, which are perceived by us as colors. Unlike the photosynthetic pigments, photoreceptors absorb a photon of a given wavelength and use this energy as a signal that initiates a photoresponse. With the exception of UVR8, all the known photoreceptors consist of a protein plus a light-absorbing prosthetic group (a nonprotein molecule attached to the photoreceptor protein) called a **chromophore**. The protein structures of the different photoreceptors vary, as does their sensitivity to light quantity (number of photons), light quality (wavelength dependency and associated action spectra), light intensity, and the duration of the exposure to light.

Among the photoreceptors that can promote photomorphogenesis in plants, the most important are those that absorb red and blue light. **Phytochromes** are photoreceptors that absorb red (600–750 nm) and far-red (710–850 nm) light most strongly, but they also absorb blue light (400–500 nm) and UV-A radiation (320–400 nm). Phytochromes mediate many aspects of vegetative and reproductive development, as will be discussed in subsequent chapters.

Three main classes of photoreceptors mediate the effects of UV-A/blue light: the **cryptochromes**, the **phototropins**, and the **ZEITLUPE** (**ZTL**) (German for "slow motion") family. All three of these receptor classes employ flavin molecules as chromophores. Cryptochromes, like the phytochromes, play a major role in plant photomorphogenesis, while phototropins primarily regulate phototropism, chloroplast movements, and stomatal opening. The ZEITLUPE photoreceptor plays roles in day-length perception and circadian (24-h cycling) rhythms. As with the hormone responses described in Chapter 12, light signaling typically involves interactions between multiple photoreceptors and their signaling intermediates.

By convention, photoreceptors are designated in lower case (e.g., phy, cry, phot) when the holoprotein (protein plus chromophore) is described, and in upper case (PHY, CRY, PHOT) when the apoprotein (protein minus chromophore) is described. To be consistent with genetic conventions, we will use uppercase italics (*PHY, CRY, PHOT*) for the genes encoding the photoreceptor apoproteins. Plants with mutations in genes encoding photoreceptor apoproteins or chromophores are designated in lowercase italics (*phy, cry, phot*).

Only the **UV RESISTANCE LOCUS 8** (**UVR8**) UV-B photoreceptor lacks a chromophore, as high-energy UV-B photons can directly excite the indole ring of exposed tryptophan residues in the photoreceptor protein to induce conformational changes and initiate photomorphogenic responses.

Photoresponses are driven by the spectral properties of the energy absorbed

It can be difficult to separate specific photoresponses in plants exposed to the full solar spectrum during normal growth, because changes in the sun's angle over the course of a day expose a plant to varying light intensities and wavelength mixtures to differentially activate multiple photoreceptors. Similarly, the process of **de-etiolation**, characterized by the production of chlorophyll in dark-grown (etiolated) (Figure 13.2) seedlings when exposed to light, results from the coaction of phytochrome absorbing red light and cryptochrome absorbing blue light from sunlight. Outputs of photosynthesis can also contribute to light responses, since photosynthetic pigments also absorb red light and blue light. How, then, can we functionally distinguish responses to individual photoreceptors?

chromophore A light-absorbing pigment molecule that is usually bound to a protein (an apoprotein).

phytochromes Plant growth-regulating photoreceptor proteins that absorb primarily red light and far-red light but also absorb blue light. The holoprotein that contains the chromophore phytochromobilin.

cryptochromes Flavoproteins implicated in many blue-light responses that have strong homology with bacterial photolyases.

phototropins Blue-light photoreceptors that primarily regulate phototropism, chloroplast movements, and stomatal opening. Phototropins are autophosphorylating protein kinases that are stimulated by blue light interactions with a flavin co-factor.

ZEITLUPE A blue-light photoreceptor that regulates day-length perception (photoperiodism) and circadian rhythms.

UV RESISTANCE LOCUS 8 (UVR8) The protein receptor that mediates various plant responses to UV-B irradiation.

de-etiolation The rapid developmental changes associated with loss of the etiolated form due to the action of light. *See* photomorphogenesis.

photoreversibility The interconversion of the Pr and Pfr forms of phytochrome.

To determine which wavelengths of light are necessary to bring about a particular plant response, photobiologists typically produce what is known as an action spectrum under carefully defined lighting conditions using precise lasers or light-emitting diodes (LEDs) and band-pass filters. Action spectra describe the wavelength specificity of a biological response within the spectral range of normal sunlight. Each photoreceptor differs in its atomic composition and arrangement and thus exhibits different absorption characteristics. In Chapter 7, the action spectrum for photosynthetic O_2 evolution is shown as a graph that plots the magnitude of a photosynthetic light response as a function of wavelength. Similarly, the action spectra of other photoresponses can then be compared with the absorption spectra of candidate photoreceptors that have been studied in isolation. The action spectra of wild-type plants can be compared to action spectra obtained from plants with mutations in genes encoding specific photoreceptors. If a particular response is missing or diminished in a particular mutant, then the response is likely to be associated with the particular photoreceptor.

These methods have been used to identify photoreceptors involved in signaling pathways. For example, red light stimulates seed germination in many species, and far-red light inhibits it (shown for lettuce in **Figure 13.5**). The action spectra for these two antagonistic effects of light on Arabidopsis seed germination are shown in **Figure 13.6A**. Stimulation shows a peak in the red region (660 nm), while inhibition has a peak in the far-red region (720 nm). Under these treatments, the purified photoreceptor phytochrome exhibits a very similar **photoreversibility** (**Figure 13.6B**), suggesting that phytochrome is the receptor for these responses. The close correspondence between the action and absorption spectra of phytochrome was used to confirm its identity as the photoreceptor involved in regulating seed germination, and also established that the red/far-red reversibility of seed germination was due to the photoreversibility of phytochrome itself.

Similarly, action spectra for blue light–stimulated phototropism, stomatal movements, and other key blue-light responses all exhibit a peak in the UV-A region (at 370 nm) and a peak in the blue region (400–500 nm) that has a characteristic "three-finger" fine structure (**Figure 13.7A**), suggesting a common photoreceptor. The absorption spectrum for the light-sensing domain of phototropin, which contains the chromophore flavin mononucleotide (FMN), is almost identical to the action spectrum for phototropism (**Figure 13.7B**), consistent with phototropin acting as the photoreceptor for these responses.

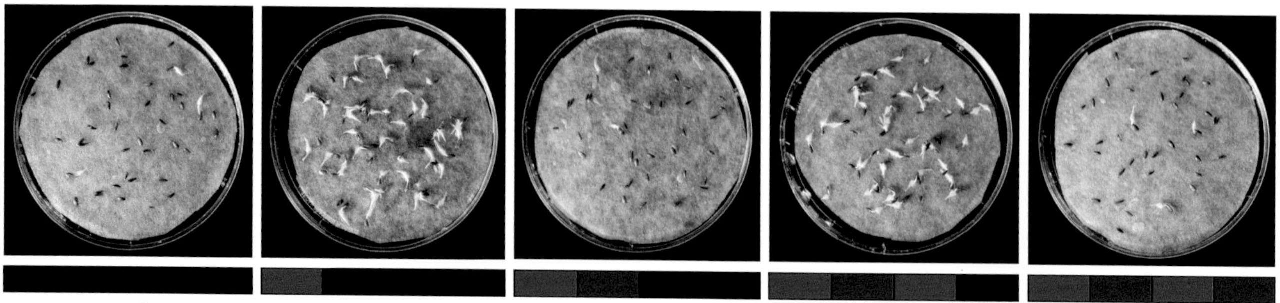

Figure 13.5 Lettuce seed germination is a typical photoreversible response controlled by phytochrome. Red light promotes lettuce seed germination, but this effect is reversed by far-red light. Imbibed (hydrated) seeds were given alternating treatments of red followed by far-red light. The effect of the light treatment depended on the last treatment given. Very few seeds germinated following the last far-red treatment.

Note: Each plate represents an independent experiment.

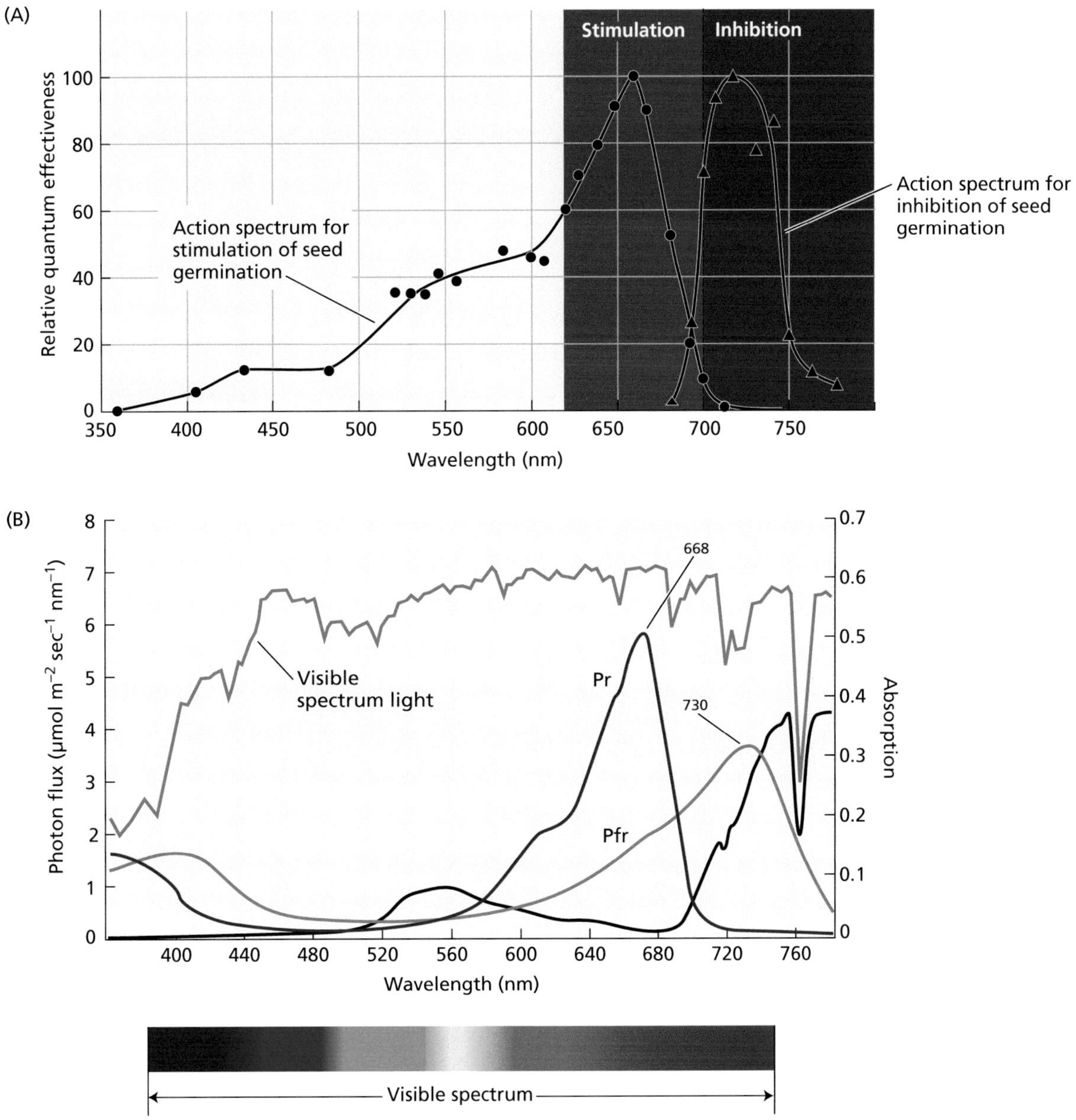

Figure 13.6 The action spectrum of phytochrome function matches its absorption spectrum. (A) Action spectra for the photoreversible stimulation and inhibition of seed germination in Arabidopsis. (B) Absorption spectra of purified oat phytochrome in the Pr (red line) and Pfr (green line) forms overlap. At the top of the canopy there is a relatively uniform distribution of visible-spectrum light (blue line), but under a dense canopy much of the red light is absorbed by plant pigments, resulting in transmittance of mostly far-red light. The black line shows the spectral properties of light that is filtered through a leaf. Thus, the relative proportions of Pr and Pfr are determined by the degree of vegetative shading in the canopy. (A after Shropshire et al. 1961; B after Kelly and Lagarias 1985, courtesy of Patrice Dubois.)

Plant responses to light can be distinguished by the amount of light required

Light responses can also be distinguished by the amount of light required to induce them. The amount of light is referred to as the **fluence**, which is defined as the total number of photons striking a unit of surface area:

$$\text{Total fluence} = \text{fluence rate} \times \text{the length of time (duration) of irradiation}$$

fluence The number of photons absorbed per unit of surface area over time ($\mu mol\ m^{-2}\ s^{-1}$).

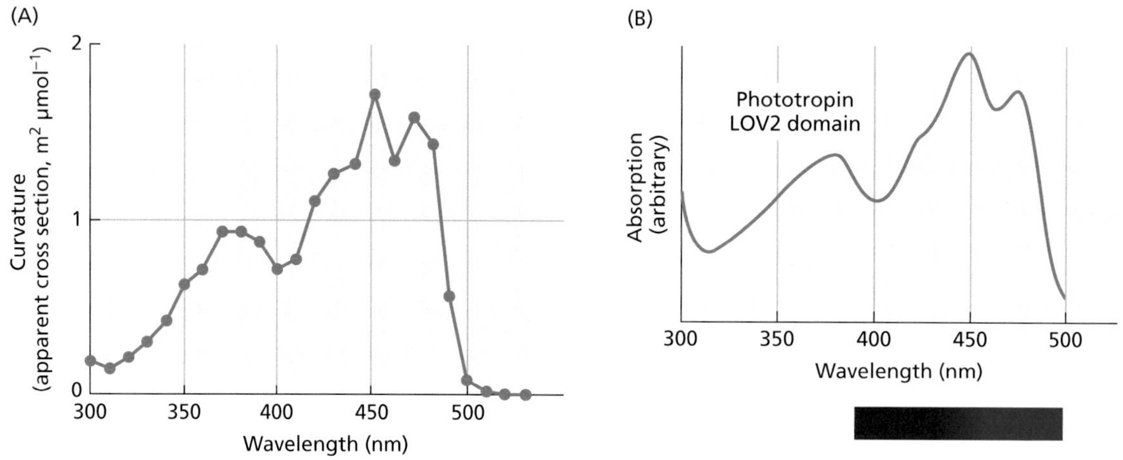

Figure 13.7 The action spectrum of phototropism matches the absorption spectrum of the light-sensing domain of phototropin. (A) Action spectrum for blue light–stimulated phototropism in alfalfa hypocotyls. The "three-finger" pattern in the 400- to 500-nm region is characteristic of many blue-light responses. (B) The absorption spectrum of the light-sensing domain of phototropin.

The standard units for fluence are micromoles of quanta (photons) per square meter (μmol m^{-2}). Some responses are sensitive not only to the total fluence but also to the **irradiance**, or fluence rate, of light. The units of irradiance (B) are micromoles of quanta per square meter per second (μmol m^{-2} s^{-1}) (see Chapter 9). From the perspective of photoreceptors, what we generally refer to as "darkness" is actually a condition of light irradiance that is close to zero.

Because photochemical responses are stimulated only when a photon is absorbed by its photoreceptor, there can be a difference between incident irradiation and absorption. For example, in photosynthesis the apparent quantum efficiency is assessed as the electron transport rate or total carbon assimilation as a function of the incident photosynthetic photon flux density (Chapters 7 and 8). However, this measure underestimates the *actual* quantum efficiency because not all the incident photons are absorbed. This caveat is also important in assessing the dose response of the photomorphogenic responses of green plants to red or blue light, because much of the light is absorbed by chlorophyll. The same principle applies to the responses to UV radiation, since the epidermis may absorb just under 100% of the incident UV radiation. Thus, the amount of radiation required to induce a photoresponse may be quite high based on the amount of incident irradiation required, and quite low based on actual photon absorption by the photoreceptor.

13.2 Phytochromes

Explain the light-induced conformation changes of phytochrome and the physical and physiological plant responses to red and far-red light.

Phytochromes were first identified in flowering plants as the photoreceptors responsible for photomorphogenesis in response to red and far-red light. However, they are members of a gene family present in the Characaean algae and all land plants (bryophytes, ferns, fern allies, and seed plants) as well as in cyanobacteria, other bacteria, fungi, and diatoms.

irradiance The amount of energy that falls on a flat surface of known area per unit of time. Expressed as watts per square meter (W m^{-2}). Time (seconds) is contained within the term watt: 1 W = 1 joule (J) s^{-1}, or as micromoles of quanta per square meter per second (μmol m^{-2} s^{-1}), also referred to as fluence rate.

Because neither red nor far-red light penetrates water to depths greater than a few meters, phytochrome would appear to be less useful as a photoreceptor for aquatic organisms. However, recent studies have shown that different algal phytochromes can sense orange, green, or even blue light, suggesting that phytochromes have been spectrally tuned during natural selection to absorb different wavelengths.

Phytochrome is the primary photoreceptor for red and far-red light

The phytochrome family of photoreceptors are red- and far-red–sensing proteins with a covalently bound bilin chromophore, termed phytochromobilin (PϕB). Purified phytochrome is a cyan-blue (midway between green and blue) or cyan-green protein with a molecular mass of about 125 kilodaltons (kDa). Phytochrome proteins function as dimers of either the red light–absorbing inactive form, called **Pr**, or the far-red light–absorbing active form, **Pfr**.

Many of the biological properties of phytochrome were established in the 1930s through studies of red light–induced morphogenic responses, especially seed germination. A key breakthrough in the history of phytochrome was the discovery that the effects of red light (620–700 nm) could be reversed by a subsequent irradiation with far-red light (710–850 nm). When the absorption spectra of each of the two forms of phytochrome are measured separately in a spectrophotometer designed to study photoreversible molecules, they correspond closely to the action spectra for the stimulation and inhibition of seed germination, respectively (Figure 13.6A). The reversibility of the red and far-red responses ultimately led to the discovery that a single photoreversible photoreceptor, phytochrome, was responsible for both activities. It was subsequently demonstrated that the two forms of phytochrome (Pr and Pfr) could be distinguished spectrophotometrically (Figure 13.6B).

Phytochrome can interconvert between Pr and Pfr forms

In dark-grown, or etiolated, seedlings, phytochrome is present in the red light–absorbing Pr form. This cyan-blue-colored inactive form is converted by red light to the far-red light–absorbing Pfr form.

$$\text{Pr} \underset{\text{Far-red light}}{\overset{\text{Red light}}{\rightleftarrows}} \text{Pfr}$$

Pfr is pale cyan-green in color and is considered to be the active form of phytochrome. Upon photoconversion to Pfr, phytochrome can be rapidly degraded by the ubiquitin pathway, the rate of degradation being dependent on the type of phytochrome (discussed later in the chapter). Alternatively, Pfr can revert back to inactive Pr in darkness, but this is a relatively slow process. However, Pfr can be rapidly converted to Pr by irradiation with far-red light. A schematic diagram of the conformational change involved in phytochrome photoreversibility is shown in **Figure 13.8**.

In natural settings, plants growing outdoors are exposed to a much broader spectrum of light, and the ratio of red light to far-red light varies with the conditions. Red light is abundant in direct sunlight, while far-red light is more abundant under foliage canopies in which chlorophyll has absorbed much of the incident red light (Figure 13.6B). Plants growing beneath a canopy can use the R:FR ratio of light to regulate such processes as seed germination and shade avoidance (discussed in Chapters 14 and 15). As we will see, phytochrome-mediated responses also play a major role in the photoperiodic control of flowering (Chapter 16).

Pr The red light–absorbing form of phytochrome. This is the form in which phytochrome is assembled. The cyan-blue colored Pr is converted by red light to the far-red light–absorbing form, Pfr.

Pfr The far-red light–absorbing form of phytochrome converted from Pr by the action of red light. The cyan-green colored Pfr is converted back to Pr by far-red light. Pfr is the physiologically active form of phytochrome.

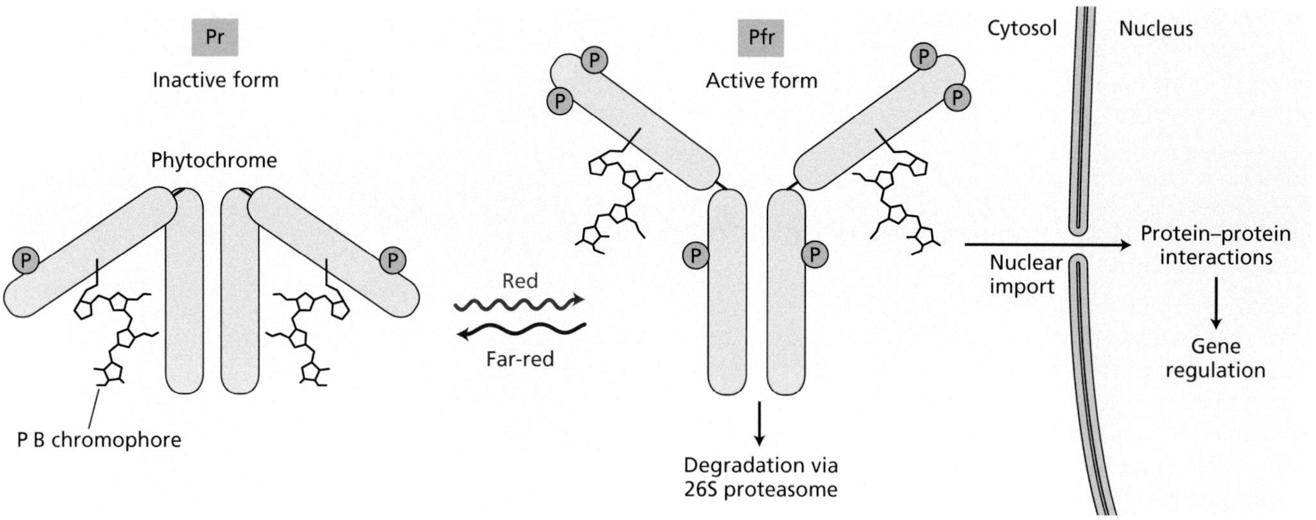

Figure 13.8 Diagram of phytochrome photoreversibility by red and far-red light. Upon perception of red light, the inactive Pr form is converted to the active Pfr form. Pfr is transported to the nucleus, where it participates in direct protein–protein interactions that result in repression or de-repression of downstream genes. Active Pfr can revert back to inactive Pr, either rapidly following far-red irradiation or slowly upon removal of red light, or it may be sent to the 26S proteasome for degradation. PϕB = the phytochrome chromophore phytochromobilin. (After art courtesy of Candace Pritchard.)

13.3 Phytochrome Responses

Describe how phytochrome-induced responses, such as germination and photomorphogenesis, are mediated by the amount of light available and how phytochrome changes gene expression to regulate photomorphogenesis.

The variety of different phytochrome responses in intact plants is extensive, in terms of both the kinds of responses (**Table 13.1**) and the quantity of light needed to induce the responses. A few responses are very rapid, such as the light-induced changes in the surface potentials (the membrane potential measured extracellularly) of oat roots or coleoptiles, which occur within seconds. Such rapid changes in surface potentials result when phytochrome interacts

Table 13.1 Typical Photoreversible Responses Induced by Phytochrome in a Variety of Higher and Lower Plants

Group	Genus	Stage of development	Effect of red light
Angiosperms	*Lactuca* (lettuce)	Seed	Promotes germination
	Avena (oat)	Seedling (etiolated)	Promotes de-etiolation (e.g., leaf unrolling)
	Sinapis (mustard)	Seedling	Promotes formation of leaf primordia, development of primary leaves, and production of anthocyanin
	Pisum (pea)	Adult	Inhibits internode elongation
	Xanthium (cocklebur)	Adult	Inhibits flowering (photoperiodic response)
Gymnosperms	*Pinus* (pine)	Seedling	Enhances rate of chlorophyll accumulation
Pteridophytes	*Onoclea* (sensitive fern)	Young gametophyte	Promotes growth
Bryophytes	*Polytrichum* (moss)	Germling	Promotes replication of plastids
Chlorophytes	*Mougeotia* (alga)	Mature gametophyte	Promotes orientation of chloroplasts to directional dim light

with cytosolic factors at or near the plasma membrane and initiates ion fluxes. However, the majority of phytochrome responses occur over a longer period of time. These slower responses affect longer-term developmental events, such as growth and organ movements.

Phytochrome responses vary in lag time and escape time

Morphological responses to the photoactivation of phytochrome are often observed visually after a *lag time*—the time between stimulation and the observed response. The lag time may be as brief as a few minutes or as long as several weeks. These differences in response times result from the multiple signal transduction pathways that function downstream of phytochrome signaling as well as interactions with other developmental mechanisms. The more rapid of these responses are usually reversible movements of organelles or reversible volume changes (swelling, shrinking) in cells, but even some growth responses are remarkably fast. For instance, red-light inhibition of the stem elongation rate of light-grown pigweed (*Chenopodium album*) and Arabidopsis is observed within minutes after the proportion of Pfr to Pr in the stem is increased. However, lag times of several weeks are observed for the induction of flowering in Arabidopsis and other species.

Red light–induced events involving phyA are reversible by far-red light for only a limited period of time, as these responses are the end result of a multistep sequence of linked biochemical reactions in the responding cells. Early stages in the sequence may be fully reversible by removing Pfr, but at some point in each sequence of events, a point of no return is reached, beyond which the reactions proceed irreversibly toward the response.

Phytochrome responses fall into three main categories based on the amount of light required

As shown in **Figure 13.9**, phytochrome responses can be classified by the amount of light they require. Responses that are dependent on light fluence levels (the number of photons striking a surface over a given time) are generally classified as very low or low fluence responses (VLFRs or LFRs). Some phytochrome responses can be initiated by fluences as low as 0.0001 μmol m^{-2} (a few seconds of starlight, or one-tenth of the amount of light emitted by a firefly in a single flash), and they become saturated (i.e., reach a maximum) at about 0.05 μmol m^{-2}. For example, Arabidopsis seeds can be induced to germinate with red light in the range of 0.001 to 0.1 μmol m^{-2}. In dark-grown oat (*Avena* spp.) seedlings, red light can stimulate the growth of the coleoptile and inhibit the growth of the mesocotyl (the elongated axis between the coleoptile and the root) at similarly low fluences.

These very low fluence responses occur only in deep-buried seeds and seedlings and are mediated by phyA, which is abundant in dark-grown tissues. VLFRs are nonreversible by far-red light because, at the extremely low light intensities involved, the amount of far-red–induced reversion of Pfr to Pr is insignificant compared with other mechanisms of Pfr degradation.

Another set of phytochrome responses cannot be initiated until the fluence reaches 1.0 μmol m^{-2}, and they are saturated at about 100 μmol m^{-2}. These LFRs include processes such as the promotion of seed germination, inhibition of

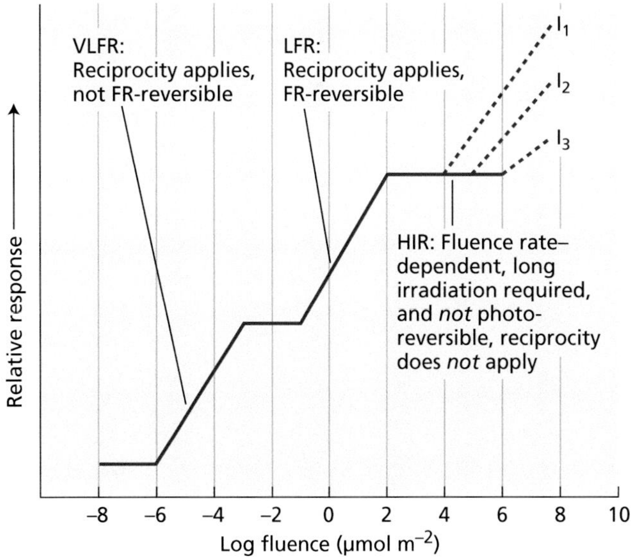

Figure 13.9 Three types of phytochrome responses, based on their sensitivities to fluence. The relative magnitudes of representative responses are plotted against increasing fluences of red light. Short light pulses activate very low fluence responses (VLFRs) and low fluence responses (LFRs). The fluence at which LFRs are initiated varies with species. Because high-irradiance responses (HIRs) are proportional to irradiance as well as to fluence, the effects of three different irradiances given continuously are illustrated by dashed lines ($I_1 > I_2 > I_3$).

reciprocity According to the Law of Reciprocity, treating plants with a brief duration of bright light will induce the same photobiological response as treating them with a long duration of dim light.

hypocotyl elongation, and regulation of leaf movements. As we saw in Figure 13.6A, the LFR action spectrum for Arabidopsis seed germination includes a main peak for stimulation in the red region (660 nm) and a major peak for inhibition in the far-red region (720 nm). The uniformity of this action spectrum in both very low fluence and low fluence light indicates that phyA is the photoreceptor involved in both responses.

Both fluence responses can be induced by brief pulses of light, provided that the total amount of light energy adds up to the required fluence. As we noted above, the total fluence is a function of two factors: the fluence rate (μmol m^{-2} s^{-1}) and the time of irradiation. Thus, a brief pulse of red light will induce a response, provided that the light is sufficiently bright; conversely, very dim light will work if the irradiation time is long enough. The magnitude of the response (e.g., percent germination or degree of inhibition of hypocotyl elongation) exhibits **reciprocity**, or dependence on the product of the fluence rate and the time of irradiation.

Phytochrome responses that require prolonged or continuous exposure to light of relatively high irradiance (fluence rate per second) and do not exhibit reciprocity are classified as high irradiance responses (HIRs). The response is proportional to the irradiance until the response saturates and additional light has no further effect. HIRs saturate at much higher fluences than LFRs—at least 100 times higher. Because neither continuous exposure to dim light nor transient exposure to bright light can induce HIRs, these responses do not show reciprocity. However, many low fluence responses are also classified as HIRs. For example, at low fluences the action spectrum for anthocyanin production in seedlings of white mustard (*Sinapis alba*) is regulated by phyA, shows a single peak in the red region of the spectrum, and is reversible with far-red light. However, if the dark-grown seedlings are instead exposed to high-irradiance light for several hours, anthocyanin production is proportional to the irradiance and is not reversible, and the action spectrum contains peaks in the far-red and blue regions.

Phytochrome A mediates responses to continuous far-red light

The different types of phytochrome responses to light dosage suggest the presence of multiple phytochrome isoforms with varying sensitivities to light. Arabidopsis contains five phytochrome isoforms (phyA–phyE). The process of Pfr degradation is conserved among all known isoforms, and phyA, which controls plant VLFRs and far-red-light–induced HIRs, is rapidly degraded as Pfr by the ubiquitin pathway.

In contrast, phyB–phyE are more light-stable. Experimental evidence has shown that phyB regulates LFRs such as photoreversible seed germination. For example, wild-type Arabidopsis seeds require light for germination, and the response shows red/far-red reversibility in the low-fluence range (Figure 13.6A). Mutants that lack phyA respond normally to red light, whereas mutants deficient in phyB are unable to respond to low-fluence red light. This experimental evidence strongly suggests that phyB mediates photoreversible seed germination.

PhyB also plays an important role in regulating plant responses to shade. Plants that are deficient in phyB often look like wild-type plants growing under dense vegetative canopies. In fact, mediating shade responses such as accelerated flowering and increased elongation growth may be one of the most ecologically important roles of phytochromes. The isoforms phyC, phyD, and phyE play unique roles in regulating other responses to red and far-red light. Responses mediated by phyD and phyE include petiole and internode elongation and the control of flowering time (described in Chapter 16).

Phytochrome regulates gene expression

All phytochrome-regulated changes in plants begin with absorption of light by the photoreceptor. After light absorption, the molecular properties of phytochrome are altered, and this affects phytochrome interactions with other cytosolic or nuclear components to ultimately bring about changes in the growth, development, or position of an organ (Table 13.1).

Although a significant portion of phytochrome remains in the cytosol, the majority of phytochrome signaling occurs in the nucleus, where it alters gene expression via a wide range of regulatory elements and transcription factors. Monitoring of gene expression profiles over time after plants have been shifted from darkness to light has led to the identification of early and late targets of *PHY* gene action. Nuclear import of phyA and phyB is highly correlated with the light quality that stimulates their activities. That is, nuclear import of phyA is activated by either red or far-red light, or low-fluence broad-spectrum light, whereas phyB import is driven by red-light exposure and is reversible by far-red light. Nuclear import of the phytochrome proteins represents a major control point in phytochrome signaling (Figure 13.8).

A family of proteins known as **phytochrome interacting factors** (**PIFs**) regulate various aspects of phytochrome-mediated activity, including photomorphogenesis, seed germination, chlorophyll biosynthesis, shade avoidance, and hypocotyl elongation. PIFs promote etiolated seedling development by activating expression of dark-induced genes and repressing genes that are expressed in the light. Red-light–induced Pfr formation initiates the degradation of PIF proteins by phosphorylation, followed by degradation via the proteasome complex (Chapter 1). The rapid degradation of PIFs may provide a mechanism for modulating light responses that is tightly coupled to the activities of phy proteins. Another group of DNA-binding, nuclear PIF-like proteins selectively interact with phytochromes in their active Pfr conformations to regulate gene transcription. Phytochrome activity is further modulated by a family of **phytochrome kinase substrate** (**PKS**) proteins that regulate phosphorylation.

As discussed in Chapter 12, the majority of plant signal transduction pathways involve the inactivation, degradation, or removal of repressor proteins. The phytochrome signaling pathway is consistent with this general principle, as rapid phyA degradation following light activation and phosphorylation is a central regulatory component of phytochrome function (Figure 13.8).

phytochrome interacting factors (PIFs) A family of phytochrome-interacting proteins that may activate and repress gene transcription; some PIFs are targets for phytochrome-mediated degradation.

phytochrome kinase substrates (PKSs) Proteins that participate in the regulation of phytochrome via direct phosphorylation or via phosphorylation by other kinases.

13.4 Blue-Light Responses and Photoreceptors

Explain the blue-light responses and the kinetics of the different blue-light photoreceptors.

Blue-light responses have been reported in seed plants, ferns, bryophytes, algae, fungi, and prokaryotes. In addition to phototropism, these responses include inhibition of seedling hypocotyl elongation, stimulation of chlorophyll and carotenoid synthesis, and activation of gene expression. Among motile unicellular organisms, including many algae and bacteria, blue light mediates *phototaxis*, the movement toward or away from light. Multiple plant responses to the environment respond to blue light. These include stomatal opening (Chapter 6), chloroplast movements and solar tracking (Chapter 9), and seedling photomorphogenesis and phototropism (Chapter 14).

Blue-light responses have characteristic kinetics and lag times

Compared to nearly instantaneous and relatively ratiometric (output that is proportional to input) photosynthetic responses, blue-light responses such as inhibition of stem elongation and stomatal opening exhibit a lag time of variable duration and proceed at maximum rates for several minutes after application of a short light pulse. For example, blue light induces a decrease in growth rate and a transient membrane depolarization in etiolated cucumber seedlings only

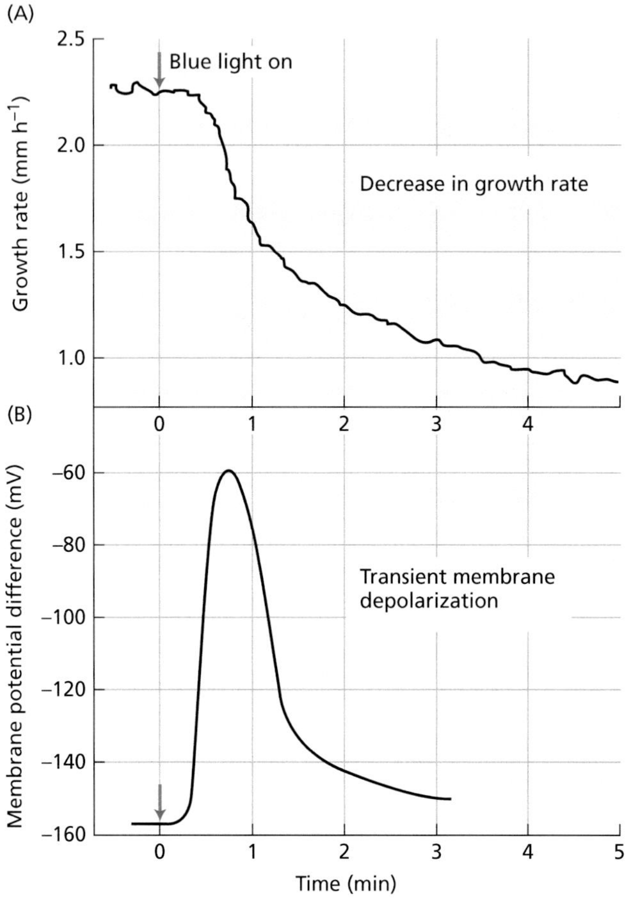

Figure 13.10 Blue light reduces hypocotyl elongation by stimulating membrane depolarization. (A) Rapid blue-light–induced changes in elongation rates of etiolated cucumber hypocotyls. (B) Blue-light–induced transient membrane depolarization of cucumber hypocotyl cells. In phototropic bending, intitial unilateral halt in growth is followed by rapid elongation on the shaded side of the hypocotyl.

after a lag time of about 25 s (**Figure 13.10**). Other blue-light responses, such as photomorphogenesis, occur over a longer time frame but also exhibit a sustained response output after receiving a threshold light exposure.

The persistence of blue-light responses in the absence of blue light has been studied using blue-light pulses. In guard cells, blue-light–induced activation of the H^+-ATPase gradually decays following a pulse of blue light, but only after several minutes have elapsed (Chapter 6). This persistence of the blue-light response after the pulse can be explained by a photochemical cycle in which the physiologically active form of the photoreceptor (in this case, phototropin), which has been converted from the inactive form by blue light, slowly reverts to the inactive form after the blue light is switched off. The rate of decay of the response to a blue-light pulse depends on the time course of the reversion of the active form of the photoreceptor back to the inactive form.

13.5 Cryptochromes

Describe domains of cryptochromes and how their activity produces physiological responses, such as flowering, and explain how different photoreceptors interact with each other in the same light-dependent response.

Cryptochromes are blue-light photoreceptors that mediate several blue-light responses in vascular plants, ferns, algae, cyanobacteria, fungi, insects, and mammals. In plants, activated cryptochromes suppress hypocotyl elongation, promote cotyledon expansion, inhibit petiole elongation, regulate anthocyanin production, mediate membrane depolarization events, and function in entrainment of the circadian clock. The cryptochrome photoreceptor was originally identified in Arabidopsis using genetic screens for mutants where hypocotyl elongation was not inhibited by white light. The long hypocotyl phenotype of one cryptochrome mutant was shown to occur in blue, but not red, light. This indicated that the loss of function did not directly involve phytochrome.

Blue-light irradiation of the cryptochrome FAD chromophore causes a conformational change

Blue-light–absorbing cryptochrome proteins have an N-terminal photolyase-related domain (PHR) that binds both the **flavin adenine dinucleotide** (**FAD**) (Figure 11.2) and the pterin 5,10-methyltetrahydrofolate (MTHF). Pterins are light-absorbing pteridine derivatives often found in pigmented cells of insects, fishes, and birds. In photolyases, blue light is absorbed by the pterin, and the excitation energy is then transferred to FAD. A similar mechanism may operate in plant cryptochromes, but is yet to be demonstrated.

Blue-light absorption alters the redox status of the bound FAD chromophore, and it is this primary event that triggers photoreceptor activation. As occurs in phytochromes and phototropins, this activation mechanism involves protein conformational changes. In the cryptochromes, light absorption by the N-terminal photolyase region appears to alter the conformation of a C-terminal extension, the cryptochrome C-terminal extension (CTE) (**Figure 13.11**). This C-terminal extension is absent from photolyase enzymes but is clearly essential

flavin adenine dinucleotide (FAD) A riboflavin-containing cofactor that undergoes a reversible two-electron reduction to produce $FADH_2$.

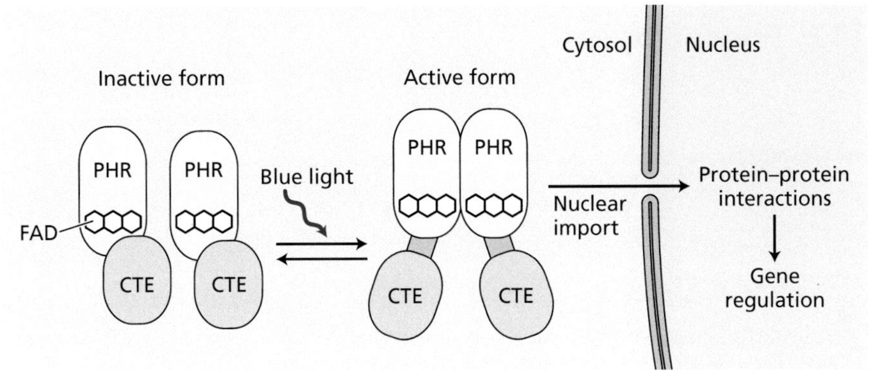

Figure 13.11 Diagram of the response of cryptochrome to blue light. Blue-light absorption by chromophores in the N-terminal photolyase-related domain (PHR) results in a conformational change of the cryptochrome C-terminal extension (CTE), enabling signaling. Following activation, the protein is transported to the nucleus, where it is involved in protein–protein interactions that initiate downstream signaling, resulting in changes such as anthocyanin accumulation, initiation of photomorphogenesis, and inhibition of elongation. (After art courtesy of Candace Pritchard.)

for cryptochrome signaling. We can therefore view plant cryptochrome as a molecular light switch whereby absorption of blue photons at the N-terminal photosensory region results in protein conformational changes at the C terminus, which initiates signaling by altering binding to specific partner proteins.

The nucleus is a primary site of cryptochrome action

Nuclear cryptochromes are involved in photomorphogenesis and act by stabilizing transcription factors and inhibiting their proteolytic degradation or directly regulating their activity. Although the cryptochrome isoform that regulates hypocotyl elongation is stable in blue light, a second isoform that regulates cotyledon expansion and floral induction is preferentially degraded under blue light. Both cryptochrome isoforms play key roles in regulating plant circadian rhythms as in insects and mammals. Rapid responses associated with growth, such as membrane depolarization, which occurs in 2 to 3 s, would be expected to occur in the cytosol, but have been shown to be regulated by cryptochrome activity in the nucleus as well.

Cryptochrome interacts with phytochrome

In Arabidopsis, continuous blue or far-red light promotes flowering, and red light inhibits flowering. Far-red light acts through phyA, and the antagonistic effect of red light is produced by phyB. One might expect cryptochrome mutants to be delayed in flowering, since blue light promotes flowering. However, cryptochrome mutants flower at the same time as the wild type under either continuous blue or continuous red light. A delay is observed only if both blue light and red light are given together, suggesting that the phytochrome and cryptochrome signaling pathways converge.

As noted previously in this chapter, several plant processes show oscillations of activity that roughly correspond to a 24-h, or circadian, cycle. This endogenous rhythm uses an oscillator that must be **entrained** (synchronized) to the daily light–dark cycles of the external environment. In experiments designed to characterize the role of photoreceptors in this process, phytochrome- and cryptochrome-deficient Arabidopsis mutants were crossed with lines carrying a reporter gene that was regulated by the circadian clock. The pace of the oscillator was slowed (i.e., the period length increased) when *phyA* mutants were grown under dim red light, whereas in *phyB* mutants, timing defects were only observed under high-irradiance red light. Similarly, cryptochrome-deficient mutants exhibited timing defects only when entrained using blue light. These studies indicate that both phytochromes and cryptochromes entrain the circadian clock in Arabidopsis.

entrain Manipulation of external controlling factors, such as light or removal of light, to synchronize biological rhythms.

LIGHT-OXYGEN-VOLTAGE (LOV) domains Highly conserved protein domains that respond to light, oxygen, or voltage changes to change receptor protein conformation. In phototropins, LOV domains are sites of binding of the FMN chromophore to phototropins and are thus the part of the protein that senses light.

13.6 Phototropins

Distinguish among the domains of phototropins, and explain how they mediate physiological responses such as phototropism and chloroplast movement.

Phototropins are the blue-light photoreceptors that mediate seedling phototropism, chloroplast movement, guard cell opening, and some leaf movements. Angiosperms contain two phototropin genes, *PHOT1* and *PHOT2*. phot1 is the primary phototropic receptor that mediates phototropism in response to both low and high fluence rates of blue light. phot2 mediates phototropism primarily in response to high light intensities. Similar overlaps in the functions of the phot1 and phot2 photoreceptors are observed for other blue-light responses, including chloroplast movements, stomatal opening, leaf movements, and leaf expansion. Together with phototropism, these processes optimize light capture and CO_2 uptake for photosynthesis. Consequently, growth of phototropin-deficient mutants is severely compromised, particularly under low light intensities.

In contrast to cryptochromes, which are predominantly localized in the nucleus, phototropin receptors are associated with the plasma membrane, where they function as light-activated serine/threonine kinases. A schematic representation of blue-light induction of conformational changes in phototropins is shown in **Figure 13.12**. Phototropins contain one or two light-sensing **LIGHT-OXYGEN-VOLTAGE (LOV) domains**, each binding a chromophore flavin mononucleotide (FMN). Only one LOV domain actively participates in photoreception. When present, a second LOV domain functions in receptor dimerization. Spectroscopic studies have shown that in the dark, one FMN molecule is noncovalently bound to each LOV domain. Photoexcitation of the FMN–LOV domain "uncages" the kinase domain and leads to autophosphorylation on multiple serine residues, which is required for phototropin kinase activity. Blue-light activation of phototropin can be reversed by a dark treatment. Dark inactivation of phototropin is mediated by a type 2A protein phosphatase that dephosphorylates the serine residues.

The ZEITLUPE family of photoreceptors also contain LOV domains with FMN cofactors, but they are F-box proteins (Chapter 12). ZEITLUPE proteins regulate transcription factor stability via protein ubiquitination in a manner similar to that of hormone receptors (Chapter 12).

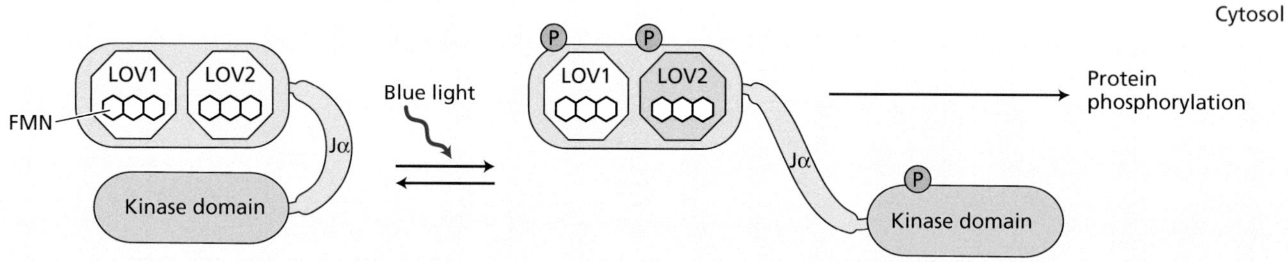

Figure 13.12 Diagram of the blue-light–induced conformational change of phototropin. Phototropin proteins consist of two light-oxygen-voltage (LOV) domains that each have a bound flavin mononucleotide (FMN) chromophore, a C-terminal serine/threonine kinase domain, and an α-helical region (Jα) that joins the N-terminal and C-terminal domains. Sensing of blue light by the LOV domains results in autophosphorylation and a dramatic conformational change that exposes the kinase domain. Direct phosphorylation of target proteins via the kinase domain results in changes such as stomatal opening, phototropism, and chloroplast relocalization. (After art courtesy of Candace Pritchard.)

Phototropins activated by directional blue light initiate seedling phototropism

Activation of phototropin kinases triggers signal transduction events that establish a variety of different responses. One of these responses is phototropism, which occurs in both mature plants and seedlings. Arabidopsis seedlings lacking phototropins fail to bend in response to unidirectional blue light (**Figure 13.13**).

As we learned in Chapter 12, seedling phototropism experiments conducted by Charles and Francis Darwin led to the discovery of the hormone auxin. Phototropic bending is a multistep process that involves an interruption of rootward auxin flows followed by lateral redistribution of auxin from the illuminated to the shaded side of the hypocotyl or coleoptile. The physiology of phototropic bending will be discussed in more detail in Chapter 14.

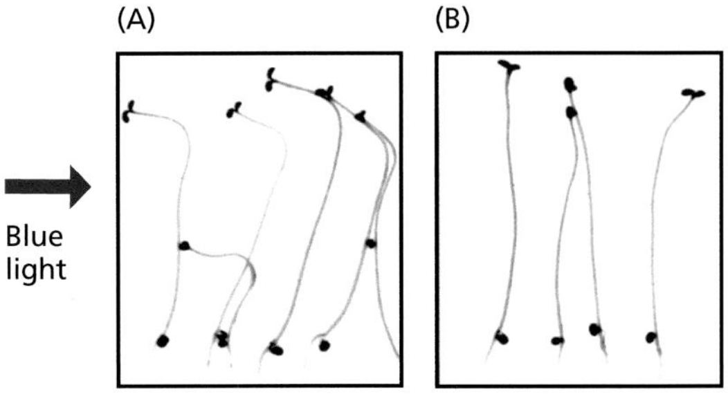

Figure 13.13 Phototropism in Arabidopsis seedlings can be used as a bioassay for phototropin activity. (A) Wild-type plants respond by bending toward the light. (B) Mutation of the LOV1 domain has little or no effect on seedling phototropism, whereas mutation in the LOV2 photoreceptor domain of phot1 abolishes the phototropic response. This indicates that activation of the LOV2 domain activates the signal transduction that triggers seedling phototropism.

Phototropins regulate chloroplast movements

Leaves can alter the intracellular distribution of their chloroplasts in response to changing light conditions. As discussed in Chapter 9, this feature is adaptive, as the redistribution of chloroplasts within the cells modulates light absorption and minimizes photodamage. Under weak illumination, chloroplasts gather near the upper and lower walls of the leaf palisade cells (accumulation), thus maximizing light absorption (**Figure 13.14**). Under strong illumination, the chloroplasts move to the lateral walls that are parallel to the incident light (avoidance), thus minimizing light absorption and photooxidative damage. In the dark, the chloroplasts move to the bottom of the cell, but the physiological function of this position is unclear. These chloroplast movements have been shown to be actin-dependent and are regulated by blue-light activation of phototropins (**Figure 13.15**). Under low light, phototropins 1 and 2 interact with CHLOROPLAST UNUSUAL POSITIONING 1 (CHUP1), a protein associated with both the chloroplast outer envelope and the plasma membrane, to maintain polymerization of actin filaments that promote chloroplast accumulations that are orthogonal to incident light. Under high light, phot2 associated with the chloroplast envelope appears to mediate the chloroplast avoidance response to high light. The importance of these chloroplast movements is demonstrated by the failure of Arabidopsis phototropin mutants to survive in the field due to observed photooxidative damage under full sunlight.

Stomatal opening is regulated by blue light, which activates the plasma membrane H^+-ATPase

Stomatal photophysiology and the impact of light on water use and photosynthesis were introduced in Chapters 3 and 6. Blue light causes rapid stomatal opening with an action spectrum similar to that of phototropism and can be shown to be distinct from the slower red-light–induced opening brought about by photosynthesis.

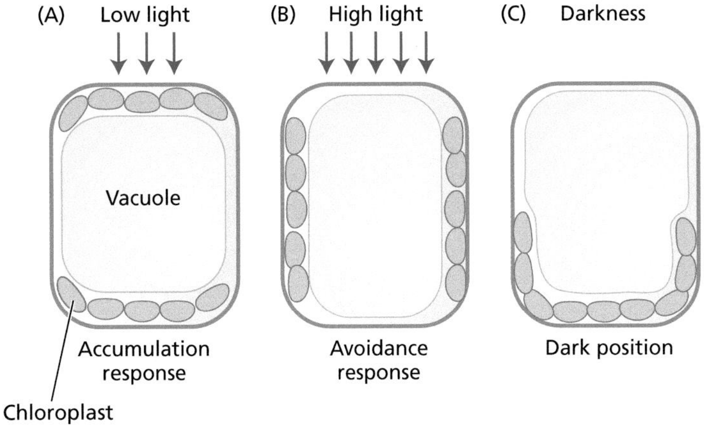

Figure 13.14 Schematic diagram of chloroplast distribution patterns in Arabidopsis leaf palisade cells in response to different light intensities. (A) Under low light conditions, chloroplasts optimize light absorption by accumulating at the upper and lower sides of palisade cells. (B) Under high light conditions, chloroplasts avoid sunlight by migrating to the side walls of palisade cells. (C) Chloroplasts move to the bottom of the cell in darkness.

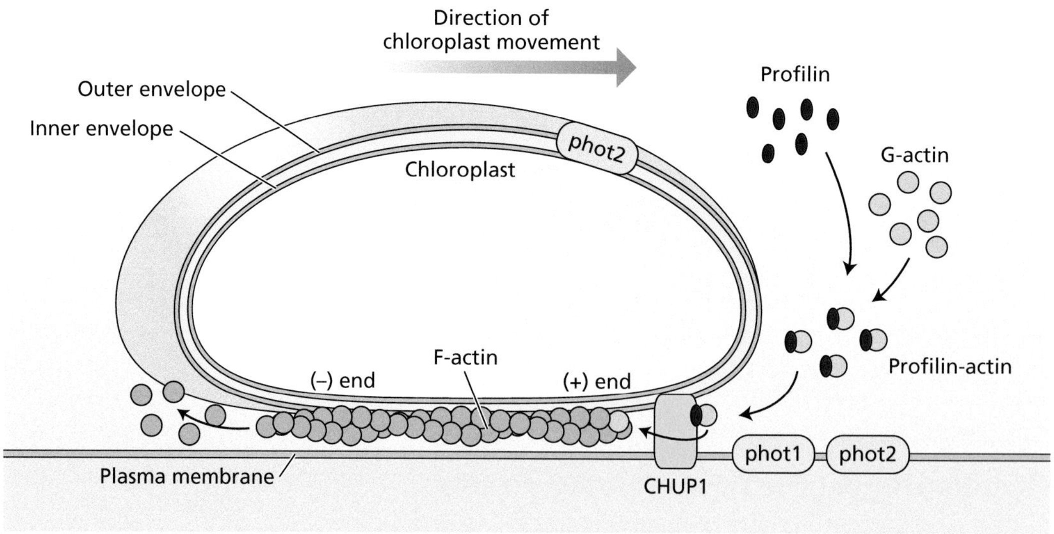

Figure 13.15 Model for phototropin-mediated chloroplast movement in *Arabidopsis thaliana*. Both phot1 and phot2 mediate the accumulation response and are localized at the plasma membrane. Under low light, phot1 and phot2 promote the activity of CHUP1, a protein that is associated with both the chloroplast outer envelope and the plasma membrane. CHUP1 is required for normal chloroplast positioning and mediates profilin-mediated actin filament polymerization (Chapter 1). Under low light conditions, phototropins 1 and 2 interact with CHUP1 to promote extension of actin filaments and chloroplast accumulation along plasma membranes that are orthogonal to incident light. The actin filaments are depolymerized at their minus ends. The green arrow shows the direction of chloroplast movement. Under high light conditions, phot2 associated with the chloroplast envelope is thought to function in avoidance alignments of chloroplasts.

As discussed in Chapter 6, the guard cell proton-pumping H^+-ATPase plays a central role in the regulation of stomatal movements. The activated H^+-ATPase transports H^+ across the membrane and increases the inside-negative electrical potential. This, in turn, drives K^+ uptake through voltage-gated inward K^+ channels. Accumulation of K^+ in the guard cells facilitates an influx of water that increases turgor pressure and stomatal opening.

Phototropins are the primary photoreceptors for blue-light–induced stomatal opening. Several key steps in the signaling pathway have been identified and are diagrammed in **Figure 13.16**. Blue light activates the H^+-ATPase via a multistep process. Initially, a membrane-associated, guard cell–specific protein kinase called BLUE LIGHT SIGNALING1 (BLUS1) is phosphorylated by phot1 and phot2 redundantly (that is, either phototropin is sufficient on its own for this step). The C terminus of the plasma membrane H^+-ATPase has an autoinhibitory domain that inhibits the activity of the enzyme. When activated by phototropin in response to blue light, BLUS1 initiates a signaling cascade that phosphorylates the C terminus of the H^+-ATPase, releasing the enzyme from inhibition. Further interactions with a 14-3-3 regulatory protein (named for the conditions under which the protein is biochemically purified) stabilize the H^+-ATPase in the active state.

Blue light also stimulates a second mechanism that contributes to increases in guard cell turgor. Activated phototropins phosphorylate cytosolic CONVERGENCE OF BLUE LIGHT AND CO_2 (CBC) proteins, which in turn inhibit plasma membrane anion channels. This increases cytosolic stores of Cl^- that can then move into the vacuole along with malate to serve as counter-ions as vacuolar K^+ increases. As CBC proteins coordinate stomatal response to blue light and CO_2 levels, and phototropins become less stable under modest increases in temperature, this control mechanism is thought to be particularly sensitive to climate change.

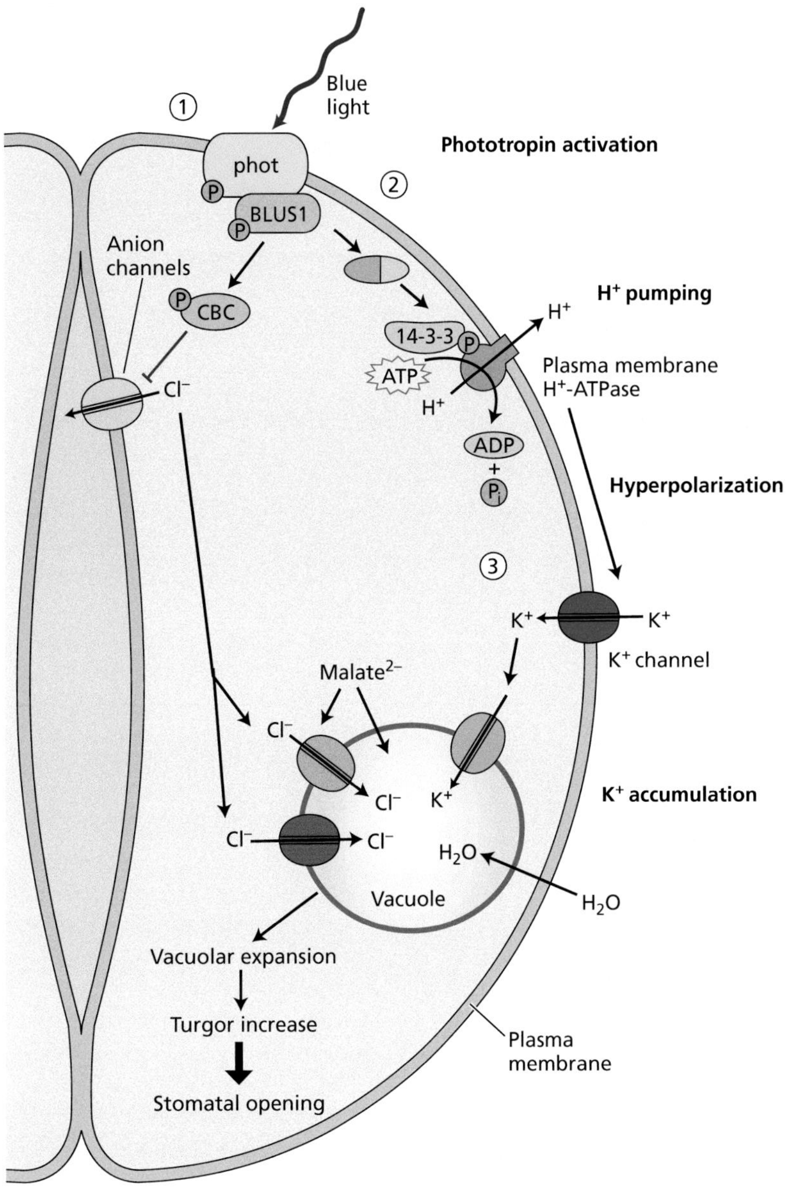

1. Blue light activates phototropin photoreceptor kinases that then activate CONVERGENCE OF BLUE LIGHT AND CO_2 (CBC) kinases. Activated CBCs inhibit transport of Cl^- out of the cell, which increases Cl^- import into the vacuole.

2. Phosphorylation of a BLUE LIGHT SIGNALING 1 (BLUS1) / kinase complex activates sequential protein phosphorylation steps that activate plasma membrane H^+-ATPases by releasing inhibitory 14-3-3 proteins.

3. Activation of H^+-ATPases hyperpolarizes the plasma membrane, which increases K^+ movement into the guard cell. Vacuolar K^+ uptake is coordinated with that of Cl^- and malate^{2-}. These ion movements lower the water potential of the vacuole and drive water uptake, vacuolar expansion, and turgor-driven stomatal opening.

Figure 13.16 Phototropin signal transduction leading to stomatal opening.

Higher fluxes of photosynthetically active photons (primarily red light) also slowly activate guard cell opening as a result of increased levels of cytosolic ATP and increased levels of photoprotective carotenoids produced in active chloroplasts (Figure 13.16). Carotenoids are also precursors in the synthesis of abscisic acid. In Chapter 19 we will discuss in detail how abscisic acid causes stomatal closure in the light.

13.7 Responses to Ultraviolet Radiation

Describe characteristics and signaling pathways of the UV-B responses by UVR8.

In addition to its cytotoxic effects, UV-B radiation can inhibit hormone-dependent growth processes and enhance synthesis and activity of stress-induced defense hormones. The photoreceptor responsible for UV-B-induced

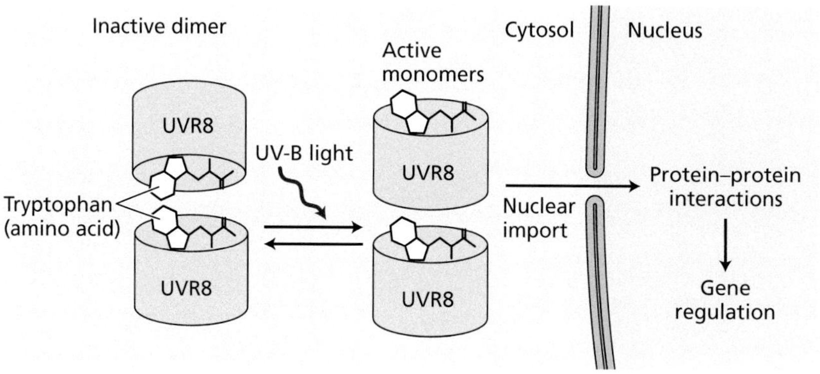

Figure 13.17 Diagram of the UV RESISTANCE LOCUS 8 (UVR8) photoreceptor response to UV-B irradiation. The inactive form of UVR8 is a dimer consisting of two monomers joined by salt bridges, with a series of tryptophan residues at the interaction surface. It is these tryptophan residues that sense UV-B, resulting in the structural changes that generate the two active monomers. Active UVR8 monomers are then transported to the nucleus where they participate in protein–protein interactions that result in downstream regulation of UV-B-responsive genes. (After art courtesy of Candace Pritchard.)

physiological responses, UVR8, is a seven-bladed β-propeller protein, which forms functionally inactive homodimers in the absence of UV-B. Unlike phytochrome, cryptochrome, and phototropin, UVR8 lacks a prosthetic chromophore. The two identical subunits of UVR8 are linked in the dimer by a network of salt bridges formed between tryptophan residues, which serve as the primary UV-B sensors, and nearby arginine residues. The indole rings of these bridging tryptophans are excited by UV-B photons, disrupting the salt bridges (**Figure 13.17**). As a result, the monomers dissociate and become functionally active. The monomers then interact with other protein complexes to activate gene expression.

Plant photoresponses observed in the field are composite activities

Much of what we know about individual photoreceptors is a result of laboratory experimentation using exquisitely precise light sources combined with mutants that lack one or more photoreceptors. Such experiments allow experimental discrimination of discrete photoresponses as well as the mechanistic pathways involved in the responses. However, it is often the case that common signaling components are identified in these experiments. In almost all cases, experimental evidence indicates that plant photoreceptor systems interact in natural conditions and that responses in the field are composites of multiple photoresponses. The increased use of wide-scale gene expression studies to analyze photoresponses in field studies overwhelmingly supports this view.

Summary

Photoreceptors, including phytochromes, cryptochromes, and phototropins, help plants regulate developmental processes over their lifetimes by sensitizing plants to incident light. Photoreceptors also initiate protective processes in response to harmful radiation.

13.1 Plant Photoreceptors

- Sunlight regulates developmental processes over the life of the plant and provides directional and nondirectional cues for growth and movement. Sunlight also contains UV and infrared radiation that can harm plant tissues (**Figures 13.1–13.4**).

(Continued)

Summary (*continued*)

- Photoreceptors undergo conformational changes after irradiation with specific light wavelengths. These conformational changes initiate downstream signaling events.
- Phytochromes (which absorb red and far-red light) and phototropins and cryptochromes (which absorb blue light and UV-A) are photoreceptors that are sensitive to light quantity, quality, and duration.
- The light-absorbing component of photoreceptors is called a chromophore. In phytochromes, cryptochromes, phototropins, and ZEITLUPE proteins, the chromophore is a small-molecule cofactor. The UVR8 photoreceptor lacks a chromophore; instead an intrinsic ring of tryptophan residues functions as a chromophore.
- Action spectra and absorption spectra help researchers determine which wavelengths of light lead to specific photoresponses (**Figures 13.5–13.7**).
- Light fluence and irradiance also govern whether a photoresponse occurs.

13.2 Phytochromes

- Phytochrome is generally sensitive to red and far-red light, and exhibits the ability to interconvert between the physiologically inactive Pr form and the active Pfr form.
- Red light triggers conformational changes in both the phytochrome chromophore and protein (**Figure 13.8**).

13.3 Phytochrome Responses

- Photoresponses can vary substantially in both their lag time (the time between exposure to light and the subsequent response) and their escape time (the time before an overall sequence of reactions becomes irreversible).
- Phytochrome-initiated responses fall into one of three main categories: very low fluence responses (VLFRs), low fluence responses (LFRs), or high irradiance responses (HIRs) (**Figure 13.9**).
- Phytochrome A mediates responses to continuous far-red light.
- Phytochrome can rapidly change membrane potentials and ion fluxes.
- Phytochrome regulates gene expression through a wide range of regulatory elements and transcription factors.
- Phytochrome itself can be phosphorylated and dephosphorylated.
- Phytochrome-induced photomorphogenesis involves protein degradation.

13.4 Blue-Light Responses and Photoreceptors

- Blue-light responses generally exhibit a lag time after irradiation, and the effect declines gradually after the disappearance of the light signal (**Figure 13.10**).

13.5 Cryptochromes

- Activation of the flavin adenine dinucleotide (FAD) chromophore causes a conformational change in cryptochrome, enabling cryptochrome to bind to other protein partners (**Figure 13.11**).
- Cryptochrome isoforms have differential developmental effects.

13.6 Phototropins

- Similar to cryptochromes, phototropins mediate photoresponses to blue light; phot1 and phot2 are sensitive to different and overlapping intensities of blue light.
- Phototropins are located at or near the plasma membrane and have flavin mononucleotide (FMN) chromophores that can induce conformational changes (**Figure 13.12**).
- When phototropins are activated by blue light, their kinase domain is "uncaged," causing autophosphorylation.
- Phototropins are required for seedling phototropism (**Figure 13.13**).
- Phototropins mediate chloroplast accumulation and avoidance responses to weak and strong light (**Figure 13.14**).
- Blue light, sensed by phototropins, causes activation of plasma membrane H^+-ATPases and inhibition of plasma membrane anion channels, and ultimately regulates stomatal opening (**Figures 13.15**, **13.16**).

13.7 Responses to Ultraviolet Radiation

- The photoreceptor involved in responses to UV-B irradiation is UVR8 (**Figure 13.17**).
- Unlike other phytochromes, cryptochromes, and phototropins, UVR8 lacks a prosthetic chromophore.
- UVR8 interacts with other protein complexes to activate the transcription of UV-B-induced genes.

Suggested Reading

Christie, J. M., and Murphy, A. S. (2013) Shoot phototropism in higher plants: New light through old concepts. *Am. J. Bot.* 100: 35–46.

Leivar, P., and Monte, E. (2014) PIFs: Systems integrators in plant development. *Plant Cell* 26: 56–78.

Nawker, G. M., Legris, M., Goyal, A., Schmid-Siegert, E., Fleury, J., Mucciolo, A., DeBellis, D., Trevisan M., Schueler, A., and Fankhauser, C. (2023) Air channels create directional light signal to regulate hypocotyl phototropism. *Science* 382: 935–940.

Rockwell, R. C., Duanmu, D., Martin, S. S., Bachy, C., Price, D. C., Bhattachary, D., Worden, A. Z., and Lagarias, J. K. (2014) Eukaryotic algal phytochromes span the visible spectrum. *Proc. Natl. Acad. Sci. USA* 111: 3871–3876.

Takemiya, A., Sugiyama, N., Fujimoto, H., Tsutsumi, T., Yamauchi, S., Hiyama, A., Tadao, Y., Christie, J. M., and Shimazaki, K.-I. (2013) Phosphorylation of BLUS1 kinase by phototropins is a primary step in stomatal opening. *Nat. Commun.* 4: 2094. DOI:10.1038/ncomms3094.

14 Seed Dormancy, Germination, and Seedling Establishment

Seeds are specialized plant structures adapted for the dispersal of the next sporophyte generation. Seeds are unique to the Spermatophyta (seed plants), which includes both the angiosperms and gymnosperms. Seeds develop from ovules, which before fertilization contain the female gametophytes and where, after fertilization, **embryogenesis** occurs (discussed in Chapter 17). Inside the seed, the embryo is surrounded by tissues that store food that the embryo uses to grow during embryogenesis and early seedling development. The seed also protects the embryo from the environment after the seed is released from the parent plant. The self-contained seed enclosing the embryo was one of many evolutionary adaptations that freed plant reproduction from a dependence on water. The evolution of seed plants thus represents an important milestone in the adaptation of plants to dry land.

In this chapter we discuss the structure and composition of various types of seeds. Next, we describe the processes of seed germination and seedling establishment—by which we mean the production of the first photosynthetic leaves and a minimal root system. Between embryogenesis and germination there is typically a period of *seed maturation* that culminates in *quiescence*, a non-germinating state characterized by a reduced metabolic rate, after which the seed is released from the parent plant. Quiescence ensures that germination is delayed until the seed reaches the soil, where it can receive the water and oxygen required for seedling growth. Although some seeds have the capacity to germinate as soon as they are mature, others remain in a state known as a *dormancy* and require an additional treatment or trigger, such as light, chilling, or physical abrasion, before they can germinate. For many agricultural crops this is

embryogenesis The formation and development of the embryo.

not a problem, because human selection for rapidly germinating seeds has resulted in the gradual loss of dormancy-inducing genes.

Once a seed begins to germinate, its food reserves are mobilized to support the growth of the emerging seedling. A wide variety of enzymes are produced that degrade the stored proteins, lipids, and starch in the seed. Hormones play an important role in coordinating the processes of germination and food mobilization.

Later in the chapter we consider the processes involved in the ongoing growth of the plant. These include photomorphogenesis (light-induced development), the development of a functioning vascular system, and the mechanisms regulating growth rate. Finally, we discuss the differential growth responses that allow the plant to navigate its environment, including gravitropism and phototropism, the processes by which plants orient themselves to gravity and light.

seed coat (or **testa)** The outer layer of the seed, derived from the integument of the ovule.

embryo The early sporophytic form of the plant formed from the zygote produced by gametic fusion or apomixis that becomes the basis of the mature plant axis. The embryo lacks leaves, stems, and reproductive structures.

radicle The embryonic root. Usually the first organ to emerge on germination.

hypocotyl The region of the seedling stem below the cotyledons and above the root.

cotyledons Also called "seed leaves," they are the main sources of nutrients during germination and post-germination photosynthesis before the first true leaves develop.

shoot apex (terminal bud) The shoot apical meristem with its associated leaf primordia and young, developing leaves.

plumule First true leaf of a growing seedling.

endosperm Tissue that surrounds and nourishes the embryo in the seeds of angiosperms (flowering plants). In angiosperms, endosperm is formed by fusion of at least one polar nucleus in the embryo sac with one of the two sperm nuclei from the pollen grain. In gymnosperms the nutritive material of the seed is haploid and present before fertilization.

double fertilization A unique feature of all angiosperms whereby, along with the fusion of a sperm with the egg to create a diploid zygote, a second male gamete fuses with the polar nuclei in the embryo sac to generate the endosperm tissue (with a triploid or higher number of chromosomes).

perisperm Storage tissue derived from the nucellus, and often consumed during embryogenesis.

14.1 Seed Anatomy

Describe the parts of the seed and be able to distinguish between eudicot and monocot seeds.

All seeds contain three basic structural features: an embryo, food storage tissue, and a protective outer layer of dead cells called the **seed coat** (or **testa**) (**Figure 14.1**). The angiosperm **embryo** consists of the embryonic axis, which is composed of the **radicle**, or embryonic root; the **hypocotyl**, to which one or two **cotyledons** are attached; and the **shoot apex** bearing the **plumule**, or first true leaf primordium. In angiosperms, the food storage tissue that nourishes the growing embryo is the triploid **endosperm** that results from **double fertilization** (Chapter 16).

Seed anatomy varies widely among different plant groups

Despite their common features, seeds range widely in size, from the dust-sized particles of orchids, weighing 1 µg, to the massive seeds of the coco de mer (*Lodoicea maldivica*), which can reach 30 cm in length and weigh 20 kg. Despite the relative simplicity of the embryo, and the limited number of tissues surrounding it, seed anatomy exhibits considerable diversity among the different plant groups.

Angiosperm seeds can be categorized broadly as endospermic or non-endospermic, depending on the presence or absence of a well-formed triploid endosperm at maturity. **Figure 14.1** shows representative examples from both categories. For example, beet seeds are non-endospermic because the triploid endosperm is largely used up during embryo development (discussed in Chapter 17). Instead, the **perisperm** and storage cotyledons serve as the main sources of nutrients during germination. Garden bean seeds (*Phaseolus vulgaris*) and legume seeds in general are also non-endospermic, relying on their large storage cotyledons, which make up the bulk of the seed, for their food reserves. In contrast, castor bean (*Ricinus communis*), onion (*Allium cepa*), bread wheat (*Triticum* spp.), and maize (corn; *Zea mays*) seeds are all endospermic. The main difference between angiosperm and gymnosperm seeds is that gymnosperms do not have triploid endosperm. Instead, gymnosperms have a haploid megagametophyte that provides stored food to the embryo.

The endosperm is typically rich in starch, lipids, and proteins. Some endosperm tissues have thick cell walls that break down during germination, releasing a variety of sugars. The outermost layer of the endosperm in some

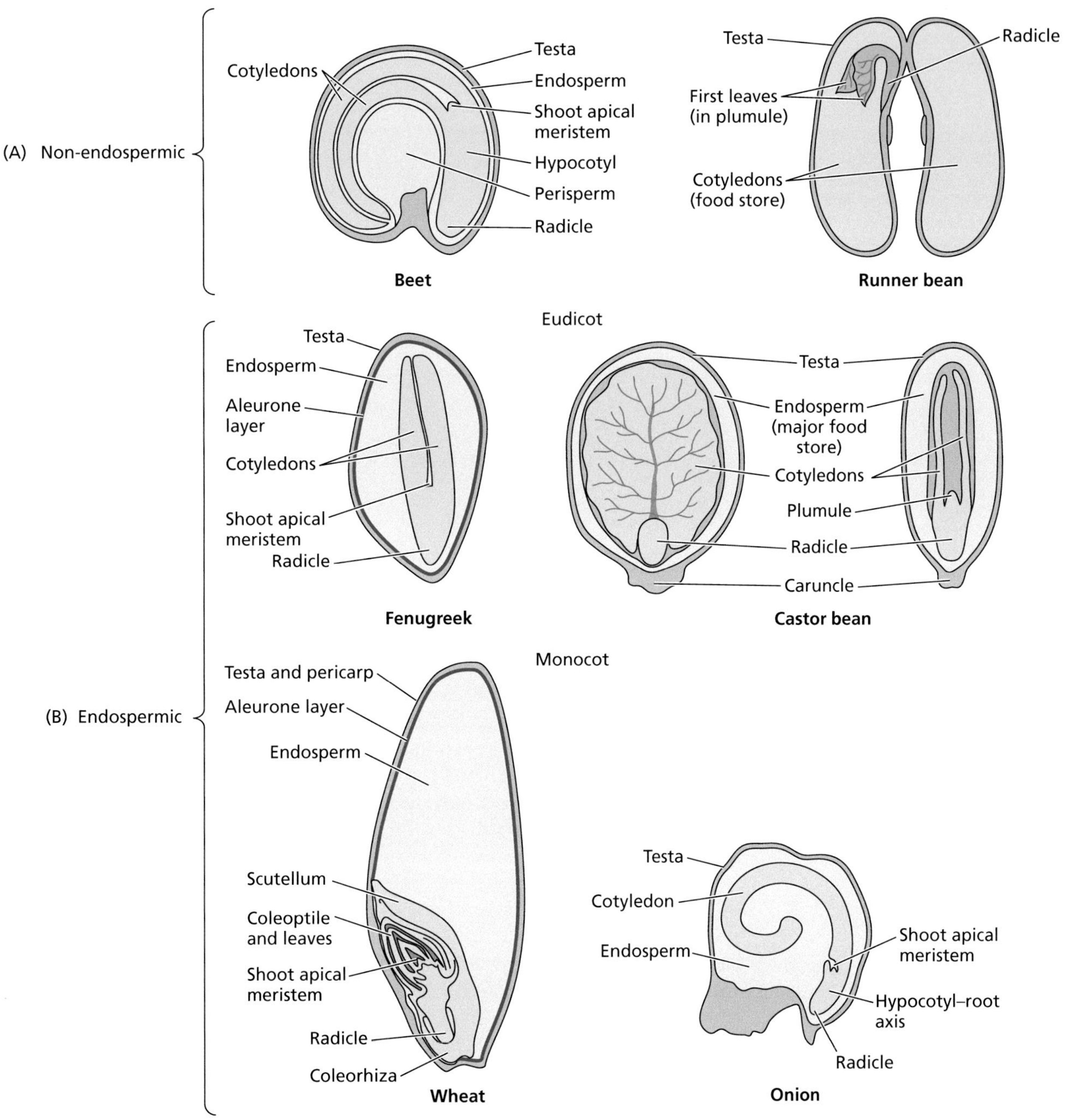

Figure 14.1 Seed structure. Structure of non-endospermic seeds (A) and endospermic seeds (B).

species differentiates into a specialized secretory tissue with thickened primary walls called the *aleurone layer*, so named because it is composed of cells filled with **protein storage vacuoles** (**PSVs**), originally called *aleurone grains*. As you will see later in the chapter, the aleurone layer plays an important role in regulating dormancy in certain eudicot seeds. In seeds of wheat and other members of the Poaceae (grass family), secretory aleurone layers are also responsible for the mobilization of stored food reserves during germination.

The embryos of cereal grains are highly specialized and merit closer examination both because of their agricultural importance and because they have been widely used as model systems to study the hormonal regulation of food

protein storage vacuoles (**PSVs**) Specialized small vacuoles that accumulate storage proteins, typically in seeds.

scutellum The single cotyledon of the grass embryo, specialized for nutrient absorption from the endosperm.

coleoptile A modified ensheathing leaf-like that covers and protects the young primary leaves of a grass seedling as it grows through the soil. Unilateral light perception, especially blue light, by the tip results in asymmetric growth and bending due to unequal auxin distribution in the lighted and shaded sides.

coleorhiza A protective sheath surrounding the embryonic radicle in members of the grass family.

mesocotyl In members of the grass family, the part of the elongating axis between the scutellum and the coleoptile.

seed dormancy A state of arrested growth of the embryo that prevents germination even when all the necessary environmental conditions for growth, such as water, O_2, and temperature, are met.

primary dormancy The failure of newly dispersed, mature seeds to germinate under normal growth conditions.

mobilization during germination. Specialized embryonic structures found in seeds from the grass family include the following structures (Figure 14.1):

- The single cotyledon has been modified by evolution to form an absorptive organ, the **scutellum**, which forms the interface between the embryo and the starchy endosperm tissue. During germination, mobilized sugars from the endosperm are absorbed by the scutellum and transported to the embryo proper.
- The basal sheath of the scutellum has been elongated to form a **coleoptile** that covers and protects the first leaves while the shoot is growing up through the soil.
- The base of the hypocotyl has been elongated to form a protective sheath around the radicle called the **coleorhiza**.
- In some members of the grass family, such as maize, the upper hypocotyl has been modified to form a **mesocotyl**. During seedling development, the growth of the mesocotyl helps raise the shoot to the soil surface, especially in the case of deeply planted seeds.

14.2 Seed Dormancy

Explain the importance of the dormancy survival mechanism and understand the types of dormancy along with the factors required to overcome each type.

During seed maturation, the embryo dehydrates and enters a quiescent phase (*seed quiescence*) where seeds do not germinate for lack of water, O_2, or proper temperature for growth. Most mature seeds typically have less than 0.1 g water g^{-1} dry weight at the time of shedding. As a consequence of dehydration, metabolism nearly comes to a halt and the seed enters a quiescent, or "resting," state. In some cases the seed becomes dormant as well. Seed germination requires rehydration and can be defined as the resumption of growth of the embryo in the mature seed. The process of germination encompasses all the events that take place between the start of *imbibition* (moistening) of the dry seed (discussed later in the context of seed germination) and the *emergence* of the embryo, usually starting with the radicle (embryonic root), from the structures that surround it. Successful completion of germination depends on the same environmental conditions as vegetative growth (Chapter 15): Water and oxygen must be available, and the temperature must be in the *physiological range* (that is, the range that does not inhibit physiological processes).

However, a viable (living) seed may not germinate even if the appropriate environmental requirements are satisfied, a phenomenon known as **seed dormancy** where additional treatments or signals are needed for germination to occur. Seed dormancy is an intrinsic temporal block to the initiation of germination that provides additional time for seed dispersal over greater distances. It also maximizes seedling survival by preventing germination under unfavorable conditions. *Seed banks* are repositories where seeds are stored under conditions that maintain both dormancy and viability. A seed bank may comprise a naturally occurring repository (such as the soil) or an artificial repository built by humans. Seeds from undomesticated plants, like wild relatives of crop plants and weeds, often have soil seed banks that span many years.

Different types of seed dormancy can be distinguished on the basis of the developmental timing of dormancy onset. Newly dispersed, mature seeds that fail to germinate under favorable conditions exhibit **primary dormancy**. Primary dormancy is typically induced by the hormone abscisic acid (ABA) during seed maturation. (ABA regulation of seed dormancy is discussed later in the chapter.)

After primary dormancy has been lost, these non-dormant seeds may acquire **secondary dormancy** if exposed to unfavorable conditions that inhibit germination over a period of time. For example, non-dormant seeds of *Avena sativa* (oat) can become dormant if exposed to temperatures higher than the maximum permissible for germination. Seeds of *Phacelia dubia* (small-flower scorpionweed) can acquire secondary dormancy at temperatures lower than the minimum for germination. The mechanisms of secondary dormancy are poorly understood but contribute to weed seed banks.

secondary dormancy Seeds that have lost their primary dormancy may become dormant again if they experience prolonged exposure to unfavorable growth conditions.

exogenous dormancy Seed dormancy that is caused by the seed coat and other surrounding tissues; may involve impermeability to water or oxygen, mechanical restraint, or the retention of endogenous inhibitors. Also called coat-imposed dormancy.

coat-imposed dormancy Seed dormancy that is caused by the seed coat and other surrounding tissues; may involve impermeability to water or oxygen, mechanical restraint, or the retention of endogenous inhibitors. Also called exogenous dormancy.

There are two basic types of seed dormancy mechanisms: exogenous and endogenous

Seed dormancy may have both physical and physiological mechanisms. According to one classification scheme, primary seed dormancy can be divided into two main types, *exogenous dormancy* and *endogenous dormancy*.

Exogenous dormancy, or **coat-imposed dormancy**, refers to the physical inhibitory effects of the seed coat or other enclosing tissues, such as the endosperm, pericarp, or extrafloral organs, on the growth of the embryo during germination. The embryos of such seeds germinate readily in the presence of water and oxygen once the seed coat and other surrounding tissues have either been removed or damaged. There are several ways that seed coats can impose dormancy on the embryo:

- *Water impermeability*: This type of dormancy, also called hard-seededness, is common in plants found in arid and semiarid regions, especially among legumes, such as clover (*Trifolium* spp.) and alfalfa (*Medicago* spp.). The classic example is Indian lotus (*Nelumbo nucifera*) seeds, which can survive up to 1,200 years because of their impermeable coats. Waxy cuticles, suberized layers, and the cell walls of palisade layers consisting of lignified sclereids all combine to restrict the penetration of water into the seed. This type of dormancy can be broken by mechanical or chemical scarification to abrupt the seed coat. In the wild, passage through the digestive tracts of animals can cause chemical scarification.
- *Interference with gas exchange*: Dormancy in some seeds can be overcome by oxygen-enriched atmospheres, suggesting that the seed coat and other surrounding tissues limit the supply of oxygen to the embryo. In wild mustard (*Sinapis arvensis*), the permeability of the seed coat to oxygen is less than that to water by a factor of 10^4. In other seeds, oxidative reactions involving phenolic compounds in the seed coat may consume large amounts of oxygen, reducing oxygen availability to the embryo.
- *Mechanical constraint*: The first visible sign of germination is typically the radicle (embryonic root) breaking through its surrounding structures, such as the endosperm, if present, and seed coat. In some cases, however, the thick-walled endosperm may be too rigid for the radicle to penetrate, as in Arabidopsis, tomato, coffee, and tobacco. For such seeds to complete germination, the endosperm cell walls must be weakened by the production of cell wall–degrading enzymes, typically where the radicle emerges.
- *Retention of inhibitors*: Dormant seeds often contain specialized metabolites, including phenolic acids, tannins, and coumarins, and repeated rinsing of such seeds with water often promotes germination. The seed coat may impose dormancy by preventing the escape of inhibitors from the seed, or the seed coat may produce inhibitors that diffuse into the embryo.

Figure 14.2 Vivipary in a wetland plant. Viviparous seeds of red mangrove (*Rhizophora mangle*).

Endogenous dormancy, or **embryo dormancy**, is intrinsic to the embryo and is not due to the other parts of the seed. Embryo dormancy is regulated by the ratio of the hormones abscisic acid (ABA) and gibberellin (GA) (discussed at the end of this section). Endogenous dormancy is typically induced by ABA at the end of embryogenesis. Fully mature dry seeds require endogenous ABA to regulate and maintain primary dormancy. ABA also regulates post-imbibition dormancy in plants such as lettuce and barley. After dormancy is broken, the seed is able to germinate over the range of conditions permissible for the particular genotype.

Non-dormant seeds can exhibit vivipary and precocious germination

In some species the mature seeds not only lack dormancy but also germinate while still on the mother plant, a phenomenon known as **vivipary**. True vivipary is extremely rare in angiosperms and is largely restricted to mangroves (**Figure 14.2**) and other plants growing in estuarine or riparian ecosystems in the tropics and subtropics. Seeds of this species germinate while still inside the attached fruit and produce an elongated, dart-like propagule that can drop from the tree and root itself in the surrounding soft mud.

The **precocious germination** of physiologically mature seeds on the mother plant is known as **preharvest sprouting** and is characteristic of some grain crops when they mature in wet weather (**Figure 14.3A**). Preharvest sprouting in cereals (e.g., wheat, barley, rice, and sorghum) reduces grain quality and causes serious economic losses. In maize, *viviparous* (*vp*) mutants have been identified in which the embryos germinate directly on the cob while still attached to the parent plant (**Figure 14.3B**). Several of these mutants are ABA deficient and one is ABA insensitive, preventing the normal ABA-mediated dormancy. Vivipary in the ABA-deficient mutants can be partially prevented by treatment with exogenous ABA. Vivipary in maize also requires GA biosynthesis early in embryogenesis as a positive signal to

(A)

(B)

Figure 14.3 Vivipary in crop plants. Preharvest sprouting in wheat (*Triticum aestivum*) (A) and maize (*Zea mays*) (B). For both species, non-sprouting heads and ears, respectively, are shown to the right for comparison.

induce germination; double mutants deficient in GA and ABA do not exhibit vivipary. This shows that germination is regulated by the ABA:GA ratio, not the actual amount of ABA.

The ABA:GA ratio is the primary determinant of embryonic seed dormancy

It has long been known that ABA exerts an inhibitory effect on seed germination, whereas GA exerts a positive effect. According to the **hormone balance theory**, the ratio of these two hormones serves as the primary determinant of seed dormancy and germination. The relative hormonal activities of ABA and GA in the seed depend on two main factors: the amounts of each hormone present in the target tissues, and the ability of the target tissues to detect and respond to each of the hormones. Hormonal sensitivity, in turn, is determined by the hormonal signaling pathways in the target tissues (**Figure 14.4**).

The amounts of ABA and GA are determined by their relative rates of biosynthesis versus inactivation (Chapter 12). The balance between the biosynthesis and inactivation pathways is regulated by the action of transcription factors. Certain environmental factors can alter the ABA:GA balance in seeds and thereby stimulate germination. Certain treatments, such as *after-ripening*, can promote germination by reducing ABA concentrations, while other treatments, such as **chilling** (or **stratification**), can promote germination by increasing GA biosynthesis (Figure 14.4).

Another factor that appears to regulate seed dormancy is the relative sensitivities of the embryo to ABA and GA. According to the present model, the hormonal sensitivity of the embryo is under both developmental and environmental control (Figure 14.4). During the early stages of seed development, ABA sensitivity is high and GA sensitivity is low, which favors dormancy over germination. Later in seed development, ABA sensitivity declines and GA sensitivity increases, favoring germination. At the same time, the seed becomes progressively more sensitive to environmental cues, such as temperature and light, that can either stimulate or inhibit germination.

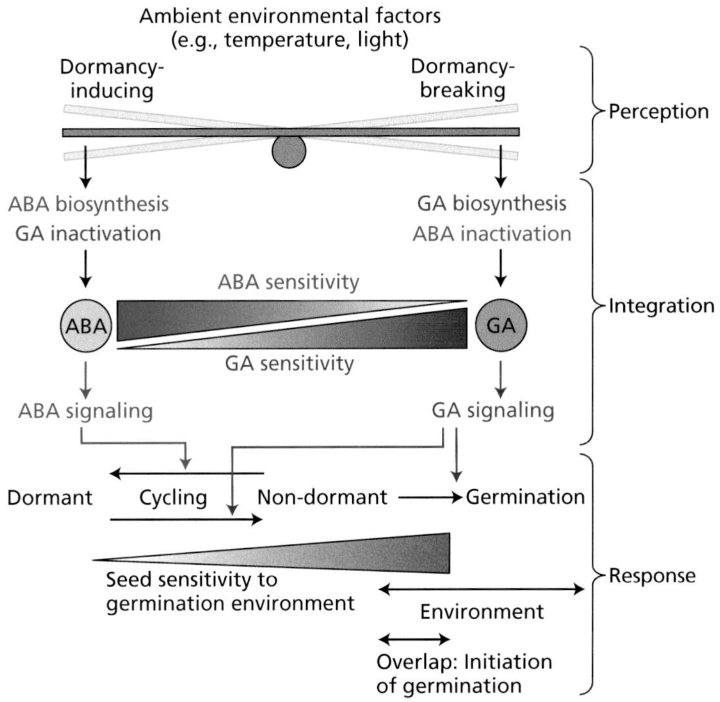

Figure 14.4 Model for ABA and GA regulation of dormancy and germination in response to environmental factors. Environmental factors such as temperature affect the ABA:GA ratios and the responsiveness of the embryo to ABA and GA. In dormancy, GA is inactivated and ABA biosynthesis and signaling predominate. In the transition to germination, ABA is inactivated and GA biosynthesis and signaling predominate. The complex interplay between ABA and GA metabolism and sensitivity in response to environmental conditions can result in cycling between dormant and non-dormant states (dormancy cycling). Germination can proceed to completion when favorable environmental conditions and non-dormancy coincide.

endogenous dormancy Seed dormancy that is intrinsic to the seed and is not due to any physical or chemical influence of the seed coat or other surrounding tissues. Endogenous dormancy is regulated by the ratio of the hormones abscisic acid (ABA) and gibberellin (GA).

embryo dormancy Seed dormancy that is caused directly by the embryo and is not due to any influence of the seed coat or other surrounding tissues.

vivipary The precocious germination of seeds in the fruit while still attached to the plant.

precocious germination Germination of viviparous mutant seeds while still attached to the mother plant.

preharvest sprouting Germination of physiologically mature wild-type seeds on the mother plant caused by wet weather.

hormone balance theory The hypothesis that seed dormancy and germination are regulated by the balance of ABA and gibberellin.

chilling (or **stratification)** Technique for breaking seed dormancy by storage at cold temperatures (1–10°C) under moist conditions for several months. The term *stratification* is derived from the former practice of breaking dormancy by allowing seeds to overwinter in small mounds of alternating layers of seeds and soil.

ABA and GA are not the only hormones regulating seed dormancy. For example, both ethylene and brassinosteroids reduce the ability of ABA to inhibit germination, apparently by negatively regulating the ABA signal transduction pathway. ABA also inhibits ethylene biosynthesis, while brassinosteroids enhance it. Thus, hormonal networks are probably involved in regulating seed dormancy, as they are in regulating most other developmental phenomena.

14.3 Release from Dormancy

Identify factors that must be present for a seed to be released from dormancy, and describe the important role that hormone ratios play in determining whether a seed will germinate or remain in dormancy.

Breaking dormancy involves a change in the metabolic state of the seed that enables the embryo to reinitiate growth. Because germination is an irreversible process that commits the seed to grow into a seedling, many species have developed sophisticated mechanisms for sensing the optimal environmental conditions for this to occur. Often there are seasonal components to the ultimate "decision" of a seed to germinate, as in the examples of secondary dormancy mentioned earlier in the chapter. In this section we discuss some of the environmental cues that bring about the release from dormancy. Although each external signal is discussed separately, seeds in nature must integrate responses to multiple environmental factors, perceived either simultaneously or in succession.

Because the ABA:GA ratio plays such a decisive role in maintaining seed dormancy, environmental conditions that break dormancy are thought ultimately to activate gene networks that affect the balance between the responses to ABA and GA. This hypothesis is consistent with the fact that treatment of seeds with GA often can substitute for a positive environmental signal in breaking dormancy.

Light is an important signal that breaks dormancy in small seeds

Many seeds have a light requirement for germination, which may involve only a brief exposure, as in the case of the "Grand Rapids" cultivar of lettuce (*Lactuca sativa*) (Chapter 13); an intermittent treatment (e.g., succulents of the genus *Kalanchoe*); or even a specific photoperiod involving short or long days. For example, birch (*Betula* spp.) seeds require long days (16 h light) to germinate, while seeds of the conifer eastern hemlock (*Tsuga canadensis*) require short days. Phytochrome, which senses red (R) and far-red (FR) wavelengths of light (Chapter 13), is the primary sensor for light-regulated seed germination. All light-requiring seeds exhibit coat-imposed dormancy, and removal of the seed coat and endosperm allows the radicle (embryonic root) to elongate in the absence of light. The effect that light has on the embryo is thus to enable the radicle to penetrate the endosperm, a process facilitated in some species by enzymatic weakening of the cell walls in the micropylar region, adjacent to the radicle.

Light is required for germination of the small seeds of numerous herbaceous and grassland species, many of which remain dormant if they are buried below the depth to which light penetrates the soil. Even when such seeds are on or near the soil surface, the amount of shading by the vegetation canopy (i.e., the R:FR light ratio the seeds receive) is likely to affect their germination. We will return to the effects of the R:FR ratio in Chapter 15 in relation to the phenomenon of shade avoidance.

Some seeds require either chilling or after-ripening to break dormancy

Many seeds require a period of low temperature (1–10°C) to germinate. For species adapted to temperate climates, this requirement has obvious survival

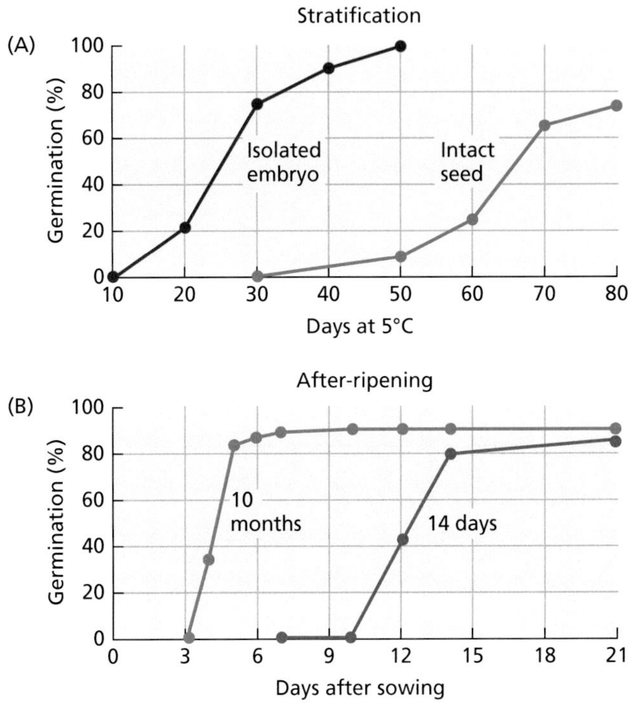

Figure 14.5 Seed dormancy can be overcome by stratification and after-ripening. (A) Release of apple seeds from dormancy by stratification, or moist chilling. Imbibed seeds were stored at 5°C and periodically removed to test the seeds or isolated embryos for germination at a permissible temperature. The germination of intact seeds (defined as emergence of the radicle) was significantly delayed compared with that of isolated embryos (defined as elongation of the radicle). (B) Effect of after-ripening on the germination of *Nicotiana plumbaginifolia seeds*. After-ripening for 10 months or longer greatly accelerated germination compared with after-ripening for only 14 days.

value, since such seeds will not germinate in the fall, but rather in the following spring. Chilling seeds to break their dormancy is called **stratification**, named for the agricultural practice of overwintering dormant seeds in layered mounds of soil or moist sand. Today seeds are simply stored moist in a refrigerator. Stratification has the added benefit that it synchronizes germination, which ensures that plants will mature at the same time. **Figure 14.5A** shows the effects of chilling on apple seed germination. Intact seeds require 80 days of chilling for maximum germination, whereas isolated embryos exhibit maximum germination after approximately 50 days. Thus, the presence of the surrounding tissues (seed coat and endosperm) increases the chilling requirement of the embryo by about 30 days.

Some seeds may require a period of **after-ripening**, meaning dry storage at room temperature, before they can germinate. The duration of the after-ripening requirement may be as short as a few weeks (e.g., barley, *Hordeum vulgare*) or as long as 5 years (e.g., curly dock, *Rumex crispus*). In the field, after-ripening may occur in winter annuals in which dormancy is broken by high summer temperatures, allowing the seeds to germinate in the fall. In contrast, moist chilling during the cold winter months is effective in many summer annuals. After-ripening of horticultural and agricultural crop seeds is usually performed in special drying cupboards that maintain the appropriate temperature, aeration, and low moisture conditions.

The effect of the duration of after-ripening on the germination of seeds of *Nicotiana plumbaginifolia* is shown in **Figure 14.5B**. Seeds after-ripened for only 14 days began to germinate after about 10 days of subsequent wetting, while seeds after-ripened for 10 months began germinating after only 3 days. Seeds are considered "dry" when their water content drops below 20%. However, if the seeds become too dry (5% water content or less), the effectiveness of after-ripening is diminished. In several species, ABA concentrations decrease during after-ripening, and even a small decline may be sufficient to break dormancy. For example, in *N. plumbaginifolia* seeds, the ABA concentration decreases by about 40% during after-ripening. In general, after-ripening promotes a decrease in ABA concentration and sensitivity, and a rise in GA concentration and sensitivity.

Seed dormancy can be broken by various chemical compounds

Numerous chemicals, such as respiratory inhibitors, sulfhydryl compounds, oxidants, and nitrogenous compounds, have been shown to break seed dormancy in specific species. However, only a few of these chemicals occur naturally in the environment. Of these, nitrate, often in combination with light, is probably the most important. Some plants, such as hedge mustard (*Sisymbrium*

stratification A process to break dormancy by chilling seeds.

after-ripening Technique for breaking seed dormancy by prolonged storage at room temperature (20–25°C) under dry conditions.

officinale), have an absolute requirement for nitrate and light for seed germination. Another chemical agent that can break dormancy is nitric oxide (NO), a signaling molecule produced by conversion of nitrite, which also functions in multiple stress responses, including hypoxia (Chapter 19). Mutants unable to synthesize NO exhibit reduced germination, and the effect can be reversed by treating the seeds with exogenous NO. Another strong chemical stimulant of seed germination in many species under natural conditions is smoke, which is produced during forest fires. Smoke is likely to contain multiple germination stimulants; one of the most active of these is **karrikinolide**, a member of the karrikin class of molecules, which structurally resemble strigolactone plant hormones that were introduced in Chapter 12.

In all three of these examples, the chemical stimulants appear to break dormancy by the same basic mechanism: by down-regulating ABA biosynthesis or signaling, and up-regulating GA biosynthesis or signaling, thus altering the ABA:GA balance.

14.4 Seed Germination

Explain the process of germination, what metabolic and developmental changes occur throughout each phase, and how water plays a vital role in the process.

Germination is the process that begins with water uptake by the dry seed and ends with the emergence of part of the embryo, usually the radicle, from its surrounding tissues. Strictly speaking, germination concludes with radicle emergence and does not include subsequent seedling growth. The rapid mobilization of stored food reserves that fuels the initial growth of the seedling is initiated during germination.

Germination requires permissive ranges of water, temperature, and oxygen, and often light and nitrate as well. Of these, water is the most critical factor. The water content of mature, air-dried seeds is in the range of 5 to 15%, well below the threshold required for fully active metabolism. In addition, water uptake is needed to generate the turgor pressure that powers cell expansion, the basis of vegetative growth and development. As we discussed in Chapter 3, water uptake is driven by the gradient in water potential (Ψ) from the soil to the seed. For example, incubating tomato seeds at high ambient water potential ($\Psi = 0$ MPa) allows 100% germination, whereas incubation at low water potential ($\Psi = -1.0$ MPa), which nullifies the gradient in water potential, completely suppresses germination (**Figure 14.6**).

Germination and postgermination can be divided into three phases corresponding to the phases of water uptake

Under normal conditions, water uptake by the seed is triphasic (**Figure 14.7**):

- *Phase I*: The dry seed takes up water rapidly by the process of imbibition.
- *Phase II*: Water uptake by imbibition declines and metabolic processes, including transcription and translation, are reinitiated. The embryo expands, and the radicle emerges from the seed coat.
- *Phase III*: Water uptake resumes because of a decrease in Ψ as the seedling grows, and the stored food reserves of the seed are fully mobilized.

The initial rapid uptake of water by the dry seed during Phase I is referred to as **imbibition**, to distinguish it from water uptake during Phase III. Although the water potential gradient drives water uptake in both cases, the causes of the gradients are different. In the dry seed, the **matric potential** (Ψ_m)

karrikinolide A component of smoke that stimulates seed germination; similar in structure to strigolactones.

germination The events that take place between the start of imbibition of the dry seed and the emergence of part of the embryo, usually the radicle, from the structures that surround it. May also be applied to other quiescent structures, such as pollen grains or spores.

imbibition The initial phase of water uptake in dry seeds, which is driven by the matric potential component of the water potential—that is, by the binding of water to surfaces, such as the cell wall and cellular macromolecules.

matric potential (Ψ_m) The sum of osmotic potential (Ψ_s) + hydrostatic pressure (Ψ_p). Useful in situations (dry soils, seeds, and cell walls) where the separate measurement of Ψ_s and Ψ_p is difficult or impossible to obtain.

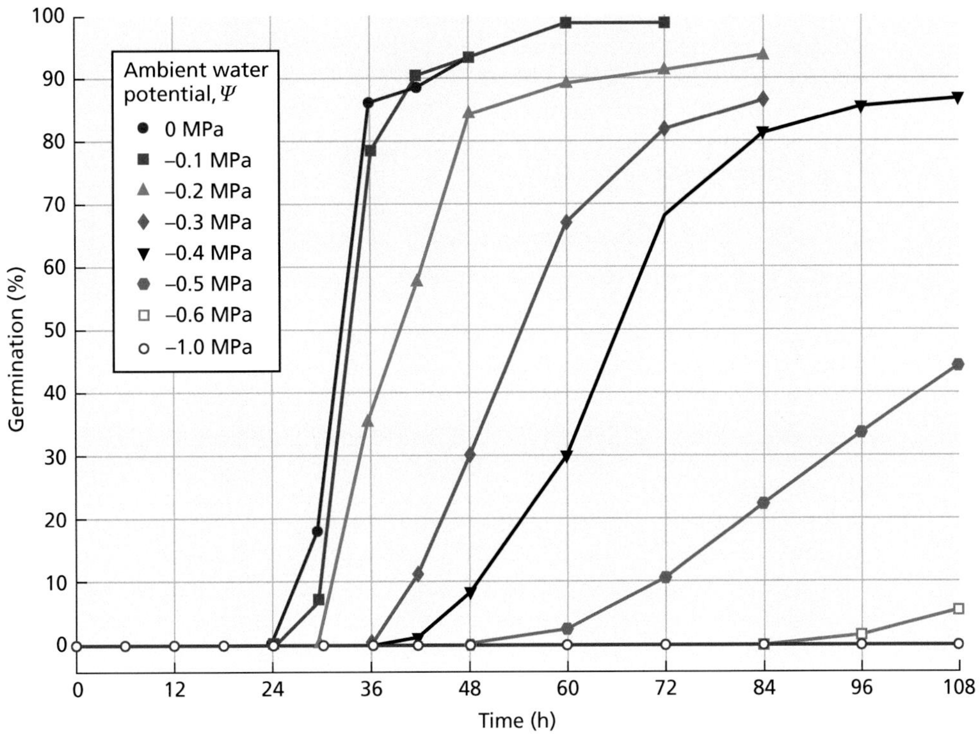

Figure 14.6 Time course of tomato seed germination at different ambient water potentials.

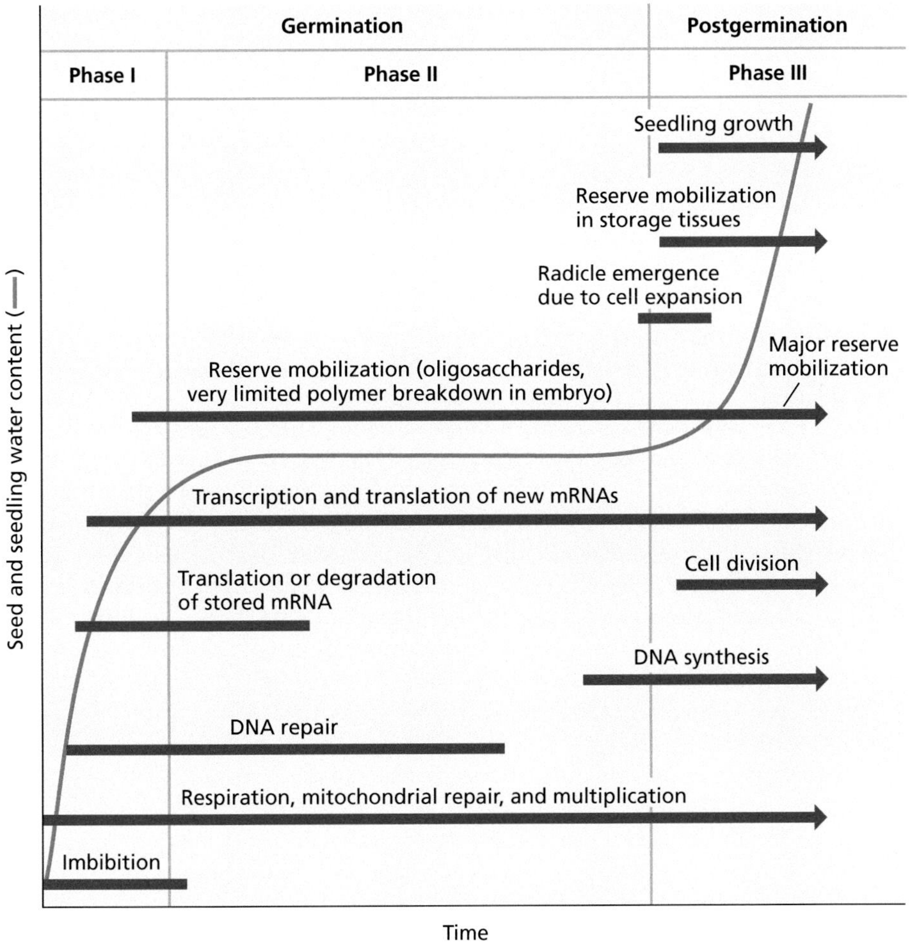

Figure 14.7 Phases of water uptake in seeds. In Phase I, dry seeds imbibe (take up) water rapidly. Since water flows from the higher to the lower water potential, the water uptake by the seed stops when the difference in water potential between the seed and the environment becomes zero. During Phase II the cells expand and the radicle emerges from the seed, completing germination. Metabolic activity increases and cell wall loosening occurs. In Phase III, water uptake resumes as the seedling becomes established.

component of the water potential equation lowers Ψ and creates the gradient. The matric potential arises from the binding of water to solid surfaces, such as the microcapillaries of cell walls and the surfaces of proteins and other macromolecules (discussed in Chapter 2). The rehydration of cellular macromolecules activates basal metabolic processes, including respiration, transcription, and translation.

Imbibition ceases when all the potential binding sites for water become saturated, and Ψ_m becomes less negative. During Phase II the rate of water uptake slows down until the water potential gradient is reestablished. Phase II can thus be thought of as the lag phase preceding growth, during which the solute potential (Ψ_s) of the embryo gradually becomes more negative because of the breakdown of stored food reserves and the liberation of osmotically active solutes. The seed volume may increase as a result, rupturing the seed coat. At the same time, additional metabolic functions come online, such as the re-formation of the cytoskeleton and the activation of DNA repair mechanisms.

The emergence of the radicle through the seed coat in Phase II marks the end of the process of germination. The cell walls of the radicle are loosened and extend in response to the increase in turgor pressure that accompanies the uptake of water, which causes cell elongation. However, in many seeds the radicle must first break through the barrier imposed by the surrounding endosperm, seed coat, or pericarp before it can emerge from the seed. Radicle emergence can be a one- or two-step process. In the one-step process, either the surrounding tissues become physically weakened during imbibition, allowing the radicle to emerge unimpeded, or the radicle expands sufficiently during imbibition to rupture the surrounding tissues. In the two-step process, the surrounding tissues must first undergo metabolic weakening before the radicle can emerge from the seed.

During Phase III the rate of water uptake increases rapidly because of the onset of cell wall loosening and cell expansion, as seedling growth begins. Thus, the water potential gradient in Phase III embryos is maintained by both cell wall relaxation and solute accumulation.

14.5 Mobilization of Stored Reserves

Explain the importance of stored food reserves, why mobilization occurs, how specific nutrients are mobilized, and the location at which nutrients are stored.

The major food reserves of angiosperm seeds are typically stored in the cotyledons or in the endosperm. Proteins, lipids, and carbohydrates (starch) are stored in specialized structures within these tissues. Starch is stored in **amyloplasts**, which are specialized plastids (discussed in Chapter 1); proteins are stored in protein storage vacuoles (PSVs); and lipids are stored in **lipid bodies**, or droplets (Chapters 1 and 11). All plants have these three types of specialized food storage structures, although in many plants one type predominates. After germination there is a massive mobilization of reserves to provide nutrients, carbon, nitrogen, minerals, and so on to the growing seedling until it becomes autotrophic.

Starch is an important form of stored carbon and is the major food reserve in cereals. The endosperm is the major tissue that stores starch in cereals, though it can also accumulate in the embryo. Starch degradation is initiated by the enzymes α-amylase and β-amylase. α-Amylase hydrolyzes starch chains internally to produce oligosaccharides consisting of α(1,4)-linked glucose residues. β-Amylase degrades these oligosaccharides from the ends to produce the disaccharide

amyloplast A starch-storing plastid found abundantly in storage tissues of shoots and roots, and in seeds. Specialized amyloplasts in the root cap and shoot also serve as gravity sensors.

lipid body Organelle that accumulates and stores triacylglycerols. It is bounded by a single phospholipid leaflet ("half–unit membrane" or "phospholipid monolayer") derived from the endoplasmic reticulum.

maltose, which is then converted to glucose by maltase. Mobilization of stored food reserves is important industrially for producing malted grains for bread, beverages, and other products.

Protein storage vacuoles are the primary source of nitrogen and amino acids for new protein synthesis in the seedling. (A micrograph of PSVs is shown in Chapter 1.) The stored proteins are hydrolyzed by proteases. Protein storage vacuoles also contain phytin, the K^+, Mg^{2+}, or Ca^{2+} salt of phytic acid (*myo*-inositol hexaphosphate, a sugar alcohol), the major storage form of phosphate in seeds. During food mobilization, the enzyme phytase hydrolyzes phytin to release phosphate and the other ions for use by the growing seedling. Legumes are an example of plants that store a large amount of protein in their seeds.

Oil bodies contain lipids, high-energy carbon sources that are converted to sucrose during germination. (Micrographs of oil bodies are shown in Chapters 1 and 11.) In addition to storing triacylglycerides and phospholipids, oil bodies also store proteins, such as oleosins, caleosins, and steroleosins, with oleosins being predominant. Oil bodies are prevalent in oilseed plants such as rapeseed, mustard, cottonseed, flax, maize, peanut, sesame, copra, palm kernel, soybean, and sunflower seed. The model plant *Arabidopsis thaliana* is also an oilseed plant. In oilseed plants, the oil bodies are synthesized in the embryo. As discussed in Chapter 11, lipids are catabolized during seed germination via the activity of lipases, β-oxidation, and the glyoxylate cycle.

starchy endosperm A starch-storing triploid endosperm tissue that makes up the bulk of the seeds of cereals and other members of the grass family.

aleurone layer The layer of aleurone cells surrounding and distinct from the starchy endosperm of cereal grains.

Cereal seeds are a model for understanding starch mobilization

Cereals have a specialized starch mobilization mechanism that has been extensively studied as a model for how ABA and GA function during germination and seedling establishment. Cereal seeds consist of three parts: the embryo, the endosperm, and the fused seed coat—pericarp (**Figure 14.8**). The embryo, which will grow into the new seedling, has a specialized absorptive organ, the scutellum. The triploid endosperm is composed of two tissues: the centrally located **starchy endosperm** and, surrounding it, a peripheral **aleurone layer**. The aleurone layer is a specialized digestive tissue; the starchy endosperm consists of thin-walled cells with amyloplasts filled with starch grains. Living cells of the aleurone layer synthesize and release α-amylase and other hydrolytic enzymes into the starchy endosperm during germination. As a consequence, the stored food reserves of the endosperm are broken down, and the solubilized sugars, amino acids, and other products are transported to the growing embryo via the scutellum.

As shown in **Figure 14.9**, GA is released from the embryo during germination to stimulate the production and release of α-amylase by the aleurone layer of cereal seeds. Once inside the aleurone cells, GA binds to its receptor and initiates a signal transduction pathway, as described in Chapter 12. GA signaling results in increased expression of a transcriptional activator, *GA-MYB*, which in turn induces expression of the α-amylase gene in the aleurone layer (Figure 14.9), leading to the production and secretion of α-amylase into the endosperm. ABA inhibits the GA-induced synthesis of hydrolytic enzymes that are essential for the breakdown of storage reserves during seedling growth. In the case of α-amylase, ABA acts by inhibiting GA-dependent transcription of α-*amylase* mRNA.

Legume seeds are a model for understanding protein mobilization

In addition to being food reserves for the developing seedling, stored proteins may also increase tolerance to desiccation and reactive oxygen species and help stabilize membranes. Seed storage proteins can accumulate in the embryo and endosperm, as well as in the aleurone layer in cereals, and different proteins may

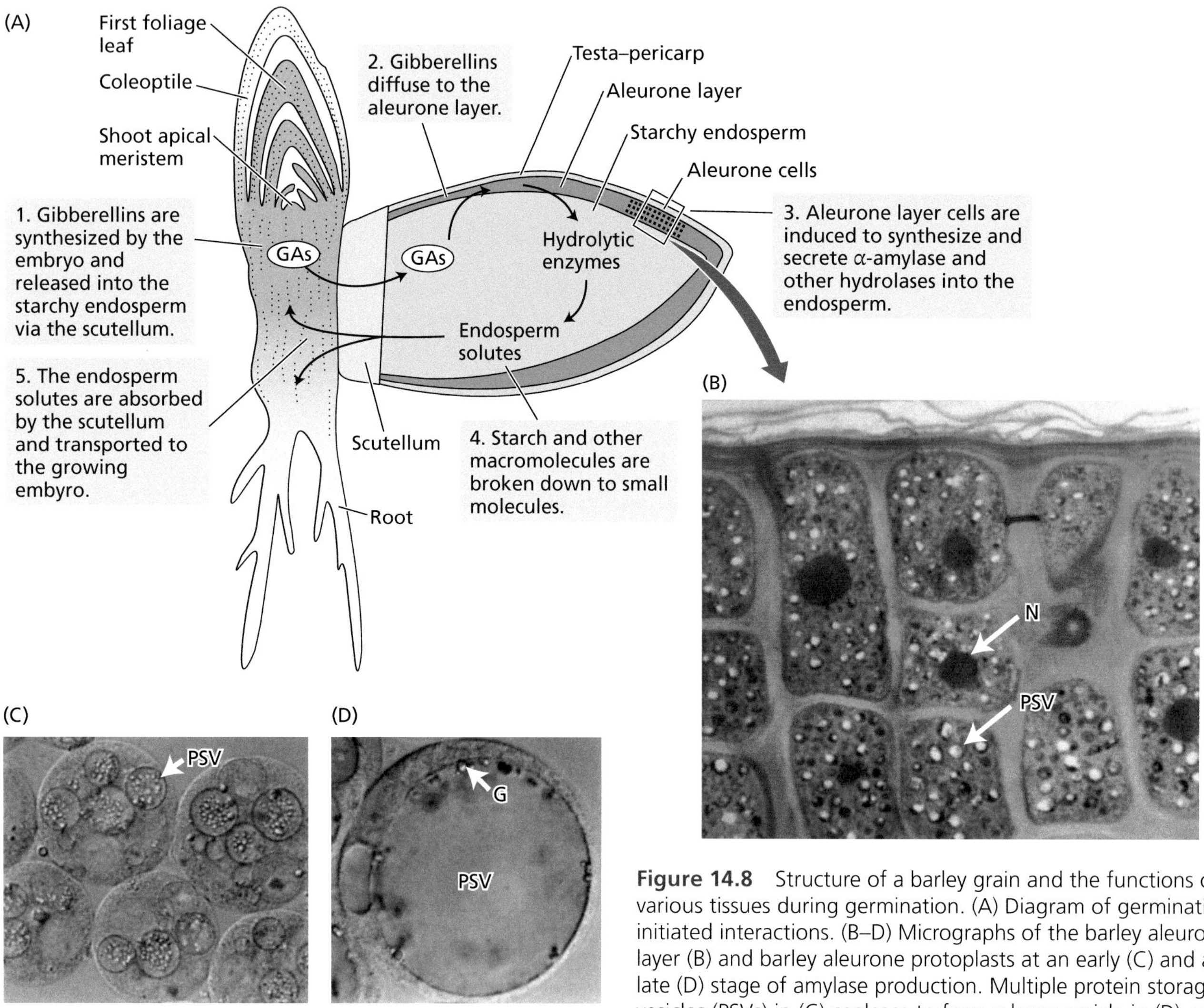
(A)
First foliage leaf
Coleoptile
Shoot apical meristem
1. Gibberellins are synthesized by the embryo and released into the starchy endosperm via the scutellum.
5. The endosperm solutes are absorbed by the scutellum and transported to the growing embyro.
GAs
GAs
2. Gibberellins diffuse to the aleurone layer.
Testa–pericarp
Aleurone layer
Starchy endosperm
Aleurone cells
Hydrolytic enzymes
Endosperm solutes
3. Aleurone layer cells are induced to synthesize and secrete α-amylase and other hydrolases into the endosperm.
4. Starch and other macromolecules are broken down to small molecules.
Scutellum
Root
(B)
N
PSV
(C)
PSV
(D)
G
PSV

Figure 14.8 Structure of a barley grain and the functions of various tissues during germination. (A) Diagram of germination-initiated interactions. (B–D) Micrographs of the barley aleurone layer (B) and barley aleurone protoplasts at an early (C) and a late (D) stage of amylase production. Multiple protein storage vesicles (PSVs) in (C) coalesce to form a large vesicle in (D), which will provide amino acids for α-amylase synthesis. G, phytin globoid that sequesters minerals; N, nucleus.

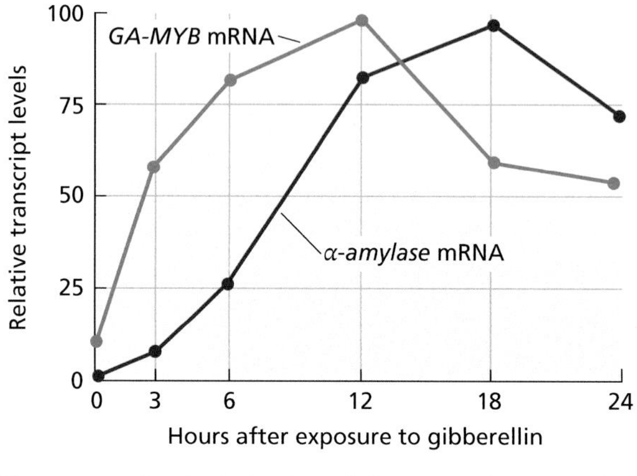
GA-MYB mRNA
α-amylase mRNA
Relative transcript levels
100
75
50
25
0
0
3
6
12
18
24
Hours after exposure to gibberellin

Figure 14.9 Time course for the induction of GA-responsive transcription factor (*GA-MYB*) and α-amylase mRNAs by GA_3. *GA-MYB* expression precedes that of α-amylase by about 3 h. In the absence of gibberellin, the concentrations of both the *GA-MYB* and α-amylase mRNAs are negligible.

be stored in different tissue types. These include water-insoluble globulins (e.g., legumins and vicilins in legumes), alcohol-soluble prolamins and alkali-soluble glutelins (in cereals), and water-soluble albumins (in many plants).

During germination, proteases degrade storage proteins to mobilize the amino acids and nitrogen they contain. Proteases that are present in dormant seeds are usually stored separately from storage proteins within the cell, and rarely within PSVs. Proteases are inactive until the seed imbibes, and seeds often contain protease inhibitors that must be degraded before the proteases can hydrolyze the stored proteins. After imbibition and activation, proteases are moved to the PSVs, via trafficking of the proteases or by fusion of the compartments. In cereals, GA increases expression of cysteine proteases in the aleurone layer needed for the mobilization of seed storage proteins. Similar proteases also function in the seasonal remobilization of storage proteins in the bark and roots of woody perennials and the roots of perennial grasses.

Oilseeds are a model for understanding lipid remobilization

Oil bodies are mainly composed of triacylglycerides, which are synthesized by the embryo. Aggregates of triacylglycerides are enclosed by a phospholipid monolayer and oleosins, structural proteins embedded in the lipid monolayer. Oleosins stabilize the oil bodies during dormancy, regulate oil body size, emulsify the oil bodies during germination to allow for lipid catabolism, regulate lipid metabolism, and provide tolerance to freezing stress. The oil body phospholipid membrane contains the proteins caleosins and steroleosins, which have peroxygenase and β-hydroxysteroid dehydrogenase activities, respectively. The Ca^{2+}-binding caleosins are also found in single-cell algae and fungi. Oleosins are found only in land plants and are thought to be derived from caleosins. The breakdown of oil bodies by proteases and lipases and the β-oxidation of the released fatty acids provide energy and raw materials for germination and seedling establishment. For oilseeds such as *Brassica napus* ssp. *napus* (canola), catabolism of lipids in the embryo oil bodies is essential for germination. If lipolysis is blocked genetically or chemically, these oilseeds cannot germinate.

14.6 Cell Expansion Mechanisms

Differentiate polar or tip growth and diffuse growth and describe the mechanisms of cell expansion.

When the cells imbibe during germination, the cell walls loosen so that the cells can expand, enabling the embryo to grow into a seedling using the mobilized seed reserves. In plants, the presence of a rigid cell wall surrounding the protoplast complicates the process of cell expansion. Cell walls have three layers: the middle lamella, the primary wall, and the secondary wall (**Figure 14.10**). After mitosis, cellulose and other components are deposited during growth, forming the primary wall. Secondary walls are formed after cell enlargement stops and thus are usually the youngest cell wall components. They are deposited between the plasma membrane and the primary cell wall. One of the functions of the cell wall is to prevent cell lysis during water uptake. Yet, for the cell to enlarge, the cell wall's resistance must somehow be overcome without compromising its integrity. The elegant mechanism plants evolved to achieve this delicate balance of yielding without breaking is described in the following sections. Both wall loosening and the directionality of cell expansion are strongly regulated by hormones.

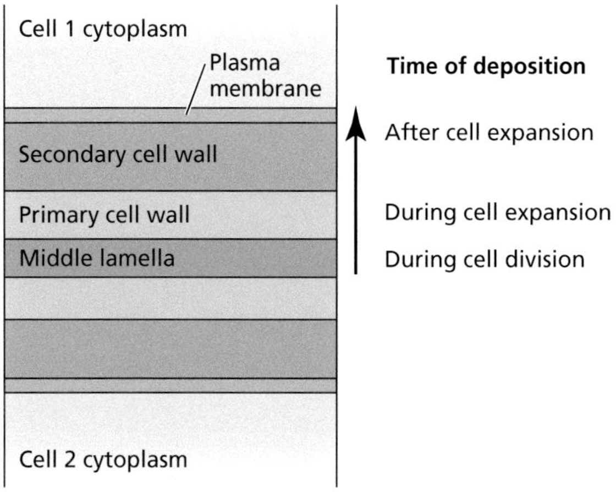

Figure 14.10 The three layers of cell walls, aligned with their time of deposition during the origin, growth, and differentiation of a cell. The middle lamella is pectin-rich and is the first layer deposited during cell plate formation in early cell division. The cell then begins to add cellulose and hemicelluloses, forming the primary cell wall. Once cell expansion has ceased, the secondary cell wall is produced.

cellulose microfibril Thin, ribbon-like structure of indeterminate length and variable width composed of 1→4-linked β-D-glucan chains tightly packed in crystalline arrays alternating with less organized amorphous regions. Provides structural integrity to the cell walls of plants and determines the directionality of cell expansion.

hemicelluloses Heterogeneous group of polysaccharides that bind to the surface of cellulose, linking cellulose microfibrils together into a network.

tip growth Localized growth at the tip of a plant cell, caused by localized secretion of new wall polymers. Occurs in pollen tubes, root hairs, some sclerenchyma fibers, and cotton fibers, as well as moss protonema and fungal hyphae. Contrast with diffuse growth.

diffuse growth A type of cell growth in plants in which expansion occurs more or less uniformly over the entire surface. Contrast with tip growth.

Polarized growth and cell expansion involve the cytoskeleton and cell wall modification

Growth is defined as the increase in cell number and cell size that occurs during an organism's life cycle. Growth in plants is qualitatively different from the growth of animals. In animals, cell divisions are evenly distributed throughout the entire body, and the organism reaches its full size by the end of the juvenile stage. Animal cells increase their size by synthesizing more cytoplasm, a process that depends primarily on the synthesis of proteins and other macromolecules.

Plant cells are surrounded by a mechanically strong cell wall that functions in regulation of cell volume, determination of cell shape, and mechanical protection of the delicate protoplast against biochemical and physical assaults. A major component of plant cell walls is the polymer cellulose. Cellulose is the most abundant polysaccharide on Earth. Its function in plants and utility for humans is based on its packaging into microfibrils. The simplest **cellulose microfibrils** are narrow structures, approximately 3 nm wide, that strengthen the cell wall, sometimes reinforcing more in one direction than in another, depending on how the microfibrils are deposited in the wall (**Figure 14.11**). The cellulose microfibrils in the primary cell wall are reinforced by pectins and **hemicelluloses**.

Early in the life of a plant cell, its primary cell wall is extensible and able to incorporate new structural material as the cell grows. This provides both

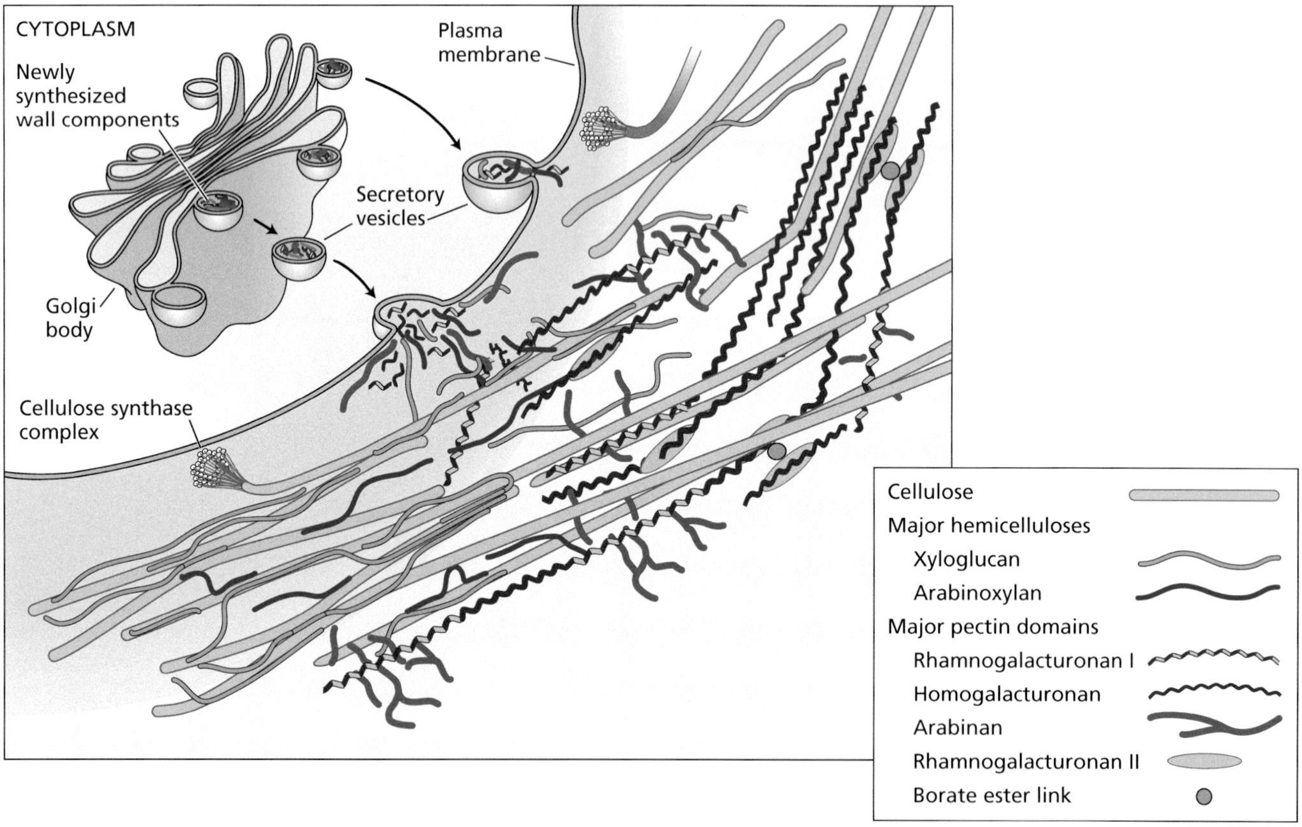

Figure 14.11 Schematic diagram of the major structural components of the primary cell wall and their possible arrangement. Cellulose microfibrils (tan rods) are synthesized at the cell surface and are partially coated with hemicelluloses (blue and purple strands), which may separate microfibrils from one another. Pectins (red, yellow, and green strands) form an interlocking matrix that controls microfibril spacing and wall porosity. Pectins and hemicelluloses are synthesized in the Golgi apparatus and delivered to the wall via vesicles that fuse with the plasma membrane and thus deposit these polymers at the cell surface. For clarity, the hemicellulose–cellulose network is emphasized on the left, and the pectin network is emphasized on the right.

flexibility and strength to the growing cell wall. During plant cell enlargement, new wall polymers are continually synthesized and secreted at the same time that the preexisting wall is expanding. Wall expansion may be highly localized (as in the case of **tip growth**) or more dispersed over the wall surface (**diffuse growth**) (**Figure 14.12**). Tip growth is characteristic of root hairs and pollen tubes; it is closely linked to cytoskeletal processes, especially those of actin microfilaments. Most of the other cells in the plant body exhibit diffuse growth, which is linked to both microtubules and actin microfilaments. Cells such as fibers, some sclereids, and trichomes grow in a pattern that is intermediate between that of diffuse growth and tip growth.

Even in cells with diffuse growth, however, different parts of the wall may enlarge at different rates or in different directions. For example, in cortical cells of the stem or epidermal cells of the young seedling hypocotyl, the end walls grow much less than the side walls. This difference in growth between cell faces may be due to structural or enzymatic variations in specific walls or to variations in the stresses borne by different walls. As a consequence of such uneven patterns of wall expansion, plant cells may assume irregular forms, often important for their functions.

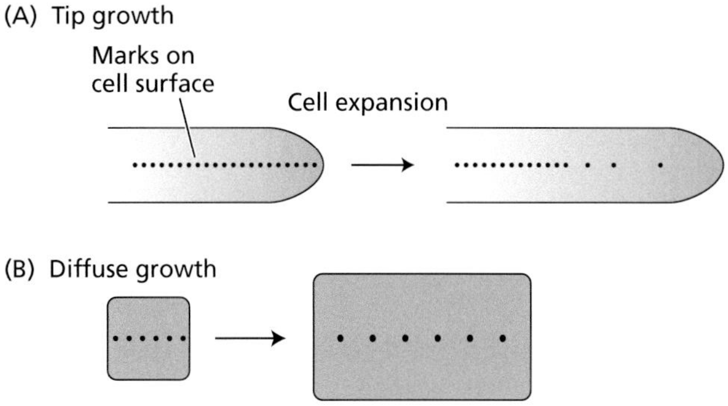

Figure 14.12 The cell surface expands differently during tip growth and diffuse growth. (A) Expansion of a tip-growing cell is confined to an apical dome at one end of the cell. If marks are placed on the cell surface and the cell is allowed to continue to grow, only the marks that were initially within the apical dome grow farther apart. Root hairs and pollen tubes are examples of plant cells that exhibit tip growth. (B) If marks are placed on the surface of a diffuse-growing cell, the distance between all the marks increases as the cell grows. Most cells in multicellular plants grow by diffuse growth.

Microfibril orientation influences growth directionality of cells with diffuse growth

During growth, the flexible cell walls are extended by physical forces generated from cell turgor pressure. Turgor pressure creates an outward-directed force, equal in all directions. The directionality of growth is determined in large part by the orientation of cellulose microfibrils within the wall.

When cells first form in the meristem, they are isodiametric; that is, they have equal diameters in all directions. If the cellulose microfibrils in the primary cell wall are randomly arranged, the cells grow isotropically (equally in all directions), expanding radially to generate a sphere (**Figure 14.13A**). In most plant cell walls, however, cellulose microfibrils are aligned in a predominant direction. The alignment of cellulose fibers allows the cell wall to hold more stress (generated by turgor) in the direction parallel to the predominant fiber alignment and yield more in the direction perpendicular to the fiber alignment. The result is **anisotropic growth** (as in stems, where cells increase much more in length than in width).

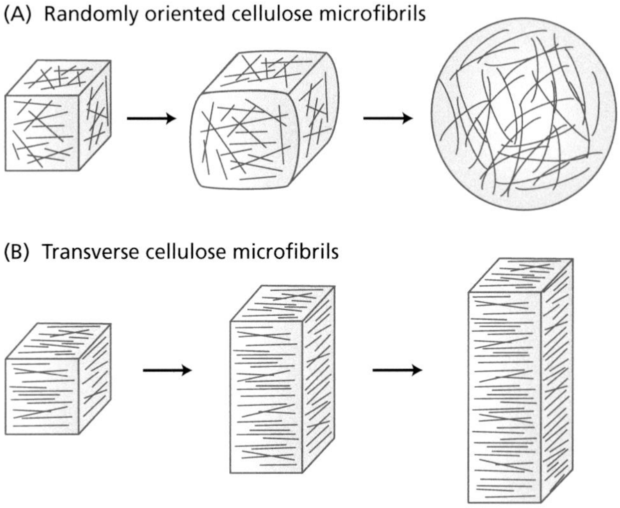

Figure 14.13 The orientation of newly deposited cellulose microfibrils determines the direction of cell expansion. (A) If the cell wall is reinforced by randomly oriented cellulose microfibrils, the cell will expand equally in all directions, forming a sphere. (B) When most of the reinforcing cellulose microfibrils have the same orientation, the cell expands at right angles to the microfibril orientation and is constrained in the direction of the reinforcement. Here the microfibril orientation is transverse, so cell expansion is longitudinal.

In the lateral walls of elongating cells such as the cortical and vascular cells of stems and roots, or the giant internode cells of the filamentous green alga *Nitella*, cellulose microfibrils are deposited circumferentially (transversely), at right angles to the long axis of the cell. The circumferential arrangement of cellulose microfibrils restricts growth in girth and promotes growth in length (**Figure 14.13B**). It is important to note that in multicellular organs, it is not necessary for all cells to display this circumferential alignment in cellulose fibers; as long as enough cells do, they can pull their neighbors with them into anisotropic

anisotropic growth Enlargement that is greater in one direction than another; for example, elongating cells in stems and roots grow more in length than in width.

initials Broadly defined as the cells of the root and shoot apical meristems. More specifically, a cluster of slowly dividing cells located within the meristem that gives rise to the more rapidly dividing cells of the surrounding meristem.

acid growth A characteristic of growing cell walls in which they extend more rapidly at acidic pH than at neutral pH.

expansins A class of wall-loosening proteins that accelerate wall yielding during cell elongation, typically with an optimum at acidic pH.

growth—something that is only possible because of the connection between neighboring cells that the wall provides.

Plant cells typically expand ten- to a thousand-fold in volume before reaching maturity. In extreme cases, cells may enlarge more than ten thousand-fold in volume compared with their meristematic **initials** (e.g., xylem vessel elements). The cell wall undergoes this massive expansion without losing its mechanical integrity and without becoming thinner. The integration of new wall polymers during cell expansion is particularly crucial for rapidly growing root hairs, pollen tubes, and other tip-growing cells, in which the region of wall deposition and surface expansion is localized to the tip of the tubelike cell.

When plant cells undergo expansion, whether by diffuse growth or tip growth, the increase in volume is due primarily to water uptake. This water ends up mainly in the vacuole, which takes up an ever-larger proportion of the cell volume as the cell enlarges.

Acid-induced growth and cell wall yielding are mediated by expansins

A common characteristic of growing cell walls is that they extend much faster at acidic pH than at neutral pH. This phenomenon is called **acid growth**. In living cells, acid growth is evident when growing cells are treated with acid buffers or with the drug fusicoccin, which induces acidification of the cell wall solution by activating an H^+-ATPase in the plasma membrane. Auxin-induced growth is also associated with wall acidification.

An example of acid-induced growth can be found in the initiation of the root hair (also called *trichoblast*), where the local wall pH drops to a value of 4.5 at the time when the epidermal cell begins to bulge outward. Wall acidification also occurs during auxin-induced growth (Chapter 13) and is an evolutionarily conserved process common to all land plants.

Acid-induced wall extension can be observed in freeze-thawed coleoptiles in which cell walls remain intact but cellular processes are disrupted. Such an observation entails using an extensometer to place the walls under tension and to measure long-term wall extension (**Figure 14.14**). When the freeze-thawed primary walls are incubated in neutral buffer (pH 7) and clamped in an extensometer, the walls extend briefly when tension is applied, but extension soon ceases. When transferred to an acidic buffer (pH 5 or less), the walls begin to extend rapidly, in some instances continuing for many hours.

Acid-induced extension is characteristic of the walls of growing cells, and it is not observed in mature (non-growing) walls. When the freeze-thawed coleoptiles are pretreated with heat, proteases, or other agents that denature proteins, they lose their ability to respond to acid. Such results indicate that acid growth is not due simply to the physical chemistry of the wall (e.g., a weakening of the pectin gel), but is catalyzed by one or more wall proteins.

The idea that proteins are required for acid growth was confirmed in reconstitution experiments in which heat-inactivated freeze-thawed coleoptiles were restored to nearly full acid-growth responsiveness by the addition of proteins extracted from the walls of growing cells (**Figure 14.15**). The active components proved to be a group of proteins that were named **expansins**. Expansins catalyze the pH-dependent yielding of cell walls. They are effective in catalytic amounts, but they do not exhibit lytic or other enzymatic activities.

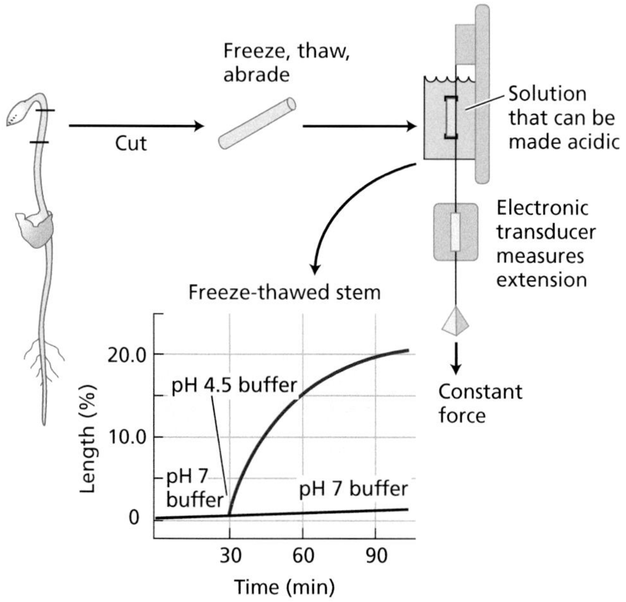

Figure 14.14 Acid-induced extension of cell walls from freeze-thawed coleoptiles, measured in an extensometer. The freeze-thawed coleoptile is clamped and put under tension in an extensometer that measures the length with an electronic transducer attached to a clamp. When the solution surrounding the wall is replaced with an acidic buffer (e.g., pH 4.5), the wall extends irreversibly in a time-dependent fashion.

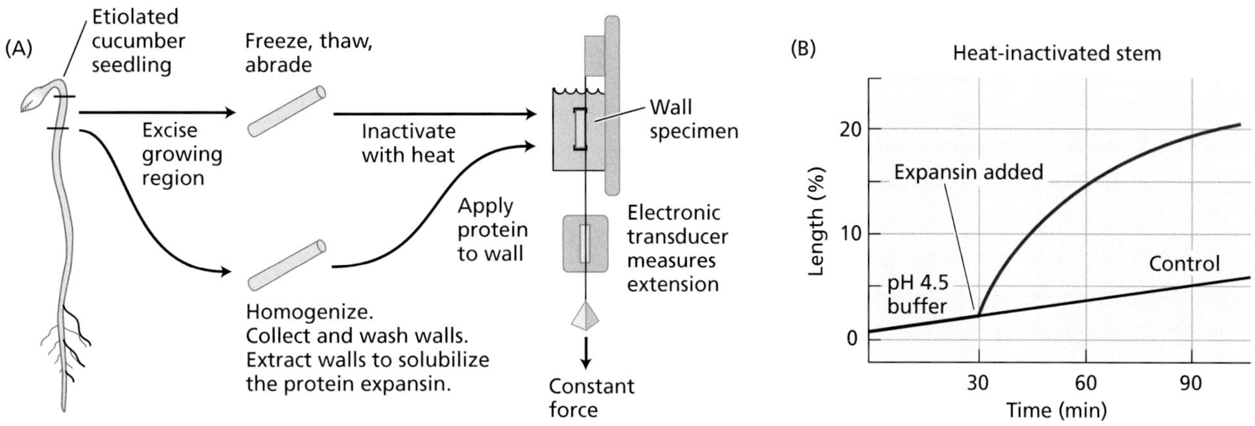

Figure 14.15 Scheme for the reconstitution of extensibility of cell walls from treated coleoptiles. (A) Tissues were prepared as in Figure 14.14 and briefly heated to inactivate the endogenous acid extension response. To restore this response, proteins are extracted from growing walls and added to the solution surrounding the wall. (B) Addition of a protein mixture containing expansins restores the acid extension properties of the wall.

14.7 Seedling Growth and Establishment

Summarize developmental changes that occur during the transition from skotomorphogenesis to photomorphogenesis, the hormones involved in this transition, and the influence of light on the development of emerging seedlings.

Broadly defined, seedling establishment is the stage when the seedling becomes competent to photosynthesize, assimilate water and nutrients from the soil, undergo normal cellular and tissue differentiation and maturation, and respond appropriately to environmental stimuli. The transition between germination and growth independent of the seed is crucial, since seedlings are highly susceptible to unfavorable biotic and abiotic factors during this stage. For example, in the field, about 10 to 55% of maize seedlings and 48 to 70% of soybean seedlings fail at this stage. Seed size is an important factor in surviving this stage because larger seeds have larger food reserves, allowing more time for seedling establishment. However, this advantage must be balanced against the larger cost to the parent plant of providing these extra reserves.

The development of emerging seedlings is strongly influenced by light

While germination is strictly defined as the emergence of the radicle from the seed, the rest of the embryo also needs to emerge from the seed, marking the beginning of seedling establishment. In monocots, the coleoptile protects the shoot apical meristem as it emerges from the seed and grows through the soil to the light. (Meristems are discussed in Chapter 15.) In eudicots, the shoot apical meristem is protected by the apical hook formed by the upper hypocotyl and the two closed cotyledons. The formation, maintenance, and opening of the apical hook are regulated by interactions between light and hormone signaling.

A key event in seedling establishment is the emergence of the shoot from the soil into the light, which triggers profound changes in shoot development.

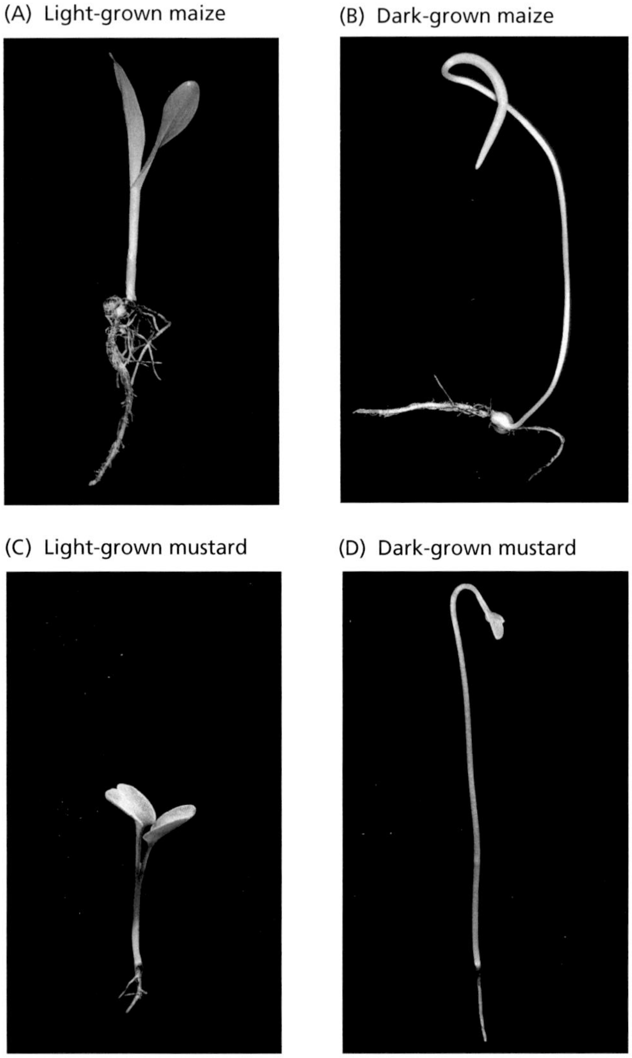

Figure 14.16 Light- and dark-grown monocot and eudicot seedlings. (A and B) Maize (corn; *Zea mays*) and (C and D) mustard (*Eruca* sp.) seedlings grown either in the light (A and C) or the dark (B and D). Symptoms of etiolation in maize, a monocot, include the absence of greening, reduction in leaf width, failure of leaves to unroll, and elongation of the coleoptile and mesocotyl. In mustard, a eudicot, etiolation symptoms include absence of greening, reduced leaf size, hypocotyl elongation, and maintenance of the apical hook.

The shoots of dark-grown seedlings are **etiolated**—that is, they have long, spindly hypocotyls, an apical hook (in the case of eudicots), closed cotyledons, and nonphotosynthetic proplastids, which causes the unexpanded leaves to have a pale-yellow color. In contrast, seedlings grown in nondirectional light have shorter, thicker hypocotyls, no apical hook, open cotyledons, and expanded leaves with photosynthetically active chloroplasts (**Figure 14.16**). Development in the dark is termed **skotomorphogenesis**, while development in the presence of light is called **photomorphogenesis**. When dark-grown seedlings are transferred to the light, photomorphogenesis takes over and the seedlings become de-etiolated.

The switch between dark- and light-grown development involves genome-wide transcriptional and translational changes triggered by the perception of light by several classes of photoreceptors. Despite the complexity of the process, the transition from skotomorphogenesis to photomorphogenesis is surprisingly rapid. Within minutes of applying a single flash of light to a dark-grown bean seedling, several developmental changes occur:

- A decrease in the rate of stem elongation
- The beginning of apical hook opening
- Initiation of the synthesis of photosynthetic pigments

Light thus acts as a signal that induces a change in the shape of the seedling, from one that facilitates growth through the soil to one that will enable the plant to efficiently harvest light energy and convert it into the essential sugars, proteins, and lipids necessary for growth.

Among the different **photoreceptors** that can promote photomorphogenic responses in plants, the most important are those that absorb red and blue light. Photoreceptors are discussed in Chapter 13. **Phytochrome** is a protein–pigment photoreceptor that absorbs red and far-red light most strongly, but it also absorbs blue light. **Cryptochromes** are flavoproteins that mediate many blue-light responses involved in photomorphogenesis, including the inhibition of hypocotyl elongation, cotyledon expansion, and petiole elongation.

Gibberellins and brassinosteroids both suppress photomorphogenesis in darkness

As we discussed in Chapter 13, photomorphogenesis is negatively regulated. In the dark, many of the transcription factors that promote photomorphogenesis are degraded in the nucleus via ubiquitination and proteolytic degradation by 26S proteasome. In the light, this degradation process is prevented, allowing photomorphogenesis to proceed. Plant hormones act to coordinate these changes throughout the plant.

In the dark, the concentration of phytochrome in the Pfr (far-red light–absorbing) form is low. Since Pfr reduces hypocotyl *sensitivity* to GA, endogenous GAs promote hypocotyl cell elongation to a greater extent in the dark than in the light, causing the spindly appearance of dark-grown seedlings.

etiolated Effects of seedling growth in the dark, in which the hypocotyl and stem are more elongated, cotyledons and leaves do not expand, and chloroplasts do not mature.

skotomorphogenesis The developmental program plants follow when seeds are germinated and grown in the dark.

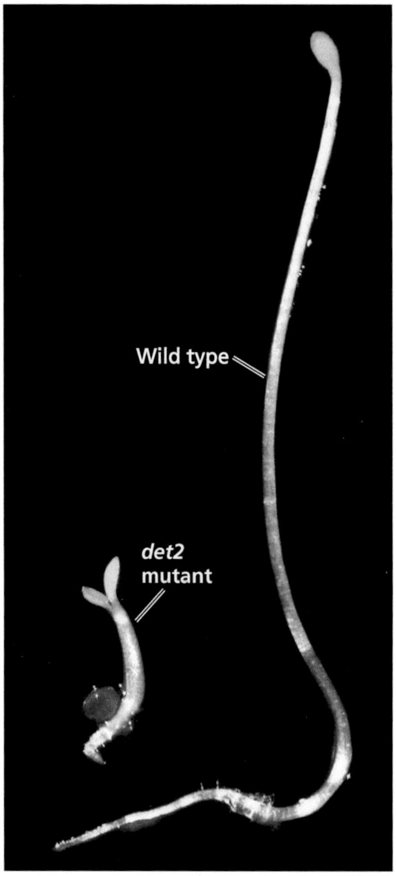

Figure 14.17 Phenotypes of dark-grown brassinosteroid-deficient mutants. The dark-grown brassinosteroid-deficient mutant (*det2*) seedling of Arabidopsis on the left has a short, thick hypocotyl and open cotyledons. A dark-grown wild type is on the right.

In the light, Pr (the red light–absorbing form of phytochrome) is converted to Pfr, which causes the hypocotyl to become less sensitive to GAs. As a result, hypocotyl elongation is greatly reduced and the seedling appears to undergo partial photomorphogenesis. For this reason, GA-deficient mutant peas grown in the dark look somewhat like light-grown seedlings, although they lack chlorophyll. Taken together, these results indicate that GAs suppress some aspects of photomorphogenesis in the dark, and the suppression is reversed by red light.

Brassinosteroids play a parallel role in suppressing photomorphogenesis in the dark. Genetic screens for mutants that appeared de-etiolated when grown in the dark led to the identification of brassinosteroid biosynthetic genes. When grown in the dark, brassinosteroid-deficient Arabidopsis mutants are short and lack the apical hooks observed in normal etiolated seedlings (**Figure 14.17**).

Hook opening is regulated by phytochrome, auxin, and ethylene

Etiolated eudicot seedlings usually form an apical hook that is located just behind the shoot apex and protects the shoot apical meristem. Hook formation and maintenance in darkness result from ethylene-induced asymmetric growth (**Figure 14.18**). In compacted soils, the gaseous hormone ethylene cannot diffuse in the soil and accumulates around the germinated seedling. This results in the triple response, in which the apical hook is tighter, the hypocotyl is shorter and thicker, and the root is very short. This change in morphology aids seedlings as they push through compacted soils and around small obstacles such as stones to reach the soil surface. Auxin-insensitive mutants do not develop an apical hook, and treatment of wild-type Arabidopsis seedlings with NPA (*N*-1-naphthylphthalamic acid), an inhibitor of polar auxin transport, also blocks apical hook formation. The closed shape of the hook is a consequence of the more rapid elongation of the outer side of the hypocotyl compared with the inner side, and auxin inhibition of cell elongation on the inner side. Therefore, auxin and ethylene function synergistically in apical hook formation. The more rapid growth of the outer tissues relative to the inner tissues could reflect an ethylene-dependent lateral redistribution of auxin analogous to what is seen during phototropic bending.

After exposure to light, the elongation rate of the inner side of the apical hook increases, equalizing the growth rates on both sides and causing the hook to open. Red light induces hook opening, and this effect is reversed by

photomorphogenesis The influence and specific roles of light on plant development. In the seedling, light-induced changes in gene expression that support aboveground growth in the light rather than belowground growth in the dark.

photoreceptors Proteins that sense the presence of light and initiate a response via a signaling pathway.

phytochromes Plant growth-regulating photoreceptor proteins that absorb primarily red light and far-red light but also absorb blue light. The holoprotein that contains the chromophore phytochromobilin.

cryptochromes Flavoproteins implicated in many blue-light responses that have strong homology with bacterial photolyases.

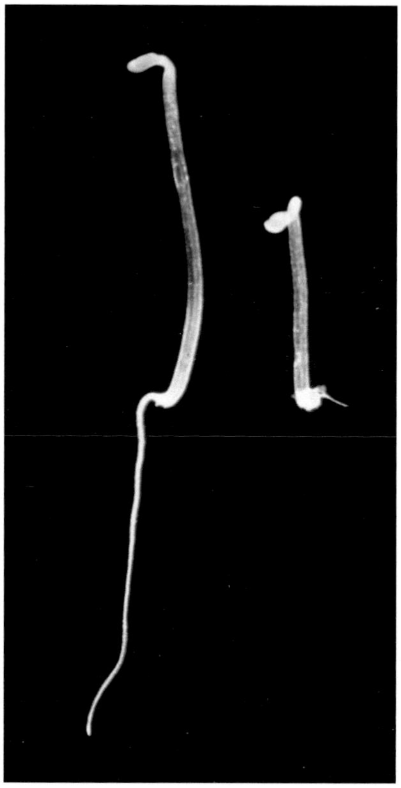

Figure 14.18 Effects of ethylene on growth in Arabidopsis seedlings. Three-day-old etiolated seedlings grown in the absence (left) or the presence (right) of 10 ppm ethylene. Note the shortened hypocotyl, reduced root elongation, and exaggeration of the curvature of the apical hook that result from the presence of ethylene; these are collectively known as the triple response.

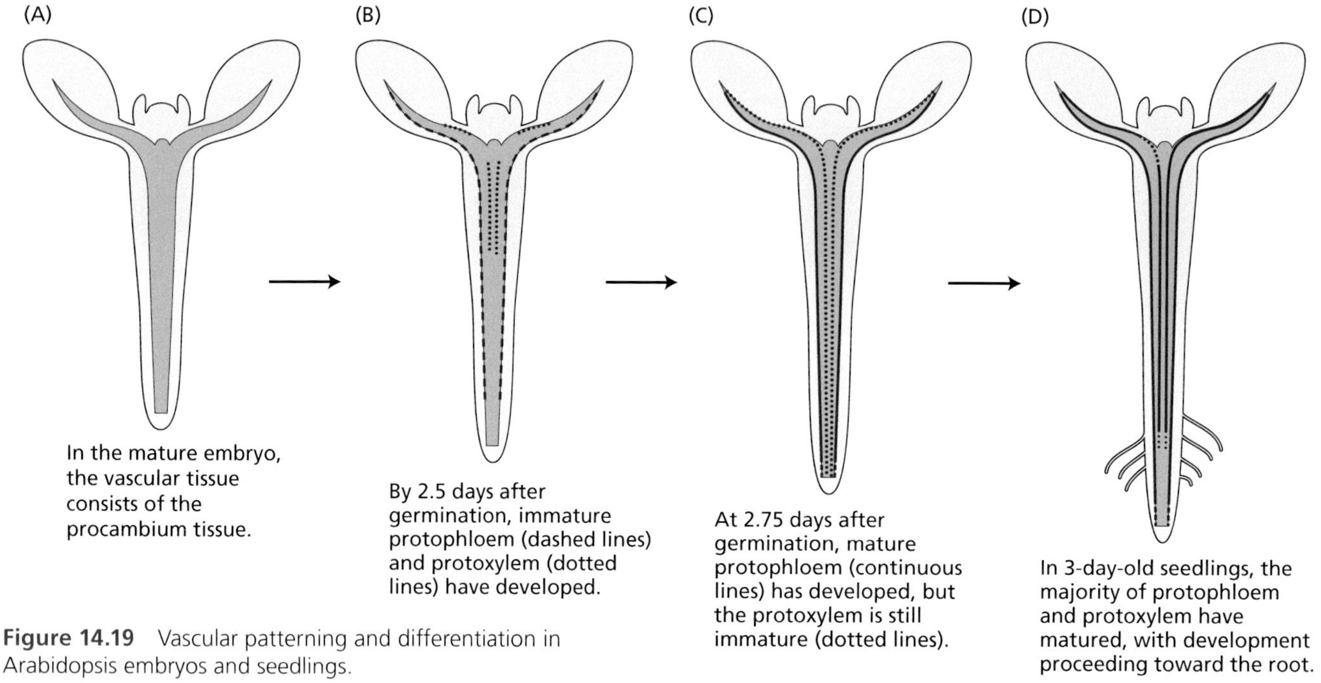

(A)
(B)
(C)
(D)
In the mature embryo, the vascular tissue consists of the procambium tissue.
By 2.5 days after germination, immature protophloem (dashed lines) and protoxylem (dotted lines) have developed.
At 2.75 days after germination, mature protophloem (continuous lines) has developed, but the protoxylem is still immature (dotted lines).
In 3-day-old seedlings, the majority of protophloem and protoxylem have matured, with development proceeding toward the root.

Figure 14.19 Vascular patterning and differentiation in Arabidopsis embryos and seedlings.

root cap Cells at the root apex that cover and protect the meristematic cells from mechanical injury as the root moves through the soil. Site for the perception of gravity and signaling for the gravitropic response in roots.

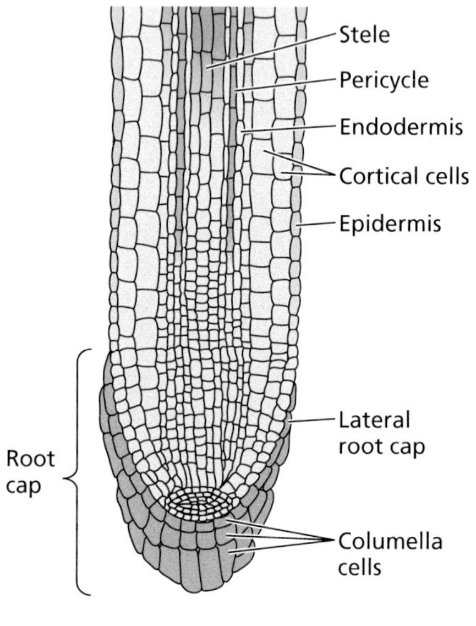

Figure 14.20 Root cell types. The root cap comprises columella cells and cells that form the lateral cap. Epidermal cells are exposed to the rhizosphere above the lateral root cap. Moving toward the center of the root, the tissue layers are the cortex, endodermis, pericycle, and stele. The stele is composed of the xylem, phloem, and their companion and supporting parenchyma cells.

far-red light, indicating that phytochrome is the photoreceptor involved in this process. A close interaction between phytochrome and ethylene controls hook opening. As long as ethylene is produced by the hook tissue in the dark, elongation of the cells on the inner side is inhibited. Red light inhibits ethylene formation, promoting growth on the inner side, thereby causing the hook to open.

Vascular differentiation begins during seedling emergence

During embryogenesis within the seed, symplasmic and apoplastic transport are sufficient to distribute water, nutrients, and signals throughout the embryo by the process of diffusion. Following germination, however, the emerging seedling requires a continuous vascular system to distribute materials quickly and efficiently throughout the plant. The vascular system of the embryo consists only of procambial strands—immature vascular tissue.

During seedling emergence, first protoxylem and protophloem cells appear, followed by the larger metaxylem and metaphloem cells (**Figure 14.19**). As the protophloem and protoxylem differentiate and mature, the vascular tissue becomes continuous between the root and the hypocotyl, which is rootlike in the organization of its vascular tissue. The pattern of differentiation and maturation is different in a germinating seedling compared with a mature plant (Chapter 15). The protophloem matures earlier than the protoxylem in the emerging radicle and during primary growth: Phloem sieve tube development occurs nearly simultaneously in the hypocotyl and root, while xylem differentiation occurs acropetally in roots.

The root tip has specialized cells

The **root cap** occupies the most distal part of the root and is composed of columella cells and cells that form the lateral cap (**Figure 14.20**). It represents a unique set of cells that are situated below the meristematic zone and cover the apical meristem and protect it from mechanical injury as the root tip is pushed through the soil. The first layer of root cap cells of emerging radicles has modified cell walls with a cuticle that protects the emerging root tip from abiotic stress. Other

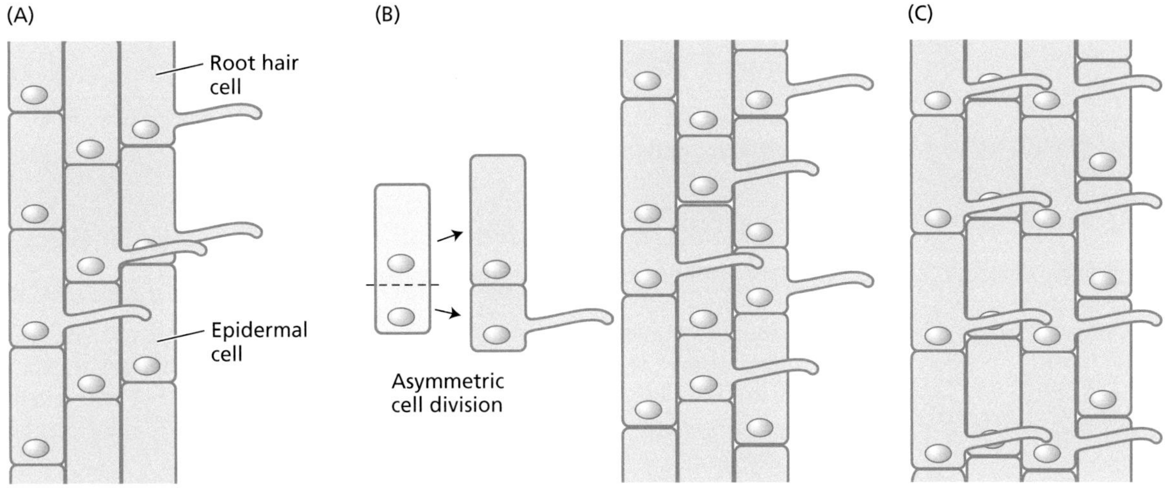

Figure 14.21 Three patterns of root hair (trichoblast) differentiation. (A) In some plants, all root epidermal cells have the potential to become root hairs (trichoblasts) and the pattern is specified by developmental cues. (B) In basal vascular plants and some monocots, the root hair results from an asymmetric cell division. (C) In the Brassicaceae, trichoblasts and atrichoblasts occur in alternating cell files.

functions of the root cap include the perception of gravity by the columella cells during gravitropism (discussed below) and the secretion of compounds, such as mucigel that help the root penetrate the soil and mobilize mineral nutrients. The root cap cells are constantly renewed by the meristematic cells and undergo developmentally programmed cell death that influences future root architecture. Epidermal cells lie beneath the root cap and become exposed to the rhizosphere above the lateral root cap.

Ethylene and other hormones regulate root hair development

Root hairs are extensions of root epidermal cells that increase the surface area of the root for water and mineral uptake and also serve a mechanical role to help anchor plants in the soil. Root hairs are also an example of tip growth. While all epidermal cells have the potential to differentiate into a root hair (trichoblast) or not (atrichoblast), developmental cues specify the pattern of trichoblasts and atrichoblasts (**Figure 14.21**). Species-specific signals and cell-to-cell communication determine the basic pattern of root hair position. Trichoblast differentiation is regulated by gene expression.

Ethylene is a positive regulator of root hair development, and ethylene-treated roots produce extra root hairs in abnormal locations (**Figure 14.22**). Conversely, fewer root hairs form in ethylene-insensitive mutants or in wild types treated with ethylene biosynthesis inhibitors. These observations indicate that ethylene acts as a positive regulator in the differentiation of root hairs.

Root hairs elongate by tip growth that responds to nutritional and water status via a combination of hormones (auxin, methyl jasmonate, brassinosteroids), and Ca^{2+} signaling. Intracellular auxin concentrations regulate root hair elongation. Methyl jasmonate has also been shown to enhance root hair growth, while brassinosteroids inhibit root hair growth.

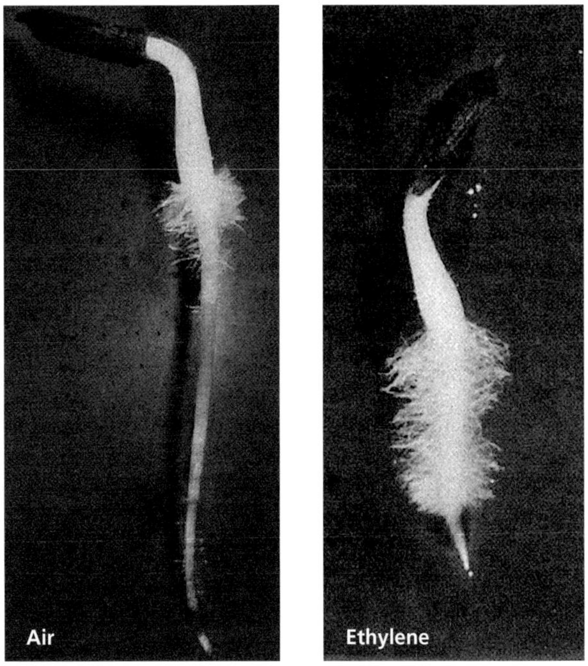

Figure 14.22 Promotion of root hair formation by ethylene in lettuce seedlings. Two-day-old seedlings were treated with air (left) or 10 ppm ethylene (right) for 24 h before the photo was taken. Note the profusion of root hairs on the ethylene-treated seedling.

14.8 Differential Growth Enables Successful Seedling Establishment

Describe the role that hormones play in cell expansion, elongation, and nutational growth in various parts of the plant.

To survive, seedlings require a functional root–soil interface, adequate gas exchange, and functional photosynthetic carbon fixation. Seedlings employ programmed and environmental responses involving preferential cell elongation to achieve establishment. Differential elongation of cortical and epidermal cells is important for penetration of soil layers. Root tip rotation during growth, also called circumnutation, is regulated by ethylene and is highly dependent on auxin reuptake into shootward flows mediated by the AUX1 auxin-uptake transporter (Chapter 12). Stems and leaves also have circumnutational growth, most notably in spiraling tendrils.

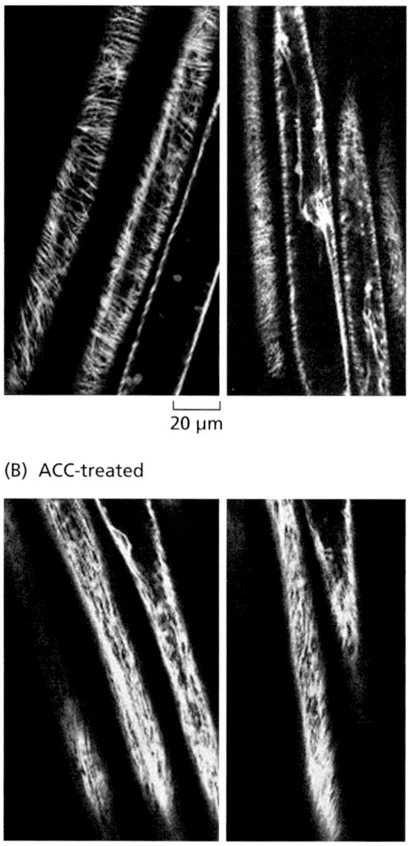

Figure 14.23 Ethylene affects microtubule orientation. (A) Microtubule orientation is transverse in hypocotyls of untreated dark-grown transgenic Arabidopsis seedlings expressing a tubulin gene tagged with green fluorescent protein. (B) Microtubule orientation is longitudinal and diagonal in hypocotyl cells from seedlings treated with the ethylene precursor 1-aminocyclopropane-1-carboxylic acid (ACC), which increases ethylene production.

Ethylene affects microtubule orientation and induces lateral cell expansion

At concentrations above 0.1 $\mu L\ L^{-1}$ (1 ppm), ethylene can reduce the rate of elongation and increase lateral expansion, leading to swelling of the hypocotyl or the epicotyl. The directionality of plant cell expansion is determined by the orientation of the cellulose microfibrils in the cell wall. Transverse microfibrils reinforce the cell wall in the lateral direction, so that turgor pressure is channeled into cell elongation. The orientation of the microfibrils is in turn determined by the orientation of the cortical array of microtubules in the cortical (peripheral) cytoplasm. In typical elongating plant cells, the cortical microtubules are arranged transversely, giving rise to transversely arranged cellulose microfibrils.

When etiolated seedlings are treated with ethylene, as shown in Figure 14.19, the microtubule alignment in the cells of the hypocotyl switches from a transverse to a diagonal or longitudinal orientation (**Figure 14.23**). This approximately 90-degree shift in microtubule orientation leads to a parallel shift in cellulose microfibril deposition in the cell wall. The newly deposited wall is reinforced in the longitudinal direction rather than the transverse direction, which promotes lateral expansion instead of elongation.

Auxin promotes growth in stems and coleoptiles, while inhibiting growth in roots

Auxin is the hormone that primarily regulates differential organ bending in seedlings. Auxin synthesized in the shoot apex is transported toward the tissues below. The steady supply of auxin arriving at the subapical region of a stem or coleoptile is required for the continued elongation of these cells. Because the concentration of endogenous auxin in the elongation region of a normal healthy plant is nearly optimal for growth, spraying the plant with exogenous auxin causes only a modest and short-lived stimulation in growth. Such spraying may even be inhibitory in the case of dark-grown seedlings, which are more sensitive to supraoptimal auxin concentrations than light-grown plants are.

However, when the endogenous source of auxin is removed by excision of stem or coleoptile sections containing the elongation zone, the growth rate rapidly decreases to a low basal rate. Such excised sections often respond to exogenous auxin by rapidly increasing their growth rate back to the concentration in the intact plant (**Figure 14.24A** and **Figure 14.24B**). Most of the elongation takes place in outer tissues and is not just a matter of tissue penetration of the auxin.

Auxin control of root elongation has been more difficult to demonstrate, perhaps because auxin induces the production of ethylene, which inhibits root growth. Elongation of root cells also requires the GA biosynthesis in epidermal

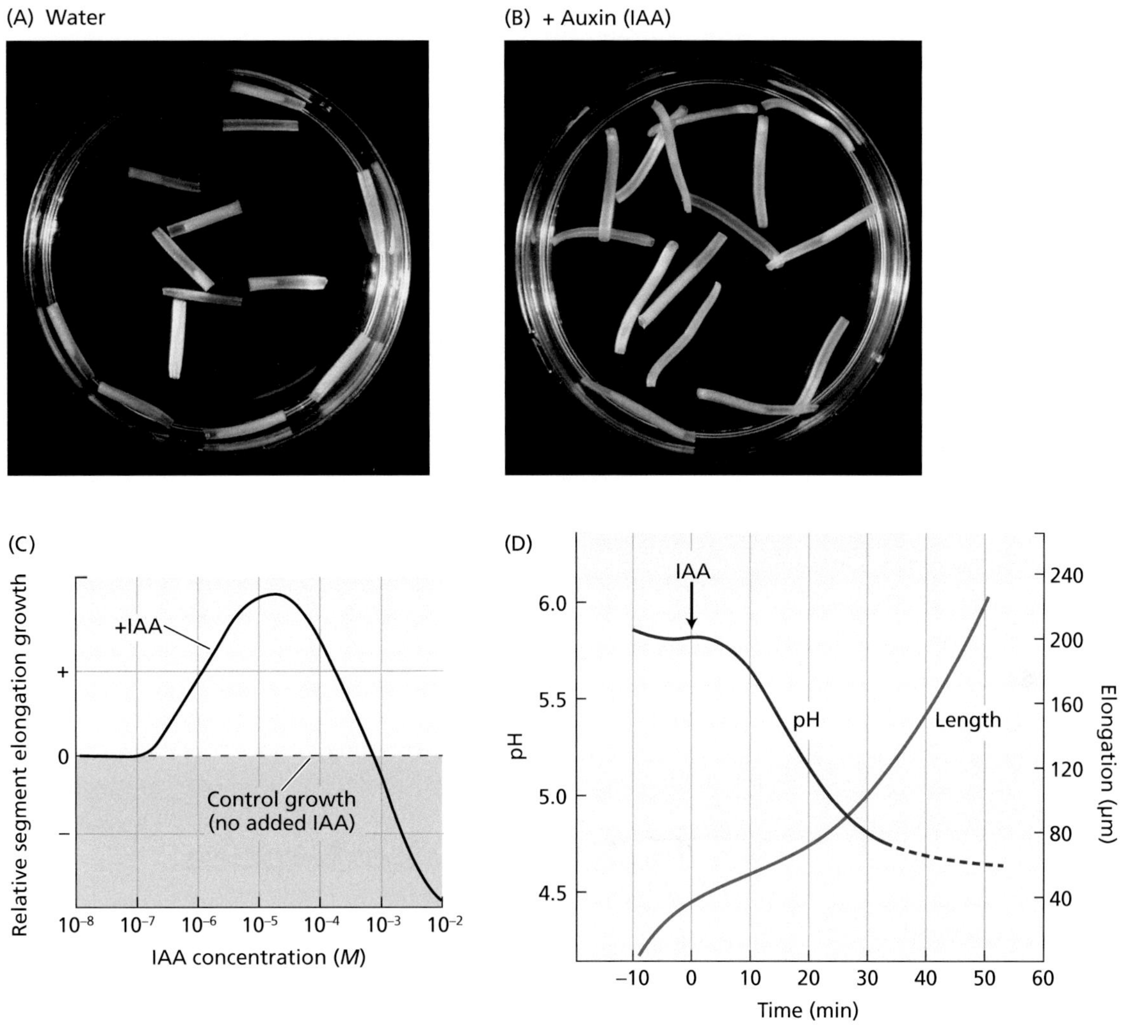

Figure 14.24 Auxin stimulates elongation. Oat coleoptile sections depleted of endogenous auxin were incubated for 18 h in either water (A) or auxin (IAA) (B). The yellow material inside the translucent coleoptile is the primary leaf tissue. (C) Typical dose-response curve for IAA-induced growth in pea stem or oat coleoptile sections. Elongation growth of excised sections of coleoptiles or young stems is plotted versus increasing concentrations of exogenous IAA. At concentrations above 10^{-5} *M*, IAA becomes less and less effective. Above about 10^{-4} *M* it becomes inhibitory, as shown by the fact that the stimulation decreases and the curve eventually falls below the dashed line, which represents growth in the absence of added IAA. (D) Kinetics of auxin-induced elongation and cell wall acidification in maize coleoptiles. The pH of the cell wall was measured with a pH microelectrode. Note the similar lag times (5–10 min) for both cell wall acidification and the increase in the rate of elongation, although the change in rate of elongation is slightly slower since it is mediated by auxin-dependent transcriptional activation.

cells. These three hormones interact differentially in root tissue to control growth. GA appears to be required for basal growth. However, even when ethylene biosynthesis is specifically blocked, low concentrations (10^{-10} to 10^{-9} *M*) of auxin promote the growth of intact roots, whereas higher concentrations (10^{-6} *M*) inhibit growth. Thus, while roots may require a minimum concentration of auxin to grow, root growth is strongly inhibited by auxin concentrations that promote elongation in stems and coleoptiles. When roots grow in well-drained or sandy soil, ethylene produced by the roots can freely diffuse in the soil, and root elongation is observed. However, in compacted soils, ethylene cannot diffuse away from the root, and the root stops elongating and increases in girth. These roots resemble the root phenotype of the ethylene-treated etiolated seedlings as seen in Figure 14.18.

The minimum lag time for auxin-induced elongation is 10 minutes

When a stem or coleoptile section is excised and inserted into a sensitive growth-measuring device, the lag time for the auxin response can be monitored with great precision. For example, addition of auxin markedly stimulates the growth rates of oat (*Avena sativa*) coleoptile and soybean (*Glycine max*) hypocotyl sections after a lag period of only 10 to 12 min. The maximum growth rate, which represents a five- to ten-fold increase over the basal rate, is reached after 30 to 60 min of auxin treatment. As is shown in **Figure 14.24C**, auxin-induced elongation is dose-dependent—that is, a threshold concentration of auxin must be reached to initiate this response. Beyond the optimal concentration, auxin becomes inhibitory. The stimulation of growth by auxin requires energy, and metabolic inhibitors inhibit the response within minutes. Auxin-induced growth is also sensitive to inhibitors of protein synthesis such as cycloheximide, suggesting that protein synthesis is required for the response. Inhibitors of RNA synthesis also inhibit auxin-induced growth after a slightly longer delay.

Auxin-induced proton extrusion loosens the cell wall

Auxin induces acidification of the apoplast by increasing the activity of plasma membrane H^+-ATPases. Cell wall acidification starts 5 to 10 min after auxin exposure, preceding the increase in growth rate by 5 to 10 min, as shown in **Figure 14.24D**. Cell wall proteins called expansins mediate wall loosening at acid pH. Once the cell wall is sufficiently loosened by expansin activity, turgor pressure initiates cell expansion.

14.9 Tropisms: Growth in Response to Directional Stimuli

Differentiate among the different tropisms, including when and why they occur, and how various hormones are involved in each response.

As sessile organisms, plants must rely on growth responses to compete for sunlight and to forage for water and minerals in the soil. Plants respond to external stimuli by altering their growth and developmental patterns. During seedling establishment, abiotic factors such as gravity, touch, and light influence the initial growth habit of the young plant. **Tropisms** are directional growth responses in relation to environmental stimuli caused by the asymmetric growth of the plant axis (stem or root) and anisotropic cell growth. Tropisms may be positive (growth toward the stimulus) or negative (growth away from the stimulus).

One of the first forces that emerging seedlings encounter is gravity. **Gravitropism**, growth in response to gravity, enables shoots to grow upward toward sunlight for photosynthesis, and roots to grow downward into the soil for water and nutrients. As soon as the shoot tip penetrates the soil surface, it encounters sunlight. **Phototropism** enables leafy shoots to grow toward sunlight, thus maximizing photosynthesis, while some roots grow away from sunlight. **Thigmotropism** is differential growth in response to touch and mechanosensing and helps roots grow around obstacles and Venus flytrap leaves to close on prey (Chapter 12). **Hydrotropism** is directional growth in response to a water or a water vapor gradient and enables roots to grow toward a source of water.

Gravitropism involves the lateral redistribution of auxin

When dark-grown *Avena* seedlings are oriented horizontally, the coleoptiles bend upward in response to gravity. According to a model independently put forward by Went and Cholodny in the early twentieth century, auxin in a horizontally

tropisms Oriented plant growth in response to a perceived directional stimulus such as light, gravity, touch, or water potential.

gravitropism Plant growth in response to gravity, enabling roots to grow downward into the soil and shoots to grow upward.

phototropism The alteration of plant growth patterns in response to the direction of incident radiation, especially blue light.

thigmotropism Plant growth in response to touch, enabling roots to grow around rocks, and shoots of climbing plants to wrap around structures for support.

hydrotropism Plant growth in response to root perception of water potential gradients in the soil, enabling root growth toward areas of greater water potential.

Cholodny–Went model Early mechanism proposed for tropisms involving stimulation of the bending of the plant axis by lateral transport of auxin in response to a stimulus, such as light, gravity, or touch. The original model has been supported and expanded by recent experimental evidence.

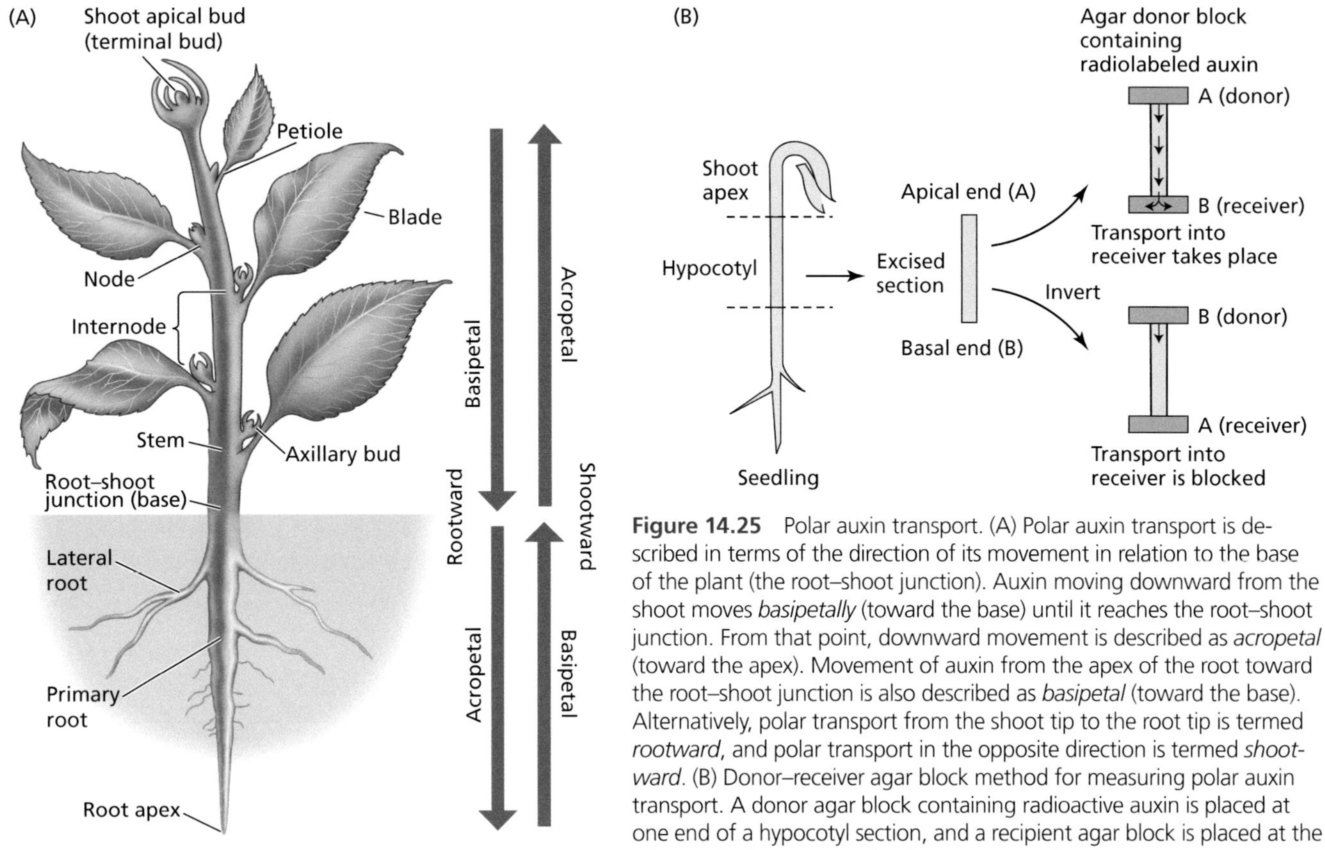

Figure 14.25 Polar auxin transport. (A) Polar auxin transport is described in terms of the direction of its movement in relation to the base of the plant (the root–shoot junction). Auxin moving downward from the shoot moves *basipetally* (toward the base) until it reaches the root–shoot junction. From that point, downward movement is described as *acropetal* (toward the apex). Movement of auxin from the apex of the root toward the root–shoot junction is also described as *basipetal* (toward the base). Alternatively, polar transport from the shoot tip to the root tip is termed *rootward*, and polar transport in the opposite direction is termed *shootward*. (B) Donor–receiver agar block method for measuring polar auxin transport. A donor agar block containing radioactive auxin is placed at one end of a hypocotyl section, and a recipient agar block is placed at the other end. The amount of radioactive auxin that accumulates in the recipient block is a measure of how much auxin is transported through the hypocotyl section. When the hypocotyl is inverted, no radiolabeled auxin is transported from the donor to the receiver block.

oriented coleoptile tip is transported laterally to the lower side, causing the lower side of the coleoptile to grow faster than the upper side (discussed below). This general **Cholodny–Went model** has subsequently been shown to apply also to phototropism. Two key features that contribute to the Cholodny–Went model are the polarity and gravity independence of long-distance auxin transport. The primary *rootward* and *shootward* polar auxin streams maintained by polar export of auxin are shown in **Figure 14.25A**. The velocity of polar auxin transport can exceed 3 mm h^{-1} in some tissues, which is faster than diffusion but slower than phloem translocation rates (Chapter 10). **Figure 14.25B** illustrates an experiment using radiolabeled auxin to demonstrate rootward polar auxin transport in a seedling hypocotyl section.

A demonstration that polar auxin transport is gravity-independent is shown in **Figure 14.26**. Grape cuttings were placed in a moist chamber, which led to the formation of adventitious roots at the basal ends of the cuttings and bud outgrowth at the apical ends. When the cuttings were inverted, the polarity of root and shoot formation was preserved. Roots formed at the basal end (now pointing upward) because root differentiation was stimulated by the auxin that accumulated there due to basipetal (rootward) polar transport. Shoots tended to form at the apical end where the auxin concentration was lowest, no matter which way the cutting was oriented with respect to gravity.

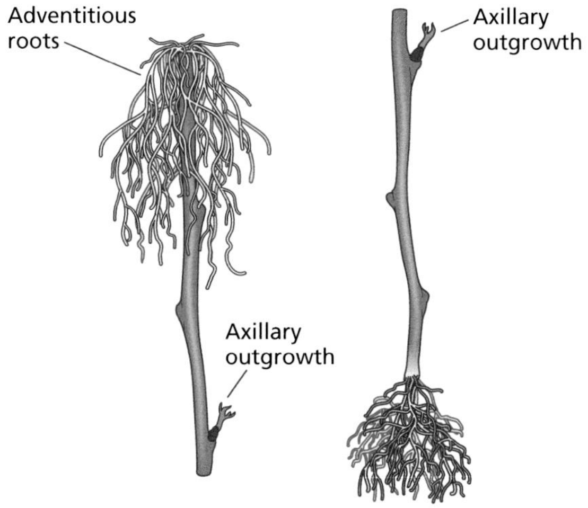

Figure 14.26 Adventitious root growth is polar. Adventitious roots grow from the basal ends of grape hardwood cuttings and axillary outgrowth occurs at the apical end, whether the cuttings are maintained in the inverted orientation (the cutting on the left) or the upright orientation (the cutting on the right). Auxin accumulates at the basal end and is depleted at the apical end because of polar transport that is independent of gravity.

The gravitropic stimulus perturbs the symmetric movements of auxin

In a young seedling, most of the auxin in the root is derived from the shoot. IAA is delivered to the root apex by a rootward directional stream directed by the efflux carrier PIN1 and maintained by the ATP-dependent activity of a group of ABCB transporters best represented by Arabidopsis ABCB19 (Chapter 12). As shown in **Figure 14.27A**, PIN proteins in the shoot apical meristem are responsible for directing auxin movement, first toward the tip of the apex, and then down the stem and into the root. In the root, PIN proteins in the cells of the vascular cylinder transport auxin to the columella (central) region of the root cap. In older plants, IAA is also synthesized in the root meristem. PIN3 efflux carriers move auxin out of the columella and provide the directional vector of auxin movement out of the columella in tropic responses in the root. Auxin is taken up into the adjacent lateral root cap cells via the AUX1 permease and then moves into a shootward epidermal stream driven by combined PIN and ABCB efflux activity. That shootward epidermal stream is primarily determined by PIN2 and maintained by ABCB4 (**Figure 14.27B**). PIN2 is localized on the upper side of root epidermal cells, conducting auxin away from the

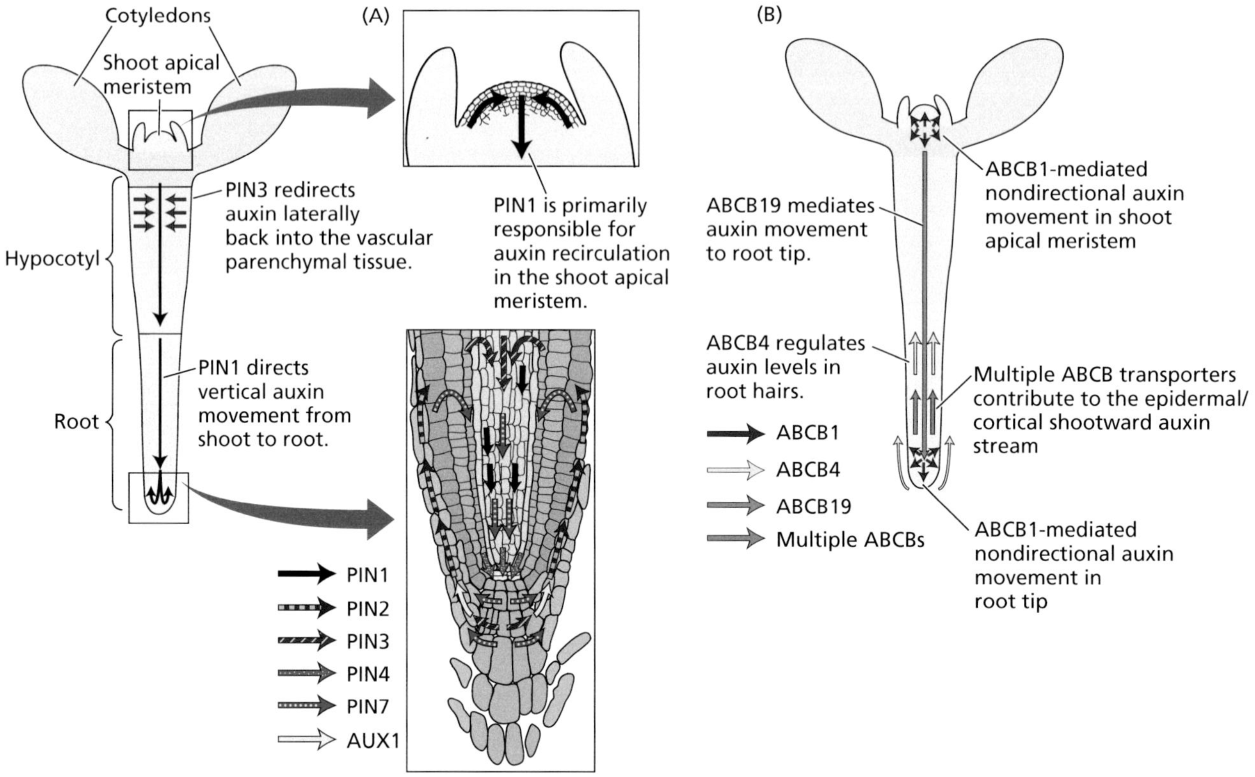

Figure 14.27 Polar auxin transport in Arabidopsis seedlings. (A) PIN transport proteins in the vascular tissue direct auxin to the root. In the root vascular cylinder PIN proteins transport auxin to the columella of the root cap. From there auxin then moves into the lateral root cap cells and is redirected by the combined uptake via AUX1 in the lateral root cap and export by PIN proteins in the epidermis. PIN proteins are also involved in redirecting auxin toward the elongation zone, after which some auxin moves back into the vascular cylinder. The term *fountain model* was suggested by the fact that the stream of auxin coming from the shoot reverses direction after reaching the root cap. The two inserts show PIN1-mediated auxin movement in the shoot apical meristem (top inset) and PIN-regulated auxin circulation in the root tip (bottom inset). (B) Auxin flow associated with ATP-dependent ABCB transport proteins. The multidirectional arrows at the shoot and root apices indicate nondirectional auxin transport. However, when combined with polarly localized PIN proteins, directional transport occurs. Multiple ABCBs function in the mobilization of auxin in cortical and epidermal tissues to the elongation and differentiation zones. ABCB4 regulates auxin concentrations in elongating root hairs.

lateral root cap toward the elongation zone, where auxin acts to stimulate cell elongation. Upon reaching the elongation zone, auxin is transported laterally back into the vascular cylinder, apparently via PIN2 localized on the inner lateral side of cortical cells (Figure 14.27A). This recirculation of auxin from the root cap to the elongation zone and back again is referred to as the *fountain model*.

Early experimental studies established that the tips of coleoptiles are the site of perception for blue-light–induced phototropic bending, and that the lateral movement of auxin to the shaded side is involved in the response (Chapters 12 and 13). The tips of coleoptiles are also able to sense gravity and redistribute auxin to the lower side. For example, if the excised tip of a coleoptile is placed on agar blocks and oriented horizontally, a greater amount of auxin diffuses into the agar block from the lower half of the tip than from the upper half, as demonstrated by a bioassay (**Figure 14.28**).

Gravitropism in roots also depends on auxin redistribution. The site of perception of gravity in roots is the root cap. When the root cap is removed from a growing root, the root no longer bends downward in response to gravity (**Figure 14.29**). In fact, the growth rate of the root actually increases slightly, suggesting that the root cap supplies an inhibitor that modulates growth in the elongation zone. Microsurgery experiments in which half of the cap was removed confirmed that the cap transports a root growth inhibitor, later identified as auxin, to the lower side of the root during gravitropic bending. Experiments with auxin transport inhibitors and auxin transporter mutants have shown that the shootward transport of auxin from the root cap to the elongation zone is required for gravitropic growth.

According to the current model for root gravitropism, shootward auxin transport in a vertically oriented root is equal on all sides. When the root is oriented horizontally, however, the signals from the cap redirect most of the auxin to the

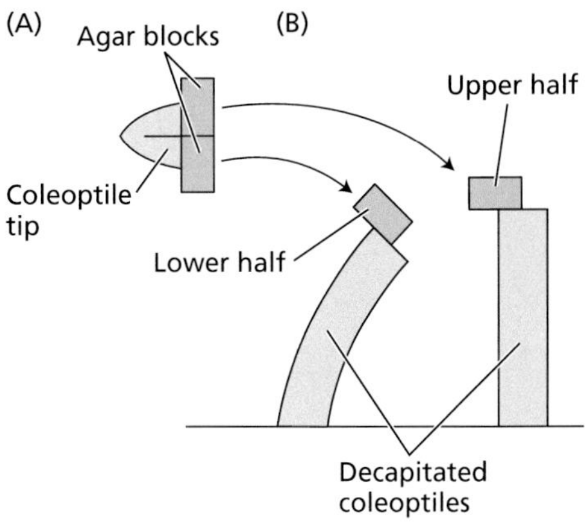

Figure 14.28 Auxin is transported to the lower side of a horizontally oriented oat coleoptile tip. (A) Auxin from the upper and lower halves of a horizontal tip is allowed to diffuse into two agar blocks. (B) The agar block from the lower half (left) induces greater curvature in a decapitated coleoptile than the agar block from the upper half (right) does.

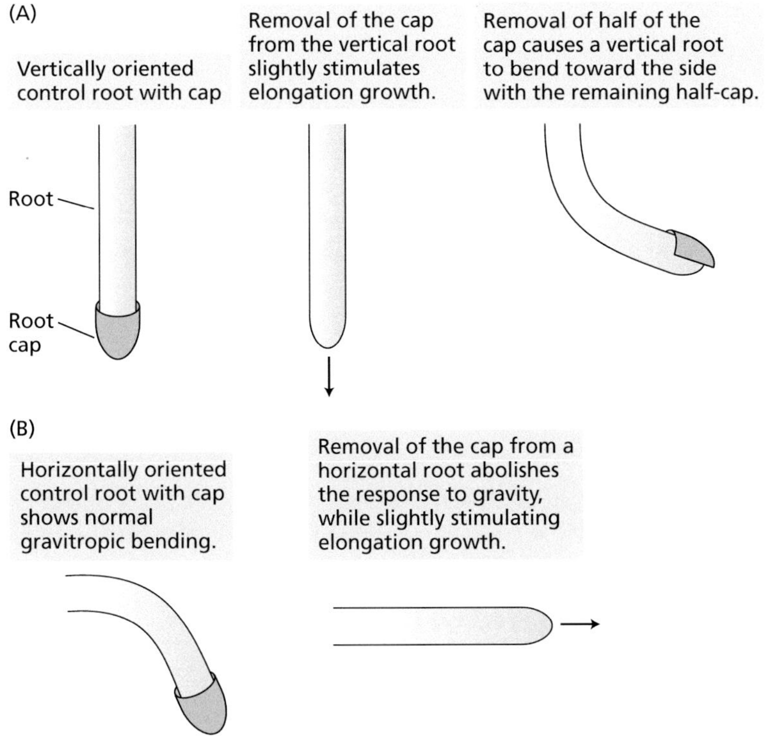

Figure 14.29 Effects of microsurgery on the direction of root growth. Microsurgery experiments demonstrate that the root cap is required for redirection of auxin and subsequent differential inhibition of elongation in root gravitropic bending. (A) Vertically growing root. (B) Horizontally growing root.

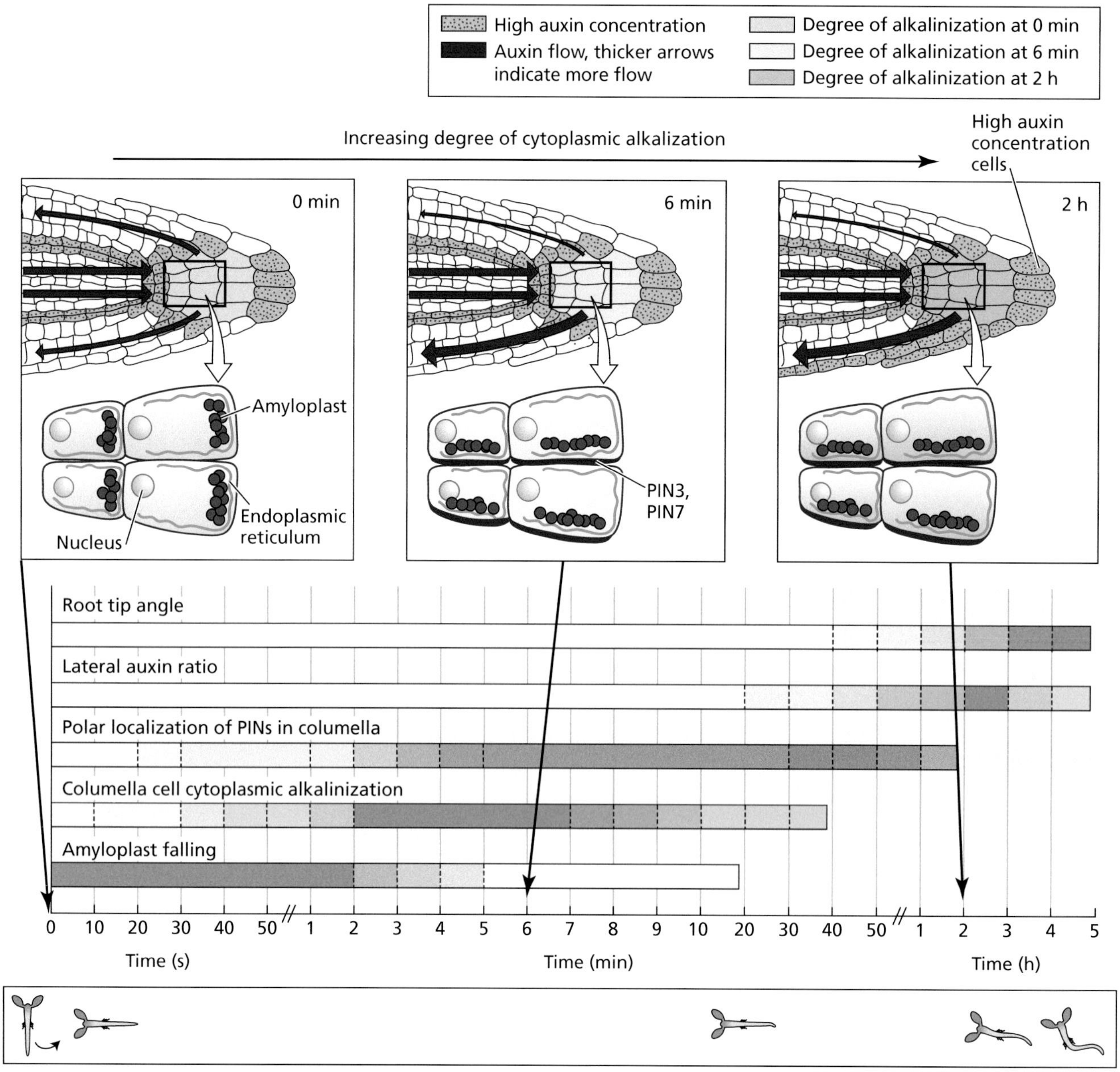

Figure 14.30 Sequence of events following gravistimulation of an Arabidopsis root. The timescale on the bottom is nonlinear. The differential growth of the shoot and roots of the seedling at different stages of the response is illustrated below the timescale. Three stages of statolith sedimentation are shown at the top. The left panel shows time zero, when the seedling is first rotated 90°. The second and third stages shown are at about 6 min and 2 h after rotation. The red arrows indicate auxin flow, with thicker arrows indicating more flow. Cells with relatively high auxin concentrations are shown in orange. Columella cells of the root tip are shown in green at time zero; the color changes to blue and then to blue-green at 2 h to indicate the degree of alkalinization of the cytoplasm. The distribution of PIN proteins is diagrammed as a purple outline on the plasma membrane of the columella cells. (After K. I. Baldwin et al. 2013. *Am. J. Bot.* 100: 126–142.)

lower side, thus inhibiting the growth of that lower side (**Figure 14.30**). Consistent with this model, IAA rapidly accumulates on the lower side of a horizontally oriented root and concentrates in the epidermal cells of the elongation zone.

Gravity perception is triggered by the sedimentation of amyloplasts

The primary mechanism by which gravity can be detected by cells is via the motion of a falling or sedimenting intracellular body. Root cap columella cells contain large, dense amyloplasts (starch-containing plastids) called **statoliths**. These statoliths readily sediment to the bottom of the cell to align with the gravity vector within 60 s (Figure 14.30). As shown in Figure 14.29, removal of the root cap from otherwise intact roots prevents root gravitropism without inhibiting growth, suggesting that the root cap columella cells function as gravity-sensing cells, or **statocytes**. Perception of the stimulus (statolith displacement by gravity) is thought to occur via membrane receptors and/or cytoskeletal interactions. Since starchless mutants still show a response to gravity (albeit reduced), additional signals must be involved in gravity perception.

In a vertically oriented root, PIN3 is uniformly distributed around the columella cells, but when the root is placed on its side, PIN3 is preferentially targeted to the lower side of these cells within 6 min (Figure 14.30). This redistribution of PIN3 occurs after sedimentation of the statoliths and before the root begins to bend and is thought to accelerate auxin transport to the lower side of the root. As a result, auxin is transported out of the columella to the lower side of the root cap. From there auxin is transported back to the elongation zone via the epidermal cells. However, as *pin3* mutants are not completely agravitropic, other asymmetric events act along with PIN3 localization to alter auxin flows. The most likely event would be an asymmetric change in apoplastic acidification, which would impose an asymmetric chemiosmotic potential to redirect auxin flow. This would cause PIN3 redistribution, which would amplify (canalize) the flow of auxin in the new direction. The activities of AUX1 and PIN2 in their respective expression domains are also necessary for gravitropic responses, as their loss-of-function mutants are agravitropic.

In eudicot stems and stemlike organs, the statoliths involved in gravity perception are located in the **starch sheath**, the innermost layer of cortical cells that surrounds the ring of vascular bundles of the shoot (**Figure 14.31**). The starch sheath is continuous with the endodermis of the root, and the amyloplasts are redistributed when the gravity vector is changed. The starch sheath cells contain ABCB19 and PIN3, which function coordinately to restrict auxin streams to the vascular system as it moves from the shoot to the root in seedlings (**Figure 14.32**). Selective regulation of the downward auxin transport stream conducted by PIN1 inside the vascular cylinder and

statoliths Cellular inclusions such as amyloplasts that act as gravity sensors by having a high density relative to the cytosol and sedimenting to the bottom of the cell.

statocytes Specialized gravity-sensing plant cells that contain statoliths.

starch sheath A layer of cells that surrounds the vascular tissues of the shoot and coleoptile and is continuous with the root endodermis. Required for gravitropism in the shoots of eudicots.

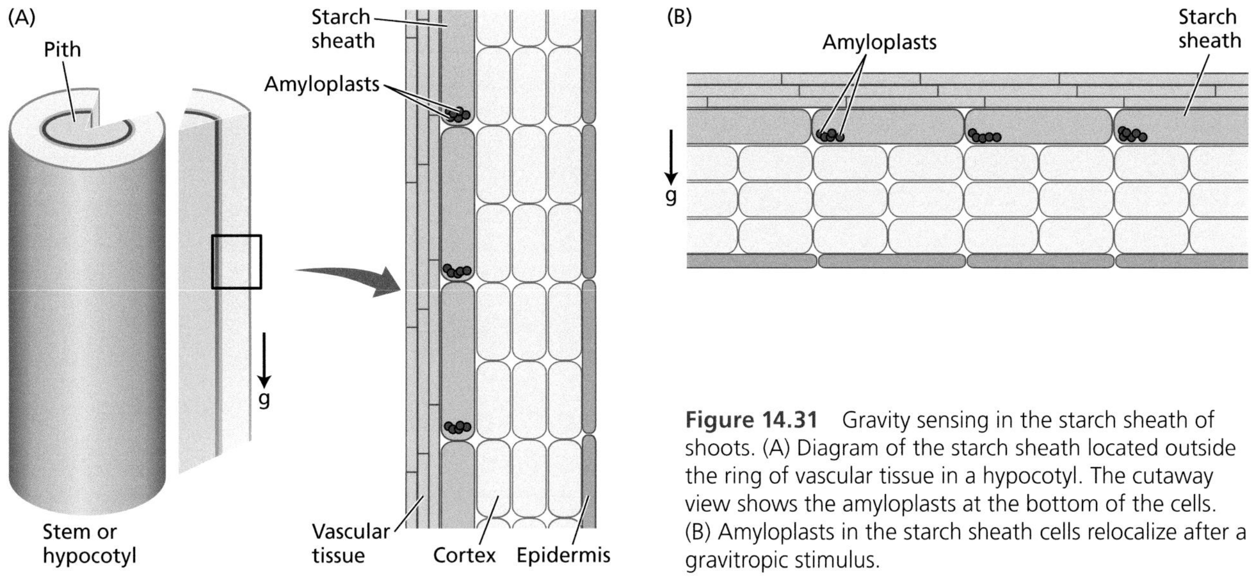

Figure 14.31 Gravity sensing in the starch sheath of shoots. (A) Diagram of the starch sheath located outside the ring of vascular tissue in a hypocotyl. The cutaway view shows the amyloplasts at the bottom of the cells. (B) Amyloplasts in the starch sheath cells relocalize after a gravitropic stimulus.

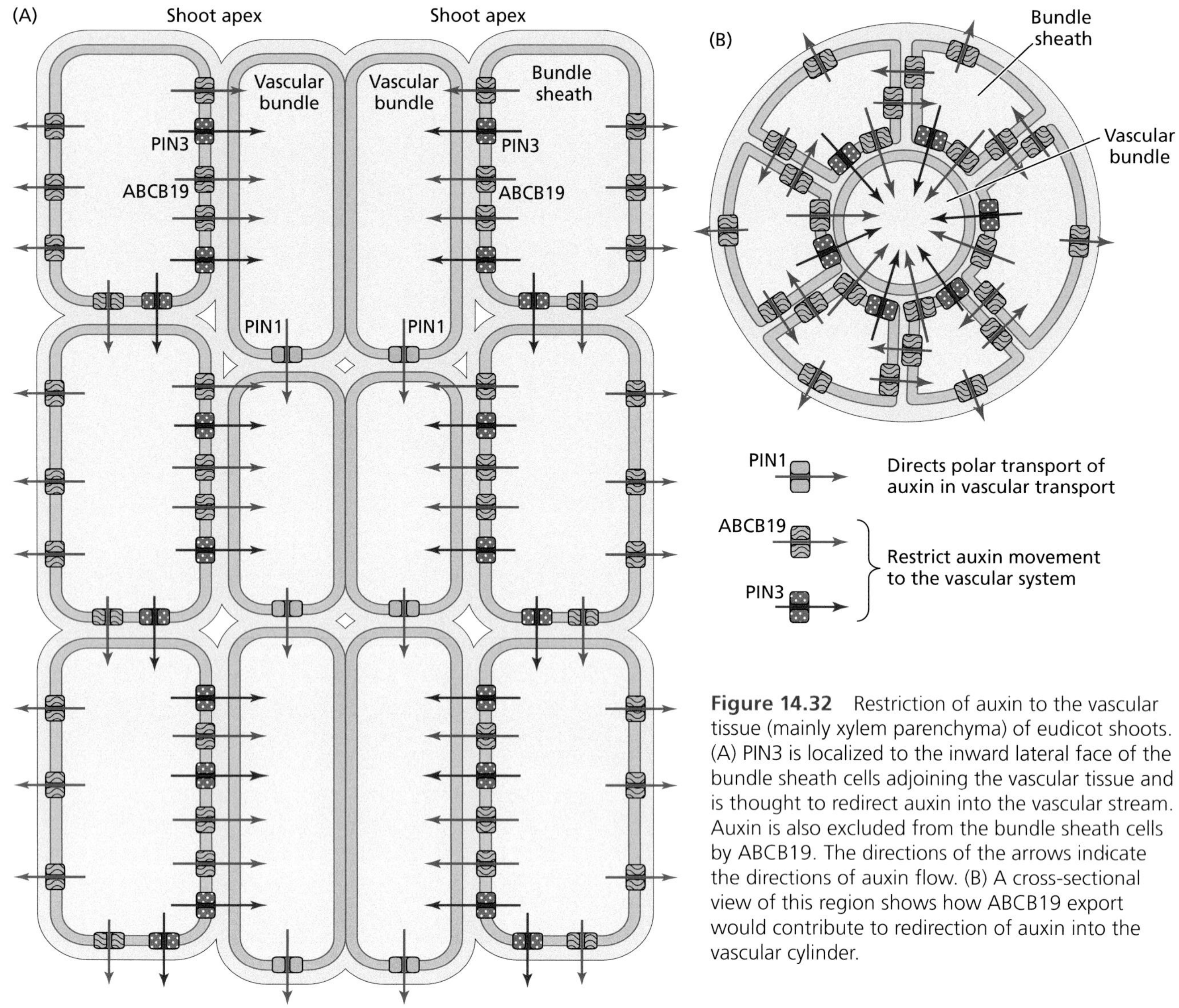

Figure 14.32 Restriction of auxin to the vascular tissue (mainly xylem parenchyma) of eudicot shoots. (A) PIN3 is localized to the inward lateral face of the bundle sheath cells adjoining the vascular tissue and is thought to redirect auxin into the vascular stream. Auxin is also excluded from the bundle sheath cells by ABCB19. The directions of the arrows indicate the directions of auxin flow. (B) A cross-sectional view of this region shows how ABCB19 export would contribute to redirection of auxin into the vascular cylinder.

selective restriction of lateral auxin movement into the starch sheath cells by ABCB19 and PIN3 appear to play a fundamental role in tropic bending. In response to a change in the gravity vector, such as occurs when heavy rain or wind causes **lodging** of a plant, the amyloplasts reorient to the bottom of the starch sheath cells and the hypocotyl will bend upward in less than 1 h. The exact sequence of subsequent events remains unresolved, but PIN3 signals are absent on the lower side of the bend in the hypocotyl starch sheath cells 4 h after gravitropic stimulation. Genetic studies have confirmed the role of the starch sheath in shoot gravitropism, and Arabidopsis and tomato mutants lacking amyloplasts in the starch sheath display reduced gravitropic shoot growth.

lodging The permanent displacement of the shoot or part of the shoot from a vertical posture. It is often caused by high wind speeds, made worse by wet conditions.

Gravity sensing may involve pH and calcium ions (Ca^{2+}) as second messengers

A variety of experiments suggest that localized changes in pH and Ca^{2+} gradients are part of the signaling that occurs during gravitropism. When pH-sensitive dyes were used to monitor both intracellular and extracellular pH in Arabidopsis roots, rapid changes were observed soon after roots were rotated to a horizontal position (**Figure 14.33A**). Within 2 min of gravistimulation, the cytoplasmic pH of the columella cells of the root cap increased from 7.2 to 7.5, while the apoplastic pH declined from 5.5 to 4.5. These changes precede any detectable tropic curvature by about 10 min. Both processes involve auxin signal amplification, as auxin perception in the gravire-sponding root is mediated by a cytosolic receptor from the TIR1/AFB family (Chapter 13) that rapidly converts ATP to a second messenger (cyclic adenosine monophosphate, cAMP). The synthesized cAMP appears to regulate a Ca^{2+} channel. Changes in extracellular pH also appear to amplify the gravitropic response and involve auxin activation of cell surface transmembrane kinase/co-receptor dimers that regulate H^+-ATPase activity.

Activation of the plasma membrane H^+-ATPase to amplify chemiosmotic gradients is one of the initial steps in activation of root gravitropism. As described in Chapter 12, the increase in chemiosmotic potential results in increased directional uptake and efflux of IAA from the affected cells. Accumulation of auxin in what becomes the lower side of the elongation zone inhibits expansion growth in those cells, resulting in the root bending in the direction of gravity.

Early physiological studies suggested that Ca^{2+} release from storage compartments might also be involved in root gravitropic signal transduction. Studies with Ca^{2+} sensors confirmed that cytosolic Ca^{2+} concentration decreases on the upper side and a Ca^{2+} spike occurs about 5 min on the lower side in the elongation zone of gravistimulated roots (**Figure 14.33B, C**). Therefore, Ca^{2+} and pH signaling appear to regulate root gravitropic bending through the propagation of a Ca^{2+}-dependent signaling pathway.

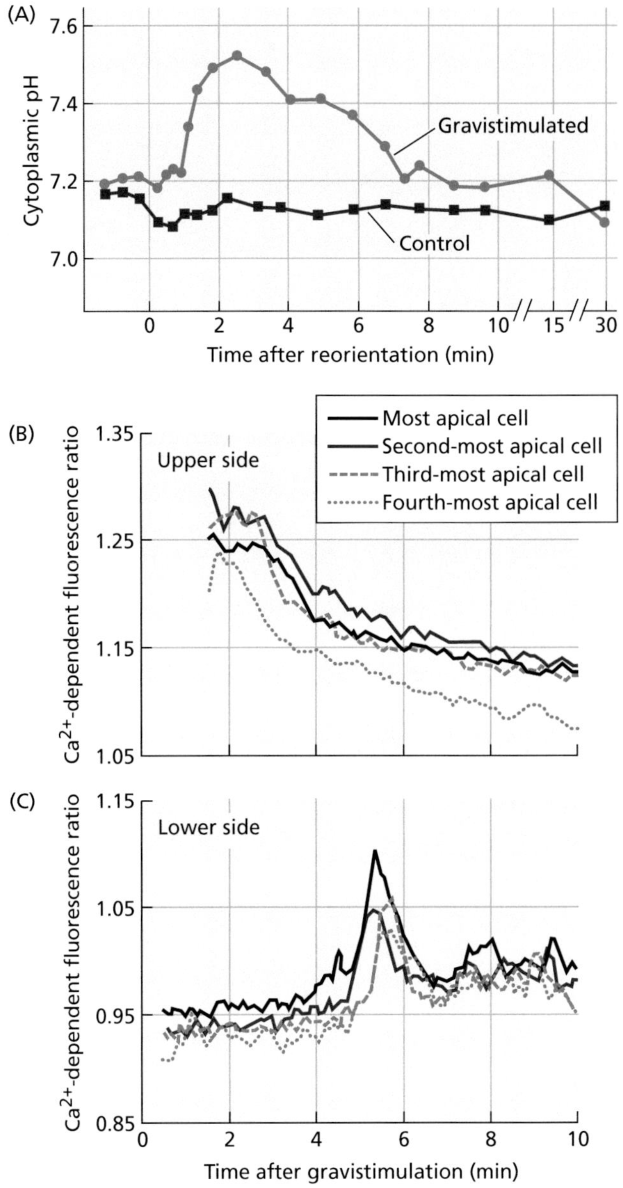

Figure 14.33 Gravitropic pH changes are regulated by a Ca^{2+}-dependent pathway. (A) Experiments with a pH-sensitive dye suggest that pH changes in columella cells of the root cap are involved in gravitropic stimulus. The cytoplasmic pH increases in less than 2 min after gravistimulation. Experiments with a Ca^{2+} sensor suggest that Ca^{2+} changes occur on the upper side (B) and lower side (C) in the elongation zone of a gravistimulated root.

Thigmotropism involves signaling by Ca^{2+}, pH, and reactive oxygen species

When the radicle emerges from the seed coat, gravity is not the only force it encounters; the radicle also immediately encounters forces imposed by the soil. Depending on where the seed ended up, the soil might be predominantly sand, clay, loam, or some other composition. The plant responds to obstacles in the soil, such as rocks, by redirecting growth around them (thigmotropism) until it can resume growth in alignment with the gravity vector. In nature, roots integrate gravitropic and thigmotropic stimuli to control their growth response.

Auxin mediates differential cell elongation growth during thigmotropism. In roots, the columella cells are the site of perception of the touch stimulus in addition to gravity stimulus, and the signal is transduced within the columella cells. The mechanical stimulus induces changes in the concentrations of the secondary messengers Ca^{2+}, pH, and reactive oxygen species. After the touch response, the concentration of cytosolic Ca^{2+} increases in the epidermal cells on the side of the root distal to the touch stimulus. This triggers an increase in reactive oxygen species in the apoplast as well as acidification of the cytosol,

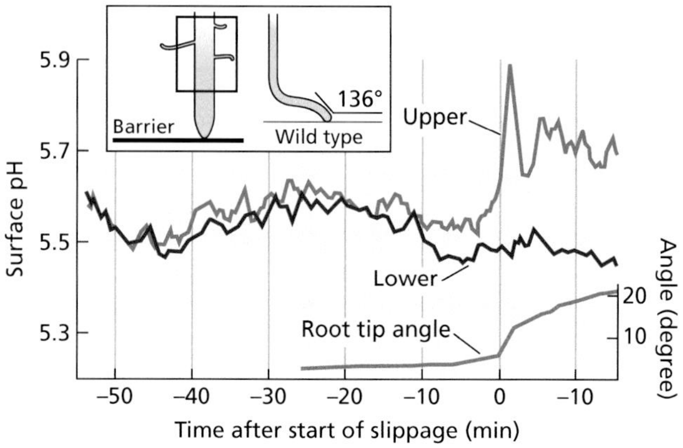

Figure 14.34 Touch sensing interacts with gravity sensing to regulate the growth of primary roots of Arabidopsis. (Inset) The primary root changes its growth direction when it encounters a barrier to downward growth. Roots grow vertically and then bend when they touch a barrier. The root tip angle of primary roots growing around a barrier is an integrated response to gravity and touch. Under normal circumstances a root encountering a horizontal barrier maintains a root tip angle of 136° until it reaches the edge of the barrier. (After G. D. Massa and S. Gilroy. 2003. *Plant J.* 33: 435–445.) pH changes were measured along the upper (blue) and lower (red) sides of the root after it encountered a barrier. The green line indicates the root tip angle. Bar = 100 μm.

and alkalization of the apoplast on the untouched side of the root (**Figure 14.34**). The asymmetric growth of the root is mediated by the redistribution of auxin that accumulates on the lower side of the root, as observed in the gravitropic response. Obstacle avoidance has two stages. Roots first bend away from the obstacle (thigmotropism) and continue to grow parallel to the obstacle. When the obstacle is no longer perceived via touch sensing, the roots bend again to align root growth with gravity (gravitropism). Therefore, touch sensing modulates gravity sensing until the touch signal is no longer perceived.

Touch responses are also triggered when a growing root encounters denser soil, resulting in root skewing or waving. This response involves the integration of gravitropism, thigmotropism, and circumnutation.

Hydrotropism involves ABA signaling and asymmetric cytokinin responses

Hydrotropism is root growth in response to asymmetric water potential of the soil (**Figure 14.35**). The adaptive response of water-foraging root growth toward areas of high water potential is essential for plant survival in soils where water is limiting. In Arabidopsis, pea, rice, and cucumber, hydrotropism still occurs when the root cap is removed using laser ablation, indicating that the root cap is not the site of perception. Instead, water gradients seem to be sensed by the cortical cells in the root elongation zone, where ABA signaling is triggered in response to a water gradient. The hydrotropism responses are dependent on ABA signaling, as the response is not observed in mutants that are insensitive to ABA, and ABA hyper-responsive mutants show rapid hydrotropism. ABA induces expression and accumulation of MIZU-KUSSEI 1 (MIZ1), named for two Japanese words meaning "water" and "tropism." MIZ1 negatively regulates the activity of the endoplasmic reticulum (ER) Ca^{2+}-ATPase 1 (ECA1). As a result, an asymmetric Ca^{2+} gradient is formed in the phloem, and Ca^{2+} diffuses down this gradient to the cortical cells in the root elongation zone on the low water potential side.

Cytokinin signaling is also involved in hydrotropism, as the response is reduced in cytokinin biosynthesis or response mutants. Asymmetric cytokinin biosynthesis and accumulation in the meristematic zone also depend on MIZ1 activity. The type A cytokinin response regulators ARR16 and ARR17, in the epidermal and cortical cells, respectively, are up-regulated, resulting in more cells available for elongation on the low water potential side. While cell elongation must occur for the root to bend toward higher moisture, the role of auxin in hydrotropic responses is species-specific. Auxin signaling or polar auxin transport resulting in the lateral redistribution of auxin is required for hydrotropism in rice and pea, but not in Arabidopsis or lotus.

Phototropins are the light receptors involved in phototropism

An emerging seedling is able to bend in any direction toward sunlight to optimize light absorption. This phenomenon is known as phototropism. As we saw in Chapter 13, blue light is particularly effective in inducing phototropism, and two flavoproteins, phototropin 1 and phototropin 2, are the photoreceptors for phototropic bending. Phototropism results from rapid signaling events that are initiated by light-activated phototropins on the illuminated side of plant organs and that result in differential elongation growth. As in the case of gravitropism,

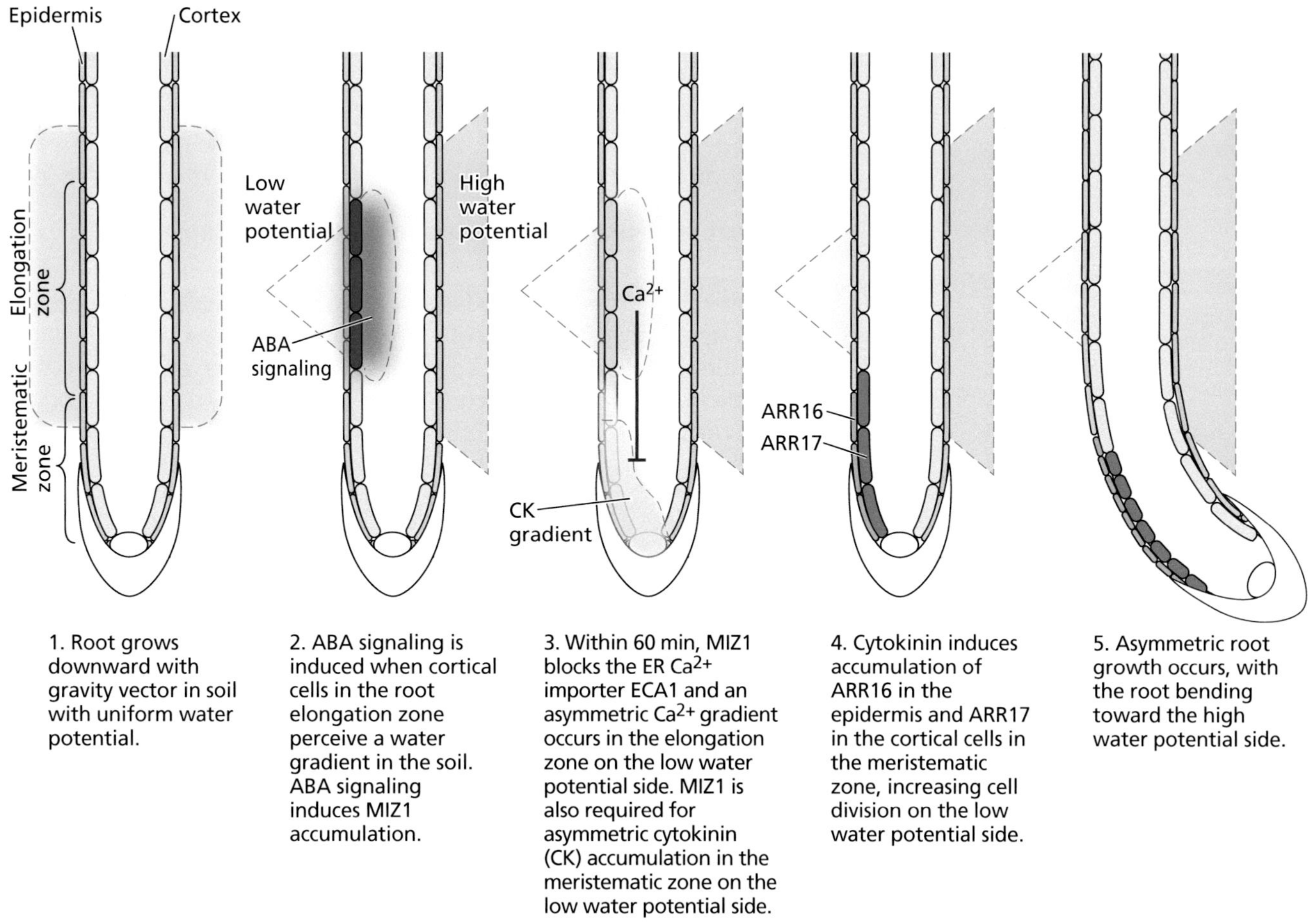
Epidermis
Cortex
Elongation zone
Meristematic zone
Low water potential
High water potential
ABA signaling
Ca2+
CK gradient
ARR16
ARR17
1. Root grows downward with gravity vector in soil with uniform water potential.
2. ABA signaling is induced when cortical cells in the root elongation zone perceive a water gradient in the soil. ABA signaling induces MIZ1 accumulation.
3. Within 60 min, MIZ1 blocks the ER Ca2+ importer ECA1 and an asymmetric Ca2+ gradient occurs in the elongation zone on the low water potential side. MIZ1 is also required for asymmetric cytokinin (CK) accumulation in the meristematic zone on the low water potential side.
4. Cytokinin induces accumulation of ARR16 in the epidermis and ARR17 in the cortical cells in the meristematic zone, increasing cell division on the low water potential side.
5. Asymmetric root growth occurs, with the root bending toward the high water potential side.

Figure 14.35 Hydrotropic growth of roots toward high water potential. Roots respond to water gradients by a pathway involving Ca^{2+} and ABA signaling events. ABA stimulates expression of MIZ1, which negatively regulates the endoplasmic reticulum Ca^{2+}-ATPase ECA1, resulting in higher Ca^{2+} concentrations on the side of the root in contact with low water potential soil. Cytokinin (CK) induces cell division on the side with low water potential via ARR16 and 17, resulting in asymmetric growth and root bending toward the high water potential side.

the bending response to directional blue light (Figure 13.13) can be explained by the Cholodny–Went model of lateral auxin redistribution.

Phototropism is mediated by the lateral redistribution of auxin

Charles and Francis Darwin provided the first clue concerning the mechanism of phototropism by demonstrating that while light is perceived at the tip of coleoptiles, bending occurs in the region below the tip. The Darwins proposed that some "influence" was transported from the tip to the growing region, thus causing the observed asymmetric growth response. This influence was later shown to be auxin.

When a shoot is growing vertically, auxin is transported polarly from the growing tip to the elongation zone. However, auxin can also be transported laterally, and this lateral diversion of auxin lies at the heart of the Cholodny–Went model for tropisms. In phototropic bending, auxin from the shoot tip is redirected laterally to the shaded side of the stem and stimulates cell elongation. The resulting differential growth causes the shoot to bend toward the light (**Figure 14.36** and **Figure 14.37**).

Although phototropic mechanisms appear to be highly conserved across plant species, the precise sites of auxin production, light perception, and lateral transport have been difficult to define. In maize coleoptiles, auxin accumulates

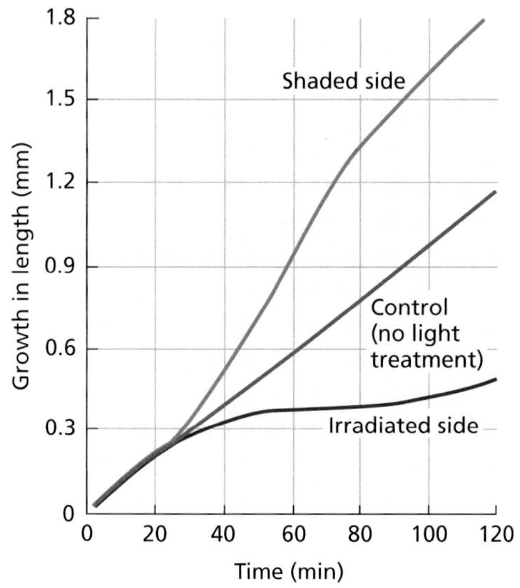

Figure 14.36 Time course of growth on the irradiated and shaded sides of a coleoptile responding to a 30-s pulse of unidirectional blue light at zero minutes. Control coleoptiles were not given a light treatment.

in the upper 1 to 2 mm of the tip. The zones of photosensing and lateral transport extend farther down, to 5 mm from the tip. The response is also strongly dependent on the light fluence (the number of photons per unit of area). Similar zones of auxin biosynthesis and accumulation, light perception, and lateral transport are seen in the true shoots of all monocots and eudicots examined to date.

Shoot phototropism occurs in a series of steps

As we mentioned earlier, the events in phototropic bending occur rapidly. Phototropism can be divided into a series of four steps: perception and early signal amplification events, rapid phototropic bending, attenuation, and elongation at the new angle (Figure 14.37). Perception and early events in phototropism, such as phosphorylation and dephosphorylation, occur within 1 s to 30 min after blue-light irradiation. Rapid phototropic bending occurs from about 30 to 150 min after perception of a blue-light signal. The signal is then attenuated from about 150 to 220 min, and finally, symmetrical elongation commences about 220 min after perception of the initial blue-light signal.

Although phototropins are hydrophilic proteins, they are primarily associated with the plasma membrane. Low-fluence blue light irradiating one side of the upper hypocotyl and shoot apex is refracted in internal air spaces to generate a gradient of activated phototropins, which then initiate a series of signal transduction events. During the first minute after irradiation, the plasma membrane is transiently depolarized, and the cytoskeleton is reoriented; existing microtubules are degraded and new, longitudinally oriented microtubules are formed.

After approximately 3 min of unilateral blue-light irradiation, phototropin 1 (phot1) undergoes autophosphorylation. Next, the activated phot1

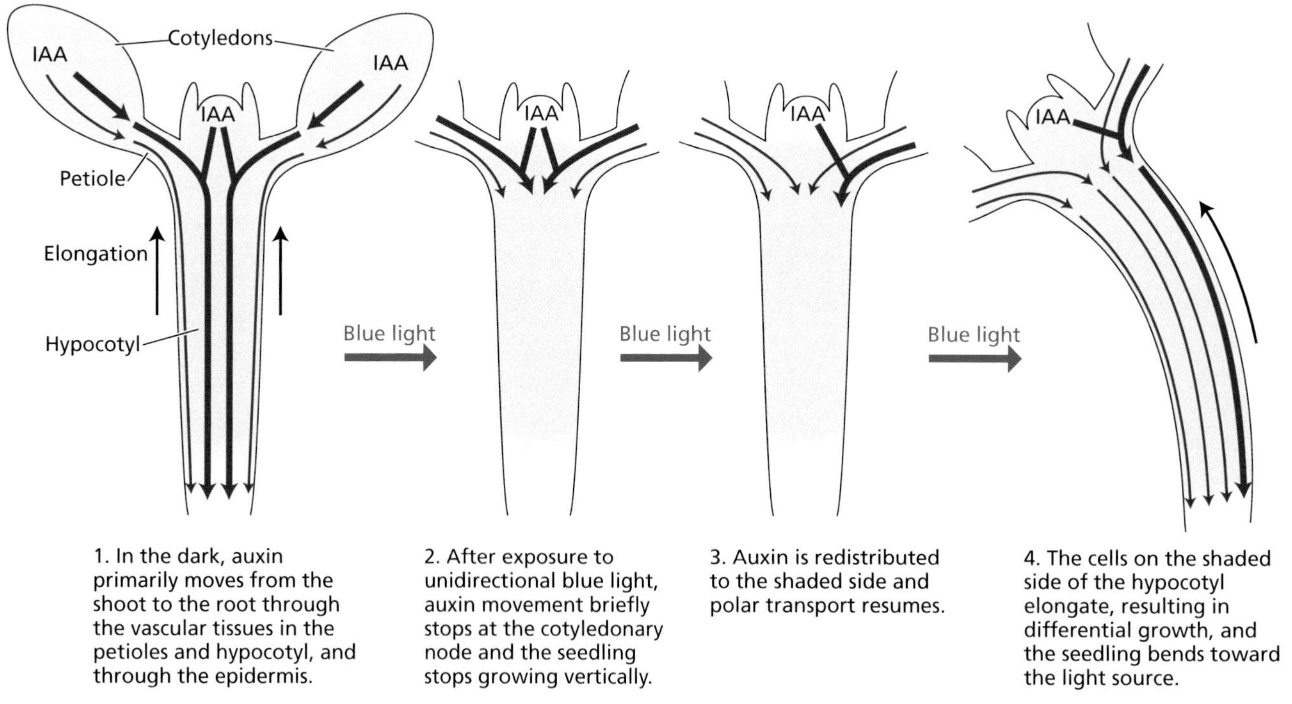

Cotyledons
IAA
IAA
IAA
Petiole
Elongation
Hypocotyl
Blue light
IAA
Blue light
IAA
Blue light
IAA
1. In the dark, auxin primarily moves from the shoot to the root through the vascular tissues in the petioles and hypocotyl, and through the epidermis.
2. After exposure to unidirectional blue light, auxin movement briefly stops at the cotyledonary node and the seedling stops growing vertically.
3. Auxin is redistributed to the shaded side and polar transport resumes.
4. The cells on the shaded side of the hypocotyl elongate, resulting in differential growth, and the seedling bends toward the light source.

Figure 14.37 Model of rootward auxin movement (red arrows within the seedlings) associated with phototropin 1–dependent phototropism in dark-acclimated seedlings of Arabidopsis.

phosphorylates the auxin transporter ABCB19. As a result of ABCB19 inhibition, auxin accumulates above the cotyledonary node and less auxin is delivered to the elongation zone, causing a rapid stop in hypocotyl elongation and circumnutation. After this pause in elongation, the pooled auxin is laterally diverted to the shaded side of the hypocotyl via a poorly understood process. Auxin accumulation on the shaded side of the upper hypocotyl can be detected after approximately 15 min of exposure to unilateral blue light. Diverted auxin is then transported to the hypocotyl elongation zone in the vascular tissues and epidermis. These later steps involve differential activation of PIN proteins and rapid auxin-dependent activation of H^+-ATPase activity on the shaded side of the hypocotyl. Bending toward the blue-light source can be observed after approximately 30 minutes.

As perception of blue light continues, phototropic signal amplification occurs via activated phot1 acting on additional substrates. Another primary substrate for activated phot1 is NONPHOTOTROPIC HYPOCOTYL 3 (NPH3), originally identified from the *nph3* mutant that did not exhibit hypocotyl bending in response to low-fluence blue light. NPH3 phosphorylation allows it to bind the 14-3-3 phospho-binding proteins that regulate multiple cellular functions, including proton pumping by plasma membrane H^+-ATPses (Chapter 6), membrane channel activities, and receptor function. Phosphorylation of NPH3 by phot1 establishes a gradient of 14-3-3 proteins from the lit to the shaded side of the seedling to modulate auxin transport functions that initiate differential bending.

Rapid auxin-dependent activation of plasma membrane H^+-ATPase activity via ABL-TMK1/4 plasma membrane receptor complexes occurs on the shaded side of the hypocotyl. Acidification of the apoplast appears to play a role in phototropic growth: The apoplastic pH is more acidic on the shaded side of phototropically bending stems or coleoptiles than on the irradiated side. Decreased pH would be expected to enhance cellular elongation and amplify auxin movement from cell to cell. Both processes would be expected to contribute to bending toward light.

Bending toward the blue-light source can be detected approximately 30 min after the initial blue-light perception and early events. After initial bending, blue-light perception continues and involves differential regulation of PIN proteins. During attenuation, the phototropic signal is diminished by degradation of phot1 and bending is completed. After bending is completed, elongation toward the light source occurs. Circumnutation resumes as phototropic and gravitropic signals are integrated during elongation. Asymmetric PIN localization is also observed 4 h after blue-light perception. Although phototropins are the primary photoreceptors for phototropism, phytochromes and cryptochromes can also contribute to the response.

Summary

Seeds require rehydration, and sometimes additional treatments, to germinate. During germination and establishment, food reserves maintain the seedling until it is autotrophic. Following emergence, the shoot responds to nondirectional cues from sunlight to undergo photomorphogenesis. At the same time, shoots are also responding to directional signals to orient themselves with respect to sunlight (phototropism), gravity (gravitropism), and a source of water (hydrotropism). The root extends downward into the soil, adjusting its growth to avoid obstacles (thigmotropism). Vascular tissue differentiates to facilitate the movement of water, minerals, and sugars from seed reserves and the environment to all parts of the seedling. Hormones play central roles as signaling agents in all of the developmental pathways associated with seedling establishment.

(Continued)

Summary (*continued*)

14.1 Seed Anatomy

- Seed anatomy varies widely in terms of the types and distribution of stored food resources and the nature of the seed coat (**Figure 14.1**).

14.2 Seed Dormancy

- Seed dormancy may be either exogenous (imposed by surrounding tissues) or endogenous (arising from the embryo itself).
- Endogenous dormancy can be due to high ABA:GA concentration ratios, or to arrested embryogenesis.
- Seeds that do not become dormant may exhibit vivipary and precocious germination (**Figures 14.2**, **14.3**).
- The primary hormones regulating seed dormancy are abscisic acid (ABA) and gibberellins (GAs) (**Figure 14.4**).

14.3 Release from Dormancy

- Light, especially red light, breaks dormancy in many small seeds, a phenomenon mediated by phytochrome.
- Some seeds require chilling or after-ripening to break dormancy (**Figure 14.5**).
- ABA and GA are not the only chemicals regulating seed dormancy. In some seeds, nitrate, nitric oxide (NO), and chemicals in smoke can break dormancy.

14.4 Seed Germination

- Germination and postgermination take place in three phases relating to water uptake (**Figures 14.6**, **14.7**).

14.5 Mobilization of Stored Reserves

- The cereal aleurone layer responds to gibberellins by secreting hydrolytic enzymes (including α-amylase) into the surrounding endosperm, making starch breakdown products available to the embryo (**Figure 14.8**).
- Gibberellins are secreted by the embryo and promote α-amylase transcription and production (**Figure 14.9**).
- ABA inhibits α-amylase transcription.
- Protein storage vacuoles (PSVs) contain phytin, the K^+, Mg^{2+}, and Ca^{2+} salt of phytic acid (*myo*-inositol hexaphosphate, a sugar alcohol), a major storage form of phosphate in seeds. Phytase releases the phosphate and other cations during germination.
- Stored proteins protect the seed against desiccation and reactive oxygen species.
- Proteases are stored in a separate compartment in the cell and are rarely packaged in the seed protein storage vacuole.
- Oil bodies store triacylglycerides and phospholipids.
- Oil bodies also store oleosins, caleosins, and steroleosins, proteins embedded in the lipid monolayer that stabilize the oil body during dormancy, regulate oil body size, and emulsify the oil bodies during germination.
- Oil bodies are broken down by proteases, lipases, and β-oxidation of the released fatty acids during germination and seedling establishment The breakdown of oil bodies by proteases and lipases and the β-oxidation of the released fatty acids provide energy and raw materials for germination and seedling establishment (discussed in Chapter 11).

14.6 Cell Expansion Mechanisms

- The cell wall has three layers: middle lamella, primary cell wall, and secondary cell wall (**Figure 14.10**).
- Cell walls are composed of many structural components that can be modified to be rigid or allow cells to grow (**Figure 14.11**).
- Cells can expand or grow nondirectionally (diffuse growth) or directionally (tip growth) (**Figure 14.12**).
- In diffuse-growing cells, the orientation of cell growth is determined by the orientation of cellulose microfibrils, which is determined by the orientation of microtubules in the cytoplasm.
- The orientation of the newly deposited cellulose microfibrils determines the direction of cell expansion (**Figure 14.13**).
- Plant cell growth and expansion is rapid at acidic pH and in the presence of expansion proteins (**Figures 14.14**, **14.15**).

14.7 Seedling Growth and Establishment

- Seedlings transition from skotomorphogenesis (development in the dark [i.e., underground]) to photomorphogenesis (development in the presence of light) at the first instance of light (**Figure 14.16**).
- In etiolated shoots, gibberellins and brassinosteroids suppress photomorphogenesis (**Figure 14.17**).
- Phytochrome, auxin, and ethylene regulate hook opening (**Figure 14.18**).
- Vascular differentiation begins during seedling emergence (**Figure 14.19**).

(*Continued*)

Summary (*continued*)

- The root cap covers the root apical meristem and protects it as the root pushes through the soil (**Figure 14.20**).
- Root hairs are specialized epidermal cells that increase the surface area of the root for water and mineral uptake and help anchor the plant in the soil (**Figure 14.21**).
- Root hair formation is regulated by ethylene (**Figure 14.22**).

14.8 Differential Growth Enables Successful Seedling Establishment

- Ethylene causes reorientation of microtubules and induces lateral cell expansion (**Figure 14.23**).
- At optimal concentrations, auxin promotes stem and coleoptile growth and inhibits root growth. However, higher concentrations of auxin can inhibit stem and coleoptile growth (**Figure 14.24**).

14.9 Tropisms: Growth in Response to Directional Stimuli

- Polarized seedling growth is directed by polar auxin streams (**Figures 14.25–14.28**).
- Most of the auxin that is redirected shootward in the apex of young seedling roots is derived from the shoot (**Figure 14.27**).
- Lateral redistribution of auxin in coleoptile tips facilitates gravitropism in coleoptiles (**Figure 14.28**).
- A horizontally oriented root redirects auxin to the lower side, inhibiting growth at the elongation zone, an activity that is mediated by the root cap (**Figures 14.29, 14.30**).
- Statoliths in the columella cells of the root cap serve as gravity sensors (**Figure 14.30**).
- The statoliths that regulate gravitropism in eudicot stems and hypocotyls are located in the starch sheath (**Figure 14.31**).
- Auxin is restricted to the vascular tissue in eudicot stems (**Figure 14.32**).
- pH and Ca^{2+} act as second messengers in the signaling that occurs during gravitropism (**Figure 14.33**).
- pH, Ca^{2+}, and reactive oxygen species act as second messengers in the signaling that occurs during thigmotropism (**Figure 14.34**).
- Auxin redistribution occurs during thigmotropism.
- The cortical cells in the root elongation zone are the site of perception of soil water potential (**Figure 14.35**).
- The shaded side of the hypocotyl elongates during phototropic bending (**Figure 14.36**).
- Blue light is refracted in air spaces in the upper hypocotyl to differentially activate phototropin light receptors
- Like gravitropism and thigmotropism, phototropism involves lateral redistribution of auxin (**Figure 14.37**).

Suggested Reading

Anderson, C. T., and Kieber, J. J. (2020) Dynamic construction, perception, and remodeling of plant cell walls. *Annu. Rev. Plant Biol.* 71(1): 39–69.

Bewley, J. D., Bradford, K. J., Hilhorst, H. W. M., and Nonogaki, H. (2013) *Seeds: Physiology of Development, Germination and Dormancy*, 3rd ed. Springer, New York.

Chang, J., Li, X., Fu, W., Wang, J., Yong, Y., Shi, H., Ding, Z., Kui, H., Gou, X., He, K., et al. (2019) Asymmetric distribution of cytokinins determines root hydrotropism in *Arabidopsis thaliana*. *Cell Res.* 29: 984–993.

Christie, J. M., Yang, H., Richter, G. L., Sullivan, S., Thomson, C. E., Lin, J., Titapiwatanakun, B., Ennis, M., Kaiserli, E., Lee, O. R., et al. (2011) phot1 inhibition of ABCB19 primes lateral auxin fluxes in the shoot apex required for phototropism. *PLOS Biol.* 9: e1001076.

Dietrich, D., Pang, L., Kobayashi, A., Fozard, J. A., Boudolf, V., Bhosale, R., Antoni, R., et al. (2017) Root hydrotropism is controlled via a cortex-specific growth mechanism. *Nat. Plants* 3: 17057.

Du, M., Spalding, E. P., and Gray, W. M. (2020) Rapid auxin-mediated cell expansion. *Annu. Rev. Plant Biol.* 71: 379–402.

Migliaccio, F., Tassone, P., and Fortunati, A. (2013) Circumnutation as an autonomous root movement in plants. *Am. J. Bot.* 100: 4–13.

Shao, Q., Liu, X., Su, T., Ma, C., and Wang, P. (2019) New insights into the role of seed oil body proteins in metabolism and plant development. *Front. Plant Sci.* 10: 1568.

Shkolnik, D., Nuriel, R., Bonza, M. C., Costa, A., and Fromm, H. (2018) MIZ1 regulates ECA1 to generate a slow, long-distance phloem-transmitted Ca^{2+} signal essential for root water tracking in Arabidopsis. *Proc. Natl. Acad. Sci. USA* 115: 8031–8036.

Wang, X., Yu, R., Wang, J., Lin, Z., Han, X., Deng, Z., Fan, L., He, H., Deng, X. W., and Chen, H. (2020) The asymmetric expression of SAUR genes mediated by ARF7/19 promotes the gravitropism and phototropism of plant hypocotyls. *Cell Rep.* 31: 107529.

15 Vegetative Growth and Senescence

root apical meristem (RAM) A group of permanently dividing cells located underneath the root cap at the tip of the root that provides cells for the primary growth of the root.

shoot apical meristem (SAM) Dome-shaped region of the shoot tip composed of meristematic cells that give rise to leaves, branches, and reproductive structures.

primary plant body The part of the plant directly derived from the shoot and root apical meristems and primary meristems.

root cap Cells at the root apex that cover and protect the meristematic cells from mechanical injury as the root moves through the soil. Site for the perception of gravity and signaling for the gravitropic response in roots.

meristematic zone The region at the tip of the root containing the meristem that generates the body of the root. Located between the root cap and the elongation zone.

The developmental polarity of the plant axis is set up during embryogenesis by the formation of the **root apical meristem (RAM)** and the **shoot apical meristem (SAM).** During early primary growth and seedling establishment, the contrasting morphologies and growth patterns of the root and shoot are elaborated. A protective root cap forms over the RAM, and continued primary growth anchors the plant in the soil and gives rise to the mature **primary plant body**. The root axis elongates, and branches (lateral roots) emerge from internal tissues. In shoots, leaf and branch primordia form on the flanks of the SAM above the elongation zone of the stem, and specialized cell types complete differentiation as elongation growth ceases. In woody perennials, secondary growth in girth is initiated by the activities of the vascular and cork cambiums (discussed in Chapter 1).

Some cells of the primary and secondary plant body, such as xylem conducting elements, are nonliving when functionally mature. The formation of such specialized cells is a developmental process that involves programmed cell death. During leaf shedding the organic constituents and mineral nutrients of the old leaves are actively recycled via the phloem to newly formed leaves. Eventually, the whole plant undergoes senescence as the result of a combination of genetic and environmental factors. In this chapter we outline many of these developmental processes and the regulatory mechanisms that underlie them.

15.1 The Root Apical Meristem (RAM)

Describe the structures and cell types that comprise the root apical meristem (RAM), and explain the genetic elements that control cell fate and the hormones that function in RAM development and maintenance.

initials Broadly defined as the cells of the root and shoot apical meristems. More specifically, a cluster of slowly dividing cells located within the meristem that gives rise to the more rapidly dividing cells of the surrounding meristem.

Roots, which anchor the plant and absorb water and mineral nutrients from the soil, display complex patterns of growth and tropisms that allow them to explore and exploit a heterogeneous environment laden with obstacles (Chapter 4). Although cells produced by the RAM divide, differentiate, and elongate as they are displaced away from the tip, lateral outgrowths such as root hairs and lateral branches emerge farther away from the root tip in regions where cell elongation is complete (Chapters 1 and 4). This spatial separation helps prevent damage to lateral structures from the shear forces generated as the root penetrates the soil.

The root tip has four developmental zones

The basic features of root development can best be described by first distinguishing zones within the root with distinct cellular behaviors. Although it is impossible to define their boundaries with absolute precision, the division of the root into the following zones provides a useful spatial framework that is relevant to our discussion of the underlying mechanisms (**Figure 15.1**).

- The **root cap** occupies the most distal part of the root. It arises from a unique set of initial cells whose derivatives are displaced distally (i.e., outward), away from the meristematic zone. The differentiated products of these divisions cover the apical meristem and protect it from mechanical injury as the root tip is pushed through the soil. Other functions of the root cap include the secretion of compounds that help the root penetrate the soil and mobilize mineral nutrients and gravity sensing and signaling that guide root gravitropism (Chapter 14).
- The **meristematic zone** is partially enclosed by the root cap. It consists of a cluster of cells that act as **initials**, dividing with characteristic polarities to produce cells that divide further and differentiate into the various mature tissues that make up the root. Cells surrounding these initials have small vacuoles and expand and divide rapidly. In contrast, the **quiescent center (QC)** of the RAM consists of a rarely dividing pool of initial cells, which provides a reservoir of new initial cells in the event of damage to the RAM.
- The **elongation zone** is the site of rapid and extensive cell elongation. Although some cells continue to divide while they elongate within this zone, the rate of division decreases progressively to zero with increasing distance from the meristem.

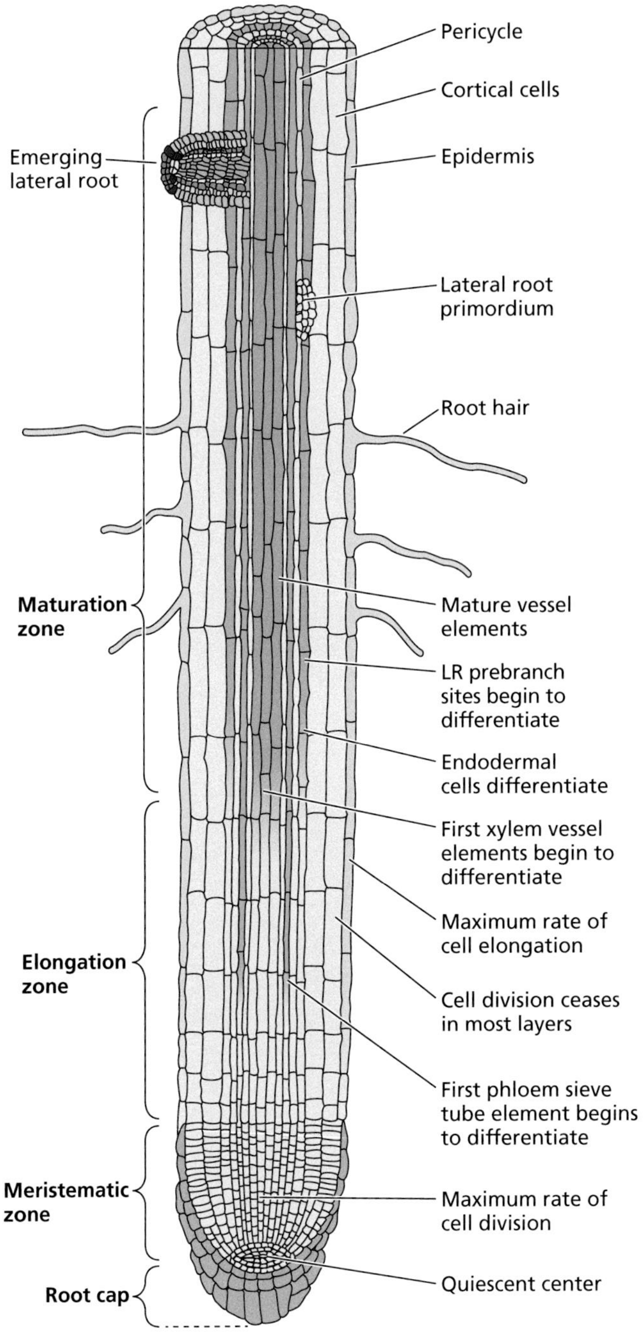

Figure 15.1 Simplified diagram of a primary root showing the root cap, meristematic zone, elongation zone, and maturation zone.

quiescent center The central region of the root meristem where cells divide more slowly than surrounding cells, or do not divide at all. Serves as a reservoir of meristematic cells for tissue regeneration in case of wounding.

elongation zone The region of rapid and extensive root cell elongation showing few, if any, cell divisions.

maturation zone The region of the root where differentiation occurs, including the production of root hairs and functional vascular tissue.

shoot apex (terminal bud) The shoot apical meristem with its associated leaf primordia and young, developing leaves.

- The **maturation zone** is the region in which cells acquire their differentiated characteristics. Cells enter the maturation zone after division and elongation have ceased, and in this region cell outgrowths called root hairs and lateral organs such as lateral roots may begin to form. Differentiation may begin much earlier, but cells do not achieve their mature state until they reach this zone.

Depending on the species, these four developmental zones can occupy little more than the first millimeter of the root or extend over a longer portion of the distal regions of the root.

Auxin and cytokinin contribute to the maintenance and function of the RAM

Auxin and cytokinin play important roles in RAM maintenance. Auxin is initially supplied to the primary root apex by rootward transport from the shoot, but is replaced by auxin produced in or around the QC as the primary root elongates and branches. Auxin accumulation and synthesis in the QC and surrounding cells maintain the expression of genes encoding a group of transcription factors that function in maintaining RAM organization.

RAM maintenance also involves the hormone cytokinin. As maturing cells move farther away from the QC in the meristematic zone, the higher auxin concentration promotes cell division and represses cytokinin responses. As progressive cell divisions gradually dilute the cellular levels of RAM-maintaining transcription factors, cytokinin signaling lowers auxin concentrations, inhibits cell division, and promotes the transition from cell division to cell elongation. A similar antagonism between auxin and cytokinin can be observed in undifferentiated callus cultures, in which varying the auxin:cytokinin ratios can be used to selectively initiate either shoots or roots.

15.2 The Shoot Apical Meristem (SAM)

Describe the structures and cell types that make up apical meristem (SAM), and explain the genetic elements that control cell fate and the hormones that function in regulating the size and cell number of SAMs.

Like the RAM, the shoot apical meristem (SAM) first forms during embryogenesis and maintains a group of undetermined cells throughout development. However, there are significant differences between the two meristem types in the way the derivatives of those cells are incorporated into lateral organs. Whereas lateral root initiation occurs well back of the root tip (discussed later in the chapter), leaves and associated axillary branches form close to the apical initials in the shoot. In place of the root cap that protects the apical initials of the root, young leaf primordia overlap and enclose the shoot tip, forming the terminal bud (**Figure 15.2**).

The term *shoot apical meristem* specifically refers to the initial cells and their undifferentiated derivatives. (The terms *initial cells* and *stem cells* are often used interchangeably by developmental biologists. To avoid confusion with the term as applied to the primary plant axis, we use the terms *initial cells, initials,* or *undifferentiated cells* throughout this.) Adjacent regions that contain cells fully committed to particular developmental fates are not included in the SAM. The more inclusive terms **shoot apex** (plural *apices*) or **terminal buds** refer to the apical meristem plus the most recently formed leaf primordia. The size, shape, and organization of the SAM vary according to plant species, developmental stage, and growth conditions. Cycads have the largest SAM among vascular plants, measuring over 3 mm in diameter; at the other extreme, the SAM of Arabidopsis is less than 50 μm in diameter and contains only a few dozen cells.

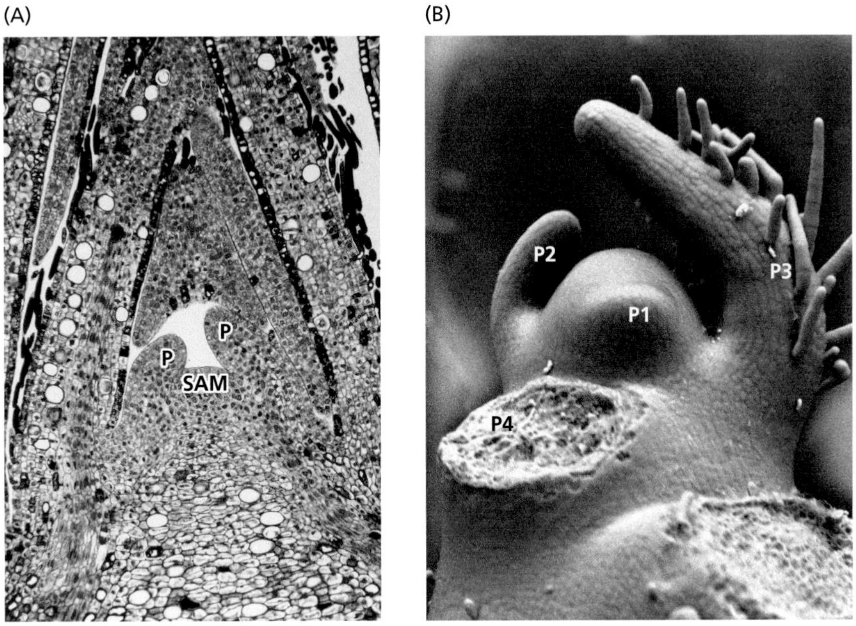

Figure 15.2 Structure of the shoot apex. (A) Stained, longitudinal section of the shoot apex of the bay laurel tree (*Laurus nobilis*). Note that the shoot apical meristem (the small mound in the center surrounded by two young leaf primordia) is enclosed and protected by the older leaf primordia. (B) The shoot apical meristem of a tomato plant. In this scanning electron microscope image, leaf primordia are labeled P1–P4 (from youngest to oldest); P4 has been removed to provide a better view of the SAM.

Significant variations in the SAM size of a given species can also occur, depending on developmental stage and environmental conditions.

The shoot apical meristem has distinct cytohistological zones and layers

The organization of the SAM has been interpreted according to two main theories, which are not mutually exclusive. For example, in **cytohistological zonation** (**Figure 15.3A**), longitudinal sections of the SAM reveal three regions distinguishable from each other by their locations, patterns of cell division, and the derivatives they form. The tip of the apical dome, called the **central zone (CZ)**, consists of a cluster of initial cells, comparable to the stem cells of animals, that gives rise to all the other cells of the SAM. These initial cells divide more slowly than the cells in the adjacent regions. A flanking region, called the **peripheral zone (PZ)**, consists of cytoplasmically dense cells that divide more frequently to produce cells that later become incorporated into lateral organs such as leaves. A centrally positioned **rib zone (RZ)** underlying the CZ (Figure 15.3A) contains dividing cells that give rise to the internal tissues of the stem. This type of organization is found in both gymnosperms and angiosperms.

The second theory of SAM organization is based on the presence of different cell layers—L1, L2, and L3—with distinct patterns of cell division and cell fates (**Figure 15.3B**). Most of the divisions are *anticlinal* (i.e., perpendicular to the surface of the SAM) in the L1 and L2 layers, which favors growth in surface area, whereas the planes of cell division are more randomly oriented in the L3 layer, which promotes three-dimensional growth. During leaf development, each of the cell layers contributes to different tissues in the leaf. The SAMs of most angiosperms exhibit a layered type of organization.

Cell identity in the SAM is determined by position rather than by cell lineage

Analyses of cell lineages show that cell divisions occasionally can cause cells of the SAM to be displaced by adjacent cells from one cell layer or cytohistological zone to another. As a result of this displacement, the developmental fate of the cell may change. For example, a cell that is displaced from the rib zone to the peripheral zone might be incorporated into a leaf primordium rather than

cytohistological zonation Regional cytological differences in cell division in the shoot apical meristems of seed plants.

central zone (CZ) A central cluster of relatively large, highly vacuolated, slowly dividing initial cells in shoot apical meristems, comparable to the quiescent center of root meristems.

peripheral zone (PZ) A doughnut-shaped region surrounding the central zone in shoot apical meristems and consisting of small, actively dividing cells with inconspicuous vacuoles. Leaf primordia are formed in the peripheral zone.

rib zone (RZ) Meristematic cells that are located beneath the central zone in shoot apical meristems and that give rise to the internal tissues of the stem.

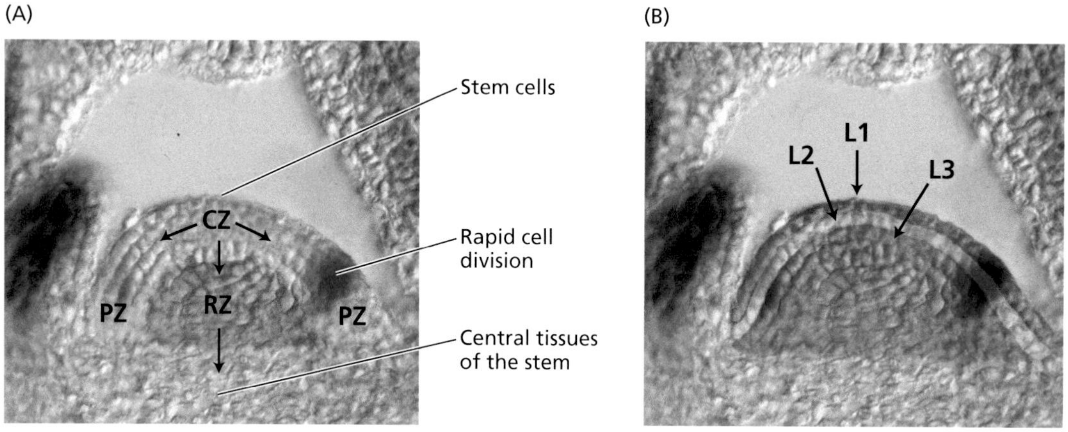

Figure 15.3 Organization of the shoot apical meristem of Arabidopsis. The organization of the shoot apical meristem can be analyzed in terms of cytohistological zones or cell layers. False color has been added to these micrographs to show the different zones and layers. (A) Cytohistological zonation. CZ, central zone (initial cells); PZ, peripheral zone (source of leaves); RZ, rib zone (source of central vascular tissues). (B) Cell layers. Most cell divisions are anticlinal in the outer, L1 and L2, layers; the planes of cell divisions are more randomly oriented in the L3 layer. The outermost (L1) layer generates the shoot epidermis; the L2 and L3 layers generate internal tissues.

the stem. Similarly, a cell that is displaced from the L2 layer to the L1 layer may become an epidermal cell rather than a mesophyll cell. This dynamic behavior indicates that the identities of cells in the SAM, including their characteristic division patterns, primarily reflect their position within the apical dome, rather than the rigidly programmed cell lineage observed in animals. This plasticity provides the opportunity for plant growth to respond to changing environmental conditions.

The gain or loss of cells from the SAM is tightly regulated during development. For example, plants that have been repeatedly wounded produce smaller leaves, a phenomenon called the "bonsai effect." The immediate cause of the smaller leaf size is the reduction in the rate of cell division in the SAM induced by the wound hormone, jasmonic acid. Interestingly, the size of the cells is either unchanged or slightly larger than normal. The smaller number of cells in the SAM available to form the leaf primordia results in smaller, though anatomically normal, leaves. The results suggest that the developmental plasticity of the cells of the SAM may be an adaptation enabling plants to reduce their leaf size in response to environmental stresses such as herbivory. Herbivory is discussed in Chapter 18.

In addition, SAMs may be either continuously meristematic (indeterminate), or may cease activity (become determinate), either by differentiating into a terminal organ such as a flower, or by undergoing growth arrest or senescence.

The size of the SAM is determined by cell number

Throughout development, the SAM maintains its size by balancing the loss of cells to differentiation in the peripheral zone with the production of more cells by division of initials in the central zone. The basic mechanism of core size control is a simple feedback loop **(Figure 15.4)**. Positive regulators are produced directly below the meristem initials in the central zone and increase the number of meristematic cells. Negative regulators are produced within the initials and suppress the positive regulators. Larger meristems have more initial cells and produce a higher dose of negative regulators, naturally limiting meristem size;

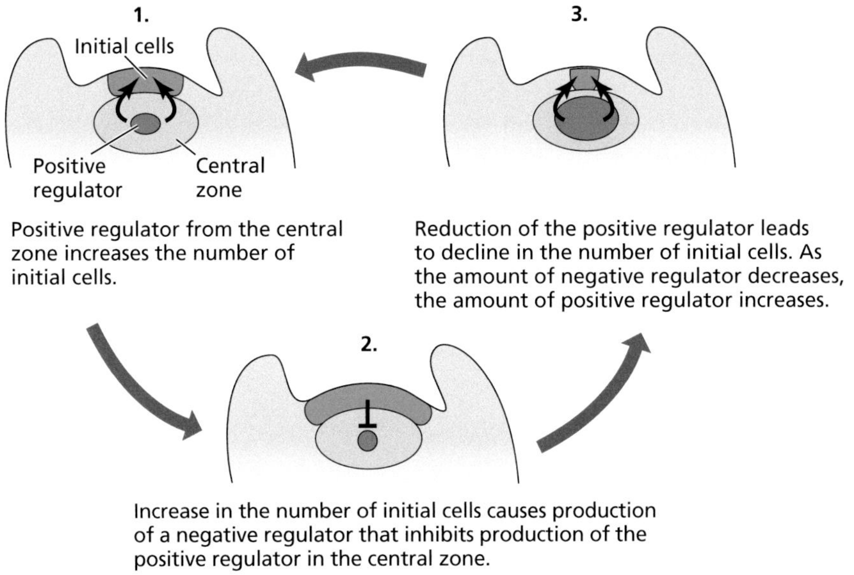

1.
Initial cells
Positive regulator
Central zone
Positive regulator from the central zone increases the number of initial cells.
3.
Reduction of the positive regulator leads to decline in the number of initial cells. As the amount of negative regulator decreases, the amount of positive regulator increases.
2.
Increase in the number of initial cells causes production of a negative regulator that inhibits production of the positive regulator in the central zone.

Figure 15.4 Model of the feedback loop that regulates the number of initial cells in the SAM.

and smaller meristems produce fewer negative regulators, allowing more positive regulators to increase meristem size. Eventually these two interacting factors reach an equilibrium and maintain constant meristem size.

15.3 Leaf Structure and Phyllotaxy

Describe the basic elements of stem morphology, leaf structure, and phyllotaxy and how hormones regulate patterning.

Morphologically, the leaf is the most variable of all the plant organs. The collective term for any type of leaf, or modified leaf, on a plant is **phyllome**. Phyllomes include the photosynthetic *foliage leaves* (what we usually mean by "leaves"), protective *bud scales, bracts* (leaves associated with inflorescences, or flowers), and *floral organs,* all of which are modified leaves. In angiosperms, the main part of the foliage leaf is expanded into a flattened structure, the **lamina** or blade. The appearance of a flat lamina in seed plants in the late Devonian was a key event in leaf evolution. A flat lamina maximizes light capture and also creates two distinct leaf domains: **adaxial** (upper surface) and **abaxial** (lower surface) (**Figure 15.5A**). Several types of leaves have evolved based on their adaxial–abaxial leaf structure.

In the majority of plants, the leaf blade is attached to the stem by a stalk called the **petiole**. However, some plants have *sessile leaves,* with the leaf blade attached directly to the stem (**Figure 15.5B**). In most monocots and certain eudicots, the base of the leaf is expanded into a sheath around the stem. Many eudicots have *stipules,* small outgrowths of the leaf primordia, located on the abaxial side of the leaf base. Stipules protect the young developing foliage leaves and are sites of auxin synthesis during early leaf development.

Leaves may be **simple** or **compound** (**Figures 15.5B** and **C**). A simple leaf has one blade, whereas a compound leaf has two or more blades, the *leaflets,* attached to a common axis, or *rachis*. Some leaves, like the adult leaves of some *Acacia* species, lack a blade and instead have a flattened petiole that functions as the blade, called the *phyllode*. In some plants the flattened stems functions as blades and are called *cladodes,* as in the cactus *Opuntia*.

phyllome The collective term for all the leaves of a plant, including structures that evolved from leaves, such as floral organs.

lamina The blade of a leaf.

abaxial Referring to the upper surface of a leaf

abaxial Referring to the lower surface of a leaf.

petiole The leaf stalk that joins the leaf blade to the stem.

simple leaf A leaf with one blade.

compound leaf A leaf subdivided into leaflets.

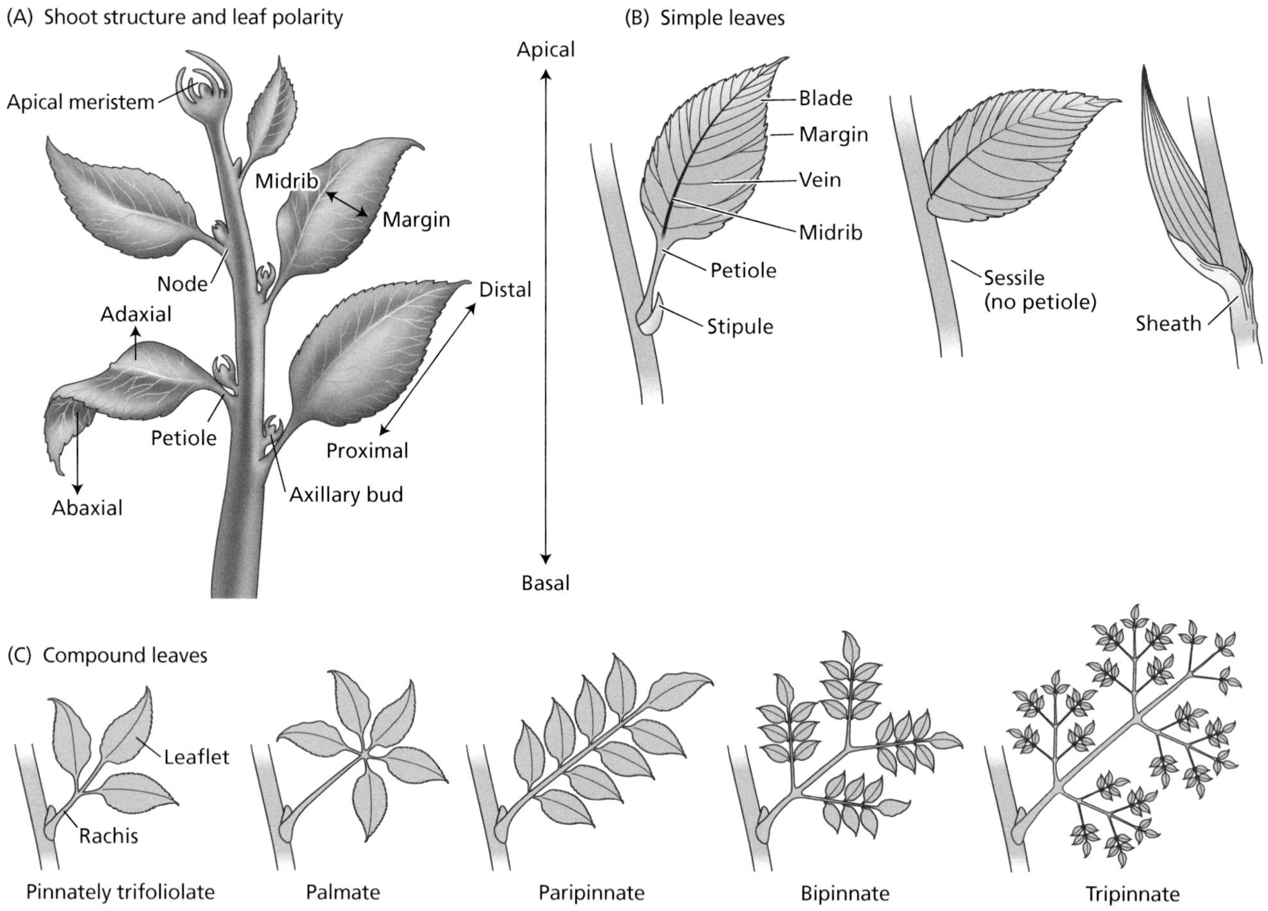

Figure 15.5 Overview of leaf structure. (A) Shoot structure, showing three types of leaf polarity: adaxial–abaxial, distal–proximal, and midrib–margin. (B) Examples of simple leaves. Variations in lower leaf structure include the presence or absence of stipules and petioles, and leaf sheaths. (C) Examples of compound leaves.

phyllotaxy (phyllotaxis) The arrangement of leaves on the stem.

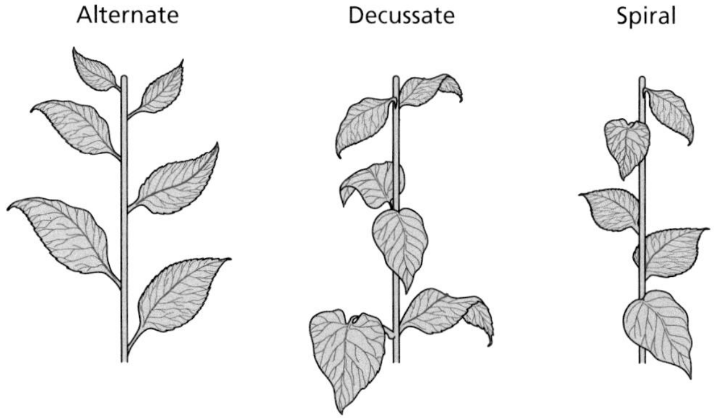

Figure 15.6 Three types of leaf arrangements (phyllotactic patterns) along the shoot axis. The three basic types of leaf arrangement (phyllotaxy) on the stem are alternate, decussate, and spiral. These patterns have their origin in the pattern of leaf primordia formation on the flanks of the SAM.

The arrangement of leaves around the shoot is called phyllotaxy

A long-standing question in plant biology is how a species' characteristic arrangement of leaves around the shoot, or **phyllotaxy**, is achieved. Three basic phyllotactic patterns, termed alternate, decussate (opposite), and spiral (**Figure 15.6**), can be directly linked to the pattern of initiation of leaf primordia on the SAM. These patterns depend on several factors, including intrinsic factors that tend to produce a phyllotaxy that is characteristic of a species. However, environmental factors or mutations that lead to changes in meristem size or shape can also affect phyllotaxy, suggesting that position-dependent mechanisms play important roles.

Auxin-dependent patterning of the shoot apex begins during embryogenesis

All leaves and modified leaves begin as small protuberances, called primordia, on the flanks of the SAM. The primordia then elongate, which establishes the proximal–distal

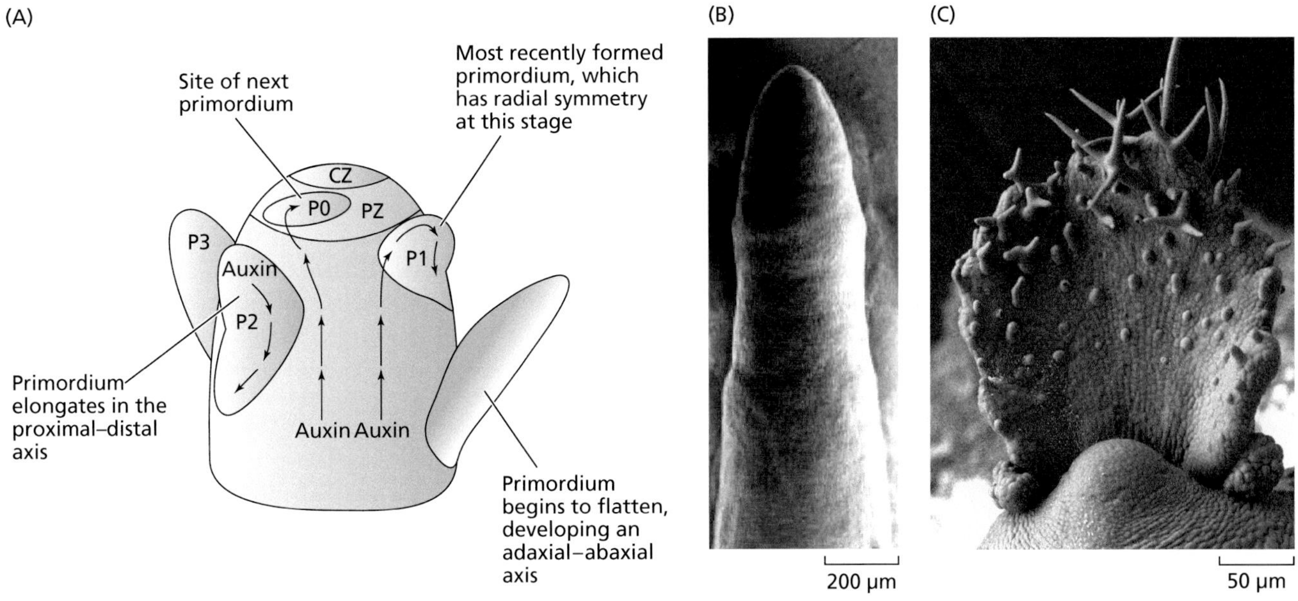

Figure 15.7 Sites of leaf formation are related to patterns of polar auxin transport. (A) Sites of leaf formation are related to patterns of polar auxin transport. P0, P1, P2, and P3 refer to the ages of leaf primordia; P0 corresponds to the stage at which the leaf begins its overt development, and P1, P2, and P3 represent increasingly older leaves. Leaf primordia are initiated where auxin accumulates. Acropetal (toward the tip) movement of auxin is blocked at the boundary separating the central and peripheral zones (CZ and PZ, respectively), leading to increased auxin levels at this position and the initiation of a leaf (P0). The newly formed leaf primordium (P1) acts as an auxin sink, thus preventing initiation of new leaves directly above it. The displacement of a more mature leaf primordium (P2) away from the PZ allows acropetal (toward the shoot tip) auxin movements to become reestablished, thus enabling the initiation of another leaf. (B) Scanning electron micrograph of a *pin1* inflorescence meristem that fails to produce leaf primordia. (C) Leaf primordium induced on the inflorescence meristem of a *pin1* mutant by placing a microdrop of IAA (indole-3-acetic acid, an auxin) in lanolin paste on the side of the meristem.

leaf axis, and flatten, which defines the adaxial and abaxial leaf surfaces. As the leaf develops, the palisade parenchyma, mesophyll, and vascular tissues differentiate. Gene expression and hormonal signaling contribute to all these processes.

The sites of leaf initiation, which ultimately give rise to the phyllotaxy of the plant, correspond to localized zones of auxin accumulation, suggesting that auxin accumulation is required for leaf initiation (**Figure 15.7A**). Support for this hypothesis comes from several experimental approaches. For example, *pin1* mutants of Arabidopsis, which are defective in polar auxin transport, fail to produce any leaf primordia on their inflorescence meristems (**Figure 15.7B**). However, application of exogenous auxin to the flank of the mutant SAM results in the production of a leaf primordium (**Figure 15.7C**). Further support for the role of auxin accumulation in leaf initiation is provided by the changes in phyllotactic patterns induced by treatment with auxin transport inhibitors.

15.4 Differentiation of Epidermal Cell Types

Identify the key components that regulate the differentiation of the leaf epidermis into specialized structures such as stomata.

In addition to the palisade parenchyma and spongy mesophyll, which are specialized for photosynthesis and gas exchange, the leaf has an epidermis that also plays vital roles in leaf function. The epidermis is the outermost layer of cells on the primary plant body, including both vegetative and

protoderm In the plant embryo and in apical meristems, the one-cell-thick surface layer that covers the young shoot and radicle of the embryo and gives rise to the epidermis.

pavement cells The predominant type of leaf epidermal cells, which secrete a waxy cuticle and serve to protect the plant from dehydration and damage from ultraviolet radiation.

trichomes Unicellular or multicellular hairlike structures that differentiate from the epidermal cells of shoots and roots. Trichomes may be structural or glandular and function in plant responses to biotic or abiotic environmental factors.

guard cells A pair of specialized epidermal cells that surround the stomatal pore and regulate its opening and closing.

reproductive structures. The epidermis usually consists of a single layer of cells derived from meristem cells known as the **protoderm**. In some plants, such as members of the Moraceae and certain species of the Begoniaceae and Piperaceae, the epidermis has two to several cell layers derived from periclinal divisions of the protoderm.

There are three main types of shoot epidermal cells found in all angiosperms: pavement cells, trichomes, and guard cells. **Pavement cells** are relatively unspecialized epidermal cells that can be regarded as the default developmental fate of the protoderm. **Trichomes** are unicellular or multicellular extensions of the shoot epidermis that take on diverse forms, structures, and functions, including protection against insect and pathogen attack, reduction of water loss, and increased tolerance of abiotic stress conditions. **Guard cells** are pairs of cells that surround the **stomata**, or pores, that are present in the photosynthetic parts of the shoot. Guard cells regulate gas exchange between the leaf and the atmosphere by undergoing tightly regulated turgor changes in response to light and other factors (Chapters 3, 6, 12, and 13). Other types of epidermal cells, such as *lithocysts*, *bulliform cells*, *silica cells*, and *cork cells* (**Figure 15.8**), play ecological roles in plant defense and drought tolerance in certain groups of plants.

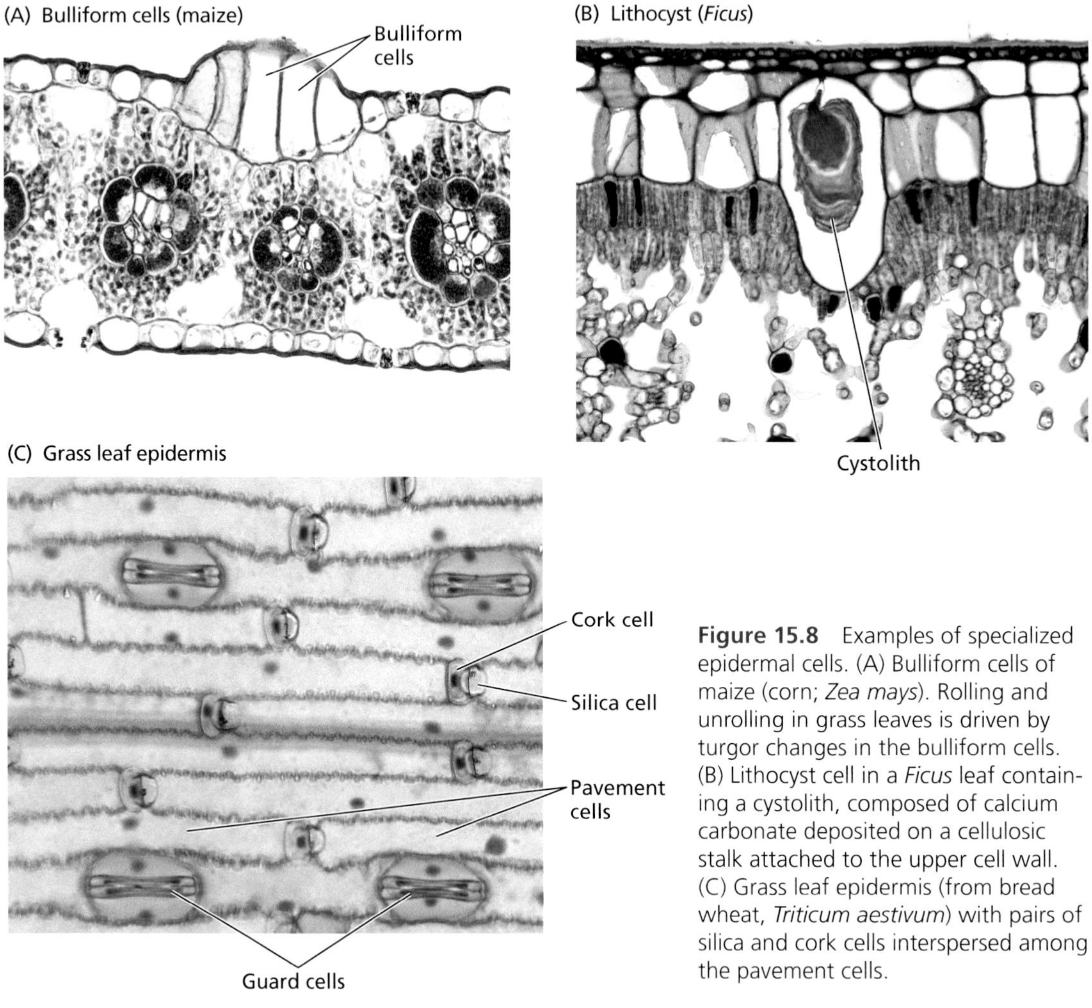

Figure 15.8 Examples of specialized epidermal cells. (A) Bulliform cells of maize (corn; *Zea mays*). Rolling and unrolling in grass leaves is driven by turgor changes in the bulliform cells. (B) Lithocyst cell in a *Ficus* leaf containing a cystolith, composed of calcium carbonate deposited on a cellulosic stalk attached to the upper cell wall. (C) Grass leaf epidermis (from bread wheat, *Triticum aestivum*) with pairs of silica and cork cells interspersed among the pavement cells.

A specialized epidermal lineage produces guard cells

In eudicots, only specific cells in a stomatal cell lineage differentiate into guard cells (**Figure 15.9**). In the developing protoderm (which gives rise to the leaf epidermis), a population of **meristemoid mother cells (MMCs)** is established. Each MMC divides asymmetrically to give rise to two morphologically distinct daughter cells—a larger stomatal lineage ground cell (SLGC) and a smaller **meristemoid**. An SLGC can either differentiate into a pavement cell or become an MMC and found secondary or satellite lineages. The meristemoid can undergo a variable number of asymmetric amplifying divisions giving rise to as many as three SLGCs, with the meristemoid ultimately differentiating into a **guard mother cell (GMC)**, which is recognizable because of its rounded morphology. The GMC then undergoes one symmetric division, forming a pair of guard cells surrounding the stomatal pore.

Stomatal densities are sensitive to abiotic conditions, especially conditions that promote leaf water deficit, such as soil water deficit, salinity, extended light exposure, and low CO_2 concentrations. Typically, moderate water deficit is associated with an increase in stomatal density, whereas severe drought is associated with a decrease in stomatal density. All other things being equal, an increase in stomatal density would be expected to *increase* both leaf CO_2 uptake and water loss. The fact that the stomatal density increases suggests that plants have evolved to enhance CO_2 fixation rather than reduce transpiration under moderate drought conditions. Under severe drought conditions a different set of signals are used to restrict transpiration by reducing stomatal density.

stomata (singular *stoma* or *stomate*) Microscopic pores in the leaf epidermis, each surrounded by a pair of guard cells and, in some species, also including subsidiary cells. Stomata regulate the gas exchange (water and CO_2) of leaves by controlling the dimension of a stomatal pore.

meristemoid mother cells (MMCs) The cells of the leaf protoderm that divide asymmetrically (the so-called entry division) to give rise to the meristemoid, a guard cell precursor.

meristemoid A small, triangular stomatal precursor cell that functions temporarily as an initial cell in a meristem.

guard mother cell (GMC) The cell that gives rise to a pair of guard cells to form a stomate.

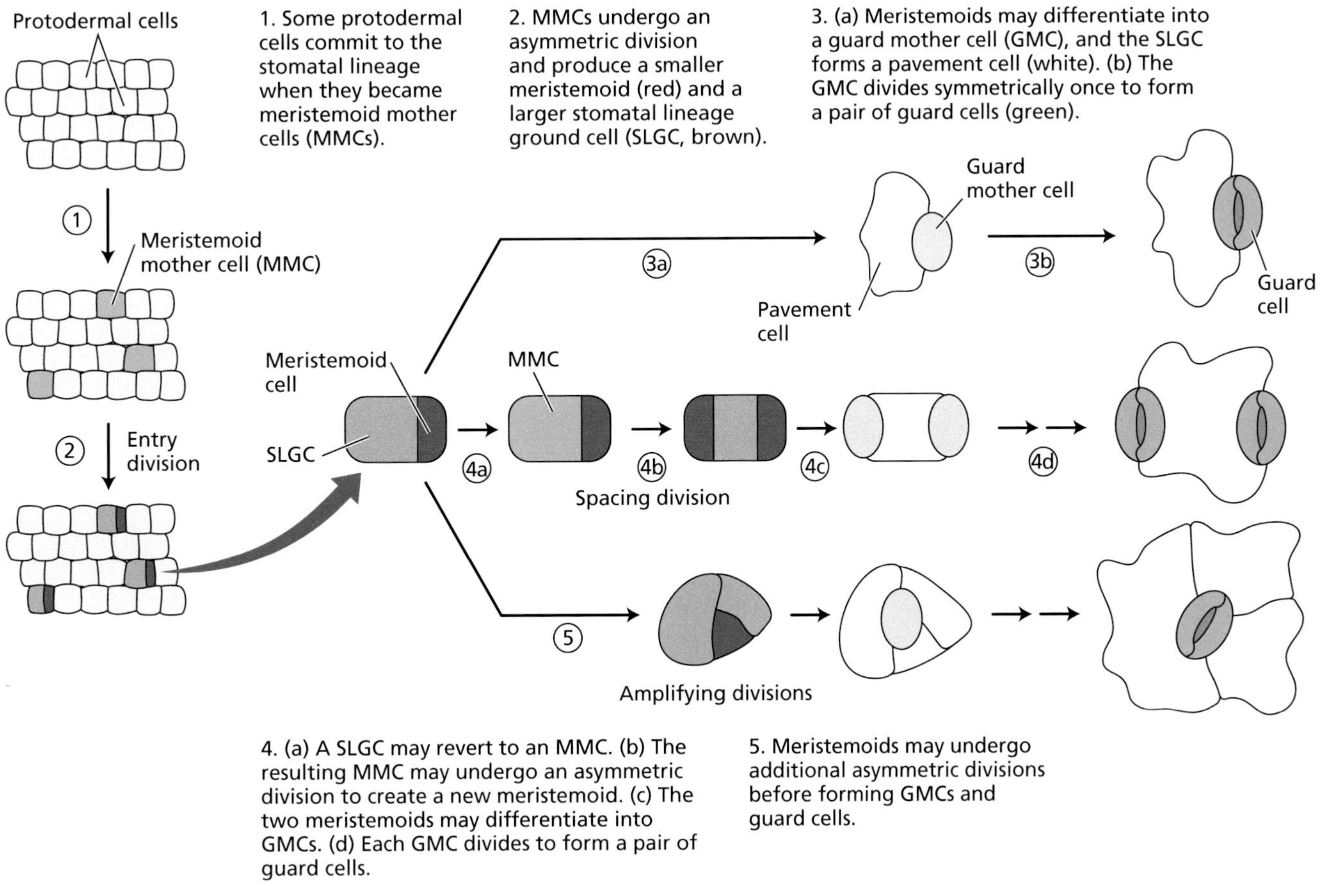
Protodermal cells
1. Some protodermal cells commit to the stomatal lineage when they became meristemoid mother cells (MMCs).
2. MMCs undergo an asymmetric division and produce a smaller meristemoid (red) and a larger stomatal lineage ground cell (SLGC, brown).
3. (a) Meristemoids may differentiate into a guard mother cell (GMC), and the SLGC forms a pavement cell (white). (b) The GMC divides symmetrically once to form a pair of guard cells (green).
1
Meristemoid mother cell (MMC)
2
Entry division
Meristemoid cell
SLGC
3a
3b
Guard mother cell
Pavement cell
Guard cell
MMC
4a
4b
4c
4d
Spacing division
5
Amplifying divisions
4. (a) A SLGC may revert to an MMC. (b) The resulting MMC may undergo an asymmetric division to create a new meristemoid. (c) The two meristemoids may differentiate into GMCs. (d) Each GMC divides to form a pair of guard cells.
5. Meristemoids may undergo additional asymmetric divisions before forming GMCs and guard cells.

Figure 15.9 Stomatal development in Arabidopsis.

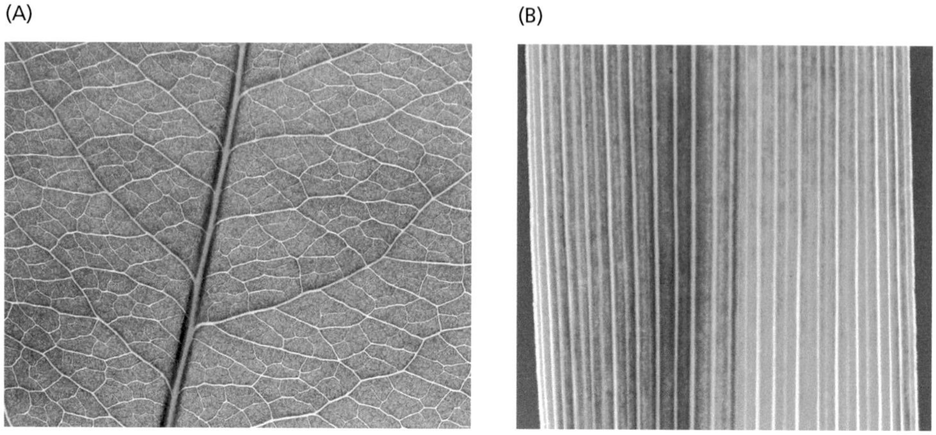

Figure 15.10 Two basic patterns of leaf venation in angiosperms. (A) Reticulate venation in *Prunus serotina*, a eudicot. (B) Parallel venation in *Iris sibirica*, a monocot.

15.5 Venation Patterns in Leaves

Describe leaf venation and the role that auxin plays in the formation of vascular tissues.

venation pattern The pattern of veins of a leaf.

bulk flow Translocation of water and solutes down a pressure gradient, as in the xylem or phloem.

phloem loading The movement of photosynthetic products into the sieve elements of mature leaves. *See also* phloem unloading.

procambium Primary meristematic tissue that differentiates into xylem, phloem, and cambium.

leaf trace The portion of the shoot primary vascular system that diverges into a leaf.

The leaf vascular system is a complex network of interconnecting veins consisting of two main conducting tissue types, xylem and phloem, as well as nonconducting cells, such as parenchyma, sclerenchyma, and fibers. The spatial organization of the leaf vascular system—its **venation pattern**—is both species- and organ-specific. Venation patterns fall into two broad categories: *reticulate venation,* found in most eudicots, and *parallel venation,* typical of many monocots (**Figure 15.10**).

Despite the diversity of leaf venation patterns, they all share a hierarchical organization. Veins are organized into distinct size classes—primary, secondary, tertiary, and so on—based on their width at the point of attachment to the parent vein (**Figure 15.11**). The smallest minor veins, called veinlets, terminate blindly in the mesophyll. The hierarchical structure of the leaf vascular system reflects the hierarchical functions of different-sized veins, with larger-diameter veins functioning in the **bulk flow** of water, minerals, sugars, and other metabolites, and smaller-diameter veins functioning in **phloem loading** (Chapter 10).

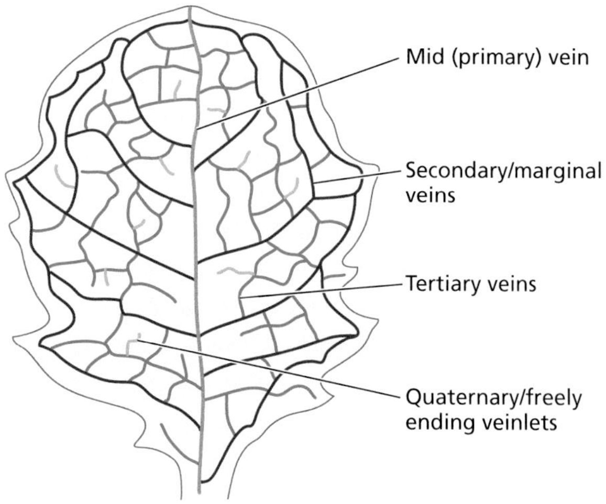

Figure 15.11 Hierarchy of venation in the mature Arabidopsis leaf based on the diameter of the veins at the site of attachment to the parent vein.

The primary leaf vein is initiated in the leaf primordium

Leaf vascular bundles arise from vascular precursor cells called the **procambium** in the emerging leaf primordia of the SAM (**Figure 15.12A**). From there the vascular bundles differentiate downward (basipetally) toward the node directly below the leaf and form a connection to the older vascular bundles that are continuous to the base of the shoot. The portion of the vascular bundle that enters the leaf is called the **leaf trace** (**Figure 15.12B**).

Auxin canalization initiates development of the leaf trace

Several lines of evidence indicate that auxin stimulates formation of vascular tissues. An example is the role of auxin in regeneration of vascular tissue after wounding (**Figure 15.13A**). Vascular regeneration is prevented by removal of the leaf and shoot above the

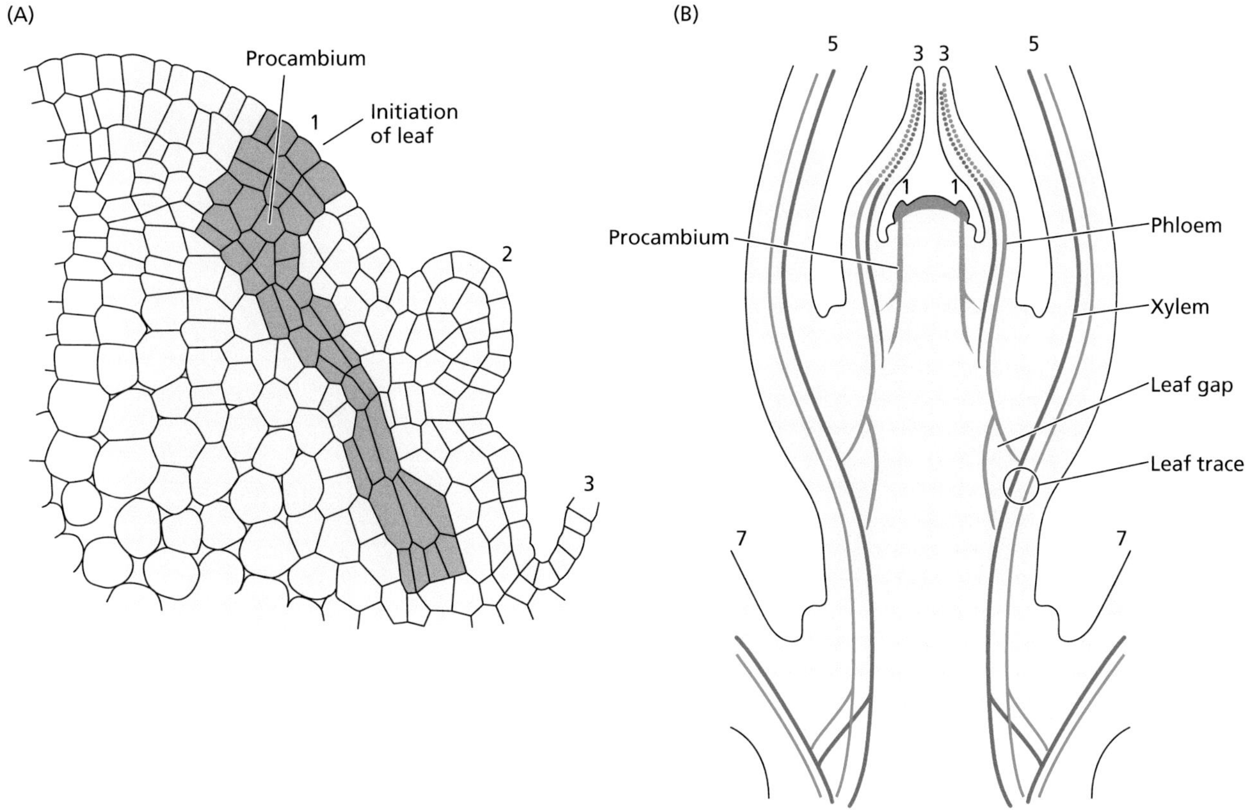

Figure 15.12 Development of the shoot vascular system. (A) Longitudinal section through the shoot tip of perennial flax (*Linum perenne*), showing the early stage in the differentiation of the leaf trace procambium at the site of a future leaf primordium. The leaf primordia and leaves are numbered, beginning with the youngest initial. (B) Early vascular development in a shoot with decussate phyllotaxy (Figure 15.6). The dark green area at the tip indicates the SAM and two young leaf primordia; procambial strands are shown in orange. Dotted lines in the pair of leaves labeled "3" represent developing xylem and phloem. Leaf traces develop basipetally to the mature vascular tissue below to form a continuous vascular bundle. Numbers correspond to the leaf order, starting with the primordia (missing leaves are in a different plane).

wound but can be restored by the application of auxin to the cut petiole above the wound, suggesting that auxin from the leaf is required for vascular regeneration. As shown in **Figure 15.13B**, the files of regenerating xylem elements originate at the source of auxin at the upper cut end of the vascular bundle, and progress basipetally until they reconnect with the cut end of the vascular bundle below, matching the presumed direction of the flow of auxin. The upper end of the cut vascular bundle thus acts as the **auxin source** and the lower cut end as the **auxin sink**.

These and similar observations in other systems, such as bud grafting, have led to the hypothesis that as auxin flows through tissues it stimulates and polarizes its own transport, which gradually becomes channeled—or canalized—into files of cells leading away from auxin sources; these cell files can then differentiate to form vascular tissue.

Consistent with this idea, local auxin application (as in the wounding experiments described in Figure 15.13) induces vascular differentiation in narrow strands leading away from the application site, rather than in broad fields of cells. New vasculature usually develops toward, and unites with, preexisting vascular strands, resulting in a connected vascular network. We would therefore predict that a developing leaf trace acts as an auxin source and the existing stem vasculature as an auxin sink. This mechanism of source/sink-based directional auxin fluxes amplified by the directional orientation of auxin efflux transporters is termed the **canalization model**. Experimental visualization of auxin transport proteins and

auxin source A cell or tissue that exports auxin to other cells or tissues by polar transport.

auxin sink A cell or tissue that takes up auxin from a nearby auxin source. Participates in auxin canalization during vascular differentiation.

canalization model The hypothesis that as auxin flows through tissues it stimulates and polarizes its own transport, which gradually becomes channeled—or canalized—into files of cells leading away from auxin sources; these cell files can then differentiate to form vascular tissue.

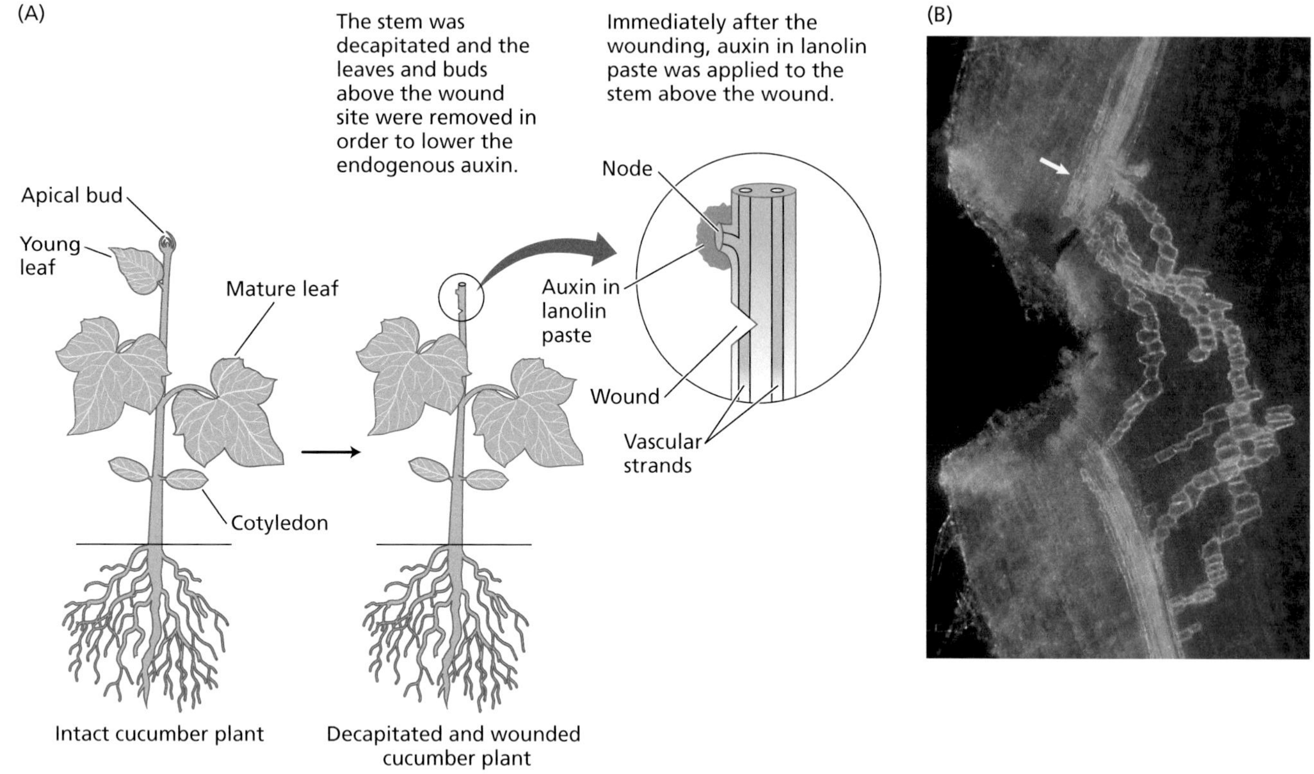
(A)
The stem was decapitated and the leaves and buds above the wound site were removed in order to lower the endogenous auxin.
Immediately after the wounding, auxin in lanolin paste was applied to the stem above the wound.
Apical bud
Young leaf
Mature leaf
Cotyledon
Node
Auxin in lanolin paste
Wound
Vascular strands
Intact cucumber plant
Decapitated and wounded cucumber plant
(B)

Figure 15.13 Auxin-induced xylem regeneration around a wound in cucumber (*Cucumis sativus*) stem tissue. (A) Method for carrying out the wound regeneration experiment. (B) Fluorescence micrograph showing regenerating vascular tissue (which is fluorescing orange) around the wound. The arrow indicates the wound site where auxin accumulates; xylem differentiation begins where auxin is perceived at the adjoining cell surface.

dynamic changes in auxin concentrations during leaf vein development suggest that both polar auxin transport and localized auxin biosynthesis are involved in vascularization during early leaf development.

15.6 Shoot Branching and Architecture

Differentiate among the patterns of shoot branching and the factors that determine their architecture.

Shoot architecture in seed plants is characterized by multiple repetitions of a basic module called the **phytomer**, which consists of an internode, a node, a leaf, and an **axillary meristem** (**Figure 15.14**). During evolution, modification of the position, size, and shape of the individual phytomer, and variations in the regulation of axillary bud outgrowth, provided the morphological basis for the remarkable diversity of shoot architecture among seed plants. Vegetative and inflorescence branches, as well as the floral primordia produced by inflorescences, are derived from axillary meristems initiated in the axils of leaves. During vegetative development, axillary meristems, like the apical meristem, initiate the formation of leaf primordia, resulting in axillary buds. These buds either become dormant or develop into lateral shoots depending on their position along the shoot axis, the developmental stage of the plant, and environmental factors. During reproductive development, axillary meristems initiate formation of inflorescence branches and flowers. Hence, the growth habit of a plant depends not only on the patterns of axillary meristem formation but also on meristem identity and its subsequent growth characteristics.

phytomer A developmental unit consisting of one or more leaves, the node to which the leaves are attached, the internode below the node, and one or more axillary buds.

axillary meristem Meristematic tissue in the axils of leaves that gives rise to axillary buds.

Auxin, cytokinins, and strigolactones regulate axillary bud outgrowth

Once the axillary meristems are formed, they may enter a phase of highly restricted growth (dormancy), or they may be released to grow into axillary branches. The "grow or don't grow" decision is determined by developmental programming and environmental responses mediated by plant hormones that act as local and long-distance signals. Interactions of hormonal signaling pathways coordinate the relative growth rates of different branches and the shoot tip, which ultimately determine shoot architecture. The main hormones involved are auxin, cytokinins, and strigolactones (Chapter 12). All three hormones are produced in varying quantities in the root and shoot, but translocation of the hormones allows them to exert effects beyond their sites of synthesis.

The role of auxin in regulating axillary bud growth is most easily demonstrated in experiments on **apical dominance**—the control exerted by the shoot tip over axillary buds and branches below. Auxin synthesized at the shoot apex is transported in a rootward polar stream (Chapters 12 and 14). Plants with strong apical dominance are typically weakly branched and show a strong branching response to decapitation (removal of the growing or expanding leaves and shoot tip) (**Figure 15.15**). Applying auxin to the decapitated stem restores apical dominance. Plants with weak apical dominance are typically highly branched and show little, if any, response to decapitation. More than a century of experimental evidence suggests that, in plants with strong apical dominance, auxin produced in the shoot tip inhibits axillary bud outgrowth. Gardeners take advantage of this phenomenon when they "pinch back" chrysanthemums and many other plants with strong apical dominance to create dense, dome-shaped bushes of blossoms.

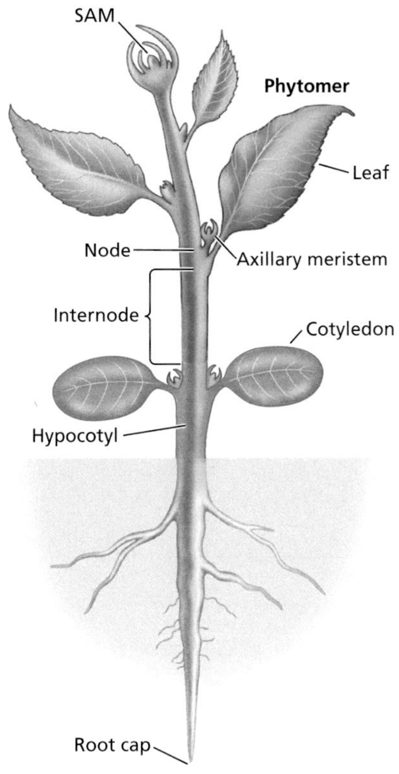

Figure 15.14 The phytomer is the basic module of shoot organization in seed plants.

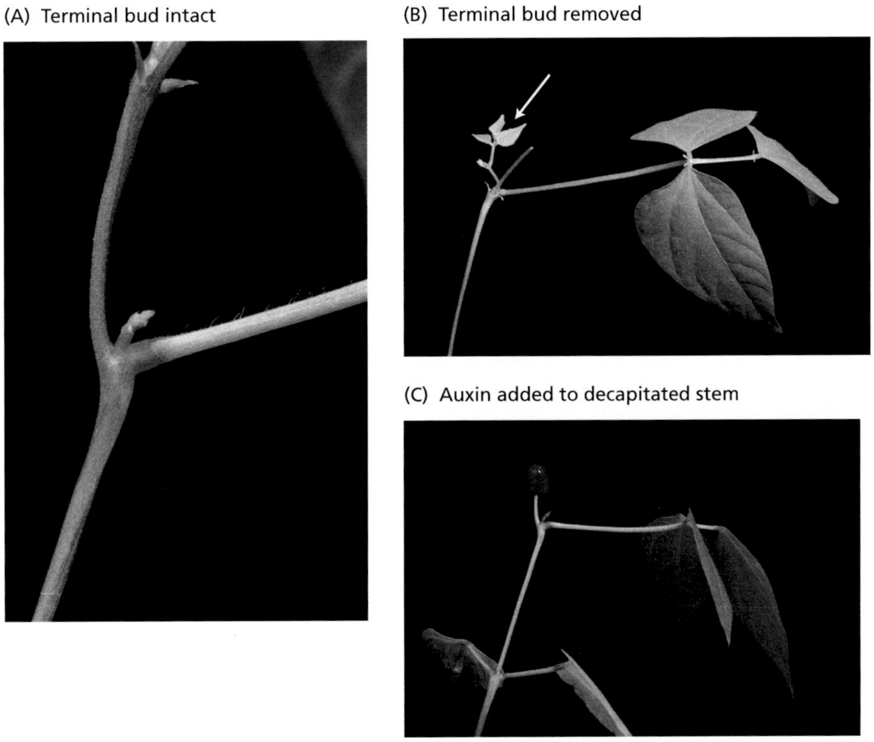

Figure 15.15 Auxin suppresses the growth of axillary buds in bean (*Phaseolus vulgaris*) plants. (A) The axillary buds are suppressed in the intact plant because of apical dominance. (B) Removal of the terminal bud releases the axillary buds from apical dominance (arrow). (C) Applying auxin in lanolin paste (contained in the red gelatin capsule) to the cut surface prevents the outgrowth of the axillary buds.

apical dominance In most higher plants, the growing apical bud's inhibition of the growth of lateral buds (axillary buds).

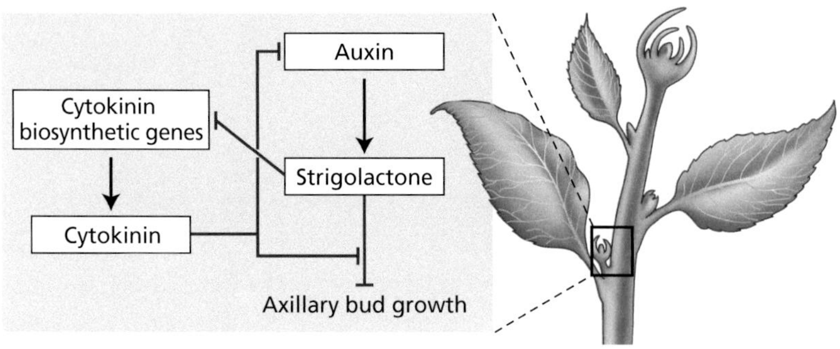

Figure 15.16 Hormonal network regulating apical dominance. Auxin from the shoot apex promotes strigolactone synthesis in the nodal region. In eudicots, strigolactone inhibits axillary bud growth and down-regulates the cytokinin biosynthetic genes. Because cytokinin blocks the inhibitory activity of strigolactone on axillary bud growth, reducing cytokinin activity enhances apical dominance.

Strigolactones act in conjunction with auxin to regulate apical dominance. Mutants defective in either strigolactone biosynthesis or signaling show increased branching without decapitation. Although strigolactone is synthesized in both the shoot and the root, grafting studies have shown that only shoot-derived strigolactone is required for apical dominance. Direct application of cytokinin to axillary buds, or stimulation of cytokinin production, promotes axillary bud growth, suggesting that cytokinins are involved in breaking apical dominance.

A simplified model for the antagonistic interactions between cytokinin and strigolactone is shown in **Figure 15.16**. Auxin maintains apical dominance by stimulating strigolactone synthesis. In eudicots, strigolactone then suppresses axillary bud growth. Strigolactone also inhibits the synthesis of cytokinin by repressing expression of cytokinin biosynthetic genes, which helps suppress axillary bud growth.

The initial signal for axillary bud growth may be an increase in sucrose availability to the bud

Recent evidence indicates that sucrose itself may serve as the initial signal in controlling bud outgrowth (**Figure 15.17**). In pea plants, axillary bud growth is initiated approximately 2.5 h after decapitation. This is 24 h prior to any detectable decline in the auxin concentration in the stem adjacent to the axillary bud, which suggests that a decrease in auxin from the tip occurs too slowly to initiate bud outgrowth.

In contrast, studies using [^{14}C]sucrose demonstrated that the concentration of leaf-derived sucrose in the stem adjacent to the bud begins to decline as early as 2 h after decapitation. This decline of sucrose in the stem is due to the uptake

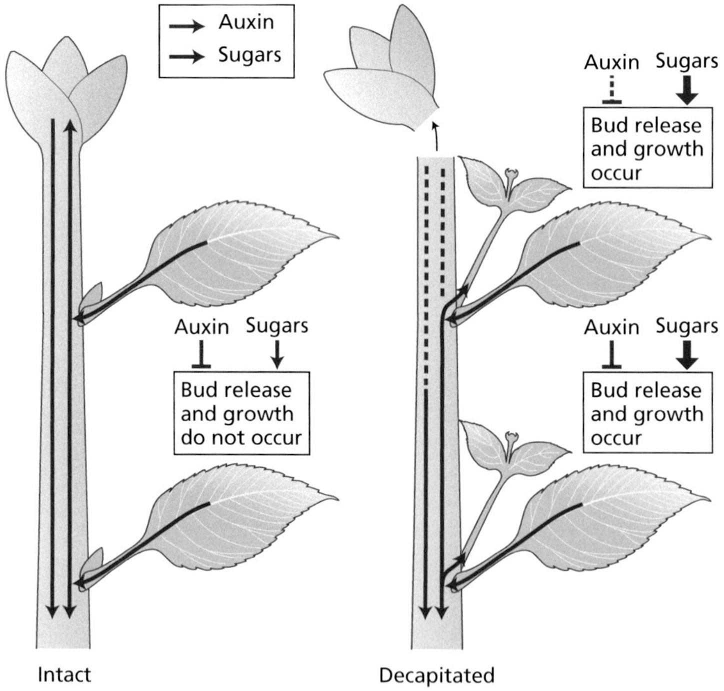

Figure 15.17 Apical dominance is regulated by sugar availability. After decapitation, sugars, which normally flow toward the shoot tip via the phloem, rapidly accumulate in axillary buds, stimulating bud outgrowth. At the same time, the loss of the apical supply of auxin results in a depletion of auxin in the stem. However, auxin depletion is relatively slow, and therefore the growing buds in the upper shoot are affected before those lower on the stem. In this model, auxin is involved predominately in the later stages of branch growth.

of sugars by the axillary bud. Thus, following decapitation, bud outgrowth is initiated *prior to* auxin depletion but *after* sucrose uptake by the axillary bud. In the intact stem, the axillary bud is starved for sucrose because the growing terminal bud is a stronger sink for sugars than the axillary bud. As a result of decapitation, the carbon supply to the axillary buds increases within the time frame sufficient to induce bud release. Apical dominance is thus regulated by the strong sink activity of the growing tip, which limits sugar availability to the axillary buds. However, sustained bud outgrowth requires the depletion of auxin in the stem adjacent to the bud as well.

shade avoidance The increased rate of stem elongation produced in certain plants in response to exposure to shade produced by green leaves.

15.7 Shade Avoidance

Describe the shade avoidance response and the signaling pathways involved.

Plants compete for sunlight to maintain their photosynthetic activity. Faster-growing plants dominate the canopy and thus receive more direct sunlight than slower-growing plants in the shade beneath the canopy. As we saw in Chapter 9, the relative spectral distribution of sunlight at the top of the canopy is qualitatively different from that of the sunlight that penetrates the canopy. **Table 15.1** compares the total fluence rate (related to light intensity) in photons (400–800 nm) and the ratio of red (R) to far-red (FR) light in eight natural conditions and environments. Compared with unfiltered daylight, there is proportionally more far-red light during sunset, under 5 mm of soil, and under the canopy of other plants (as on the floor of a forest). The reduction in the R:FR ratio results from the fact that green leaves absorb red light because of their high chlorophyll content, but are relatively transparent to far-red light. Exposure of certain plants to the shade of other plants increases the rate of stem elongation, a phenomenon termed **shade avoidance**. The response is specific for shade produced by green leaves, which act as filters for red and blue light, and is not induced by other types of shade. Plants that increase stem extension in response to shading are called "sun plants."

The photoreceptor that controls shade avoidance in sun plants is phytochrome, which has two interconvertible forms: One is inactive and absorbs red light (Pr), and the other is active and absorbs far-red light (Pfr) (Chapter 13). As shading increases, the R:FR ratio decreases. A higher proportion of far-red light converts more Pfr to Pr, and the ratio of Pfr to total phytochrome (Pfr:P_{total}) decreases. When sun plants (plants adapted to growing in direct sunlight) are grown in natural light with shade cloths to control R:FR, stem extension rates increase in response to a higher far-red content (i.e., a lower Pfr:P_{total}) (**Figure 15.18**). In other words, simulated canopy shading (high levels of far-red light, low Pfr:P_{total}) induces

Table 15.1 Ecologically Important Light Parameters

	Fluence rate (μmol $m^{-2}s^{-1}$)	R:FR[a]
Daylight	1900	1.19
Sunset	26.5	0.96
Moonlight	0.005	0.94
Beneath ivy canopy	17.7	0.13
Soil, at a depth of 5 mm	8.6	0.88
Lakes, at a depth of 1 m		
Black Loch	680	17.2
Loch Leven	300	3.1
Loch Borralie	1200	1.2

Source: Smith 1982, p. 493.

Note: The light intensity factor (400–800 nm) is given as the photon fluence rate, and phytochrome-active light is given as the R:FR ratio.

[a] Absolute values taken from spectroradiometer scans; the values should be taken to indicate the relationships between the various natural conditions and not as actual environmental means.

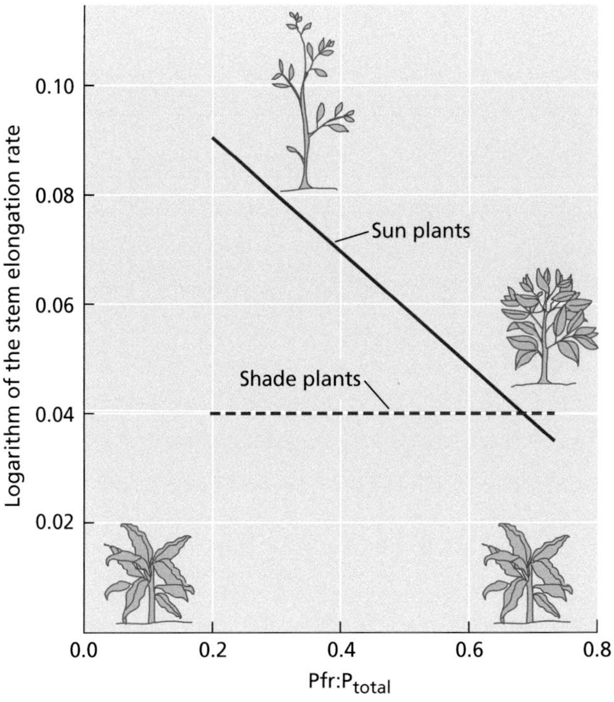

Figure 15.18 Phytochrome plays a predominant role in controlling stem elongation rate in sun plants (solid line) but not in shade plants (dashed line).

these plants to allocate more of their resources to growing taller. This correlation is absent for "shade plants," which are adapted to grow under a leaf canopy. Shade plants exhibit almost no reduction in their stem extension rate when they are exposed to higher R:FR values. Thus, there appears to be a systematic relationship between phytochrome-controlled growth and species habitat. Such results are taken as an indication of the involvement of phytochrome in shade perception.

For a sun plant or "shade-avoiding plant," there is a clear adaptive value in allocating its resources toward more rapid extension growth when it is shaded by another plant. The shade avoidance response enhances its chances of growing above the canopy and acquiring a greater share of unfiltered, photosynthetically active light. The price for increased internode elongation is usually reduced leaf area and reduced branching, but at least in the short term, this adaptation to canopy shade increases plant fitness. When the plant grows above the canopy or a canopy gap occurs when a tree falls in the forest, the plant is released from shade avoidance and reverts to it normal growth habit.

Reducing shade avoidance responses can improve crop yields

Shade avoidance responses may be highly adaptive in a natural setting to help plants outcompete neighboring vegetation. But for many crop species, a reallocation of resources from reproductive to vegetative growth can reduce crop yield. In recent years, yield gains in crops such as maize (corn; *Zea mays*) have come largely through the breeding of new varieties with a higher tolerance to crowding (which induces shade avoidance responses) rather than through increases in basic yield per plant. As a consequence, today's maize crops can be grown at higher densities than older varieties without suffering decreases in plant yield (**Figure 15.19**).

(A)

(B)

Figure 15.19 High-density planting and crop yield. (A) Maize field from the 1930s. (B) Modern maize varieties are planted at high density. Traditionally, Native Americans grew maize on small hills or mounds, which were separated by more than a meter. The three sisters, maize, beans, squash were co-planted in the mounds. The plants were short and often produced multiple small ears. In contrast, modern hybrids are machine-planted in dense rows with little space between them (typically 74,000–94,000 plants per hectare). Although yield per plant has not increased dramatically for many years in commercial hybrids, overall yields have continued to increase, largely because of better performance of plants at high planting density. As shown in this image from upstate New York, modern varieties of maize have upright leaves that help the plants capture sunlight energy under crowded conditions.

15.8 Root Branching and Architecture

Describe the different types of root system architecture, and explain how developmental and environmental factors control it.

Plant root systems, through natural selection, have evolved the ability to adapt to variations in soil composition, competition with other root systems, abiotic stress, and biotic interactions in the rhizosphere. Diversity and phenotypic plasticity are therefore key features of root architecture and are essential to the plant's survival. *Root system architecture* refers to the three-dimensional arrangement of individual roots within the plant's root system. Root systems are composed of different root types, and plant species differ in their ability to control the types of roots they produce as well as their growth rates, degree of branching, and growth angles. Post-embryonic roots arising from existing roots are termed *lateral* or *branch roots*, while roots arising from non-root tissues are termed *adventitious roots*. In general, monocots such as the grasses have fibrous root systems consisting mainly of adventitious roots that in turn produce abundant lateral roots. By contrast, eudicots such as Arabidopsis typically produce a single dominant primary root, or taproot, that gives rise to all of the lateral roots and their branches. Whereas in shoots, axillary bud primordia are periodically initiated on the flanks of apical meristems in precise patterns, lateral root primordia are initiated in the zone of differentiation of the growing primary root. The arrangement of lateral roots along the surface of a root is determined by a combination of endogenous and environmental factors. New cells are constantly produced at the growing tip, and the older cells continue to develop and mature (Figure 15.1). As a result, the youngest lateral root primordia are found just behind the growing tip while the oldest lateral roots are located toward the root base.

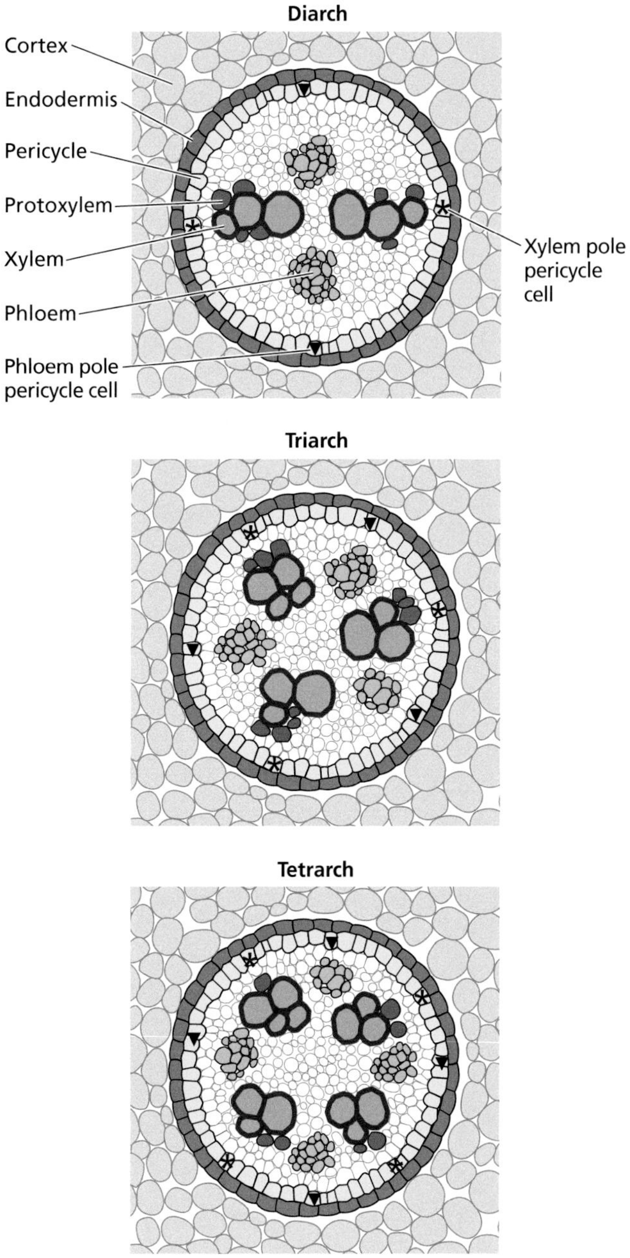

Figure 15.20 Examples of different protoxylem and protophloem arrangements in eudicot roots.

Lateral root primordia arise from initial cells in the pericycle

In Arabidopsis and most other eudicots and nongrass monocots, lateral root initiation occurs exclusively in the pericycle cells adjacent to xylem arms, while in the grasses (e.g., maize and barley) lateral root initiation is restricted to the pericycle cells adjacent to the phloem. The arrangement of vascular tissue in the root is characterized as diarch (as in Arabidopsis), triarch, tetrarch, and so on, depending on the number of xylem arms, which corresponds to the number of protoxylem strands (**Figure 15.20**). Because most of the research in this area has been conducted on Arabidopsis, we will focus on the development of lateral roots from pericycle initial cells adjacent to the xylem.

Lateral root formation can be divided into four distinct stages

Lateral root primordia arise in the pericycle from specialized initial cells adjacent to the xylem arm. These specialized initial cells only achieve full competence to differentiate

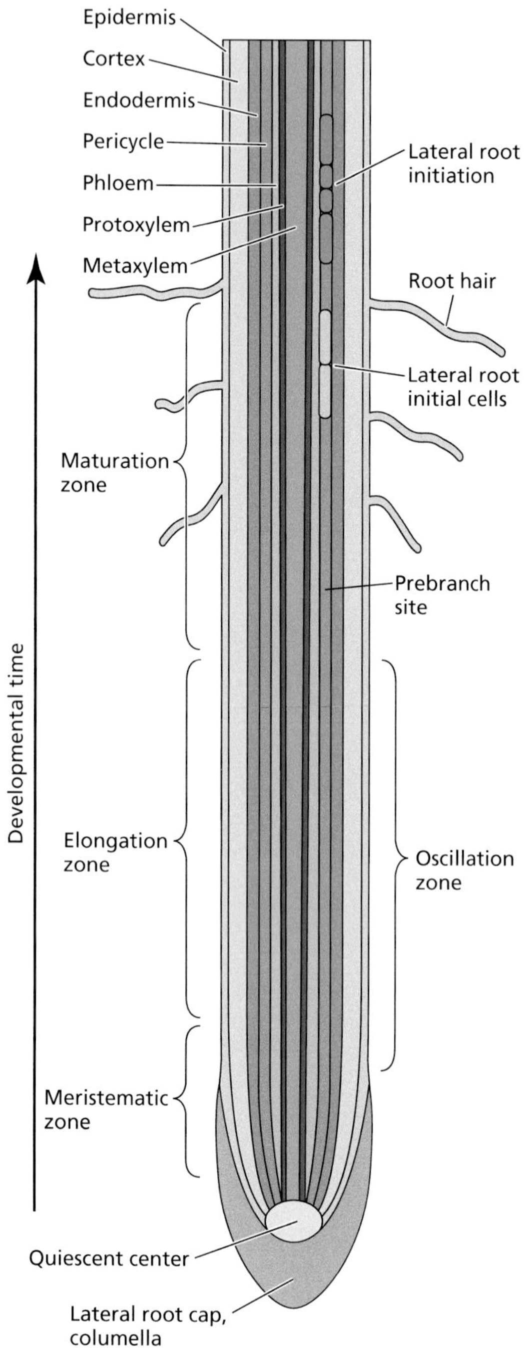

Figure 15.21 Lateral root development in Arabidopsis. Longitudinal section of the primary root showing its developmental zones. The prebranch site (where a branch root will develop if given the appropriate signals) forms after an oscillatory phase of auxin-responsive gene expression within the oscillation zone. Within the oscillation zone, xylem pole pericycle cells acquire competence to differentiate into lateral root founder cells (light blue). (Drawing not to scale.)

into lateral root cells as they enter the maturation zone of the root tip. The precise location where the initial cells become competent to form lateral root cells is called a prebranch site (**Figure 15.21**). It is here that some xylem pole pericycle cells undergo priming, which conditions them to become lateral root initial cells.

Lateral root formation can be divided into four distinct stages:

I. Formation of the prebranch sites along the root
II. Lateral root initiation
III. Formation of a lateral root primordium with a new root apical meristem
IV. Emergence of the new lateral root emergence through the cortex and epidermis of the root.

There is now abundant evidence suggesting that specification of prebranch sites requires a pulse of auxin. According to the **auxin clock model**, the spacing of prebranch sites at the root tip is regulated by periodic oscillations of auxin. According to the model, only those initial cells of the pericycle that are exposed to an auxin concentration maximum go on to form a prebranch site.

Plants can modify their root system architecture to optimize water and nutrient uptake

Root system architecture refers to the geometric arrangement of individual roots within the plant's root system in the three-dimensional soil space. Root systems are composed of different root types, and plants are able to modify and control the types of roots they produce, the root angles with respect to gravity, the rates of root growth, and the degree of branching. In addition to having a primary root system, some plants can produce additional roots from shoot tissues, called **adventitious roots**, as an adaptation to a particular environment or in response to nutritional or water stress. Variations in root system architecture within and across species have been linked to resource acquisition and growth. Root system architecture varies widely among species, even among those living in the same habitat (**Figure 15.22**).

auxin clock model A model for the spacing of prebranch sites in roots based on the periodic fluctuation of auxin activity or concentration in the oscillation zone.

adventitious roots Roots arising from any organ other than a root.

seminal roots Adventitious roots that arise during embryogenesis from the stem tissue between the scutellum and the coleoptile.

Monocots and eudicots differ in their root system architecture

Monocot and eudicot root systems are roughly similar in structure, consisting of an embryonically derived primary root (the radicle), lateral roots, and adventitious roots. However, there are significant differences between monocot and eudicot root systems. Monocot root systems are generally more fibrous and complex than the root systems of eudicots, especially in the cereals. For example, the maize seedling root system consists of a primary root that develops from the radicle, lateral roots, **seminal roots** (adventitious roots that branch from the hypocotyl

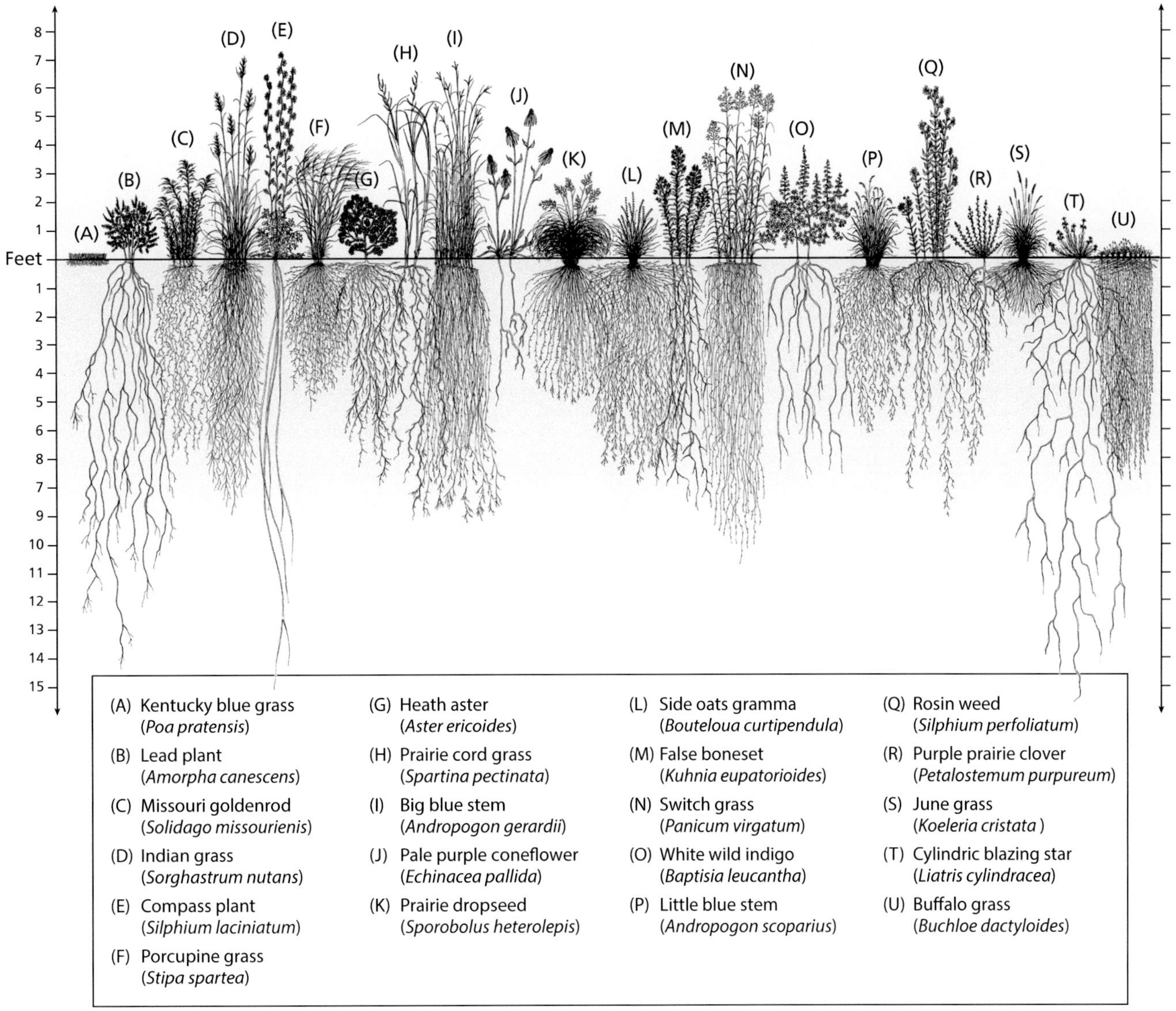

Figure 15.22 Diversity of root systems in prairie plants.

of the embryo), and post-embryonically derived **crown roots** and **brace roots** (**Figure 15.23**). The primary and seminal roots are highly branched and fibrous. The crown roots derive from the lowermost node of the stem. The brace roots, also called "prop roots," are adventitious roots that grow from the stem nodes. Although crown roots are relatively unimportant in seedlings (Figure 15.23A) in contrast to the primary and seminal roots, crown and brace roots continue to form, develop, and branch throughout vegetative growth. Thus, the crown and brace root system make up the major part of the root system in adult maize plants (Figure 15.23B).

The root system of a young eudicot consists of the primary root (or taproot) and its branch roots. As the root system matures, basal roots arise from the base of the taproot. In addition, adventitious roots can arise from subterranean stems or from the hypocotyl, and can be considered to be loosely analogous to the adventitious crown roots in the cereals. **Figure 15.24** depicts the root system of a soybean plant, a representative eudicot.

crown roots Adventitious roots that emerge from the lowermost nodes of a stem.

brace roots Roots derived from aerial nodes of some monocot plants that penetrate the soil to anchor large shoots.

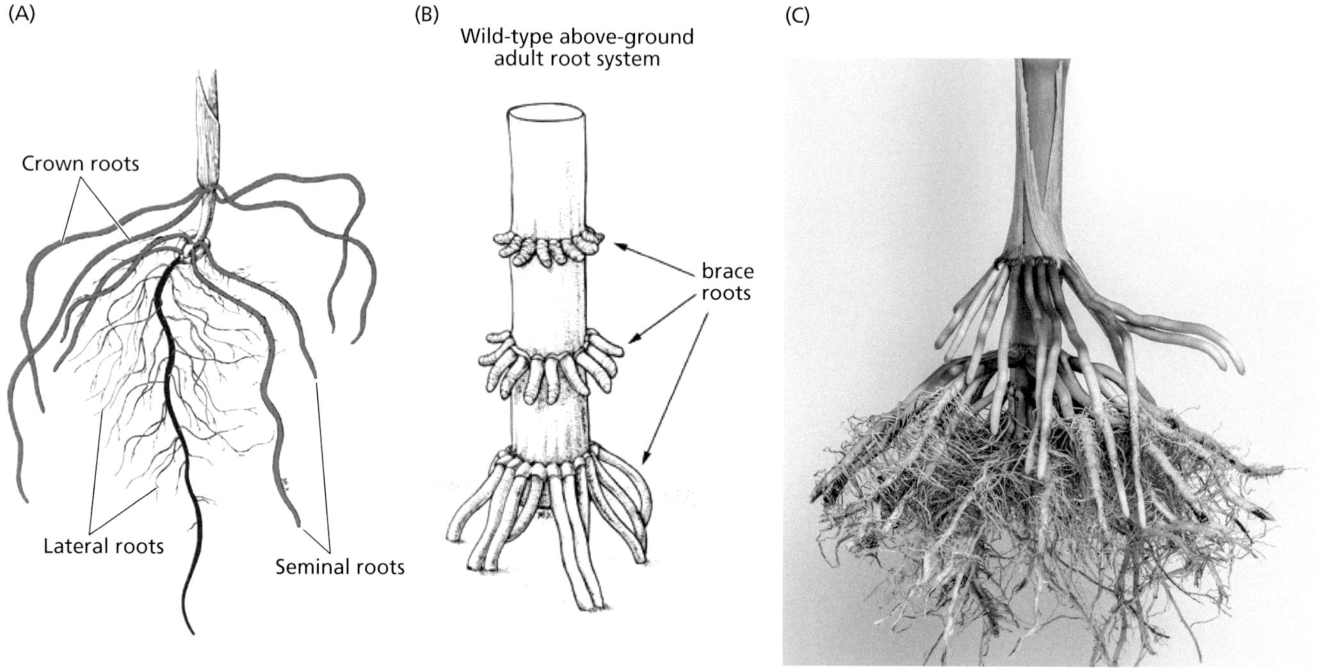

Figure 15.23 Root system of a 14-day-old maize seedling. (A) Root system of a 14-day-old maize seedling composed of the primary root derived from the embryonic radicle, the seminal roots derived from the scutellar node, post-embryonically formed crown roots that arise at nodes above the mesocotyl, and lateral roots. (B) Brace roots from the stem nodes of a 6-week-old plant. (C) Mature maize root system.

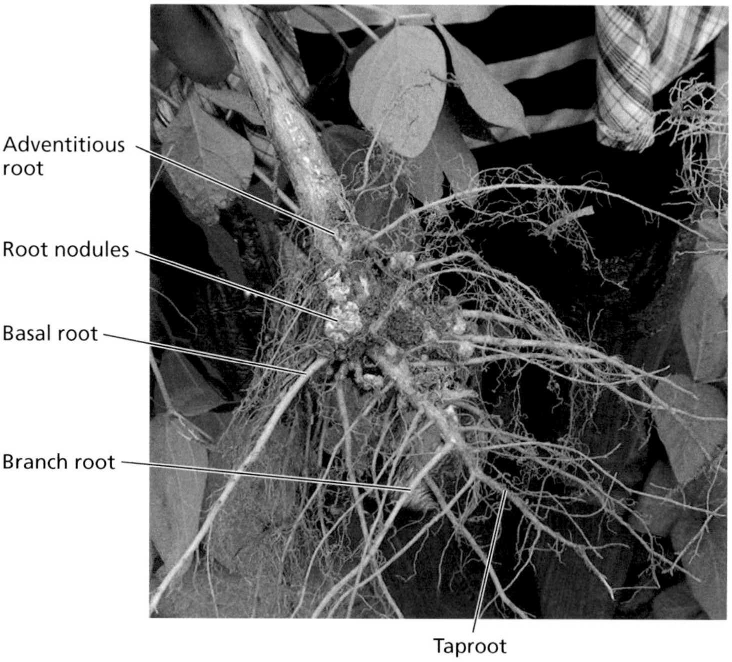

Figure 15.24 Soybean root system, showing the primary root (taproot), branch roots, basal roots, and adventitious roots.

Root system architecture changes in response to phosphorus deficiencies

Phosphorus is, along with nitrogen, the most limiting mineral nutrient for crop production (Chapters 4 and 5). Phosphorus limitation is a particular problem in tropical soils, where the highly weathered acidic soils tend to bind phosphorus tightly, rendering much of it unavailable for acquisition by roots. Plant root systems undergo well-documented morphological alterations in response to phosphorus deficiency (**Figure 15.25**). These responses can vary somewhat from species to species, but in general they include a reduction in primary root elongation, an increase in lateral root proliferation and elongation, and an increase in the number of root hairs.

As described in Chapter 4, soil phosphorus is bound tightly to clay particles or organic structures and is more enriched in the surface horizons (layers) of the soil. Phosphorus deficiency can trigger "topsoil foraging" by plants. For example, some bean varieties (phosphorus-efficient genotypes) respond to phosphorus deficiency by producing more adventitious lateral roots that grow at a more horizontal angle so that they are shallower. There is also an overall increase in the number of lateral roots and root hairs. These bean varieties have been used extensively to breed more phosphorus-efficient beans and soybeans for growth in low-phosphorus soils.

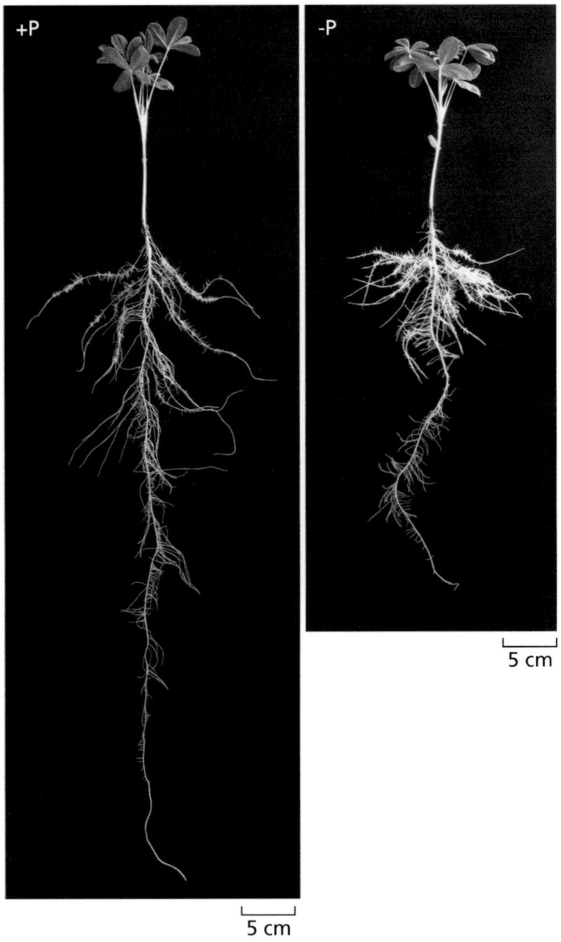

Figure 15.25 In response to phosphorus deficiency, plants may alter their root architecture in ways that increase their ability to forage for the nutrient. These white lupines (*Lupinus albus*) were grown hydroponically in a nutrient solution with (+P) or without (–P) phosphate. The phosphorus-deficient plant has a shallower root system and many more lateral roots in the upper part of the system. These lateral roots in turn are covered with short, densely packed lateral roots called cluster roots; these features increase the surface area of the root system in the upper layers of the soil. Other morphological responses of plants to phosphorus deficiency include the formation of longer, denser root hairs, an increased root:shoot ratio, and formation of aerenchyma (air spaces in the cortex).

15.9 Secondary Growth

Explain the process of secondary growth (radial growth and wood formation) and the cell types involved.

All gymnosperms and most eudicots—including large herbaceous plants, woody shrubs, and trees—develop a lateral meristem called the vascular cambium that gives rise to secondary growth. Secondary growth provides a mechanism for radially thickening and strengthening the primary plant axis. Wood from secondary growth makes up more than half the world's biomass, and forest biomes provide habitats for about 80% of the world's biodiversity. The protective outer layer of the secondary plant body that replaces the epidermis and cortical tissues is the bark, composed of secondary phloem and the periderm. The periderm results from the activity of a second type of lateral meristem called the cork cambium.

The vascular cambium produces secondary xylem and phloem

Vascular cambia have arisen independently multiple times during the evolution of land plants (**Figure 15.26**). Although secondary growth in extant taxa is restricted to seed plants, studies of the fossil record indicate the presence of many arborescent non-seed plants (e.g., giant lycopods and horsetails) in the coal swamps of the Carboniferous Period. The secondary growth of these extinct

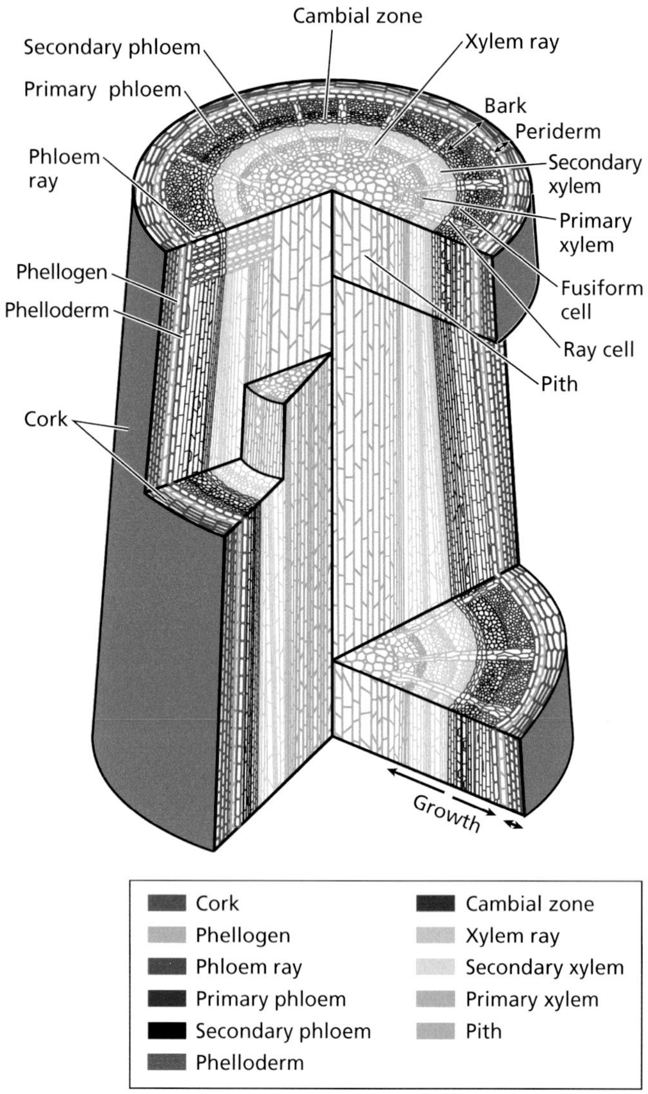

Figure 15.26 Internal anatomy of a woody stem. The vascular cambium (red region) consists of a single layer of cambial initial cells and its immediate derivatives on either side, and is surrounded by an outer layer of secondary phloem cells (black) and an inner layer of secondary xylem cells (light green). The primary phloem (dark blue), primary xylem (dark green), and pith (light blue) are also shown. The periderm includes the phellogen (tan cell layer), phelloderm (pink), and phellem (cork cells, in brown). The bark layer includes all tissues external to the vascular cambium, including the primary and secondary phloem. Most angiosperm and gymnosperm tree species also contain radial files of ray cells that play a role in nutrient transport and storage.

treelike forms consisted only of secondary xylem, in contrast to the secondary growth of contemporary seed plants, which includes both secondary xylem and secondary phloem. The vascular cambium of the former produces internal derivative cells exclusively, while the vascular cambium of the latter produces derivative cells in both directions: xylem cells inwardly and phloem cells outwardly (**Figure 15.27**). Wood consists of secondary xylem tissue with thickened secondary cell walls. Physiologically, wood serves transport, storage, and mechanical functions.

The cork cambium produces the periderm

Secondary growth is generated by the activities of two types of lateral meristems: the vascular cambium, which produces secondary vascular tissues, and the cork cambium, or phellogen, which produces the outer protective layers of the secondary plant body called the periderm (Figure 15.26). The periderm produced by the cork cambium replaces the outer tissues of the primary plant axis, the epidermis and cortex, and protects the plant from biotic and abiotic stresses.

The transition from primary to secondary growth in gymnosperms and eudicots is readily visible along the shoot axis. In poplar (*Populus* spp.), for example, primary growth occurs in the top eight internodes, up to approximately 15 cm from the shoot apical meristem (SAM). Primary growth then gives way to secondary (woody) growth that produces secondary xylem and secondary phloem (Figure 15.27A). The primary and secondary growth zones are separated both spatially and temporally, are easily discernible, and develop rapidly (within 1–2 months) in fast-growing species such as poplar. The vascular cambium produces secondary xylem and phloem. In contrast to the terminally positioned SAM and RAM, the vascular cambium presents a very distinct meristem organization that extends along nearly the entire length of the apical–basal axis of the plant and functions to produce vascular tissues along the radial axis. The vascular cambium consists of a single layer of initial cells (cambial initials) organized in radial files that form a continuous cylinder around the stem. Cambial initials divide to produce mother cells for xylem, phloem, and rays (discussed shortly), which in turn undergo several rounds of division to form the cambium zone, a zone of relatively undifferentiated cells with variable numbers of cells per radial cell file. These cells then differentiate into several different cell types as they progress away from the cambium zone.

The vascular cambium displays two main division patterns—**anticlinal** (perpendicular to the stem surface) and **periclinal** (parallel to the stem surface) (Figure 15.27B). Anticlinal divisions add more cells to the cambium to accommodate the increasing girth of the stem. The peak of anticlinal divisions (typically within the first or second cell file proximal to the phloem) is usually used to pinpoint the location of the vascular cambium. While other meristems produce cells only

anticlinal Pertaining to the orientation of cell division such that the new cell plate forms perpendicular to the tissue surface.

periclinal Pertaining to the orientation of cell division such that the new cell plate forms parallel to the tissue surface.

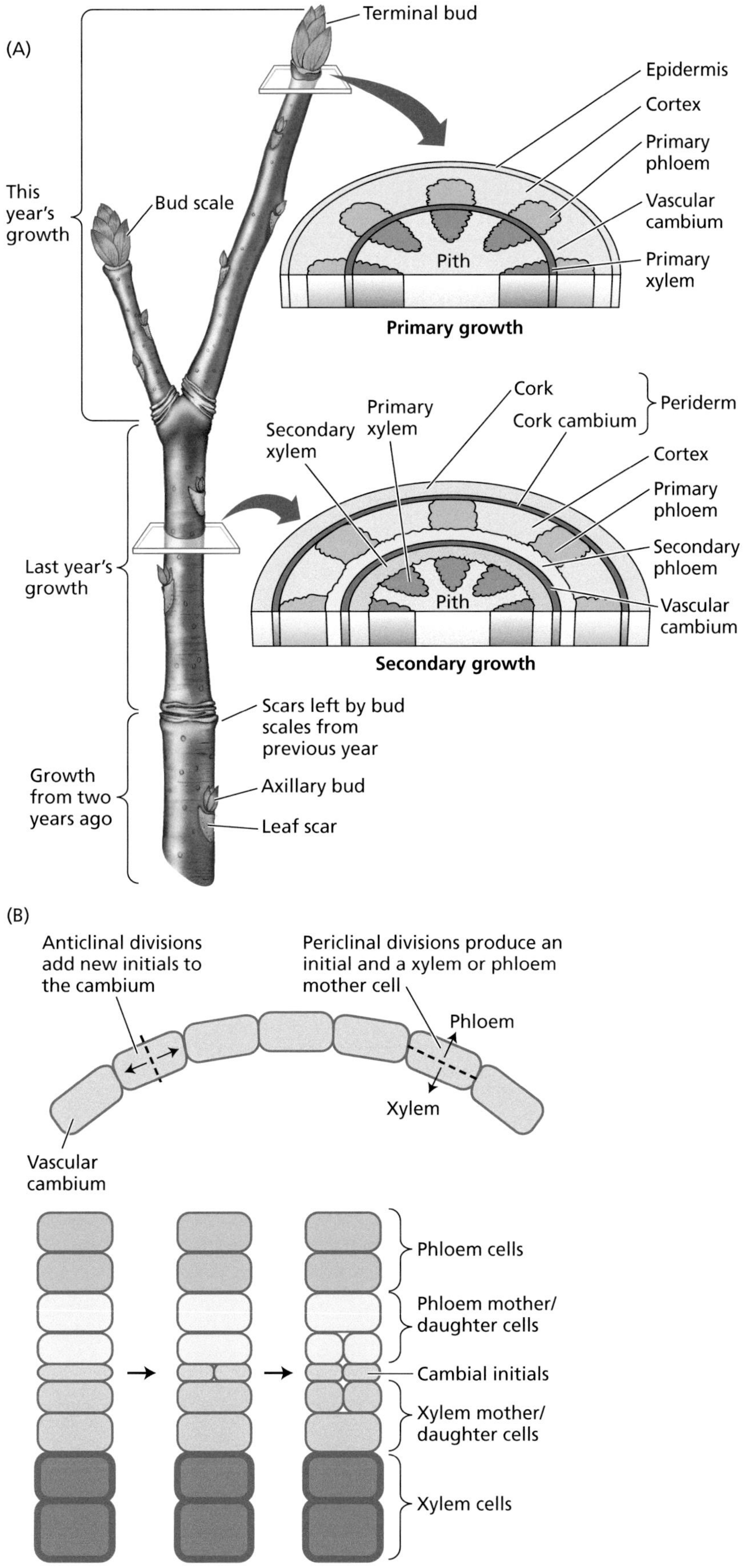

Figure 15.27 Development of secondary vascular tissue. (A) Primary growth in woody stems occurs in the spring, followed by secondary growth. (B) Orientations of the cell division planes in the cambial zone maintain the proper balance between growth in diameter versus circumference. Cambial cells initially divide anticlinally to produce new initials and increase the circumference of the cambium. The same initials also divide periclinally to produce xylem and phloem mother cells, always leaving behind another initial.

on one side, both of the lateral meristems involved in secondary growth (the vascular and cork cambiums) are bifacial, forming new cells on both sides. The vascular cambium produces xylem inward and phloem outward, while the cork cambium produces phellem outward and phelloderm inward (Figure 15.26).

In addition to phloem and xylem cells, the vascular cambium produces ray cells, parenchyma cells that serve as conduits for lateral transport in the stem and for storage during unfavorable conditions such as winter dormancy. Ray cells can be arranged in one file (uniseriate) or multiple files (multiseriate) to form a tissue known as rays that traverses the phloem, cambium, and xylem (Figure 15.26). One distinct challenge that trees face is climate seasonality, which poses risks to their survival during prolonged (seasonal) unfavorable or potentially lethal conditions, such as those encountered during winter months in temperate and boreal regions. To endure dehydration and freezing stress during the winter, trees enter a period of dormancy. The vascular cambium's annual alternation between active growth and dormancy results in the formation of tree rings that record the amount of lateral growth of the tree each year. The molecular mechanisms that control vascular cambium growth during growth–dormancy cycles are poorly understood, though phytohormones such as auxin and gibberellins are thought to play major roles.

15.10 Plant Senescence

Identify the types and processes of plant senescence, and explain how and when it is initiated and regulated.

Senescence refers to the energy-dependent, autolytic (self-digesting) process that leads to the death of targeted cells. As is the case with most features of plant development, senescence is controlled by the interaction of environmental factors with genetically regulated developmental programs.

In temperate climates, the leaves of deciduous trees (trees that shed their leaves annually) turn yellow, orange, or red and fall from their branches in response to shorter day lengths and cooler temperatures, which trigger two related developmental processes: senescence and abscission. **Abscission** refers to the separation of cell layers that occurs at the bases of leaves, floral parts, and fruits, which allows them to be shed easily without damaging the plant. Before the senescing leaves are shed from the plant, however, nitrogen and other nutrients are transported from them back to the branches.

There are three types of plant senescence, based on the level of structural organization of the senescing unit: *programmed cell death*, *organ senescence*, and *whole plant senescence*. **Programmed cell death (PCD)** is a general term referring to the genetically regulated death of individual cells. During PCD, the protoplasm, and sometimes the cell wall, undergoes autolysis. In the case of the development of xylem tracheary elements and fibers (Chapter 3), however, secondary wall layers are deposited prior to cell death. **Organ senescence**—the senescence of whole leaves, branches, flowers, or fruits—occurs at various stages of vegetative and reproductive development and typically includes abscission of the senescing organ. As previously noted, leaf senescence is strongly influenced by photoperiod and temperature. Finally, **whole plant senescence** involves the death of the entire plant. In this chapter we focus on leaf senescence (a type of organ senescence) and whole plant senescence.

senescence An active, genetically controlled, developmental process in which cellular structures and macromolecules are broken down and translocated away from the senescing organ (typically leaves) to actively growing regions that serve as nutrient sinks. Initiated by environmental cues and regulated by hormones.

abscission The shedding of leaves, flowers, and fruits from a living plant. The process whereby specific cells in the leaf petiole (stalk) differentiate to form an abscission layer, allowing a dying or dead organ to separate from the plant.

programmed cell death (PCD) A process whereby individual cells activate an intrinsic senescence program accompanied by a distinct set of morphological and biochemical changes similar to mammalian apoptosis.

organ senescence The developmentally regulated senescence of individual organs, such as leaves, flowers, fruits, or roots.

whole plant senescence The death of the entire plant, as opposed to the death of individual cells, tissues, or organs.

Programmed cell death (PCD) is a normal aspect of development in all eukaryotic organisms

PCD can be initiated by specific developmental signals or by potentially lethal events, such as pathogen attacks or errors in DNA replication during cell division. It involves the expression of a characteristic set of genes that orchestrates the dismantling of cellular components, ultimately causing cell death. PCD, sometimes referred to as apoptosis or autolysis, involves the controlled activities of various hydrolytic enzymes that break down the nucleic acids, proteins, fatty acids, and polysaccharides that make up the cell.

Normal plant development involves two main types of PCD. *Developmental PCD* functions to eliminate senescing or no-longer-required cells, or to generate tissues composed of modified cell "corpses" that retain structural or storage functions when dead. This process occurs during gametophyte maturation, degeneration of embryo **suspensor** cells, formation of xylem tracheary elements, root cap growth, and leaf senescence (**Figure 15.28**). Leaf senescence differs

suspensor In seed plant embryogenesis, the structure that develops from the basal cell following the first division of the zygote. It supports, but is not part of, the embryo.

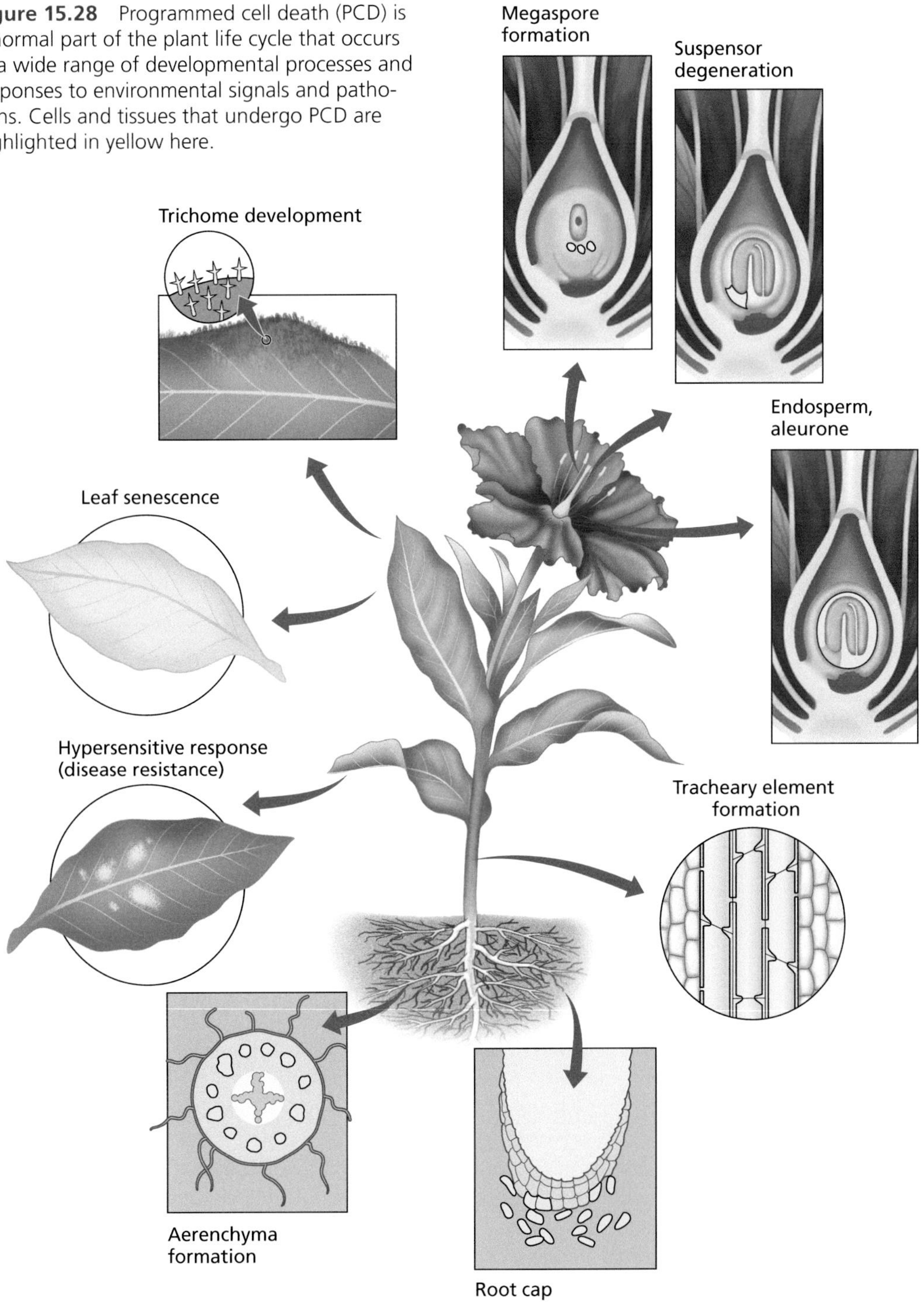

Figure 15.28 Programmed cell death (PCD) is a normal part of the plant life cycle that occurs in a wide range of developmental processes and responses to environmental signals and pathogens. Cells and tissues that undergo PCD are highlighted in yellow here.

from other types of developmental PCD because it develops over a broad area of the organ and at a relatively slow rate. The slow progression of PCD in senescing leaf cells allows sufficient time for remobilization of nutrients and their translocation to other parts of the plant. A second type of PCD occurs during the hypersensitive response and other forms of pathogen-triggered cell death. *Pathogen-triggered PCD* plays a prominent role in plant defense responses, as will be discussed further in Chapter 18.

During leaf senescence, nutrients are remobilized from the source leaf to vegetative or reproductive sinks

All leaves, including those of evergreens, undergo senescence, whether in response to age-dependent factors, environmental cues, biotic stress, or abiotic stress. During developmental senescence, cells undergo genetically programmed changes in cell structure and metabolism. The earliest structural change in leaf cells is the breakdown of the chloroplast, which contains up to 70% of the leaf protein. Carbon assimilation is replaced by the breakdown and conversion of chlorophyll, proteins, and other macromolecules to exportable nutrients that can be translocated to growing vegetative organs or developing seeds and fruits. The resulting sugars, nucleosides, and amino acids are then transported via the phloem back into the main body of the plant, where they are reused for biosynthesis. Many minerals are also transported out of senescing organs back into the plant.

Since senescence redistributes nutrients to growing parts of the plant, it can serve as a survival mechanism during environmentally adverse conditions, such as drought or temperature stress (Chapter 19). However, leaf senescence occurs even under optimal growth conditions and is therefore part of the plant's normal developmental program. As new leaves are initiated at the SAM, older mature leaves may become shaded and lose the ability to function efficiently in photosynthesis, triggering their senescence. In eudicots, senescence is usually followed by abscission, the process that enables plants to shed senescent leaves from the plants. Together, the coupled programs of leaf senescence and abscission help optimize the photosynthetic and nutrient efficiency of the plant.

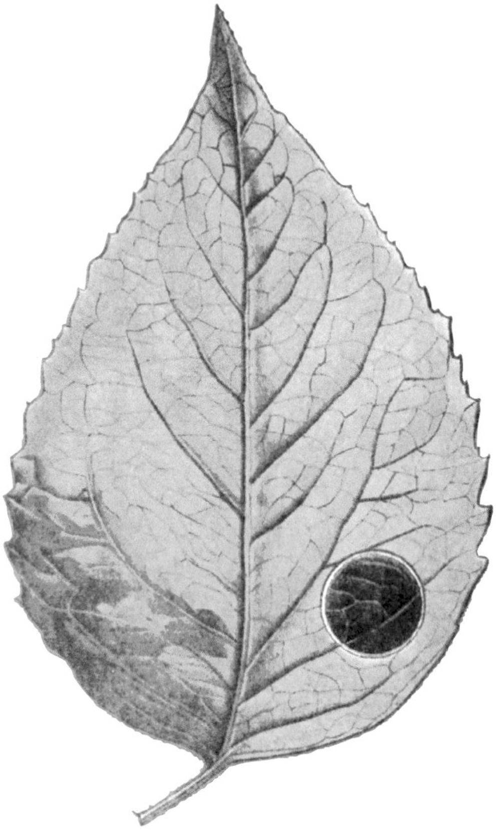

Figure 15.29 Early leaf senescence experiment. This experiment shows the delayed senescence of a leaf disc cultured in the laboratory in a dilute mineral nutrient solution compared with the leaf of mock orange (*Philadelphus grandiflora*) from which the disc was excised, which remained on the bush.

The developmental age of a leaf may differ from its chronological age

Both internal and external cues influence the developmental age of leaf tissue, which may or may not correspond to the leaf's chronological age. The distinction between developmental and chronological age is nicely illustrated by a simple experiment carried out by the German plant physiologist Ernst Stahl in 1909. Stahl excised a small disc from a green leaf of the deciduous shrub mock orange (*Philadelphus grandiflora*). He then incubated the disc on a simple nutrient solution in the laboratory until the autumn, by which time the leaf attached to the plant had turned yellow. The illustration in **Figure 15.29** shows the incubated leaf disc superimposed on the leaf from which it was removed, both at the end of the experiment. Although the chronological ages of the leaf and the disc were the same, the leaf was now developmentally much older than the disc tissue. The leaf that remained on the shrub was subjected to a variety of internal signals coming from the surrounding leaf tissue and other parts of the plant, while the disc was physically cut off from these influences. Furthermore, the leaf attached to the plant remained outdoors exposed to the changing seasons, while the disc was cultured indoors under more or less constant conditions. Shielded from both internal and external cues, the leaf disc remained at the same developmental age as at the start of the experiment, while the attached leaf became developmentally older.

Leaf senescence may be sequential, seasonal, or stress-induced

Leaf senescence under normal growth conditions is governed by the developmental age of the leaf, which is a function of hormones and other regulatory factors. Under these conditions, there is usually a senescence gradient from the oldest leaves located near the base of the shoot to the youngest leaves located near the growing tip—a pattern known as **sequential leaf senescence** (**Figure 15.30**). In contrast, the leaves of deciduous trees in temperate climates senesce all at once in response to the shorter days and cooler temperatures of autumn, a pattern known as **seasonal leaf senescence** (**Figure 15.31**). Both sequential and seasonal leaf senescence are variations of developmental senescence, since they occur under normal growing conditions.

Sequential and seasonal leaf senescence can be divided into three distinct phases: initiation, degeneration, and termination (**Figure 15.32**). During the *initiation phase,* the leaf receives developmental and environmental signals that initiate a decline in photosynthesis and a transition from being a nitrogen sink to a nitrogen source. Most of the degradation of cellular organelles and macromolecules occurs during the *degenerative phase* of leaf senescence. The solubilized mineral and organic nutrients are then remobilized via the phloem to growing sinks, such as young leaves, underground storage organs, or reproductive structures. The abscission layer forms during the degenerative phase

Figure 15.30 Sequential leaf senescence of barley stems. The photograph shows a gradient of senescence from the older leaves at the base to the younger leaves near the tip.

(A) September 8 (B) September 13 (C) September 18

(D) September 25 (E) October 3 (F) October 8

Figure 15.31 Seasonal leaf senescence in an aspen tree (*Populus tremula*). All of the leaves begin to senesce in late September and undergo abscission in early October.

sequential leaf senescence The pattern of leaf senescence in which there is a gradient of senescence from the growing tip of the shoot to the oldest leaves at the base.

seasonal leaf senescence In temperate climates, the pattern of leaf senescence in deciduous trees in which all of the leaves undergo senescence and abscission in the autumn.

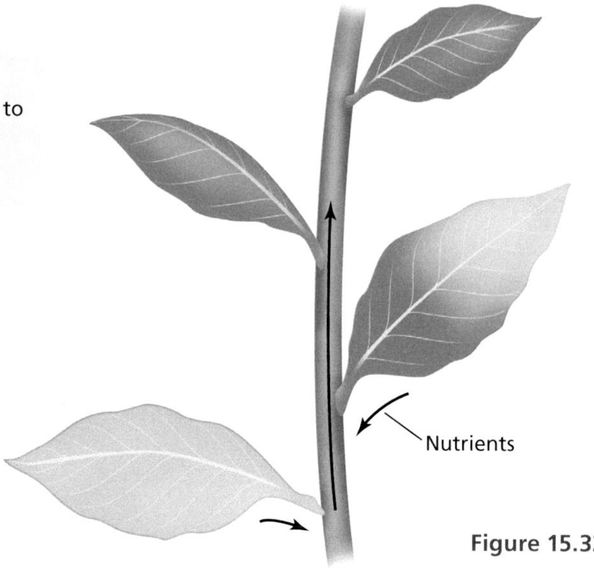

Figure 15.32 The three stages of leaf senescence.

of leaf senescence. During the *terminal phase*, autolysis is completed and cell separation takes place at the abscission layer, resulting in leaf abscission.

The earliest cellular changes during leaf senescence occur in the chloroplast

Chloroplasts contain about 70% of the total leaf protein, most of which consists of ribulose 1,5-bisphosphate carboxylase/oxygenase (Rubisco) localized in the stroma, and light-harvesting chlorophyll-binding protein II associated with the thylakoid membranes (Chapters 7 and 8). Catabolism and remobilization of chloroplast proteins thus provide the primary source of amino acids and nitrogen for sink organs, and represent the earliest changes that occur during leaf senescence. Unlike chloroplasts, the nucleus and mitochondria, which are required for gene expression and energy production, remain intact until the later stages of senescence.

The transition from a mature, photosynthetically active leaf to a senescing leaf involves increased expression of senescence associated genes (SAGs) and decreased expression of metabolic maintenance-related genes. SAGs include genes regulating many processes associated with abiotic and biotic stress (**Figure 15.33**).

Reactive oxygen species serve as internal signaling agents in leaf senescence

There is growing evidence that reactive oxygen species (ROS), especially H_2O_2, play important roles as signals during leaf senescence. ROS are toxic chemicals that cause oxidative damage to DNA, proteins, and membrane lipids (Chapter 18). They are produced primarily as byproducts of normal metabolic processes, such as respiration and photosynthesis, in chloroplasts, mitochondria, peroxisomes, and at the plasma membrane. Plants use ROS-scavenging systems, such as enzymes (catalase, superoxide dismutase, ascorbate peroxidase) and antioxidant molecules (e.g., ascorbate and glutathione), to protect themselves from oxidative damage. However, the plant's antioxidant concentrations decrease during leaf

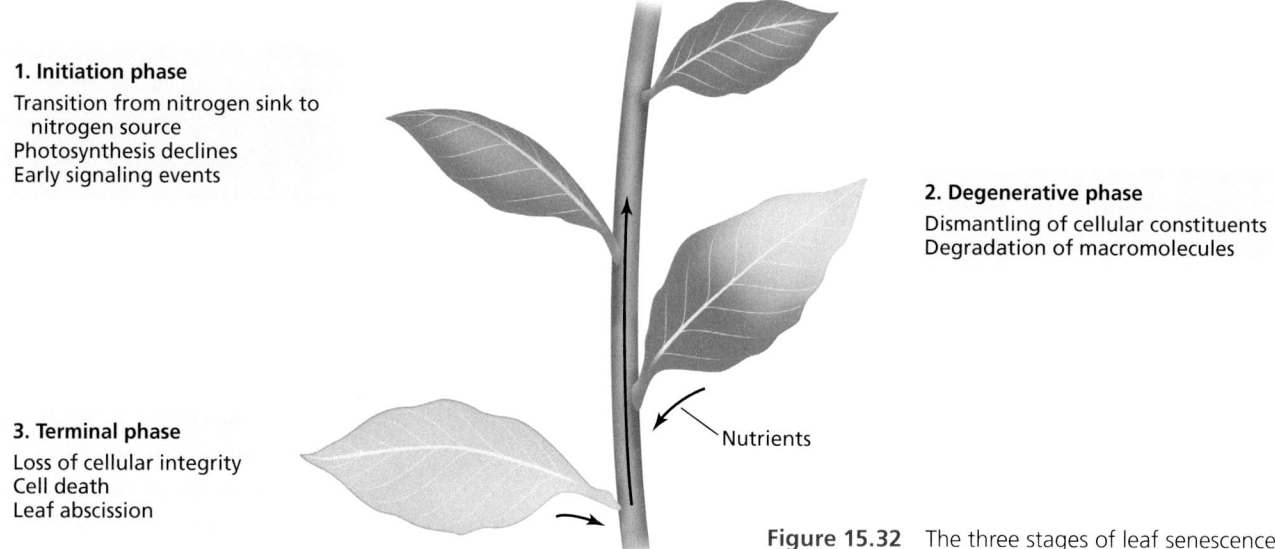

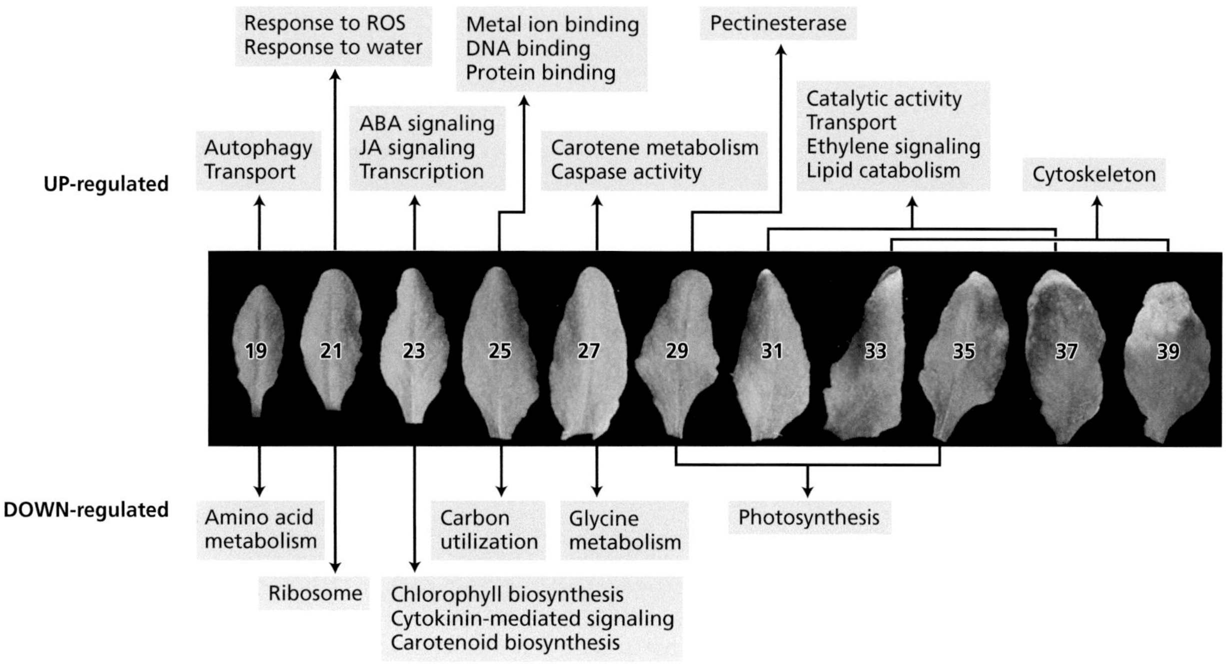

Figure 15.33 Metabolic pathways that are either up-regulated or down-regulated during leaf senescence in Arabidopsis. The seventh leaf was sampled from plants at different stages of senescence. The numbers 19 through 39 refer to the ages of the plants, expressed as the number of days after sowing, from which the seventh leaf was taken. Caspase is a type of cysteine protease.

senescence, while ROS levels increase. In senescence, ROS act as signals that activate genetically programmed cell death events.

Plant hormones interact in the regulation of leaf senescence

Leaf senescence is an evolutionarily selected, genetically regulated process that ensures efficient remobilization of nutrients to vegetative or reproductive sink organs. No mutation, treatment, or environmental condition has yet been found that abolishes the process completely, suggesting that leaf senescence is ultimately governed either by developmental or chronological age. However, both the timing and progression of senescence are flexible, and hormones are key developmental signals that accelerate or delay the timing of leaf senescence.

The senescence-repressing role of cytokinins appears to be universal in plants and has been demonstrated in many types of studies. Although applied cytokinins do not prevent senescence completely, their effects can be dramatic, particularly when the cytokinin is sprayed directly on an intact plant. If only one leaf is treated, it remains green after other leaves of similar age have yellowed and dropped off the plant. If a small spot on a leaf is treated with cytokinin, that spot will remain green after the surrounding tissues on the same leaf begin to senesce. This "green island" effect can also be observed in leaves infected by some fungal pathogens, as well as in those hosting galls produced by insects. Such green islands have higher levels of cytokinins than surrounding leaf tissue. Unlike young leaves, mature leaves produce little, if any, cytokinin. During senescence, the transcript abundance of genes involved in cytokinin biosynthesis declines, whereas transcripts of

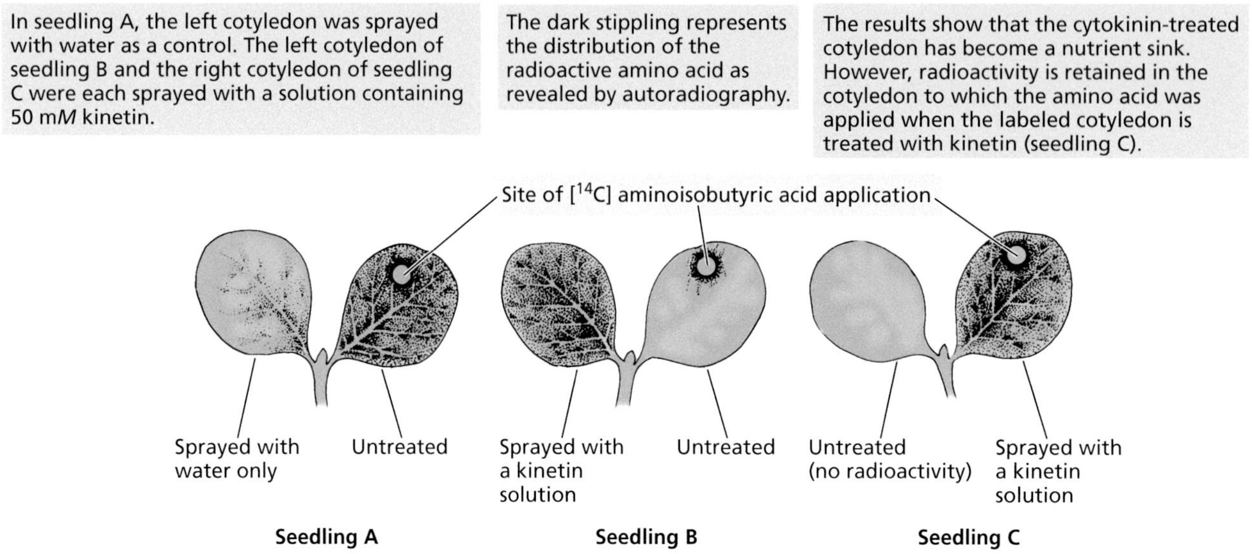

In seedling A, the left cotyledon was sprayed with water as a control. The left cotyledon of seedling B and the right cotyledon of seedling C were each sprayed with a solution containing 50 mM kinetin.
The dark stippling represents the distribution of the radioactive amino acid as revealed by autoradiography.
The results show that the cytokinin-treated cotyledon has become a nutrient sink. However, radioactivity is retained in the cotyledon to which the amino acid was applied when the labeled cotyledon is treated with kinetin (seedling C).
Site of [14C] aminoisobutyric acid application
Sprayed with water only
Untreated
Sprayed with a kinetin solution
Untreated
Untreated (no radioactivity)
Sprayed with a kinetin solution
Seedling A
Seedling B
Seedling C

Figure 15.34 Effect of cytokinin (kinetin) on the movement of an amino acid in cucumber seedlings. Radioactively labeled α-aminoisobutyric acid, a non-metabolizable amino acid that cannot be used in protein synthesis, was applied as a discrete spot on the right cotyledon of each of these seedlings. After a set amount of time, the seedlings were placed on X-ray film to detect the movement of the radioactive amino acid. The black stippling indicates the distribution of radioactivity outside the application spot.

genes involved in degrading cytokinins, such as cytokinin oxidase, increase during senescence. Mature leaves may therefore depend on root-derived cytokinins to postpone their senescence.

Thus far, the molecular mechanism of cytokinin action in delaying leaf senescence remains unclear. According to a long-standing hypothesis, cytokinin represses leaf senescence by regulating nutrient mobilization and source–sink relations. This phenomenon can be observed when nutrients (sugars, amino acids, and so on) radiolabeled with ^{14}C or ^{3}H are fed to plants after one leaf or part of a leaf is treated with a cytokinin (**Figure 15.34**). Subsequent autoradiography of the whole plant reveals the pattern of movement and the sites at which the labeled nutrients accumulate. Experiments of this nature have demonstrated that nutrients are preferentially transported to and accumulated in the cytokinin-treated tissues, which retain the nutrient sink status associated with young, growing tissues.

Gibberellins (GAs) are also senescence-repressing hormones. The abundance of active forms of gibberellin declines in leaves as they age. For example, the senescence of excised leaf discs of *Taraxacum* and *Rumex* is delayed by treatment with GA. Auxin also plays a more limited role in delaying senescence and is associated with delays in leaf senescence and decreases in SAG expression. As described below, auxin production in leaves inhibits the initiation of leaf abscission.

Ethylene, by contrast, is regarded as a senescence-promoting hormone because ethylene treatment accelerates leaf and flower senescence, and inhibitors of ethylene synthesis and action can delay senescence. As we discuss in the next section, ethylene plays an important role in abscission as well. Abscisic acid (ABA) levels increase in senescing leaves, and exogenous application of ABA rapidly promotes leaf senescence and expression of several SAGs, which is consistent with ABA's effects on leaf senescence. However, like ethylene, ABA is considered an enhancer rather than a triggering factor of leaf senescence.

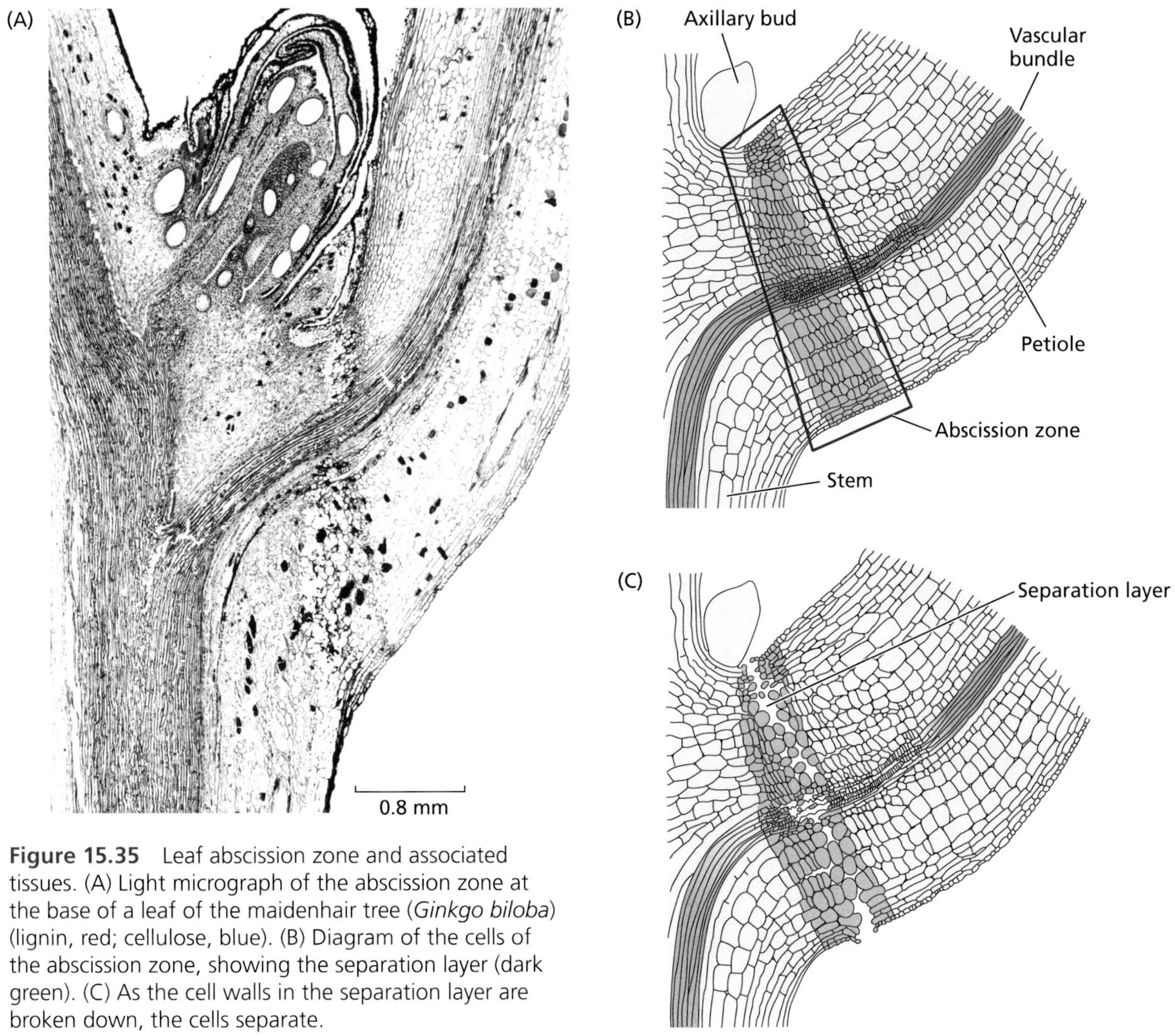

Figure 15.35 Leaf abscission zone and associated tissues. (A) Light micrograph of the abscission zone at the base of a leaf of the maidenhair tree (*Ginkgo biloba*) (lignin, red; cellulose, blue). (B) Diagram of the cells of the abscission zone, showing the separation layer (dark green). (C) As the cell walls in the separation layer are broken down, the cells separate.

15.11 Leaf Abscission

Describe how a leaf is abscised and how hormones interact in the process.

The shedding of leaves, fruits, flowers, and other plant organs is termed *abscission*. Abscission takes place within specific layers of cells called the **abscission zone**, located near the base of the petiole (**Figure 15.35**). The abscission zone becomes morphologically and biochemically differentiated during organ development, many months before organ separation actually takes place. Often the abscission zone can be morphologically identified as one or more layers of isodiametrically flattened cells (Figure 15.35B).

The timing of leaf abscission is regulated by the interaction of ethylene and auxin

Ethylene plays a key role in activating the events leading to cell separation within the abscission zone. The ability of ethylene gas to cause defoliation in young birch trees is shown in **Figure 15.36**. The wild-type tree on the left has lost most of its leaves; only the younger leaves at the top fail to abscise. The tree on the

abscission zone The region that contains the abscission layer and is located near the base of the petiole of leaves.

Figure 15.36 Effect of ethylene on abscission in the birch *Betula pendula*. The tree on the left is the wild type; the tree on the right has a dominant mutation than knocks out all activity of the ethylene receptor. Both trees had been fumigated for 3 days with 50 ppm ethylene. One of the characteristics of these mutant trees is that they do not drop their leaves, as wild-type plants do.

right has been genetically transformed to be insensitive to ethylene and retains its leaves after ethylene treatment.

The process of leaf abscission can be divided into three distinct developmental phases during which the cells of the abscission zone become competent to respond to ethylene (**Figure 15.37**).

1. *Leaf maintenance phase*. Prior to the perception of any signal (internal or external) that initiates the abscission process, the leaf remains healthy and fully functional. A gradient of auxin from the leaf blade to the stem maintains the abscission zone in an insensitive state. Removal of the leaf blade (the site of auxin production) promotes petiole abscission. Application of exogenous auxin to petioles from which the leaf blade has been removed delays the abscission process.
2. *Abscission induction phase*. A reduction or reversal in the auxin gradient from the leaf blade, normally associated with leaf senescence, causes the abscission zone to become sensitive to ethylene. Treatments that enhance leaf senescence do so by promoting abscission through interference with auxin synthesis or transport in the leaf. In the abscission induction phase, typically associated with leaf senescence, the amount of auxin from the leaf blade decreases and the ethylene level rises. Ethylene appears to decrease the concentration of auxin both by reducing its synthesis and transport and by increasing its degradation. The reduction in the concentration of free auxin increases the response of specific target cells in the abscission zone to ethylene.

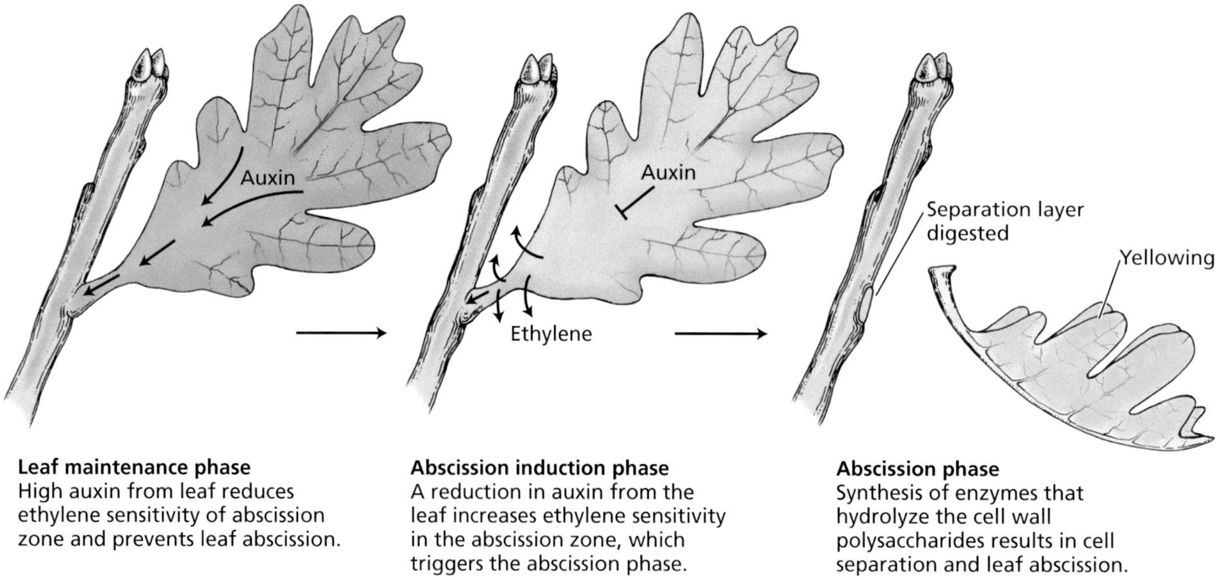

Auxin
Auxin
Ethylene
Separation layer digested
Yellowing
Leaf maintenance phase
High auxin from leaf reduces ethylene sensitivity of abscission zone and prevents leaf abscission.
Abscission induction phase
A reduction in auxin from the leaf increases ethylene sensitivity in the abscission zone, which triggers the abscission phase.
Abscission phase
Synthesis of enzymes that hydrolyze the cell wall polysaccharides results in cell separation and leaf abscission.

Figure 15.37 Schematic view of the roles of auxin and ethylene during leaf abscission. In the abscission induction phase, the level of auxin decreases and the level of ethylene increases. These changes in the hormonal balance increase the sensitivity of the target cells to ethylene.

3. *Abscission phase*. The sensitized cells of the abscission zone respond to low concentrations of endogenous ethylene by synthesizing and secreting cell wall–degrading enzymes and cell wall–remodeling proteins, including β-1,4-glucanase (cellulase), polygalacturonase, **xyloglucan** endotransglucosylase/hydrolase, and expansin, resulting in cell separation and leaf abscission.

15.12 Whole Plant Senescence

Explain whole plant senescence in annuals and biennials.

The programmed death of individual leaves is an adaptation that benefits the plant as a whole by increasing its evolutionary fitness. However, the death of whole plants cannot easily be rationalized in evolutionary terms, even though the life spans of individual plants are, to a large extent, genetically determined.

Angiosperm life cycles may be annual, biennial, or perennial

Individual plant life spans vary from a few weeks in the case of desert ephemerals, which grow and reproduce rapidly in response to brief episodes of rain, to about 4600 years in the case of bristlecone pine. In general, **annual plants** grow, reproduce, senesce, and die in a single season. **Biennial plants** devote their first year to vegetative growth and food storage, and their second year to reproduction, senescence, and death. Because annual and biennial plants undergo whole plant senescence following fruit and seed production, both are termed **monocarpic** because they reproduce only once (**Figure 15.38**).

Perennial plants live for 3 years or longer and may be herbaceous or woody. The range of maximum life spans for perennial plants is given in **Table 15.2**. Perennial plants are usually **polycarpic**, producing fruits and seeds over multiple seasons. However, there are also examples of monocarpic perennials, such as the century plant (*Agave americana*) and Japanese timber bamboo (*Phyllostachys bambusoides*). The century plant grows vegetatively for 10 to 30 years before flowering, fruiting, and senescing, while Japanese timber bamboo can grow vegetatively for 60 to 120 years before reproduction and death. Remarkably, all clones from the same bamboo stock flower and senesce simultaneously, regardless of geographic location or climatic condition, which suggests the presence of a long-term biological clock of some kind.

Many perennial plants that form clones by asexual reproduction can proliferate into community-sized interconnected "individuals" that achieve astounding ages, such as King's lomatia (*Lomatia tasmanica*), a Tasmanian shrub in the Proteaceae family that can be over 43,000 years old. Each individual lomatia plant lives only about 300 years, but

Figure 15.38 Monocarpic senescence in soybean (*Glycine max*). The entire plant on the left underwent senescence after flowering and producing fruit (pods). The plant on the right remained green and vegetative because its flowers were continually removed.

xyloglucan A hemicellulose with a backbone of 1→4-linked β-D-glucose residues and short side chains that contain xylose, galactose, and sometimes fucose.

annual plant A plant that completes its life cycle from seed to seed, senesces, and dies within 1 year.

biennial plant A plant that requires two growing seasons to flower and produce seed.

monocarpic Referring to plants, typically annuals, that produce fruits only once and then die.

perennial plants Plants that live for more than 2 years.

polycarpic Referring to perennial plants that produce fruit many times.

Table 15.2 Longevity of Various Individual and Clonal Perennial Plants

Species	Age (yr)
Individual plants	
Bristlecone pine (*Pinus longaeva*)	4,600
Giant sequoia (*Sequoiadendron giganteum*)	3,200
Stone pine (*Pinus cembra*)	1,200
European beech (*Fagus sylvatica*)	930
Blackgum (*Nyssa sylvatica*)	679
Red oak (*Quercus rubra*)	326
English ivy (*Hedera helix*)	200
Bigtooth aspen (*Populus grandidentata*)	113
Scots heather (*Calluna vulgaris*)	42
Spring heath (*Erica carnea*)	21
Clonal plants	
King's lomatia (*Lomatia tasmanica*)	43,000+
Creosote (*Larrea tridentata*)	11,000+
Bracken (*Pteridium aquilinum*)	1,400
Ground pine (*Lycopodium complanatum*)	850
Reed grass (*Calamagrostis epigeios*)	400+
Wood sage (*Teucrium scorodonia*)	10

because it does not transfer any senescence signal to its clones, the clonal community apparently grows and proliferates indefinitely.

Nutrient or hormonal redistribution may trigger senescence in monocarpic plants

A diagnostic feature of monocarpic senescence is the ability to delay senescence well beyond the plant's normal life span by the removal of the reproductive structures. For example, repeated depodding (removing flowers and pods) enables soybean plants to remain vegetative for many years under favorable growing conditions, leading to a treelike appearance. Monocarpic senescence is thought to result from the redistribution of vital nutrients via the phloem from vegetative sources to reproductive sinks. However, the critical redistributed compound that triggers monocarpic senescence is not likely to be a carbohydrate, as the carbohydrate content of leaves actually *increases* during senescence. This observation is consistent with the ability of exogenous sugars to trigger leaf senescence. Rather than carbohydrate loss, it may be alterations in source–sink relations caused by floral development that induce a global shift in the hormonal or nutrient balance of the vegetative organs. A loss of nitrogen coupled with a simultaneous accumulation of carbohydrate would cause an increase in the C:N ratio, which has been associated with autolysis in senescing leaves.

Summary

After seedling establishment, development of vegetative organs occurs primarily from the meristem tissues. Vegetative growth is controlled by developmental processes involving molecular interactions and regulatory feedback. These mechanisms create root and shoot polarity, allowing plants to produce lateral organs (e.g., leaves and branching systems), which form an overall vegetative architecture. Senescence at the cellular level, called programmed cell death, is an integral part of plant development. Senescence also occurs at the organ level, as in the case of leaf senescence, during which the leaf undergoes a genetically programmed sequence of macromolecular turnover and nutrient recycling, followed by abscission. The regulation of plant senescence differs in annuals, biennials, and perennials. In monocarpic species, senescence may be triggered by nutrient or hormonal redistribution during fruit production. Many perennial species can live for thousands of years.

15.1 The Root Apical Meristem (RAM)

- The RAM gives rise to the primary root axis.
- The root tip has four developmental zones (**Figure 15.1**).
- Auxin and cytokinin help regulate and maintain the RAM.

15.2 The Shoot Apical Meristem (SAM)

- After seedling germination, vegetative shoot organs are derived from a compact SAM (**Figure 15.2**).
- Organogenesis originates in discrete SAM zones (**Figure 15.3**).

15.3 Leaf Structure and Phyllotaxy

- The size of the SAM is regulated by a feedback loop of positive and negative regulators (**Figure 15.4**).

(Continued)

Summary (continued)

- The development of flat laminae in seed plants was a key evolutionary event; since then, phyllome morphology has diversified dramatically (**Figure 15.5**).
- The three basic types of leaf arrangement (phyllotaxy) are alternate, decussate, and spiral (**Figure 15.6**).
- Leaf phyllotactic patterns are established at the shoot apex by localized zones of auxin accumulation resulting from polar auxin transport (**Figure 15.7**).

15.4 Differentiation of Epidermal Cell Types

- The epidermis is derived from the protoderm and has three main cell types: pavement cells, trichomes, and stomatal guard cells, as well as other specialized cell types, depending on the species (**Figure 15.8**).
- Not only guard cells but also the majority of leaf epidermal cells arise from specialized meristemoid mother cells (MMCs), stomatal lineage ground cells (SLGCs), meristemoids, and guard mother cells (GMCs) (**Figure 15.9**).

15.5 Venation Patterns in Leaves

- Leaf venation patterns differ in eudicots and monocots (**Figure 15.10**).
- Leaf veins exhibit a hierarchy based on their diameter at the site of attachment to the parent vein (**Figure 15.11**).
- Leaf vascular bundles arise from the procambium and differentiate downward, forming a connection to the older vascular bundles that are continuous to the base of the shoot (**Figure 15.12**).
- The rootward transport of auxin in shoots induces the differentiation of new xylem cells after wounding (**Figure 15.13**).
- Both localized auxin biosynthesis and polar auxin transport are involved in vascularization during early leaf development (**Figure 15.13**).
- The canalization, or source–sink, model helps to explain the venation patterns of plant leaves.

15.6 Shoot Branching and Architecture

- Plant shoot architecture is based on a repeating unit called the phytomer, consisting of an internode, a node, a leaf, and an axillary bud (**Figure 15.14**).
- Auxin, cytokinins, and strigolactones regulate apical dominance (**Figures 15.15**, **15.16**).
- Cytokinins are involved in breaking apical dominance and stimulating axillary bud growth (**Figure 15.16**).
- Sucrose also serves as the initial signal for axillary bud growth (**Figure 15.17**).

15.7 Shade Avoidance

- Plants compete for sunlight and respond to shading from other plants by detecting decreased R:FR ratios (**Table 15.1**; **Figure 15.18**).
- Phytochrome is the photoreceptor that senses the R:FR ratio during shade avoidance (**Figure 15.18**).
- Plants react to shading by increasing elongation growth (**Figure 15.18**).
- Genetic modification of shade avoidance responses can increase crop yields (**Figure 15.19**).

15.8 Root Branching and Architecture

- Branch root initials form opposite the protoxylem poles in pericycle lateral root initial cells (**Figure 15.20**).
- The prebranch site is located in the zone of maturation of the root (**Figure 15.21**).
- According to the auxin clock model, the spacing of prebranch sites is regulated by periodic oscillations of auxin.
- Species-specific root system architecture optimizes water and nutrient uptake (**Figure 15.22**).
- Monocot root systems, as exemplified by maize roots, are composed of the primary root, seminal and crown roots, and lateral roots (**Figures 15.23**); eudicot root systems, such as those of soybean, include the primary root (taproot) and branch, basal, and adventitious roots (**Figures 15.24**).
- Phosphorus availability can alter root system architecture (**Figure 15.25**).

15.9 Secondary Growth

- Secondary growth results from the activities of two types of lateral meristems: the vascular and cork cambiums (**Figure 15.26**).
- The vascular cambium produces secondary xylem cells inwardly and secondary phloem cells outwardly (**Figure 15.27**).
- The cork cambium produces the outer protective layers of the secondary plant body called the periderm (**Figure 15.27**).
- Bark is composed of the periderm plus the older layers of secondary phloem.

15.10 Plant Senescence

- There are three types of plant senescence, based on the level of structural organization of the senescing unit: programmed cell death (PCD), organ senescence, and whole plant senescence.

(Continued)

Summary *(continued)*

- PCD is a general term referring to the genetically regulated death of individual cells (**Figure 15.28**).
- Leaf senescence is a type of organ senescence. It involves the genetically regulated autolysis of cellular proteins, carbohydrates, and nucleic acids and the redistribution of their components from the leaf back into the main body of the plant, into the actively growing tissues within the main body of the plant. Minerals are also transported out of senescing leaves back into the plant.
- Normal leaf senescence is regulated by internal, developmental signals as well as external, environmental signals (**Figure 15.29**).
- Leaf senescence may exhibit a sequential or seasonal pattern (**Figures 15.30**, **15.31**).
- Leaf senescence can be divided into three phases: initiation, degeneration, and termination (**Figure 15.32**).
- Leaf senescence is preceded by the increased expression of senescence-associated genes (**Figure 15.33**).
- There is growing evidence that reactive oxygen species (ROS), especially H_2O_2, can serve as internal signals to promote senescence.
- Plant hormones interact to regulate leaf senescence.
- Cytokinin may delay senescence by causing leaf cells to become sinks for nutrients (**Figure 15.34**).

15.11 Leaf Abscission

- Abscission is the shedding of leaves, fruits, flowers, or other plant organs, and is caused by the separation of cell layers within the abscission zone (**Figure 15.35**).
- Ethylene plays a fundamental signaling role in leaf senescence and abscission (**Figure 15.36**).
- High levels of auxin keep leaf tissue in an ethylene-insensitive state, but as auxin levels decrease, the abscission-promoting and auxin-repressing effects of ethylene become stronger (**Figure 15.37**).

15.12 Whole Plant Senescence

- In general, annuals and biennials reproduce only once before senescing, while perennials can reproduce multiple times before senescing.
- There is wide variation in the longevities of different plant species (**Table 15.2**).
- Nutrient or hormonal redistribution from vegetative structures to reproductive sinks may trigger whole plant senescence in monocarpic plants (**Figure 15.38**).

Suggested Reading

Bayer, I., Smith, R. S., Mandel, T., Nakayama, N., Sauer, M., Prusinkiewicz, P., and Kuhlemeier, C. (2009) Integration of transport-based models for phyllotaxis and midvein formation. *Genes Dev.* 23: 373–384.

Breeze, E., Harrison, E., McHattie, S., Hughes, L., Hickman, R., Hill, C., Kiddle, S., Kim,Y.-S., Penfold, C. A., Jenkins, D., et al. (2011) High-resolution temporal profiling of transcripts during Arabidopsis leaf senescence reveals a distinct chronology of processes and regulation. *Plant Cell* 23: 873–894.

Caño-Delgado, A., Lee, J.Y., and Demura,T. (2010) Regulatory mechanisms for specification and patterning of plant vascular tissues. *Annu. Rev. Cell Dev. Biol.* 26: 605–637.

Domagalska, M. A., and Leyser, O. (2011) Signal integration in the control of shoot branching. *Nat. Rev. Mol. Cell Biol.* 12: 211–221.

Lau, S., and Bergmann, D. C. (2012) Stomatal development: A plant's perspective on cell polarity, cell fate transitions and intercellular communication. *Development* 139: 3683–3692.

Lucas, W. J., Groover, A., Lichtenberger, R., Furuta, K., Yadav, S. R., Helariutta, Y., He, X. Q., Fukuda, H., Kang, J., Brady, S. M., et al. (2013) The plant vascular system: Evolution, development and functions. *J. Integr. Plant Biol.* 55: 294–388.

Mason, M. G., Ross, J. J., Babst, B. A., Wienclaw, B. N., and Beveridge, C. A. (2014) Sugar demand, not auxin, is the initial regulator of apical dominance. *Proc. Natl. Acad. Sci. USA* 111: 6092–6097.

Nunes, T. D. G., Zhang, D., and Raissig, M. T. (2020) Form, development and function of grass stomata. *Plant J.* 101: 780–799.

Risopatron, J. P. M., Sun, Y., and Jones, B. J. (2012) The vascular cambium: Molecular control of cellular structure. *Protoplasma* 247: 145–161.

Smetana, O., Mäkilä, R., Lyu, M., Amiryousefi, A., Sánchez Rodríguez, F., Wu, M.-F., Solé-Gil, A., Leal Gavarrón, M., Siligato, R., Miyashima, S., et al. (2019) High levels of auxin signaling define the stem-cell organizer of the vascular cambium. *Nature* 565: 485–489.

Woo, H. R., Kim, H. J., Lim, P. O., and Nam, H. G. (2019) Leaf senescence: Systems and dynamics aspects. *Ann Rev Plant Biol.* 70(1): 347–376.

Zhang, Z., Liao, H., and Lucas, W. J. (2014) Molecular mechanisms underlying phosphate sensing, signaling, and adaptation in plants. *J. Integr. Plant Biol.* 56: 192–220.

16 Flowering and Double Fertilization

Most people look forward to the spring season and the profusion of flowers it brings. Many people plan their vacations to coincide with specific blooming seasons: *Citrus* along the Blossom Trail in southern California and tulips in the Netherlands. In Japan, Taiwan, and many parts of the United States, the cherry blossoms are received with spirited ceremonies. As spring transitions into summer, summer into autumn, and autumn into winter, wildflowers bloom at their appointed times. Flowering at the correct time of year is crucial for the reproductive fitness of the plant; plants that are cross-pollinated must flower in synchrony with other individuals of their species as well as with their pollinators at the time of year that is optimal for seed set.

Although the strong correlation between flowering and seasons is common knowledge, the phenomenon poses fundamental questions that we address in this chapter:

- How do plants keep track of the seasons of the year and the time of day?
- Which environmental signals influence flowering, and how are those signals perceived?
- What is the mechanism of floral organ formation?
- How are the male and female gametophytes formed within the flower?
- What happens during fertilization?

The transition to flowering involves major changes in the pattern of morphogenesis and cell differentiation at the shoot apical meristem (SAM). Ultimately, as we will see, this process leads to the production of the floral organs—sepals, petals, stamens, and carpels—as well as the male and female gametophyte generations.

16.1 Floral Evocation: Integrating Environmental Cues

Describe how the integration of autonomous (developmental) and environmental factors allows the synchronization of flowering favoring reproductive success.

The developmental decision to flower is an important event in the plant life cycle. The process by which the shoot apical meristem (SAM) becomes committed to forming flowers is called **floral evocation**. Delaying this commitment to flower increases the carbohydrate reserves that will be available for mobilization, allowing more and better-provisioned seeds to mature. Delaying flowering, however, also potentially increases the danger that the plant will be eaten, killed by abiotic stress, or outcompeted by other plants before it reproduces. Therefore, plants have evolved an extraordinary range of reproductive adaptations—for example, annual versus perennial life cycles.

Annual plants such as groundsel (*Senecio vulgaris*) may flower within a few weeks after germinating, but trees may grow for 20 or more years before they begin to produce flowers. Across the plant kingdom, different species flower at a wide range of ages, indicating that the age, or perhaps the size, of the plant is an internal factor controlling the switch to reproductive development.

Autonomous flowering is regulated by developmental programming and is mostly independent of environmental conditions. In species that exhibit an absolute requirement for a specific set of environmental cues to bloom, flowering is described as *obligate* or *qualitative*. If flowering is promoted by certain environmental cues but will eventually occur in the absence of such cues, the flowering response is *facultative* or *quantitative*.

Photoperiodism and *vernalization* are two of the most important processes that regulate seasonal responses. Photoperiodism is a response to the length of day or night; vernalization is the promotion of flowering by prolonged cold temperature. Other signals, such as light quality, ambient temperature, and abiotic stress, are also important external cues for plant development.

The evolution of both internal (autonomous) and external (environment-sensing) control systems enables plants to precisely regulate flowering so that it occurs at the optimal time for reproductive success. For example, in many populations of a particular species, such as bamboo, flowering is synchronized, which favors crossbreeding. Pruning can also synchronize flowering in trees. Flowering in response to environmental cues also helps ensure that seeds are produced under favorable conditions, particularly with respect to water and temperature. However, this makes plants especially vulnerable to rapid climate change, such as global warming, which can alter the regulatory networks that govern floral timing. Several studies have shown that many plant species, such as the highbush blueberry (*Vaccinium corymbosum*), are now flowering several days to weeks earlier than they did in the nineteenth century, while other plants are no longer synchronized with pollinators, thus impacting plant reproduction and availability of food (nectar, pollen) for pollinators.

16.2 The Shoot Apex and Phase Changes

Describe developmental phase changes and the influence of hormones and nutritional and other signals on them.

All multicellular organisms pass through a series of more or less defined developmental stages, each with its characteristic features. In humans, infancy, childhood, adolescence, and adulthood represent four general stages of development, with

floral evocation The events occurring in the shoot apex that specifically commit the apical meristem to produce flowers.

phase change The phenomenon in which the fates of the meristematic cells become altered in ways that cause them to produce new types of structures.

puberty as the dividing line between the nonreproductive and the reproductive phases. Similarly, plants pass through distinct developmental phases. However, unlike animals, plants produce new organs from the SAM more or less continuously. This ability allows plants to synchronize their development with a changing environment by controlling the timing of phase transitions. In addition, the plant must integrate information from the environment as well as autonomous signals within the plant to maximize its reproductive fitness.

Plant development has three phases

Postembryonic development in plants can be divided into three phases:

1. The juvenile phase
2. The adult vegetative phase
3. The adult reproductive phase

The transition from one phase to another is called **phase change**.

The primary distinction between the juvenile and the adult vegetative phases is that the latter has the ability to form reproductive structures (flowers in angiosperms, cones in gymnosperms); the adult vegetative phase can be extremely brief in some species, with plants transitioning from juvenile to reproductive very quickly. Flowering, the manifestation of reproductive competence of the adult phase, often depends on specific environmental and developmental signals. Thus, the absence of flowering itself is not a reliable indicator of juvenility.

The transition from juvenile to adult is frequently accompanied by changes in vegetative characteristics, such as leaf morphology, phyllotaxy (the arrangement of leaves on the stem), thorniness, rooting capacity, and leaf retention in deciduous plants such as English ivy (*Hedera helix*) (**Figure 16.1**). These changes are most evident in woody perennials but can be seen in many herbaceous species. Unlike the abrupt transition from the adult vegetative phase to the reproductive phase, the transition from juvenile to adult vegetative is usually gradual, involving intermediate forms.

Juvenile tissues are produced first and are located at the base of the shoot

The timing of the three developmental phases results in a spatial gradient of juvenility along the shoot axis. Because growth in height is restricted to the apical meristem, the juvenile tissues and organs form first at the base of the shoot. In rapidly flowering herbaceous species, the juvenile phase may last only a few days, and few juvenile structures are produced. In contrast, woody species have a more prolonged juvenile phase, in some cases lasting 30 to 40 years (**Table 16.1**). In these cases the juvenile structures can account for a significant portion of the mature plant.

Once the meristem has switched to the adult phase, only adult vegetative structures are produced, culminating in flowering. The adult and reproductive phases are therefore located in the upper and peripheral regions of the shoot.

Figure 16.1 Juvenile and adult forms of English ivy (*Hedera helix*). The juvenile form has lobed palmate leaves arranged alternately, a climbing growth habit, and no flowers. The adult form (projecting out to the right) has entire ovate leaves arranged in spirals, an upright growth habit, and flowers that develop into fruits.

Table 16.1 Length of Juvenile Period in Some Woody Plants

Species	Length of juvenile period
Rose (*Rosa* [hybrid tea])	20–30 days
Grape (*Vitis* spp.)	1 year
Apple (*Malus* spp.)	4–8 years
Citrus spp.	5–8 years
English ivy (*Hedera helix*)	5–10 years
Redwood (*Sequoia sempervirens*)	5–15 years
Sycamore maple (*Acer pseudoplatanus*)	15–20 years
English oak (*Quercus robur*)	25–30 years
European beech (*Fagus sylvatica*)	30–40 years

Source: Clark 1983.

In most species, growing to a sufficiently large size appears to be more important than the plant's chronological age in determining the transition to the adult phase. Conditions that retard growth, such as mineral deficiencies, low light, water stress, defoliation, and low temperature, prolong the juvenile phase or even cause reversion to juvenility of adult shoots. In contrast, conditions that promote vigorous growth accelerate the transition to the adult phase. When growth is accelerated, exposure to the correct flower-inducing treatment can induce flowering.

Although plant size seems to be the most important factor, it is not always clear which specific component associated with size is critical. In some *Nicotiana* species, it appears that plants must produce a certain number of leaves to transmit enough of the floral stimulus to the apex.

The adult phase is relatively stable and is maintained during vegetative propagation or grafting. For example, cuttings taken from the basal region of mature plants of English ivy develop into juvenile plants, while those taken from the tip region develop into adult plants. When scions were taken from the base of a flowering silver birch (*Betula verrucosa*) and grafted onto seedling rootstocks, there were no flowers on the grafts for the first 2 years. In contrast, grafts taken from the top of the mature tree flowered immediately.

The term *juvenility* has different meanings for herbaceous and woody species. Whereas juvenile herbaceous meristems flower readily when grafted onto flowering adult plants, juvenile woody meristems generally do not. Juvenile woody meristems are thus said to lack the competence to flower.

Phase changes can be influenced by nutrients, gibberellins, and epigenetic regulation

The transition from the juvenile to the adult phase can be affected by factors translocated from the rest of the plant to the shoot apex. In many plants, exposure to low-light conditions prolongs juvenility or causes reversion to juvenility. A major consequence of a low-light regime is a reduction in the supply of carbohydrates to the apex; thus, carbohydrate supply, especially sucrose, affects the size of the shoot apex and may play a role in the transition between juvenility and maturity. For example, in hybrid chrysanthemums (*Chrysanthemum morifolium*), flower primordia are not initiated until a minimum apex size has been reached. In some species, flowering is induced when sucrose is added exogenously.

In some plants gibberellin signaling is involved in these responses. In pines and some other conifers, gibberellins accumulate under conditions that promote cone production (e.g., root removal, water stress, and nitrogen starvation), and application of gibberellins is used to stimulate production of reproductive structures in juvenile trees, such as birch, citrus, and cherry.

In many herbaceous species, juvenile status is maintained by **epigenetic changes** in gene expression that are associated with transition to the adult phase. A common epigenetic mechanism for temporary repression of a large number of genes is the production of short, noncoding RNA sequences called microRNAs (miRNAs). These miRNAs contain sequences that are complementary to regions of mRNAs that are transcribed from target genes. Cellular mechanisms that recognize double-stranded RNAs degrade the target mRNA, thus preventing the synthesis of the protein. In Arabidopsis, two specific miRNAs that target adult/reproductive-phase genes are produced during juvenility. At the onset of the transition to the adult phase, the decreased expression of these miRNAs stabilizes the target mRNAs, resulting in increased expression of the proteins encoded by the adult/reproductive-phase genes.

epigenetic changes Chemical modifications to DNA and histones that cause heritable changes in gene activity without altering the underlying DNA sequence.

16.3 Photoperiodism: Monitoring Day Length

Describe photoperiodism, the types of photoperiodism, and how it interacts with the circadian clock to promote flowering.

photoperiodism A biological response to the length and timing of day and night, making it possible for an event to occur at a particular time of year.

Photoperiodism is the ability of an organism to detect day length to ensure that events occur at the appropriate time of year, thus allowing for a seasonal response. Photoperiodic phenomena are found in both animals and plants. In the animal kingdom, day length controls such seasonal activities as hibernation, development of summer and winter coats, and reproductive activity. Plant responses controlled by day length are numerous; they include the initiation of flowering, asexual reproduction, the formation of storage organs, and the onset of dormancy. In a natural environment, light and dark periods change seasonally according to latitude (**Figure 16.2**), and plants must have mechanisms to adjust to these changes to ensure survival and reproduction.

Plants can be classified according to their photoperiodic responses

Numerous plant species flower during the long days of summer, and for many years plant physiologists believed that the correlation between long days and flowering was a consequence of the accumulation of photosynthetic products synthesized during long days.

This hypothesis was shown to be incorrect by the work of Wightman Garner and Henry Allard, conducted in the 1920s at the U.S. Department of Agriculture

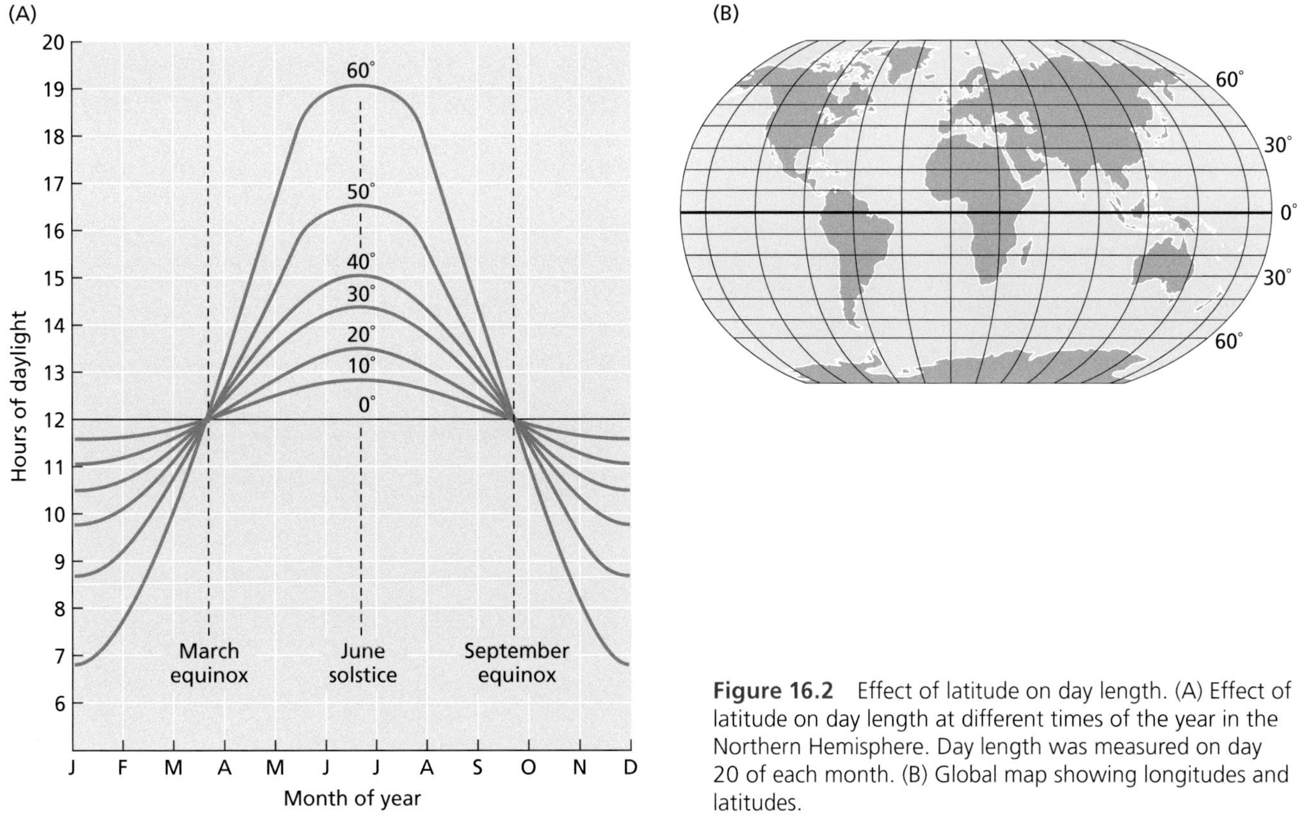

Figure 16.2 Effect of latitude on day length. (A) Effect of latitude on day length at different times of the year in the Northern Hemisphere. Day length was measured on day 20 of each month. (B) Global map showing longitudes and latitudes.

Figure 16.3 "Maryland Mammoth" mutant of tobacco (right) compared with wild-type tobacco (left). Both plants were grown during summer in the greenhouse. (University of Wisconsin graduate students used for scale.)

short-day plant (SDP) A plant that flowers only in short days (i.e., shorter than a critical day length) or whose flowering is accelerated by short days (quantitative SDP).

long-day plant (LDP) A plant that flowers only in long days (qualitative LDP) or whose flowering is accelerated by long days (quantitative LDP).

critical day length The minimum length of the day required for flowering of a long-day plant; the maximum length of day that will allow short-day plants to flower. However, studies have shown that it is the length of the night, not the length of the day, that is important.

laboratories in Beltsville, Maryland. Garner and Allard found that a mutant variety of tobacco, "Maryland Mammoth," grew profusely to about 5 m in height, but failed to flower in the summer (**Figure 16.3**). However, the plants flowered in the greenhouse during the winter under natural light conditions.

These results ultimately led Garner and Allard to test the effect of artificially shortened days by covering plants grown during the long days of summer with a light-tight tent from late in the afternoon until the following morning. These artificial short days also caused the plants to flower (like the short days in winter). Garner and Allard concluded that day length, rather than the accumulation of photosynthate, was the determining factor in flowering. They were able to confirm their hypothesis in many different species and conditions. This work laid the foundations for the extensive subsequent research on photoperiodic responses.

Although many other aspects of plants' development may also be affected by day length, flowering is the response that has been studied the most. Flowering species tend to fall into one of two main photoperiodic response categories: short-day plants and long-day plants.

- **Short-day plants (SDPs)** flower only in short days (*qualitative* SDPs), or their flowering is accelerated by short days (*quantitative* SDPs).
- **Long-day plants (LDPs)** flower only in long days (*qualitative* LDPs), or their flowering is accelerated by long days (*quantitative* LDPs).

The essential distinction between long-day and short-day plants is that flowering in LDPs is promoted only when the day length *exceeds* a certain duration, called the **critical day length**, in every 24-h cycle, whereas promotion of flowering in SDPs requires a day length that is *less than* the critical day length. The absolute value of the critical day length varies widely among species, and only when flowering is examined for a range of day lengths can the correct photoperiodic classification be established (**Figure 16.4**).

LDPs can effectively measure the lengthening days of spring or early summer and delay flowering until the critical day length is reached. Many varieties of bread wheat (*Triticum aestivum*) behave in this way. SDPs often flower in the fall when the days shorten below the critical day length, as in many varieties of *Chrysanthemum morifolium*. However, day length alone is an ambiguous signal, because the plant cannot distinguish between spring and autumn.

Plants exhibit several adaptations for avoiding the ambiguity of the day-length signal. One is the presence of a juvenile phase that prevents the plant from responding to day length during the spring. Another mechanism is the

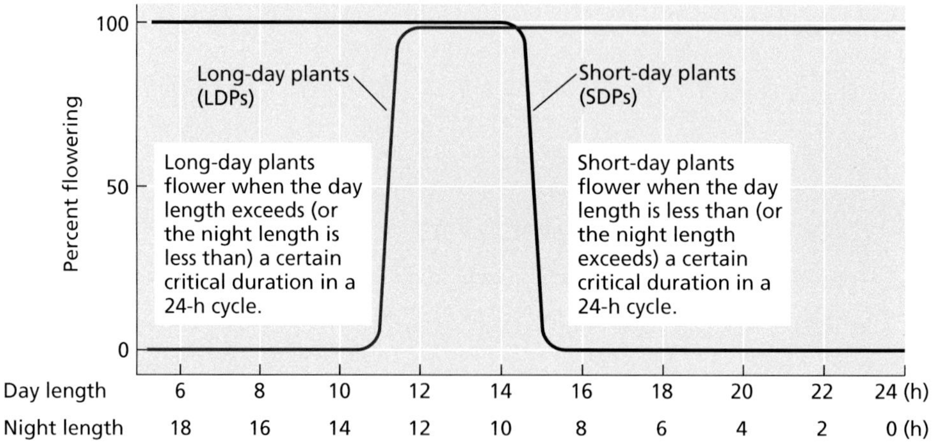

Figure 16.4 Photoperiodic response in long- and short-day plants. The critical day length varies among species. In this example, both the SDPs and the LDPs would flower in photoperiods between 12 and 14 h long.

coupling of a temperature requirement to a photoperiodic response. Certain plant species, such as winter wheat, a variety of bread wheat, do not respond to photoperiod until after a cold period (vernalization or overwintering) has occurred. (We discuss vernalization later in this chapter.)

Other plants avoid seasonal ambiguity by distinguishing between *shortening* and *lengthening* days. Such "dual–day length plants" fall into two categories:

- **Long–short-day plants (LSDPs)** flower only after a sequence of long days followed by short days. LSDPs, such as *Bryophyllum*, *Kalanchoe*, and night-blooming jasmine (*Cestrum nocturnum*), flower in the late summer and fall, when the days are shortening.
- **Short–long-day plants (SLDPs)** flower only after a sequence of short days followed by long days. SLDPs, such as white clover (*Trifolium repens*), Canterbury bells (*Campanula medium*), and echeveria (*Echeveria harmsii*), flower in the early spring in response to lengthening days.

Finally, species that flower under any photoperiodic condition are referred to as day-neutral plants. **Day-neutral plants (DNPs)** are insensitive to day length. Flowering in DNPs is typically under autonomous regulation—that is, internal developmental control. Some day-neutral species, such as maize (*Zea mays*), evolved near the equator where the day length is virtually constant throughout the year (Figure 16.2). Many desert annuals, such as desert paintbrush (*Castilleja chromosa*) and desert sand verbena (*Abronia villosa*), evolved to germinate, grow, and flower quickly whenever sufficient water is available. These are also DNPs.

long–short-day plant (LSDP) A plant that flowers in response to a shift from long days to short days.

short–long-day plant (SLDP) A plant that flowers only after a sequence of short days followed by long days.

day-neutral plant (DNP) A plant whose flowering is not regulated by day length.

circadian rhythm A physiological process that oscillates endogenously with a cycle of approximately 24 hours.

period In cyclic (rhythmic) phenomena, the time between comparable points in the repeating cycle, such as peaks or troughs.

Photoperiodism is one of many plant processes controlled by a circadian rhythm

Organisms are normally subjected to daily cycles of light and lack of light, usually referred to as darkness. Both plants and animals often exhibit rhythmic behavior in association with these changes. Examples of such rhythms include leaf and petal movements (day and night positions) (Chapter 13), stomatal opening and closing (Chapters 3, 6, and 13), growth and sporulation patterns in fungi (e.g., *Pilobolus* and *Neurospora*), time of day of pupal emergence (the fruit fly *Drosophila*), and activity cycles in rodents, as well as daily changes in the rates of metabolic processes such as photosynthesis and respiration.

When organisms are transferred from daily light–dark cycles to continuous darkness or continuous light, many of these rhythms continue to be expressed, at least for several days. Under such uniform conditions the period of the rhythm is close to 24 h, a **circadian rhythm** (from the Latin *circa*, "about," and *diem*, "day"). Because the rhythms continue under constant light or darkness, these circadian rhythms cannot be direct responses to the presence or absence of light, but must be based on an internal mechanism called an *endogenous oscillator*. A single oscillator mechanism can be tied to multiple downstream processes at different times. Endogenous oscillators are thought to be regulated by the interactions of four sets of genes expressed in the dawn, morning, afternoon, and evening hours. Light may augment the amplitude of the oscillation by activating the morning and evening genes.

Circadian rhythms exhibit characteristic features

Circadian rhythms arise from cyclic phenomena that can be depicted as wave forms and are defined by three parameters:

1. **Period** is the time between comparable points in the repeating cycle. Typically the period is measured as the time between consecutive maxima (peaks) or minima (troughs) (**Figure 16.5A**).

phase In cyclic (rhythmic) phenomena, any point in the cycle recognizable by its relationship to the rest of the cycle, for example the maximum and minimum positions.

amplitude In a biological rhythm, the distance between peak and trough; it can often vary while the period remains unchanged.

entrain Manipulation of external controlling factors, such as light or removal of light, to synchronize biological rhythms.

Zeitgebers Environmental signals such as light-to-dark or dark-to-light transitions that synchronize the endogenous oscillator to a 24-h periodicity.

2. **Phase** is any point in the cycle that is recognizable by its relationship to the rest of the cycle. The most obvious phase points are the peak and trough positions. (The term *phase* in this context should not be confused with the term *phase change* in meristem development discussed earlier.)
3. **Amplitude** is usually considered to be the distance between peak and trough. The amplitude of a biological rhythm can often vary while the period remains unchanged (as, for example, in **Figure 16.5B**).

In constant light or darkness, rhythms depart from an exact 24-h period. The rhythms then drift in relation to solar time, either gaining or losing time depending on whether the endogenous period is shorter or longer than 24 h. Plants can be **entrained** (synchronized) to a 24-h period (or any other period) by environmental signals; the most important of these are the light-to-dark transition at dusk and the dark-to-light transition at dawn (**Figure 16.5C**).

Such environmental signals are termed **Zeitgebers** (German for "time givers"). When such signals are removed—for example, by transfer to continuous

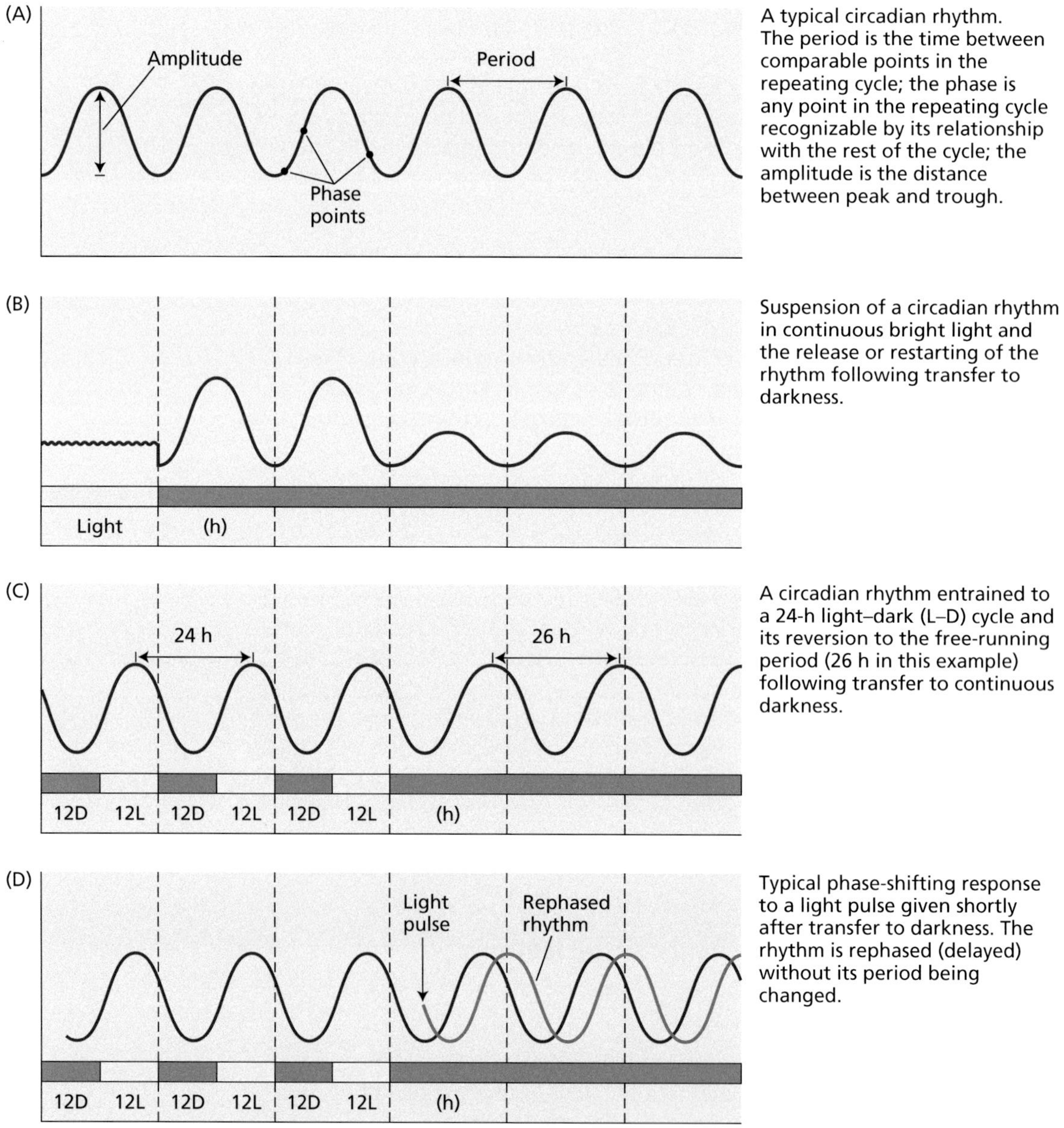

(A)
Amplitude
Period
Phase points
A typical circadian rhythm. The period is the time between comparable points in the repeating cycle; the phase is any point in the repeating cycle recognizable by its relationship with the rest of the cycle; the amplitude is the distance between peak and trough.
(B)
Light
(h)
Suspension of a circadian rhythm in continuous bright light and the release or restarting of the rhythm following transfer to darkness.
(C)
24 h
26 h
12D 12L 12D 12L 12D 12L (h)
A circadian rhythm entrained to a 24-h light–dark (L–D) cycle and its reversion to the free-running period (26 h in this example) following transfer to continuous darkness.
(D)
Light pulse
Rephased rhythm
12D 12L 12D 12L 12D 12L (h)
Typical phase-shifting response to a light pulse given shortly after transfer to darkness. The rhythm is rephased (delayed) without its period being changed.

Figure 16.5 Some characteristics of circadian rhythms.

darkness—the rhythm is said to be **free-running**, and it reverts to the circadian period that is characteristic of the particular organism. Although the rhythms are generated internally, they normally require an environmental signal, such as exposure to light or a change in temperature, to synchronize their expression. In addition, many rhythms "damp out" (i.e., the amplitude decreases) when the organism is subjected to a constant environment for several cycles. When this occurs, a Zeitgeber, such as a transfer from light to dark or a change in temperature, is required to restart the rhythm (Figure 16.5B). Note that the clock itself need not damp out; only the coupling between the molecular clock (endogenous oscillator) and the physiological function is affected.

The circadian clock would be of no value to the organism if it could not keep accurate time under the fluctuating temperatures experienced in natural conditions. Indeed, temperature has little or no effect on the period of the free-running rhythm. The feature that enables the clock to keep time at different temperatures is called **temperature compensation**.

free-running Designation of the biological rhythm that is characteristic for a particular organism when environmental signals are removed, as in total darkness. *See* Zeitgeber.

temperature compensation A characteristic of circadian rhythms, which can maintain their circadian periodicity over a broad range of temperatures within the physiological range.

photoperiodic induction The photoperiod-regulated processes that occur in leaves resulting in the transmission of a floral stimulus to the shoot apex.

Circadian rhythms adjust to different day–night cycles

How do circadian rhythms remain constant when the daily durations of light and darkness change with the seasons? Investigators typically test the response of the endogenous oscillator by placing the organism in continuous darkness and examining the response to a short pulse of light (usually <1 h) given at different time points in the free-running rhythm. If a light pulse is given during the first few hours of the original night period, the phase of the rhythm is delayed; the organism interprets the light pulse as the end of the previous day (**Figure 16.5D**). As would be expected, a light pulse given toward the end of the original night period advances the rhythm phase; now the organism interprets the light pulse as the beginning of the following day.

The biochemical mechanism that enables a light signal to cause these phase shifts is not yet known, but photoreceptor studies (Chapter 13) have improved our understanding of how light regulates the process. The low levels and specific wavelengths of light that can induce phase-shifting indicate that the light response must be mediated by specific photoreceptors rather than by photosynthetic rate. Phytochromes are the primary photoreceptors influencing circadian rhythms, but cryptochromes also participate in blue-light entrainment of the clock in plants.

The leaf is the site of perception of the photoperiodic signal

The photoperiodic stimulus in both LDPs and SDPs is perceived by the leaves. For example, treatment of a single leaf of the SDP *Xanthium* (cocklebur) with short photoperiods is sufficient to cause the formation of flowers, even when the rest of the plant is exposed to long days. Thus, in response to photoperiod the leaf transmits a signal that regulates the transition to flowering at the shoot apex. The photoperiod-regulated processes that occur in the leaves resulting in the transmission (translocation) of a floral stimulus to the shoot apex are referred to collectively as **photoperiodic induction**.

Photoperiodic induction can take place in a leaf that has been separated from the plant. For example, in the SDP *Perilla crispa* (a member of the mint family), an excised leaf exposed to short days can cause flowering when subsequently grafted to a noninduced plant maintained in long days (**Figure 16.6**). This result indicates that photoperiodic induction depends on events that take place exclusively in the leaf.

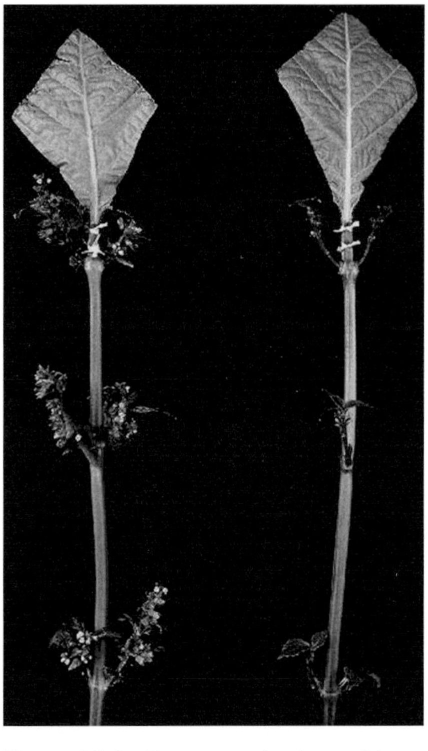

Figure 16.6 Demonstration by grafting of a leaf-generated floral stimulus in the SDP *Perilla crispa*. (Left) Grafting an induced leaf from a plant grown under short days onto a noninduced shoot causes the axillary shoots to produce flowers. The donor leaf has been trimmed to facilitate grafting, and the upper leaves have been removed from the stock to promote phloem translocation from the scion to the receptor shoots. (Right) Grafting a noninduced leaf from a plant grown under long days results in the formation of vegetative branches only. The graft junction is below the leaf.

Plants monitor day length by measuring the length of the night

Under natural conditions, day and night lengths configure a 24-h cycle of light and darkness. In principle, a plant could perceive a critical day length by measuring the duration of either light or darkness. Flowering of SDPs is primarily

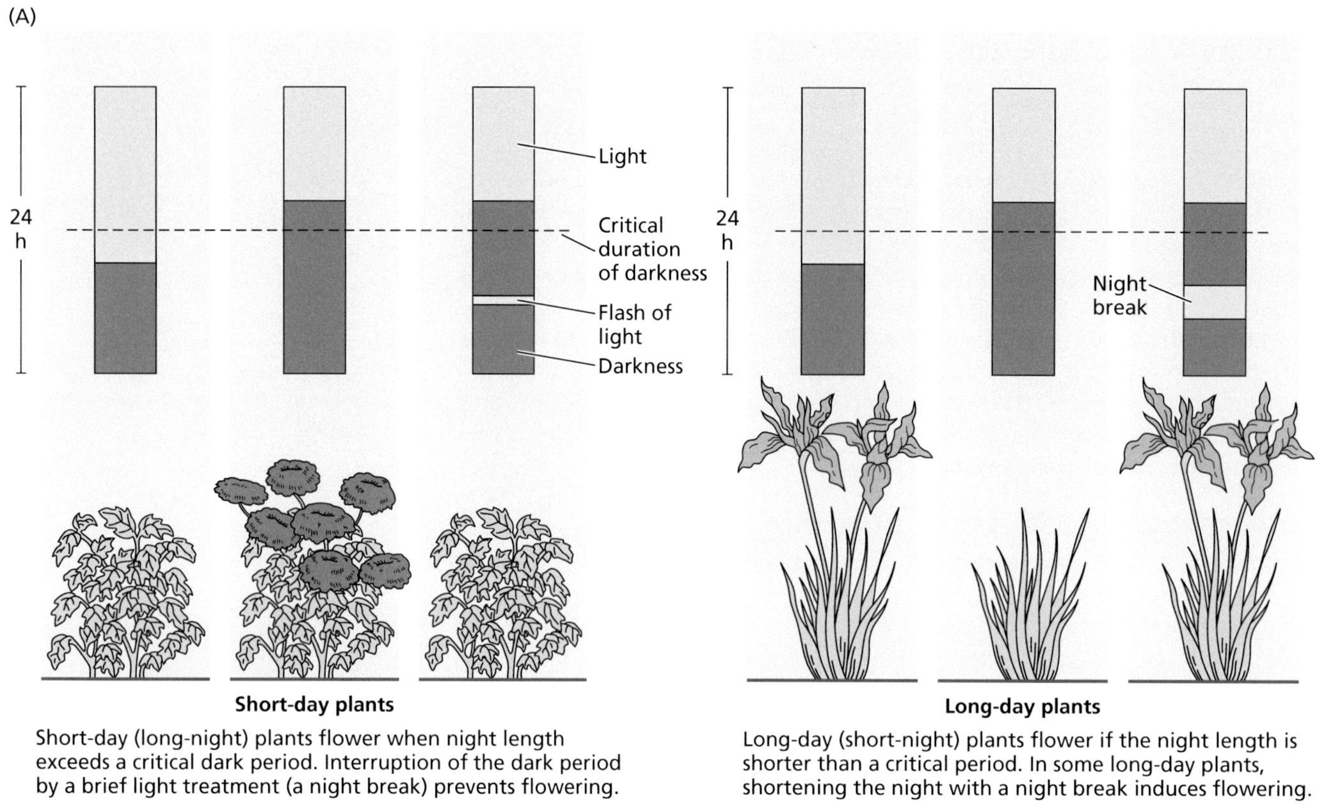
(A)
24 h
Light
Critical duration of darkness
Flash of light
Darkness
24 h
Night break
Short-day plants
Short-day (long-night) plants flower when night length exceeds a critical dark period. Interruption of the dark period by a brief light treatment (a night break) prevents flowering.
Long-day plants
Long-day (short-night) plants flower if the night length is shorter than a critical period. In some long-day plants, shortening the night with a night break induces flowering.

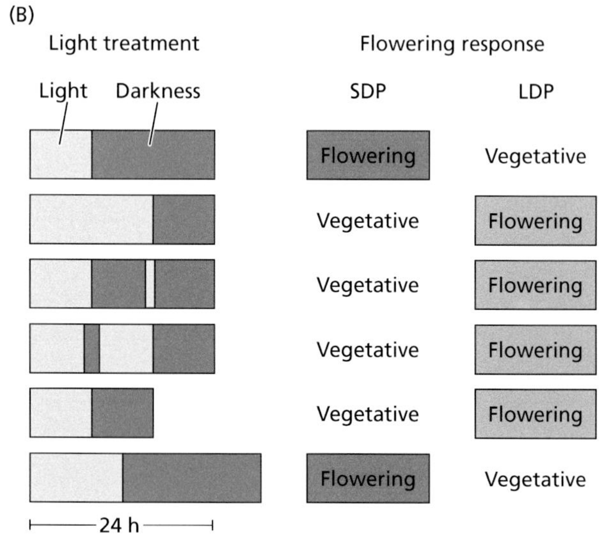
(B)
Light treatment
Light
Darkness
Flowering response
SDP
LDP
Flowering
Vegetative
Vegetative
Flowering
Vegetative
Flowering
Vegetative
Flowering
Vegetative
Flowering
Flowering
Vegetative
24 h

Figure 16.7 Photoperiodic regulation of flowering. (A) Effects on SDPs and LDPs. (B) Effects of the duration of the dark period on flowering. Treating SDPs and LDPs with different photoperiods clearly shows that the critical variable for flowering is actually the length of the dark period rather than the day length.

determined by the duration of darkness (**Figure 16.7A**). Flowering can be induced in SDPs with light periods longer than the critical value, provided that these were followed by sufficiently long nights (**Figure 16.7B**). Similarly, SDPs did not flower when short days were followed by short nights.

More detailed experiments demonstrated that the mechanism of photoperiodic timekeeping in SDPs is based on the duration of darkness. For example, flowering occurred only when the dark period exceeded 8.5 h in cocklebur (*Xanthium strumarium*) or 10 h in soybean (*Glycine max*). The duration of darkness was also shown to be important in LDPs (Figure 16.7B). These plants flowered in short days, provided that the accompanying night length was also short; however, a regime of long days followed by long nights was ineffective.

Night breaks can cancel the effect of the dark period

A feature that underscores the importance of the dark period is that it can be made ineffective by interruption with a short exposure to light, called a **night break** (Figure 16.7A). In contrast, interrupting a long day with a brief dark period does not cancel the effect of the long day (Figure 16.7B). Night-break treatments of only a few minutes are effective in *preventing* flowering in many SDPs, including *Xanthium* and *Pharbitis*, but much longer exposures are often required to *promote* flowering in LDPs. Consistent with the involvement of a circadian rhythm, the effect of a night break varies greatly according to the time when it is given. For both LDPs and SDPs, a night break was found to be most effective when given near the middle of a dark period of 16 h (**Figure 16.8**).

night break An interruption of the dark period with a short exposure to light that makes the entire dark period ineffective.

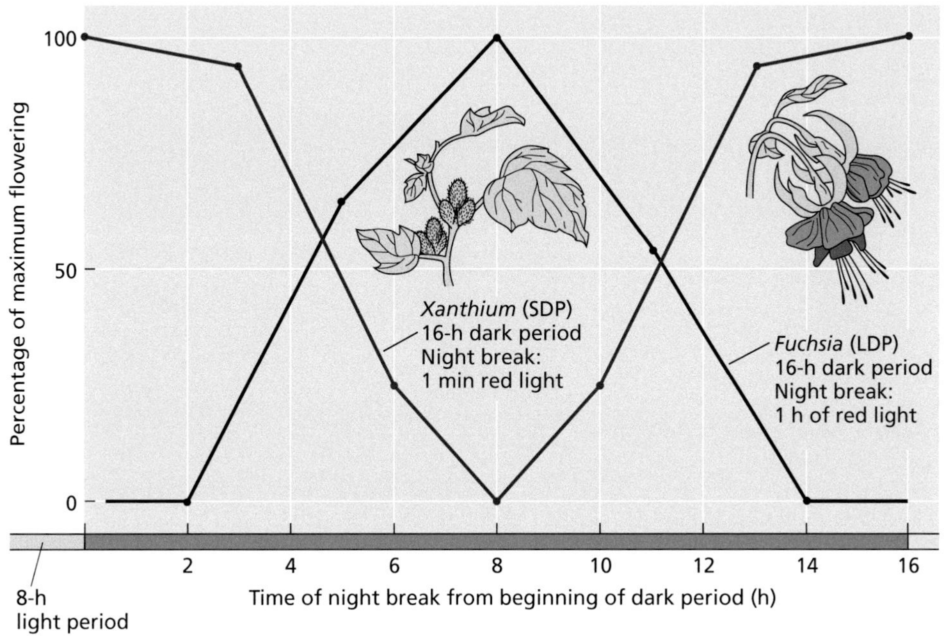

Figure 16.8 The time at which a night break is given determines the flowering response. When given during a long dark period, a night break promotes flowering in LDPs and inhibits flowering in SDPs. In both cases, the greatest effect on flowering occurs when the night break is given near the middle of the 16-h dark period. The LDP *Fuchsia* was given a 1-h exposure to red light in a 16-h dark period. The SDP *Xanthium* was exposed to red light for 1 min in a 16-h dark period.

The discovery of the night-break effect, and its time dependence, had several important consequences. It established the central role of the dark period and provided a valuable probe for studying photoperiodic timekeeping. Because only small amounts of light are needed, it became possible to study the action and identity of the photoreceptor without the interfering effects of photosynthesis and other nonphotoperiodic phenomena. This discovery has also led to the development of commercial methods for regulating the time of flowering in horticultural species, such as *Kalanchoe*, chrysanthemum, and poinsettia (*Euphorbia pulcherrima*).

Photoperiodic timekeeping during the night depends on a circadian clock

Measuring the passage of time in darkness is important for photoperiodic timekeeping and appears to depend on an endogenous circadian oscillator (Figure 16.5). The role of circadian rhythms in photoperiodic timekeeping can be investigated by measuring flowering after interrupting an extended night period with a short night break. For example, when short-day soybean plants are transferred from an 8-h light period to a 64-h dark period, the flowering response to 4-h night breaks given at different times during the dark period shows a circadian rhythm (**Figure 16.9**). Soybean plants would normally flower in response to 8-h day lengths. *Inhibition* of flowering by a night break (Figure 16.7A) therefore corresponds to a period of light sensitivity. As shown in Figure 16.9, during a 64-h night period, the peaks of light inhibition of flowering occur at 24-h intervals, indicating that the phases of light sensitivity and insensitivity show circadian periodicity even during continuous darkness.

The coincidence model links oscillating light sensitivity and photoperiodism

How does an oscillation with a 24-h period measure a critical duration of darkness of, say, 8 to 9 h, as in the SDP *Xanthium*? In 1936 Erwin Bünning proposed that the control of flowering by photoperiodism is achieved by an oscillation of phases with different sensitivities to light. This proposal has evolved into the **coincidence model**, whereby the endogenous oscillator controls the timing of light-sensitive and light-insensitive phases.

coincidence model A model for flowering in photoperiodic plants in which the circadian oscillator controls the timing of light-sensitive and light-insensitive phases during the 24-h cycle.

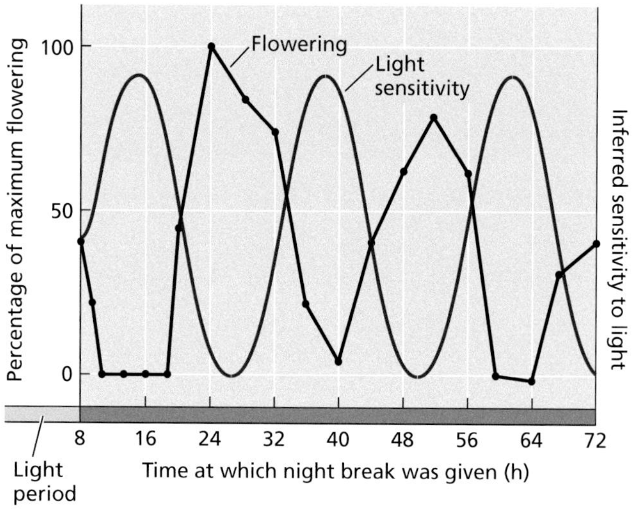

Figure 16.9 Rhythmic flowering in response to night breaks. In this experiment, the SDP soybean (*Glycine max*) received cycles of an 8-h light period followed by a 64-h dark period. A 4-h night break (light treatment) was given at various times during the long inductive dark period. The flowering response, plotted as the percentage of the maximum, was then plotted for each night break given. Note, for example, that a night break given at 26 h resulted in maximum flowering, while no flowering was obtained when the night break was given at 40 h. Since in SDPs light acts as an inhibitor of flowering during the long night period, we can infer that 26 h corresponds to a minimum for light sensitivity, while 40 h corresponds to a maximum for light sensitivity. The flowering data can thus be used to infer the periodicity of the plants' sensitivity to the effect of the night break over time (the red curve), which exhibits a circadian rhythm. These data support a model in which flowering in SDPs is induced only when dawn (or a night break) occurs after the completion of the light-sensitive phase. Since in LDPs light stimulates flowering during the dark period, the light break must coincide with the light-sensitive phase for flowering to occur in LDPs.

The ability of light either to promote or to inhibit flowering depends on the phase in which the light is given. When a light signal is administered during the light-sensitive phase of the rhythm, the effect is either to *promote* flowering in LDPs or to *prevent* flowering in SDPs. As shown in Figure 16.9, the phases of sensitivity and insensitivity to light continue to oscillate in darkness. Flowering in SDPs is induced only when exposure to light during a night break or at dawn occurs after completion of the light-sensitive phase of the rhythm.

If a similar experiment is performed with an LDP, flowering is induced only when the night break occurs *during* the light-sensitive phase of the rhythm. In other words, *flowering in both SDPs and LDPs is induced when the light exposure is coincident with the appropriate phase of the rhythm*. This continued oscillation of sensitive and insensitive phases in the absence of dawn and dusk light signals is characteristic of a variety of processes controlled by the circadian oscillator.

Molecular evidence from monocot and eudicot species supports the function of a coincidence model in the control of floral induction in both SDPs and LDPs. The mechanisms are highly conserved and involve the expression in the leaf of a conserved floral induction gene coding for a transcriptional activator that serves as a master switch. The expression of this master switch gene is controlled by the circadian clock, with the peak of gene expression occurring about 12 to 16 h after dawn (**Figure 16.10**). In LDPs such as Arabidopsis, the master switch protein acts as an inducer of flowering, while in SDPs such as rice it acts as an inhibitor. Because the master switch mRNA reaches its peak around 16 h after dawn, it will be coincident with light exposure only under long days. Therefore, under short-day conditions, when the master switch protein is not coincident with light, the master switch protein does not accumulate and LDPs will remain vegetative (Figure 16.10A). In contrast, SDPs will flower under short-day conditions because the master switch protein acts as an inhibitor (Figure 16.10C). Under long-day conditions, when the master switch mRNA is coincident with light, the master switch protein accumulates and LDPs will flower (Figure 16.11B), whereas SDPs will remain vegetative (Figure 16.11D).

Phytochrome is the primary photoreceptor in photoperiodism

Night-break experiments led to the identification of the photoreceptors that perceive light signals during the photoperiodic response. The inhibition of flowering in SDPs by night breaks was one of the first physiological processes shown to be under the control of phytochrome (**Figure 16.11**).

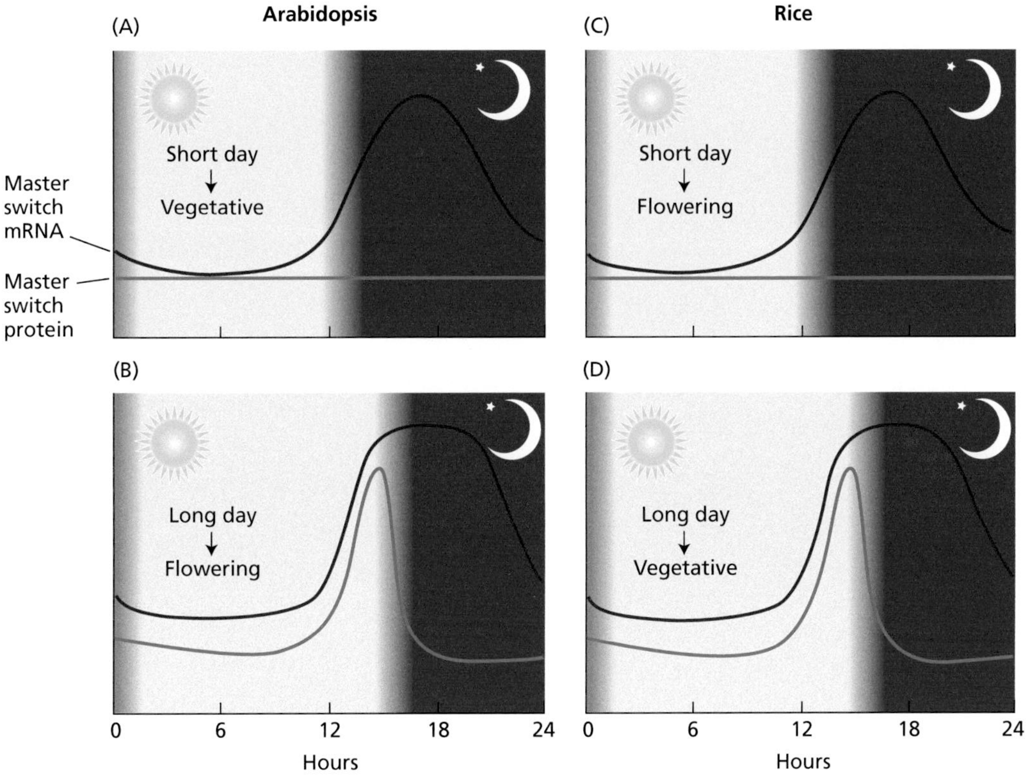

Figure 16.10 The coincidence model of photoperiodic induction in LDP Arabidopsis (A and B) and SDP rice (C and D). (A) In Arabidopsis under short days, there is little overlap between the expression of the master switch gene and daylight, and the plant remains vegetative. (B) Under long days, the peak of expression of the master switch gene (at hours 12–16) overlaps with daylight (sensed by phytochrome), allowing the protein to accumulate. (C) In rice under short days, the lack of coincidence between the master switch gene mRNA expression and daylight prevents the accumulation of the protein, which in this case is an inhibitor of flowering. In the absence of the inhibitor, the plants flower. (D) Under long days (sensed by phytochrome), the peak of the master switch gene expression overlaps with the day, allowing the accumulation of the repressor protein. As a result, the plant remains vegetative.

In many SDPs, a night break becomes effective only when the supplied dose of light is sufficient to saturate the photoconversion of **Pr** (phytochrome that absorbs red light) to **Pfr** (phytochrome that absorbs far-red light) (Chapter 13). A subsequent exposure to far-red light, which photoconverts the pigment back to the physiologically inactive Pr form, restores the flowering response.

Action spectra for the inhibition and restoration of the flowering response in dark-grown and green SDPs are shown in **Figure 16.12**. The action spectrum shows a peak at 660 nm (the absorption maximum of Pr) when dark-grown *Pharbitis* seedlings are used to avoid interference from chlorophyll. In contrast, the spectrum for *Xanthium* shows the response in green plants, and that the presence of chlorophyll can cause some discrepancy between the action spectrum and the absorption spectrum of Pr. These action spectra plus the red/far-red reversibility of the night-break responses confirm the role of phytochrome as the photoreceptor involved in photoperiod measurement in SDPs. Confirming the function of phytochrome, flowering is promoted by red-light night breaks, and a subsequent exposure to far-red light prevents this response (Figure 16.11).

A circadian rhythm in the promotion of flowering by far-red light has been observed in the LDPs darnel ryegrass (*Lolium temulentum*), barley (*Hordeum vulgare*), and Arabidopsis. The response is proportional to the irradiance and duration of far-red light and is therefore a high-irradiance response (HIR; Chapter 13). As in other HIRs, phyA is the phytochrome that mediates the response to far-red light.

Pr The red light–absorbing form of phytochrome. This is the form in which phytochrome is assembled. The cyan-blue colored Pr is converted by red light to the far-red light–absorbing form, Pfr.

Pfr The far-red light–absorbing form of phytochrome converted from Pr by the action of red light. The cyan-green colored Pfr is converted back to Pr by far-red light. Pfr is the physiologically active form of phytochrome.

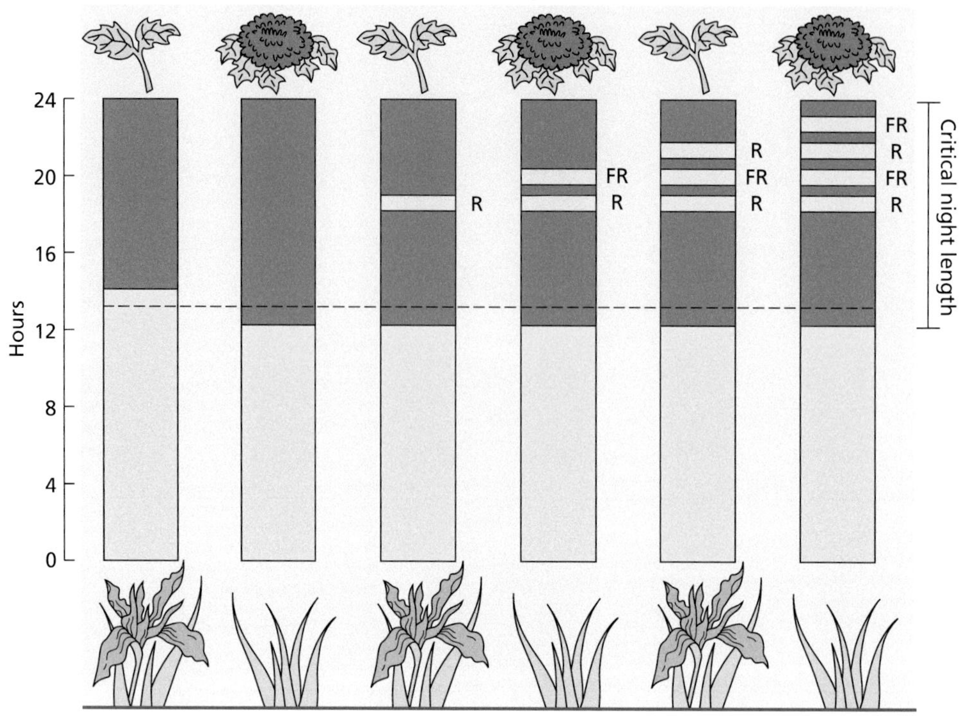

Figure 16.11 Phytochrome control of flowering by red (R) and far-red (FR) light. A flash of red light during the dark period induces flowering in an LDP, and the effect is reversed by a flash of far-red light. This response indicates the involvement of phytochrome. In SDPs, a flash of red light prevents flowering, and the effect is reversed by a flash of far-red light.

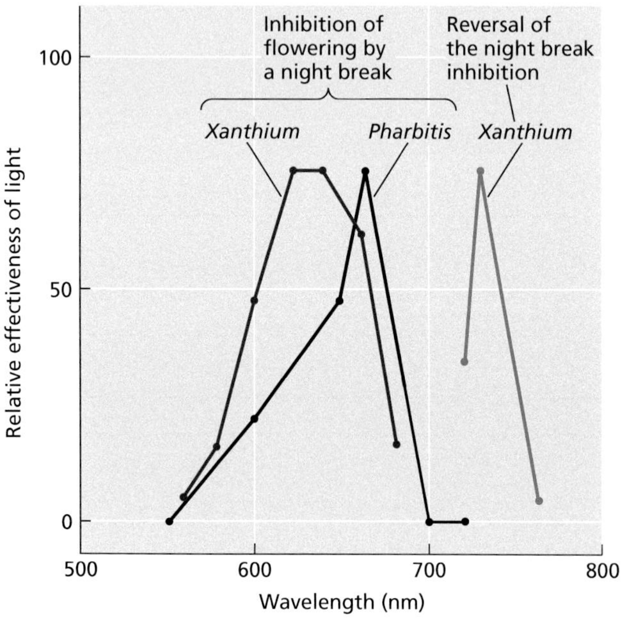

Figure 16.12 Action spectra for the control of flowering by night breaks implicate phytochrome. Flowering in SDPs grown under inductive short-day conditions is inhibited by a short light treatment (night break) during the long night. In the SDP *Xanthium strumarium*, red-light night breaks of 620 to 640 nm are the most effective wavelengths. Reversal of the red-light effect is maximal at 725 nm. In the dark-grown SDP *Pharbitis nil*, which is devoid of chlorophyll and its interference with light absorption, night breaks of 660 nm are the most effective. This 660-nm maximum coincides with the absorption maximum of phytochrome. The data were normalized relative to the dark control at each wavelength.

16.4 Vernalization: Promoting Flowering with Cold Treatment

Describe vernalization and how it mediates epigenetic changes that allow flowering.

Vernalization is the process whereby repression of flowering is alleviated by a cold treatment given to a hydrated seed (i.e., a seed that has imbibed water) or to a growing plant (dry seeds do not respond to the cold treatment because vernalization is an active metabolic process). Without the cold treatment, plants that require vernalization show delayed flowering or remain vegetative, and they are not competent to respond to floral signals such as inductive photoperiods. This requirement is important for production of overwintering cereal crops such as barley (**Figure 16.13**) and in overwintering annual eudicots that grow as rosettes with no elongation of the stem until longer days signal the end of the frost period

vernalization In some species, the cold temperature requirement for flowering. The term is derived from the Latin word for "spring."

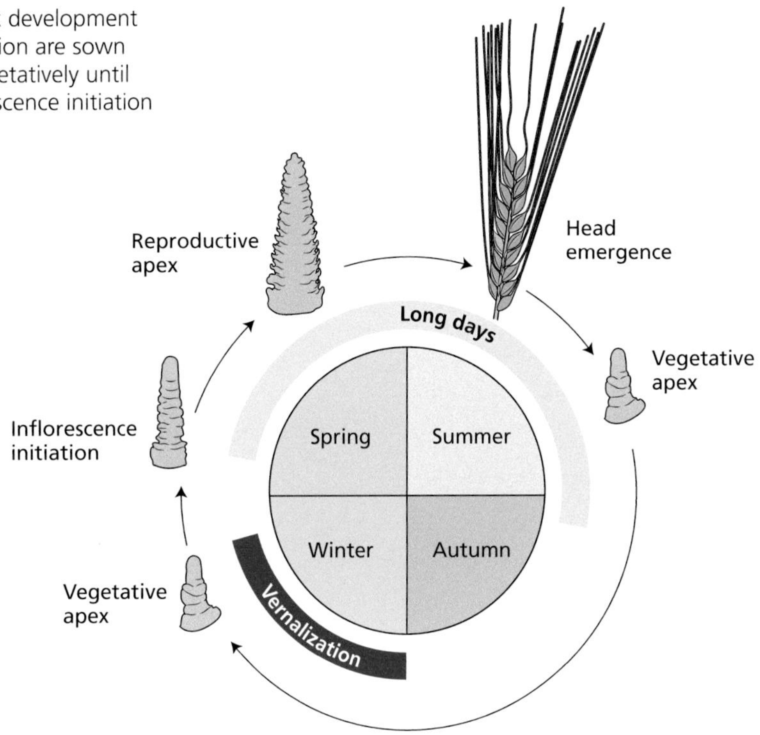

Figure 16.13 Influence of seasonal cues on shoot apex development in the temperate cereals. Varieties that require vernalization are sown in late summer or autumn. The shoot apex develops vegetatively until winter, when vernalization occurs. This promotes inflorescence initiation as temperatures increase in spring.

(**Figure 16.14**). (Chapter 14 describes *stratification,* whereby hydrated seeds are sometimes chilled in order to break seed dormancy.)

Plants differ considerably in the age at which they can become vernalized. Winter annuals, such as the winter forms of cereals (sown in the fall and flower in the following summer), respond to low temperature very early in their life cycle. In fact, many winter annuals can be vernalized before germination (i.e., radicle emergence from the seed) if the seeds have imbibed water and become metabolically active. Other plants, including most biennials (which grow as rosettes during the first season after sowing and flower in the following summer), must reach a minimum size before they become sensitive to low temperature for vernalization.

The effective temperature range for vernalization is from just below freezing to about 10°C, with an optimum usually between 1 and 7°C. The effect increases with the duration of the cold treatment until the response is saturated. The response usually requires several weeks of exposure to low temperature, but the precise duration varies widely with species and variety. High temperature and other stresses can "devernalize" overwintering annuals, but the longer the exposure to low temperature, the more irreversible the vernalization effect (**Figure 16.15**).

Figure 16.14 Vernalization induces flowering in the winter-annual types of *Arabidopsis thaliana*. The plant on the left is a winter-annual type that has not been exposed to cold. The plant on the right is a genetically identical winter-annual type that was exposed to 40 days of temperatures slightly above freezing (4°C) as a seedling. It flowered 3 weeks after the end of the cold treatment with about nine leaves on the primary stem.

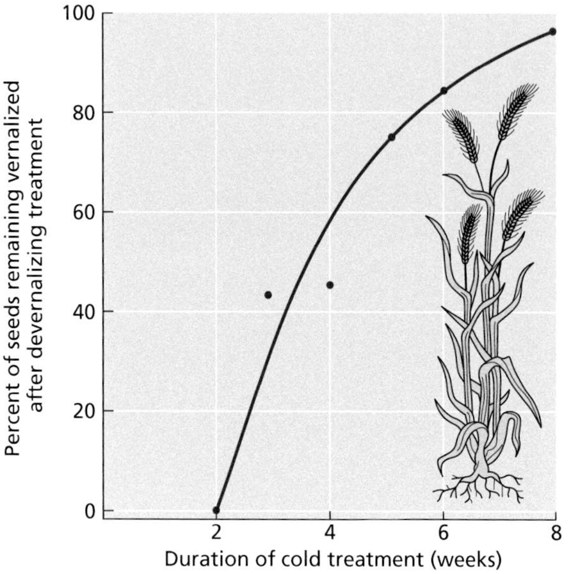

Figure 16.15 Duration of exposure to low temperature increases the stability of the vernalization effect. The longer that winter rye (*Secale cereale*) is exposed to a cold treatment, the greater the number of plants that remain vernalized when the cold treatment is followed by a devernalizing treatment. In this experiment, seeds of rye that had imbibed water were exposed to 5°C for different lengths of time, then immediately given a devernalizing treatment of 3 days at 35°C.

Vernalization appears to take place primarily in the SAM. Localized cooling of the shoot apex causes flowering, and this effect appears to be largely independent of the temperature experienced by the rest of the plant. This effect is clearly visualized in winter cereals, where experimental temperature treatments of isolated stem apices are relatively easy.

In developmental terms, vernalization results in the acquisition of competence of the meristem to undergo the floral phase transition. However, competence to flower does not guarantee that flowering *will* occur: A vernalization requirement is often linked to a requirement for a particular inductive photoperiod. The most common combination is a requirement for cold treatment *followed* by a requirement for long days—a combination that leads to flowering in early summer at high latitudes.

One model for how vernalization stably affects competence is that there are changes in the pattern of gene expression in the meristem after cold treatment that persist into the spring and throughout the remainder of the life cycle. These epigenetic changes are stable even after the signal (in this case cold) that induced the change is no longer present.

16.5 Long-Distance Signaling Involved in Flowering

Summarize the signals that travel from the leaves to the meristems to promote the transition to the flowering phase.

Although floral evocation occurs at the apical meristems of shoots, in photoperiodic plants inductive photoperiods are sensed by the leaves. A long-range signal must move from the leaves to the apex, as demonstrated experimentally through extensive grafting experiments in many different plant species (Figure 16.7).

In some plant species, noninduced plants can be stimulated to flower by having a leaf or shoot from a photoperiodically induced donor plant grafted to them. For example, in the SDP *Perilla crispa*, grafting a leaf from a plant grown under inductive short days onto a plant grown under noninductive long days causes the latter to flower (Figure 16.7). The movement of the floral stimulus from a donor leaf to the rest of the plant requires establishment of vascular continuity across the graft union and correlates closely with the translocation of ^{14}C-labeled assimilates from the donor. Disruptions of the phloem also prevent flowering, and experimental measurements and modeling show that floral induction correlates with phloem translocation rates.

The floral stimulus seems to be the same in plants with different photoperiodic requirements. Thus, grafting an induced shoot from the LDP *Nicotiana sylvestris*, grown under long days, onto the SDP "Maryland Mammoth" tobacco caused the latter to flower under noninductive (long-day) conditions. The leaves of DNPs have also been shown to produce a graft-transmissible floral stimulus (**Table 16.2**). For example, grafting a single leaf of a day-neutral variety

Table 16.2 Transmission of the Flowering Signal Occurs Through a Graft Junction

Donor plants maintained under flower-inducing conditions	Photoperiod type	Vegetative receptor plant induced to flower	Photoperiod type
Helianthus annus	DNP in LD	*H. tuberosus*	SDP in LD
Nicotiana tabacum "Delcrest"	DNP in SD	*N. sylvestris*	LDP in SD
Nicotiana sylvestris	LDP in LD	*N. tabacum* "Maryland Mammoth"	SDP in LD
Nicotiana tabacum "Maryland Mammoth"	SDP in SD	*N. sylvestris*	LDP in SD

Note: The successful transfer of a flowering induction signal by grafting between plants of different photoperiodic response groups shows the existence of a transmissible floral hormone that is effective. DNPs = day-neutral plants; LD = long days; LDPs = long-day plants; SD = short days; SDPs = short-day plants.

of soybean, "Agate," onto the short-day variety, "Biloxi," caused flowering in "Biloxi" even when the latter was maintained in noninductive long days. Similarly, a shoot from a day-neutral variety of tobacco (*Nicotiana tabacum*, cv. Trapezond) grafted onto the LDP *Nicotiana sylvestris* induced the latter to flower under noninductive short days.

In Arabidopsis and other species, the transmissible flowering stimulus has been shown to be a small globular protein called FLOWERING LOCUS T (FT), which is expressed in companion cells and is translocated to the SAM, where FT interacts with transcription factors to induce flowering. Transcription of the gene *CONSTANS* (or *HEADING* (*flowering*) *DATE 1* in cereal plants) is regulated by the circadian clock. *CONSTANS* acts as a light-inducible master switch in leaves and regulates the movement of FT into the phloem, where it is translocated along the source–sink gradient to the shoot apex. Once in the floral meristem, the FT protein enters the nuclei of the cells and forms an active complex with the transcription factor FLOWERING LOCUS D (FD). The FT–FD complex then activates various floral identity genes (discussed later) that convert the vegetative meristem into a floral meristem. This process is diagrammed in **Figure 16.16**.

Gibberellins and ethylene can induce flowering

Gibberellins promote the transition from vegetative to reproductive phases in many plants. Exogenous gibberellin can induce flowering when applied either to rosette LDPs such as Arabidopsis, or to dual–day-length plants such as the long–short-day plant *Bryophyllum*, when grown under short days. Gibberellin can also induce flowering in cold-requiring plants that have not been vernalized and promote cone formation in juvenile plants of several gymnosperm families. Thus, in some plants exogenous gibberellins can bypass the endogenous trigger of age in autonomous flowering, as well as the primary environmental signals of day length and temperature.

Gibberellin metabolism in the plant is strongly affected by day length. For example, in the LDP spinach (*Spinacia oleracea*), the levels of gibberellins are relatively low in short days, and the plants maintain a rosette form. After the plants are transferred to long days, there is a fivefold increase in the level of physiologically active gibberellin, GA_1, which causes the marked stem elongation that accompanies flowering.

In addition to gibberellins, other growth hormones can either inhibit or promote flowering. One commercially important example is the promotion of flowering in pineapple (*Ananas comosus*) by ethylene and ethylene-releasing compounds—a response that appears to be restricted to members of the pineapple family (*Ananas comosus*).

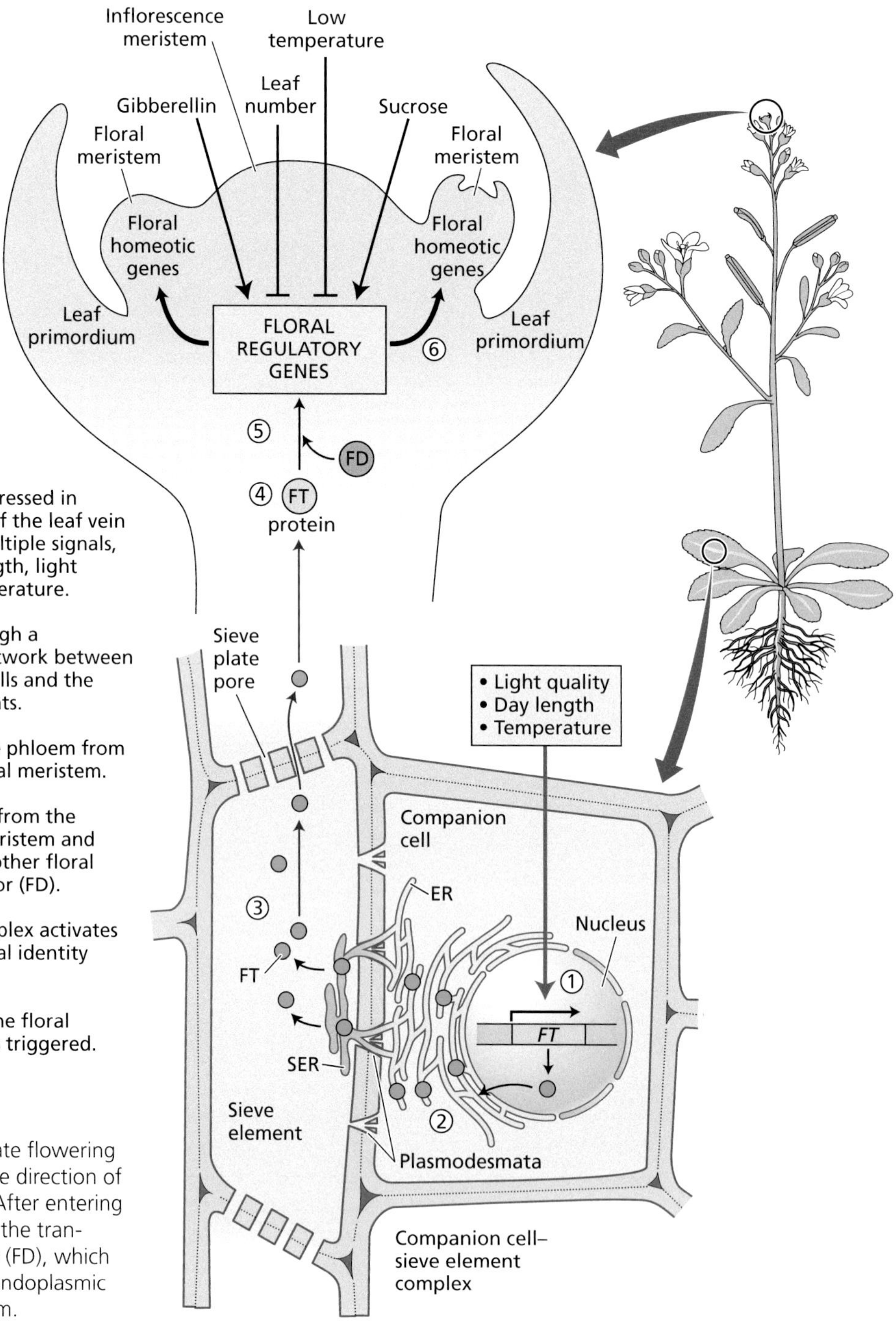

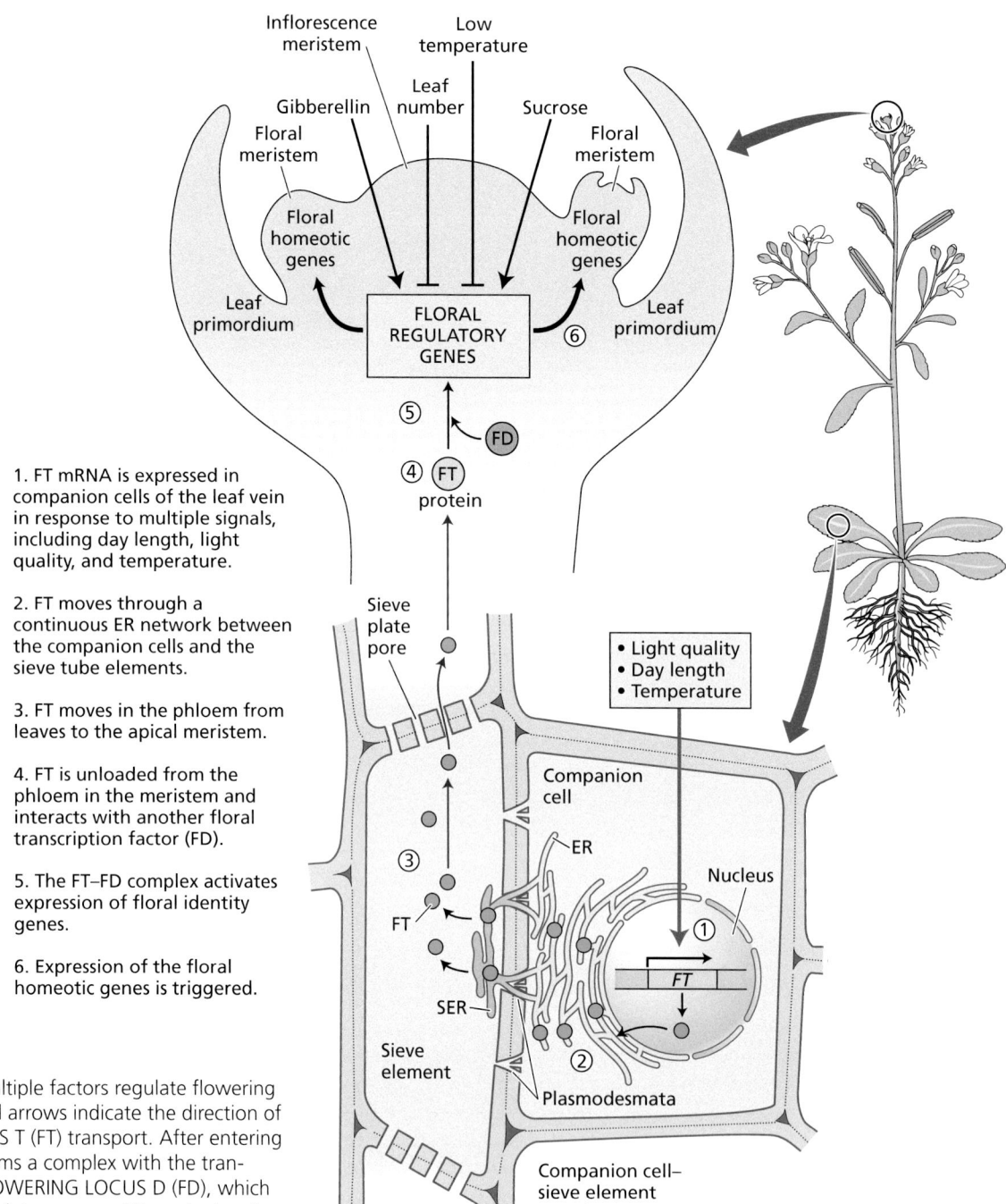

1. FT mRNA is expressed in companion cells of the leaf vein in response to multiple signals, including day length, light quality, and temperature.

2. FT moves through a continuous ER network between the companion cells and the sieve tube elements.

3. FT moves in the phloem from leaves to the apical meristem.

4. FT is unloaded from the phloem in the meristem and interacts with another floral transcription factor (FD).

5. The FT–FD complex activates expression of floral identity genes.

6. Expression of the floral homeotic genes is triggered.

Figure 16.16 Multiple factors regulate flowering in Arabidopsis. Red arrows indicate the direction of FLOWERING LOCUS T (FT) transport. After entering the nucleus, FT forms a complex with the transcription factor FLOWERING LOCUS D (FD), which activates floral regulatory genes. ER, endoplasmic reticulum; SER, sieve element reticulum.

16.6 Floral Meristems and Floral Organ Development

Describe the floral meristem identity genes and the floral organ identity genes, and how the floral organ identity genes interact.

Once flowering has been induced, the process of building flowers begins. The shapes of flowers are extremely diverse, reflecting adaptations to protect developing gametophytes, attract pollinators, promote self-pollination or cross-pollination as appropriate, and produce and disperse fruits and seeds. Despite this diversity, genetic and molecular studies have identified a network of genes that control floral morphogenesis in flowers across species. We will focus on floral development in Arabidopsis, which has been studied extensively.

The SAM in Arabidopsis changes with development

Floral meristems can usually be distinguished from vegetative meristems by their larger size. In the vegetative meristem, the cells of the central zone complete their division cycles slowly. The transition from vegetative to reproductive development is marked by an increase in the frequency of cell divisions within the central zone of the SAM. The increase in the size of the meristem is largely a result of the increased division rate of these central cells.

During the vegetative phase of growth, SAMs give rise to leaf primordia on the flanks of the shoot apex (**Figure 16.17A**). When reproductive development is initiated, the vegetative meristems are transformed either directly into a floral meristem, or indirectly into a primary inflorescence meristem, depending on the species. In rosette plants such as Arabidopsis, the **primary inflorescence meristem** produces an elongated inflorescence axis bearing two types of lateral organs: cauline (stem-borne) leaves and flowers. The axillary buds of the cauline leaves develop into **secondary inflorescence meristems** (not shown in figure), and their activity repeats the pattern of development of the primary inflorescence meristem. The Arabidopsis inflorescence meristem has the potential to grow indefinitely and thus exhibits **indeterminate growth**. Flowers arise from **floral meristems** that form on the flanks of the inflorescence meristem (**Figure 16.17B**). In contrast to the inflorescence meristem, the floral meristem is determinate.

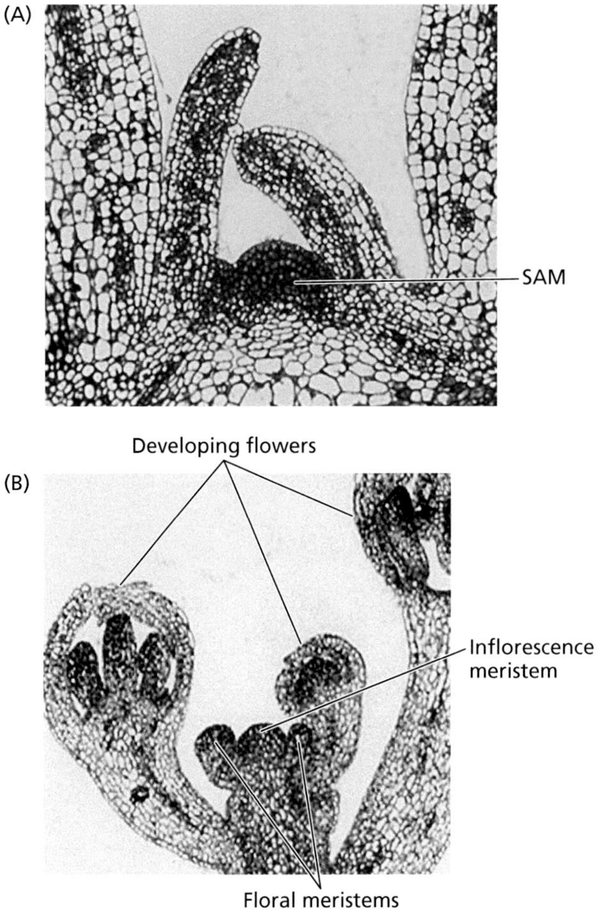

Figure 16.17 Longitudinal sections through a vegetative (A) and a reproductive (B) shoot apical region of Arabidopsis.

The four different types of floral organs are initiated as separate whorls

Floral meristems initiate four different types of floral organs: sepals, petals, stamens, and carpels. These sets of organs are initiated in concentric rings, called **whorls**, around the flanks of the meristem (**Figure 16.18**). In Arabidopsis, the whorls are arranged as follows:

- The first (outermost) whorl consists of four sepals, which are green at maturity.
- The second whorl is composed of four petals, which are white at maturity.
- The third whorl contains six stamens (the male reproductive structures), two of which are shorter than the other four.
- The fourth (innermost) whorl is a single complex organ, the gynoecium or pistil (the female reproductive structure), which is composed of an ovary with two fused carpels, each containing numerous ovules, and a short style capped with a stigma.

primary inflorescence meristem The meristem that produces stem-bearing flowers; it is formed from the shoot apical meristem.

secondary inflorescence meristem The inflorescence meristem that develops from the axillary buds of stem-borne leaves of the primary inflorescence.

indeterminate growth The ability to keep growing and developing until the onset of senescence.

floral meristem The meristem that forms floral (reproductive) organs: sepals, petals, stamens, and carpels. May form directly from a vegetative meristem or indirectly via an inflorescence meristem.

whorl Pertaining to the concentric pattern of a set of organs that are initiated around the flanks of the meristem.

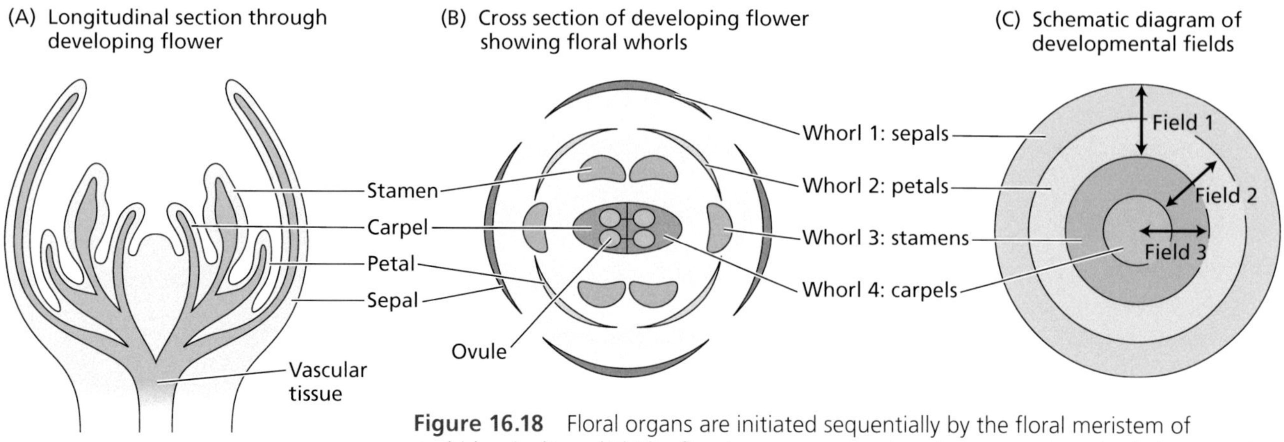

Figure 16.18 Floral organs are initiated sequentially by the floral meristem of Arabidopsis. (A and B) The floral organs are produced as successive whorls (concentric circles), starting with the sepals and progressing inward. (C) According to the combinatorial model, the functions of each whorl are determined by three overlapping developmental fields. These fields correspond to the expression patterns of specific floral organ identity genes.

floral meristem identity genes Two classes of genes, one required for the conversion of a shoot apical meristem into a floral meristem, the other that maintains the identity of an inflorescence meristem (as opposed to a floral meristem).

floral organ identity genes Three classes of genes that control the specific locations of floral organs in the flower. *See also* ABC model.

Two major categories of genes regulate floral development

Studies of mutations have enabled identification of two key categories of genes that regulate floral development: floral meristem identity genes and floral organ identity genes.

1. **Floral meristem identity genes** encode transcription factors that are necessary for the initial induction of floral organ identity genes. In general, these are the positive regulators of floral organ identity in the developing floral meristem.
2. **Floral organ identity genes** directly control floral organ identity. The proteins encoded by these genes are transcription factors that interact with other protein cofactors to control the expression of downstream genes whose products are involved in the formation or function of floral organs. Mutations of these genes often result in dramatic changes in flower structures, with sepals, petals, stamens, and carpels formed in the wrong whorl. Many of these mutants are used extensively in flower breeding to introduce flowers with unique and striking forms. Developmental mutations of this type, which cause one organ to be replaced by another, are called *homeotic mutations*, and the associated genes are called *homeotic genes*.

While certain genes fit neatly within these two general categories, it is important to keep in mind that floral development involves complex, nonlinear gene networks. In these networks, individual genes often play multiple roles.

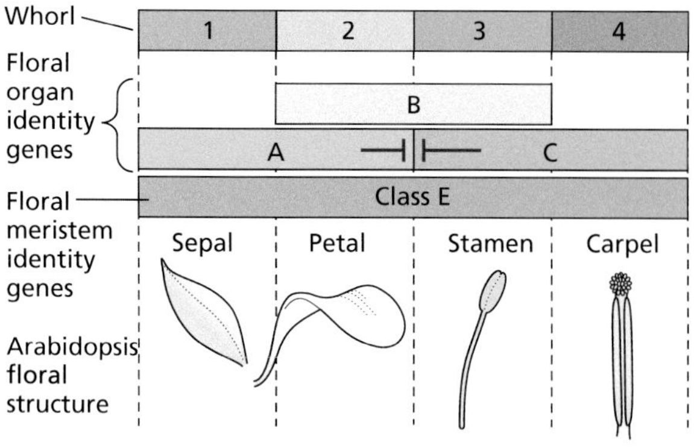

Figure 16.19 The ABCE model of floral development.

The ABC model partially explains the determination of floral organ identity

Floral organ identity genes fall into three classes (A, B, and C), which define three different kinds of activities encoded by three distinct types of genes (**Figure 16.19**).

- Class A activity controls organ identity in the first and second whorls. Loss of Class A activity results in the formation of carpels instead of sepals in the first whorl, and of stamens instead of petals in the second whorl.

- Class B activity controls organ determination in the second and third whorls. Loss of Class B activity results in the formation of sepals instead of petals in the second whorl, and of carpels instead of stamens in the third whorl.
- Class C activity controls events in the third and fourth whorls. Loss of Class C activity results in the formation of petals instead of stamens in the third whorl. Moreover, in the absence of Class C activity, the fourth whorl (normally a carpel) is replaced by a *new flower.* As a result, the fourth whorl of a Class C gene mutant flower is occupied by sepals. The floral meristem is no longer determinate. Flowers continue to form *within* flowers, and the pattern of organs (from outside to inside) is: sepal, petal, petal; sepal, petal, petal; and so on.

The **ABC model** provides a way of understanding how relatively few key regulators can provide a complex outcome. The ABC model postulates that organ identity in each whorl is determined by a unique combination of the three organ identity gene activities:

- Class A activity alone specifies sepals.
- Class A and B activities are required for the formation of petals.
- Class B and C activities form stamens.
- Class C activity alone specifies carpels.

The model further proposes that Class A and C activities mutually repress each other; that is, both A- and C-Class genes exclude each other from their expression domains, in addition to their function in determining organ identity.

Although the patterns of organ formation in wild-type flowers and most of the mutants are predicted and explained by this model, not all observations can be accounted for by the ABC genes alone. For example, expression of the ABC genes throughout the plant does not transform vegetative leaves into floral organs. Thus, the ABC genes, while necessary, are not sufficient to impose floral organ identity onto a leaf developmental program.

A fourth class of genes, the Class E genes, have been shown to be required for the other three floral homeotic genes to be active. E-Class genes can therefore be described as floral meristem identity genes. The augmented ABC model is referred to as the ABCE model (Figure 16.19). The ABCE model was formulated based on genetic experiments in Arabidopsis and snapdragon. Flowers of different species have evolved diverse structures by modifying the regulatory networks described by the ABCE model.

ABC model A proposal for the way in which floral homeotic genes control organ formation in flowers. According to the model, organ identity in each whorl is determined by a unique combination of the three organ identity gene activities.

16.7 Pollen Development in the Anther

Describe male gametophyte development.

As described in Chapter 1, the plant life cycle consists of two entirely separate haploid individuals called the male and female gametophytes. Strictly speaking, the stamens and carpels of flowers are spore-producing structures rather than sexual structures. The spores produced in these structures develop into the male and female gametophytes. The role of the male and female gametophytes is to produce the gametes (sperm and egg, respectively), making them the true sexual phases of the plant life cycle.

Male gametophytes, or pollen grains, are formed in the stamen of the flower. Typically, the stamen is made up of a delicate filament attached to an anther composed of four microsporangia arranged in opposite pairs (**Figure 16.20**). Each pair of microsporangia is separated from the other by a central region of sterile tissue surrounding a vascular bundle. Development of the male gametophyte, or

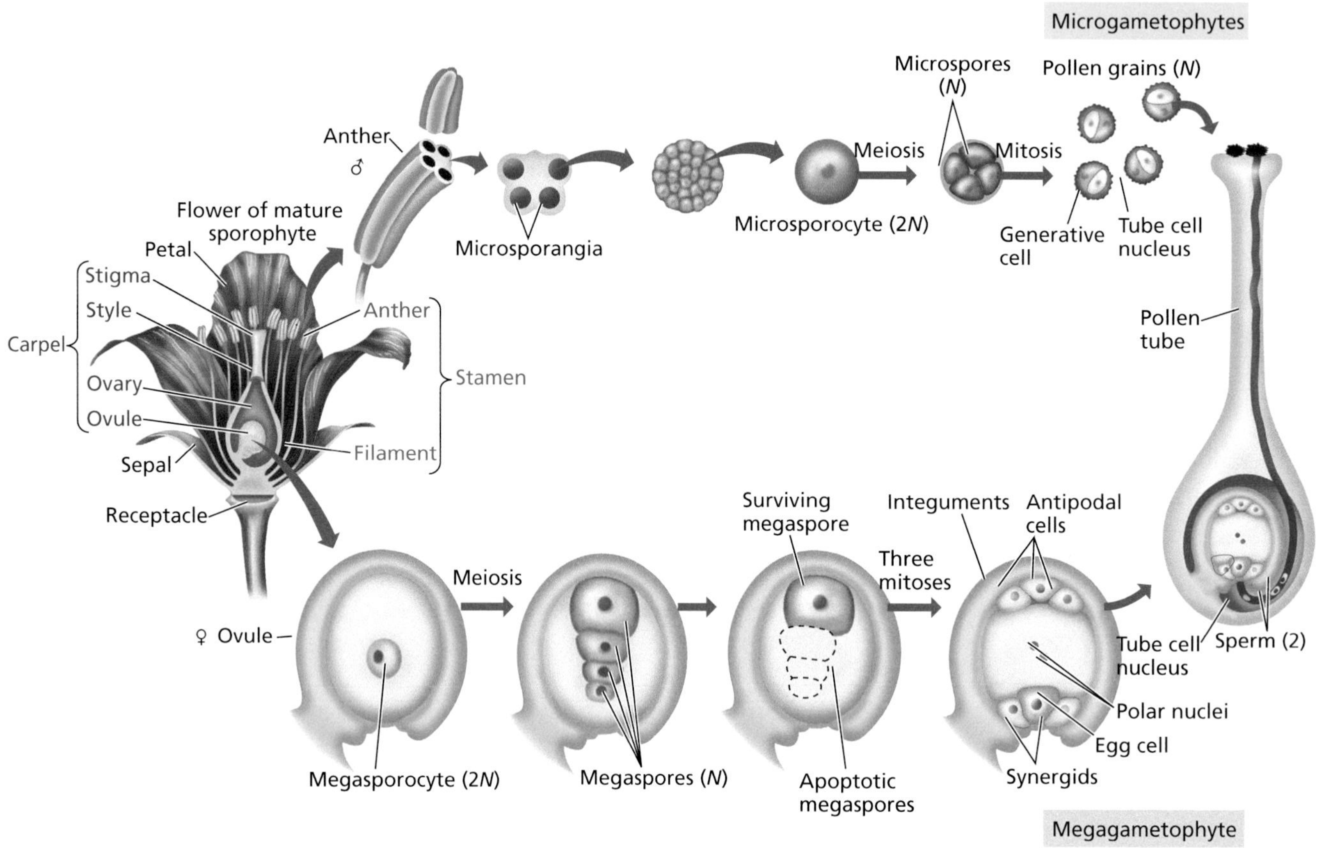

Figure 16.20 Angiosperm life cycle.

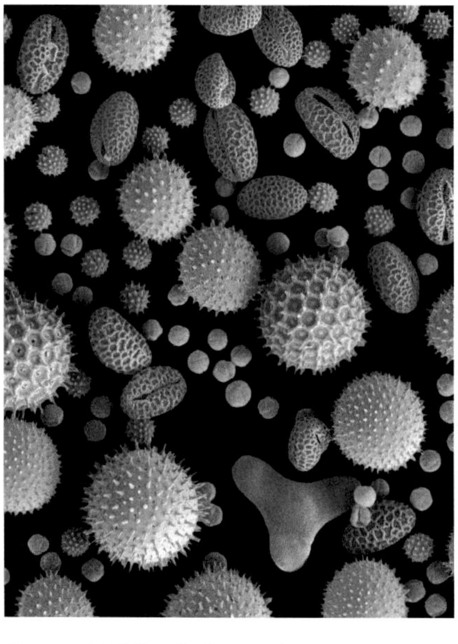

Figure 16.21 Pollen grain structure. Scanning electron microscope image of pollen grains from different species exhibiting distinct ornamentation.

pollen grain, begins with the process of microsporogenesis. During this process cells in the anther divide to form diploid pollen mother cells (microsporocytes) that subsequently undergo meiosis to produce haploid microspores (Figure 16.20). The microspores further divide and differentiate to form pollen grains with highly sculpted cell walls, and play important ecological roles in transferring the pollen from flower to flower (**Figure 16.21**).

Pollen grains consist of two sperm cells (containing DNA) surrounded by a vegetative cell that later forms the pollen tube (**Figures 16.22**). The pollen tube functions in sperm delivery during fertilization.

16.8 Embryo Sac Development in the Ovule

Describe female gametophyte development.

Ovule primordia arise in a specialized ovary tissue called the *placenta*. The female gametophyte, or *embryo sac*, develops within the ovule. This process is more complex and more diverse than the development of the male gametophyte within the anther. According to one classification scheme, there are more than 15 different patterns of embryo sac development in angiosperms. The most common pattern was first described in the genus *Polygonum* (knotweed) and is therefore called the *Polygonum* type of embryo sac. In angiosperms, *ovules* are located within the *ovary* of the *gynoecium*, the collective term for the carpels in flowers with multiple carpels (Figure 16.20). Upon fertilization of the female gamete, or egg, by a sperm cell, embryogenesis is initiated, and the ovule develops into a seed. Simultaneously, the ovary enlarges and becomes a fruit (Chapter 17).

Functional megaspores undergo a series of free nuclear mitotic divisions followed by cellularization

Female gametogenesis begins with the formation of the megaspore mother cell (megasporocyte) that undergoes meiosis to form four haploid megaspores in *Polygonum*-type embryo sacs (Figure 16.20). Three of the megaspores subsequently undergo programmed cell death, leaving only one functional megaspore. Functional megaspores then undergo three rounds of free nuclear mitotic divisions (mitoses without cytokinesis) to produce a multinucleate cell formed by nuclear divisions. The result is an eight-nucleate, immature embryo sac. Four of the nuclei then migrate to one pole, and the other four migrate to the opposite pole. Three of the nuclei at each pole undergo cellularization, while the remaining two nuclei, called polar nuclei, migrate toward the central region of the embryo sac (Figure 16.20 and **Figure 16.23**). The cytoplasm and the two polar nuclei develop their own plasma membrane and cell wall, giving rise to a large binucleate cell. The fully cellularized embryo sac represents the mature female gametophyte or embryo sac. At maturity, the *Polygonum*-type embryo sac consists of seven cells and eight nuclei and a large central vacuole.

The **egg cell** (the female gamete that combines with a sperm cell to form the zygote) and the two **synergid cells** are located at one end of the embryo sac adjacent to the **micropyle**, a small pore in the ovule that allows pollen tube entry during pollination (**Figure 16.24**). The synergid cells are involved in the final stages of pollen tube attraction, the discharge of pollen tube contents into the embryo sac, and gamete fusion.

The large binucleate cell in the middle of the embryo sac, composed of two **polar nuclei**, is called the **central cell**. Although its developmental fate is quite different from that of the egg, the central cell is also regarded as a gamete because it fuses with one of the sperm cells during a process that is unique to angiosperms called **double fertilization** (Figure 16.24).

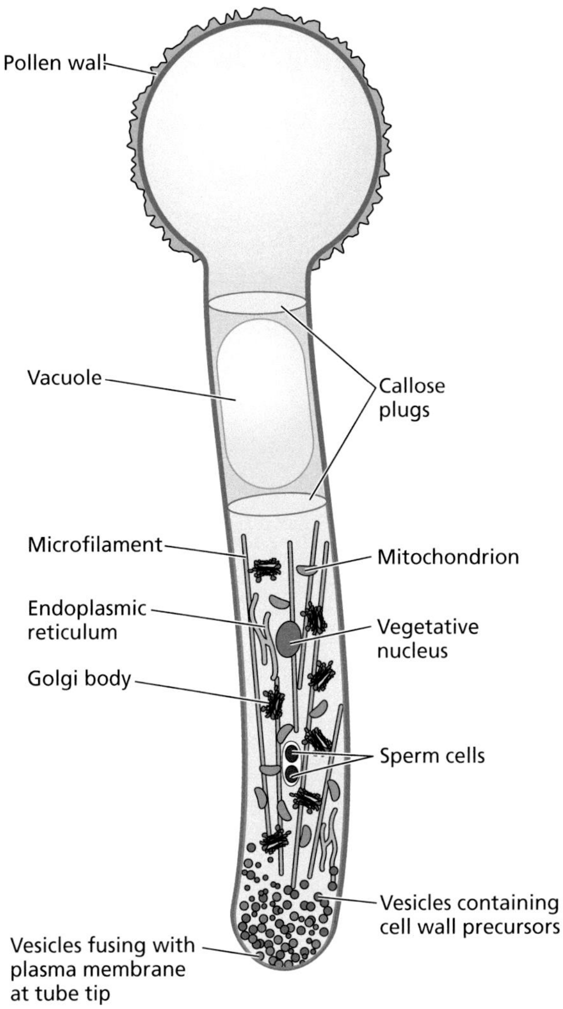

Figure 16.22 Pollen tubes elongate by tip growth. The cytoplasm is concentrated in the growing region of the tube by large vacuoles and callose partitions.

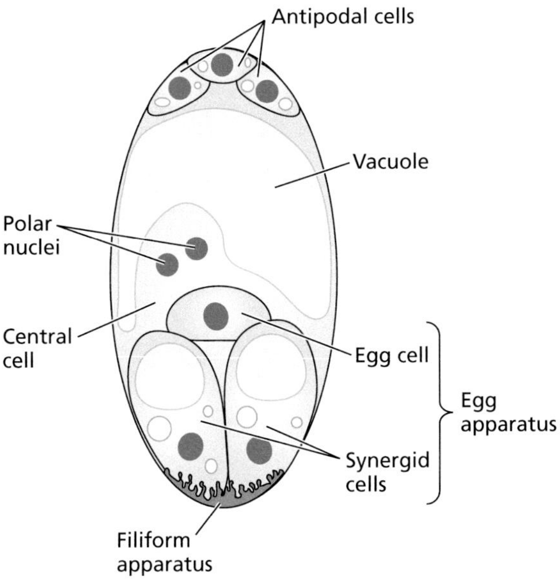

Figure 16.23 Diagram of the egg apparatus and filiform apparatus of the *Polygonum*-type embryo sac.

egg cell The female gamete.

synergid cells Two cells adjacent to the egg cell of the embryo sac, one of which is penetrated by the pollen tube upon entry into the ovule.

micropyle The small opening at the distal end of the ovule, through which the pollen tube passes prior to fertilization.

polar nuclei The two haploid nuclei at the center of the embryo sac that normally fuse to form the diploid nucleus of the central cell.

central cell The cell in the embryo sac that fuses with the second sperm cell, giving rise to the primary endosperm cell.

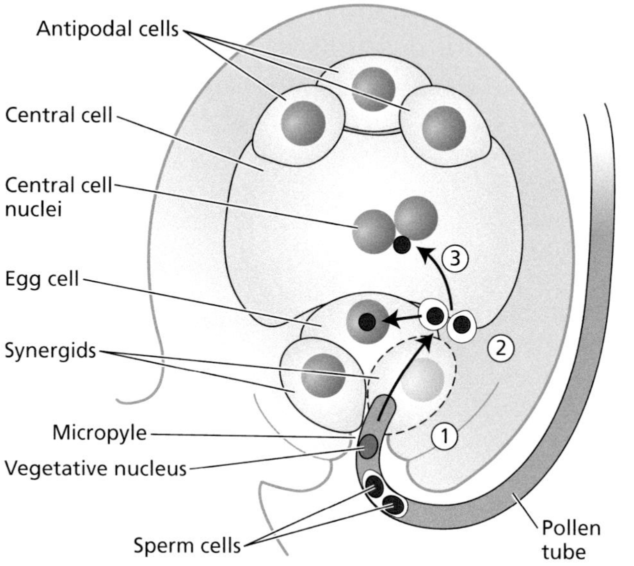

1. The pollen tube bursts and discharges. Sperm cells are delivered rapidly from the pollen tube into the female gametophyte. The receptive synergid cell is likely to break down just after the start of pollen tube discharge.

2. Two sperm cells remain at the boundary region between the egg cell and the central cell for several minutes.

3. One sperm cell fuses with the egg cell and the other fuses with the central cell, and their nuclei move toward the target gamete nuclei.

Figure 16.24 Sperm cell behavior during double fertilization in Arabidopsis can be divided into three stages.

16.9 Pollination and Double Fertilization in Flowering Plants

Describe pollination and fertilization in flowering plants.

Pollination in angiosperms is the process of transferring pollen grains from the anther of the stamen (the male organ of the flower) to the stigma of the pistil (the female organ of the flower). In some species, such as Arabidopsis and rice, reproduction typically occurs via self-pollination, or selfing—that is, the pollen and the stigma belong to the same individual sporophyte. In other species, either **cross-pollination** or **outcrossing** is the norm—when the male parent and the female parent are separate sporophytic individuals. Many species can reproduce by either self- or cross-pollination; other species have various mechanisms for promoting cross-pollination, and may even be incapable of reproducing by self-pollination.

In the case of cross-pollination, pollen might travel great distances before landing on a suitable stigma. Produced in excess, pollen grains are dispersed by wind, insects, birds, and mammals, which carry the nonmotile male gametes of angiosperms much farther than the motile sperm of lower plants could ever swim.

Successful pollination depends on several factors, including ambient temperature, timing, and the receptivity of the stigma of a compatible flower. Many pollen grains can tolerate desiccation and high temperatures during their journey to the stigma. However, some pollen grains, such as those of tomato, are damaged by heat. Understanding how some pollen grains tolerate

double fertilization A unique feature of all angiosperms whereby, along with the fusion of a sperm with the egg to create a diploid zygote, a second male gamete fuses with the polar nuclei in the embryo sac to generate the endosperm tissue (with a triploid or higher number of chromosomes).

cross-pollination Pollination of a flower by pollen from the flower of a different plant.

outcrossing The mating of two plants with different genotypes by cross-pollination.

periods of high temperature will help ensure our food supply as the global climate changes.

Two sperm cells are delivered to the female gametophyte by the pollen tube

The female gametes are protected from the environment by the ovary tissues. Consequently, to reach an unfertilized egg, sperm cells must be delivered by a pollen tube that grows from the stigma to the ovule. Successful delivery of the two sperm cells to the two female gametes (egg and central cell) depends on extensive interactions and communication between the pollen tube, the pistil, and the female gametophyte.

Pollination begins with adhesion and hydration of a pollen grain on a compatible flower

Angiosperm reproduction is highly selective. The surfaces of stigmas can discriminate among diverse pollen grains, accepting those from the appropriate species and rejecting others from unrelated species. When pollen lands on a compatible stigma, the grains physically adhere to the surface *papillae cells*, named for their fingerlike projections, due to biophysical and chemical interactions between pollen proteins and lipids and stigma surface proteins. Most pollen grains adhere poorly to the stigmas of plants of other families.

Flowers have either wet or dry stigmas. Pollen grains become hydrated by default on wet stigmas, whereas hydration of pollen on dry stigmas is regulated by the formation of a capillary system through which water can flow out onto the stigmatic surface and hydrate the pollen grain.

During hydration, the pollen grain becomes physiologically activated. Calcium ion influx into the vegetative cell triggers reorganization of the cytoskeleton and causes the cell to become physiologically and structurally polarized. The source of the Ca^{2+} may be either the cytoplasm or the cell wall of the papillar cell.

Pollen tubes grow by tip growth

Following germination on the stigma surface, the pollen tube begins to grow downward through the style by *tip growth* (Figure 16.22; Chapter 14). The pollen tube growth rate of angiosperms ranges from approximately 10 μm per h to more than 20,000 μm (20 mm) per h, about 100 times faster than the growth rate of gymnosperm pollen tubes. Some pollen tubes can reach up to 40 cm in length, as in the case of maize (corn; *Zea mays*) pollen tubes growing through the threadlike styles ("silk") to reach the ovaries.

Growing pollen tubes restrict the cytoplasm, the two sperm cells, and the vegetative nucleus to the growing apical region by forming large vacuoles and callose partitions that seal off the rear portion of the tube (Figure 16.22). The apical end of the pollen tube is packed with small secretory vesicles delivering wall materials and new membranes to the growing tip.

For successful fertilization to occur, the pollen tube must find its way to the micropyle of an ovule. In fact, there is often competition among pollen tubes to arrive at the micropyle first and thus be the one to fertilize the egg. The surrounding maternal tissues may even influence the outcome of this "race"—a type of female mate selection.

Double fertilization results in the formation of the zygote and the primary endosperm cell

When the pollen tube senses chemical attractants secreted by the synergids, the tube grows through the micropyle, penetrates the embryo sac, and enters one of the synergid cells. Once inside the synergid, the pollen tube stops growing

primary endosperm cell A diploid cell formed by fusion of two polar cells in the angiosperm embryo sac prior to fertilization.

and the tip bursts, releasing the two sperm cells. During double fertilization in the *Polygonum*-type embryo sac, one sperm cell fuses with the egg to produce the zygote, and the other fuses with the binucleate central cell (which includes the two polar nuclei) to produce the triploid **primary endosperm cell** that divides mitotically to give rise to the nutritive endosperm of the seed (Figure 16.24). Because different types of embryo sacs contain different numbers of polar nuclei, the ploidy level of the endosperm ranges from 2*N* in *Oenothera* to 15*N* in *Peperomia*. As is described in Chapter 17, the processes of embryogenesis and seed development begin rapidly after double fertilization occurs.

Summary

At the appropriate stage in development the shoot apical meristem undergoes a massive reorganization from the adult vegetative phase to the reproductive phase to give rise to the floral organs: sepals, petals, stamens, and carpels. A network of genes that control floral morphogenesis has been identified in several species. These genes are activated in response to a range of internal and external cues that must be integrated to effect flowering. The male and female gametophyte stages develop in the anther and ovary, respectively. Sexual reproduction is initiated by pollination and culminates in the process of double fertilization.

16.1 Floral Evocation: Integrating Environmental Cues

- Internal (autonomous) and external (environment-sensing) control systems enable plants to precisely regulate and time flowering for reproductive success.
- Two of the most important seasonal responses that affect floral development are photoperiodism (response to changes in day length) and vernalization (response to prolonged cold).
- Synchronized flowering favors crossbreeding and helps ensure seed production under favorable conditions.

16.2 The Shoot Apex and Phase Changes

- Postembryonic development in plants can be divided into three phases with different physiological and morphological features: juvenile, adult vegetative, and reproductive (**Figure 16.1**).
- The transition from one phase to another is called phase change.
- Juvenile tissues are produced first and are located at the base of the shoot.
- Gibberellins play a role in regulating phase changes in some species.

16.3 Photoperiodism: Monitoring Day Length

- Plants can detect seasonal changes in day length that occur at latitudes away from the equator (**Figure 16.2**).
- Flowering in LDPs requires that the day length exceed a certain duration, called the critical day length. Flowering in SDPs requires a day length that is less than the critical day length (**Figure 16.4**).
- Circadian rhythms are defined by three parameters: period, phase, and amplitude (**Figure 16.5**).
- Temperature compensation prevents temperature changes from affecting the period of the circadian clock.
- Phytochromes and cryptochromes entrain the circadian clock.
- Leaves are the sites of perception of the photoperiodic stimulus in both LDPs and SDPs.
- Leaf grafting experiments have shown that a photoperiodically induced leaf can induce an uninduced plant to flower (**Figure 16.6**).
- Plants monitor day length by measuring the length of the night; flowering in both SDPs and LDPs is determined primarily by the duration of darkness (**Figure 16.7**).
- For both LDPs and SDPs, the dark period can be made ineffective by interruption with a short exposure to light (a night break) (**Figure 16.8**).
- The flowering response to night breaks shows a circadian rhythm, supporting the clock hypothesis (**Figure 16.8**).
- Flowering in LDPs is promoted when the inductive light treatment coincides with a peak in light sensitivity, which follows a circadian rhythm (**Figure 16.9**).

(Continued)

Summary (*continued*)

- Flowering in SDPs is inhibited when the inductive light treatment coincides with a peak in light sensitivity, which follows a circadian rhythm (**Figure 16.9**).
- In the coincidence model, flowering is induced in both SDPs and LDPs when light exposure is coincident with the appropriate phase of the oscillator (**Figure 16.10**).
- The effects of red and far-red night breaks implicate phytochrome in the control of flowering in SDPs and LDPs (**Figures 16.11**, **16.12**).

16.4 Vernalization: Promoting Flowering with Cold Treatment

- In vernalization-requiring plants, a cold treatment is required for plants to respond to floral signals such as inductive photoperiods (**Figures 16.13**, **16.14**).
- High temperatures can cause devernalization in vernalized plants.
- The longer the cold treatment, the more stable the vernalized state of the plant is (**Figure 16.15**).
- For vernalization to occur, active metabolism is required during the cold treatment.
- Vernalization takes place primarily in the SAM.

16.5 Long-Distance Signaling Involved in Flowering

- In photoperiodic plants, a long-range signal is transmitted via the phloem from the leaves to the apex, permitting floral evocation (**Figure 16.16**).
- A transcriptional activator, the small protein FT, serves as a master switch controlling the expression of floral stimulus genes (**Figure 16.16**).
- FT moves via the phloem from the leaves to the SAM under inductive photoperiods.
- In the meristem, FT forms a complex with the transcription factor FD to activate floral identity genes (**Figure 16.16**).

16.6 Floral Meristems and Floral Organ Development

- Floral meristems may develop directly from vegetative meristems or indirectly from inflorescence meristems (**Figure 16.17**).
- The four different types of floral organs are initiated sequentially in separate, concentric whorls (**Figure 16.18**).
- Formation of floral meristems requires active floral meristem identity genes, while development of the floral organs requires active floral organ identity genes.
- Mutations in floral organ identity genes alter the types of floral organs produced in each of the whorls.
- The ABC model states that organ identity in each whorl is determined by the combined activity of three floral organ identity genes (**Figure 16.19**).
- According to the ABCE model, a fourth gene is required for the three floral organ identity genes to be active (**Figure 16.19**).

16.7 Pollen Development in the Anther

- Plants undergo both a diploid and a haploid generation in order to make gametes and reproduce (**Figure 16.20**).
- The diverse shapes, sizes, and surface features of pollen grains reflect adaptations that facilitate pollination (**Figure 16.21**).

16.8 Embryo Sac Development in the Ovule

- Eggs are formed in the female gametophyte (embryo sac) first by megaspore formation and then by the development of the female gametophyte (**Figure 16.20**).
- Most angiosperms exhibit *Polygonum*-type embryo sac development, in which meiosis of a diploid megaspore mother cell produces four haploid megaspores, only one of which goes on to form the female gametophyte (embryo sac) (**Figure 16.23**).
- Embryo sac development begins with three mitotic divisions without cytokinesis, followed by cellularization (**Figure 16.20**).

16.9 Pollination and Double Fertilization in Flowering Plants

- Pollen tubes grow by tip growth (**Figure 16.22**).
- Once the pollen tube has reached the ovule, two sperm cells are released into the embryo sac (**Figure 16.24**).
- During double fertilization, one sperm cell fertilizes the egg and the other fuses with the binucleate central cell, forming the zygote and primary endosperm cell, respectively (**Figure 16.24**).
- The primary endosperm cell divides mitotically to form triploid endosperm tissue, the main nutritive tissue of angiosperm seeds.

Suggested Reading

Amasino, R. (2010) Seasonal and developmental timing of flowering. *Plant J.* 61: 1001–1013. DOI:10.1111/j.1365-313X.2010.04148.x.

Dresselhaus, T., and Franklin-Tong, N. (2013) Male-female crosstalk during pollen germination, tube growth and guidance, and double fertilization. *Mol. Plant* 6: 1018–1036.

Fattorini, R., and Glover, B. J. (2020) Molecular mechanisms of pollination biology. *Annu. Rev. Plant Biol.* 71(1): 487–515.

Huijser, P., and Schmid, M. (2011) The control of developmental phase transitions in plants. *Development* 138: 4117–4129. DOI:10.1242/dev.063511.

Krizek, B. A., and Fletcher, J. C. (2005) Molecular mechanisms of flower development: An armchair guide. *Nat. Rev. Genet.* 6: 688–698.

Song, Y. H., Ito, S., and Imaizumi, T. (2013) Flowering time regulation: Photoperiod- and temperature-sensing in leaves. *Trends Plant Sci.* 18: 575–583.

Song, Y. H., Kubota, A., Kwon, M. S., Covington, M. F., Lee, N., Taagen, E. R., Cintrón, D. L., Hwang, D. W., Akiyama, R., Hodge, S. K., et al. (2018) Molecular basis of flowering under natural long-day conditions in Arabidopsis. *Nat. Plants* 4: 824–835.

Twell, D. (2010) Male gametophyte development. In *Plant Developmental Biology—Biotechnological Perspectives*, Vol. 1, E. C. Pua and M. R. Davey, eds., Springer-Verlag, Berlin, pp. 225–244.

Yang, W.-C., Shi, D.-Q., and Chen, Y.-H. (2010) Female gametophyte development in flowering plants. *Annu. Rev. Plant Biol.* 61: 89–108.

17 Seed and Fruit Development

In Chapter 16, we discussed the processes of flowering and sexual reproduction that precede embryogenesis and the formation of seeds and fruits. Land plants comprise a monophyletic group called the embryophytes—plants that make embryos. What sets them apart from their aquatic, algal ancestors is the ability to develop a multicellular sporophyte following fertilization of the egg cell by the sperm cell.

Vascular plant (tracheophyte) zygotes (fertilized egg cells) divide to produce a diploid sporophyte that lives independently of the gametophytes that produced it. Furthermore, cells of vascular plant sporophytes differentiate into the three main tissue types—dermal, ground, and vascular tissues—that become further specialized into a wide variety of cell and tissue types that form the major organ systems of vegetative and reproductive structures. In seed plants (spermatophytes, which include gymnosperms and angiosperms), embryogenesis terminates with the formation of a dormant (or quiescent) seed. Due to the tremendous importance of seed plants to food production, we know much more about embryogenesis in seed plants than in non-seed plants. Accordingly, this chapter focuses on embryogenesis and seed/fruit development of seed plants, with an emphasis on the angiosperms.

In seed plants, embryogenesis transforms the single-celled zygote into a complex, multicellular individual contained within a mature seed. As such, embryogenesis provides many examples of developmental processes by which the basic architecture of the plant is established, including the elaboration of forms (**morphogenesis**), the associated formation of functionally organized structures (**organogenesis**), and the **differentiation** of cells to produce anatomically and functionally distinct tissues

morphogenesis The developmental processes that give rise to biological form.

organogenesis The development of organs such as roots, shoots, and flowers from embryonic or meristematic tissue.

differentiation Process by which a cell acquires metabolic, structural, and functional properties that are distinct from those of its progenitor cell. In plants, differentiation is frequently reversible, when excised differentiated cells are placed in tissue culture.

histogenesis The differentiation of cells to produce various tissues.

dormancy A living condition in which growth does not occur under conditions that are normally favorable to growth.

germination The events that take place between the start of imbibition of the dry seed and the emergence of part of the embryo, usually the radicle, from the structures that surround it. May also be applied to other quiescent structures, such as pollen grains or spores.

seed coat (or **testa)** The outer layer of the seed, derived from the integument of the ovule.

(**histogenesis**). The development of the embryo also features complex changes in physiology that enable the embryo to withstand prolonged periods of reduced metabolic activity (**dormancy**) and to recognize and respond to environmental cues that signal the plant to resume growth (**germination**) (discussed in Chapter 14).

The association of seed and fruit development with an embryo derived from sexual reproduction only began to gain acceptance in the late seventeenth century. Prior to that time, seeds were thought to be produced by an asexual, vegetative process similar to bud formation.

We begin with a discussion of the development of the embryo and associated seed structures that protect and nourish the embryo to ensure its viability and distribution to a suitable site to germinate.

17.1 Seed Structure

Describe the parts of the seed, and differentiate among seeds, seedlike fruits, and fruits.

All seeds contain three basic structural features: an embryo, food storage tissue, and a protective outer layer of dead cells called the **seed coat** (or **testa**). The angiosperm

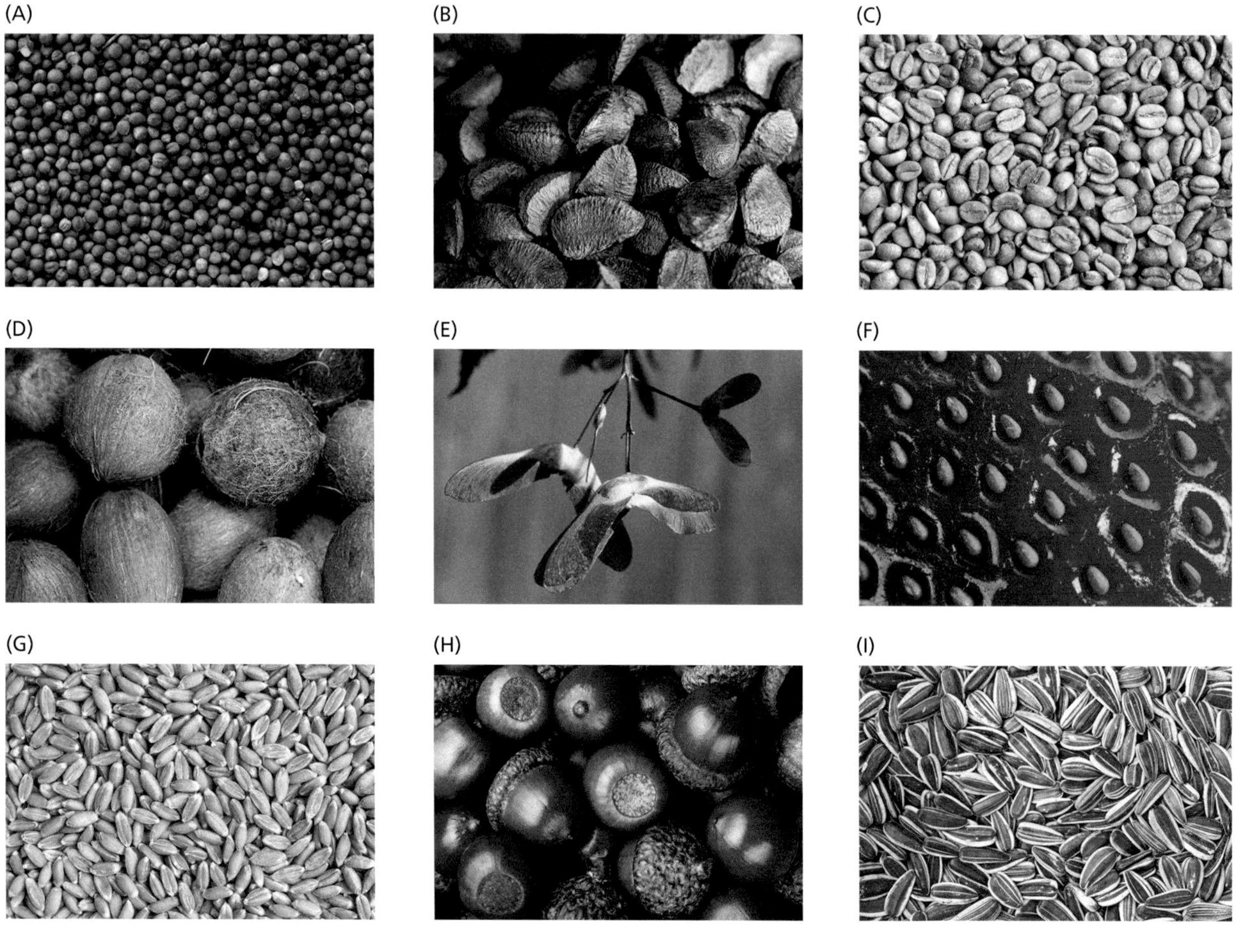

Figure 17.1 Seeds and seedlike fruits. (A–D) True seeds. (A) Rapeseed (*Brassica napus*). (B) Brazil nut (*Bertholletia excelsa*). (C) Coffee bean (*Coffea* sp.). (D) Coconut (*Cocos nucifera*). (E–I) Simple dried indehiscent fruits. (E) Samara, a winged achene, from maple (*Acer* sp.). (F) Achene from strawberry (*Fragaria* sp.). (G) Caryopses from wheat (*Triticum* spp.) and other cereals. (H) Nut from oak (*Quercus* sp.). (I) Cypselae, an achene with two carpals, from sunflower (*Helianthus* spp.) and other composites. Monocot caryopses and Asteraceae cypselae are routinely referred to as seeds.

embryo consists of the embryonic axis, which is composed of the **radicle**, or embryonic root; the **hypocotyl**, to which one or two **cotyledons** are attached; and the **shoot apex** bearing the **plumule**, or first true leaf primordium. In angiosperms, the food storage tissue that nourishes the growing embryo is the triploid **endosperm** that results from **double fertilization**. In some angiosperm species, the seed coat is fused to the fruit wall, or **pericarp**, which is derived from the ovary wall. In cereals, the testa and pericarps are fused, technically making these "seeds" fruits, but they will be referred to as seeds throughout the book. **Figure 17.1** shows a variety of familiar true seeds as well as fruits with a seedlike appearance.

17.2 Establishment of the Embryonic Axis

Distinguish between endosperm and embryo development and the mechanisms that regulate their formation.

A characteristic feature of seed plants is a polarity in which tissues and organs are arrayed in a highly conserved order along an axis that extends from the shoot apical meristem to the root apical meristem. An early manifestation of this apical–basal axis is seen in the zygote itself, which elongates approximately threefold and becomes polarized with respect to its intracellular composition. The apical end of the zygote is densely cytoplasmic, in contrast to the basal end, which contains a large central vacuole. These differences in cytoplasmic density are fixed in place when the zygote divides asymmetrically to yield a short, cytoplasmically dense **apical cell** and a longer, vacuolated **basal cell** (**Figure 17.2**). This first cell division is a critical step in embryogenesis because it defines the polar axis of the plant. A similar asymmetric division is observed in many (but not all) flowering plants.

Much of what we know about the establishment of the apical–basal embryonic axis has been derived from the model plant Arabidopsis, which displays simple and highly stereotyped cell divisions during the early stages of its embryonic development. A graphic diagrammatic depiction of the earliest cell divisions in Arabidopsis, provided in Figure 17.2, offers a convenient guide for the discussion that follows.

The reproducible patterns of cell division during early embryogenesis in Arabidopsis might suggest that a fixed sequence of cell division is essential to this phase of development. Such a *lineage-dependent* mechanism is found in animals and can be likened to assembling a structure from a standard set of parts according to self-contained instructions. However, plant development is more robust when challenged by environmental or genetic perturbation and exhibits more variation in cell division patterning.

In part, this flexibility in polar development is a result of the role of the plant hormone auxin in maintaining both programmed and plastic development (Chapter 12). Molecular genetic and physiological analyses have shown that polar auxin transport is required for maintenance of the apical–basal axis and organogenesis (Chapter 15). Polar auxin streams are also essential to morphological changes in response to environmental stimuli (Chapter 14).

17.3 Seed Endosperm Development

Describe endosperm development.

The ovule that contains the developing embryo is connected to the mother plant via a threadlike structure called the *funiculus*. The funiculus supplies nutrients from the vascular system to a structure known as the *chalaza*, where a tissue connecting to the embryo sac called the *nucellus* adjoins two integuments that

embryo The early sporophytic form of the plant formed from the zygote produced by gametic fusion or apomixis that becomes the basis of the mature plant axis. The embryo lacks leaves, stems, and reproductive structures.

radicle The embryonic root. Usually the first organ to emerge on germination.

hypocotyl The region of the seedling stem below the cotyledons and above the root.

cotyledons Also called "seed leaves," they are the main sources of nutrients during germination and post-germination photosynthesis before the first true leaves develop.

shoot apex (terminal bud) The shoot apical meristem with its associated leaf primordia and young, developing leaves.

plumule First true leaf of a growing seedling.

endosperm Tissue that surrounds and nourishes the embryo in the seeds of angiosperms (flowering plants). In angiosperms, endosperm is formed by fusion of at least one polar nucleus in the embryo sac with one of the two sperm nuclei from the pollen grain. In gymnosperms the nutritive material of the seed is haploid and present before fertilization.

double fertilization A unique feature of all angiosperms whereby, along with the fusion of a sperm with the egg to create a diploid zygote, a second male gamete fuses with the polar nuclei in the embryo sac to generate the endosperm tissue (with a triploid or higher number of chromosomes).

pericarp The fruit wall surrounding a fruit, derived from the ovary wall.

apical cell The smaller, cytoplasm-rich cell formed by the first division of the zygote.

basal cell In embryogenesis, the larger, vacuolated cell formed by the first division of the zygote. Gives rise to the suspensor.

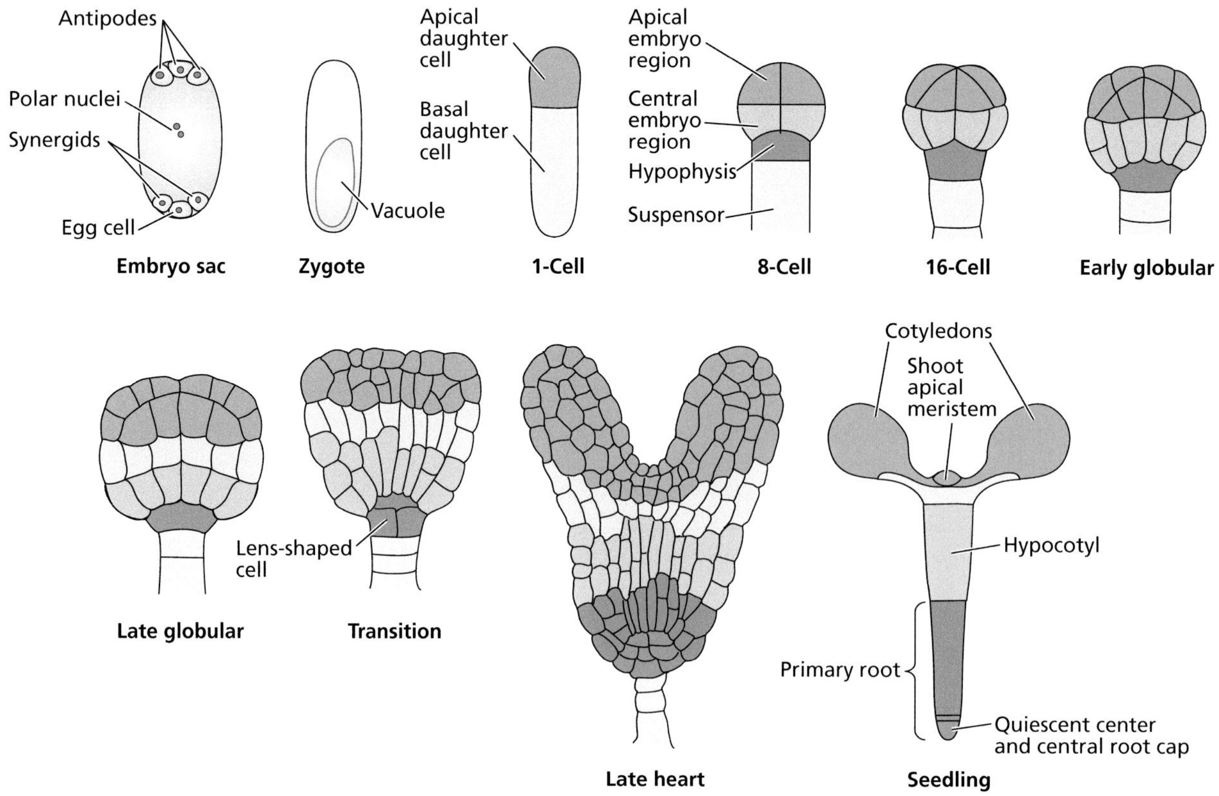

Figure 17.2 Pattern formation during Arabidopsis embryogenesis. A series of successive stages are shown to illustrate how specific cells in the young embryo contribute to specific anatomically defined features of the seedling. Clonally related groups of cells (cells that can be traced back to a common progenitor) are indicated by distinct colors. Following the asymmetric division of the zygote, the smaller, apical daughter cell divides to form an eight-cell embryo consisting of two tiers of four cells each. The upper tier gives rise to the shoot apical meristem and most of the flanking cotyledon primordia. The lower tier produces the hypocotyl and some of the cotyledons, the embryonic root, and the upper cells of the root apical meristem. The basal daughter cell produces a single file of cells that make up the suspensor. The uppermost cell of the suspensor becomes the **hypophysis**, which is part of the embryo. The hypophysis divides to form the quiescent center and the initial cells that form the root cap.

eventually associate with the endosperm or seedcoat. During seed morphogenesis, the endosperm provides nutrition to the developing embryo. In some species there is sufficient endosperm remaining to provide nutrition to the germinated seedling as well. In Arabidopsis and many other species, the endosperm is almost entirely reabsorbed (solubilized and absorbed) during embryogenesis, and the reserves that will support early seedling growth are stored in the embryo's cotyledons (**Figure 17.3**). The fleshy cotyledons of legumes are highly specialized for food storage. In cereals and other grasses, the endosperm persists during seed development and becomes the major site for the storage of starch and protein (**Figure 17.4**). Mobilization of these reserves for transport to the embryo is the final function of the endosperm before it undergoes programmed cell death as the seedling becomes established.

The endosperm develops from mitotic divisions of the primary endosperm nucleus resulting from double fertilization. The most common form of endosperm development in angiosperms is termed *nuclear,* since it results from multiple post-fertilization nuclear divisions to form a multinucleate coenocyte before cellularization and cell wall formation are initiated.

hypophysis The cell located directly below the octant stage of the embryo that gives rise to the root cap and part of the root apical meristem.

somatic embryogenesis The process by which somatic cells (i.e., non-germline cells) de-differentiate and undergo an autonomous embryogenesis process that recapitulates all the zygotic embryogenesis steps and results in viable embryos.

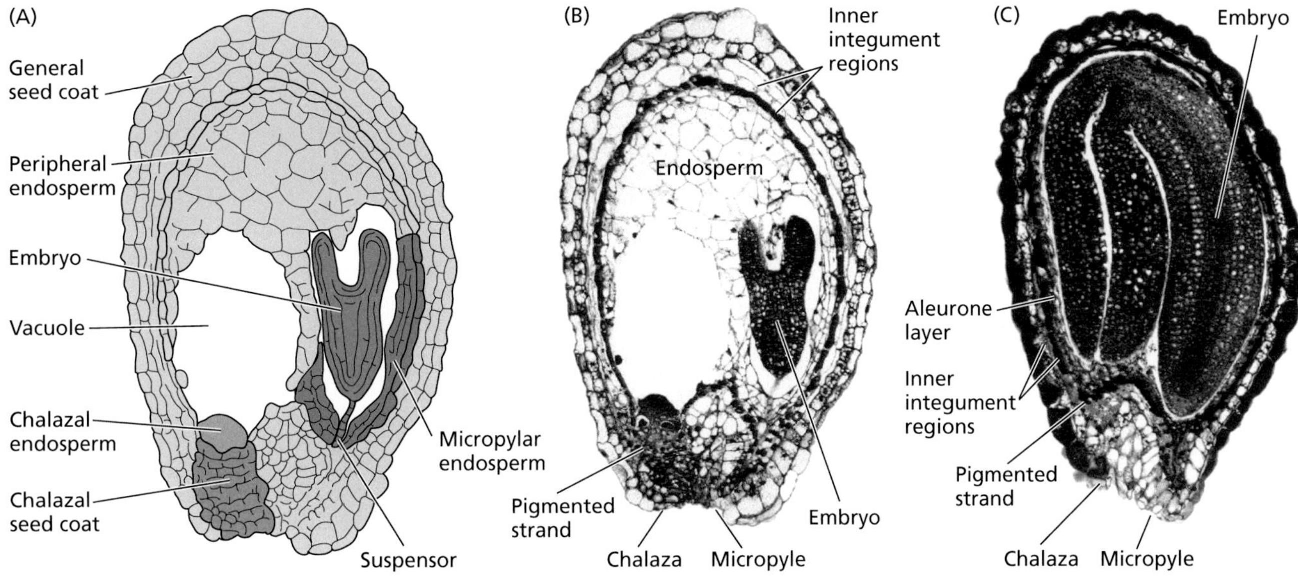

B and C from I. Debeaujon et al. 2003. *Plant Cell* 15: 2514–2531

Figure 17.3 Arabidopsis seed structure. (A) Diagram of an Arabidopsis seed with the embryo at the torpedo stage of development. (B) Light micrograph of a stained section of an Arabidopsis seed at the same stage as in (A). The embryo is embedded in mature endosperm tissue. The seed is covered by a seed coat derived from the inner and outer integument tissues of the ovule. (C) Mature seed. The endosperm has mostly been reabsorbed, and the embryo fills the seed. The cotyledons contain stored reserves that will support early seedling growth following germination.

Many aspects of endosperm development are a result of the parental origin of a particular cell type. Most commonly, *maternal effects* arc associated with the genetic background of the female genotype. Some maternal effects are particularly evident in hybrids, where the pollen and ovary are derived from two different species.

Endosperm development and embryogenesis can occur autonomously

Although embryogenesis and endosperm formation occur concurrently and in close proximity in seeds, these two processes are not obligately coupled. Endosperm development can continue to a limited extent if separated from the embryo. By contrast, embryogenesis can take place in vegetative tissue without endosperm development during **somatic embryogenesis** in *Bryophyllum* and *Kalanchoe*, which can reproduce asexually by producing plantlets on their leaf margins (**Figure 17.5A**). Somatic embryos can also be generated experimentally via tissue culture, a routine procedure in many biotechnology laboratories (**Figure 17.5B**). Once initiated, somatic embryos produced via tissue culture pass through the same developmental stages as normal embryos do and can grow into otherwise normal plants, demonstrating that a nutrient medium can substitute for the nutritive endosperm during in vitro embryogenesis. Despite this apparent morphological recapitulation of zygotic embryogenesis, somatic embryos are formed through different processes. For example, rather than developing from a single-celled zygote, they develop from small groups of cells, and the somatic embryos are typically larger than zygotic embryos.

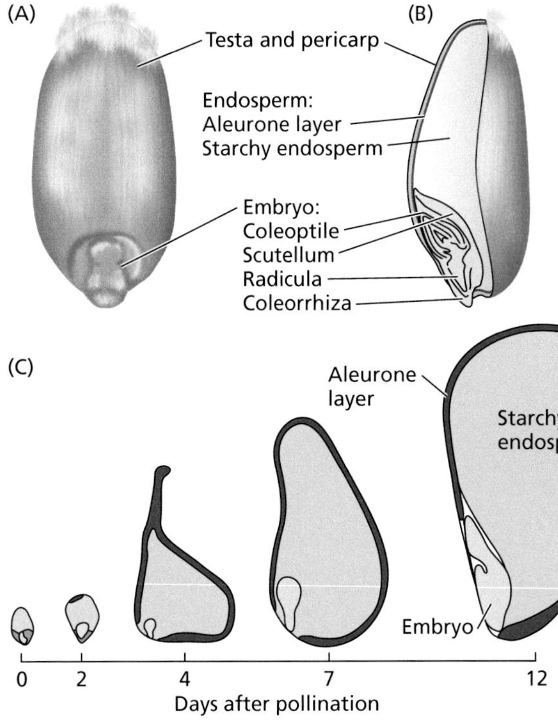

Figure 17.4 Cereal seed structure, as illustrated by wheat (*Triticum aestivum*). (A) Surface view of the seed, showing the location of the embryo in relation to the endosperm. (B) Longitudinal section through a seed. (C) Development of the embryo.

Figure 17.5 Examples of asexual embryogenesis. (A) In *Kalanchoe*, small embryos, sometimes referred as propagules, differentiate from the border of the leaves. They eventually fall off the leaf and start growing as a germinated seed. (B) Somatic embryos can be generated in tissue culture from many species, using young tissue or anthers. These explants are first subjected to a strong osmotic, hormonal, or temperature shock, and afterward grown in a recovery medium as calli or cells in suspension. Some clusters of cells then assume an "egg-like" fate and start recapitulating all the morphogenetic steps of zygotic embryogenesis. Here, a heart-stage embryo is seen emanating from a callus of undifferentiated cells.

Cells of the starchy endosperm and aleurone layer follow divergent developmental pathways

Whereas seeds of many species store reserves as proteins and oils, the endosperm of cereals stores large quantities of starch. The **starchy endosperm** is a unique tissue, accounting for the bulk of the endosperm in cereal grains (Figure 17.4). The starchy endosperm of cereals also contains storage proteins, which are deposited in protein storage vacuoles.

starchy endosperm A starch-storing triploid endosperm tissue that makes up the bulk of the seeds of cereals and other members of the grass family.

polyploidy Describes cells that are polyploid, or have more than two sets of chromosomes.

Endoreduplication—that is, chromosomal replication without mitosis, resulting in high DNA content and **polyploidy**—appears to play an important role in starchy endosperm development. In maize, for example, up to five rounds of endoreduplication may lead to a DNA content per nucleus of 96C—that is, 96 times the amount (content) present in the haploid nucleus. Endoreduplication begins during reserve deposition, and the accumulation of DNA prevents further nuclear or cell division.

The typical size of cells seems to be limited by the amount of DNA in the nucleus. Therefore, endoreduplication enables certain cells of plants and other organisms to expand to larger-than-normal volumes, maximizing their capacity to store food reserves (e.g., endosperm, up to 96C) or secretory compounds (e.g., leaf trichomes, up to 16C, Chapter 18). The starchy endosperm of cereals is dead at maturity due to programmed cell death, an event linked to the ethylene signaling pathway (Chapter 15).

As discussed in Chapter 14, the aleurone layer (the outermost layer[s] of the endosperm) functions during early seedling growth by mobilizing starch and storage protein reserves in the starchy endosperm through the production of an α-amylase, protease, and other hydrolases in response to gibberellins (GA) produced by the embryo. Maize and wheat have one layer of aleurone cells, rice has one to several layers, and barley has three layers. In cereal grains, the aleurone layer is the only part of the endosperm that may become pigmented.

17.4 Seed Coat Development

Describe seed coat development.

By serving as the outer protective layer of the seed, the seed coat is the first line of defense for the embryonic plant against adverse abiotic and biotic environmental factors. However, as discussed in Chapter 14, seed coats can also regulate dormancy, ensuring that external conditions are favorable for growth before germination starts. Following fertilization, the seed coat differentiates from the cells of the maternally derived ovule integuments as they undergo a dramatic period of growth in the first few days after fertilization through both cell division and expansion. Cells of the innermost layer synthesize **proanthocyanidin** flavonoid compounds, also known as **condensed tannins**, which accumulate in the central vacuole of the endothelial cells during the first week after fertilization and later become oxidized, imparting color to the pigmented cell layer and the whole seed coat. Because the seed coat surrounds the seed, its growth in surface area must be coordinated with the growth of the embryo and endosperm for the seed to reach its mature size. If the seed coat fails to expand, seed size is reduced.

Seed coat development seems to depend directly on the endosperm. In crosses between wild-type Arabidopsis and mutants producing only one sperm cell (and which were thus able to fuse with only one of the female gametes—discussed in Chapter 16), seeds in which only the central cell was fertilized produced normal seed coats, whereas seeds in which only the egg cell was fertilized did not, suggesting that seed coat development depends on endosperm development, and not on embryo development. The reason that the endosperm is required for seed coat development is that, following fertilization, the endosperm, which synthesizes several hormones, supplies auxin to the outer tissues of the ovule to trigger coat development.

17.5 Seed Maturation and Desiccation Tolerance

Explain seed maturation and the role of LEA proteins in desiccation tolerance.

The final phase of seed development is termed **maturation.** Abscisic acid (ABA) functions in the later stages of seed maturation and regulates seed dormancy (Chapters 12 and 14). Maturation is accompanied by the evaporative loss of water to produce a dry seed, a prerequisite for the quiescent state that precedes germination in many species of plants. For many species, maturation also includes the acquisition of **desiccation tolerance**. Desiccation tolerance is often correlated with **seed longevity**, the ability of the seed to remain viable in the dry state over long periods of time.

The term **orthodox seed** denotes seeds that can tolerate desiccation down to a water content as low as 5%. They are typically storable in a dry state for extensive periods of time, depending on the species. Examples of orthodox seeds include cereals, legumes, and Arabidopsis. By contrast, **recalcitrant seeds** are released from the plant with a relatively high water content and active metabolism. Recalcitrant seeds therefore tend to deteriorate and lose viability upon drying, unlike orthodox seeds. Mango and avocado are examples of plants with recalcitrant seeds.

The timing of the development of desiccation tolerance and seed longevity in relation to the attainment of mature size and seed dispersal varies across species. For most species, the acquisition of desiccation tolerance occurs during seed filling. Subsequently, during late maturation, seeds progressively acquire longevity. Genetic studies have shown that this progression involves multiple metabolic processes at each stage.

proanthocyanidin A group of condensed tannins that are present in many plants that serve as defensive chemicals against plant pathogens and herbivores.

condensed tannins Tannins that are polymers of flavonoid units. Require use of strong acid for hydrolysis.

maturation The final phase of seed development.

desiccation tolerance A plant's ability to function while dehydrated.

seed longevity The length of time a seed can remain dormant without losing viability.

orthodox seed A seed that can tolerate desiccation and remain viable after storage in a dry state.

recalcitrant seeds Seeds that are released from the plant with a relatively high water content and active metabolism and, as a consequence, deteriorate upon dehydration and do not survive storage.

For example, four stages of seed growth and development (embryogenesis, seed filling, late maturation, and pod abscission) for seeds of barrel clover (*Medicago truncatula*) are shown in **Figure 17.6A**. Embryogenesis proceeds during the first 10 days after pollination, after which seed filling begins, as indicated by the increase in seed dry weight. Simultaneously, the water content of the seed declines (**Figure 17.6B**). The acquisition of desiccation tolerance begins approximately 24 days after pollination, and overlaps with the seed-filling and dehydration phases. From 28 days after pollination onward, seeds gradually acquire longevity (**Figure 17.6C**). Freshly harvested seeds acquire the ability to germinate at approximately 16 days after pollination (about 20%). The ability to complete germination increases to 50% from 22 to 32 days after pollination, after which it declines to 10% due to the onset of dormancy (**Figure 17.6D**) (seed dormancy is discussed in Chapter 14). However, this dormancy can be overcome by dry storage for 6 months (after-ripening), after which the fully mature seeds germinate within 24 h.

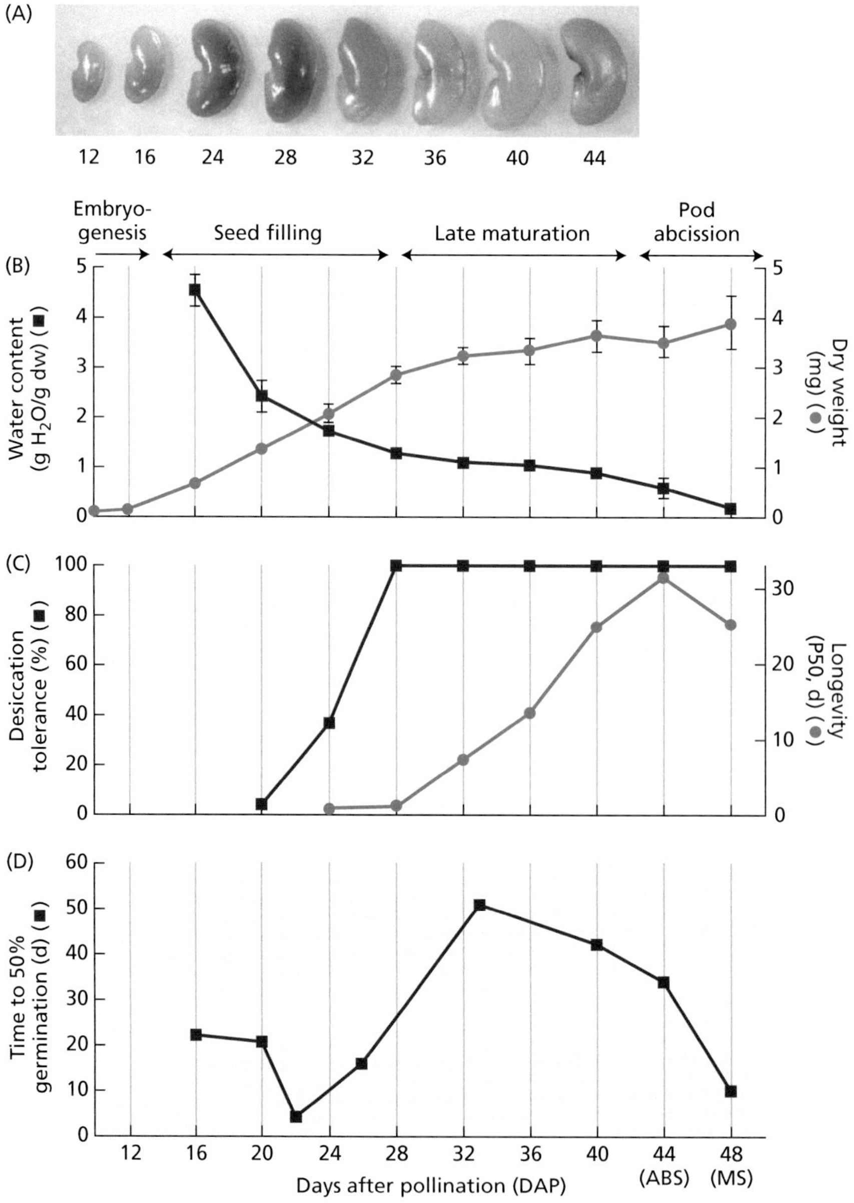

Figure 17.6 Metabolic and physiological changes during seed maturation of *Medicago truncatula*. Seed development is divided into four major phases: embryogenesis, seed filling, late maturation, and pod abscission. (A) Time course of seed development. (B) Water content and dry weight (dw) changes. (C) Acquisition of desiccation tolerance, measured as the percentage germination after rapid drying to 43% relative humidity, and longevity, determined as the time to reduce viability to 50% under storage at 75% relative humidity and 35°C. (D) Changes in germination speed or dormancy, determined as the time required for 50% of seeds to complete germination at 20°C. Error bars represent ± standard error of the mean. ABS, abscission; MS, mature seed.

Embryos accumulate proteins and sugars during desiccation

Desiccation can severely damage membranes and other cellular constituents (Chapter 19). Mature seeds have as little as 0.1 g water g^{-1} dry weight, with water potentials between −350 and −50 MPa. As seeds begin to dehydrate, embryos accumulate nonreducing sugars (primarily sucrose, raffinose, and stachyose) as well as **late-embryogenesis-abundant (LEA) proteins**.

The accumulation of these molecules transforms the embryo to a *glassy state*. In general, a glass is defined as an amorphous, metastable state that resembles a solid, brittle material but retains the disorder and physical properties of the liquid state. Biological glasses are highly viscous liquids with very low molecular diffusion rates, and therefore can participate only in limited chemical reactions.

LEA proteins are small, hydrophilic, largely disordered, and thermostable proteins that are synthesized in orthodox seeds during mid- to late maturation and in vegetative tissues in response to osmotic stress. They are thought to have a range of protective functions against desiccation with different efficiencies, including ion binding, antioxidant activity, hydration buffering, and membrane and protein stabilization. Since LEA proteins were first described in the early 1980s in cotton seeds, related proteins have been identified in seeds and pollen grains of other plant species, as well as in bacteria, cyanobacteria, and some invertebrates. The ability of "resurrection plants" (e.g., *Craterostigma plantagineum*) to survive extreme desiccation has been linked to the accumulation of LEA proteins in the vegetative organs. In addition, LEA proteins may play a role in the response to freezing and salinity stress, both of which involve cellular dehydration (Chapter 19).

late-embryogenesis-abundant (LEA) proteins Proteins involved in desiccation tolerance. They interact to form a highly viscous liquid with very slow diffusion and therefore limited chemical reactions. Encoded by a group of genes that are regulated by osmotic stress, first characterized in desiccating embryos during seed maturation.

Coat-imposed dormancy correlates with long-term seed viability

The seeds of many herbs and garden vegetables, such as onion, okra, and soybean, can remain viable under storage for only 1 to 2 years. Others, such as those of cucumber and celery, remain viable for up to 5 years. In 1879, W. J. Beal initiated the longest-running experiment on seed longevity by burying the seeds of 20 different species in unstoppered bottles in a sandy hilltop near the Michigan Agricultural College in East Lansing. After 120 years (in the year 2000), only one species, moth mullein (*Verbascum blattaria*), remained viable, with half of the 50 seeds in the bottle germinating. In 2021 another bottle was dug up and the seeds were tested for viability; only 11 seeds germinated. However, 142 years is by no means the maximum longevity for seeds. Seeds of canna lily (*Canna compacta*) apparently can live for at least 600 years. The oldest authenticated surviving seeds are those of sacred Indian lotus (*Nelumbo nucifera*), at nearly 1,300 years old, and a single date palm (*Phoenix dactylifera*) seed found buried at Masada in Israel, at a remarkable 2,000 years old. These two species have highly impermeable seed coats, suggesting that coat-imposed dormancy is associated with long-term viability. However, many orthodox seeds of lesser longevity can be stored for a long time under seed-bank conditions (either 1–2°C or −18°C), depending on the plant species.

17.6 Fruit Development and Ripening

Distinguish among the variety of mechanisms involved in fruit development and ripening.

True fruits are found only in the flowering plants. In fact, fruits are a defining feature of the angiosperms (*angio* means "vessel" or "container" in Greek and *sperm* means "seed"). Fruits are typically derived from a seed-containing mature ovary; ovary-derived fruit is often referred to as *botanical fruit*. Examples of botanical fruits are peaches, tomatoes, and Arabidopsis siliques. However, fruits

fruit In angiosperms, one or more mature ovaries containing seeds and sometimes adjacent attached parts.

dehiscent fruit A dry fruit that spontaneously splits, releasing its seeds.

indehiscent Lacking spontaneous opening of a mature anther or fruit.

berry A simple, fleshy fruit produced from a single ovary and consisting of an outer pigmented exocarp, a fleshy, juicy mesocarp, and a membranous inner endocarp.

drupe A structure similar to a berry, but with a hardened, shell-like endocarp (pit or stone) that contains a seed.

pome A fruit, such as an apple, composed of one or more carpels surrounded by accessory tissue derived from the receptacle.

fruit set The commitment to initiate fruit development. It is a key developmental switch that normally occurs following successful fertilization and/or pollination.

can also develop from a variety of non-ovary tissues, and in these cases they are referred to as *accessory fruit* (or *false fruit*). Strawberry is an example of accessory fruit. The fleshy part of the strawberry is actually the receptacle, the stem tissue to which the floral organs are attached, while the true fruits are the dry achenes embedded in the receptacle.

Fruits are seed-dispersal units, and they can be grouped according to several features. For example, they can be either dry or fleshy based on their structure and moisture content. Some dry fruits, such as soybean pods, split to release the seeds and are called **dehiscent fruits**. Other dry fruits, such as the winged seeds (samaras) of maple and cypselae of dandelion, do not undergo this process and are thus **indehiscent**. The fleshy fruits we are most familiar with are indehiscent and occur in a variety of forms. Tomatoes, bananas, and grapes are defined botanically as **berries**, in which the seeds are embedded in a fleshy mass, while peaches, plums, and apricots are classified as **drupes**, in which the seed is enclosed in a stony endocarp. Apples and pears are **pome** fruits, in which the edible tissue is derived from an accessory structure called a *hypanthium*, the fused bases of the sepals, petals, and sometimes stamens. Fruits can also be defined as *simple*, with a mature single or compound ovary, as in hazelnut and tomato. Alternatively, they may be *aggregate*, like raspberry, composed of multiple carpels that are not joined together. Finally, they may be *multiple fruits* formed by the fusion of fruits produced by clusters of flowers, as in the case of pineapple. Some examples of fleshy and dry fruit types are illustrated in **Figure 17.7**. It is interesting to note that even within a single plant family, a wide variety of fruit types can be found, including fleshy fruits such as berries, pomes, and drupes and dry fruits such as achenes, capsules, and follicles. This developmental plasticity in fruit type probably reflects rapid evolutionary adaptation in modes of seed protection and dispersal.

Auxin and gibberellic acid (GA) regulate fruit set and parthenocarpy

In most angiosperms, the floral organs senesce and die if unfertilized. The developmental switch that turns a floral tissue into a growing fruit is called **fruit set**. The mechanisms underlying this induction of fruit set are evolutionarily conserved and have been studied extensively in strawberry as well as other commercially valuable fruits. A strawberry fruit consists of a receptacle dotted

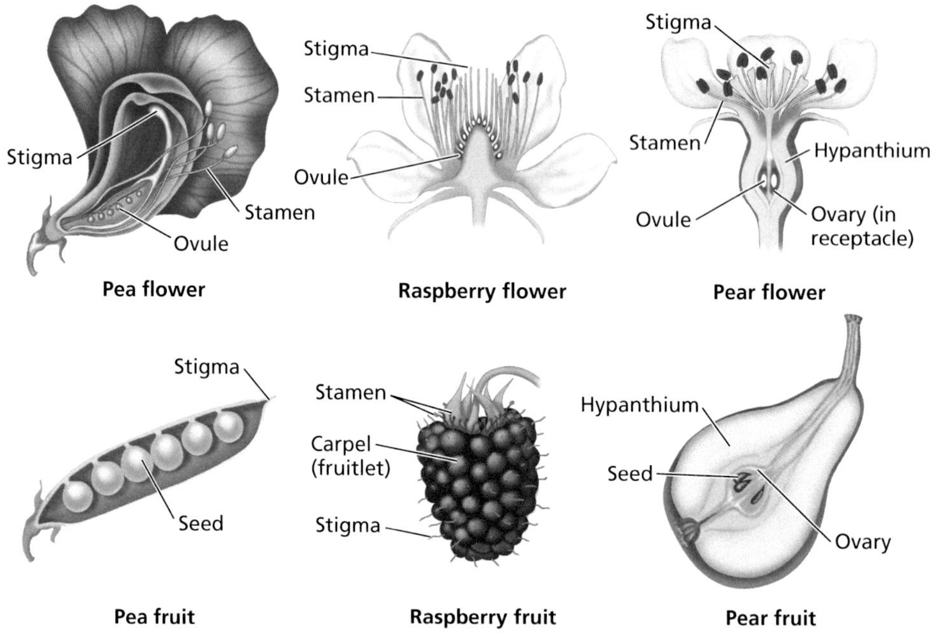

Figure 17.7 Three types of fruit and their flowers: pea, raspberry, and pear.

with about 200 seed-containing ovaries (achenes) on its surface, which makes the achenes (the true fruits) easily accessible for experimental manipulation. In 1950, Nitsch showed that removal of the achenes prevented development of the receptacle into a fruit. However, application of auxin to the receptacle could substitute for the achenes and stimulate receptacle enlargement (**Figure 17.8**). This simple experiment suggested that achenes were the source of auxin that stimulated the development of the receptacle. This and later experiments showed that the phytohormones auxin and GA are produced by the seed upon successful fertilization and that the production or exogenous application of these phytohormones is required to promote fruit set and subsequent fruit enlargement.

The molecular mechanisms regulating fruit set have been extensively investigated using genetic and mutational analyses in conjunction with techniques to analyze gene expression and hormone levels in seedless (*parthenocarpic*) cultivars. Induction of parthenocarpy is used to enhance fruit yield and quality by producing seedless fruits for human consumption.

As shown in **Figure 17.9A** and **Figure 17.9B**, wild bananas are seeded, but commercial varieties are *triploid* (three rather than two sets of chromosomes) and don't produce seeds. Because of this, they are propagated asexually, and the resultant lack of genetic diversity underlies the threat to global banana production by only three species of Shigatoka fungi from the genus *Mycosphaerella*.

Increased production or perception of the hormones auxin and GA in early fruit development also can produce parthenocarpic fruit. For example, spraying with GA produces large Thompson Seedless grapes (Chapter 12). In tomato and eggplant, transgenic plants that overproduce auxin or an auxin receptor develop parthenocarpic fruits (**Figure 17.9C–Figure 17.9F**). Mutations in genes

Figure 17.8 Nitsch's 1950 experiment on strawberry fruit development. (A) From left to right: a small fruit flesh induced by the removal of all but a single fertilized achene; a fully fertilized and enlarged fruit; a fruit that failed to develop any fruit flesh due to the removal of all achenes; and a fruit with no achenes that still enlarged to form fruit flesh due to the exogenous application of auxin. (B) Selective removal of most of the fertilized ovaries in a strawberry flower resulted in the development of two separate miniature "fleshy red fruits" induced by the two remaining achenes (black arrowheads). (C) Aggregation of five separate "tiny red fruits," each of which developed below a fertilized achene (white and black arrowheads).

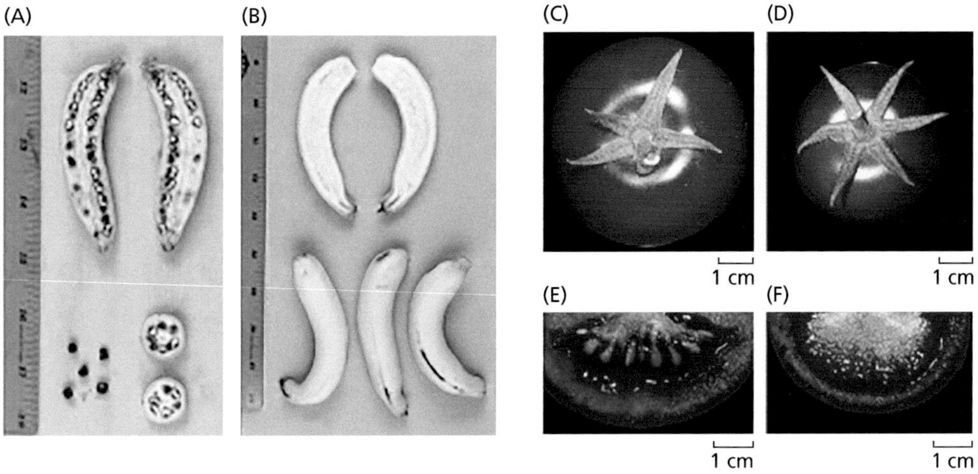

Figure 17.9 Parthenocarpic fruits compared with fertilized fruits. (A) Fruits of wild, seeded banana (*Musa acuminata banksia*). (B) Domesticated parthenocarpic banana fruits. (C) Fertilization-induced wild-type tomato fruit of the "Monalbo" variety. (D) The "Monalbo" wild-type fruit cut in half; the locule is filled with pulp and seeds. (E) Parthenocarpic "Monalbo" fruit containing an abnormal auxin response factor. (F) The parthenocarpic "Monalbo" fruit cut in half to show the seedless endocarp.

stone A hardened endocarp (the inner layer of the ovary) in drupe fruits such as peach.

pit A microscopic region where the secondary wall of a tracheary element is absent and the primary wall is thin and porous

coding for repressors of auxin responses also lead to parthenocarpy. Similarly, the tomato *procera* mutant, which has reduced responses to GA, also develops parthenocarpic fruit.

Mature dehiscent fruits open spontaneously to release seeds

The dry, dehiscent fruits of Arabidopsis serve as a model for the study of fruit dehiscence. In Arabidopsis the gynoecium arises from the fusion of two carpels, referred to collectively as the pistil, at the center of the flower. The Arabidopsis gynoecium consists of two ovary walls or pericarps (also known as valves), a central replum, a septum, and valve margins (**Figure 17.10A**). After fertilization, the valve margins differentiate into a dehiscence zone that consists of a lignified layer and a separation layer (**Figure 17.10B** and **Figure 17.10C**). The separation layer contains small cells held together weakly by the extracellular matrix; at maturity the springlike tension in the lignified layer and adjoining tissues splits open the fruit along the separation layer, causing the valve to detach from the replum during pod shattering. In seed crops such as legumes (e.g., soybean) and canola (rapeseed), loss of fruit dehiscence (pod shattering resistance) is a key agronomical trait selected to prevent significant yield loss.

Fleshy fruits have extensive mesocarp layers

The pericarp of fleshy fruits is typically divided into three layers: exocarp (skin), mesocarp (intermediate layer), and endocarp (innermost layer) (**Figure 17.11**). The soft edible part of the fruit is often the mesocarp. In peach, pistachio, and other drupe fruits, the hardened endocarp is referred to as the **stone** or **pit**. The stone provides a physical barrier around the seed to protect seeds from pathogens and herbivores.

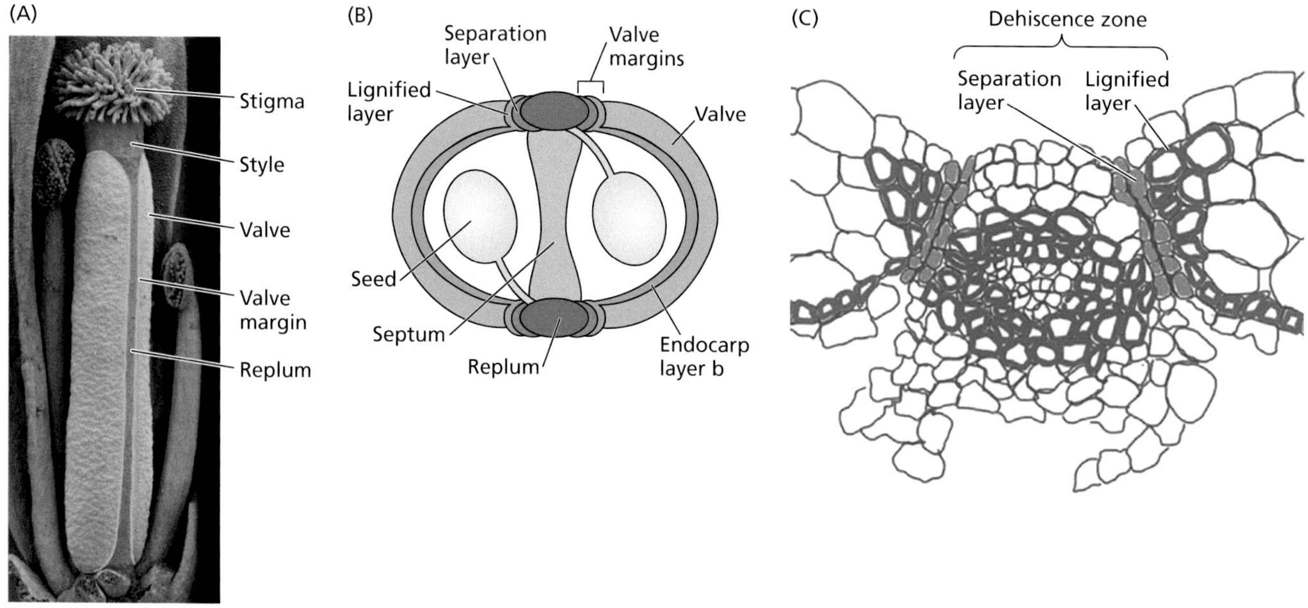

Figure 17.10 The structure and regulation of the dehiscence zone. (A) False-color scanning electron micrograph of the Arabidopsis gynoecium showing the stigma, style, valves, replum, and valve margins. (B) Diagram of a transverse cross section of the ovary region of the gynoecium, showing the septum and seeds and the two layers of the dehiscence zone. The endocarp layer b, which is interior to the epidermal layer on the inside surface, is lignified and contributes to the tension mechanism of pod shatter. (C) Drawing of a cross section through the dehiscence zone in Arabidopsis. The separation and lignified layers are indicated by different colors.

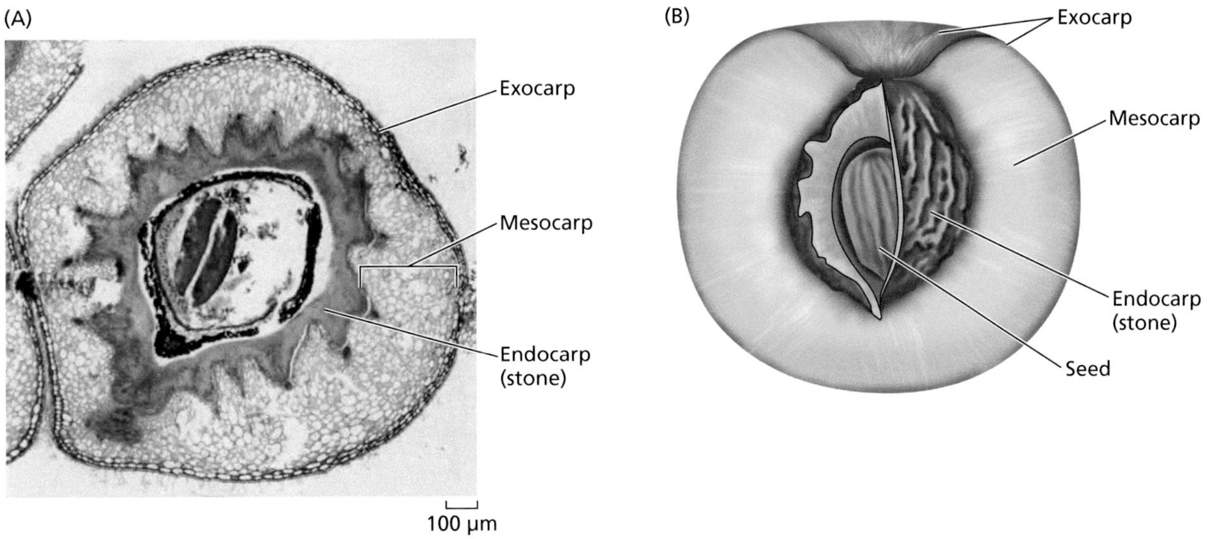

Figure 17.11 Examples of tissue differentiation in fleshy fruits. (A) Cross section of a developing raspberry drupelet. The lignin in the thickening endocarp (stone) is stained red. (B) Drawing of a peach fruit illustrating the three tissue layers and a seed inside the endocarp.

Most fleshy fruits are indehiscent and are derived from the fusion of carpels with seeds that are attached to the placenta. In fleshy fruits, cell division is usually followed by massive cell expansion (**Figure 17.12**). In some varieties of tomato, pericarp cell diameters may reach 0.5 mm. Cell division and expansion in fleshy fruits are controlled by highly regulated staging of gene expression.

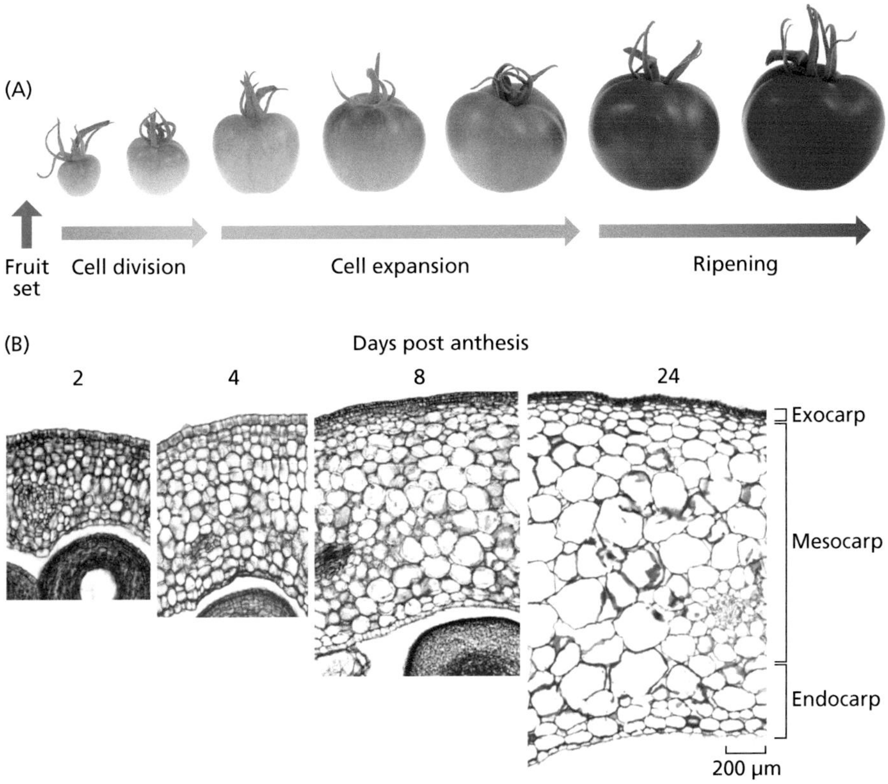

Figure 17.12 Tomato fruit growth. (A) Photographs of developmental stages in a miniature tomato. (B) Light microscope images of a cross section of tomato pericarp at 2, 4, 8, and 24 days after flower opening (anthesis).

ripening The process that causes fruits to become more palatable, including softening, increasing sweetness, loss of acidity, and changes in coloration.

Fleshy fruits undergo ripening

Ripening in fleshy fruits refers to the changes that make them attractive (to humans and other animals) and ready to eat. Such changes typically include color development, softening, starch hydrolysis, sugar accumulation, production of aroma compounds, and the disappearance of organic acids and phenolic compounds, including tannins. Dry fruits do not undergo a true ripening process, and many of the same families of genes that control dehiscence in dry fruits seem to have been recruited to new functions in ripening in fleshy fruits. Because of the importance of fruits in agriculture and their health benefits, the vast majority of studies on fruit ripening have focused on edible fruits. Tomato is the established model for studying fruit ripening, as it has proved highly amenable to biochemical, molecular, and genetic ripening studies.

Ripening involves changes in the color of fruit

Fruits ripen from green to a range of colors, including red, orange, yellow, purple, and blue. The pigments involved affect not only the visual appeal of the fruit, but also its taste and aroma, and are known to have health benefits for humans. Fruits typically contain a mixture of pigments, including the green chlorophylls; yellow, orange, and red carotenoids; red, blue, and violet anthocyanins; and yellow flavonoids. The loss of the green pigment at the onset of ripening is caused by the degradation of chlorophyll and the conversion of chloroplasts to chromoplasts, which act as the site for the accumulation of carotenoids.

Carotenoids are responsible for the red color of tomato fruits. During ripening in tomato, the concentration of carotenoids increases between 10- and 14-fold, mainly due to the accumulation of the deep red pigment lycopene. Fruit ripening involves the active biosynthesis of carotenoids, the chemical precursors for which are synthesized in the plastids. The first committed step is the formation of the colorless molecule phytoene by the enzyme phytoene synthase. In tomato, phytoene is then converted to the red pigment lycopene through a series of further reactions.

Anthocyanins are the pigments responsible for the blue and purple color in some berries and red in strawberry (**Figure 17.13**). Anthocyanins are made through the phenylpropanoid pathway; that is, they are derived from the amino acid phenylalanine. Phenylpropanoids constitute some of the most important sets of specialized metabolites in plants. They contribute not only to the characteristic color and flavor of fruits but also to unfavorable traits, such as browning of fruit tissues via enzymatic oxidation of phenolic compounds by polyphenol oxidases.

(A) (B)

Figure 17.13 Examples of pigment formation during fruit ripening. (A) Blueberries accumulate more than a dozen different anthocyanins during ripening, including malvidin-, delphinidin-, petunidin-, cyanidin-, and peonidin-glycosides, which give them a deep purple color. (B) A pelargonidin-based anthocyanin contributes to the red color of strawberry fruit.

Fruit softening involves the coordinated action of many cell wall–degrading enzymes

Softening of fruit involves changes in the fruit's cell walls. In most fleshy fruits, the cell walls consist of a semirigid composite of cellulose microfibrils, thought to be tethered by a xyloglucan network, which is embedded in a gel-like pectin matrix. In tomato, more than 50 genes related to cell wall structure show changes in expression during ripening, indicating a highly complex set of events connected with cell wall remodeling during the ripening process.

Experiments in transgenic plants have demonstrated that no single cell wall–degrading enzyme can account for all aspects of softening in tomato or other fruits. It appears that texture changes result from the synergistic action of a range of cell wall–degrading enzymes and that suites of texture-related genes give different fruits their unique melting, crisp, or mealy textures. However, even in tomato, the precise contribution of each type of enzyme to fruit texture is still poorly understood. Changes in the cuticle of the fruit that affect water loss also affect texture and shelf life.

ACC synthase The enzyme that catalyzes the synthesis of 1-aminocyclopropane-1-carboxylic acid (ACC) from *S*-adenosylmethionine.

ACC oxidase The enzyme that catalyzes the conversion of 1-aminocyclopropane-1-carboxylic acid (ACC) to ethylene, the last step in ethylene biosynthesis.

Taste and flavor reflect changes in acids, sugars, aroma, and other compounds

Most fleshy fruits that are consumed by humans undergo changes that make them especially palatable to eat when they are ripe. These chemical changes include alterations in sugars and acids, the release of aroma compounds, and the reduction of unpalatable metabolites.

One example of human selection for more palatable fruits is the selection against bitterness in cucumber. Wild cucumbers produce cucurbitacins, triterpenoids that confer a bitter taste that repels herbivores. Cultivated nonbitter varieties of cucumber that have been historically selected by humans for taste and digestibility contain mutations in transcription factors that regulate cucurbitacin synthesis.

In many fruits, starch is converted at the onset of ripening to glucose and fructose, and citric and malic acids are also abundant. However, although sugars and acids are vital for taste, volatiles are what really determine the unique flavors of fruits. Flavor volatiles arise from a wide range of compounds. Some of the most detailed studies have been undertaken in tomato. They show that of the 400 or so volatiles produced by tomato, only a small number have a positive effect on flavor. The most important flavor volatiles in tomato (hexenal and hexenol isomers) are derived from the catabolism of the fatty acids linoleic acid and linolenic acid via lipoxygenase activity. Other important volatiles are derived from amino acids and include 2- and 3-methylbutanal, 3-methylbutanol, phenylacetaldehyde, 2-phenyl-ethanol, and methyl salicylate. A third class of volatiles are the apocarotenoids, which are derived via the oxidative cleavage of carotenoids. Apocarotenoids such as β-damascenone are important in tomato, apple, and grape.

Ethylene regulates the ripening of many fruits

Ethylene has long been recognized as the hormone that can accelerate the ripening of many edible fruits. However, the definitive demonstration that ethylene is required for fruit ripening was provided by experiments in which ethylene biosynthesis was blocked by inhibiting expression of the genes that encode either **ACC synthase** (ACS) or **ACC oxidase** (ACO). ACS and ACO are the enzymes involved in the second-to-last and last steps in ethylene synthesis, respectively (Chapter 12). These two steps in the pathway are normally tightly regulated. Genome editing of *ACC* inhibits ripening in transgenic tomatoes (**Figure 17.14**). Exogenous ethylene restores normal ripening in the transgenic tomato fruits. A similar effect is observed with inhibition of *ACO* gene expression.

Further demonstration of the requirement for ethylene in fruit ripening came from analysis of the *Never-ripe* mutation in tomato commonly found in supermarket tomatoes that are shipped over long distances. As the name implies, this mutation completely blocks the ripening of tomato fruit until it is treated with ethylene. The *Never-ripe* phenotype is caused by a mutation in one of the ethylene receptors in tomato fruits to

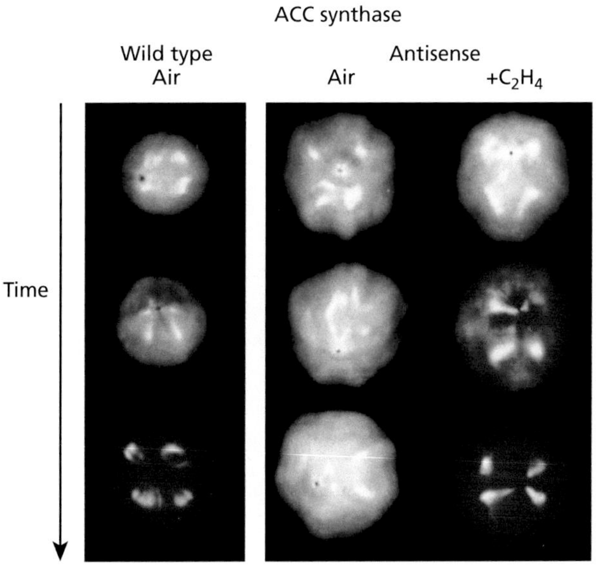

Figure 17.14 Genome editing of ACC synthase inhibits ripening. Fruit wherein levels of ACC synthase have been decreased by genome editing compared with controls (wild type). Note that in air the antisense fruit did not ripen, but did senesce after 70 days (yellow); ripening could be restored by adding external ethylene (C_2H_4).

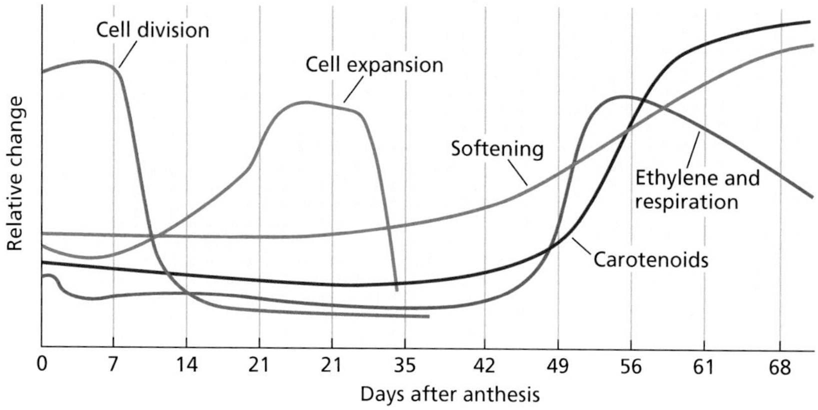

Figure 17.15 Growth and development of tomato fruits in relation to the effects of ethylene and ripening. Climacteric fruits show a characteristic increase in respiration and ethylene production, which signals the onset of ripening.

reduce ethylene binding. Treatment with high concentrations of ethylene during tomato shipment is required to activate ethylene perception and ripening.

The role of plant hormones other than ethylene in controlling ripening is much less well understood, although auxin, ABA, and GA are known to have an effect on this important developmental process.

Climacteric and non-climacteric fruit differ in their ethylene responses

Fleshy fruits have traditionally been placed in two groups as defined by the presence or absence of a characteristic respiratory rise called a **climacteric** at the onset of ripening. Climacteric fruits show this respiratory increase along with a preceding or coincident spike of ethylene production (**Figure 17.15**). Apple, banana, avocado, and tomato are examples of climacteric fruits. In contrast, fruits such as citrus, strawberry, and grape do not exhibit such large changes in respiration and ethylene production and are termed **non-climacteric** fruits.

In plants with climacteric fruits, two systems of ethylene production operate, depending on the stage of development:

- In System 1, which acts in immature climacteric fruit, ethylene inhibits its own biosynthesis by negative feedback.
- In System 2, which occurs in mature climacteric fruit and in senescing petals in some species, ethylene stimulates its own biosynthesis.

The positive feedback loop for ethylene biosynthesis in System 2 ensures that the entire fruit ripens evenly once ripening has commenced.

When mature climacteric fruits are treated with ethylene, the onset of the climacteric rise and the changes associated with ripening are hastened. In contrast, when immature climacteric fruits are treated with ethylene, the respiration rate increases gradually as a function of the ethylene concentration, but the treatment does not trigger production of endogenous ethylene or induce ripening. Ethylene treatment of non-climacteric fruits, such as citrus, strawberry, and grape, does not cause an increase in respiration and is not required for ripening. However, it can alter ripening characteristics in some species, such as the enhancement of color in citrus.

Although the distinction between climacteric and non-climacteric fruits is a useful generalization, some non-climacteric fruits may also respond to ethylene; for example, citrus fruits de-green in response to exogenous ethylene. Indeed, the distinction between climacteric and non-climacteric fruits may be less dramatic than previously thought, with some species showing contrasting behavior depending on the cultivar. For example, melon (*Cucumis melo*) can be climacteric or non-climacteric depending on the variety.

Considerably less is known about the molecular control of the ripening process in non-climacteric fruits. Accumulating evidence suggests that abscisic acid and sucrose promote fruit ripening, while auxin inhibits strawberry fruit ripening, despite its early role in fruit set and enlargement.

climacteric Marked rise in respiration at the onset of ripening that occurs in all fruits that ripen in response to ethylene, and in the senescence process of detached leaves and flowers.

non-climacteric Referring to a type of fruit that does not undergo a climacteric, or respiratory burst, during ripening.

Commercial control of fruit ripening

Elucidation of the role of ethylene in the ripening of climacteric fruits has resulted in many practical applications aimed at causing either uniform ripening or a delay in ripening. For example, banana bunches are picked when they are still green and hard, which helps them survive the journey from the fields of Central and South America to their final destinations around the world. The clusters of unripe fruits are cut from the bunch, treated with fungicide, packed in cartons, and shipped overseas. On arrival at their destination, the bananas are placed in temperature-controlled rooms and treated with low concentrations of ethylene gas to initiate ripening. This mirrors the natural ripening process but ensures that fruits at different stages of maturity will initiate ripening at the same time, making them easier to market.

In the case of fruits such as apples, ripening can be delayed using controlled-atmosphere (i.e., ~2.5% oxygen and 2.5% carbon dioxide) storage and refrigeration, thus extending the marketable period for the crop.

It is also possible to manipulate the nutritional quality of fruit. For example, anthocyanins, like carotenoids, are thought to protect against heart disease and certain cancers, as they are strong antioxidants that can scavenge excessive damaging reactive oxygen species. Anthocyanin levels in fruits can be manipulated by transgenic approaches, even to the extent of introducing high levels of these compounds in tomato flesh, where they do not normally occur (**Figure 17.16**). A greater understanding of the molecular determinants of other aspects of fruit development, such as the production of volatiles, will presumably offer other opportunities to improve fruit quality.

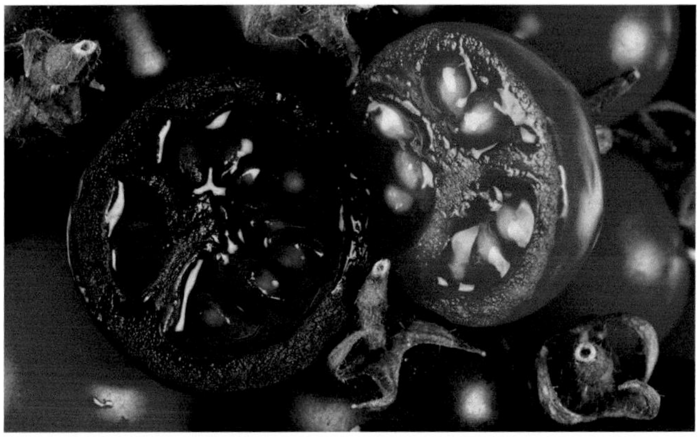

Figure 17.16 Anthocyanin production can be induced in tomato by introduction of transcription factors that control the biosynthesis of these compounds in snapdragon (*Antirrhinum*).

Summary

Seeds are a hallmark of spermatophyte land plants. The tremendous importance of seed plants to food production has led to extensive study of seed and fruit development. In seed plants, embryogenesis transforms the single-cell zygote into a complex, multicellular individual contained within a mature seed. This developmental process proceeds from maturation to dormancy and culminates in germination. Fruits play an important role in sustaining seeds and ensuring their dispersal. Seed and fruit development and ripening are regulated by gene expression and hormonal control.

17.1 Seed Structure

- Seeds are surrounded by a seed coat (testa), whereas fruits are enclosed by the pericarp (**Figure 17.1**).
- Many "seeds" are actually seedlike fruits (**Figure 17.1**).

17.2 Establishment of the Embryonic Axis

- Among seed plants, the apical–basal polarity is established early in embryogenesis by programmed cell division and differentiation as well as polar auxin flows (**Figure 17.2**).

17.3 Seed Endosperm Development

- Seed structures share common elements, but vary across phyla (**Figures 17.3** and **17.4**).
- Embryos can be asexually produced from vegetative tissues (**Figure 17.5**).
- After fertilization, the diploid endosperm, which will provide nutrition to the embryo, becomes multinucleate (coenocytic). Cellularization of the coenocytic endosperm in Arabidopsis proceeds from the micropylar to the chalazal region, while cellularization of cereal endosperms proceeds centripetally (**Figures 17.3** and **17.4**).

(*Continued*)

Summary (*continued*)

- Endosperm development is generally controlled primarily by maternally expressed genes, not by the embryo.
- Endosperm development and embryogenesis can occur autonomously.
- Cells of the starchy endosperm and aleurone layer follow divergent developmental pathways.

17.4 Seed Coat Development

- The seed coat arises from maternal integuments, but its development is regulated by the endosperm.

17.5 Seed Maturation and Desiccation Tolerance

- Seed filling and the acquisition of desiccation tolerance overlap in most species.
- The acquisition of desiccation tolerance is aided by LEA proteins, which form hydrogen bonds with nonreducing sugars, allowing embryo cells to acquire a glassy state that makes them more stable than cells that are simply dehydrated.
- Impermeable seed coats and low temperatures can increase seed longevity, which otherwise is highly variable across species.
- Embryos accumulate proteins and sugars during desiccation (**Figure 17.6**).
- Coat-imposed dormancy correlates with long-term seed viability.

17.6 Fruit Development and Ripening

- Fruits are seed-dispersal units that either arise from the pistil and contain the seed(s) or are derived from accessory floral tissues (**Figure 17.7**).
- Fruit initiation is triggered by fertilization of the ovules and pollination of the stigma. This developmental switch, also called *fruit set*, initiates cell division and expansion that can also be induced by auxin and GA treatment (**Figure 17.8**).
- Induction of parthenocarpy—the development of fruit without fertilization—can improve fruit quality and yield (**Figure 17.9**).
- Dehiscent fruits release their seeds through pod shattering. Proper morphogenesis and differentiation of the dehiscence zone is tightly regulated (**Figure 17.10**).
- Fleshy fruits undergo ripening, which involves color changes, highly coordinated fruit softening, and other changes (**Figures 17.11–17.14**).
- Acids, sugars, and volatiles determine the flavor of ripe and unripe fleshy fruits.
- Ethylene accelerates ripening, particularly in climacteric fruits (**Figures 17.14** and **17.15**).
- Fruit color and nutrient quality can be modified by genome engineering (**Figure 17.16**).

Suggested Reading

Angelovici, R., Galili, G., Fernie, A. R., and Fait, A. (2010) Seed desiccation: A bridge between maturation and germination. *Trends Plant Sci.* 15: 201–208.

Dresselhaus, T., and Jürgens G. (2021) Comparative embryogenesis in angiosperms: Activation and patterning of embryonic cell lineages. *Annu. Rev. Plant Biol.* 72: 641–676.

Knapp, S., and Litt, A. (2013) Fruit—An angiosperm innovation. In *The Molecular Biology and Biochemistry of Fruit Ripening*, G. B. Seymour, G. A. Tucker, M. Poole, and J. J. Giovannoni, eds., Wiley-Blackwell, Oxford, UK, p. 206.

Li, J., and Berger, F. (2012) Endosperm: Food for humankind and fodder for scientific discoveries. *New Phytol.* 195: 290–305.

Palovaara, J., de Zeeuw, T., and Weijers, D. (2016) Tissue and organ initiation in the plant embryo: A first time for everything. *Annu. Rev. Cell Dev. Biol.* 32: 47–75.

Radoeva, T., Vaddepalli, P., Zhang, Z., and Weijers, D. (2019) Evolution, initiation and diversity in early plant embryogenesis. *Dev. Cell* 50: 533–543.

Seymour, G. B., Østergaard, L., Chapman, N. H., Knapp, S., and Martin, C. (2013) Fruit development and ripening. *Annu. Rev. Plant Biol.* 64: 209–241.

Wilkinson, J. Q., Lanahan, M. B., Yen, H.-C., Giovannoni, J. J., and Klee, H. J. (1995) An ethylene-inducible component of signal transduction encoded by Never-ripe. *Science* 270: 1807–1809.

18 Biotic Interactions

In natural habitats, plants live in complex, diverse environments in which they interact with a wide variety of organisms (**Figure 18.1**). Some of these interactions are clearly beneficial, if not essential, to both the plant and the other organism. Such a mutually beneficial biotic interaction is termed a **mutualism**. Examples of mutualism include the symbiotic relationship between nitrogen-fixing bacteria (rhizobia) and leguminous plants described in Chapter 5, the mycorrhizal associations observed between most plants and fungi (Chapter 4), and plant–pollinator interactions. Other types of biotic interactions, including **herbivory**, infection by **microbial pathogens** or **parasites** (organisms living in or on the plant and gaining their nutrition at the expense of their hosts), and **allelopathy** (chemical warfare between plants), are detrimental. Plants have evolved intricate defense mechanisms to protect themselves against harmful organisms, and these harmful organisms have in turn evolved counter-mechanisms to defeat the plants' defenses. Such tit-for-tat evolutionary processes are examples of **coevolution**, which accounts for the complex interactions between plants and other organisms that we observe today.

However, it would be an oversimplification to characterize all organisms that interact with plants as either beneficial or harmful. For example, the grazing of flowers by mammals decreases fitness in some plant species, but in others it can lead to an increase in the number of flowering stalks, thus increasing fitness. There are also organisms that benefit from their interaction with the plant without causing any harmful effects. Such neutral interactions (from the plant's perspective) are termed **commensalism**. Commensal organisms can become beneficial if they protect the plant from a second, harmful organism. For example, nonpathogenic soil bacteria

mutualism A symbiotic relationship in which both organisms benefit.

herbivory Consumption of plants or parts of plants as a food source.

microbial pathogens Bacterial or fungal organisms that cause disease in a host plant.

parasite An organism that lives on or in an organism of another species, known as the host, from the body of which the parasite obtains nutriment.

allelopathy Release by plants of substances into the environment that have harmful effects on neighboring plants.

coevolution Linked adaptations of two or more organisms.

commensalism A relationship between two organisms in which one organism benefits without negatively affecting the other.

Figure 18.1 Virtually every part of the plant is adapted to coexist with the organisms in its immediate environment.

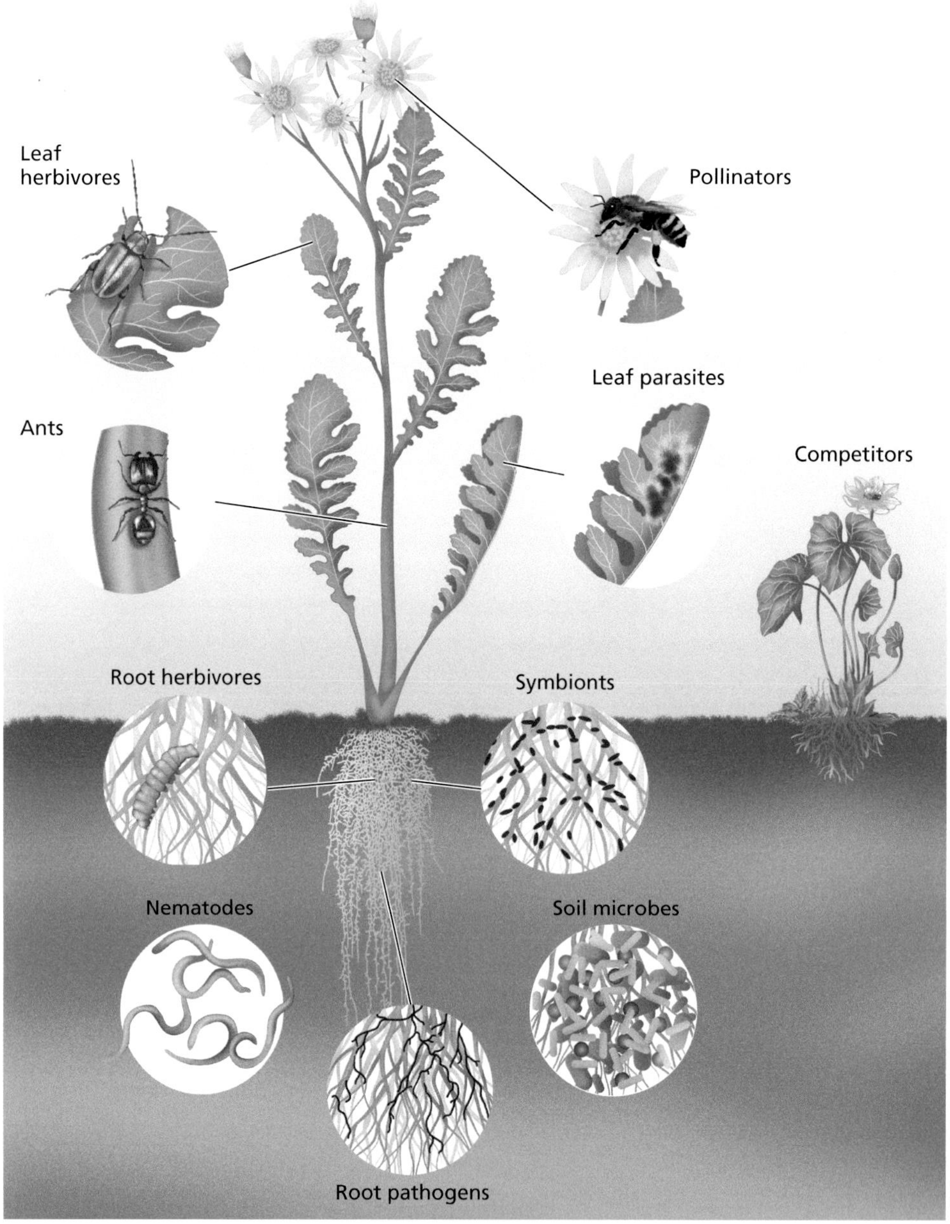

and fungi, which cause no harm to the plant, may nevertheless stimulate the plant's innate immune system and thus protect the plant from other, pathogenic microorganisms.

Preexisting **constitutive defenses** are the first line of defense deployed by plants against potentially harmful organisms. Structures such as spines, thorns, and hairs filled with toxins can deter animals from eating the plant. The cuticle (a waxy outer layer), the periderm, and other mechanical barriers help block bacterial, fungal, and insect entry. Plants also accumulate toxic specialized metabolites to deter herbivores or defend against pathogen invasion. The second line of defense typically involves **inducible defenses** such as the production of one or more toxic chemicals in response to attack. Inducible defenses require the presence of

constitutive defenses Plant defenses that are always immediately available or operational; that is, defenses that are not induced.

inducible defenses Defense responses that exist at low levels until a biotic or abiotic stress is encountered.

specific detection systems and signal transduction pathways that can sense the presence of an herbivore or pathogen and alter defense-related gene expression and metabolism accordingly.

We begin our discussion of these biotic interactions with examples of beneficial associations between plants and microorganisms. Next we consider various types of harmful interactions between plants, herbivores, and pathogens. We then survey the wide range of inducible defenses that plants have evolved to fend off insect herbivores, and the signaling molecules and signal transduction pathways that regulate them. We follow with a discussion of plant responses to microbial pathogens, for which some of the same themes of constitutive and inducible defenses are employed. Some of these responses can confer both local and **systemic** (plant-wide) resistance to pathogens. Finally, we discuss two other types of plant biotic interactions, nematodes and parasitic plants, as well as the ecological role that toxic root exudates play during plant–plant competition.

systemic Movement of a signal or molecule from one location in a plant to distal locations in the plant.

18.1 Plant Interactions with Beneficial Microorganisms

Describe the relationship between plants and beneficial microorganisms and the signaling mechanisms that mediate it.

The first land plants appear in the fossil record about 450 to 500 million years ago (mya), with the first fossil plants that also show fungal associations with roots (i.e., mycorrhizal associations) dated at approximately 400 mya. Thus, a mutualistic symbiosis between plant and fungus appears to have occurred very early in the land plant lineage and was probably an important factor in helping plants invade the terrestrial environment. Indeed, land plants are colonized by a wide variety of beneficial microorganisms: endophytic and mycorrhizal fungi, bacteria in the form of biofilms on the surfaces of roots and leaves, endophytic bacteria within the plant, and nitrogen-fixing bacteria contained in specialized root or stem nodules.

In Chapter 4, we discussed the structure and function of mycorrhizae in mineral uptake and, in Chapter 5, the structure and function of nitrogen-fixing nodules. The fungal partner in arbuscular mycorrhizal associations belongs to the ancient phylum Glomeromycota. The interaction of Glomeromycota with plant roots is so finely tuned that these fungi have lost the ability to complete their life cycles outside the plant. In part, the fungal dependence on plant partners results from a need to incorporate plant membrane lipids in periarbuscular membrane structures formed by mycorrhizal fungi within plant root cells. Because arbuscular mycorrhizas can be grown only in the presence of their host, progress on dissecting the signaling between the mycorrhizas and their host initially proceeded slowly. However, the rhizobium–legume association proved to be much easier to approach because the bacterial partner in this symbiosis could be cultured independently, allowing the chemicals it produced to be more easily defined, isolated, and then tested to determine their effects on the plant. Once the molecular details of this plant–bacterium interaction were characterized, researchers could ask whether similar processes also occurred during plant–mycorrhizal associations. The similarities have led to the proposal that the rhizobium–legume signaling pathway actually evolved from the more ancient arbuscular mycorrhizal–plant interaction pathway. This idea fits well with the timing of the evolution of the rhizobium–legume symbiosis, which is thought to have formed relatively recently, approximately 60 mya.

common symbiotic pathway A sequence of common cellular events in plant roots that occurs in both mycorrhizal formation and root nodulation.

extrafloral nectar Nectar produced outside the flower and not involved in pollination events.

plant growth promoting rhizobacteria (PGPR) Soil bacteria associated with root surfaces that promote plant growth by producing plant growth regulators and/or fixing nitrogen.

In legumes, a number of core symbiotic genes have been shown to be required both for legume nodulation and for arbuscular mycorrhizal association. Similarly, the chemical structures of the bacterially-secreted factors that trigger nodulation resemble those of signaling molecules that elicit similar events in response to mycorrhizal fungi. Taken together, these parallels have led to the idea that the interaction between legumes and nitrogen-fixing rhizobia probably evolved from the more ancient interaction between plants and mycorrhizal fungi, with shared signaling elements forming a **common symbiotic pathway**. Non-nodulating or non-mycorrhizal plants appear to have lost some elements of this core symbiotic pathway, but they do have close relatives of some components that function in defenses against fungi. This suggests that a defense response system was recruited first for initiation of mycorrhizal associations and subsequently for legume–rhizobium signaling. Some plant lineages then lost the symbiotic associations for this pathway, which has been repurposed for the detection of pathogenic fungi.

Some plants have evolved strategies to attract microbes or animals that defend the plant from herbivores. An example of plant structures that serve this defense function is **extrafloral nectaries** that primarily attract ants and other animals that, in turn, repel or consume herbivores. Another example is protein-rich bodies produced on ant acacia trees (*Acacia cornigera*) to attract *Pseudomyrmex ferruginea* ants that fiercely defend the trees from animal herbivores.

Rhizobacteria can increase nutrient availability, stimulate root branching, and protect against pathogens

Plant roots provide a nutrient-rich habitat for the proliferation of soil bacteria that thrive on root exudates and lysates, which can represent as much as 40% of the total carbon fixed during photosynthesis. Population densities of bacteria in the rhizosphere can be up to 100-fold higher than in bulk soil, and up to 15% of the root surface may be covered by microcolonies of a variety of bacteria. While these bacteria use the nutrients that are released from the host plant, they also secrete metabolites into the rhizosphere.

A loosely defined group of **plant growth-promoting rhizobacteria** (**PGPR**) provides several beneficial services to growing host plants (**Figure 18.2**). For example, volatile chemicals (i.e., molecules that easily vaporize into the air) produced by the bacterium *Bacillus subtilis* alter root architecture by changing root length and lateral root density. In addition to having these developmental effects, the same volatiles have been shown to enhance proton release by Arabidopsis roots in iron-deficient growth media. These protons acidify the soil, thereby facilitating increased iron uptake and leading to plants with higher chlorophyll content, higher photosynthetic efficiency, and increased size.

PGPR can also control the buildup of harmful soil organisms. For example, root-associated *Pseudomonas* species synthesize the antifungal compound 2, 4-diacetylphloroglucinol, suppressing the growth of the

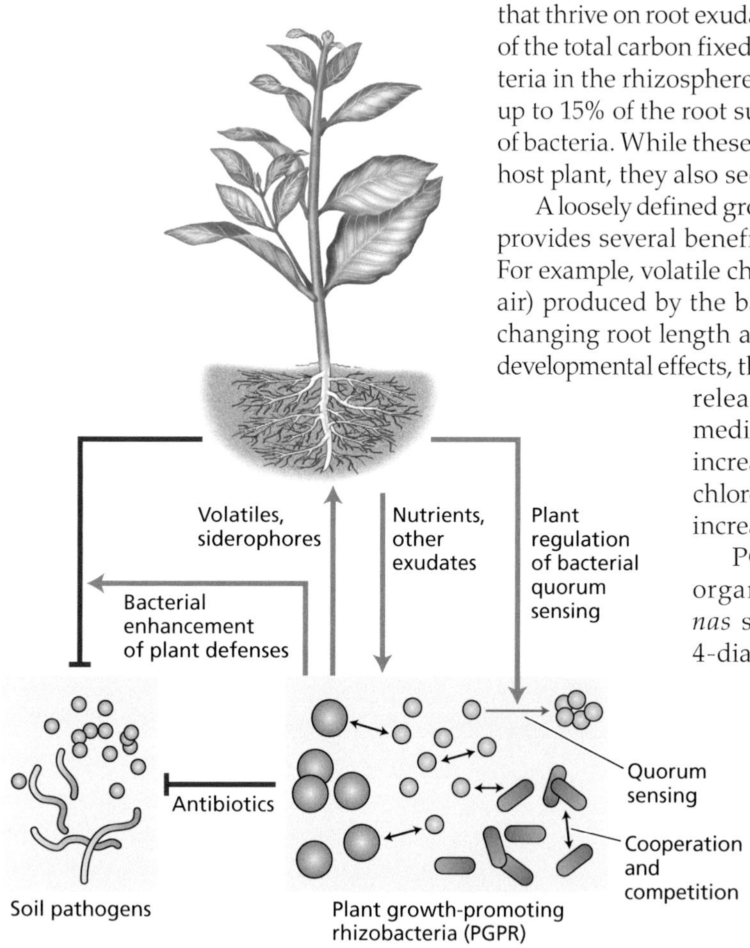

Figure 18.2 Diagram of the interactions between plants and plant growth-promoting rhizobacteria, such as *Pseudomonas aeruginosa*, which is thought to release antibiotics or siderophores into the soil that alleviate plant abiotic or biotic stress. The plant exerts control over the bacterial population by regulating bacterial quorum sensing signaling pathways through the release of exudates by the roots.

pathogenic fungus *Gaeumannomyces graminis*. In addition, several studies have suggested that *Pseudomonas aeruginosa* can alleviate the symptoms of biotic and abiotic stress by the release of antibiotics or iron-scavenging **siderophores** (Chapters 5 and 6). The amounts of the compounds released by *P. aeruginosa* are controlled by the bacterium's own **quorum sensing** signaling pathways, which are activated when the bacterium population density reaches a certain level. Plants influence the amount of antibiotics or siderophores released by rhizobacteria through the production of root exudates that regulate these bacterial quorum sensing pathways. In addition to producing chemicals that act to alter the microbial populations or characteristics of the soil around the root, PGPR microbes can also provide protection against pathogenic organisms by altering plant growth via production of plant growth regulators (Chapter 12) or activating the *systemic acquired resistance* (*SAR*) pathway described later in this chapter.

18.2 Herbivore Interactions That Harm Plants

Describe the mechanical barriers and chemical compounds that plants use for defense against herbivores and pathogens.

Along with microbial pathogens, vertebrate and invertebrate herbivores represent a major cause of biotic plant damage. Numerous mammals, reptiles, and mollusks feed on plants, as do about half of the more than 5 million insect species, with 20 to 40% of global crop productivity being lost annually to these pests. In more than 350 million years of plant–herbivore coevolution, herbivores have developed diverse feeding styles and behaviors, while plants have evolved mechanisms to defend themselves against herbivory. Plants employ an array of strategies to resist being eaten, including mechanical barriers, constitutive chemical defenses, and both direct and indirect inducible defenses. These defense mechanisms are quite effective, since most plant species are resistant to a wide range of herbivores. For example, about 90% of insect herbivores are restricted to consuming a single family of plants or just a few closely related plant species, while only 10% are able to eat a wide range of plants. This observation suggests that the vast majority of plant–herbivore interactions involved coevolution, whereby the insect has developed behavioral or metabolic strategies that circumvent the defenses of its particular family of plants, but then cannot cope with the defenses employed by other plants.

Mechanical barriers provide a first line of defense against insect pests and pathogens

Mechanical barriers, including surface structures, mineral crystals, and thigmonastic (touch-induced) leaf movements, often provide a first line of defense against pests and pathogens for many plant species. The most common surface structures are thorns, spines, prickles, and trichomes (**Figure 18.3**). **Thorns** are modified branches, as in citrus and acacia; **spines** are modified leaves, as in cacti; and **prickles** are derived mainly from the epidermis, as in roses. All of these structures possess sharp, pointed tips that physically protect plants from larger herbivores, such as mammals. However, they are less effective against smaller, insect herbivores, which can easily avoid such defenses and reach the edible parts of the shoot. **Trichomes**, or hairs, provide another layer of defense that can be effective against insect pests. These structures occur in a variety of forms, either as simple hairs that can mechanically interfere with insect locomotion, or as glandular trichomes that can wage chemical warfare against a potential herbivore. Glandular trichomes store noxious specialized metabolites such as terpenoids and phenolics in a pocket formed between the cell wall and the cuticle. These pockets burst and release their contents upon herbivore contact, and

siderophores Small molecules secreted by non-grass plants and some microbes to chelate iron, which is then taken up into cells at the root surface.

quorum sensing A system of coordinated signals and responses by which populations regulate growth and environmental responses. This is a common mechanisms in microbial organisms.

thorns Sharp plant structures that physically deter herbivores and are derived from branches.

spines Sharp, stiff plant surface structures, thought to be modified leaves, that physically deter herbivores and may assist in water conservation.

prickles Pointed plant structures that physically deter herbivory and are derived from epidermal cells.

trichomes Unicellular or multicellular hairlike structures that differentiate from the epidermal cells of shoots and roots. Trichomes may be structural or glandular and function in plant responses to biotic or abiotic environmental factors.

Figure 18.3 Examples of mechanical barriers developed by plants. (A) Thorns on a lemon tree (*Citrus* sp.) are modified branches, as can be seen by their position in the axil of a leaf. (B) Spines, which are characteristic of cacti (*Opuntia* spp.) in the New World, are modified leaves. (C) Prickles can be found on the stem and petioles of roses (*Rosa* spp.) and are formed by the epidermis. (D) Trichomes on shoot and leaves of tomato (*Solanum lycopersicum*) are also derived from epidermal cells.

the strong smell and bitter taste of these compounds often repel insect pests; in some cases, the sticky exudate from the trichome can actually immobilize the attacker. In addition to serving as barriers to insect herbivory, trichomes—when bent or damaged—may also act as herbivore sensors, sending electrical or chemical signals to surrounding cells. As discussed later in this chapter, such signals can trigger the formation of inducible defense compounds.

The leaves of the stinging nettle (*Urtica dioica*) possess highly specialized "stinging hairs" that form an effective physical and chemical barrier against larger herbivores. The cell walls of these hollow, needlelike trichomes are reinforced with silicates (glass) and filled with a nasty "cocktail" of acids such as oxalic acid, tartaric acid, formic acid, the neurotransmitter serotonin, and the inflammatory agent histamine (**Figure 18.4**). Prior to contact, the tip of the stinging hair is covered by a tiny glass bulb that readily snaps off when touched by an herbivore (or an unlucky human who happens to brush against it), leaving an extremely sharp point to puncture the animal's skin. The pressure of contact also pushes the needlelike trichome shaft down onto the spongy plant tissue at its base, which acts like the plunger of a syringe, injecting the cocktail and causing severe irritation and inflammation.

A different type of mechanical obstacle to herbivory is provided by mineral crystals that are present in many plant species. For example, silica crystals, called **phytoliths**, form in the epidermal cell walls, and sometimes in the vacuoles, of many grass species. Phytoliths add toughness to the cell walls and make the leaves of grasses difficult for herbivores to chew. The cell walls of the horsetail *Equisetum hyemale* contain so much abrasive silica that the indigenous peoples

phytoliths Discrete cells that accumulate silica in leaves or roots.

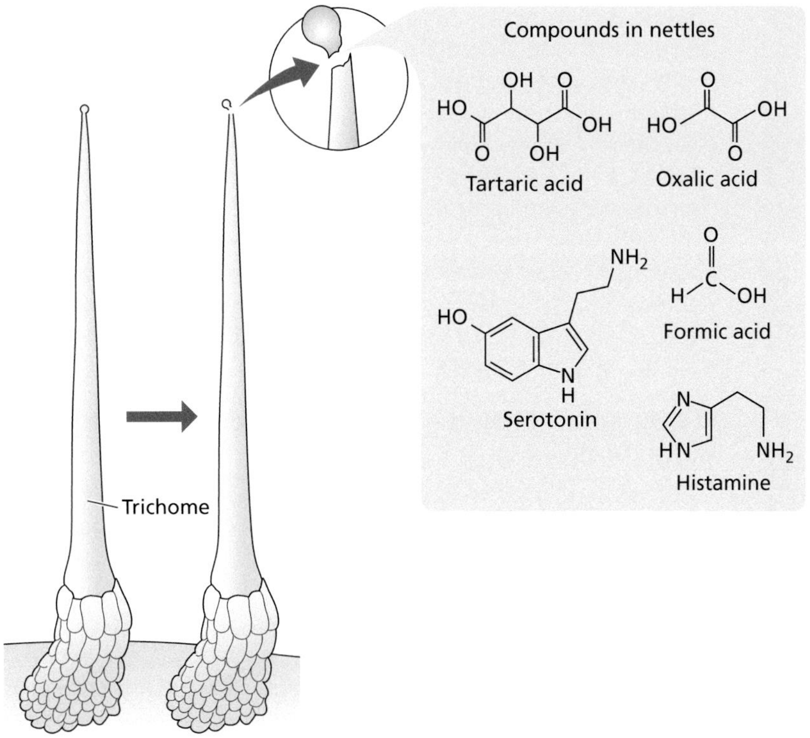

Figure 18.4 Trichomes of nettles (*Urtica dioica*) have a multicellular base with a single stinging cell protruding. The cell wall of this single cell is reinforced by silicates and breaks easily on contact, releasing a "cocktail" of specialized metabolites that can cause severe irritation of the skin of animals.

calcium oxalate crystals Crystals of calcium oxalate that form in the vacuoles of some plant species to deter herbivory by insects and mammals.

idioblast A "special" cell that differs markedly in form, content, or size from the other cells in the same tissue.

raphides Needles of calcium oxalate or carbonate that function in plant defense.

of North America and Mexico used the stems of these plants to clean their cooking pots.

Calcium oxalate crystals provide another, widespread mechanical deterrent to eating certain plants. These crystals are present in the vacuoles of more than 200 families of plants, including species in the genera *Vitis* (grapes), *Agave*, and *Medicago* (medick, or burclover). Much as with phytoliths, these crystals act as physical deterrents, for example having abrasive effects on the mouthparts of insect herbivores, especially the mandibles. They therefore serve as a mechanical deterrent for attack by insects, mollusks, and other herbivores. These crystals may be distributed evenly throughout the leaf or restricted to specialized cells called **idioblasts**, with some also forming bunches of needlelike structures called **raphides** (**Figure 18.5**). Raphides have extremely sharp tips that can penetrate the soft tissue in the throat and esophagus of an herbivore. The raphide-rich *Dieffenbachia*, a popular tropical houseplant, is called "dumb cane" because chewing the leaves reputedly renders the victim speechless due to the inflammation caused by consuming the oxalate crystals. Besides inflicting mechanical damage, raphides may also allow other toxic compounds produced by the plant to penetrate through the wounds they produce.

A very different means of avoiding herbivory is employed by sensitive plant (*Mimosa* spp.). *Mimosa* species possess composite leaves consisting of many individual leaflets that are connected to the midrib by a joint-like structure called the pulvinus. This pulvinus acts as a turgor-driven hinge, swelling or deflating and so causing each pair of attached leaflets to move. The pulvinus is responsive to various

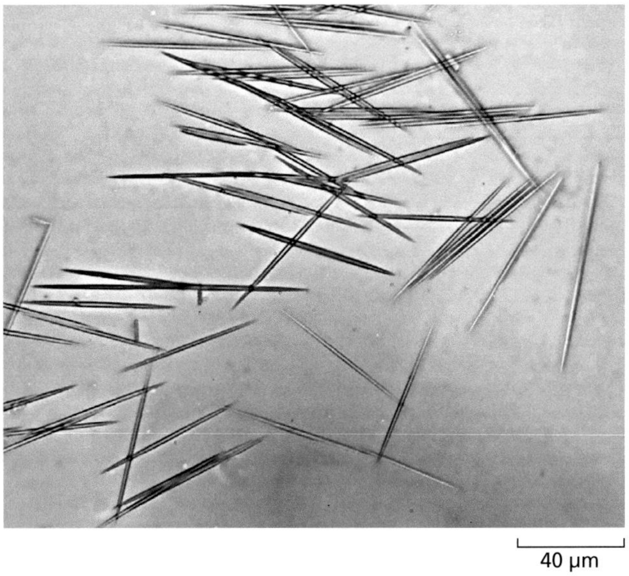

Figure 18.5 Calcium oxalate crystals (raphides) from the leaf of an agave (*Agave weberi*). These raphides are tightly packed in specialized cells called idioblasts and are released upon damage. Note the size and pointed ends of these structures.

Figure 18.6 Leaves of the sensitive plant (here *Mimosa pudica*) respond rapidly to touch by folding in their individual leaflets within seconds. This rapid movement may deter potential insect herbivores. (A) Untouched (control) leaves. (B) Leaves 5 s after touch.

stimuli, including heat, diurnal cycles (in what is called *nyctinasty*, or sleep movements), solar tracking, water deficiency, and, most important in the context of escaping being eaten, touch and damage. If an insect herbivore attempts to chew on a *Mimosa* leaflet, the pulvinus at the base of the damaged leaflet almost immediately deflates, causing the attached leaflets to fold together. Over the course of just a few seconds, this response then spreads to the other undamaged parts of the leaf, causing a wave of leaflet movements. If the stress signal is strong enough, the entire leaf collapses downward due to the action of another pulvinus located at the base of the petiole. Such rapid movements of leaflets and leaves may deter feeding insects and grazing herbivores by startling them (**Figure 18.6**).

Plants' specialized metabolites can deter insect herbivores

Plants are amazing biochemical factories, so it should not be surprising that chemical defense mechanisms provide a further line of defense against plant pests and pathogens. Plants produce a wide array of chemicals that can be categorized into primary metabolites and specialized metabolites. **Primary metabolites** are those compounds that all plants produce and that are directly involved in growth and development. They include sugars, amino acids, fatty acids and lipids, and nucleotides, as well as the larger compounds that are made of these building blocks, such as proteins, polysaccharides, DNA, and RNA. In contrast, **specialized metabolites** (also called *secondary metabolites*) are often species-specific and usually belong to one of three major classes of chemicals: terpenoids, phenolics, or alkaloids.

Plants store constitutive toxic compounds in specialized structures

Plants can synthesize a broad range of specialized metabolites that have negative effects on the growth and development of other organisms. Classic examples of plants that are toxic to humans because of their specialized metabolites are hemlock (*Cicuta* spp.) and foxglove (*Digitalis* spp.), which produce the highly toxic cicutoxin (a neurotoxin) and digitoxin (a cardiac toxin), respectively (**Figure 18.7**). In some cases these compounds have proven useful for medicinal purposes. For example, digitoxin is a **cardiac glycoside** that acts by inhibiting the Na^+/K^+-ATPase pump in the plasma membranes of heart cells. This inhibition leads to increased myocardial contraction. Although fatal at high levels, in therapeutic doses digitoxin has been used to treat congestive heart failure and cardiac arrhythmia. Indeed, plants continue to be a rich source for the discovery of such biomedicinal compounds. However, it is important to remember that the reason these compounds exist is largely because plants use them to prevent themselves from being eaten, by conducting chemical warfare on their herbivores.

Some of these constitutively produced defensive specialized metabolites can also have deleterious effects on the plant itself. To avoid such autotoxicity, these compounds must be safely stored within the cell, and they must also be relatively isolated from susceptible tissues in the event of release due to cellular damage. Plants therefore tend to accumulate toxic specialized metabolites in storage organelles such as vacuoles, or in specialized anatomical structures such as *resin ducts*, *laticifers* (latex-producing cells), or glandular trichomes. Following an attack by herbivores or pathogens, the toxins are released and become active at the site of damage, without adversely affecting vital growing areas. **Conifer resin ducts**, which are found in the cortex and phloem, contain a mixture of diverse terpenoids, as well as resin acids (**Figure 18.8**). This cocktail

primary metabolites Metabolites associated with basic cellular functions (e.g., sugars, amino acids, lipids).

specialized metabolites (also secondary metabolites) Plant compounds that have no direct role in plant growth and development, but function as defenses against herbivores and microbial infection by microbial pathogens, attractants for pollinators and seed-dispersing animals, and agents of plant–plant competition.

cardiac glycosides Glycosylated organic plant defense compounds—similar to oleandrin from oleander bushes—that are poisonous to animals and inhibit Na^+/K^+ channels to cause contractions in cardiac muscles.

conifer resin ducts Ducts or channels in conifer leaves and woody tissue that conduct terpenoid defense compounds. They may be constitutive, or their formation may be induced by wounding or defense responses.

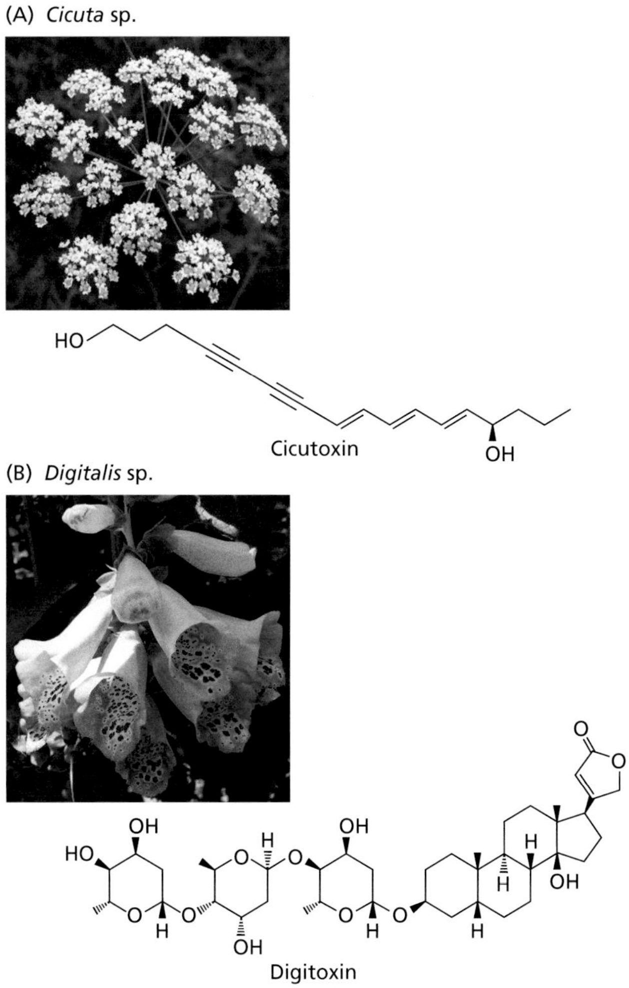

Figure 18.7 Constitutive chemical defenses are effective against many different herbivores, including insects and mammals. Hemlock (*Cicuta* sp.) produces cicutoxin, a diacetylene that prolongs the repolarization of neuronal action potentials. The active principle in foxglove (*Digitalis* sp.) is digitoxin, a cardiac glycoside that inhibits ATPase activity in the plasma membrane of heart muscle cells and can increase myocardial contraction.

is released immediately upon damage to the plant, acting either to poison the attacking herbivore or, because the resin is sticky, to glue the herbivore's mouthparts together. In extreme cases the resin may even engulf the whole insect or pathogen, effectively killing it. However, some herbivores have evolved strategies to avoid these effects. For example, bark beetles circumvent this defense by attacking water-stressed or weakened trees where the flow of resin is much reduced.

Most of the resin ducts in coniferous plants are considered constitutive defenses, but they can also be induced upon herbivore damage. These adventitious structures, sometimes called *traumatic resin ducts,* form and produce resin in response to the hormone jasmonate.

Laticifers are composed of cells that produce **latex**, a milky fluid of emulsified components that coagulate upon exposure to air. Compared with resins, latex is usually chemically much more complex and may also contain proteins and sugars besides toxic or repelling specialized metabolites. Laticifers may consist of a series of fused cells (articulated laticifers) or one long syncytial cell (non-articulated laticifers) (**Figure 18.9**). Most notable among the latex-producing plants is the rubber plant (*Hevea brasiliensis*), which has been grown commercially as a source of natural rubber. Upon wounding, this plant releases huge amounts of latex, which is collected by humans and later converted into rubber. Under natural conditions, the rubber released by wounded trees defends the plant against herbivores and pathogens, either by repelling or engulfing them.

While specialized metabolites can be effective at deterring herbivores, some animals have evolved adaptations to avoid them (e.g., the bark beetles described earlier) or actually capitalize on their production for their own defense. For example, milkweed (*Asclepias curassavica*) and related genera, such as oleander

laticifers In many plants, an elongated, often interconnected network of separately differentiated cells that contain latex (hence the term laticifer), rubber, and other secondary metabolites.

latex A complex, often milky solution exuded from cut surfaces of some plants that represents the cytoplasm of laticifers and may contain defensive substances.

Figure 18.8 Resin stored in its resin ducts is released when herbivores damage the plant. (A) Resin duct in the wood of a conifer (*Araucaria* sp.). Note that the resin duct is surrounded by secretory cells that release resin components into the duct system. (B) Upon wounding, resin is released at the damage site. There it seals off the damage and serves as a repellent against further herbivory. (C) Two terpenoids that are common components of resin.

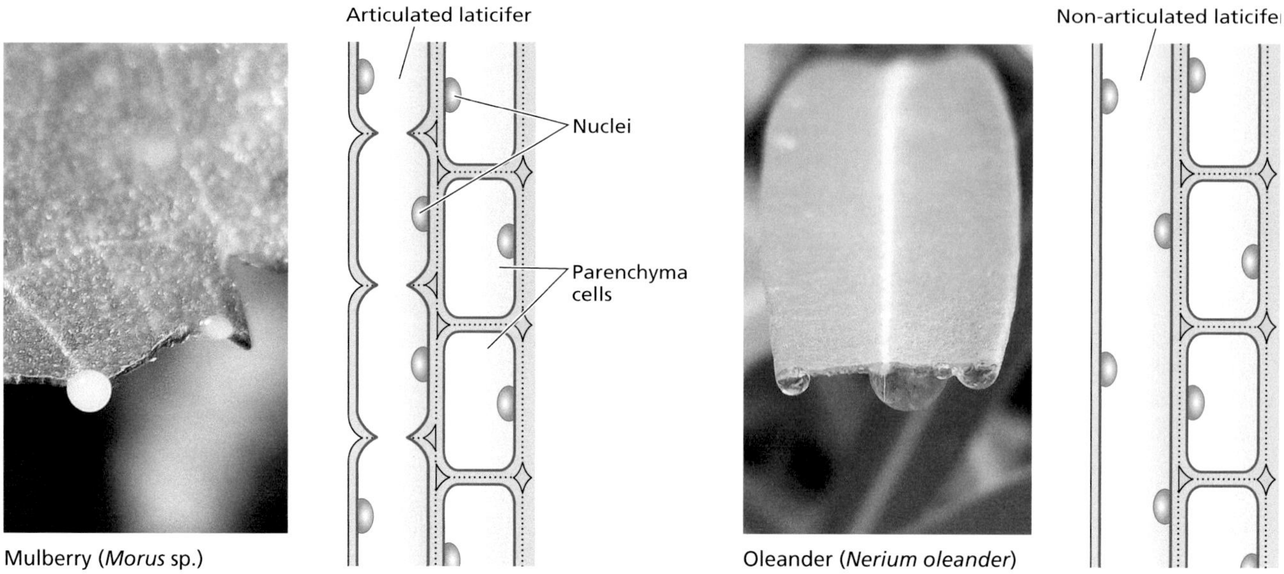

Figure 18.9 Laticifers are made of individual cells and can occur either as articulated systems (single cells connected by a small tube) or non-articulated systems (one large syncytial cell). Latex in the laticifers is released upon damage and often contains cardiac glycosides, which fend off herbivores. While mulberry (*Morus* sp.) produces a milky latex in its articulated laticifers, oleander (*Nerium oleander*) releases a clear latex from non-articulated laticifers.

cardenolides Steroidal glycosides that taste bitter and are extremely toxic to higher animals through their action on Na^+/K^+-activated ATPases. Extracted from foxglove (*Digitalis*) for treatment of human heart disorders.

(*Nerium oleander*), produce latex that contains significant amounts of poisonous steroids called **cardenolides**. Generalist insect herbivores respond to these compounds either by being repelled or by suffering spasms that lead to death. In contrast, the specialist caterpillars of the monarch butterfly (*Danaus plexippus*) are insensitive to the toxins. They feed on milkweed leaves and retain the cardenolides in their bodies for their own defense, a trick also adopted by the large milkweed bug (*Oncopeltus fasciatus*) and the milkweed aphid (*Aphis nerii*) (**Figure 18.10**). As a result, most insectivorous birds quickly learn to avoid eating these insects. Indeed, the bright and distinctive coloration of the monarch caterpillars and butterflies serves to advertise their toxic nature. Other insects have further adapted to recognize volatile defense compounds as feeding attractants or pheromones to attract potential mates.

In a further embellishment to the milkweed saga, the parasitic fly *Zenillia adamsoni* can obtain its cardenolide secondhand from the monarch caterpillar. When the female fly is ready to lay eggs, she seeks out a monarch caterpillar and deposits her eggs on its surface. Upon hatching, the fly larvae bore into the caterpillar and consume it from within. Besides using the caterpillar for nourishment, the fly larvae are able to store the caterpillar's toxic cardenolide and retain it as a defense throughout their adult stage.

Plants often store defense chemicals as nontoxic water-soluble sugar conjugates in the vacuole

In addition to accumulation within specialized structures, another common mechanism used by plants to safely store potentially dangerous specialized metabolites is to conjugate the toxic compound with a sugar such as glucose to produce a glycoside. This chemical modification produces a nontoxic precursor as well as making the specialized metabolite more water soluble. To become active, the glycosidic linkages often need to be enzymatically hydrolyzed. Uncontrolled activation is prevented by the spatial separation of the activating hydrolases and their respective toxic substrates, which come together only when an herbivore starts chewing on the plant.

(A) Monarch butterfly (*Danaus plexippus*)

(B) Milkweed bug (*Oncopeltus fasciatus*)

(C) Milkweed aphid (*Aphis nerii*)

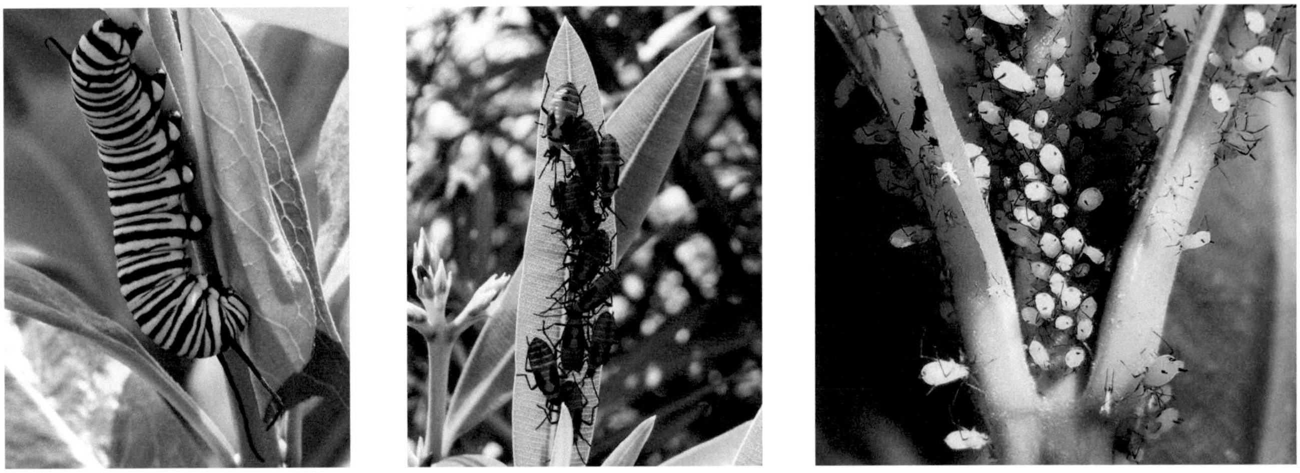

Figure 18.10 Insect specialists "borrow" toxic specialized compounds for defense. While most herbivores are very susceptible to the toxic metabolites in the latex of milkweed and oleander plants, some insect herbivores incorporate these compounds into their bodies and show this to potential predators by presenting bright colors. Shown here are three specialist insect herbivores that feed on these latex-producing plants: the caterpillar of the monarch butterfly (A), the milkweed bug (B), and the milkweed aphid (C). Of these, the milkweed bug and the milkweed aphid will use oleander as a food source if milkweed plants are not available.

Vacuolar storage of glycosylated chemical defenses is used to deter both insects and pathogens. In the order Brassicales (cabbages and their relatives), many species produce glucosinolates—sulfur-containing organic compounds derived from glucose and an amino acid—as their major defensive specialized metabolites. The hydrolyzing enzyme, myrosinase, is stored in different cells than its substrates. Thus, while myrosinase-containing cells are mostly glucosinolate-free, the so-called sulfur-rich cells contain high concentrations of glucosinolates. When the tissue is damaged, the released myrosinase and glucosinolates mix, resulting in the production of a variety of biologically active compounds that deter most generalist insect herbivores (**Figure 18.11A**). The flavors of mustard, wasabi, radish, brussels sprout, and other related vegetables are due to isothiocyanates produced in this manner. Indeed, humans often prize plants, or parts of them, for the culinary flavors imparted by these compounds. However, the high content of the specialized metabolites in these plants is primarily there not for our use in cooking, but to provide the noxious chemical defenses for protection from damage by pests and pathogens.

Cyanogenic glycosides represent a particularly toxic class of specialized metabolites. Upon tissue damage, these glycosides break down and release hydrogen cyanide (HCN). Cyanide inhibits mitochondrial function, so ultimately the herbivore's cells run out of energy and die. Several plant species of economic and nutritional importance, including sorghum (*Sorghum bicolor*) and cassava (*Manihot esculenta*), produce different types of cyanogenic glycosides (**Figure 18.11B**). Cassava roots are a major source of dietary calories for people in tropical regions, but they must be carefully prepared by grating and extensive washing before cooking to remove the glycosides and so avoid cyanide toxicity when they are eaten.

In some members of the Poaceae (grass) family, benzoxazinoids (toxic alkaloids) are constitutively produced as defensive specialized metabolites. These compounds are usually stored in the vacuole as glycosides coupled to D-glucose (Glc) and are hydrolyzed to form toxic products after tissue disruption by insect herbivores or pathogens.

cyanogenic glycosides Nonalkaloid, nitrogenous protective compounds that break down to give off the poisonous gas hydrogen cyanide when the plant is crushed.

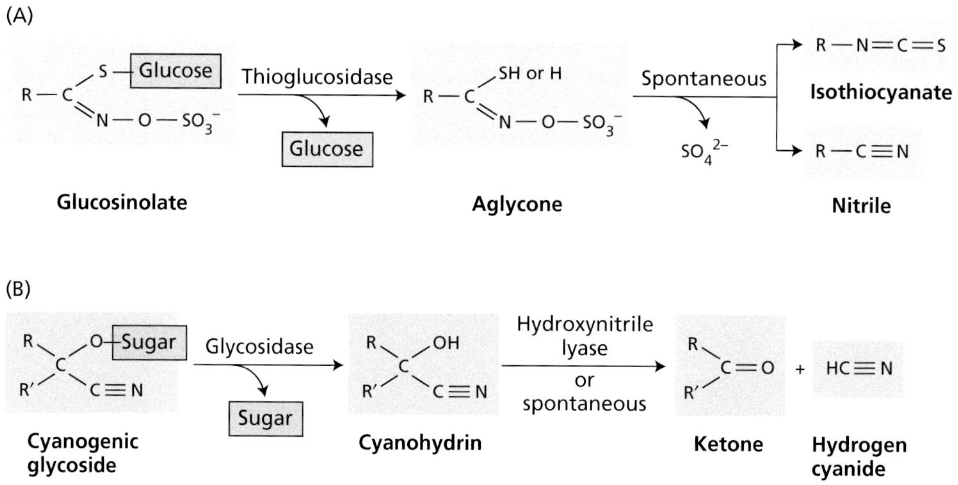

Figure 18.11 Hydrolysis of glycosides into active defense compounds. (A) Glucosinolates are hydrolyzed to form mustard-smelling volatiles. (B) Enzyme-catalyzed hydrolysis of cyanogenic glycosides releases hydrogen cyanide. R and R′ represent various alkyl or aryl substituents.

18.3 Inducible Defense Responses to Insect Herbivores

Describe how plants identify an herbivore attack, the subsequent signaling mechanisms at the cellular and whole-plant levels, and the differences between the immediate and systemic responses to herbivory.

Constitutive chemical defenses provide plants with protection against many pests and pathogens and are common among plants in nature. However, there are disadvantages to this type of defense strategy. First, constitutive defenses are costly to the plant. The production of specialized metabolites requires a significant energy investment derived from primary metabolism, which is then unavailable for use in growth and reproduction. This trade-off is most obvious in agricultural crops, in which yield is increased, in part, by reducing the capacity of the plant to defend itself. Second, pests and pathogens can adapt to the plant's constitutive chemical defenses, as we saw in the case of the monarch caterpillar and milkweed, even using these compounds to defend themselves against their own predators or parasites. Accordingly, most plants have evolved inducible defense systems in addition to whatever constitutive defenses they may have. Inducible defense systems enable plants to respond more flexibly to the full panoply of threats presented by pests and pathogens.

Inducible defenses require systems to monitor for signs of damage or of an attack and then trigger a response. Plants have many such systems, and below we discuss some of the mechanisms by which plants recognize insect herbivores and how they then mount their inducible defenses. These plant responses include not only the de novo synthesis of toxic specialized metabolites and proteins but also the recruitment of natural enemies of the attacker along with the sending of signals to nearby plants to prepare them against impending herbivory.

Based on feeding behaviors, three major types of insect herbivores can be distinguished:

1. *Phloem feeders*, such as aphids and whiteflies, cause little direct tissue damage. Phloem feeders insert their narrow *stylet*, which is an elongated mouthpart, between the cells of the epidermis and mesophyll and into

the phloem sieve tubes of leaves and stems. The plant defense response to phloem feeders more closely resembles the response to pathogens than to other herbivores. Although the amount of direct tissue injury to the plant from insertion of the stylet is low, these insects can serve as vectors for plant viruses that can then cause great damage.

2. *Cell-content feeders*, such as mites and thrips, are piercing-and-sucking insects that cause an intermediate amount of physical damage to plant cells.
3. *Chewing insects*, such as caterpillars (the larvae of moths and butterflies), grasshoppers, and beetles, cause significant damage to plants. In the discussion that follows, our definition of "insect herbivory" mostly relates to this type of insect damage.

Plants can recognize specific components of insect saliva

To mount an effective inducible defense against pests or pathogens, the host plant must be able to sense that damage has occurred and determine that the source is a biotic attack. Mechanical damage, be it from an insect chewing on a leaf or from hailstones damaging the leaf surface, releases plant compounds that are not normally present in the apoplast. These include contents of the cytoplasm of the damaged cells, as well as fragments of cell wall such as oligogalacturonides. These compounds can then trigger responses to mechanical damage. However, most plant responses to insect herbivory involve both this general mechanical wound response and the recognition of certain compounds abundant in the insect's saliva or regurgitant that trigger further levels of defense. These insect compounds belong to a broad group of chemicals called **elicitors**, which trigger defense responses in plants to a wide variety of herbivores and pathogens. Although repeated mechanical wounding can induce responses similar to those caused by insect herbivory in some plants, certain molecules in insect saliva can serve as enhancers of this stimulus. Many of these are fatty acid–amino conjugates (fatty acid amides) and may induce responses in a broad or narrow range of plants. All of these responses involve intermediary Ca^{2+} and phosphorylation signaling events. Not surprisingly, some compounds present in insect saliva can actually suppress plant defense responses. In addition, wounding and insect-derived elicitors can trigger signaling pathways *systemically* (that is, throughout the plant), thereby initiating defense responses that can minimize further damage in regions of the plant that are not yet being directly attacked.

Jasmonate activates defense responses against insect herbivores

A major signaling pathway involved in most plant defenses against insect herbivores is triggered by the hormone jasmonate (primarily in the form of the amino acid conjugate jasmonic acid-isoleucine, JA-Ile). Jasmonate levels rise steeply in response to damage and trigger the production of many proteins involved in plant defenses. Direct demonstration of the role of jasmonate in insect resistance has come from research with jasmonate-deficient mutant lines of Arabidopsis, tomato, and maize. Such mutants are easily killed by insect pests that normally cannot damage wild-type plants. Application of exogenous JA restores resistance nearly to the levels of the wild-type plant.

In plants, jasmonate is synthesized from the fatty acid linolenic acid (Chapter 11), which is released from membrane lipids and then converted to JA-Ile, as outlined in **Figure 18.12**. Two organelles participate in jasmonate biosynthesis, chloroplasts and peroxisomes. Jasmonate triggers a wide range of defense processes, including inducing the transcription of multiple genes for enzymes in all major pathways for specialized metabolite biosynthesis. Jasmonate also shuts down growth, allowing the plant to reallocate its resources toward metabolic pathways

elicitors Specific pathogen molecules or cell wall fragments that bind to plant proteins and thereby act as signals for activation of plant defense against a pathogen.

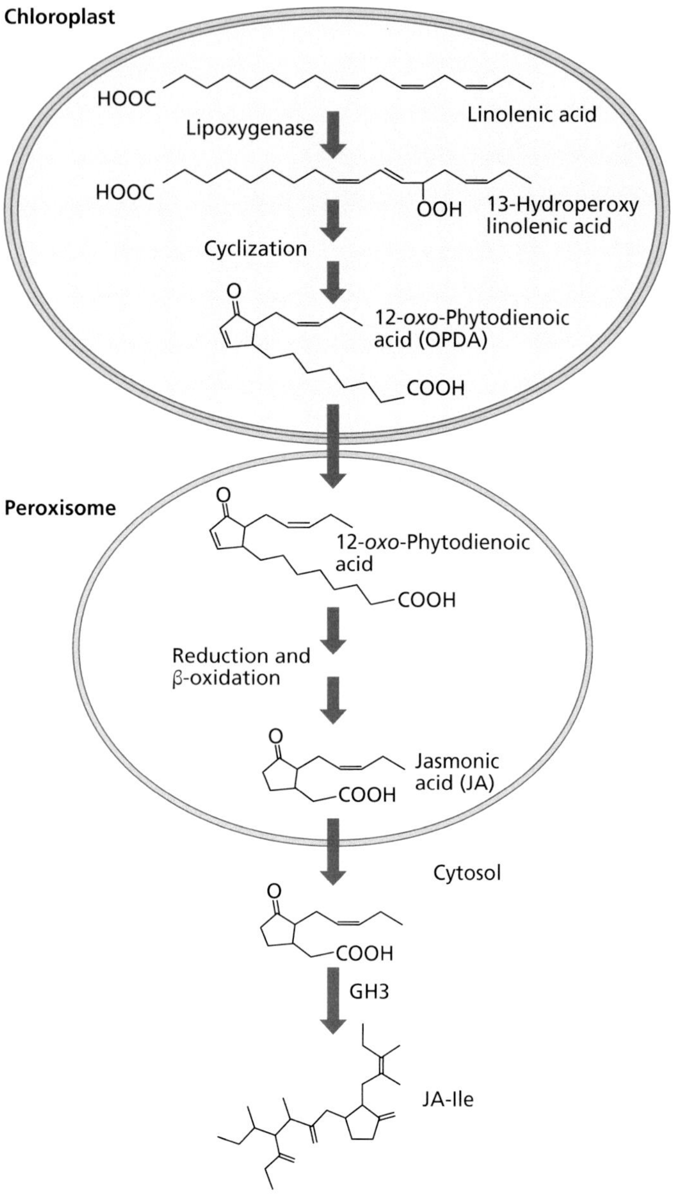

Figure 18.12 Steps in the pathway for conversion of linolenic acid into jasmonate. The first enzymatic steps occur in the chloroplast, resulting in the cyclized product 12-*oxo*-phytodienoic acid (OPDA). This intermediate is transported to the peroxisome, where it is first reduced and then converted into jasmonic acid by β-oxidation. The amino acid leucine is enzymatically conjugated to jasmonic acid by GH3.

involved in defense. As introduced in Chapter 12, JA-Ile acts through a conserved ubiquitin ligase–based signaling mechanism that bears close resemblance to those described for auxin and gibberellin.

Hormonal interactions contribute to plant–insect herbivore interactions

In addition to jasmonate, several other signaling agents—including ethylene, salicylic acid, and methyl salicylate—are often induced by insect herbivory. In particular, ethylene appears to play an important role in this context. When applied alone to plants, ethylene has little effect on defense-related gene activation. However, when applied together with JA it seems to enhance jasmonate-related responses. Similarly, when plants are treated with elicitors such as fatty acid amides (which by themselves do not induce the production of significant amounts of ethylene) in combination with ethylene, defense responses are significantly increased. Results such as these demonstrate that a concerted action of these signaling compounds is required for the full activation of induced defense responses. Such control through multiple factors allows plants to integrate numerous environmental signals in modulating the defense response.

Jasmonate initiates the production of defense proteins that inhibit herbivore digestion

Besides activating pathways for the production of toxic or repelling specialized metabolites, JA also initiates the biosynthesis of defense proteins. Most of these proteins interfere with the herbivore digestive system. For example, some legumes synthesize **α-amylase inhibitors**, which block the action of the starch-digesting enzyme α-amylase in the insect gut. Other plant species produce **lectins**, defense proteins that bind to carbohydrates or carbohydrate-containing proteins. After ingestion by an herbivore, lectins bind to the epithelial cells lining the digestive tract and interfere with nutrient absorption.

A more direct attack on the insect herbivore's digestive system is performed by some plants through the production of a specific cysteine protease that disrupts the membrane that protects the gut epithelium of many insects. Plants also produce **proteinase inhibitors**. Found in legumes, tomato, and other plants, these substances block the action of herbivore proteolytic enzymes. After entering the herbivore's digestive tract, they hinder protein digestion by binding tightly and specifically to the active site of protein-hydrolyzing enzymes such as trypsin and chymotrypsin. Insects that feed on plants containing proteinase inhibitors suffer reduced rates of growth and development. Experiments have shown that this effect can be offset by adding supplemental amino acids to their diet, directly linking their impaired growth to disruption of their ability to digest proteins in their diet.

Genetic engineering of crop plants with a proteinase inhibitor from *Bacillus thuringiensis* has been a highly successful strategy used to prevent crop damage

α-amylase inhibitors Substances synthesized by some legumes that interfere with herbivore digestion by blocking the action of the starch-digesting enzyme α-amylase.

lectins Defensive plant proteins that bind to carbohydrates; or carbohydrate-containing proteins inhibiting their digestion by a herbivore.

from the European corn borer (*Ostrinia nubilalis*). However, coevolution is a powerful force in plant–insect interactions, and just as has occurred with native plant proteinase inhibitors over longer periods of time, borer resistance to this transgenic proteinase began to emerge only a few decades after its introduction.

Herbivore damage induces systemic defenses

In tomato, insect feeding leads to the rapid accumulation of proteinase inhibitors throughout the plant, even in undamaged areas far from the initial feeding site. Similarly, in Arabidopsis the wounding of one leaf leads within minutes to the production of jasmonates in unwounded leaves. These observations indicate that plants possess a system capable of rapid signaling throughout the plant body to trigger preemptive defenses even in undamaged parts. Systemic responses based on transmission of species-specific peptide hormones have been demonstrated in multiple plant species.

Several lines of evidence suggest a role for electrical signaling in systemic defense responses. For example, the feeding of Egyptian cotton leafworm (*Spodoptera littoralis*) on bean leaves induces a wave of plasma membrane depolarizations that spreads to undamaged areas of the leaf. Similarly, measurements of Arabidopsis responses have now confirmed the role of electrical signaling in the spread of the jasmonate-induced defenses to undamaged leaves in response to both purely mechanical damage and insect herbivory. During feeding of *S. littoralis* larvae, electrical signals induced near the site of attack subsequently spread to neighboring leaves at a maximum speed of 9 cm per min (**Figure 18.13**). These changes are also accompanied by propagating waves of Ca^{2+} and reactive oxygen species (ROS, another ubiquitous cellular signaling molecule). These signals propagate rapidly through the vasculature and then spread throughout the undamaged leaves, probably moving from cell to cell through plasmodesmata to initiate JA-Ile–mediated defense responses. A family of glutamate receptor-like (GLR) genes has been identified as playing an important role in this process. These GLRs encode Ca^{2+}-permeable ion channels, and both the electrical and Ca^{2+} waves no longer propagate after wounding when these genes are mutated. Similarly, JA-Ile response gene expression and herbivore defense in the unwounded leaves are reduced. The GLRs have also previously been implicated in microbe-related defense responses, suggesting a potentially widespread role for the pathways leading to inducible defenses in the plant.

Herbivore-induced volatiles can repel herbivores and attract natural enemies

Volatile organic compounds (VOCs), or volatiles, are plant-derived organic molecules that are induced or released upon damage. The response of insects to these volatiles provides an excellent example of how coevolution can shape plant–insect interactions. The combination of emitted molecules is often specific for each insect herbivore species and typically includes representatives from the three major pathways of specialized metabolism: the terpenoids, alkaloids, and phenolics. All plants also emit lipid-derived products, such as **green-leaf volatiles** (a mixture of six-carbon aldehydes, alcohols, and esters) in response to mechanical damage. The ecological functions of these volatiles are manifold. Many of these compounds, although volatile, remain attached to the surface of the leaf and serve as feeding deterrents because of their taste. In other cases they can act almost like a cry for help from the plant, attracting natural enemies of the attacking insect herbivores (predators or parasites), which use the volatile cues to find their prey or a host for their offspring. For example, in maize the elicitor present in the saliva of beet armyworm larvae can induce the synthesis of volatiles that attract parasitoids. Maize seedlings that are treated with very low concentrations of the armyworm elicitor release relatively large amounts of terpenoids, which attract the tiny parasitoid wasp *Microplitis croceipes*, which then lays its eggs on the invading armyworms.

proteinase inhibitors Compounds that inhibit the enzymatic activity of proteases.

green-leaf volatiles A mixture of lipid-derived six-carbon aldehydes, alcohols, and esters released by plants in response to mechanical damage.

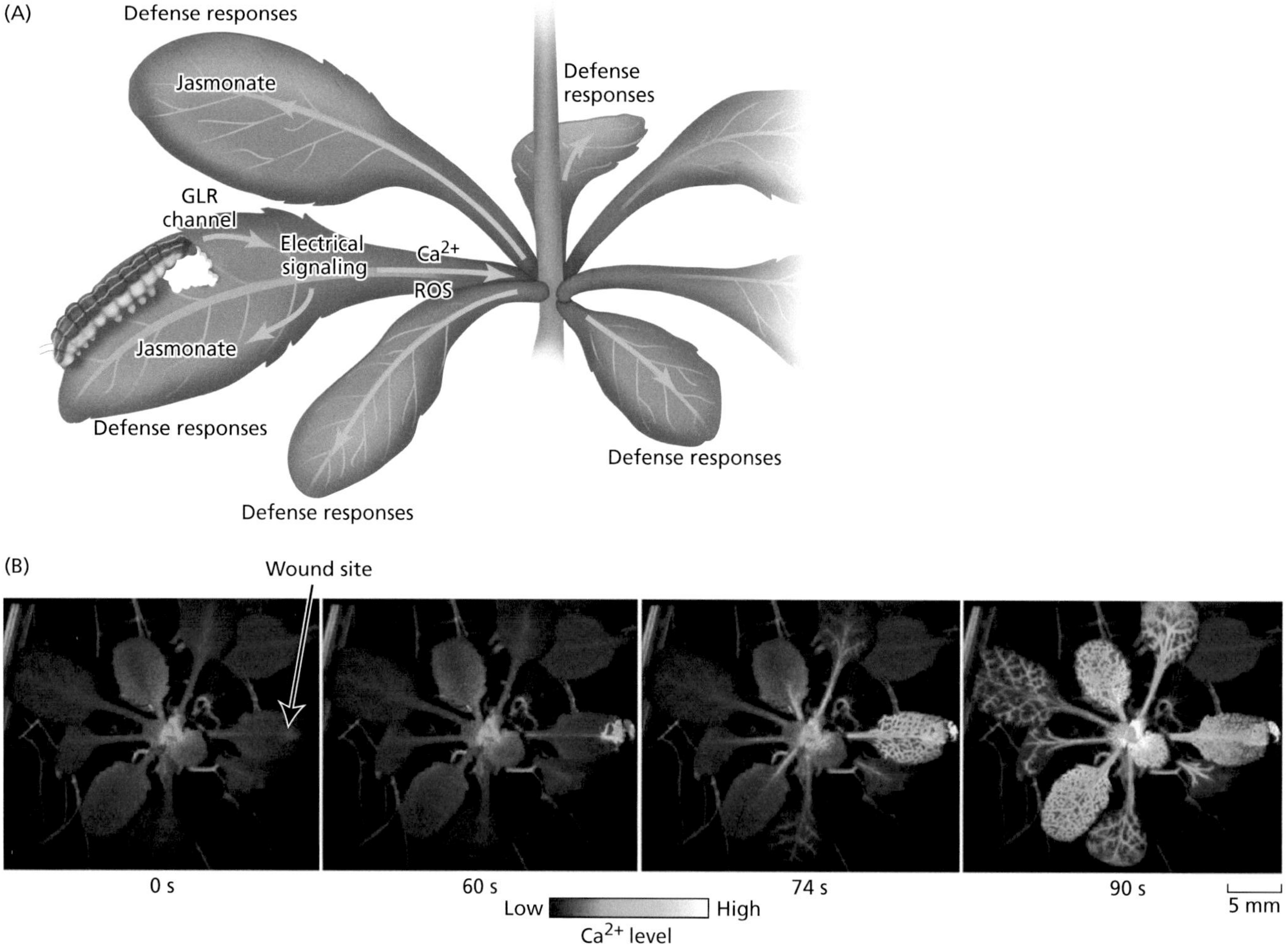

Figure 18.13 Systemic response of Arabidopsis to herbivore attack. (A) Injury to the leaf caused by herbivory activates glutamate receptor-like (GLR) ion channels. The GLRs trigger electrical signals that are thought to travel through the vascular system alongside waves of chemical signaling such as changes in the concentration of Ca^{2+} and reactive oxygen species (ROS), thereby stimulating jasmonate production both locally and in other leaves. Jasmonate then initiates defense responses that discourage further herbivory. (B) Spread of Ca^{2+} signal through a plant in response to local application of a wound elicitor. The plant is expressing a Ca^{2+}-sensitive fluorescent reporter protein.

In contrast, some herbivores actually capitalize on these volatile emissions. For example, volatiles released by leaves during moth oviposition (egg laying) can act as repellents to other female moths, thereby preventing further egg deposition and reducing subsequent competition among the larvae that emerge. In some cases, female insects ingest these volatiles and utilize them as pheromones to signal mating status (**Figure 18.14**).

Plants have the ability to distinguish among various insect herbivore species and to respond differentially. For example, following herbivory, *Nicotiana attenuata*, a wild tobacco that grows in the deserts of the Great Basin in the western United States, typically produces higher levels of nicotine, which poisons the insect central nervous system. However, when wild tobacco plants are attacked by nicotine-tolerant caterpillars, the plants show no increase in nicotine. Instead, they release volatile terpenes that attract insect predators of the caterpillars. Clearly, wild tobacco and other plants can discriminate between insect herbivores by the manner in which their foliage is damaged or the distinctive elicitors they produce.

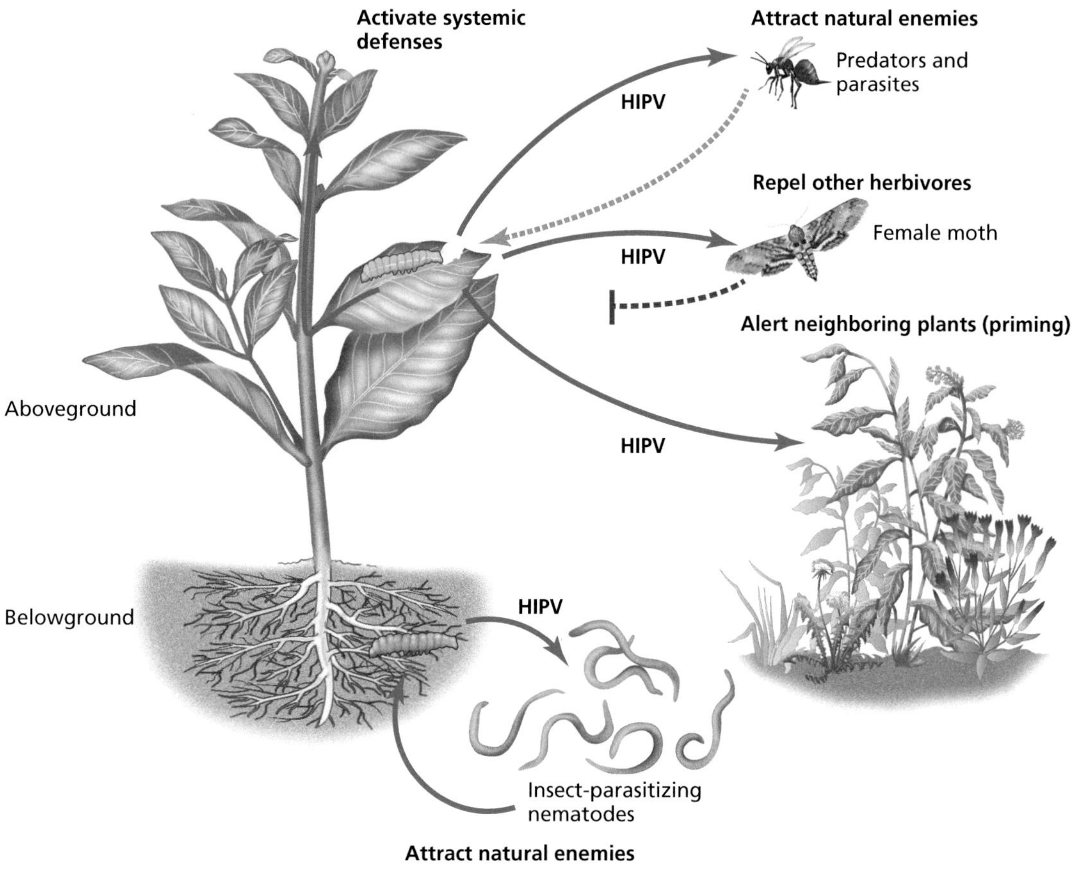

Figure 18.14 Ecological functions of herbivore-induced plant volatiles (HIPV). Many plants release a specific bouquet of volatile organic compounds when attacked by insect herbivores. These volatiles can consist of compounds from all major pathways for specialized metabolite biosynthesis, including terpenoids (mono- and sesquiterpenes), alkaloids (indole), and phenylpropanes (methyl salicylate), as well as green-leaf volatiles. These volatiles can act as cues for natural enemies of the insect herbivore, for example parasitic wasps. Belowground parts of plants can also release volatiles when attacked by herbivores. It has been shown that these volatiles attract insect-parasitizing nematodes, which then attack the herbivore. Volatiles may also serve as a repellant for female moths, thereby reducing further egg deposition. Volatiles have also been found to act as a systemic defense signal between plants over short distances and in highly sectorial plants with interrupted vascular connections. There, these volatile signals prepare the receiving plant against impending herbivory by priming (preparing) defense responses, resulting in a stronger and faster response when the receiving plant is actually attacked.

Herbivore-induced volatiles can serve as long-distance signals between plants

The role of herbivore-induced plant volatiles is not limited to the interactions between plants and insects: Certain volatiles emitted by infested plants can also serve as signals for neighboring plants to initiate preemptive anti-herbivore defenses (Figure 18.14). In addition to several terpenoids, green-leaf volatiles act as potent signals in this process. Green-leaf volatiles are the major components of the familiar scent of freshly cut grasses and have been shown to prime or sensitize the defense mechanisms of a variety of other plant species. They can also activate the production of antimicrobial compounds.

Herbivore-induced volatiles can also act as systemic signals within a plant

Besides providing a signal for neighboring plants, infested plants may well send a volatile signal to other parts of themselves (Figure 18.14). From an evolutionary point of view, this may have been the original function of those volatiles.

pattern recognition receptors (PRRs) Innate immune system proteins that are associated with microbe-associated molecular patterns (MAMPs) and damage-associated molecular patterns (DAMPs).

effector-triggered immunity Immune responses that are mediated by a class of intracellular R proteins.

hypersensitive response A common plant defense following microbial infection, in which cells immediately surrounding the infection site die rapidly, depriving the pathogen of nutrients and preventing its spread.

systemic acquired resistance (SAR) The increased resistance throughout a plant to a range of pathogens following infection by a pathogen at one site.

Volatiles have been shown to act as inducers of herbivore resistance in plants that are adapted to water deficits by having vascular systems that are divided into discrete sectors. In these plants, volatiles are used in place of defense systems spreading through the vasculature to provide systemic signaling.

Insects have evolved mechanisms to defeat plant defenses

This section has demonstrated how plants have evolved an array of physical and chemical mechanisms to protect themselves. These defenses are generally highly effective, but as we have seen, some herbivorous insects have evolved mechanisms for circumventing or overcoming these plant defenses by the process of *reciprocal evolutionary change between plant and insect*, a type of coevolution. These insect adaptations, like plant defense responses, can be either constitutive or induced. Constitutive adaptations are more widely distributed among specialist herbivorous insects, which can feed on only a few plant species, whereas induced adaptations are more likely to be found among insects that are dietary generalists. Although it is not always obvious, in most natural environments plant–insect interactions have led to a standoff where there are effective defenses and countermeasures to elude these defenses and in which plant and insect can each develop and survive, albeit under suboptimal conditions.

18.4 Plant Defenses against Pathogens

Describe the mechanisms of plant immunity and systemic acquired resistance against pathogens and the mechanism of resistance-gene recognition of strain-specific pathogen effectors.

Plant pathology is the study of plant diseases. Pathogens that cause infectious diseases in plants include viruses, bacteria, fungi, oomycetes, and nematodes. Oomycetes are filamentous organisms that are evolutionarily closer to brown algae than to fungi and include some of the most destructive plant pathogens in history. Oomycetes include the genus *Phytophthora*, cause of the disastrous potato blight of the Great Irish Famine (1845–1849). Collectively, these pathogens lead to an estimated reduction in global crop production of 15% annually, despite massive use of pesticides that are costly to both farmers and the environment. A major goal of plant pathology is to understand the molecular and cellular basis of the plant immune system so that this knowledge can be applied to developing crop plants with more robust immune systems, which would then enable a reduction in pesticide use.

The diverse array of mechanisms that plants have evolved to resist infection locally include **pattern recognition receptor (PRR)–triggered immunity (PTI)**, **effector-triggered immunity (ETI)**, the production of antimicrobial agents, and a type of programmed cell death called the **hypersensitive response**. Plant defenses also include a form of systemic plant immunity referred to as **systemic acquired resistance (SAR)**.

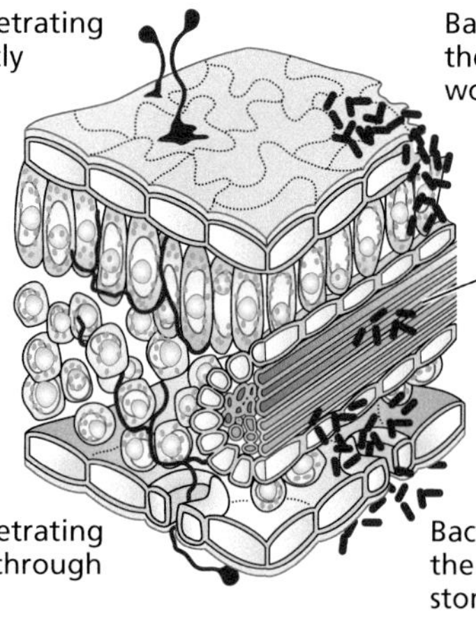

Figure 18.15 Bacterial and fungal plant pathogens have developed various methods for invading plants. Some fungi have mechanisms that allow them to directly penetrate the cuticle and cell wall of the plant. Other fungi, and also pathogenic bacteria, enter through natural openings such as stomata or through existing wounds caused by herbivores. Bacteria movement through the plant is restricted to the xylem, although a few species move through the phloem. Viruses are transmitted to plants by herbivores and viruses may spread throughout the plant via plasmodesmata, xylem, or phloem.

Microbial pathogens have evolved various strategies to invade host plants

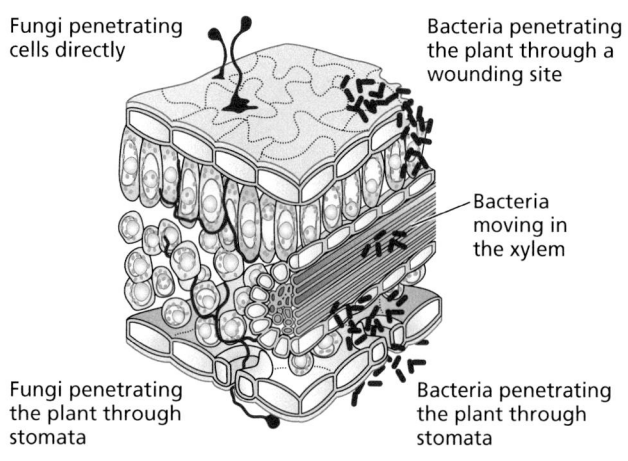

Throughout their lives, plants are continuously exposed to a diverse array of pathogens. Successful pathogens have evolved various mechanisms to invade their host plant and cause disease (**Figure 18.15**). Some penetrate the cuticle and cell wall directly by secreting lytic enzymes, which digest these mechanical barriers. Others enter the plant through natural openings such as stomata, hydathodes, and lenticels. Pathogens in a third category invade the plant through wound sites, for example those caused by insect herbivores. Additionally, many viruses, as well as other types of pathogens, are transferred by insect herbivores, which serve as vectors, and invade the plant from the insect feeding site. Phloem feeders such as whiteflies and aphids deposit these pathogens directly into the vascular system, from which they can easily spread throughout the plant.

Once inside the plant, pathogens generally employ one of three main attack strategies to subdue the host plant and obtain nutrients for their own proliferation. **Necrotrophic pathogens** attack their host by secreting cell wall–degrading enzymes or toxins, which eventually kill the affected plant cells, leading to extensive tissue maceration (softening of the tissues after death by autolysis). This dead tissue is then colonized by the pathogens and serves as a food source. A different strategy is used by **biotrophic pathogens**; after infection, most of the plant tissue remains alive and only minimal cell damage can be observed, as the pathogens feed on substrates that are stolen from their host. **Hemibiotrophic pathogens** are characterized by an initial biotrophic stage, in which the host cells are kept alive as described for biotrophic pathogens. This phase is followed by a necrotrophic stage, in which the pathogens can cause extensive tissue damage.

Although these invasion and infection strategies are individually successful, plant disease epidemics are rare in natural ecosystems. This is because plants have evolved effective defense strategies against a diverse array of pathogens.

Pathogens produce effector molecules that aid in the colonization of their plant host cells

Plant pathogens can produce a wide array of effectors that support their ability to successfully colonize their host and gain nutritional benefits. **Effectors** are secreted molecules that can suppress the host immune system, and/or alter the plant's structure, metabolism, or hormonal regulation to the advantage of the pathogen. Because invasion of a suitable host is often the most difficult step for a pathogen, many pathogens produce enzymes that can degrade the plant cuticle and cell wall. Among those enzymes are cutinases, cellulases, xylanases, pectinases, and polygalacturonases. These enzymes can compromise the integrity of the cuticle as well as the primary and secondary cell walls.

Degradation of cutin and host cell walls by pathogen enzymes leads to release of molecules that signal the plant that it is under attack. These **damage-associated molecular patterns (DAMPs)** are sensed by cell surface **pattern recognition receptors (PRRs)** that then induce a strong immune response—PRR-triggered immunity, or PTI—which we describe in more detail shortly. To be successful, pathogens must suppress this PTI response (**Figure 18.16**). This is most often accomplished by translocation of protein effectors from the pathogen into host cells, where the proteins bind to and inactivate key components of the PTI signaling pathway. A single bacterial pathogen strain such as *Pseudomonas syringae* pathovar *tomato* is known to secrete more than 30 different protein effectors that target several different steps in the PTI pathway. Even more striking, oomycete pathogens such as *Phytophthora infestans* are believed to translocate several hundred different protein effectors. Such effectors are proving to be very useful tools for identifying proteins in the host that contribute to immune signaling.

In addition to suppressing PTI, some effectors promote host cell susceptibility by inducing cellular responses that promote pathogen growth, such as the release of sugars from cells. A particularly interesting example of this comes from bacterial pathogens in the genus *Xanthomonas*, which secrete effectors known as transcription activator-like factors that mimic host plant transcription factors and activate the expression of genes encoding sugar transporters.

necrotrophic pathogens Pathogens that attack their host plant first by secreting cell wall–degrading enzymes or toxins, which will lead to massive tissue laceration and plant death.

biotrophic pathogens Pathogens that leave infected tissue alive and only minimally damaged while the pathogen continues to feed on host resources.

hemibiotrophic pathogens Pathogens that show an initial biotrophic stage, which is followed by a necrotrophic stage in which the pathogen causes extensive tissue damage.

effector A molecule that binds a protein to change its activity. Bacterial effectors are secreted by pathogens to act on proteins within a host cell.

damage-associated molecular patterns (DAMPs) Molecules within cells that are released from damaged or dying cells due to injury or pathogen infection to activate innate immune responses.

pattern recognition receptors (PRRs) Innate immune system proteins that are associated with microbe-associated molecular patterns (MAMPs) and damage-associated molecular patterns (DAMPs).

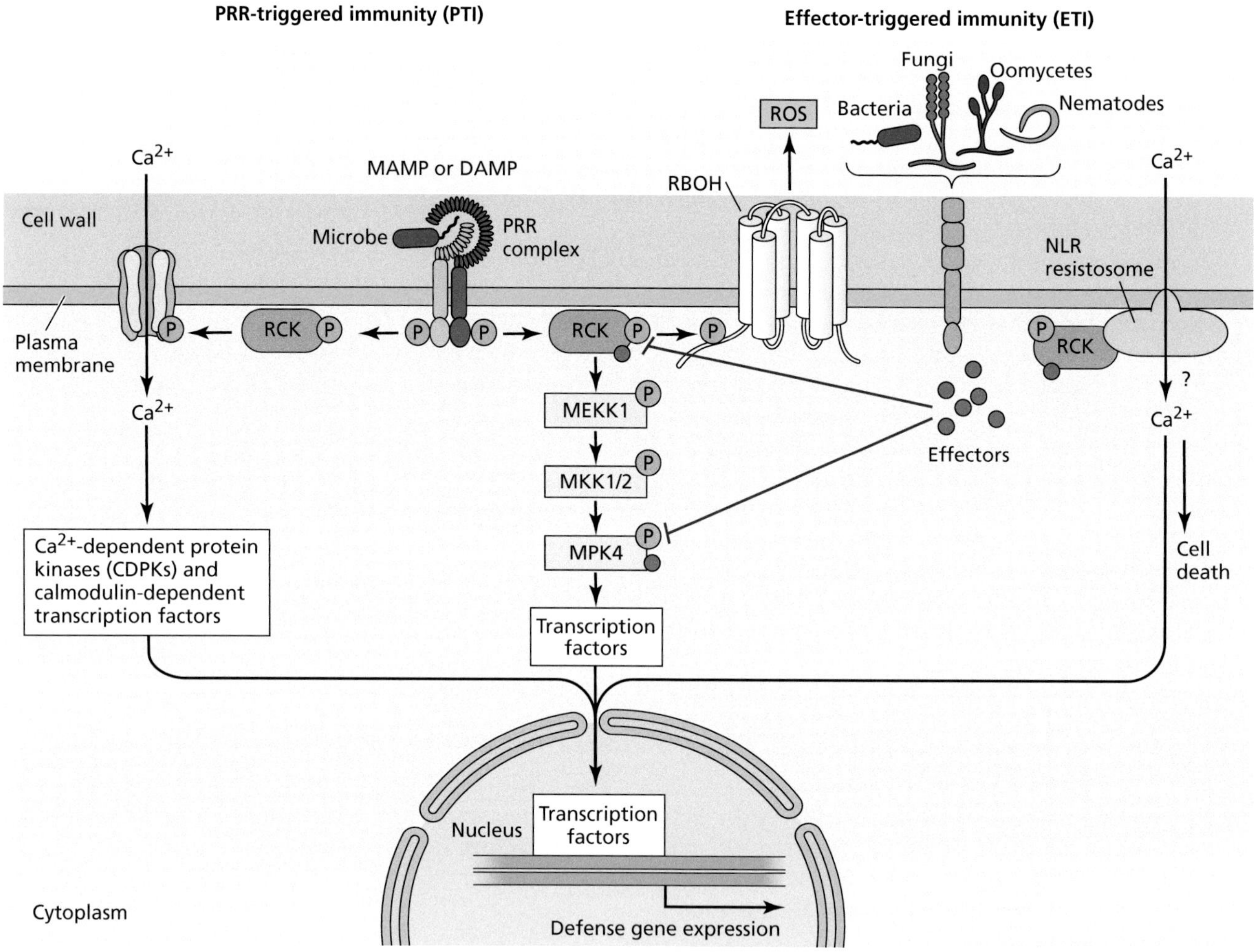

Figure 18.16 Plants have evolved defense responses to a variety of danger signals of biotic origin. These danger signals include microbe-associated molecular patterns (MAMPs), damage-associated molecular patterns (DAMPs), and effectors. Extracellular MAMPs produced by microbes, and DAMPs released by microbial enzymes, bind to pattern recognition receptors (PRRs) on the cell surface. As plants coevolved with pathogens, the pathogens acquired effectors as virulence factors that mainly function to suppress PRR signaling. When MAMPs, DAMPs, and effectors bind to their PRRs and resistance (R) proteins, two types of defense responses are induced: PRR-triggered immunity and effector-triggered immunity. A discrete group of receptor cytoplasmic kinases (RCK) are phosphorylated by receptor kinase components of PRR complexes. MEKK1, MKK1/2, MPK4, components of MAP kinase signal amplification cascade; NLR, nucleotide-binding site–leucine rich repeat protein; RCK, RECEPTOR CYTOPLASMIC KINASE; RBOH, respiratory burst oxidase homolog.

HC-toxin A cell permeant cyclic tetrapeptide produced by the maize pathogen *Cochliobolus carbonum* that inhibits histone deacetylases.

fusicoccin A fungal toxin that induces acidification of plant cell walls by activating H^+-ATPases in the plasma membrane. Fusicoccin stimulates rapid acid growth in stem and coleoptile sections. It also stimulates stomatal opening by stimulating proton pumping at the guard cell plasma membrane.

As we discussed in Chapter 12, the class of plant hormones known as gibberellins (GA) were discovered as a result of the production of GA_3 as a compound produced by a fungal pathogen to alter the growth of rice seedlings. Fungal spores released from the taller, infected plants are more likely to spread to surrounding plants because of their height advantage. Other pathogens produce effector molecules that significantly interfere with the hormonal balance of the plant host.

In addition to secreting protein-based effectors, many pathogens secrete toxins that target specific proteins in the plant (**Figure 18.17**). For example, the **HC-toxin** from the fungus *Cochliobolus carbonum,* which causes northern leaf blight disease, inhibits specific histone deacetylases in maize. In general, decreased deacetylation of histones, which are essential in the organization of the chromatin, tends to increase the expression of associated genes. However, it is not yet known whether this is how HC-toxin causes disease in maize.

Fusicoccin, shown in Figure 18.17, is a toxin from the fungus *Fusicoccum amygdali.* **Fusicoccin** constitutively activates the plant plasma membrane H^+-ATPase by first binding to a specific protein of the 14-3-3 group of regulators. This complex then binds to the C-terminal region of the H^+-ATPase and activates it irreversibly, leading to cell wall overacidification and plasma membrane hyperpolarization that aids infection and growth of the fungus.

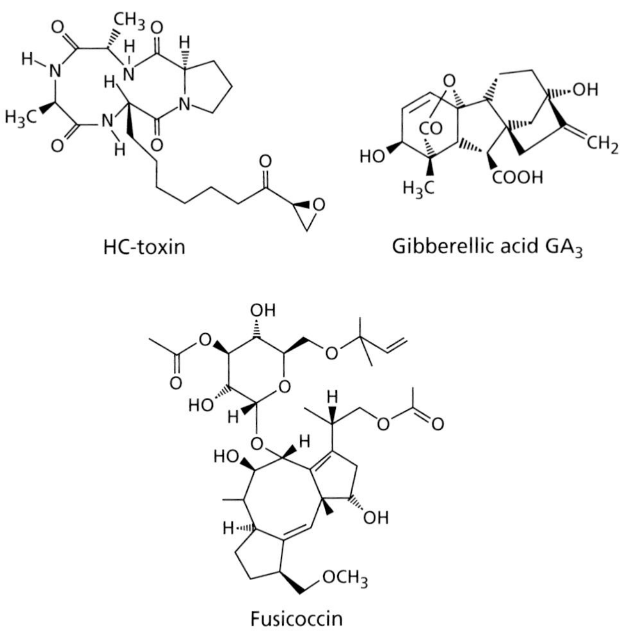

Figure 18.17 Toxins and hormones produced by pathogens help pathogens infect plants. Some pathogens produce specific toxins that significantly alter the physiology of the plant. The HC-toxin, a cyclic peptide, inhibits the enzyme histone deacetylase in the nucleus, and may have a compromising effect on the expression of genes involved in defense. Fusicoccin binds to plant plasma membrane H^+-ATPases, in particular those in stomata, and activates them irreversibly. Other pathogens produce compounds identical to plant hormones, like gibberellic acid, GA_3.

Plants can detect pathogens through perception of pathogen-derived "danger signals"

As mentioned earlier, immune signaling in plants is initiated by cell surface receptors called PRRs. Based on transcriptome sequences, we know that plants express hundreds of different PRRs. In addition to sensing host cell–derived DAMPs, these PRRs detect **microbe-associated molecular patterns (MAMPs)**, which are conserved among a specific class of microorganisms (e.g., chitin fragments for fungi, flagellin for bacteria) but are absent in the host. A subset of the receptor kinases (RKs) introduced in Chapter 12 are key PRRs for microbe- and plant-derived molecular signals associated with pathogen infection (Figure 18.16). In addition to RKs, some PRRs include receptor-like proteins that contain an extracellular domain and a transmembrane domain but lack an intracellular kinase domain. These typically associate with RKs to form heterodimers, with the kinase domain of the RK providing the intracellular signaling function.

Among the well-studied microorganism-derived MAMPs are flg22 from flagellin (Chapter 12); Pep-13, a 13–amino acid peptide from a cell wall–localized transglutaminase of the oomycete *Phytophthora*, the cause of the infamous potato blight in Ireland; and elf18, an 18–amino acid fragment of a bacterial elongation factor. Because these molecules are common to many, if not all, species within groups of microorganisms, their detection allows the plant to distinguish entire classes of potentially pathogenic organisms, such as fungi versus bacteria.

microbe-associated molecular patterns (MAMPs) Microbially produced molecules that are recognized by host cells.

Perception of MAMPs or DAMPs by cell surface PRRs initiates PTI, which inhibits the growth and activity of non-adapted pathogens or pests. The signaling pathways from cell surface PRRs to activation of diverse defense responses is complex, and invariably includes activation of a discrete set of receptor cytoplasmic kinases (RCK) (Figure 18.16) that are phylogenetically related to the kinase domains of PRRs. Once activated by PRRs, these kinases phosphorylate and activate multiple target proteins, including plasma membrane–localized NADPH/respiratory burst oxidase homologs (RBOHs), anion channels, initiators of kinase signaling cascades, and plasma membrane–localized Ca^{2+} channels. Opening of the last leads to an influx of Ca^{2+} ions that then activate Ca^{2+}-dependent protein kinases (CDPKs). CDPKs constitute a plant-specific family of protein kinases with 34 members in *Arabidopsis thaliana*. These various signaling pathways ultimately lead to a dramatic change in gene expression, which consequently induces a major shift in cellular resources away from growth and toward defense. Included among the up-regulated genes are those encoding hydrolytic enzymes that attack the cell walls of pathogens, such as glucanases, chitinases, and other hydrolases. The resulting breakdown of pathogen cell walls, in turn, releases additional MAMPs, thus producing a self-amplifying loop that produces a robust defense response.

The activation of RBOH enzymes is especially significant, as this leads to dramatic increases in extracellular ROS such as the superoxide anion (O_2^-), hydrogen peroxide (H_2O_2), and the hydroxyl radical (•OH), as is seen in some

resistance (R) proteins Proteins that function in plant defense against fungi, bacteria, and nematodes by binding to specific pathogen molecules, elicitors.

nucleotide-binding domain (NBD) Protein domain that binds nucleotides such as ATP. Found in NLR proteins.

leucine rich repeats (LRRs) Amino acid sequence motifs in proteins that provide a structural framework for protein–protein interactions.

nucleotide-binding site–leucine rich repeat proteins (NLRs) Proteins found in plants and animals that function regulate innate immune responses. NLRs function in the formation of protein complexes that function in cellular defenses.

hypersensitive response A common plant defense following microbial infection, in which cells immediately surrounding the infection site die rapidly, depriving the pathogen of nutrients and preventing its spread.

abiotic stress responses (Chapter 19). The hydroxyl radical is the strongest oxidant of these ROS and can initiate radical chain reactions with a range of organic molecules, leading to lipid peroxidation and enzyme inactivation, which are directly toxic to many microbes.

The ROS burst also contributes to strengthening of the plant cell wall, as certain proline-rich proteins of the wall become oxidatively cross-linked after pathogen attack in an H_2O_2-mediated reaction. Cell wall strengthening is further enhanced by deposition of lignin and callose at the infection site. These polymers are thought to serve as barriers, walling off the pathogen from the rest of the plant and physically blocking its spread.

Resistosomes recognize strain-specific effectors

The evolution of effector proteins by pathogens that can suppress PTI placed plants under tremendous evolutionary pressure to evolve additional resistance mechanisms. Plants have responded by evolving specialized **resistance (R) proteins** that recognize these intracellular effectors and then trigger the opening of Ca^{2+} channels by an independent pathway, which then activates many of the same defense systems as PTI. This second pathway is called effector-triggered immunity (ETI) (Figure 18.16).

Most R proteins contain a **nucleotide-binding domain (NBD)** and **leucine rich repeats (LRRs)** and hence are commonly referred to as **nucleotide-binding site–leucine rich repeat proteins (NLRs)**. NLRs are distinct oligomeric structures that form when activated by effectors. One class of NLRs detects pathogen effectors directly via a physical association of the effector with two NLR subunits. Binding of the effector induces a conformational change that enables oligomerization. This, in turn, forms a binding pocket for nicotinamide adenine dinucleotide (NAD), which is broken down, releasing a cyclic nucleotide that functions as a second messenger. This second messenger appears to then activate a subfamily of second messenger proteins. The other type of NLRs recognize pathogen effectors indirectly by binding to plant proteins that are the targets of pathogen effectors. In the absence of pathogen effectors, component NLR proteins of this type are distributed in the cytoplasm and associate with effector targets in a monomeric inactive state. However, when the effector interacts with its target, the NLR protein undergoes a conformational change that enables oligomerization and activity.

The hypersensitive response is a common defense against pathogens

Although there is considerable overlap between ETI- and PTI-induced defense responses, ETI differs from PTI in that ETI commonly leads to a localized form of cell death called the **hypersensitive response (HR)**, in which the attacked cell and cells immediately surrounding the infection site die rapidly. In the case of biotrophic pathogens, especially viruses, this cell death is highly effective at walling off the pathogen and preventing its spread. After a successful hypersensitive response, a small region of dead tissue is left at the site of the attempted invasion, but the rest of the plant is unaffected.

Exactly how ETI leads to cell death is not well understood but appears to require Ca^{2+} influx from the extracellular space mediated by one type of NLR complexes. Regardless of the precise mechanism, a hallmark of ETI-induced cell death is disruption of the plasma membrane and leakage of electrolytes into the apoplastic space.

A single encounter with a pathogen may increase resistance to future attacks

In addition to triggering defense responses locally, microbial pathogens also induce the production of signals such as salicylic acid, methyl salicylate, and other compounds that lead to systemic expression of the antimicrobial **pathogenesis-related (PR) genes**. **Salicylic acid (SA)** is produced from the nonenzymatic breakdown of isochorismate-9-glutamate in the cytosol. Isochorismate is synthesized by isochorismate synthase from pools of chorismate in the chloroplast and then exported to the cytosol, where conjugation to glutamate occurs. Isochorismate levels increase as the day progresses, which is thought to contribute to the enhanced resistance of some plants to pathogenic bacteria where infection tends to occur in the early morning as opposed to the evening. Pathogen infection increases isochorismate export to the cytosol, where it is conjugated to glutamate that rapidly degrades to form SA.

PR genes encode low-molecular-weight proteins (6–43 kDa) composed of a diverse group of hydrolytic enzymes, wall-modifying enzymes, antifungal agents, and components of signaling pathways. PR proteins are localized either in vacuoles or in the apoplast and are most abundant in leaves, where they are presumed to confer protection against secondary infections. This phenomenon of local pathogen challenge enhancing resistance to secondary infection is called **systemic acquired resistance (SAR)** and normally develops over a period of several days.

Although the phenomenon of SAR has been known for decades, the identity of the molecule responsible for systemic signaling was not uncovered until 2018, when N-hydroxy-pipecolic acid (NHP) was shown to move systemically in plants and to induce defense genes. NHP is an amino acid derived from L-lysine, and its biosynthesis is dramatically up-regulated locally in response to pathogen attack (**Figure 18.18**). Exogenously applied NHP is rapidly converted to N-OGlc-pipecolic acid, which is conjugated to a sugar and then rapidly accumulates in

pathogenesis-related (PR) genes Genes that encode small proteins that function either as antimicrobials or in initiating systemic defense responses.

salicylic acid (SA) A benzoic acid derivative that functions as a local endogenous signal in coordination with systemically distributed pipecolates for systemic acquired resistance.

systemic acquired resistance (SAR) The increased resistance throughout a plant to a range of pathogens following infection by a pathogen at one site.

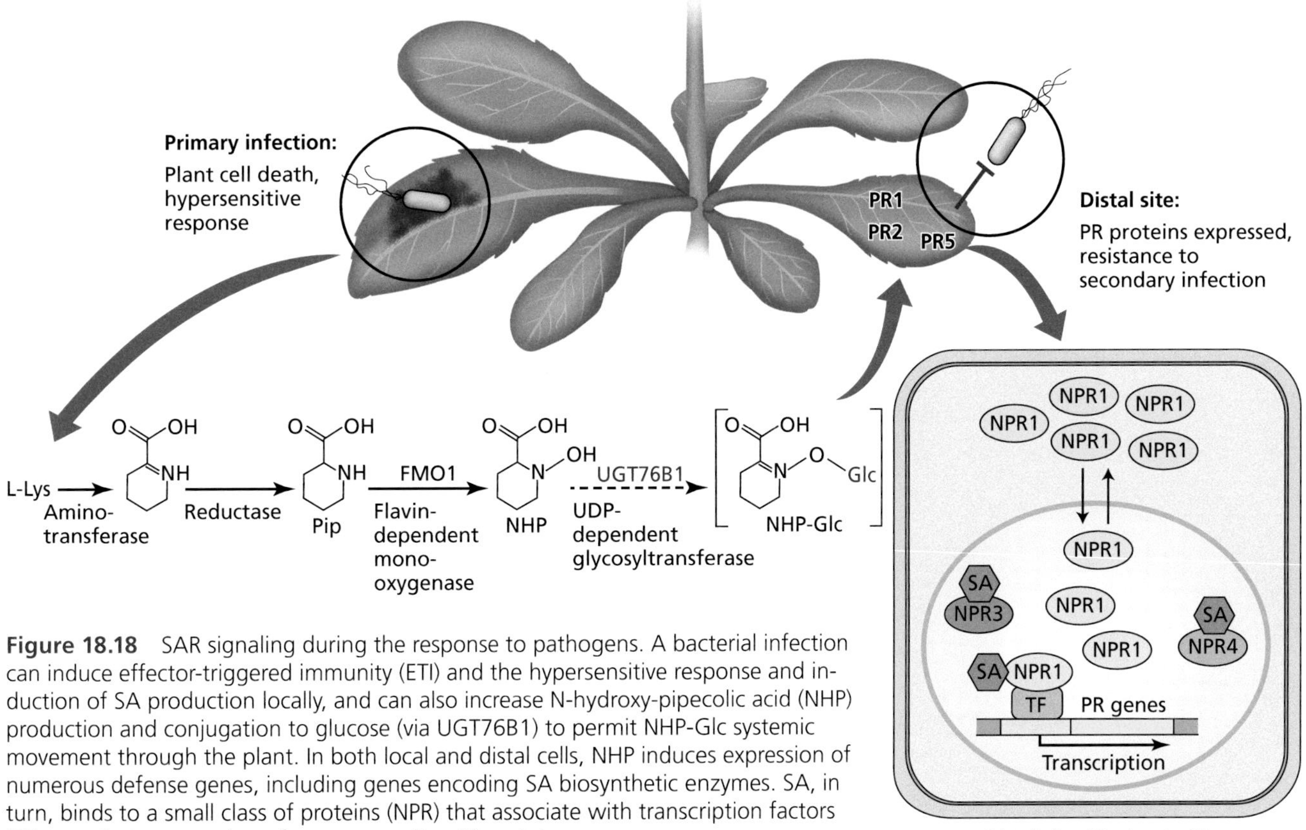

Figure 18.18 SAR signaling during the response to pathogens. A bacterial infection can induce effector-triggered immunity (ETI) and the hypersensitive response and induction of SA production locally, and can also increase N-hydroxy-pipecolic acid (NHP) production and conjugation to glucose (via UGT76B1) to permit NHP-Glc systemic movement through the plant. In both local and distal cells, NHP induces expression of numerous defense genes, including genes encoding SA biosynthetic enzymes. SA, in turn, binds to a small class of proteins (NPR) that associate with transcription factors (TF) to activate expression of genes encoding PR proteins.

phytoalexins A chemically diverse group of secondary metabolites with strong antimicrobial activity that are synthesized following infection and that accumulate at the site of infection.

cyst nematodes Parasitic nematodes that invade roots and transform into a non-motile cyst. The soybean cyst nematode *Heterodera glycines* is a major threat to soybean production.

root knot nematodes Plant parasites from the genus *Meloidogyne* found in tropical and subtropical soils. Root-knot nematode larvae infect plant roots to form root-knot galls and are a major cause of crop losses.

distal leaves. Importantly, the distal accumulation occurs after NHP application even in plants that lack FLAVIN-DEPENDENT MONOOXYGENASE1 (FMO1), which is required for the synthesis of NHP. This observation demonstrates that the exogenously added NHP is transported and that NHP produced by the plant after attack probably moves in a similar manner. Significantly, NHP application causes rapid global changes in defense gene expression and metabolic pathways that lead to enhanced resistance to pathogens, including induction of SA production and downstream induction of defense-related genes.

Phytoalexins with antimicrobial activity accumulate after pathogen attack

Phytoalexins are a chemically diverse group of specialized metabolites with strong antimicrobial activities that accumulate around the infection site. Phytoalexin production appears to be a common mechanism of resistance to pathogenic microbes in a wide range of plants. However, different plant families employ different types of specialized products as phytoalexins. For example, in leguminous plants, such as alfalfa and soybean, isoflavonoids are common phytoalexins, whereas in solanaceous plants, such as potato, tobacco, and tomato, various sesquiterpenes are produced as phytoalexins.

Phytoalexins are generally undetectable in the plant prior to infection, but they are synthesized rapidly after microbial attack. The point of control is usually the expression of genes encoding enzymes for phytoalexin biosynthesis. Plants do not appear to store any of the enzymatic machinery required for phytoalexin synthesis. Instead, soon after microbial invasion they begin transcribing the appropriate genes into mRNA and translating the mRNAs into enzymes that synthesize phytoalexins de novo.

RNA interference plays a central role in antiviral immune responses in plants

The mechanisms of RNA-mediated gene silencing are important components of plant antiviral immune responses. A key breakthrough in our understanding of the molecular mechanisms underlying RNA interference (RNAi) came when it was discovered that infection of plants with a positive-strand RNA virus (potato virus X) induces accumulation of small RNAs homologous to the infecting virus, which are now known as small-interfering RNAs (siRNAs). This discovery indicated that RNAi might be associated with an RNA-based immune response that occurs even in non-transgenic plants. This was based on the observation that plants frequently recover from a virus infection, with newly emerging leaves on a virus-infected plant displaying greatly enhanced resistance to reinfection by the same virus—as well as to infection by viruses with closely related sequences, but not to more distantly related viruses. This finding indicates that some sort of sequence-specific immune response must be occurring. In support of this idea, inserting an endogenous plant gene's antisense cDNA sequence into an RNA virus can induce silencing of that gene upon viral infection. Such virus-induced gene silencing (VIGS) has become a useful tool in plant biotechnology, as it provides a rapid method for systemic silencing of genes in plants.

Some plant parasitic nematodes form specific associations through the formation of distinct feeding structures

Nematodes, or roundworms, are water and soil inhabitants that often outnumber all other animals in their respective environments. Many nematodes exist as parasites relying on other living organisms, including plants, to complete their life cycle. Nematodes can cause severe losses of agricultural crops and ornamental plants. Plant parasitic nematodes can infect all parts of a plant, from roots to leaves, and may even live in the bark of forest trees. Nematodes feed through

Figure 18.19 Free-living juvenile nematodes are attracted to secretions by the roots. After penetration, the nematode starts feeding on cells in the vasculature. (A) Cyst nematodes cause the formation of a specific feeding structure (syncytium) in the vasculature but do not cause other morphological changes. After fertilization, the female cyst nematode dies, thereby forming a cyst containing the fertilized eggs on the outside of the root, from which a new generation of infective juveniles hatches. (B) Infection by root knot nematodes causes the formation of giant cells via cell expansion and repeated rounds of mitosis without cytokinesis, resulting in the typical root knots. Upon maturation, the female nematode releases an egg mass from which new infective juveniles hatch and cause further infestations of plants. N indicates a female nematode head.

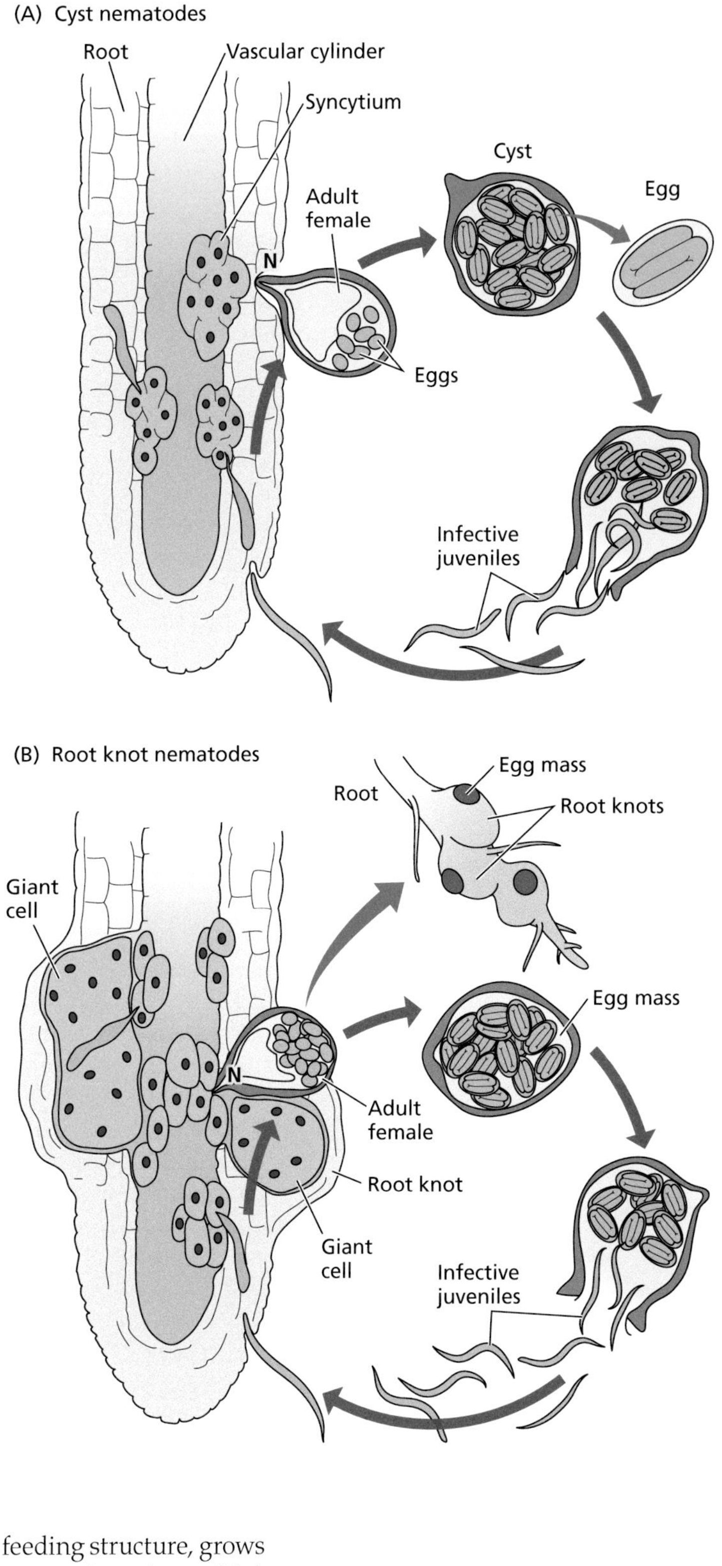

a hollow stylet that can easily penetrate plant cell walls. In the soil, nematodes can move from plant to plant, thereby causing extensive damage. Arguably the best studied among the plant parasitic nematodes are the **cyst nematodes** and those causing gall formation on infected roots, the so-called **root knot nematodes**. Both are endoparasites that depend on a living plant as host to complete their life cycles and are therefore categorized as biotrophs. The life cycle of parasitic nematodes begins when dormant eggs recognize specific compounds secreted by the plant root (**Figure 18.19**). Once hatched, the juvenile nematodes swim to the root and penetrate it. There they migrate to the vascular tissue, where they begin feeding on the cells of the vascular system.

At the permanent feeding site, usually in the root cortex, a cyst nematode larva pierces a cell with its stylet and injects saliva. The saliva contains numerous effector proteins produced by the salivary glands of the nematode. Gene silencing experiments indicate that these effector proteins are required for formation of a **syncytium** (Figure 18.19A). The syncytium is a large, metabolically active feeding site that becomes multinucleate as neighboring plant cells are incorporated into it by cell wall dissolution and cell fusion. The syncytium continues to spread centripetally toward the vascular tissue, incorporating pericycle cells and xylem parenchyma. The outer walls of the syncytium adjacent to the conducting elements form protuberances resembling those of transfer cells (Chapter 10), indicating that the syncytium now functions as a nutrient sink.

The cyst nematode, after establishing itself in such a feeding structure, grows and undergoes three molting stages while becoming a vermiform (wormlike) adult. At maturity, the female produces eggs internally, swells, and protrudes from the root surface. The mature male nematodes are released from the root into the soil and are attracted by pheromones to protruding females on the root surface. After fertilization, the female dies, forming a cyst containing the fertilized eggs.

syncytium A multinucleate cell that can result from multiple cell fusions of uninuclear cells, usually in response to viral infection.

root exudates Sugars and other compounds secreted into the soil by roots.

hemiparasitic plants Photosynthetic plants that are also parasites.

holoparasitic plants Nonphotosynthetic plants that are obligate parasites.

Roots infected by root knot nematodes form large cells, resulting in the establishment of the characteristic knot or gall, which also remains in close contact with the vasculature and provides the nematode with nutrients (Figure 18.19B).

As already mentioned, plant parasitic nematodes secrete a large number of effector proteins that affect the morphology and physiology of the plant. Among those effector proteins are some that are specifically recognized by plants and activate defense responses through recognition by NLR proteins, as described earlier for plant–pathogen interactions. Several of these plant NLR proteins have been identified and have been shown to participate in plant resistance to microbial pathogens as well. This suggests that nematode and microbial effectors may target some of the same host proteins.

Plants compete with other plants by secreting allelopathic specialized metabolites into the soil

Plants release compounds (**root exudates**) into their environment that change soil chemistry, thus increasing nutrient uptake or protecting against metal toxicity. Plants also secrete chemical signals that are essential for mediating interactions between plant roots and nonpathogenic soil bacteria, including nitrogen-fixing bacterial symbionts. However, microbes are not the only organisms that are influenced by specialized metabolites released by plant roots; some of these chemicals are also involved in direct communication between plants. Plants release specialized metabolites to the soil to inhibit the roots of other plants, a phenomenon known as allelopathy.

Interest in allelopathy has increased in recent years because of the problem of invasive species that outcompete native species and take over natural habitats. A devastating example is spotted knapweed (*Centaurea maculosa*), an invasive exotic weed introduced to North America that releases phytotoxic specialized metabolites into the soil. Spotted knapweed, a member of the aster family (Asteraceae), is native to Europe, where it is not a dominant or problematic species. However, in the northwestern United States it has become one of the worst invasive weeds, infesting more than 1.8 million ha (~4.4 million acres) in Montana alone. Spotted knapweed often colonizes disturbed areas in North America, but it also invades rangelands, pastures, and prairies, where it displaces native species and establishes dense monocultures.

The phytotoxic specialized metabolites that spotted knapweed roots release into the soil have been identified as a racemic mixture of highly phytotoxic toxic −catechin (**Figure 18.20**) with less toxic +catechin. Catechin acts as a phytotoxin by triggering a wave of ROS initiated at the root meristem, which leads to a Ca^{2+} signaling cascade that, in turn, triggers genome-wide changes in gene expression. In Arabidopsis, catechin doubled the expression of about 1,000 genes within 1 h of treatment. By 12 h, many of these same genes were repressed, which may reflect the onset of cell death. Laboratory experiments examining the effects of catechin on plant germination and growth showed that native North American grassland species vary considerably in their sensitivity to catechin. Resistant species may produce root exudates that detoxify this allelochemical.

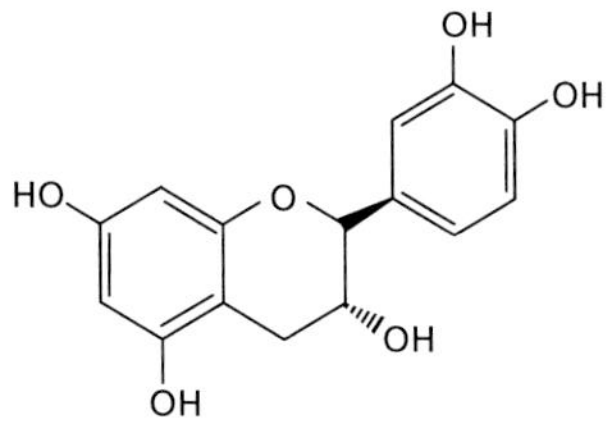

(–)-Catechin

Figure 18.20 Phytotoxic allelopathic compound −catechin produced by spotted knapweed (*Centaurea maculosa*).

Some plants are parasites of other plants

While most plants are autotrophic, some plants have evolved into parasites that rely on other plants to provide essential nutrients for their own growth and development. Parasitic plants can be divided into two main groups depending on the degree of parasitism. **Hemiparasitic plants** retain the ability to perform at least some photosynthesis, while **holoparasitic plants** are completely parasitic on their host plants and have lost the ability to carry out photosynthesis. For example, mistletoe (genus *Viscum*), which has green leaves and is able to perform photosynthesis, is a hemiparasite (**Figure 18.21A, B**).

In contrast, dodder (genus *Cuscuta*), which has lost the ability to photosynthesize and depends entirely on the host for sugars, is a holoparasite (**Figure 18.21C, D**).

Parasitic plants have developed a specialized structure, the **haustorium**, which is a modified root (**Figure 18.22A**). After establishing contact with its host plant, the haustorium penetrates the epidermis or bark and then the parenchyma to grow into the vascular tissue and absorb nutrients from the host (**Figure 18.22B–D**). To reach the host plant, seeds of parasitic plants are either directly deposited by birds or are more randomly distributed by wind or other means. After germination, the seedlings must rely for a time on their seeds for their food supply, until they can find a suitable host. Recent research has shown that low amounts of species-specific plant volatiles may serve as cues for dodder seedlings and direct their growth toward the host. Alternatively, in the case of root parasites, such as *Striga*, compounds secreted by the host root guide the growth of the seedling roots toward the host. Upon contacting the host root, the *Striga* seedling root develops into a haustorium. The haustorium then penetrates the host root and grows directly into the vascular system through the pits of xylem vessels, where it absorbs the necessary nutrients through tubelike protoplasmic structures not covered by a cell wall. The haustorium are thought to absorb nutrients at the xylem–phloem interface.

The mechanisms of these interactions between parasitic plants and their hosts have been studied mostly at the morphological level, and little is known about the signaling mechanisms involved. Emerging data, however, indicate that RNA molecules, especially microRNAs, move from the parasitic plant into the host plant, where they affect gene expression in the host. There is also little known about the defense mechanisms of the host plant. It is likely that common defense signaling pathways, including jasmonate, salicylic acid, and ethylene, may play an important role in the defense against parasitic plants, but much more research is needed.

haustorium The hyphal tip of a fungus or root tip of a parasitic plant that penetrates host plant tissue.

Figure 18.21 Parasitic plants. (A) Mistletoe (*Viscum* sp.) on a mesquite tree (genus *Prosopis*). (B) Clearly visible is the green stem of the mistletoe growing through the bark of the host plant. (C) Dodder (*Cuscuta* sp.) (yellow) growing on a patch of sand verbenas (*Abronia umbellata*) on dunes at the Pacific coast in California. (D) Close-up showing the high density of infestation of dodder on its host plant.

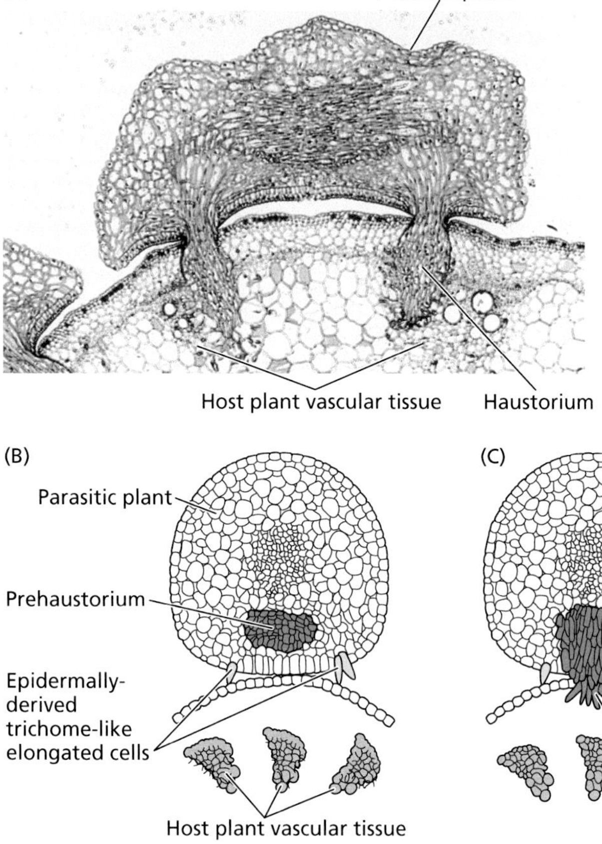

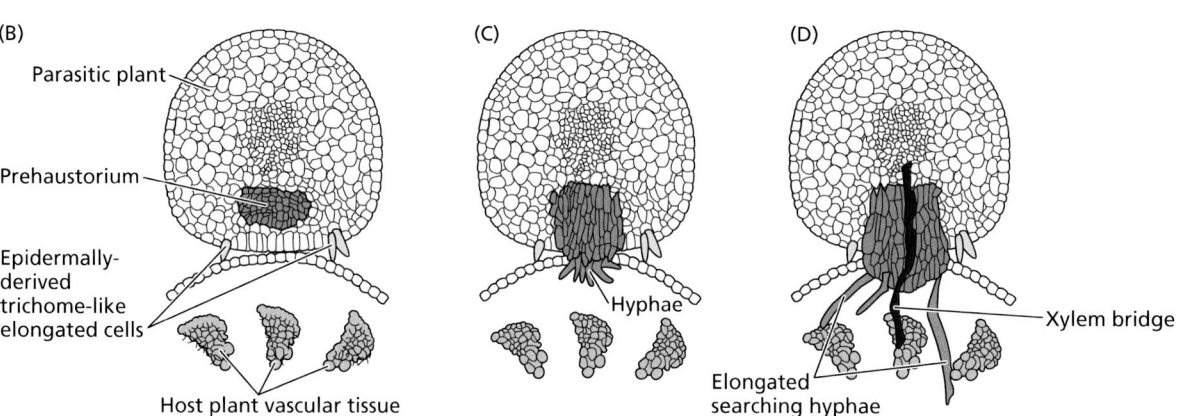

Figure 18.22 Dodder (*Cuscuta* sp.) haustorium. (A) Micrograph showing the haustorium of dodder penetrating the tissues of its host plant. (B–D) Diagram of the penetration process. (B) The disclike meristem appears in the prehaustorium (green), and epidermally derived trichome-like elongated cells (yellow) form attachment. (C) The cortex-derived haustorium penetrates the host tissue. (D) Inside the host tissue, the elongated searching hyphae (orange) grow toward the host vasculature. The searching hyphae that come into contact with the host xylem establish the xylem bridge (red).

Summary

Plants have evolved many strategies to cope with the threats posed by pests and pathogens. These strategies include sophisticated detection mechanisms and the production of toxic and repelling specialized metabolites. While some of these strategies are constitutive, others are inducible. Overall, these defenses have led to a standoff in the coevolutionary arms race between plants and their enemies and plant competitors.

18.1 Plant Interactions with Beneficial Microorganisms

- In native environments, almost every part of a plant coexists with other organisms (**Figure 18.1**).
- Interactions with symbiotic nitrogen-fixing bacteria are evolutionarily derived from interactions with pathogenic bacteria. Rhizobia release factors that set off a series of reactions leading to infection and the formation of nodules.
- Rhizobacteria can release metabolites that assist plant growth by increasing nutrient availability and pathogen protection (**Figure 18.2**).

18.2 Herbivore Interactions That Harm Plants

- Mechanical barriers that provide a first line of defense against pests and pathogens include thorns, spines, prickles, trichomes, and raphides (**Figures 18.3–18.6**).
- Specialized metabolites that serve defense functions are stored in structures that release their contents only upon damage (**Figures 18.7–18.10**).
- Some specialized metabolites are stored in the vacuole as water-soluble sugar conjugates that are spatially separated from their activating enzymes (**Figure 18.11**).

18.3 Inducible Defense Responses to Insect Herbivores

- Rather than producing defensive specialized metabolites continuously, plants can save energy by producing defense compounds only when induced by mechanical damage or specific components of insect saliva (elicitors).

(Continued)

Summary (*continued*)

- The jasmonate concentration increases rapidly in response to insect damage and induces transcription of genes involved in plant defense (**Figure 18.12**).
- Herbivore damage can induce systemic defenses by causing the synthesis of polypeptide signals or by rapidly propagating electrical and chemical signals to initiate defense responses in as-yet-undamaged tissues (**Figure 18.13**).
- Plants may release volatile compounds to attract natural enemies of herbivores, or to signal neighboring plants to initiate defense mechanisms (**Figure 18.14**).

18.4 Plant Defenses against Pathogens

- Pathogens can invade plants through cell walls by secreting lytic enzymes, through natural openings such as stomata and lenticels, and through wounds. Insect herbivores may also be pathogen vectors (**Figure 18.15**).
- Pathogens generally use one of three attack strategies: necrotrophism, biotrophism, or hemibiotrophism.
- Pathogens often produce effector proteins that aid in initial infection (**Figure 18.16**).
- All plants have pattern recognition receptors (PRRs) that set off defense responses when activated by evolutionarily conserved microbe-associated molecular patterns (MAMPs; e.g., flagellin, chitin) or host-derived damage-associated molecular patterns (DAMPS; e.g., cell wall fragments, extracellular ATP, specific peptides) (**Figure 18.16**).
- Plant resistance genes encode cytosolic receptors (NLR proteins) that recognize pathogen-derived protein effectors in the cytosol. NLR proteins can recognize pathogen effectors by direct binding, or indirectly by sensing conformational changes in effector targets. Upon activation, NLR proteins induce a hypersensitive response, in which cells surrounding the infected site die rapidly, thereby limiting the spread of infection (**Figure 18.16**).
- Pathogens can produce toxins and hormones to facilitate infection of host plants (**Figure 18.17**).
- A plant that survives local pathogen infection often develops increased resistance to subsequent attack, a phenomenon called systemic acquired resistance (SAR) (**Figure 18.18**).
- In response to infection, many plants produce phytoalexins, specialized metabolites with strong antimicrobial activity.
- RNA interference, a process that eliminates specific RNA transcripts, probably evolved as an antiviral defense response.
- Nematodes (roundworms) are parasites that can move between hosts and that induce formation of feeding structures and galls from vascular plant tissues. In response, plants use defense signaling pathways similar to those used for pathogen infection (**Figure 18.19**).
- Some plants produce allelopathic specialized metabolites that enable them to outcompete nearby plant species (**Figure 18.20**).
- Some plants are parasitic on other plants. Parasitic plants can be divided into two main groups (hemiparasites and holoparasites) depending on their ability to perform some photosynthesis (**Figure 18.21**).
- Parasitic plants use a specialized structure, the haustorium, to penetrate their host, grow into the vasculature, and absorb nutrients (**Figure 18.22**).

Suggested Reading

Chen, C. Y., Liu, Y. Q., Song, W. M., Chen, D. Y., Chen, F. Y., Chen, X. Y., Chen, Z. W., Ge, S. X., Wang, C. Z., Zhan, S., et al. (2019) An effector from cotton bollworm oral secretion impairs host plant defense signaling. *Proc. Natl. Acad. Sci. USA* 116: 14331–14338.

Farmer, E. E., Gao, Y.-Q., Lenzoni, G., Wolfender, J.-L., and Wu, Q. (2020) Wound- and mechanostimulated electrical signals control hormone responses. *New Phytol.* 227: 1037–1050.

Guo, Q., Major, I. T., and Howe, G.A. (2018) Resolution of growth defense conflict: Mechanistic insights from jasmonate signaling. *Curr. Opin. Plant Biol.* 44: 72–81.

Lolle, S., Stevens, D., and Coaker, G. (2020) Plant NLR-triggered immunity: From receptor activation to downstream signaling. *Curr. Opin. Immunol.* 62: 99–105.

Rosa, C., Kuo, Y. W., Wuriyanghan, H., and Falk, B. W. (2018) RNA interference mechanisms and applications in plant pathology. *Annu. Rev. Phytopathol.* 56: 581–610.

Sharifi, R., Lee, S.-M., and Ryu, C.-M. (2018) Microbe-induced plant volatiles. *New Phytol.* 220: 684–691.

Wang, J., Hu, M., Wang, J., Qi, J., Han, Z., Wang, G., Qi, Y., Wang, H. W., Zhou, J. M., and Chai, J. (2019) Reconstitution and structure of a plant NLR resistosome conferring immunity. *Science* 364: eaav5870.

Zhou, J. M., and Zhang, Y. (2020) Plant immunity: Danger perception and signaling. *Cell* 181: 978–989.

19 Abiotic Stress

Plants grow and reproduce in environments characterized by a multitude of abiotic (nonliving) factors that vary both with time and geographic location. The predominant abiotic environmental parameters that affect plant growth are light (intensity, quality, and duration), water (soil availability and humidity), carbon dioxide, oxygen, soil nutrients (composition and availability), temperature, and toxins (e.g., heavy metals and salinity). Fluctuations of these abiotic factors outside their normal ranges usually have negative biochemical and physiological consequences for plants. Being sessile, plants are unable to avoid abiotic stress simply by moving to a more favorable environment. Instead, plants have evolved the ability to compensate for stressful conditions by altering physiological and developmental processes to maintain growth and reproduction.

In the past several decades, the introduction of greenhouse gases into our atmosphere has resulted in a cumulative warming of air, land, and ocean (global warming). This warming has become a major contributor to changes in our climate, causing an increase in the frequency and intensity of weather events that subject plants to severe abiotic stress conditions, such as prolonged droughts, heat waves, flooding, and salinity. As we describe in this chapter, these conditions can have a negative impact on plant growth and yield, threatening our food supply and highlighting the importance of studying abiotic stress responses in different crops, herbaceous plants, and trees.

In this chapter we provide an integrated view of how plants adapt and respond to abiotic stresses in the environment. Like all living organisms, plants are complex biological systems containing thousands of different genes, proteins, regulatory molecules, signaling agents, and chemical compounds that form hundreds of interlinked pathways and networks.

These interlinked pathways are physiologically calibrated to the particular set of environmental conditions under which the species evolved. Under normal growing conditions, the different biochemical pathways and signaling networks act in a coordinated manner to balance environmental inputs with the plant's genetic imperative to grow and reproduce. When exposed to unfavorable environmental conditions, this complex interactive system adjusts homeostatically to minimize the negative impacts of stress and maintain metabolic equilibrium (**Figure 19.1**).

For example, one way plants can adapt to abiotic stress, such as global warming, is to shift their seasonal flowering time, or *phenology*. Comparisons of modern-day flowering times with historical data collected by the nineteenth-century American philosopher and naturalist Henry David Thoreau have shown that during the past 170 years, plants growing around Walden Pond in Concord, Massachusetts, USA have shifted their flowering times an average of seven days earlier relative to the nineteenth century. Those species best able to adjust their flowering times in response to global warming have persisted in the landscape, while those least able to adapt phenologically have either succumbed or are in decline. Presumably, the loss of these non-adapted species will have negative impacts on the populations of insects, birds, and herbivores that depend on them for sustenance.

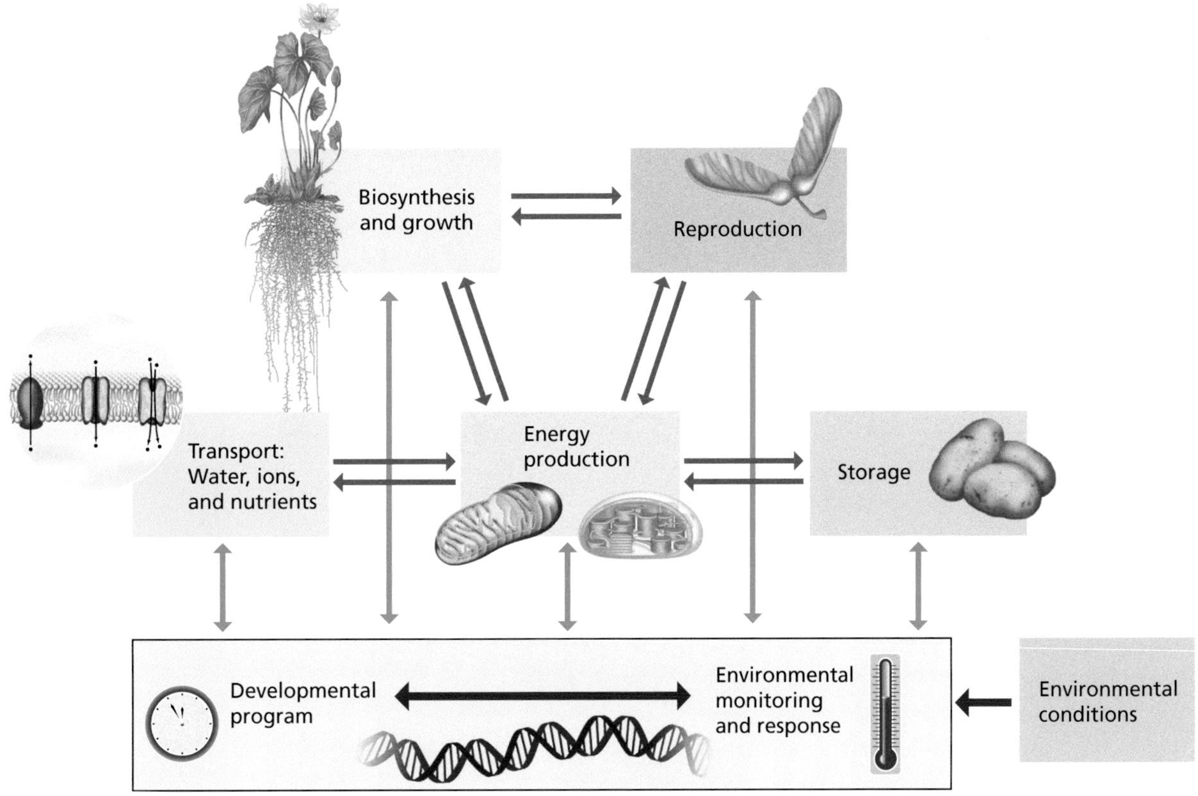

Figure 19.1 Interactions between environmental conditions and plant development, growth, energy production, and ion and nutrient balance and storage. The balance between these processes is controlled by the plant genome (lower blue box), which encodes sensors and signal transduction pathways that monitor and adjust for environmental parameters. Based on the different environmental stress signals, the plant genome can thus direct the flow of energy between the different processes (brown arrows) to establish a new homeostatic state matched to the specific stress conditions.

stress Disadvantageous influences exerted on a plant by external abiotic or biotic factor(s), such as herbivory, infection, heat, water, or anoxia. Measured in relation to plant survival, crop yield, biomass accumulation, or CO_2 uptake.

19.1 Defining Plant Stress

Explain what plant stress is and how plants respond to it.

The ideal growth conditions for a given plant can be defined as the conditions that allow the plant to achieve its maximum growth and reproductive potential as measured by plant weight, height, and seed number, which together make up the *total biomass* of the plant. **Stress** can be defined as any environmental condition that prevents the plant from achieving its full genetic potential. For example, a decrease in light intensity would cause a reduction in photosynthetic activity, with a concomitant decrease in the energy supply to the plant. Under these conditions, the plant could compensate either by slowing down biosynthesis, thus reducing its growth rate or yield, or by drawing on its stored food reserves in the form of starch or oil (Figure 19.1).

As anyone who has ever forgotten to water their garden can attest, water-deficit stress can cause severe wilting (**Figure 19.2A** and **Figure 19.2B**). A decrease in water availability can also have a deleterious effect on growth. One way that plants compensate for a decrease in water potential is by closing their stomata, which reduces water loss by transpiration (Chapter 3). However, stomatal closure also decreases CO_2 uptake by the leaf, thereby reducing photosynthesis and suppressing growth. An example of the effects of water deficit on the growth of rice plants is shown in **Figure 19.2C**. Rice is able to tolerate moderate water deficit without any measurable effect on growth, but severe water deficit strongly inhibits vegetative growth.

In addition to suppressing vegetative growth, abiotic stresses can reduce seed production due to limited photosynthetic resources and cause deleterious

Figure 19.2 Comparisons of control (well-watered) and water-deficit–stressed squash (*Cucurbita pepo*) and rice (*Oryza sativa*) plants. (A) Well-watered squash plant. (B) Water-deficit–stressed squash plant. (C) Control, moderately water-deficit–stressed, and severely water-deficit–stressed rice plants.

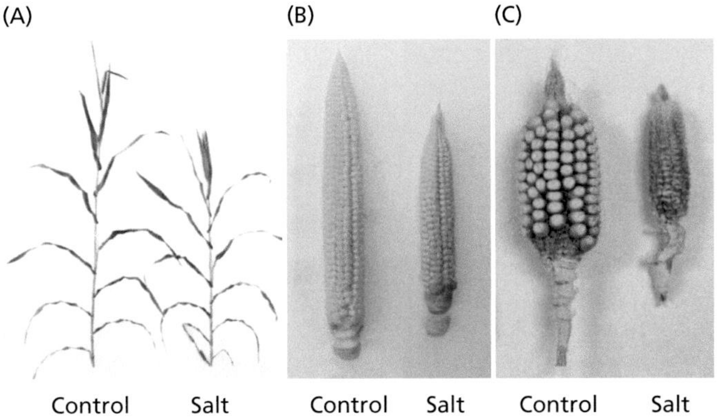

Figure 19.3 Effects of salt stress on maize production. Effect of salt stress on (A) maize ear size; (B) maize kernel size; (C) maize kernel number.

effects on reproductive processes. For example, salinity decreases shoot height, ear size, and kernel size in maize (corn; *Zea mays*) (**Figure 19.3**).

Physiological adjustment to abiotic stress involves trade-offs between vegetative and reproductive development

How do changes in environmental conditions affect seed production? Under optimal growing conditions, the competition for resources among the different plant organs or developmental phases is minimal. The transition to reproductive growth occurs only after the vegetative adult phase completes its genetically determined developmental program (Chapter 16). Under abiotic stress conditions, however, the vegetative growth program may terminate prematurely, and the plant may advance immediately to the reproductive phase. In this case the plant undergoes a transition to flowering, fertilization, and seed set before reaching its full size, producing a smaller plant with fewer leaves to provide photosynthates (Figure 19.3A). As a result, the seeds of plants grown under such suboptimal conditions may be fewer and smaller than normal seeds (Figure 19.3 B and C).

The particular developmental pathway that a plant uses to maximize its reproductive potential under abiotic stress depends to a large extent on the plant's life cycle. For example, *annual plants* complete their life cycle in a single season. It is thus advantageous for annual plants to adjust their metabolism and developmental programs so as to produce the maximum number of viable seeds under whatever environmental conditions are encountered during the season. By contrast, *perennial plants*, which have multiple seasons in which to produce seeds, tend to adjust their metabolism and developmental programs to ensure the optimal storage of food resources that will enable the plants to survive to the next season, even at the expense of seed production. Faced with a severe abiotic stress period, perennial plants might therefore abort flowering altogether and go dormant until conditions improve.

19.2 Acclimation versus Adaptation

Distinguish between acclimation and adaptation to environmental conditions.

Individual plants respond to changes in the environment by directly altering their physiology or morphology to enhance survival and reproduction. Such responses require no new genetic modifications. If the response of the individual plant improves with repeated exposure to the environmental stress, then the

acclimation The increase in plant stress tolerance due to exposure to prior stress. May involve changes in gene expression. Contrast with *adaptation*.

adaptation An inherited level of stress resistance acquired by a process of selection over many generations. Contrast with *acclimation*.

response is termed **acclimation**. Acclimation represents a nonpermanent change in the physiology or morphology of the individual that can be reversed if the prevailing environmental conditions change. Epigenetic mechanisms that alter the expression of genes without changing the genetic code of an organism can extend the duration of acclimation responses and make them heritable.

One example of acclimation, from gardening, is a process known as *hardening off*. To speed up the growth of plants, gardeners often start by growing them indoors in pots under optimal growth conditions. The gardeners then move the plants outdoors for part of the day over a period long enough to acclimate, or "harden," the plants to outdoor weather before moving them outdoors permanently.

Another example of acclimation is the response of salt-sensitive plants, termed glycophytic plants, to salinity. Although glycophytic plants are not genetically adapted to growth in saline environments, when exposed to elevated salinity they can activate several stress responses that allow the plants to cope with the physiological perturbations imposed by elevated salinity in their environment. Salt exposure stimulates enhanced efflux of Na^+ from cells, which reduces salinity-induced toxicity. This response is not sufficient to protect plants under hypersaline conditions, but it can afford short-term protection when salt concentrations temporarily increase in drying soils.

Therefore, acclimation is the result of transient changes in an individual plant. In contrast, morphological or physiological changes in a population that have become genetically fixed over many generations by natural selection are referred to as **adaptations**.

A remarkable example of adaptation to an extreme abiotic environment is the growth of plants in serpentine soils. Serpentine soils are characterized by low moisture, low concentrations of macronutrients, and elevated levels of nickel, cobalt, and chromium ions. These conditions would result in severe stress conditions for most plants. However, it is not unusual to find populations of plants that have become genetically adapted to serpentine soils growing not far from closely related non-adapted plants growing on "normal" soils. Simple transplant experiments have shown that only the adapted populations can grow and reproduce on the serpentine soil, and genetic crosses reveal the stable genetic basis of this adaptation.

The evolution of adaptive mechanisms in plants to a particular set of environmental conditions generally involves processes that allow the avoidance of the potentially damaging effects of these conditions. For example, the predominant forms of the toxic metalloid arsenic taken up by plant cells are arsenate (As^{IV}) and the reduced form arsenite (As^{III}). As noted in Chapter 6, As^{IV} resembles phosphate and can enter plant cells via the phosphate transporter. Around arsenic-contaminated mine sites in southwestern England, populations of the weed Yorkshire fog grass (*Holcus lanatus*) have become adapted to arsenic toxicity by means of a specific genetic modification of the phosphate transporter. This mutation reduces the uptake of both phosphate and arsenate, enabling the plants to avoid arsenic toxicity. In contrast, populations of Yorkshire fog grass growing on uncontaminated soils are less likely to contain this genetic modification.

Both genetic adaptation and physiological acclimation can contribute to the plants' overall tolerance of extremes in their abiotic environment. For example, genetic adaptation in the arsenic-tolerant Yorkshire fog grass population reduced arsenate uptake but did not block it entirely. Once inside the plant cell the arsenate ion is rapidly reduced to arsenite, an even more toxic form of arsenic. To mitigate the toxic effects of arsenite, adapted plants use the same biochemical mechanism that non-adapted plants use to respond to the toxic effects of arsenic accumulation. This mechanism involves the biosynthesis of low-molecular-weight, metal-binding molecules called phytochelatins (discussed below), which leads to the sequestration and detoxification of the phytochelatin–arsenite complex in the vacuole. Thus, the ability of Yorkshire fog grass to thrive on arsenic-contaminated mine waste depends on both a genetic adaptation specific to the tolerant population (arsenate exclusion) and on an acclimation response (arsenite sequestration), which is common to all plants that respond to arsenic by producing phytochelatins.

reactive oxygen species (ROS) These include the superoxide anion ($O_2{\bullet}^-$), hydrogen peroxide (H_2O_2), the hydroxyl radical (HO•), and singlet oxygen. They are generated in several cell compartments and can act as signals or cause damage to cellular components.

19.3 Environmental Stressors

Distinguish among the effects of different environmental stressors on physiological processes.

As with every biological system, plant survival and growth depend on complex networks of coupled anabolic and catabolic pathways that direct the flow of energy and resources within and among cells. Disruption of these networks by environmental factors can result in uncoupling of the pathways. For example, metabolic enzymes can, and often do, have different temperature optima. An increase or decrease in ambient temperature can inhibit or increase the activity of a subset of enzymes without affecting other enzymes in the same or connected pathways. Such functional uncoupling of metabolic pathways could result in the accumulation of intermediate compounds that could be converted to toxic byproducts.

One of the most common groups of toxic intermediates produced by stress are **reactive oxygen species** (**ROS**), which are highly reactive forms of oxygen, often possessing at least one unpaired electron in their orbitals. ROS are capable of rapidly reacting with, and oxidizing, a wide variety of cellular constituents, including proteins, DNA, RNA, and lipids. The most common forms of ROS in plant cells are superoxide ($O_2^{\bullet-}$), singlet oxygen ($^1O_2$), hydrogen peroxide (H_2O_2), and hydroxyl radicals ($OH^{\bullet}$). ROS can also trigger an autocatalytic process of membrane oxidation that can result in the degradation of organelles and the plasma membrane, and cell death (Chapter 15). Despite their mechanistic differences, most abiotic stresses result in the production of ROS, and ROS are used by plants as signal transduction molecules for stress sensing (**Figure 19.4**). ROS signaling can also regulate aspects of normal development as well as stress responses (Chapter 12).

Environmental stresses can also disrupt the *compartmentation* of metabolic processes that separates them from other cellular components. The same temperature extremes that can inhibit enzyme activity can also affect membrane fluidity: High temperatures increase membrane fluidity, and low temperatures decrease membrane fluidity. Changes in membrane fluidity can disrupt the coupling between different protein complexes in chloroplast or mitochondrial membranes, resulting in the uncontrolled transfer of electrons to oxygen and the formation of ROS.

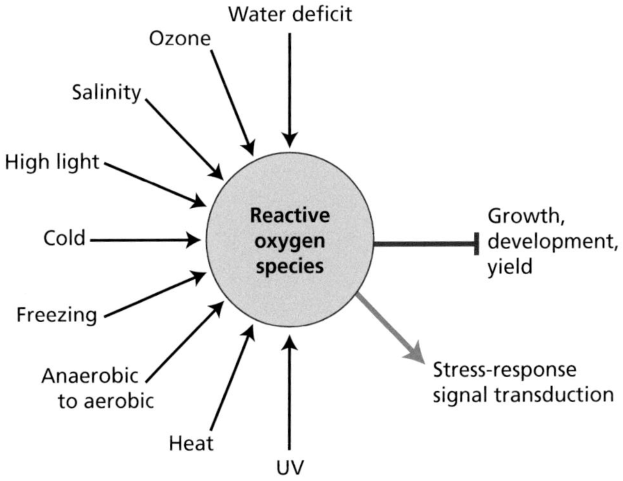

Figure 19.4 Disruption of metabolic processes during abiotic stress can produce reactive oxygen species (ROS). A variety of abiotic stresses result in the accumulation of ROS in cells. ROS accumulation has both positive and negative effects. Negative effects include decreased plant growth, development, and yield. Cellular ROS accumulation also activates signal transduction pathways (Chapter 12) that induce acclimation mechanisms. These, in turn, counteract the negative effects of stress (including ROS accumulation).

Water deficit decreases turgor pressure, increases ion toxicity, and inhibits photosynthesis

As in most other organisms, water makes up the largest proportion of the cellular volume in plants and is the most limiting resource. About 97% of water taken up by plants is lost to the atmosphere (mostly by transpiration). About 2% is used for volume increase or cell expansion, and 1% for metabolic processes, predominantly photosynthesis. Water deficit (insufficient water availability) occurs in most natural and agricultural habitats and is caused mainly by intermittent to continuous periods without precipitation. *Drought* is the meteorological term for an extended period of suboptimal precipitation that ultimately results in plant water deficit. *Water-deficit stress* (rather than "drought stress") is the amount of stress a plant experiences under drought conditions and depends on the soil's water-holding capacity and the depth of the water table.

Water deficit can affect plants differently during vegetative versus reproductive growth. When plant cells experience water deficit, cell dehydration occurs. Cell dehydration adversely affects many basic physiological processes (**Table 19.1**). For example, during water deficit the water potential (Ψ) of the apoplast becomes more negative than that of the symplasm, causing reductions in pressure potential (turgor) (Ψ_p) and volume (Chapter 3). A secondary effect of cell dehydration is that ions become more concentrated and may become cytotoxic. Water deficit also induces the accumulation of abscisic acid (ABA), which promotes stomatal closure, reducing gas exchange and inhibiting photosynthesis (**Figure 19.5**). As a result of dehydration-induced uncoupling of the photosystems, free electrons produced by the reaction centers are not transferred to $NADP^+$, leading to the generation of ROS that could cause cellular damage if not removed.

Table 19.1 Physiological and Biochemical Perturbations in Plants Caused by Fluctuations in the Abiotic Environment

Environmental factor	Primary effects	Secondary effects
Water deficit	Water potential (Ψ) reduction Cell dehydration Hydraulic resistance	Reduced cell/leaf expansion Reduced cel lular and metabolic activities Stomatal closure Phot osynthetic inhibition Leaf abscission Altered carbon partitioning Cavitation Membrane and protein destabilization ROS production Cell death
Salinity	Water potential (Ψ) reduction Cell dehydration Ion cytotoxicity	Same as for water deficit (see above)
Flooding and soil compaction	Hypoxia Anoxia	Reduced respiration Fermentative metabolism Inadequate ATP production Production of toxins by anaerobic microbes ROS production Stomatal closure
High temperature	Membrane and protein destabilization	Photosynthetic and respiratory inhibition ROS production Cell death
Chilling	Membrane destabilization	Membrane dysfunction
Freezing	Water potential (Ψ) reduction Cell dehydration Symplastic ice crystal formation	Same as water deficit (see above) Physical destruction
Trace element toxicity	Disturbed cofactor binding to proteins and DNA ROS production	Disruption of metabolism
High light intensity	Photoinhibition ROS production	Inhibition of PSII repair

Temperature stress affects a broad spectrum of physiological processes

Temperature stress disrupts plant metabolism because of its differential effect on protein stability and enzymatic reactions, causing the uncoupling of different reactions and the accumulation of toxic intermediates and ROS. As mentioned earlier, heat stress increases membrane fluidity, whereas cold stress decreases membrane fluidity, causing the uncoupling of different multiprotein complexes, disruption of electron flow and energetic reactions, and disruption of ion homeostasis and regulation. Heat stress can also destabilize (melt) RNA and DNA secondary structures, and cold stress can overstabilize (harden) them, disrupting transcription, translation, and RNA processing and turnover. In addition, temperature stress can affect protein structure and block protein degradation, causing the buildup of protein aggregates. Such protein clumps can disrupt normal cellular functions by interfering with the function of the cytoskeleton and associated organelles. Heat stress is also known to negatively affect plant reproduction by causing decreased pollen viability, fertilization, and seed filling, as well as increased flower and fruit abortion. Because of the effects of heat stress on plant reproductive processes, global warming is expected to cause an overall decrease in the yield of major grain crops such as wheat, maize, and rice.

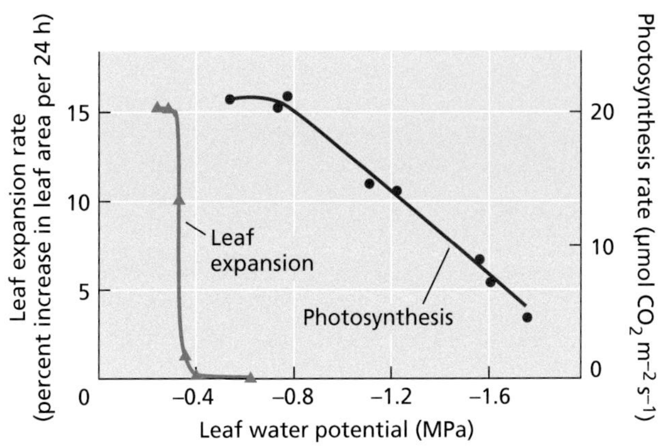

Figure 19.5 Effects of water-deficit stress on photosynthesis and leaf expansion of sunflower (*Helianthus annuus*). In this species, leaf expansion is completely inhibited under mild water-deficit stress levels that minimally affect photosynthetic rates.

Flooding causes anaerobic stress to roots

When a field is flooded, the O_2 levels at the root surface decrease dramatically because most of the air in the soil is displaced by water, which has a significantly lower O_2 concentration than air does. For example, the atmosphere contains about 20% O_2 or 200,000 ppm, compared with less than 10 ppm dissolved O_2 in flooded soil. Under these conditions, respiration in roots is suppressed and fermentation is enhanced (Chapter 11). This metabolic shift can cause energy depletion, acidification of the cytosol, and toxicity from ethanol accumulation. As a consequence of energy depletion, many processes such as protein synthesis are suppressed. Anaerobic stress can cause cell death within hours or days, depending on the degree of genetic adaptation of the species. Even if the O_2-deprived plant is returned to normal O_2 levels, the recovery process itself can pose a hazard. While the roots are under anaerobic stress, the absence of O_2 prevents the formation of ROS. But if the O_2 level in the soil is rapidly increased, much of it may be used to form ROS, causing oxidative damage to the root cells.

Salinity stress has both osmotic and cytotoxic effects

Excess soil salinity brought about by a combination of over-irrigation and poor soil drainage affects large regions of the world's land mass and has a severe impact on agriculture. About 20% of all cultivated land is currently under irrigation and contributes to about 40% of global food production; roughly 20% of all irrigated land is affected by salinity stress. It is likely that the percentage of irrigated land will increase significantly in areas most affected by global warming.

Salinity stress has two components: nonspecific **osmotic stress** that causes water deficits, and specific ion effects resulting from the accumulation of toxic ions, which interfere with nutrient uptake and cause cytotoxicity. Salt-tolerant plants genetically adapted to salinity are termed **halophytes** (from the Greek *halo*, "salty"), while less salt-tolerant plants that are not adapted to salinity are

osmotic stress Stress imposed on cells or whole plants when the osmotic potential of external solutions is more negative than that of the solution inside the plant.

halophytes Plants that are native to saline soils and complete their life cycles in that environment. Contrast with *glycophytes*.

glycophytes Plants that are not able to resist salts to the same degree as halophytes. Show growth inhibition, leaf discoloration, and loss of dry weight at soil salt concentrations above a threshold. Contrast with *halophytes*.

termed **glycophytes** (from the Greek *glyco*, "sweet"). Under non-saline conditions, the cytosol of higher plant cells contains about 100 m*M* K^+ and less than 10 m*M* Na^+, an ionic environment in which enzymes are optimally functional. In saline environments, cytosolic Na^+ and Cl^- increase to more than 100 m*M*, and these ions become cytotoxic. One reason that Na^+ is toxic at high concentrations is that it is chemically similar to K^+ and can competitively inhibit K^+-dependent cellular processes. Several enzymes, including phosphofructokinase involved in glycolysis, require K^+ for their function and can be inhibited by high cytosolic Na^+ levels. At high concentrations, apoplastic Na^+ can also compete for sites on transport proteins that are necessary for high-affinity uptake of K^+, an essential macronutrient.

The effects of high salinity in plants occur through a two-phase process: a fast response to the high osmotic pressure at the root–soil interface and a slower response caused by the accumulation of Na^+ (and Cl^-) in the leaves. In the osmotic phase there is a reduction in shoot growth, with reduced leaf expansion and inhibition of lateral bud formation. The second phase starts with the accumulation of toxic amounts of Na^+ in the leaves, leading to the inhibition of photosynthesis and other biosynthetic processes. Although in most species Na^+ reaches toxic concentrations before Cl^-, some plant species, such as citrus, grapevine, and soybean, are highly sensitive to excess Cl^-.

During freezing stress, extracellular ice crystal formation causes cell dehydration

Plants subjected to freezing temperatures must contend with the formation of ice crystals, either extracellularly or intracellularly. Intracellular ice crystal formation nearly always proves lethal to the cell. However, the solute concentration in the apoplast is relatively low and the water therefore has a higher freezing point than that of the more concentrated symplasm. As a result, ice crystals tend to form in the apoplast and in the xylem tracheids and vessels, along which the ice can quickly propagate. The formation of ice crystals lowers the apoplastic water potential (Ψ), which becomes more negative than that of the symplasm. Unfrozen water within the cell moves down this gradient out of the cell toward the ice crystals in the intercellular spaces. As water leaves the cell, the plasma membrane contracts and pulls away from the cell wall. During this process the plasma membrane, rigidified by the low temperature, may become damaged. The lower the temperatures, the more water travels down the gradient toward the frozen water. For example, at –10°C the symplasm loses about 90% of its osmotically active water to the apoplast. In this respect, freezing stress has much in common with water-deficit stress. As with water-deficit stress, the apoplasts of cells that are already dehydrated, like those in seeds and pollen, are less likely to undergo further dehydration by extracellular ice crystal formation and therefore less likely to be damaged. Thus, the low water content in seeds and pollen is a survival mechanism.

Heavy metals can both mimic essential mineral nutrients and generate ROS

The uptake of heavy metal ions such as Cd^{2+} by plant cells can lead to the accumulation of ROS, inhibition of photosynthesis, disruption of membrane structure and ion homeostasis, inhibition of enzymatic reactions, and activation of programmed cell death (Chapter 15). One of the reasons that heavy metals are so toxic is that they can mimic other essential metals (e.g., Cd^{2+} can mimic Zn^{2+}), take their place in essential reactions, and disrupt these reactions. The mimicking of essential elements can also explain the uptake of Cd^{2+} and other heavy metal ions into cells via channels that evolved to transport essential elements. Cd^{2+} can also inhibit the uptake of Zn^{2+}, Fe^{2+}, and Mn^{2+} by competing with these ions for the same transport systems. (Note that Fe^{3+} is transported by a different mechanism, as discussed in Chapter 6.) In addition to having these toxic effects, some heavy metals can directly interact with oxygen to form ROS.

Combinations of abiotic stresses can induce unique signaling and metabolic pathways

In the field, plants are often subjected to a combination of different abiotic stresses simultaneously. Water-deficit stress and heat stress are examples of two different abiotic stresses that almost always occur together in the environment, with devastating results. Between 1980 and 2012 in the United States, the cost of crop damage due to droughts combined with heat waves was four times greater than that due to droughts alone (**Figure 19.6A**), and these extreme weather events are forecast to increase in frequency and intensity because of climatic changes driven by global warming.

Plants respond differently to different abiotic stresses, but the physiological responses to the different stresses are not additive when they occur in combination. **Figure 19.6B** shows the effects of heat and water deficit, applied separately, on four different physiological parameters of Arabidopsis: photosynthesis, respiration, stomatal conductance, and leaf temperature. The physiological profiles under the two stresses applied individually are quite different. For example, during heat stress alone, plants *increase* their stomatal conductance, which cools their leaves by transpiration. However, if heat stress is combined with water-deficit stress, the stomata close to conserve water, causing the leaf temperature to rise by 2 to 5°C. In other words, water-deficit stress exacerbates the effects of heat stress on leaf temperature, and the combination can be lethal.

The combination of heat stress plus water-deficit stress also induces different patterns of gene expression and metabolite turnover than either stress alone. Differences in physiological parameters, transcript accumulation, and metabolites could be a result of conflicting physiological responses to the two stresses. Because global warming is likely to be accompanied by more frequent droughts

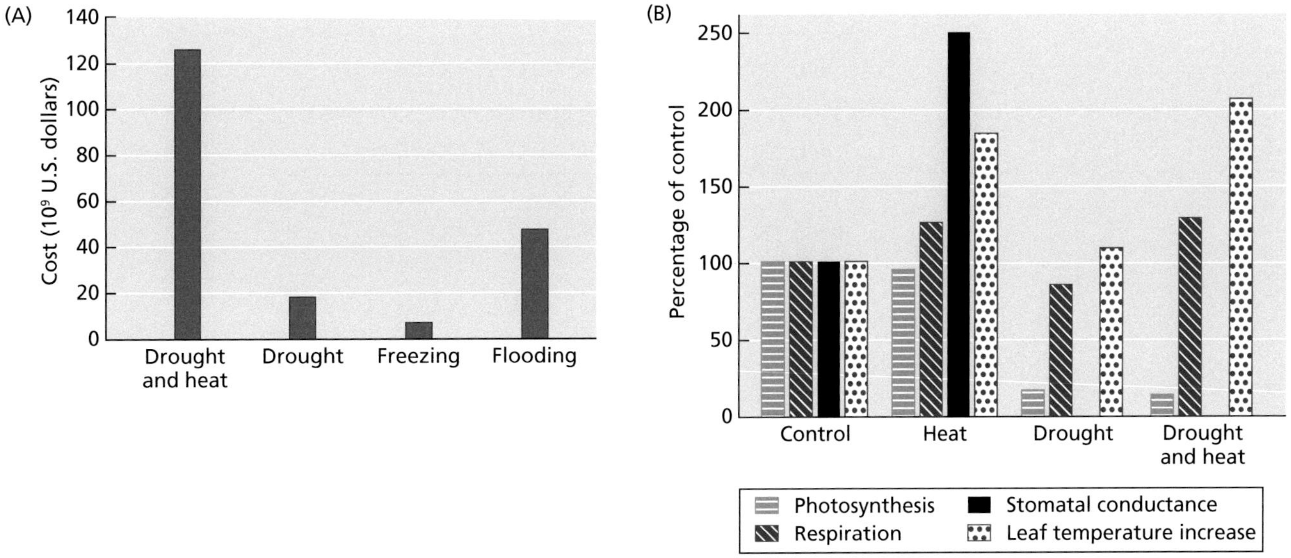

Figure 19.6 Effect of combined abiotic stresses on plant productivity, physiology, and molecular responses. (A) Losses to U.S. agriculture resulting from a combination of water-deficit stress and heat stress were much higher than losses caused by water-deficit stress, freezing, or flooding alone between 1980 and 2019. (B) Effect of combined water-deficit stress and heat stress on plant physiology. Note the complete closure of plant stomata (zero conductance), which results in a higher leaf temperature.

cross-protection A plant response to one environmental stress that confers resistance to another stress.

in many regions of the world, both agriculture and natural ecosystems are likely to be negatively affected.

A similar problem can arise when heat stress occurs in combination with salinity or heavy metal stress, because enhanced transpiration can lead to greater uptake of salt or heavy metals. Among the several stress interactions that could have a deleterious effect on crop productivity are water deficit and heat, salinity and heat, nutrient deficiency stress and water deficit, and nutrient deficiency and salinity. On the other hand, some stress combinations could have beneficial effects on plants compared with the individual stresses applied separately. For example, water-deficit stress, which causes stomatal closure, could potentially enhance tolerance to ozone.

Sequential exposure to different abiotic stresses sometimes confers cross-protection

Several studies have reported that the application of a particular abiotic stress condition can enhance the tolerance of plants to a subsequent exposure to a different type of abiotic stress. This phenomenon is called **cross-protection**. It occurs because many stresses result in the accumulation of the same general stress-response proteins and metabolites, for example ROS-scavenging enzymes, molecular chaperones, and osmoprotectants. These persist in plants for some time even after the stress conditions have subsided. The application of a second stress to the same plants that experienced the initial stress may therefore have a decreased effect because the plants are already primed and ready to deal with several different aspects of the new stress condition.

19.4 Stress-Sensing Mechanisms in Plants

Describe the different stress-sensing mechanisms, how they are integrated, and the signaling pathways activated in response to abiotic stress.

Environmental stress disrupts or alters many physiological processes in the plant by affecting protein and RNA stability, ion transport, the coupling of reactions, or other cellular functions. Any of these primary disruptions can signal to the plant that a change in environmental conditions has occurred and that it's time to respond by altering existing pathways or by activating stress-response pathways. At least five different types of stress-sensing mechanisms can be distinguished:

1. *Physical sensing* refers to the mechanical effects of stress on the plant or cell structure—for example, the contraction of the plasma membrane from the cell wall during water-deficit stress.
2. *Biophysical sensing* might involve changes in protein structure or enzymatic activity, such as the inhibition of different enzymes during heat stress.
3. *Metabolic sensing* usually results from the detection of byproducts that accumulate in cells due to the uncoupling of enzymatic or electron transfer reactions, such as the accumulation of ROS during stress caused by too much light.
4. *Biochemical sensing* often involves the presence of specialized proteins that were selected during evolution for their ability to sense a particular stress—for example, Ca^{2+} channels that can sense changes in osmotic stress and alter Ca^{2+} homeostasis.
5. *Epigenetic sensing* refers to modifications of DNA or RNA structure that do not alter genetic sequences, such as the changes in chromatin structure that occur during temperature stress.

Each of these stress-sensing mechanisms can act individually or in combination to activate downstream signal transduction pathways.

Plants use a variety of mechanisms to sense abiotic stress

Abiotic stress, besides affecting various physical parameters of the cell and basic metabolic processes, can also activate receptors and downstream second messenger or phosphorylation signaling to activate physiological responses (Chapter 12). Nonspecific stress activation of receptors can sometimes add to damage, especially in a plant species that is not adapted to a particular form of stress. As noted above, stress-sensing mechanisms can act individually or in combination to activate downstream signal transduction pathways.

Initial stress-sensing mechanisms trigger downstream responses involving multiple signal transduction pathways. As diagrammed in **Figure 19.7**, these pathways involve receptors, calcium ions, protein kinases, protein phosphatases, ROS signaling, transcriptional regulators, and plant hormones. The signals that emerge from these pathways, in turn, activate or suppress various networks that may either allow growth and reproduction to continue under stress conditions, or enable the plant to survive the stress until more favorable conditions return.

Chloroplasts and mitochondria respond to abiotic stress by sending stress signals to the nucleus

We usually think of the nucleus as the master organelle of the cell, controlling the activities of the other organelles by regulating nuclear gene expression. However, retrograde, or reverse, signaling from the chloroplast or the mitochondria to the nucleus also plays a key role in abiotic stress perception. Many abiotic stress conditions affect chloroplasts and mitochondria, either directly or indirectly, and can potentially generate signals that can influence nuclear gene expression and acclimation responses. Exposure to excess light, for example, can cause over-reduction of the electron transport chain and enhanced accumulation of ROS.

During the acclimation to stress caused by excess light, which can damage the photosynthetic apparatus, the levels of light-harvesting complex II (LHCII) decline because of the down-regulation of the *LHCB* gene, which encodes the apoprotein of the LHCII complex (Chapter 7). Because *LHCB* is a nuclear

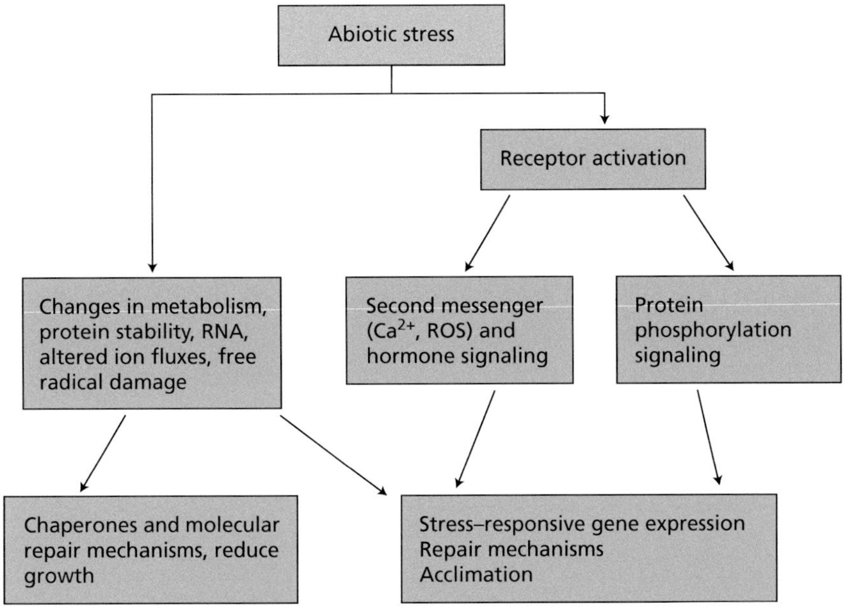

Figure 19.7 Signal transduction and acclimation pathways activated by abiotic stress in plants.

systemic acquired acclimation (SAA) A system whereby exposure of one part of a plant to an abiotic stress generates signals that can initiate acclimation to that stress in other, unexposed parts of the plant.

gene, the chloroplast sends a stress signal to the nucleus, which results in the down-regulation of *LHCB* gene expression.

Plant-wide waves of Ca^{2+} and ROS mediate systemic acquired acclimation

As in systemic acquired resistance (SAR) during biotic stress (Chapter 18), abiotic stress applied to one part of the plant can generate signals that are transported to the rest of the plant, initiating acclimation even in parts of the plant that have not been subjected to the stress. This process is called **systemic acquired acclimation (SAA)**. Rapid SAA responses to different abiotic stress conditions, including heat, cold, salinity, and high light intensity, have been demonstrated to be mediated by systemic plant-wide waves of ROS production and Ca^{2+} signaling, which travel at rates of 1 to 4 cm min^{-1} and are dependent on the presence of specific NADPH oxidases, such as **respiratory burst oxidase homolog D**, and various Ca^{2+} channels, such as glutamate-like receptors (**Figure 19.8**). The rapid movement of abiotic

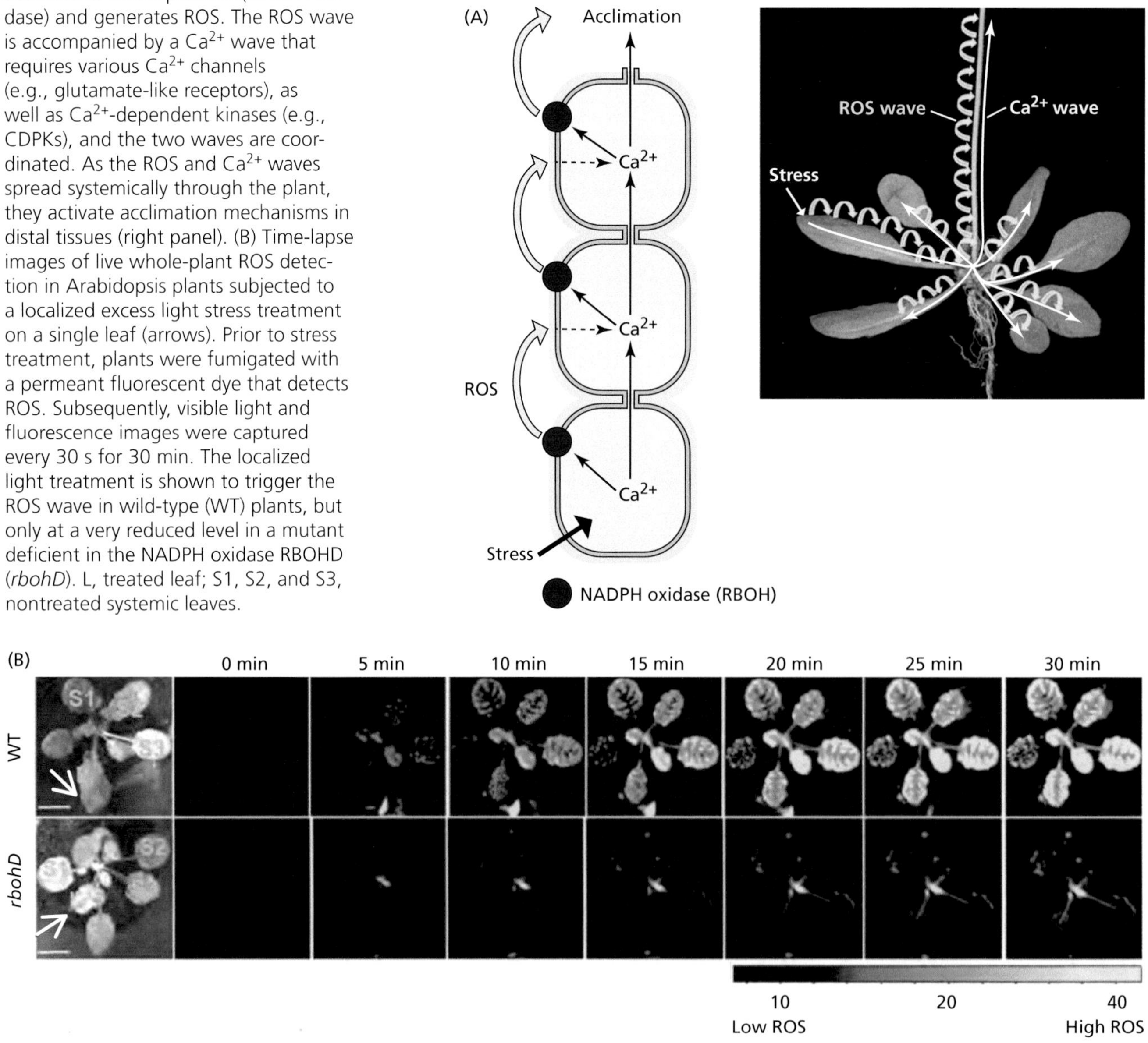

Figure 19.8 Rapid systemic signaling in response to local sensing of abiotic stress. (A) Schematic model of the reactive oxygen species (ROS) and Ca^{2+} waves that are required to mediate rapid systemic signaling in response to abiotic stress. The ROS wave is generated by an active, self-propagating wave of ROS production that starts at the initial tissue subjected to stress and spreads to the entire plant. As shown in the left panel, each cell along the path of the signal activates its RBOH proteins (NADPH oxidase) and generates ROS. The ROS wave is accompanied by a Ca^{2+} wave that requires various Ca^{2+} channels (e.g., glutamate-like receptors), as well as Ca^{2+}-dependent kinases (e.g., CDPKs), and the two waves are coordinated. As the ROS and Ca^{2+} waves spread systemically through the plant, they activate acclimation mechanisms in distal tissues (right panel). (B) Time-lapse images of live whole-plant ROS detection in Arabidopsis plants subjected to a localized excess light stress treatment on a single leaf (arrows). Prior to stress treatment, plants were fumigated with a permeant fluorescent dye that detects ROS. Subsequently, visible light and fluorescence images were captured every 30 s for 30 min. The localized light treatment is shown to trigger the ROS wave in wild-type (WT) plants, but only at a very reduced level in a mutant deficient in the NADPH oxidase RBOHD (*rbohD*). L, treated leaf; S1, S2, and S3, nontreated systemic leaves.

stress–triggered ROS and Ca^{2+} signals through the plant suggests that many abiotic stress responses occur much more quickly than previously thought. In fact, changes in gene expression and metabolism can occur within minutes of abiotic stress application.

Hormonal interactions regulate a wide range of abiotic stress responses

Plant hormones mediate a wide range of acclimation responses and are essential for the ability of plants to adapt to abiotic stresses. ABA biosynthesis is among the most rapid responses of plants to abiotic stress. As shown in **Figure 19.9**, ABA concentrations in leaves can increase up to 20-fold under water-deficit conditions—the most dramatic change in concentration reported for any hormone in response to an environmental signal. Redistribution or biosynthesis of ABA is very effective in causing stomatal closure, and ABA accumulation in stressed leaves plays an important role in the reduction of water loss by transpiration under water-stress conditions (Chapter 12; Figure 19.9). Increases in humidity reduce ABA concentrations by increasing ABA breakdown, thereby permitting stomata to reopen. ABA-biosynthesis or -response mutants are unable to close their stomata under water-deficit conditions and wilt rapidly.

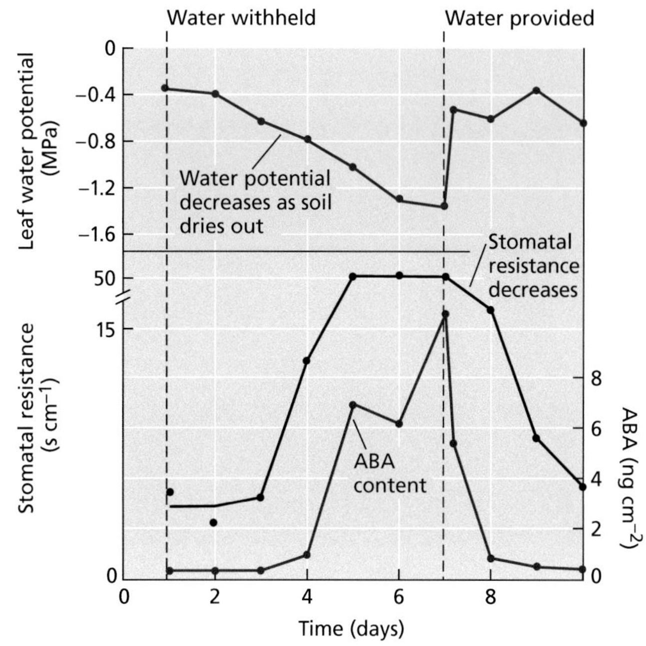

Figure 19.9 Changes in leaf water potential, stomatal resistance (the inverse of stomatal conductance), and ABA content in maize in response to water-deficit stress. As the soil dries out, the water potential of the leaf decreases, and the ABA content and stomatal resistance increase. Rewatering reverses the process.

Within the limits set by the plant's genetic potential, a shoot tends to grow until water uptake by the roots becomes limiting to further growth; conversely, roots tend to grow until their demand for photosynthate from the shoot exceeds the supply. This functional balance is shifted if the water supply decreases. When water to the shoot becomes limiting, leaf expansion is reduced before photosynthetic activity is affected. Inhibition of leaf expansion reduces the consumption of carbon and energy, and a greater proportion of the plant's assimilates can be allocated to the root system, where they can support further root growth. This root growth is sensitive to the water status of the soil microenvironment; the root apices in dry soil lose turgor, while roots in the soil zones that remain moist continue to grow.

ABA plays an important role in regulating the root:shoot ratio during water stress. As shown in **Figure 19.10**, under water-stress conditions the root-to-shoot biomass ratio increases, because the roots grow at the expense of the leaves. However, ABA-deficient mutants are unable to change their root:shoot ratio in response to water stress. Thus, ABA is required for the change in the root:shoot ratio to occur. ABA also interacts with the hormone auxin to regulate root branching under water-deficit stress.

Another plant hormone that plays a key role in acclimation to various abiotic stresses is cytokinin. Cytokinin and ABA have antagonistic effects on stomatal opening, transpiration, and photosynthesis. Water deficit results in decreased cytokinin levels and increased ABA levels. Although ABA is normally required for stomatal closure, preventing excessive water loss, water-deficit stress conditions can also inhibit photosynthesis and cause premature leaf senescence. Cytokinins appear to be able to ameliorate the effects of water deficit. As shown in **Figure 19.11**, transgenic plants that overexpress the gene encoding the enzyme isopentenyl transferase, the enzyme that catalyzes the rate-limiting step in cytokinin synthesis (Chapter 12), exhibit remarkable water-deficit tolerance compared with wild-type plants. The experiment shows that cytokinins are able to protect biochemical processes associated with photosynthesis and delay senescence during water-deficit stress.

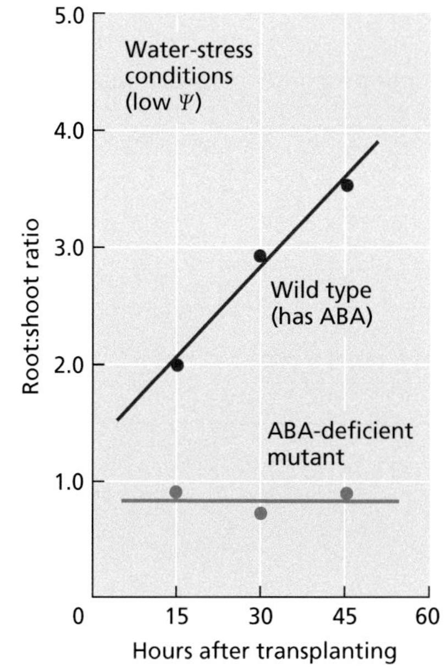

Figure 19.10 The root:shoot ratio of maize seedlings grown under water-stress conditions (i.e., at low Ψ). The ratio of root growth to shoot growth is much higher in the wild type, which synthesizes its own endogenous ABA, than in the ABA-deficient mutant. Note that the root:shoot ratio of the wild type was already higher than that of the mutant at the beginning of the experiment.

Figure 19.11 Effects of water deficit on wild-type and transgenic tobacco plants expressing isopentenyl transferase (a key enzyme in the production of cytokinin) under the control of a stress-induced promoter. Shown are wild-type (A) and transgenic (B) plants after 15 days of water deficit followed by 7 days of rewatering. The stress-inducible promoter in the transgenic plants ensures that cytokinins are only produced during water stress.

respiratory burst oxidase homolog D (RBOHD) An enzyme that generates superoxide using NADPH as an electron donor.

epigenome Heritable chemical modifications to DNA and chromatin, including DNA methylation, histone methylation, and acetylation.

In addition to ABA and cytokinin, other hormones play important roles in the response of plants to abiotic stress, including gibberellic acid, auxin, salicylic acid, ethylene, jasmonic acid, strigolactones, and brassinosteroids (Chapter 12).

Epigenetic mechanisms are also involved in sensing abiotic stress

Thus far we have discussed responses to abiotic stress in terms of signaling cascades and altered gene expression—acclimation processes that can be reversed when more favorable conditions arise. Recently, attention has focused on epigenetic changes, which can potentially provide long-term adaptations to abiotic stress. Because some chromatin modifications are mitotically and meiotically heritable, stress-induced epigenetic changes might even have evolutionary implications. Chromatin immunoprecipitation of DNA cross-linked to modified histones, coupled with modern sequencing technologies, has opened the door to genome-wide analyses of changes in the **epigenome**. Stable or heritable DNA methylation and histone modifications have now been linked with specific abiotic stresses (**Figure 19.12**).

19.5 Developmental and Physiological Mechanisms That Protect Plants against Abiotic Stress

Define the major physiological and developmental changes in response to abiotic stress.

Terrestrial plants first colonized land more than 500 million years ago, ample time to evolve mechanisms to cope with various types of abiotic stresses. These mechanisms include the abilities to accumulate protective metabolites and proteins and to regulate growth, morphogenesis, photosynthesis, membrane transport, stomatal aperture, and resource allocation. The effects of these and other changes are to attain cellular homeostasis so that the plant's life cycle can be completed under the new environmental regime.

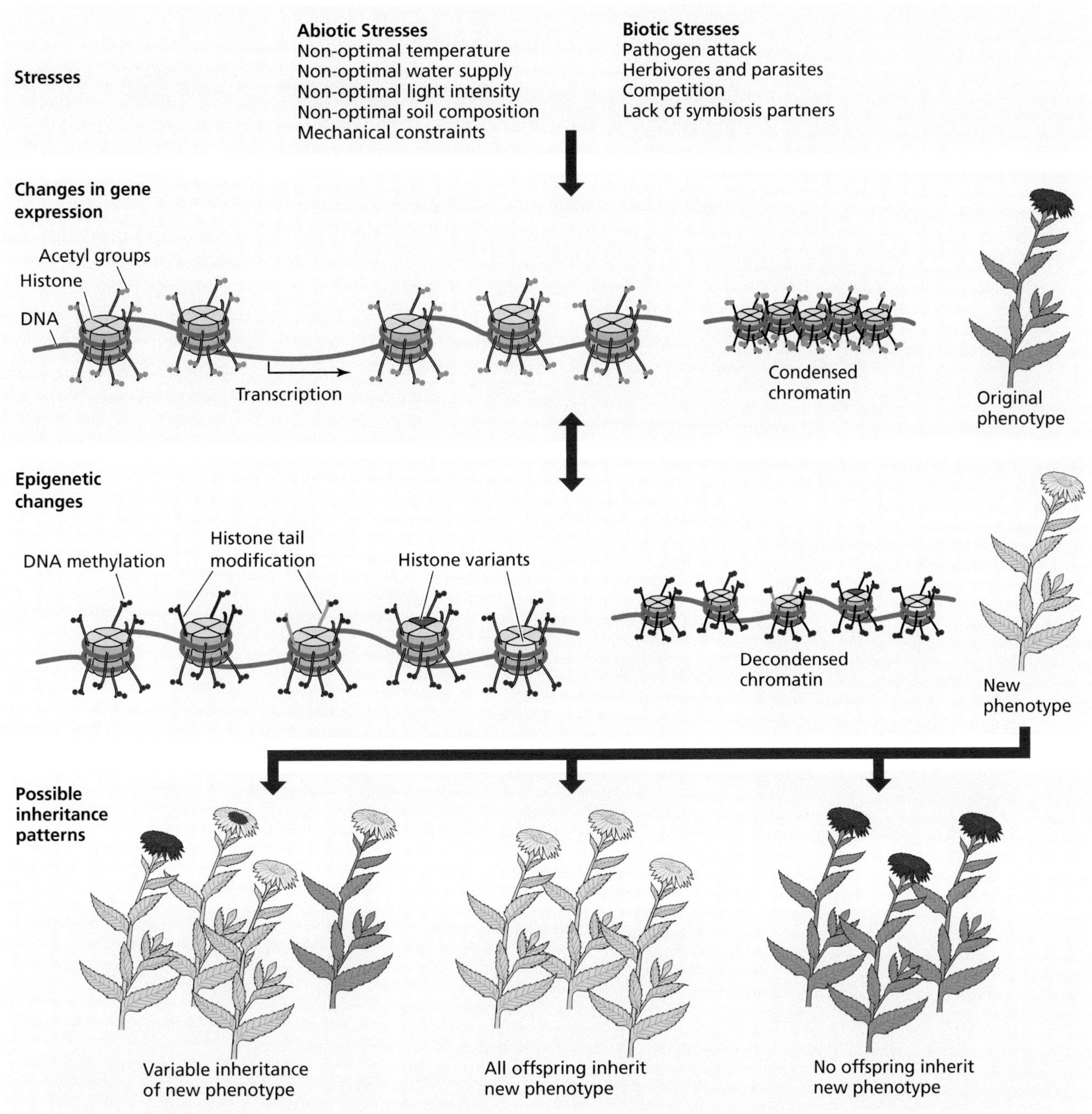

Figure 19.12 Stress-induced changes in gene expression. Stress-induced changes can be mediated by protein, lipid, or nucleic acid modification, second messengers, or hormones (e.g., abscisic acid, salicylic acid, jasmonic acid, and ethylene). Transcriptional changes or stress factors can affect chromatin via DNA methylation, histone tail modifications, histone variant replacements, or nucleosome loss and chromatin decondensation. These changes are reversible and can modify plant metabolism or morphology under stress conditions. Usually, the new phenotypes are not transmitted to progeny; however, chromatin-associated changes have the potential to be heritable, which could result in maintenance of the new features and epigenetic diversity in the next generation.

Plants can activate developmental programs that alter their phenotype

In response to abiotic stress, plants can activate developmental programs that alter their phenotype, a phenomenon known as **phenotypic plasticity**. Phenotypic plasticity can result in adaptive anatomical changes that enable plants to avoid some of the harmful effects of abiotic stress.

phenotypic plasticity Physiological or developmental responses of a plant to its environment that do not involve genetic changes.

An important example of phenotypic plasticity is the ability to alter leaf shape and the root-to-shoot ratio. As biological solar collectors, leaves must be exposed to sunlight and air, which makes them vulnerable to environmental extremes. Plants have thus evolved the ability to modify leaf morphology and leaf production in ways that enable them to avoid or mitigate the effects of abiotic extremes.

LEAF AREA Large, flat leaves provide optimal surfaces for the production of photosynthates. However, they can be detrimental to crop growth and survival under stress conditions because they have a large surface area for evaporation of water that can lead to rapid depletion of soil water or damage from excessive absorption of solar energy. Plants can reduce their overall leaf area by reducing leaf cell division and expansion, altering leaf shapes (**Figure 19.13A**), and initiating the senescence and abscission of leaves (**Figure 19.13B**). This phenomenon can lead to certain types of heterophylly (variation in leaf shape), as in aquatic plant.

Figure 19.13 Morphological changes in response to abiotic stress. (A) Altered leaf shape can occur in response to environmental changes. The oak (*Quercus* sp.) leaf on the left is from the outside of a tree canopy, where temperatures are higher than they are inside the canopy. The leaf on the right is from the inside of the canopy. The deeper sinuses between the lobes of the leaf on the left result in a thinner boundary layer, which allows for better evaporative cooling. (B) The leaves of young cotton (*Gossypium hirsutum*) plants abscise in response to water-deficit stress. The plants at left (control) were watered throughout the experiment; those in the middle and at right were subjected to moderate water-deficit stress and severe water-deficit stress, respectively, before being watered again. Only a tuft of leaves at the top of the stem is left on the severely stressed plants. The pressure/water potential units of −5 bar, −12 bar, and −24 bar marked on the cards in the photo are equivalent to −0.5, −1.2, and −2.4 MPa, respectively, in SI (International Symbol) pressure/water potential units (Chapter 2). (C) Leaf movements in soybean in response to osmotic stress. Leaflet orientation of field-grown soybean (*Glycine max*) plants in the well-watered, unstressed position; during mild water-deficit stress; and during severe water-deficit stress. The large leaf movements induced by mild stress are quite different from wilting, which occurs during severe stress. Note that during mild stress the terminal leaflet has been raised, whereas the two lateral leaflets have been lowered; each is almost vertical.

LEAF ORIENTATION As described in Chapter 9, the leaves of some plants may orient themselves parallel or perpendicular to the sun's rays to reduce damage. Wilting also changes the angle of the leaf, and leaf rolling minimizes the leaf area exposed to the sun (**Figure 19.13C**).

TRICHOMES Many leaves and stems have hairlike epidermal cells known as trichomes or hairs (Chapters 1 and 15). Trichomes can either be ephemeral or persist throughout the life of the organ. Some persisting trichomes remain alive, while others undergo programmed cell death, leaving only their cell walls. Densely packed trichomes on a leaf surface keep leaves cooler by reflecting radiation and reducing evaporation by creating a thicker unstirred surface layer. Leaves of some plants have a silvery white appearance because the densely packed trichomes reflect a large amount of light. However, pubescent leaves are a disadvantage in the cooler spring months because the trichomes also reflect the visible light needed for photosynthesis.

CUTICLE The cuticle is a multilayered structure of waxes and related hydrocarbons deposited on the outer cell walls of the leaf epidermis. The cuticle, like trichomes, can reflect light, thereby reducing heat load. The cuticle appears to also restrict the diffusion of water and gases, as well as the entrance of pathogens. A developmental response to water deficit in some plants is the production of a thick, waxy cuticle, which decreases apoplastic evaporation.

ROOT-TO-SHOOT RATIO The root-to-shoot ratio is another important example of phenotypic plasticity. The root:shoot ratio appears to be governed by a functional balance between water uptake by the root and photosynthesis by the shoot. Within the limits set by the plant's genetic potential, a shoot tends to grow until water uptake by the roots becomes limiting to further growth; conversely, roots tend to grow until their demand for photosynthates from the shoot exceeds the supply. This functional balance is shifted if the water supply decreases.

Plants adjust osmotically to drying soils by accumulating solutes

Water can only move through the soil–plant–atmosphere continuum if water potential decreases along that path (Chapters 2 and 3). Recall from Chapter 2 that $\Psi = \Psi_s + \Psi_p$, where Ψ = water potential, Ψ_s = osmotic potential, and Ψ_p = pressure potential (turgor). When the water potential of the rhizosphere (the microenvironment surrounding the root) decreases because of water deficit or salinity, plants can continue to absorb water only as long as Ψ is lower (more negative) in the plant than it is in the soil water. **Osmotic adjustment** is the capacity of plant cells to accumulate solutes and use them to lower Ψ during periods of osmotic stress. This adjustment involves a net increase in solute content per cell that is independent of the volume changes that result from loss of water. The decrease in Ψ_s is typically limited to about 0.2 to 0.8 MPa, except in plants adapted to extremely dry conditions.

There are two main ways by which osmotic adjustment can take place, one involving the vacuole and the other the cytosol. A plant may take up ions from the soil, or transport ions from other plant organs to the root, so that the solute concentration of the root cells increases. For example, increased uptake and accumulation of K^+ will lead to decreases in Ψ_s, because of the effect of the K^+ on the osmotic pressure within the cell. This response is common in plants growing in saline soils, where ions such as K^+ and Ca^{2+} are readily available to the plant. The uptake of K^+ and other cations must be electrically balanced either by the uptake of inorganic anions, such as Cl^-, or by the production and vacuolar accumulation of organic acids such as malate or citrate.

osmotic adjustment The ability of the cell to accumulate compatible solutes and lower water potential during periods of osmotic stress.

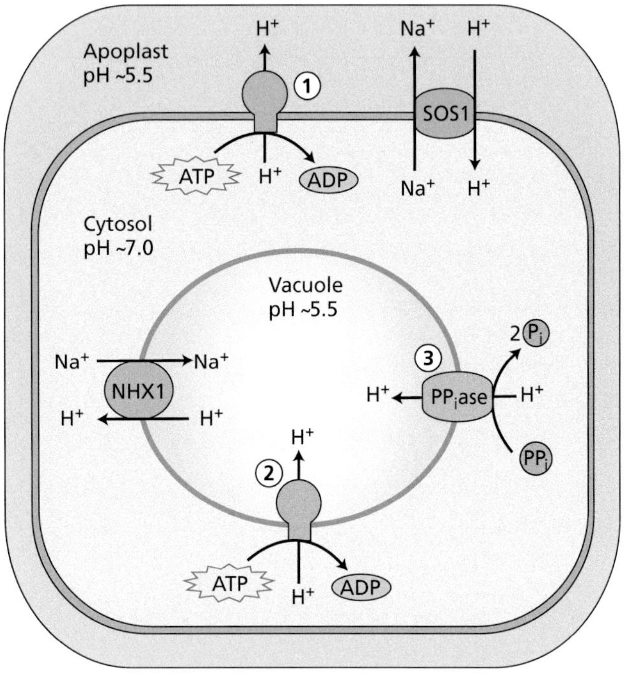

Figure 19.14 Primary and secondary active transport mechanisms function in salt stress responses. The plasma membrane–localized H^+-pumping ATPase (P-type ATPase) (1) and the tonoplast-localized H^+-pumping ATPase (V-type ATPase) (2), and pyrophosphatase (PPiase) (3) are primary active transport systems that energize the plasma membrane and the tonoplast, respectively. By coupling the energy released by hydrolysis of ATP or pyrophosphate, these proton pumps are able to transport H^+ across the plasma membrane and the tonoplast against an electrochemical gradient. The H^+–Na^+ antiporters SOS1 and NHX1 are secondary active transport systems that couple the transport of Na^+ against its electrochemical gradient with that of H^+ down its electrochemical gradient. SOS1 transports Na^+ out of the cell, whereas NHX1 transports Na^+ into the vacuole (Chapter 6).

There is a potential problem, however, when ions are used to decrease Ψ_s. Some ions, such as Na^+ or Cl^-, are essential to plant growth in low concentrations, but at higher concentrations can have a detrimental effect on cellular metabolism. Other ions, such as K^+, are required in larger quantities, but at high concentrations can still have a detrimental effect on the plant, mostly through disruption of plasma membranes or denaturation of proteins. The accumulation of ions during osmotic adjustment is predominantly restricted to the vacuoles, where the ions are kept out of contact with cytosolic enzymes or organelles. For example, many halophytes (plants adapted to salinity) use vacuolar compartmentation of Na^+ and Cl^- to facilitate osmotic adjustment (**Figure 19.14**) that sustains or enhances growth in saline environments.

When the ion concentration in the vacuole increases, other solutes must accumulate in the cytosol to maintain water potential equilibrium between the two compartments. These solutes are called compatible solutes (or compatible osmolytes). **Compatible solutes** are organic compounds that are osmotically active in the cell, but do not destabilize the membranes or interfere with enzyme function at high concentrations, as ions do. Plant cells can tolerate high concentrations of these compounds without detrimental effects on metabolism. Common compatible solutes include amino acids such as proline, sugar alcohols such as **sorbitol**, and quaternary ammonium compounds such as **glycine betaine** (**Figure 19.15**). Some of these solutes, such as proline, also seem to have an osmoprotectant function, whereby they protect plants from toxic byproducts produced during periods of water shortage and provide a source of carbon and nitrogen to the cell when conditions return to normal. Each plant family tends to use one or two compatible solutes in preference to others. Because the synthesis of compatible solutes is an active metabolic process, energy is required. The amount of carbon used for the synthesis of these organic solutes can be rather large, and for this reason the synthesis of these compounds tends to reduce crop yield.

Submerged organs develop aerenchyma tissue in response to hypoxia

How do plants cope with too much water? Flooded soil is depleted of oxygen and is therefore anaerobic. Water-saturated (flooded) soils or submergence of the plant in water can lead to **hypoxia**, low levels of oxygen, and produce anaerobic stress. In most wetland plants, such as rice, and in many plants that acclimate well to wet conditions, the stem and roots develop longitudinally interconnected, gas-filled channels that provide a low-resistance pathway for the movement of O_2 and other

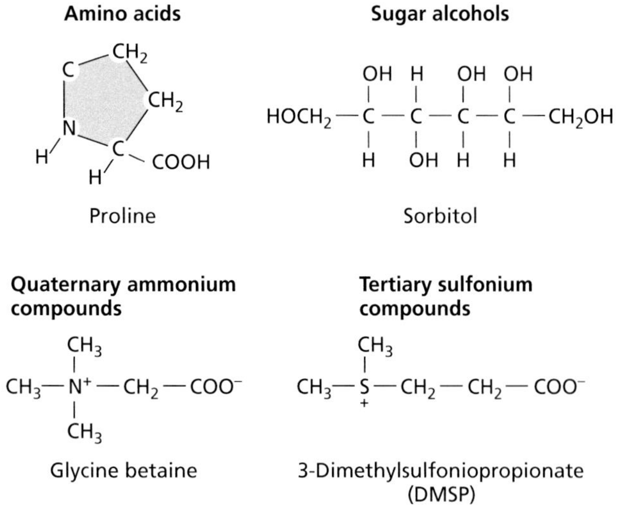

Figure 19.15 Compatible solutes. Four groups of molecules frequently serve as compatible solutes: amino acids, sugar alcohols, quaternary ammonium compounds, and tertiary sulfonium compounds. Note that these compounds are small and have no net charge.

gases. The gases (air) enter through stomata, or through lenticels (porous regions of the periderm that allow gas exchange) on woody stems and roots, and travel by molecular diffusion or by convection driven by small pressure gradients. In many plants adapted to wetland growth, root cells are separated by prominent, gas-filled spaces that form a tissue called **aerenchyma**. These cells develop in the roots of wetland plants independently of environmental stimuli. In some non-wetland monocots and eudicots, O_2 deficiency induces the formation of aerenchyma in the stem base and newly developing roots.

An example of induced aerenchyma occurs in maize (**Figure 19.16**). Hypoxia stimulates the activity of **ACC synthase** and **ACC oxidase** in the root tips of maize and causes ACC and ethylene to be produced faster. Ethylene triggers programmed cell death and disintegration of cells in the root cortex. The spaces formerly occupied by these cells provide gas-filled voids that facilitate movement of O_2. Ethylene-triggered cell death is highly selective; only some cells have the potential to initiate the developmental program that creates the aerenchyma.

When aerenchyma formation is induced, a rise in cytosolic Ca^{2+} concentration is thought to be part of the ethylene signal transduction pathway leading to cell death. Signals that elevate cytosolic Ca^{2+} concentration can promote cell death in the absence of hypoxia. Conversely, signals that lower cytosolic Ca^{2+} concentration block cell death in hypoxic roots that would normally form aerenchyma.

In roots of rice and other typical wetland plants, structural barriers composed of suberized and lignified cell walls prevent O_2 diffusion outward to the soil. The O_2 thus retained supplies the apical meristem and allows growth to extend 50 cm or more into anaerobic soil. In contrast, roots of non-wetland species, such as maize, leak O_2. Internal O_2 becomes insufficient for aerobic respiration in the root apex of these non-wetland plants, and this lack of O_2 severely limits the depth to which such roots can extend into anaerobic soil.

compatible solutes Organic compounds that are accumulated in the cytosol during osmotic adjustment. Compatible solutes do not inhibit cytosolic enzymes, unlike high concentrations of ions. Examples of compatible solutes include proline, sorbitol, mannitol, and glycine betaine.

sorbitol A sugar alcohol formed by reduction of aldehyde glucose.

glycine betaine *N,N,N*-trimethylglycine, which functions in drought stress protection, and was originally identified in sugar beet (*Beta vulgaris*).

hypoxia Low levels of oxygen in tissue.

aerenchyma An anatomical feature of roots found in hypoxic conditions, showing large, gas-filled intercellular spaces in the root cortex.

ACC synthase The enzyme that catalyzes the synthesis of 1-aminocyclopropane-1-carboxylic acid (ACC) from *S*-adenosylmethionine.

ACC oxidase The enzyme that catalyzes the conversion of 1-aminocyclopropane-1-carboxylic acid (ACC) to ethylene, the last step in ethylene biosynthesis.

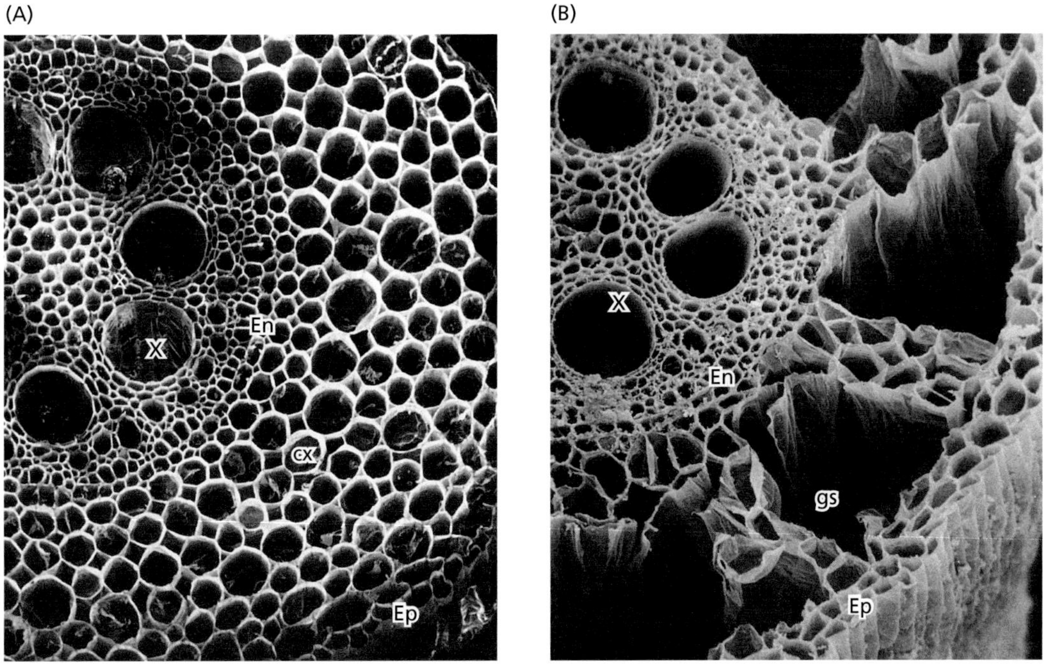

Figure 19.16 Scanning electron micrographs of transverse sections through roots of maize, showing changes in structure with O_2 supply. (A) Control root, supplied with air, with intact cortical cells (cx). (B) O_2-deficient root growing in a nonaerated nutrient solution. Note the prominent gas-filled spaces (gs) in the cortex formed by degeneration of cells. The stele (all cells interior to the endodermis, En) and the epidermis (Ep) remain intact. X, xylem. (150× magnification)

ROS scavenging Detoxification of reactive oxygen species via interactions with proteins and electron acceptor molecules.

antioxidative enzymes Proteins that detoxify reactive oxygen species.

superoxide dismutase An enzyme that converts superoxide radicals to hydrogen peroxide.

ascorbate peroxidase An enzyme that converts peroxide and ascorbate to dehydroascorbate and water.

catalase An enzyme that breaks hydrogen peroxide down into water. When it is abundant in peroxisomes, it may form crystalline arrays.

Antioxidants and ROS-scavenging pathways protect cells from oxidative stress

Reactive oxygen species accumulate in cells during many different types of environmental stresses and are detoxified by specialized enzymes and antioxidants, a process referred to as **ROS scavenging**. Biological antioxidants are small organic compounds or small peptides that can accept electrons from ROS, such as superoxide or H_2O_2, and neutralize them. Common antioxidants in plants (**Figure 19.17**) include the water-soluble ascorbate (vitamin C), the reduced tripeptide glutathione (GSH in reduced form, GSSG in oxidized form), the lipid-soluble α-tocopherol (vitamin E), and β-carotene (vitamin A). Ascorbate and glutathione function in a complementary manner in the cytoplasm, while the two vitamins function similarly in the membranes. To maintain an adequate supply of these compounds in the reduced state, cells rely on various reductases, such as glutathione reductase, dehydroascorbate reductase, and monodehydroascorbate reductase, which use the reducing power of NADPH or NADH produced by respiration or photosynthesis.

Some ROS may react spontaneously with cellular antioxidants, and some are unstable and decay before they cause cellular damage. However, plants have evolved several different **antioxidative enzymes** that increase the efficiency of these processes. For example, **superoxide dismutase** is an enzyme that simultaneously oxidizes and reduces the superoxide anion to produce hydrogen peroxide and oxygen according to the reaction: $2\ O_2^- + 2\ H^+ \rightarrow O_2 + H_2O_2$. Variants of superoxide dismutase are found in chloroplasts, peroxisomes, mitochondria, the cytosol, and the apoplast. Different forms of **ascorbate peroxidase** are present in the same cellular compartments as superoxide dismutase. Ascorbate peroxidase catalyzes the reduction of hydrogen peroxide using ascorbic acid as a reducing agent in the following reaction: 2 L-ascorbate + H_2O_2 + 2 H^+ $\rightarrow$ 2 monodehydroascorbate + 2 H_2O. **Catalase** catalyzes the detoxification of hydrogen peroxide in peroxisomes by converting it to water and oxygen according to this reaction: $2\ H_2O_2 \rightarrow 2\ H_2O + O_2$.

Molecular chaperones and molecular shields protect proteins and membranes during abiotic stress

Protein structure is sensitive to disruption by changes in temperature, pH, or ionic strength associated with different types of abiotic stresses. Plants have several mechanisms to limit or avoid such problems, including osmotic adjustment

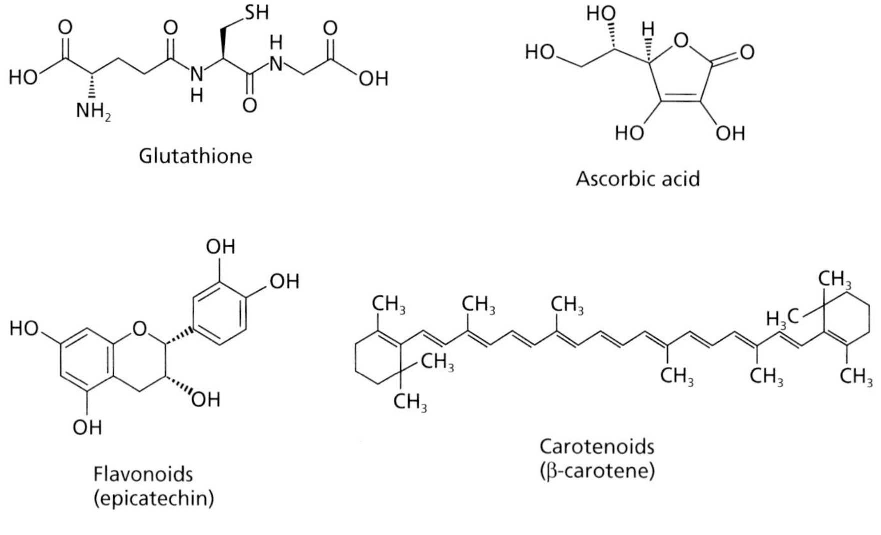

Figure 19.17 Antioxidant ROS scavengers.

for maintenance of hydration, proton pumps to maintain pH homeostasis, and **molecular chaperone proteins** that physically interact with other proteins to facilitate protein folding, reduce misfolding, stabilize tertiary structure, and prevent aggregation or disaggregation. A unique set of chaperones, called **heat shock proteins** (**HSPs**), are synthesized in response to a variety of environmental stresses. Cells that synthesize HSPs in response to heat stress show improved thermal tolerance and can tolerate subsequent exposures to higher temperatures that would otherwise be lethal. HSPs are induced by widely different environmental conditions, including water deficit, wounding, low temperature, and salinity. In this way, cells that have previously experienced one stress may gain cross-protection against another stress.

HSPs were discovered in the fruit fly (*Drosophila melanogaster*) and appear to be ubiquitous in plants, animals, fungi, and microorganisms. The heat shock response appears to be mediated by one or more signal transduction pathways, one of which involves a specific set of transcription factors, called **heat shock factors**, that regulate the transcription of HSP mRNAs. There are several different classes of HSPs, each with different mechanisms to correctly fold proteins or refold misfolded proteins into their active forms, as shown in **Figure 19.18**. These different classes of HSPs function together as a molecular network to alleviate the impact of heat stress.

Numerous other proteins have been identified that act in a similar fashion to stabilize proteins and membranes during dehydration, temperature extremes, and ion imbalance. These include the Late Embryogenesis Abundant (LEA) protein family. LEA proteins accumulate in response to dehydration during the latter stages of seed maturation (Chapter 17). **Dehydrins** (DHNs) also accumulate in plant tissues in response to a variety of abiotic stresses, including salinity, dehydration, cold, and freezing stress. Dehydrins, like LEA proteins, are highly hydrophilic and are intrinsically disordered proteins. Their ability to serve as molecular shields and as cryoprotectants has been attributed to their flexibility and minimal secondary structure. Both LEAs and DHNs are often induced by abscisic acid.

Plants can alter their membrane structure in response to temperature and other abiotic stresses

As temperatures decrease, membranes may go through a phase transition from a flexible liquid-crystalline structure to a solid gel structure. The phase transition temperature varies depending on the lipid composition of the membranes. In general, saturated fatty acids (lipids with no double bonds) solidify at lower temperatures than lipids containing polyunsaturated fatty acids. Polyunsaturated fatty acids have kinks in their hydrocarbon chains and do not pack as closely as saturated fatty acids because of their double bonds. Chilling-resistant plants tend to have membranes with more unsaturated fatty acids that increase their fluidity, whereas chilling-sensitive plants have a high percentage of saturated fatty acid chains that tend to solidify at low temperatures (**Table 19.2**).

Prolonged exposure to extreme temperatures may result in an altered composition of membrane lipids, a form of acclimation. Certain transmembrane enzymes can alter lipid saturation, by introducing one or more double bonds into fatty acids. For example, during acclimation to cold temperatures, the activities of **desaturase enzymes** increase and the proportion of unsaturated lipids rises. This modification lowers the temperature at which the membrane lipids begin a gradual phase change from fluid to semicrystalline form and allows membranes to remain fluid at lower temperatures, thus protecting the plant against damage from chilling. Conversely, a greater degree of saturation of the fatty acids in membrane lipids makes the membranes less fluid. Certain Arabidopsis mutants have reduced activity of omega-3 fatty acid desaturases. These mutants show an increased tolerance of photosynthesis to high temperatures, presumably because the degree of saturation of chloroplast lipids is increased.

molecular chaperone proteins Proteins that maintain or restore the active three-dimensional structures of other macromolecules.

heat shock proteins (HSPs) A specific set of proteins that are induced by a rapid rise in temperature, and by other factors that lead to protein denaturation. Most act as molecular chaperones.

heat shock factors Transcription factors that regulate expression of heat shock proteins.

dehydrins Hydrophilic plant proteins that accumulate in response to drought stress and cold temperatures.

desaturase enzymes Enzymes that remove hydrogens in a carbon chain to create a double bond between carbons.

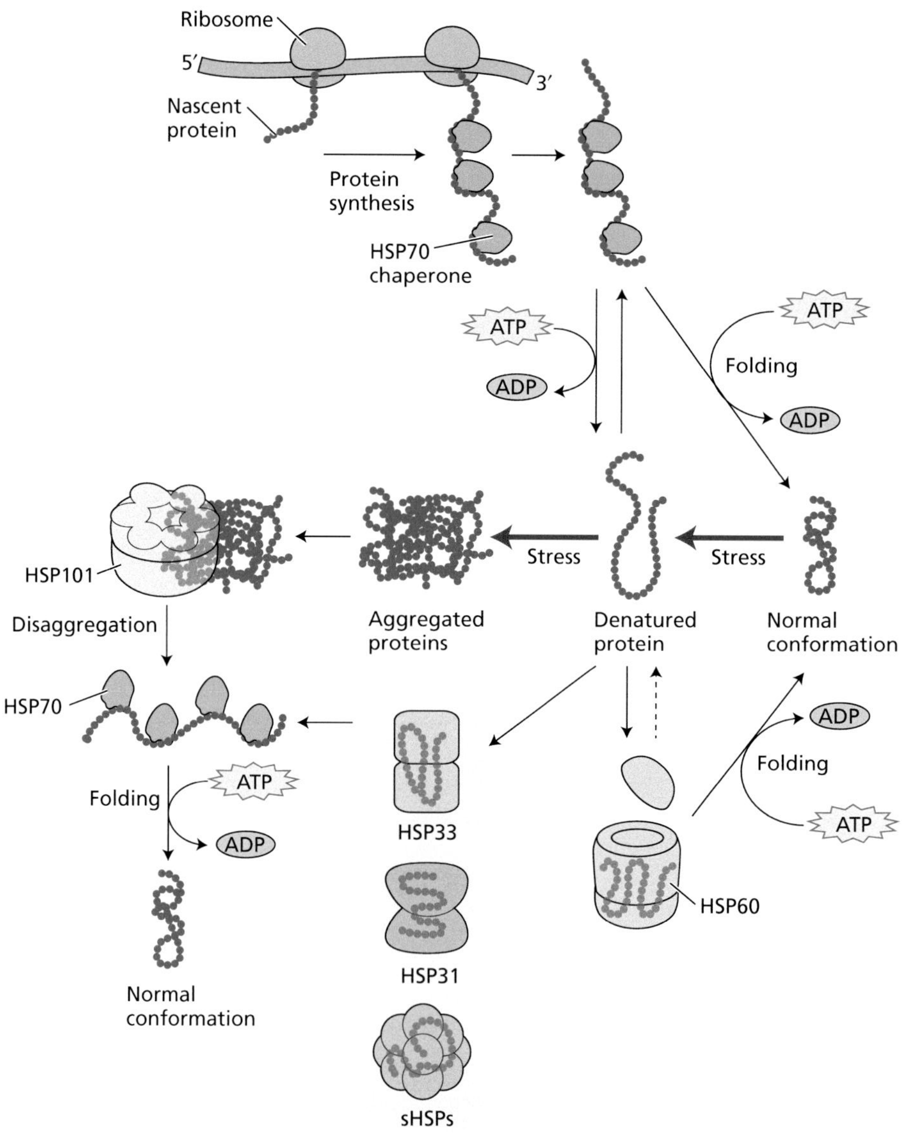

Figure 19.18 Molecular chaperone network in cells. Nascent proteins requiring the assistance of molecular chaperones to reach a proper conformation are associated with HSP70 chaperones (top). Native proteins that undergo denaturation during stress (right) associate with HSP70s that bind and release misfolded proteins (top right) and HSP60s that produce huge barrel-like complexes that are used as chambers for protein folding (bottom right). If protein aggregates are formed (left middle), they are disaggregated by HSP101s that mediate the disaggregation of protein aggregates and HSP70s (left). Additional stress-related chaperones such as HSP31, HSP33, and sHSPs can also associate with denatured proteins during stress.

Table 19.2 Fatty Acid Composition of Mitochondria Isolated from Chilling-Resistant and Chilling-Sensitive Species

	Percent weight of total fatty acid content					
	Chilling-resistant species			Chilling-sensitive species		
Major fatty acids[a]	Cauliflower bud	Turnip root	Pea shoot	Bean shoot	Sweet potato	Maize shoot
Palmitic (16:0)	21.3	19.0	17.8	24.0	24.9	28.3
Stearic (18:0)	1.9	1.1	2.9	2.2	2.6	1.6
Oleic (18:1)	7.0	12.2	3.1	3.8	0.6	4.6
Linoleic (18:2)	16.1	20.6	61.9	43.6	50.8	54.6
Linolenic (18:3)	49.4	44.9	13.2	24.3	10.6	6.8
Ratio of unsaturated to saturated fatty acid	3.2	3.9	3.8	2.9	1.7	2.1

Source: After Lyons et al. 1964

[a] Shown in parentheses are the number of carbon atoms in the fatty acid chain and the number of double bonds.

Exclusion and internal tolerance mechanisms allow plants to cope with toxic ions

Some metal ions, such as zinc and copper, are cofactors for some enzymes, and are required in small amounts. However, in excess, metal and metaloid ions can be toxic to plants. Two basic mechanisms are employed by plants to tolerate the presence of high concentrations of toxic ions in the environment, including aluminum, arsenic, cadmium, copper, nickel, selenium, and zinc ions: exclusion and internal tolerance. *Exclusion* refers to the ability to block the uptake of toxic ions into the cell, thereby preventing the concentrations of these ions from reaching a toxic threshold level. Exclusion mechanisms often involve activation of plasma membrane ion transport proteins. **Internal tolerance** typically involves biochemical adaptations that enable the plant to tolerate, compartmentalize, or chelate elevated concentrations of potentially toxic ions. Typical detoxifying chelators are organic acids such as citric acid and the specialized small peptides known as *phytochelatins* (**Figure 19.19**).

Chelation is the binding of an ion with at least two ligating atoms within a chelating molecule. Chelating molecules can have different atoms available for ligation, such as sulfur (S), nitrogen (N), or oxygen (O), and these different atoms have different affinities for the ions they chelate. The chelating molecule renders the ion less chemically active by wrapping itself around the ion it binds to form a complex, thereby reducing the potential toxicity of the ion. The complex is then usually translocated to other parts of the plant, or stored away from the cytoplasm (typically in the vacuole).

Phytochelatins are low-molecular-weight thiols consisting of the amino acids glutamate, cysteine, and glycine, with the general form of $(\gamma\text{-Glu-Cys})_n\text{Gly}$ (Figure 19.19B). Phytochelatins are synthesized by the enzyme phytochelatin synthase. The thiol groups act as ligands for ions of trace elements such as cadmium and arsenic. Once formed, the phytochelatin–metal complex is transported into the vacuole for storage.

Synthesis of phytochelatins has been shown to be necessary for resistance to cadmium and arsenic. In addition to chelation, active transport of metal ions into the vacuole and out of the cell also contributes to internal tolerance to metals.

Hyperaccumulation is an extreme example of internal tolerance to toxic ions. Hyperaccumulating plants can tolerate foliar concentrations of various trace elements, such as arsenic, cadmium, nickel, zinc, and selenium, of up to 1% of their shoot dry weight (10 mg per gram dry weight). Hyperaccumulation is a

internal tolerance Tolerance mechanisms that function in the symplast (as opposed to exclusion mechanisms).

hyperaccumulation Accumulation of metals in a healthy plant to levels that are much higher than those found in the soil and that are generally toxic to non-accumulators.

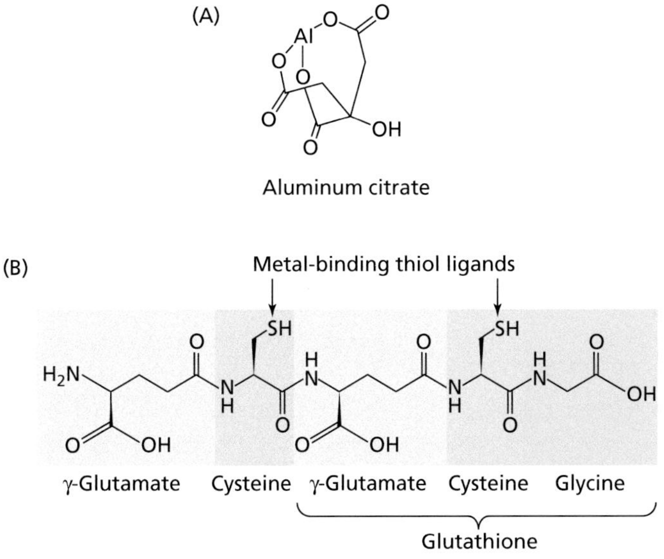

Figure 19.19 Molecular structure of metal chelators. (A) Citrate, shown coordinated with aluminum. (B) Phytochelatin, which uses the sulfur in cysteine to bind ions such as cadmium, zinc, and arsenic.

supercool The condition in which cellular water remains liquid because of its solute content, even at temperatures several degrees below its theoretical freezing point.

antifreeze proteins Proteins that lower the temperature at which aqueous solutions freeze. When induced by cold temperatures, these plant proteins bind to the surfaces of ice crystals to prevent or slow further crystal growth, thereby limiting or preventing freeze damage. Some antifreeze proteins are similar to pathogenesis-related proteins.

relatively rare plant adaptation to potentially toxic ions that requires heritable genetic changes that enhance the expression of the ion transporters involved in the uptake and vacuolar compartmentation of these ions.

Plants use cryoprotectant molecules and antifreeze proteins to prevent ice crystal formation

During freezing, the protoplast, including the vacuole, may **supercool**—that is, the cellular water may remain liquid even at temperatures several degrees below its theoretical freezing point. Supercooling is common in many species of the hardwood forests of southeastern Canada and the eastern United States. Cells can supercool only to about −40°C, the temperature at which ice forms spontaneously. Spontaneous ice formation sets the *low-temperature limit* at which many alpine and subarctic species that undergo deep supercooling can survive. It may also explain why the altitude of the timberline in mountain ranges is at or near the −40°C minimum isotherm.

Several specialized plant proteins, termed **antifreeze proteins**, limit the growth of ice crystals through a mechanism independent of lowering of the freezing point of water. Synthesis of these antifreeze proteins is induced by freezing temperatures. The proteins bind to the surfaces of ice crystals to prevent or slow further crystal growth. Sugars, polysaccharides, osmoprotectant solutes, dehydrins, and other cold-induced proteins can also have cryoprotective effects.

Summary

Plants sense changes in their environment and respond to them by means of dedicated stress-response pathways. These pathways involve gene networks, regulatory proteins, and signaling intermediates, as well as proteins, enzymes, and molecules that act to protect cells from the detrimental effects of abiotic stress. Together these anti-stress mechanisms enable plants to acclimate or adapt to stresses such as water deficit, heat, cold, and salinity, as well as their possible combinations. A major research goal is to use some of these acclimation or adaptation mechanisms to protect crops from the adverse climatic conditions that are expected to result from global warming.

19.1 Defining Plant Stress

- Stress can be defined generally as any environmental condition that prevents the plant from achieving its full genetic potential measured under ideal growth conditions (**Figures 19.1**, **19.2**).
- Plant responses to abiotic stress involve trade-offs between vegetative and reproductive growth, which may differ depending on whether the plant is an annual or a perennial (**Figure 19.3**).

19.2 Acclimation versus Adaptation

- Adaptation is characterized by genetic changes in an entire population that have been fixed by natural selection over many generations.
- Acclimation is the process whereby individual plants respond to periodic changes in the environment by directly altering their morphology or physiology. The physiological changes associated with acclimation require no genetic modifications, and many are reversible.

19.3 Environmental Stressors

- Environmental stresses can disrupt plant metabolism through a variety of mechanisms, most of which result in the accumulation of reactive oxygen species (ROS) (**Figure 19.4**).
- Cell dehydration leads to decreased turgor pressure, increased ion toxicity, and inhibition of photosynthesis (**Figure 19.5**, **Table 19.1**).
- Temperature stress affects protein stability, enzymatic reactions, membrane fluidity, and the secondary structures of RNA and DNA (**Table 19.1**).
- Flooded soil is depleted of oxygen, leading to anaerobic stress to the root (**Table 19.1**).
- Salinity stress causes cell dehydration coupled with inhibition of enzymatic activity due to ion toxicity, thereby reducing aboveground plant growth and inhibiting photosynthesis (**Table 19.1**).
- Freezing stress, like water-deficit stress, causes cell dehydration, but can also disrupt cellular integrity due to membrane damage (**Table 19.1**).

(Continued)

Summary (*continued*)

- Heavy metals can replace other essential metals in enzymes and disrupt key reactions (**Table 19.1**).
- Ozone and ultraviolet light induce the formation of ROS, which in turn induce the formation of leaf lesions and programmed cell death (**Figure 19.4**).
- Combinations of abiotic stresses can have effects on plant physiology and productivity that are different than those of individual stresses (**Figure 19.6**).
- Plants can gain cross-protection when they are sequentially exposed to different abiotic stresses.

19.4 Stress-Sensing Mechanisms in Plants

- Plants use physical, biophysical, metabolic, and biochemical mechanisms to detect stresses and activate response pathways (**Figure 19.7**).
- Chloroplasts and mitochondria can send status and distress signals to the nucleus.
- Stress-sensing pathways involve various signaling intermediates (**Figure 19.7**).
- A systemic plant-wide wave of ROS production and Ca^{2+} signaling alerts as-yet-unstressed parts of the plant of the need for a response (**Figure 19.8**).
- Hormones act separately and together to regulate abiotic stress responses (**Figure 19.9**, **19.10**, **19.11**).

19.5 Developmental and Physiological Mechanisms That Protect Plants against Abiotic Stress

- Plants can be protected from water-deficit stress by overexpression of a gene for cytokinin biosynthesis (**Figure 19.11**).
- Stress-induced phenotypic changes can be mediated by epigenetic mechanisms (**Figure 19.12**)
- Epigenetic stress-response mechanisms may lead to heritable protection (**Figure 19.12**).
- Abiotic stress can induce changes in leaf shape, leaf abscission, or leaf angle. (**Figure 19.13**)
- Plants may lower root Ψ to continue to absorb water in drying soil (**Figures 19.14**, **19.15**).
- Aerenchyma tissue allows O_2 to diffuse down to submerged organs (**Figure 19.16**).
- ROS can be detoxified through scavenging pathways that reduce oxidative stress (**Figure 19.17**).
- Molecular chaperone proteins protect sensitive proteins and membranes during abiotic stress (**Figure 19.18**).
- Prolonged exposure to extreme temperatures can alter the composition of membrane lipids, thereby allowing plants to maintain membrane fluidity (**Table 19.2**).
- Plants cope with toxic ions through exclusion and internal tolerance mechanisms (**Figure 19.19**).

Suggested Reading

Bailey-Serres, J., Parker, J. E., Ainsworth, E. A., Oldroyd, G. E. D., and Schroeder, J. I. (2019) Genetic strategies for improving crop yields. *Nature* 575: 109–119.

Cheung, A. Y., Qu, L. J., Russinova, E., Zhao, Y., and Zipfel, C. (2020) Update on receptors and signaling. *Plant Physiol.* 182: 1527–1530.

Kollist, H., Zandalinas, S. I., Sengupta, S., Nuhkat, M., Kangasjärvi, J., and Mittler, R. (2019) Rapid responses to abiotic stress: Priming the landscape for the signal transduction network. *Trends Plant Sci.* 24: 25–37.

Lamers, J., van der Meer, T., and Testerink, C. (2020) How plants sense and respond to stressful environments. *Plant Physiol.* 182: 1624–1635.

Liang, Y., Liu, H.-J., Yan, J., and Tian F. (2021) Natural variation in crops: Realized understanding, continuing promise. *Annu. Rev. Plant Biol.* 72(1): 357–385.

Mhamdi, A., and Van Breusegem, F. (2018) Reactive oxygen species in plant development. *Development* 145: dev164376.

Mittler, R. (2017) ROS are good. *Trends Plant Sci.* 22: 11–19.

Ohama, N., Sato, H., Shinozaki, K., and Yamaguchi-Shinozaki, K. (2017) Transcriptional regulatory network of plant heat stress response. *Trends Plant Sci.* 22: 53–65.

Saijo, Y., and Loo, E. P. (2020) Plant immunity in signal integration between biotic and abiotic stress responses. *New Phytol.* 225: 87–104.

Zandalinas, S. I, Fritschi, F. B., and Mittler, R. (2021) Global warming, climate change, and environmental pollution: Recipe for a multifactorial stress combination disaster. *Trends Plant Sci.* 26: 588–599.

Zhu, J. K. (2016) Abiotic stress signaling and responses in plants. *Cell* 167: 313–324.

Appendix Plant Breeding and Genome Engineering

Generation of **polyploid** plants is widely used in crop breeding. Plants can undergo additional rounds of DNA replication without undergoing cytokinesis to generate polyploid genomes. When polyploidy is restricted to vegetative organs such as leaves, it is described as *endopolyploidy*. If whole-genome duplication passes into the germ line (gametes), a uniformly polyploid next generation can result. Polyploidy is not a rare event; it is a common phenomenon that has occurred at least once in all angiosperm lineages and occasionally in gymnosperms. **Autopolyploid plants** contain multiple complete genomes of a single species, while **allopolyploid plants** contain multiple complete genomes derived from two or more separate species.

Both types of polyploidy can result from incomplete meiosis during gametogenesis. During normal meiosis, the chromosomes of a diploid germ cell undergo DNA replication followed by two rounds of meiosis to produce four haploid (*1N*) cells (Chapter 17). If the normal two rounds of cell division during meiosis does not occur after chromosome duplication, then the formation of an autopolyploid is the result. If interspecies fertilization occurs and results in genome duplication, a viable allopolyploid is often formed. Diploid interspecies hybrids also occur naturally, but they are frequently sterile.

Humans have modified crop plants through selective breeding for many centuries to produce varieties that have higher yields, are better adapted to specific climates, or are resistant to plant pathogens. A classic example from nature where multiple species in the same genus have produced allopolyploid offspring that have been selected for food production comes from the mustard family, Brassicaceae (**Figure A.1**). Aside from the natural

polyploid Cell with more than two sets of chromosomes.

autopolyploid plants Polyploids containing multiple complete genomes of a single species.

allopolyploid plants Polyploids with multiple complete genomes derived from two separate species.

Figure A.1 Three common species of plants in the mustard family (Brassicaceae) have interbred with one another in nature to form new allotetraploid species. Their relationship is depicted in the so-called triangle of U, named after the Korean scientist Nagaharu U. The three corners of the triangle show diploid species of *Brassica*. Each of the three species can interbreed with the two others to form new allopolyploids.

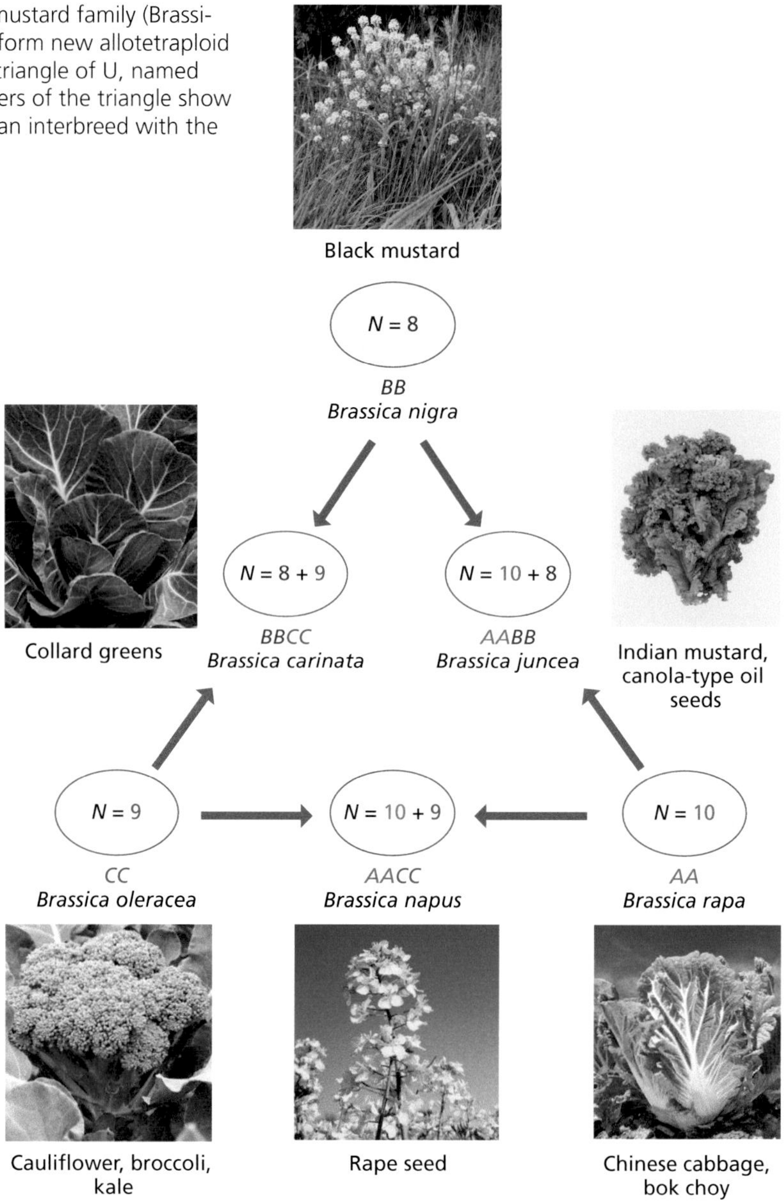

subfunctionalization The process by which evolution acts on duplicate genes, causing one copy to be either lost or changed in function, while the other retains its original function.

hybrid vigor (heterosis) The increased vigor often observed in the offspring of crosses between two inbred varieties of the same plant species.

occurrence of genome duplication, polyploidy can also be induced artificially by treatment with the natural toxin colchicine, which is derived from the autumn crocus (*Colchicum autumnale*). Colchicine prevents cell division but does not interfere with DNA replication. Treatment with colchicine therefore results in an undivided nucleus containing multiple copies of the genome.

As polyploidy leads to multiple, redundant copies of genes in the genome, one copy may be either lost or changed in function over multiple generations, a process known as **subfunctionalization**. Polyploids may also revert to diploid status, and many diploid species exhibit evidence of genome duplication in their evolutionary history.

Allopolyploids often exhibit **hybrid vigor**, or **heterosis**, the increased vigor often (but not always) observed in the offspring of crosses between two inbred varieties of the same plant species. Heterosis can contribute to larger plants, greater biomass, and higher yields in agricultural crops. Plant breeders have used heterosis as a tool to increase crop plant biomass and productivity for centuries.

The generation of allopolyploids plays a prominent role in breeding of agricultural crops. Breeding has even produced entirely new species, such as the common bread wheat *Triticum aestivum*, which is allohexaploid and arose from the cross-pollination of three different progenitor species. Monoculture has reduced their genetic diversity, and food production has become increasingly subject to losses from climate volatility, diminished natural resources, and pest pressures. Unlike modern crop varieties that have been selected for yield under ideal agronomic conditions, wild relatives of crop plants retain useful resistance and resilience **traits** that are otherwise undesirable for food production and consumption. Historically, the process of crop domestication involved selection of variants that perform reliably due to loss of traits that provide selective advantages in the wild. Examples include loss of free shattering (seed dispersal after ripening) in wheat, barley, rice, and sorghum; apical dominance in maize; and grain quality and threshability in bread wheat (*Triticum aestivum*). These traits are controlled by a small number of genes and were eliminated by crop breeding, with the result that a "domestication bottleneck" reduced the genetic diversity that underpins complex traits such as disease resistance, heat, and water-deficit stress tolerance.

trait A specific character of an organism that is determined by genetic factors (genotype) and environment.

Marker-assisted breeding, advances in DNA sequencing, and automated genotyping have revolutionized the breeding process and have reduced the time and costs required to transfer useful genes and alleles from wild species of crop plants. Introgression of useful genetic material from wild relatives to cultivated varieties has traditionally been performed by direct introduction of wild pollen into the emasculated or female flowers of domesticated crops. However, successful crossing is often challenging and time-consuming because of the incompatibilities between the two species.

In many cases, interspecific hybrids do not form viable seeds or healthy embryos. Technically difficult approaches that artificially insert genetic material from one species into cells from another are sometimes employed when sexual hybridization is not feasible. Breeders often increase ploidy (genome) numbers to achieve necessary genetic stability and viability after hybridization. Chromosome doubling using colchicine is employed for this purpose. Synthetic hexaploid wheat lines have been developed using this technique by crossing tetraploid wheat with a diploid progenitor of modern hexaploid wheat, and then doubling the genomes of the triploid interspecific hybrid using colchicine.

Other techniques include "bridge" crossing, which involves the introduction of genes from a wild relative into an intermediate species that is compatible with both the parents followed by crosses into the target crop species and the use of γ-radiation to fragment chromosome before crossing to introduce portions of chromosomes into the genome of a cultivar. Both processes require a great deal of subsequent crossing and selection to eliminate undesirable genes and gene products.

Figure A.2 Classic breeding and domestication of the wild grass teosinte (left) over hundreds of years have led to the crop plant *Zea mays* (maize, right).

In addition to hybridization and allopolyploidy, selection of naturally occurring mutations has played a prominent role in the development of food and ornamental crops. For example, modern maize varieties are the domesticated descendants of wild relatives in the genus *Zea*, known as teosinte (**Figure A.2**). Mesoamerican breeders selected what we would recognize as maize 4,500 years ago. Modern DNA sequencing has shown that these ancient agriculturalists selected plants with mutations that increase yield and seed mass at the expense of some survivability traits.

In the 1920s, H. J. Muller and L. J. Stadler independently experimented with the effects of X-rays on the stability of chromosomes in barley. Both researchers reported heritable changes in the treated organisms. In the following years, other techniques for inducing mutations were developed. These techniques include the use of other forms of radiation and mutagenic chemicals. Mutagenesis with either radiation or chemicals induces nucleotide changes randomly throughout the genome. Breeders then select desirable traits that are associated with a gene mutation to develop valuable agricultural or ornamental varieties.

In the twentieth century, mutagenesis was achieved in a number of species, particularly maize, by crossing of a plant of interest to a member of the same or similar species that carries a DNA fragment of an ancestral virus (transposon) that randomly inserts itself in other genes. This is followed by screening the offspring for mutant phenotypes that are of potential value. Because the sequence of the transposon is known, these mutations are "tagged"; thus, the DNA sequences adjacent to the transposon can be readily found and analyzed to identify the mutated gene. This technique is called **transposon tagging**. Using a technique very similar to transposon tagging, plant genes can also be tagged with known sequences from specific bacteria or viruses. Most common is the use of the soil-borne mutagenic *Agrobacterium* to transfer a piece of its genome into a host plant's genome, creating a mutation and leaving behind a "tag," similar to that of transposons. This bacterium causes infected plants to overproduce growth hormones, which induce the formation of a tumor called a **crown gall** (Chapter 13). Crown gall disease is a serious problem for certain agricultural crops, such as fruit trees, as it can reduce crop yield and decrease overall plant health.

transposon tagging The technique of inserting a transposon into a gene and thereby marking that gene with a known DNA sequence.

crown gall A tumor-forming plant disease resulting from wound infection of the stem or trunk by the soil-dwelling bacterium *Agrobacterium tumefaciens*. A tumor resulting from the disease.

protoplast fusion A technique for incorporating foreign genes into plant genomes by fusion of two genetically different cells from which the cell walls have been removed.

biolistics (also **gene gun)** A procedure, also called the "gene gun" technique, in which tiny gold particles coated with the genes of interest are mechanically shot into cultured cells. Some of the DNA is randomly incorporated into the genome of the targeted cells.

A.1 Genetic Modification of Plants

Agrobacterium is sometimes referred to as a natural genetic engineer because it has the ability to transform plant cells with a small subset of its own genes. The genes transferred to the plant genome are part of a circular extrachromosomal piece of DNA called the tumor-inducing (Ti) plasmid. The Ti plasmid contains genes that are required for initiating and conducting the transfer of the T-DNA into the plant cell. Once transferred, the T-DNA inserts itself randomly into the plant nuclear genome. The genes on the Ti plasmid that induce grown gall formation and further bacterial growth are removed and replaced with the DNA encoding the *transgenes* to be inserted in their place. As *Agrobacterium* is attracted to wounded cells at sites of entry that produce the compound acetosyringone, this compound can be used to enhance growth and *transformation* (insertion of the plasmid DNA) into target plants. A gene encoding an enzyme that provides resistance to an antibiotic or herbicide is

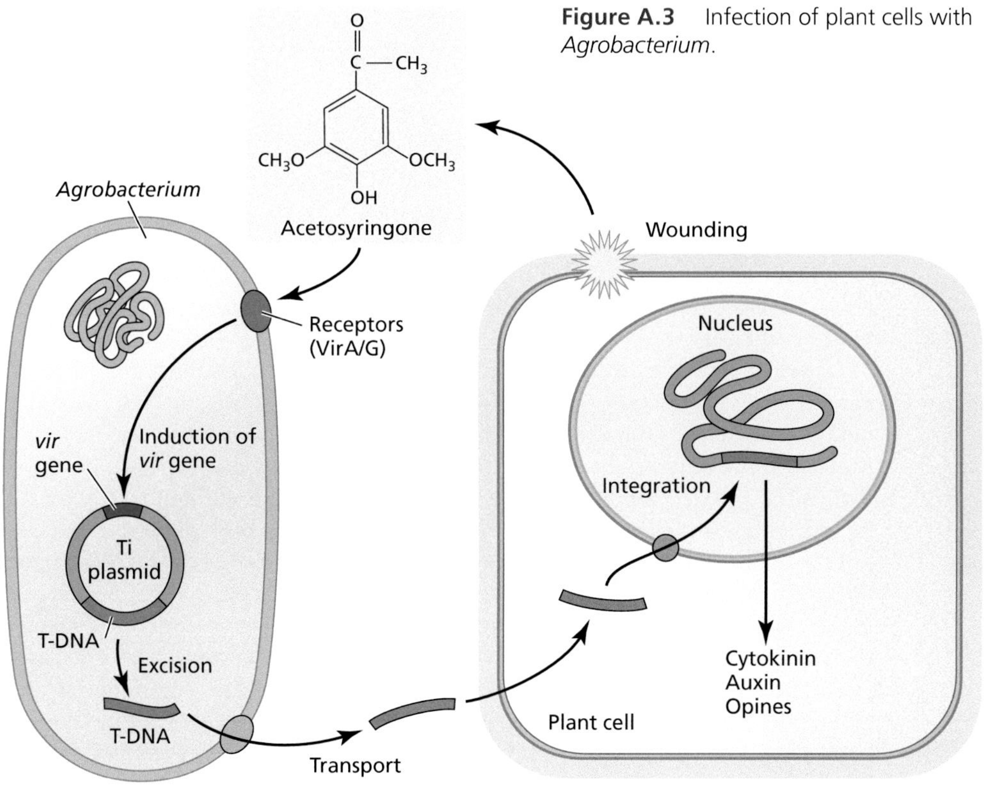

Figure A.3 Infection of plant cells with *Agrobacterium*.

included in the plasmid to allow for selection of plants that have been successfully transformed. An overview of the steps involved in the transformation of plant cells by *Agrobacterium* is shown in **Figure A.3**.

Besides *Agrobacterium*-mediated transformation, several other techniques have been developed to incorporate foreign genes into plant genomes. One such technique is the fusion of two plant cells with different genomic information, called **protoplast fusion**. Another is **biolistics**, sometimes referred to as the **gene gun** technique, in which tiny particles of gold coated with the genetic construct of interest are shot into cells growing in culture dishes (**Figure A.4**). The genetic material is randomly incorporated into the cells' genomes. The cells can then be transferred to solid culture medium and grown into mature transgenic individuals.

Two of the types of transgenes that have been most commonly used in commercial crops today are genes that allow plants to resist herbicide applications and genes that allow plants to withstand attack by certain insects. Weed invasion and insect infestation are two of the main causes of crop yield reductions in agriculture.

Plants carrying a transgene for **glyphosate resistance** will survive a field application of glyphosate (the commercial herbicide Roundup), which kills weeds but does not harm resistant crop plants. Glyphosate inhibits the enzyme enolpyruvalshikimate-3-phosphate synthase (EPSPS), which catalyzes a key reaction in the shikimic acid pathway, a plant-specific metabolic pathway necessary for the production of many specialized compounds, including auxin and aromatic amino acids. Glyphosate-resistant plants carry either a gene encoding a bacterial form of EPSPS that is insensitive to the

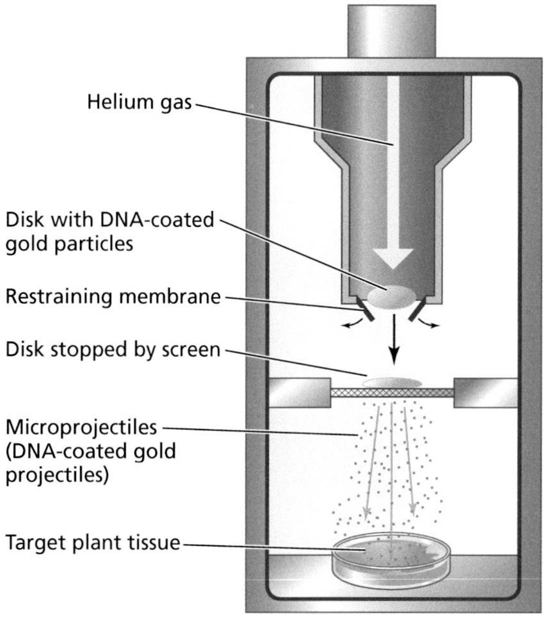

Figure A.4 Recombinant DNA can be inserted into plant cells using biolistics with a "gene gun."

glyphosate resistance Genetic capacity to survive a field application of the commercial herbicide Roundup, which kills weeds but does not harm resistant crop plants.

***Bacillus thuringiensis* (Bt)** A soil bacterium that is the source for a commonly used transgene encoding an insecticidal toxin.

herbicide, or transgene constructs that fuse high-activity promoters with the wild-type EPSPS gene to achieve herbicide resistance by overproduction of the enzyme.

Another commonly used transgene encodes an insecticidal toxin from the soil bacterium ***Bacillus thuringiensis*** (**Bt**). Bt toxin interferes with a receptor found only in the larval gut of certain insects, eventually killing them. Plants expressing Bt toxin are toxic to susceptible insects but harmless to most other organisms, including nontarget insect species.

More recently, overexpression of an oxalate oxidase gene from maize in American chestnut (*Castanea dentata*) has been used to confer resistance to chestnut blight in this keystone North American forest species that is threatened with extinction.

Transgenic plants are also being developed that have enhanced nutritional value. Every year, according to the World Health Organization, dietary vitamin A deficiency causes as many as 500,000 children in developing nations to go blind. Many of these children live in Southeast Asia, where rice is the main part of the diet. Although rice plants synthesize abundant levels of β-carotene (provitamin A) in their leaves, rice endosperm, which makes up the bulk of the grain, does not normally express the genes required for three of the steps in the β-carotene biosynthesis pathway. To overcome this block, Ingo Potrykus, Peter Beyer, and their colleagues developed new strains of rice that carry genes from other species that can complete the β-carotene biosynthesis pathway (**Figure A.5**). The most efficient strain uses two transgenes: a phytoene synthase gene from maize and a bacterial carotene desaturase gene. Together, these two genes allow the rice plant to accumulate large amounts of β-carotene. Facing many regulatory and intellectual property hurdles, this "golden rice" was field tested for more than ten years and was finally approved for use in the United States in 2018. This was not the first time that the β-carotene content of a crop plant was altered by agriculturalists. Carrots, for example, were either red or yellow before the seventeenth century, when a Dutch horticulturist selected the first orange-colored varieties.

Other researchers are developing transgenic plants that express vaccines in their edible fruit as an alternative, more convenient way to vaccinate people in parts of the world where medical facilities are insufficient for the administration of conventional vaccines. However, concerns about the deployment of transgenic crops and their potential impact on the environment have made such deployment extremely costly and have limited their use in many countries.

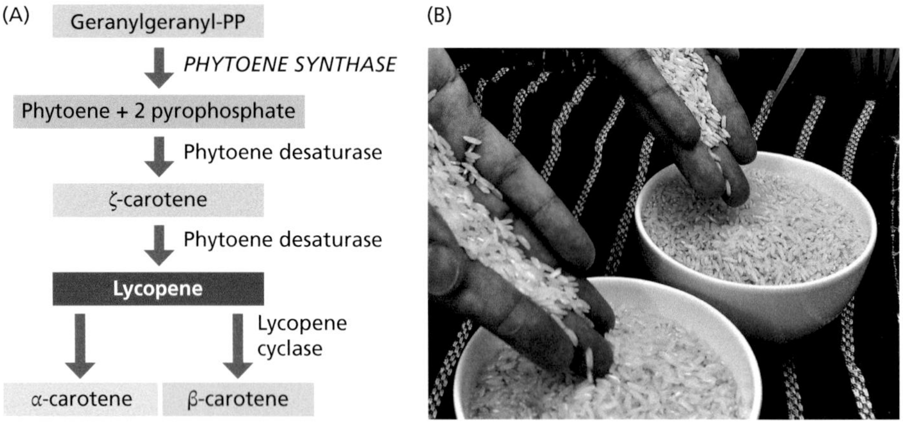

Figure A.5 Golden rice was produced by inserting a gene encoding a phytoene synthase from maize and another gene encoding a phytoene desaturase enzyme from the soil bacterium *Erwinia uredevora* to catalyze lycopene synthesis in rice grains. A rice lycopene cyclase converts lycopene to α- and β-carotene. (A) The β-carotene biosynthesis pathway in golden rice. (B) Normal white rice (left) compared with golden rice (right).

A.2 Editing Plant Genomes with Sequence-Specific Nucleases

To introduce specific mutations into the plant genome, one can engineer sequence-specific nucleases, which generate double-strand breaks at the target DNA sequence(s). Transcription activator-like effector nucleases (TALENs) were early sequence-specific nucleases used in biotechnology. They are derived from *Xanthomonas* bacterial pathogens that secrete transcription activator-like (TAL) effectors into plant cells to disrupt host metabolism and cause disease. TAL effectors bind to the promoters of some host genes and activate their expression, thereby modifying host processes to benefit the bacterial pathogens.

The most popular and powerful sequence-specific nuclease technologies in current use are based on CRISPR (Clustered *R*egularly *I*nterspaced *S*hort *P*alindromic *R*epeats)-Cas (CRISPR-*as*sociated protein) systems. The CRISPR-Cas system was originally identified in the immune system of the bacterium *Streptococcus thermophilus*, which is used in the dairy industry to fend off bacteriophages (viruses that attack bacteria) that can spoil the yogurt-making process. A functional CRISPR-Cas system can snip off a piece of DNA from the genome of an invading phage and insert it between CRISPR repeats in the bacterial genome, creating a "memory" of that phage. This phage-derived memory DNA is called a protospacer and can be transcribed to produce CRISPR RNA (crRNA), which includes both phage and CRISPR sequences. A trans-activating CRISPR RNA (tracrRNA) that can hybridize with the CRISPR sequence in the crRNA is also required by the system. Cas proteins are nonspecific DNA nucleases that are guided by the crRNA-tracrRNA to target and cleave the invading phage DNA at a site adjacent to the protospacer motif (**Figure A.6A**). In 2012, Jennifer Doudna and Emmanuelle Charpentier reengineered a CRISPR-Cas system to generate an RNA-guided gene-editing system. Importantly, a guide RNA (gRNA) was engineered by fusing crRNA and tracrRNA together, which reduced the CRISPR-Cas system to two components, Cas (in this case Cas9) and gRNA (**Figure A.6B**). Soon after, CRISPR-Cas9 was being used for genome editing in eukaryotes, including plants. Later, additional Cas components such as Cas12a and Cas12b were demonstrated to work in plants. As RNA-guided nucleases, CRISPR-Cas systems are easy to engineer; this enables targeting of nearly any sequence in a genome of interest. For their work on adapting the CRISPR system into a programmable gene-editing tool, Doudna and Charpentier were awarded the 2020 Nobel Prize in Chemistry.

The CRISPR-Cas system has now been further engineered to:

1. Introduce a frame shift in the DNA with the result that no viable protein is produced
2. Replace a specific DNA sequence within a gene so that the encoded protein has altered function and/or cellular distribution
3. Remodel the promoters of a gene so that it is expressed at higher or lower levels under certain environmental conditions or in a different tissue at a different developmental stage
4. Edit a single nucleotide to alter post-translational modifications or interactions with other cellular components (Chapters 1 and 12). This approach can also be used to introduce single nucleotide polymorphisms (changes; SNPs) that are associated with enhanced pest resistance or yield but are difficult to introduce via standard breeding approaches.

Given the promise of precise gene replacement for improving crop resilience and yield in suboptimal climatic conditions, innovative solutions to current limitations of these technologies are expected. The perceived advantages of

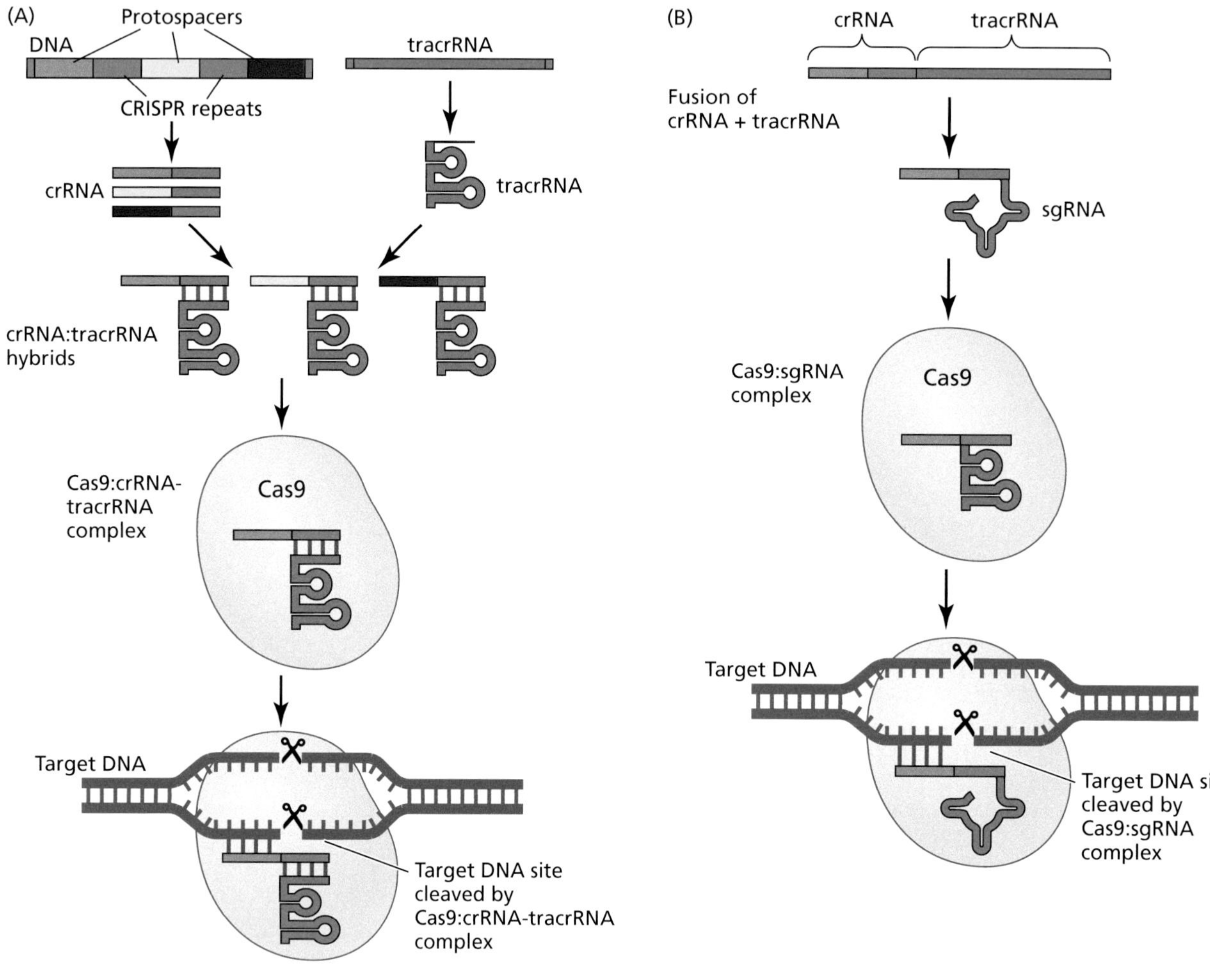

Figure A.6 CRISPR-Cas system for gene editing. (A) The CRISPR regions in bacterial genomes contain snippets of DNA (protospacers, depicted as colored segments) that are remnants of invasive DNA from a bacterial or viral attack in the past that did not lead to the host cell's death. Incorporation of these foreign DNA pieces into the bacterial host genome creates a memory bank of DNA sequences to be potentially targeted for destruction in the event of a future attack. Each DNA snippet can be transcribed into RNA, called CRISPR-RNA (or crRNA), each of which is complementary to one specific foreign DNA sequence. The second major component of the host's CRISPR system is the transactivating CRISPR RNA (tracr RNA). crRNA and tracrRNA hybridize and bind to the Cas9 protein, also encoded in the host's genome. This complex can then target complementary invading DNA and cleave it using the nuclease function of the Cas9 protein. Although the cleaved DNA can be repaired, such repairs are imperfect and create mutations in the invading DNA, which often impairs its function. (B) CRISPR systems engineered for genome editing use a synthetic piece of RNA, called the guide RNA (gRNA), which is a fusion of the tracrRNA and the target sequence. A synthetic construct made in the lab that encodes the gRNA is transformed into the target host together with a copy of the Cas9 gene. Inside the target cell, the Cas9:gRNA complex can create mutations in a gene of interest by cleavage via Cas9 followed by imperfect repair of the damage by the host cell's DNA repair machinery.

gene editing over the use of transgenics has broadened their appeal and has led to preferential use of this approach in generating herbicide, insect, and disease resistant cultivars (**Figure A.7**).

A.3 Genetic Engineering of Plants Remains Controversial

The engineering of crop plants has not been greeted with universal enthusiasm and support. In spite of the enormous humanitarian potential of accelerated breeding of sustainable agricultural traits, many individuals, as well as the governments of some countries, look on genetic engineering with suspicion and

Figure A.7 Generation of herbicide-resistant rice plants by CRISPR-Cas9 engineering. The plants are shown 36 days after being sprayed with 100 µM bispyribac sodium at the five-leaf stage. On the left is a rice line with the edited acetolactate synthase (*ALS*) gene involved in the synthesis of branched amino acids and on the right is a wild-type (WT) rice line. The herbicide bispyribac sodium inhibits the WT ALS protein function but fails to inhibit the modified ALS protein. Note that the wild-type plants died whereas the gene-edited plants grew normally.

concern. In particular, opponents to the introduction of genes from other organisms into food plants cite the possibility of inadvertently producing crops that express allergens introduced from another species. They also worry that the overuse of genes encoding Bt toxin might select for insects that develop resistance to the toxin or that windblown pollen from transgenic herbicide-resistant crops could cross-pollinate nearby wild species, thereby producing weeds with herbicide resistance or contaminating organic crops with transgenes. According to the World Health Organization, however, the concern that the ingestion of currently grown and marketed genetically engineered foods poses risks to human health has so far been unsubstantiated.

Crop plants produced by introduction of transgenes or spliced DNA from other organisms are commonly referred to as Genetically Modified Organisms (GMOs). As the sophistication of plant genome-editing techniques have become more refined, plant biotechnologists are able to avoid the release of plants containing transgenes into the field. The term Genetically Engineered Organisms (GEOs) is applied to gene-edited crops. Many countries have adopted guidelines that differentiate GMO and GEO crops. For example, a 2016 report from the U.S. National Academies of Sciences, Engineering, and Medicine recommended regulatory differentiation of GMO and GEO approaches, and the U.S. Department of Agriculture (USDA) has adopted the 2020 Sustainable, Ecological, Consistent, Uniform, Responsible, Efficient (SECURE) rule, which specifies that most genetically engineered crops will not be regulated if they do not contain transgenes.

Overall, genetic engineering holds considerable promise for safely modernizing crop breeding to meet the challenges of global climate change. Many of the targeted mutations generated by gene-editing technologies already occur in nature or have already arisen during conventional nontargeted mutagenesis or crossbreeding. Given the potential contributions of genome editing to future crop breeding, it is anticipated that these new technologies will be widely used to produce crops that have higher yields, reduced reliance on chemicals to prevent losses to pest and disease, and more resilience to climate-dependent stress.

Research is ongoing to monitor the effects of the new technologies on human health and the environment. In the end, the controversy may come down to this question: How much risk is acceptable in an attempt to satisfy the needs of an ever-increasing world population for food, clothing, and shelter?

Glossary

1*N* (haploid) Having a single unpaired set of chromosomes; the gametophyte generation is characteristically haploid.

26S proteasome A large proteolytic complex that degrades intracellular proteins marked for destruction by the attachment of one or more copies of the small protein, ubiquitin.

A

abaxial Referring to the lower surface of a leaf.

ABC model A proposal for the way in which floral homeotic genes control organ formation in flowers. According to the model, organ identity in each whorl is determined by a unique combination of the three organ identity gene activities.

abscisic acid (ABA) A plant hormone that functions in regulation of seed dormancy as well as in stress responses such as stomatal closure during water deficit and responses to cold and heat. ABA is derived from carotenoid precursors.

abscission The shedding of leaves, flowers, and fruits from a living plant. The process whereby specific cells in the leaf petiole (stalk) differentiate to form an abscission layer, allowing a dying or dead organ to separate from the plant.

abscission zone The region that contains the abscission layer and is located near the base of the petiole of leaves.

absorption spectrum A graphic representation of the amount of light energy absorbed by a substance plotted against the wavelength of the light.

accessory pigments Light-absorbing molecules in photosynthetic organisms that work with chlorophyll *a* in the absorption of light used for photosynthesis. They include carotenoids, other chlorophylls, and phycobiliproteins.

acclimation The increase in plant stress tolerance due to exposure to prior stress. May involve changes in gene expression. Contrast with *adaptation*.

ACC oxidase The enzyme that catalyzes the conversion of 1-aminocyclopropane-1-carboxylic acid (ACC) to ethylene, the last step in ethylene biosynthesis.

ACC synthase The enzyme that catalyzes the synthesis of 1-aminocyclopropane-1-carboxylic acid (ACC) from *S*-adenosylmethionine.

acid growth A characteristic of growing cell walls in which they extend more rapidly at acidic pH than at neutral pH.

acropetal From the base to the tip of an organ, such as a stem, root, or leaf.

actin A major ATP-binding cytoskeletal protein. The monomeric, globular form of actin is called G-actin; the polymerized form in microfilaments is F-actin.

actinorhizal Pertaining to several woody plant species, such as alder trees, in which symbiosis occurs with soil bacteria of the nitrogen-fixing genus *Frankia*.

actin-related protein 2/3 (Arp 2/3) Actin-related proteins 2 and 3, which bind to the side of a pre-existing actin filament and form a complex with actin to initiate growth of an actin filament branch.

action potential A transient event in which the membrane potential difference rapidly decreases (depolarizes) and abruptly increases (hyperpolarizes). Action potentials, which are triggered by the opening of ion channels, can be self-propagating along linear files of cells, especially in the vascular systems of plants.

action spectrum A graphic representation of the magnitude of a biological response to light as a function of wavelength.

active transport The use of energy to move a solute across a membrane against a concentration gradient, a potential gradient, or both (electrochemical potential gradient). Uphill transport.

adaptation An inherited level of stress resistance acquired by a process of selection over many generations. Contrast with *acclimation*.

adaxial Referring to the upper surface of a leaf.

adenosine triphosphate (ATP) The major carrier of chemical energy in the cell, which by hydrolysis is converted to adenosine diphosphate (ADP) or adenosine monophosphate (AMP) with release of energy.

adhesion The attraction of water to a solid phase such as a cell wall or glass surface, due primarily to the formation of hydrogen bonds.

adventitious roots Roots arising from any organ other than a root.

aerenchyma An anatomical feature of roots found in hypoxic conditions, showing large, gas-filled intercellular spaces in the root cortex.

aerobic respiration The complete oxidation of carbon compounds to CO_2 and H_2O, using oxygen as the final electron acceptor. Energy is released and conserved as ATP.

aeroponics The technique by which plants are grown with their roots suspended in air while being sprayed continuously with a nutrient solution.

after-ripening Technique for breaking seed dormancy by prolonged storage at room temperature (20–25°C) under dry conditions.

AGC kinase A member of a large subfamily of protein kinases that are activated by environmental stimuli of second messengers. AGC kinases share a conserved catalytic domain and a regulatory C-terminal domain.

albuminous cells Sieve element–associated cells in the phloem of gymnosperms. Although similar to companion cells in angiosperms, they have a different developmental origin. Also called *Strasburger cells*.

aleurone layer The layer of aleurone cells surrounding and distinct from the starchy endosperm of cereal grains.

allelopathy Release by plants of substances into the environment that have harmful effects on neighboring plants.

allocation The regulated diversion of photosynthate into storage, utilization, and/or transport.

allopolyploid plants Polyploids with multiple complete genomes derived from two separate species.

alternation of generations The presence of two genetically distinct multicellular stages, one haploid and one diploid, in the plant life cycle. The haploid gametophyte generation begins with meiosis, while the diploid sporophyte generation begins with the fusion of sperm and egg.

alternative oxidase An enzyme in the mitochondrial electron transport chain that reduces oxygen to water and oxidizes ubiquinol (ubihydroquinone).

amplitude In a biological rhythm, the distance between peak and trough; it can often vary while the period remains unchanged.

α-amylase inhibitors Substances synthesized by some legumes that interfere with herbivore digestion by blocking the action of the starch-digesting enzyme α-amylase.

amyloplast A starch-storing plastid found abundantly in storage tissues of shoots and roots, and in seeds. Specialized amyloplasts in the root cap and shoot also serve as gravity sensors.

anaphase The stage of mitosis during which the two chromatids of each replicated chromosome are separated and move toward opposite poles.

anchored proteins Proteins that are bound to the membrane surface via lipid molecules, to which they are covalently attached.

angiosperms The flowering plants. With their innovative reproductive organ, the flower, they are a more advanced type of seed plant and dominate the landscape. Distinguished from gymnosperms by the presence of a carpel that encloses the seeds.

anisotropic growth Enlargement that is greater in one direction than another; for example, elongating cells in stems and roots grow more in length than in width.

annual plant A plant that completes its life cycle from seed to seed, senesces, and dies within 1 year.

annual rings Alternating rings of spring and summer wood (formed by secondary growth of the xylem) seen in cross sections of stems of woody species.

antenna complex A group of pigment molecules that cooperate to absorb light energy and transfer it to a reaction center complex.

anticlinal Pertaining to the orientation of cell division such that the new cell plate forms perpendicular to the tissue surface.

antifreeze proteins Proteins that lower the temperature at which aqueous solutions freeze. When induced by cold temperatures, these plant proteins bind to the surfaces of ice crystals to prevent or slow further crystal growth, thereby limiting or preventing freeze damage. Some antifreeze proteins are similar to pathogenesis-related proteins.

antioxidative enzymes Proteins that detoxify reactive oxygen species.

antiport A type of secondary active transport in which the passive (downhill) movement of protons or other ions drives the active (uphill) transport of a solute in the opposite direction.

apical–basal axis An axis that extends from the shoot apical meristem to the root apical meristem.

apical cell The smaller, cytoplasm-rich cell formed by the first division of the zygote.

apical dominance In most higher plants, the growing apical bud's inhibition of the growth of lateral buds (axillary buds).

apical meristems Localized regions made up of undifferentiated cells undergoing cell division without differentiation at the tips of shoots and roots.

apoplast The mostly continuous system of cell walls, intercellular air spaces, and xylem vessels in a plant.

apoplastic transport Movement of molecules through the cell wall continuum that is called the apoplast. Molecules may move through the linked cell walls of adjacent cells, and in that way move throughout the plant without crossing a plasma membrane.

aquaporins Integral membrane proteins that form channels across a membrane, many of which are selective for water (hence the name). Such channels facilitate water movement across a membrane.

arbuscular mycorrhizae Symbioses between fungi in the phylum Glomeromycota and the roots of a broad range of angiosperms, gymnosperms, ferns, and liverworts. The hyphae of arbuscular mycorrhizae penetrate the cortical cells of the root.

arbuscules Branched structures formed by mycorrhizal fungi inside cortical cells of the host plant root; the sites of nutrient transfer between the fungus and the plant.

ascorbate peroxidase An enzyme that converts peroxide and ascorbate to dehydroascorbate and water.

asparagine synthetase (AS) An enzyme that transfers nitrogen as an amino group from glutamine to aspartate, forming asparagine.

aspartate aminotransferase An aminotransferase that transfers the amino group from glutamate to the carboxyl atom of oxaloacetate to form aspartate.

ATPase The multi-subunit protein complex that synthesizes ATP from ADP and phosphate (P_i). F_oF_1 and CF_0-CF_1 types are present in mitochondria and chloroplasts, respectively. Also called *ATP synthase* and *CF_0-CF_1 ATP synthase*.

ATP synthase The multi-subunit protein complex that synthesizes ATP from ADP and phosphate (P_i). F_oF_1 and CF_0-CF_1 types are present in mitochondria and chloroplasts, respectively. Also called *ATPase* and *CF_0–CF_1 ATP synthase*.

autophagy A catabolic mechanism that conveys cellular macromolecules and organelles via autophagosomes to lytic vacuoles where they are degraded and recycled.

autopolyploid plants Polyploids containing multiple complete genomes of a single species.

auxin clock model A model for the spacing of prebranch sites in roots based on the periodic fluctuation of auxin activity or concentration in the oscillation zone.

AUXIN/INDOLE-3-ACETIC ACID (AUX/IAA) A family of short-lived small proteins that combine with the TIR1/AFB proteins to form the primary auxin receptor. This family of short-lived small proteins in Arabidopsis regulates auxin-induced gene expression by binding to ARF protein that is bound to DNA. If the specific ARF is a transcriptional activator, the Aux/IAA binding represses transcription.

auxin response element (AuxRE) A DNA promoter sequence that modulates gene expression when bound by auxin-responsive transcription factors.

auxin response factors (ARFs) A family of proteins that regulate the transcription of specific genes involved in auxin responses; they are inhibited by association with specific Aux/IAA repressor proteins, which are degraded in the presence of auxin.

auxins Major plant hormones involved in numerous developmental processes, including cell elongation, organogenesis, apical–basal polarity, differentiation, apical dominance, and tropic responses. The most abundant chemical form is indole-3-acetic acid.

auxin sink A cell or tissue that takes up auxin from a nearby auxin source. Participates in auxin canalization during vascular differentiation.

auxin source A cell or tissue that exports auxin to other cells or tissues by polar transport.

axillary buds Secondary meristems that are formed in the axils of leaves.

axillary meristem Meristematic tissue in the axils of leaves that gives rise to axillary buds.

B

***Bacillus thuringiensis* (Bt)** A soil bacterium that is the source for a commonly used transgene encoding an insecticidal toxin.

bacteriochlorophylls Light-absorbing pigments active in photosynthesis in anoxygenic photosynthetic organisms.

bacteroids Endosymbiotic bacteria that have differentiated into a nondividing nitrogen-fixing state.

bark Collective term for all the tissues outside the cambium of a woody stem or root, and composed of phloem, periderm and dead tissue.

basal cell In embryogenesis, the larger, vacuolated cell formed by the first division of the zygote. Gives rise to the suspensor.

basipetal From the growing tip of a shoot or root toward the base (junction of the root and shoot).

berry A simple, fleshy fruit produced from a single ovary and consisting of an outer pigmented exocarp, a fleshy, juicy mesocarp, and a membranous inner endocarp.

β-oxidation Oxidation of fatty acids into fatty acyl-CoA, and the sequential breakdown of the fatty acids into acetyl-CoA units. NADH is also produced.

biennial plant A plant that requires two growing seasons to flower and produce seed.

biolistics A procedure, also called the "gene gun" technique, in which tiny gold particles coated with the genes of interest are mechanically shot into cultured cells. Some of the DNA is randomly incorporated into the genome of the targeted cells.

biosphere The parts of the surface and atmosphere of Earth that support life as well as the organisms living there.

biotrophic pathogens Pathogens that leave infected tissue alive and only minimally damaged while the pathogen continues to feed on host resources.

bleached The loss of chlorophyll's characteristic absorbance due to its conversion into another structural state, often by oxidation.

bordered pit A pit pair in which the pit chamber is over-arched by the cell wall, creating a larger pit chamber and smaller pit aperture.

boundary layer resistance (r_b) The resistance to the diffusion of water vapor, CO_2, and heat due to the layer of unstirred air next to the leaf surface. A component of diffusional resistance.

Bowen ratio The ratio of sensible heat loss to evaporative heat loss, the two most important processes in the regulation of leaf temperature.

brace roots Roots derived from aerial nodes of some monocot plants that penetrate the soil to anchor large shoots.

brassinolide A plant steroid hormone with growth-promoting activity, first isolated from *Brassica napus* pollen. One of a group of plant hormones with similar structures and activities, called brassinosteroids.

brassinosteroids (BRs) A group of plant steroid hormones that play important roles in many developmental processes, including cell division and cell elongation in stems and roots, photomorphogenesis, reproductive development, leaf senescence, and stress responses.

bryophyte *See* nonvascular plants.

bulk flow Translocation of water and solutes down a pressure gradient, as in the xylem or phloem.

bundle sheath One or more layers of closely packed cells surrounding the small veins of leaves and the primary vascular bundles of stems.

C

C_4 photosynthesis Photosynthetic carbon metabolism in which the initial fixation of CO_2 is catalyzed by phospho*enol*pyruvate carboxylase (not by Rubisco as in C_3 photosynthesis), producing a four-carbon compound (oxaloacetate). The fixed carbon is subsequently released and refixed by the Calvin–Benson cycle.

calcium oxalate crystals Crystals of calcium oxalate that form in the vacuoles of some plant species to deter herbivory by insects and mammals.

callose A β-1,3-glucan synthesized in the plasma membrane and deposited between the plasma membrane and the cell wall. Synthesized by sieve elements in response to damage or stress, or as part of a normal developmental process.

calmodulin A Ca^{2+}-binding protein that regulates many cellular processes in a Ca^{2+}-dependent manner.

Calvin–Benson cycle The biochemical pathway for the reduction of CO_2 to carbohydrate. The cycle involves three phases: the carboxylation of ribulose 1,5-bisphosphate with atmospheric CO_2, catalyzed by Rubisco; the reduction of the formed 3-phosphoglycerate to triose phosphates; and the regeneration of ribulose 1,5-bisphosphate through the concerted action of ten enzymatic reactions.

CAM *See* Crassulacean acid metabolism.

cambium Layer of meristematic cells between the xylem and phloem that produces cells of these tissues and results in the lateral (secondary) growth of the stem or root.

canalization model The hypothesis that as auxin flows through tissues it stimulates and polarizes its own transport, which gradually becomes channeled—or canalized—into files of cells leading away from auxin sources; these cell files can then differentiate to form vascular tissue.

capillarity The movement of water for small distances up a glass capillary tube or within the cell wall, due to water's cohesion, adhesion, and surface tension.

carbon fixation reactions The synthetic reactions occurring in the stroma of the chloroplast that use the high-energy compounds ATP and NADPH for the incorporation of CO_2 into carbon compounds.

carbon reactions *See* carbon fixation reactions.

cardenolides Steroidal glycosides that taste bitter and are extremely toxic to higher animals through their action on Na^+/K^+-activated ATPases. Extracted from foxglove (*Digitalis*) for treatment of human heart disorders.

cardiac glycosides Glycosylated organic plant defense compounds—similar to oleandrin from oleander bushes—that are poisonous to animals and inhibit Na^+/K^+ channels to cause contractions in cardiac muscles.

carotenoids Linear polyenes arranged as a planar zigzag chain, with conjugated double bonds. These orange pigments serve both as antenna pigments and photoprotective agents.

carriers Membrane transport proteins that bind to a solute, undergo conformational change, and release the solute on the other side of the membrane.

Casparian strip A band in the cell walls of the endodermis that is impregnated with lignin. Prevents apoplastic movement of water and solutes into the stele.

catalase An enzyme that breaks hydrogen peroxide down into water. When it is abundant in peroxisomes, it may form crystalline arrays.

cation exchange The replacement of mineral cations adsorbed to the surface of soil particles by other cations.

cavitation The collapse of tension in a column of water resulting from the formation and expansion of tiny gas bubbles.a particular cell.

cell plate A wall-like structure that separates newly divided cells. Formed by the phragmoplast and later becomes the cell wall.

cellulose A linear chain of (1,4)-linked β-D-glucose. The repeating unit is cellobiose.

cellulose microfibril Thin, ribbon-like structure of indeterminate length and variable width composed of 1→4-linked β-D-glucan chains tightly packed in crystalline arrays alternating with less organized amorphous regions. Provides structural integrity to the cell walls of plants and determines the directionality of cell expansion.

cellulose synthase Enzyme that catalyzes the synthesis of individual (1,4)-linked β-D-glucans that make up cellulose microfibrils.

cell wall The rigid cell surface structure external to the plasma membrane that supports, binds, and protects the cell. Composed of cellulose and other polysaccharides and proteins.

central cell The cell in the embryo sac that fuses with the second sperm cell, giving rise to the primary endosperm cell.

central zone (CZ) A central cluster of relatively large, highly vacuolated, slowly dividing initial cells in shoot apical meristems, comparable to the quiescent center of root meristems.

centromere The constricted region on the mitotic chromosome where the kinetochore forms and to which spindle fibers attach.

CF_0CF_1-ATP synthase The multi-subunit protein complex that synthesizes ATP from ADP and phosphate (P_i). F_oF_1 and CF_0-CF_1 types are present in mitochondria and chloroplasts, respectively. Also called ATP synthase and ATPase.

channels Transmembrane proteins that function as selective pores for passive transport of ions or water across the membrane.

charge-separated state State produced in photosynthetic reaction centers after excitation by light in which an electron is moved from the lumen side to the stromal side. This movement of electrons forms an electric field across the thylakoid.

checkpoint One of several key regulatory points in the cell cycle. One checkpoint, late in G_1, determines if the cell is committed to the initiation of DNA synthesis.

chelator A carbon compound (e.g., malic acid or citric acid) that can form a soluble noncovalent complex with certain cations, thereby facilitating their uptake.

chemical fertilizers Fertilizers that provide nutrients in inorganic form.

chemical potential The free energy associated with a substance that is available to perform work.

chemiosmotic mechanism The mechanism whereby the electrochemical gradient of protons established across a membrane by an electron transport process is used to either drive energy-requiring ATP synthesis (mitochondria and chloroplasts) or carrier-mediated anion efflux at the plasma membrane (as in polar auxin transport).

chilling (or stratification) Technique for breaking seed dormancy by storage at cold temperatures (1–10°C) under moist conditions for several months. The term *stratification* is derived from the former practice of breaking dormancy by allowing seeds to overwinter in small mounds of alternating layers of seeds and soil.

chlorophyll *a/b* antenna proteins Chlorophyll-containing proteins associated with one or the other of the two photosystems in eukaryotic organisms. Also known as light-harvesting complex proteins (LHC proteins).

chlorophylls A group of light-absorbing green pigments active in photosynthesis.

chloroplast The organelle that is the site of photosynthesis in photosynthetic eukaryotic organisms.

chloroplast envelope The double-membrane system surrounding the chloroplast.

chlorosis The yellowing of plant leaves such as occurs as a result of mineral deficiency. The leaves affected and the parts of the leaves that yellow can be diagnostic for the type of deficiency.

Cholodny–Went model Early mechanism proposed for tropisms involving stimulation of the bending of the plant axis by lateral transport of auxin in response to a stimulus, such as light, gravity, or touch. The original model has been supported and expanded by recent experimental evidence.

chromatin The DNA–protein complex found in the interphase nucleus. Condensation of chromatin occurs during cell division to form the rod-shaped mitotic and meiotic chromosomes.

chromophore A light-absorbing pigment molecule that is usually bound to a protein (an apoprotein).

chromoplasts Plastids that contain high concentrations of carotenoid pigments, rather than chlorophyll.

chromosome A threadlike or rod-shaped structure of DNA and protein found in the nucleus of most living cells, encoding genetic information in the form of genes.

chronic photoinhibition Photoinhibition of photosynthentic activity in which both quantum efficiency and the maximum rate of photosynthesis are decreased. Occurs under high levels of excess light.

circadian rhythm A physiological process that oscillates endogenously with a cycle of approximately 24 hours.

cisternae (singular *cisterna*) A network of flattened saccules and tubules that compose the endoplasmic reticulum.

clathrin Proteins that have a unique *triskelion* structure that spontaneously assemble into 100-nm cages that coat vesicles associated with endocytosis at the plasma membrane and other cellular trafficking events.

climacteric Marked rise in respiration at the onset of ripening that occurs in all fruits that ripen in response to ethylene, and in the senescence process of detached leaves and flowers.

CO_2 compensation point The CO_2 concentration at which the rate of respiration balances the photosynthetic rate.

coat-imposed dormancy Seed dormancy that is caused by the seed coat and other surrounding tissues; may involve impermeability to water or oxygen, mechanical restraint, or the retention of endogenous inhibitors. Also called exogenous dormancy.

coevolution Linked adaptations of two or more organisms.

cohesion The mutual attraction between water molecules due to extensive hydrogen bonding.

coincidence model A model for flowering in photoperiodic plants in which the circadian oscillator controls the timing of light-sensitive and light-insensitive phases during the 24-h cycle.

coleoptile A modified ensheathing leaf-like that covers and protects the young primary leaves of a grass seedling as it grows through the soil. Unilateral light perception, especially blue light, by the tip results in asymmetric growth and bending due to unequal auxin distribution in the lighted and shaded sides.

coleorhiza A protective sheath surrounding the embryonic radicle in members of the grass family.

collection phloem Sieve elements of sources.

collenchyma A specialized parenchyma with irregularly thickened, pectin-rich, primary cell walls.

commensalism A relationship between two organisms in which one organism benefits without negatively affecting the other.

common symbiotic pathway A sequence of common cellular events in plant roots that occurs in both mycorrhizal formation and root nodulation.

companion cells In angiosperms, metabolically active cells that are connected to their sieve element by large, branched plasmodesmata and that take over many of the metabolic activities of the sieve element. In source leaves, they function in the transport of photosynthate into the sieve elements.

compatible solutes Organic compounds that are accumulated in the cytosol during osmotic adjustment. Compatible solutes do not inhibit cytosolic enzymes, unlike high concentrations of ions. Examples of compatible solutes include proline, sorbitol, mannitol, and glycine betaine.

compound leaf A leaf subdivided into leaflets.

condensed tannins Tannins that are polymers of flavonoid units. Require use of strong acid for hydrolysis.

conifer resin ducts Ducts or channels in conifer leaves and woody tissue that conduct terpenoid defense compounds. They may be constitutive, or their formation may be induced by wounding or defense responses.

conifers Cone-bearing trees.

constitutive defenses Plant defenses that are always immediately available or operational; that is, defenses that are not induced.

contact angle A quantitative measure of the degree to which a water molecule is attracted to a solid phase versus to itself.

cork cambium A layer of lateral meristem that develops within mature cells of the cortex and the secondary phloem. Produces the secondary protective layer, the periderm. Also called phellogen.

cortex Ground tissue in the region of the primary stem or root located between the vascular tissue and the epidermis, mainly consisting of parenchyma.

cortical ER The network of endoplasmic reticulum that lies just under the plasma membrane and is associated with the plasma membrane at specific contact points.

cotyledons Also called "seed leaves," they are the main sources of nutrients during germination and post-germination photosynthesis before the first true leaves develop.

crassulacean acid metabolism (CAM) A biochemical process for concentrating CO_2 at the carboxylation site of Rubisco. Found in the family Crassulaceae (*Crassula*, *Kalanchoe*, *Sedum*) and numerous other families of angiosperms. In CAM, CO_2 uptake and initial fixation take place at night, and decarboxylation and reduction of the internally released CO_2 occur during the day.

cristae Folds in the inner mitochondrial membrane that project into the mitochondrial matrix.

critical day length The minimum length of the day required for flowering of a long-day plant; the maximum length of day that will allow short-day plants to flower. However, studies have shown that it is the length of the night, not the length of the day, that is important.

cross-pollination Pollination of a flower by pollen from the flower of a different plant.

cross-protection A plant response to one environmental stress that confers resistance to another stress.

cross-regulation The interaction of two or more signaling pathways.

crown gall A tumor-forming plant disease resulting from wound infection of the stem or trunk by the soil-dwelling bacterium *Agrobacterium tumefaciens*. A tumor resulting from the disease.

crown roots Adventitious roots that emerge from the lowermost nodes of a stem.

cryptochromes Flavoproteins implicated in many blue-light responses that have strong homology with bacterial photolyases.

cyanogenic glycosides Nonalkaloid, nitrogenous protective compounds that break down to give off the poisonous gas hydrogen cyanide when the plant is crushed.

cyclic electron flow In photosystem I, the flow of electrons from the electron acceptors through the cytochrome b_6f complex and back to P700, coupled to proton pumping into the lumen. This electron flow energizes ATP synthesis but does not oxidize water or reduce $NADP^+$.

cyclin-dependent kinases (CDKs) Protein kinases that regulate the transitions from G_1 to S, from S to G_2, and from G_2 to mitosis, during the cell cycle.

cyclins Regulatory proteins associated with cyclin-dependent kinases that play a crucial role in regulating the cell cycle.

cyst nematodes Parasitic nematodes that invade roots and transform into a non-motile cyst. The soybean cyst nematode *Heterodera glycines* is a major threat to soybean production.

cytochrome b_6f complex A large multi-subunit protein complex containing two *b*-type hemes, one *c*-type heme (cytochrome *f*), and a Rieske iron–sulfur protein. A relatively immobile complex distributed equally between the grana and the stroma regions of the thylakoid membranes.

cytochrome *c* A peripheral, mobile component of the mitochondrial electron transport chain that oxidizes complex III and reduces complex IV.

cytochrome *f* A subunit in the cytochrome b_6f complex that plays a role in electron transport between photosystems I and II.

cytochrome P450 monooxygenase (CYP) A generic term for a large number of related, but distinct, mixed-function oxidative enzymes localized on the endoplasmic reticulum. CYPs participate in a variety of oxidative processes, including steps in the biosynthesis of gibberellins and brassinosteroids.

cytohistological zonation Regional cytological differences in cell division in the shoot apical meristems of seed plants.

cytokinesis In plant cells, following nuclear division, the separation of daughter nuclei by the formation of new cell wall.

cytokinins A class of plant hormones that function in cell division and differentiation, as well as axillary bud growth, and leaf senescence. Cytokinins are adenine derivatives, the most common form of which is zeatin.

cytoplasm The cellular matter enclosed by the plasma membrane exclusive of the nucleus.

cytoplasmic streaming The coordinated movement of particles and organelles through the cytosol.

cytoskeleton Composed of polarized microfilaments of actin or microtubules of tubulin, the cytoskeleton helps control the organization and polarity of organelles and cells during growth.

cytosol The aqueous phase of the cytoplasm containing dissolved solutes but excluding supramolecular structures, such as ribosomes and components of the cytoskeleton.

D

damage-associated molecular patterns (DAMPs) Molecules within cells that are released from damaged or dying cells due to injury or pathogen infection to activate innate immune responses.

day-neutral plant (DNP) A plant whose flowering is not regulated by day length.

de-etiolation The rapid developmental changes associated with loss of the etiolated form due to the action of light. *See* photomorphogenesis.

deficiency zone In plant tissue, the range of concentrations of a mineral nutrient below the critical concentration, the highest concentration where reduced plant growth or yield is observed.

dehiscent fruit A dry fruit that spontaneously splits, releasing its seeds.

dehydrins Hydrophilic plant proteins that accumulate in response to drought stress and cold temperatures.

dermal tissue The tissue system that covers the outside of the plant body; the epidermis or periderm.

desaturase enzymes Enzymes that remove hydrogens in a carbon chain to create a double bond between carbons.

desiccation tolerance A plant's ability to function while dehydrated.

diacylglycerol (DAG) A molecule consisting of the three-carbon glycerol molecule to which two fatty acids are covalently attached by ester linkages.

difference in water vapor concentration (Δc_{wv}) The difference between the water vapor concentration of the air spaces inside the leaf and that of the air outside the leaf.

differentiation Process by which a cell acquires metabolic, structural, and functional properties that are distinct from those of its progenitor cell. In plants, differentiation is frequently reversible, when excised differentiated cells are placed in tissue culture.

diffuse growth A type of cell growth in plants in which expansion occurs more or less uniformly over the entire surface. Contrast with tip growth.

diffusion coefficient (D_s) The proportionality constant that measures how easily a specific substance *s* moves through a particular medium. The diffusion coefficient is a characteristic of the substance and depends on both the medium and the temperature.

diffusion The movement of substances due to random thermal agitation from regions of high free energy to regions of low free energy (e.g., from high to low concentration).

diffusional resistance (*r*) The restriction posed by the boundary layer and the stomata to the free diffusion of gases from and into the leaf.

diffusion potential The potential (voltage) difference that develops across a semipermeable membrane as a result of the differential permeability of solutes with opposite charges (for example, K^+ and Cl^-).

diploid Cell with two copies of each chromosome and abbreviated as having 2*N* chromosomes.

dioecious Refers to plants in which male and female flowers are found on different individuals, such as spinach (*Spinacia*) and hemp (*Cannabis sativa*). Contrast with *monoecious*.

directed organelle movement Movement of an organelle in a particular direction, which can be driven by the interaction with molecular motors associated with the cytoskeleton.

DNA synthesis phase (S, or S phase) In the cell cycle, the stage during which DNA is replicated; it follows G_1 and precedes G_2.

dormancy A living condition in which growth does not occur under conditions that are normally favorable to growth.

double fertilization A unique feature of all angiosperms whereby, along with the fusion of a sperm with the egg to create a diploid zygote, a second male gamete fuses with the polar nuclei in the embryo sac to generate the endosperm tissue (with a triploid or higher number of chromosomes).

drupe A structure similar to a berry, but with a hardened, shell-like endocarp (pit or stone) that contains a seed.

dynamic photoinhibition Photoinhibition of photosynthesis in which quantum efficiency decreases but the maximum photosynthetic rate remains unchanged. Occurs under moderate, not high, excess light.

E

ectomycorrhizae Symbioses in which the fungus typically forms a thick sheath, or mantle, around the roots. The root cells themselves are not penetrated by the fungal hyphae, and instead are surrounded by a network of hyphae called the Hartig net.

effector A molecule that binds a protein to change its activity. Bacterial effectors are secreted by pathogens to act on proteins within a host cell.

effector-triggered immunity Immune responses that are mediated by a class of intracellular R proteins.

egg cell The female gamete.

electrochemical potential The chemical potential of an electrically charged solute.

electrochemical proton gradient The sum of the electrical charge gradient and the pH gradient across the membrane, resulting from a concentration gradient of protons.

electrogenic transport Active ion transport involving the net movement of charge across a membrane.

electronegative Having the capacity to attract electrons and thus producing a slightly negative electric charge.

electroneutral transport Active ion transport that involves no net movement of charge across a membrane.

elicitors Specific pathogen molecules or cell wall fragments that bind to plant proteins and thereby act as signals for activation of plant defense against a pathogen.

elongation zone The region of rapid and extensive root cell elongation showing few, if any, cell divisions.

embryo The early sporophytic form of the plant formed from the zygote produced by gametic fusion or apomixis that becomes the basis of the mature plant axis. The embryo lacks leaves, stems, and reproductive structures.

embryo dormancy Seed dormancy that is caused directly by the embryo and is not due to any influence of the seed coat or other surrounding tissues.

embryogenesis The formation and development of the embryo.

embryophyte The plant group, including all land plants, characterized by the ability of the gametophyte to contain and nurture the young sporophyte within its tissues through the earliest stages of development.

endocytosis The formation of small vesicles from the plasma membrane, which detach and move into the cytosol, where they fuse with elements of the endomembrane system.

endodermis A specialized layer of cells surrounding the vascular tissue in roots and some stems.

endogenous dormancy Seed dormancy that is intrinsic to the seed and is not due to any physical or chemical influence of the seed coat or other surrounding tissues. Endogenous dormancy is regulated by the ratio of the hormones abscisic acid (ABA) and gibberellin (GA).

endoplasmic reticulum (ER) A continuous membrane system within the cytoplasm of eukaryotic cells that serves multiple functions, including the synthesis, modification, and intracellular transport of proteins.

endoreduplication Cycles of nuclear DNA replication without mitosis resulting in polyploidization.

endosperm Tissue that surrounds and nourishes the embryo in the seeds of angiosperms (flowering plants). In angiosperms, endosperm is formed by fusion of at least one polar nucleus in the embryo sac with one of the two sperm nuclei from the pollen grain. In gymnosperms the nutritive material of the seed is haploid and present before fertilization.

energy conversion efficiency The fraction of energy from photons that is stored in photochemical reactions. Energy is lost in making reactions go forward, which prevents back reactions. This is distinct from the quantum yield, which describes the fraction of light that induces stable photochemistry.

energy transfer In the light reactions of photosynthesis, the direct transfer of energy from an excited molecule, such as β-carotene, to another molecule, such as chlorophyll. Energy transfer can also take place between chemically identical molecules, as in chlorophyll-to-chlorophyll transfer.

entrain Manipulation of external controlling factors, such as light or removal of light, to synchronize biological rhythms.

envelope The double-membrane system surrounding the chloroplast or the nucleus. The outer membrane of the nuclear envelope is continuous with the endoplasmic reticulum.

epidermis The outermost layer of plant cells, typically one cell thick.

epigenetic changes Chemical modifications to DNA and histones that cause heritable changes in gene activity without altering the underlying DNA sequence.

epigenome Heritable chemical modifications to DNA and chromatin, including DNA methylation, histone methylation, and acetylation.

essential element A chemical element that is an intrinsic component of the structure or metabolism of a plant. When the element is in limited supply, a plant suffers abnormal growth, development, or reproduction.

ethylene A gaseous plant hormone involved in fruit ripening, abscission, and the growth of etiolated seedlings. The chemical formula of ethylene is C_2H_4.

etiolated Effects of seedling growth in the dark, in which the hypocotyl and stem are more elongated, cotyledons and leaves do not expand, and chloroplasts do not mature.

etioplast Photosynthetically inactive form of chloroplast found in etiolated seedlings.

euchromatin The dispersed, transcriptionally active form of chromatin. Compare with *heterochromatin*.

eudicot Abbreviated name for the eudicotyledons, one of two major classes of the angiosperms, and refers to the fact that plants in this class have two seed leaves (cotyledons).

exocytosis The process by which vesicle contents are distributed to the plasma membrane or apoplast.

exodermis A specialized layer of cells in most angiosperms, but not gymnosperms, containing a Casparian strip, in the outer layer of the cortex of the root.

exogenous dormancy Seed dormancy that is caused by the seed coat and other surrounding tissues; may involve impermeability to water or oxygen, mechanical restraint, or the retention of endogenous inhibitors. Also called coat-imposed dormancy.

expansins A class of wall-loosening proteins that accelerate wall yielding during cell elongation, typically with an optimum at acidic pH.

export The movement of photosynthate in sieve elements away from the source tissue.

extracellular space In plants, the space continuum outside the plasma membrane made up of interconnecting cell walls through which water and mineral nutrients readily diffuse.

extrafloral nectar Nectar produced outside the flower and not involved in pollination events.

F

F_1 The ATP-binding, matrix-facing part of the F_oF_1-ATP synthase.

facilitated diffusion Passive transport across a membrane using a carrier.

Fe–S centers Prosthetic groups consisting of inorganic iron and sulfur that are abundant in proteins in respiratory and photosynthetic electron transport.

fermentation The metabolism of pyruvate in the absence of oxygen, leading to the oxidation of the NADH generated in glycolysis to NAD^+. Allows glycolytic ATP production to function in the absence of oxygen.

ferredoxin (Fd) A small, water-soluble iron–sulfur protein involved in electron transport in photosystem I.

ferredoxin–$NADP^+$ reductase (FNR) A membrane-associated flavoprotein that receives electrons from photosystem I and reduces $NADP^+$ to NADPH.

ferredoxin–thioredoxin system Three chloroplast proteins (ferredoxin, ferredoxin–thioredoxin reductase, thioredoxin). The three proteins use reducing power from the photosynthetic electron transport chain to reduce protein disulfide bonds in a cascade of thiol–disulfide exchanges. As a result, light controls the activity of several enzymes of the Calvin–Benson cycle.

ferritin A protein that functions in cellular iron storage in multiple compartments, including plastids and mitochondria.

fertilization The formation of a diploid (2*N*) zygote from the cellular and nuclear fusion of two haploid (1*N*) gametes, the egg and the sperm. In angiosperms, fertilization also involves fusion of a second sperm nucleus with the haploid nuclei (usually two) of the central cell to form the endosperm (usually triploid).

FeS_R An iron- and sulfur-containing subunit of the cytochrome b_6f complex, involved in electron and proton transfer. *See also* Rieske iron–sulfur protein.

FeS_X, FeS_A, FeS_B Membrane-bound iron–sulfur proteins that transfer electrons between photosystem I and ferredoxin.

fiber An elongated, tapered sclerenchyma cell that provides support in vascular plants.

fimbrin An actin-binding protein that bundles F-actin filaments together into larger filamentous bundles.

flavin adenine dinucleotide (FAD) A riboflavin-containing cofactor that undergoes a reversible two-electron reduction to produce $FADH_2$.

flavin mononucleotide (FMN) A riboflavin-containing cofactor that undergoes a reversible one- or two-electron reduction to produce FMNH or $FMNH_2$. FAD and FMN occur as cofactors in flavoproteins.

flippase Enzyme that "flips" newly synthesized phospholipids across the bilayer from the outer (cytoplasmic) face of the membrane to the inner leaflet, thereby ensuring symmetrical lipid composition of the membrane.

floral evocation The events occurring in the shoot apex that specifically commit the apical meristem to produce flowers.

floral meristem The meristem that forms floral (reproductive) organs: sepals, petals, stamens, and carpels. May form directly from a vegetative meristem or indirectly via an inflorescence meristem.

floral meristem identity genes Two classes of genes, one required for the conversion of a shoot apical meristem into a floral meristem, the other that maintains the identity of an inflorescence meristem (as opposed to a floral meristem).

floral organ identity genes Three classes of genes that control the specific locations of floral organs in the flower. *See also* ABC model.

flowering plant *See* angiosperm.

fluence The number of photons absorbed per unit of surface area over time ($\mu mol\ m^{-2}\ s^{-1}$).

fluid mosaic The common molecular lipid–protein structure for all biological membranes. A double layer (bilayer) of polar lipids (phospholipids or, in chloroplasts, glycosylglycerides) has a hydrophobic, fluid-like interior. When structural sterols and sphingolipids are present in the bilayer, the fluidity of the interior is decreased. Membrane proteins are embedded in the bilayer and may move laterally.

fluorescence Following light absorption, the emission of light at a slightly longer wavelength (lower energy) than the wavelength of the absorbed light.

fluorescence resonance energy transfer (FRET) The physical mechanism by which excitation energy is conveyed from a molecule that absorbs light to an adjacent molecule.

flux density (J_s) The rate of transport of a substance *s* across a unit of area per unit of time. J_s may have units of moles per square meter per second (mol m^{-2} s^{-1}).

F_o The integral membrane part of the F_oF_1-ATP synthase.

F_oF_1-ATP synthase A multi-subunit protein complex associated with the inner mitochondrial membrane that couples the passage of protons across the membrane to the synthesis of ATP from ADP and phosphate. The subscript "o" in F_o refers to the binding of the inhibitor oligomycin. Similar to CF_0CF_1-ATP synthase in photophosphorylation, to which oligomycin does not bind and inhibit (hence the subscript is "0").

foliar application The application, and subsequent absorption, of mineral nutrients or other chemical compounds to leaves as sprays.

formins Proteins that bind actin and actin–profilin complexes to initiate polymerization of the actin filament.

free-running Designation of the biological rhythm that is characteristic for a particular organism when environmental signals are removed, as in total darkness. *See* Zeitgeber.

frequency (ν) A unit of measurement that characterizes waves, in particular light energy. The number of wave crests that pass an observer in a given time.

fruit In angiosperms, one or more mature ovaries containing seeds and sometimes adjacent attached parts.

fruit set The commitment to initiate fruit development. It is a key developmental switch that normally occurs following successful fertilization and/or pollination.

fusicoccin A fungal toxin that induces acidification of plant cell walls by activating H^+-ATPases in the plasma membrane. Fusicoccin stimulates rapid acid growth in stem and coleoptile sections. It also stimulates stomatal opening by stimulating proton pumping at the guard cell plasma membrane.

G

G_1 The phase of the cell cycle preceding the synthesis of DNA.

G_2 The phase of the cell cycle following the synthesis of DNA.

GABA shunt A pathway supplementing the tricarboxylic acid cycle with the ability to form and degrade GABA.

gamete A haploid (1*N*) reproductive cell.

gametophyte The haploid (1*N*) multicellular structure that produces haploid gametes by mitosis and differentiation. Contrast with sporophyte.

gametophyte generation The stage, or generation, in the life cycle of plants that produces gametes. It alternates with the sporophyte generation, in a process called alternation of generations.

gap 1 (G_1) The phase of the cell cycle preceding the synthesis of DNA.

gap 2 (G_2) The phase of the cell cycle following the synthesis of DNA.

gate A structural domain of the channel protein that opens or closes the channel in response to external signals such as voltage changes, hormone binding, or light.

gene gun *See* biolistics.

genome Refers to all the genes in a haploid complement of eukaryotic chromosomes, in an organelle, a microbe, or the DNA or RNA content of a virus.

genotype The genetic constitution of an individual organism. Sometimes used to indicate a variant, cultivar, or ecotype carrying a unique set of genomic sequence variations.

germination The events that take place between the start of imbibition of the dry seed and the emergence of part of the embryo, usually the radicle, from the structures that surround it. May also be applied to other quiescent structures, such as pollen grains or spores.

GIBBERELLIN INSENSITIVE DWARF 1 (GID1) Gibberellin receptor protein in rice.

gibberellins (GAs) A large group of chemically related plant hormones synthesized by a branch of the terpenoid pathway and associated with the promotion of stem growth (especially in dwarf and rosette plants), seed germination, and many other functions.

glucan A polysaccharide made from glucose units.

gluconeogenesis The synthesis of carbohydrates through the reversal of glycolysis.

glucose 6-phosphate dehydrogenase A cytosolic and plastidic enzyme that catalyzes the initial reaction of the oxidative pentose phosphate pathway.

glutamate dehydrogenase (GDH) An enzyme that catalyzes a reversible reaction that synthesizes or deaminates glutamate as part of the nitrogen assimilation process.

glutamate synthase (GOGAT) An enzyme that transfers the amide group of glutamine to 2-oxoglutarate, yielding two molecules of glutamate. Also known as glutamine:2-oxoglutarate aminotransferase.

glutamine synthetase (GS) An enzyme that catalyzes the condensation of ammonium and glutamate to form glutamine. The reaction is critical for the assimilation of ammonium into essential amino acids. Two forms of GS exist—one in the cytosol and one in chloroplasts.

glyceroglycolipids Glycerolipids in which sugars form the polar head group. Glyceroglycolipids are the most abundant glycerolipids in chloroplast membranes.

glycosides Compounds containing an attached sugar or sugars.

glycosylphosphatidylinositol anchor In plants, a phosphoglyceride protein modification that anchors proteins to ordered plasma membrane domains.

glycerophospholipids Polar glycerolipids in which the hydrophobic portion consists of two 16-carbon or 18-carbon fatty acid chains esterified to positions 1 and 2 of a glycerol backbone. The phosphate-containing polar head group is attached to position 3 of the glycerol.

glycine betaine *N*,*N*,*N*-trimethylglycine, which functions in drought stress protection, and was originally identified in sugar beet (*Beta vulgaris*).

glycolysis A series of reactions in which a sugar is oxidized to produce two molecules of pyruvate. A small amount of ATP and NADH is produced.

glycophytes Plants that are not able to resist salts to the same degree as halophytes. Show growth inhibition, leaf discoloration, and loss of dry weight at soil salt concentrations above a threshold. Contrast with *halophytes*.

glycoprotein Protein that has covalently attached sugar oligomers or polymers.

glyoxylate cycle The sequence of reactions that convert two molecules of acetyl-CoA to succinate in the glyoxysome.

glyoxysome An organelle found in the oil-rich storage tissues of seeds in which fatty acids are oxidized. A type of microbody.

glyphosate resistance Genetic capacity to survive a field application of the commercial herbicide Roundup, which kills weeds but does not harm resistant crop plants.

Goldman equation An equation that predicts the diffusion potential across a membrane, as a function of the concentrations and permeabilities of all ions (e.g., K^+, Na^+, and Cl^-) that permeate the membrane.

Golgi apparatus Endomembrane organelle that modifies and packages secreted proteins and some cell wall components. A site of key steps in protein glycosylation. Named for its discoverer, Camillo Golgi.

grana lamellae Stacked thylakoid membranes within the chloroplast. Each stack is called a granum.

granum (plural *grana*) In the chloroplast, a stack of thylakoids.

gravitational potential (Ψ_g) The part of the water potential caused by gravity. It is only of a significant size when considering water transport into trees and drainage in soils.

gravitropic response The growth initiated by the root cap's perception of gravity and the signal that directs the roots to grow downward.

gravitropism Plant growth in response to gravity, enabling roots to grow downward into the soil and shoots to grow upward.

greenhouse effect The warming of Earth's climate, caused by the trapping of long-wavelength radiation by CO_2 and other gases in the atmosphere. The term is derived from the heating of a greenhouse that results from the penetration of long-wavelength radiation through the glass roof, the conversion of the long-wave radiation to heat, and the blocking of the heat escape by the glass roof.

green-leaf volatiles A mixture of lipid-derived six-carbon aldehydes, alcohols, and esters released by plants in response to mechanical damage.

ground tissue The internal tissues of the plant, other than the vascular tissues.

growth respiration The respiration that provides the energy needed for converting sugars into the building blocks that make up new tissue. Contrast with *maintenance respiration*.

guard cells A pair of specialized epidermal cells that surround the stomatal pore and regulate its opening and closing.

guard mother cell (GMC) The cell that gives rise to a pair of guard cells to form a stomate.

guttation An exudation of liquid from the leaves due to root pressure.

gymnosperms An early group of seed plants. Distinguished from angiosperms by having seeds borne unprotected (naked) in cones.

H

halophytes Plants that are native to saline soils and complete their life cycles in that environment. Contrast with *glycophytes*.

haploid Cell with only one copy of each chromosome, abbreviated as 1*N*.

Hartig net A network of fungal hyphae that surround, but do not penetrate, the cortical cells of roots in ectomycorrhizal symbioses.

haustorium The hyphal tip of a fungus or root tip of a parasitic plant that penetrates host plant tissue.

HC-toxin A cell permeant cyclic tetrapeptide produced by the maize pathogen *Cochliobolus carbonum* that inhibits histone deacetylases.

heat shock factors Transcription factors that regulate expression of heat shock proteins.

heat shock proteins (HSPs) A specific set of proteins that are induced by a rapid rise in temperature, and by other factors that lead to protein denaturation. Most act as molecular chaperones.

heliotropism The movements of leaves toward or away from the sun.

hemibiotrophic pathogens Pathogens that show an initial biotrophic stage, which is followed by a necrotrophic stage in which the pathogen causes extensive tissue damage.

hemicelluloses Heterogeneous group of polysaccharides that bind to the surface of cellulose, linking cellulose microfibrils together into a network.

hemiparasitic plants Photosynthetic plants that are also parasites.

herbivory Consumption of plants or parts of plants as a food source.

heterochromatin Chromatin that is densely packed and transcriptionally modified or suppressed. Compare with *euchromatin*.

hexose phosphates Six-carbon sugars with phosphate groups attached.

histogenesis The differentiation of cells to produce various tissues.

histones A family of proteins that interact with DNA and around which DNA is wound to form a nucleosome.

Hoagland solution A type of nutrient solution for plant growth, originally formulated by Dennis R. Hoagland.

holoparasitic plants Nonphotosynthetic plants that are obligate parasites.

hormone balance theory The hypothesis that seed dormancy and germination are regulated by the balance of ABA and gibberellin.

H^+-pyrophosphatase (H^+-PPase) An electrogenic pump that moves protons into the vacuole, energized by the hydrolysis of pyrophosphate. H^+-PPases are also thought to function in reverse to regenerate pyrophosphate when cytosolic levels are depleted.

hybrid vigor (heterosis) The increased vigor often observed in the offspring of crosses between two inbred varieties of the same plant species.

hydraulic conductivity A measure of how readily water can move across a membrane; it is expressed in terms of volume of water per unit of area of membrane per unit of time per unit of driving force (i.e., $m^3\ m^{-2}\ s^{-1}\ MPa^{-1}$).

hydrogen bonds Weak chemical bonds formed between a hydrogen atom and an oxygen or nitrogen atom.

hydroponics A technique for growing plants with their roots immersed in nutrient solution without soil.

hydrostatic pressure Pressure generated by compression of water into a confined space. Measured in units called pascals (Pa) or, more conveniently, megapascals (MPa).

hydrotropism Plant growth in response to root perception of water potential gradients in the soil, enabling root growth toward areas of greater water potential.

hyperaccumulation Accumulation of metals in a healthy plant to levels that are much higher than those found in the soil and that are generally toxic to non-accumulators.

hypersensitive response A common plant defense following microbial infection, in which cells immediately surrounding the infection site die rapidly, depriving the pathogen of nutrients and preventing its spread.

hyphal coils Coiled structures formed by mycorrhizal fungi within the root cortical cells of its host plant; the sites of nutrient transfer between the fungus and the plant. Also called arbuscules.

hypocotyl The region of the seedling stem below the cotyledons and above the root.

hypophysis The cell located directly below the octant stage of the embryo that gives rise to the root cap and part of the root apical meristem.

hypoxia Low levels of oxygen in tissue.

I

idioblast A "special" cell that differs markedly in form, content, or size from the other cells in the same tissue.

imbibition The initial phase of water uptake in dry seeds, which is driven by the matric potential component of the water potential—that is, by the binding of water to surfaces, such as the cell wall and cellular macromolecules.

import The movement of photosynthate in sieve elements into sink organs.

indehiscent Lacking spontaneous opening of a mature anther or fruit.

indeterminate growth The ability to keep growing and developing until the onset of senescence.

inducible defenses Defense responses that exist at low levels until a biotic or abiotic stress is encountered.

induction period The period of time (time lag) between the perception of a signal and the activation of the response. In the Calvin–Benson cycle, the time elapsed between the onset of illumination and the full activation of the cycle.

infection thread An internal tubular extension of the plasma membrane of root hairs through which rhizobia enter root cortical cells.

initials Broadly defined as the cells of the root and shoot apical meristems. More specifically, a cluster of slowly dividing cells located within the meristem that gives rise to the more rapidly dividing cells of the surrounding meristem.

inner mitochondrial membrane The inner of the two mitochondrial membranes, containing the electron transport chain, the F_oF_1-ATP synthase, and numerous transporters.

integral membrane proteins Proteins that are embedded in the lipid bilayer of a membrane via at least one transmembrane domain.

intercellular air space resistance The resistance or hindrance that slows down the diffusion of CO_2 inside a leaf, from the substomatal cavity to the walls of the mesophyll cells.

interface light scattering The randomization of the direction of photon movement within plant tissues due to the reflecting and refracting of light from the many air–water interfaces. Greatly increases the probability of photon absorption within a leaf.

intermediary cell A type of companion cell with numerous plasmodesmatal connections to surrounding cells, particularly to the bundle sheath cells.

intermembrane space The fluid-filled space between the two mitochondrial membranes or between the two chloroplast envelope membranes.

internal tolerance Tolerance mechanisms that function in the symplast (as opposed to exclusion mechanisms).

internode The portion of a stem between nodes.

interphase Collectively the G_1, S, and G_2 phases of the cell cycle.

inwardly rectifying Refers to ion channels that open only at potentials more negative than the prevailing Nernst potential for a cation, or more positive than the prevailing Nernst potential for an anion, and thus mediate inward current.

ionome The composition of all inorganic elements, including nutrients, in an organism.

irradiance The amount of energy that falls on a flat surface of known area per unit of time. Expressed as watts per square meter ($W\ m^{-2}$). Time (seconds) is contained within the term watt: 1 W = 1 joule (J) s^{-1}, or as micromoles of quanta per square meter per second ($\mu mol\ m^{-2}\ s^{-1}$), also referred to as fluence rate.

J

JA-ILE ZIM-DOMAIN (JAZ) A transcriptional repressor that serves as a switch for jasmonate signaling. In the presence of JA, JAZ is degraded, allowing positive transcriptional regulators to activate JA-induced genes.

jasmonates Plant hormones that function in plant defenses to biotic and abiotic stresses as well as some other aspects of development. Jasmonic acid is derived from the octadecanoid pathway. Can function as a volatile signal when methylated (methyl jasmonate) and is primarily active when conjugated to the amino acid isoleucine (JA-Ile).

K

karrikinolide A component of smoke that stimulates seed germination; similar in structure to strigolactones.

kinases Enzymes that have the capacity to transfer phosphate groups from ATP to other molecules.

kinetochore The site of spindle fiber attachment to the chromosome in anaphase.

Kranz anatomy (German *kranz*, "wreath") The wreathlike arrangement of mesophyll cells around a layer of bundle sheath cells. The two concentric layers of photosynthetic tissue surround the vascular bundle. This anatomical feature is typical of leaves of many C_4 plants.

L

lamina The blade of a leaf.

land plants (or embryophytes) All the families of plants inclusive of the non-seed, nonvascular plants.

late-embryogenesis-abundant (LEA) proteins Proteins involved in desiccation tolerance. They interact to form a highly viscous liquid with very slow diffusion and therefore limited chemical reactions. Encoded by a group of genes that are regulated by osmotic stress, first characterized in desiccating embryos during seed maturation.

latent heat of vaporization The energy needed to separate molecules from the liquid phase and move them into the gas phase at constant temperature.

lateral roots Arise from the pericycle in mature regions of the root through establishment of secondary meristems that grow out through the cortex and epidermis, establishing a new growth axis.

latex A complex, often milky solution exuded from cut surfaces of some plants that represents the cytoplasm of laticifers and may contain defensive substances.

laticifers In many plants, an elongated, often interconnected network of separately differentiated cells that contain latex (hence the term laticifer), rubber, and other secondary metabolites.

leaf blade The broad, expanded area of the leaf; also called the lamina.

leaf stomatal resistance (r_s) The resistance to CO_2 diffusion imposed by the stomatal pores.

leaf trace The portion of the shoot primary vascular system that diverges into a leaf.

leaves The main lateral appendages radiating out from stems and branches. Green leaves are usually the major photosynthetic organs of the plant.

lectins Defensive plant proteins that bind to carbohydrates; or carbohydrate-containing proteins inhibiting their digestion by a herbivore.

leghemoglobin An oxygen-binding heme protein produced by legumes during nitrogen-fixing symbioses with rhizobia. Found in the cytoplasm of infected nodule cells, it facilitates the diffusion of oxygen to the respiring symbiotic bacteria.

leucine rich repeats (LRRs) Amino acid sequence motifs in proteins that provide a structural framework for protein–protein interactions.

leucoplasts Nonpigmented plastids, the most important of which is the amyloplast.

light channeling In photosynthetic cells, the propagation of some of the incident light through the central vacuole of the palisade cells and through the air spaces between the cells.

light compensation point The amount of light reaching a photosynthesizing leaf at which photosynthetic CO_2 uptake exactly balances respiratory CO_2 release.

light energy The energy associated with photons.

light-harvesting complex I (LHCI) One of a large family of the most abundant antenna proteins, associated primarily with photosystem II.

light-harvesting complex II (LHCII) One of a large family of the most abundant antenna proteins, associated primarily with photosystem II.

LIGHT-OXYGEN-VOLTAGE (LOV) domains Highly conserved protein domains that respond to light, oxygen, or voltage changes to change receptor protein conformation. In phototropins,

LOV domains are sites of binding of the FMN chromophore to phototropins and are thus the part of the protein that senses light.

light reactions *See* thylakoid reactions.

lignin Highly branched phenolic polymer made up of phenylpropanoid alcohols that is deposited in secondary cell walls.

limiting nutrient A nutrient that is not sufficiently available to support plant growth. Its application will normally increase growth as long as other factors (abiotic or biotic) are not also limiting.

lipid body (also known as **oleosomes** or **oil bodies**) Organelle that accumulates and stores triacylglycerols. It is bounded by a single phospholipid leaflet ("half–unit membrane" or "phospholipid monolayer") derived from the endoplasmic reticulum. =

lodging The permanent displacement of the shoot or part of the shoot from a vertical posture. It is often caused by high wind speeds, made worse by wet conditions.

long-day plant (LDP) A plant that flowers only in long days (qualitative LDP) or whose flowering is accelerated by long days (quantitative LDP).

long-distance transport Translocation through the phloem to the sink.

long–short-day plant (LSDP) A plant that flowers in response to a shift from long days to short days.

lowest excited state The excited state with the lowest energy attained when a chlorophyll molecule in a higher energy state gives up some of its energy to the surroundings as heat.

lytic vacuoles Analogous to lysosomes in animal cells, plant lytic vacuoles contain hydrolytic enzymes that break down cellular macromolecules during senescence.

M

maintenance respiration The respiration needed to support the function and turnover of existing tissue. Contrast with *growth respiration*.

malic enzyme An enzyme that catalyzes the oxidation of malate to pyruvate, permitting plant mitochondria to oxidize malate or citrate to CO_2 without involving pyruvate generated by glycolysis.

MAP (Mitogen-Activated Protein) kinase cascade A series of protein kinases that transmit an activation signal from a cell surface receptor to DNA in the nucleus.

margo A porous and relatively flexible region of the pit membranes in tracheids of conifer xylem, surrounding a central thickening, the torus.

mass transfer rate The quantity of material passing through a given cross section of phloem or sieve elements per unit of time.

matric potential (Ψ_m) The sum of osmotic potential (Ψ_s) + hydrostatic pressure (Ψ_p). Useful in situations (dry soils, seeds, and cell walls) where the separate measurement of Ψ_s and Ψ_p is difficult or impossible to obtain.

matrix The aqueous, gel-like phase of a mitochondrion that occupies the internal space into which the cristae extend. Contains the DNA, ribosomes, and soluble enzymes required for the tricarboxylic acid cycle, oxidative phosphorylation, and other metabolic reactions.

matrix polysaccharides Polysaccharides comprising the matrix of plant cell walls. In primary cell walls they consist of pectins, hemicelluloses, and proteins.

maturation The final phase of seed development.

maturation zone The region of the root where differentiation occurs, including the production of root hairs and functional vascular tissue.

maximum quantum yield The ratio between photosynthetic product and the number of photons absorbed by a photosynthetic tissue. In a graphic plot of photon flux and photosynthetic rate, the maximum quantum yield is given by the slope of the linear portion of the curve.

megaspore The haploid (1*N*) spore that develops into the female gametophyte.

megastrobili The strobili or cones that contain the female gametophytic tissue.

meiosis The "reduction division" whereby two successive cell divisions produce four haploid (1*N*) cells from one diploid (2*N*) cell. In plants with alternation of generations, spores are produced by meiosis. In animals, which don't have alternation of generations, gametes are produced by meiosis.

membrane permeability The extent to which a membrane permits or restricts the movement of a substance.

meristematic zone The region at the tip of the root containing the meristem that generates the body of the root. Located between the root cap and the elongation zone.

meristemoid A small, triangular stomatal precursor cell that functions temporarily as an initial cell in a meristem.

meristemoid mother cells (MMCs) The cells of the leaf protoderm that divide asymmetrically (the so-called entry division) to give rise to the meristemoid, a guard cell precursor.

meristems Localized regions of ongoing cell division that enable growth during postembryonic development.

mesocotyl In members of the grass family, the part of the elongating axis between the scutellum and the coleoptile.

mesophyll Leaf tissue found between the upper and lower epidermal layers.

mesophyll resistance The resistance to CO_2 diffusion imposed by the liquid phase inside leaves. The liquid phase includes diffusion from the intercellular leaf spaces to the carboxylation sites in the chloroplast.

metabolic redundancy A common feature of plant metabolism in which different pathways serve a similar function. They can therefore replace each other without apparent loss in function.

metaphase A stage of mitosis during which the nuclear envelope breaks down and the condensed chromosomes align in the middle of the cell.

microbe-associated molecular patterns (MAMPs) Microbially produced molecules that are recognized by host cells.

microbial pathogens Bacterial or fungal organisms that cause disease in a host plant.

microfibril The major fibrillar component of the cell wall composed of layers of cellulose molecules packed tightly together by extensive hydrogen bonding.

microfilament A component of the cell cytoskeleton made of actin; it is involved in organelle motility within cells.

micropyle The small opening at the distal end of the ovule, through which the pollen tube passes prior to fertilization.

microspores The haploid (1*N*) cell that develops into the pollen tube or male gametophyte.

microstrobili The strobili or cones that contain the male gametophytic tissue.

microtubule A component of the cell cytoskeleton and the mitotic spindle, and a player in the orientation of cellulose microfibrils in the cell wall. Made of tubulin.

middle lamella A thin layer of pectin-rich material at the junction where the primary walls of neighboring cells come into contact.

mineralization The process of breaking down organic compounds, usually by soil microorganisms, and thereby releasing mineral nutrients in forms that can be assimilated by plants.

mineral nutrition The study of how plants obtain and use mineral nutrients.

mitochondrion (plural *mitochondria*) The organelle that is the site for most reactions in the respiratory process in eukaryotes.

Mitogen Activated Protein (MAP) kinase A large family of eukaryotic serine/threonine protein kinases that function in diverse environmental and developmental processes. MAP kinases generally initiate protein kinase cascades that rapidly amplify signals to activate cellular responses.

mitosis The ordered cellular process by which replicated chromosomes are distributed to daughter cells formed by cytokinesis.

mitotic phase (M, or M phase) Includes all the stages of mitosis, from metaphase to telophase.

mitotic spindle The mitotic structure involved in chromosome movement. Polymerized from α- and β-tubulin monomers formed by the disassembly of the preprophase band in early metaphase.

model organism Organism that is particularly accessible and convenient for research and that provides information for hypothesis testing in other organisms.

molecular chaperone proteins Proteins that maintain or restore the active three-dimensional structures of other macromolecules.

monocarpic Referring to plants, typically annuals, that produce fruits only once and then die.

monocot One of the two classes of flowering plants, characterized by a single seed leaf (cotyledon) in the embryo.

monoecious Refers to plants in which male and female flowers are found on the same individuals, such as cucumber (*Cucumis sativus*) and maize (corn; *Zea mays*). Contrast with *dioecious*.

morphogenesis The developmental processes that give rise to biological form.

movement proteins Nonstructural proteins encoded by a virus's genome that facilitate movement of that virus through the symplast.

M phase The phase of the cell cycle in which the replicated chromosomes condense, move to opposite poles, and come to reside in the nuclei of two identical cells.

multivesicular bodies Part of the prevacuolar sorting compartment that functions in the degradation of vacuoles and their membranes.

mutualism A symbiotic relationship in which both organisms benefit.

mycorrhiza (plural *mycorrhizae*) The symbiotic (mutualistic) association of certain fungi and plant roots. Facilitates the uptake of mineral nutrients by roots.

mycorrhizal fungi Fungi that can form mycorrhizal symbioses with plants.

N

NADH dehydrogenase (complex I) A multi-subunit protein complex in the mitochondrial electron transport chain that catalyzes oxidation of NADH and reduction of ubiquinone linked to the pumping of protons from the matrix to the intermembrane space.

NAD(P)H dehydrogenases A collective term for membrane-bound enzymes that oxidize NADH or NADPH, or both, and reduce quinone. Several are present in the electron transport chain of mitochondria; for example, the proton-pumping complex I but also simpler non–proton-pumping enzymes.

necrotic spots Small spots of dead leaf tissue.

necrotrophic pathogens Pathogens that attack their host plant first by secreting cell wall–degrading enzymes or toxins, which will lead to massive tissue laceration and plant death.

Nernst potential The electrical potential described by the Nernst equation.

night break An interruption of the dark period with a short exposure to light that makes the entire dark period ineffective.

nitrate reductase An enzyme located in the cytosol that reduces nitrate (NO_3^-) to nitrite (NO_2^-). It catalyzes the first step by which nitrate absorbed by roots is assimilated into organic form.

nitrogenase enzyme complex The two-component protein complex that catalyzes the biological nitrogen fixation reaction in which ammonia is produced from molecular nitrogen.

nitrogen fixation The natural or industrial processes by which atmospheric nitrogen N_2 is converted to ammonia (NH_3) or nitrate (NO_3^-).

N-linked glycoprotein Glycan linked via a nitrogen atom to a protein. Formed by transfer of a 14-sugar glycan from the ER membrane–embedded dolichol diphosphate to the nascent polypeptide as it enters the lumen of the ER.

nodal roots In monocots, adventitious roots that form after the emergence of primary roots.

node Position on the stem where leaves are attached.

Nod factors Lipochitin oligosaccharide signal molecules active in regulating gene expression during nitrogen-fixing nodule formation. All Nod factors have a chitin β-1,4-linked *N*-acetyl-D-glucosamine backbone (varying in length from three to six sugar units) and a fatty acid chain on the C-2 position of the nonreducing sugar.

nodulation (*nod*) genes Rhizobial genes, the products of which participate in nodule formation.

nodules Specialized organs of a plant host that contain symbiotic nitrogen-fixing bacteria.

nodulin genes Plant genes specific to nodule formation.

non-climacteric Referring to a type of fruit that does not undergo a climacteric, or respiratory burst, during ripening.

non-seed plants Plant families that do not produce a seed.

nonvascular plants Plants that do not have vascular tissues, such as xylem and phloem.

nuclear envelope The double membrane surrounding the nucleus.

nuclear genome The entire complement of DNA found in the nucleus.

nuclear pores Sites where the two membranes of the nuclear envelope join, forming a partial opening between the interior of nucleus and the cytosol.

nucleolus (plural *nucleoli*) A densely granular region in the nucleus that is the site of ribosome synthesis.

nucleoplasm The soluble matrix of the nucleus in which the chromosomes and nucleolus are suspended.

nucleosome A structure consisting of eight histone proteins around which DNA is coiled.

nucleotide-binding domain (NBD) Protein domain that binds nucleotides such as ATP. Found in NLR proteins.

nucleotide-binding site–leucine rich repeat proteins (NLRs) Proteins found in plants and animals that function regulate innate immune responses. NLRs function in the formation of protein complexes that function in cellular defenses.

nucleus (plural *nuclei*) The organelle that contains the genetic information primarily responsible for regulating cellular metabolism, growth, and differentiation.

nutrient assimilation The incorporation of mineral nutrients into carbon compounds such as pigments, enzyme cofactors, lipids, nucleic acids, or amino acids.

nutrient depletion zone The region surrounding the root surface showing diminished nutrient concentrations due to uptake into the roots and slow replacement by diffusion.

nutrient solution A solution containing only inorganic salts that supports the growth of plants in sunlight without soil or organic matter.

nyctinasty Sleep movements of leaves. Leaves extend horizontally to face the light during the day and fold together vertically at night.

O

oil body *See* lipid body

oleosome *See* lipid body.

oligomer-trapping model A model that explains the active accumulation of tri-, tetra-, and pentasaccharides in sieve element–companion cell complexes of symplasmically loading species.

O-linked oligosaccharide Small polysaccharide that is covalently linked to the side chain hydroxyl of serine or threonine residues in a subgroup of plant glycoproteins. O-linked glycosylation occurs in the Golgi apparatus.

ordinary companion cell A type of companion cell with a variable number of plasmodesmata connecting it to neighboring cells other than its associated sieve element.

organic fertilizer A common usage describing fertilizer that contains nutrient elements derived from natural sources without any synthetic additions.

organogenesis The development of organs such as roots, shoots, and flowers from embryonic or meristematic tissue.

organ senescence The developmentally regulated senescence of individual organs, such as leaves, flowers, fruits, or roots.

orthodox seed A seed that can tolerate desiccation and remain viable after storage in a dry state.

osmolarity A unit of concentration expressed as moles of total dissolved solutes per liter of solution (mol L^{-1}). In biology, the solvent is usually water.

osmosis The net movement of water across a selectively permeable membrane toward the region of more negative water potential, Ψ (lower concentration of water).

osmotic adjustment The ability of the cell to accumulate compatible solutes and lower water potential during periods of osmotic stress.

osmotic potential (Ψ_s) The effect of dissolved solutes on water potential. Also called solute potential.

osmotic stress Stress imposed on cells or whole plants when the osmotic potential of external solutions is more negative than that of the solution inside the plant.

outcrossing The mating of two plants with different genotypes by cross-pollination.

outer mitochondrial membrane The outer of the two mitochondrial membranes, which appears to be freely permeable to all small molecules.

outwardly rectifying Refers to ion channels that open only at potentials more positive than the prevailing Nernst potential for a cation, or more negative than the prevailing Nernst potential for an anion, and thus mediate outward current.

oxidative pentose phosphate pathway A cytosolic and plastidic pathway that oxidizes glucose and produces NADPH and several sugar phosphates.

oxidative phosphorylation The transfer of electrons to oxygen in the mitochondrial electron transport chain that is coupled to ATP synthesis from ADP and phosphate by ATP synthase.

oxygen-evolving complex (OEC) The complex associated with photosystem II that oxidizes water and produces molecular oxygen (O_2).

P

P680 The chlorophyll of the photosystem II reaction center that absorbs maximally at 680 nm in its neutral state. The P stands for pigment.

P700 The chlorophyll of the photosystem I reaction center that absorbs maximally at 700 nm in its neutral state. The P stands for pigment.

palisade cells Below the leaf upper epidermis, the top one to three layers of pillar-shaped photosynthetic cells.

parasite An organism that lives on or in an organism of another species, known as the host, from the body of which the parasite obtains nutriment.

parenchyma Metabolically active ground tissue consisting of thin-walled cells.

partitioning The differential distribution of photosynthate to multiple sinks within the plant.

passive transport Diffusion across a membrane. The spontaneous movement of a solute across a membrane in the direction of a gradient of (electro)chemical potential (from higher to lower potential). Downhill transport.

pathogenesis-related (PR) genes Genes that encode small proteins that function either as antimicrobials or in initiating systemic defense responses.

pattern recognition receptors (PRRs) Innate immune system proteins that are associated with microbe-associated molecular patterns (MAMPs) and damage-associated molecular patterns (DAMPs).

pavement cells The predominant type of leaf epidermal cells, which secrete a waxy cuticle and serve to protect the plant from dehydration and damage from ultraviolet radiation.

pectins A heterogeneous group of complex cell wall polysaccharides that form a gel in which the cellulose–hemicellulose network is embedded.

PEP carboxylase A cytosolic enzyme that forms oxaloacetate by the carboxylation of phospho*enol*pyruvate.

perennial plants Plants that live for more than 2 years.

perforation plate The perforated end wall of a vessel element in the xylem.

pericarp The fruit wall surrounding a fruit, derived from the ovary wall.

periclinal Pertaining to the orientation of cell division such that the new cell plate forms parallel to the tissue surface.

pericycle Meristematic cells forming the outermost layer of the vascular cylinder in the stem or root, interior to the endodermis.

periderm Tissue produced by the cork cambium that contributes to the outer bark of stems and roots during secondary growth of woody plants. Also replaces epidermis after wounding and in abscission layers after the shedding of plant parts.

period In cyclic (rhythmic) phenomena, the time between comparable points in the repeating cycle, such as peaks or troughs.

peripheral membrane proteins Proteins that are bound to the membrane surface by noncovalent bonds, such as ionic bonds or hydrogen bonds.

peripheral zone (PZ) A doughnut-shaped region surrounding the central zone in shoot apical meristems and consisting of small, actively dividing cells with inconspicuous vacuoles. Leaf primordia are formed in the peripheral zone.

perisperm Storage tissue derived from the nucellus, and often consumed during embryogenesis.

permanent charge Ionic charge of soil minerals caused by substitution of ions (for example, Al^{3+} for Si^{4+}) in the crystal structure during mineral formation.

peroxisome Organelle in which organic substrates are oxidized by O_2. These reactions generate H_2O_2 that is broken down to water by the peroxisomal enzyme catalase.

petiole The leaf stalk that joins the leaf blade to the stem.

Pfr The far-red light–absorbing form of phytochrome converted from Pr by the action of red light. The cyan-green colored Pfr is converted back to Pr by far-red light. Pfr is the physiologically active form of phytochrome.

phase In cyclic (rhythmic) phenomena, any point in the cycle recognizable by its relationship to the rest of the cycle, for example the maximum and minimum positions.

phase change The phenomenon in which the fates of the meristematic cells become altered in ways that cause them to produce new types of structures.

phellogen *See* cork cambium.

phenotype A set of observed or measured characteristics of an individual genotype found under a specific set of environmental conditions.

phenotypic plasticity Physiological or developmental responses of a plant to its environment that do not involve genetic changes.

pheophytin A chlorophyll in which the central magnesium atom has been replaced by two hydrogen atoms.

phloem The tissue that transports the products of photosynthesis from mature leaves (or storage organs) to areas of growth and storage, including the roots.

phloem fiber Elongated, tapering sclerenchyma cell associated with the other cells in the phloem.

phloem loading The movement of photosynthetic products into the sieve elements of mature leaves. *See also* phloem unloading.

phloem unloading The movement of photosynthates from the sieve elements to neighboring cells that store or metabolize them or pass them on to other sink cells via short-distance transport. *See also* phloem loading.

phosphatases Enzymes that remove a phosphate group from a protein.

phosphate transporter Protein in the plasma membrane specific for the uptake of phosphate by the cell.

phosphatidic acid (PA) A diacylglycerol that has a phosphate on the third carbon of the glycerol backbone.

phosphatidylinositol 4,5-bisphosphate (PIP_2) A group of phosphorylated derivatives of phosphatidylinositol.

photoassimilation The coupling of nutrient assimilation to photosynthetic electron transport.

photochemistry The very rapid chemical reactions in which light energy absorbed by a molecule causes a chemical reaction to occur.

photoinhibition The inhibition of photosynthesis by excess light.

photomorphogenesis The influence and specific roles of light on plant development. In the seedling, light-induced changes in gene expression that support aboveground growth in the light rather than belowground growth in the dark.

photon A discrete physical unit of radiant energy.

photonasty Plant movements in response to nondirectional light.

photon flux density (PFD) The amount of energy striking a leaf per unit of time, expressed as moles of quanta per square meter per second (mol m^{-2} s^{-1}). Also referred to as fluence rate.

photoperiodic induction The photoperiod-regulated processes that occur in leaves resulting in the transmission of a floral stimulus to the shoot apex.

photoperiodism A biological response to the length and timing of day and night, making it possible for an event to occur at a particular time of year.

photophosphorylation The formation of ATP from ADP and inorganic phosphate (P_i), catalyzed by the CF_0F_1-ATP synthase and using light energy stored in the proton gradient across the thylakoid membrane.

photoreceptors Proteins that sense the presence of light and initiate a response via a signaling pathway.

photorespiration Uptake of atmospheric O_2 with a concomitant release of CO_2 by illuminated leaves. Molecular oxygen serves as substrate for Rubisco, producing 2-phosphoglycolate that enters the photorespiratory cycle (the oxidative photosynthetic carbon cycle). The activity of the cycle recovers some of the carbon found in 2-phosphoglycolate, but some is lost to the atmosphere.

photoreversibility The interconversion of the Pr and Pfr forms of phytochrome.

photosynthate A carbon-containing product of photosynthesis.

photosystem I (PSI) A system of photoreactions that absorbs maximally far-red light (700 nm), oxidizes plastocyanin, and reduces ferredoxin.

photosystem II (PSII) A system of photoreactions that absorbs maximally red light (680 nm), oxidizes water, and reduces plastoquinone. Operates very poorly under far-red light.

phototropins Blue-light photoreceptors that primarily regulate phototropism, chloroplast movements, and stomatal opening. Phototropins are autophosphorylating protein kinases that are stimulated by blue light interactions with a flavin co-factor.

phototropism The alteration of plant growth patterns in response to the direction of incident radiation, especially blue light.

phragmoplast An assembly of microtubules, membranes, and vesicles that forms during late anaphase or early telophase and precedes fusion of vesicles to form the cell plate.

phyllome The collective term for all the leaves of a plant, including structures that evolved from leaves, such as floral organs.

Phyllotaxy (phyllotaxis) The arrangement of leaves on the stem.

phytoalexins A chemically diverse group of secondary metabolites with strong antimicrobial activity that are synthesized following infection and that accumulate at the site of infection.

phytochrome interacting factors (PIFs) A family of phytochrome-interacting proteins that may activate and repress gene transcription; some PIFs are targets for phytochrome-mediated degradation.

phytochrome kinase substrates (PKSs) Proteins that participate in the regulation of phytochrome via direct phosphorylation or via phosphorylation by other kinases.

phytochromes Plant growth-regulating photoreceptor proteins that absorb primarily red light and far-red light but also absorb blue light. The holoprotein that contains the chromophore phytochromobilin.

phytoliths Discrete cells that accumulate silica in leaves or roots.

phytomer A developmental unit consisting of one or more leaves, the node to which the leaves are attached, the internode below the node, and one or more axillary buds.

PIN auxin efflux carrier proteins Plasma membrane transport proteins that amplify localized, directional auxin streams associated with embryonic development, organogenesis, and tropic growth.

pipecolates hormones derived from pipecolic acid that function as long distance signals in systemic acquired resistance responses

pit A microscopic region where the secondary wall of a tracheary element is absent and the primary wall is thin and porous

pith The ground tissue in the center of the stem or root.

pit membrane The porous layer in the xylem between pit pairs, consisting of two thinned primary walls and a middle lamella.

pit pair Two pits occurring opposite one another in the walls of adjacent tracheids or vessel elements. Pit pairs constitute a low-resistance path for water movement between the conducting cells of the xylem.

plant growth promoting rhizobacteria (PGPR) Soil bacteria associated with root surfaces that promote plant growth by producing plant growth regulators and/or fixing nitrogen.

plant tissue analysis In the context of mineral nutrition, the analysis of the concentrations of mineral nutrients in a plant sample.

plants All the families of plants inclusive of the non-seed, nonvascular plants.

plasma membrane (plasmalemma) A bilayer of polar lipids (phospholipids or glycosylglycerides) and embedded proteins that together form a selectively permeable boundary around a cell.

plasma membrane H^+-ATPase An ATPase that pumps H^+ across the plasma membrane energized by ATP hydrolysis.

plasmodesmata (singular *plasmodesma*) Microscopic membrane-lined channel connecting adjacent cells through the cell wall and filled with cytoplasm and a central rod derived from the ER called the desmotubule.

plasticity The ability to adjust morphologically, physiologically, and biochemically in response to changes in the environment.

plastids Cellular organelles found in eukaryotes, bounded by a double membrane, and sometimes containing extensive membrane systems. They perform many different functions: photosynthesis, starch storage, pigment storage, and energy transformations.

plastocyanin (PC) A small (10.5 kDa), water-soluble, copper-containing protein that transfers electrons between the cytochrome b_6f complex and P700. This protein is found in the lumenal space.

plastohydroquinone (PQH_2) The fully reduced form of plastoquinone.

plumule First true leaf of a growing seedling.

polar A small difference in charge between two atoms in a molecule. An example is the water molecule where the oxygen atom has a partial negative charge relative to the two hydrogen atoms.

polar auxin transport Directional auxin movement that functions in programmed development and plastic growth responses. Long-distance polar auxin transport maintains the overall polarity of the plant apical–basal axis and supplies auxin for direction into localized streams.

polar glycerolipids The main structural lipids in membranes, in which the hydrophobic portion consists of two 16-carbon or 18-carbon fatty acid chains esterified to positions 1 and 2 of a glycerol.

polarity (1) Property of some molecules, such as water, in which differences in the electronegativity of some atoms result in a partial negative charge at one end of the molecule and a partial positive charge at the other end. (2) Refers to the distinct ends and intermediate regions along an axis. Beginning with the single-celled zygote, the progressive development of distinctions along two axes: an apical–basal axis and a radial axis.

polar nuclei The two haploid nuclei at the center of the embryo sac that normally fuse to form the diploid nucleus of the central cell.

pollen Small structures (microspores) produced by anthers of seed plants. Contain haploid male nuclei that will fertilize the egg in the ovule.

polycarpic Referring to perennial plants that produce fruit many times.

polyploid Cell with more than two sets of chromosomes.

polyploidy Describes cells that are polyploid, or have more than two sets of chromosomes.

pome A fruit, such as an apple, composed of one or more carpels surrounded by accessory tissue derived from the receptacle.

pore–plasmodesma contacts Specific symplasmic contacts between a sieve element and its companion cell, consisting of sieve pores and branched plasmodesmata on the sieve element and companion cell side, respectively.

P-proteins Phloem proteins that act to seal damaged sieve elements by plugging the sieve element pores. Abundant in the sieve elements of most angiosperms, but absent from gymnosperms.

Pr The red light–absorbing form of phytochrome. This is the form in which phytochrome is assembled. The cyan-blue colored Pr is converted by red light to the far-red light–absorbing form, Pfr.

precocious germination Germination of viviparous mutant seeds while still attached to the mother plant.

preharvest sprouting Germination of physiologically mature wild-type seeds on the mother plant caused by wet weather.

pre-phloem transport The movement of photosynthates from to mesophyll to bundle sheath and phloem parenchyma, prior to their uptake by sieve element–companion cell complexes.

preprophase band A circular array of microtubules and microfilaments formed in the cortical cytoplasm just prior to cell division that encircles the nucleus and predicts the plane of cytokinesis following mitosis.

preprophase In mitosis, the stage just before prophase during which the G_2 microtubules are completely reorganized into a preprophase band.

pressure-flow model A widely accepted model of phloem translocation in angiosperms. It states that transport in the sieve elements is driven by a pressure gradient between source and sink. The pressure gradient is osmotically generated and results from the loading at the source and unloading at the sink.

pressure potential (Ψ_p) The hydrostatic pressure of a solution in excess of ambient atmospheric pressure.

prevacuolar compartment A membrane compartment equivalent to the late endosome in animal cells where sorting occurs before cargo is delivered to the lytic vacuole. *See also* multivesicular body.

prickles Pointed plant structures that physically deter herbivory and are derived from epidermal cells.

primary active transport The direct coupling of a metabolic energy source such as ATP hydrolysis, oxidation–reduction reaction, or light absorption to active transport by a carrier protein.

primary cell walls The thin (< 1 μm), unspecialized cell walls that are characteristic of young, growing cells.

primary cross-regulation Involves distinct signaling pathways regulating a shared transduction component in a positive or a negative manner.

primary dormancy The failure of newly dispersed, mature seeds to germinate under normal growth conditions.

primary endosperm cell A diploid cell formed by fusion of two polar cells in the angiosperm embryo sac prior to fertilization.

primary growth The phase of plant development that gives rise to new organs and to the basic plant form.

primary inflorescence meristem The meristem that produces stem-bearing flowers; it is formed from the shoot apical meristem.

primary metabolites Metabolites associated with basic cellular functions (e.g., sugars, amino acids, lipids).

primary plant axis The longitudinal axis of the plant defined by the positions of the shoot and root apical meristems.

primary plant body The part of the plant directly derived from the shoot and root apical meristems and primary meristems.

primary root In monocots, a root generated directly by growth of the embryonic root or radicle.

proanthocyanidin A group of condensed tannins that are present in many plants that serve as defensive chemicals against plant pathogens and herbivores.

procambium Primary meristematic tissue that differentiates into xylem, phloem, and cambium.

profilin Actin binding protein that keeps the depolymerized globular G-actin monomers charged with ATP, so that they can be rapidly re-integrated into F-actin. This protein also binds formins and thereby accelerates the formation of F-actin from formins.

programmed cell death (PCD) A process whereby individual cells activate an intrinsic senescence program accompanied by a distinct set of morphological and biochemical changes similar to mammalian apoptosis.

prophase The first stage of mitosis (and meiosis) prior to disassembly of the nuclear envelope, during which the chromatin condenses to form distinct chromosomes.

proplastid Type of immature, undeveloped plastid found in meristematic tissue.

proteinase inhibitors Compounds that inhibit the enzymatic activity of proteases.

protein storage vacuoles (PSVs) Specialized small vacuoles that accumulate storage proteins, typically in seeds.

protoderm In the plant embryo and in apical meristems, the one-cell-thick surface layer that covers the young shoot and radicle of the embryo and gives rise to the epidermis.

protofilament A column of polymerized tubulin monomers (α- and β-tubulin heterodimers) or a chain of polymerized actin subunits.

proton motive force (PMF) The energetic effect of the electrochemical H^+ gradient across a membrane, expressed in units of electrical potential.

protoplast fusion A technique for incorporating foreign genes into plant genomes by fusion of two genetically different cells from which the cell walls have been removed.

PRR-triggered immunity (PTI) Immune response activated by PRRs.

pulvinus (plural *pulvini*) A turgor-driven organ found at the junction between the blade and the petiole of the leaf that provides the mechanical force for leaf movements.

pumps Membrane proteins that carry out primary active transport across a biological membrane. Most pumps transport ions, such as H^+ or Ca^{2+}.

pyruvate dehydrogenase An enzyme in the mitochondrial matrix that decarboxylates pyruvate, producing NADH (from NAD^+), CO_2, and acetic acid in the form of acetyl-CoA (acetic acid bound to coenzyme A).

Q

Q cycle A mechanism for oxidation of plastohydroquinone (reduced plastoquinone, also called plastoquinol) in chloroplasts and of ubihydroquinone (reduced ubiquinone, also called ubiquinol) in mitochondria.

quantum (plural *quanta*) A discrete packet of energy contained in a photon.

quantum yield (ϕ) The ratio of the yield of a particular product of a photochemical process to the total number of quanta absorbed.

quiescent center The central region of the root meristem where cells divide more slowly than surrounding cells, or do not divide at all. Serves as a reservoir of meristematic cells for tissue regeneration in case of wounding.

quorum sensing A system of coordinated signals and responses by which populations regulate growth and environmental responses. This is a common mechanisms in microbial organisms.

R

radial axis An axis that extends from the center of a root or stem to its outer surface.

radicle The embryonic root. Usually the first organ to emerge on germination.

raphides Needles of calcium oxalate or carbonate that function in plant defense.

rays Tissues of various height and width that radiate through the secondary xylem and phloem, and are formed from ray initials in the vascular cambium.

reaction center complex A group of electron transfer proteins that receive energy from the antenna complex and convert it into chemical energy using oxidation–reduction reactions.

reactive oxygen species (ROS) These include the superoxide anion ($O_2^{\bullet -}$), hydrogen peroxide (H_2O_2), the hydroxyl radical (HO•), and singlet oxygen. They are generated in several cell compartments and can act as signals or cause damage to cellular components.

recalcitrant seeds Seeds that are released from the plant with a relatively high water content and active metabolism and, as a consequence, deteriorate upon dehydration and do not survive storage.

receptor kinase A protein in a signaling pathway that detects the presence of a ligand, such as a hormone, by phosphorylating itself or another protein.

reciprocity According to the Law of Reciprocity, treating plants with a brief duration of bright light will induce the same photobiological response as treating them with a long duration of dim light.

release phloem Sieve elements of sinks where sugars and other photosynthetic products are unloaded into sink tissues.

resistance (R) proteins Proteins that function in plant defense against fungi, bacteria, and nematodes by binding to specific pathogen molecules, elicitors.

respiratory burst oxidase homolog D (RBOHD) An enzyme that generates superoxide using NADPH as an electron donor.

respiratory quotient (RQ) The ratio of CO_2 evolution to O_2 consumption.

response regulator One component of the two-component regulatory systems that are composed of a histidine kinase sensor protein and a response regulator protein. Response regulators have a *receiver domain*, which is phosphorylated by the sensor protein, and an *output domain*, which carries out the response.

rhizobia A collective term for the genera of soil bacteria that form symbiotic (mutualistic) relationships with members of the plant family Fabaceae (Leguminosae).

rhizosphere The immediate microenvironment surrounding the root.

ribosome The site of cellular protein synthesis; consists of RNA and protein.

ribulose-1,5-bisphosphate carboxylase/oxygenase *See* Rubisco.

ribulose 5-phosphate In the pentose phosphate pathway, the initial five-carbon product of the oxidation of glucose 6-phosphate; in subsequent reactions, it is converted into sugars containing three to seven carbon atoms.

rib zone (RZ) Meristematic cells that are located beneath the central zone in shoot apical meristems and that give rise to the internal tissues of the stem.

Rieske iron–sulfur protein A protein subunit in the cytochrome b_6f complex, in which two iron atoms are bridged by two sulfur atoms, with two histidine and two cysteine ligands.

ripening The process that causes fruits to become more palatable, including softening, increasing sweetness, loss of acidity, and changes in coloration.

root The organ, usually underground, that serves to anchor the plant in the soil, and to absorb water and mineral ions and conduct them to the shoot. In contrast to shoots, roots lack buds, leaves, or nodes.

root apical meristem (RAM) A group of permanently dividing cells located underneath the root cap at the tip of the root that provides cells for the primary growth of the root.

root cap Cells at the root apex that cover and protect the meristematic cells from mechanical injury as the root moves through the soil. Site for the perception of gravity and signaling for the gravitropic response in roots.

root exudates Sugars and other compounds secreted into the soil by roots.

root hairs Microscopic extensions of root epidermal cells that greatly increase the surface area of the root for absorption.

root knot nematodes Plant parasites from the genus *Meloidogyne* found in tropical and subtropical soils. Root-knot nematode larvae infect plant roots to form root-knot galls and are a major cause of crop losses.

root pressure A positive hydrostatic pressure in the xylem of roots that typically occurs at night in the absence of transpiration.

ROS scavenging Detoxification of reactive oxygen species via interactions with proteins and electron acceptor molecules.

rough ER The endoplasmic reticulum to which ribosomes are attached. Rough ER synthesizes proteins that are transported by vesicles either to internal organelles or the plasma membrane.

Rubisco The acronym for the chloroplast enzyme *ri*bulose 1,5-*bis*phosphate carboxylase/oxygenase. In a carboxylase reaction, Rubisco uses atmospheric CO_2 and ribulose 1,5-bisphosphate to form two molecules of 3-phosphoglycerate. It also functions as an oxygenase that adds O_2 to ribulose 1,5-bisphosphate to yield 3-phosphoglycerate and 2-phosphoglycolate. The competition between CO_2 and O_2 for ribulose 1,5-bisphosphate limits net CO_2 fixation.

S

S phase The phase in the cell cycle during which DNA is replicated; it follows G_1 and precedes G_2.

salicylic acid (SA) A benzoic acid derivative that functions as a local endogenous signal in coordination with systemically distributed pipecolates for systemic acquired resistance.

salt stress The adverse effects of excess minerals on plants.

salt-tolerant plants Plants that can survive or even thrive in high-salt soils. *See also* halophytes.

sclerenchyma Plant tissue composed of cells (sclereids and fibers), often dead at maturity, with thick, lignified secondary cell walls.

scutellum The single cotyledon of the grass embryo, specialized for nutrient absorption from the endosperm.

seasonal leaf senescence In temperate climates, the pattern of leaf senescence in deciduous trees in which all of the leaves undergo senescence and abscission in the autumn.

secondary active transport Active transport that uses energy stored in the proton motive force or other ion gradient and operates by symport or antiport.

secondary cell wall Cell wall synthesized by nongrowing cells. Often multilayered and containing lignin, it differs in composition and structure from the primary wall.

secondary cross-regulation Regulation by the output of one signal pathway of the abundance or perception of a second signal.

secondary dormancy Seeds that have lost their primary dormancy may become dormant again if they experience prolonged exposure to unfavorable growth conditions.

secondary growth Growth, generally circumferential, that occurs after primary growth (stem and root elongation) is complete. It involves the vascular cambium (producing the secondary xylem and phloem) and the cork cambium (producing the periderm).

secondary inflorescence meristem The inflorescence meristem that develops from the axillary buds of stem-borne leaves of the primary inflorescence.

secondary signal A signaling component that is produced or released by activation of a primary signaling pathway. Often a hormone that acts on sites distal to the site of activation.

second messenger A small intracellular molecule (e.g., cyclic AMP, cyclic GMP, calcium, IP_3, or diacylglycerol) whose concentration increases or decreases in response to the activation of a receptor by an external signal, such as hormones or light. It then diffuses intracellularly to the target enzymes or intracellular receptor to produce and amplify the physiological response.

seed coat The outer layer of the seed, derived from the integument of the ovule.

seed dormancy A state of arrested growth of the embryo that prevents germination even when all the necessary environmental conditions for growth, such as water, O_2, and temperature, are met.

seed longevity The length of time a seed can remain dormant without losing viability.

seed plants Plants in which the embryo is protected and nourished within a seed; gymnosperms and angiosperms.

seed quiescence A state of suspended growth of the embryo due to a lack of water, O_2, or proper temperature for growth. Germination of quiescent seeds proceeds immediately once these conditions are met.

selective permeability The membrane property that allows diffusion of some molecules across the membrane to a different extent than other molecules.

seminal roots Adventitious roots that arise during embryogenesis from the stem tissue between the scutellum and the coleoptile.

senescence An active, genetically controlled, developmental process in which cellular structures and macromolecules are broken down and translocated away from the senescing organ (typically leaves) to actively growing regions that serve as nutrient sinks. Initiated by environmental cues and regulated by hormones.

sensor proteins Bacterial receptor proteins that perceive external or internal signals as part of two-component regulatory systems. They consist of two domains, an *input domain*, which receives the environmental signal, and a *transmitter domain*, which transmits the signal to the response regulator.

sequential leaf senescence The pattern of leaf senescence in which there is a gradient of senescence from the growing tip of the shoot to the oldest leaves at the base.

shade avoidance The increased rate of stem elongation produced in certain plants in response to exposure to shade produced by green leaves.

shoot apex (terminal bud) The shoot apical meristem with its associated leaf primordia and young, developing leaves.

shoot apical meristem (SAM) Dome-shaped region of the shoot tip composed of meristematic cells that give rise to leaves, branches, and reproductive structures.

shoots The organ, usually aboveground, that includes the stem, leaves, buds, and reproductive structures. Function in photosynthesis and reproduction.

short-day plant (SDP) A plant that flowers only in short days (i.e., shorter than a critical day length) or whose flowering is accelerated by short days (quantitative SDP).

short–long-day plant (SLDP) A plant that flowers only after a sequence of short days followed by long days.

siderophores Small molecules secreted by non-grass plants and some microbes to chelate iron, which is then taken up into cells at the root surface.

sieve area A depression in the cell wall of a sieve tube element that contains a field of plasmodesmata.

sieve cells The relatively unspecialized sieve elements of gymnosperms.

sieve effect The penetration of photosynthetically active light through several layers of cells due to the gaps between chloroplasts permitting the passage of light.

sieve elements Cells of the phloem that conduct sugars and other organic materials throughout the plant. Refers to both sieve tube elements (angiosperms) and sieve cells (gymnosperms).

sieve plates Sieve areas found in angiosperm sieve tube elements; they have larger pores (sieve plate pores) than other sieve areas and are generally found in end walls of sieve tube elements.

sieve tube Tube formed by the joining together of individual sieve tube elements at their end walls.

sieve tube elements The highly differentiated sieve elements typical of the angiosperms.

signal An environmental input that initiates one or more plant responses.

signal peptide A hydrophobic sequence of 18 to 30 amino acid residues at the amino-terminal end of a chain; it is found on all secretory proteins and most integral membrane proteins and permits their transit across the membrane of the rough ER.

signal recognition particle (SRP) A ribonucleoprotein (protein–RNA complex) that recognizes and targets specific proteins to the endoplasmic reticulum in eukaryotes.

signal transduction pathway A sequence of biochemical processes by which an extracellular signal (typically light or a hormone) interacts with a receptor, causing a change in the level of a second messenger and ultimately a change in cell function.

simple leaf A leaf with one blade.

sink Any organ that imports photosynthate, including nonphotosynthetic organs and organs that do not produce enough photosynthetic products to support their own growth or storage needs, such as roots, tubers, developing fruits, and immature leaves. Contrast with *source*.

sink activity The rate of uptake of photosynthate per unit of weight of sink tissue.

sink size The total weight of the sink.

sink strength The ability of a sink organ to mobilize assimilates toward itself. Depends on two factors: sink size and sink activity.

size exclusion limit (SEL) The restriction on the size of molecules that can be transported via the symplast. It is imposed by the width of the cytoplasmic sleeve that surrounds the desmotubule in the center of the plasmodesma.

skotomorphogenesis The developmental program plants follow when seeds are germinated and grown in the dark.

smooth ER The endoplasmic reticulum lacking attached ribosomes and usually consisting of tubules. Functions in lipid synthesis.

SNF1-Related Kinase2 (SnRK2) A family of kinases that includes ABA-activated protein kinases or stress-activated protein kinases.

soil analysis The chemical determination of the nutrient content in a soil sample, typically from the root zone.

soil hydraulic conductivity A measure of the ease with which water moves through a soil.

solar tracking The movement of leaf blades throughout the day so that the planar surface of the blade remains perpendicular to the sun's rays.

solute potential The effect of dissolved solutes on water potential. Also called osmotic potential (Ψ_s).

solution culture A technique for growing plants with their roots immersed in nutrient solution without soil. Also called hydroponics.

somatic embryogenesis The process by which somatic cells (i.e., non-germline cells) de-differentiate and undergo an autonomous embryogenesis process that recapitulates all the zygotic embryogenesis steps and results in viable embryos.

sorbitol A sugar alcohol formed by reduction of aldehyde glucose.

source Any organ that is capable of exporting photosynthetic products in excess of its own needs, such as a mature leaf or a storage organ. Contrast with *sink*.

specialized metabolites (also secondary metabolites) Plant compounds that have no direct role in plant growth and development, but function as defenses against herbivores and microbial infection by microbial pathogens, attractants for pollinators and seed-dispersing animals, and agents of plant–plant competition.

specific heat capacity Ratio of the heat capacity of a substance to the heat capacity of a reference substance, usually water. Heat capacity is the amount of heat needed to change the temperature of a unit mass by 1°C. The heat capacity of water is 1 calorie (4.184 Joule) per gram per degree Celsius.

S-phase kinase associated protein 1 (Skp1)/Cullin/F-box (SCF) complexes Large protein complexes that function as E3 ubiquitin ligases in the signaling pathways of several plant hormones.

spines Sharp, stiff plant surface structures, thought to be modified leaves, that physically deter herbivores and may assist in water conservation.

spongy mesophyll Mesophyll cells of very irregular shape located below the palisade cells and surrounded by large air spaces. They function in photosynthesis and gas exchange.

spores Reproductive cells formed in plants by meiosis in the sporophyte generation. They give rise by mitotic divisions to the gametophyte generation.

sporophyte The diploid (2*N*) multicellular structure that produces haploid spores by meiosis. Contrast with *gametophyte*.

SRP receptor A receptor protein on the ER membrane that binds to the ribosome–SRP complex, permitting the ribosome to dock with the translocon pore through which the elongating polypeptide will enter the lumen of the ER.

starch A polyglucan consisting of long chains of 1,4-linked glucose molecules and branch points where 1,6-linkages are used. Starch is the carbohydrate storage form in most plants.

starch sheath A layer of cells that surrounds the vascular tissues of the shoot and coleoptile and is continuous with the root endodermis. Required for gravitropism in the shoots of eudicots.

starchy endosperm A starch-storing triploid endosperm tissue that makes up the bulk of the seeds of cereals and other members of the grass family.

statocytes Specialized gravity-sensing plant cells that contain statoliths.

statoliths Cellular inclusions such as amyloplasts that act as gravity sensors by having a high density relative to the cytosol and sedimenting to the bottom of the cell.

stele In the root, the tissues located interior to the endodermis. The stele contains the vascular elements of the root: the phloem and the xylem.

stem The typically aboveground primary axis of the shoot that bears leaves and buds. May also occur underground in the form of rhizomes, corms, and tubers.

stomata (singular *stoma* or *stomate*) Microscopic pores in the leaf epidermis, each surrounded by a pair of guard cells and, in some species, also including subsidiary cells. Stomata regulate the gas exchange (water and CO_2) of leaves by controlling the dimension of a stomatal pore.

stomatal complex The guard cells, subsidiary cells, and stomatal pore, which together regulate leaf transpiration.

stomatal resistance A measurement of the limitation to the free diffusion of gases from and into the leaf posed by the stomatal pores. The inverse of stomatal conductance.

stone A hardened endocarp (the inner layer of the ovary) in drupe fruits such as peach.

stratification A process to break dormancy by chilling seeds.

stress Disadvantageous influences exerted on a plant by external abiotic or biotic factor(s), such as herbivory, infection, heat, water, or anoxia. Measured in relation to plant survival, crop yield, biomass accumulation, or CO_2 uptake.

strigolactones Carotenoid-derived plant hormones that inhibit shoot branching. They also play roles in the soil by stimulating the growth of arbuscular mycorrhizae and the germination of parasitic plant seeds, such as those of *Striga* (the source of their name).

stroma The fluid component surrounding the thylakoid membranes of a chloroplast.

stroma lamellae Unstacked thylakoid membranes within the chloroplast.

subfunctionalization The process by which evolution acts on duplicate genes, causing one copy to be either lost or changed in function, while the other retains its original function.

subsidiary cells Specialized epidermal cells that flank the guard cells and work with the guard cells in the control of stomatal apertures.

substrate-level phosphorylation A process that involves the direct transfer of a phosphate group from a substrate molecule to ADP to form ATP.

sucrose A disaccharide consisting of a glucose and a fructose molecule linked via an ether bond between C-1 on the glucosyl subunit and C-2 on the fructosyl unit. Sucrose is the carbohydrate transport form (e.g., in the phloem between source and sink).

sunflecks Patches of sunlight that pass through openings in a forest canopy to the forest floor. A major source of incident radiation for plants growing under the forest canopy.

supercool The condition in which cellular water remains liquid because of its solute content, even at temperatures several degrees below its theoretical freezing point.

superoxide dismutase An enzyme that converts superoxide radicals to hydrogen peroxide.

surface tension A force exerted by water molecules at the air–water interface, resulting from the *cohesion* and *adhesion* properties of water molecules. This force minimizes the surface area of the air–water interface.

suspensor In seed plant embryogenesis, the structure that develops from the basal cell following the first division of the zygote. It supports, but is not part of, the embryo.

symbiosis The close association of two organisms in a relationship that may or may not be mutually beneficial. Often applied to beneficial (mutualistic) relationships.

symplasm The continuum of cellular cytoplasm connected by plasmodesmata.

symplast The continuous system of cell protoplasts interconnected by plasmodesmata.

symplasmic transport The intercellular transport of water and solutes through plasmodesmata.

symport A type of secondary active transport in which two substances are moved in the same direction across a membrane.

syncytium A multinucleate cell that can result from multiple cell fusions of uninuclear cells, usually in response to viral infection.

synergid cells Two cells adjacent to the egg cell of the embryo sac, one of which is penetrated by the pollen tube upon entry into the ovule.

systemic Movement of a signal or molecule from one location in a plant to distal locations in the plant.

systemic acquired acclimation (SAA) A system whereby exposure of one part of a plant to an abiotic stress generates signals that can initiate acclimation to that stress in other, unexposed parts of the plant.

systemic acquired resistance (SAR) The increased resistance throughout a plant to a range of pathogens following infection by a pathogen at one site.

T

taproot In eudicots, the main single root axis from which lateral roots develop.

telomeres Regions of repetitive DNA that form the chromosome ends and protect them from degradation.

telophase Prior to cytokinesis, the final stage of mitosis (or meiosis) during which the chromatin decondenses, the nuclear envelope re-forms, and the cell plate extends.

temperature coefficient (Q_{10}) The increase in the rate of a process (e.g., respiration) for every 10°C increase in temperature.

temperature compensation A characteristic of circadian rhythms, which can maintain their circadian periodicity over a broad range of temperatures within the physiological range.

tensile strength The ability to resist a pulling force. Water has a high tensile strength.

tension Negative hydrostatic pressure.

tertiary cross-regulation Regulation that involves the outputs of two distinct signaling pathways exerting influences on one another.

testa The outer layer of the seed, also called the seed coat, derived from the integument of the ovule.

thigmotropism Plant growth in response to touch, enabling roots to grow around rocks, and shoots of climbing plants to wrap around structures for support.

thorns Sharp plant structures that physically deter herbivores and are derived from branches.

thylakoid reactions The chemical reactions of photosynthesis that occur in the specialized internal membranes of the chloroplast (called thylakoids). Include photosynthetic electron transport and ATP synthesis.

thylakoids The specialized, internal chlorophyll-containing membranes of the chloroplast where light absorption and the chemical reactions of photosynthesis take place.

tip growth Localized growth at the tip of a plant cell, caused by localized secretion of new wall polymers. Occurs in pollen tubes, root hairs, some sclerenchyma fibers, and cotton fibers, as well as moss protonema and fungal hyphae. Contrast with *diffuse growth*.

tonoplast The vacuolar membrane.

torus A central thickening found in the pit membranes of tracheids in the xylem of most gymnosperms.

tracheary elements Water-transporting cells of the xylem.

tracheids Spindle-shaped, water-conducting cells with tapered ends and pitted walls without perforations found in the xylem of both angiosperms and gymnosperms.

tracheophyte A plant that has xylem and phloem. Also called a vascular plant.

trait A specific character of an organism that is determined by genetic factors (genotype) and environment.

transcription The process by which the base sequence information in DNA is copied into an RNA molecule.

transcytosis Redirection of a secreted protein from one membrane domain within a cell to another, polarized domain.

transfer cell A type of companion cell similar to an ordinary companion cell, but with fingerlike projections of the cell wall that greatly increase the surface area of the plasma membrane and increase the capacity for solute transport across the membrane from the apoplast.

trans **Golgi network (TGN)** in plants, a collective compartment of vesicles derived from the Golgi apparatus that are sorted to different subcellular destinations. The TGN also receives extracellular materials and recycled molecules from endocytic compartments.

translation The process whereby a specific protein is synthesized on ribosomes according to the sequence information encoded by the mRNA.

translocation The movement of photosynthate from sources to sinks in the phloem.

translocons Pores in the rough endoplasmic reticulum that enable proteins synthesized on ribosomes to enter the ER lumen.

transpiration The evaporation of water from the surface of leaves and stems.

transpiration ratio The ratio of water loss to photosynthetic carbon gain. Measures the effectiveness of plants in moderating water loss while allowing sufficient CO_2 uptake for photosynthesis.

transport Molecular or ionic movement from one location to another; may involve crossing a diffusion barrier such as one or more membranes.

transport phloem Sieve elements of the pathway connecting source and sinks such as the phloem of midribs and stem internodes.

transport proteins Transmembrane proteins that are involved in the movement of molecules or ions from one side of a membrane to the other side.

transposon tagging The technique of inserting a transposon into a gene and thereby marking that gene with a known DNA sequence.

treadmilling During interphase, a process by which microfilaments or microtubules in the cortical cytoplasm appear to migrate around the cell periphery due to addition of G-actin or tubulin heterodimers, respectively, to the plus end at the same rate as their removal from the minus end.

triacylglycerols Three fatty acyl groups in ester linkage to the three hydroxyl groups of glycerol. Fats and oils.

tricarboxylic acid (TCA) cycle A cycle of reactions catalyzed by enzymes localized in the mitochondrial matrix that leads to the oxidation of pyruvate to CO_2. ATP and NADH are generated in the process. Also called the citric acid cycle.

trichomes Unicellular or multicellular hairlike structures that differentiate from the epidermal cells of shoots and roots. Trichomes may be structural or glandular and function in plant responses to biotic or abiotic environmental factors.

triose phosphate A three-carbon sugar phosphate.

tropisms Oriented plant growth in response to a perceived directional stimulus such as light, gravity, touch, or water potential.

tubulin A family of cytoskeletal GTP-binding proteins with three members, α-, β-, and γ-tubulin. α-tubulin forms heterodimers with β-tubulin, which polymerize to form microtubules.

turgor pressure Force per unit of area in a liquid. In a plant cell, turgor pressure pushes the plasma membrane against the rigid cell wall and provides a force for cell expansion.

turnover The balance between the rate of synthesis and the rate of degradation, usually applied to protein or RNA. An increase in turnover typically refers to an increase in degradation.

two-component regulatory systems Signaling pathways common in prokaryotes. They typically involve a membrane-bound histidine kinase sensor protein that senses environmental signals and a response regulator protein that mediates the response. Although rare in eukaryotes, systems resembling bacterial two-component systems are involved in both ethylene and cytokinin signaling.

U

ubiquinone A mobile electron carrier of the mitochondrial electron transport chain. Chemically and functionally similar to plastoquinone in the photosynthetic electron transport chain.

ubiquitin A small polypeptide that is covalently attached to proteins, and that serves as a recognition site for a large proteolytic complex, the proteasome.

ubiquitin–proteasome pathway A mechanism for the specific degradation of cellular proteins involving two discrete steps: the polyubiquitination of proteins via the E3 ubiquitin ligase and the degradation of the tagged protein by the 26S proteasome.

uncoupled A process in which protons translocated into the lumen as part of the light reactions can escape to the stroma and bypass the ATP synthase so that ATP is not synthesized. Similar uncoupled reactions occur in the mitochondria when protons that are translocated out of the matrix as a result of the electron transport activity of complexes I, III and IV return to the matrix and bypass the F_oF_1-ATP synthase.

uncoupler A chemical compound that increases the proton permeability of membranes and thus uncouples the formation of the proton gradient from ATP synthesis.

uncoupling protein A protein that increases the proton permeability of the inner mitochondrial membrane and thereby decreases energy conservation.

unfolded protein response A cellular response that eliminates misfolded proteins in the lumen of the endoplasmic reticulum.

UV RESISTANCE LOCUS 8 (UVR8) The protein receptor that mediates various plant responses to UV-B irradiation.

V

vacuolar H^+-ATPase (V-ATPase) A large, multi-subunit enzyme complex, related to F_oF_1-ATPases, present in endomembranes (tonoplast, Golgi). Acidifies the vacuole and provides the proton motive force for the secondary transport of a variety of solutes into the lumen. V-ATPases also function in the regulation of intracellular protein trafficking.

vacuolar sap The fluid contents of a vacuole, which may include water, inorganic ions, sugars, organic acids, and pigments.

vacuole A membrane-bound cellular organelle containing water and dissolved substances such as salts, sugars, enzymes, and amino acids. Vacuoles function in maintaining water balance and cellular turgor. Lytic vacuoles are characterized by a more acidic pH and can often occupy a majority of the space within a cell. Specialized protein storage vacuoles are found primarily in seeds and very young seedlings.

vascular cambium A lateral meristem consisting of fusiform and ray stem cells, giving rise to secondary xylem and phloem elements, as well as ray parenchyma.

vascular plant A plant that has xylem and phloem. Also called a tracheophyte.

vascular tissues Plant tissues specialized for the transport of water (xylem) and photosynthetic products (phloem).

velocity The linear distanced traveled per unit time.

venation pattern The pattern of veins of a leaf.

vernalization In some species, the cold temperature requirement for flowering. The term is derived from the Latin word for "spring."

vessel A stack of two or more vessel elements in the xylem.

vessel elements Nonliving water-conducting cells with perforated end walls found only in angiosperms and a small group of gymnosperms.

villin An actin binding protein that bundles F-actin filaments.

vivipary The precocious germination of seeds in the fruit while still attached to the plant.

W

water potential (Ψ) A measure of the free energy associated with water per unit of volume (J m^{-3}). These units are equivalent to pressure units such as pascals. Ψ is a function of the solute potential, the pressure potential, and the gravitational potential: $\Psi = \Psi_s + \Psi_p + \Psi_g$. The term Ψ_g is often ignored because it is negligible for heights under 5 m.

wavelength (λ) A unit of measurement for characterizing light energy. The distance between successive wave crests. In the visible spectrum, it corresponds to a color.

whole plant senescence The death of the entire plant, as opposed to the death of individual cells, tissues, or organs.

whorl Pertaining to the concentric pattern of a set of organs that are initiated around the flanks of the meristem.

wilting Loss of rigidity, leading to a flaccid state, due to turgor pressure falling to zero.

wound callose Callose deposited in the sieve pores of damaged sieve elements that seals them off from surrounding intact tissue. As sieve elements recover, the callose disappears from pores.

X

xylem The vascular tissue that transports water and ions from the root to the other parts of the plant.

xylem loading The process whereby ions exit the symplast and enter the conducting cells of the xylem.

xyloglucan A hemicellulose with a backbone of 1→4-linked β-D-glucose residues and short side chains that contain xylose, galactose, and sometimes fucose.

Z

Zeitgebers Environmental signals such as light-to-dark or dark-to-light transitions that synchronize the endogenous oscillator to a 24-h periodicity.

ZEITLUPE A blue-light photoreceptor that regulates day-length perception (photoperiodism) and circadian rhythms.

Illustration Credits

CHAPTER 1

[Chapter 1 Opener]Biophoto Associates / Science Source, p. 1; [Figure 1.1a-j] A:Johnmartindavies/Wikipedia/CC BY-SA 3.0, B: Kristian Peters/Wikipedia/ CC BY-SA 3.0, C: Olivier_s/Wikipedia/ CC BY-SA 3.0, D: © Sikth | Dreamstime.com, E: Velela/Wikipedia, F: Fir0002/ Flagstaffotos/Wikipedia/CC BY-NC, G: Photo by Osman Rana on Unsplash, H: DavetheMage/Wikipedia/CC BY 3.0, I: Photo by REGINE THOLEN on Unsplash, J: Esculapio/Wikipedia/CC BY-SA 3.0, p. 3; [Figure 1.3a, e]A: Courtesy of David McIntyre., E: David McIntyre., p. 7; [Figure 1.4b-c]B: Robinson-Beers, K., and Evert, R. F. (1991) Fine structure of plasmodesmata in mature leaves of sugar cane. Planta 184: 307–318, C: Bell, K., and Oparka, K. (2011) Imaging plasmodesmata. Protoplasma 248: 9–25, p. 8; [Figure 1.6b-e]B: B.E.S. Gunning. 2009. Plant Cell Biology on DVD: Information for students and a resource for teachers. Springer: New York. Courtesy of Brian Gunning., C: Courtesy of R. Heady, D: B.E.S. Gunning. 2009. Plant Cell Biology on DVD: Information for students and a resource for teachers. Springer: New York. Courtesy of Brian Gunning., E: Courtesy of A. Cleary, p. 11; [Figure 1.7]Both micrographs from B.E.S. Gunning. 2009. Plant Cell Biology on DVD: Information for students and a resource for teachers. Springer: New York. Courtesy of Brian Gunning., p. 12; [Figure 1.9b]B: Daniel R. Froelich and others, Phloem Ultrastructure and Pressure Flow: Sieve-Element-Occlusion-Related Agglomerations Do Not Affect Translocation, The Plant Cell, Volume 23, Issue 12, December 2011, Pages 4428–4445., p. 14; [Figure 1.12c] L. A. Staehelin and E. H. Newcomb. 2000. In Biochemistry and Molecular Biology of Plants. B. B. Buchanan et al. [eds.]. American Society of Plant Biologists, Rockville, MD., p. 17; [Figure 1.15a-b]A: Courtesy of R. Evert., B: Fiserova, J., Kiseleva, E., and Goldberg, M. W. (2009) Nuclear envelope and nuclear pore complex structure and organization in tobacco BY-2 cells. Plant J. 59: 243–255., p. 22; [Figure 1.16]Alberts, B., Johnson, A., Lewis, J., Raff, M., Roberts, K., and Walter, P. (2007) Molecular Biology of the Cell. 5th ed. Garland Science, New York., p. 23; [Figure 1.19a-c]All photos from B. E. S. Gunning and M. W. Steer. 1996. Plant Cell Biology: Structure and Function of Plant Cells. Jones and Bartlett, Boston., p. 27; [Figure 1.20]B. E. S. Gunning and M. W. Steer. 1996. Plant Cell Biology: Structure and Function of Plant Cells. Jones and Bartlett, Boston., p. 28; [Figure 1.21] Both photos courtesy of Ulla Newmann, p. 29; [Figure 1.22]Micrograph by S. E. Frederick, courtesy of E. H. Newcomb., p. 30; [Figure 1.23]Micrographs by W. P. Wergin, courtesy of E. H. Newcomb., p. 31; [Figure 1.24]Gunning, B. E. S., and Steer, M. W. (1996) Plant Cell Biology: Structure and Function of Plant Cells. Jones and Bartlett, Boston., p. 32; [Figure 1.26]© Sebastian Kaulitzki/Shutterstock., p. 35; [Figure 1.27]T. Higaki et al. 2008. BMC Plant Biol. 8: 80. CC BY 2.0, p. 36; [Figure 1.28]J. M. Seguí-Simarro et al. 2004. Plant Cell 16: 836–856, p. 38

CHAPTER 2

[Chapter 2 Opener]namaki/ Shutterstock, p. 41; [Figure 2.1]H. Jones. 2013. Plants and Microclimate: A Quantitative Approach to Environmental Plant Physiology (3rd ed.). Cambridge, UK: Cambridge University Press. Data from W. Day et al. 1978. J. Agric. Sci. 91: 599–623, and P. Innes and R.D. Blackwell. 1981. J. Agric. Sci. 96: 603–610., p. 42; [Figure 2.2]Schuur, E. A. (2003) Productivity and Global Climate Revisited: The Sensitivity of Tropical Forest Growth to Precipitation. Ecology 84: 1165-1170., p. 42; [Figure 2.8]Nobel, P. S. (1991) Physicochemical and Environmental Plant Physiology,. Academic Press, San Diego, CA., p. 48; [Figure 2.11]Hsiao, T. C., and Xu, L. K. (2000) Sensitivity of growth of roots versus leaves to water stress: Biophysical analysis and relation to water transport. J. Exp. Bot. 51: 1595–1616., p. 55; [Figure 2.12]Courtesy of David McIntyre, p. 56; [Figure 2.15] Hsiao, T. C. (1973) Plant responses to water stress. Ann Rev. Plant Physiol. 24: 519-70, p. 58

CHAPTER 3

[Chapter 3 Opener]Nomad_Soul/ Shutterstock, p. 61; [Figure 3.3]Graph from Kramer, P. J. (1983) Water Relations of Plants. Academic Press, San Diego, CA., p. 64; [Figure 3.5]David McIntyre., p. 65; [Figure 3.6b-e]B: © Steve Gschmeissner/Science Source., C: Zimmermann, M. H. (1983) Xylem Structure and the Ascent of Sap. Springer, Berlin., D: Zimmermann, M. H. (1983) Xylem Structure and the Ascent of Sap. Springer, Berlin., E: Pittermann J, Sperry JS, Hacke UG, Wheeler JK, Sikkema EH. Torus-margo pits help conifers compete with angiosperms. Science. 2005 Dec 23;310(5756):1924. doi: 10.1126/ science.1120479. PMID: 16373568., F: Pittermann J, Sperry JS, Hacke UG, Wheeler JK, Sikkema EH. Torus-margo pits help conifers compete with angiosperms. Science. 2005 Dec 23;310(5756):1924. doi: 10.1126/ science.1120479. PMID: 16373568., p. 68; [Figure 3.8]Micrograph from B. E. S. Gunning and M. W. Steer. 1996. Plant Cell Biology: Structure and Function of Plant Cells. Jones and Bartlett, Boston., p. 72; [Figure 3.9]Sperry, J. S. (2000) Hydraulic constraints on plant gas exchange. Agric. For. Meteorol. 104: 13–23., p. 74; [Figure 3.11]Bange, G. G. J. (1953) On the quantitative explanation of stomatal transpiration. Acta Bot. Neerl. 2: 255–296., p. 78; [Figure 3.12a-c]A: Ray Simons/Science Source, B: agefotostock/Alamy Stock, C: Meidner, H., and Mansfield, D. (1968) Physiology of Stomata. McGraw-Hill, London., p. 79; [Figure 3.13a-b]A: Palevitz, B. A. (1981) The structure and development of guard cells. In Stomatal Physiology, P. G. Jarvis and T. A. Mansfield, eds., Cambridge University Press, Cambridge, pp. 1–23, B: Sack, F. D. (1987) The development and structure of stomata. In Stomatal Function, E. Zeiger, G. Farquhar, and

I. Cowan, eds., Stanford University Press, Stanford, CA, pp. 59–90, p. 80

CHAPTER 4

[Chapter 4 Opener]std/Moment/Getty Images., p. 85; [Figure 4.1]E. Epstein and A. J. Bloom. 2005. Mineral Nutrition of Plants: Principles and Perspectives, 2nd ed. Oxford University Press/ Sinauer, Sunderland, MA, p. 89; [Figure 4.2]Adapted from Vukosav P. et al., Revision of iron(III)–citrate speciation in aqueous solution. Voltammetric and spectrophotometric studies, Analytica Chimica Acta, Vol. 745, 2012, pp. 85-91., p. 91; [Figure 4.3]Lucas, R. E., and Davis, J. F. (1961) Relationships between pH values of organic soils and availabilities of 12 plant nutrients. Soil Sci. 92: 177–182., p. 99; [Figure 4.5]J. E. Weaver. 1926. Root Development of Field Crops. McGraw-Hill, New York. In the Public Domain, p. 104; [Figure 4.6]From B. Péret et al. 2014. Plant Physiol. 166: 1713–1723. Photographs by Thirumurugen Kuppusamy, School of Plant Biology, University of Western Australia, p. 105; [Figure 4.8]After A. J. Bloom et al. 1993. Plant Cell Environ. 16: 199–206, p. 108; [Figure 4.9]After J. P. Lynch. 2007. Aust. J. Bot. 55: 493–512, p. 109; [Figure 4.10a-b]A: From T. Remans et al. 2006. Proc. Natl. Acad. Sci. USA 103: 19206–19211. © 2006 National Academy of Sciences, B: B after Y. Matsubayashi. 2018. Proc. Jpn. Acad. Ser. B. Phys. Biol. Sci. 94: 59–74, p. 110; [Figure 4.11]Manuela Giovannetti, Luciano Avio, Paola Fortuna, Elisa Pellegrino, Cristiana Sbrana & Patrizia Strani. At the Root of the Wood Wide Web Plant Signaling & Behavior Vol. 1, Iss. 1, 2006, p. 111; [Figure 4.12]At the Root of the Wood Wide Web, p. 112; [Figure 4.13]Plant Signaling & Behavior Vol. 1, Iss. 1, 2006, p. 112

CHAPTER 5

[Chapter 5 Opener]tamu1500/ Shutterstock, p. 117; [Figure 5.2]Bloom, A. J. (1997) Nitrogen as a limiting factor: Crop acquisition of ammonium and nitrate. In Ecology in Agriculture, L. E. Jackson, ed., Academic Press, San Diego, CA, pp. 145–172., p. 120; [Figure 5.6]Pate, J.S. 1973. Uptake, assimilation and transport of nitrogen compounds by plants. Soil Biol. Biochem. 5, 109-119., p. 123; [Figure 5.11]David McIntyre., p. 131; [Figure 5.12]© Dr. Peter Siver/ Alamy Photo Stock, p. XXX; [Figure 5.13]Stokkermans, T. J. W., Ikeshita, S., Cohn, J., Carlson, R. W., Stacey, G., Ogawa, T., and Peters, N. K. (1995) Structural requirements of synthetic and natural product lipo-chitin oligosaccharides for induction of nodule primordia on Glycine soja. Plant Physiol. 108: 1587–1595., p. 132; [Figure 5.16]Dixon, R. O. D., and Wheeler, C. T. (1986) Nitrogen Fixation in Plants. Chapman and Hall, New York; Buchanan, B., Gruissem, W., and Jones, R., eds. (2000) Biochemistry and Molecular Biology of Plants. American Society of Plant Physiologists, Rockville, MD., p. 137

CHAPTER 6

[Chapter 6 Opener]George Bernard / Science Source, p. 143; [Figure 6.4] Higinbotham, N., Graves, J. S., and Davis, R. F. (1970) Evidence for an electrogenic ion transport pump in cells of higher plants. J. Membr. Biol. 3: 210–222., p. 150; [Figure 6.7a-b]A: Q. Leng et al. 2002. Plant Physiol. 128: 400–410. B: D. Sanders and P. Bethke. 2000. In Biochemistry and Molecular Biology of Plants, B. B. Buchanan et al. [eds.], pp. 110-158. American Society of Plant Physiologists: Rockville, MD., p. 152; [Figure 6.8b]B: L. Perfus-Barbeoch and S. M. Assmann, unpublished data, p. 154; [Figure 6.12]Lin, W., Schmitt, M. R., Hitz, W. D., and Giaquinta, R. T. (1984) Sugar transport into protoplasts isolated from developing soybean cotyledons. Plant Physiol. 75: 936–940., p. 157; [Figure 6.14a-d]A: Lebaudy, A., Véry, A., and Sentenac, H. (2007) K+ channel activity in plants: Genes, regulations and functions. FEBS Lett. 581: 2357–2366, B-D: Very, A. A., and Sentenac, H. (2002) Cation channels in the Arabidopsis plasma membrane. Trends Plant Sci. 7: 168–175., p. 160; [Figure 6.17]M. L. Guerinot, Y. Yi, Iron: Nutritious, Noxious, and Not Readily Available, Plant Physiology, Volume 104, Issue 3, March 1994, pp. 815–820, p. 163; [Figure 6.20]D. Focht et al. Front. Physiol., 11 April 2017 Sec. Membrane Physiology and Membrane Biophysics Volume 8 - 202. CC BY 4.0, p. 166; [Figure 6.21]Kluge, C., Lahr, J., Hanitzsch, M., Bolte, S., Golldack, D., and Dietz, K. J. (2003) New insight into the structure and regulation of the plant vacuolar H+-ATPase. J. Bioenerg. Biomembr. 35: 377–388., p. 167; [Figure 6.22]Courtesy of E. Raveh., p. 168

CHAPTER 7

[Chapter 7 Opener]Hans/Pixabay, p. 178; [Figure 7.13]Courtesy of J. Swafford., p. 187; [Figure 7.14]Becker, W. M. (1986) The World of the Cell. Benjamin/Cummings, Menlo Park, CA., p. 187; [Figure 7.15]Nelson, N., and Ben-Shem, A. (2004) The complex architecture of oxygenic photosynthesis. Nat. Rev. Mol. Cell Biol. 5: 971–982., p. 189; [Figure 7.17]Blankenship, R. E., and Prince, R. C. (1985) Excited-state redox potentials and the Z scheme of photosynthesis. Trends Biochem. Sci. 10: 382–383., p. 192; [Figure 7.19] Blankenship, R. E., and Prince, R. C. (1985) Excited-state redox potentials and the Z scheme of photosynthesis. Trends Biochem. Sci. 10: 382–383., p. 194; [Figure 7.20]RCSB PDB Core Operations are funded by the National Science Foundation (DBI-1832184), the US Department of Energy (DE-SC0019749), and the National Cancer Institute, National Institute of Allergy and Infectious Diseases, and National Institute of General Medical Sciences of the National Institutes of Health under grant R01GM133198 hence public domain., p. 195; [Figure 7.22]Marcin Sarewicz et al. High-resolution cryo-EM structures of plant cytochrome b6f at work.Science Advances 9, eadd9688 (2023)., p. 197; [Figure 7.24]Buchanan, B. B., Gruissem, W., and Jones, R. L., eds. (2000) Biochemistry and Molecular Biology of Plants. American Society of Plant Physiologists, Rockville, MD, p. 199

CHAPTER 8

[Chapter 8 Opener]jipatafoto89/ Shutterstock, p. 205; [Figure 8.8b-c]B: Courtesy of Athena McKown, C: S. D. X. Simon D.X. Chuong, Vincent R. Franceschi, Gerald E. Edwards, The Cytoskeleton Maintains Organelle Partitioning Required for Single-Cell C4 Photosynthesis in Chenopodiaceae Species, The Plant Cell, Volume 18, Issue 9, September 2006, pp. 2207–2223., p. 224

CHAPTER 9

[Chapter 9 Opener]Christian Hutter/ imageBROKER/Shutterstock, p. 233; [Figure 9.1]Courtesy of T. Vogelman., p. 234; [Figure 9.3]Smith, H. (1986). The perception of light quality. In Photomorphogenesis in Plants, R. E. Kendrick and G. H. M. Kronenberg, eds., Nijhoff, Dordrecht, Netherlands, pp. 187–217., p. 236; [Figure 9.4]Smith, H. (1994) Sensing the light environment: The functions of the phytochrome family. In Photomorphogenesis in Plants, 2nd ed., R. E. Kendrick and G. H. M. Kronenberg, eds., Nijhoff, Dordrecht, Netherlands, pp. 377–416., p. 237; [Figure 9.5]Vogelmann, T. C., and Björn, L. O. (1983) Response to directional light by leaves of a sun-tracking lupine (Lupinus succulentus). Physiol. Plant. 59: 533–538., p. 238; [Figure 9.7]Harvey, G. W. (1979) Photosynthetic performance of isolated leaf cells from sun and shade plants. Carnegie Inst. Wash. Yb. 79: 161–164. Courtesy, Carnegie Institution for Science., p. 241; [Figure 9.8]Björkman, O. (1981) Responses to different quantum flux densities. In Encyclopedia of Plant Physiology, New Series, Vol. 12A, O. L. Lange, P. S. Nobel, C. B. Osmond, and H. Zeigler, eds., Springer, Berlin, pp. 57–107., p. 242; [Figure 9.9] After P.G. Jarvis and J.W. Leverenz. 1983. In Physiological Plant Ecology IV. Encyclopedia of Plant Physiology (New Series), vol 12D, O.L. Lange et al. [eds], pp. 233-280. Springer: Berlin, Heidelberg., p. 242; [Figure 9.10] Osmond, C. B. (1994) What is photoinhibition? Some insights from comparisons of shade and sun plants. In Photoinhibition of Photosynthesis: From Molecular Mechanisms to the Field, N. R. Baker and J. R. Bowyer, eds., BIOS Scientific, Oxford, pp. 1–24., p. 243; [Figure 9.12]Adams, W.W. & Demmig-Adams, B. (1992) Operation of the xanthophyll cycle in higher plants in response to diurnal changes in incident sunlight. Planta 186: 390., p. 244; [Figure 9.13]Tlalka, M., and Fricker, M. (1999) The role of calcium in blue-light-dependent chloroplast movement in Lemna trisulca L. Plant J. 20: 461-473. Images courtesy of M. Tlalka and M. Fricker., p. 244; [Figure 9.14]Demming-Adams, B., and Adams, W. (2000) Harvesting sunlight safely. Nature 403: 371–372., p. 245; [Figure 9.16] Björkman, O., Mooney, H.A., Ehleringer, J. 1975. Photosynthetic responses of plants from habitats with contrasting thermal environments: comparison of photosynthetic characteristics of intact plants. Carnegie Inst. Washington Yearb. 74:743-48. Courtesy, Carnegie Institution for Science., p. 247; [Figure 9.17] Ehleringer, J. and Björkman, O. 1977. Quantum Yields for CO2 Uptake in C3 and C4 Plants. Plant Physiology Jan 1977, 59 (1) 86-90., p. 249; [Figure 9.18] Ehleringer, J. R. (1978) Implications of quantum yield differences on the distributions of C3 and C4 grasses. Oecologia 31: 255–267., p. 249; [Figure 9.19a-c]A: Barnola, J. M., Raynaud, D., Lorius, C., and Barkov, N.I. (2003) Historical CO2 record from the Vostok ice core. In Trends: A Compendium of Data on Global Change, T. A. Boden, D. P. Kaiser, R. J. Sepanski, and F. W. Stoss, eds., Carbon Dioxide Information Analysis Center, Oak Ridge National Laboratory, U.S. Dept. of Energy, Oak Ridge, TN, pp. 7–10., B: Etheridge, D.M., Steele, L.P., Langenfelds, R.L., Francey, R.J., Barnola, J.-M. and Morgan, V.I.. 1998. Historical CO2 records from the Law Dome DE08, DE08-2, and DSS ice cores. In Trends: A Compendium of Data on Global Change. Carbon Dioxide Information Analysis Center, Oak Ridge National Laboratory, U.S. Department of Energy, Oak Ridge, Tenn., U.S.A., C: Keeling, C. D., and Whorf, T. P. (1994) Atmospheric CO2 records from sites in the SIO air sampling network. In Trends '93: A Compendium of Data on Global Change, T. A. Boden, D. P. Kaiser, R. J. Sepanski, and F. W. Stoss, eds., Carbon Dioxide Information Center, Oak Ridge National Laboratory, Oak Ridge, TN, pp. 16–26; updated using data from Dr. Pieter Tans, NOAA/ESRL (www.esrl.noaa.gov/gmd/ccgg/trends/) and Dr. Ralph Keeling, Scripps Institution of Oceanography (scrippsco2.ucsd.edu/)., p. 250; [Figure 9.21]Berry, J. A., and Downton, J. S. (1982) Environmental regulation of photosynthesis. In Photosynthesis: Development, Carbon Metabolism and Plant Productivity, Vol. 2, Govindjee, ed., Academic Press, New York, pp. 263–343., p. 252; [Figure 9.22] Ehleringer, J. R., Cerling, T. E., and Helliker, B. R. (1997) C4 photosynthesis, atmospheric CO2, and climate. Oecologia 112: 285–299., p. 253; [Figure 9.23]Min-Wha Jeon, Mohammad Babar Ali, Eun-Joo Hahn, Kee-Yoeup Paek, Photosynthetic pigments, morphology and leaf gas exchange during ex vitro acclimatization of micropropagated CAM Doritaenopsis plantlets under relative humidity and air temperature, Environmental and Experimental Botany, Volume 55, Issues 1–2, 2006, pp. 183-194, p. 254; [Figure 9.24a-c]A: Courtesy of David F. Karnosky, B: Courtesy of the USDA/ARS., C: Long, S. P., Ainsworth, E. A., Leakey, A. D., Nosberger, J., and Ort, D. R. (2006) Food for thought: Lower-than-expected crop stimulation with rising CO2 concentrations. Science 312: 1918–1921., p. 256

CHAPTER 10

[Chapter 10 Opener]© VisionsPictures/Minden Pictures., p. 259; [Figure 10.2a-b]A: Joy, K. W. (1964) Translocation in sugar beet. I. Assimilation of 14CO2 and distribution of materials from leaves. J. Exp. Bot. 15: 485–494., B: Courtesy of R. Aloni., p. 261; [Figure 10.3]Courtesy of A. Schulz, p. 262; [Figure 10.4]Courtesy of A. Schulz, p. 262; [Figure 10.6a-c]A: A. Schulz and G. A. Thompson. 2009. In Encyclopedia of Life Sciences. John Wiley & Sons, New York., B: Courtesy of A. Schulz, C: A. Schulz and G. A. Thompson. 2009. In Encyclopedia of Life Sciences. John Wiley & Sons, New York., p. 263; [Figure 10.7a-d]A and B: Evert, R. F. (1982) Sieve-tube structure in relation to function. Bioscience 32: 789–795, C: A. Schulz and G. A. Thompson. 2009. In Encyclopedia of Life Sciences. John Wiley & Sons, New York., D: K. H. Jensen et al. 2012. Frontiers in Plant Science. 3: 151. CC BY 3.0, p. 264; [Figure 10.8]A. Schulz and G. A. Thompson. 2009. In Encyclopedia of Life Sciences. John Wiley & Sons, New York., p. 266; [Figure 10.10]Fondy, B. R. (1975) Sugar selectivity of phloem loading in Beta vulgaris, vulgaris L. and Fraxinus americanus, americana L. Ph.D. diss., University of Dayton, Dayton, OH, courtesy of D. Geiger., p. 268; [Figure 10.12a-b]A: From B. Bentwood and J. Cronshaw. 1978. Planta 140: 111-120., B: From R. Turgeon et al. 1993. Planta 191: 446–456., p. 269; [Figure 10.13]van Bel, A. J. E. (1992) Different phloem-loading machineries correlated with the climate. Acta Bot. Neerl. 41: 121–141., p. 270; [Figure 10.14]Nobel, P. S. (2005) Physicochemical and Environmental Plant Physiology, 3rd ed., Academic Press, San Diego, CA., p. 271; [Figure 10.16a-c]A: Hunziker, P., Schulz, A. (2019). Transmission Electron Microscopy of the Phloem with Minimal Artifacts. In: Liesche, J. (eds) Phloem. Methods in Molecular Biology, vol 2014. Humana, New York, NY., B: Schulz, A. Wound phloem in transition to bundle phloem in primary roots of Pisum sativum L. Protoplasma 130, 27–40 (1986), C: From A.Furch, A. C., Hafke, J. B., Schulz, A., & van Bel, A. J. (2007). Ca2+-mediated remote control of reversible sieve tube occlusion in Vicia faba. Journal of Experimental Botany, 58(11), 2827-2838., p. 281; [Figure 10.18]All photos from Werner, D., Gerlitz, N. & Stadler, R. A dual switch in phloem unloading during ovule development in Arabidopsis. Protoplasma 248, 225–235 (2011)., p. 284; [Figure 10.19]Turgeon, R., and Webb, J. A. (1973) Leaf development and phloem transport in Cucurbita pepo: Transition from import to export. Planta 113: 179–191., p. 285; [Figure 10.20] Turgeon, R. (2006) Phloem loading: How leaves gain their independence. BioScience 56: 15–24., p. 285; [Figure 10.21a-c]A: Schneidereit, A., Imlau, A., and Sauer, N. (2008) Conserved cis-regulatory elements for DNA-binding-with-one-finger and homeo-domain-leucine-zipper transcription factors regulate companion cell-specific expression of the Arabidopsis thaliana SUCROSE TRANSPORTER 2 gene. Plant 228: 651–662, B and C: Ruth Stadler, Kathryn M. Wright, Christian Lauterbach, Gabi Amon, Manfred Gahrtz, Andrea Feuerstein, Karl J. Oparka, Norbert Sauer (2005). Expression of GFP-fusions in Arabidopsis companion cells reveals non-specific protein trafficking into sieve elements and identifies a novel post-phloem domain in roots, Plant J. 41: 319–331, p. 286

CHAPTER 11

[Chapter 11 Opener]Photo by M. H. Siddall, p. 297; [Figure 11.5a-c]A: Perkins, G., Renken, C., Martone, M. E., Young, S. J., Ellisman, M., and Frey, T. (1997) Electtron tomography of neuronal mitochondria: Three-dimensional structure and organization of cristae and membrane contacts. J. Struct. Biol. 119: 260–272, B: B. E. S. Gunning and M. W. Steer. 1996. Plant Cell Biology: Structure and Function of Plant Cells. Jones and Bartlett, Boston., C: Time lapse pictures courtesy of David C. Logan., p. 329; [Figure 11.16b]B: Courtesy of R. N. Trelase., p. 333

CHAPTER 12

[Chapter 12 Opener]Carol Dembinsky/Dembinsky Photo Associates/Alamy Stock Photo., p. 337; [Figure 12.1a-e]A: Nigel Cattlin / Alamy Stock Photo, B: blickwinkel / Alamy Stock Photo, C: Steven Sheppardson / Alamy Stock Photo, D: Courtesy of David McIntyre, E: Courtesy of David McIntyre, p. 339; [Figure 12.4b]B: Xiangxiu Liang and Jian-Min Zhou Annual Review of Plant Biology 2018 69:1, 267-299, p. 341; [Figure 12.10a-c]A: © Sylvan Wittwer/Visuals Unlimited., B: Courtesy of B. Phinney., C: © Sylvan Wittwer/Visuals Unlimited., p. 348; [Figure 12.11a-b]A: Catherine Riou-Khamichi, Rachel Huntley, Annie Jacqmard, and James A.H. Murray. (1999), Cytokinin Activation of Arabidopsis Cell Division Through a D-Type Cyclin, Science 283: 1541-1544., B: Aloni R, Wolf A, Feigenbaum P, Avni A, Klee HJ. The never ripe mutant provides evidence that tumor-induced ethylene controls the morphogenesis of agrobacterium tumefaciens-induced crown galls on tomato stems. Plant Physiol. 1998 Jul; 117(3):841-9. Photos courtesy of R. Aloni., p. 349; [Figure 12.12a-b]A: © Ray Simons/Science Source., B: © Ray Simons/Science Source., p. 349; [Figure 12.13]Hartwig, T. et al. (2011) Brassinosteroid control of sex determination in maize PNAS 108: 19814-19819, p. 350; [Figure 12.17a-b]A: Images 1, 2 Courtesy of L. Gaelweiler and K. Palme; Image 3 Custom Life Science Images / Alamy Stock Photo, B: Multani, D. S., Briggs, S. P., Chamberlin, M. A., Blakeslee, J. J., Murphy, A. S., & Johal, G. S. (2003). Loss of an MDR transporter in compact stalks of maize br2 and sorghum dw3 mutants. Science, 302(5642), 81-84., p. 350; [Figure 12.21a-b]A: blickwinkel / Alamy Stock Photo, B: Jennifer Böhm, Sönke Scherzer, Signaling and transport processes related to the carnivorous lifestyle of plants living on nutrient-poor soil, Plant Physiology, Volume 187, Issue 4, December 2021, pp. 2017–2031. Reprinted by permission of Oxford University Press on behalf of the American Society of Plant Biologists., p. 365

CHAPTER 13

[Chapter 13 Opener]Anest/Shutterstock, p. 370; [Figure 13.1]© Shosei/Corbis., p. 371; [Figure 13.2]© Nigel Cattlin/Alamy., p. 371; [Figure 13.3]Courtesy of M. A. Quiñones., p. 372; [Figure 13.5] Courtesy of David McIntyre., p. 374; [Figure 13.6a-b]A: Shropshire, W., Jr., Klein, W. H., and Elstad, V. B. (1961) Action spectra of photomorphogenic induction and photoinactivation of germination in Arabidopsis thaliana. Plant Cell Physiol. 2: 63–69, B: Kelly, J. M., and Lagarias, J. C. (1985) Photochemistry of 124-kilodalton Avena phytochrome under constant illumination in vitro. Biochemistry 24: 6003–6010, courtesy of Patrice Dubois., p. 375; [Figure 13.7a-b]A: Baskin, T. I. and Iino, M. (1987), An Action Spectrum in the Blue and Ultraviolet for Phototropism in Alfalfa. Photochemistry and Photobiology, 46: 127-136., B: Briggs, W.R. and Christie, J.M. (2002) Phototropins 1 and 2: versatile plant blue-light receptors. Trends in Plant Science 7(5): 204–210., p. 376; [Figure 13.8]After art courtesy of Candace Pritchard, p. 378; [Figure 13.9]Briggs, W. R., Mandoli, D. F., Shinkle, J. R., Kaufman, L. S., Watson, J. C., and Thompson, W. F. (1984) Phytochrome regulation of plant development at the whole plant, physiological, and molecular levels. In Sensory Perception and Transduction in Aneural Organisms, G. Colombetti, F. Lenci, and P.-S. Song, eds., Plenum, New York, pp. 265–280., p. 379; [Figure 13.10] Spalding, E. P., and Cosgrove, D. J. (1989) Large membrane depolarization precedes rapid blue-light induced growth inhibition in cucumber. Planta 178: 407–410., p. 382; [Figure 13.11]Art courtesy of Candace Pritchard, p. 383; [Figure 13.12]Art courtesy of Candace Pritchard, p. 384; [Figure 13.13] Christie, J. M., Swartz, T. E., Bogomolni, R. A. and Briggs, W. R. (2002) Phototropin LOV domains exhibit distinct roles in regulating photoreceptor function. The Plant Journal 32: 205-219., p. 385; [Figure 13.14]Wada, M. (2013) Chloroplast movement. Plant Sci. 210: 177–182., p. 385; [Figure 13.15]Wada, M. (2013) Chloroplast movement. Plant Sci. 210: 177–182., p. 386; [Figure 13.16]Inoue, S., and Kinoshita, T. (2017) Blue light regulation of stomatal opening and the plasma membrane H+-ATPase. Plant Physiol. 174: 531-538. CC BY 4.0, p. 387; [Figure 13.17]Art courtesy of Candace Pritchard, p. 388

CHAPTER 14

[Chapter 14 Opener]Allexxandar/Shutterstock, p. 391; [Figure 14.2]Larry Larsen / Alamy Stock Photo, p. 396; [Figure 14.3a-c]A: Jaswinder Singh/McGill University, B: Photo by Jenny Rees, C: Branex/Dreamstime.com, p. 396; [Figure 14.4]Finch-Savage, W. E. and Leubner-Metzger, G. (2006) Seed dormancy and the control of germination. New Phytol. 171: 501–523., p. 397; [Figure 14.5a-b]A: Bewley J.D., Bradford K.J., Hilhorst H.W.M., Nonogaki H. (2013) Dormancy and the Control of Germination. In: Seeds: Physiology of Development, Germination and Dormancy, 3rd Edition. Springer, New York, NY, B: Grappin, P., Bouinot, D., Sotta, B., Migniac, E., and Julien, M. (2000) Control of seed dormancy in Nicotiana plumbaginifolia: post-imbibition abscisic acid synthesis imposes dormancy maintenance. Planta 210: 279–285., p. 399; [Figure 14.6]Liptay, A., and Schopfer, P. (1983) Effect of water stress, seed coat restraint, and abscisic acid upon different germination capabilities of two tomato lines at low temperature. Plant Physiol. 73: 935–938., p. 401; [Figure 14.7]J. D. Bewley. 1997. The Plant Cell 9: 1055–1066; H. Nonogaki et al. 2007. In Annual Plant Reviews, Vol. 27: Seed Development, Dormancy, and Germination; H. Nonogaki et al. 2010. Plant Sci. 179: 574–581, p. 401; [Figure 14.8b-d]Paul C.

Bethke, Robert Schuurink, Russell L. Jones, Hormonal signalling in cereal aleurone, Journal of Experimental Botany, Volume 48, Issue 7, July 1997, pp. 1337–1356, courtesy of P. C. Bethke., p. 404; [Figure 14.9]Gubler, F., Kalla, R., Roberts, J. K., and Jacobsen, J. V. (1995) Gibberellin-regulated expression of a myb gene in barley aleurone cells: Evidence of Myb transactivation of a high-pI alpha-amylase gene promoter. Plant Cell 7: 1879–1891., p. 404; [Figure 14.11]Cosgrove D. J. Growth of the plant cell wall. Nature Reviews Molecular Cell Biology 2005 Nov; 6(11):850-861., p. 406; [Figure 14.14] Cosgrove, D. J. (1997) Assembly and enlargement of the primary cell wall in plants. Annu. Rev. Cell Dev. Biol. 13: 171–201., p. 408; [Figure 14.15]Cosgrove, D. J. (1997) Assembly and enlargement of the primary cell wall in plants. Annu. Rev. Cell Dev. Biol. 13: 171–201., p. 409; [Figure 14.16a-d]A and B: Courtesy of Patrice G. Dubois., C and D: Courtesy of David McIntyre, p. 410; [Figure 14.17] Courtesy of S. Savaldi-Goldstein., p. 411; [Figure 14.18]Courtesy of Joe Kieber, UNC., p. 411; [Figure 14.19a-d] Busse, J. S., and Evert, R. F. (1999) Pattern of differentiation of the first vascular elements in the embryo and seedling of Arabidopsis thaliana. Int. J. Plant Sci. 160: 1–13., p. 412; [Figure 14.21]Bibikova, T. and S. Gilroy. 2003. Root Hair Development. Journal of Plant Growth Regulation. 21: 383–415, p. 413; [Figure 14.22]Frederick B. Abeles, Page W. Morgan, Mikal E. Saltveit, CHAPTER 5 - Roles and Physiological Effects of Ethylene in Plant Physiology: Dormancy, Growth, and Development, Editor(s): Frederick B. Abeles, Page W. Morgan, Mikal E. Saltveit, Ethylene in Plant Biology (Second Edition), Academic Press, 1992, pp. 120-181., p. 413; [Figure 14.23a-b] Le, J., Vandenbussche, F., De Cnodder, T., Van Der Straeten, D., and Verbelen, J. P. (2005) Cell elongation and microtubule behavior in the Arabidopsis hypocotyl: Responses to ethylene and auxin. J. Plant Growth Regul. 24: 166–178., p. 414; [Figure 14.24a-d]A and B:© M. B. Wilkins., C and D: Mark Jacobs, Peter M. Ray, Rapid Auxin-induced Decrease in Free Space pH and Its Relationship to Auxin-induced Growth in Maize and Pea, Plant Physiology, Volume 58, Issue 2, August 1976, pp. 203–209, p. 415; [Figure 14.26]Adapted from Hartmann, H. T., and Kester, D. E. 1983. Plant Propagation: Principles and Practices, 4th ed. Prentice-Hall, Inc., NJ., p. 417; [Figure 14.27]Blilou, I., Xu, J., Wildwater, M., Willemsen, V., Paponov, I., Friml, J., Heidstra, R., Aida, M., Palme, K., and Scheres, B. (2005) The PIN auxin efflux facilitator network controls growth and patterning in Arabidopsis roots. Nature 433: 39–44., p. 418; [Figure 14.29]Shaw, S., and Wilkins, M. B. (1973) The source and lateral transport of growth inhibitors in geotropically stimulated roots of Zea mays and Pisum sativum. Planta 109: 11–26., p. 419; [Figure 14.30]Baldwin, K.I. et al. 2013. Gravity Sensing and Signal Transduction in Vascular Plant Primary Roots. American Journal of Botany. 100: 126–142, p. 420; [Figure 14.31]Palmieri, M and Kiss, J.Z. 2007. The Role of Plastids in Gravitropism. In The Structure and Function of Plastids. Advances in Photosynthesis and Respiration, vol. 23, R. R. Wise and J. K. Hoober [eds.], pp. 507–525. Springer, Dordrecht, p. 421; [Figure 14.33]Fasano, J. M., Swanson, S. J., Blancaflor, E. B., Dowd, P. E., Kao, T. H., and Gilroy, S. (2001) Changes in root cap pH are required for the gravity response of the Arabidopsis root. Plant Cell 13: 907–921., p. 423; [Figure 14.34] Monshausen, G. B. et al. 2009. Ca2+ Regulates Reactive Oxygen Species Production and pH during Mechanosensing in Arabidopsis Roots. Plant Cell 21: 2341–2356, p. 424; [Figure 14.36]Iino, M. and Briggs, W. R. 1984. Growth distribution during first positive phototropic curvature of maize coleoptiles. Plant, Cell & Environment. 7: 97–104, p. 426; [Figure 14.37] Christie, J. M. et al. 2011. phot1 Inhibition of ABCB19 Primes Lateral Auxin Fluxes in the Shoot Apex Required For Phototropism. PLoS Biol 9(6): e1001076., p. 426

CHAPTER 15

[Chapter 15 Opener]imageBROKER. com GmbH & Co. KG / Alamy Stock Photo, p. 430; [Figure 15.2a-b]A: © Biology Pics/Science Source., B: Kuhlemeier, C., and Reinhardt, D. (2001) Auxin and phyllotaxis. Trends Plant Sci. 6: 187-189. courtesy of D. Reinhardt., p. 433; [Figure 15.3] Bowman, J. L., and Eshed, Y. (2000) Formation and maintenance of the shoot apical meristem. Trends Plant Sci. 5: 110–115., p. 434; [Figure 15.7a-c]A: Reinhardt, D., Pesce, E. R., Stieger, P., Mandel, T., Baltensperger, K., Bennett, M., Traas, J., Friml, J., and Kuhlemeier, C. (2003) Regulation of phyllotaxis by polar auxin transport. Nature 426: 255–260, B: Vernoux, T., Kronenberger, J., Grandjean, O., Laufs, P., and Traas, J. (2000) PIN-FORMED 1 regulates cell fate at the periphery of the shoot apical meristem. Development 127: 5157–5165., C: Reinhardt, D., Pesce, E. R., Stieger, P., Mandel, T., Baltensperger, K., Bennett, M., Traas, J., Friml, J., and Kuhlemeier, C. (2003) Regulation of phyllotaxis by polar auxin transport. Nature 426: 255–260, p. 436; [Figure 15.8a-c]A: © Dr. Ken Wagner/Visuals Unlimited, Inc. B: © Jon Bertsch/Visuals Unlimited, Inc. C: Garry DeLong/Science Source, p. 437; [Figure 15.9]Lau, S., and Bergmann, D. C. (2012) Stomatal development: A plant's perspective on cell polarity, cell fate transitions and intercellular communication. Development 139: 3683–3692., p. 439; [Figure 15.10]Courtesy of David McIntyre, p. 440; [Figure 15.11]Lucas, W. J., Groover, A., Lichtenberger, R., Furuta, K., Yadav, S.-R., Helariutta, Y., He, X.-Q., Fukuda, H., Kang, J., Brady, S. M., et al. (2013) The plant vascular system: Evolution, development and functions. J. Integr. Plant Biol. 55: 294–388., p. 440; [Figure 15.13b]Courtesy of R. Aloni., p. 442; [Figure 15.15]Courtesy of David McIntyre, p. 443; [Figure 15.17]Mason, M. G., Ross, J. J., Babst, B. A., Wienclaw, B. N., and Beveridge, C. A. (2014) Sugar demand, not auxin, is the initial regulator of apical dominance. Proc. Natl. Acad. Sci. USA 111: 6092–6097., p. 444; [Figure 15.19a-b]A: Courtesy of T. Brutnell, B: Glasshouse Images / Alamy Stock Photo, p. 446; [Figure 15.20]M. Megías et al. 2019. Atlas of Plant and Animal Histology. Dept. Func. Biol. Hlth. Sci., University of Vigo, Spain. https://mmegias.webs.uvigo.es/02-english/2-organos-v/guiada_o_v_rprimario.php, p. 447; [Figure 15.23b-c]B: Hochholdinger, F., and Tuberosa, R. (2009) Genetic and genomic dissection of maize root development and architecture. Curr. Opin. Plant Biol. 12: 172–177., C: J.J. Gouin/Shutterstock, p. 450; [Figure 15.24]Courtesy of Leon Kochian., p. 450; [Figure 15.25]Péret, B., Desnos, T., Jost, R., Kanno, S., Berkowitz, O., and Nussaume, L. (2014) Root architecture responses: in search of phosphate. Plant Physiology 166: 1713–1723., p. 451; [Figure 15.26] Risopatron M. JP., Sun Y., Jones BJ. The vascular cambium: molecular control of cellular structure. Protoplasma. 2010 Dec; 247(3-4):145-161, p. 452; [Figure 15.27b] B: R. Spicer and A. Groover. 2010. New Phytologist 186: 577-592., p. 453; [Figure 15.30]Courtesy of Andreas M. Fischer., p. 457; [Figure 15.31a-f] Keskitalo, J., Bergquist, G., Gardestrom, P., and Jansson, S. (2005) A cellular timetable of autumn senescence. Plant Physiol. 139: 1635–1648., p. 457; [Figure 15.33]Breeze, E., Harrison, E., McHattie, S., Hughes, L., Hickman, R., Hill, C., Kiddle, S., Kim, Y.-S., Penfold, C. A., Jenkins, D., et al. (2011) High-resolution temporal profiling of transcripts during Arabidopsis leaf senescence reveals a distinct chronology of processes and regulation. Plant Cell 23: 873–894., p. 459; [Figure 15.36] Vahala, J., Ruonala, R., Keinänen, M., Tuominen, H., and Kangasjärvi, J. (2003) Ethylene insensitivity modulates ozone-induced cell death in birch (Betula pendula). Plant Physiol. 132: 185–195., p. 462; [Figure 15.37]Morgan, P. W. (1984) Is ethylene the natural regulator of abscission? In Ethylene: Biochemical, Physiological and Applied Aspects, Y. Fuchs and E. Chalutz, eds., Martinus Nijhoff, The Hague, Netherlands, pp. 231–240., p. 462; [Figure 15.38] Courtesy of L. Nooden., p. 463

CHAPTER 16

[Chapter 16 Opener]tinnko/ Shutterstock, p. 467; [Figure 16.1] Courtesy of Luigi Rignanese., p. 469; [Figure 16.3]Courtesy of Richard Amasino., p. 472; [Figure 16.6]Courtesy of J. A. D. Zeevaart., p. 475; [Figure 16.8]Vince-Prue, D. (1975) Photoperiodism in Plants. McGraw-Hill, London; Salisbury, F. B. (1963) Biological timing and hormone synthesis in flowering of Xanthium. Planta 49: 518–524; Papenfuss, H. D., and Salisbury, F. B. (1967) Aspects of clock resetting in flowering of Xanthium. Plant Physiol. 42: 1562–1568., p. 477; [Figure 16.9]Coulter, M. W., and Hamner, K. C. (1964) Photoperiodic flowering response of Biloxi soybean in 72 hour cycles. Plant Physiol. 39: 848–856., p. 478; [Figure 16.10]Hayama, R., and Coupland, G. (2004) The molecular basis of diversity in the photoperiodic flowering responses of Arabidopsis and rice. Plant Physiol. 135: 677–684., p. 479; [Figure 16.12]Hendricks, S. B., and Siegelman, H. W. (1967) Phytochrome and photoperiodism in plants. Comp. Biochem. 27: 211–235; Saji, H., Vince-Prue, D., and Furuya, M. (1983) Studies on the photoreceptors for the promotion and inhibition of flowering in dark-grown seedlings of Pharbitis nil choisy. Plant Cell Physiol. 67: 1183–1189., p. 480; [Figure 16.13]Trevaskis, B., Hemming MN, Dennis ES, Peacock WJ (2007) The molecular basis of vernalization-induced flowering in cereals. Trends Plant Sci. 12: 352-357., p. 481; [Figure 16.15]Purvis, O. N., and Gregory, F. G. (1952) Studies in vernalization of cereals. XII. The reversibility by high temperature of the vernalized condition in Petkus winter rye. Ann. Bot. 1: 569–592, p. 482; [Figure 16.16]Liu, L., Zhu, Y., Shen, L., and Yu, H. (2013) Emerging insights into florigen transport. Curr. Opin. Plant Biol. 16: 607–613, p. 484; [Figure 16.17]Courtesy of V. Grbic and M. Nelson., p. 485; [Figure 16.18]Bewley, J. D., Hempel, F. D., McCormick, S., and Zambryski, P. (2000) Reproductive Development. In Biochemistry and Molecular Biology of Plants, B. B. Buchanan, W. Gruissem, and R. L. Jones, eds., American Society of Plant Biologists, Rockville, MD, pp. 988–1034., p. 486; [Figure 16.19]Krizek, B. A., and Fletcher, J. C. (2005) Molecular mechanisms of flower development: An armchair guide. Nat. Rev. Genet. 6: 688–698., p. 486; [Figure 16.21]© Scientifica/RMF/Visuals Unlimited, Inc., p. 488; [Figure 16.22]Jones, R. L., Ougham, H., Thomas H., Waaland, S. (2013) The Molecular Life of Plants. Wiley-Blackwell., p. 489

CHAPTER 17

[Chapter 17 Opener]Alena Brozova/ Shutterstock, p. 495; [17.1a-I]A: Roman Nerud/Shutterstock.com, B: iStock.com/ sdstockphoto., C: Jiri Hera/Shutterstock.com., D: iStock.com/kickers., E: iStock.com/SweetpeaAnna., F: Suslik1983/ Shutterstock.com., G: BW Folsom/ Shutterstock.com., H: Vania Zhukevych/ Shutterstock.com., I: iStock.com/ surabky, p. 496; [Figure 17.2]Thomas Laux and others, Genetic Regulation of Embryonic Pattern Formation, The Plant Cell, Volume 16, Issue suppl_1, June 2004, S190–S202, p. 498; [Figure 17.3]I. Debeaujon et al. 2003. Plant Cell 15: 2514–2531 Proanthocyanidin-Accumulating Cells in Arabidopsis Testa: Regulation of Differentiation and Role in Seed Development, The Plant Cell, Volume 15, Issue 11, November 2003, Pages 2514–2531nd C from I. Debeaujon et al. 2003. Plant Cell 15: 2514–2531, p. 499; [Figure 17.4a-c]A and B: W. Troll. 1937. Vergleichende Morphologie der hoheren Pflanzen, Gebruder Borntrager, Berlin. C: M. Losségal et al. 2007. In Endosperm: Plant Cell Monographs, vol 8. Olsen, O.A. [ed.], pp. 57-71. Springer: Berlin, Heidelberg., p. 499; [Figure 17.5a-b]A: Patrice 785000/Wikimedia CC-By-SA 3.0, B: P. P. Kumar and C. S. Loh, 2012. In Plant Biotechnology and Agriculture. Arie Altman, Paul Michael Hasegawa (eds.) pp. 131-138. Academic Press: Amersterdam. Photo by Wendy Shu, p. 500; [Figure 17.6a-d]A: J. Verdier et al. 2013. Plant Physiol. 163: 757–774. CC BY 4.0. B-D: J. Verdier et al. 2013. Plant Physiol. 163: 757–774. CC BY 4.0, p. 502; [Figure 17.8]R. Aloni. 2021. Vascular Differentiation and Plant Hormones, p. 174. Springer Nature: Cham, Switzerland. Courtesy of R. Aloni., p. 505; [Figure 17.9b-d]B: Irish (USDA-ARS) for the Taxonomic Reference Collection exercise., C, D: from M. Goetz et al. 2007. Plant Physiol. 145: 351–366. CC BY 4.0, p. 505; [Figure 17.10 a-c]A: G. B. Seymour et al. 2013. Annu. Rev. Plant Biol. 64: 219–241, B: J. R. Dinneny and M. F. Yanofsky. 2004. BioEssays 27: 42–49, C: P. Ballester and C. Ferrándiz. 2017. Curr. Opin. Plant Biol. 35: 68–75, p. 506; [Figure 17.11a-b] A: Courtesy of Zhongchi Liu, B: Drawing by Zhongchi Liu, p. 507; [Figure 17.12a-b]A: © brozova/istock., B: G. B. Seymour et al. 2013. Annu. Rev. Plant Biol. 64: 219–241, p. 507; [Figure 17.13a-b]A: Courtesy of David McIntyre., B: Courtesy of Zhongchi Liu, p. 508; [Figure 17.14a-c]A: P. Oeller et. al. 1991. Science. 254: 437-439. B, C: D. Grierson. 2013. In The Molecular Biology and Biochemistry of Fruit Ripening. Seymour, G. B. et al. [eds.], pp. 43-73. Wiley-Blackwell: New York, p. 509; [Figure 17.15]After J. J. Giovannoni. 2004. Plant Cell 16: S170–S180., p. 510; [Figure 17.16]Courtesy of Andy Davis, John Innes Centre., p. 511

CHAPTER 18

[Chapter 18 Opener]Oenz/Shutterstock, p. 513; [Figure 18.1]van Dam, N. M. (2009) How plants cope with biotic interactions. Plant Biol. 11: 1–5., p. 514; [Figure 18.2]Goh, C.-H., Veliz Vallejos, D. F., Nicotra, A. B., and Mathesius, U. (2013) The impact of beneficial plant-associated microbes on plant phenotypic plasticity. J. Chem. Ecol. 39: 826–839., p. 516; [Figure 18.3a-d]© J. Engelberth., p. 518; [Figure 18.5]Courtesy of Agong1/Wikipedia., p. 519; [Figure 18.6]© J. Engelberth., p. 520; [Figure 18.7a-b]A: © J. Engelberth., B: © J. Engelberth., p. 521; [Figure 18.8b]© J. Engelberth., p. 521; [Figure 18.10a-c]© J. Engelberth., p. 523; [Figure 18.13a-b] A: Christmann, A., and Grill, E. (2013) Electric defence. Nature 500: 404–405., B: M. Toyota et al. 2018. Science 361: 1112–1115, p. 528; [Figure 18.16]Boller, T., and Felix, G. (2009) A Renaissance of elicitors: Perception of microbe-associate molecular patterns and danger signals by pattern-recognition receptors. Annu. Rev. Plant Biol. 60: 379–406.,

p. 532; [Figure 18.21a-d]© J. Engelberth., p. 539; [Figure 18.22a-d]A: © Biodisc/Visuals Unlimited, Inc., B-D: S. Yoshida et al. 2016. Annu. Rev. Plant Biol. 67: 643–647, p. 540

CHAPTER 19

[Chapter 19 Opener]janews/ Shutterstock, p. 542; [Figure 19.2a-c]A and B: iStock.com/PlazacCameraman, C: Courtesy of E. Blumwald., Differential Role for Trehalose Metabolism in Salt-Stressed Maize. Clémence Henry, Samuel W. Bledsoe, Cara A. Griffiths, Alec Kollman, Matthew J. Paul, Soulaiman Sakr, L. Mark Lagrimini Plant Physiology Oct 2015, 169 (2) 1072-1089; DOI: 10.1104/ pp.15.00729, p. 544; [Figure 19.5]Boyer, J. S. (1970) Leaf enlargement and metabolic rates in corn, soybean, and sunflower at various leaf water potentials. Plant Physiol. 46: 233–235., p. 549; [Figure 19.6]Mittler, R. (2006) Abiotic stress, the field environment and stress combination. Trends Plant Sci. 11: 15–19., p. 551; [Figure 19.8]Mittler, R., Vanderauwera, S., Suzuki, N., Miller, G., Tognetti, V. B., Vandepoele, K., Gollery, G., Shulaev, V., and Van Breusegem, F. (2011) ROS signaling: The new wave? Trends Plant Sci. 16: 300–309, p. 554; [Figure 19.9]Beardsell, M. F., and Cohen, D. (1975) Relationships between leaf water status, abscisic acid levels, and stomatal resistance in maize and sorghum. Plant Physiol. 56: 207–212., p. 555; [Figure 19.11]Courtesy of E. Blumwald., p. 556; [Figure 19.12] Gutzat, R., and Mittelsten-Scheid, O. (2012) Epigenetic responses to stress: Triple defense? Curr. Opin. Plant. Biol. 15: 568–573., p. 557; [Figure 19.16] Courtesy of J. L. Basq and M. C. Drew., p. 558

APPENDIX

[Appendix Chapter Opener] International Rice Research Institute; photo licensed under CC BY 2.0, p. A-569; [Figure A.1]© Jim Mills/istock., p. A-570; [Figure A.5]F. Wu et al. 2021. Proc Natl Acad Sci 118: e2120901118. Image: International Rice Research Institute; photo licensed under CC BY 2.0., p. A-574; [Figure A.6] J. D. Sander and J. K. Joung. 2014. Nat Biotechnol. 32: 347–355, p. A-576; [Figure A.7]Y. Sun et al. (2016). Engineering Herbicide-Resistant Rice Plants through CRISPR/Cas9-Mediated Homologous Recombination of Acetolactate Synthase. Mol Plant. Apr 4; 9(4):628-631., p. A-577

Index

Tables, figures, and boxes are indicated by an italic *t*, *f*, and *b* following the page/paragraph number.